T0139317

Handbook of Finite Translation Planes

Norman L. Johnson

University of Iowa
Iowa City, IA, U.S.A.

Vikram Jha

Caledonian University
Glasgow, Scotland, U.K.

Mauro Biliotti

University of Leece
Italy

 Chapman & Hall/CRC
Taylor & Francis Group

Boca Raton   London   New York

Chapman & Hall/CRC is an imprint of the
Taylor & Francis Group, an informa business

PURE AND APPLIED MATHEMATICS

A Program of Monographs, Textbooks, and Lecture Notes

MONOGRAPHS AND TEXTBOOKS IN PURE AND APPLIED MATHEMATICS

Recent Titles

W. J. Wickless, A First Graduate Course in Abstract Algebra (2004)

R. P. Agarwal, M. Bohner, and W-T Li, Nonoscillation and Oscillation Theory for Functional Differential Equations (2004)

J. Galambos and I. Simonelli, Products of Random Variables: Applications to Problems of Physics and to Arithmetical Functions (2004)

Walter Ferrer and Alvaro Rittatore, Actions and Invariants of Algebraic Groups (2005)

Christof Eck, Jiri Jarusek, and Miroslav Krbec, Unilateral Contact Problems: Variational Methods and Existence Theorems (2005)

M. M. Rao, Conditional Measures and Applications, Second Edition (2005)

A. B. Kharazishvili, Strange Functions in Real Analysis, Second Edition (2006)

Vincenzo Ancona and Bernard Gaveau, Differential Forms on Singular Varieties: De Rham and Hodge Theory Simplified (2005)

Santiago Alves Tavares, Generation of Multivariate Hermite Interpolating Polynomials (2005)

Sergio Macías, Topics on Continua (2005)

Mircea Sofonea, Weimin Han, and Meir Shillor, Analysis and Approximation of Contact Problems with Adhesion or Damage (2006)

Marwan Moubachir and Jean-Paul Zolésio, Moving Shape Analysis and Control: Applications to Fluid Structure Interactions (2006)

Alfred Geroldinger and Franz Halter-Koch, Non-Unique Factorizations: Algebraic, Combinatorial and Analytic Theory (2006)

Kevin J. Hastings, Introduction to the Mathematics of Operations Research with *Mathematica*®, Second Edition (2006)

Robert Carlson, A Concrete Introduction to Real Analysis (2006)

John Dauns and Yiqiang Zhou, Classes of Modules (2006)

N. K. Govil, H. N. Mhaskar, Ram N. Mohapatra, Zuhair Nashed, and J. Szabados, Frontiers in Interpolation and Approximation (2006)

Luca Lorenzi and Marcello Bertoldi, Analytical Methods for Markov Semigroups (2006)

M. A. Al-Gwaiz and S. A. Elsanousi, Elements of Real Analysis (2006)

Theodore G. Faticoni, Direct Sum Decompositions of Torsion-free Finite Rank Groups (2006)

R. Sivaramakrishnan, Certain Number-Theoretic Episodes in Algebra (2006)

Aderemi Kuku, Representation Theory and Higher Algebraic K-Theory (2006)

Robert Piziak and P. L. Odell, Matrix Theory: From Generalized Inverses to Jordan Form (2007)

Norman L. Johnson, Vikram Jha, and Mauro Biliotti, Handbook of Finite Translation Planes (2007)

Chapman & Hall/CRC
Taylor & Francis Group
6000 Broken Sound Parkway NW, Suite 300
Boca Raton, FL 33487-2742

International Standard Book Number-10: 1-58488-605-6 (Hardcover)
International Standard Book Number-13: 978-1-58488-605-1 (Hardcover)

Visit the Taylor & Francis Web site at
http://www.taylorandfrancis.com

and the CRC Press Web site at
http://www.crcpress.com

Contents

Preface and Acknowledgments.

The authors' previous work *Foundations of Translation Planes* (see Biliotti, Jha, Johnson [**123**]) provides the basics of translation planes and a general theory of translation planes written for the student of the area of translation planes, the interested incidence geometer and, as a general reference, to combinatoralists.

Furthermore, there were a considerable number of construction techniques considered in the *Foundations* text, including the methods of 'derivation', 's-square' and 's-inversion', 'T-extension', 'T-distortion', 'transpose' and 'lifting'.

In the text on subplane covered nets (Johnson [**753**]) by one of the authors, there is also the construction technique of multiple direct products corresponding to Desarguesian parallelisms and partial parallelisms of projective spaces.

However, we mentioned in the *Foundations* text that although there were a varied and complex variety of examples given within the two texts previously mentioned, the set of examples was not to be considered the last word on the nature of examples. Although it is presumptuous to claim that herein lies that last word on the set of examples of translation planes, at least up to this date, we would like to think that we have provided a comprehensive listing of all the translation planes that arise using a fundamental construction technique or are otherwise of interest in that they appear as examples within a classification scheme. The translation planes of orders 4, 9, 16, 25, 27 and 49 are all completely determined, many classes of which by computer. These listings are very important for at least two reasons: First sporadic examples often point to various heretofore unknown infinite classes and second, certain arguments tend to reduce to analysis of a few special orders and knowledge of special orders then becomes vital. So, we are interested in all translation planes but, for reasons of space, planes known mostly by computer shall not be listed in complete detail.

As the reader will most certainly determine, there is such a tremendous variety and type of translation planes that it might appear that there is no possible reasonable manner of providing a classification. We therefore feel that it is more appropriate to provide a combination of description and methods of construction to give an explication of the classes of translation planes. Furthermore, as the validity for the interest in certain translation planes arises primarily from their place within certain major classification results and their consequent use in combinatorial geometries, we also will be interested in providing at least comprehensive sketches of these major theorems.

So, we provide here both classification results for translation planes and the general outlines of their proofs, a complete review of all of the known construction techniques for translation planes and a comprehensive treatment of the known examples.

We would like to acknowledge contributions to this book.

From one of the authors, Norman L. Johnson:

It seems that a set of strong women greatly influenced my life and although this set properly contains the following subset of three, it is these three to whom I basically owe everything and about whom, I feel I need to amplify.

Let me begin with my mother Catherine Elizabeth Lamb (Gabriel), to whom I owe whatever talent I may have in mathematics. The funny thing about my mother is that although she was denied a university education by my grandfather, who had rather old-fashioned ideas about 'girls' attending a university, she did acquire some knowledge of mathematics in high school by use of a 'zero-sum' game. Her father wanted her to take either 'lady-like' courses or business courses (perhaps not mutually exclusive) but my mother said she would not take any course that was not matched, as it were, by a course in some sort of mathematics. So, the family arranged for my mother to take special courses in mathematics, where usually she was the only one in the course and always the only one in the course who knew what was going on—and that number would probably include the teacher of the course. Years later, she was able to cope with my total lack of interest in mathematics but somehow perhaps subtly transferred her love of the game to me. As such an occurrence was extremely unlikely, I am still not quite sure how she was able to accomplish that.

Ida Boquist grew up in Haugesund, Norway, the daughter of a couple who ran a local inn. Working for her father at the inn, Ida learned to speak the various languages of those who came through and since this was wartime, there were a variety of languages heard. Eventually, Ida came to the United States, matriculated through the University of Oregon and came to teach English, German, and Latin in a small town on the Oregon coast called "Tillamook," which means "land of many waters" in the original indigenous Indian language. This is where I come in. Mrs. B (Boquist) was the favorite of many of the students at Tillamook High School, but she seemed to have a very soft spot in her heart for all mischievous and rebellious students, a category in which I fit perfectly.

Upon entering the service (army) midway through my senior year—in those days a somewhat honorable exit for even the most rebellious—I started getting interested in languages and studied German on my own while also taking a variety of elementary courses first at the University of San Francisco and then later at the University of Washington.

After the service, I asked Mrs. B if she would be willing to tutor me in German. Of course, that soft spot was still there and she taught me German one year, then French the next, not to mention pushing some English literature as well.

Anyway, she made the difference as I set out for university life where I would study languages and mathematics and 'Bonnie', my singular passions—more or less in that sequential order—sometimes it just takes a gentle push.

I began my studies at Portland State University in Portland, Oregon, where I met the love of my life, Bonnie Lynn Hemenover. Somehow, this connection was meant to be as Bonnie was from Pendleton, Oregon, and we were not likely to meet. But, she did study at Kinman Business University in Spokane and made her way to Portland to work for an insurance company in the heart of the city. To make ends meet, she shared a room with a 'Tillamook girl' and after walking several miles from the center of the city toward their apartment near Portland State, Bonnie and her roommate would stop at the apartment that I shared with my 'ol buddy' Clark

Ferry from Tillamook (another mischievous and rebellious type), and still another unlikely occurrence made its appearance.

Bonnie has this wonderful ability to bring out the positive in everyone she meets and has encouraged me tremendously in everything that I have been involved in—languages and music (maybe the same thing), in mathematics (again maybe the same thing)—and gave all of us continuous encouragement to finish these three books. She is the one 'constant' in my life of ever changing variables.

It is a mysterious fact that this book would never have happened but for these three strong women, so I gratefully acknowledge the very many hidden contributions of Catherine Lamb, Ida Boquist and my wife Bonnie Lynn Hemenover.

In general, the authors are greatly indebted to Brian Treadway of the University of Iowa for his considerable help with these three volumes, *Subplane Covered Nets*, *Foundations of Translation Planes*, and *Handbook of Finite Translation Planes*. As with the previous volumes, the layout and typesetting and considerable editing was accomplished by Brian with great skill and we are very thankful for this help.

Finally, the authors would like to thank our teachers, friends and collaborators L.A. Rosati and A. Barlotti, D.A. Foulser and Dan Hughes, and T.G. Ostrom both singly and collectively.

In addition, the authors are indebted to the Universities of Lecce, Caledonian and Iowa for their invaluable contributions to the writing of this book.

The authors would like to acknowledge and thank the International Programs of the University of Iowa, which supported one of the authors (V. Jha) during the summer of 2002 and further provided assistance for another of the authors (N.L. Johnson) to visit Scotland and Italy on several occasions to allow collaboration on this text.

In addition, the authors are grateful for the support of the London Mathematical Society, which supported one of the authors (N.L. Johnson) during the summer of 2004, and also thank the MIUR of Italy for essentially continuous summer support.

Norman L. Johnson
University of Iowa,
Iowa City, Iowa, U.S.A.

Vikram Jha
Caledonian University,
Glasgow, Scotland

Mauro Biliotti,
University of Lecce, Italy
Lecce, Italy

CHAPTER 1

An Overview.

There are only a few texts that deal with the subject of translation planes. Undoubtedly, this is due both to the great difficulty in presenting the theory in a cohesive manner and to the lack of a uniform model with which to study the structure.

This book originated with the goal to try to unify the various diverse approaches to the subject and to present the material in a way which could be utilized by other mathematicians in related areas.

Since there are combinatorial and coding, graph- and group-theoretical components to this topic, as well as rather standard connections to projective geometry and linear and multilinear algebra, mathematicians who are interested in any of these areas will probably also be interested in this text.

As the material for this text developed, we realized the need for a detailed text on fundamentals which emphasized technique and the interaction between translation planes and their coordinatizing structures such as spreads, matrix spread sets and quasifields. But also, it became clear to us that the general geometer is not aware of the many and varied examples and techniques available in the general theory of translation planes. As these incidence structures are of vital importance in the area of finite incidence geometry, we decided to have a separate text detailing essentially all translation planes either by example, classification, or construction method. In our text *Foundations of Translation Planes*, the reader will find all of the basic material required for research in this area and/or to appreciate the thrust of the examples. As there is certain material from the *Foundations* text which we have included in the present text, the interested reader will find certain necessary duplications. In addition, material on parallelisms from one author's (Johnson's) text on *Subplane Covered Nets* is also included, as this relates directly to an important class of translation planes. Indeed, fundamental material on derivation and derivable nets is also taken from the subplanes book. However, we have written the present text more as a compendium of examples, processes, construction techniques and models than as a true textbook on translation planes. Thus, we see this book as a suitable resource for the finite geometer who might contribute to this area or who would seek a source for the existence of an isomorphic copy of any given translation plane.

Hence, we have written this book for the general geometer. Most of the material presented here was done initially in the finite case. Occasionally, we have avoided the use of the finite argument in favor of a general argument over arbitrary/infinite structures. In this way, there are also some connections to topological and general projective geometries so that the text might appeal to those working both in finite and infinite incidence structures.

Here, we are interested in presenting the material in a way which could be used by students and the general finite geometer, so we pay particular attention to the various models which may be used to study translation planes. These include the methods of André which utilize vector spaces presented in the text mentioned above, the studies of Bose–Bruck who define translation planes within projective spaces. We show how to visualize one of these processes within the other and generalize the construction techniques so as to connect constructions such as "retraction" of vector space spreads and the geometric "lifting" procedures of projective space partitions.

We also develop coodinate procedures based upon linear transformation sets and on coordinatizing ternary rings. We also consider the procedures which identify points as field or skewfield elements such as those of Oyama and Ebert. It might be mentioned that various of these methods are explicated in the *Foundations* book or the *Subplane* text; we revisit them here for completeness but in a more concentrated and deliberate manner.

Also, certain translation planes may be best visualized as certain sets of points ('ovoids') within the Klein quadric in five dimensional projective space. We, therefore, present the necessary classical background which allows the student to transfer the point of view from points and lines as vectors and certain subspaces and translates to sets of points within quadrics. Ovoids in 7- or 8-dimensional finite vector spaces 'slice' to ovoids of the Klein quadric, so we provide a description of all known examples. For example, there are the unitary and Ree–Tits ovoids, all of whose slices have been constructed and studied by Kantor. Further, there are the binary and ternary ovoids of Conway, Kleidman and Wilson, of which only the binary slices have been studied. Finally, we have the r-ary ovoids of Moorhouse, which provide an enormous variety of translation planes of which very little is known. We also consider the methods and procedures of Kantor on slicing, spreading and extending within projective spaces with partitioning of quadrics as the principal starting point.

We also pay particular attention to construction techniques. These include the general procedures of net replacement particularly multiple derivation. Since duals of certain ubiquitous translation planes permit derivation and produce semi-translation planes, we provide a more general theory of derivation and the sorts of geometric structures which can be realized or inherited using derivation. There is a very important class of examples, "generalized André" systems/planes to which we pay particular attention. Here we show important new developments by the use of "hyper-reguli." We also discuss the structure of derivable nets and their generalizations. We include the related construction techniques of transpose, t-extension, t-distortion and (algebraic) lifting.

A principal theme of the text will also be to show how the collineation group can affect the structure of the plane and what information about the plane can be obtained by imposing group-theoretic conditions on the plane. We shall present the important studies of Foulser on Baer collineations and shall be particularly interested in multiply transitive group operations on various related geometric incidence structures and the action of $GL(n, V)$ on various geometries.

We present various approaches to the study of semifield or division ring planes. These include both coordinate and linear algebra approaches. Furthermore, we present a unified approach to the cyclic semifield planes and to certain of their related translation planes which admit $GL(2, K)$ as collineation groups.

We include in this volume an introduction to the geometries which are connected to two-dimensional translation planes. These include flocks of quadric sets in three-dimensional projective space, certain generalized quadrangles and projective planes with restricted point-line transitivity. In addition, we include material on generalizations of flocks to higher-dimensional spaces and the consequent import for the theory of translation planes.

All of these geometries are related to sets of reguli that are interwoven in particular ways. We present some structure theory for translation planes and geometries which are related to various sets of reguli. These also include the chains of Bruen and the nests of reguli of Ebert and Baker.

Finally, but perhaps most important, we include an atlas of planes. This is, in fact, our guiding principle; to develop the material, models and methods and processes so that the examples can be understood with a minimum of difficulty.

With the few possible exceptions, all known finite projective (or affine) planes are intimately connected with translation planes. So, we provide an atlas of the known finite translation planes. However, as certain planes are known only in the sense that they may be constructed with a certain procedure, the atlas will refer back to the construction procedures developed in the text. An example of planes of this sort are the myriad planes which may be constructed from a given finite Desarguesian plane by the construction techniques of 'lifting' and 'contracting'. We would list only a parent plane in this case and provide some examples of lifting and contracting, which are in the literature. We also pay more attention to the planes that are included in one of the classification results, which we shall be discussing, as say to a list by computer enumeration of all planes of a certain order. Also, we make some subjective judgements as to what planes are 'interesting' and explicated or merely listed by process or computer. We furthermore include in this text an atlas of all finite projective (or affine planes).

The set of all known finite projective planes may be briefly partitioned in the following manner. There are the infinite classes of projective planes of Hughes of order q^2 and those derived from the Hughes planes, the Ostrom–Rosati planes, the planes of Figueroa of order q^3 and the planes of Coulter–Matthews of order q^2, all of which are not translation planes. There are the unusual projective planes of order 16 of Mathon and of Moorhouse of order 25, also not translation planes. Apart from these projective planes, all other known finite projective planes are either translation planes or their duals or constructible from translation planes by sequences of dualization and derivation constructions. All of these planes are either translation planes or their duals or semi-translation or their duals (in fact, the Hughes planes are also semi-translation planes).

Thus, we would hope that this handbook would provide the geometer with a search tool to find a particular plane and a stimulus to show that the 'new' plane is, in fact, not on the list.

Also, the authors appreciate the support of the Universities of Lecce, Caledonian and Iowa for providing the sort of supportive atmosphere which permitted the writing of this text.

Remarks on the Bibliography. The text is intended to serve as a handbook for those working in the area of finite translation planes and touches on a variety of areas of finite geometry that connect with translation planes. However, it is an easy matter to find areas that we have not covered, simply for lack of time and space. In light of this, we have included a bibliography which purports to list every reference that we have found in our work which is connected to translation planes, either directly or through various of the many geometries that are connected to translation planes. For example, there are references to unitals in Desarguesian planes, sharply transitive sets of permutations, and groups acting abstractly on point-line incidence geometries, ovals, generalized quadrangles, flocks of quadric sets, ovoids and spreads in projective space, and so on. A bibliography like this is certainly eclectic and most certainly incomplete. However, it is our hope that a list of references connected to translation planes even obliquely will serve as a valuable resource for the geometer looking for strings to this area. So, our list of approximately 1,200 references serves notice again that geometers working in one particular area need to be aware and cognizant of those working in related areas, and that those connecting areas continue to grow.

Remarks on the Index. The index is intended to be a search source and we have organized this into three parts, listed in the following order: Theorems, Models, general index, each of which is listed alphabetically. For example, where to find the "Thas, Bader–Lunardon Theorem," the reader would look at the "Theorems" part of the index, under the t's or to search for the "Penttila–Law Flocks," the reader would look at the "Models" part of the index under the p's. Ordinary terms such as "flock of a quadratic cone," would be found by looking through the general index under the f's.

Finally, we offer a few words about models. Because of its heterogenous and multidisciplinary nature, the study of translation planes used the terminology of the varied disciplines to which it is linked: areas from which it borrows tools and results and areas to which it contributes. Moreover, translation planes themselves are represented by a range of equivalent models: as affine planes, as spreads on vector spaces, as partitions of projective spaces, as quasifields and so on.

Hence, fixing notation by rigidly adopting a specific model would have the tendency of locking readers out of many published works which might adopt a different model. So no attempt has been made at consistency of notation; our excuse is that this is to the benefit of the reader, for the reasons just indicated. Moreover, to emphasize that any particular result be viewed in a number of alternative ways, we often go as far as to tacitly switch points of view 'midstream' within theorems and proofs.

The Design is a Puzzle and a Pun. For all puzzle lovers: The design on the cover of the book is intended to be a puzzle and a pun of sorts. Can you figure out what it is trying to say? (Hint: What is the significance of the "gold line"?) See the last page of Chapter 105 for the solution.

CHAPTER 2

Translation Plane Structure Theory.

DEFINITION 2.1. A 'translation plane' is an affine plane that admits a group of translations that acts transitively on its points.

We have mentioned that essentially any affine or projective plane is intrinsically connected to some translation plane. To see why this might be so, we mention the structure theory for the set of translation planes.

It is well known that the points of each translation plane may be viewed as the vectors of some vector space over a skewfield K and that the lines may be considered as translates of a particular set of 'half-dimensional' vector subspaces the union of which forms a cover of the vector space (of the points). Of course, since our emphasis here is on finite translation planes, the skewfield K becomes a field isomorphic to $GF(p^t)$, where p is a prime and t is a positive integer. However, the particular field K sometimes becomes an issue, so we retain the notation $PG(2n - 1, K)$ as opposed to the more common $PG(2n - 1, q)$, when considering projective spaces and the 'spread' associated with a translation plane.

NOTATION 2.2. We shall use the notation K_q when referring to a field isomorphic to $GF(p^t)$, where $q = p^t$.

REMARK 2.3. A 'translation plane' is necessarily an affine plane. When we wish to consider the associated projective plane, we shall use the notation 'projective translation plane'.

DEFINITION 2.4. A partition of a vector space V of dimension $2r$ over a field K_q (isomorphic to $GF(p^t)$) by a collection $\mathcal{T}$ of pairwise trivially intersecting subspaces of dimension r is called a 'spread' on V. The associated collection of all cosets $v+T$, for $v \in V$ and $T \in \mathcal{T}$, becomes the set of lines of a translation plane whose point-set is V.

We note that the 'order' of the translation plane then is q^r.

DEFINITION 2.5. We define a 'Desarguesian' translation plane or simply a 'Desarguesian plane' as a translation plane given by a partition of dimension 2 over a field K_q.

2.1. Group Partitions.

DEFINITION 2.6. Let G be a group. A 'partition' of G is a set $\mathfrak{N}$ of trivially pairwise intersecting disjoint proper subgroups of G such that $G = \bigcup \mathfrak{N}$. The members of $\mathfrak{N}$ are the 'components' of the partition. If all the components in $\mathfrak{N}$ are normal in G then $\mathfrak{N}$ is said to be a 'normal partition' of G.

A vector space of dimension at least 2, regarded as an Abelian group, admits a partition by all its dimension-one subspaces. The components of a partition, even a normal one, need not have the same size:

EXAMPLE 2.7. Let V be a vector space of dimension r over $K = GF(q^d) \supset GF(q) - F$. Let $\mathfrak{K}$ be the partition of V by its dimension-one K-spaces. Choose $A \in \mathfrak{K}$ and let $\mathfrak{A}$ be the set of dimension-one F-spaces in A. Then $(\mathfrak{K} \setminus \{A\}) \cup \mathfrak{A}$ is a normal partition of $(V, +)$ with some components of size q^d and others of size q.

More exaggerated examples of normal partitions of vector spaces are easily obtainable by modifying the strategy of the example. However, if we impose the additional condition that a normally partitioned group G is generated by any pair of distinct components, then the possibilities are much more restricted.

DEFINITION 2.8. A normal partition $\mathfrak{N}$ of a group G is called a 'splitting normal partition' of G if $G = \langle N_1, N_2 \rangle \; (= N_1 N_2)$, whenever N_1 and N_2 are distinct components of G. (The term 'congruence partition' is sometimes used for 'splitting normal partition'.)

We shall show in two stages that splitting normal partitions lead to translation planes. The first stage is to show that the group of a splitting normal partition is an Abelian group that is a direct sum of any pair of distinct components.

THEOREM 2.9. Let G be a group that admits a splitting normal partition $\mathfrak{N}$:
$$G = \langle N_1, N_2 \rangle \; \forall N_1, N_2 \in \mathfrak{N}, N_1 \neq N_2.$$
Then G is an Abelian group such that $G = N_1 N_2$ whenever $N_1, N_2 \in \mathfrak{N}$ are distinct components. Moreover, all the components $N \in \mathfrak{N}$ are mutually isomorphic groups.

THEOREM 2.10. Let $\mathfrak{N}$ be the components of a splitting normal partition (see Definition 2.8) of a group G. Then the following hold:

(1) G is an Abelian group and $G = N_1 \oplus N_2$ whenever $N_1 \cong N_2$ are distinct components.

(2) Then the set-theoretical incidence structure on G is defined as follows:
$$\pi(\mathfrak{N}) := (G, \{ x + N \mid x \in G, \; N \in \mathfrak{N} \}),$$
whose 'points' are the elements of G and whose 'lines' are the cosets $x + N$ of the components $N \in \mathfrak{N}$, is an affine plane. Two lines of $\pi(\mathfrak{N})$ are 'parallel' if and only if they are cosets of the same $N \in \mathfrak{N}$.

(3) The 'translation group' of $\pi(\mathfrak{N})$ is simply the translation group of the Abelian group $(G, +)$:
$$\tau_G := \{ \tau_g : x \mapsto x + g \mid g \in G \}.$$

(4) The translation group τ_G is regular on the affine points and hence $\pi(\mathfrak{N})$ is a translation plane by definition.

The set of endomorphisms of an Abelian group G forms a ring under standard addition and composition. We shall define a subring of this ring that is associated with any partition of G.

DEFINITION 2.11. Let $(G, +)$ be an Abelian group admitting a partition $\mathfrak{N}$ and let $\mathrm{Hom}(G, +)$ denote the ring of additive endomorphisms of $(G, +)$ under addition and map composition. Then the 'kernel' $(\mathcal{K}, +, \circ)$ of the partition, or the 'kernel' of $(G, \mathfrak{N})$, is the subring of the ring of $\mathrm{Hom}(G, +)$ that fixes every component $N \in \mathfrak{N}$. The members of the kernel of any partition will be called the 'kernel endomorphisms' of the partition.

We refrain from fixing one of the two possible interpretations for the kernel composition operation in $\text{Hom}(G, +)$. When G is finite, the most common situation we encounter, the interpretations coincide because the following result forces finite kernels to be commutative.

THEOREM 2.12. Let G be a group that admits a splitting normal partition of $\mathfrak{N}$ (Definition 2.8), so G is Abelian. Then the kernel $\mathcal{K}$ of $(G, \mathfrak{N})$ (Definition 2.11) is a skewfield and G is a vector space over $\mathcal{K}$ under its standard action on G. Moreover, every component of $\mathfrak{N}$ is a $\mathcal{K}$-subspace of V.

We now have a better picture of translation planes associated with splitting normal partitions of a group. Later we show that all translation planes are of this form.

COROLLARY 2.13. Let the group $(G, +)$ admit a splitting normal partition $\mathfrak{N}$, and let $\mathbf{o}_G \in G$ denote the identity element. Then $(G, +)$ is an Abelian group that becomes a vector space under the standard action of the kernel $\mathcal{K}$ of $\mathfrak{N}$. Furthermore, the translation plane $\pi_{\mathfrak{N}}$ admits $\mathcal{K}^*$ as a group of homologies with center $\mathbf{o}_G$. In addition, the lines through $\mathbf{o}_G$ are the members of $\mathfrak{N}$ and $\mathcal{K}$ is the largest subgroup of $\text{Hom}(G, +)$ that leaves invariant each of the lines through $\mathbf{o}_G$.

The corollary above characterizes splitting normal partitions of a group as being equivalent to partitions of vector spaces by a collection of subspaces such that the direct-sum of any two is the whole space. Such partitions of a vector space are called 'spreads', and from now on, we shall replace the terminology based on group partitions with the essentially equivalent terminology of spreads, based on vector spaces. This agreement permits convenient access to the notions of linear and projective spaces in the study of translation planes, since all translation planes turn out to be equivalent to spreads (and splitting normal partitions). Furthermore, since combinatorial incidence geometries tend to be realizable from classical structures such as vector spaces, affine or projective spaces, it becomes reasonable to presume that most affine or projective planes could be connected in some manner to a translation plane.

If one wishes to pursue the study of geometries which admit translations but which are not affine planes (or affine spaces), this may be done by relaxing the requirement that an Abelian partition be a congruence partition. In particular, translation 'Sperner spaces' may be obtained in this way.

2.2. André's Characterization of Spreads.

We begin this section by describing some of the terminology associated with spreads.

DEFINITION 2.14. Let $(V, +)$ be a vector space over any skewfield F. Then a collection $\mathcal{S}$ of subspaces of V is called a 'spread' , or an 'F-spread', if (1) $A, B \in \mathcal{S}$ and $A \neq B$ then $V = A \oplus B$ and (2) every $x \in V^*$ lies in a unique member of $\mathcal{S}$. The members of $\mathcal{S}$ are the 'components' of the spread and V is the 'ambient space' for the spread. The zero element $\mathbf{o}_V$ is also the 'origin' of the spread.

REMARK 2.15. We note that the use of the term 'spread' is not consistent in the literature. We emphasize that not all spreads encountered elsewhere will necessarily have the direct sum property. Note that it is important to note that by definition, the subspaces of a spread are mutually disjoint as vector subspaces.

By Corollary 2.13, it is obvious that spreads and splitting normal partitions of groups are the same objects:

COROLLARY 2.16. Let $(G, +)$ be a group that admits a splitting normal partition $\mathfrak{N}$ and let $\mathcal{K}$ be the kernel of $\mathfrak{N}$. Then $(G, +)$ is a vector space for every subskewfield $\mathcal{F} \subseteq \mathcal{K}$ (under its standard action). Furthermore, regarding $(G, +)$ as a vector space over $\mathcal{F}$, the set $\mathfrak{N}$ is an $\mathcal{F}$-spread. Conversely, every $\mathcal{F}$-spread $\mathfrak{S}$ on $(V, +)$, a vector space over a skewfield $\mathcal{F}$, is a splitting normal partition on $(V, +)$ and the kernel $\mathcal{K}$ of this spread is the subring of all endomorphisms in $\mathrm{Hom}(V, +)$ that leaves every component invariant. Hence, $\mathcal{K}$ is a skewfield that contains $\mathcal{F}$ as a subskewfield (in the sense that a subfield of $\mathcal{K}$ induces the same action on $(V, +)$ as $\mathcal{F}$).

We have already noted that every splitting normal partition of a group yields a translation plane and we have introduced some terminology relating the partition to the plane (see Theorem 2.10). In view of the corollary above, we may assume a parallel terminology relating spreads $(V, \mathfrak{S})$ to their translation planes $\pi_{\mathfrak{S}}$.

However, in view of the paramount importance of spreads in translation plane theory, and to avoid being distracted by 'normal partitions', which are no longer needed, we summarize, in the following result, some of the above work on normal partitions explicitly in terms of 'vector-space' partitions.

THEOREM 2.17. Let V denote a vector space over a skewfield F admitting a spread $\mathfrak{S}$, consisting of F-subspaces of V. Thus, $\mathfrak{S}$ is an F-spread, by definition, and $\Pi(\mathfrak{S})$, the set-theoretic incidence structure associated with $\mathfrak{S}$, has V as its point set and its lines are the cosets of the components $S \in \mathfrak{S}$. The following hold:

(1) $\Pi(\mathfrak{S})$ is a translation plane and its translation group is $\{ t_a \colon x \mapsto x + a \mid a \in V \}$, the translation group of $(V, +)$ viewed as an Abelian group.

(2) The group $\{ h_f \colon v \mapsto vf \mid f \in F \}$, the group of scalar multiplication on V induced by F^*, is a group of kernel homologies (this is, by definition, the homology group with axis the line at infinity, of $\Pi(\mathfrak{S})$ (see Theorem 2.17)).

(3) The skewfield F lies in the unique maximal ring R of $\mathrm{Hom}(V, +)$ that leaves every component of $\mathfrak{S}$ invariant and R is a skewfield such that V is a vector space over R. Moreover, $\mathfrak{S}$ is also an R-spread and $\{ h_r \colon v \mapsto xr \mid r \in R \}$, the action of R on V, is the full group of kernel homologies of the translation plane $\Pi(\mathfrak{S})$.

(4) The components of $\mathfrak{S}$ are isomorphic as vector subspaces of V regarded as an F-space.

The following definition is unambiguous because of part (4) in Theorem 2.17 above.

DEFINITION 2.18. Let $\pi = (V, \mathfrak{S})$ be an F-spread and V a vector space over the skewfield F. Then the 'F-dimension' of π is the common value of the dimension of the components of $\mathfrak{S}$, regarded as F-spaces. In particular, the 'dimension' of π is its K-dimension when V is viewed as a vector space over K, the kernel of π.

We may now classify the translation planes associated with dimension-one spreads as being Desarguesian.

PROPOSITION 2.19. $AG(2, K)$ is the translation plane associated with a spread of K-dimension one. Conversely, every affine plane associated with a dimension-one spread over a skewfield K is isomorphic to $AG(2, K)$.

Recall that for an arbitrary affine plane the kernel homologies, by definition, are the central collineations with the line at infinity as axis and with a given proper center (see Theorem 2.17).

REMARK 2.20. A translation plane admits a transitive group of kernel homologies if and only if the plane is Desarguesian.

Our next objective is to describe all possible isomorphisms between the affine translation 'planes' associated with spreads in terms of the structure of the spreads themselves. Thus, we need to define in what sense two spreads are isomorphic and then to describe the isomorphisms among the associated planes in terms of the isomorphisms of the spreads.

DEFINITION 2.21. Let $(A, \mathfrak{A})$ and $(B, \mathfrak{B})$ be spreads, with associated ambient Abelian groups $(A, +)$ and $(B, +)$, respectively. Then a 'spread isomorphism', from $(A, \mathfrak{A})$ to $(B, \mathfrak{B})$, is a bijective additive group isomorphism $f \colon A \to B$ such that f induces a bijection from the components of $\mathfrak{A}$ onto the components of $\mathfrak{B}$.

It turns out that this concept of an isomorphism between spreads is much more restrictive than it might appear as the isomorphisms are necessarily vector-space isomorphisms when A and B are viewed as vector spaces over their respective kernels. To clarify this notion, it is desirable to make explicit what we mean by an isomorphism among vector spaces over isomorphic skewfields that are not necessarily identical.

DEFINITION 2.22. Assume that $(A, +)$ and $(B, +)$ are vector spaces over skewfields K and L, respectively—with the skewfields operating from, say, the left. Then a bijective additive isomorphism $\Psi \colon A \to B$ is a '(K, L)-isomorphism' if there is a bijective ring isomorphism $\psi \colon K \to L$ such that $\forall k \in K, a \in A : \Psi(ka) = k^{\psi} \Psi(a)$. Note that when $A = B$ and $K = L$ then the isomorphisms are simply the members of $\Gamma L(A, K)$, the group of K-semilinear automorphisms of A.

Assume that the skewfield K, in Definition 2.22 above, is a subring of $\mathrm{Hom}(A, +)$ and $\Phi \colon A \to B$ is a bijective additive isomorphism, from the additive group $(A, +)$ onto the additive group $(B, +)$. Under these conditions $L := K^{\Phi} = \Phi K \Phi^{-1}$ is a skewfield in the ring $\mathrm{Hom}(B, +)$ and conjugation by Φ induces a bijective ring isomorphism from the skewfield K onto the skewfield L. That is, we obtain that $k \mapsto \Phi k \Phi^{-1} := k^{\phi}$ is a ring isomorphism from the skewfield K onto the skewfield $L := \Phi K \Phi^{-1}$, and moreover since Φ and k are in $\mathrm{Hom}(A, +)$, $\Phi k(a) = \Phi(ka)$ so: $\Phi(ka) = \Phi(k\Phi^{-1}\Phi(a)) = k^{\phi}\Phi(a)$, implying that Φ is a (K, L) isomorphism from A to $\Phi(A) = B$, in the sense of Definition 2.22 above.

In the context of 'spreads', $(A, \mathfrak{A})$ and $(B, \mathfrak{B})$, a spread isomorphism $\Phi \colon A \to B$ conjugates the kernel skewfield $\mathcal{A}$ of $(A, \mathfrak{A})$ onto the kernel skewfield $\mathfrak{B}$ of $(B, \mathfrak{B})$. In particular, this is because these are, by definition, the maximal subrings of $\mathrm{Hom}(A, +)$ and $\mathrm{Hom}(B, +)$ that, respectively, leave invariant each of the components of $\mathfrak{A}$ and $\mathfrak{B}$. Thus, an additive map is a spread isomorphism if and only if it is a vector-space isomorphism, in the sense of Definition 2.22, relative to the kernels.

THEOREM 2.23. Let $(A, \mathfrak{A})$ and $(B, \mathfrak{B})$ be spreads, with kernels $\mathcal{K}$ and $\mathcal{L}$, respectively. Then an additive map $\Phi \in A \to B$ induces a spread isomorphism from $(A, \mathfrak{A})$ to $(B, \mathfrak{B})$ if and only if Φ is a $(\mathcal{K}, \mathcal{L})$-semilinear bijective isomorphism from A to B that maps components onto components.

The special case when the two spreads coincide is worth stressing as the spread isomorphisms from $(A, \mathfrak{A})$ onto itself are thus 'automorphisms':

DEFINITION 2.24. Let $(V, \mathfrak{S})$ be a spread on the additive group $(V, +)$. Then $\mathrm{Aut}(V, +)$, the 'automorphism group' of $(V, \mathfrak{S})$, is the subgroup of $GL(V, +)$ that maps the components of $\mathfrak{S}$ onto components of $\mathfrak{S}$.

In this context, Theorem 2.23 may be stated as:

THEOREM 2.25. Let $(V, \mathfrak{S})$ be a spread with kernel $\mathcal{K}$. Then the largest subgroup of $GL(V, +)$ that permutes the members of $\mathfrak{S}$ among themselves is in $\Gamma(V, \mathcal{K})$, when V is viewed as a vector space under the given action of $\mathcal{K}$.

Thus, additive bijections of a vector space that leave a spread invariant are, in fact, semilinear. So, we observe that when the kernel is large, the groups tend to get 'boring'. In fact, to find exotic groups acting on spreads, it is necessary to consider spreads that have 'large' dimensions over their kernels whatever that may mean in the present context. On the other hand, boring groups might act on 'interesting' translation planes. One possible boring group would be the identity group. Is it possible that a translation plane might have its full collineation complement simply the group of translations that define its structure? Since we have the kernel homology group as a collineation group, such a translation plane must have kernel isomorphic to $GF(2)$ and then would not admit further collineations that leave the zero vector invariant. We note that such translation planes do, in fact, exist. Of course, the existence of interesting groups might force the translation plane itself to be known or classical in some sense.

2.2.1. André's Spread Isomorphism Theorem.

THEOREM 2.26. Let $\Pi_{(A,\mathfrak{A})}$ and $\Pi_{(B,\mathfrak{B})}$ denote the translation planes associated with the spreads $(A, \mathfrak{A})$ and $(B, \mathfrak{B})$, respectively. Let K and L denote the kernels of the spreads $\mathfrak{A}$ and $\mathfrak{B}$, respectively. Then the following are equivalent:

(1) $\Psi \colon A \to B$ induces an isomorphism from the affine plane $\Pi_{(A,\mathfrak{A})}$ onto the affine plane $\Pi_{(B,\mathfrak{B})}$;

(2) $\Psi = \tau\Phi \colon A \to B$ where $\tau \colon B \to B$ is a translation of $\Pi_{(B,\mathfrak{B})}$ and $\Phi \colon A \to B$ is an 'additive' bijection mapping the components of $\mathfrak{A}$ onto the components of $\mathfrak{B}$;

(3) $\Psi = \tau\Phi \colon A \to B$ where $\tau \colon B \to B$ is a translation of $\Pi_{(B,\mathfrak{B})}$ and $\Phi \colon A \to B$ is a (K, L)-semilinear bijective isomorphism from A onto B that maps the components of $\mathfrak{A}$ onto the components of $\mathfrak{B}$.

Note that the theorem implies that non-isomorphic spreads yield non-isomorphic planes. It also means that the isomorphisms between any two affine 'planes' associated with spreads can be completely described in terms of the isomorphisms between the 'spreads'. In particular, the automorphism group of a spread completely determines the collineation group of the associated plane.

The importance of these results stems from the second fundamental theorem of André that states that every affine translation plane is associated with a spread (which is unique up to isomorphism by the above theorem).

Before turning to this result, we explicitly state the special case of the above theorem for a single spread. We describe the 'collineation group' of the translation 'plane' arising from a spread in terms of the automorphism group of the spread. This corollary is the most frequent form in which the above theorem is encountered.

THEOREM 2.27. Let π be the translation plane associated with a spread $(V, \mathfrak{S})$, so that the translation group of π is: $\tau_V := \{\tau_v \colon x \mapsto x + v,\ x \in V \mid v \in V\}$. Then τ_V is normal in the collineation group of the plane π and $\mathrm{Aut}(\pi)$—the 'full' collineation group of the *plane* π—is given by the following semidirect product: $\mathrm{Aut}(\pi) = \tau_V\, \mathrm{Aut}(V, \mathfrak{S})$. Furthermore, the subgroup $\mathrm{Aut}(V, \mathfrak{S})$ is the 'stabilizer of $\mathbf{o}$ in $\mathrm{Aut}(\pi)$'.

In view of Remark 2.25, the theorem may be restated as follows.

COROLLARY 2.28. Let $(V, \mathfrak{S})$ be a spread, with V viewed as a vector space over the kernel $\mathcal{K}$ of $\mathfrak{S}$. Then the collineation group of the associated translation plane $\pi_{(V, \mathfrak{S})}$ is $G = TG_{\mathbf{o}}$, where T is the translation group of $(V, +)$ and $G_{\mathbf{o}}$ is the largest subgroup of $\Gamma(V, \mathcal{K})$ that permutes the components of the spread among themselves. This group coincides with the subgroup of $GL(V, +)$ that permutes the components of the spread among themselves.

Thus, we have complete knowledge of isomorphisms and collineations of translation planes, 'associated with spreads', in terms of the isomorphisms and automorphisms of spreads themselves. We now demonstrate that all translation planes are associated with spreads. So, the study of translation planes and their collineation groups reduces in principle to the corresponding study of spreads themselves.

2.2.2. The Fundamental Theorem.

THEOREM 2.29. Let π be a translation plane with translation group $(T, +)$, written additively, and let $\mathcal{M}$ denote the set of parallel classes of π. Let T_μ denote the subgroup of T fixing all the lines of μ, for $\mu \in \mathcal{M}$. Then:

(1) $\Gamma = \{T_\mu \mid \mu \in \mathcal{M}\}$ is a spread on T and hence T is a vector space over the associated kernel $\mathcal{K}$.

(2) π is isomorphic to π_Γ, the translation plane associated with the spread (T, Γ).

André's description of translation planes means that every translation plane may be identified with a spread $\mathfrak{S}$ on a vector space V, over some skewfield K. So subsets of spreads naturally have a major impact on investigations concerned with spreads and translation planes. These are considered within the framework of partial spreads introduced in Chapters 3 and 4. However, first we pause for some examples of translation planes/spreads. We do not, however, consider the question of whether such examples are possibly isomorphic.

CHAPTER 3

Partial Spreads and Translation Nets.

Partial spreads are generalizations of spreads. They are essentially collections τ of subspaces of a vector space V such that any two distinct members of τ direct-sum to V. Since subsets of spreads are the motivating examples of partial spreads, it is not surprising that partial spreads are fundamental to the study of spreads. That is, any subset of a spread is a partial spread but there may be partial spreads that cannot be contained in (or extended to) a spread.

Hence, of related interest are the 'maximal partial spreads' and these occur in a variety of ways, some of which will be considered in this text. Some questions occur in the following manner:

(1) If a partial spread $\mathcal{P}$ is given via a vector space over a skewfield K such that the partial spread components are K-subspaces, is there a partial spread $\mathcal{P}^+$ consisting of K-subspaces that properly contains $\mathcal{P}$?

Also, does there exist a spread with kernel containing K that properly contains $\mathcal{P}$?

(2) If a partial spread $\mathcal{P}$ is given via a vector space over a skewfield K, is there a subskewfield F of K such that there is a partial spread $\mathcal{P}^+$ consisting of F-subspaces that properly contains $\mathcal{P}$?

In this chapter, we consider the basics of partial spreads. We generalize the standard theory associated with spreads and their connection with translation planes to a corresponding theory applicable to partial spreads and their associated incidence structures, 'translation nets'.

A third question related to partial spreads and their corresponding translation nets is both interesting and important:

(3) Can a given translation net be extended to an affine plane and if so, how many such planes contain the net and do the planes necessarily all need to be translation planes?

Note that various of these questions and ideas of the theory also involve clarifying the meaning of isomorphisms and automorphisms. Furthermore, these notions, which depend partly on André theory, need to be defined with some care since André theory does not generalize easily to partial spreads.

After dealing with these introductory notions, we turn to the notion of a 'subsystem' of a partial spread and, hence, also of a spread. In the most important case, a subsystem of a spread π on a vector space V, essentially involves a subspace W of V on which π 'induces' a spread.

This is followed by establishing the elementary but basic connections between partial spreads and translation nets. In particular, we consider the connection between subsystems of a partial spread and the subincidence structures of the associated translation net. This leads to and yields the basic lattice-like intersection

property of subsystems of partial spreads, particularly when the subsystems correspond to subplanes.

This is followed by a specialized investigation of 'rational translation nets':
these are essentially the nets that 'lie across' some subplane of a partial spread, and
are absolutely fundamental to the study of translation planes. Special attention is
given to the existence of a 'Foulser cover' for rational 'Desarguesian' nets. Roughly
speaking, a Foulser cover is a partition of the points covered by a partial spread
consisting of subspaces that are subplanes. There is a comprehensive and complete
theory of Foulser-covers for completely arbitrary nets, based on the classification
of derivation arising from Prohaska and Cofman, the more recent work of Johnson,
Thas and De Clerck, and culminating in the beautiful and surprising theorem of
Johnson [**753**].

3.1. Isomorphisms of Partial Spreads.

In the most general sense, a partial spread might be taken to be a collection
of subgroups of a group G any two of which must direct-sum to the whole group.
However, in this setup, André theory no longer applies as the ambient group G
need not be Abelian. For example, if $G = H \oplus H$, with H any group, then G
admits a 'partial spread' with at least three components (consisting of H-isomorphic
subgroups).

In translation plane theory, the partial spreads of most importance are the
subsets of a spread π, on some ambient space G. Hence partial spreads are defined
only when the ambient space G may be regarded as a vector space over some
skewfield K.

DEFINITION 3.1. Let V be a vector space over a skewfield K and suppose that
$\mathfrak{T}$ is a non-empty collection of K-subspaces of V. Then $(V, \mathfrak{T})$, or $\mathfrak{T}$, is a 'K-partial
spread', or simply a 'partial spread', with ambient space V, if $V = A \oplus B$ for every
pair of distinct $A, B \in \mathfrak{T}$, and if $|\mathfrak{T}| \leq 2$ assume explicitly that $V/A \cong A$ for $A \in \mathfrak{T}$.

The members of $\mathfrak{T}$, are the 'components' of the partial spread $\mathfrak{T}$, and the
common cardinal number $|\mathfrak{T}|$ is the 'degree' of $\mathfrak{T}$. The 'dimension' of the partial
spread $\mathfrak{T}$ is the common K-dimension of the 'components' in τ, regarded as K-
spaces (cf. Definition 2.18).

The legitimacy of the final sentence is easily checked:

REMARK 3.2. All the components of the partial spread $\mathfrak{T}$ are isomorphic as
vector spaces, over K (or any subskewfield of K).

This suggests a more general approach to partial spreads than the one we have
adopted. This becomes particularly relevant when the ambient space V is infinite-
dimensional. Instead of insisting, as we have, that V is a direct sum of any two
distinct components—thereby forcing all components to have 'half the dimension'
of V—we might instead have defined 'half-dimensional' subspaces by insisting that
a collection of pairwise trivially intersecting subspaces $\mathfrak{S}$ of V is a (partial) spread
if every $S \in \mathfrak{S}$ is isomorphic as a vector space to some X such that $V = X \oplus X$.

Although such 'X-partial spreads' arise inevitably in the study of infinite nets,
we shall not consider them further since the definition of a partial spread that we
have adopted leads to less technical-sounding results.

If $\mathfrak{T}$ is a partial spread, on the ambient vector space V, then so is $\mathfrak{U}$ whenever
$\mathfrak{U} \subset \mathfrak{T}$. Thus we introduce the following terminology:

DEFINITION 3.3. A subcollection $\mathfrak{U}$, of a partial spread $\mathfrak{T}$, is a 'subpartial spread' of $\mathfrak{T}$.

Note, however, that to the term 'subpartial spread' will be assigned a broader and more significant meaning in Definition 4.7.

Although subsets of spreads are always partial spreads, as we suggested, there are many partial spreads that cannot be extended to spreads and we shall see that it will emerge later that there are maximal partial spreads that are 'not' spreads, so André theory applies only in a much weaker form to partial spreads than it does to spreads.

Thus, some care is needed when considering isomorphisms between 'partial' spreads. In particular, for 'spreads', virtually all reasonable definitions are forced to be mutually equivalent by the restrictions imposed by André theory. However, for partial spreads, it becomes necessary to guard against potential pathologies.

DEFINITION 3.4. Let $(V_i, \mathfrak{T}_i)$, $i = 1, 2$, be partial spreads, where V_1 and V_2 are the associated ambient vector spaces, over a common skewfield K. Then a K-linear bijection Ψ from $\mathfrak{T}_1$ to $\mathfrak{T}_2$ is a 'K-linear isomorphism' if and only if Ψ bijectively maps the components of $\mathfrak{T}_1$ onto those of $\mathfrak{T}_2$. More generally, an 'isomorphism' from $\mathfrak{T}_1$ onto $\mathfrak{T}_2$ is an additive isomorphism from V_1 onto V_2 that maps components bijectively onto components.

The definition of an automorphism of a partial spreads, implied by the above definition, could be simplified for the finite case: an automorphism of a spread $\mathfrak{S}$, on a finite vector space V, is a map $\phi \in GL(V, +)$ that maps the components of $\mathfrak{S}$ into itself: this automatically induces a bijection on $\mathfrak{S}$. However, in the infinite case this might fail as the following exercise illustrates.

However, a linear bijection ϕ that maps the components from a 'spread' into some (partial) spread forces the latter to be a spread and ϕ to be a spread isomorphism:

REMARK 3.5. Let $(V, \mathfrak{S})$ and $(W, \mathfrak{T})$ be (partial) K-spreads, where V and W denote vector spaces over a skewfield K. Let $\phi \colon V \to W$ be a semilinear bijection that maps $\mathfrak{S}$ into $\mathfrak{T}$. Then

(1) If $\mathfrak{S}$ is a spread then $\mathfrak{T}$ is a spread and ϕ is a spread isomorphism, mapping $\mathfrak{S}$ onto $\mathfrak{T}$.

(2) $\mathfrak{S}$ is a spread if and only if $\mathfrak{T}$ is a spread.

The following corollary provides a helpful technique for deciding when a linear bijection is a spread automorphism and shows that it is sufficient to check that it maps components to components.

COROLLARY 3.6. Let $\phi \in GL(V, +)$ map a spread $\mathfrak{S}$ on V into itself. Then $\phi \in \operatorname{Aut}(\mathfrak{S})$, and hence also $\phi \in \Gamma L(V, K)$, where K is the largest subfield of $GL(V, +)$ leaving every member of $\mathfrak{S}$ invariant, i.e., the kernel of the spread.

3.2. Translation Nets.

To provide a framework for considering the geometry of partial spreads, we introduce the terminology of nets. The motivating examples of a net are the incidence substructures of an affine plane A, whose points are those of A and whose lines are the union of a collection of parallel classes of A. However, here we only consider *translation* nets, which are those associated with the parallel classes of a (partial) spread.

DEFINITION 3.7. Let $\pi = (V, \mathcal{S})$ be a (partial) spread. The associated 'net' $\mathcal{N}(\mathcal{S})$ is the set-theoretic incidence structure whose point set is V and whose line set $\mathcal{L}$ consists of all the cosets of the components of the partial spread $\mathcal{S}$, thus $\mathcal{L} = \{\, v + S \mid v \in V, \ S \in \mathcal{S} \,\}$. A 'parallel class' of any component $S \in \mathcal{S}$ consists of the set of all cosets of S: so parallel classes are the maximal pairwise disjoint sets of lines. An incidence structure isomorphic to a net associated with a partial spread is a 'translation net'. The 'subnets' of $\mathcal{N}(\mathcal{S})$ are the translation nets of form $\mathcal{N}(\mathcal{T})$, for $\mathcal{T} \subset \mathcal{S}$.

So translation planes are examples of translation nets. Note also that it has become common practice to ignore the distinction between partial spreads and the associated translation nets. However, we shall maintain the distinction for some time.

Some extensions of the above terminology are included in:

(1) Any two distinct points of a translation net lie on at most one line. Hence the terminology uses the word 'lines' rather than the word 'blocks'.

(2) Any translation net admits a regular translation group fixing each parallel class globally, which is the standard action of the associated vector space on itself.

(3) A partial spread $\pi = (V, \mathcal{S})$ is a spread if and only if the associated net $\mathcal{N}(\mathcal{S})$ is a translation plane.

CHAPTER 4

Partial Spreads and Generalizations.

In this chapter, we give some generalizations to Chapter 3 and also endeavor to provide sufficient background for discussion of 'quasi-subgeometry' partitions.

DEFINITION 4.1. A 'partial vector-space spread' of a vector space V is merely a set of vector subspaces any two distinct elements of which direct sum to the vector space and are each isomorphic to a fixed subspace. In a similar manner, one may define a 'partial projective spread'. The elements of a partial spread are called 'components'. From a partial vector-space spread $\mathcal{P}$, we may form a 'translation net' $T(\mathcal{P})$ by taking the 'points' of the net to be the vectors and the 'lines' of the net to be the vector translates of the partial spread components.

The analysis of André versus Bruck–Bose works for partial spreads just as for spreads. Hence, a partial vector-space spread produces a partial projective spread and conversely.

DEFINITION 4.2. Let π be a translation plane defined using the André method. The group $\mathcal{T}$ of translations of the underlying vector space V over a finite field K are collineations of π. This group is called the 'translation group' of π. Let $\mathcal{Z}$ denote the vector-space spread defining π, so that $\pi = \pi_{\mathcal{Z}}$. Let $\Gamma L(V, K)$ denote the semi-linear group of V over K. If σ is a collineation of π then σ is an element of the semi-direct product of $\Gamma L(V, K)$ by $\mathcal{T}$. Let $\mathcal{F}$ denote the full collineation group of π. Then, $(\Gamma L(V, K) \cap \mathcal{F})\mathcal{T} = \mathcal{F}$. We note that $\Gamma L(V, K) \cap \mathcal{F}$ may be regarded as the stabilizer subgroup of the zero vector 0: $\mathcal{F}_0$.

The group $\Gamma L(V, K) \cap \mathcal{F}$ is called the 'translation complement of π with respect to K'.

If $GL(V, K)$ is the general linear subgroup of $\Gamma L(V, K)$, then $GL(V, K) \cap \mathcal{F}$ is called the 'linear translation complement' of π with respect to K'.

If K^+ is the maximal field containing K such that the spread components are also K^+-subspaces, we shall say that K^+ is the 'kernel' of π. In this case, if $\mathcal{F}$ is the full collineation group of π then also $(\Gamma L(V, K^+) \cap \mathcal{F})\mathcal{T} = \mathcal{F}$. We then use the terms 'translation complement' and 'linear translation complement' when referring to the translation and linear translation complements with respect to the kernel K^+. Note that the translation complement is the subgroup of $\Gamma L(V, K^+)$ that permutes the spread components.

We may extend the previous definitions to collineation groups of partial vector-space spreads and speak of the translation complement of a translation net as the subgroup of $\Gamma L(V, K)$ that permutes the partial spread components. However, it may also be when considering the action on partial spreads that the associated vector space V may be allowed to be a vector space over a field L different from K and L need not be a superfield of K and a collineation might be in $\Gamma L(V, L)$ instead of $\Gamma L(V, K)$.

Generalized Partial Spreads.

DEFINITION 4.3. (1) Let V be a vector space over a field K. Let $\{\, T_i;\ i \in \Omega \,\} = \mathcal{T}$ be a set of mutually disjoint K-vector subspaces, then $\mathcal{T}$ shall be called a 'generalized partial spread'. If $V = \bigcup_{i \in \Omega} T_i$, then $\{\, T_i;\ i \in \Omega \,\}$ shall be called a 'generalized spread' of V. Note that this allows that the individual subspaces can be of different dimensions.

(2) Assume that V is isomorphic to the external direct sum $\sum_{i \in \lambda} W_{o,i}$, of dimension $|\lambda|\, w_o$, where all $W_{o,i}$ are K-vector spaces that are K-isomorphic to a fixed subspace W_o over K and all T_i are K-isomorphic to W_o, and $\dim_K W_o = w_o$, we shall say that $\{\, T_i;\ i \in \Omega \,\}$ is a '$|\lambda|$-generalized spread of size (w_o, K)'.

In all cases, the subspaces T_i are called the 'components' of the generalized spread. For example, if V is a dw_o-dimensional over K for d finite and W_o is w_o-dimensional, we obtain a 'd-generalized spread of size (w_o, K)'.

(3) If $d = 2$ and all components of the generalized spread pairwise generate V, then we have a spread. If K is finite of order q, and w_o is also finite, we shall say that we have a 'd-generalized spread of size (w_o, q)'. We shall also use the terminology 'd-spread' to indicate this same structure.

We shall be interested in the case when K is a field.

DEFINITION 4.4. We shall say that an extension field D of K 'acts' on a generalized spread if and only if (i) the multiplicative group D^* is in $GL(V, K)$, (ii) D^* contains the scalar group K^*, (iii) permutes the spread components and (iv) acts fixed-point-free (each non-identity element fixes only the zero vector). Hence, a field D acting on a generalized spread maps components of a given dimension onto components of the same dimension. Any orbit of components of a generalized spread under a field D shall be called a 'fan'. If the subfield D_L of D is induced from the stabilizer of a component L, we shall call the orbit $O(L)$ under D a 'D_L-fan'.

DEFINITION 4.5. Let $\mathcal{S}$ be a d-spread of size (s, K) of a vector space V over a field K. If we define 'points' as vectors and 'lines' as translates of spread components, we obtain an incidence geometry of points and lines such that the line set is partitioned into a set of parallel classes. If $d = 2$, then distinct lines from different parallel classes must intersect, whereas if $d > 2$, this is not the case. Such an incidence geometry is called a 'translation space' and is an example of a 'Sperner space'.

REMARK 4.6. We shall be interested in the collineation group of a translation space as well as the collineation group of a generalized spread. We define the 'linear' translation complement of a generalized spread over the field to be the subgroup of $GL(V, K)$ that permutes the components of the spread.

4.1. Subspreads and Planar Groups.

Given a partial spread $\pi = (V, \mathcal{S})$, we consider what is meant by a subsystem *induced* on a subspace $W \leq V$, by $\mathcal{S}$. Basically, the subsystem induced on W consists of the non-trivial intersections of W with $\mathcal{S}$, and the most important case is when these intersections induce a partial spread on W. In this case, W is a '*sub*partial spread' of $\mathcal{S}$.

DEFINITION 4.7. Let $\pi := (V, \mathcal{S})$ be a (partial) spread and suppose W is a non-zero additive subspace of $(V, +)$. Then $\mathcal{S}(W) := \{\, s \in \mathcal{S} \mid s \cap W \neq \mathbf{O} \,\}$ is the subpartial spread of $\mathcal{S}$ 'across' W. Moreover, the subsystem of $\pi := (V, \mathcal{S})$ 'induced' on W, is $\mathcal{S}_W = \{\, s \cap W \mid s \in \mathcal{S}(W) \,\}$, the set of non-zero intersections of W with the components of $\mathcal{S}$ across it.

The induced subsystem $\mathcal{S}_W$, also written $\pi_W := (W, \mathcal{S}_W)$, is a 'sub(partial) spread' of the (partial) spread $\pi = (V, \mathcal{S})$ if π_W turns out to be a (partial) spread on W; π_W is the (partial) spread 'induced' on W; in this case W itself is often called: (i) a (partial) subspread of π; (ii) a subnet of π; (iii) a subplane of π, if W is a subspread of π.

Note that we now have essentially two distinct usages of the term 'subpartial spread'. It might simply mean:

(1) A subcollection of components of a partial spread, Definition 3.3, and this usage includes $\mathcal{S}(W)$, the subpartial spread across W; or it might mean

(2) A sub(partial) spread induced on some subspace W. In this case, this means that the components $\mathcal{S}(W)$ across W induce a (partial) spread on W, Definition 4.7 above. The context will determine the intended meaning of the terminology.

Also note that the substructure induced by π on a subspace $W \leq V$, viz., $\mathcal{S}_W$, depends only on the components of $\mathcal{S}$ that are *across* W: the other components of $\mathcal{S}$ have no influence on the structure of $\mathcal{S}_W$.

We summarize these observations.

REMARK 4.8. Let $\pi := (V, \mathcal{S})$ be a partial spread. (1) A subpartial spread ψ of π means: (a) $\psi := (V, \mathcal{U})$, for some non-empty collection $\mathcal{U} \subset \mathcal{S}$ (Definition 3.3); or (b) $\psi := \pi_W$ where W is a subspace of V on which π induces a partial spread (Definition 4.7). Explicitly, $\pi_W = (W, \mathcal{S}_W)$ where $\mathcal{S}_W$ is a partial spread consisting of all the intersections of W with the components of $\mathcal{S}$ that lie across W. (2) If W is a subspace of V then for all $\mathcal{T}$ satisfying $\mathcal{S} \supset \mathcal{T} \supset \mathcal{S}(W)$, the system $\pi_W := (W, \mathcal{T}_W)$ coincides with $\pi_W := (W, \mathcal{S}_W)$. In particular, if a subspace W that lies within a partial spread π is a (partial) subspread of π, then all partial spreads across W induce the same (partial) spread π_W on W.

Of course, not all subspaces $W \leq V$, of a partial spread $\pi = (V, \mathcal{S})$, correspond to a partial spread. For example, W might be part of a component or the intersections with W of components that meet W might not always pairwise direct-sum to W. This happens if the components that meet W non-trivially do so in subspaces of varying sizes, or if they meet W in non-zero subspaces that are too small to direct-sum to W. But, if W is a sub(partial) spread of $\pi := (V, \mathcal{S})$, then the translation net associated with W coincides with the incidence structure that it inherits from π as is noted in the following:

Let $\pi = (V, \mathcal{S})$ be a (partial) spread, and suppose that W is a non-zero additive subspace of $(V, +)$, on which a partial spread is induced. Then the translation net defined by the partial spread induced on W coincides with the subincidence structure induced on W by the translation net associated with π.

The following propositions stated without proof on subplanes highlight important cases of the remark above.

PROPOSITION 4.9. Let $\pi = (V, \mathcal{S})$ be a (partial) spread and Π the associated translation net on V. For any subspace W of V, use the notation $\Pi(v, W)$ to denote the subincidence structure of Π whose point set is $v + W$.

PROPOSITION 4.10. Suppose W is a non-zero additive subspace of $(V, +)$, on which a **spread** is induced. Then the following hold:

(1) The translation net $\Pi(\mathbf{o}, W)$, induced by the translation net Π, coincides with the translation plane associated with the spread induced on W by π. Furthermore, the translation group of W is W itself and this coincides with the subgroup of the translation group of Π leaving W invariant.

(2) The subplanes of Π that are translation planes and that admit W as their translation subgroup are precisely the cosets of W, viz., the incidence structures of form $\Pi(v, W)$, $v \in V$.

(3) Every subplane of Π that is a translation plane is of the form $\Pi(v, W)$, for some subspread $W \leq V$ of π.

Thus, the above exercises imply that *results concerning subspreads of a spread readily convert to corresponding results concerning subplanes of the associated translation planes.* Similarly, results concerning subpartial spreads have an interpretation for translation nets.

Generally speaking, such conversions are left to the reader. However, the case of *planar groups* deserves special attention.

DEFINITION 4.11. Let $\pi = (V, \mathfrak{T})$ be a spread with associated translation plane Π. If $G \leq \mathrm{Aut}(\pi)$ is an automorphism group of the spread π such that $\mathrm{Fix}(G)$ is a subspread ψ of π then G is a 'planar group' of π that is often denoted by π_G. Furthermore, Π_G denotes the subincidence structure of Π associated with π_G.

The following remark lists the obvious connections between π_G and Π_G. We shall take these for granted.

REMARK 4.12. Let π be a spread, Π the associated translation plane, and suppose that $G \leq \mathrm{Aut}(\pi)$. Then the following statements are equivalent:

(1) The group G is a planar group of the spread π with fixed subspread π_G.

(2) The group G is a planar group of the translation plane Π such that its fixed subplane Π_G includes $\mathbf{o}$, the zero of the ambient space of π.

(3) The group G is a planar group of π and Π such that Π_G, the subplane of Π fixed by G, coincides with the translation plane associated with the spread π_G, the subspread of π fixed by G.

Hence, π_G, as well as Π_G, will be called the 'fixed plane of G'; we shall usually not distinguish between π_G and Π_G.

4.2. Kernel Subplanes of a Spread.

Recall that a 'kernel-homology' of an arbitrary affine plane Π is a homology with the line at infinity as axis. Below, we introduce the kernel subplane of a translation plane.

DEFINITION 4.13. Let $\pi = (V, \mathfrak{T})$ be a spread with kernel K. Thus, K is the largest subfield in $\mathrm{Hom}(V, +)$ such that the components are subspaces. Let $\sigma = (S, \mathfrak{S})$ be a subspread of π across some subspace over the prime field $S \leq V$. Then the subspread σ, or S, is a 'kernel subspread' (or a 'kernel subplane'), if S is a dimension-2 vector space over some subskewfield F of K. Furthermore, σ, and also S, are 'F-subspreads' of π.

The following proposition lists other versions of the definition. Clearly, these characterizations, stated for spreads, may be reformulated to apply to translation planes.

PROPOSITION 4.14 (Kernel Subspreads). Let $\pi = (V, \mathfrak{T})$ and $\sigma = (S, \mathfrak{S})$ be a subspread induced on a subspace S of V. Then the following are equivalent:

(1) σ is a kernel subplane.

(2) σ is Desarguesian and the kernel homologies of σ extend to kernel homologies of π.

(3) The kernel homologies of π that leave σ invariant induce a transitive group on Z^*, for every component Z of σ.

The following remark shows that any 2-dimension subspace over a kernel field is automatically a kernel plane, provided it is not contained in a component of the spread. Hence, any spread has a large number of kernel subspreads.

REMARK 4.15. Let $(V, \mathfrak{T})$ be a spread with kernel K, and suppose $F \leq K$ is a subskewfield of K. Then an F-subspace W of V is a kernel subplane, in fact an F-subspread, Definition 4.13, if and only if W has dimension 2 over F and meets more than one component non-trivially.

The following remark shows that a unique kernel subspread is specified, relative to a kernel field F, once a pair of 'axes' and a 'unit point' off of the axes are given.

REMARK 4.16. Let $\pi = (V, \mathfrak{T})$ be a spread with kernel K, and suppose $F \leq K$ is a subskewfield of K. Consider the set $\{e, X, Y\}$ where $X, Y \in \mathfrak{T}$ are distinct components of π, and $e \in V \setminus (X \cup Y)$. Then there is a unique F-kernel subspread $\sigma = (S, \mathfrak{S})$ of π such that X and Y are components of σ (equivalently they both meet S non-trivially) and $e \in S$.

The use of the term kernel spread is sometimes used in a different sense than from the meaning assigned to it in Definition 4.13. To cover these alternative usages, we introduce the following terminology.

DEFINITION 4.17. Let $\pi = (V, \mathfrak{T})$ be a spread with kernel K. Let $\sigma = (S, \mathfrak{S})$ be a subspread of π across some subspace $S \leq V$. (1) The subspread σ is 'kernel-invariant' if K^* leaves σ invariant. (2) The subspread σ is a 'maximal kernel subspread' if K^* induces a transitive homology group on σ (this is always taken to mean transitive on the non-zero points of each component).

Thus all maximal kernel subspreads are kernel subspreads, as defined above, and a kernel subspread is a maximal kernel subspread if and only if it is kernel-invariant. Moreover, F-spreads are simply the minimal F-invariant subspreads.

As remarked earlier, we shall *freely use planar versions of these results* without further justification.

The following situation arises very frequently.

REMARK 4.18. Let $\pi = (V, \mathfrak{T})$ be a spread of finite order q^2 with kernel $K = GF(q)$. Then every 2-dimensional subspace of V over K is either a line or a Baer subplane of $\Pi(V, \mathfrak{T})$, in the sense that the 1-dimensional subspaces of a 2-dimensional space which is not a line is a spread for the associated affine Baer subplane.

4.3. Intersections of Partial Spreads.

We are concerned with the intersection of spreads and partial spreads and are particularly interested in the associated lattice structure that this implies. Perhaps the most natural way to consider this is to step back and look at the corresponding results for affine planes. So we first review the lattice structure of affine subplanes of an affine plane that share a non-degenerate subincidence structure.

REMARK 4.19. Let Π be an affine plane. Consider a collection of affine sub-planes $(\Pi_i)_{i\in I}$ that share a quadrangle in the projective closure of Π. Each Π_i includes at least two non-parallel lines, say X and Y, and an affine point $u \notin (X \cup Y)$. Let P_i denote the points of Π_i, $i \in I$. Then the intersection $\bigcap(P_i)_{i\in I}$ is the set of points of an affine subplane of Π, denoted by $\bigcap(\Pi_i)_{i\in I}$.

The usual notion of "generation" can thus be justified. However, we shall only apply the concept in the following restricted sense.

DEFINITION 4.20. Let W be a collection of points and lines of an affine plane Π that includes two non-parallel affine lines, X and Y and an affine point $u \notin (X \cup Y)$. Then the affine plane 'generated' by W is the unique smallest affine subplane Π_0 of Π that includes W as a subconfiguration.

Hence, we immediately obtain:

REMARK 4.21. Let Π be an affine plane and W a collection of points and lines that includes two non-parallel affine lines X and Y, and an affine point $u \notin (X \cup Y)$ (note that the point $X \cap Y$ is contained in Π). Then the set of subplanes of Π containing W forms a lattice $\mathcal{L}_W$, where the sum of two subplanes is the plane generated by their union and where the 'join' is defined by the standard intersection operation.

The following result is essentially a special case of the previous one considered in the context of spreads and subspreads.

REMARK 4.22. Let $\pi = (V, \mathfrak{T})$ be a spread. Let $\{\pi_i = (V_i, \mathfrak{T}_i) \mid i \in I\}$ be a collection of subspreads such that every subspread π_i, $i \in I$, shares at least two distinct components $X, Y \in \mathfrak{T}$ and there is a point $u \in V \setminus (X \cup Y)$ contained in some subspread component of each π_i. Then π induces a subspread π_W on $W = \bigcap\{V_i \mid i \in I\}$, the 'intersection subspread' of the family $(\pi_i)_{i\in I}$.

Although the above may be proved by converting to a result concerning affine planes, a direct vector-space argument also works. This alternative argument also works for partial spreads, as shown below. By slightly modifying these arguments we can obtain a simple alternative proof of the above based on linear algebra.

PROPOSITION 4.23. Let $\pi = (V, \mathfrak{T})$ be a partial spread. Let $(W_i)_{i\in I}$ be a family of subspaces of V on each of that π induces a subpartial spread, and let $W = \bigcap(W_i)_{i\in I}$. Then W is either a subspace of some component of π or $W = \bigcap(W_i)_{i\in I}$ is a subpartial spread of π.

COROLLARY 4.24. A collection of subpartial spreads $(W_i)_{i\in I}$ of a partial spread $\pi = (V, \mathcal{S})$ that share at least two common components, say G and H, and a common affine point $u \notin G \cup H$, must intersect in a subpartial spread of π. Moreover, if every W_i, for $i \in I$, is a *subspread* of π then $W := \bigcap(W_i)_{i\in I}$ is also a *subspread*.

Note that the result concerning *subspreads* in the corollary cannot legitimately be deduced from Proposition 4.23 by simply deleting all the 'partials'. When such deletions may be automatically applied, they may be written parenthetically '(partial)'; the brackets are intended to signify that the reader is encouraged to delete all partials to obtain (often) a more interesting result. Unparenthesized partials indicate that such deletions might not be permissible. Thus, the following remark is really a two-in-one remark.

REMARK 4.25. Let $\pi = (V, \mathfrak{T})$ be a (partial) spread and S a non-empty subset of V not contained in a component. Then the intersection of all (partial) subspreads of V that contain S is a sub(partial) spread $\langle S \rangle$—'generated' by S.

The above statement may be regarded as summing up the main point of this section that the set $\mathcal{L}_S$ of subspreads $W \leq V$, of a spread $(V, \mathfrak{T})$, that share a subset $S \subset V$, not contained in a component have a natural lattice structure. This structure may be identified with the lattice structure of subplanes of a translation plane that contain a set of non-collinear points that may be identified with $S \cup \{o\}$.

4.4. Rational Partial Spreads and Foulser-Covers.

If a partial spread $\pi = (V, \mathfrak{S})$ induces a subspread on some subspace $W \leq V$ (cf. Definition 4.7(iii)) then the subset of components of π that meet W non-trivially is a *rational* partial spread. Hence, a rational partial spread is one having a subplane 'across' it.

As rational partial spreads play a paramount role in translation plane theory, we record the definition explicitly, even repeating the meaning of some earlier terminology.

DEFINITION 4.26. Let $\pi = (V, \mathcal{S})$ be a (partial) spread, and suppose that W is a non-zero subspace of V on which a **spread** is induced. Thus, $\pi_W := (W, \mathcal{T})$, where $\mathcal{T} = \mathcal{S}_W \subset \mathcal{S}$ denotes the set of components of π that meet W non-trivially, induces a spread on W. Then π_W is a 'rational partial spread' that lies 'across' W and is 'determined' by W. That is, W is a subspread of π 'across' π_W. A 'rational partial spread' in π is any subspread $\psi \subset \pi$ such that it is across some subplane W of π (cf. Definition 4.7(iii)). In particular, a rational partial spread $\pi_W := (W, \mathcal{T})$, across a subplane W of π, is a 'rational Desarguesian partial spread' if additionally $\mathcal{T} \subset \mathcal{D}$, where $(V, \mathcal{D})$ is a Desarguesian spread.

Thus a rational partial spread $\pi_W := (W, \mathcal{T})$, across the subplane W of a partial spread π, is Desarguesian only if the planes across π_W are all themselves Desarguesian.

However, the fact that the subspace W has a Desarguesian plane π_W across it is not in itself sufficient to force the partial spread across W to be a (rational) Desarguesian partial spread. This will be evident later, but is strongly suggested by the following remark, and Remark 4.31 ahead.

REMARK 4.27. Let $(V, \mathfrak{T})$ be any spread over a skewfield K and let W be any 2-dimensional subspace of V, but not lying in a component. Then the partial spread across W has at least one Desarguesian subspread.

Thus, the set of components *in any spread* meeting any dimension-two subspace W that is not contained in a component is a rational partial spread and has W as a Desarguesian subplane across it.

REMARK 4.28. In spreads of order p^2, p prime, *all* rational subspreads, including the non-Desarguesian ones, have Desarguesian planes across them: All the proper subspreads of a spread of order p^2 are Desarguesian.

REMARK 4.29. It turns out that all spreads of order p^2 contain many rational partial spreads, each with at least one Desarguesian spread across it. In general, these rational partial spreads will fail to be Desarguesian because they fail to have the following *covering* property that Desarguesian spreads clearly always possess:

DEFINITION 4.30. A rational partial spread (V, ψ) has a 'Foulser-cover' $(W_i)_{i \in I}$ if the W_i, for $i \in I$, are pairwise disjoint additive subspaces of V each of which is across ψ (that is, ψ induces a spread on every W_i and every component of ψ meets each W_i non-trivially) and $\bigcup(W_i)_{i \in I} = \bigcup \psi$.

To show that rational partial subspreads of a Desarguesian spread always admit a Foulser-cover, we first record the following frequently used fact.

REMARK 4.31. (1) The subplanes across any rational partial spread have pairwise trivial intersection.

(2) Any rational partial spread admits at most one Foulser-cover.

(3) Every rational partial spread across a Desarguesian spread admits a unique Foulser-cover.

(4) Subplanes of a Desarguesian spread are kernel subplanes.

(5) The Foulser-covers of a Desarguesian spread consist of kernel subplanes.

Thus, a rational partial spread ρ, even if it has *a Desarguesian spread across it*, will *not* be a rational *Desarguesian* partial spread if ρ does not admit a Foulser-cover, or if some plane across ρ is not a kernel subplane. The reader interested in infinite analogues of the above ideas might note the following to see how the difficulties can arise when we have skewfields K that are not fields.

REMARK 4.32. Let K be a skewfield and F a subskewfield such that K has left dimension 2 over F. Consider $K \oplus K$ as a left K-vector space whose spread of 1-dimensional left K-subspaces form a Desarguesian spread. Now consider the partial spread $\mathcal{D}$ arising from the subskewfield F. If we represent the associated Desarguesian affine plane with spread set

$$x = 0, y = xm \text{ for all } m \in K$$

and juxtaposition represents multiplication in K, then the partial spread $\mathcal{D}$ has the form

$$x = 0, y = x\alpha \text{ for all } \alpha \in F.$$

Now we note that there is a set of Desarguesian affine subplanes of the associated net corresponding to $\mathcal{D}$ that have the following equations (represent $x = (x_1, x_2)$, $y = (y_1, y_2)$ over F as left 2-vectors):

$$\pi_{a,b} = \{ (a\alpha, b\alpha, a\beta, b\beta); \ \alpha, \beta \in F \}$$

where $(a, b) \neq (0, 0)$, $a, b \in F$. We note that $\pi_{a,b}$ intersects $y = x\delta$ in

$$\{ (a\alpha, b\alpha, a\alpha\delta, b\alpha\delta); \ \alpha \in F \}.$$

Now consider (c, e) as an element of K for $c, e \in F$ and define the 'conjugate of F by (c, e)' as follows:

$$F^{(a,b)} = \{ d \in K; \ d \in (c, e)F(c, e)^{-1} \}.$$

Noting that $(c, e)\rho = (c\rho, e\rho)$ for $\rho \in F$, let $(c, e)^{-1}d(c, e) = \alpha_d$ for $\alpha_d \in F$. Hence, we obtain

$$d(c, e) = (c\alpha_d, e\alpha_d) \text{ so that}$$
$$d(c\alpha, e\alpha) = d(c, e)\alpha = (c\alpha_d\alpha, e\alpha_d\alpha) \text{ for all } \alpha \in F.$$

Now it follows that the mappings:

$$(x, y) \longmapsto (dx, dy) \text{ for all } d \in F^{(a,b)}$$

map

$$\{ (a\alpha, b\alpha, a\alpha\delta, b\alpha\delta); \ \alpha \in F \}.$$

onto

$$\{ (a\alpha_d\alpha, b\alpha_d\alpha, a_d\alpha\delta, b\alpha_d\alpha\delta); \ \alpha \in F \}.$$

In other words, $F^{(a,b)}$ acts as the kernel of $\pi_{a,b}$. Hence, we have a set of subspreads $\pi_{a,b}$ and a set of 2-dimensional subskewfields $F^{(a,b)}$ of K such that $\pi_{a,b}$ is a left 2-dimensional $F^{(a,b)}$-subspace. Thus, when a net isomorphic to $\mathcal{D}$ lies within a Desarguesian affine plane, then every subplane incident with the zero vector becomes a 'kernel' subplane. Now consider that $\mathcal{D}$ lies within a translation plane with kernel F but not K. Then we would not have available the subskewfields $F^{(a,b)}$ and the 'only' 'kernel' subplanes would be those when there is a 'natural' coordinate system such that $F^{(a,b)} = F$ or equivalently where $(a, b) \in Z(F)$ (center of F). For example, $\pi_{0,1}$, $\pi_{1,0}$ and $\pi_{1,1}$ would always be kernel subplanes but the only way that all such subplanes would be kernel subplanes is for $Z(F) = F$.

Another approach to the problem of constructing rational partial spreads of various types is based on *quasifields*.

Further investigation of spreads requires the introduction of coordinate based methods.

4.5. André versus Bruck–Bose.

There are two well-known approaches to the study of translation planes: one using vector spaces, due to André [31], and another using projective spaces due to Bruck–Bose [**182**]. We shall employ a combination of these two methods, noting that it is often more expedient when working with translation planes to work in the associated vector space.

We recall that using the André method or vector-space approach, a 'finite (vector-space) spread' is a set of mutually disjoint half-dimensional vector subspaces which cover the vector space as above. Actually, it is also the associated affine space corresponding to the vector space that is of importance in interconnecting the methods under discussion. For these interconnections, we adopt some notation which is useful in these contexts.

NOTATION 4.33. Let V be a vector space over a field K. Then, we denote $AG(V, K)$ to be the affine space whose 'points' are the vectors of V and whose 'subspaces' are the vector translates of the vector subspaces of V. Now there are two projective spaces associated with V and/or $AG(V, K)$:

(i) Extend $AG(V, K)$ to a projective space by the method of adjunction of a 'hyperplane at infinity'. We shall call this projective space $PG(V, K)$.

(ii) Form the projective space obtained from V by taking the 'points' to be the 1-dimensional K-subspaces and the set of 'projective subspaces' to be the lattice of vector subspaces. Let $V = W \oplus K$, and denote W by V^- and V by W^+. We shall use the notation $PG(V^-, K)$ to denote this projective space. Note that if V is k-dimensional over K, $AG(V, K)$ and $PG(V, K)$ are denoted by $AG(k, K)$ and $PG(k, K)$, respectively. Furthermore, $PG(V^-, K)$ is then denoted by $PG(k-1, K)$.

(iii) If $\mathcal{S}$ is a set of vector subspaces of V, we write $P(\mathcal{S})$ to denote the set $\{ PG(W^-, K); W \in \mathcal{S} \}$ in $PG(V^-, K)$.

DEFINITION 4.34. Let K be a field and let $\mathcal{P}$ be a projective space isomorphic to $PG(2n-1, K)$. A 'finite-dimensional projective spread' of $\mathcal{P}$ is a set $\mathcal{S}_{\mathcal{P}}$ of mutually disjoint $(n-1)$-dimensional projective subspaces such that $\bigcup \{ W_{\mathcal{P}}; W_{\mathcal{P}} \in \mathcal{S}_{\mathcal{P}} \} = \mathcal{P}$.

DEFINITION 4.35. Let K be a field, V a vector space which is the external direct sum $W_o \oplus W_o$, where W_o is a K-vector space. Let $\mathcal{P}$ denote the lattice of projective subspaces of V. A 'projective spread' of $\mathcal{P}$ is a set $\mathcal{S}_{\mathcal{P}}$ of mutually disjoint projective subspaces of $PG(V^-, K)$ such that:

(i) Each element of $\mathcal{P}$ is isomorphic to $PG(W_o^-, K)$, any two distinct elements of $\mathcal{P}$ generate $PG(V^-, K)$ (in the sense that all points are on lines joining pairs of points from the two projective subspaces), and

(ii) $\bigcup \{ W_{\mathcal{P}}; W_{\mathcal{P}} \in \mathcal{S}_{\mathcal{P}} \} = \mathcal{P}$.

In either setting an element of a spread (vector-space or projective) is called a 'component'.

Let Σ denote the projective space (isomorphic to) $PG(V^-, K)$. Now embed Σ into a projective space Σ^+ so that Σ is a (projective) hyperplane of $\pm^+$. This is accomplished as follows: Let V^+ denote the associated vector space over K such that $V^+ = V \oplus Q$, where Q is a 1-dimensional K-vector space and the sum is considered an 'external direct sum'. Hence, $\pm$ is isomorphic to $PG(V^-, K)$ and $\pm^+$ is isomorphic to $PG(V^{+-}, K)$. Furthermore, let $W_o^+ = W_o \oplus Q$ (again, an external direct sum).

We now define an affine translation plane $\pi_{\mathcal{S}_{\mathcal{P}}}$ as follows: the 'points' are the points of $\mathcal{P}^+ - \mathcal{P}$ and the 'lines' are the projective subspaces of $\mathcal{P}^+$ isomorphic to $PG(W_o^{+-}, K)$ that intersect $\mathcal{P}$ in an element of $\mathcal{S}_{\mathcal{P}}$. When an affine translation plane is obtained as above using a projective spread, we shall say that the plane is obtained using the 'Bruck–Bose method'.

So, using the Bruck–Bose method or projective-space approach, a finite projective spread is a set of mutually skew $n - 1$-dimensional projective spaces covering the points of $PG(2n - 1, q)$.

The essential difference between the André and Bruck–Bose approaches then depends on the methods used in constructing the affine translation plane corresponding to the spread whether it be a vector-space or a projective spread.

In the following discussion, we shall give the interconnections between these two methods, generally considering an arbitrary vector space over a field K of the form the external direct sum $W_o \oplus W_o$, for W_o a K-vector space.

We begin with a fundamental lemma which shows that the projective space $PG(V^{+-}, K)$ obtained from the vector space V^+ over K by defining the incidence geometry as the lattice structure of vector subspaces of V^+ is isomorphic to $PG(V, K)$, the projective space obtained by adjunction of a hyperplane at infinity to the affine space $AG(V, K)$ associated with V. In the following, we adopt the notation developed in the previous definitions.

LEMMA 4.36. (1) $PG(V^{+-}, K)$ is isomorphic to $PG(V, K)$.

(2) Bases may be chosen for V and V^+ so that

  (a) vectors of V^+ may be represented in the form:

    $((x_i), x_\infty)$ for all $i \in \rho$, ρ an index set, where $x_i, x_\infty \in K$,

    $((x_i), 0)$ for all $i \in \rho$, ρ an index set, where $x_i \in K$ represent vectors in V,

  (b) regarding two non-zero 'tuples' above to be equal if and only if they are K-scalar multiples of each other produces the 'homogeneous coordinates' of the associated projective spaces $PG(V^-, K)$ and $PG(V^+ - 1, K)$, and

  (c) $((x_i), 1)$ for all $i \in \rho$, ρ an index set, where $x_i \in K$ represents homogeneous coordinates for a subset isomorphic to $AG(V, K)$.

(3) Furthermore, we may consider $PG(V, K)$ as the adjunction of $PG(V^-, K)$ as the hyperplane at infinity of $AG(V, K)$.

4.5.1. André Implies Bruck–Bose.

THEOREM 4.37. André implies Bruck–Bose: Hence, from a vector space V over a field K and spread S, we obtain a translation plane π_S and a corresponding projective partial spread $P(S)$ of $PG(V^-, K)$.

Note that although it is immediate that from a vector-space spread S a projective-space spread $P(S)$ is obtained, we have not determined whether the translation plane π_S obtained using the André method is isomorphic to the translation plane $\pi_{P(S)}$ obtained using the Bruck–Bose method. We shall now show this to be the case while considering the more complicated converse to the above theorem.

4.5.2. Bruck–Bose Implies André.

THEOREM 4.38. Bruck–Bose implies André: Let V be a vector space over a field K, which is the external direct sum $W_o \oplus W_o$, where W_o is a K-vector space. If Q is any 1-dimensional K-subspace, form the external direct sum $V \oplus Q = V^+$ and let the lattice of K-subspaces be denoted by $PG(V^{+-}, K)$.

(1) Then, $PG(V^-, K)$ may be considered a hyperplane of $PG(V^{+-}, K)$ ('co-dimension 1-subspace'), which we call the 'hyperplane at infinity'.

(2) Let S be a projective spread in $PG(V^-, K)$ and let $\mathcal{L}$ be an element of S. Now $\mathcal{L}$ is a subspace of V that is K-isomorphic to W_o. Consider $\mathcal{L}^+ = \mathcal{L} \oplus Q$. Then, in $PG(V^{+-}, K)$, (a) $PG(\mathcal{L}^{+-}, K)$ is isomorphic to $PG(W_o, K)$ and (b) $PG(\mathcal{L}^{+-}, K)$ intersects $PG(V^-, K)$ in $\mathcal{L}$.

(3) From $PG(V^{+-}, K)$, remove the hyperplane at infinity $PG(V^-, K)$ to produce (a) an affine space isomorphic to $AG(V, K)$ and (b) a corresponding vector-space spread $V(S)$ obtained by taking

$$V(S) = \left\{ PG(\mathcal{L}^{+-}, K) - PG(V^-, K) \cap PG(\mathcal{L}^{+-}, K); \ \mathcal{L} \in S \right\}.$$

The proof to the above result will become clear once we compare the translation planes obtained from the two processes.

4.5.3. The Translation Planes Are Isomorphic.

NOTATION 4.39. If S is a projective spread, we shall use the notation introduced in the theorem of the previous subsection to obtain a vector space spread $V(S)$. If Z is a vector-space spread, we shall use the notation $P(Z)$ to denote the corresponding projective spread obtained using lattices.

THEOREM 4.40. (1) The translation plane π_S obtained from the projective spread S by using the Bruck–Bose method is isomorphic to the translation plane $\pi_{V(S)}$ obtained from the vector-space spread $V(S)$ using the André method.

(2) The translation plane π_Z obtained from a vector-space spread Z using the André method is isomorphic to the translation plane $\pi_{P(Z)}$ using the Bruck–Bose method.

(3) $\pi_S \simeq \pi_{V(S)} \simeq \pi_{P(V(S))}$ and $\pi_Z \simeq \pi_{P(Z)} \simeq \pi_{V(P(Z))}$, using the notations of (1) and (2).

The model that we want to emphasize to the reader when going from vector space to affine space to projective spaces is summarized in the following remark.

REMARK 4.41. Take a vector-space spread Z and form the associated affine space. Extend the affine space to a projective space and extend the affine spaces corresponding to spread components to projective subspaces. Construct a projective spread on the hyperplane at infinity by intersection of the projective 'component' subspaces. Thus, abusing language somewhat we have:

$$\text{Bruck–Bose is André at infinity.}$$

CHAPTER 5

Quasifields.

In this chapter, we review the coordinate systems of translation planes, called quasifields. Many of the important classes of translation planes were originally constructed using quasifields, which often were obtained by varying the multiplication of an associated field. However, two quasifields can produce the same, or isomorphic translation planes. This material is from the authors' text [**123**] in a more compact form.

5.1. Coordinates: Spread Set.

Let π be a finite spread. In this section, we shall consider various ways in which π might be coordinatized in terms of concrete structures: spread sets and pre-quasifields. The most primitive form of coordinatization depends on choosing a pair of *axes*—any ordered pair (X, Y) of distinct components of π. Relative to this choice of axes, the corresponding spread set of π consists of a set of additive maps $\tau \subset \mathrm{Hom}(X \to Y, +)$ such that the components ($\neq Y$) of π may be expressed in the form $y = xT$, for $T \in \tau$. The set τ is the *slope set* for π relative to the chosen axes. Subsets $\sigma \subset \mathrm{Hom}(X \to Y, +)$ that are the slope sets of some spread on $X \oplus Y$ are called *spread sets* and the associated spread is denoted by $\pi(\sigma)$.

Spread sets may be characterized axiomatically.

REMARK 5.1. If $V = X \oplus Y$ is an elementary Abelian group, of order p^{2r}, with subgroups $X \cong Y$, then a subset $\sigma \subset \mathrm{Hom}(X \to Y, +)$ is a spread set iff: (1) $|\sigma| = p^r$; (2) if $S_1, S_2 \in \sigma$ then the difference $S_1 - S_2 \colon X \to Y$ is a bijection; and σ includes the zero.

It follows immediately that the non-zero maps in the spread set σ are bijective and that there is an associated spread on $V = X \oplus Y$ specified by $\pi(\sigma) = \{\, y = xS : S \in \sigma \,\} \cup \{Y\}$, and, moreover, $\pi(\sigma)$ actually has σ as its slope set. Thus, the slope sets of spreads yield spread sets, and conversely every spread set may be regarded as a slope set of a given spread. So we may regard slope sets and spread sets as being conceptually synonymous: slope sets correspond to a generic construction for sets of linear maps that satisfy the axioms for spread sets.

We may get new spread sets, from the given spread set σ, in several ways. For example, if we simply choose a pair of $GF(p)$-linear bijections, $A \colon S \to X$ and $B \colon Y \to T$, then $A\sigma B$ is a spread set: we shall consider spread sets as being equivalent—by basis-change—if they are related in this way. We shall regard such equivalent spread sets as coordinatizing the given initial spread π, relative to the axes (X, Y). This equivalence has a basic geometric interpretation:

Fundamental Equivalence of Spread Isotopisms and Spread Sets.
Given spreads π and ψ with chosen axes (X, Y) and (S, R), respectively, the spread

sets coordinatizing these spreads, relative to the chosen axes, are equivalent iff there is a spread isomorphism from π to ψ that preserves the axis.

The isotopism terminology anticipates the algebraic structures 'coordinatizing' spread sets. These algebraic structures contribute, internally, to simplifying the type of spread sets coordinatizing a spread π, relative to some chosen axes (X, Y). The idea is to replace the 'intrinsic' spread sets in $\mathrm{Hom}(X, Y)$ by the less clumsy, but somewhat ad hoc, spread sets in $\mathrm{Hom}(Z, +)$ obtained by identifying X and Y with a common isomorphic vector space Z, via a pair of additive bijections $(\lambda_X \colon X \to Z, \lambda_Y \colon Y \to Z)$, which we call a *labeling*. Every such labeling defines a spread set $\zeta \subset \mathrm{Hom}(Z, +)$, equivalent to the spread sets above.

Such a spread set ζ also provides a convenient route to the algebraic structures associated with the classical planar ternary rings that coordinatize arbitrary affine translation planes, viz., *(pre)-quasifields*. Thus, upon choosing a bijection $\nu \colon Z \to \zeta$, we define $z_1 \circ z_2 = (z_1)(z_2)^\nu$, for all $z_1, z_2 \in Z$. This makes $(Z, +, \circ)$ into a *pre-quasifield* that *labels* π, relative to the choices made. Such pre-quasifields are considered isotopic if there is a multiplicative isotopism between them (as defined in the theory of non-associative binary systems) such that the three component maps of the isotopism are also additive bijections.

Fundamental Equivalence of Spread Isotopisms and Pre-Quasifields. Given spreads π and ψ with chosen axes (X, Y) and (S, R), respectively, the pre-quasifields labeling these spreads, relative to the chosen axes, etc., are isotopic iff there is a spread isomorphism from π to ψ that preserves the axes.

In particular, all the pre-quasifields labeling π relative to fixed axes (X, Y) are mutually isotopic.

Like spread sets, pre-quasifields may be characterized axiomatically, and there is a natural (but complex!) correspondence between spread sets and pre-quasifields and hence also between spreads and pre-quasifields. Thus, just as many essentially distinct spread sets may be associated to a given spread, so many non-isomorphic pre-quasifields yield the same spread set, and hence also the same spread.

The pre-quasifields with identity are *quasifields*, and these are precisely the planar ternary rings coordinatizing affine translation planes. This reflects the fact that any pre-quasifield may be converted to a quasifield that defines the same spread.

It is also desirable to introduce an intermediate structure between pre-quasifields and quasifields: pre-quasifields with *at least a left identity*. This is because although any pre-quasifield Q, with associated spread set σ, directly converts to a pre-quasifield Q_ℓ with a *left* identity e, e cannot be a 2-sided identity unless σ contains the identity map: in such cases it becomes necessary to first 'shift' σ to an equivalent spread set σ' with identity. Moreover, pre-quasifields with left identity are used directly in defining several major classes of spreads in their simplest form. Accordingly, to give it some status, we shall regard a pre-quasifield Q with a left identity as *coordinatizing*, rather than just *labeling*, the associated spread.

5.2. Coordinatizing Spreads by Spread Sets.

Let π be a rank-r spread over $GF(q)$, with ambient vector space V. Thus π consists of $q^r + 1$ pairwise skew rank-r $GF(q)$-spaces that induce a partition of the non-zero points of V. We consider how π may be described ('coordinatized') in terms of its various *spread sets*.

DEFINITION 5.2. Let X and Y be n-dimensional vector spaces over $F = GF(q)$, where $q = p^r$, p prime. Then a set τ consisting of q^n linear maps from X to Y is an X-Y SPREAD SET, over $GF(q)$ if: (1) $\mathbf{0} \in \tau$; (2) If $A, B \in \tau$ then $A - B$ is a linear bijection from X to Y. The default choice for F is often the prime field $GF(p)$.

Every spread set is *associated* to a spread as follows.

REMARK 5.3. Let X and Y be n-dimensional vector spaces over $GF(q)$, and suppose τ is an X-Y spread set. Then, on the vector space $V = X \oplus Y$, the collection of subspaces of V defined by:

$$\pi(\tau) := \{\, \{\, x \oplus T(x) : x \in X \,\} : T \in \tau \,\} \cup \{\mathbf{0} \oplus Y\}$$

is a $GF(q)$-spread; $\pi(\tau)$ is the spread set, ASSOCIATED TO the spread set τ, relative to the AXES (X, Y).

We turn to the converse: every spread π may be expressed in the form $\pi(\tau)$, for various spread sets τ. Such τ will be regarded as *coordinatizing* π.

DEFINITION 5.4. Choose distinct components $X, Y \in \pi$. Then $\mu \in \mathrm{Hom}(X, Y)$ is the SLOPE MAP of a component $M \in \pi \setminus \{Y\}$ if: $M = \{\, x \oplus \mu(x) : x \in X \,\}$. Let τ_M denote the slope map of the component $M \in \pi \setminus \{Y\}$. Then $\tau_\pi := \{\, T_M : M \in \pi \setminus \{Y\} \,\}$ is the SLOPE SET of π, that COORDINATIZES π relative to AXES (X, Y).

It is straightforward to verify that the slope set τ of a spread π is a spread set and the spread $\pi(\tau)$, assigned to π, coincides with π itself.

REMARK 5.5. Let π be a spread and $X, Y \in \pi$ be distinct components. Then the slope set τ_π coordinatizng π, relative to (X, Y), is a spread set. Moreover $\pi = \pi(\tau_\pi)$—the spread assigned to the spread set τ_π.

Thus, summarizing Remarks 5.3 and 5.5, *every spread π is coordinatized by a spread set τ_π, for each choice of axis pair (X, Y), and conversely, given a spread set τ from X to Y, there is a spread $\pi(\tau)$ on $X \oplus Y$ whose slope set is τ itself.*

The following observation emphasizes the uniqueness of the coordinatizing spread set once the axes (X, Y) are fixed.

REMARK 5.6. Let π be a spread. Then to each pair of distinct components (X, Y) of π, there corresponds a unique $(X$-$Y)$-spread-set τ that coordinatizes π relative to (X, Y).

The above discussion establishes a natural many-one correspondence between spread sets and spreads: every spread set specifies a unique spread, but every spread π may be obtained by a range of distinct spread sets τ based on the choice of axes (X, Y) of π. In fact, the various spread sets are not generally 'equivalent by basis change', cf. Theorem 5.11. To make all this precise, we prove:

PROPOSITION 5.7. Suppose (V, π) and (W, ψ) are $GF(q)$-spreads of order q^r with ambient spaces V and W, respectively. Let τ and σ be, respectively, the spread sets of π and ψ relative to the axis pairs (X, Y) and (A, B). Then there is a $GF(q)$-linear spread isomorphism $\Theta \colon \pi \to \psi$ such that $(X)\Theta = A, (Y)\Theta = B$ iff $\sigma = G\tau H$ for non-singular maps $G \colon A \to X$ and $H \colon Y \to B$.

DEFINITION 5.8. Let $\tau \subset \mathrm{Hom}_{GF(q)}(X,Y)$ $\sigma \subset \mathrm{Hom}_{GF(q)}(A,B)$ be $GF(q)$-linear sets of the same size q^r, and let $GF(p)$ be the prime subfield in $GF(q)$. Then σ and τ are $GF(q)$-EQUIVALENT BY BASIS CHANGE, if there are $GF(q)$-linear bijections $G \colon A \to X$, $B \colon Y \to B$ such that $\sigma = A\tau B$.

The spread sets σ and τ are EQUIVALENT BY BASIS CHANGE if they are $GF(p)$-equivalent by basis change.

Thus, Proposition 5.7 might be interpreted as asserting that a linear spread isomorphism mapping the ordered pair of components (X,Y) onto the ordered pair of components (A,B) is possible iff the associated spread sets are obtainable from each other by a change of basis:

COROLLARY 5.9. Spread sets τ and σ are equivalent by basis change iff there is an isomorphism of the associated spreads, $\Theta \colon \pi(\tau) \to \pi(\sigma)$, that preserves both the axes. Moreover, Θ can be chosen to be induced by a $GF(q)$-linear bijection of the associated ambient spaces iff τ and σ may both be regarded as $GF(q)$-linear spread sets.

The following special case of the corollary, when the two spreads coincide, is of paramount importance: it is more important than the corollary itself.

THEOREM 5.10 (Axes Orbits Correspond to Equivalent Spread Sets). Let V denote a vector space over a field $K = GF(q)$, and let π be a spread of K-subspaces of V. Let (X,Y) and (A,B) be two pairs of axes for π; so $X \neq Y$ and $A \neq B$ are all components of π. Let $\theta \subset \mathrm{Hom}_K(X,Y)$ and $\phi \subset \mathrm{Hom}_K(A,B)$ denote the slope set of π relative to (X,Y) and (A,B), respectively; in particular, θ and ϕ are K-spread-sets relative to (X,Y) and (A,B), respectively. Then $\Phi \in GL(V,K)$ is in $\mathrm{Aut}(\pi)$ iff θ and ϕ are equivalent under basis-change.

THEOREM 5.11. Let π be a spread on a vector space V with characteristic p. Then all the spread sets associated with π are all mutually equivalent by basis change, over $GF(p)$ iff $\mathrm{Aut}(\pi)$ induces a 2-transitive group on the components of π.

The Desarguesian and the Lüneburg–Tits spreads admit 2-transitive groups on their components. In fact these are the only spreads admitting such groups. This is a consequence of the classification of the finite 2-transitive groups. This result, due to Kallaher (unpublished manuscript), generalizes the Czerwinski–Schulz [**268, 1151**] theorem, proved well before the classification of finite simple groups was available.

COROLLARY 5.12. (Kallaher; Czerwinski and Schulz [**268, 1151**]). All the spread sets associated with a spread π are equivalent under basis change iff the spread is Desarguesian or Lüneburg–Tits.

Thus, the existence of inequivalent spread sets, for a given spread π, is the rule rather than the exception.

5.3. Labeled Coordinatization.

When coordinatizing a spread π, relative to axes (X,Y), it is usually desirable to identify the isomorphic spaces X and Y with a common vector space Z, via linear bijections $(\lambda_X \colon X \to Z, \lambda_Y \colon Y \to Z)$. Thus, π may be identified with a spread π_Z, and π_Z *coordinatizes* π relative to the *labeling* $(\lambda_X \colon X \to Z, \lambda_Y \colon Y \to$

Z). Similarly, we refer to the slope set of π_Z, relative to the natural axes, as *coordinatizing* the slope set of π relative to (X, Y). We record this terminology:

DEFINITION 5.13. Let π be a spread on V with components of rank r over $GF(q)$. Choose a rank-r $GF(q)$ vector space Z, and linear bijections $\lambda_X \colon X \to Z$ and $\lambda_Y \colon Y \to Z$. Then $\lambda_X \oplus \lambda_Y \colon X \oplus Y \to Z \oplus Z$, defines a linear bijection from V to the $Z \oplus Z$ such that the image of the spread π is an isomorphic spread π_Z on $Z \oplus Z$.

The spread π_Z COORDINATIZES π relative to AXES (X, Y), based on the LABELING $(\lambda_X \colon X \to Z, \lambda_Y \colon Y \to Z)$; the map $\lambda_X \oplus \lambda_Y \colon \pi \to \pi_Z$ is the COORDINATIZING SPREAD ISOMORPHISM OF π BY π_Z.

The slope set $\tau \subset \operatorname{Hom}(Z, +)$ of π_Z relative to axes $(Z \oplus 0, 0 \oplus Z)$, Definition 5.4, is the SLOPE SET OF π COORDINATIZED BY THE LABELING $(\lambda_X \colon X \to Z, \lambda_Y \colon Y \to Z)$.

REMARK 5.14. Choosing $q = p$, gives an additive version of the definition.

Returning to Definition 5.13, the slope set τ_π of π *coordinatized by a given labeling* is identified with the slope set of π_Z, the spread defined by the labeling isomorphism, relative to the standard axis of π_Z. The slope set of π_Z is easily expressed in terms of the slope set of π.

REMARK 5.15. Let the slope set of the spread π, relative to axes (X, Y), be $\tau_\pi \subset \operatorname{Hom}_{GF(q)}(X, Y)$. Suppose π is coordinatized by the spread π_Z, based on axes (X, Y) and labeling $(\lambda_X \colon X \to Z, \lambda_Y \colon Y \to Z)$. Then the slope set for π, relative to this labeled coordinatization, is $\lambda_X^{-1} \tau_\pi \lambda_Y \subset GL_{GF(q)}(Z, +) \cup \{\mathbf{0}\}$.

Thus, when considering a single spread π with axes (X, Y) and components isomorphic to a vector space Z, we may identify π with a spread π_Z on $Z \oplus Z$, while simultaneously identifying the components X and Y with components $Z \oplus \mathbf{0}$ and $\mathbf{0} \oplus Z$, respectively, of π_Z; the slope set τ of π, relative to (X, Y), corresponds to the slope set τ_Z of π_Z, relative to the natural axes $(Z \oplus \mathbf{0}, \mathbf{0} \oplus Z)$; in particular, τ_Z is a spread set that satisfies the condition: $\mathbf{0} \in \tau \subset GL(Z, +) \cup \{\mathbf{0}\}$. In this context, the following special case of Proposition 5.7 is worth stressing.

COROLLARY 5.16. Suppose π is a $GF(q)$-spread of order q^r with ambient space $Z \oplus Z$, and includes the components $X := Z \oplus \mathbf{0}$ and $Y := \mathbf{0} \oplus Z$. Let τ be the slope set of π, relative to the axis pair (X, Y), and let σ be the slope set of π relative to axes (A, B), for any $A, B \in \pi$, $A \neq B$. Then there is a $GF(q)$-linear spread isomorphism $\Theta \in \operatorname{Aut}(\pi)$ such that $(X)\Theta = A, (Y)\Theta = B$ iff $\sigma = G\tau H$ for non-singular maps $G \colon A \to X$ and $H \colon Y \to B$.

Specializing further by taking $(X, Y) = (A, B)$, we have a useful method for computing the full stabilizer of any two components, in the linear automorphism group of a spread.

COROLLARY 5.17. Let π be any $GF(q)$-spread of order q^r with ambient space $Z \oplus Z$, and such that $X, Y \in \pi$, where $X := Z \oplus \mathbf{0}$ and $Y := \mathbf{0} \oplus Z$. Let τ be the slope set of π relative to the axis pair (X, Y). Then a $GF(q)$-linear isomorphism $\Theta \in \operatorname{Aut}(\pi)$ fixes X and Y iff $\tau = G\tau H$ where $\Theta^X := G$, $\Theta^Y := H$.

REMARK 5.18. Since any spread π may be regarded as a spread on a vector space V over a prime field $GF(p)$, Corollary 5.17 actually enables a determination of the full group fixing any ordered pair of components of π. Such groups will be called *autotopism groups* because of their algebraic interpretation, considered ahead.

We return to the identification of the general spread π with a spread π_Z on $Z \oplus Z$. We have seen that a pair of distinct components $X, Y \in \pi$ may be identified with $Z \oplus \mathbf{0}$ and $\mathbf{0} \oplus Z$, respectively. Actually, this can be done so that any third component $U \in \pi$ becomes the unit line $I = \{\, z \oplus z : z \in Z \,\}$. In this situation, a single linear bijection $\lambda_U \to I$ uniquely determines λ_X and λ_Y. Thus, we have a further refinement of 'coordinatization' of spreads by spread sets, Definition 5.13.

DEFINITION 5.19. [Coordinatization Based on Unit Line]. Let π be a spread on V with components of rank r over $GF(q)$. Choose an ordered triple (X, Y, U) consisting of three distinct components of π, and a linear bijection $\lambda \colon U \to Z$, where Z is any rank r vector space over $GF(q)$. Then there are unique linear bijections $\lambda_X \colon X \to Z \oplus \mathbf{0}$ and $\lambda_Y \colon Y \to \mathbf{0} \oplus Z$ such that the linear bijection $\lambda_X \oplus \lambda_Y \colon V \to Z \oplus Z$ coincides with λ on U and maps the ordered pair (X, Y) onto $(Z \oplus \mathbf{0}, \mathbf{0} \oplus Z)$.

Thus the image of π under $\lambda_X \oplus \lambda_Y \colon V \to Z \oplus Z$, is a spread π_Z isomorphic to the spread π, and the slope set $\tau \subset \mathrm{Hom}(Z, +)$ of π_Z relative to $(Z \oplus \mathbf{0}, \mathbf{0} \oplus Z)$, as in Definition 5.13, is the SLOPE SET OF π COORDINATIZED by the (UNIT-LINE) LABELING $\lambda \colon U \to Z$.

REMARK 5.20. As indicated in Remark 5.14, the definition includes the case based on choosing $\lambda \colon U \to Z$ to be merely an additive bijection.

Thus, the definition forces a unique choice of (λ_X, λ_Y) in terms of $\lambda_Z \colon U \to Z$, hence the coordinatization based on $\lambda_Z \colon U \to Z$, corresponds to coordinatization by (λ_X, λ_Y), in the sense of Definition 5.13. So we are coordinatizing a spread, using Definition 5.13, but simultaneously ensuring that the unit line $y = x$ is included in the coordinatizing spread.

Expressed intuitively, the entire situation may be summarized as follows. We have demonstrated that given an ordered triple of distinct components (X, Y, J), of a spread π over $K = GF(q)$, we may coordinatize π by any vector space Z isomorphic to the components of π, such that the components X, Y and J correspond, respectively, to the subspaces of $Z \oplus Z$ specified by $y = 0$, $x = 0$ and $y = x$; these components are often referred to as the x-axis, the y-axis and the unit line, respectively. The representation of the entire spread, by an isomorphic spread π_Z on $Z \oplus Z$, has as its component set $\{\, y = xT : T \in \tau \,\} \cup \{y = 0\}$, where $\tau \subset GL(Z, K) \cup \{\mathbf{0}\}$ is a spread set containing zero and the *identity* map.

Summary of Terminology. We summarize the various types of 'labeling' and 'coordinatization' that convert a spread into a spread set, and hence define an isomorphism from the spread onto a standard type of spread. We assume π is a spread.

(1) Let $\tau \subset \mathrm{Hom}(X, Y)$ be a spread set. Then the *associated spread* $\pi(\tau)$, Remark 5.3, on $V = X \oplus Y$ is specified by

$$\pi(\tau) = \{\, (x, xT) : T \in \tau \,\} \cup \{Y\}.$$

(2) Let the ordered pair of distinct components (X, Y) of π be chosen as axes of π. Then the slope set of π, Definition 5.4, is

$$\tau_\pi = \{\, T \in \mathrm{Hom}(X, Y) : (y = xT) \in \pi \,\};$$

τ_π is the spread set *coordinatizing* π relative to (X, Y).

(3) Let π be a spread on V with all its components additively isomorphic to a vector space Z, and fix additive bijections $\lambda_X \colon X \to Z$ and $\lambda_Y \colon Y \to Z$. Then, under the additive bijection $\Lambda \colon V \to Z \oplus Z$ defined by

$$\Lambda \colon \lambda_X \oplus \lambda_Y \colon X \oplus Y \to Z \oplus Z,$$

the spread $\pi_Z := \Lambda(\pi)$, on $Z \oplus Z$, COORDINATIZES π relative to AXES (X, Y), based on the LABELING $(\lambda_X \colon X \to Z, \lambda_Y \colon Y \to Z)$; the map $\lambda_X \oplus \lambda_Y \colon \pi \to \pi_Z$ is the COORDINATIZING SPREAD ISOMORPHISM OF π BY π_Z, cf. Definition 5.13 and Remark 5.14.

The slope set $\tau \subset \mathrm{Hom}(Z, +)$ of π_Z relative to axes $(Z \oplus \mathbf{0}, \mathbf{0} \oplus Z)$, Definition 5.4, is the *slope set*, or *spread set*, of π coordinatized by the labeling $(\lambda_X \colon X \to Z, \lambda_Y \colon Y \to Z)$.

(4) Choose an ordered triple (X, Y, U) consisting of three distinct components of π, and a linear bijection $\lambda \colon U \to Z$, where Z is isomorphic to the components of π qua vector spaces. Then there are unique linear bijections, $\lambda_X \colon X \to Z \oplus \mathbf{0}$ and $\lambda_Y \colon Y \to \mathbf{0} \oplus Z$, such that the linear bijection $\lambda_X \oplus \lambda_Y \colon V \to Z \oplus Z$ coincides with λ on U and maps the ordered pair (X, Y) onto $(Z \oplus \mathbf{0}, \mathbf{0} \oplus Z)$.

Then the spread isomorphism $\lambda_X \oplus \lambda_Y \colon \pi \to \pi_Z$, from π by π_Z and corresponding to the coordinatization of the spread π by π_Z, cf. Definition 5.13 or the case above, is also defined to be the coordinatizing isomorphism of π relative to *the axes* (X, Y) *and* UNIT-LINE LABELING $\lambda \colon U \to Z$, Definition 5.19. Similarly, the spread set coordinatized by (λ_X, λ_Y) is also regarded as coordinatizing π relative to the axes (X, Y) and *unit-line labeling* $\lambda \colon U \to Z$.

5.4. Quasigroups and Isotopisms.

The binary systems associated with the multiplication of pre-quasifields—the structures coordinatizing spreads—are quasigroups. We introduce the terminology, notation and basic properties of quasigroups viewed as binary systems.

DEFINITION 5.21. Let $(B, \circ)$ be a binary system. Then for each $a \in B$, the associated RIGHT MULTIPLIER and LEFT MULTIPLIER are, respectively, the maps $R_a \colon B \to B$ and $L_a \colon B \to B$, specified by the conditions:

$$\forall x \in B : R_a \colon x \mapsto x \circ a, \quad L_a \colon x \mapsto a \circ x.$$

The system $(B, \circ)$ is a QUASIGROUP if every right multiplier R_a and left multiplier L_a is a bijection of B. A quasigroup with an identity is a LOOP.

An isomorphism between binary systems is a special case of an *isotopism*. This corresponds to a relabeling process which might for instance be used to demonstrate the existence of large numbers of non-isomorphic quasigroups of the same size.

DEFINITION 5.22. Let $\mathcal{Z} := (Z, \circ)$ and $\mathcal{W} := (W, *)$ be binary systems. A triple of bijections (α, β, γ), each from Z onto W, is an ISOTOPISM from $\mathcal{Z}$ to $\mathcal{W}$ if:

$$\forall x, y \in Z : (x \circ y)\beta = (x\alpha) * (y\gamma);$$

the isotopism is denoted by $A \to B : (\alpha, \beta, \gamma)$ or $(\alpha, \beta, \gamma) \colon A \to B$.

Composition and inverses of isotopisms between binary systems are defined in the obvious way.

DEFINITION 5.23. Let $\mathcal{A}$, $\mathcal{B}$, $\mathcal{C}$, be three binary systems, and suppose that

$$\sigma := (\alpha_1, \beta_1, \gamma_1) \colon \mathcal{A} \to \mathcal{B}, \quad \tau := (\alpha_2, \beta_2, \gamma_2) \colon \mathcal{B} \to \mathcal{C},$$

are isotopisms. Then their COMPOSITION is the isotopism $\sigma \circ \tau$ (also written $(\ )\sigma\tau$, $\tau\sigma(\ )$, etc.) is the isotopism from $A \to C$ specified by the triple of bijections:

$$((\ )\alpha_1 \circ \alpha_2, (\ )\beta_1 \circ \beta_2, (\ )\gamma_1 \circ \gamma_2).$$

Similarly, σ^{-1} is the isotopism specified by $(\alpha_1^{-1}, \beta_1^{-1}, \gamma_1^{-1})$, from $\mathcal{B}$ to $\mathcal{A}$.

The AUTOTOPISM GROUP, of any binary system $\mathcal{Z}$, is the group of all isotopisms of $\mathcal{Z}$ onto itself, under composition.

We note that isotopisms preserve quasigroups, isotopic quasigroups need not be isomorphic, and every quasigroup is isotopic to a loop.

5.5. Pre-Quasifields and Isotopisms.

We define pre-quasifields to be the algebraic structures coordinatizing *finite* spreads, and hence translation planes. The finiteness condition permits a simplification of the axioms for pre-quasifields that are necessary for the general case. The notion of isotopisms among quasigroups is extended to pre-quasifields.

DEFINITION 5.24. A system $\mathcal{Q} = (V, +, \circ)$, with $|V|$ *finite*, is a PRE-QUASIFIELD if the following axioms hold:

 (1) $(V, +)$ is an Abelian group, with additive identity 0;
 (2) $(V^*, \circ)$ is a quasigroup;
 (3) $\forall x, y, z \in V : (x + y) \circ z = x \circ z + y \circ z$;
 (4) $x \circ 0 = 0 \forall x \in V$.

A pre-quasifield with a multiplicative identity is a QUASIFIELD.

We have already dealt with isotopisms (α, β, γ) among quasigroups. We extend the concept to pre-quasifields: by insisting that α and β be *additive* bijections of the ambient vector space V, but γ is required only to be an *arbitrary* 0-fixing bijection of V.

DEFINITION 5.25. Let $\mathcal{Z} := (Z, +, \circ)$ and $\mathcal{W} = (W, +, *)$ be pre-quasifields. Then a triple of bijections $(\alpha \colon Z \to W, \; \beta \colon Z \to W, \; \gamma \colon Z \to W)$ is a PRE-QUASIFIELD ISOTOPISM if the triple $(\alpha|Z^*, \beta|Z^*, \gamma|Z^*)$ is an isotopism from the multiplicative quasigroup $(Z^*, \circ)$ onto the multiplicative quasigroup $(W^*, *)$, and additionally the following conditions hold: (1) $\alpha \colon Z \to W$ and $\beta \colon Z \to W$ are additive, so $\alpha(0) = \beta(0) = 0$; (2) $\gamma(0) = 0$.

Notation such as $(\alpha, \beta, \gamma) \colon \mathcal{Z} \to \mathcal{W}$ denotes an isotopism from $\mathcal{Z}$ to $\mathcal{W}$: in this case the two pre-quasifields are mutually ISOTOPIC. If $(\mathcal{W}, +, *) = (\mathcal{Z}, +, \circ)$, then an isotopism is an AUTOTOPISM.

It is unnecessary to explicitly define the notion of composition and inverses of isotopisms for pre-quasifields, since they have already been defined for arbitrary binary systems, Definition 5.23. Hence, the autotopism group of a pre-quasifield Q is the group of all autotopisms of Q under the standard composition for autotopisms.

By default an isotopism (α, β, γ) among pre-quasifields is assumed to be a pre-quasifield isotopism: α and β are required to be additive isomorphisms.

The prevalence of isotopes of a quasigroup, and the existence of an isotope with multiplicative identity, are reflected in corresponding properties for isotopisms among pre-quasifields.

PROPOSITION 5.26. (1) Suppose $(Z, +, \circ)$ is a pre-quasifield and the additive group $(Z, +) \cong (W, +)$. Choose $\alpha, \beta \in \mathrm{Hom}(Z, W)$ to be isomorphisms, and $\gamma \colon Z \to W$ to be an arbitrary bijection such that $\gamma(0) = 0$. Suppose $(Z, +, \circ)$ is a pre-quasifield and define $*$ by:

$$\forall x, y \in Z : (x \circ y)\beta = (x\alpha) * (y\gamma).$$

Then $(W, +, *)$ is a pre-quasifield isotopic to $(Z, +, \circ)$ and (α, β, γ) is an isotopism from $(Z, +, \circ)$ to $(W, +, *)$.

(2) Let $(Z, +, \circ)$ be a pre-quasifield. Choose $e \in Z$ and define $*$ by

$$\forall x, y \in X : x \circ y = (x \circ e) * (e \circ y).$$

Then $(Z, +, *)$ is a pre-quasifield that has identity $e \circ e$ and is isotopic to $(Z, +, \circ)$.

PROOF. Straightforward: since the multiplicative groups of pre-quasifields are quasigroups, straightforward arguments yield the desired properties for pre-quasifields. □

Let $\mathcal{A} = (A, +, \circ)$, $\mathcal{B} = (B, +, *)$ be pre-quasifields. Then any pre-quasifield isomorphism $\phi \colon A \to B$ is an isotopism of form (ϕ, ϕ, ϕ). Hence the automorphism group of a pre-quasifield is a subgroup of its autotopism group.

5.6. The Spread of a Pre-Quasifield.

A pre-quasifield Q is essentially a planar ternary ring, although possibly without a multiplicative identity. Hence by the familiar process of coordinatization, applicable to arbitrary affine planes, any pre-quasifield yields an affine plane $\Pi(Q)$, and the right distributive law forces $\Pi(Q)$ to be a translation plane. The details of this coordinatization may be viewed as a multi-stage process: (i) Q is assigned a spread set τ_Q on $(Q, +)$: technically this requires $(Q, +)$ to be an *elementary* Abelian group, so this needs to be established; (ii) $\pi(Q)$ is simply $\pi(\tau_Q)$, the spread assigned to τ_Q on $Q \oplus Q$, cf. Remark 5.3; (iii) the standard translation plane on $Q \oplus Q$ associated to $\pi(Q)$ is denoted by $\Pi(Q)$: so the lines of $\Pi(Q)$ are the additive cosets of the subspaces $S \in \pi(Q)$. We develop the details of this process.

DEFINITION 5.27. Let $(Q, +, \circ)$ be a pre-quasifield, in particular $(Q, +)$ is a finite Abelian group. The SLOPE MAP for each $m \in Q$ is the map $T_m \in \mathrm{Hom}(Q, +)$ specified by: $T_m \colon x \mapsto x \circ m$, and the SLOPE SET of Q is the set $\tau_Q = \{ T_m : m \in Q \} \subset \mathrm{Hom}(Q, +)$. The ASSOCIATED SPREAD $\pi(Q)$, COORDINATIZED BY Q, and τ_Q, is $\pi(\tau_Q)$, cf. Remark 5.3; explicitly, $\pi(Q) = \pi(\tau_Q)$ is the following set of additive subspaces of $Q \oplus Q$: $\{ y = x \circ m : m \in V \} \cup \{x = \mathbf{0}\}$. The ASSOCIATED TRANSLATION PLANE $\Pi(Q)$ has $Q \oplus Q$ as its points and the lines are the cosets of the members of $\pi(Q)$.

We note the tacit claims in Definition 5.27 are validated if $(Q, +)$ is elementary Abelian: this is necessary for τ_Q to technically satisfy the definition of a spread set.

PROPOSITION 5.28. Suppose $(Q, +, \circ)$ is a pre-quasifield. Then the following hold:

(1) $\pi(Q)$ is a genuine spread and $\Pi(Q)$ is its associated translation plane.

(2) $(Q, +)$ is a vector space over some field.

(3) τ_Q is a spread set on $(Q, +)$ and the associated spread, in the sense of Remark 5.3, coincides with $\pi(Q)$.

5.7. Coordinatization by Pre-Quasifields.

We have seen that any pre-quasifield Q yields a spread $\pi(Q)$. We now consider the converse problem of describing all the pre-quasifields Q such that the associated spreads $\pi(Q)$ are isomorphic to a given spread π.

Actually, we shall mostly restrict ourselves to considering pre-quasifields Q that possess at least a left identity, since the general case is essentially equivalent to this case—indeed, reduces to this case, cf. Proposition 5.31. Also note that the standard pre-quasifields used to define translation planes virtually always possess at least a left identity.

DEFINITION 5.29. Let $\tau \subset \operatorname{Hom}(Z, +)$ be a spread set on the elementary Abelian group $(Z, +)$. Then for any $e \in Z^*$ let T_y be the unique member of τ such that $(e)T_y = y$, and define:

$$\forall x, y \in Z : x \circ y = (x)T_y.$$

Then $Q_e := (Q, +, \circ)$ is the pre-quasifield COORDINATIZING τ RELATIVE TO e.

The assumption in the definition, that Q_e turns out to be pre-quasifield, is among the following list of routinely verified properties of Q_e.

REMARK 5.30. Let τ be a spread set on $(Z, +)$, and let Q_e be as in Definition 5.29. Then the following hold:

(1) Q_e is a pre-quasifield and has e as multiplicative left identity: $\forall y \in Z :$ $e \circ y = y$.

(2) e is a 2-sided multiplicative identity for Q_e iff $\mathbf{1} \in \tau$.

(3) τ is the slope set of Q_e.

(4) If Q is a pre-quasifield with left identity e such that its slope set $\tau_Q = \tau$ then $Q = Q_e$.

Thus, we have found all pre-quasifields with a left-identity that coordinatize a given spread set. In particular, Remark 5.30 implies that *every* spread set τ on an elementary Abelian group $(Z, +)$ is of form τ_Q, where $Q = (Z, +, \circ)$ is a pre-quasifield with left identity $e \in Z^*$. We may generalize this to describe all the pre-quasifields (including those not possessing a left identity) that have as their slope set a given spread set τ.

PROPOSITION 5.31. Let τ be a spread set on an elementary Abelian group $(V, +)$. Choose any bijection $\psi \colon V \to \tau$ such that $\psi(0) = \mathbf{0}$. Then the following hold:

(1) $Q_\psi := (V, +, \circ)$ defined by: $\forall x, y \in V : x \circ y = (x)\psi(y)$, is a pre-quasifield such that its slope set is τ, and conversely any pre-quasifield with slope set τ has the form Q_ψ for some ψ. The pre-quasifield Q_ψ is LABELED by ψ.

(2) Q_e, as in Definition 5.29, is a pre-quasifield labeling τ.

Hence, *labeling* a spread set by a pre-quasifield, as defined in Proposition 5.31 above, slightly generalizes the notion of *coordinatizing* a spread set, Definition 5.29: coordinatizing, unlike labeling, requires the pre-quasifield to include a left identity.

We assign to any pre-quasifield Q the attributes associated to its slope set τ_Q, viewed as a spread set. For example, Q is associated to the same spread as τ_Q, viz., $\pi(\tau_Q)$, and, in the reverse direction, if a spread set τ 'coordinatizes' (in one of the acceptable senses) a spread π then so do all the pre-quasifields Q that have τ as their slope set. The main possibilities are listed in the following:

DEFINITION 5.32. Let π be a spread of order p^r, with ambient space an elementary Abelian group $(V, +)$ of order p^{2r}. Fix an ordered pair of additive bijections $\Lambda := (\lambda_X : X \to Z, \lambda_Y : Y \to Z)$, where Z is an elementary Abelian group of order p^r, and (X, Y) is an ordered pair of distinct components of π. Let τ be the slope set of π coordinatized by the labeling Λ, Definition 5.13; thus π_τ, the spread associated to τ, coordinatizes π on $Z \oplus Z$ relative to the labeling $\lambda_X \oplus \lambda_Y$. (Hence the map $\lambda_X \oplus \lambda_Y V \to Z \oplus Z$, is the coordinatizing spread-isomorphism, Definition 5.13, from π onto π_τ.)

(1) Let Q_ψ be the pre-quasifield labeled by a given bijection $\psi \colon Z \to \tau$, thus $\psi(0) = \mathbf{0}$. Then Q_ψ LABELS π relative to labeled axes $\Lambda = (\lambda_X : X \to Z, \lambda_Y : V \to Z)$ and SLOPE LABELING ψ.

(2) Let I be any point on V such that its image, under $\Lambda = (\lambda_X : X \to Z, \lambda_Y : V \to Z)$, has form $e \oplus e \in Z \oplus Z$, for some $e \in Z^*$. Then the PRE-QUASIFIELD COORDINATIZING π, relative to the AXES (X, Y), LABELING Λ and UNIT POINT I, is the unique pre-quasifield Q_e, Definition 5.29, that coordinatizes τ relative to e.

Since the slope set of Q_ψ above was designed to coincide with the spread set τ, it follows that $\pi(Q_\psi) = \pi_\tau$, whence it follows:

THEOREM 5.33. Suppose π is a spread on V and that $\Lambda := (\lambda_X : X \to Z, \lambda_Y : Y \to Z)$, is an ordered pair of linear bijections from the distinct components $X, Y \in \pi$ onto a vector space Z. Then for any labeling of π, based on the spread set determined by Λ (Definition 5.13) by a pre-quasifield Q_ψ (Definition 5.32) there is a spread isomorphism $\Theta \colon \pi \to \pi(Q_\psi)$ such that $\Theta(X) = Z \oplus \mathbf{0}$ and $\Theta(Y) = \mathbf{0} \oplus Z$.

The following is easily verified:

REMARK 5.34. Conversely, suppose there is an isomorphism $\Theta \colon \pi \to \pi(Q)$, for some pre-quasifield $Q = (Z, +, \circ)$, and let (X, Y) be the ordered pair of components of π that map onto the ordered pair of components $(Z \oplus \mathbf{0}, \mathbf{0} \oplus Z)$ of $\pi(Q)$. Then Q is a labeling[1] of π based on the axes choice (X, Y). In fact, Q may be fully specified by the labeling $\lambda_X \oplus \lambda_Y \colon X \oplus Y \to Z \oplus Z$ where $\lambda_X = \Theta^X$, and $\lambda_Y = \Theta^Y$.

Thus, the pre-quasifields Q *labeling* (i.e., no left identity required) π are precisely the pre-quasifields such that $\pi(Q) \cong \pi$. Similarly, we may describe all the pre-quasifields Q that *coordinatize* π: i.e., describe all the Q, with at least a *left identity* such that $\pi(Q) \cong \pi$.

---

[1]Recall the convention that for the labeling Q to *coordinatize* π, Q must possess at least a left identity.

PROPOSITION 5.35. Let π be coordinatized, Definition 5.32, by a pre-quasifield Q_e with left identity e, and let τ be the slope set for Q_e.

(1) Then e is the unique left multiplicative identity of Q_e, and there is a spread-isomorphism $\Theta \colon \pi \to \pi(Q_e)$ such that Θ preserves the axes and maps I to (e, e).

(2) $\pi(Q_e)$ contains the unit line $y = x$ iff e is a 2-sided identity.

(3) τ contains the identity map iff e is a 2-sided identity.

5.8. Geometric Interpretations.

Any isotopism between pre-quasifields corresponds to an isomorphism of the associated spreads that preserves the coordinate axes.

PROPOSITION 5.36 (Standard Isomorphism from Isotopism). Let $(Q, +, \circ)$ be a pre-quasifield. Suppose that (α, β, γ) is an isotopism from Q to a pre-quasifield $(R, +, *)$, in particular $\forall x, y \in V : \beta(x \circ y) = \alpha(x) * \gamma(y)$. Then the additive bijection $\alpha \oplus \beta \colon Q \oplus Q \to R \oplus R \colon (x, y) \mapsto (\alpha(x), \beta(y))$ is an isomorphism from the spread $\pi(Q)$ onto the spread $\pi(R)$ that maps the component $y = x \circ m$ onto the component $y = x * (m)\gamma$, for $m \in Q$. The isomorphism maps the X and Y axes of $\pi(Q)$ onto the corresponding axes of $\pi(R)$.

We refer to the spread isomorphism induced by a pre-quasifield isotopism, as in Proposition 5.36 above, as the STANDARD ISOMORPHISM associated with the pre-quasifield isotopism. Hence standard spread isomorphisms, among pre-quasifield defined spreads, preserve the axes.

We turn to the converse: among pre-quasifield-defined spreads, isomorphisms preserving chosen axes are standard, i.e., are based on pre-quasifield isotopisms.

PROPOSITION 5.37. Let $(Q, +, \circ)$ and $(R, +, *)$ be pre-quasifields. Suppose $g \colon Q \oplus Q \to R \oplus R$ is an additive bijection that induces a spread isomorphism from $\pi(Q)$ onto $\pi(R)$ such that g maps the X and Y-axis of $\pi(Q)$ onto, respectively, the X and Y-axis of $\pi(R)$; so g^X and g^Y may be identified, respectively, with the maps $\alpha \colon Q \to R$ and $\beta \colon Q \to R$. Similarly, define the map $\gamma \colon Q \to R$ by the condition that g maps the component $y = x \circ m$ onto the component $y = x * (m)\gamma$. Then the following hold:

(1) The maps $\alpha \colon Q \to R$ and $\beta \colon Q \to R$ are additive bijections and $\gamma \colon Q \to R$ is a bijection, with $(0)\gamma = 0$. Thus, g may be expressed in the form:
$$g \colon x \oplus y \mapsto (x)\alpha \oplus (y)\beta; \quad g \colon \text{`} y = x \circ m\text{'} \mapsto \text{`} y = x * ((m)\gamma)\text{'}.$$

(2) The triple (α, β, γ) is a pre-quasifield isotopism from Q onto R, in particular it is a multiplicative isotopism: $\forall x, m \in Q : (x)\alpha * (m)\gamma = (x \circ m)\beta$.

(3) If Q possesses a multiplicative left identity e then: $\forall m \in Q : A * (m)\gamma = (m)\beta$, where $A = (e)\alpha$.

We emphasize special cases of the above, when both the pre-quasifields possess at least a left identity.

COROLLARY 5.38. Let $(Q, +, \circ)$ and $(R, +, *)$ be pre-quasifields possessing left multiplicative identities e and f, respectively. Suppose that $\alpha \colon Q \to R$ and $\beta \colon Q \to R$ are additive isomorphisms such that the additive bijection $\alpha \oplus \beta \colon Q \oplus Q \to R \oplus R$, induces a spread isomorphism from $\pi(Q)$ onto $\pi(R)$. Then the following hold whenever $\alpha(e) = f$:

(1) (α, β, β) is an isotopism from $(Q, +, \circ)$ to $(R, +, *)$.

(2) If e and f are two-sided identities, i.e., Q and R are quasifields, then the isotopism (α, β, γ) is simply a quasifield isomorphism, i.e., $\alpha = \beta = \gamma$, which means (cf. (2) of Proposition 5.37): $\forall x, m \in Q : (x)\alpha * (m)\alpha = (x \circ m)\alpha$.

Conversely, an isotopism (α, β, β) from $(Q, +, \circ)$ to $(R, +, *)$ induces a spread isomorphism from $\pi(Q)$ to $\pi(R)$, viz., the standard isomorphism of Proposition 5.36, that preserves the axes and maps the left identity (e, e) to the left identity (f, f).

Another specialization of Proposition 5.37 and its converse, to the case when the two quasifields coincide, we obtain *an identification of the autotopisms of a single quasifield Q with the axes-fixing automorphisms of the associated spread $\pi(Q)$.* This time we make no assumption concerning the existence of a left identity.

COROLLARY 5.39. Let $(Q, +, \circ)$ be any pre-quasifield. Then the subgroup $\mathrm{Aut}(\pi(Q))$ of the spread $\pi(Q)$ that leaves invariant the components X and Y may be identified with the autotopism group G of the pre-quasifield Q. Explicitly, if (α, β, γ) is an autotopism of $(Q, +, \circ)$, then it induces the spread automorphism of $\pi(Q)$ given by $\forall x, y \in Q: x \oplus y \mapsto \alpha(x) \oplus \beta(y)$, forcing the components $y = x \circ m$ to map onto the components $y = x \circ (m)\gamma$.

Conversely, suppose $\alpha, \beta \in \mathrm{Aut}(Q, +)$ is such that $\alpha \oplus \beta \in \mathrm{Aut}(\pi(Q))$ and let γ denote the action induced by $\alpha \oplus \beta$ on the components of $\pi(Q)$, thus, '$y = x \circ m$' $\mapsto$ '$y = x \circ (m)\gamma$'. Then: $\forall x, y, m \in Q : \beta(x \circ m) = (x)\alpha \circ (m)\gamma$, thus (α, β, γ) is an autotopism of Q.

In view of Corollary 5.39, if Q is any pre-quasifield, then the subgroup of $\mathrm{Aut}(\pi(Q))$ that leaves invariant X-axis and Y-axis will be called the *autotopism group* of the spread $\pi(Q)$, *relative to the axes* $(X = Q \oplus 0, Y = 0 \oplus Q)$. We extend the terminology:

DEFINITION 5.40. Let (X, Y) be an ordered pair of distinct components of a spread π. Then $\mathrm{Aut}(\pi)_{\{\{X\}\{Y\}\}}$, the global stabilizer of X and Y in $\mathrm{Aut}(\pi)$, is the AUTOTOPISM GROUP of π relative to (X, Y).

In this terminology, Corollary 5.39 may be interpreted as follows.

REMARK 5.41. Let $X, Y \in \pi$ be distinct components of a spread π. Then the autotopism group of π, relative to (X, Y), may be identified with the autotopism group of any pre-quasifield Q that coordinatizes π, relative to the axis pair (X, Y), for any choice of labeling functions $(\lambda_X : X \to Z, \lambda_Y : Y \to Z)$.

A more general version of this result, dealing with isotopisms rather than autotopisms is the following basic result.

PROPOSITION 5.42. Let $\Theta : V \to W$ be an isomorphism between additive spaces V and W such that Θ maps a spread π, with ambient space V, onto a spread ψ with ambient space W. Let $(S, T) = (\Theta(X), \Theta(Y))$ where $X, Y \in \pi$ are distinct components. Let $\mathcal{Q}$ (respectively $\mathcal{R}$) be the class of pre-quasifields labeling π (respectively ψ) based on the choice of axes (X, Y) (respectively (S, T)). Then the following hold:

(1) Every $Q \in \mathcal{Q}$ is isotopic to every $R \in \mathcal{R}$.
(2) The isomorphism Θ may be expressed in terms of an isotopism between any $Q \in \mathcal{Q}$ and $R \in \mathcal{R}$. More precisely, there are labelings $A : \pi \to \pi(Q)$

and $B\colon \pi \to \pi(R)$, mapping (X,Y), respectively (S,T), onto the standard axes of $\pi(Q)$, respectively $\pi(R)$, and an isotopically defined isomorphism $\Xi\colon \pi(Q) \to \pi(R)$, in the sense of Proposition 5.37, such that $\Theta = (\ )\,\alpha\Xi\beta$.

PROOF. By Theorem 5.33, corresponding to any $Q \in \mathcal{Q}$ (respectively $R \in \mathcal{R}$), there are spread isomorphisms $A\colon \pi \to \pi(Q)$ and $B\colon \psi \to \pi(R)$, mapping (X,Y) (respectively (S,T)) onto the standard axes of $\pi(Q)$ (respectively $\pi(R)$). Now $(\ )A^{-1}\Theta B = \Xi$ is an isomorphism from $\pi(Q)$ onto $\pi(R)$ that preserves the axes. Hence by Proposition 5.37 the isomorphism Ξ may be represented in terms of an isotopism from Q onto R. The result follows.   □

5.9. Semifield Spreads.

Recall that a binary system $(B,\circ)$ is a *quasigroup* if for all $b \in B$, the maps $L_a\colon x \mapsto a \circ x$ and $R_a\colon x \mapsto x \circ a$ are both bijections on B, and B is a *loop* if it has an identity. Indeed, any quasigroup is isotopic with a loop. A ring $\mathcal{D} := (D,+,\circ)$ in which the non-zero elements D^* form a multiplicative quasigroup is a pre-*division ring*, and $\mathcal{D}$ is a *division ring* if its multiplicative quasigroup is a loop. Thus, Wedderburn's celebrated theorem asserts that finite division rings are (Galois) fields, and hence completely known. By extending the notion of isotopisms concepts, from quasigroups to pre-semifields, it may be seen that any associative pre-quasifield is isotopic to an associative quasifield, thus Wedderburn's theorem leads to the description of all finite associative (pre-)division rings.

We shall only be concerned with *finite* pre-division rings and division rings. To emphasize this, we shall refer to them collectively as pre-*semifields*.

DEFINITION 5.43. A system $\mathcal{D} := (D,+,\circ)$ is a pre-SEMIFIELD if $(D,+)$ is a *finite* Abelian group such that both the distributive laws are valid:

$$z \circ (x+y) = z \circ x + z \circ y; \quad (x+y) \circ z = x \circ z + y \circ z,$$

and $(D^*,\circ)$ is a quasigroup. A pre-semifield $\mathcal{D} := (D,+,\circ)$ is a SEMIFIELD if it possesses a multiplicative identity, or equivalently, if $(D^*,\circ)$ is a loop.

We introduce some terminology and simple facts associated with the duals and multipliers of any (pre)-semifield.

REMARK 5.44. The DUAL of a (pre)-semifield $(D,+,\circ)$ is the (pre)-semifield $(D,+,\star)$, where the operation $\star$ is defined by: $\forall x,y \in D\colon \quad x \star y = y \circ x$.

For each $d \in D$, the left and right multipliers are, respectively, the maps $L_d\colon D \to D$ and $R_d\colon D \to D$, defined by $L_d\colon x \mapsto d \circ x$, $R_d\colon x \mapsto d \circ x$.

For any $d \neq 0$, we have $L_d, R_d \in GL(D,+)$. Moreover, R_d and L_d associated with $(D,+,\circ)$ become, respectively, L_d and R_d when d is regarded as an element of the dual: $(D,+,\star)$.

Note that (pre-)semifields are simply finite (pre-)quasifields that obey *both* distributive laws. In particular, the spreads associated with (pre-)semifields dualize to yield new spreads. In fact, recalling the connection between (pre-)quasifields and spreads yields:

REMARK 5.45. Let π be the translation plane associated with the spread defined by a (pre)-semifield $\mathcal{D} := (D,+,\circ)$. Let $\mathcal{D}'$ be the dual (pre)-semifield $\mathcal{D}$. Then π', the plane coordinatized by the spread associated with $\mathcal{D}'$ is the dual of the translation plane π.

Thus, every (pre)-semifield yields two translation planes, and they need not be isomorphic, as we shall see. Since (pre)-semifields are (pre)-quasifields we may consider their slope sets. The fact that both distributive laws apply to (pre)-semifields implies that their slope sets are additive. The converse also is equally trivial: finite (pre)-quasifields with additive slope sets are (pre)-semifields. However, to impose a relatively flat structure on this section, we define *additive* spread sets explicitly, without reference to spread sets or pre-quasifields.

DEFINITION 5.46. An ADDITIVE SPREAD SET is a subset $\mathcal{M} \leq GL(n,q) \cup \{\mathbf{0}\}$ such that $|\mathcal{M}| = q^n$ and

$$A, B \in \mathcal{M} \implies A - B \in \mathcal{M}.$$

Any additive spread set yields several pre-semifields.

REMARK 5.47. Let V be a vector space over $GF(q)$ and suppose $\mathcal{M} \subset GL(V,q) \cup \{\mathbf{0}\}$ is an additive spread set. Then for $e \in V^*$ define $\forall a, b \in V : a \circ b = (a)M_b$, where M_b is the unique element of $\mathcal{M}^*$ such that $eM_b = b$. Then the system $\mathcal{D}_e := (V, +, \circ)$ is a pre-semifield such that its right multipliers $R_b = M_b$, for all $b \in V$; and $\mathcal{D}_e$ is a semifield iff the spread set $\mathbf{1} \in \mathcal{M}$.

Any field of linear maps $\mathcal{F} \cong GF(q^n)$ contained in $GL(q,n) \cup \{\mathbf{0}\}$ is obviously an additive spread set. Hence Remark 5.47 above shows how such fields convert to pre-semifields that are not semifields and hence not fields.

REMARK 5.48. Let V be a rank n vector space over a field $F \cong GF(q)$. Choose a field $\mathcal{F} \cong GF(q^n)$ in $GL(V, F) \cup \{\mathbf{0}\}$, and choose $e \in V^*$. Then (1) The pre-semifield $\mathcal{F}_e$ is a field, (2) for any choice of non-singular $A \in GL(V, F)$, $\Delta := A\mathcal{F}$ is an additive spread set and the pre-semifield Δ_e is not a semifield.

Thus, from the point of view of semifield theory, even without reference to the associated geometry, it is desirable to regard pre-semifields as 'equivalent' if their spreads can be forced to coincide by a basis change of the underlying vector space. We note that such basis changes convert any additive spread sets to another additive spread sets, and that the additive spread set of any pre-semifield is equivalent to the additive spread set of a semifield.

REMARK 5.49. If $\mathcal{M} \subset GL(n,q) \cup \{\mathbf{0}\}$ an additive spread set then is $A\mathcal{M}B$ is also an additive spread set, for any $L, R \in GL(n,q)$. In particular, L and R may be chosen so that $L\mathcal{M}$ and $\mathcal{M}R$ are both additive spread sets with identity.

The significance of this remark becomes clear in the context of the following result, which specializes the standard connection, between spread sets and spreads, to finite additive spread sets.

PROPOSITION 5.50. Suppose V is a vector space of rank n over $GF(q)$ and $\mathcal{M} \subset GL(V,q) \cup \{\mathbf{0}\}$ an additive spread set. Then the collection of subspaces of $V \oplus V$:

$$\nabla M := \{\, 'y = xM' : M \in \mathcal{M} \,\} \cup \{\mathbf{0} \oplus V\}$$

is a spread on $V \oplus V$ that admits as a transitive $\mathbf{0} \oplus V$-shears group: $\{\, x \oplus y \mapsto x \oplus xA + y : A \in \mathcal{M} \,\}$.

Conversely, if a spread on $V \oplus V$ defined by **any** spread set $\mathcal{M}$, via the standard formula ∇M, admits a transitive shears group, with axis $\{\mathbf{0} \oplus V\}$, then $\mathcal{M}$ is additive spread set, linear over $GF(q)$.

In view of Proposition 5.50 above, and Remark 5.49, it follows:

COROLLARY 5.51. The spreads of order q^n that are coordinatized by additive $GF(q)$-spread-sets may always be coordinatized by additive spread sets that include identity. Hence, a finite spread coordinatized by a pre-semifield may always be coordinatized by a semifield.

There are many variants of this connection between pre-semifields and semifields. Since pre-semifields are pre-quasifields, we may regard $GF(q)$ as being in the kern of a pre-semifield if the associated spread has $GF(q)$ in its kern. In this case the spread sets defined by the pre-semifield are linear maps over $GF(q)$. Thus we easily obtain:

COROLLARY 5.52. If P is a pre-semifield with $GF(q)$ in its kern, then there is a semifield D with $GF(q)$ in its kern such that the associated spreads are isomorphic.

Thus, pre-semifields may 'in principle' be dispensed with. However, many major classes of semifields are best defined via an associated pre-semifield, which are often tidier than any semifield coordinatizing the associated plane. Hence, we continue to consider pre-semifields along with semifields.

The construction of pre-semifields is equivalent to the construction of bilinear maps, on finite vector spaces, that do not admit 'zero divisors':

PROPOSITION 5.53. Let $(D, +)$ be a finite vector space over $F = GF(q)$. Then the following are equivalent:

(1) $\mathcal{D} := (D, +, \circ)$ is a pre-semifield and F is in the kern of the associated spread: that is, the additive spread set for $\mathcal{D}$ is in $GL(D, F) \cup \{\mathbf{0}\}$.
(2) The $\circ\colon D \times D \to D$ is bilinear over F, such that $x \circ y = \mathbf{0}$ if and only if $\{x, y\} \supseteq \{\mathbf{0}\}$.

As an immediate application of Proposition 5.53, we obtain a major class of pre-semifields, and hence semifields coordinatizing their spreads.

CONSTRUCTION 5.54. **(Generalized Twisted Fields).** Let $(V, +)$ denote the additive group of $K := GF(q^n) > GF(q) := F$ and suppose $S, T \in \operatorname{Gal}(K/F)$ such that

(1) $1 \neq S \neq T \neq 1$; and
(2) $\operatorname{Fix}(S, T) = F$.

Let $N = K^{S-1} K^{T-1}$, so N^* is a subgroup of K^*. Suppose $c \in K^* \subset N^*$. Then define the Albert product:

$$\forall x, y \in K : x \circ y := \langle x, y \rangle_c := xy - x^T y^S c.$$

Then $\mathcal{A}_c := (K, +, \circ)$ is a pre-semifield: a (GENERALIZED) TWISTED FIELD; the associated spreads are ALBERT SPREADS, and these contain $GF(q)$ in their kern.

In view of Remark 5.48, it is still necessary to rule out the putative possibility that *all* the semifields associated with the generalized twisted fields are merely fields. It will be convenient to consider such matters in the context of certain subfields of semifields fundamental to their study: their various nuclear and central fields.

DEFINITION 5.55. Let $\mathcal{D} := (D, +, \circ)$ be a semifield. Then its LEFT NUCLEUS N_ℓ, MIDDLE NUCLEUS N_m, and RIGHT NUCLEUS N_r are specified, respectively, by:

$$N_\ell := \{ \ell \in D : \forall a, b \in D : \ell \circ (a \circ b) = (\ell \circ a) \circ b \},$$
$$N_m := \{ m \in D : \forall a, b \in D : a \circ (m \circ b) = (a \circ m) \circ b \},$$
$$N_r := \{ r \in D : \forall a, b \in D : a \circ (b \circ r) = (a \circ b) \circ r \}.$$

The NUCLEUS of $\mathcal{D}$, written $N := N(\mathcal{D})$, or $N(D)$, etc., is specified by

$$N(D) := N_\ell \cap N_m \cap N_r,$$

and the CENTER of $\mathcal{D}$, written $Z(D)$, $Z(\mathcal{D})$, etc., is specified by

$$Z(D) := \{ z \in N : \forall d \in D : z \circ d = d \circ z \}.$$

A subfield of any of the fields N_ℓ, N_m and N_r, is a SEMINUCLEAR field: in particular, N_ℓ, N_m and N_r are themselves seminuclear fields. Finally, subfields of the nucleus (respectively, center) are NUCLEAR (respectively, CENTRAL) fields.

We have tacitly assumed that the various seminuclear and other 'structures' defined above are fields. All this may be routinely verified:

REMARK 5.56. If D is a semifield then $N_\ell(D)$, $N_m(D)$, $N_r(D)$, $N(D)$, $Z(D)$, as defined above, are all fields. Moreover, D is a left (respectively, right) vector space over its left (respectively, right) nucleus and a two-sided vector space over its nuclear fields and central fields.

By Wedderburn's theorem, a *proper* semifield $\mathcal{D}$, cannot have associative multiplication. Thus if D is a proper semifield, then its rank $r > 1$ over any of its seminuclear fields, in particular, over its center. A standard exercise is to verify that $r = 2$ is also impossible over the center, or see the proof of Proposition 12.19, on p. 102.

REMARK 5.57. A semifield $\mathcal{D}$ of rank ≤ 2 over a central field Z must be a field.

Thus, there are no proper semifields D of rank $r \leq 2$ over a central field F. However, there do exist proper semifields that are rank 2 over their nucleus. They are obtainable as a subclass of semifields that are rank 2 over some subfield F lying in two or more coincident seminuclei. In fact, all such semifields may be described in terms of a construction of Hughes and Kleinfeld, which we only consider for the finite case.

LEMMA 5.58. Let $V = F \oplus F\lambda$ be a rank two vector space over a field F (operating from the left) with basis $(1, \lambda)$, where $1 := 1_F$, the identity of F, and $\lambda \notin F$ is an 'indeterminate' symbol. Choose $\alpha, \beta \in F$, and $\sigma \in \mathrm{Gal}(F)$. Then there is at most one semifield $\mathcal{V}_{\alpha, \beta, \sigma} := (V, +, \circ)$, where $(V, +)$ is the additive group of V, such that:

(1) $F \subseteq N_m(\mathcal{V}) \cap N_\ell(\mathcal{V})$;
(2) $\lambda^2 = \lambda\alpha + \beta$;
(3) For all $x \in F$: $\lambda \circ x = x^\sigma \circ \lambda$.

Moreover, a semifield $\mathcal{V} := \mathcal{V}_{\alpha, \beta, \sigma}$ satisfies the above conditions iff the equation $x^{1+\sigma} + x\alpha - \beta = 0$ has roots $x \in F$. The multiplication $\circ$ for this semifield is given by

$$(5.9.1) \qquad (x + y\lambda) \circ (u + v\lambda) = (xu + yv^\sigma \beta) + (xv + yu^\sigma + yv^\sigma \alpha)\lambda,$$

hence the associated additive spread set must have as its components

$$(5.9\,?) \qquad\qquad R_{u+\lambda v} := \begin{pmatrix} u & v \\ v^\sigma \beta & u^\sigma + v^\sigma \alpha \end{pmatrix}, \text{ for } u, v \in F.$$

5.10. Shears and Spread Sets.

Let $\pi := \pi_S$ be a spread specified by a spread set $S \subset \operatorname{Hom}(Z, Z)$, where Z is a finite vector space over $GF(q)$. We have seen that any finite spread may be specified in this way, on the ambient space $V = Z \oplus Z$ and such that two of the components are $X := Z \oplus 0$ and $Y := 0 \oplus Z$. When necessary, we may assume that $1 \in \operatorname{Hom}(Z, Z)$, or equivalently, that $y = x$ is a component of π.

Any linear map of V may be expressed in the form

$$\alpha \colon x \oplus y \mapsto (xA + yC) \oplus xB \oplus yD, \text{ where } A, B, C, D \in \operatorname{Hom}(Z, Z),$$

or, more conveniently, in matrix notation:

$$\alpha \colon (x, y) \mapsto (x, y) \begin{pmatrix} A, B \\ C, D \end{pmatrix} = (xA + yC, xB + yD).$$

We consider the case when α is an elation of π, with axis Y. Evidently, e.g., since all vertical lines must be invariant, α leaves the X-coordinate unchanged thus forcing $A = 1, C = 0$. Since α is the identity on Y, we similarly have $D = 1$, thus $\alpha = \begin{pmatrix} 1, B \\ 0, 1 \end{pmatrix}$. This maps the component $y = xS$, $S \in S$, onto the subspace $y = x(S + B)$. Hence α is in $\operatorname{Aut}(\pi)$ iff $B + S = S$. So we have shown:

PROPOSITION 5.59. Suppose $V = Z \oplus Z$ is the ambient space of a finite spread $\pi := \pi_S$, specified by a spread set $S \subset \operatorname{Hom}(Z, Z)$, with components including $X := Z \oplus 0$ and $Y := 0 \oplus Z$. Then the shears group $\mathfrak{G}$ of π with axis Y is isomorphic with the additive subgroup σ of S specified by $\Sigma := \{ T \in S : T + S \subseteq S \}$. In fact, $\mathfrak{G} = \{ \sigma_B \colon x \oplus y \mapsto x \oplus (xB + y) : B \in \Sigma \}$, and σ_B maps the general component $y = xM$ onto the component $y = x(M + B)$. In particular, the shears group with axis Y is transitive iff S, the slope set of π, is an additive group.

A finite spread is a *shears spread* iff it admits a transitive shears group, with affine axis. Since we may always choose the shears axis as the Y-axis, the proposition implies:

COROLLARY 5.60. A finite spread is a shears spread (semifield spread) iff it may be coordinatized by an additive spread set.

It might be remarked that we shall be discussing a very general method of representation of spreads due to Oyama. However, we postpone this discussion until we have defined generalized André planes and quasifields.

CHAPTER 6

Derivation.

The concept of a finite derivable affine plane was conceived by T.G. Ostrom in the early 1960s (see [**1037**] and [**1039**]) and has been arguably the most important construction procedure of affine planes developed in the last thirty-five years. Certain finite affine planes may be 'derived' to produce other affine planes of the same order. For example, the Hall planes of order q^2 originally constructed by Marshall Hall Jr. [**424**] by coordinate methods were shown by Albert [**26**] to be constructible from any Desarguesian affine plane of order q^2 by the method of derivation. The Hughes planes [**556**] of order q^2 were shown to be derivable and the projective planes constructed were the first examples of finite projective planes of Lenz–Barlotti class II-1 (there is a single, incident, point-line transitivity). The planes obtained were independently discovered by T.G. Ostrom and L.A. Rosati [**1140**] and are called the 'Ostrom–Rosati planes'.

The description of a finite derivable affine plane is as follows:

DEFINITION 6.1. Let π denote a finite affine plane of order q^2 and let π^E denote the projective extension of π by the adjunction of the set ℓ_∞ of parallel classes as a line. Let $\mathcal{D}_\infty$ denote a subset of $q + 1$ points of ℓ_∞. $\mathcal{D}_\infty$ is said to be a 'derivation set' if and only if it satisfies the following property:

If A and B are any two distinct points of π whose join in π^E intersects ℓ_∞ in $\mathcal{D}_\infty$ then there is an affine subplane $\pi_{A,B}$ of order q containing A and B and whose $q + 1$ infinite points are exactly those of $\mathcal{D}_\infty$.

Given any derivation set $\mathcal{D}_\infty$, there is a corresponding set $\mathcal{B}$ of $q^2(q + 1)$ affine subplanes of order q each of which has $\mathcal{D}_\infty$ as its set of infinite points. π is said to be 'derivable' if it contains a derivation set.

The main result of Ostrom is as follows:

THEOREM 6.2. (Ostrom [**1039**]). Let π be a finite derivable affine plane of order q^2 with derivation set $\mathcal{D}_\infty$. Let $\mathcal{B}$ denote the associated set of $q^2(q + 1)$ subplanes of order q each of which has $\mathcal{D}_\infty$ as its set of infinite points. Form the following incidence structure $\pi(\mathcal{D}_\infty)$: The 'points' of $\pi(\mathcal{D}_\infty)$ are the points of π and the 'lines' of $\pi(\mathcal{D}_\infty)$ are the lines of π which do not intersect $\mathcal{D}_\infty$ in the projective extension and the subplanes of $\mathcal{B}$. Then $\pi(\mathcal{D}_\infty)$ is an affine plane of order q^2.

PROOF. To prove this, we need only show that two distinct points are incident with a unique line and two lines either uniquely intersect or are disjoint.

Let P and Q be distinct points of π and let PQ denote the unique line of π containing P and Q. If PQ is not a line of $\mathcal{D}_\infty$ then PQ is also a line of $\pi(\mathcal{D}_\infty)$. If PQ is a line of $\mathcal{D}_\infty$, there is a subplane $\pi_{P,Q}$ containing P and Q whose parallel classes are exactly those of $\mathcal{D}_\infty$. Note that the infinite points together with P and Q generate a 'unique' subplane containing P and Q. Hence, given two distinct points of π, there is a unique line of $\pi(\mathcal{D}_\infty)$ containing the two points.

The lines of $\pi(\mathcal{D}_\infty)$ are of two types; lines of π and the subplanes of $\mathcal{B}$. Two distinct lines must either be disjoint or share a unique point. By the above remark on the generation of subplanes, we need only consider the situation where there is a line of each type. A subplane $\pi_{P,Q}$ and a line ℓ which does not lie in $\mathcal{D}_\infty$ must share at least one common point by a simple counting argument and cannot share more than two points since otherwise ℓ would not appear in the construction process. $\quad\square$

DEFINITION 6.3. The affine plane $\pi(\mathcal{D}_\infty)$ is called the plane 'derived' from π by the derivation of $\mathcal{D}_\infty$ or merely the 'plane derived from π'.

From here it is not difficult to show that:

THEOREM 6.4. Any affine plane $AG(2, q^2)$ is derivable.

DEFINITION 6.5. We 'define' the 'Hall planes' as the affine planes obtained by the derivation of $AG(2, q^2)$.

6.1. $PG(3, q)$ and the Regulus.

In the previous example of a derivable plane $AG(2, q^2)$, we first realized $AG(2, q^2)$ as a 2-dimensional $GF(q^2)$-vector space and then decomposed the space into a 4-dimensional $GF(q)$-vector space. From the 4-dimensional $GF(q)$-vector space, we may then construct the associated lattice of subspaces in $PG(3, q)$.

DEFINITION 6.6. A 'regulus' $\mathcal{R}$ is a set of $q + 1$ lines in $PG(3, q)$ with the property that any line ℓ which intersects $\mathcal{R}$ in at least three lines of $\mathcal{R}$ must then intersect each line of $\mathcal{R}$. Any such line ℓ is called a 'transversal' to $\mathcal{R}$. The set of $q + 1$ of transversals to $\mathcal{R}$ form a regulus $\mathcal{R}^{\mathrm{opp}}$ called the 'opposite regulus'. Note that the opposite regulus to the opposite regulus of $\mathcal{R}$ is $\mathcal{R}$.

6.1.1. The Standard Vector-Space Form of a Regulus.

PROPOSITION 6.7. Let $\mathcal{R}$ be a regulus in $PG(3, q)$ and let $V_{\mathcal{R}}$ denote the set of 2-dimensional $GF(q)$-subspaces corresponding to $\mathcal{R}$. We may choose a basis for $V_{\mathcal{R}}$ so that vectors are represented in the form (x_1, x_2, y_1, y_2) where $x = (x_1, x_2)$ and $y = (y_1, y_2)$ with scalar multiplication $(z_1, z_2)\delta = (z_1\delta, z_2\delta)$. Then the 2-dimensional $GF(q)$-subspaces of $V_{\mathcal{R}}$ are $x = 0, y = x\delta \; \forall \delta \in GF(q)$. The opposite regulus in vector-space form is the set of subplanes $\pi_{(a,b)} = \{ (a\alpha, b\alpha, a\beta, b\beta); \alpha, \beta \in GF(q) \}$ where not both a and b are 0.

PROOF. Choose a basis so that three lines of $\mathcal{R}$ as 2-dimensional $GF(q)$-subspaces are represented as $x = 0, y = 0, y = x$ where the vectors have the general form (x_1, x_2, y_1, y_2). Now take any 2-dimensional $GF(q)$-subspace W which shares a 1-dimensional subspace with each of $x = 0, y = 0, y = x$. Note that this is the vector-space version of the requirement for a transversal.

Hence, W is $\langle (0, 0, y_1^*, y_2^*), (x_1^*, x_2^*, 0, 0) \rangle$ where in order to intersect $y = x$, we must have that $(y_1^*, y_2^*) = (x_1^*, x_2^*)\delta$ for some $\delta \in GF(q)$. Hence, W is $\pi_{(x_1^*, x_2^*)}$. That is, the transversals have the form maintained in the theorem.

Once we have $\mathcal{R}^{\mathrm{opp}}$ represented in the form maintained, we may use the analogous argument showing that $AG(2, q^2)$ is derivable to show that $\mathcal{R}$ may be represented as maintained. $\quad\square$

Hence, in our example of a derivable affine plane, $AG(2, q^2)$, given any affine point O, the set of lines incident with O and intersecting $GF(q) \cup \{\infty\} = PG(1, q)$

projectively defines a regulus $\mathcal{R}$ in the associated 3-dimensional projective space $PG(3,q)$. The set of subplanes of order q incident with O each of whose set of infinite points is $GF(q) \cup \{\infty\}$ is the opposite regulus $\mathcal{R}^{\mathrm{opp}}$ to $\mathcal{R}$. As we mentioned previously, it is immediate that the subplanes of $\mathcal{R}$ are Desarguesian subplanes of order q.

6.1.2. The Regulus Question. We have seen that $AG(2,q^2)$ is a derivable affine plane. In fact, Albert [**26**] used this to construct the Hall planes by derivation.

The derivation set used for the derivation corresponds to a regulus in the associated projective space $PG(3,q)$. Perhaps the most important question dating from the 1960s is:

Does every derivation set in a finite affine plane correspond to a regulus?

Actually, Johnson (see [**753**]) was able to prove this by embedding a derivable net into a 3-dimensional projective space as follows:

6.2. The Fundamental Theorem on Derivation and Projective Space.

Formally, we recall the definition of a derivable net.

DEFINITION 6.8. A derivable net $N = (P, L, B, C, I)$ is an incidence structure with a set P of points, a set L of lines, a set B of Baer subplanes of the net, a set C of parallel classes of lines and a set I which is called the incidence set such that the following properties hold:

(i) Every point is incident with exactly one line from each parallel class, each parallel class is a cover of the points and each line of L is incident with exactly one of the classes of C.

(ii) Two distinct points are incident with exactly one line of L or are not incident.

(iii) If we refer to the set C as the set of infinite points, then the subplanes of B are affine planes with infinite points exactly those of the set C.

(iv) Given any two distinct points a and b of P that are incident with a line of L, there is a Baer subplane $\pi_{a,b}$ of B containing (incident with) a and b.

REMARK 6.9. In the following lemmas, we assume that $D = (P, L, B, C, I)$ is a derivable net satisfying properties (i), (ii), (iii) and (iv). If $\beta \in C$ and Q a point of P, $Q\beta$ shall denote the unique line of β incident with Q. Furthermore, QC shall denote the set of all net lines incident with Q.

THEOREM 6.10. (Johnson [**753**]). Let $D = (P, L, B, C, I)$ be a derivable net where the indicated sets denote the sets of points P, lines L, Baer subplanes B, parallel classes C and incidence I.

(I) Then there exists a skewfield K and a 3-dimensional projective geometry Σ isomorphic to $PG(3,K)$ together with a fixed line N of Σ such that:

(1) The points P of D are the lines of Σ which are skew to a fixed line N,

(2) The lines L of D are the points of $\Sigma - N$,

(3) The parallel classes C of D are the planes of Σ which contain N, and

(4) The subplanes B of D are the planes of Σ which do not contain N.

(II) Conversely, if $\Sigma \simeq PG(3,K)$ is a 3-dimensional projective geometry over the skewfield K and N is any fixed line of Σ, define sets of points P lines L parallel classes C, subplanes B to agree with the correspondence above where incidence

I is relative incidence in Σ. Then $D = (P, L, B, C, I)$ is a derivable net. Hence, derivable nets are equivalent to 3-dimensional projective geometries.

Now returning to the question is given a finite derivable net D and hence a 3-dimensional projective geometry $PG(3, K)$ where $K \simeq GF(q)$, is there a second 3-dimensional projective geometry $PG(3, q)$ such that the derivable net corresponds to a regulus in the second geometry?

In other words: **Is a finite derivable net a regulus net?**

6.3. The Collineation Group of a Derivable Net.

Using the embedding theorem, the following may be obtained:

THEOREM 6.11. Let $D = (P, L, B, C, I)$ be a derivable net and let $\Sigma \simeq PG(3, K)$ denote the corresponding 3-dimensional projective geometry with designated line N. Then the full collineation group G_D of D is isomorphic to the stabilizer of the line N in the full collineation group of Σ. Hence, the full collineation group of D is isomorphic to $P\Gamma L(4, K)_N$.

6.3.1. The Translation Group of a Derivable Net.

DEFINITION 6.12. Let V be a vector space over a skewfield K. A 'transvection' τ of $\Gamma L(V, K)$ such that there exists a vector d such that $\tau(v) - v$ is in $\langle d \rangle$ for all $v \in V$. In this case, v is said to be 'in the direction of d'. We shall also use the term 'transvection' to refer to the element in $P\Gamma L(V, K)$ induced by a transvection in $\Gamma L(V, K)$. A transvection of $PG(3, K)$ will fix a plane pointwise. We shall call the pointwise fixed plane, the 'axis of the transvection'. We shall be considering transvections with axes planes containing the distinguished line N of the 3-dimensional projective geometry Σ corresponding to the derivable net and in direction d a point of N. We shall use the notation τ^* to denote the quotient group action by an element τ of $\Gamma L(V, K)$.

DEFINITION 6.13. A collineation σ of the derivable net D shall be called a 'translation' if and only if σ fixes each parallel class in C and fixes some parallel class of lines linewise. The subgroup of D which is generated by the set of translations of D shall be called the translation group of D and denoted by T. Note that not every element of a translation group is necessarily a translation.

Actually, it turns out that if τ is a transvection of Σ with axis a plane A_α^+ and in the direction d, a point of N, then τ^* is a translation of the derivable net D.

REMARK 6.14. We denote by T the translation group of the derivable net D

$$\langle \tau^*; \tau^* \text{ is a transvection with axis } A_\alpha^+ \text{ and some in } N, \forall \alpha \in C \rangle.$$

We denote by T_α the subgroup of transvections with fixed axis A_α^+ with direction in N. Clearly, T_α is a normal subgroup of T as T leaves each parallel class invariant. Note that any transvection with direction d will leave invariant any plane of Σ which contains d.

THEOREM 6.15. (1) The translation group T of a derivable net D is transitive on the points P of D.

(2) The translation group T is normal in the full collineation group G_D of D and $G_D = TG_0$ where G_0 is the full collineation group which fixes a given point 0 of P.

Hence, T is Abelian and generated by any two transvection groups T_α, T_β for $\alpha, \beta \in C, \alpha \neq \beta$.

PROOF. Let (x_1, x_2, x_3, x_4) represent a vector (point) of V (of Σ) and consider the matrix action by a 4×4 K-matrix M to be $(x_1, x_2, x_3, x_4)M$.

We see that the indicated group in the statement of the theorem is the product of the following two transvection groups:

$$T_4 = \left\{ \tau_{(0,0,c_3,c_4)} = \begin{bmatrix} 1 & 0 & 0 & 0 \\ 0 & 1 & 0 & 0 \\ 0 & 0 & 1 & 0 \\ c_3 & c_4 & 0 & 1 \end{bmatrix} \right\}, \quad c_i \in K, \, i = 1, 2, 3, 4,$$

which has as its axis the plane given by the 3-dimensional vector space whose equation is $x_4 = 0$ in the representation of V, and

$$T_3 = \left\{ \tau_{(c_1,c_2,0,0)} = \begin{bmatrix} 1 & 0 & 0 & 0 \\ 0 & 1 & 0 & 0 \\ c_1 & c_2 & 1 & 0 \\ 0 & 0 & 0 & 1 \end{bmatrix} \right\}, \quad c_i \in K, \, i = 1, 2, 3, 4,$$

which has as its axis the plane given by equation x_3 in the representation of V.

Note that $T_3 T_4$ is Abelian and $\tau_{(c_1,c_2,0,0)} \tau_{(0,0,c_3,c_4)} = \tau_{(c_1,c_2,c_3,c_4)}$.

Now let τ be any transvection with direction $d = (y_1, y_2, 0, 0) \in N$ for $y_i \in K$ and $i = 1, 2$ with axis A_α^+.

Since τ fixes N pointwise, it follows directly that τ is linear, as it fixes a K-subspace pointwise and must be represented in the general form

$$\tau = \begin{bmatrix} 1 & 0 & 0 & 0 \\ 0 & 1 & 0 & 0 \\ c_1 & c_2 & a & b \\ c_3 & c_4 & c & d \end{bmatrix}.$$

(Note that if the $(1,1)$ and $(2,2)$ entries are not 1, we may multiply by an appropriate scalar matrix to obtain the required form.)

However, $\tau(x_1, x_2, x_3, x_4) - (x_1, x_2, x_3, x_4) = \lambda(y_1, y_2, 0, 0)$ for $\lambda \in K$.

This means that

$$(x_3 a + x_4 c - x_3) = 0 = (x_3 b + x_4 d - x_4).$$

Since this is valid for any x_i for $i = 1, 2, 3, 4$, it follows that $a = d = 1$ and $c = d = 0$.

This shows that any transvection may be written as a product of transvections from T_3 and T_4, respectively. $\square$

COROLLARY 6.16. (1) Any line of Σ which is skew to N has a basis of the form

$$\{(d_1, d_2, 1, 0), (d_3, d_4, 0, 1)\}.$$

(2) The element $\tau_{(c_1,c_2,c_3,c_4)}$ of T maps the line with basis

$$\{(d_1, d_2, 1, 0), (d_3, d_4, 0, 1)\}$$

onto the line with basis

$$\{(d_1 + c_1, d_2 + c_2, 1, 0), (d_3 + c_3, d_4 + c_4, 0, 1)\}.$$

(3) The translation group T acts regularly on the derivable net D.

PROOF. (1) Any line W of Σ which is skew to N intersects the planes $x_4 = 0$ and $x_3 = 0$ in different points with uniquely determined representations $(c_1, c_2, 1, 0)$ and $(c_3, c_4, 0, 1)$, respectively. These constitute the unique basis of the form required in the statement of (1).

(2) is immediate and (3) follows since the points of the derivable net are the lines of Σ skew to N.   □

REMARK 6.17. To distinguish between points of the net and skew lines with basis

$$\{(d_1, d_2, 1, 0), (d_3, d_4, 0, 1)\},$$

we shall use the notation $P(d_1, d_2, d_3, d_4)$ to represent points.

In this case, the element $\tau_{(c_1, c_2, c_3, c_4)}$ maps $P(d_1, d_2, d_3, d_4)$ onto $P(d_1 + c_1, d_2 + c_2, d_3 + c_3, d_4 + c_4)$.

Hence, T may be represented acting in D by the following:

$$\{P(x_1, x_2, x_3, x_4) \longrightarrow P(x_1 + c_1, x_2 + c_2, x_3 + c_3, x_4, c_4)\}$$

$$\forall c_i \in K \text{ and } i = 1, 2, 3, 4.$$

6.3.2. The Group of Skew Perspectivities.

DEFINITION 6.18. An element which fixes two skew lines of $PG(3, K)$ is called a skew perspectivity.

Let 0 be a fixed point of the derivable net D, M_0 the corresponding line of Σ considered also as a 2-dimensional K-vector subspace skew to the line N.

Let S_0 denote the group generated by the set of skew perspectivities that fix the lines N and M_0 pointwise.

LEMMA 6.19. With the choice of basis above, S_0 may be represented as follows:

$$\{\sigma_\delta \colon (x_1, x_2, x_3, x_4) \longmapsto (\delta x_1, \delta x_2, x_3, x_4); \delta \in K - \{0\}\}$$

The action on the P-points is as follows:

$$P(x_1, x_2, x_3, x_4) \longmapsto (P(\delta x_2, \delta x_2, \delta x_3, \delta x_4).$$

PROOF. We first note that the mappings

$$g_{\alpha, \beta} \colon (x_1, x_2, x_3, x_4) \longmapsto (\alpha x_1, \alpha x_2, \beta x_3, \beta x_4)$$

for $\alpha, \beta \in K - \{0\}$ will fix each 1-dimensional K-subspace of M_0 and each 1-dimensional K-subspace of N. Since any skew perspectivity fixes a K-subspace pointwise, it is linear and multiplication by a scalar shows that the $(1, 1)$ and $(2, 2)$ entries may be taken to be 1. Hence, the form of the skew perspectivity is as maintained.

We have

$$\langle (x_1, x_2, 1, 0) \rangle \sigma_\delta = \langle (x_1, x_2, \delta^{-1}, 0) \rangle = \langle (\delta x_1, \delta x_2, 1, 0) \rangle$$

and

$$\langle (x_3, x_4, 0, 1) \rangle \sigma_\delta = \langle (x_3, x_4, 0, \delta^{-1}) \rangle = \langle (\delta x_3, \delta x_4, 0, 1) \rangle .$$

Using the P-notation introduced above, the action on the P-points is as stated in the lemma.   □

Recall that we are trying to obtain a representation of a derivable net algebraically close to that of a regulus. In order to do this, we require that we identify the points of the derivable net with the vectors of a 4-dimensional left K-vector space and the lines of the net incident with the zero vector with a set of 2-dimensional left K-subspaces corresponding to the lines of the regulus in $PG(3,K)$. We see that we do have 4-dimensional left K-vector space defined by $P(x_1, x_2, x_3, x_4)$ with scalar multiplication

$$\alpha P(x_1, x_2, x_3, x_4) = P(\alpha x_1, \alpha x_2, \alpha x_3, \alpha x_{43}).$$

However, we shall see that in order that the lines of the derivable net incident with the point 0 become subspaces over a field L, we must restrict the scalars to the center of K, $Z(K)$.

Using the general setup, the theorem presented in the next section is obtained.

6.4. Structure Theorem for Derivable Nets.

The following result is due to Johnson (see [**753**]).

THEOREM 6.20. Let $D = (P, L, B, C, I)$ be a derivable net.

(1) Then, there exists a skewfield K such that the points of D may be considered the vectors of a left K-subspace represented in the form (x_1, x_2, y_1, y_2) and let $x = (x_1, x_2)$, $y = (y_1, y_2)$.

(2) The lines of the net which are incident with the zero vector 0 are left $Z(K)$-subspaces which may be represented by the following equations: $x = 0, y = \delta x$ $\forall \delta \in K$.

(3) The Baer subplanes of the net which are incident with the zero vector are left K-subspaces which may be represented by the following equations: $\pi_{d_1, d_2} = \{ (\alpha d_1, \alpha d_2, \beta d_1, \beta d_2); \alpha, \beta \in K \}$, where $(d_1, d_2) \neq (0, 0)$.

(4) The remaining lines and Baer subplanes are translates of the lines and Baer subplanes incident with the zero vector.

PROOF. Note that in the statement of the theorem, we are suppressing the P-notation. We also note that the π_{d_1, d_2} are clearly left K-subspaces and the mappings $\sigma_\delta\colon (x_1, x_2, y_1, y_2) \longmapsto (\delta x_1, \delta x_2, \delta y_1, \delta y_2)$ for $\delta \in K$ leave each such Baer subplane invariant. $\qquad\square$

Now we may answer the 'regulus' question for finite derivable nets.

COROLLARY 6.21. Every finite derivable net corresponds to a regulus in some 3-dimensional finite projective geometry.

6.5. Derivation of Spreads in $PG(3,q)$.

As an application of the ideas, we note the forms that derivable affine translation planes take when the derivable net is a regulus net in $PG(3,q)$ and then more generally when the derivable net does not correspond to a regulus in this particular space (note again that every finite derivable net is a regulus net is some projective space by Johnson's structure theorem).

THEOREM 6.22. Let S be a spread represented in the form:

$$x = 0, y = x \begin{pmatrix} g(u,t) & f(u,t) \\ t & u \end{pmatrix} \forall t, u \in K.$$

Assume that $\mathcal{S}$ contains a regulus $\mathcal{R}$ in standard form

$$x = 0, y = x \begin{pmatrix} u & 0 \\ 0 & u \end{pmatrix} \forall u \in K \simeq GF(q).$$

Hence, $g(u,0) = u$ and $f(u,0) = 0$ for all $u \in K$.
Then,

$$x = 0, y = x \begin{pmatrix} u & 0 \\ 0 & u \end{pmatrix} \forall u \in K \simeq GF(q)$$

together with

$$y = x \begin{pmatrix} -g(u,t)t^{-1} & f(u,t) - g(u,t)t^{-1}u \\ t^{-1} & ut^{-1} \end{pmatrix} \forall t \neq 0, u \in K$$

is also a spread in $PG(3,q)$.

This spread is called the 'derived spread' of the original. (We shall revisit such spreads later in the text and formulate some corresponding theory.)

A regulus partial spread is then a derivable partial spread (again, the reader is referred to Johnson [753] for more details on derivation and derivable nets, derivable partial spreads).

In general a derivable net within a spread in $PG(3,q)$ may be given the following representation:

$$x = 0, y = x \begin{pmatrix} u & 0 \\ 0 & u^\sigma \end{pmatrix} \forall u \in K \simeq GF(q), \sigma \in \operatorname{Aut} GF(q).$$

When this occurs and σ is not 1, then the derived translation plane does not have its spread in $PG(3,q)$, so this is a simple way of creating large dimensional spreads.

An easy example of spreads in $PG(3,q)$ are the Kantor–Knuth spreads:

$$x = 0, y = x \begin{bmatrix} u & \gamma t^\sigma \\ t & u \end{bmatrix} ; u, t \in GF(q),$$

q odd, σ an automorphism of $GF(q)$, γ a non-square.

So, from the Knuth spreads, we obtain:

COROLLARY 6.23. The following represents a spread in $PG(3,q)$:

$$x = 0, y = x \begin{pmatrix} v & 0 \\ 0 & v \end{pmatrix}, y = x \begin{pmatrix} -ut^{-1} & \gamma t^\sigma - u^2 t^{-1} \\ t^{-1} & ut^{-1} \end{pmatrix}$$

$\forall t \neq 0, v, u \in GF(q)$ for q odd, γ a non-square and $\sigma \in \operatorname{Gal} GF(q)$.

CHAPTER 7

Frequently Used Tools.

In this small chapter, we gather various important and frequently used tools. These appear elsewhere in the text but we list these as a sort of small tool-kit for the geometer.

7.1. p-Primitive Elements.

THEOREM 7.1. (Zsigmondy [**1219**]; see Lüneburg [**961**, Theorem 6.2, p. 27]). Consider a prime power p^n, where n is an integer. Then there is a prime u dividing $p^n - 1$ but not dividing $p^k - 1$ for k a positive integer $< n$ or one of the following occurs: (1) $n = 2$ and $p + 1 = 2^a$, i.e., p is a Mersenne prime or (2) $p = 2$ and $n = 6$.

The element u mentioned above is a prime 'p-primitive divisor' of $p^n - 1$. For the more general situation of divisors of $q^n - 1$, where q is itself a prime power, u is said to be a 'q-primitive divisor' of $q^n - 1$.

7.2. Schur's Lemma.

THEOREM 7.2. Let V be a vector space over a field K. Let S be a set of linear transformations of V/K which act irreducibly on V. Then the centralizer of S in $GL(V, K)$ is a field.

7.3. Maschke Complements.

THEOREM 7.3. Let G be a group of $GL(V, K)$ acting on a vector space V of finite order over K. Assume that $(|G|, |K|) = 1$ and G leaves invariant a subspace W of V. Then there is a complement C of W that is left invariant by G; $W \oplus C = V$. The subspace C is said to be a 'Maschke Complement'.

Assume that τ is a collineation of a finite translation plane in the translation complement. Assume that τ leaves invariant a subspace W of order p^n. If the order of τ is a p-primitive divisor of $p^n - 1$, then τ acts irreducibly on W or trivially. Hence, if τ fixes a proper subspace of W, then τ fixes W pointwise.

7.4. Ostrom Phantom.

THEOREM 7.4. (See Johnson [**700**]). Let V be a vector space of dimension $2n$ over a field $F \simeq GF(q)$, for $q = p^r$, p a prime. Assume that a collineation σ of $GL(2n, q)$ has order dividing $q^n - 1$ but not dividing $q^t - 1$ for $t < n$. If σ fixes at least three mutually disjoint n-dimensional F-subspaces, then there is an associated Desarguesian spread Σ admitting σ as a kernel homology. Furthermore, the normalizer of $\langle \sigma \rangle$ is a collineation group of Σ.

In the situation above, Σ is called an 'Ostrom phantom'.

7.5. Hering–Ostrom.

THEOREM 7.5. (Hering [**476**], Ostrom [**1047, 1052**]). Let π be a finite translation plane of order p^r, for p a prime. Let E denote the group that is generated by all affine elations in the translation complement (that fix the zero vector). Then one of the following situations occurs:

(1) E is elementary Abelian, and there is a unique elation axis.

(2) E is isomorphic to $SL(2, p^t)$, and there are exactly $p^t + 1$ elation axes, and $p^t + 1$ elementary Abelian elation groups of order p^t.

(3) E is isomorphic to $SL(2, 5)$, $p = 3$, and the number of elation axes is 10.

(4) E is isomorphic to $S_z(2^{2t+1})$ (a Suzuki group), and the number of elation axes is $2^{2(2t+1)} + 1$.

(5) The order of E is $2d$ where d is an odd number and $p = 2$.

7.6. Structure of Baer Subplanes.

THEOREM 7.6. (Foulser [**375**]). Let π be a translation plane of order q^2, for q a prime power p^n. Let B be a Baer group (fixes a Baer subplane π_0 of order q pointwise).

(1) Then $B = PQ$, where P is an elementary Abelian normal Sylow p-subgroup of B and Q is cyclic subgroup of order dividing $|P| - 1$ and $|B|$ divides $q(q - 1)$. In addition, the order of P divides the kernel subgroup of π_0.

(2) (Jha [**570**]). If q or $q - 1$ divides $|B|$, then π_0 is Desarguesian.

7.7. Planar p-Bound of Jha.

THEOREM 7.7. (Jha [**570**]). Let π be a translation plane of order p^n, for p a prime and n an integer, admitting a planar automorphism p-group of order p^r. Then $r \leqq n - 1$ unless $r = n/2$ and the fixed point plane is Baer. When $n = r - 1$ then $p = 2$ and $p^n = 16$.

7.8. Incompatibility and Structure of Baer Subplanes.

THEOREM 7.8. (Foulser [**375**], Jha and Johnson [**588**]). Let π be a translation plane of order p^n, for p an odd prime and n an even positive integer.

(1) Assume that p is odd. (a) Then Baer p-collineations and elations are incompatible. (b) If $p > 3$, then all Baer subplanes pointwise fixed by Baer p-collineations lie in the same net of degree $\sqrt{p^n} + 1$.

(2) For p even or odd, if there are at least three Baer subplanes fixed pointwise by Baer p-collineation groups of order $p^{n/2}$, then the net containing the Baer axes is a rational Desarguesian net.

(3) For $p = 2$, let P be a Baer p-group and let E denote an elation group (both groups in the translation complement). (a) If $|B| > \sqrt[4]{p^n}$, then E centralizes B and $|E| \leqq 2$. (b) If $|E| = q$ and E centralizes B, then $|B| \leqq 2$.

7.9. Foulser–Johnson.

The following result is an extension of the Foulser–Johnson theorem on translation planes of order q^2 admitting a collineation group isomorphic to $SL(2,q)$. The version given is due to Biliotti–Jha–Johnson.

The reader is referred to Chapter 47 for all details regarding the planes listed in the theorem as well as a sketch of the proof.

THEOREM 7.9. Let π denote an affine translation plane of order q^2. Assume that π admits a collineation group G containing a normal subgroup N, such that G/N is isomorphic to $PSL(2,q)$. Then π is one of the following planes: (1) Desarguesian, (2) Hall, (3) Hering, (4) Ott–Schaeffer, (5) one of three planes of Walker of order 25, or (6) the Dempwolff plane of order 16.

CHAPTER 8

Sharply Transitive Sets.

In this chapter, we show that sharply transitive sets are closely connected to translation planes.

THEOREM 8.1. Let π be a finite translation plane of order p^n, where p is a prime and n a positive integer. Choose a basis so that a spread may be represented as follows:

$$x = 0, y = 0, y = xM_i, \ i = 1, 2, \ldots, p^n - 1$$

$$\text{for } M_i, \ M_i - M_j \text{ non-singular, } \forall \, i \neq j = 1, 2, \ldots, p^n - 1.$$

(1) Then $\{ M_i; \ i = 1, \ldots, p^n - 1 \}$ acts sharply transitively on the non-zero points of an underlying n-dimensional vector space over $GF(p)$.

(2) Conversely, a set of linear transformations that acts sharply transitively on the non-zero points of an n-dimensional vector space over a field isomorphic to $GF(p)$ constructs a finite translation plane whose spread representation is as in part (1).

PROOF. Let x_0 be any non-zero vector of a vector space upon which the matrices M_i act. Assume that $x_0 M_i = x_0 M_j$. Then $M_i - M_j$ is singular, a contradiction if i is not j. Hence, there are exactly $p^n - 1$ images under the set of matrices, which is the number of points in the vector space. This proves part (1); the converse is similar. □

In this context, we note a connection with finite sharply two-transitive groups, which have been completely determined by Zassenhaus [**1218**].

THEOREM 8.2. Let π be a finite translation plane of order p^n, for p a prime and n a positive integer with spread defined as above:

$$x = 0, y = 0, y = xM_i, \ i = 1, 2, \ldots, p^n - 1$$

$$\text{for } M_i, \ M_i - M_j \text{ non-singular, } \forall \, i \neq j = 1, 2, \ldots, p^n - 1.$$

Let T denote the translation group of the underlying vector space V_n of dimension n over $GF(p)$. (1) Then $T \{ M_i; \ i = 1, \ldots, p^n - 1 \}$ is a sharply doubly transitive set of permutations of V_n, which contains a transitive normal subgroup T. (2) Conversely if S is a sharply doubly transitive set of permutations of V_n, containing a transitive normal subgroup, then there is a corresponding translation plane with spread defined exactly as above.

PROOF. $T \{ M_i; \ i = 1, \ldots, p^n - 1 \}$ is clearly transitive and the stabilizer of the zero vector is transitive on the non-zero vectors, which is equivalent to the set being doubly transitive. Since T has order p^n it follows that the set is sharply doubly transitive and the normality of T follows easily. The converse is similar. □

Now assume that we have a finite sharply doubly transitive group on a set X. Since we now have the classification theorem of finite simple groups and hence a listing of all finite doubly transitive group actions (see, for example, Chapter 104), it follows that there is a normal elementary Abelian normal subgroup. This implies that $|X| = p^n$ for some prime p and positive integer n. We then obtain a translation plane by the above theorem. Furthermore, since the stabilizer of a point (the 'zero vector') is a group, it will now follow that there is an affine homology group of order $p^n - 1$ and by considering the mapping $(x, y) \to (y, x)$, then we will obtain a second affine homology group of order $p^n - 1$, as this mapping now becomes a collineation of the translation plane.

DEFINITION 8.3. A finite translation plane of order p^n which admits an affine homology group of order $p^n - 1$ with axis L and coaxis M also admits an affine homology group of order $p^n - 1$ with axis M and coaxis M. Any such plane is said to be a 'nearfield plane'.

8.1. The Nearfield Planes.

As mentioned finite translation planes of order p^n, for p a prime, are equivalent to sharply transitive sets within $GL(2n, p)$. When the sharply transitive set is a sharply transitive group acting on the non-zero vectors, we have a nearfield plane. Since we are considering now doubly transitive groups with an elementary Abelian normal subgroup, represented in the form $x \longmapsto x + a$, for all a in $GF(p^n)$. In what are called the 'regular nearfield planes', the group which fixes 0, has elements of the form $x \longmapsto x^{p^{\lambda(m)}} m$, where λ is a function from $GF(p^n)^*$ to the set of integers $\{0, 1, 2, \ldots, m - 1\}$ and is hence a generalized André plane. Dickson [**317**] completely determined these groups and these shall be listed below.

8.2. The Irregular Nearfield Planes.

However, there are other sharply transitive groups, all of which are defined as 2-dimensional matrix groups over $GF(p)$, produce translation planes (nearfield planes) of order p^2 and involve either $SL(2, 5)$ or $SL(2, 3)$. These planes are called the 'irregular nearfield planes'. Furthermore, all of these groups are in $SL(2, t)Z$, where Z is $GF(p)^*$, of order $p - 1$.

In the $SL(2, 5)$ case, the central involution fixes all 1-dimensional $GF(p)$-subspaces. Since Z fixes all 1-dimensional $GF(p)$-subspaces, it follows that $p + 1$ must divide 60, so that $p = 11, 19, 29, 59$. For $p = 19$, however, $p^2 - 1 = 60 \cdot 3$, and it is impossible for $SL(2, 5)Z^i$, to act sharply transitively. For $p^2 - 1 = 60 \cdot 2, 60 \cdot 14, 60 \cdot 58$, respectively, for $p = 11, 19, 59$, the sharply transitive groups are, respectively, $SL(2, 5)Z^{11-1} = SL(2, 5)$, $SL(2, 5)Z^4$, $SL(2, 5)Z^2$. The corresponding nearfield planes are the 'irregular nearfield planes with non-solvable groups'.

In the $SL(2, 3)$ case, there is a subgroup N containing $SL(2, 3)$ as an index 2 subgroup. In this setting $p+1$ must divide 24, implying that $p = 3, 5, 7, 11, 23$. When $p = 3$, the corresponding plane is the Hall plane, which is a regular nearfield plane. For $p^2 - 1 = 24, 24 \cdot 2, 24 \cdot 5, 24 \cdot 22$, respectively for $p = 5, 7, 11, 23$, the corresponding sharply transitive groups are $SL(2, 3)$, N, $SL(2, 3)Z^2$, NZ^2, respectively. These planes are called the 'irregular nearfield planes with solvable groups'. It might also be pointed out that Dickson also determined the irregular nearfield planes. The general argument for sharply 2-transitive groups is due to Zassenhaus [**1218**].

The specific descriptions are as follows, with the type number, order and generators of the nearfield generator groups M listed.

$$I.\ 5^2,\ A\ =\ \begin{bmatrix} 0 & -1 \\ 1 & 0 \end{bmatrix}, B = \begin{bmatrix} 1 & -2 \\ -1 & -2 \end{bmatrix}, M \simeq SL(2,3)$$

$$II.\ 11^2,\ A\ =\ \begin{bmatrix} 0 & -1 \\ 1 & 0 \end{bmatrix}, B = \begin{bmatrix} 1 & 5 \\ -5 & -2 \end{bmatrix}, C = B = \begin{bmatrix} 4 & 0 \\ 0 & 4 \end{bmatrix},$$
$$M\ \simeq\ SL(2,3) \times \langle C \rangle$$

$$III.\ 7^2,\ A\ =\ \begin{bmatrix} 0 & -1 \\ 1 & 0 \end{bmatrix}, B = \begin{bmatrix} 1 & 3 \\ -1 & -2 \end{bmatrix}, M \simeq N$$

$$IV.\ 23^2,\ A\ =\ \begin{bmatrix} 0 & -1 \\ 1 & 0 \end{bmatrix}, B = \begin{bmatrix} 1 & -6 \\ 12 & -2 \end{bmatrix}, C = B = \begin{bmatrix} 2 & 0 \\ 0 & 2 \end{bmatrix},$$
$$M\ \simeq\ N \times \langle C \rangle$$

$$V.\ 11^2,\ A\ =\ \begin{bmatrix} 0 & -1 \\ 1 & 0 \end{bmatrix}, B = \begin{bmatrix} 2 & 4 \\ 1 & -3 \end{bmatrix}, M \simeq SL(2,5)$$

$$VI.\ 29^2,\ A\ =\ \begin{bmatrix} 0 & -1 \\ 1 & 0 \end{bmatrix}, B = \begin{bmatrix} 1 & -7 \\ -12 & -2 \end{bmatrix}, C = B = \begin{bmatrix} 16 & 0 \\ 0 & 16 \end{bmatrix},$$
$$M\ \simeq\ SL(2,5) \times \langle C \rangle$$

$$VII.\ 59^2,\ A\ =\ \begin{bmatrix} 0 & -1 \\ 1 & 0 \end{bmatrix}, B = \begin{bmatrix} 9 & 15 \\ -10 & -10 \end{bmatrix}, C = B = \begin{bmatrix} 4 & 0 \\ 0 & 4 \end{bmatrix},$$
$$M\ \simeq\ SL(2,5) \times \langle C \rangle$$

The Hall Planes of Order 9. The Hall plane of order 9 is also a nearfield plane and admits a collineation group acting transitively on the line at infinity. Furthermore, there are five symmetric pairs of components admitting affine homologies of order 2 (arising from Baer involutions of the associated Desarguesian plane), which are permuted by the full collineation group. The group induced on these pairs turns out to be S_5 and the kernel of the action has order 2^5, so that the full collineation group is of order $2^5 \cdot 5!$.

Noting that we have defined the Hall planes as constructed from a Desarguesian affine plane by derivation of a regulus. The collineation groups of all Hall planes of orders $q^2 > 9$, however, have the full collineation group inherited from the Desarguesian plane and hence fixes a regulus. There is a group isomorphic to $GL(2,q)$ acting on the Hall planes, where the p-elements, $p^r = q$, are Baer collineations.

8.2.1. The Regular Nearfield Planes; Dickson Planes.

DEFINITION 8.4. Let $q = p^r$, where p is a prime and r is a positive integer. Let n be a positive integer. If every prime divisor of n also divides $q - 1$ and if $n \not\equiv 0 \bmod 4$ when $q \equiv 3 \bmod 4$, then $\{q, n\}$ is said to be a 'Dickson pair'. It follows that n divides $q^n - 1$.

THEOREM 8.5. (Dickson [**317**]). Let $\{q, n\}$ be a Dickson pair. Let U be a subgroup of $GF(q^n)^*$ of order $(q^n - 1)/n$. Let wU be a generator of the cyclic group $GF(q^n)^*/U$.

(1) Then the following is a nearfield spread, also called a 'Dickson' nearfield spread (Dickson nearfield plane):

$$x = 0, y = 0, y = x^{q^i} m; \; m \in w^{\left(\frac{q^i-1}{q-1}\right)}, \; i = 1, 2, 3, \ldots, n,$$

and all regular nearfield spreads (corresponding to regular nearfield planes) may be represented in this form (Zassenhaus [**1218**], Lüneburg [**961**]).

(2) (Lüneburg [**961**]). Let φ denote the Euler function and f the order of $p \bmod n$ (the smallest integer k such that $w^{(p^k-1)} \in U$). Then the number of Dickson nearfield planes is $\frac{\varphi(n)}{f}$.

CHAPTER 9

$SL(2,p) \times SL(2,p)$-**Planes.**

The irregular nearfield planes of order p^2, admit affine homology groups isomorphic to either $SL(2,5)$ or $SL(2,3)$ in their corresponding sharply transitive groups. Furthermore, by taking the product G of the two affine homology groups with symmetric axes and coaxes, we obtain a collineation group with the property that given any non-axis (of an affine homology) N, then G_N is transitive on the non-zero vectors of N. We note that in the irregular nearfield planes there is a collineation of the form $(x,y) \longmapsto (y,x)$ if the axes are called $x = 0, y = 0$. In particular, this means that the irregular nearfield planes are of rank 3. Indeed, if a rank 3 translation plane has an orbit of length 2 on the line at infinity (i.e., interchanges say $x = 0, y = 0$), usually the translation plane is a generalized André and only is not when possibly the order is in $\{5, 7, 11, 23, 29, 59\}$. Since the irregular nearfield planes are not generalized André planes, we see the restriction on orders is strict. But, in particular, rank 3 translation planes with an orbit of length 2 are interesting and actually may be classified. The classification involves the analysis of some exceptional planes. The work of Kallaher, Ostrom [**844**] and Lüneburg [**960**] (together with the computer analysis of all translation planes of order 7^2) gives the following result:

THEOREM 9.1. (Kallaher–Ostrom [**844**], Lüneburg [**960**] + translation planes of order 7^2). Let π be a rank 3 translation plane with an orbit of length 2 on the line at infinity. Then either the plane is a generalized André plane, one of the irregular nearfield planes, or the exceptional Lüneburg plane of order 7^2 (see below for the description).

We note that the statement that the plane is a generalized André plane does not specify the generalized André planes which are rank 3. We note this as an open problem.

PROBLEM 9.2. Determine all rank 3 generalized André planes.

There are two exceptional classes of translation planes of order p^2 that are 2/3-triangle transitive; there is automorphism group G in the translation complement such that G fixes two components L and M and acts transitively on the non-zero vectors on each of L and M. It is known that non-solvable affine homology groups much involve $SL(2,5)$. However, it is also possible to have affine homology groups isomorphic to $SL(2,3)$. The exceptional Lüneburg planes are precisely the translation planes of order p^2 with the following properties:

(i) There is an autotopism collineation group that acts transitively on the non-zero vectors of each affine axis L and M.

(ii) The subgroup $G_{[L]}$ fixing L pointwise is isomorphic to the subgroup $G_{[M]}$ pointwise, and this group is isomorphic to $SL(2,t)$, where either $t = 3$ or 5 and the component orbits of each group are the same.

(iii) Given any component N not equal to L or M, the stabilizer subgroup G_N is transitive on the non-zero vectors of N.

Recalling again the theorem of Lüneburg [**961**, Theorem 9.2] which characterizes finite generalized André planes, which is the existence of an Abelian autotopism group A with property (iii), the planes under consideration are exceptional indeed to exist at all. Recalling again the classification theorem of finite doubly transitive groups (see Chapter 104):

(B) G has a regular normal subgroup N which is elementary Abelian of order $v = h^a$, where h is a prime. Identify G with a group of affine transformations $x \longmapsto x^g + c$ of $GF(h^a)$, where $g \in G_0$. Then one of the following occurs:
(1) $G \leq A\Gamma L(1,v)$, (2) $G_0 \trianglerighteq SL(n,z)$, $z^n = h^a$, (3) $G_0 \trianglerighteq Sp(n,z)$, $z^n = h^a$, (4) $G_0 \trianglerighteq G_2(z)'$, $z^6 = h^a$, z even, (5) $G_0 \trianglerighteq A_6$ or A_7, $v = 2^4$, (6) $G_0 \trianglerighteq SL(2,3)$ or $SL(2,5)$, $v = h^2$, $h = 5, 7, 11, 19, 23, 29$, or 59 or $v = 3^4$, (7) G_0 has a normal extraspecial subgroup E of order 2^5 and G_0/E is isomorphic to a subgroup of S_5, where $v = 3^4$, (8) $G_0 = SL(2,13)$, $v = 3^6$.

Hence, we will then isolate on case (6), which will imply in the $SL(2,5)$ case that $p \in \{11, 19, 29, 59\}$. Similarly, in the $SL(2,3)$ case, the restriction will turn out to be $p \in \{5, 7, 11, 23\}$. However, there is another way to see the numerical restriction. The kernel homology group of order $p-1$ fixes each 1-dimensional $GF(p)$ subspace and acts transitively on the $p-1$ points of that space. Moreover, by the structure of affine homology groups, we have that $SL(2,t)$ is a normal subgroup of $G \mid L$. Consider the group induced in $PSL(2,p)$. The structure is known, and since we have either A_5 or A_4, we see that in either case this implies that the group induced on the set of 1-dimensional $GF(p)$-subspaces is in fact the normalizer of A_5 or A_4 in $PGL(2,p)$. In the A_5 case, A_5 is its own normalizer, and in the A_4 case, A_4 has index 2 in its normalizer. Hence, we have either $p + 1$ divides 60 or $p + 1$ divides 24. Furthermore, since $SL(2,t)$ is an affine homology group, we know that 120 or 24 divides $p^2 - 1$. In the $SL(2,5)$-case, we then obtain $p = 11, 19, 29, 59$ and in the $SL(2,3)$-case, we have $p = 5, 7, 11, 23$. Note that in terms of numbers we now include the possibility that $p = 19$, apart from the numbers involved in irregular nearfield planes. Hence, in all but one of the orders, we will find examples of nearfield planes.

9.1. $SL(2,5)$-Exceptional Lüneburg Planes.

For more details, the reader is directed to Lüneburg [**961**] (note also our notation is slightly different from Lüneburg's). Take a group isomorphic to $SL(2,5)$ in $SL(2,p)$, where $p = 11, 19, 29$, or 59, acting on a 2-dimensional $GF(p)$-space. Let Z denote the scalar group of order $p-1$ of $GF(p)^*$. Then $SL(2,5)Z^2$ acts transitively on the associated 2-dimensional vector space V. Form $V \oplus V$ and consider $SL(2,5)Z^2 \times SL(2,5)Z^2$ acting in the natural manner. We intend to construct a spread in $PG(3,p)$, so we let the putative kernel group with elements $(x,y) \longmapsto (kx, ky)$, for all $k \in GF(p)^*$ of order $p-1$, be denoted by K^*. The analysis of the emerging spreads will deal with the group $SL(2,5) \times SL(2,5)K^{*2}$. Form the following partial spread (a regulus in $PG(3,p)$):

$$\{x = 0, y = 0, y = xk; \; k \in GF(p)^*\} = R,$$

and note that the diagonal group D with elements

$$d_g : \begin{bmatrix} g & 0 \\ 0 & g \end{bmatrix} ; g \in SL(2,5)$$

fixes each $y = xk$.

The following sets of 2-dimensional $GF(p)$-subspaces, become partial spreads:

$$\mathcal{P}_{(y=xk)} = \mathcal{P}_k = \{ (y = xk)h; \ h \in SL(2,5) \times SL(2,5)K^* \},$$

and other than $(0,0)$, the union as a set of points is a point orbit under $SL(2,5) \times SL(2,5)K^{*2}$. It turns out $SL(2,5) \times SL(2,5)D$ is the stabilizer of each partial spread $\mathcal{P}_k$. So, the only other action involves subgroups of $Z^2 \times Z^2$. Furthermore, the set of point orbit lengths under $SL(2,5) \times SL(2,5)K^{*2}$ of $V \oplus V$ of points not on either $x = 0$ or $y = 0$ are

$$p = 11, \qquad \text{one orbit of length } 120^2,$$

$$p = 19, \qquad \text{three orbits of length } 120^2 \cdot 3,$$

$$p = 29, \qquad \text{seven orbits of length } 120^2 \cdot 7,$$

$$p = 59, \qquad \text{twenty-nine orbits of length } 120^2 \cdot 29.$$

Actually, the number of elements $y = xk$ (other than $(0,0)$) which lie within orbits of $SL(2,5) \times SL(2,5)K^{*2}$ is the same, say a, within each orbit and may also be enumerated as follows:

$$a = 10 \qquad \text{for } p = 11,$$

$$a = 6 \qquad \text{for } p = 19,$$

$$a = 4 \qquad \text{for } p = 29,$$

$$a = 2 \qquad \text{for } p = 59.$$

Note that $Z^2 \times Z^2$ will permute transitively the point orbits off of $x = 0, y = 0$ of $SL(2,5) \times SL(2,5)K^{*2}$. Now concerning the partial spreads $\mathcal{P}_k$, it will turn out that there are exactly two components of R within this partial spread. Hence, there are exactly $a/2$ partial spreads obtained in this way, for each of the a orbits and recall that the set of non-zero points is a point orbit under $SL(2,5) \times SL(2,5)K^{*2}$. Hence, for each of the point orbits, there are a components $y = xk$ (other than $(0,0)$) with each orbits, and within each of the partial spreads $\mathcal{P}_k$ there are two components $y = xt$, so there are $a/2$ possible choices for a partial spread of this type.

For example, for $p = 19$, there are three point orbits, each of which contains 6 regulus components, defining three possible partial spreads. If we choose a partial spread for each point orbit union $x = 0, y = 0$, this becomes a spread. Hence, there are 3^3 possible constructed translation planes of order 19^2.

The number of spreads constructed is then k_p, of order p^2:

$$k_p = 1, \qquad p = 11;$$

$$k_p = 3^3, \qquad p = 19;$$

$$k_p = 2^7, \qquad p = 29;$$

$$k_p = 1, \qquad p = 59.$$

Hence, in particular, we see that there are unique spreads constructed in this manner of order 11^2 and 59^2, namely the irregular nearfield planes. Using the group, however, it turns out that there are exactly three mutually non-isomorphic planes of order 19^2 and nine mutually non-isomorphic planes of order 29^2, of which one is the irregular nearfield plane.

The main results of Lüneburg [**961**] (see Chapter III) are as follows:

THEOREM 9.3. The translation planes of order p^2 with the following properties are completely determined:

(*i*) There is an autotopism collineation group that acts transitively on the non-zero vectors of each affine axis L and M.

(*ii*) The subgroup $G_{[L]}$ fixing L pointwise is isomorphic to the subgroup $G_{[M]}$ pointwise, and this group is isomorphic to $SL(2,5)$, and the component orbits of each group are the same.

(*iii*) Given any component N not equal to L or M, the stabilizer subgroup G_N is transitive on the non-zero vectors of N. Then $p = 11, 19, 29, 59$ and

(*a*)$_{11}$, there is a unique translation plane of order 11^2, the irregular nearfield plane,

(*b*)$_{19}$, there are three translation planes of order 19^2, the exceptional Lüneburg planes of order 19^2,

(*c*)$_{29}$, there are nine translation planes of order 29^2, the irregular nearfield plane and eight other planes, the exceptional Lüneburg planes of order 29^2,

(*d*)$_{59}$, there is a unique translation plane of order 59^2, the irregular nearfield plane.

9.2. $SL(2,3)$-Exceptional Lüneburg Planes.

In this case, we are concerned with planes of order p^2, where $p = 5, 7, 11, 23$. Actually, Lüneburg [**961**](Chapter III) shows that planes of order 5^2 or 23^2 satisfying the conditions (*i*), (*ii*), and (*iii*) with $t = 3$ are exactly the irregular nearfield planes. For planes of orders 7^2 and 11^2, there are exceptional planes obtained by an analysis of the group $SL(2,3) \times SL(2,3)$ similar to the $SL(2,5)$-case. We consider $SL(2,3)$ within $SL(2,p)$ acting on a 2-dimensional $GF(p)$-vector space.

In both cases, we consider the regulus R in $PG(3,p)$, represented as before,

$$x = 0, y = 0, y = xk; \ k \in GF(p)^*$$

and define what turn out to be partial spreads as follows:

$$\mathcal{P}_k = \{ (y = xk)g; \ g \in SL(2,3) \times SL(2,3) \}.$$

Just as in the $SL(2,5)$ case, these partial spreads each contain exactly two components of the regulus (namely $y = xk$ and $y = -xk$).

$p = 7$: First take $p = 7$. Let B_1 denote the union set of points of $\mathcal{P}_1$. All components $y = xk$ of the regulus R are actually contained within B_1. Hence, B_1 contains three partial spreads $\mathcal{P}_k$. There is a subgroup N containing $SL(2,3)$ as an index two subgroup. We are then considering the group $N \times NK^*$, where K^* is the scalar group of order $p - 1$, as a putative collineation group of the constructed translation plane. Let $M = \{ (w,w); \ w \in N \}$, so that $K^* \subset M$.

Take $B_2 = B_1(1,v)$, where $(1,v) \in N \times N - (SL(2,3) \times SL(2,3)K^*)M(\langle 1 \rangle \times GF(p)^*)$, then it turns out that B_2 and B_1 intersect precisely at $(0,0)$, showing that there are also three partial spreads within B_2. Furthermore, B_1 and B_2 partition the underlying vector space $V \oplus V$. Choose any of the three partial spreads from within B_1 and any of the three within B_2. Together with $x = 0, y = 0$, this becomes a spread. Hence, there are 3^2 possible spreads so constructed. In fact, up to isomorphisms, there are exactly two, one being the irregular nearfield plane. The other is a rank 3 translation plane.

$p = 11$: Here let $B_k = \bigcup_{T \in \mathcal{P}_k} \mathcal{P}_k$. Then $\bigcup B_k - \{(0,0)\}$ contains all points not on $x = 0, y = 0$ and $B_k = B_{-k}$ but $B_k \cap B_j = \{(0,0)\}$ if k is not j or $-j$. Take $v \in NGF(p)^* - SL(2,3)GF(p)^*$. It then follows that $\mathcal{P}_k(1,v)$ is a partial spread within $W = \bigcup B_k$. Hence, there are five sets B_k and in each there are two partial spreads. For each B_k (modulo $\pm k$) choose either $\mathcal{P}_k$ or $\mathcal{P}_k(1,v)$. The union of these partial spreads together with $x = 0, y = 0$ form a spread. Hence, there are 2^5 spreads. Up to isomorphisms, there are exactly four spreads, one corresponding to the irregular nearfield plane of order 11^2 with solvable group.

THEOREM 9.4. (Lüneburg [**961**, Chapter III]). The translation planes of order p^2 with the following properties are completely determined:

(i) There is an autotopism collineation group that acts transitively on the non-zero vectors of each affine axis L and M.

(ii) The subgroup $G_{[L]}$ fixing L pointwise is isomorphic to the subgroup $G_{[M]}$ pointwise, and this group is isomorphic to $SL(2,3)$, and the component orbits of each group are the same.

(iii) Given any component N not equal to L or M, the stabilizer subgroup G_N is transitive on the non-zero vectors of N. Then $p = 5, 7, 11, 23$ and

$(a)_5$, there is a unique translation plane of order 5^2, the irregular nearfield plane,

$(b)_7$, there are two translation planes of order 7^2, the irregular nearfield plane and the exceptional Lüneburg plane of order 7^2,

$(c)_{11}$, there are four translation planes of order 11^2, the irregular nearfield plane with solvable group and three other planes, the exceptional Lüneburg planes of order 11^2,

$(d)_{23}$, there is a unique translation plane of order 23^2, the irregular nearfield plane.

9.3. $SL(2,5) \times SL(2,5)$-**Planes of Hiramine.**

Now assume that a translation plane of prime square order p^2 admits a collineation group isomorphic to $SL(2,5) \times SL(2,5)$ fixing components $x = 0$ and $y = 0$. Then $SL(2,5) \times \langle 1 \rangle$ becomes an affine homology group and we see that our group becomes the product of affine homology groups with symmetric axes. The exceptional Lüneburg planes have this property, in addition to the assumption that each orbit of one affine homology group is the same as the other. If we do not assume this condition, then other translation planes arise, as determined by Hiramine [**507**], who assumed instead that all but possibly one orbit on the line at infinity has length 120. Under these conditions, the spread has the following form:

$$x = 0, y = 0, y = xg_1 H \cup y = xg_2 H \cup \cdots \cup y = xg_r H \cup y = xHsH$$

where H is $\langle 1 \rangle \times SL(2,5)$, and g_i are appropriate elements of $GL(2,p)$. The length of the remaining orbit $y = xHsH$ is $n120$, where $n \in \{5, 6, 10, 12, 20, 30, 60\}$ (the 15-case cannot occur).

For planes of order p^2, the first basic requirement is that 120 divides $p^2 - 1$. Actually, Hiramine notices that if $y = xg_1$ has an orbit of length 120 then $y = xk$, for some $k \in GF(p)^*$. Moreover, if $y = xk_1 H$ and $y = xk_2 H$, for k_1 and k_2 in

$GF(p)^*$, form a partial spread, then $k_1^{-1} k_2$ cannot have order 2, 3, or 5. Using this idea, it is possible to restrict the possible orders p^2. Indeed, if (n, p) denotes the size of the 'long' orbit of length $120n$ for $n \geq 1$, where the possible order of the translation plane is p^2. Then, with the use of the computer, the spreads are completely determined. Actually, when all orbits have length 120, it is shown that only the irregular nearfield and exceptional Lüneburg show up.

We denote the spread by representatives of the orbits of lengths 120 by $\{k_1, \ldots, k_r\}$ (i.e., $y = x k_i H$) and the representative of $y = x H s H$ of the order of length $120n$ by $s = \begin{bmatrix} a & b \\ c & d \end{bmatrix}$, a 2×2 matrix in $GF(p)$. For convenience, we represent the spread by $\left\{ 1, 3, 11, \begin{bmatrix} 4 & 5 \\ 6 & 27 \end{bmatrix} \right\}$. That is, this notation represents the following spread:

$$x = 0, y = 0, y = xH \cup y = x3H \cup y = x11H \cup y = xH \begin{bmatrix} 4 & 5 \\ 6 & 27 \end{bmatrix} H,$$

of order 31^2, where $n = 5$.

THEOREM 9.5. (Hiramine [**507**]). If π is a translation plane of order p^2, where p is a prime admitting the collineation group $SL(2, 5) \times SL(2, 5)$, generated by affine homology groups isomorphic to $SL(2, 5)$, assume that there is exactly one component orbits of length > 120, say $120n$. Then π is one of the following spreads:

(*i*) $p = 31, n = 5$ and the four spreads are as follows:

$$\left\{ 1, 3, 11, \begin{bmatrix} 4 & 5 \\ 6 & 27 \end{bmatrix} \right\}$$
$$\left\{ 3, 4, 5, \begin{bmatrix} 4 & 5 \\ 6 & 27 \end{bmatrix} \right\}$$
$$\left\{ 1, 7, 13, \begin{bmatrix} 4 & 5 \\ 6 & 27 \end{bmatrix} \right\},$$
$$\left\{ 4, 6, 13, \begin{bmatrix} 4 & 5 \\ 6 & 27 \end{bmatrix} \right\}$$

(*ii*)$_5$ $p = 59$, $n = 5$ and one spread

$$\left\{ k \in GF(59)^* - \{7, 8, 11, 19, 25\}, \begin{bmatrix} 1 & 32 \\ 48 & 58 \end{bmatrix} \right\},$$

(*iii*)$_6$ $p = 59$, $n = 6$ and one spread

$$\left\{ k \in GF(59)^* - \{5, 11, 14, 16, 17, 19\}, \begin{bmatrix} 28 & 2 \\ 3 & 31 \end{bmatrix} \right\},$$

(*iv*)$_7$ $p = 59$, $n = 7$ and one spread

$$\left\{ 1, 2, 4, 5, 6, 7, 10, 14, 17, 18, 20, 22, 24, 25, 26, 27, \begin{bmatrix} 8 & 1 \\ 31 & 39 \end{bmatrix} \right\},$$

Let K_{p-1}^* denote the associated kernel homology group of order $p-1$. Hiramine also shows that the full translation complement is $SL(2, 5) \times SL(2, 5) K_{30}^*$, for the first two spreads listed in (*i*) and is $SL(2, 5) \times SL(2, 5) K_p^* \langle \tau \rangle$, where $\tau^2 \in K_p^*$, where τ invert $x = 0$ and $y = 0$ for the remaining five spreads. Finally, we mention the following open problem.

PROBLEM 9.6. Determine all translation planes of order p^2, for p a prime, admitting $SL(2,5) \times SL(2,5)$ as a collineation generated by affine homologies.

CHAPTER 10

Classical Semifields.

Recall that a distributive quasifield is called a *semifield*. Equivalently, a semifield is a 'non-associative [skew]field' as seen in the following characterization. The aim of this chapter is to address the following question: what are the possible sizes of finite *non-associative* semifields? We shall see that semifields that are of order p^2 are always fields. Also all translation planes of order 8 are known to be Desarguesian. But the twisted fields of A.A. Albert and the even-order commutative semifields of D.E. Knuth, taken together, demonstrate that *for all other prime-power orders n* at least one non-associative semifield plane of order n exists. The main goal of this chapter is to introduce these planes and demonstrate that they are non-associative.

The following theorem is an analogue of the elementary result: finite [associative] integral domains are fields. Here we prove that finite 'non-associative' integral domains are semifields. Many important constructions of finite [pre]semifields are based on this principle.

REMARK 10.1. A system $(D, +, \circ)$ is a semifield iff the following axioms hold:

(1) $(D, +)$ is an Abelian group;
(2) The distributive laws are valid for $x, y, z \in D$:
   (a) $x \circ (y + z) = x \circ y + x \circ z$;
   (b) $(y + z) \circ x = y \circ x + z \circ x$.
(3) $(D^*, \circ)$ is a loop.

A semifield that is not a [skew]field is called a proper semifield. We shall be concerned with *finite* semifields from now on. Thus the basic question is what are the possible orders of proper non-associative semifields? This question has a complete answer, but first we draw attention to some elementary facts.

Let $(D, +, \circ)$ be a finite semifield. Then its three seminuclei $N\ell$, N_m and N_r are all fields, in particular its kern coincides with N_ℓ and $(D, +)$ is a vector space over each of these nuclei, as well as over its nucleus and center (both of which are also fields).

A semifield two-dimensional over a field in its center is a field, Remark 5.57. Hence all semifields of order p^2 are known. Thus all semifield planes of order p^2 are known. A spectacular extension of this result follows from a theorem of Menichetti [**979**]: all semifield planes of order p^3 are known. They are forced to be coordinatized by the generalized twisted fields of Albert, see Definition 10.13.

10.1. The Knuth Commutative Semifields.

Finite commutative semifields (that are not associative) appear to be quite hard to find. The following construction due to Knuth [**897**] established the existence of commutative semifields of *even* order N, where $N > 8$ is not a power of 2.

THEOREM 10.2 (The Binary Knuth Semifields). Let

$$K = GF(2^{nm}) \supset GF(2^m) = K_0,$$

where $n > 1$ is odd. Let $f : K \to K_0$ be any non-zero linear functional of K as a K_0 vector space. Define a new multiplication as follows: $a \circ b = ab + (f(a)b + f(b)a)^2$. The algebraic system $(K, +, \circ)$ is a pre-semifield.

We note that the usual procedure for converting a pre-quasifield to a quasifield '$(a \circ b) = (a \circ e) * (e \circ b)$', where e is an arbitrary non-zero element. However, to ensure that the resulting commutative semifield is not a field f needs to be chosen with some care. Such an f is introduced in the following theorem.

The theorem also demonstrates that in converting a pre-semifield to a semifield it is desirable to choose the identity 'e' with care, to avoid creating a semifield with a more opaque structure than the pre-semifield used to construct it.

THEOREM 10.3 (The Binary Knuth Semifields). Let

$$K = GF(2^{nm}) \supset GF(2^m) = K_0,$$

where $n > 3$ is odd. Fix a K_0-basis of K of type $(1, \alpha, \alpha^2, \ldots, \alpha^{n-1})$, and choose the K_0-valued functional $f : K \to K_0$ such that $f(\alpha^i) = 0$ for $0 \leq i \leq n - 2$, and $f(\alpha^{n-1}) = 1$. Define new multiplications $\circ$ and $\odot$ on K as follows for all $a, b \in K$: $a \circ b = ab + (f(a)b + f(b)a)^2$ and $a \circ b = (a \circ 1) \odot (1 \circ b)$. The algebraic system $(K, +, \circ)$ is a commutative pre-semifield and $(K, +, \odot)$ is a commutative semifield (but not a field) such that they both coordinatize the same semifield plane.

Perhaps the most important feature of the theorem above is that it ensures the existence of non-Desarguesian projective planes of order 2^p, p any *prime* > 3.

10.2. Twisted Fields.

Let $c \in K = GF(q^n)$ such that $c \notin K^{q-1}$. Then $GF(q)$-linear maps of $K = GF(q^n)$ defined by

(10.2.1) $P^{-1} \quad : \quad x \mapsto x - cx^q,$

(10.2.2) $Q^{-1} \quad : \quad x \mapsto x^q - cx$

are bijective (thus justifying the inverse notation) because $x = xc^q$ or $x^q = cx$ both contradict the assumption $c \notin K^{q-1}$.

Since P^{-1} and Q^{-1} both map 1 to $1 - c = f$, we also have $P(f) = Q(f) = 1$. We now define the semifield associated with (P, Q); the above equation will establish the multiplicative identity.

THEOREM 10.4. Define $\odot$ by $x \odot y = xP(yQ)^q - (xP)^q(yQ)c$, and let $f = 1 - c$. Then $(K, +, \odot)$ is a division algebra with identity $f = 1 - c$ and center $F \odot f$ where $F = GF(q) \subset GF(q^n)$.

As indicated by Albert, the product $\odot$ cannot be regarded as explicitly known until the Vaughan polynomials for P *and* Q are explicitly known. However, in view of the close connection between the definitions of P^{-1} and Q^{-1}, cf. (10.2.1) and (10.2.2), it is possible to deduce the Vaughan polynomial of Q from that of P, so we only compute P explicitly.

10.2.1. Polynomial for P; Non-Commutative of a Semifield. In this section we adopt the following:

NOTATION 10.5. Regarding $K = GF(q^n) \supset F = GF(q)$ as a rank n vector space over F, and define the F-linear maps of K: (1) $S : x \mapsto x^q$; (2) $R_a : x \mapsto xa$, for $a \in K$.

We regard members of $\mathrm{Hom}_F(K, +)$ as acting on K from the right. The associative ring $\sum_{i=0}^{n-1} S^i R_{a_i}$, for $a_i \in K$, forms an F-algebra; F may be identified with the central field $\{R_f \mid f \in F\}$. By Vaughan polynomials the S^i's in the expression are linearly independent over F and hence the expressions account for $|K|^n$ K-linear maps in $\mathrm{Hom}_F(K, +)$, but since this set has size $|F|^{n^2}$, we have a fundamental fact concerning Vaughan polynomials.

RESULT 10.6 (Fundamental Theorem of Vaughan Polynomials). The K-algebra $\mathrm{Hom}_F(K, +)$ is the K-algebra $\left\{ \sum_{i=0}^{n-1} S^i R_{a_i} \mid a_i \in F, \ \forall i \in [0, n-1) \right\}$.

We now compute P using the equation (10.2.1), which may be written as $P^{-1} = x - xSR_c$, and the elementary ring identity $(1 - \theta)(1 + \theta + \theta^2 + \cdots + \theta^{n-1}) = 1 - \theta^n$, by noticing that $\theta := SR_c$ implies: $1 - \theta = P^{-1}$. Thus we have:

$$(10.2.3) \qquad P^{-1} \left(1 + SR_c + (SR_c)^2 + \cdots + (SR_c)^{n-1} \right) = 1 - (SR_c)^n$$

and now $(SR_c)^i$ may be expressed in the following notation, $(SR_c)^i = S^i R_{c_i}$, where $c_i \in F^*$ is uniquely defined by the above requirement. In particular, we need to record:

REMARK 10.7. Define $c_i \in F$ in terms of c by: $\forall i \in [1, n] : (SR_c)^i = S^i R_{c_i}$. Then

(1) P^{-1} commutes with all terms of type $S^i R_{c_i}$;
(2) $c_n \in GF(q)$, $c_{i+1} = (c_i)Sc$; and
(3) $c_n \in GF(q)^*$, but $c_n \neq 1$.

THEOREM 10.8. Assume $n > 2$. Then $(D, +, \odot)$ is commutative iff $c = -1 \neq 1$ and $P = Q - (1 + S)$.

10.3. Generalized Twisted Fields.

The twisted fields of Albert, discussed in the previous section, are important partly because they help to demonstrate that non-associative semifields of odd order p^r exist, for p prime, iff $r > 2$. The generalized twisted fields, introduced in this section, have proven to be of importance because they arise in several major classification theorems: Menichetti's classification of the semifields of order p^3 and in the Cordero–Figueroa–Liebler classification of semifield planes admitting large autotopism groups of various types. In all these cases the associated planes are shown to be among the class of generalized twisted fields of Albert, rather than in the class of planes coordinatized by just the ordinary twisted fields of the previous section. We begin with an elementary result from arithmetic that has wide applications in the exploitation of finite fields.

THEOREM 10.9. Let q be a prime power. Then

$$\gcd \left(q^a - 1, q^b - 1 \right) = q^{\gcd(a,b)} - 1.$$

Throughout the section we adopt the following hypothesis:

NOTATION 10.10. The integer $q = p^s > 1$ is a power of the prime p. We set

$$K = GF(q^n)$$

and Aut K denotes the associated Galois group generated by $\rho : x \mapsto x^q$. Assume $S, T \in$ Aut K such that (1) $1 \neq S \neq T \neq 1$; and (2) Fix$(S, T) = GF(q)$.

Note that any finite field with two distinct non-trivial automorphisms, S and T, can be viewed as satisfying all the above conditions if we define $GF(q)$ to be the fixed field of the group $\langle S, T \rangle$. Write $N = K^{S-1}K^{T-1}$, so N^* is a multiplicative subgroup of K^*. Fix an element $c \in N - K$. The *Albert product* on K, written $\langle x, y \rangle_c$ and abbreviated to $x \circ y$ is defined by:

(10.3.1) $\forall x, y \in K : x \circ y := \langle x, y \rangle_c := xy - x^T y^S c.$

Clearly, $\langle x, y \rangle_c = 0 \iff x = 0 \vee y = 0$.

Since S and T are additive, $(K, +, \circ)$ must also satisfy both distributive laws: so we have a finite 'non-associative integral domain' and, as in the associative case, this means that multiplication defines a quasigroup on the non-zero elements. Thus we have:

LEMMA 10.11. Suppose the triple $(D, +, \circ)$ is such that $(D, +)$ is a *finite* Abelian group such that both the distributive laws hold. Then $(D^*, \circ)$ is a quasigroup, or equivalently, $(D, +, \circ)$ is a pre-semifield if and only if

$$x \circ y = 0 \iff x = 0 \vee y = 0.$$

THEOREM 10.12. Let $\mathcal{A}_c := (K, +, \circ)$, where $\circ = \langle \cdot, \cdot \rangle_c$ is an Albert product on $K = GF(q^n)$ and $(K, +)$ is the additive group of the field. Then $\mathcal{A}_c$ is a presemifield.

DEFINITION 10.13. The planes coordinatized by the pre-semifields $\mathcal{A}_c$ will be called the *Albert planes*. The pre-semifields $\mathcal{A}_c$ will be called *generalized twisted fields*.

The following proposition yields the list of orders that Albert planes have.

PROPOSITION 10.14. Let $K = GF(q^n)$, Fix$(\langle S, T \rangle) = GF(q)$, where $S \neq T$ are distinct non-trivial $GF(q)$-linear field automorphisms in Aut K such that Fix$(\langle S, T \rangle) = GF(q)$. Let $N = K^{S-1}K^{T-1}$, then $K - N \neq \emptyset$ iff

(1) $q > 2$ and $n > 2$; now any pair of distinct non-trivial S and T will yield $K - N \neq \emptyset$;

(2) If $q = 2$ and n is not a prime; now, without loss of generality $1 \leq a < b < n$, the pair $\left(S : x \mapsto x^{p^a}, T : x \mapsto x^{p^b} \right)$ yields $K - N \neq \emptyset$ iff and gcd$(a, b) > 1$ shares a non-trivial factor with n.

Hence, a generalized twisted field of order 2^n exists iff n is a non-prime ≥ 6.

10.3.1. Menichetti's Characterization of Albert Systems. Considering a semifield as a vector space n over its center K, the spectacular result of Menichetti [**979, 980**] shows that when n is prime and K is large enough relative to n then the only possible semifields are generalized twisted fields.

THEOREM 10.15. (Menichetti [**979, 980**]). Let S be a finite semifield of prime dimension n over the center K of order q. Then there is an integer $\nu(n)$ such that if $q \geqslant \nu(n)$ then S is a generalized twisted field.

We note that in the theorem of Menichetti, the right, left and middle are equal to the center. Also, there are strict counterexamples to the theorem without $q \geqslant \nu(n)$. For example, the Knuth binary of order 2^p, for p prime, is not generalized twisted, so it would be a counterexample to a putative general result on semifields of prime dimension.

10.4. Dickson Commutative Semifields.

In this subsection we mention two classes of semifields whose planes admit geometric characterizations. They are also associated with tangentially transitive planes. We use the following notation. Let F be a finite field of odd order and $a \in F^*$ a non-square in F. Let λ be an indeterminate over F, and θ a non-trivial field automorphism of F. Let $D = F \oplus \lambda F$.

THEOREM 10.16 (Dickson's Commutative Semifields). Suppose $a \in F^*$ is non-square, so F is odd. Then

$$(x + \lambda y) \circ (z + \lambda t) = \left(xz + a(yt)^\theta\right) + \lambda(yz + xt)$$

is a commutative semifield such that: (1) F is the middle nucleus of $(D, +, \circ)$; (2) $K = \text{Fix}(\theta) \cap F$ is the right nucleus, the left nucleus and hence also the center of D.

10.5. Hughes–Kleinfeld Semifields

The semifields of order q^2 with two nuclei equal and isomorphic to $GF(q)$ are known by Hughes and Kleinfeld. These are defined as follows:

THEOREM 10.17. (Hughes–Kleinfeld Semifields). Suppose $a = x^{1+\theta} + xb$ has no solution for x in F. Then

$$(x + \lambda y) \circ (z + \lambda t) = (xz + aty^\theta) + \lambda \left(yz + (x^\theta + y^\theta b)t\right)$$

is a semifield and F is its right and middle nucleus. Conversely, if D is a semifield that is a finite two-dimensional over a field F such that the middle and right nucleus of D coincide then D is a Hughes–Kleinfeld semifield.

In Chapter 12, any semifield of order q^2 having two nuclei isomorphic to $GF(q)$ becomes a Hughes–Kleinfeld semifield by 'fusing' the nuclei.

CHAPTER 11

Groups of Generalized Twisted Field Planes.

In the late 1950s and early 1960s, A.A. Albert wrote several seminal papers on semifield planes and, more particularly, on twisted and generalized twisted field planes (see Albert [**22, 23, 24**]). In particular, Albert determined the solvability of the autotopism groups of twisted field planes, completely determined the generalized twisted fields whose associated isotopes with unity are commutative and gave exact conditions when two generalized twisted fields coordinatize isomorphic semifield planes.

Using group theory, it is possible to show that generalized twisted field planes which admit an autotopism group acting transitively on the non-vertex points of at least one side of the autotopism triangle always have solvable autotopism groups. Moreover, Biliotti–Jha–Johnson [**121**] determined the collineation groups acting relative to a given field of reference K and conclude that, in particular, the twisted field planes admit autotopism collineation groups acting transitively on two but not always on three of the sides of the autotopism triangle and more generally the autotopism collineation groups always sit as subgroups of $\Gamma L(1, K) \times \Gamma L(1, K)$.

However, not all generalized twisted field planes admit a collineation group acting transitively on the non-vertex points of a side of the autotopism triangle. So, in order to completely determine the autotopism groups of generalized twisted field planes, we considered a more general method which would be applicable to all such planes.

As noted by Albert, any autotopism collineation may be represented in terms of three additive mappings and is of the general form $(x, y) \longmapsto (f(x), h(y))$ and $y = x \circ m \longmapsto y = x \circ g(m)$, where $y = x \circ m$ represents the components of the generalized twisted field plane. Writing these additive functions f, g and h over $GF(p)$ as polynomial functions of degrees $\leq p^{r-1}$, when the order is p^r, and using the pre-semifield components as opposed to the semifield components, Biliotti, Jha and Johnson are able to completely determine the autotopism collineation group of any non-Desarguesian generalized twisted field plane. The analysis ultimately uses fundamental aspects of group theory results. Once the autotopism groups have been determined, it is a simple matter to extend these results to the analysis of isotopisms.

The main results in this regard are due to Biliotti, Jha and Johnson and are as follows:

THEOREM 11.1. (Biliotti, Jha, Johnson [**121**]). Let $\pi = \pi(K, \alpha, \beta, c)$ be a non-Desarguesian generalized twisted field plane of order p^r with components

$$x = 0, \ y = xm - cx^\alpha m^\beta \text{ for all } m \in K \simeq GF(p^r),$$

where α and β are automorphisms of K and c is a constant such that $c \neq x^{\alpha-1} m^{\beta-1}$ for all $x, m \in K$.

(1) The full autotopism collineation group is a subgroup of $\Gamma L(1, K) \times \Gamma L(1, K)$.
(2) The autotopism collineations are exactly the maps of the form $(x, y) \longmapsto$ $(x^\sigma e, y^\rho d)$, where $e, d \in K - \{0\}$, $\sigma, \rho \in \text{Aut} K$, such that $\rho = \sigma$ and $e^{\alpha - \beta} d^{\beta - 1} = c^{\sigma - 1}$.
(3) Every isotopism of π onto another generalized twisted field plane may be assumed to be within $\Gamma L(1, K) \times \Gamma L(1, K)$.

There have been quite a few studies involving twisted and generalized twisted field planes for various reasons, but there are several results in the literature which can be completed only by a deep knowledge of what occurs in generalized twisted field planes.

Ganley and Jha [392] studied finite translation planes that admit a collineation group that fixes an infinite point (∞) and acts doubly transitively on the remaining infinite points. The main theorem of Ganley and Jha is the following.

THEOREM 11.2. (Ganley and Jha [392]). Let π denote a finite translation plane which admits a collineation group that fixes an infinite point (∞) and acts doubly transitively on the remaining infinite points. Then π is a semifield plane and the full collineation group is solvable.

Dempwolff [304] used the coordinate methods of Oyama [1079] to analyze those semifield planes that admit a cyclic and transitive autotopism group.

THEOREM 11.3. (Dempwolff [304]). Let π denote a finite semifield plane which admits a cyclic and transitive autotopism group on the non-vertex infinite points. Then π is a generalized twisted field plane.

More generally, in a series of papers concerning characterizations of generalized twisted field planes, Cordero and Figueroa [253] show that the main ingredient which forces a semifield plane to become a generalized twisted field plane is the presence of an autotopism collineation which induces on a side of the autotopism triangle a permutation of order a p-primitive divisor of $p^r - 1$. Then, with a few exceptions, the plane is a generalized twisted field plane. In particular, they prove the following

THEOREM 11.4. (Cordero and Figueroa [253]). Let π be a semifield plane of order $p^r \neq 2^6$. If the autotopism group acts transitively on the non-vertex points of at least one side of the autotopism triangle, then π is a generalized twisted field plane.

So, the main open question is to establish when these transitive autotopism groups exist in generalized twisted field planes. The main results of Biliotti–Jha–Johnson [121] is as follows:

THEOREM 11.5. Let $\pi = \pi(K, \alpha, \beta, c)$ be a generalized twisted field plane and let G be the full autotopism group. Let $\alpha = p^a$ and $\beta = p^b$.
(1) G admits a cyclic subgroup acting transitively on the non-zero vectors of $(y = 0)$ if and only if $(r, a, b) = (r, b)$.
(2) G admits a cyclic subgroup acting transitively on the non-zero vectors of $(x = 0)$ if and only if $(r, a, b) = (r, a - b)$.
(3) G admits a cyclic subgroup acting transitively on $\ell_\infty - \{(\infty), (0)\}$ if and only if $(r, a, b) = (r, a)$.

Hence, there are cyclic groups acting transitively on the non-vertex points of each side of the autotopism triangle if and only if the right, middle, and left nuclei are equal.

More generally, it might be asked the nature of generalized twisted field planes that admit only transitive groups, but not necessarily cyclic transitive groups, on the non-vertex points of a side of the autotopism triangle. By utilizing the methods of Foulser [366], Biliotti–Jha–Johnson [121] are able to show:

THEOREM 11.6. (Biliotti–Jha–Johnson [121]). Let π be a generalized twisted field plane and let G be the full autotopism group. If G acts transitively on the non-vertex points of a side of the autotopism triangle, then there is a cyclic subgroup of G which also acts transitively on the non-vertex points of that side.

Hence, of particular interest is the following:

COROLLARY 11.7. (Biliotti–Jha–Johnson [121]) Let π be a generalized twisted field plane. Then π admits an autotopism collineation group acting transitively on the non-vertex points of each side of the autotopism triangle if and only if the left, right and middle nuclei are equal.

Actually, it is known that twisted field planes admit such cyclic groups acting transitively on the non-vertex points of the infinite side of the autotopism triangle, but Dempwolff was not able to show that the generalized twisted field planes with such a group must be twisted field planes.

It is immediate from our results that not every generalized twisted field plane with cyclic group acting transitively on the non-vertex points of the infinite side of the autotopism triangle must be twisted.

Precisely, the result may be given as follows:

THEOREM 11.8. (Biliotti–Jha–Johnson [121]). Let $\pi = \pi\,(K, \alpha, \beta, c)$ be a generalized twisted field plane of order p^r. Then π admits a cyclic collineation group transitive on the non-vertex points of the infinite side of the autotopism triangle if and only if $(r, a) = (r, a, b)$.

In a similar manner, combining our results with those of Ganley–Jha and Cordero–Figueroa, it is possible to obtain several characterization theorems which are given in the last part of this chapter.

There is an intrinsic connection within generalized twisted field planes and certain distance-transitive graphs with imprimitive automorphism groups. In particular, Gardiner [394] develops connections between certain semifield planes and distance-transitive graphs. In fact, Gardiner proves that distance-transitive q-fold antipodal coverings of the complete bipartite graph $K_{q,q}$ are equivalent to self-dual semifield planes of order q which admit an autotopism group that acts transitively on the non-vertex points of each side of the autotopism triangle. Chuvaeva and Pasechnik [233] show that the commutative twisted field planes are self-dual and admit such an autotopism group, thus producing examples of such distance-transitive covering graphs. Here the early definition of "twisted field" meant distorting the multiplication of a field $K \cong GF(q^s)$, q a prime power, as follows $x \circ m = x^q m - cxm^q$, for c a suitable constant in K. It follows directly from the previously mentioned work that the given examples are slightly more general than those considered by Chuvaeva and Pasechnik. Moreover, Liebler and his colleagues

[**923**] were able to completely determine the possible distance-transitive q-fold antipodal covers of $K_{q,q}$ by showing that the self-dual semifield planes which admit an
autotopism group transitive on the non-vertex points of each side of the autotopism
triangle are exactly the self-dual twisted field planes.

Here we consider the self-dual generalized twisted field planes. In particular,
combining the works listed about of Biliotti, Jha, Johnson and Liebler, we obtain
the following result:

THEOREM 11.9. Distance-transitive q-fold antipodal covers of $K_{q,q}$ with $q =$
p^r are in one-to-one correspondence with $PG(2,q)$ and the twisted field planes
$\pi\left(K, \alpha, \alpha^{-1}, c\right)$ of order q such that $r/(r,a)$ is odd and $c^{\rho+1} = z^{1-\alpha}$ for some
$z \in K^*$ and $\rho \in \operatorname{Aut} K$.

From a slightly different angle, Enea and Korchmáros [**353**] give a fundamental
contribution to the determination of two-transitive parabolic ovals in translation
planes of odd order.

THEOREM 11.10. (Enea and Korchmáros [**353**]). Let π be a translation plane of
odd order which admits a 2-transitive parabolic oval. Then π may be coordinatized
by a commutative semifield.

As already remarked by Enea and Korchmáros, the results of Cordero–Figueroa
apply in this situation, so that π may be coordinatized by a generalized twisted
field. Since Biliotti, Jha and Johnson have analyzed the isotopisms of any generalized twisted field plane, it is then possible to make a slight improvement of this
theorem. We mention this theorem here and then also provide a sketch of how this
is accomplished in the next two sections.

THEOREM 11.11. (Biliotti–Jha–Johnson [**121**], Enea and Korchmáros [**353**]).
Let π be a translation plane of odd order p^r which admits a 2-transitive parabolic
oval. Then π is a commutative twisted field plane and may be coordinatized by a
commutative twisted field with multiplication: $x \circ m = x^\alpha m - x m^\alpha$, for all x, m
in $K \cong GF(p^r)$. Hence, if $\alpha = p^a$ then $r/(r,a)$ is odd. Thus, a translation plane
of odd order admits a 2-transitive parabolic oval if and only if it is a commutative
twisted field plane.

Albert [**22**] originally considered a particular case of generalized twisted fields,
under the name of "twisted fields". Precisely Albert assumes that $\beta = \alpha^{-1}$ and
$(r,a) = a$ and considers the following isotopic form (see [**290**]):

$$x * y = x \circ y^\alpha = x y^\alpha - c x^\alpha y = x y^q - c x^q y$$

where $p^a = q$ and $c \neq x^{q-1}$ for all $x \in K^*$. Subsequently, in a note of [**23**], Albert
refers to twisted fields without the assumption $(r,a) = a$, which is actually an
unnecessary condition. In general throughout this text we use the term "twisted
field" in the broad meaning.

An autotopism of $A(K, \alpha, \beta, c)$ is a triple (f,g,h) of non-singular additive functions over K such that for all $x, m \in K$. The associated autotopism collineation of
π acts as follows $(x,y) \longrightarrow (f(x), h(x)), y = x \circ m \longrightarrow y = x \circ g(m)$.

Every generalized twisted field plane possesses the class of autotopism collineations given in the following:

THEOREM 11.12. *Let $\pi = \pi(K, \alpha, \beta, c)$ be a non-Desarguesian generalized twisted field plane. Then the map $(x, y) \longrightarrow (x^\sigma e, y^\rho d)$ with $e, d \in K^*$ and $\sigma, \rho \in$ Aut K is an autotopism collineation of π if and only if $\sigma = \rho$ and $e^{\alpha-\beta} d^{\beta-1} = c^{\sigma-1}$.*

In the treatment given here we are working with pre-semifields. We make the following assumptions.

(a) The right nucleus (the kernel) of $\pi = \pi(K, \alpha, \beta, c)$ consists of those elements $e \in K^*$ such that $a(e, e, I)$ is a collineation of π (together with 0). It is the subfield of K of order $p^{(r,a)}$ since by Theorem 11.12 we must have $e^{\alpha-\beta} e^{\beta-1} = e^{\alpha-1} = 1$.

(b) The middle nucleus of π consists of those elements $e \in K^*$ such that $a(e, 1, I)$ is a collineation of π (together with 0). It is the subfield of K of order $p^{(r,a-b)}$ since we must have $e^{\alpha-\beta} = 1$.

(c) The left nucleus of π consists of those elements $d \in K^*$ such that $a(1, d, I)$ is a collineation of π (together with 0). It is the subfield of K of order $p^{(r,b)}$ since we must have $d^{\beta-1} = 1$.

Our assumptions agree with Albert's results. Clearly these assumptions are correct only when $\pi(K, \alpha, \beta, c)$ is non-Desarguesian, that is $\alpha \neq \beta$ and neither α or β is I.

In the following, with the term "generalized twisted field plane" we refer only to non-Desarguesian planes.

REMARK 11.13. It is easily seen that the generalized twisted field $A(K, \alpha, \beta, c)$ is isotopic to the generalized twisted field $A(K, \alpha^{-1}, \beta^{-1}, c^{-1})$. So, if $\alpha : k \longrightarrow k^{p^a}$, $\beta : k \longrightarrow k^{p^b}$ we may suppose, up to an isotopism, $a > b$ when necessary.

Group-Theoretic Methods in Twisted Field Planes. Using results of permutation group theory, Biliotti, Jha, and Johnson [121] are able to show:

PROPOSITION 11.14. (Biliotti, Jha, Johnson [121]). *Let G be the full autotopism group of π. If $N \mid l$ is normal in $G \mid l$ for at least one side l of the autotopism triangle.*

THEOREM 11.15. (Biliotti, Jha, Johnson [121]). *Let G be the full autotopism group of π. If G is transitive on the non-vertex points of at least one side of the autotopism triangle, then G is a subgroup of $\Gamma L(1, K) \times \Gamma L(1, K)$.*

In the following, we give a sketch of how to determine those generalized twisted field planes which admit a cyclic autotopism group acting transitively on the non-vertex points of at least one side of the autotopism triangle. We note that a variation of the following lemma occurs in Liebler [921, Corollary 3.5]; the proof is technical and is omitted.

LEMMA 11.16. *Let $K \cong GF(p^r)$ and let $\alpha : x \longrightarrow x^{p^a}$, $\beta : x \longrightarrow x^{p^b}$ be automorphisms of K. Then the following hold:*

(1) *The equation $e^{\alpha-\beta} d^{\beta-1} = 1$ admits a solution (e, d) if and only if $e^{\alpha-\beta}$ or $d^{\beta-1}$ lie in the subgroup of K^* of order $\dfrac{(p^r-1)(p^{(r,a,b)}-1)}{(p^{(r,a-b)}-1)(p^{(r,b)}-1)}$.*

(2) *The equation $e^{\alpha-1} f^{\beta-1} = 1$ admits a solution (e, f) if and only if $e^{\alpha-1}$ or $f^{\beta-1}$ lie in the subgroup of K^* of order $\dfrac{(p^r-1)(p^{(r,a,b)}-1)}{(p^{(r,a)}-1)(p^{(r,b)}-1)}$.*

COROLLARY 11.17. *Let $\pi = \pi(K, \alpha, \beta, c)$ be a generalized twisted field plane.*

(i) π admits autotopism collineations of the form $a(e, d, I)$ for all $e \in K^*$ if and only if $(r, a, b) = (r, b)$.

(ii) π admits autotopism collineations of the form $a(e, d, I)$ for all $d \in K^*$ if and only if $(r, a, b) = (r, a - b)$.

(iii) π admits autotopism collineations of the form $a(e, d, I)$ for any value of de^{-1} in K^* if and only if $(r, a, b) = (r, a)$.

Then using these ideas, Biliotti, Jha, and Johnson prove:

THEOREM 11.18. (Biliotti, Jha, Johnson [**121**]). Let $\pi = \pi(K, \alpha, \beta, c)$ be a generalized twisted field plane and let G be the full autotopism group of π.

(1) G admits a cyclic subgroup which acts transitively on the non-zero vectors of $(y = 0)$ if and only if $(r, a, b) = (r, b)$.

(2) G admits a cyclic subgroup which acts transitively on the non-zero vectors of $(x = 0)$ if and only if $(r, a, b) = (r, a - b)$.

(3) G admits a cyclic subgroup which acts transitively on $\ell_\infty - \{(\infty), (0)\}$ if and only if $(r, a, b) = (r, a)$.

COROLLARY 11.19. Let $\pi = \pi(K, \alpha, \alpha^{-1}, c)$ be a twisted field plane and let G be the full autotopism group of π. Then:

(1) $G \leq \Gamma L(1, K) \times \Gamma L(1, K)$ and G admits cyclic subgroups acting transitively on the non-zero vectors of $(y = 0)$ and on $\ell_\infty - \{(\infty), (0)\}$, respectively.

(2) G admits cyclic subgroups acting transitively on the non-vertex points for each side of the autotopism triangle if and only if $r / (r, a)$ is odd.

Assume that $\alpha = \beta^k$ for some integer $k > 1$.

We have that $a(e, d, I)$ is a collineation of the plane $\pi = \pi(K, \alpha, \beta, c)$ if and only if $e^{\alpha - \beta} d^{\beta - 1} = 1$. Since, $\beta - 1 \mid \alpha - \beta$, we may choose directly $d = e^{(\beta - \alpha)/(\beta - 1)}$.

11.1. General Transitivity.

Strengthened results on the transitivity of the autotopism group on the non-vertex points of the sides of the autotopism triangle may be obtained using the classification of doubly transitive subgroups of $A\Gamma L(1, K)$, $K \cong GF(p^r)$, given by Foulser in [**366**, Section 15], and also by Woltermann [**1217**].

THEOREM 11.20 (Foulser [**366**]). Let the elements of $A\Gamma L(1, K)$ represented by the mappings $x \longmapsto gx^\mu + b$ over K with $b \in K$, $g \in K^*$ and $\mu \in \mathrm{Aut}\, K$. Let w denote a primitive root of $GF(p^r)$. Let w, ρ and τ_a denote the mappings

$$x \longmapsto wx, \quad x \longmapsto x^p, \quad x \longmapsto x + a,$$

respectively. Let $T = \langle \tau_a : a \in K \rangle$. Let F be a doubly transitive subgroup of $A\Gamma L(1, K)$. Then:

(1) $T \subset F$ and $F = T \cdot H$ is the split extension of T by $H = F_0$.

(2) H may be represented in the form $\langle w^d, w^e \rho^s \rangle$ where

  (i) $d > 0$ and $d \mid p^r - 1$,

  (ii) $s > 0$ and $r \equiv 0 \pmod{sd}$,

  (iii) $(d, e) = 1$,

  (iv) the prime divisors of d divide $p^s - 1$ and if $4 \mid p^s + 1$, then $4 \nmid d$.

We need the following:

LEMMA 11.21. Let $\pi = \pi\,(K, \alpha, \beta, c)$ be a generalized twisted field plane. There exists $\sigma_A \in \operatorname{Aut} K$ defined by $\sigma_A : x \longrightarrow x^{p^{s_A}}$ such that

(i) Every autotopism $a\,(e, d, \sigma)$ of π is of the form $a\left(e, d, (\sigma_A)^i\right)$ for $0 < i \leq r/s_A$, and

(ii) $s_A \mid (r, a, b)$.

PROOF. Let σ_0 be an element of maximal order in $\operatorname{Aut} K$ such that $a\,(e, d, \sigma_0)$ is an autotopism of π for suitable $e, d \in K^*$. If $a\,(e', d', \sigma)$ is any other autotopism of π, then $\sigma \in \langle \sigma_0 \rangle$ by the maximality of the order of σ_0. Let $\sigma_0 : x \longrightarrow x^{p^{s_0}}$. Put $s_A = (r, s_0)$ and let $\sigma_A : x \longrightarrow x^{p^{s_A}}$. We have (i) since $\langle \sigma_0 \rangle = \langle \sigma_A \rangle$. Now $\left|\langle e^{\alpha - \beta} d^{\beta - 1}\rangle\right| = \frac{p^r - 1}{p^{(r,a,b)} - 1}$. Hence $c^{p^{(r,a,b)} - 1} \in \langle e^{\alpha - \beta} d^{\beta - 1}\rangle$. By Theorem 11.12 we have (ii). $\qquad\square$

Then using these ideas, Biliotti, Jha, and Johnson show:

THEOREM 11.22. Let $\pi = \pi\,(K, \alpha, \beta, c)$ be a generalized twisted field plane and let G be the full autotopism group of π. The following hold.

(1) G is transitive on the non-zero vectors of $(y = 0)$ if and only if $(r, b) = (r, a, b)$.

(2) G is transitive on the non-zero vectors of $(x = 0)$ if and only if $(r, a - b) = (r, a, b)$.

(3) G is transitive on $\ell_\infty - \{(\infty), (0)\}$ if and only if $(r, a) = (r, a, b)$.

When G is transitive on the non-vertex points of a side of the autotopism triangle, then G possesses a cyclic subgroup transitive on the non-vertex points of that side.

COROLLARY 11.23. Let $\pi = \pi\,(K, \alpha, \beta, c)$ be a generalized twisted field plane and let G be the full autotopism group of π. Then G acts transitively on the non-vertex points of each side of the autotopism triangle if and only if the right, middle, and left nuclei are all equal.

PROOF. By Theorem 11.22 G acts transitively on the non-vertex points of each side of the autotopism triangle if and only if $(r, a, b) = (r, a) = (r, b) = (r, a - b)$. By what we have seen in Section 11.1, this is equivalent to the assertion that the right, middle and left nuclei are all equal. $\qquad\square$

11.2. Autotopisms as Triples of Additive Functions.

We include some remarks on the use of autotopisms in generalized twisted field planes. Let $A = A(K, \alpha, \beta, c)$, $K \cong GF(p^r)$, be a generalized twisted field and let (f, g, h) be an autotopism of A. Then f, g and h are non-singular additive functions over K such that $f(x) \circ g(m) = h(x \circ m)$. Put

$$f(x) = \sum_{s=0}^{r-1} f_s x^{p^s}, \quad g(m) = \sum_{s=0}^{r-1} g_s m^{p^s}, \quad h(y) = \sum_{s=0}^{r-1} h_s y^{p^s};$$

then this yields $f(x)g(m) - cf(x)^\alpha g(m)^\beta = h(xm - cx^\alpha m^\beta)$ if and only if

$$\left(\sum_{s=0}^{r-1} f_s x^{p^s}\right) g(m) - c \left(\sum_{s=0}^{r-1} f_s^\alpha x^{\alpha p^s}\right) g(m)^\beta = \sum_{s=0}^{r-1} h_s x^{p^s} m^{p^s} - \sum_{s=0}^{r-1} h_s c^{p^s} x^{\alpha p^s} m^{\beta p^s}$$

for all x and for all m in K.

Using such equations, Biliotti–Jha–Johnson [121] prove the following result mentioned previously.

THEOREM 11.24. Let $\pi - \pi(K, \alpha, \beta, c)$ be a generalized twisted field plane. Then the full autotopism collineation group G of π is a subgroup of $\Gamma L(1, K) \times \Gamma L(1, K)$.

11.3. Isotopism.

Similarly, isotopism may be handled as follows:

THEOREM 11.25. (Biliotti, Jha, Johnson [121]). Let $\pi = \pi(K, \alpha, \beta, c)$ be a generalized twisted field plane. Then we have that

(1) If $\pi(K, \alpha', \beta', c')$ is a generalized twisted field plane isomorphic to π, then $A(K, \alpha, \beta, c)$ and $A(K, \alpha', \beta', c')$ correspond by an isotopism $\mathbf{h} = (F, G, H)$ with $F, G, H \leq \Gamma L(1, K)$.

(2) Two generalized twisted fields $A(K, \alpha, \beta, c)$ and $A(K, \alpha', \beta', c')$ are isotopic if and only if either
   (i) $\alpha' = \alpha$, $\beta' = \beta$, $c'^\rho = c\left(f^{\alpha-1}g^{\beta-1}\right)$ for some $\rho \in \operatorname{Aut} K$, $f, g \in K^*$
   or
   (ii) $\alpha' = \alpha^{-1}$, $\beta' = \beta^{-1}$, $c'^\rho = c^{-1}\left(f^{1-\alpha}g^{1-\beta}\right)$ for some $\rho \in \operatorname{Aut} K$, $f, g \in K^*$.

(3) The isomorphism induced by $\mathbf{h}$ between the planes π and

$$\pi_2 = \pi(K, \alpha', \beta', c')$$

may be written in the form

$$(x, y) \longrightarrow (x^\rho f, y^\rho fg); \ (m) \longrightarrow (m^\rho g)$$

where $f, g \in K^*$, $\rho \in \operatorname{Aut} K$ in the case (2.i) and

$$(x, y) \longrightarrow \left(x^{\rho\alpha^{-1}}f, -y^\rho cf^\alpha g^\beta\right); \ (m) \longrightarrow \left(m^{\rho\beta^{-1}}g\right)$$

where $f, g \in K^*$, $\rho \in \operatorname{Aut} K$ in the case (2.ii).

To each pre-semifield $A(k, \alpha, \beta, c)$ is associated an isotopic semifield $D(k, \alpha, \beta, c)$ where the multiplication rule "$\cdot$" is defined by $x \cdot m = (x - cx^\alpha) \circ (m - cm^\beta)$.

The semifield $D(k, \alpha, \beta, c)$ has the unity element $1 - c$, where 1 is the unity of K. It is not difficult to see that our Theorem 11.25 claims the same as [24, Theorem 2], in terms of pre-semifields instead of semifields. Theorem 11.25 can be reformulated for twisted field planes as follows.

THEOREM 11.26. Let $\pi = \pi(K, \alpha, \alpha^{-1}, c)$ be a twisted field plane. Then we have that

(1) If $\pi(K, \alpha', \alpha'^{-1}, c')$ is a twisted field plane isomorphic to π, then the two pre-semifields $A(K, \alpha, \alpha^{-1}, c)$ and $A(K, \alpha', \alpha'^{-1}, c')$ correspond by an isotopism $\mathbf{h} = (F, G, H)$ with $F, G, H \leq \Gamma L(1, K)$.

(2) Two twisted fields $A(K, \alpha, \alpha^{-1}, c)$ and $A(K, \alpha', \alpha'^{-1}, c')$ are isotopic if and only if either
   (i) $\alpha' = \alpha$, $c'^\rho = cf^{\alpha-1}$ for some $\rho \in \operatorname{Aut} K$, $f \in K^*$ or
   (ii) $\alpha' = \alpha^{-1}$, $c'^\rho = c^{-1}z^{1-\alpha}$ for some $\rho \in \operatorname{Aut} K$, $z \in K^*$.

Actually, Purpura [**1121**] is able to use this result to give a number of non-isotopic generalized twisted fields of odd order. Let $g(p^n)$ denote the number of non-isotopic generalized twisted fields of orders p^n, where p is an odd prime. In particular, Purpura proves the following theorems:

THEOREM 11.27. (Purpura [**1121**]). The number of non-isotopic generalized twisted fields of order p^n is at least $\binom{n-2}{2}(p-2)$ and the bound is sharp if and only if n is prime or $n = 4$.

THEOREM 11.28. (Purpura [**1121**]). The number $g(p^n)$ is a polynomial in p of degree $\mu(n)$, where $\mu(n)$ is defined as the largest divisor of n which is strictly less than $n/2$.

We can also prove the following:

PROPOSITION 11.29. A generalized twisted field plane $\pi(K, \alpha, \beta, c)$ is self-dual if and only if $\beta = \alpha^{-1}$ and $c^{\rho+1} = z^{1-\alpha}$ for some $z \in K^*$ and $\rho \in \mathrm{Aut}\, K$.

That is, the only self-dual generalized twisted field planes are the self-dual twisted field planes.

PROOF. Assume that $\pi(K, \alpha, \beta, c)$ is isomorphic to the dual plane. This means that if the original pre-semifield multiplication rule is $x \circ m = xm - cx^\alpha m^\beta$, then a multiplication rule in the dual may be defined by $x \circ_d m = m \circ x = xm - cx^\beta m^\alpha$.

Since then the two pre-semifields are isotopic, it follows from Theorem 11.25(2) that $\beta = \alpha^{-1}$ and $c^\rho = c^{-1}z^{1-\alpha}$.   □

REMARK 11.30. The correlations in a self-dual twisted field plane are induced by the isotopisms between the two pre-semifields considered in the above proposition. Thus any correlation Φ is represented by three non-singular additive maps Φ_1, Φ_2 and Φ_3 such that $(y \circ x)\Phi_1 = x\Phi_2 \circ y\Phi_3$. It operates on the plane as follows: $(x, y) \longmapsto [x\Phi_3, y\Phi_1]$; $[m, k] \longmapsto (m\Phi_2, k\Phi_1)$ (see [**385**]). These maps may be determined by combining the results of Theorem 11.25 and Proposition 11.29. Precisely, we must have

$$\Phi_1 : x \longrightarrow -x^\rho c f^\alpha g^{\alpha^{-1}}; \quad \Phi_2 : x \longrightarrow x^{\rho\alpha^{-1}} f; \quad \Phi_3 : x \longrightarrow x^{\rho\alpha} g$$

with $c^{\rho+1} = f^{1-\alpha}g^{1-\alpha^{-1}}$.

A plane $\pi(k, \alpha, \beta, c)$ is called a *commutative* generalized twisted field plane if there is a commutative isotope of $A(k, \alpha, \beta, c)$.

PROPOSITION 11.31. A plane $\pi(k, \alpha, \beta, c)$ is a commutative generalized twisted field plane if and only if $\beta = \alpha^{-1}$ and $c = -z^{\alpha-1}$ for some $z \in K^*$, that is if and only if it may be coordinatized, up to an isotopism, by a commutative twisted field.

PROOF. Let $S(+, *)$ be a commutative isotope of $A(k, \alpha, \beta, c)$. Then $S(+, \star)$ with multiplication rule $x \star m = (x * e) * (e * m)$ is isotopic to $S(+, *)$ and is commutative. Indeed

$$x \star m = (x * e) * (e * m) = (e * x) * (m * e) = (m * e) * (e * x) = m \star x.$$

By [**24**, Theorem 1], $S(+, \star)$ is isomorphic to a semifield $D(k, \alpha, \beta, d)$ for some d such that $d = cf^{\alpha-1}g^{\beta-1}$ for some $f, g \in K^*$. Hence $D(k, \alpha, \beta, d)$ must be commutative. By [**23**, Theorem 2], it must be $\beta = \alpha^{-1}$ and $d = -1$. Thus $c = -f^{\alpha-1}g^{\alpha^{-1}-1}$ for some $f, g \in K^*$, which is equivalent to $c = -z^{\alpha-1}$ for some $z \in K^*$.   □

It is now not difficult to characterize generalized twisted field planes admitting polarities.

PROPOSITION 11.32. A generalized twisted field plane $\pi(K, \alpha, \beta, c)$ possesses a polarity if and only if either

(1) $\pi = \pi\left(K, \alpha, \alpha^{-1}, -1\right)$ is a commutative twisted field plane, or
(2) $\pi = \pi\left(K, \alpha, \alpha^{-1}, c\right)$ where r is even, $r/(r, a)$ is odd and c is an element of K^* such that $o(c) \nmid \frac{p^r-1}{p^{(r,a)}-1}$, but $o(c) \mid \frac{p^r-1}{p^{(r,a)/2}-1}$.

REMARK 11.33. Note that from Proposition 11.31 we infer that commutative (generalized) twisted field planes must have odd order. Furthermore, $K = GF(p^r)$ with r odd. Indeed, $-1 \notin \left\langle z^{\alpha-1} : z \in K^* \right\rangle$. Since $\left| \left\langle z^{\alpha-1} : z \in K^* \right\rangle \right| = (p^r - 1)/\left(p^{(r,a)} - 1\right)$ we have that $(p^r - 1)/\left(p^{(r,a)} - 1\right) = \sum_{i=1}^{\overline{(r,a)}} p^{r-i(r,a)}$ must be odd, that is $\frac{r}{(r,a)}$ must be odd. Self-dual twisted field planes exist also for even order. Indeed suppose that (r, a) is even and let u be a primitive root of K. Put $c = u^{p^{(r,a)/2}-1}$. Then $\pi\left(K, \alpha, \alpha^{-1}, c\right)$ is a self-dual twisted field plane. Not all self-dual twisted field planes possess an autotopism group acting transitively on each side of the autotopism triangle. For example $\pi\left(GF(p^{12}), p^2, p^{10}, u^{p-1}\right)$, u a primitive root of $GF(p^{12})$, is such that $r/(r, a) = 6$ is even. Not all self-dual twisted field planes with $r/(r, a)$ odd possess a polarity. Take for example $\pi\left(GF(p^{24}), p^8, p^{16}, u^{(p^4+1)(p^2-1)}\right)$, u a primitive root of $GF(p^{24})$. In this case $r/(r, a) = 3$, but $o(u^{(p^4+1)(p^2-1)}) = \frac{p^{24}-1}{(p^4+1)(p^2-1)} \nmid \frac{p^{24}-1}{p^4-1}$.

Dempwolff [304] proved that if a semifield plane π has a cyclic group that acts transitively on the non-vertex points of the infinite side of the autotopism triangle, then π is a Desarguesian or a generalized twisted field plane. Furthermore, as mentioned previously, Cordero and Figueroa extended Dempwolff's result (see [253]).

Using the previously mentioned Theorem 11.22, Biliotti, Jha and Johnson provide a characterization theorem which completes the results of both Dempwolff and Cordero–Figueroa. Note that we may always assume $\Delta = \{(x = 0), (y = 0), l_\infty\}$.

THEOREM 11.34. Let π be a semifield plane of order $p^r \neq 2^6$ and let G be the autotopism group relative to the autotopism triangle Δ. G acts transitively on the non-vertex points of some side of Δ if and only if π is a Desarguesian plane or a generalized twisted field plane $\pi(K, \alpha, \beta, c)$ satisfying at least one of the following conditions:

(i) $(r, b) = (r, a, b)$ (in this case G possesses a cyclic subgroup transitive on the non-zero vectors of $(y = 0)$).
(ii) $(r, a - b) = (r, a, b)$ (in this case G possesses a cyclic subgroup transitive on the non-zero vectors of $(x = 0)$).
(iii) $(r, a) = (r, a, b)$ (in this case G possesses a cyclic subgroup transitive on $l_\infty - \{(\infty), (0)\}$).

Also, we have:

THEOREM 11.35. Let π be a semifield plane of order $p^r \neq 2^6$. Then π is of rank 3 if and only if π is a generalized twisted field plane $\pi(K, \alpha, \beta, c)$ such that the left, middle and right nuclei are all equal.

11.3.1. Theorem of Ganley–Jha. Ganley and Jha [**392**] proved that if a translation plane π admits a collineation group acting doubly transitively on $\ell_\infty - \{L\}$, where L is a point on the line at infinity ℓ_∞, then π is a semifield plane. Combining this result with Theorem 11.34 we obtain:

THEOREM 11.36. A translation plane π of order $p^r \neq 2^6$ admits a collineation group acting doubly transitively on $\ell_\infty - \{L\}$, where L is a point on the line at infinity ℓ_∞, if and only if π is a Desarguesian plane or a generalized twisted field plane $\pi(K, \alpha, \beta, c)$ satisfying the condition $(r, a) = (r, a, b)$.

11.3.2. Theorem of Enea–Korchmáros. The theorem of Enea and Korchmáros [**353**] mentioned above shows that a translation plane of odd order admitting a transitive parabolic oval is a commutative semifield plane. Furthermore, Enea and Korchmáros exhibited an example of a 2-transitive parabolic oval in any commutative twisted field plane $\pi(K, \alpha, \alpha^{-1}, -1)$ with $a \mid r$. In a subsequent work of Abatangelo, Enea, Korchmáros and Larato [**8**] the same example, in a slightly different form, is presented again and investigated in detail. Actually Enea and Korchmáros consider the isotopic form of $A(K, \alpha, \alpha^{-1}, -1)$ with multiplication rule "$*$" given by $x*y = xR_1^{-1} \circ yL_1^{-1}$ where $R_1 : x \longrightarrow x \circ 1$ and $L_1 : x \to 1 \circ x$. Nevertheless, they prefer to write $xR_1 * y^\alpha L_1 = x \circ y^\alpha$, that is, $(x^\alpha + x) * (y^\alpha + y) = xy^\alpha + x^\alpha y$, representing each element $z \in K$ in the form $z = x^\alpha + x$ for $x \in K$. Note that x is uniquely determined.

Then they prove that the set of points of π defined as $\left\{ (c, 2c^{\alpha+1}) : c \in K \right\}$ is a 2-transitive parabolic oval. It is easily seen that the assumption $a \mid r$ is unnecessary since the arguments of [**353**] work for any α. Hence, also in this case, the term "commutative twisted field" must be considered in the broad meaning. The projective extension of Ω has exactly one point (∞) on ℓ_∞ and the group G, doubly transitive on Ω, acts doubly transitive also on the points of $\ell_\infty - \{(\infty)\}$. Then π is a Desarguesian plane or a generalized twisted field plane $\pi(K, \alpha, \beta, c)$ by Theorem 11.36.

In a final remark of [**353**], Enea and Korchmáros ask what the combined conditions "π commutative semifield plane" and "π generalized twisted field plane" imply. An answer is given by Proposition 11.31, so that Biliotti, Jha and Johnson essentially answer this question.

THEOREM 11.37. Let π be a translation plane of odd order which admits a parabolic oval Ω and a collineation group G which fixes Ω and acts doubly transitive on its points. Then π is a Desarguesian plane or a commutative twisted field plane.

11.3.3. Chuvaeva–Pasechnik/Liebler Theorems. Gardiner [**394**] proved a theorem which gives a one-to-one correspondence between distance-transitive q-fold antipodal covers of the complete bipartite graph $K_{q,q}$ and the planes of a class of rank-three affine planes admitting a correlation. By the well-known results about rank-three affine planes, the result of Gardiner may be expressed as follows (see [**923**]).

THEOREM 11.38. Distance-transitive q-fold antipodal covers of $K_{q,q}$ are in one-to-one correspondence with semifield planes π of order q satisfying the conditions:
   (i) π is self-dual.
   (ii) The autotopism group of π is transitive on the non-vertex points of each side of the autotopism triangle.

Liebler [**923**] proved that only Desarguesian planes and self-dual twisted field planes satisfy the conditions of Theorem 11.38. Actually Theorem 11.34 and Proposition 11.29 show that the following characterization of self-dual twisted field planes may be given.

THEOREM 11.39. A self-dual translation plane of order $p^r \neq 2^6$ is a self-dual twisted field plane if and only if it admits an autotopism collineation group transitive on the non-vertex points of at least one side of the autotopism triangle.

Using Proposition 11.29 and Corollary 11.19 an explicit characterization may be given.

THEOREM 11.40. Distance-transitive q-fold antipodal covers of $K_{q,q}$ with $q = p^r$ are in one-to-one correspondence with $PG(2, q)$ and the twisted field planes $\pi\left(K, \alpha, \alpha^{-1}, c\right)$ of order q such that $r/\left(r, a\right)$ is odd and $c^{\rho+1} = z^{1-\alpha}$ for some $z \in K^*$ and $\rho \in \operatorname{Aut} K$.

Chuvaeva and Pasechnik [**233**] assert that condition (ii) of Theorem 11.38 may be replaced by the strengthened condition that π possesses a polarity. This assertion seems to be inconsistent. In any case the generalized twisted field planes admitting polarities are characterized in our Proposition 11.32. Note that the existence of a polarity forces $r/\left(r, a\right)$ to be odd.

11.3.4. Ganley, Jha, Dempwolff, Liebler, Cordero, Figueroa Theorem. The combined work of Ganley and Jha [**392**], Dempwolff [**301**] Liebler [**923**] and Cordero–Figueroa [**253**] proves the following important theorem, referred to as the GJDLCF theorem.

Suppose that we have a finite semifield plane of order q^n that admits a collineation group fixing the elation axis and acting doubly transitively on the non-axis components, is it possible to determine the plane? For example, various of the generalized twisted field planes certainly admit such groups as is seen previously. Ganley and Jha [**392**] showed that translation planes admitting a group fixing a component and acting doubly transitive on the remaining components is always a semifield plane. With the contributions of Dempwolff [**301**] and Cordero–Figueroa [**253**], we have the following important theorem. Furthermore, many of the results are intrinsic in the work of Liebler, who is instrumental in developing this material.

THEOREM 11.41. (Ganley, Jha [**392**], Dempwolff [**301**], Liebler [**923**], Cordero, Figueroa [**253**]). Let π be a finite translation plane of order $q^n \neq 2^6$ that admits a collineation group fixing a component and acting doubly transitively on the remaining components. Then π is a generalized twisted field plane.

11.3.5. p-Primitive Groups; Figueroa's Theorem. Actually, more can be said regarding groups forcing a semifield plane to be generalized twisted. In particular, Figueroa [**357**] proves the following result:

THEOREM 11.42. (Figueroa [**357**]). Let π be a finite translation plane of order p^n, for p odd and $n \geqslant 3$. Assume that the autotopism group of π has a collineation of order $(p^k - 1)(\frac{p^n - 1}{p^m - 1})$, then π is a generalized twisted field plane, with the possible exceptions of $p = 3$, $k = 1$ and $n = 2m$.

We note that when a finite translation plane of order p^n admits a group that fixes one parallel class and is doubly transitive on the others, the stabilizer of a second parallel class admits a group of order $(p^n - 1)$. If this group admits an element of order $p^n - 1$, then Dempwolff [301] concludes that the plane is a generalized twisted field plane. So, the above result may be regarded as an odd-order variant of the theorem of Dempwolff. Note that generalized twisted field planes of characteristic larger than 5 do admit such collineations as in the statement of the above theorem. Furthermore, it might be hoped that Figueroa's result could be generalized to having simply a collineation of order a prime p-primitive divisor, which may be true for planes of characteristic at least 5. However, this cannot be true for characteristic 3, due to the sporadic semifield plane of order 3^6, due to Cordero and Figueroa [253].

CHAPTER 12

Nuclear Fusion in Semifields.

In this chapter, we consider a powerful technique available in semifields of "fusing" selected nuclei. In particular, we recall that the affine planes $\pi(D)$, coordinatized by a semifield D, have maximal homology groups of three distinct types, each isomorphic to the multiplicative group of certain subfields of D, that may be characterized as being the left, middle and right associators of D. This material follows Jha and Johnson [**634**] but here we give either no proof or a brief sketch of the proof.

Recalling the definitions of the left, middle and right nuclei, respectively defined as follows:

$$N_\ell(D) := \{k \in D : k \circ (d_1 \circ d_2) = (k \circ d_1) \circ d_2, \forall d_1, d_2 \in D\};$$
$$N_m(D) := \{k \in D : d_1 \circ (k \circ d_2) = (d_1 \circ k) \circ d_2, \forall d_1, d_2 \in D\};$$
$$N_r(D) := \{k \in D : d_1 \circ (d_2 \circ k) = (d_1 \circ) d_2 \circ k, \forall d_1, d_2 \in D\}.$$

It is easily seen that $N_\ell(D)$, $N_m(D)$, and $N_r(D)$, are fields and their multiplicative groups are isomorphic to one of the three types of maximal homologies that $\pi(D)$ admits. Thus, $N_\ell(D)^*$, $N_m(D)^*$ and $N_r(D)^*$ are isomorphic respectively to the maximum group of kernel homologies, the maximal group of homologies with the elation axis as axis, and maximal groups of affine homologies with axis not the shears axis.

Furthermore, since the order of the nuclei (collectively called the 'seminuclei') is an invariant, then although a non-Desarguesian semifield plane may always be coordinatized by a variety of mutually non-isomorphic semifields D, the order of each of the associated seminuclear fields, $N_\ell(D)$, $N_m(D)$ and $N_r(D)$, does *not* depend on the choice of D.

However, there are a number of other subfields of D, with natural algebraic definitions, which one may examine for geometric invariance, i.e., their dependence, up to isomorphism, on the choice of the coordinatizing semifield D. The most important of these are the *nucleus* and the *center* defined, respectively, by the notation

$$N(D) := N_\ell(D) \cap N_m(D) \cap N_r(D);$$
$$Z(D) := \{z \in N : z \circ d = d \circ z \ \forall d \in D\}.$$

The geometric invariance of the center and the three seminuclei of a finite semifield are immediate consequences of a more general result concerning loops due to Bruck, [**173**, p. 57]. On the other hand, in Jha and Johnson [**604**] it has been shown that the nucleus $N(D)$ is not in general an invariant. In fact, if $N(D)$ is *not* in the center, then the spread can *always* be recoordinatized by a semifield E such that $|N(E)| \neq |N(D)|$.

This results in what might be considered a 'splitting' of the nucleus. This is most clearly seen in the case when D has prime dimension k over its nucleus N, with N non-central. In this situation, all the seminuclear fields clearly coincide with the nucleus. In such a case, clearly $N(D) = N_\ell(D) = N_m(D) - N_r(D)$.

Since some recoordinatization is guaranteed to change the size of the nucleus, and cannot change the size of the seminuclei, it is possible that using an appropriate recoordinatization E, the three seminuclei 'split', that is $N_\ell(E)$, $N_m(E)$ and $N_r(E)$ no longer coincide—hence they 'split'.

Similar splitting results actually hold for all non-central fields that are in the intersection of at least two seminuclear fields (see Jha and Johnson [604]).

Such splitting is important for a variety of applications, for example, in the derivation of a semifield plane. If a semifield plane of order q^2 is of dimension two over either its right or middle nucleus, then any net invariant under an associated affine homology group becomes derivable. Hence, by splitting, it is possible to construct various derivable nets, which produce potentially non-isomorphic translation planes. This is the case, for example, in the construction of the translation planes of order 16 that may be derived from a unique semifield plane of order 16 and kernel $GF(4)$; it is possible to split the right and middle nuclei to produce various translation planes. The Johnson–Walker planes and the Lorimer–Rahilly planes of order 16 may be so constructed (see Johnson [684]).

Suppose one considers the reverse process, which might be called 'fusion'. In the simplest case, one might ask: If a finite plane is coordinatized by a semifield D such that all three of its seminuclei contain $GF(q)$, can the spread be recoordinatized by a semifield E such that $N(E) \supseteq GF(q)$? Thus, one asks whether one can guarantee that three seminuclei fields of the same size can be 'fused' to a common nucleus. These questions are considered by Jha and Johnson [634], who answer this question in the affirmative.

THEOREM 12.1 (Jha–Johnson Theorem on Nuclear Fusion). Let π be a finite plane coordinatized by semifield D such that its seminuclei, $N_\ell(D)$, $N_m(D)$, and $N_r(D)$, each contain a subfield $\cong GF(q)$. Then π can be recoordinatized by a semifield E such that $N(E)$ contains a field isomorphic to $GF(q)$.

Actually, there it is possible to establish similar results concerning 'fusion' of two seminuclei (rather than all three as in the nuclear case), and these form the initial stages in arriving at the fusion theorem above. It should be noted that the converses of all types of fusion results that we establish here (viz., 'splitting' theorems) follow from Jha and Johnson [604].

A basic theorem concerning semifield spreads π asserts that either π is Desarguesian or there are guaranteed non-isomorphic semifields coordinatizing π, and all the semifields coordinatizing π may be obtained by choosing the elation axis Y as the y-axis and any other component X as the x-axis. This raises fundamental issues concerning the algebraic properties of semifields coordinatizing a given semifield spread.

Which algebraically defined substructures of a semifield remain isomorphic for all semifields coordinatizing the same spread? If an algebraically defined substructure of a semifield varies with the semifields coordinatizing a spread π, then will there be 'canonical' versions of the structure, for all such semifields?

The main results of Jha and Johnson [604] address invariance questions of this type for various subfields associated with seminuclear intersections of a semifield.

Perhaps the most basic result of this type is the invariance of the order of the three seminuclei, $N_\ell(D)$, $N_m(D)$ and $N_r(D)$, of all the semifields D coordinatizing a semifield spread π. This may be established, as indicated earlier, by noting that the multiplicative group of each of these seminuclei is isomorphic to one of the three types of maximal homology groups of π: respectively, those with infinite axis (the kernel homology group), with the shears axis as axis, and with affine axis $\neq Y$, the shears axis.

However, the size of the nucleus $N(D) = N_\ell(D) \cap N_m(D) \cap N_r(D)$ can, and does, vary with the choice of D, unless some D (or equivalently all D) has its nucleus isomorphic to $N(D)$. More generally, it has been determined precisely when the intersection of two or more seminuclear fields, of all the semifields D coordinatizing the same spread π, are always mutually isomorphic, and also when they are always central in D:

THEOREM 12.2 (Jha and Johnson [604]). Let Δ be the class of all non-isomorphic semifields coordinatizing a semifield spread π that has at least one seminuclear field $\cong GF(q)$, thus π has order q^r. Suppose that at least one of the following conditions holds:

(1) $N_m(D) \cap N_r(D) \supseteq GF(q) \forall D \in \Delta$;
(2) $N_\ell(D) \cap N_r(D) \supseteq GF(q) \forall D \in \Delta$;
(3) $N_\ell(D) \cap N_m(D) \supseteq GF(q) \forall D \in \Delta$.

Then all three conditions hold and $Z(D) \supseteq GF(q)$ for every $D \in \Delta$.

Thus, if two or more seminuclear fields of a semifield D have an intersection in $F \cong GF(q)$, then, provided F is not in the center, we may recoordinatize the spread π associated with D such that $GF(q)$ is no longer in the intersection of the seminuclei involved: the seminuclei have split. One of the main results of this paper is that the reverse, fusion, also holds: if D has two or more seminuclear fields of order q then, by recoordinatization, if necessary, all these seminuclei must coincide.

Actually there is an interesting combinatorial problem involved in the ideas of fusion and splitting, which is to determine the combinatorial configurations associated with all possible seminuclear intersections, in particular the nuclei, of all the semifields coordinatizing a given semifield plane π; the fusion results indicated above are a by-product of this analysis.

In fact, the study of seminuclear intersections of a semifield spread is described in terms of what are called Desarguesian *hyperbolic covers* $\mathcal{E}$ of the components of the semifield spread π under consideration. Thus, we shall show that each seminuclear field (other than the kernel) of order q is associated with a hyperbolic cover $\mathcal{E}$ consisting of rational Desarguesian nets of order q that share exactly two components (the shears axis Y and another component $X \in \pi \setminus \{X\}$) and that various 'fusions' correspond to the set of planes shared by the two nets and also to the kernel invariance of such planes and their intersections.

This analysis will enable us to describe precisely and explicitly, in terms of the hyperbolic coverings, all planes of order q that are coordinatized by any chosen type of seminuclear intersections. For example, suppose π is a semifield spread with $GF(q)$ in all its seminuclear fields, thus π has order q^t, and for simplicity assume $GF(q)$ is not central (in which case our earlier results apply). We shall show not only that such a plane may be recoordinatized by a semifield D with $GF(q) \subseteq N(D)$ ('fusion'), but also that π has exactly t^2 subplanes (through the shears axis Y and a

second component X) coordinatizable by nuclear planes of order q and we describe
their exact configuration in terms of certain hyperbolic coverings.

As a further application to our results, it is possible to obtain a complete
classification of all semifield spreads of order q^2 that admit homology groups of
order $q - 1$ with different axes, Theorem 12.22. We shall show that these are the
spreads coordinatized by semifields of types II, III and IV according to Knuth's
taxonomy [**898**, (7.17)]. In fact, Theorem 12.22 may be regarded as a geometric
generalization of a result of Knuth [**898**, Theorem 7.4.1] which arrives at the same
conclusions as Theorem 12.22. We also note that Theorem 12.22, our generalization
of Knuth's result, may also be used to generalize the basic fact that finite semifields
of order q^2, with $GF(q)$ in the center, are fields. We show that this conclusion holds
under a weaker hypothesis on $GF(q)$.

THEOREM 12.3. *Let D be a semifield of order q^2 with subfields F and G in its
left and right nucleus, respectively, such that $F \cong G \cong GF(q)$. Then D is a field if
and only if at least one of F and G centralizes D multiplicatively.*

It should be noted that the assumptions in the theorem above are weaker than
assuming that D contains $GF(q)$ in its center, and that the result becomes false if
the centralizing condition adopted is dropped.

12.1. The Seminuclear Homology Groups.

In this section we recall basic results concerning the seminuclei of semifields,
geared to our needs. These will be expressed in terms of the standard vector space
structures on the line at infinity, which also plays a prominent role in our analysis.

If D is a finite semifield then $\pi(D)$, the associated translation plane, has its set
of 'infinite' points described by the notation $\Delta := \{(d) : d \in D\} \cup \{(\infty)\}$, and if
$W \subset D$ we write $(W) := \{(w) : w \in D\}$.

Thus, Δ is identified with D and hence has the identical vector space structures
that D has over the subfields of the seminuclei, nucleus and center of D. The
action of these (semi)nuclei corresponds to the action induced on Δ by the standard
homology groups. Since the exact form of these connections will be repeatedly
required, and are determined partly by the coordinatization conventions, we state
them explicitly.

RESULT 12.4. *Let D be any finite semifield and K^* a multiplicative subgroup of
a (sub)field of the left, middle, or right nucleus of D. Then, depending on whether
K is a left-, middle-, or right-nucleus semifield, K^* is isomorphic to a homology
group of $\pi(K)$ of at least one of the following types.*

(X-axis, Y-coaxis): For $K \subseteq N_r(D)$:

(12.1.1) $(x, y) \mapsto (x, yk); (z) \mapsto (zk), \forall k \in K^*, x, y, z \in D.$

Such a group is also called right-nucleus homology group.
(Y-axis, X-coaxis): For $K \subseteq N_m(D)$:

(12.1.2) $(x, y) \mapsto (xk, y); (z) \mapsto (k^{-1}z), \forall k \in K^*, x, y, z \in D.$

Such a group is also called middle-nucleus homology group.
Kernel Homologies Group: For $K \subseteq N_\ell(D)$:

(12.1.3) $(x, y) \mapsto (kx, ky); (z) \mapsto (z), \forall k \in K^*, x, y, z \in D.$

We focus mainly on the right and middle nuclei which are related by a sort of 'duality'; the kernel (left nucleus) is related to the middle nucleus by a similar 'duality' in the opposite (dual) semifield, but we shall mostly ignore this.

COROLLARY 12.5. Let D be a finite semifield, and let F, K be subfields of $N_m(D)$, $N_r(D)$, respectively. Then the set of non-vertical infinite points, viz., $\Delta = \{(d) : d \in D\}$ is a 2-sided vector space over F and K under natural addition and with the multiplicative actions specified by

$$f(d)k = (fdk)\forall f \in F, k \in K, d \in D.$$

Moreover, assuming these actions, the following hold: The *group* induced on Δ by the left action (respectively right action) of F^* (respectively K^*) coincides with the *group* induced by the Y-axis, X-coaxis (respectively X-axis, Y-coaxis) homologies associated with F^* (respectively K^*). This group action on Δ is linear when D is regarded as a right (respectively left) vector space over the right-nucleus field K (respectively middle-nucleus field F).

12.2. Combinatorics of Seminuclei; Hyperbolic Covers.

We now describe the set of all middle or left nuclear subplanes of a semifield plane in a coordinate-free manner, using combinatorial structures, hyperbolic covers, that are associated with the spreads of semifield planes. In fact, we shall find it convenient to always work with *spreads* π rather than the more cumbersome affine translation planes obtained by taking their cosets; for a systematic treatment of spreads, semifields and rational Desarguesian partial spreads.

DEFINITION 12.6. Let π be a finite spread. **A hyperbolic cover** of π is a collection $\mathcal{E}$ of isomorphic rational **Desarguesian** partial subspreads ('nets') of π all of which share exactly two components $Y, X \in \pi$, the **carriers** of $\mathcal{E}$, and such that any component $\ell \in \pi \setminus \{X, Y\}$ lies in exactly one member of $\mathcal{E}$. The subplanes **across** $\mathcal{E}$ are the subplanes across the nets in $\mathcal{E}$, and the order of these planes is the **order** of $\mathcal{E}$.

Since a rational Desarguesian net of degree $q + 1$, by definition, consists of the partial spread of components of a Desarguesian subplane of order q, contained in some larger Desarguesian plane, we may easily count the number of subplanes involved in any hyperbolic covering of a spread:

Let π be a spread associated with a translation plane of order n that admits a hyperbolic cover $\mathcal{E}$ of order q. Then $n = q^r$, for some integer r, and $\mathcal{E}$ has exactly $(q^r - 1/q - 1)^2$ subplanes across the rational Desarguesian nets in $\mathcal{E}$. The following lemma will lead to hyperbolic covers associated with middle-nucleus and left-nucleus planes. In this lemma, and throughout this section, it will prove convenient to work with spreads rather than the equivalent translation planes. The proof is straightforward (but also see Jha and Johnson [**634**]).

LEMMA 12.7. Let ψ be a subspread of a semifield spread π such that the shears axis $Y \in \pi$ is in ψ, and let $X \in \pi$ be any other components of π that also lies in ψ. Let the degree of ψ be $q+1$. Suppose the subgroup of $(Y$-axis, X-coaxis$)$-homologies (respectively $(X$-axis, Y-coaxis$)$-homologies) of π that leaves ψ invariant induces a transitive homology group on ψ. Then the following hold.

(1) The coordinatization of π with the unit point in ψ, the y-axis and x-axis as Y and X, respectively, yields a semifield $(D, +, \circ)$ such that $N_m(D)$ (respectively $N_r(D)$) contains a subfield $K \simeq GF(q)$ that coordinatizes ψ, and the largest subgroup of $(Y$-axis, X-coaxis)-homologies (respectively $(X$-axis, Y-coaxis)-homologies) of π that leaves ψ invariant is precisely the group defined in Equation (12.1.2) (respectively Equation (12.1.1)).

(2) The subspread ψ defines a rational Desarguesian net, hence this net is covered by Desarguesian subplanes that have pairwise trivial intersection. Moreover, each of these subplanes are middle-nucleus (respectively right-nucleus) subplanes of order q. The slopes of all these planes are coordinatized by $(K) \cup (\infty)$.

(3) Suppose $\pi(D)$ is recoordinatized by **any** semifield $(E+, \bullet)$, with Y as the shears axis and X as x-axis. Then the non-vertical slopes of the order-$|K|$ middle-nucleus (respectively right-nucleus) planes in $\pi(N_m(E))$ (respectively $\pi(N_r(E))$) range over the dimension one left (respectively right) K-subspaces of (D): that is, they have the form $(K \circ t)$ (respectively $(t \circ K)$), for $t \in K^*$.

A second elementary result concerning vector spaces will be used repeatedly.

LEMMA 12.8. *Let V be a vector space of dimension r over $GF(q)$. Suppose that H is a cyclic linear group of V such that $|H| \, | \, q - 1$.*

(1) *Then, H fixes r 1-dimensional $GF(q)$ subspaces that form a direct sum.*

(2) *Represent a generator of H as follows:*

$$(x_1, x_2, \ldots, x_r) \longmapsto (\lambda_1 x_1, \lambda_2 x_2, \ldots, \lambda_r x_r) \text{ for } x_i \in GF(q)^*.$$

Assume that there are exactly s eigenvalues $\lambda_1, \ldots, \lambda_s$ by re-indexing, if necessary. Let the dimension of the λ_j-eigenspace be d_i, so that $\sum_{i=1}^s d_i = r$. Then H leaves invariant exactly

$$\sum_{j=1}^s (q^{d_j} - 1)/(q - 1)$$

dimension one subspaces of V.

(3) *Hence, if H fixes $> (q^{r-1} - 1)/(q + 1) + 1$ 1-dimensional subspaces then H fixes all 1-dimensional subspaces.*

PROOF. Since $|H|$ divides $q - 1$, all the eigenvalues of H are in $GF(q)$. Hence, H must fix a dimension one subspace C of V and thus also leave invariant a Maschke complement M of C. By repeating the argument on H^M, we eventually see that V is a direct sum of r dimension one subspaces each of which is H invariant. This proves part (1). Now we decompose the vector space as a direct sum of mutually distinct λ_j-eigenspaces. In any eigenspace, there are exactly $(q^{d_j} - 1)/(q - 1)$ 1-dimensional $GF(q)$-subspaces and these are all fixed by H. Since $\lambda_j \neq \lambda_i$ for $i \neq j$, it follows that we have exactly

$$\sum_{j=1}^s (q^{d_j} - 1)/(q - 1)$$

1-dimensional $GF(q)$-subspaces fixed by H. $\square$

We now apply Lemma 12.7 to show that any middle-/right-nucleus subfield of order q, in a semifield D, is associated with a hyperbolic covering of order q for

$\pi(D)$. Note that for left nuclear fields the corresponding situation turns out to be more complicated: a 'natural' hyperbolic cover is only guaranteed to exist in the dual of $\pi(D)$.

The following count of the number of middle-nucleus (right-nucleus) subplanes is fundamental.

PROPOSITION 12.9. Let π be a semifield plane with shears axis Y and a second component $X \in \pi \setminus \{Y\}$. Suppose π admits a middle- (respectively right-) nucleus field of order q; so π has order q^r, for some integer r. Then the following hold.

(1) π admits a unique Desarguesian hyperbolic cover $\mathcal{E}_M$ (respectively $\mathcal{E}_R$) of order q with carriers $\{X, Y\}$ such that the set of planes across $\mathcal{E}_M$ (respectively $\mathcal{E}_R$) is the set of all the middle-nucleus (respectively right-nucleus) subplanes of π of order q, whose components include $\{X, Y\}$.

(2) If H is a (X-axis, Y-coaxis) (respectively (Y-axis, X-coaxis)) homology group of order dividing $q - 1$, then there exist positive integers d_i for $i = 1, 2, \ldots, s$ such that $\sum_{i=1}^{s} d_i = r$ and H leaves invariant

$$\sum_{j=1}^{s} (q^{d_j} - 1)/(q - 1)$$

Desarguesian nets of $\mathcal{E}_M$ (respectively $\mathcal{E}_R$), and all the planes across them, or H fixes every plane across $\mathcal{E}_M$ (respectively $\mathcal{E}_R$).

Now specializing the argument to the case where H corresponds to the full multiplicative group of $GF(q)$, we can count the exact number of subplanes of a semifield plane that are simultaneously middle and right nuclear planes of order q.

THEOREM 12.10. Let π be a semifield spread, with shears axis Y and a second component X, that admits middle- and right-nucleus fields $\cong GF(q)$. MAKE THE CONVENTION THAT ANY SUBSPREAD OR SUBPLANE OF π MUST INCLUDE X AND Y AMONG ITS COMPONENTS.

Hence, π has order q^r, for some integer r, that admits hyperbolic covers of order q, $\mathcal{E}_M$ and $\mathcal{E}_R$, both with carriers $\{X, Y\}$, such that the set of planes of the hyperbolic covers $\mathcal{E}_M$ and $\mathcal{E}_R$ are, respectively, the set of all the middle-nucleus subplanes of π that have order q, and the set of all the right-nucleus subplanes of π that have order q.

Thus there are $(q^r - 1/q - 1)^2$ middle-nucleus planes of order q and this is also the number of right-nucleus planes of order q. Moreover, exactly one of the following possibilities holds:

(1) The hyperbolic covers $\mathcal{E}_M$ and $\mathcal{E}_R$ of order q share exactly $\sum_{j=1}^{s} (q^{d_j} - 1)/(q-1) = R$ Desarguesian nets $\mathcal{N}_r$ (any two intersecting in $\{X, Y\}$) and the simultaneously middle and right nuclear subplanes of π, of order q, are the subplanes across the R Desarguesian nets in $\mathcal{N}_r$, where $\sum_{i=1}^{s} d_i = r$ and $s \neq r$.

Thus there are exactly $(\sum_{j=1}^{s} (q^{d_j} - 1)/(q - 1)) (q^r - 1/q - 1)$ subplanes that are simultaneously middle and right planes of order q.

(2) All the middle and right nuclear planes of order q coincide, in which case they constitute the set of central subplanes of order q in π. Thus, $\mathcal{E}_M = \mathcal{E}_R$ consists of the set of all central subplanes of order q; in particular, there are exactly $(q^r - 1/q - 1)^2$ central subplanes of π that are of order q.

We summarize some consequences of the above.

COROLLARY 12.11. Suppose π is a semifield spread of order q^r that admits a middle-nucleus plane of order q and a right-nucleus plane of order q. Let $\mathcal{E}_M$ and $\mathcal{E}_R$ denote, respectively, the associated middle-nucleus and right-nucleus hyperbolic covers, relative to the same carriers $\{X, Y\}$. Then the following are equivalent.

(1) The hyperbolic covers $\mathcal{E}_M$ and $\mathcal{E}_R$ share $> (q^{r-1} - 1)/(q-1) + 1$ rational Desarguesian nets.
(2) $\mathcal{E}_M = \mathcal{E}_R$.
(3) Some plane across a net in at least one of the hyperbolic covers $\mathcal{E}_M$ or $\mathcal{E}_R$ is coordinatized by the center of a semifield D coordinatizing π.
(4) All planes across the hyperbolic covers $\mathcal{E}_M$ and $\mathcal{E}_R$ are central planes of order q (and these are the only central planes through the carriers of these hyperbolic covers).

This corollary immediately implies the invariance of the size of the center of all the semifields coordinatizing a given semifield plane. This may be phrased slightly more generally as follows.

COROLLARY 12.12 (Invariance of Central Fields). Let π be a finite semifield plane that may be coordinatized by a semifield D of dimension r over a central field $F \cong GF(q)$. Then every semifield coordinatizing π has a unique central field $\cong GF(q)$.

Of course, our results imply that the corresponding result concerning the nucleus is false.

12.3. Kernel and Fusion.

So far we have considered the configuration of subplanes, of a semifield spread π, that are simultaneously middle- and right-nucleus planes. This analysis used the fact that if $GF(q)$ is in $N_m(D)$ and in $N_r(D)$, for some semifield D coordinatizing π, then the spread π admits hyperbolic covers of order q: one associated with the middle nucleus and one with the right nucleus.

We now consider configurations involving the remaining seminucleus $N_\ell(D)$, viz., the kernel. Thus, in terms of the standard notions of spread theory [123] we are concerned with a combinatorial description of the subplanes of order q, of a semifield spread π in $PG(2r-1, q)$, that are simultaneously kernel planes, i.e., lines of $PG(2r-1, q)$ that are not in components of π, and, simultaneously, middle- (respectively right-) nucleus subplanes. The situation is somewhat different from before in that $N_\ell(D)$, the left nucleus of a coordinatizing semifield D is no longer associated with a hyperbolic cover in $\pi(D)$. However, the kernel of any spread has special properties that we are able to exploit to complete our analysis. In particular, quasifields that have central kernels may be identified geometrically as follows.

THEOREM 12.13. Let π be a spread of order q^n with $GF(q)$ in its kernel. Suppose that π has a kernel subplane ψ of order q; thus ψ is Desarguesian and the kernel homologies of ψ extend to the kernel homologies of π. Then the following are equivalent.

(1) The components of π through ψ form a rational Desarguesian net such that all planes across the net are kernel planes.

(2) For some choice of the axes and unit point in ψ, the corresponding quasifield Q has a kernel subfield $K \cong GF(q)$ that centralizes Q multiplicatively and $\pi(K) = \psi$.

(3) If ξ is any plane across the rational Desarguesian net, then coordinatizing π in any way with axes and unit point in ξ yields a quasifield Q' such that its kernel subfield K' of order q centralizes Q' multiplicatively.

(4) If π is viewed as a spread in $PG(2n+1, q)$, then the components through ψ form a regulus $\mathcal{R}$ and the set of transversals τ across $\mathcal{R}$ consists of the kernel planes sharing the same slope set as ψ.

The following theorem considers the situation when the kernel and one of the other seminuclear fields, of a finite semifield spread, contain subfields $\cong GF(q)$. Notice that part (1) of the result describes completely the configuration of all subplanes of order q that are simultaneously kernel planes and in one of the other seminuclei.

THEOREM 12.14. Let π be a semifield spread with a kernel field $\cong GF(q)$, and hence π is of order q^r, for some integer r. Assume Y is the shears axis and fix $X \in \pi \setminus \{Y\}$. Suppose π admits a coordinatization by a semifield with $GF(q)$ in its middle (respectively right) nucleus, or equivalently, π admits a hyperbolic cover $\mathcal{E}_M$ (respectively $\mathcal{E}_R$) of order q, with carriers $\{X, Y\}$. Then the following hold.

(1) There is a non-trivial partition of $r = \sum_{j=1}^{t} k_j$ for $t \neq r$ such that any rational Desarguesian net in $\mathcal{E}_M$ (respectively $\mathcal{E}_R$), contains exactly $(\sum_{j=1}^{t}(q^{k_j} - 1)/(q - 1))$ kernel subplanes across it, or the set of all the planes of $\mathcal{E}_M$ (respectively $\mathcal{E}_R$) is also the set of all central subplanes (containing $\{X, Y\}$) of π that have order q.

(2) The spread π may be coordinatized by a semifield D such that $D \supset F \cong GF(q)$ where F lies in $N_m(D) \cap N_\ell(D)$ (respectively $N_r(D) \cap N_\ell(D)$).

(3) The following conditions are equivalent: (1) The spread π may be co-ordinatized by some semifield E such that the intersection field $F \subseteq N_m(E) \cap N_\ell(E)$ (respectively $F \subseteq N_r(E) \cap N_\ell(E)$) **does not** contain $GF(q)$; (2) some semifield coordinatizing π does not have $GF(q)$ in its center; (3) every semifield coordinatizing π does not have $GF(q)$ in its center.

As an immediate consequence we are able to count exactly the number of subplanes of order q that are simultaneously in the middle (respectively right) nucleus and the kernel.

COROLLARY 12.15. Let π be a semifield spread which may be represented projectively in $PG(2r-1, q)$ and that additionally has $GF(q)$ in its middle (respectively right) nucleus. Let Y be the shears axis and X another component of π. Then one of the following holds:

(1) There exists a non-trivial partition $\sum_{i=1}^{s} d_i = r$, $s \neq r$, such that π has exactly

$$(\sum_{j=1}^{s}(q^{d_j} - 1)/(q - 1)) (q^r - 1/q - 1)$$

line transversals to $\{X, Y\}$, in $PG(2r - 1, q)$, that are simultaneously middle- (respectively right-) nucleus subplanes of π.

(2) The middle- (respectively right-) nucleus subplanes of order q define q-reguli and the central planes of π of order q are precisely the planes across these reguli. Hence, there are exactly $(q^r - 1/q - 1)^2$ central planes sharing two components (including the shears axis).

The following result summarizes the configurations arising when two or *more* seminuclear fields of a semifield spread coincide. An important consequence is *nuclear fusion*: a finite semifield plane may always be recoordinatized so that *all* its seminuclear subfields of the *same size* coincide.

THEOREM 12.16 (Nuclear Fusion). Let π a finite semifield spread, with shears axis Y and let $X \in \pi \setminus \{Y\}$. For any semifield D coordinatizing π, use the notation $N_1 = N_\ell(D)$, $N_2 = N_m(D)$, and $N_3 = N_r(D)$. Then the following hold.

(1) If two fields N_i and N_j, for $i, j \in \{1, 2, 3\}$, each contain a subfield isomorphic to $GF(q)$ then π may be recoordinatized by a semifield E such that $N_i(E) \cap N_j(E) \supseteq F \cong GF(q)$.

(2) Suppose each N_i, $i = 1, 2, 3$, contains a subfield $\cong GF(q)$, but some semifield coordinatizing π does *not* have $GF(q)$ in its center.[1]

    (a) Then π has order-q hyperbolic covers $\mathcal{E}_M$ and $\mathcal{E}_R$, associated respectively with the middle and right nucleus and with carriers $\{Y, X\}$, and there is a non-trivial partition of $r = \sum_{i=1}^s d_i$, for $s \neq r$ and the hyperbolic covers share a set of exactly $(\sum_{j=1}^s (q^{d_j} - 1)/(q - 1))$ rational Desarguesian nets $\mathcal{R}$.

    (b) Furthermore, there is another non-trivial partition $\sum_{i=1}^t k_i$, $t \neq r$ and each of these nets contain exactly $(\sum_{j=1}^t (q^{k_j} - 1)/(q - 1))$ nuclear planes of order q (and these constitute the set of all the nuclear planes of order q that contain $\{Y, X\}$).

(3) Suppose each N_i, $i = 1, 2, 3$, contains a subfield $\cong GF(q)$. Then either all the seminuclei of order q in every coordinatizing semifield are in the center or π has no central planes of order q and there are two non-trivial partitions of $r = \sum_{j=1}^s d_j$, $s \neq r$, and $r = \sum_{i=1}^t k_i$, $t \neq r$, and there are exactly

$$\left(\sum_{j=1}^s (q^{d_j} - 1)/(q - 1)\right)\left(\sum_{j=1}^t (q^{k_j} - 1)/(q - 1)\right)$$

nuclear planes of order q passing through the axes $\{X, Y\}$.

(4) If each of the three seminuclei, N_1, N_2 and N_3, contains subfields $\cong GF(q)$, then π may be coordinatized by a semifield E such that the nucleus of E contains a field $F \cong GF(q)$, that is

$$GF(q) \cong F \subseteq N(E) = N_\ell(E) \cap N_m(E) \cap N_r(E).$$

PROOF. We recall that if π is coordinatizable by even one semifield that has a central field $Z \cong GF(q)$, then all seminuclear fields (and the nucleus) of order q, of any semifield coordinatizing Z, coincide with the central field of order q, and the configuration of the associated central planes is completely known, e.g., see Theorem 12.10. Hence, we shall assume throughout that *none* of the coordinatizing semifields of the considered plane has $GF(q)$ in its center.

---

[1] Recall that the configuration of all central planes of order q is fully known if one such central plane is known.

Since part (1) is covered by the earlier result and part (3) and part (4) are immediate consequences of part (2), it is sufficient to prove part (2). Now, by hypothesis, N_m and N_r both contain $GF(q)$, and hence, by Theorem 12.10, the associated hyperbolic covers of order q, viz., $\mathcal{E}_M$ and $\mathcal{E}_R$, must share exactly $(\sum_{j=1}^{s}(q^{d_j} - 1)/(q-1))$ rational Desarguesian nets $\mathcal{R}$; the planes on the nets across $\mathcal{R}$ are the simultaneously middle and right nuclear subplanes of order q.

However, the group of kernel homologies of order $q-1$ must fix either $(\sum_{j=1}^{t}(q^{k_j} - 1)/(q-1))$ or all of the planes across each such net, Theorem 12.14(1). So these planes of order q are precisely the subplanes of order q that are in each of the three seminuclear planes. Thus, coordinatizing with unit point in these nets, always yields semifields D with the nucleus $N(D) \cong GF(q)$ and all nuclear plane of order q (containing the carriers of the hyperbolic covers) are of this type. ☐

The above is a 'fusion' theorem; the converse is a 'splitting' theorem: if two or more seminuclei overlap in a field $GF(q)$ when can we recoordinatize so that the overlap no longer includes $GF(q)$. The answer is always unless one, and hence all, coordinatizing semifields have $GF(q)$ in their center. Thus we observe:

COROLLARY 12.17. If each seminucleus of a finite semifield plane π contains a subfield $\cong GF(q)$, then π may be coordinatized by a semifield D such that the nucleus F of D contains $GF(q)$. Moreover, π may also be recoordinatized by a semifield such that its nucleus does **not** contain $GF(q)$ unless $GF(q)$ is in the center of one (and hence all) coordinatizing semifields.

12.4. The Semifield Planes with Two Isomorphic Baer Seminuclei.

We begin by stating a well-known characterization of fields of square order.

THEOREM 12.18. If D has a Baer subfield Z in its center, then D is a field.

PROOF. Choose an element $t \in D \setminus Z$, so $t^2 = t\alpha + \beta$, $\alpha, \beta \in Z$. Now use the defining identities for a semifield and its center to verify that multiplication in D is indeed associative. Hence D is a skewfield, so by Wedderburn's theorem D is a field. ☐

Note that counterexamples arise if $GF(q)$ above is merely assumed to be in the nucleus, rather than the center, of a semifield of order q^2. The aim of this section is to classify all semifield spreads of order q^2 that admits homology groups of order $q - 1$ relative to at least two distinct axes.

A key role in all this involves the Hughes–Kleinfeld semifields, and their characterization in the finite case as being the semifields D of order q^2 such that $N_\ell(D) = N_m(D) \cong GF(q)$. We shall show that if a semifield spread of order q^2 has two Baer seminuclear field (which might not coincide), then the associated spread is either a Hughes–Kleinfeld spread, its dual or its transpose. The main step is to show that two of the seminuclei of the same size q can be forced to coincide, possibly after recoordinatization, and then reducing to the Hughes–Kleinfeld's classification.

PROPOSITION 12.19. Let π be a non-Desarguesian spread of order q^2 coordinatized by a semifield D of order q^2 such that its left and middle nucleus are both isomorphic to $GF(q)$. Then π may be coordinatized by a Hughes–Kleinfeld semifield. In fact, *every* net $\mathcal{H}$ across a middle-nucleus subplane of order q contains exactly two middle-nucleus planes $\psi_1, \psi_2 \in \mathcal{H}$ that are also kernel planes: thus if the spread is coordinatized by a semifield S based on axes and unit points in a plane $\psi \in \mathcal{H}$ then S is a Hughes–Kleinfeld semifield if and only if $\psi \in \{\psi_1, \psi_2\}$. Furthermore, there are exactly $2(q+1)$ kernel and middle-nucleus planes.

PROOF. Applying Theorem 12.16(1), either the result holds or the spread is coordinatizable by a semifield K of order q^2 with $GF(q)$ in its center, but this forces K to be a field, Theorem 12.18.   □

We remark that the result includes the well-known fact that a Hughes–Kleinfeld plane may always be coordinatized by semifields that are *not* Hughes–Kleinfeld semifields. More interestingly, the above describes precisely the configuration of all the coordinatizations that lead to Hughes–Kleinfeld semifields. We use this to show that a Hughes–Kleinfeld spread has no regulus containing the shears axis, a fact we shall require. It is convenient to begin with a few remarks concerning reguli in arbitrary line spreads.

If a spread π is dimension 4 over a kernel field $K = GF(q)$, then π may be viewed as a spread of lines in $PG(3, K)$, and the kernel subplanes of π are simply the lines of $PG(3, K)$ that are not components. Suppose ρ is any such line (kernel plane) and let $\mathcal{R}$ be the set of components meeting ρ non-trivially. Then ρ determines a regulus if and only if the set of transversals τ to the $q+1$ components in $\mathcal{R}$ forms a covering of the point set $\bigcup \mathcal{R}$. Moreover, it is well known from elementary projective geometry that three skew lines lie in a unique regulus.

Now specialize again to a Hughes–Kleinfeld plane ξ, and consider any set of three components $\tau = \{Y, X, I\}$, where Y is the shears axis. Thus the kernel planes that include τ are transversals to τ and hence lie in the regulus determined by τ assuming it exists. However, by Proposition 12.19 there are exactly two kernel planes that meet non-trivially each member of τ, so τ cannot lie in a regulus. In particular, if D coordinatizes ξ, then its kernel $K = N_\ell(D) \cong GF(q)$ cannot centralize D multiplicatively. Thus we have shown:

COROLLARY 12.20. A Hughes–Kleinfeld semifield of order q^2 cannot admit a regulus of $q+1$ components that includes the shears axis. In particular, if D is a semifield coordinatizing a Hughes–Kleinfeld semifield, then its kernel $N_\ell(D) = GF(q)$ cannot centralize D multiplicatively.

As a simple application we mention the following result:

COROLLARY 12.21. Suppose D is a semifield of order q^2 with its left and right nuclei containing, respectively, subfields F and G, both $\cong GF(q)$ and such that F or G centralizes D multiplicatively. Then D is a field.

We next classify all semifield planes π of order q^2 such that of the three types of seminuclear planes (left, middle, right) at least two are Baer-sized. We state these conditions in terms of the homology groups of the plane.

THEOREM 12.22. Let π^{ℓ_∞} be a semifield plane of order q^2 that admits two homology groups of order $q-1$ with distinct axes and coaxes. Then π is one of the following:

   (1) A Hughes–Kleinfeld plane (Knuth Type II, [**898**, (7.17)]);
   (2) A dual of a Hughes–Kleinfeld plane (Knuth Type III, [**898**, (7.17)]);
   (3) A transpose of a Hughes–Kleinfeld plane (Knuth Type IV, [**898**, (7.17)]);

(4) A Desarguesian plane.

We note that the above result generalizes the algebraic characterization of semi-fields of Knuth types II, III, IV, [**898**, Theorem 7.4.1], by providing a geometric characterization of the planes associated with these Knuth types: we have essentially replaced the seminuclear coincidence assumptions of Knuth [**898**, Theorem 7.4.1] by the weaker assumption that the seminuclear fields involved have the same size.

12.5. The Numbers of Nuclear Subplanes.

The configurational theorems that we have established were used in the previous section to planes coordinatized by semifields with at least two Baer seminuclei. The question arises as to whether there are non-coincident seminuclei $\cong GF(q)$ in semifield spreads of order q^r, particularly when r is odd, to enable us to apply the configurational theorems established above.

In fact, the class of semifields called *cyclic semifields* provides a large supply of planes coordinatized by semifields whose dimensions over various *non-central* seminuclear fields can more or less be controlled at will. The results in this paper impose limitations on the transitivity groups of such planes. For example, consider a spread π coordinatized by a semifield D of order q^r with a *non-central* field $N \cong GF(q)$ in the nucleus $N(D)$. Then our results show that $G = \text{Aut}(\pi)$ must leave invariant a set of at most $((q^{r-1} - 1)/(q+1) + 1)^2$ subplanes (associated with nuclear planes).

Furthermore, the partition numbers in the statement of our results can be made much finer. When we discussed the eigenspaces of the middle-nucleus homology group acting on the right-nucleus homology group space on the line at infinity, we did not make use of the fact that the middle nucleus must act transitively on each right-nucleus orbit and furthermore, the middle-nucleus group union the zero mapping forms a field.

Hence, if we are to let the middle-nucleus homology group act with generator as follows: $(x_1, \ldots, x_r) \longmapsto (\lambda_1 x_1, \ldots, \lambda_r x_r)$, then since the group is cyclic of order $q - 1$ and fixed-point-free (acting on the line at infinity) $-(\infty)$, it follows that we may assume that $\lambda_i = \lambda_1^{p^{\lambda(i)}}$, for all $i = 2, 3, \ldots, r$, where $q = p^u$, p prime so $\lambda(i) = 1, 2, \ldots, u$. Thus, in the partition of r described in our theorems, we have only partitions of at most u distinct elements d_i.

Hence, when u is small relative to r, we have severe restrictions on the automorphism group.

For example, assume that we fuse three seminuclei isomorphic to $GF(p^2)$ but right and middle nuclei do not commute over the semifield. Then, $\lambda(i) = 0$ or 1 and we have at most two distinct eigenspaces. Hence, there exist integers a, b, c, d all strictly less than r such there are exactly $((q^a - 1)(q^b - 1)(q^c - 1)(q^d - 1))/(q-1)^4$ nucleus subplanes and hence this group must be permuted by the full automorphism group.

More generally, the same statement holds for three seminuclei fused to $GF(q^2)$ where the center is $GF(q)$. This situation turns out to occur in certain Hughes–Kleinfeld semifields of order q^4 with left, right and middle nuclei isomorphic to $GF(q^2)$ with center $GF(q)$. Hence, we obtain a regulus hyperbolic cover of $GF(q)$-reguli.

THEOREM 12.23. Let π be a Hughes–Kleinfeld semifield plane of order q^4 and kernel, right and middle nuclei isomorphic to $GF(q^2)$ where the center is $GF(q)$. Then the following hold.

(1) For any subfield $GF(q')$ of $GF(q)$, there is a q'-regulus hyperbolic cover of the associated spread.
(2) There exists a right hyperbolic $GF(q^2)$-cover and a middle hyperbolic $GF(q^2)$-cover.
(3) There are exactly four nucleus subplanes.

When considering the 'infinite' points of a plane or spread $\pi(D)$, coordinatized by semifield D, we shall adopt the notation $(W) := \{(w) : w \in W\}$ to denote the points on the line at infinity with slope set $W \subset D$; so (D) denotes the set of all points $\neq (\infty)$ on the line at infinity.

We recall that, for any finite semifield D, if K^* denotes the multiplicative group of a (sub)field of the left, middle, or right nucleus of D then in each case K^* is associated with an appropriate homology group, the 'K-homology group' of $\pi(D)$. Specifically the K-homology groups, associated with K, are at least one of the following types:

X-**axis, Y-coaxis:** For $K \subseteq N_r(D)$:

(12.5.1) $\qquad (x, y) \mapsto (x, yk); (z) \mapsto (zk), \ \forall k \in K^*, x, y, z \in D.$

Y-**axis, X-coaxis:** For $K \subseteq N_m(D)$:

(12.5.2) $\qquad (x, y) \mapsto (xk, y); (z) \mapsto (k^{-1}z), \ \forall k \in K^*, x, y, z \in D.$

Kern Homologies: For $K \subseteq N_\ell(D)$:

(12.5.3) $\qquad (x, y) \mapsto (kx, ky); (z) \mapsto (z), \ \forall k \in K^*, x, y, z \in D.$

We focus mainly on the right and middle nucleus which are related by a sort of 'duality'; the kern (left homology) is related to the middle nucleus by a similar 'duality' in the opposite (dual) semifield, but we shall mostly ignore this.

COROLLARY 12.24. Let D be a finite semifield, and let F and K be any subfields of $N_m(D)$ and $N_r(D)$, respectively. Then the set of non-vertical infinite points, viz., $(D) := \{(d) : d \in D\}$ is a 2-sided vector space over F and K under natural addition and with the multiplicative actions specified by

$$f(d)k = (fdk) \forall f \in F, k \in K, d \in D.$$

Moreover, assuming these actions, the following holds: The *group* induced on (D) by the left action (respectively right action) of F^* (respectively K^*) coincides with the *group* induced by the Y-axis, X-coaxis (respectively X-axis, Y-coaxis) homologies associated with F^* (respectively K^*). This group action on (D) is linear when D is regarded as a right (respectively left) vector space over the right-nucleus field K (respectively middle-nucleus field F).

We turn to the coordinate-free description (up to duality) of the subplane and net structures suggested by the above results: doing this enables genuine additional information to be extracted from the setup above.

It will be useful to begin by introducing the notion of an elliptic cover of a spread, consisting of rational Desarguesian nets.

DEFINITION 12.25. Let π be a finite spread. A **Desarguesian elliptic cover** of π is a collection of rational Desarguesian nets $\mathcal{E}$ of π all of which share exactly two components $Y, X \in \pi$ and such that any component $\ell \in \pi \setminus \{X, Y\}$ lies in exactly one member of $\mathcal{E}$. The **order** of $\mathcal{E}$ is the order of the subplanes of the nets across $\mathcal{E}$.

The following remark provides a complete description of the middle and right nuclear subplanes of a semifield plane in terms of a coordinatizing semifield. After verifying the result, we provide a compact reformulation in terms of elliptic covers.

REMARK 12.26. Let ψ be a subspread of a semifield spread π such that the shears axis $Y \in \pi$ also lies in ψ, and let $X \in \pi$ be any other components of π that lies in ψ. Suppose the subgroup of (Y-axis, X-coaxis)-homologies (respectively (X-axis, Y-coaxis)-homologies) of π that leaves ψ invariant induces a transitive homology group on ψ. Then the following hold.

(1) The coordinatization of π with the unit point in ψ, the y-axis and x-axis as respectively Y and X, yields a semifield D such that $N_m(D)$ (respectively N_r) contains a subfield K that coordinatizes ψ, and the largest subgroup of (Y-axis, X-coaxis)-homologies (respectively (X-axis, Y-coaxis)-homologies) of π that leaves ψ invariant is precisely the group defined in Equation (12.5.2) (respectively Equation (12.5.1)).

(2) The subspread ψ defines a rational Desarguesian net, and this net is covered by Desarguesian subplanes that have pairwise trivial intersection and each of these subplanes are middle-nucleus (respectively right-nucleus) subplanes of order $|K|$. The slopes of all these planes are coordinatized by $(K) \cup (\infty)$.

(3) Suppose π is coordinatized by a semifield D, when Y is the shears axis and X is chosen as the x-axis. Then the slope set of middle-nucleus (respectively right-nucleus) planes of order $|K|$ that include X and Y among their components are the rank one left (respectively right) K-spaces of D: that is, they have the form (Kt) (respectively (tK)), for $t \in K^*$.

PROOF. We only consider the (Y-axis, X-coaxis)-homologies case, since the other case is very similar. It is straightforward to verify that the coordinatization chosen leads to a coordinatization of π by a semifield D such that ψ is coordinatized by a middle-nucleus field K. Hence D is a 2-sided vector space over K, *a fortiori* a right vector space, so the slope maps associated with K form a field, and hence ψ defines a rational Desarguesian net. The homology group of π that induces a transitive homology group of ψ obviously induces a transitive group on the slopes of all the Desarguesian planes across the net defined by ψ. Thus, the arguments applied to ψ apply equally well to all the planes across its components. Thus all subplanes across ψ are middle-nucleus subplanes, with the same slope set as ψ.

Let θ be any middle-nucleus subplane of order $|K|$, associated with (Y-axis, X-coaxis)-homologies. Since $\pi(D)$ has a unique group of (Y-axis, X-coaxis)-homologies of order K^* and this must leave invariant any middle-nucleus subplane (which includes X and Y), it follows that the slopes of $\theta \notin \{(\infty), (0)\}$ must form an orbit under the group of (Y-axis, X-coaxis)-homologies of $\pi(D)$ of order K^*. But these slope orbits have the form K^*t, $t \in D^*$.

The result has now been established. $\qquad\square$

We may now state the above in terms of elliptic covers.

COROLLARY 12.27. Let π be a semifield plane with shears axis Y and a second component $X \in \pi \setminus \{Y\}$. Suppose π admits a right (respectively middle-) nucleus field of order q. Then π admits a unique Desarguesian elliptic cover $\mathcal{E}_M$ (respectively $\mathcal{E}_R$) of order q supported by $\{X, Y\}$ such that the planes across the rational Desarguesian nets in $\mathcal{E}_M$ (respectively $\mathcal{E}_R$) are precisely all the middle-nucleus (respectively right-nucleus) subplanes of π supported by $\{X, Y\}$ and having order q.

Suppose $\mathcal{E}_M$ of order q exists, so π has order q^r for some integer r. If H is a (X-axis, Y-coaxis) homology group of order dividing $q - 1$, then either H leaves invariant exactly r Desarguesian nets of $\mathcal{E}_M$, and all the planes across them, or H fixes every plane across $\mathcal{E}_M$.

Now specialize the argument to the case where H corresponds to the full multiplicative group of a field. This shows that either exactly r of the nets have planes that are right-nucleus planes, or $\mathcal{E}$ is also the right-nucleus elliptic cover associated with q. It turns out that every member of $\mathcal{E}$ consists of planes in the center.

The most important case of this argument is when a middle and right nucleus have the same order q. For this, we have established:

THEOREM 12.28. Let π be a semifield spread, with shears axis Y and a second component X, that admits middle- and right-nucleus fields $\cong GF(q)$; thus π has order q^r for some integer r. Then the following hold.

(1) There are exactly $(q^r - 1/q - 1)^2$ middle (right) nuclear planes (containing $\{X, Y\}$) of order q.

(2) Either all the middle and right nuclear planes of order q coincide, in which case they are precisely the set of central planes (containing $\{X, Y\}$) of order q in π or there are either exactly $r(q^r - 1/q - 1)$ planes that are simultaneously middle and right planes of order q (that are required to contain $\{X, Y\}$).

In any semifield spread $\pi(D)$ the group of kern homologies, Equation (12.5.3), evidently centralizes the group of middle-nucleus homologies, Equation (12.5.2), and also the group of right-nucleus homologies, Equation (12.5.1). Consider the kern-middle-nucleus case in further detail (the kern-right-nucleus case may be similarly treated). Suppose $F \cong GF(q)$ is in $N_m(D)$ and $K \cong GF(q)$ is in $N_\ell(D)$ (a kern field). It is evident that the X-axis as an F-space admits a faithful K-linear representation of the kern homology group H of order $q - 1$. Thus, since the primitive element in H has an eigenvalue in F^*, we get a splitting of H into r F-spaces on X, and hence each of the derivable middle-nucleus nets associated with F has exactly r planes that are kern planes or all the middle-nucleus planes are kern planes. In the latter event, the kern slopes correspond to reguli on all nets, hence every middle-nucleus plane is also a left-nucleus plane and all middle-nucleus planes of order q are coordinatized by a field in the center of the corresponding semifield.

A similar analysis holds if $GF(q)$ is in the kern and the right nucleus.

In particular, recall that if N_m and N_r both contain $GF(q)$, then some of the nets in their elliptic cover are common and on these nets, if the *kern* also contains $GF(q)$, then exactly r of these or all of them are kern planes as well as. In the latter case we have the central situation, as indicated.

Hence we may summarize *some* of our conclusions as follows.

THEOREM 12.29 (Nuclear Fusion). Let π a finite semifield spread such that at least two of the three seminuclei contain subfields $\cong GF(q)$, so π has order q^r. Then the following hold.

(1) π may be coordinatized by a semifield D such that D contains a subfield $F \cong GF(q)$ such that F lies in two of the three seminuclei in the triple $(N_\ell(D), N_m(D), N_r(D))$.

(2) If all three seminuclei listed in $(N_\ell(D), N_m(D), N_r(D))$ contain subfields $\cong GF(q)$, then π may be coordinatized by a semifield D such that the nucleus of D contains a field $F \cong GF(q)$ and F lies in every member of the triple $(N_\ell(D), N_m(D), N_r(D))$.

Note that by the above arguments we can also list exactly the number of nuclear planes of order q and also the number of central planes of order q.

CHAPTER 13

Cyclic Semifields.

In Chapter 33, we consider extensions of derivable partial spreads which are necessarily Desarguesian. In this chapter, we consider a similar study of extensions of rational Desarguesian partial spreads but here the partial spread is not necessarily derivable. Although our manner of construction is completely independent from that of Sandler [**1147**], the constructed 'semifield spreads' bear a resemblance to those given by Sandler so we suspect there is more than a coincidental intersection. We shall be constructing semifields of order q^n whose spreads contain a Desarguesian partial spread of degree $q + 1$. However, when $n = 3$, we here also describe some definitely non-semifield translation planes of order q^3 that admit a collineation group isomorphic to $SL(2, q)$. These $SL(2, q)$-spreads and their parent or related semifields have also been described in a completely different manner in the work of Glynn [**402**].

Let W be a finite n-dimensional vector space, $n > 1$ over a field F and suppose that $T \in \Gamma L(W, F) \setminus GL(W, F)$ is a strictly semi-linear bijection of W, where W is regarded as an F-space. Also, let K be a subfield of F such that $T \in GL(W, K)$. For example, K might be chosen to be the prime subfield of F. We are interested here in the case when T is F-irreducible, that is, when T does not leave invariant any non-trivial proper F-subspace of W. Examples of such T are easily constructed, for instance on choosing $S \in GL(W, F)$ to correspond to a Singer cycle of $PG(n-1, F)$, $\sigma \in \text{Gal}(F)^*$, we might define $T = S\hat{\sigma}$; it is also not hard to see that S^k, for many values of k, work as well as S itself. We now observe that the F-subspace of $\text{Hom}(W, F)$, generated by the powers of T, form an additive spread set and thus yields a semifield; the strict F-semilinearity of T ensures that none of these semifields will be a field. We shall call these semifields *cyclic*.

PROPOSITION 13.1. Suppose that W is a finite n-dimensional vector space, $n > 1$, over a field F and assume that $T \in GL(W, K)$, where K is a proper subfield of F. If $T \in \Gamma L(W, F) \setminus GL(W, F)$ is F-irreducible, then viewing T and $f \in F$ as elements of $GL(W, K)$, the set:

$$\Delta(T, F) := \{ \mathbf{1} a_0 + T a_1 + \ldots + T^{n-1} a_{n-1} \mid a_0, a_1, \ldots a_{n-1} \in F \}$$

is an additive spread set over the field K. Such spread sets will be called 'cyclic semifield' spread sets.

The kernel of $\Delta(T, F)$ is isomorphic to the centralizer of $\{T\} \cup F$ in $\text{Hom}(W, +)$.

The Sandler semifields and the finite Hughes–Kleinfeld semifields are cyclic semifields. Thus, cyclic semifields, considered generally, may be regarded as providing a uniform characterization of the finite Hughes–Kleinfeld and Sandler semifields, in slightly generalized form. Since there are a variety of ways to construct cyclic semifield planes, the question certainly arises as to the number of mutually non-isomorphic planes so obtained. Kantor and Liebler in unpublished work have shown the following.

THEOREM 13.2. (Kantor and Liebler). The number of mutually non-isomorphic cyclic semifield planes of order q^n is less than $n^2 q^n$.

In the following we shall discuss certain particular classes of cyclic semifield planes. In terms of particular orders, for example, it turns out that the number of non-isomorphic cyclic semifield of order q^6, where the middle and right nuclei are isomorphic to $GF(q^2)$ and the left nucleus is $GF(q^3)$ is at least $q(q+1)/6e$, where $p^e = q$, for p a prime.

13.1. Jha–Johnson Cyclic Semifields.

Jha and Johnson [**606**] give various constructions, which we describe here.

DEFINITION 13.3. Let $\ell = \operatorname{lcm}(m,n)$, $\ell > 1$, where $m, n > 1$ are integers. Define the following mapping T_ω

$$T_\omega : GF(q^\ell) \to GF(q^\ell), \quad xT = \omega x^{q^n},$$

where ω is a primitive element of $GF(q^\ell)$. Then a cyclic semifield is obtained (see Jha and Johnson [**606**], p. 16). It turns out that the middle and right nucleus share a subfield isomorphic to $GF(q^m)$ and the left nucleus (the kernel of the semifield plane) is isomorphic to $GF(q^n)$. This class of semifields is called the 'Jha–Johnson cyclic semifields of type $S(\omega, m, n)$'.

DEFINITION 13.4. Let $GF(p^f) \subset GF(p^m)$, and let $\lambda = m/f > 1$, where p is a prime. Assume that $n > m$ is chosen so that $\gcd(n, \lambda) = 1$ and $p^{nf} - 1$ has a p-primitive divisor u. Assume that $B \in GL(n, p^f) \subset GL(n, p^m)$ has order u and let $1 \neq \sigma \in \operatorname{Gal}_{GF(p^f)} GF(p^m)$. Then $B\sigma$ defines a cyclic semifield. This class is called the 'Jha–Johnson cyclic semifields of p-primitive type 1' (see Jha and Johnson [**606**], p. 20).

DEFINITION 13.5. Assume that p is a prime and m, n integers > 1 such that $p^{nm} \neq 2^6$. Let $q = p^m$ and u be any prime q-primitive divisor of $q^n - 1$. Let $B \in GL(mn, p) \subset GL(mn, q)$ of order u. Then $V = GF(q^{mn})$ is a $\langle B \rangle GF(q)$-module. Let $1 \neq \sigma \in \operatorname{Gal} GF(q)$, of order k. Then V contains an irreducible $\langle B\sigma \rangle GF(q)$-submodule W such that $|W| = q^{nk}$ and $B\sigma \mid W$ produces a cyclic semifield. This class of semifields is called the 'Jha–Johnson cyclic semifields of q-primitive type 2'.

REMARK 13.6. In the following, we shall be interested in cyclic semifields of order h^3. Hence,

(1) When $\operatorname{lcm}(m,n)$ is divisible by 3, there are a variety of Jha–Johnson cyclic semifields of type $S(\omega, m, n)$ of order h^3, where $h = q^{\ell/3}$,

(2) Take m or n to be 3 to obtain a Jha–Johnson cyclic semifield of p-primitive type 1 of order p^{3j}, or

(3) Choose $n = 3$ and obtain a Jha–Johnson cyclic semifield of q-primitive type 2 of order q^{3k}.

For an easy summary of the types of orders obtained, Jha and Johnson [**606**] show the following:

THEOREM 13.7. A cyclic semifield plane of order p^{mn} exists where the associated $T \in \Gamma L(n, p^m) - GL(n, p^m)$ for all primes p and for all integers $m, n > 1$.

13.2. The Cofman Problem.

The 'Cofman question' is where there exist finite affine planes that contain more subplanes than does a Desarguesian plane? It turns out that any cyclic semifield plane has more affine subplanes than does the corresponding Desarguesian plane of the same order. Hence, the Cofman problem of finding all orders of planes for which such affine planes exist is completely determined by the previous theorem of Jha and Johnson.

CHAPTER 14

T-Cyclic $GL(2, q)$-Spreads.

We now define spreads that are *never* semifield spreads, but still based on a field $\mathcal{F}$ of K-linear mappings of an n-dimensional K-vector space W, K any finite field. All of this material arises primarily from the work of Jha and Johnson [**611**] and related work.

The construction is best described directly, as constructing a spread on $V = W \oplus W$, rather than via a spread set, so it becomes convenient to work with matrices, relative to a chosen K-basis of W, and we make the identifications $W = K^n$, $V = K^n \oplus K^n$. Now the field of linear mappings associated with the scalar action of F on W, viz., $\hat{f} : x \mapsto xf$, becomes identified with a field $\mathcal{F}$ of $n \times n$ matrices over K, acting on K^n, and $T \in GL(n, K)$ is still required to be strictly $\mathcal{F}$-semilinear on K^n, or equivalently: $T \in N_{GL(n,K)}(\mathcal{F}) - C_{GL(n,K)}(\mathcal{F})$. Furthermore, we shall insist that T does not leave invariant any non-trivial $\mathcal{F}$-subspace of rank ≤ 2, rather than insisting that T acts irreducibly, as in the previous case.

We shall demonstrate that the orbit τ of the subspace $y = xT$ of V, under the standard action of $\mathcal{G} = GL(2, \mathcal{F})$ on V, forms a partial spread that extends to a larger $\mathcal{G}$-invariant partial spread $\pi(T, \mathcal{F}) := \pi_{\mathcal{F}} \cup \tau$, where $\pi_{\mathcal{F}}$ is the (rational Desarguesian) partial spread associated with $\mathcal{F}$. On specializing to the case $dim_F W = 3$, the partial spread $\pi(T, \mathcal{F})$ becomes a non-Desarguesian spread of order q^3 admitting $GL(2, q)$, where $\mathcal{F} \cong GF(q)$.

THEOREM 14.1. Let $W = K^n$ be the standard n-dimensional vector space over a finite field $K = GF(q)$, for $n > 3$. Suppose that $\mathcal{F} \subset GL(n, K)$ is a field, containing the scalar field K, and $T \in N_{GL(n,K)}(\mathcal{F}) - C_{GL(n,K)}(\mathcal{F})$. So, there is a non-trivial field automorphism $\sigma \in \mathrm{Gal}(\mathcal{F}/K)^*$ such that $\forall X \in \mathcal{F} : X^\sigma = T^{-1}XT$. Let $\pi_{\mathcal{F}}$ be the rational Desarguesian partial spread determined on $V := W \oplus W$ by the spread set $\mathcal{F}$, and let τ be the orbit of the K-subspace $y = xT$, of V, under the group:

$$\mathcal{G} := \left\{ \begin{pmatrix} a & b \\ c & d \end{pmatrix} \mid a, b, c, d \in \mathcal{F}, ad - bc \neq 0 \right\} \cong GL(2, \mathcal{F}),$$

in its standard action on V.
Put: $\pi(T, \mathcal{F}) := \tau \cup \pi_{\mathcal{F}}$. Suppose that T does not leave invariant any non-zero $\mathcal{F}$-subspace of W that has rank ≤ 2. Then the following hold:

(1) τ is a partial spread containing $q(q^2 - 1)$ components and the global stabilizer of $y = xT$ in $\mathcal{G}$ is the diagonal group

$$\{\mathrm{Diag}[A, A^\sigma] \mid A \in \mathcal{F}^*\}.$$

(2) The rational Desarguesian partial spread $\pi_{\mathcal{F}}$ is a $\mathcal{G}$-orbit, and $\mathcal{G}$ acts triply transitively on its components.

(3) The $\mathcal{G}$-orbits, τ and $\pi_{\mathcal{F}}$, do not share any components and $\pi(T, \mathcal{F})$ is also a partial spread.

(4) $\pi(T, \mathcal{F})$ is a spread if and only if $dim_{\mathcal{F}}W = 3$. In this case, the spread admits $\mathcal{G} = GL(2, \mathcal{F})$ so that this group partitions the components of $\pi(T, \mathcal{F})$ into two orbits, viz., τ and $\pi_{\mathcal{F}}$, and $\mathcal{G}$ acts triply transitively on the orbit $\pi_{\mathcal{F}}$ and transitively on the orbit τ. The kernel of $\pi(T, \mathcal{F})$ is isomorphic to the centralizer of $\{T\} \cup \mathcal{F}$ in $\mathrm{Hom}(W, +)$; hence, $K = GF(q)$ is always in the kernel, and $\mathcal{F}$ is not: so the spread is non-Desarguesian.

Instructive Diversion. *This is the case of a simple but useful principle:*
(a) *If a dimension* r *subspace* A *of a vector space* V *of dimension* $2r$ *has an orbit* $\mathcal{A}$ *under a subgroup* $G \leq GL(V, +)$ *such that* $A - A^g$ *is non-singular or zero, for all* g, *then* $\mathcal{A}$ *is a partial spread that is* G-*invariant;*
(b) *If the subspace* A *is disjoint from all the members of a* G-*invariant partial spread* $\mathcal{B}$ *then* $\mathcal{A} \cup \mathcal{B}$ *is also a partial spread.* $\square$

Next, to apply the second part of the above principle, consider the possibility that $y = xT$ meets $\pi_{\mathcal{F}}$, the rational Desarguesian spread coordinatized by $\mathcal{F}$. If $T - A$ is singular for $A \in \mathcal{F}^*$, then $xT = xA$, for some non-zero $x \in W$. Thus, $y = xT$ and $y = xA$ are disjoint subspaces of V, for $A \in \mathcal{F}^*$: otherwise T leaves invariant $x\mathcal{F}$, contrary to the hypothesis. Moreover, $y = xT$ is certainly disjoint from $\mathbf{O} \oplus W$. Hence $y = xT$ is disjoint from the rational Desarguesian partial spread coordinatized by the spread set $\mathcal{F}$. However, this partial spread, viz., $\pi_{\mathcal{F}} := \{y = xA \mid A \in \mathcal{F}\} \cup \{Y\}$ is also invariant under $\mathcal{G}$. To see this, note that

$$(\mathbf{0}, u) \begin{pmatrix} a & b \\ c & d \end{pmatrix} = (uc, ud)$$

shows that Y is left invariant when $c = \mathbf{0}$, and otherwise, when $cu \neq \mathbf{0}$, Y maps to $y = x(uc)^{-1}ud$, which is a component of type $y = xf$, $f \in \mathcal{F}$.

Similarly, we can determine that $y = xf$, $f \in \mathcal{F}$, maps under $\mathcal{G}$ into the rational Desarguesian partial spread $\pi_{\mathcal{F}}$:

$$(y = xf) \begin{pmatrix} a & b \\ c & d \end{pmatrix} \mapsto \left\{ \begin{array}{ll} (y = x(a + fc)^{-1}(b + fd)) & \text{if } a + fc \neq \mathbf{O}; \\ (x = \mathbf{0}) & \text{otherwise.} \end{array} \right\}.$$

In particular, Y is not $\mathcal{G}$ invariant, and the global stabilizer $\mathcal{G}_{\{Y\}}$ of Y is doubly transitive on all the other components of $\pi_{\mathcal{F}}$. For example, note that $\mathcal{G}_{\{Y\}}$ does not leave X invariant and the global stabilizer of X in $\mathcal{G}_{\{Y\}}$ is transitive on the components in $\pi_{\mathcal{F}} \setminus \{X, Y\}$. Hence, $\mathcal{G}$ leaves $\pi_{\mathcal{F}}$ invariant and acts 3-transitively on its components.

Thus, recalling that the members of $\pi_{\mathcal{F}}$ are disjoint from $y = xT$, we see that the orbit $(y = xT)\mathcal{G}$ is a partial spread such that its members all have trivial intersection with the members of $\pi_{\mathcal{F}}$.

Now specialize to the case $\mathcal{F} = GF(q)$ and $dim_{\mathcal{F}}W = 3$. The partial spreads $\pi_{\mathcal{F}}$ and τ together contribute $q + 1 + q(q^2 - 1) = q^3 + 1$ components of the partial spread $\pi(T, \mathcal{F})$, and this is the size needed to make it into a spread. Since the $\mathcal{G}$-orbit τ now has the size of $\pi(T, \mathcal{F}) \setminus \pi_{\mathcal{F}}$, we conclude that $\mathcal{G}$ is transitive on the components of the spread outside $\pi_{\mathcal{F}}$. This spread is coordinatized by a spread set $\mathcal{S} \supset \mathcal{F} \cup \{T\}$ that includes the identity and yet $\mathcal{S}$ is not a field because T does not centralize $\mathcal{F}$. The slope set of $\pi(T, \mathcal{F})$ is clearly in $\mathrm{Hom}(W, +)$, so its kernel is as claimed.

By varying T, for a fixed choice of $\mathcal{F}$, it is possible to ensure that the dimension of the spread $\pi(T, \mathcal{F})$, over its kernel, can be made arbitrarily large. In particular, this means that non-Desarguesian translation planes of order q^3 that admit $SL(2, q)$ can be chosen to have arbitrarily large dimension.

14.1. Known $GL(2, q)$-Planes of Order q^3.

Prior to the cyclic semifield connection, there were several known classes of generalized Desarguesian planes.

There are as follows:

(1) **The Kantor planes of order q^6 and kernel $GF(q^3)$ admitting $GL(2, q^2)$ (Kantor [862, 864]).**

(2) **There are similar planes of Liebler [922].**

(3) **Also, there are the planes of Bartolone and Ostrom [102].**

All of these planes arise from cyclic semifields of order h^3 and may be referred to as 'Jha–Johnson $GL(2, h)$, h^3-planes'.

Hence, we now may add

(4) **The $GL(2, h)$, h^3 planes arising from the Jha–Johnson cyclic semifields of type $S(\omega, m, n)$, of p-primitive type 1, or of q-primitive type 2.**

CHAPTER 15

Cone Representation Theory.

The theory of cone representations of geometries may be viewed an alternative model to representations using orthogonal arrays. This model is due to D. Glynn in a series of articles [**402, 403, 404, 405**]. Although not mined thoroughly, this is a very general representation method, which should provide significant constructions and interconnections. In particular, there are interconnections with cyclic semifields of order q^3 and kernel containing $GF(q)$, their associated generalized Desarguesian planes with kernels containing $GF(q)$ and certain cone representations.

First we consider a finite affine plane π of order N. Since we are primarily interested in translation planes, we assume that N is a prime power q^i, where $q = p^z$, for p a prime and z a positive integer. The affine plane π has q^{2i} points, $q^i + q^{2i}$ lines and $q^i + 1$ parallel classes. Consider the points of the Desarguesian affine plane of order q^i coordinatized by $GF(q^i)$. Choose any two parallel classes (∞) and (0) with lines $x = c$, $y = d$, for $c, d \in GF(q^i)$. Represent the other lines by a ternary function $y = T(x, m, b)$, where the parallel classes are designated by (m) and (∞) for $m \in GF(q^i)$. Hence, $y = T(x, m, b)$ is the unique line incident with the point $(0, b)$ in the parallel class (m).

Fix a parallel class (m) and form a $q^i \times q^i$-matrix such that the (k, j)-entry is t if and only if $T(k, m, t) = j$, where context determines whether k and j represent integers or elements of $GF(q^i)$. That is, we form a bijection between the elements of $GF(q^i)$ and the integers $0, 1, \ldots, q^i$, and use the integer context when considering dimensions and entries of the matrix. In this way, we determine both a 'Latin square' and/or a $q^i \times q^i$-matrix with entries from $GF(q^i)$, such each row and column contains all elements of $GF(q^i)$. We can construct such matrices for each of the $q^i - 1$ parallel classes (m), $m \neq 0$. As matrices, if we choose any two Latin squares, and superimpose these, we will obtain all possible 2-vectors (a, b) for $a, b \in GF(q^i)$.

Now consider these Latin squares as q^{2i}-row vectors with entries from $GF(q^i)$. Then we will have a $(q^i - 1) \times q^{2i}$-matrix such that any two rows will have all possible 2-vectors $\begin{bmatrix} a \\ b \end{bmatrix}$, $a, b \in GF(q^i)$, in the set of columns of the two rows. Actually, viewed in this way, we may also attach the following matrices

$$
\begin{bmatrix}
a_1 & a_1 & . & . & a_1 \\
a_2 & a_2 & . & . & a_2 \\
. & . & . & . & . \\
. & . & . & . & . \\
a_{q^i} & a_{q^i} & . & . & a_{q^i}
\end{bmatrix}
,
\begin{bmatrix}
a_1 & a_2 & . & . & a_{q^i} \\
a_1 & a_2 & . & . & a_{q^i} \\
. & . & . & . & . \\
. & . & . & . & . \\
a_1 & a_2 & . & . & a_{q^i}
\end{bmatrix}
,
$$

to obtain a $(q^{2i} + 1) \times q^{2i}$ matrix with this same property. This is called an 'orthogonal array $OA(q^i + 1, q^{2i})$'. Conversely, it is straightforward to reverse this procedure and construct from an orthogonal array with these parameters, an affine plane of order q^{2i}. More generally, a net of degree k constructs an orthogonal array $OA(k, q^{2i})$.

Similarly, for Laguerre planes, an analogous construction will construct similarly orthogonal arrays, except that 2 is replaced by 3.

We consider generalizations of these concepts.

DEFINITION 15.1. Let S be a set of cardinality g and let k, λ, s be positive integers, $k \geq 3$, $s \geq 2$, $\lambda \geq 1$. An orthogonal array $OA(g, k, s, \lambda)$ of 'order g, degree k , index λ and strength s' is a $k \times g^2\lambda$-matrix with entries from S such that every s rows admit every possible $s \times 1$ column vector over S exactly λ times.

In this chapter, we shall assume that $\lambda = 1$, and S is $GF(q^i)$. We shall only be concerned with $s = 2$, for which orthogonal arrays corresponds to nets of degree $k + 2$.

We begin with an $A = OA(q^i, k, s)$ (alternating the notation for $\lambda = 1$). For each row of q^{2i} elements of $GF(q^i)$, consider each element as an i-vector over $GF(q)$. Basically, we choose a basis for $GF(q^i)$ over $GF(q)$. We may choose different bases for each row of A. For each j of the possible i-components of each i-vector, let A_j denote the associated matrix with entries the jth component of each element of the matrix A. In this way, there are i-matrix constructed, each of which is of dimension $k \times q^{2i}$.

Now assuming that each element of the matrix A is an i-vector over $GF(q)$, let the rank of A be r. This is the same as the dimension of the subspace of a k-dimensional vector space V_k over $GF(q)$, generated by the iq^{2i} columns of the set of matrices A_j for $j = 1, 2, \ldots, i$.

Factor: Factor $A_j = XY_j$, where X is of dimension $k \times r$, Y_j is of dimension $r \times q^{si}$ and both of rank r and the columns of X generate a subspace of V_k over $GF(q)$ of dimension r. To associate an orthogonal array $OA(q^i, k, s)$ to a cone in $PG(n - 1, q)$, we form a vector space of dimension $n = r + i$ as follows. Define the vector space $V_n = V_r \oplus V_i$ as the set of vectors (x, y), such that x in V_r and y in V_i, vector spaces of dimensions r and i over $GF(q)$.

DEFINITION 15.2. The 'vertex V' isomorphic to V_i of the cone will be the i-dimensional vector space $x = 0$. The 'base space B' isomorphic to V_r is the r-dimensional space $y = 0$. The 'base curve' P is the set

$$\{(x_t, 0); x_t \text{ is the } t\text{th row of } X; t = 1, 2 \ldots, k.\}.$$

The 'generator spaces' are the $i + 1$-dimensional spaces

$$V_{i+1}^t = V \oplus \langle (x_t, 0) \rangle,$$

for $i = 1, 2, \ldots, k$. The 'cone with vertex V' over the base curve P is the set of points of V_n taking projectively on the generator spaces.

Consider a standard cone over an oval in $PG(3, q)$. Choose an oval in a plane and choose an exterior point V called the vertex. For the set of generator lines by joining V to the points of the oval. The set of points on these lines is called an 'oval cone'. If we think of the base curve in place of the oval and the vertex (space) in place of the point vertex, and the generator spaces in place of the generator lines, we have a mental image of this generalization.

For oval cones, we are interested in covering the non-vertex points of the cone by ovals of intersection that lie in planes of intersection. In the generalization, we consider r-dimensional vector spaces disjoint from the vertex space V, which cover the non-vertex cone points, such that given any s non-vertex points on s different generator spaces lie on exactly one intersection r-space.

DEFINITION 15.3. For each $w = 1, 2, \ldots, q^{si}$, form a matrix B_w by taking the j-th column of B_w as the w-th column of Y_j in V_r over $GF(q)$. Hence, B_w is of dimension $r \times i$. Now form $B_w^+ : y = xB_w$, as x varies over the r-vectors of V_r over $GF(q)$, as the set of points $(x, y) \in V_r \oplus V_i$. The subspaces B_w^+ are called the 'intersection spaces' and are r-dimensional $GF(q)$-subspaces of $V_r \oplus V_i$, disjoint from V, since Y_j is of rank r.

REMARK 15.4. The vector in the (t, w)-entry of A is $x_t B_w$ as an i-vector. The point $(x_t, x_t B_w)$ is on the generator space and on $B_w^+ \equiv y = xB_w$. Take s rows of

$$A, t_1, t_2, \ldots, t_s, \text{ choose any } s \times 1 \text{ column vector } \begin{bmatrix} a_1 \\ \cdot \\ a_s \end{bmatrix} \text{ of } A. \text{ Each } a_i \text{ is an } i\text{-vector}$$

over $GF(q)$, there exists a unique w between 1 and q^{si} such that a_i is the entry in the (t_i, w)-entry. Hence, there is a unique intersection space $y = xB_w$ that contains these s vectors.

Hence, we obtain the following theorem.

THEOREM 15.5. (Glynn [403, 1.4]). Orthogonal arrays $OA(q^i, k, s)$ are equivalent to cones in $PG(n-1, q)$, $n = r + i$, with vertex a $PG(i-1, q)$, and k generator spaces $PG(i, q)$'s containing V, which is cut by q^{si} intersection spaces $PG(r-1, q)$, with the property that each s non-vertex points on s different generator spaces are on a unique intersection space.

DEFINITION 15.6. Any cone structure defined as in the previous theorem is called a 'cone representation' of the associated geometry determined by the orthogonal array. If $n = r + i$, the cone representation is said to be of 'rank r' and 'index i'.

15.1. Cone Representations of Affine Planes.

Now consider orthogonal arrays with strength $s = 2$ and $k = q^i - 1$, which correspond to affine plane of order q^i. The associated cone representation of rank r of the $(q^i + 1) \times q^{2i}$ matrix has a base curve of $q^i + 1$ points $(x_t, 0)$ and this set of points is a subset of the base space of vector dimension r. Hence, $q^i + 1 \le q^{r-1} + q^{r-2} + \cdots + q + 1$, so that $r \ge i + 1$.

REMARK 15.7. Assume that $r = i + 1$. Choose any point D not on the base curve but on the base space of dimension $i + 1$ and assume that no line (2-dimensional vector space) incident with D is tangent to the base curve. Adjoin D to the base curve. This does not change the intersections subspaces, but does add a parallel class to the affine plane considered as a net. Hence, any such point D lies on a tangent line to the base curve.

LEMMA 15.8. If $i = 2$, then the base curve is $AG(2, q)$ union a point.

PROOF. The base curve contains $q^{i=2} + 1$ $r = 3$-vectors in a 3-dimensional vector space. Hence, the base space, projectively is $PG(2, q)$, a set of $1 + q + q^2$

points containing the $q^2 + 1$ points of the base curve. Take any point D of the base space not on the base curve. Then there is a tangent to the base curve C incident with D. The lemma now follows. $\square$

Assume that there are two distinct intersection vector subspaces of dimension $i + 1$ that intersect in a i-dimensional vector subspace R. Note that any two distinct intersection subspaces in a $n = r + i = 2i + 1$ certainly non-trivially intersect. The following lemma is immediate.

LEMMA 15.9. *The projection of R from the vertex to the base is a vector subspace T of vector dimension i which is tangent to the base curve.*

PROOF. In the vector setting R is a subspace of $V \oplus B$, where V is the vertex and B is the base space. Since R is disjoint with V and every element r of R is a sum $v + b$ of $V \oplus B$, it follows that the set $\{b \in B; v + b \in R, \text{ for some } v \in V\}$ is an i-dimensional subspace T, tangent to the base curve. $\square$

$V \oplus T$ is a vector subspace of dimension $2i$. Any pair of intersection subspaces, which then intersect $V \oplus T$ non-trivially will then intersect in a i-dimensional $GF(q)$ subspaces and these intersections are mutually disjoint. It then follows that the subspace intersections together with V produce a spread of $V \oplus T$ (Glynn [**403**, p. 798]). Hence, we obtain the following result. We give the vector-space version of the theorem.

THEOREM 15.10. (Glynn [**403**, Theorem (2.2) and following]). *Assume that there is a cone representation of an affine plane of order q^i of rank $i + 1$ and index i. If there exist two intersection subspaces that intersect in an i-dimensional vector subspace R of the ambient $2i + 1$ vector space, then the affine plane is a translation plane. If V is the vertex of the cone and T is the projection of R from V to the base space B, then there is a spread induced on the $2i$-dimensional $GF(q)$-vector space $V \oplus T$. The spread is the i-space intersections by pairs of intersection subspaces union V. Hence, translation planes of order q^2 with spreads in $PG(3, q)$ correspond to cone representations of rank 3 and index 2.*

As noted in Chapter 13 on cyclic semifields of Jha–Johnson of order q^n, when $n = 3$, there are associated generalized Desarguesian planes of order q^3 admitting $GL(2, q)$, acting canonically. However, note that the kernel of these spreads, cyclic of order q^3 and generalized Desarguesian can be chosen in a variety of ways. In particular, the kernel can be considerably smaller that $GF(q)$. However, when the kernel is $GF(q)$, there are translation planes due to Ostrom and Bartolone which again may be constructed using cyclic semifields of order q^3. In any case, in the special case of cyclic semifields of order q^3 and left nucleus $GF(q)$ and the accompanying generalized Desarguesian planes, both show up in terms of cone representations of rank 4 and index 3. However, it is the dual of the projective extension of the generalized Desarguesian plane that appears in this cone representation, and furthermore this cone representation is not a spread representation, even in the Desarguesian case (which is both a cyclic semifield plane and a generalized Desarguesian plane).

THEOREM 15.11. (Glynn [**405**]). *A cone representation of rank 4 and index 3 of a projective plane produces either a spread representation or exactly one non-spread representation.*

Indeed, the non-spread representation is of interest. Glynn shows that a projective plane π of order q^3 with the non-spread cone representation of rank 4 and index 3 has a certain set L of $q+1$ lines incident with a point W. There is a collection of $q^6 + q^5 + q^4$ projective subplanes of order q each of which contains V and every line of L. These subplanes form in a well-defined manner the structure of $PG(6, q)$, corresponding to almost all of the hyperplanes of $PG(6, q)$. Furthermore, there is a construction and characterization of a dual translation planes admitting this cone representation.

If the set L incident with a point W in a intersection space isomorphic to $PG(3, q)$, then associated with the cone representation is a fixed-point-free collineation T of $PG(2, q)$, which then lies in $\Gamma L(3, q)$. Furthermore, it turns out that the associated dual translation plane of order q^3 admits $GL(2, q)$. Actually, once it is known that there is an associated fixed-point-free collineation, the constructions of Jha and Johnson apply as follows. Consider $GL(2, q)$ acting canonically on an affine Desarguesian plane of order q^2 as

$$\left\{ (x, y) \longmapsto \begin{bmatrix} a & b \\ c & d \end{bmatrix} (x, y); a, b, c, d \in GF(q); ab - bc \neq 0 \right\}.$$

We know that T has the property that $\beta T = T \beta^\sigma$, where $\sigma \in \operatorname{Gal} GF(q)$. In this case, the subgroup of $GL(2, q)$ which fixes $y = xT$, considering x as a k-vector, for suitable k, is

$$\left\{ \begin{bmatrix} a & 0 \\ 0 & a^{-\sigma} \end{bmatrix} a \in GF(q)^*; \right\},$$

of order $q-1$. Then it follows that $(y = xT)GL(2, q)$ is a partial spread of cardinality $q(q^2 - 1)(q - 1)/(q - 1) = q^3 - q$. It is now not difficult to verify that

$$\{x = 0, y = x\beta; \beta \in GF(q)\} \cup (y = xT)GL(2, q)$$

is a translation plane of order q^3 and admitting $GL(3, q)$, which is now recognizable as a generalized Desarguesian plane. Actually, the construction of such planes may be done more generally by taking a fixed-point-free collineation T in $\Gamma L(3m, p)$, which normalizes but does not centralize a field K isomorphic to $GL(q)$ contained in $GL(3n, p)$, where $q = p^n$, p a prime. We note that it then follows that

$$\{y = x(\beta_0 + \beta_1 T + \beta_2 T^2); \beta_i \in GF(q); i = 0, 1, 2\}$$

is a semifield spread (a cyclic semifield of Jha and Johnson).

When $T \in \Gamma L(3, q)$, consider the set $\mathcal{F} = \{x = 0, y = x\beta; \beta \in GF(q)\}$. The parallel classes of $\mathcal{F}$ is a set of $q + 1$ points isomorphic to $PG(1, q)$. Also, the set $\mathcal{F}$ contains at least two Baer subplanes which are $GF(q)$-subspaces. Consider the dual translation plane where the line ℓ_∞ becomes the point W above. There we have a set L of $q+1$ lines incident with W with the property that there are at least two transversal lines.

THEOREM 15.12. (Glynn [**405**]). A dual translation plane π of order q^3 having the second kind of cone representation of rank 4 and 3, where the set of $q+1$ special lines in a intersection space $PG(3, q)$ has at least two transversal lines is associated with a fixed-point-free collineation T of $PG(2, q)$. This collineation and $GL(2, q)$ completely determines π.

Glynn also points out that at any point of the dual translation plane π not on one of the $q + 1$ lines of L, a semifield plane of order q^3 may be constructed. So, we have cone representation of the duals of generalized Desarguesian planes.

CHAPTER 16

André Net Replacements and Ostrom–Wilke Generalizations.

André translation planes may be regarded as a far-reaching generalization of the Hall planes and the Desarguesian planes. They exist for any strict prime power $n = p^d > p$, and although they are never (proper) semifield planes, the class of of André planes overlaps the class of nearfield planes—neither class includes the other. Thus, for more than two decades the André planes included the only known class of projective planes (up to duality) of non-square order that are not nearfield or semifield planes. This remained the situation until Foulser in 1967 [**369**] constructed his λ-planes, which include all the André planes as well as all the regular nearfield planes. We discuss Foulser's λ-planes in Chapter 17 below. The λ-planes, like all André and nearfield planes, are disjoint from the class of non-Desarguesian planes.

An important distinction between André planes and the λ-planes generalizing them is that whereas the André planes are completely known, there remain potentially interesting λ-planes to discover. But there is a well-developed theory for λ-planes, including many construction techniques, that we describe in Chapter 17.

The aim of the present chapter is to focus on André nets in Desarguesian spreads rather than the corresponding André planes themselves. The André planes arise by making standard or André replacements of pairwise disjoint André collection nets in a Desarguesian spread. Thus, André nets generalize derivable nets and André planes are the planes obtained by taking arbitrary André replacements of pairwise disjoint generalized André planes.

Using Ostrom–Wilke modifications of this process, we obtain further replaceable nets and further translation planes of 'multiple-André' type. These turn out to be examples of λ-planes, more or less by definition, and hence contribute to the theory of λ-planes. However, in this chapter, we consider these constructions with only minimal reference to λ-planes; cf. Chapter 17 for a detailed investigation of λ-planes. We end the present chapter by drawing attention to a putatively powerful generalization of the replacement techniques, due to Ostrom, that has great potential for constructing finite planes.

DEFINITION 16.1. Let Σ_F be the standard Desarguesian spread on $F \oplus F$, $F = GF(q^n) > GF(q) = K$, with component set $\{ y = fx : f \in F \} \cup \{(x = 0)\}$. Then the subnet $\mathcal{A}_\alpha := \{ y = xm : m^{(q^n-1)/(q-1)} = \alpha \}$, $\alpha \in GF(q)^*$, is an **André net**, or an **André q-net**.

The collection of subspaces $\mathcal{A}_\alpha^\rho := \{ y = x^\rho m : m^{(q^n-1)/(q-1)} = \alpha \}$, for any $\rho \in \mathrm{Gal}(F/K)$, forms a net replacement for the André net $\mathcal{A}_\alpha$, and is called an **André replacement** for $\mathcal{A}_\alpha$, or an **André ρ-replacement** for $\mathcal{A}_\alpha$; if $\rho \colon x \mapsto x^{q^r}$, then $\mathcal{A}_\alpha^{q^r}$, or $\mathcal{A}_\alpha^\rho$, denotes the André ρ-replacement.

We summarize some immediate consequences of the definitions:

REMARK 16.2. Assume the notation of Definition 16.1, and let $C_{\frac{q^n-1}{q-1}}$ be the cyclic subgroup of order $\frac{q^n-1}{q-1}$ in F^*. Then:

(1) For each $\alpha \in GF(q)^*$, the André net $\mathcal{A}_\alpha$ is one of the $q-1$ cosets of the group $C_{\frac{q^n-1}{q-1}}$ in the multiplicative group F^*.

(2) The map $(x,y) \mapsto (x^\rho, y)$ is a net isomorphism from the André net $\mathcal{A}_\alpha$ to its André replacement $\mathcal{A}_\alpha^\rho$.

(3) The scalar group on $F \oplus F$, $\{ \hat{f} : x \oplus y \mapsto fx \oplus fy : f \in F^* \}$, leaves invariant each André net $\mathcal{A}_\alpha$, and also each André replacement $\mathcal{A}_\alpha^\rho$.

Continuing with the notation of Definition 16.1, we define the André spreads on $F \oplus F$ as the spreads obtained obtained by taking André replacements of some collection of André nets, with common size $(q^n-1)/(q-1)$, in the Desarguesian spread Σ_F: this procedure always yields a legitimate spread since distinct André nets of fixed size are pairwise disjoint, Remark 16.2(1). (If convenient, we assume α runs over all $GF(q)^*$—unchanged $\mathcal{A}_\alpha$'s are self-replaced.) In 'λ-notation':

DEFINITION 16.3. (André Planes). Let Σ_F be the Desarguesian spread on $F \oplus F$, $F = GF(q^n) > GF(q) = K$. Let $\mathcal{A}_q := \{ \mathcal{A}_\alpha : \alpha \in K^* \}$ be the set of André q-nets in Σ_F, and let $\lambda \colon K^* \mapsto \mathrm{Gal}(F/K)$ be any mapping. Then $\Sigma_F^\lambda := \{ \mathcal{A}_\alpha^{\lambda(\alpha)} : \alpha \in K^8 \}$, the spread obtained by replacing each André net $\mathcal{A}_\alpha$, $\alpha \in GF(q)^*$, by $\mathcal{A}_\alpha^{\lambda(\alpha)}$, is an **André spread** and the corresponding translation plane is an **André plane**, more precisely, an **André q-plane**.

Since André net replacements are invariant under the kernel homologies of the underlying Desarguesian spread, Remark 16.2(3), we have:

COROLLARY 16.4. Let Σ_F be the standard Desarguesian spread on $F \oplus F$, $F = GF(q^n)$. Then any associated André plane on $F \oplus F$ admits as a collineation group the full kernel homology group of order $q^n - 1$ of the underlying Desarguesian spread Σ_F.

Thus, non-Desarguesian André spreads are invariant under a field F not contained in its kernel; the component orbits of F^* are called 'fans'. We shall explore fans in detail in several parts of this book.

First, we consider generalizations of André planes based on André replacements of André nets of *different* sizes. André replacements of pairwise disjoint André nets of different sizes may be used to construct planes that are not André planes. Large numbers of such nets may be constructed by the recursive application of the following principle.

LEMMA 16.5. If d is any divisor of n, then the André q-nets, Definition 16.1, in the Desarguesian spread Σ_F, $F = GF(q^n) > GF(q) = K$, are partitioned by the André q^d nets in Σ_F.

The partition of any André q-net A_i, $1 \leq i \leq (q^n-1)/(q-1)$, into a collection of $(q^n-1)/(q^d-1)$ André q^d-nets will be called an **André partition of** A_i; the trivial case $d = 1$ permitted.

PROOF. If $A > B > C$ is a chain of Abelian groups, then the A-cosets of C refine the partition of A by the A-cosets of B. The result follows, since André nets of a fixed size may be identified with the cosets of a cyclic subgroup of F^*.   □

The lemma legitimizes the following *recursive* construction of multiple André partitions of *varying sizes*, hence simultaneously replacing them by their André replacements yields further planes.

CONSTRUCTION 16.6. Let Σ_F be the Desarguesian spread associated with $F = GF(q^n)$, $n > 1$. A **multiple André partition** of an André q-net $A := A_q < \Sigma_{F'}$ is an André partition of $A < \Sigma_F$, cf. Lemma 16.5, or the partition of A into André nets obtained by applying arbitrary (possibly trivial) André partitions to every André net in a multiple André partition of A_q.

A multiple André replacement A', of an André net A, is the net obtained by applying arbitrary (and possibly trivial) André replacements, Definition 16.1, to every André net in a multiple André partition of an André net.

A **multiple André plane** is defined to be one of the André planes obtained by multiply replacing some of the André q-nets in Σ_F.

Since the multiple André nets are pairwise disjoint unions of André nets of varying sizes, the multiple André planes are André planes in the 'degenerate' case when all the replaced André nets replaced have the same size. However, in all other cases, which are clearly far more frequent, the associated multiple André planes are guaranteed to be non-isomorphic to André planes by a result of Foulser, Corollary 17.28. However, all multiple André planes are by definition instances of the λ-planes, cf. Theorem 17.3.

The idea of multiple André replacements is due to Ostrom: see [**1046**], where he provides a variety of ways to construct mutually disjoint André replacement nets of different sizes. Now relative to our general problem on the determination of replacements of a given replaceable net, we formulate the specific problem that we shall consider.

PROBLEM 16.7. Find replacements of André nets that are not André-type replacements.

We now describe a class of non-André nets that replace André nets due to Wilke. The principle underlying Wilke's constructions has been generalized by Ostrom to provide powerful techniques for the construction of replaceable nets in finite affine planes.

The Wilke Replacements. The following elementary principle enables non-André nets to be constructed from André nets. It is also the basis for many net replacement techniques due to Ostrom.

REMARK 16.8. Let $B > C$ be partial spreads on a vector space V. Suppose $\sigma, \tau \in GL(V, +)$ are such that B^σ and C^τ are replacements for B and C, respectively. Then B admits as a replacement: $(B^\sigma \setminus C^\sigma) \cup (C^\sigma)^{\sigma^{-1}\tau\sigma}$, hence if τ and σ commute then $(B^\sigma \setminus C^\sigma) \cup (C^\sigma)^\tau$ is a replacement for B.

PROOF. Let $[X]$ denote the points covered by any partial spread. Thus

$$([C]^\sigma)^{\sigma^{-1}\tau\sigma} = ([C])^{\tau\sigma} = [C]^\sigma < [B]^\sigma.$$

Hence $([C]^\sigma)^{\sigma^{-1}\tau\sigma} < B^\sigma$ is a replacement for $C^\sigma < B^\sigma$, and the result follows. □

This has an obvious recursive extension. Moreover, the linear maps τ and σ commute if they are automorphisms for some V-based field $F = (V, +, \cdot)$. Thus, we clearly have:

THEOREM 16.9. (Wilke Replacement-Nets; Wilke [**1213**]). For a strictly decreasing sequence of fields

$$F = GF(q^n) \succ F_0 - CF(q^{d_0}) \succ F_1 = GF(q^{d_1}), \ldots, F_k = GF(q^{d_k}), \qquad k \geq 1,$$

let $C_0 < C_1 < \cdots < C_k$ be the corresponding decreasing chain of multiplicative subgroups of F^* such that $|C_i| = (q^n - 1) / (q^{d_i} - 1)$, for all i. Then the collection of subspaces of the standard Desarguesian spread Σ_F, on $F \oplus F$, specified by

$$
y = \begin{cases}
x^{q^{d_k}} m : m \in C_k \setminus C_{k-1}; \\
\vdots \\
x^{q^{d_{i+1}}} m : m \in C_{d_i+1} \setminus C_{d_i}; \\
\vdots \\
x^{q^{d_1}} m : m \in C_1 \setminus C_0; \\
x^{q^{d_0}} m : m \in C_0
\end{cases}
$$

forms a replacement for the André net $\{\, y = xm : m^{\frac{q^n - 1}{q^{d_0} - 1}} = 1 \,\}$. The translation planes obtained by applying such replacements are the **Wilke planes**.

The Wilke planes turn out to be λ-planes but are never André planes, cf. Corollary 17.29. We remark there are several variants of the Wilke planes, for instance, the planes obtained by applying Wilke replacements to André nets that are pairwise disjoint and not necessarily of the same size. As indicated already, Ostrom greatly extended the scope of Wilke's idea. We consider this now.

16.1. Ostrom's Replacement Theorems.

The ideas for the general replacements in the section on multiple André replacement are due to Ostrom [**1046**]. In this section, we prove an extremely general replacement theorem, which incorporates essentially all of the replacements previously listed. The idea is as follows: Consider an André net $A_\alpha = \{\, y = xm;\ m^{(q^n-1)/(q-1)} = \alpha \,\}$ in a Desarguesian affine plane of order q^n, for q a prime power p^r. Choose any integer i, $1 \leq i \leq n-1$. Then the mapping $\sigma_i \colon (x, y) \longmapsto (x^{q^{n-i}}, y)$, maps $y = xm$ to $y = x^{q^i} m$ and hence transforms the André net into one of the basic $n-1$ André replacements $A_\alpha^i = \{\, y = xq^i m;\ m^{(q^n-1)/(q-1)} = \alpha \,\}$. Now if σ_j is applied to A_α^i, we simply obtain A_α^{i+j}, another replacement (reducing $i + j$ modulo n).

Now suppose we start with an André net A_α and apply any replacement procedure we wish on A_α to construct say the net A_α^*. Then consider $A_\alpha^* \sigma_i$, we claim that this is another replacement for A_α. To see this, we note that $A_\alpha \sigma_i$ is an André net and a replacement for A_α. Two points P and Q joined in A_α by a line, map to points $P\sigma_i$ and $Q\sigma_i$ joined by a line of $A_\alpha \sigma_i$. But, P and Q are also joined by a line of A_α^*, which implies that $P\sigma_i$ and $Q\sigma_i$ are also joined by a line of $A_\alpha^* \sigma_i$. By symmetry, $A_\alpha^* \sigma_i$ replaces the original André net. Of course, when A_α^* is another André replacement net, we simply obtain another André replacement; however, in the more general case, we obtain a potentially quite different replacement. We note that the mappings considered σ_i are not collineations of the associated Desarguesian affine plane, so while the nets A_α^* and $A_\alpha^* \sigma_i$ are isomorphic as to themselves, sitting within a translation plane, the existence of these nets will usually force the translation planes to be non-isomorphic. Of course, such constructions may be done independently on various André nets. All of this generalizes for arbitrary mappings τ and arbitrary nets N such that $N\tau$ becomes a replacement for N.

Ostrom's 1st replacement theorem gives the appropriate generalization.

16.1.1. Ostrom's 1st Replacement Theorem.

THEOREM 16.10. (Ostrom [**1046**, (3.17)]). Let N be a finite net $N = \bigcup_{i=1}^{k} N_i \cup M$, where the notation indicates nets on the same affine points with disjoint partial spreads.

(1) If τ_i is a mapping on points such that $N\tau_i$ is a replacement for N_i and τ is a mapping on points such that $N\tau$ is a replacement for N, then $\bigcup_{i=1}^{k} N_i\tau_i\tau \cup M\tau$ is a replacement for N.

(2) If N_i' is a replacement for N_i and $N\tau$ is a replacement for N, then $\bigcup_{i=1}^{k} N_i'\tau \cup M\tau$ is a replacement for N.

PROOF. The proof is essentially immediate from our previous discussion. Note that (2) implies (1) and allows André mapping replacements, Johnson generalized replacements and Foulser replacements. □

We now provide a slightly different interpretation of certain replacements. Consider the André net

$$A_1 = \left\{ y = xm; \, m^{(q^n-1)/(q-1)} = 1 \right\}.$$

Suppose that k_1 divides n. Decompose A_1 into $(q^{k_1}-1)/(q-1)$ smaller André nets of degree $(q^n-1)/(q^{k_1}-1)$. Now basically we are considering cosets of $GF(q)^*$ in $GF(q^{k_1})^*$, as there are $(q^{k_1}-1)/(q-1)$ such cosets. Suppose that we replace all of these coset André nets of degree $(q^{k_1}-1)/(q-1)$ '**except**' the coset $GF(q)^*$ by components of the form $y = x^{q^{k_1 t_1}} m$. What this means is that if $m^{(q^n-1)/(q^{k_1}-1)} \neq 1$, we use the replacement as above, where $f_1(k_1)$ is a fixed integer. This amounts to saying that if $m \in GF(q^n)^*/GF(q)^*$ but not in $GF(q^n)^*/GF(q^{k_1})^*$, we use the replacement as above. In general, use the notation $G_i = GF(q^n)^*/GF(q^{k_i})^*$. Note that G_i is a cyclic group of order $(q^n-1)/(q^{k_i}-1)$, and if $k_i \mid k_{i+1}$, with $k_0 = 1$, and since $(q^n-1)/(q^{k_{i+1}}-1)$ then divides $(q^n-1)/(q^{k_i}-1)$, we may consider that $G_{i+1} \subset G_i$. We now write A_1^{q-1} as $R_1 \cup A_1^{q^{k_1}-1}$, where the notation is to indicate that $A_1^{q^{k_1}-1}$ is the André subnet that is unreplaced above. Now partition $A_1^{q^{k_1}-1}$ as $R_2 \cup A_1^{q^{k_2}-1}$.

Let $N_3 = A_1^{q^{k_2}-1}$ and let $N_3\sigma_3$ replace N_3, where σ_3 is any corresponding André replacement mapping. Let $N_2 = R_2 \cup A_1^{q^{k_2}-1}$ and note that $N_2 - N_3 = R_2$. Considering for a moment that R_2 can be replaced by components of the form $y = x^{q^{k_1 t_1}} m$, this means that N_2 is a replacement for $A_1^{q^{k_1}-1}$, so if σ_2 is a corresponding replacement mapping for the k_1-André net then $N_2\sigma_2 = (N_2 - N_3)\sigma_2 \cup N_3\sigma_3\sigma_2$ replaces the k_1-André net. Let $N_1 = R_1 \cup A_1^{q^{k_1}-1}$ is a replacement for A_1^{q-1}, so if σ_1 is a corresponding replacement mapping for the 1-André net then $N_1 = (N_1 - N_2) \cup N_2$ has replacement $(N_1 - N_2) \cup (N_2 - N_3)\sigma_2 \cup N_3\sigma_3\sigma_2$, so the original André net has replacement $(N_1 - N_2)\sigma_1 \cup (N_2 - N_3)\sigma_2\sigma_1 \cup N_3\sigma_3\sigma_2\sigma_1$.

As indicated earlier, this idea is due to Wilke [**1213**], cf. Theorem 16.9. These Wilke translation planes are generalized by Ostrom in [**1046**]. Generalizing of these ideas yields the following theorem.

16.1.2. Ostrom's 2nd Replacement Theorem.

THEOREM 16.11. (Ostrom [**1046**]). Let N_i, $i = 0, 1, 2, \ldots, t$, be a finite net such that $N_t \subset N_{t-1} \subset N_{t-2} \subset \cdots \subset N_0$. Assume that $N_j \sigma_j$ is a replacement for N_j. Then

$$N_t^{\sigma_t \sigma_{t-1} \cdots \sigma_1 \sigma_0} \cup (N_{t-1} - N_t)^{\sigma_{t-1} \sigma_{t-2} \cdots \sigma_1} \cup (N_{t-2} - N_{t-1})^{\sigma_{t-2} \cdots \sigma_1} \cup \cdots \cup (N_1 - N_2)^{\sigma_1}$$

is a replacement for N_1.

This theorem has really never been mined for the most general form and as such has been essentially overlooked. However, this is really an omnibus theorem, from which a tremendous number of different translation planes may be constructed using Ostrom's 2nd replacement. We offer here a few applications illustrating the many variations of replacements of André nets. We call the first theorem the 'all except' theorem, reflecting the ideas of the previous analysis.

THEOREM 16.12. ('All Except Theorem'). Consider any field $GF(q^n)$, where q is a prime power p^r. Let $\{1, k_1, k_2, \ldots, k_t\}$ be a set of divisors of n, where $k_i \mid k_{i+1}$, for $i = 0, 1, 2, \ldots, t$, and let $k_0 = 1$. Define $\sigma_i := \sigma_i^{s_i} : (x, y) \longmapsto (x^{q^{n-k_i s_i}}, y)$, where s_i is an integer $1 \leq s_i < n/k_i$.

Consider the André net $N_i = A_1^{k_i} = \left\{ xm;\ m^{(q^n-1)/(q^{k_i}-1)} = 1 \right\}$, $i = 0, 1, \ldots, t$.

Note that $N_i \sigma_i^{s_i}$ replaces N_i and that $N_t \subset N_{t-1} \subset N_{t-2} \subset \cdots \subset N_0$. Then we obtain the following spread, which is a generalized André spread:

$$x = 0, y = 0,$$

$$y = x^{q^{k_t s_t + k_{t-1} s_{t-1} + \cdots + k_1 s_1 + s_0}} m;\ m^{\frac{(q^n-1)}{(q^{k_t}-1)}} = 1,$$

$$y = x^{q^{k_{t-1} s_{t-1} + \cdots + k_1 s_1 + s_0}} m;\ m^{\frac{(q^n-1)}{(q^{k_{t-1}}-1)}} = 1,\ m^{\frac{(q^n-1)}{(q^{k_t}-1)}} \neq 1,$$

$$y = x^{q^{k_{t-2} s_{t-2} + \cdots + k_1 s_1 + s_0}} m;\ m^{\frac{(q^n-1)}{(q^{k_{t-2}}-1)}} = 1,\ m^{\frac{(q^n-1)}{(q^{k_{t-1}}-1)}} \neq 1,$$

$$\vdots$$

$$y = x^{q^{s_0}} m;\ m^{\frac{(q^n-1)}{(q-1)}} = 1,\ m^{\frac{(q^n-1)}{(q^{k_1}-1)}} \neq 1.$$

PROOF. Starting from the last listed subspace, we noted that this considers the largest André net of degree $\frac{(q^n-1)}{(q-1)}$, decomposes this net into $\frac{(q^{k_1}-1)}{(q-1)}$ André nets of degree $\frac{(q^n-1)}{(q^{k_1}-1)}$ and replaces all except the André net $y = xm; m^{\frac{(q^n-1)}{(q-1)}} = 1$. The theorem now follows immediately by iteration of these ideas together with the indicated replacement mappings. $\square$

We note that the 'all except' theorem basically replaces all of the smaller coset André nets 'except' the net corresponding to $y = xm; m^{\frac{(q^n-1)}{(q^{l_i}-1)}} = 1$, for $i = 0, 1, \ldots, t$ by the same replacement mapping. However, we may choose 'any' replacement for the coset André nets, so that each coset André net may be replaced independently. These replacements may be obtained from replacement mappings or from Foulser replacement or from Johnson replacements or so on. Furthermore, except to obtain a sequence of nets, there is no particular reason to single out the nets $y = xm; m^{\frac{(q^n-1)}{(q^{l_i}-1)}} = 1$, not to replace until the next stage, for we may select any subset of coset André nets to postpone replacement until the next stage. Hence, the variety of generalized André planes, as well as more general translation planes, that may be obtained by variations of these themes is extraordinarily large.

We also might mention that we may begin with a set of $q - 1$ 'big' André nets and apply Ostrom's 1st and 2nd replacement theorems on each of the $q - 1$ nets independently to construct generalized André planes. For example, the net $y = xm; m^{\frac{(q^n-1)}{(q-1)}} = \beta_0$ is in $GF(q)^*$. Decompose into $(q^{k_1} - 1)/(q - 1)$ nets of degree $\frac{(q^n-1)}{(q^{k_1}-1)}$. If $(x, y) \longmapsto (x, yb)$, such that $b^{\frac{(q^n-1)}{(q-1)}} = \beta_0^{-1}$, then, for example, apply 'all except' to the image and then map back to the net in question. As an example of the complexity, consider the example below.

EXAMPLE 16.13. Choose $n = 2^2 \cdot 3 \cdot 5$ and an arbitrary prime power q. We may take sequences of divisors $\{1, 2, 6, 12\}$, $\{1, 3, 6, 12\}$, $\{1, 2, 10, 20\}$, $\{1, 2, 4, 12\}$ and so on. We have $q - 1$ André nets of degree $\frac{q^n-1}{q-1}$. For each of these we may choose the 'all except' procedure or any desired variation. Furthermore, for each of the $q - 1$ nets, we may independently choose any sequence of divisors.

For example, take $q = 5$. There are four big André nets, A_i, $i = 1, 2, 3, 4$, $A_i = \left\{ y = xm; m^{(5^n-1)/(5-1)} = i \right\}$. Take the sequences

$$\{1, 2, 6, 12\}, \ \{1, 3, 6, 12\}, \ \{1, 2, 10, 20\}, \ \{1, 2, 4, 12\},$$

for $i = 1, 2, 3, 4$, respectively, and apply 'all except' to each net. Let $s_{i,j}$ denote the requisite divisors. Let b_i be an element of $GF(5^n)^*$ such that $b_i^{(5^n-1)/(5-1)} = i$ and $b_1 = 1$. Then we obtain the following spread:

$$x = 0, y = 0,$$

$$y = x^{q^{12s_{1,3}+6s_{1,2}+2s_{1,1}+s_{1,0}}} m; m^{\frac{(q^n-1)}{(q^{12}-1)}} = 1,$$

$$y = x^{q^{6s_{1,2}+2s_{1,1}+s_{1,0}}} m; m^{\frac{(q^n-1)}{(q^6-1)}} = 1, m^{\frac{(q^n-1)}{(q^{12}-1)}} \neq 1,$$

$$y = x^{q^{2s_{1,1}+s_{1,0}}} m; m^{\frac{(q^n-1)}{(q^2-1)}} = 1, m^{\frac{(q^n-1)}{(q^6-1)}} \neq 1,$$

$$y = x^{q^{s_{1,0}}} m; m^{\frac{(q^n-1)}{(q-1)}} = 1, m^{\frac{(q^n-1)}{(q^2-1)}} \neq 1,$$

$$y = x^{q^{12s_{2,3}+6s_{2,2}+3s_{2,1}+s_{2,0}}} mb_2; m^{\frac{(q^n-1)}{(q^{12}-1)}} = 1,$$

$$y = x^{q^{6s_{2,2}+3s_{2,1}+s_{2,0}}} mb_2; m^{\frac{(q^n-1)}{(q^6-1)}} = 1, m^{\frac{(q^n-1)}{(q^{12}-1)}} \neq 1,$$

$$y = x^{q^{3s_{2,1}+s_{2,0}}} m; mb_2^{\frac{(q^n-1)}{(q^3-1)}} = 1, m^{\frac{(q^n-1)}{(q^6-1)}} \neq 1,$$

$$y = x^{q^{s_{2,0}}} mb_2; m^{\frac{(q^n-1)}{(q-1)}} = 1, m^{\frac{(q^n-1)}{(q^3-1)}} \neq 1,$$

$$y = x^{q^{20s_{3,3}+10s_{3,2}+2s_{3,1}+s_{3,0}}} mb_3; m^{\frac{(q^n-1)}{(q^{20}-1)}} = 1,$$

$$y = x^{q^{10s_{3,2}+2s_{3,1}+s_{3,0}}} mb_3; m^{\frac{(q^n-1)}{(q^{10}-1)}} = 1, m^{\frac{(q^n-1)}{(q^{20}-1)}} \neq 1,$$

$$y = x^{q^{2s_{3,1}+s_{3,0}}} mb_3; m_3^{\frac{(q^n-1)}{(q^2-1)}} = 1, m^{\frac{(q^n-1)}{(q^{10}-1)}} \neq 1,$$

$$y = x^{q^{s_{3,0}}} mb_3; m^{\frac{(q^n-1)}{(q-1)}} = 1, m^{\frac{(q^n-1)}{(q^2-1)}} \neq 1,$$

$$y = x^{q^{12s_{4,3}+4s_{4,3}+2s_{4,1}+s_{4,0}}} mb_4; m^{\frac{(q^n-1)}{(q^{12}-1)}} = 1,$$

$$y = x^{q^{4s_{4,3}+2s_{4,1}+s_{4,0}}} mb_4; m^{\frac{(q^n-1)}{(q^4-1)}} = 1, m^{\frac{(q^n-1)}{(q^{12}-1)}} \neq 1,$$

$$y = x^{q^{2s_{4,1}+s_{4,0}}} mb_4; m_3^{\frac{(q^n-1)}{(q^2-1)}} = 1, m^{\frac{(q^n-1)}{(q^4-1)}} \neq 1,$$

$$y = x^{q^{s_{4,0}}} mb_4; m^{\frac{(q^n-1)}{(q-1)}} = 1, m^{\frac{(q^n-1)}{(q^2-1)}} \neq 1.$$

The reader will note that the construction in these examples involves components of the form $y = x^\lambda M$, where $x \mapsto x^\lambda$ is not a field automorphism; this contrasts with Wilke's construction, Theorem 16.9, where all 'λ' maps are field automorphisms. The latter condition guarantees that the Wilke planes are instances of generalized André planes, which we now investigate.

CHAPTER 17

Foulser's λ-Planes.

We indicated earlier that for a long time the only all known projective planes of non-square order were, up to duality, semifield, nearfield or André planes, Definition 16.3. In this chapter we consider the λ-planes of Foulser, usually called generalized André planes. The generalized André planes are also never (proper) semifield planes, and include all the André planes as well as the non-exceptional nearfield planes as a proper subset. In particular, the λ-planes furnish examples of planes of non-square order that are neither André nor nearfield planes.

Briefly, if $(F, +, \cdot)$ is a finite field, then the corresponding generalized André plane is, by definition, the plane obtained when the associated spread set consists of a subset of $\mathrm{Gal}(F)$. Thus, the problem of finding such sets is equivalent to finding functions $\lambda \colon GF(q^n) \to \mathrm{Aut}\, GF(q)$ such that $a \circ b = a^{\lambda(b)} b$ defines a quasifield. Foulser showed that this is equivalent to a number-theoretic definition of λ. We introduce this next, and examine in this chapter the corresponding theory and construction techniques. First, we study the λ-quasifields from mainly an algebraic standpoint, then we consider the corresponding translation planes from a geometric point of view.

17.1. λ-Quasifields.

Unless the contrary is clearly indicated, we usually assume implicitly:

NOTATION 17.1. $q = p^r$ is any power of the prime p. For any positive integer m: $I_m := \{0, 1, \ldots, m - 1\}$.

DEFINITION 17.2. (Foulser λ-function [**369**]). A function $\lambda \colon I_{q^d-1} \to I_{d-1}$ is a Foulser λ-FUNCTION if the FOULSER λ-CONDITION holds:

$$t := \gcd(d, \lambda(i) - \lambda(j)), i \equiv j \pmod{q^t - 1} \implies i \equiv j \pmod{q^d - 1}.$$

If $\lambda(0) = 0$, then the λ-function is NORMAL.

THEOREM 17.3. (Foulser λ-Condition [**369**]). Let $F = GF(q^d) > GF(q) = K$, q any prime power and $d \geq 1$ any integer. Fix a generator ω of the cyclic group F^*. Let $\lambda \colon I_{q^d-1} \to I_d$ be any map, and define $Q_\lambda := (F, +, \circ)$ by

$$\forall x \in F, i \in I_{q^d-1} : x \circ \omega^i = x^{q^{\lambda(i)}} \omega^i, x \circ 0 = 0.$$

Then the following hold.

(1) Q_λ is a pre-quasifield iff the function λ is a Foulser λ-function, Definition 17.2; the associated spread, or plane, coordinatized by Q_λ is called a GENERALIZED ANDRÉ SPREAD π_λ, or a λ-PLANE. Explicitly, π_λ is defined

on the K-space $V = F \oplus F$ and its component set consists of the K-subspaces

$$x - 0, y = 0, y = x^{\mu^{\lambda(i)}}\omega^i, i \in I_{q^d-1}.$$

(2) If Q_λ is a pre-quasifield, then it is a quasifield iff $\lambda(0) = 0$. Such quasifields are called λ-SYSTEMS or λ-QUASIFIELDS. The kernel of Q_λ is the subfield of F specified by $K = \text{Fix}(\langle \rho \rangle)$.

(3) If $\lambda \colon I_{q^d-1} \to I_d$ is a λ-function, then the function $\lambda_0 = (\lambda - \lambda(0)) \bmod d$ is a normalized λ-function, thus $\lambda_0(0) = 1$, which also coordinatizes the generalized André plane π_λ.

PROOF. We prove only the first part, the rest follow easily. We obtain a spread iff $\omega^{kq^{\lambda(i)}}\omega^i = \omega^{kq^{\lambda(j)}}\omega^j$ forces $i \equiv j \pmod{q^d - 1}$. Thus, we have a spread iff for $\lambda(i) \geq \lambda(j)$ the following condition holds:

$$kq^{\lambda(j)}(q^{\lambda(i)-\lambda(j)} - 1) + (i - j) \equiv 0 \pmod{q^d - 1} \implies i - j \equiv 0 \pmod{q^d - 1}.$$

Put $t = \gcd(d, \lambda(i) - \lambda(j))$, and note that this means $q^t - 1$ divides both $q^d - 1$ and $q^{\lambda(i)-\lambda(j)} - 1$, hence also $i - j$. Thus the condition above is precisely the Foulser condition, Definition 17.2, yielding the desired result. $\square$

The kernel of the λ-planes above evidently contains $GF(q)$. This follows upon noting that the associated λ-quasifields have kernel containing $GF(q)$. However, we note that the kernel is the fixed field of the automorphisms corresponding to the image of λ.

REMARK 17.4. Assume $F = GF(q^d) > GF(q) = K$. Choose any map $\Lambda \colon F^* \to \text{Gal}(F, K)$, subject to $\Lambda(1) = \mathbf{1}$. Define $Q_\Lambda := F(+, \circ)$ by

$$\forall\, x, m \in F : x \circ m = x^{\Lambda(m)}m, x \circ 0 = 0.$$

Then Q_Λ is a quasifield iff it is a generalized André system Q_λ, based on some map $\lambda \colon I_{q^d-1} \to I_{q-1}$, dependent on the choice of the primitive generator ω of F^*. If Q_Λ is a quasifield its kernel contains K and coincides with the fixed field of $\text{Fix}(\Lambda(F^*))$.

PROOF. Suppose Λ is a quasifield and fix ω. Since $\Lambda < \text{Gal}(F/GF(q))$, we may identify Λ with a unique map $\lambda \colon I_{q^d-1} \to I_d$, $\lambda(0) = 0$, such that the product on Q_Λ is specified by $\omega^j \circ \omega^i = (\omega^j)^{q^{\lambda(j)}}\omega^i$; now Theorem 17.3 applies. The converse is trivial, and obviously $GF(q)$ is in the kernel. To show that the kernel is precisely $GF(q)$ requires some manipulations; see [**369**, Lemma 2.2]. $\square$

If the choice of the primitive ω is changed then the (pre)-quasifield Q_λ changes in general to another (pre)-quasifield, say Q'. It is evident, however, that Q' may be represented using ω and changing instead λ to another λ-function λ'. Similar arguments show that any generalized André quasifield Q_λ, defined on the field $F = GF(q^d)$ by a λ-function $\lambda \colon I_{q^d-1} \to I_d$, has $GF(q)$ as its full kernel: if not, one may choose a new λ-function $\mu \colon I_{q^d-1} \to I_{q^m-1}$, where $GF(q^m)$ is the kernel of Q_λ, such that $Q_\lambda = Q_\mu$.

Hence, unless otherwise indicated, all generalized André systems of order q^d with kernel $GF(q)$ are specified by λ-functions of the form $\lambda \colon I_{q^d} \to I_d$, cf. Theorem 17.3, such that $GF(q)$ is the full kernel and the same primitive element ω is used to specify all these generalized André systems.

REMARK 17.5. The obvious examples of generalized André systems are fields, by Remark 17.4 above, and of course the André systems. The latter holds since the collection of André nets defining an André plane may be collectively specified by a λ-function, cf. Remark 17.4. Thus

(1) A λ-system is a field iff the corresponding λ-function is identically zero.
(2) Let $F = GF(q^d)$. Define $\nu: I_{q^d-1} \to I_{q-1}$ by $\nu(i) = i \bmod q - 1$ and $\lambda = \mu\nu: I_{q^d-1} \to I_d$, where $\mu: I_{q-1} \to I_d$ is an arbitrary function such that $\mu(0) = 0$. Then λ is a λ-function, and the associated λ-systems Q_λ coordinatize André planes; such Q_λ are called ANDRÉ SYSTEMS. Conversely, every André plane is coordinatized by an André system.

There is an integer v, defining the 'periodicity' of λ, that is helpful in recognizing generalized André systems that are not André systems.

DEFINITION 17.6. Let $\lambda: I_{q^d-1} \to I_d$ define a λ-system on $GF(q^d)$. Then the period of λ is the unique integer $v := v_\lambda$, in I_{q^d-1}, such that v_λ is a generator of the ideal $J = \{ m : \lambda(i) = \lambda(j) \pmod{m}, \ i, j \in I_{q^d-1} \}$.

We now have a criterion for recognizing 'proper' generalized André systems.

REMARK 17.7. An arbitrary λ-system, based on $\lambda: I_{q^d-1} \to I_d$, is an André system iff $v_\lambda | q - 1$.

All the regular nearfields may be viewed as λ-systems.

REMARK 17.8. Let q be a prime power and $d > 1$ any integer such that the prime divisors of d divide $q-1$, and let $d \not\equiv 0 \bmod 4$ if $q \equiv 3 \bmod 4$. Then d divides $(q^n - 1)/(q-1)$, and $(q^i - 1)/(q-1)$, $1 \le i \le n$, runs over a complete set of residues mod n, [**961**, Theorem 6.4]. Hence, we may define $\lambda: I_{q^d-1} \to I_d$ so that

$$\frac{q^{\lambda(i)} - 1}{q - 1} \equiv i \pmod{d}, \qquad i \in I_{q^d-1}.$$

The putative generalized André systems on $GF(q^d)$, with kernel $GF(q)$ and based on λ, are actual generalized André systems, viz., the regular nearfields. Hence such nearfields have 'period' $v_\lambda = d$.

PROOF. If λ satisfying the construction exists, it may be regarded as an additive epimorphism $\lambda: I_{q^d-1} \to I_d$, and this implies the corresponding λ-system is a nearfield. The existence of such λ follows from the definition and existence of the Dickson nearfields, Theorem 44.13, specified by $\lambda_0(i) = i$. (The fact that λ_0 is a λ-function directly follows from the Foulser condition, Theorem 17.3.) Hence the λ defined above are the λ-systems, defining nearfields, that may be identified with the class of additive epimorphisms $\lambda_{a,b}: I_{q^d-1} \to I_d$, $i \mapsto a\lambda_0(b^{-1}i)$, a, b invertible in I_{q^d-1}. This holds since all these maps are λ-systems, because λ_0 is, and they form the full set of additive additive epimorphisms from I_{q^d-1} to I_d. However, all these λ-systems include the nearfields obtained by changing the primitive ω, generating $GF(q^d)^*$ and defining the above Dickson nearfields based λ_0, to ω^b. By definition, these are just the regular nearfields, and the value of v_λ follows by inspection. □

Comparing v_λ for nearfields and André systems yields:

COROLLARY 17.9. There are infinitely many nearfields that are generalized André systems but not André systems, and there are infinitely many regular nearfields that are André systems.

PROOF. By Remark 17.7 a regular nearfield is an André system iff $v_\lambda | q - 1$, by Remark 17.8 v_λ divides $d - 1$. □

Thus, the next question to resolve is whether there are generalized André systems that are neither nearfields nor André systems.

We construct an important class of λ-system using the following number-theoretic fact, a special case of a more comprehensive result, for instance see [**961**, Theorem 6.3, p. 28].

LEMMA 17.10. Let u be any prime, and suppose $u^b | n$. Then for any integer $q > 1$, the condition $u^a || q - 1$ implies $u^{a+b} | q^n - 1$.

PROOF. We induct on b, noting the result is trivial if $b = 0$. Assuming, for all possible q and n that $u^a || q - 1, u^b || n \implies u^{a+b} | q^n - 1$, consider the situation when $u^{b+1} || n = n_1 u^b$, $u^a || q - 1$. Thus

$$q^n - 1 = (q^{n_1})^u - 1 = (q^{n_1} - 1) \left(\sum_{i=0}^{u-1} (q^{n_1})^i \right),$$

and since the inductive hypothesis implies $u^{a+b} | q^{n_1} - 1$, a fortiori $(q^{n_1})^i \equiv 1 \pmod{u}$, the final right-hand-side term is divisible by $u^{a+b} u$, as required. □

CONSTRUCTION 17.11. (The λ-systems Q_g). Let $q = p^r$, p prime, and g be any divisor of d such that all the prime divisors of g divide $q - 1$. Assume $(g, m) = (g, r)$ and $(d, a) = 1$, for positive integers $m, n, a \geq 1$. Then $\lambda \colon I_{q^d - 1} \to I_d$ specifies a generalized André plane, called a Q_g-SYSTEM, if it is defined as follows:

$$\forall i \in I_{q^d - 1}, \nu(i) = (mi) \bmod g, \lambda(i) = \nu(i) \bmod d.$$

The λ-system Q_g has period $v_\lambda = g$.

PROOF. In general, if λ defines a generalized André system then evidently so does $a\lambda$, provided $(a, d) = 1$. Hence, we assume $a = 1$. For the given λ, the λ-test condition $i \equiv j \pmod{q^{\gcd(d, \lambda(i) - \lambda(j))} - 1}$ becomes

$$i - j \equiv 0 \pmod{q^{\gcd(d, m(i-j)+gx)} - 1}.$$

We claim that this implies $g | i - j$. If not, then for some prime power u^a we have $u^a || i - j$ and $u^b | g$ for $b > a$. Now, as $\gcd(m, g) = 1$ and $g | d$, we have $u^a || \gcd(d, m(i - j) + gx)$. But then the number-theoretic Lemma 17.10 applied to the congruence above, yields $u^{a+1} || (i - j)$, contradicting $u^a || i - j$. Hence $i \equiv j \pmod{g}$. This forces $\lambda(i) = \lambda(j)$ since λ is defined to be constant on the g-classes. Thus the Foulser λ-test, Theorem 17.3, forces $i \equiv j \pmod{q^d - 1}$, as required. The value v_λ is immediate. □

Comparing the periods of Q_g, nearfields, Remark 17.8, and André systems, Remark 17.7, shows that infinitely many λ-systems are neither André nor nearfields.

COROLLARY 17.12. The λ-system Q_g, Construction 17.11, is an André system iff $g | q - 1$ and is a nearfield iff $g = d$ and $g | q - 1$.

The class of all λ-systems of order q^d, with d an arbitrary prime power and kernel containing $GF(q)$, have been explicitly listed by Foulser [**369**]:

PROPOSITION 17.13. (Foulser [**367**, **368**, **369**]). Suppose $d = c^t$, c prime. There are exactly c^f possible definitions of λ-functions $\lambda\colon I_{q^d-1} \to I_d$ where

$$f = t(q-2) + \sum_{k=1}^{t-1}(t-k)(q^{c^k} - q^{c^{k-1}}).$$

Proposition 17.13 follows from Foulser's formula [**369**] for a large class of λ-systems based on $\lambda\colon I_{q^d-1} \to I_{d-1}$, whenever d is divisible by a prime power c^t: the number constructed is an upper bound for the number of possible λ when $d = c^t$.

However, if d is prime, the formula for the number of λ in Proposition 17.13 coincides with the number of distinct André systems, hence:

COROLLARY 17.14 (Foulser [**367**, **368**, **369**]). Let d be a prime. Then any generalized André system of order q^d with $GF(q)$ in kernel is an André system (possibly a field).

Every generalized André plane, based on an arbitrary normalized $\lambda\colon I_{q^d-1} \to I_d$, admits a certain large homology group that depends on the field $GF(q^d)$, but is otherwise independent of the choice of λ.

DEFINITION 17.15. The SOCLE of $GF(q^d)^*$ is the integer $D := \operatorname{lcm}\{\, q^k - 1 : 0 < k < d, \ k \mid d \,\}$, and the SOCLE INDEX of $GF(q^d)^*$ is its cyclic subgroup of order $(q^d - 1)/D$.

LEMMA 17.16. Let N_D be the socle index of $GF(q^d)$, Definition 17.15, and let $\lambda\colon I_{q^d-1} \to I_d$ be any λ-function, $\lambda(0) = 0$. Let Q_λ be a corresponding generalized André plane, thus:

$$\omega^i \circ \omega^j = \omega^{iq^{\lambda(j)}+j}.$$

Then:

    (1) If $i \equiv j \pmod{D}$ then $\lambda(i) = \lambda(j)$.
    (2) The socle D divides the period v_λ, Definition 17.6.
    (3) $\lambda(m) = 0$, whenever $\omega^m \in N_D$.

PROOF. Case (1) is an immediate consequence of Definition 17.2 and Theorem 17.3. Case (2) follows from case (1) and the definition of v_λ, Definition 17.6. Case (3) holds because of case (1) and the fact that ω^D generates N_D, with $\lambda(0) = 0$. □

The following corollary lists elementary properties associated with N_D, and indicates their fundamental role in the *geometry* of λ-planes. In particular, any λ-system admits large affine homology groups with natural axis–coaxis pairs, and demonstrates the presence of irreducible collineation groups stabilizing any given component.

COROLLARY 17.17. Let Q_λ and $\pi_\lambda = \pi(Q_\lambda)$ be the quasifield and the corresponding generalized André plane coordinatized by a normalized λ-function $\lambda\colon I_{q^d-1} \to I_d$, N_D the corresponding socle index, Definition 17.15. Then the following hold.

    (1) The collection of maps of π_λ specified by

$$\Gamma_\infty := \{\, (x,y) \mapsto (x, y \circ b) : b \in N_D \,\},$$

    is a cyclic homology group with axis-coaxis pair $(y = 0, x = 0)$ of order $(q^d - 1)/D$, cf. Definition 17.15.

(2) The collection of maps of π_λ specified by

$$\Gamma_0 := \{\, (x, y) \mapsto (x \circ a, y) : a \in N_D \,\},$$

is a cyclic homology group with axis-coaxis pair $(x = 0, y = 0)$ of order $(q^d - 1)/D$.

(3) The subgroup $N_D < GF(q^d)$ is a subgroup of the index of $N_m(Q_\lambda) \cap N_r(Q_\lambda)$.

(4) Hence

$$\Gamma := \{\, (x, y) \mapsto (x \circ a, y \circ b) : a, b \in N_D \,\} \cong \Gamma_0 \times \Gamma_\infty$$

is an Abelian autotopism group that is the direct product of the cyclic homology groups $\Gamma_0 \cong \Gamma_\infty$, each of order $(q^d - 1)/D$.

(5) The slope orbits of Γ_0 and Γ_∞ coincide.

(6) The stabilizer of any component $M \in \pi_\lambda$, in the group $\Gamma_0 \times \Gamma_\infty$ acts $GF(q)$-irreducibly on M.

PROOF. The first two cases may be verified using the properties listed in Lemma 1; case (3) is the algebraic version of cases (1) and (2) considered together, as does case (4). Case (5) follows, for instance, by computing the orbit of $(1, m)$, $m \neq 0$, defining the component $y = xm$, under Γ_0 and under Γ_∞. By case (5) the stabilizer of any component in Γ (a fortiori, the stabilizer of the axes) is divisible by q-primitive elements. Hence case (6) holds.   □

To consider when the planes coordinatized by these quasifields are non-isomorphic we require several classical theorems of a fundamental nature due to Foulser [369], Ostrom, and Lüneburg [961], dealing with group-theoretic characterizations of various classes of (generalized) André planes. We consider these next.

17.2. Collineation Groups of λ-Planes.

We characterize André and generalized André planes in terms of their Abelian autotopism groups, due to Ostrom and Lüneburg, respectively, and state earlier fundamental results due to Foulser [369]. An important consequence, due to Foulser [369] is that an André plane cannot be isomorphic to a plane coordinatized by a non-André generalized André system. To fix notation, we recall the usual properties of the field-norm required to prove Ostrom's theorem.

REMARK 17.18. Let Γ be an automorphism group of order n of a finite field F; thus $K = \text{Fix}(\Gamma) = GF(q)$, $F = GF(q^n)$, and the corresponding norm function $\nu_\Gamma \colon f \mapsto \prod_{\gamma \in \Gamma} f^\gamma$ is a multiplicative homomorphism of F^* such that for all $f \in F^*$

$$\nu_\Gamma(f) = \prod_{i=0}^{n-1} f^{q^i} = f^{(q^n-1)/(q-1)} \in K^*.$$

Hence: (1) $\nu_\Gamma(F^*) = K^*$ and $\ker(\nu_\Gamma)$ is the cyclic subgroup $C_{q^n-1/q-1} < F^*$, of order $(q^n - 1)/(q - 1)$; (2) If M is a multiplicative subgroup of F^*, then Γ induces the trivial group on F^*/M iff $M \geq \ker(\nu_\Gamma)$.

PROOF. Case (1) is a familiar exercise, and case (2) follows by noting that $x = f^\gamma/f \in M$ holds, for all $f \in F^*$, $\gamma \in \Gamma$, iff $M \geq \{\, f^{q^i-1} : f \in F^*, \ i \geq 1 \,\} = C_{q^n-1/q-1}$.   □

COROLLARY 17.19. (André Construction). As in Remark 17.18 above, let $F = GF(q^n) > GF(q) = K$, $\Gamma = \mathrm{Gal}(F, K)$, and suppose $M < F^*$ is any subgroup such that Γ induces a trivial group on F^*/M.

Let $\lambda\colon F^* \to \Gamma$ be any mapping constant on the multiplicative cosets of M^*, with $\lambda(K^*) = 1$. Define $\circ\colon F \times F \to F$ by $x \circ 0 = 0$ and $x \circ y = (x)^{\lambda(y)}y$, for $y \in F^*$, $x \in F$. Then $(F, +, \circ)$ is an André quasifield and conversely every André quasifield may be expressed in this form. The field $K = \mathrm{Fix}(\Gamma)$ is the kernel of Γ.

PROOF. Apply Remark 17.18, and hence observe that the cosets of K^* are André nets by definition: so the cosets of M^*, which are unions of K^*-cosets, are unions of disjoint André nets. The result follows from the definition of André quasifields/spreads and André nets. $\square$

To prove a fundamental theorem of Ostrom, we will use case (3) of the following lemma; the previous cases are merely steps toward its proof.

LEMMA 17.20. (Cf. [**961**, Theorem 12.1]). Suppose $\lambda\colon F^* \to \mathrm{Aut}(F)$ is any map, F a finite field, with $\lambda(1) = \mathbf{1}$, and let Γ be the subgroup of $\mathrm{Aut}(F^*)$ generated by the set of automorphisms $\lambda(F^*)$, and define the subgroup $\Phi = \langle\, f^{\gamma-1} : f \in F^*, \gamma \in \Gamma \,\rangle < F^*$. Then the following hold.

(1) The group $\Phi = \langle\, f^{(\lambda(f_1)-1)\cdots(\lambda(f_i)-1)\cdots(\lambda(f_n)-1)} : n \geq 1,\ f, f_i \in F^* \,\rangle$, hence also $\Phi = \langle\, f^{(\lambda(g)-1)} : f, g \in F^* \,\rangle$.

(2) The subgroup $\Phi = \langle\, f^{\gamma-1} : f \in F^*, \gamma \in \Gamma \,\rangle < F^*$ is invariant under Γ and Γ acts trivially on F^*/Φ.

(3) If λ agrees on $\alpha, \beta \in F^*$ whenever $\beta = \alpha f^{\lambda(g)-1}$, for some $f, g \in F^*$, then λ is constant on the cosets of Φ, hence λ defines an André plane, by the standard construction, cf. Corollary 17.19.

PROOF. Since any member of Γ has the form $\gamma = \lambda(f_1)\cdots\lambda(f_n)$, every $f_i \in F^*$, Φ is generated by the elements of the form $f^{\lambda(f_1)\cdots\lambda(f_n)-1}$, $f \in F^*$. We show this may be expressed in the required form $g^{(\lambda(f_1')-1)\cdots(\lambda(f_\nu')-1)}$, by inducting on n, assuming the result always holds for $n-1$, f arbitrary. For $n = 1$ there is nothing to prove. For $\gamma = \lambda(f_1)\cdots\lambda(f_n)$, we have:

$$f^{\gamma-1} = f^{\lambda(f_1)\cdots\lambda(f_i)\cdots\lambda(f_n)-1} = \left(f^{\lambda(f_1)}\right)^{\lambda(f_2)\cdots\lambda(f_n)-1} f^{\lambda(f_1)-1},$$

hence by the inductive hypothesis we have a product of two terms of type $f^{\gamma-1}$, as desired. Case (2) follows immediately, because Φ is Γ-invariant since F^* is cyclic, and the condition $f^\gamma/f \in \Phi$ means Γ acts trivially on F^*/Φ^*.

Case (3). By the hypothesis and transitivity of equality, the assumed condition for the equality $\lambda(\alpha) = \lambda(\beta)$ generalizes, for $a, b \in F^*$, to:

$$b = aa_1 f^{\lambda(g_1)-1}\cdots a_n f^{\lambda(g_n)-1}, \exists n \geq 1, f_i, g_i \in F^* \implies \lambda(b) = \lambda(a).$$

Hence, by case (1), λ is constant on the coset $a\Phi$. So by Corollary 17.19 and case (2), λ defines an André plane with λ constant on the cosets of Φ. $\square$

Generalized André spreads π may be characterized by the fact that they admit Abelian autotopism groups A such that for every component m, the stabilizer A_m of m induces an irreducible group on m. This theorem of Lüneburg implies, via Lemma 17.20(3) above, a corresponding characterization of André planes due to Ostrom: if for each m, the stabilizer A_m acts transitively on m^*, then π is an André spread. The proof below is based on Lüneburg [**961**], where the treatment is more involved because the the infinite case is not excluded.

THEOREM 17.21. Let π be a finite spread, with kernel containing a field K. Then the following hold.

(1) (LÜNEBURG [**960**, Theorem 11.5]). π is a generalized André plane iff π admits an Abelian K-linear autotopism group A, fixing at least two distinct components $X, Y \in \pi$, such that for every other component $Z \in \pi \setminus \{X, Y\}$, the global stabilizer A_Z induces an irreducible group on Z.

(2) (OSTROM [**1042**]). π is an André plane iff π admits an Abelian K-linear autotopism group A, fixing at least two distinct components $X, Y \in \pi$, such that for every other component $Z \in \pi \setminus \{X, Y\}$, the global stabilizer A_Z induces a transitive group on $Z \setminus \{0\}$.

PROOF. We only consider the 'if' parts, since generalized André planes and André planes evidently satisfy the conditions claimed. Hence, for both cases, we have the condition on A, stated in case (1).

Coordinatize π with any quasifield $Q = (X, +, \circ)$ based on axes X and Y, the components fixed by A. Thus, $\pi = \pi(Q)$, $kern(Q) > K \cong GF(q)$, such that the components consist of the subspaces $V(m) := y = x \circ m, m \in Q$, of the vector K-space $V = Q \oplus Q$, and the subspace $Y := V(\infty) = \{\{0\} \times Q\}$, such that the component $X = V(0)$. Now A is a K-linear Abelian autotopism group of π, fixing $V(0)$ and $V(\infty)$, such that A_m, the global stabilizer of each component $m \neq V(0), V(\infty)$ acts irreducibly on the K-space V_m; of course, the triad principle implies that A_m is also irreducible on the axes $V(0)$ and $V(\infty)$.

So for $m \in Q^*$, Schur's lemma (plus the fact that A_m is Abelian) implies that the ring $K_m := K[A_m]$ is in a field, hence it is a finite integral domain, hence a field. Moreover, the irreducibility of $K_m := K[A_m]$ prevents it from being a subfield of a larger field, hence $V(m)$ is evidently a one-space over the field $K_m|V(m)$. Since a group fixing three components of a partial spread acts identically on all components it fixes, this 'triad principle' forces K_m to act faithfully and irreducibly on each of the subspaces $V(m)$, $V(0)$, and $V(\infty)$. Thus, V becomes a rank-two vector space over the field K_m.

Consider the field K_1. Since the components $V(1) = \{ (x, x) : x \in K \}$, $V(0) = \{ (x, 0) : x \in K \}$, and $V(\infty) = \{ (0, x) : x \in K \}$ are among the fixed components of K_1, we evidently have a field $\mathfrak{F} < \mathrm{Hom}(X, +)$ such that every element in K_1 has the form $(x, y) \mapsto (x^a, y^a) : a \in F$. Since $|\mathfrak{F}| = |X|$, the field of maps $\mathfrak{F}$ forms a spread set on X, hence we have a field $F = (X, +, \cdot) \cong \mathfrak{F}$ with identity 1 by defining $x \cdot m = (1)T_m$ where $T_m \in \mathfrak{F}$ is such that $T_m(1) = x$, etc. Hence, every member of K_1 has the form $(x, y) \mapsto (xa, ya)$, denoting field product by juxtaposition.

Now consider the action of K_m, $m \in F^*$, on the x-axis $X = V(0)$. Since X is a rank-one K_m-space, $K_m|X$ must coincide with the centralizer, in $\mathrm{Hom}(X, K)$, of the irreducible Abelian group $A|X$. However, since $A[K]$ is a commutative ring, the field of maps $\{(x, 0) \mapsto (xa, 0), a \in F\}$ centralizes $A|X$. Hence $K_m|X$ is the field $\{(x, 0) \mapsto (xa, 0), a \in F\}$, and this field is independent of the choice of $m \in X^*$. By considering $K_m|Y = V(\infty)$, we similarly establish that any map in the restriction $K_m|Y$, $Y = V(\infty)$, has the form $(x, 0) \mapsto (x, xb)$, $b \in F$. Hence every member $g \in K_m$ has the form $(x, y) \mapsto (xa_m^g, yb_m^g)$, for $a_m^g, b_m^g \in K^*$. Thus, $g \mapsto a_m^g$ and $g \mapsto b_m^g$ are bijections from K_m^* onto X^*. For convenience, we drop the superscript 'g' and instead make the convention that a_m and b_m are related by some (clearly unique) $g \in K_m$, as just indicated. Now the projections $\pi_1 \colon (a_m, b_m) \mapsto a_m$ and $\pi_2 \colon (a_m, b_m) \mapsto b_m$ are field isomorphisms, since the domain and images are fields of the same size, hence the map $\pi_1^{-1} \circ \pi_2 := a_m \mapsto b_m$ is in $\mathrm{Gal}(F/K)$.

Since any $\alpha_m \in K_m$ fixes $y = x \circ m$ and has the form $(x, y) \mapsto (xa_m, yb_m)$, it follows, on comparing $(xa_m, (x \circ m)b_m) = (xa_m, (xa_m) \circ m)$ and putting $x = 1$, that $b_m = (a_m \circ m)m^{-1}$. But since we have established that $\pi_1^{-1} \circ \pi_2 \colon a_m \mapsto b_m :=$ $(a_m)^{\lambda_m}$ is a field isomorphism, in $\mathrm{Gal}(F/K)$, we have $(a_m)^{\lambda_m} = (a_m \circ m)m^{-1}$, for all $a_m \in F$. Recall that for fixed m, $a_m \ (= a_m^g)$ ranges over all $a \in F^*$. Thus, $a^{\lambda_m} m = a \circ m$, $m \in F^*$, $a \in F$, hence $\circ$ defines a λ-system, with multiplication $\circ$, assuming the convention $a^{\lambda(0)} = 0$. Hence case (1), Lüneburg's theorem, has been established.

For case (2), Ostrom's theorem, we must consider the special situation when A_m is transitive on $y = x \circ m$, for each $m \in X^*$. Thus A_m coincides with K_m, for all $m \in X^*$. So, recalling that $b_m = a_m^{\lambda_m}$, the members of A_m have the form $g_m \colon (x, y) \mapsto (xa, ya^{\lambda_m})$, where a ranges over all F^*.

The collineation $\hat{f} \colon (x, y) \mapsto (xf, yf^{\lambda_m})$ maps the component $y = x \circ M$ onto the subspace $\{ (xf, (x \circ M)f^{\lambda_m}) : x \in X \} = \{ (x, ((xf^{-1}) \circ M)f^{\lambda_m}) : x \in X \}$. Since this component contains the point $(1, (f^{-1} \circ M)f^{\lambda_m})$, we have the identity $((xf^{-1}) \circ M)f^{\lambda_m} = x \circ ((f^{-1} \circ M)f^{\lambda_m})$ and converting $\circ$ to λ:

$$((xf^{-1})^{\lambda_M} M)f^{\lambda_m} = (x)^{\lambda}(f^{-\lambda_M}M)f^{\lambda_m} \, (f^{-1})^{\lambda_M} \, M \, f^{\lambda_m}$$

yielding $(x)^{\lambda_M} = (x)^{\lambda}f^{\lambda_m - \lambda_M}M = (x)^{\lambda}f^{1-(\lambda_M/\lambda_m)}M$. Letting m range over F^*, this shows λ is constant on the multiplicative cosets of $\langle f^{1-\gamma} : f \in F^*, \gamma \in \Gamma \rangle$, hence by Lemma 17.20 (3), it follows that λ defines an André plane.   □

Lüneburg's theorem implies that quasifields isotopic to generalized André systems are themselves generalized André systems. In fact, a slightly more general theorem holds—we get nothing new if the coordinate axes are interchanged:

COROLLARY 17.22. Let π be a spread coordinatized by a quasifield Q, based on both axes selected from two components $\{X, Y\} \subset \pi$. Then Q is a generalized André quasifield iff every quasifield Q', coordinatizing π, based on axes $\{X, Y\} \subset \pi$, is a generalized André system.

Ostrom's theorem yields an analogous result for André planes.

COROLLARY 17.23. Let π be a spread coordinatized by a quasifield Q, based on both axes selected from two components $\{X, Y\} \subset \pi$. Then Q is an André quasifield iff every quasifield Q', coordinatizing π, based on axes $\{X, Y\} \subset \pi$, is an André system.

The above raises the question regarding the uniqueness of the axis pair $\{X, Y\}$ for generalized André planes. In fact, this and several related fundamental questions were completely settled earlier in Foulser's pioneering paper on generalized André planes. We list some of these results [369], referring the reader to Lüneburg [961, 14.10–14.11] for complete proofs.

THEOREM 17.24. (Foulser [367, 368, 369], [961, Corollary 14.10]). Let π^{ℓ_∞} be an affine generalized André plane, with full collineation group G. Suppose π is not a Desarguesian or a Hall plane. Then the following hold.

(1) G leaves invariant a set $\{P, Q\} \subset \ell_\infty$ of two distinct points.

(2) If $B \subset \ell_\infty$ is a G-orbit then either $B \subseteq \{P, Q\}$ or $|B| > 2$.
(3) Hence π cannot admit an affine elation group of order > 2, so π is not a semifield plane.

The set $\{P, Q\}$, of the generalized André planes as above, corresponds to the axes of the largest irreducible Abelian autotopism group A associated with Lüneburg's theorem mentioned above, and A is index 2 in the full collineation group of the spread. This is reflected in:

THEOREM 17.25. Let $(L, +, \cdot)$ be a finite field defining a generalized André system $Q_\lambda = (L, +, \circ)$, based on the λ-function $\lambda\colon L \to \mathrm{Gal}(L)$. Assume that the spread $\pi_\lambda := \pi(Q_\lambda)$, on $L \oplus L$, is neither a Desarguesian nor a Hall plane. Then the full collineation group of π_λ is a subgroup of $(\Gamma(1, L) \times \Gamma(1, L))\langle \rho \rangle$, where $\rho\colon (x, y) \mapsto (y, x) \, \forall \, (x, y) \in L \oplus L$.

COROLLARY 17.26. (Foulser [**367**], [**368**], [**961**, Corollary 14.12]). The plane coordinatized by a λ-system Q_λ is isomorphic to an André plane iff Q_λ is an André system.

PROOF. Let π be the spread of generalized André, hence for a suitable choice of axis components $\{X, Y\} \subset \pi$, there corresponds a generalized λ-system Q_λ, which is not an André system. It follows that π is not Desarguesian or Hall, since Q_λ must have dimension > 2 over its kernel to prevent it from being an André plane. Hence, $\mathrm{Aut}(\pi)$ leaves only one 2-set of components invariant, Theorem 17.24; so a quasifield R that coordinatizes π and is a generalized André plane can only arise when the coordinate axes are selected from $\{X, Y\}$, since this set is invariant under $\mathrm{Aut}(\pi)$. Hence since Q_λ is not an André quasifield, then as Corollary 17.23 shows, neither is R. □

COROLLARY 17.27. Let π be a translation plane coordinatized by a generalized André system Q_g, of order $q^d(> 9)$ with kernel $GF(q)$. Then the following hold.
(1) The plane π is an André plane iff $g | q - 1$.
(2) The plane π is a nearfield plane iff $g | d$.

PROOF. Suppose π is an André plane. Then Corollary 17.26 applies. If π is isomorphic to a plane coordinatized by a nearfield N, then by Theorem 17.24, the coordinatizing axes $\{X, Y\}$ yielding Q_g and N must coincide. Thus Q_g also has associative multiplication. So Q_g is a nearfield. □

The proof of Corollary 17.27 above may essentially be repeated to yield the following further examples of λ-planes guaranteed to be non-André planes.

COROLLARY 17.28. Any multiple André plane π from a Desarguesian spread Σ, defined in Construction 16.6, is a generalized André plane. It is not an André plane iff at least two of those André nets of Σ to which non-trivial André replacements are applied to obtain π have different degrees.

COROLLARY 17.29. The Wilke planes, Theorem 16.9, are generalized André planes but not André planes.

We pause to summarize briefly the contents of this chapter. Although there is a complete classification of André planes, the generalized André systems may be too numerous to classify. However, in some cases generalized André planes have been completely classified: the λ-planes of prime dimension over their kernels are all André planes, and Foulser has construction technique for all λ-planes that are of prime-power dimension over their kernels. In particular, his description implies a lower bound for the number of non-isomorphic generalized André planes of prime-power dimension over their kernels. Finally, although the λ-planes themselves may be too numerous to classify, the structures of their collineation groups are quite well understood, due to the work of Foulser, Ostrom, and Lüneburg. In particular, André planes and generalized André planes have striking characterizations in terms of their autotopism groups, as established by Ostrom and generalized by Lüneburg.

CHAPTER 18

Regulus Lifts, Intersections over Extension Fields.

We have already analyzed the structure of a *single* regulus, in the Desarguesian spread Σ_{q^2}. Relative to a suitable coordinatization any such regulus may be identified with the STANDARD or SCALAR regulus $\{ (y = mx) : m \in GF(q) \} \cup \{ (x = 0) \}$. However, the problem remains of finding a nice form for the remaining reguli. We shall resolve this problem, in Section 18.1, by showing that all reguli are in the y-axis central collineation orbits of the scalar regulus or of an André regulus (which also have nice forms, as we have seen). Thus, we have explicit computationally useful forms for all reguli in Σ_{q^2} which enable us to study algebraically the intersections (set of shared components) of any two reguli in Σ_{q^2}.

The reguli in Σ_{q^2} extend naturally to reguli in $\Sigma_{q^{2t}}$, using the general notion of 'tensor extensions' of spreads, Section 18.2. This prepares us to consider the intersection of reguli, $\overline{R_1} > R_1$ and $\overline{R_2} > R_2$, in $\Sigma_{q^{2t}} > \Sigma_{q^2}$, that extend the reguli R_1 and R_2 in Σ_{q^2}. We show that $|R_1 \cap R_2| = |\overline{R_1} \cap \overline{R_2}|$ for *odd t*. This result forms the basis of subregular towers due to Johnson. We end by defining the Johnson towers and investigating them in detail, as a prelude to the study of subregular spreads.

Finally, note that since the study of reguli in Desarguesian spreads is equivalent to the study of circles in Miquellian inversive planes, e.g., [**333**], much of the material in this chapter might alternatively have been developed within this elegant framework.

18.1. General Regulus Coordinates in Σ_{q^2}.

We determine a computationally useful form (i.e., coordinatization) of the *general* regulus in Σ_{q^2}. The coordinatization of the reguli through the y-axis is obtained by noting that they form an orbit of the scalar regulus under the group of y-axis central collineations, and the remaining reguli will be seen to be images of André reguli. We recall some notation for regular (Desarguesian) spreads.

CONVENTION. If $L = GF(q^\ell)$, then $\Sigma_L = \Sigma_{q^\ell}$ is the standard Desarguesian spread of order q^ℓ on the vector space $L \oplus L$, with component set specified by:

$$\{ y = x\lambda : \lambda \in L \} \cup \{ (x = 0) \}.$$

If $L = GF(q^\ell) \supset GF(q^j) = J$, then Σ_J is identified with its standard embedding in Σ_L, specified as follows.

DEFINITION 18.1. Let $L = GF(q^\ell) \supset GF(q^j) = J$, then the spread induced by the Desarguesian spread $(L \oplus L, \Sigma_L)$ on the additive subgroup $J \oplus J < L \oplus L$, is the Desarguesian spread Σ_J, with component set $\{ y = xj : j \in J \} \cup \{ (x = 0) \}$.

We recall the terminology for the STANDARD or SCALAR regulus in Σ_{q^2}.

REMARK 18.2. Let $GF(q^2) = F \supset K = GF(q)$. Then the set of components of Σ_K in Σ_F forms a regulus, the SCALAR or STANDARD regulus, specified by

$$\Delta_F := \{\, y - fx : f \in F \,\} \cup \{x - 0\}.$$

The general formula for the reguli in the y-axis elation orbit of the standard regulus will be needed.

REMARK 18.3. Let $GF(q^2) = F \supset K = GF(q)$. Then the orbit of the scalar (standard) regulus $\Sigma_K < \Sigma_F$, under the full elation group with axis $(x = 0)$, is the collection of q distinct reguli:

$$\Delta(K_t) := \{\, y = (k + t)x : k \in K \,\} \cup \{(x = 0)\}, \qquad t \in F;$$

these q reguli form a parabolic cover of the Σ_F: so any two distinct reguli of type $\Delta(K_t)$ share precisely $(x = 0)$.

We proceed to similarly coordinatize y-axis elation-shifts of André reguli:

CONVENTION. If $m | q - 1$, q a prime power, then C_m denotes the unique multiplicative (cyclic) group of order m in some $GF(q)$ under consideration.

Recall that the two-dimensional André nets are reguli and their (unique nontrivial) André replacement corresponds to derivation:

REMARK 18.4. Suppose $F = GF(q^2) \supset GF(q) = K$. Then the André net

$$\mathcal{A}_\alpha = \{\, y = xm\alpha : m \in C_{q+1} \,\} < \Sigma_F, \text{ for } \alpha \in GF(q)^*,$$

is a regulus and the corresponding André replacement is the opposite regulus:

$$\mathcal{A}_\alpha^q := \{\, y = x^q m\alpha : m \in C_{q+1} \,\},$$

and each component of the replacement is a K-subspace of Σ_F.

The André reguli lie in distinct orbits of the full y-axis elation group of Σ_{q^2}:

LEMMA 18.5. In the standard Desarguesian spread Σ_{q^2}, let E denote the elation group of order q^2 with axis $(x = 0)$. Then for each $\alpha \in GF(q)^*$, the $q - 1$ André reguli $\mathcal{A}_\alpha := \{\, y = xm\alpha : m^{q+1} = 1 \ \forall m \in C_{q+1} \,\}$ lie in $q - 1$ distinct regulus orbits under E.

PROOF. Any elation η of Σ_{q^2} with axis $x = 0$ maps each component $y = xM$, $M \in GF(q^2)$, to a component of the form $y = x(M + T)$, for fixed $T \in GF(q^2)$, cf. Remark 18.3. Suppose if possible that some André regulus $\mathcal{A}_\alpha$ satisfies the condition $\eta(\mathcal{A}_\alpha) = \mathcal{A}_\beta$; hence, for all $m \in GF(q^2)$ satisfying $m^{q+1} = 1$ we have:

$$(m\alpha + T)^{q+1} = \beta \implies \left(\frac{1}{m}\alpha^q + T^q\right)(m\alpha + T) = \beta,$$

and since also $m^{-1} = m^q$, we have the quadratic condition on the $q + 1$ values of m of the form

$$m^2 \alpha T^q + m(TT^q + \alpha\alpha^q - \beta) + \alpha^q T = 0,$$

a contradiction unless $T = 0$. $\square$

Since the elation axis $x = 0$ does not lie in the André reguli, the E-orbits of the André reguli have size q^2. Thus, applying all possible elations to $\mathcal{A}_\alpha$, as in the proof above, yields:

COROLLARY 18.6. *The orbit of the André net $\mathcal{A}_\alpha$ under the full $(x = 0)$-elation group of Σ_F, $F = GF(q^2)$, is the set of q^2 reguli of the form $\mathcal{A}_\alpha(t)$, for $t \in F$, ranging over the additive cosets of F, specified by*

$$\mathcal{A}_\alpha(t) := \{\, y = x(m + t)\alpha : m^{q+1} = 1 \;\forall m \in C_{q+1} \,\}, \qquad \alpha \in C_{q-1} = GF(q)^*.$$

Moreover, an André net $\mathcal{A}_\beta = \mathcal{A}_\alpha(t)$, for some t, iff $\beta = \alpha$ and $t = 0$.

COROLLARY 18.7. *For the Desarguesian spread extension $\Sigma_{q^2} \supset \Sigma_q$, the full elation group of order q^2, with axis $(x = 0)$, has $q - 1$ regulus orbits, each of size q^2, such that each orbit contains exactly one André regulus, $\mathcal{A}_\alpha = \mathcal{A}_\alpha(0)$.*

To show that the reguli listed in Corollary 18.7 above account for all the reguli in $\Sigma_{q^2} \setminus \{(x = 0)\}$, we count the reguli not meeting any selected component.

REMARK 18.8. *Any Desarguesian spread Σ in $PG(3, q)$ contains $q(q^2 + 1)$ reguli and $q(q + 1)$ of these share any chosen component $\ell \in \Sigma$. Hence there are $q^2(q - 1)$ reguli, contained in Σ, that do not share any specified component $\ell \in \Sigma$.*

PROOF. Since any 3-set of components specifies a unique regulus, the number of reguli in a Desarguesian spread in $PG(3, q)$ is $\binom{q^2+1}{3}$. Similarly, the number of reguli containing a fixed component ℓ is $\binom{q^2}{2} / \binom{q}{2}$. Thus, the number of reguli not through ℓ is the difference between these counts. $\square$

COROLLARY 18.9. *For any choice of a component ℓ of a Desarguesian spread Σ, the spread may be coordinatized by $F = GF(q^2) \supset GF(q) = K$ so that ℓ gets identified with $(x = 0)$, and the reguli of Σ not through ℓ are all (shifted) André reguli, viz., $A_\alpha(t)$, $\alpha \in C_{q+1}$, $t \in GF(q^2)$, cf. Corollary 18.6.*

The remaining reguli contain ℓ and form an orbit of $q(q + 1)$ reguli under the group of central collineations with axis $(x = 0)$:

$$\{\, y = (tk + s)x : k \in K \,\} \cup \{(x = 0)\}, \qquad t \in F^*, \; s \in F.$$

PROOF. Apply Remark 18.8, Corollary 18.7; for the final sentence note that the y-axis central collineations are transitive on reguli through the y-axis. $\square$

18.2. Field Extensions of (Partial) Spreads.

A spread π of K subspaces of a finite vector space V may be canonically identified with a *partial* spread '$F \otimes \pi$' of F-subspaces of $F \otimes V$, for any finite field $F > K$. There are many known cases when π_F may be extended to an F-spread, and many cases where such extensions are not known to be possible. Here, we list the basic notions concerning such putative extensions.

DEFINITION 18.10. Suppose π is a (partial) spread on a vector space V over a field K. Then for any field $F > K$ we define a corresponding partial spread $F \otimes \pi := \{\, F \otimes \omega : \omega \in \pi \,\}$ on $F \otimes V$.

The fact that $F \otimes \pi$ is a partial spread follows from:

REMARK 18.11. *If W is a vector space over a field K, then for any extension field $F > K$ the K-independent subsets of W are F-independent subspaces of $F \otimes W$.*

Remark 18.11 also yields:

REMARK 18.12. Suppose $F > K$ are fields and W is a two-dimensional vector space over K. Then π_K denotes the Desarguesian spread consisting of the one-dimensional K-subspaces of W, and $\pi_F > \pi_K$ is the Desarguesian spread of one-dimensional F-subspaces of $V = F \otimes W$.

The EXTENSION, or π_F-extension, of any component $\gamma \in \pi_K$, is the π_F-component $\hat{\gamma} := F \otimes \gamma \supset \gamma$. So if $\mathcal{R} \leq \pi_K$ is any (rational) net then the corresponding (rational) net in π_F is $F \otimes \mathcal{R} \leq \pi_K := \{ \overline{\gamma} : \gamma \in \mathcal{R} \}$.

Hence for any rational net $R \subseteq \pi_K$, the net $F \otimes R \subset \pi_F$ is a rational net of π_F, the EXTENSION of π_R in π_F. Moreover disjoint (not necessarily rational) nets in π_K extend to disjoint nets in π_F, thus:

$$R_1 \cap R_2 = \varnothing \implies (F \otimes R_1) \cap (F \otimes R_2) = \varnothing.$$

If g is a K-linear map on the ambient space W of π_K, then $\overline{g} = F \otimes g$ is the unique F-linear map of $V = F \otimes W$, such that $\overline{g}|W = g$.

Remark 18.12 above, and the fact that a regulus is uniquely determined by any three of its components, immediately yields:

REMARK 18.13. Let $F = GF(q^{2n}) > GF(q^2) = K$; so $\pi_F > \pi_K$ are Desarguesian spreads, and $\pi_F \supset F \otimes \pi_K$. Then for any regulus $\mathcal{R} \subset \pi_K$, the corresponding rational Desarguesian net $F \otimes \mathcal{R} \subset \pi_F$ lies in a unique regulus $\overline{\mathcal{R}} \supset F \otimes \mathcal{R}$ of π_F, and $\overline{\mathcal{R}}$ is the **regulus extension**, in π_F, of the π_K-regulus $\mathcal{R}$.

Moreover, if R_1 and R_2 are reguli in π_K, then $\mathcal{R}_1 \cap \mathcal{R}_2 \subseteq \overline{\mathcal{R}_1} \cap \overline{\mathcal{R}_2}$.

PROOF. The final sentence follows from Remark 18.11.   □

WARNING. If R_1 and R_2 are disjoint reguli in π_K, then so are their rational extensions $F \otimes R_1$ and $F \otimes R_2$, in $\pi_F > \pi_K$. This does **not** imply that **regulus** extensions of R_1 and R_2—uniquely determined as extensions of $F \otimes R_1$ and $F \otimes R_2$—are themselves disjoint.

However, by Johnson, Lemma 18.19, for n odd, disjoint reguli have disjoint regulus extensions. This implies, cf. Theorem 18.21, that any *subregular* plane extends recursively to yield an infinite ascending (non-Baer) chain of subregular planes.

REMARK 18.14. Let $F = GF(q^n) > GF(q) = K$. Then, for the Desarguesian planes $\pi_F > \pi_K$, the following hold.

(1) If g is an elation (respectively, homology) of π_K, on the K-space W, then $F \otimes g$, the unique extension of g to an F-linear map on $V = F \otimes W$ is an elation (respectively, homology) of π_F, of the same order.

(2) The collineation group $\Gamma L(2, q)$ of π_K is the restriction of the collineation group $F \otimes \Gamma L(2, q^n)$, acting on π_F, to π_K. This restriction corresponds to the embedding $g \mapsto g \otimes F$, which defines a faithful isomorphism $\Gamma L(2, K) \to \Gamma L(2, F)$.

(3) In particular, the linear collineation group $GL(2, K)$ of π_K is the restriction of the linear collineation group $F \otimes GL(2, K)$, such that $g \mapsto F \otimes g$ specifies a faithful isomorphism $GL(2, K) \to GL(2, F)$.

PROOF. This follows from the fact that $\text{Fix}(g^F)$ has an F-basis containing a K-basis of $\text{Fix}(g)$, and (1) follows from Remark 18.11. The other parts follow from the definition of $F \otimes g$.   □

COROLLARY 18.15. For $K = GF(q) < GF(q^n) = F$, any planar collineation group $G < \Gamma L(2, K)$ of the Desarguesian spread π_K extends to a planar subgroup $\overline{G} = F \otimes G < \Gamma L(2, F)$ of $\pi_F > F \otimes \pi_K$ that has the same order as G. In particular, Baer involutions extend to Baer involutions.

In studying subregular spreads obtained by deriving sets of disjoint reguli in Desarguesian spreads, we require the fact that the collineation groups of a Desarguesian spread π_K that permute a given set of disjoint reguli Θ (the corresponding 'inherited group') extend to the inherited group of the corresponding reguli in appropriate extensions of Π_K.

REMARK 18.16. Suppose $F = GF(q^{2n}) > GF(q^2) = K$, $n > 1$ odd. Suppose Θ is a collection of pairwise disjoint reguli in the Desarguesian spread π_K, and $F \otimes \Theta = \{ F \otimes T : T \in \Theta \}$. Let $\mathrm{Aut}(\Theta)$ be the subgroup of $\mathrm{Aut}(\pi_K)$ that permutes the members of Θ among themselves. Then $F \otimes \mathrm{Aut}(\Theta)$ is a subgroup of $\mathrm{Aut}(\pi_F)$, $\pi_F > F \otimes \pi_K$, that permutes the members of the pairwise disjoint reguli that form the unique reguli in π_F extending the rational nets in $F \otimes \Theta$.

PROOF. Apply Remark 18.14 to $F \otimes \mathrm{Aut}(\Theta)$, noting that any three components of a regulus of π_K extend uniquely to a regulus of $F \otimes \pi_K$. $\square$

18.3. Field Extensions of André Reguli.

We consider the number of intersections of two distinct reguli in Σ_K, when one A_K is a (shifted) André net and the other is the scalar regulus S_K. If F is a larger field then $S_1 = F \otimes S$ is a scalar regulus in Σ_F and there is a natural shifted André net $A_1 = A_F > F \otimes A_K$. Our primary goal is to count the number of shared components $A_1 \cap F_1$. The case when $A_1 \cap F_1 = \varnothing$ will be of crucial importance in the extension of subregular spreads.

REMARK 18.17. Let $F = GF(q^2) \supset GF(q) = K$, so

$$\mathcal{K} := \{ y = kx : k \in K \} \cup \{ (x = 0) \}$$

is the scalar regulus in Σ_F, associated with the standard embedding $\Sigma_K < \Sigma_F$. Consider any elation shifted André net in Σ_F, viz.,

$$\mathcal{A}_\alpha^{(t)} := \{ y = x(m + t) : m^{q+1} = \alpha \} \text{ for some } t \in F, \ \alpha \in K^*.$$

This net shares with the scalar regulus $\mathcal{K}$ the components $(y = kx)$ that are specified by $\{ k \in K : k^2 - (t^q + t)k + (t^q t - \alpha) = 0 \}$, the set of K-roots of the K-quadratic equation $k^2 - (t^q + t)k + (t^q t - \alpha) = 0$.

PROOF. The reguli meet where $(k - t)^{q+1} = \alpha$, equivalently $(k - t)^q (k - t) = \alpha$, and since $k = k^q$ the result follows. $\square$

We review some conventions concerning a Desarguesian spread Σ_B embedded in a Desarguesian spread Σ_A.

CONVENTION. If A is a subfield of the field B, then the standard embedding of Σ_A in Σ_B corresponds to the rational net $B \otimes \Sigma_A < \Sigma_B$. If $G \leq L \leq F$ is any chain of fields, then any component $\gamma \in \Sigma_G$ is identified with the component of $L \otimes \gamma \in \Sigma_L$. Hence, if γ is specified by $y = xg$, it is identified with the component $L \otimes \gamma$ of Σ_L, and this coincides with $\{ y = xg|_{x \in L} \} := \{ (x, xg) : x \in L \}$. Similarly $x = 0|_{x \in L}$ means $L \otimes (x = 0)$.

Any André regulus A in Σ_{q^2} 'induces' a corresponding André net in $\Sigma_{q^{2n}}$ with a similar formula but over a larger field.

DEFINITION 18.18. Let $F = GF(q^{2n}) > GF(q^n) = K \supset GF(q) = P$. Let $L = GF(q^2) < GF(q^{2n}) = F$. Then for any (shifted) André regulus for L/P:

$$\mathcal{A}_\alpha^{(t)} := \{\, y = x(m+t)|_{x \in L} : m^{q+1} = \alpha,\ m \in L \,\}, \alpha \in P = GF(q), t \in L,$$

the (shifted) André regulus, over F/K,

$$\overline{\mathcal{A}}_\alpha^{(t)} := \{\, y = x(m+t)|_{x \in F} : m^{q^n+1} = \alpha,\ m \in F \,\},$$

is INDUCED by $\mathcal{A}_\alpha^{(t)}$.

Since for odd n, and $m \in L$, we have $m^{q^n+1} = m^{q+1}$, it follows that:

NOTE. Assume the conditions of Definition 18.18 above. Then for odd n, the André F-regulus $\overline{\mathcal{A}}_\alpha^{(t)} \supset F \otimes \mathcal{A}_\alpha^{(t)}$. Thus, $\overline{\mathcal{A}}_\alpha^{(t)}$ is an F-regulus extension of the André L-regulus $\mathcal{A}_\alpha^{(t)}$.

The following result will show that if $\pi_F > \pi_K$ is an odd-dimensional extension of a Desarguesian spread π_K, then disjoint reguli in π_K extend to disjoint reguli in π_F.

LEMMA 18.19 (Johnson). Let $F = GF(q^{2n}) > GF(q) = P$, n odd, and choose the subfields $K = GF(q^n) > P$, $L = GF(q^2) > P$. Hence the scalar regulus in Σ_L, viz., $L \otimes \Sigma_P$, extends to the following rational net in the regular spread Σ_F:

$$\mathcal{P} := F \otimes \Sigma_P = \{\, y = kx|_{x \in F} : k \in P \,\} \cup \{\, x = 0|_{x \in F} \,\};$$

similarly the scalar regulus in Σ_F is specified by

$$\mathcal{K} := F \otimes \Sigma_K = \{\, y = kx|_{x \in F} : k \in K \,\} \cup \{x = 0|_{x \in F}\}.$$

Then, in the notation of Definition 18.18, the following hold.

(1) The (shifted) André regulus $\overline{\mathcal{A}}_\alpha^{(t)}$ in Σ_F, induced by the André L/P regulus $\mathcal{A}_\alpha^{(t)} < \Sigma_L$, hence $t \in L$ and $\alpha \in P$, is the unique regulus in Σ_F that F-extends the components of the (shifted) André regulus $\mathcal{A}_\alpha^{(t)}$ in Σ_L, that is, $\overline{\mathcal{A}}_\alpha^{(t)} \supset F \otimes \mathcal{A}_\alpha^{(t)}$.

(2) $\overline{\mathcal{A}}_\alpha^{(t)} \cap \mathcal{K} = F \otimes \left(\mathcal{A}_\alpha^{(t)} \cap \Sigma_P \right)$.

PROOF. Since case (1) follows from Note 18.3, we only consider case (2). The first intersection is the component set of Σ_F specified by

(18.3.1) $\{\, y = x(m+t) : m^{q^n+1} = \alpha,\ m \in F \,\} \cap \{\, y = xk : k \in K \,\}$,

which is the same as the component set

$$\{\, y = xk : (k-t)^{q^n+1} = \alpha,\ k \in K \,\},$$

and the number of solutions is the same as the solutions for $k \in K$ satisfying the quadratic $k^2 - k(t^{q^n} + t) + (t^{q^n}t - \alpha) = 0$, for $\alpha \in P$, $t \in L$.

But since is n odd, this becomes

(18.3.2) $k^2 - k(t^q + t) + (t^q t - \alpha) = 0;\ \alpha \in P,\ t \in L.$

However, since $t \in L$ *this is a P-quadratic equation,* so its K-solutions are merely its P-solutions, since the degree $[K : P]$ is odd. Hence we have shown:

$$(18.3.3) \qquad \left| \overline{\mathcal{A}}_\alpha^{(t)} \cap \mathcal{K} \right| = \#P\text{-solutions of } P\text{-quadratic equation (18.3.2)}.$$

Similarly, the intersection $(F \otimes \mathcal{A}_\alpha^{(t)}) \cap \mathcal{P}$, corresponds to the component set in Σ_L specified by

$$\{\, y = x(m + t) : m^{q+1} = \alpha, \ m \in L \,\} \cap \{\, y = xk : k \in P \,\},$$

which is the same as the component set

$$\{\, y = xk : (k - t)^{q+1} = \alpha, \ k \in P \,\},$$

and the solution for $k \in P$ satisfying this condition is the set of $k \in P$, satisfying the previous P-quadratic equation (18.3.2) above. So the quadratic equation (18.3.2) does not admit any further solutions other than the $k \in P$ just mentioned, since K is an odd-dimensional extension of P. The result follows, since $(F \otimes \mathcal{A}_\alpha^{(t)}) \cap \mathcal{P} :=$ $(F \otimes \mathcal{A}_\alpha^{(t)}) \cap (F \otimes \Sigma_P) = F \otimes \left(\mathcal{A}_\alpha^{(t)} \cap \Sigma_P \right).$ □

We deduce that the number of components shared by any two reguli, in a Desarguesian spread Σ_{q^2}, n odd, are the same as the number of components shared by their (regulus) extensions in Σq^{2n}.

COROLLARY 18.20. *Let $\Delta_{q^{2n}}$ be any Desarguesian spread extension of order q^{2n}, n odd, of a Desarguesian spread Δ_{q^2}, of order q^2. If $\mathcal{R}_i$, $i = 1, 2$, are distinct reguli in Δ_{q^2}, then the number of components in $\mathcal{R}_1 \cap \mathcal{R}_2$ is the same as the number of components in $\overline{\mathcal{R}}_1 \cap \overline{\mathcal{R}}_2$, where $\overline{\mathcal{R}}_i$, $i = 1, 2$, denotes the unique regulus in $\Delta_{q^{2n}}$ that extends the rational net $GF(q^{2n}) \otimes \mathcal{R}_i$.*

PROOF. Coordinatize with axial components $\{(x = 0), (y = 0), (y = x)\}$ chosen in $\mathcal{R}_1$ such that $(x = 0) \notin \mathcal{R}_2$. Now the decreasing chain $\Delta_{q^{2n}} > \Delta_{q^n} > \mathcal{R}_1$ becomes identified, after coordinatization with axes and unit line in $\mathcal{R}_1$, with the 'standard' chain $\Sigma_{q^2} > \Sigma_q > \mathcal{R}_1$, where $\mathcal{R}_1$ is the scalar regulus in Σ_q. Hence, by Corollary 18.9, $\mathcal{R}_2$ may be chosen to be a shifted André spread. Now Lemma 18.19(2) yields the result. □

Specializing to the case of pairwise disjoint reguli, we have:

THEOREM 18.21 (Johnson's Tower). *Let $\Delta_{q^{2n}} > \Delta_{q^2}$ be Desarguesian spreads, for n odd, and suppose*

$$\Theta := \{\pi, \psi \dots\} \subset \Delta_{q^2},$$

is any collection of pairwise disjoint reguli in Δ_{q^2}. Then their unique extension reguli

$$\overline{\Theta} := \{\overline{\pi}, \overline{\psi} \dots\} \subset \Delta_{q^{2n}},$$

are also distinct pairwise disjoint reguli in $\Delta_{q^{2n}}$.

PROOF. Apply Corollary 18.20. □

COROLLARY 18.22. *Assume the conditions of Theorem 18.21, and write $F = GF(q^{2n}) > GF(q^2) = K$, n odd. The group $\mathrm{Aut}(\Theta)$, the subgroup of $\Gamma L(2, K) = \mathrm{Aut}(\Delta_{q^2})$ permuting the members of Θ, extends to the group $F \otimes \mathrm{Aut}(\Theta)$ in $\Gamma L(2, F) < \mathrm{Aut}(\Delta_{q^{2n}})$, and permutes the members of $F \otimes \Theta$, i.e., $\mathrm{Aut}(F \otimes \Theta) > \mathrm{Aut}(\Theta)$.*

If $\theta \in \text{Aut}(\Theta)$ and θ is a Baer collineation, elation, etc., then $F \otimes \Theta \in \text{Aut}(F \otimes \Theta)$ is respectively a Baer collineation, elation, etc.

PROOF. By the above and Remark 18.16 The final assertion follows from the fact that $F \otimes g$, has the same 'geometric' properties as g, cf. Remark 18.14 and Corollary 18.15. □

Thus Theorem 18.21 provides a rich source of planes that are multiply derivable from a given plane. That is:

COROLLARY 18.23 (Field-Lift of Reguli). For any infinite ascending chain of Desarguesian spreads Σ_{q^i} of the form:

$$\Sigma_{q^{2r_0}} < \Sigma_{q^{2r_1}} < \cdots < \Sigma_{q^{2r_i}} < \cdots , \qquad \text{with } r_0 = 1, \ r_i \in \{\, 2n+1 : n \geq 1 \,\}, \ i \geq 1,$$

any set of r pairwise disjoint reguli $\Lambda_{r_0} = \{\, \mathcal{R}_k : 1 \leq k \leq r \,\}$ in the square-order plane $\Sigma_{q^{2r_0}}$ uniquely specifies in every Desarguesian spread $\Sigma_{q^{2r_i}}$, in the ascending chain, a corresponding set of r disjoint reguli Λ_{r_i} such that the r disjoint reguli $\Lambda_{r_{i+1}}$ in $\Sigma_{q^{2r_{i+1}}}$ are the unique reguli extending the r disjoint reguli Λ_{r_i} in $\Sigma_{q^{2r_i}}$.

PROPOSITION 18.24. Assume the conditions of Theorem 18.21, and write $F = GF(q^{2n}) > GF(q^2) = K$, n odd. The group $\text{Aut}(\Theta)$, the subgroup of $\Gamma L(2, K)$ permuting the members of Θ, extends to the group $F \otimes \text{Aut}(\Theta)$ in $\Gamma L(2, F)$, and permutes the members of $F \otimes \Theta$, i.e., $\text{Aut}(F \otimes \Theta) > \text{Aut}(\Theta)$.

REMARK 18.25. By the above and Corollary 18.22.

If a Desarguesian plane of square order admits a net which may be expressed as a disjoint union of reguli, then the net may be replaced by a net obtained by deriving all the reguli in the net. Such multiply derived planes, or subregular planes, yield many line spreads with interesting properties, and contribute to the set of translation planes which are neither generalized André planes nor semifield planes.

The field-lift Corollary 18.23 above demonstrates that even a *single example of a multiply derived line spread, possibly even computer-generated, yields an infinite ascending class of such planes.* Moreover, by Remark 18.14, the entire sequence of these multiply derived planes 'inherit' the subgroup of the base Desarguesian plane that permutes its selected reguli. Thus, there is some control, a lower bound, for the collineation group of any chain of lifted multiply derived planes.

As one would expect, this type of 'increasing chain' property also applies equally well to the multiply derived spreads arising from a Desarguesian chain as above. For example, this follows from:

REMARK 18.26. In any Desarguesian spread $\pi_F > F \otimes \pi_K$, for $K = GF(q^2)$, each regulus $R < \pi_K$ extends to a rational net $F \otimes \pi$, and this specifies a unique regulus $R_F := F \otimes \pi_K$. The identity mapping, embedding π_K into π_F, maps R into R_F, hence also R' into R'_F.

If n is odd and R_i are given pairwise disjoint reguli of π_K, then the identity map the R_i into R'_i's. So the spreads multiply derived from π_K are embedded in the spreads multiply derived from π_F.

PROOF. The K-linear identity embedding $\mathbf{1}_F \colon W \to F \otimes W$ maps reguli R of π_K into rational subnet of π_F that define reguli R_F, and hence also the opposite reguli R' into the corresponding R'_F (e.g., apply Remark 18.11 to see partial spreads map into partial spreads, etc.). Thus, if a disjoint set of reguli in π_K are replaced by their opposites, then $\mathbf{1}_F$ continues to be an embedding from the multiply derived spreads of π_K into the corresponding multiply derived spreads in π_F: these will be disjoint if the degree $[F : K]$ is odd, Corollary 18.20. □

COROLLARY 18.27 (Johnson's Multiply Derived Chains of Planes). For any infinite ascending chain of Desarguesian spreads, as in Corollary 18.23, of the form

$$\left(\Sigma_{q^{2r_i}} \right)_{i \geq 0}, r_i \equiv 1 \pmod 2, \qquad r_i = 1 \Leftrightarrow i = 0,$$

any set of r pairwise disjoint reguli $\Lambda_{r_0} = \{ \mathcal{R}_k : 1 \leq k \leq r \}$ in $\Sigma_{q^{2r_0}}$ uniquely specifies in every Desarguesian spread $\Sigma_{q^{2r_i}}$, in the ascending chain, a corresponding set of r disjoint reguli Λ_{r_i} such that the corresponding infinite chain of multiply derived planes form an increasing sequence of planes

$$\Sigma'_{q^{2r_0}} < \Sigma'_{q^{2r_1}} < \cdots < \Sigma'_{q^{2r_i}} < \cdots, \qquad \text{with } r_0 = 1, r_i \in \{ 2n+1 : n \geq 1 \}, \ i \geq 1,$$

where each $\pi'_i = \Sigma'_{q^{2r_i}}$ is a subplane of $\pi'_{i+1} = \Sigma'_{q^{2r_{i+1}}}$.

The maximal subgroup of $G < \Gamma L(2, q^2) < \Sigma_{q^{2r_0}} \ (= \Sigma_{q^2})$ that permutes the selected set of pairwise disjoint reguli Λ_{r_0} in Σ_{q^2} extends to the subgroup $GF(q^{2r_i}) \otimes G$ of the spread $\Sigma_{q^{2r_i}}$, and this permutes the reguli in Λ_{r_i} and hence is simultaneously a collineation group of the multiply derived plane $\pi'_i = \Sigma'_{q^{2r_1}}$.

CHAPTER 19

Hyper-Reguli Arising from André Hyper-Reguli.

In this chapter, we begin by isolating attention to the following problem:

PROBLEM 19.1. Given a finite André net, does the set of replacements consist of the André-type replacements?

19.1. The Generalized André Planes of Johnson.

DEFINITION 19.2. A 'hyper-regulus' in a vector space V_{2n} of dimension $2n$ over $GF(q)$ is a set of $(q^n - 1)/(q - 1)$ mutually disjoint n-dimensional subspaces over $GF(q)$, which is covered by another set of $(q^n - 1)/(q - 1)$ n-dimensional subspaces over $GF(q)$.

When $n = 2$, a hyper-regulus is simply a regulus. It is well known that replacement of a regulus that sits as a partial spread in an affine Desarguesian plane of order q^2 by its opposite regulus produces a Hall plane. If there is a set of mutually line-disjoint reguli that lie in an affine Desarguesian plane of order q^2, then the translation plane obtained by replacement of each regulus in the set by its opposite regulus is called a 'subregular' plane.

In [1070], Ostrom asked whether vector-space replacement or vector-space $\mathcal{B}$-derivation of a hyper-regulus vector-space net implies that the associated translation plane becomes an André plane, whether the replacement was a standard André replacement. Furthermore, there was an answer provided to this question by Ostrom by giving an example of a hyper-regulus net of order q^6 and degree $(q^6 - 1)/(q - 1)$ in a Desarguesian affine plane which did not give rise to an André plane.

These ideas may be generalized to answer the question in the most general possible manner.

19.2. The Existence of Hyper-Reguli of Non-André Type.

Following the ideas implicit in Ostrom's result [1070], the following is proved by Johnson [760]. These examples may be obtained by variations of the ideas on Wilke replacements as presented in Chapter 16, but here we vary the emphasis somewhat with more focus on the collineation groups possibly obtained. One of the rationals for this variation is to later apply these ideas for the construction of subgeometry and quasi-subgeometry partitions of projective spaces.

THEOREM 19.3. (Johnson [760]). Let q be any prime power and let n be any composite integer. Then, there exists a hyper-regulus which is not an André hyper-regulus of order q^n and degree $(q^n - 1)/(q - 1)$. Consider

$$y = x^q m; \ m^{(q^n - 1)/(q^d - 1)} = 1,$$
$$y = x^{q^{n-(d-1)}} m; \ m^{(q^n - 1)/(q - 1)} = 1 \text{ but } m^{(q^n - 1)/(q^d - 1)} \neq 1.$$

This net is replacement for the André net.

COROLLARY 19.4. The translation plane obtained from a Desarguesian plane by the replacement of an André net by a non-André hyper-regulus of the previous type is not an André plane.

DEFINITION 19.5. The replacements that use subspaces of the form $y = x^{q^{\rho(i)}} m$ as above are called 'generalized André replacements' or the 'generalized André replacements of Johnson'.

It is now possible to construct a tremendous variety of 'new' translation planes by generalizations of the previous constructions. We shall not be concerned with obtaining strictly hyper-regulus replacements, although there will be some of these, as we are more interested in developing a general type of replacement procedure that produces translation planes admitting field groups. To point the direction, we note that it is possible to find hyper-regulus replacements admitting fixed-point-free groups F_d^*.

DEFINITION 19.6. A vector-space net of order q^n and degree $(q^d - 1)/(q - 1)$, where d divides n that also admits a collineation group F_d^* isomorphic to $GF(q^d)^*$ acting transitively on the components of the net is said to be a 'q^d-fan'.

It turns out that these are easy to construct as follows (see [760]).

THEOREM 19.7. Let q be any prime power and let n be any composite integer and let d denote the smallest prime properly dividing n.

(1) If $(n/d, (q^d - 1)/(q - 1)) \neq 1$, then there exists a non-André hyper-regulus of order q^n and degree $(q^n - 1)/(q - 1)$ that admits F_d^*, as a fixed-point-free group with orbits of length $(q^d - 1)/(q - 1)$.

(2) The translation plane of order q^n obtained by replacement of any such non-André hyper-regulus has a spread that is a union of q-fans or q^d-fans.

19.2.1. Replacements Admitting F_d^*; Fans. We now show that we can find new generalized André planes of order q^{ds}, that admit F_d^* as a fixed-point-free collineation group not all of whose orbits of components are trivial. In this section, there is no restriction on d. The following results may be found in Johnson [760]. We now let Σ be a Desarguesian affine plane of order q^{ds} defined by the field F isomorphic to $GF(q^{ds})$ and let F_d denote the subfield isomorphic to $GF(q^{ds})^*$. Let F^* and F_d^*, respectively, denote the associated multiplicative groups. Let e be any divisor of d.

PROPOSITION 19.8. Let $\mathcal{A}_\alpha$ denote the q^e-André net

$$\{ y = xm; \ m^{(q^{ds}-1)/(q^e-1)} = \alpha \},$$

let $\alpha = 1$ and consider cosets of $F_d^{*s(q^e-1)}$ in $F_d^{*(q^e-1)}$. Let

$$\left\{ \alpha_i; \ i = 1, \ldots, (s, \frac{(q^d-1)}{(q^e-1)}) \right\}$$

be a set of coset representatives for $F_d^{*s(q^e-1)}$. Let

$$\mathcal{A}_1^{q^{e\lambda(i)}} : \{ y = x^{q^{e\lambda(i)}} m; \ m^{(q^{ds}-1)/(q^d-1)} \in \alpha_i F_d^{*s(q^e-1)} \},$$

$$\text{where } (d/e, \lambda(i)) = 1, \qquad i = 1, \ldots, \left(s, \frac{(q^d-1)}{(q^e-1)} \right).$$

Then, the kernel homology subgroup of Σ corresponding to F_d^* leaves $\mathcal{A}_1^{q^{e\lambda(i)}}$ invariant and has

$$\frac{(q^{ds}-1)}{(q^d-1)} \bigg/ \left(s, \frac{(q^d-1)}{(q^e-1)} \right)$$

orbits of length

$$(q^d-1)/(q^e-1).$$

Then considering standard André constructions, the following provides the necessary and sufficient conditions.

PROPOSITION 19.9. Assuming the conditions as above, then $\left(s, \frac{(q^d-1)}{(q^e-1)} \right) = 1$ if and only if $\left| \mathcal{A}_1^{q^{e\lambda(1)}} \right| = \frac{(q^{ds}-1)}{(q^e-1)}$ if and only if (for arbitrary $\lambda(i)$) we have a standard q^e-André replacement.

For hyper-regulus replacements, the situation is more complex.

PROPOSITION 19.10. Again assuming the previous conditions.
If $(\lambda(i) - \lambda(j), d/e) = d/e$ for all $i, j = 1, 2, \ldots, \left(s, \frac{(q^d-1)}{(q^e-1)} \right)$, then

$$\bigcup_{i=1}^{\frac{(q^{ds}-1)}{(q^d-1)}} \mathcal{A}_1^{q^{e\lambda(i)}}$$

forms a generalized André replacement, admitting F_d^* as a fixed-point-free collineation group. Hence, we have a generalized André replacement that is a union of q^e-fans. If, in addition, $(\lambda(i), ds/e) = 1$, for all $i = 1, \ldots, \left(s, \frac{(q^d-1)}{(q^e-1)} \right)$, we obtain a hyper-regulus replacement.

Then we have this result:

PROPOSITION 19.11. The generalized André replacement of Proposition 19.10 is not André if and only if we choose at least two of the $\lambda(i)$'s to be distinct, which is possible if and only if there are at least two distinct integers i and j involved in the construction, if and only if $\left(s, \frac{(q^d-1)}{(q^e-1)} \right) \neq 1$.

COROLLARY 19.12. Choose any fixed integer f less than d/e such that $(f, d/e) = 1$. For each i, $i = 1, \ldots, \left(s, \frac{(q^d-1)}{(q^e-1)} \right)$, choose any integer k_i such that $k_i d/e + f \leq ds$ and let $\lambda(i) = k_i d/e + f$. Call the corresponding generalized André replacement $\Sigma_{k_i, f}$. If $\left(s, \frac{q^d-1}{q^e-1} \right) \neq 1$ and at least two of the integers k_i are distinct, then we obtain a generalized André replacement which is not André. Furthermore, the translation plane obtained by this replacement of a single André net is not an André plane.

COROLLARY 19.13. If A_α is any q^e-André net, for $\alpha \in GF(q^e) - \{0\}$, there is a generalized André replacement isomorphic to the replacement for A_1, considered as a q^e-André net.

Although we have a wide variety of generalized André planes which may be constructed by unusual replacements of André nets, it is possible to generalize this even further.

19.2.2. Multiple Generalized André Replacement. Let Σ be a Desarguesian affine plane of order $q^{ds} = q^{d_\alpha s_\alpha}$, for all $\alpha \in GF(q)^*$, where $ds = d_\alpha s_\alpha$. There are $(q-1)$ mutually disjoint q-André nets A_α where

$$A_\alpha = \left\{ y = xm; \, m^{(q^{d_\alpha s_\alpha}-1)/(q-1)} = \alpha \right\}, \qquad \text{for } \alpha \in GF(q)^*.$$

Let $\left\{ \alpha_i; \, i = 1, \ldots, (s, \frac{(q^{d_\alpha}-1)}{(q-1)}) \right\}$ be a coset representative set for $F_d^{*s_\alpha(q-1)}$ in $F_d^{*(q-1)}$. Let

$$\mathcal{A}_\alpha^{q^{\lambda_\alpha(i)}} : \left\{ y = x^{q^{\lambda_\alpha(i)}} m b_\alpha; \, m^{(q^{d_\alpha s_\alpha}-1)/(q^{d_\alpha}-1)} \in \alpha_i F_{d_\alpha}^{*s_\alpha(q-1)} \right\},$$

where $(d_a, \lambda_\alpha(i)) = 1$ and $(\lambda_\alpha(i) - \lambda_\alpha(j), d_\alpha) = d_\alpha$.

REMARK 19.14. (1) The groups $F_{d_\alpha}^{*s_\alpha(q-1)}$ are not necessarily equal for all d_α, as this depends on the integers d_α and s_α. The order of such a group is $\frac{(q^{d_\alpha}-1)}{(q-1)} / \left(s_\alpha, \frac{(q^{d_\alpha}-1)}{(q-1)} \right)$.

(2) If $\lambda_\alpha(i) = \lambda_\alpha(j)$ for all i, j, then we have a standard André replacement.

(3) Let $\eta(ds)$ denote the number of divisors of ds not equal to ds but including 1. For each α, choose any divisor d_α of ds not equal to ds and let $s_\alpha = ds/d_\alpha$. There are at least $\left(s_\alpha, \frac{(q^{d_a}-1)}{(q-1)} \right)^{(s_\alpha-1)}$ possible sets $\left\{ \lambda_\alpha(i); \, i = 1, 2, \ldots, \left(s_\alpha, \frac{(q^{d_a}-1)}{(q-1)} \right) \right\}$, leading to at least this number of distinct replacements for each André net.

Hence, there are at least $\left(s_\alpha, \frac{(q^{d_a}-1)}{(q-1)} \right)^{(s_\alpha-1)\eta(ds)}$ possible generalized André replacements for each André net.

Now we may choose any number of the $(q-1)$ André nets to replace. If we replace exactly one André net, it is clear that the associated plane is isomorphic to the translation plane obtained from any other André net. However, except for such a choice, it is not completely clear what sorts of replacements provide isomorphic translation planes. Hence, we consider that there might be $1 + \sum_{k=2}^{q-1} \binom{q-1}{k}$ possible ways of producing mutually non-isomorphic translation planes.

THEOREM 19.15. (Johnson [**760**]). Let Σ be an affine Desarguesian plane of order q^{ds}. For each André net A_α for $\alpha \in GF(q)^*$, choose a divisor $d_\alpha \neq ds$ of ds and let $s_\alpha = ds/d_\alpha$. Let

$$\mathcal{S}_\alpha^{\lambda_a} = \bigcup \mathcal{A}_\alpha^{q^{\lambda_\alpha(i)}} : \left\{ y = x^{q^{\lambda_\alpha(i)}} m b_\alpha; \, m^{(q^{d_\alpha s_\alpha}-1)/(q^{d_\alpha}-1)} \in \alpha_i F_{d_\alpha}^{*s_\alpha(q-1)} \right\},$$

where $(d_\alpha, \lambda_\alpha(i)) = 1$ and $(\lambda_\alpha(i) - \lambda_\alpha(j), d_\alpha) = d_\alpha$, $i = 1, \ldots, \left(s_\alpha, \frac{(q^{d_\alpha}-1)}{(q-1)} \right)$, $b_\alpha^{(q^{ds}-1)/(q-1)} = \alpha$.

(1) Then $\mathcal{S}_\alpha^{\lambda_a} \cup \{x = 0, y = 0\}$ is a spread for a generalized André plane.

(2) Let $\eta(ds) + 1$ equal the number of divisors of ds. There are at least

$$\left(s_\alpha, \frac{(q^{d_\alpha}-1)}{(q-1)} \right)^{(s_\alpha-1)\eta(ds)k} \left(1 + \sum_{k=2}^{q-1} \binom{q-1}{k} \right)$$

different ways of choosing André nets and generalized André replacements that possibly lead to mutually non-isomorphic translation planes.

We certainly have not determined the exact number of mutually non-isomorphic translation planes that may be obtained from the previous. We leave this as an open problem.

PROBLEM 19.16. How many of the new generalized André planes of order q^{ds} are mutually non-isomorphic?

COROLLARY 19.17. If $d_\alpha = d$ for all choices then the group F_d^* acts as a fixed-point-free collineation group of the associated translation plane. In this case, any such spread is a union of orbits of length either $(q^d - 1)/(q - 1)$ or 1. Hence, the spread is a union of q-fans or components fixed by F_d^* (q^d-fans).

We now consider mixing the replacements by allowing the basic André nets to be of different cardinalities.

19.2.3. Mixed Multiple Hyper-Regulus Replacement. We have previously considered partitioning André nets into parts of sizes $(q^{ds} - 1)/(q^d - 1)$ for various values of d, for up to $(q - 1)$ mutually disjoint André nets. If we take different values of d per André net, we have effectively partitioned the spread into a set of different sized André nets and considered replacements of each of these nets. However, our replacements restricted to the nets of the partition are never André replacements on the nets of the partitions of the q^d-André nets, since in the André case the replacement components would necessarily be of the form $y = x^{q^d \rho(m)} m$, where our components have the form $y = x^{q^{\lambda(i)}} m$, where $(\lambda(i), d) = 1$, for various functions λ.

The previous multiple replacement method may be generalized as follows: Let $q^{ds} = q_\beta^{d_\beta s_\beta}$, where q_α is not necessarily equal to q. Although, we use a similar notation as in the previous, d_β need not be a divisor of d in this setting. Choose any set of mutually disjoint q_β-André nets of $(q^{ds} - 1)/(q_\beta - 1)$ components each for various q_β's. Choose generalized André replacements for each of these nets. Any such translation plane constructed will be a potentially new generalized André plane.

To construct spreads that are unions of fans, we formally present a variation of the basic construction method.

THEOREM 19.18. (Johnson [760]). Let Σ be a Desarguesian affine plane of order q^{ds}. For each q-André A_α, $\alpha \in GF(q)$, choose a divisor e_α of d and partition the components of A_α into $\frac{(q^{e_\alpha} - 1)}{(q - 1)}$ pieces each of cardinality $\frac{(q^{ds} - 1)}{(q^{e_\alpha} - 1)}$.

(1) Each of the partition pieces will be q^{e_α}-André nets admitting the group F_d^*. Let $A_{\alpha,i}$, for $i = 1, \ldots, \frac{(q^{e_\alpha} - 1)}{(q - 1)}$, denote the set of q^{e_α}-André nets arising from A_α.

(2) Consider the subgroup $F_d^{*s(q^{e_\alpha} - 1)}$ of $F_d^{*(q^{e_\alpha} - 1)}$ and let

$$\left\{ \beta_{\alpha,i,j}; j = 1, \ldots, (s, \frac{q^d - 1}{q^{e_\alpha} - 1}) \right\}, \ i = 1, \ldots, \frac{(q^{e_\alpha} - 1)}{(q - 1)}, \ \alpha \in GF(q) - \{0\}$$

be a coset representative set for $F_d^{*s(q^{e_\alpha} - 1)}$ of $F_d^{*(q^{e_\alpha} - 1)}$, depending on α and i.

Let b_α be any element of F_{ds}^* such that $b_\alpha^{(q^{ds} - 1)/(q - 1)} = \alpha$. Let

$$A_{\alpha,i}^{e_\alpha \lambda_{\alpha,i}(j)} = \left\{ y = x^{q^{e_\alpha \lambda_{\alpha,i}(j)}} m b_\alpha; \ m^{(q^{ds} - 1)/(q^d - 1)} \in \beta_{\alpha,i,j} F_d^{*s(q^{e_\alpha} - 1)} \right\},$$

where the functions $\lambda_{\alpha,i}$ are chosen to satisfy the conditions $(\lambda_{\alpha,i}(j), d/e_\alpha) = 1$ and $(\lambda_{\alpha,i}(j) - \lambda_{\alpha,i}(j), d/e_\alpha) = d/e_\alpha$.

Then $A_{\alpha,i}^{e_\alpha \lambda_{\alpha,i}(j)}$ is a union of $\left(\frac{(q^{ds}-1)}{(q^d-1)} \Big/ \left(s, \frac{(q^d-1)}{(q^{e_\alpha}-1)} \right) \right) F_d^*$ orbits of length $\frac{(q^d-1)}{(q^{e_\alpha}-1)}$;
a set of $\left(\frac{(q^{ds}-1)}{(q^d-1)} \Big/ \left(s, \frac{(q^d-1)}{(q^{e_\alpha}-1)} \right) \right) q^{e_\alpha}$-fans.

Hence, $\bigcup_{i=1}^{\frac{(q^{e_\alpha}-1)}{(q-1)}} A_{\alpha,i}^{e_\alpha \lambda_{\alpha,i}(j)}$ consists of $\frac{(q^{e_\alpha}-1)}{(q-1)} \left(\frac{(q^{ds}-1)}{(q^d-1)} \Big/ \left(s, \frac{(q^d-1)}{(q^{e_\alpha}-1)} \right) \right) q^{e_\alpha}$-fans.

(3) $\left\{ x=0, y=0, \bigcup_{\alpha \in GF(q)^*} \bigcup_{i=1}^{\frac{(q^{e_\alpha}-1)}{(q-1)}} A_{\alpha,i}^{e_\alpha \lambda_{\alpha,i}(j)} \right\} = S^{(e_\alpha;\, \alpha \in GF(q)^*)}$ is a spread

defining a generalized André plane that admits F_d^* as a fixed-point-free collineation
group. The spread is the union of two q^d-fans ($x=0, y=0$) and

$$\frac{(q^{e_\alpha}-1)}{(q-1)} \left(\frac{(q^{ds}-1)}{(q^d-1)} \Big/ \left(s, \frac{(q^d-1)}{(q^{e_\alpha}-1)} \right) \right)$$

q^{e_α}-fans, for each of $q-1$ possible divisors e_α of d.

For $\alpha \neq \delta$, the divisors e_α and e_δ of d are independent. For $i \neq k$, the functions
$\lambda_{\alpha,i}$ and $\lambda_{\delta,k}$ are independent subject to the prescribed conditions.

PROBLEM 19.19. Completely determine the isomorphism classes of translation
planes that arise from the mixed multiple generalized André replacements.

CHAPTER 20

Translation Planes with Large Homology Groups.

In Chapter 16, we considered generalized André planes and listed a variety of examples. In this chapter, we consider translation planes admitting large affine homology groups and basically show that we return usually to generalized André planes.

In order to appreciate how translation planes admitting large homology groups connect with generalized André planes, we begin by reminding the reader that for a translation plane of order q^t, the largest affine homology group has order $q^t - 1$. When this maximal order occurs, there are actually two affine homology groups of orders $q^t - 1$ and the plane is a nearfield plane. These groups fix the same two parallel classes and so the axis of one group is the coaxis of the other (the coaxis is the affine line containing the center of the group on the line at infinity). When two affine homology groups have the sort of symmetry, we call the two groups 'symmetric homology groups'. If q^t is odd, suppose there is an affine homology group of order $(q^t - 1)/2$, then it turns out that there is not necessarily a second corresponding affine homology group so that the union is a symmetric set of homology groups. In Chapter 51 on sharply k-transitive sets, we mention the connection with affine homology groups and flocks of hyperbolic quadrics (a translation plane of order q^2 with spread in $PG(3, q)$ admitting an affine homology group of order $q-1$ with the property that the axis and coaxis union one component orbit is a regulus corresponds to a flock of a hyperbolic quadric). When one assumes the existence of two affine homology groups of order $(q^t - 1)/2$, then it is not clear that the two groups are symmetric. However,

THEOREM 20.1. (Hiramine and Johnson [**521**]). Let π be a translation plane of order p^m which admits at least two affine homology groups of order $(p^m - 1)/2$. Then π is one of the following planes: (i) Desarguesian, (ii) a generalized André plane, (iii) the irregular nearfield plane of order 23^2, (iv) the irregular nearfield plane of order 7^2, (v) the exceptional Lüneburg plane of order 7^2.

The proof uses combinatorial group theory to establish that if the plane is not Desarguesian, then the two affine homology groups are symmetric. Then the application of Lüneburg's theorem, on the existence of Abelian groups G fixing two components, such that G_L is irreducible on L, for any remaining component L, shows that the plane is a generalized André plane with two exceptional orders.

In a similar analysis, Draayer [**327**] considers translation planes with pairs of groups of order $(q^t - 1)/3$.

THEOREM 20.2. (Draayer [**327**, (1.1)]). Assume that π is a translation plane of order p^t, for p a prime, that admits a pair of distinct affine homology groups of order $(p^t - 1)/3$ in the translation complement. Then one of the following two situations occurs: (i) π is a generalized André plane (or Desarguesian), (ii) $t = 2$ and $p \in \{5, 7, 11, 17, 19, 23, 47, 71, 89, 179\}$, or (iii) $p^t = 2^t$.

In this setting, there are exceptional cases, which are as yet unresolved. However:

THEOREM 20.3. (Draayer [**327**, (5.5)]). The irregular nearfield planes of orders 5^2, 7^2, 11^2 (with solvable group) and 23^2 are non-generalized André planes of order p^2 admitting two affine homology groups of orders $(p^2 - 1)/3$.

Now that the focus is back on generalized André planes, there is no particular need to assume that there are two affine homology groups of order $(q^t - 1)/2$.

However, Chapter 69 on the so-called triangle transitive translation planes, translation planes of order q^2 that admit affine homology groups of order $(q^t - 1)/2$ are considered from a different viewpoint. There and here the following theorem is of importance.

THEOREM 20.4. (see Hiramine and Johnson [**520**] and Draayer [**326**, (4.2) and (5.2)]). Let π be a generalized André plane of order q^n that admits symmetric homology groups of order $(q^n - 1)/2$ (the axis and coaxis of one group is the coaxis and axis, respectively, of the second group). Assume that the right and middle nuclei of the corresponding quasifield coordinatizing π are equal. Then the spread may be represented as follows:

$$x = 0, y = x^{q^i} w^{s(q^i-1)/(q-1)} \alpha, y = x^{p^t q^j} w^{v q^j} w^{s(q^j-1)/(q-1)} \beta$$

where $\alpha, \beta \in \mathcal{A}$ a subgroup of $GF(q^n)^*$ of order $(q^n - 1)/2n$, where $i, j = 0, 1, \ldots, n-1$ and where t and v have the following restrictions:

$$\begin{aligned} \text{for } q &\equiv& 1 \bmod 4 \text{ or } n \text{ odd, we may choose } s = 2, \, v = 1, \, t \geq 0, \\ \text{for } q &\equiv& -1 \bmod 4 \text{ and } n \equiv 2 \bmod 4, \text{ we may choose } s = 1, \\ v &=& n, \, t \geq 0 \text{ and even, or} \\ \text{for } q &\equiv& 3 \bmod 8 \text{ and } n \equiv 0 \bmod 4, \text{ we may choose } s = 1, \\ v &=& 2[n]_{2'}, \, t \geq 0 \text{ and even.} \end{aligned}$$

Furthermore, we have the congruence:

$$s(p^t - 1) \equiv v(q - 1) + 2nz \bmod (q^n - 1),$$

$$\text{for some } z \text{ so that } s(p^t - 1) \equiv v(q - 1) \bmod 2n.$$

Conversely, generalized André planes with the above conditions have two symmetric homology groups corresponding to equal right and middle nuclei in the associated quasifield.

When there is exactly one affine homology group of order $(q^t - 1)/2$ is a generalized André plane, it is still possible to give general requirements for the associated spread. For example,

THEOREM 20.5. (Draayer [**326**]). Let π be a translation plane of order q^n that admits an affine homology group H_y of order $(q^n - 1)/2$ with axis $y = 0$ and co-axis $x = 0$. Then the spread for π may be represented as follows:

$$x = 0, y = x^{q^i} w^{s(q^i-1)/(q-1)} \alpha, \quad y = x^{p^t q^j} w^{v q^j} w^{s(q^j-1)/(q-1)} \beta$$

where $\alpha, \beta \in \mathcal{A}$ a subgroup of $GF(q^n)^*$ of order $(q^n-1)/2n$, where $i, j = 0, 1, \ldots, n-1$ and where t and v have the following restrictions:

for $q \equiv 1 \bmod 4$ or n odd, let $s = 2$, $v = 1$, $t \geq 0$, or

for $q \equiv -1 \bmod 4$ and $n \equiv 2 \bmod 4$, let $s = 1$, $v = n$, $t \geq 0$ and even, or

for $q \equiv 3 \bmod 8$ and $n \equiv 0 \bmod 4$, let $s = 1$, $v = 2[n]_{2'}$ $t \geq 0$ and even.

Furthermore, the group H_y has the following form:

$$\left\langle (x,y) \longmapsto (x, y^{q^i} w^{s(q^i-1)/(q-1)} \alpha); \ \alpha \in \mathcal{A}, \text{ where } i = 0, 1, \ldots, n-1 \right\rangle.$$

We note that it is clear that the regular nearfield planes are generalized André planes. We also note that it is possible to have generalized André planes of order q^t with one but not two affine homology groups of order $(q^t-1)/2$ as noted above. The classification of such generalized André is accomplished in Hiramine and Johnson [**520**] with the exception that one case was overlooked. This was, however, corrected by Draayer [**326**].

20.1. The Generalized André Planes with Index 2 Homology Groups.

An affine homology group corresponds to either a right or middle nucleus of the associated coordinate system. If a finite translation plane admits a right nucleus R, then in the translation plane associated with the dual spread of the original plane, R is a middle nucleus.

REMARK 20.6. If π is a finite generalized André plane with spread S and right nucleus R and π^d is the translation plane corresponding to the dual of S, then π^d is a generalized André with middle nucleus R.

PROOF. The group of π^d is isomorphic to the group of π, implying that π^d is a generalized André plane. $\qquad\square$

Hence, in the following, we assume that π is a generalized André of order p^t with a right nucleus whose non-zero elements have order $(p^t - 1)/2$.

THEOREM 20.7. (Hiramine–Johnson [**520**, (4.1)], Draayer [**326**, (5.2)]). Let $(F, +, \cdot)$ be $GF(p^t)$, where p is an odd prime. Let $q^n = p^t$ and assume that every prime factor of n divides $q - 1$.

(a) If q is not congruent to $-1 \bmod 4$, let $s = 2, r_1 = 1$, and $t_1 \geq 0$.

(b) If $q \equiv -1 \bmod 4$ and $n \equiv 2 \bmod 4$, let $s = 1, r_1 = n$, and $t_1 \geq 0$, t_1 even.

(c) If $q \equiv 3 \bmod 8$ and $n \equiv 0 \bmod 4$, let $s = 1, r_1 = 2[n]_{2'}$ and $t_1 \geq 0$, t_1 even.

Let ω be a primitive element of the multiplicative group of $(F, +, \cdot)$. Then

$$F^* = \bigcup_{0 \leq k < 2} \bigcup_{0 \leq i < n} \omega^{kr_1 q^i} \cdot \omega^{s(q^i-1)/(q-1)} \cdot \omega^{2n\mathbb{Z}}.$$

Furthermore, the following collections of subspaces including $x = 0, y = 0$,

$$y = x^{q^i} \omega^{s(q^i-1)/(q-1)+2nj}, \ y = x^{p^{t_1} q^i} \omega^{r_1 q^i + s(q^i-1)/(q-1)+2nj},$$

$$0 \leq i < n, 0 \leq j < (q^n - 1)/2n$$

is a spread for a generalized André plane of order p^t admitting an affine homology group of order $(p^t - 1)/2$ with axis $y = 0$ and coaxis $x = 0$.

We now investigate how such translation planes may be constructed by considering net replacements in certain associated nearfields.

20.2. Net Replacements in the Dickson Nearfield Planes.

The reader is directed to Section 8.1 for definitions and construction of the nearfield planes. We next consider net replacements in the regular nearfield planes due to Hiramine and Johnson [**520**] for planes of order q^t admitting affine homology groups of order $(q^t - 1)/2$ and of Draayer [**326**] for planes of admitting affine homology groups of order $(q^t-1)/3$. In all cases, the resulting translation planes will be generalized André planes; however, it is probable that these cannot be obtained using Ostrom's 1st or 2nd replacement theorems, see Section 16.1. These planes were found from translation planes of order q^n that admit affine homology groups of orders $(q^n - 1)/i$, where $i = 1$,or 2. Since certain of the regular nearfield planes admit affine homology groups of such orders, it is possible to make net replacements to construct other such planes (the nearfield planes admit affine homology groups of orders $(q^n - 1)$ but not all admit subgroups on index 2).

Let $\{q, n\}$ be a Dickson pair, where q is odd and let π be a Dickson nearfield plane with coordinate system $(F^*, +, \circ)$. Let F be isomorphic to $GF(q^n)$ and assume that ω is a primitive element of F^*. Let Z denote the subgroup of order $(q^n - 1)/n$.

Then

$$F^* = \omega^s Z \cup \omega^{s(q^2-1)/(q-1)} Z \cup \cdots \cup \omega^{s(q^i-1)/(q-1)} \cup \cdots \cup \omega^{s(q^n-1)/(q-1)},$$

where $(s, n) = 1$ and $x \circ m = x^{q^i} m; m \in \left\langle \omega^{s(q^i-1)/(q-1)} \right\rangle$.

Then in Hiramine and Johnson [**520**], there is a development of net replacements that isolates groups of order $(q^t - 1)/2$. The main idea is in the following lemma.

LEMMA 20.8. (Hiramine and Johnson [**520**]). Represent a Dickson nearfield plane as above. Then $\left\langle x \longmapsto x^{q^i} \omega^{s(q^i-1)/(q-1)} \alpha;\ \alpha \in Z,\ i = 1, 2, \ldots, n \right\rangle$ represents the associated nearfield group N. If N^- is a subgroup of order $(q^n - 1)/2$, then N is one of the following two types:

$$\text{Type 1: } H_1 : \left\langle x \longmapsto x^{q^i} \omega^{s(q^i-1)/(q-1)} \alpha;\ \alpha \in Z^-,\ i = 1, 2, \ldots, n \right\rangle,$$

where Z^- is a cyclic subgroup of order $(q^n - 1)/2n$, or

$$\text{Type 2: } H_2 : \left\langle x \longmapsto x^{q^i} \omega^{s(q^i-1)/(q-1)} \alpha;\ \alpha \in Z,\ i = 1, 2, \ldots, n/2 \right\rangle.$$

20.2.1. Type 1 Replacements.

THEOREM 20.9. (Hiramine and Johnson [**520**, (3.2)]). Consider a Dickson nearfield plane represented as above. Let Z denote the subgroup of F^* of order $(q^n - 1)/n$ and Z^- the subgroup of F^* of order $(q^n - 1)/2n$ (note for any pair $2n$ divides $(q^n - 1)$, for q odd). Let the spread be represented as follows:

$$x = 0, y = 0, x^{q^i} \omega^{s(q^i-1)/(q-1)} \alpha;\ \alpha \in Z,\ i = 1, 2, \ldots, n,\ \langle \omega \rangle = F^*, (s, n) = 1.$$

(i) If n is odd and s even or n even and $q \equiv -1 \bmod 4$, then

$$P_0 = \left\{ y = x^{q^i} \omega^{s(q^i-1)/(q-1)} \beta;\ \beta \in Z^-,\ i = 1, 2, \ldots, n \right\}$$

and

$$P_1 = \left\{ y = x^{q^i} \omega^{s(q^i-1)/(q-1)} b;\ b \in Z - Z^-,\ i = 1, 2, \ldots, n \right\}$$

are partial spreads of degree $(q^n - 1)/2$ invariant under the affine homology group

$$G = \left\langle (x,y) \longmapsto \left(x, y^{q^i} \omega^{s(q^i-1)/(q-1)}\beta\right); \beta \in Z^-, \ i = 1,2,\dots,n \right\rangle$$

of order $(q^n - 1)/2$. Furthermore, the spread for π is $\{x = 0, y = 0\} \cup P_0 \cup P_1$.

(ii) When the spread does have the form $\{x - 0, y = 0\} \cup P_0 \cup P_1$, then

$$\overline{P_{1,t}} = \left\{ y = x^{p^t q^i} \omega^{n q^i} \omega^{s(q^i-1)/(q-1)}\beta; \ \beta \in Z^-, \ i = 1,2,\dots,n \right\},$$

where $q = p^r$ and t any fixed integer, is a partial spread which is invariant under the group G.

(a) If n is even (and hence s is odd), then $\overline{P_{1,t}}$ is a replacement for P_1, for any t.

(b) If n is odd and s even, then $\overline{P_{1,t}}$, for $t = 2s$, is a replacement for P_1.

(iii) Thus, we obtain a set of translation planes of order q^n which admits an affine homology group $(q^n - 1)/2$. The kernel of the translation plane obtained using $p^{2t} = h$ is the intersection of the fixed field of $x \longmapsto x^h$ with $GF(q)$.

20.2.2. Type 2 Replacements. Here we again follow Hiramine and Johnson [**520**], who point out that many of generalized André planes of order q^n that admit affine homology groups may be obtained from a Dickson nearfield plane by appropriate net replacement. Draayer [**327**] generalizes this construction to produce from Dickson nearfield planes generalized André planes of order q^n that admit affine homology groups of order $(q^n - 1)/d$, where d divides n. Draayer shows that the kernel of the associated constructed translation plane may be any subfield of $GF(q^d)$, so in the case of $d = 2$, we may construct planes of order q^n, where the kernel is any subfield of $GF(q^2)$.

THEOREM 20.10. (Hiramine and Johnson [**520**, (3.4)] but note that the kernel is a subfield of $GF(q^2)$). Let π be a Dickson nearfield plane of order q^n, where $\{q,n\}$ is a Dickson pair and n is even. Let $F \simeq GF(q^n)$, $\langle \omega \rangle = F^*$ and $Z \le F^*$ of order $(q^n - 1)/n$ and let $(s,n) = 1$.

(i) Then

$$P_0 = \left\{ y = x^{q^{2i}} \omega^{s(q^{2i}-1)/(q-1)}\beta; \ \beta \in Z, \ i = 1,2,\dots,n/2 \right\}$$

and

$$P_1 = \left\{ y = x^{q^{2i+1}} \omega^{s(q^{2i+1}-1)/(q-1)}\alpha; \ b \in Z, \ i = 1,2,\dots,n/2 \right\}$$

are partial spreads invariant under the affine homology group

$$G = \left\langle (x,y) \longmapsto \left(x, y^{q^{2i}} \omega^{s(q^{2i}-1)/(q-1)}\beta\right); \beta \in Z, \ i = 1,2,\dots,n/2 \right\rangle$$

of order $(q^n - 1)/2$. The spread for π is

$$\{x = 0, y = 0\} \cup P_0 \cup P_1.$$

(ii) Let

$$\overline{P_{1,t}} = \left\{ y = x^{p^t q^{2i}} \omega^{q^{2i}} \omega^{s(q^{2i+1}-1)/(q-1)}\alpha; \ \alpha \in Z, \ i = 1,2,\dots,n/2 \right\},$$

where $q = p^r$ and t any fixed integer.

Then $\overline{P_{1,t}}$ is a replacement for P_1 and the corresponding spread

$$\{x = 0, y = 0\} \cup P_0 \cup \overline{P_{1,t}}$$

is a translation plane admitting an affine homology group G of order $(q^n - 1)/2$.

RESULT 20.11. There is a direct generalization of the previous theorem due to Draayer [**326**, (6.3)]. If d is any divisor of n there are corresponding replacements $\overline{P_{1,t}}$, where the corresponding requirement is that d divides $p^t - 1$, and where 2 is replaced by d in the above statement.

THEOREM 20.12. (Draayer [**520**, (6.3)]). Let π be a Dickson nearfield plane of order q^n, where $\{q, n\}$ is a Dickson pair and n is even. Let $F \simeq GF(q^n)$, $\langle \omega \rangle = F^*$ and $Z \le F^*$ of order $(q^n - 1)/n$.

For each k, $0 \le k < d$, select an automorphism $\rho_k \colon x \longmapsto x^{p^{t_k}}$ of F, with $\rho_0 = 1$.

(i) Then, for each k,

$$P_k = \left\{ y = x^{q^{k+di}} \, \omega^{(q^{k+di}-1)/(q-1)} \beta; \ \beta \in Z, \ i = 1, 2, \ldots, n/d \right\}$$

is a partial spread invariant under the affine homology group

$$G = \left\langle (x, y) \longmapsto \left(x, y^{q^{di}} \omega^{(q^{di}-1)/(q-1)} \beta \right); \ \beta \in Z, \ i = 1, 2, \ldots, n/d \right\rangle$$

of order $(q^n - 1)/d$. The spread for π is $\{x = 0, y = 0\} \cup \{ P_k; 0 \le k < d \}$.

(ii) If $d \mid p^{t_k} - 1 \ \forall k$, $0 \le k < d$, then

$$\{x = 0, y = 0\} \cup \left\{ \overline{P_{t_k}}; 0 \le k < d \right\},$$

$$\overline{P_{t_k}} = \left\{ y = x^{p^{t_k} q^{di+k}} \, \omega^{(q^{di+k}-1)/(q-1)} \alpha; \ \alpha \in Z, \ i = 1, 2, \ldots, n/2 \right\},$$

where $q = p^r$, is a translation plane admitting an affine homology group G of order $(q^n - 1)/d$.

The kernel of this generalized André plane has order p^g, where

$$g = \gcd \left\{ t_k + rd; 1 \le k \le d \right\}$$

(where $t_d = 0$), so that the kernel is a subfield of $GF(q^d)$.

The above theorem of Draayer thus constructs a variety of interesting and important generalized André planes of order q^n admitting affine homology groups of order $(q^n - 1)/d$. In particular, Draayer is interested in constructing planes admitting affine homology groups of order $(q^n - 1)/3$, since his theorem on symmetric homology groups of this order normally implies that the associated translation plane is generalized André.

CHAPTER 21

Derived Generalized André Planes.

The fundamental question arises, when is a translation plane of order q^n constructed from a Desarguesian plane by replacement of a set of mutually disjoint hyper-reguli? We have previously discussed theorems that force a translation plane to be subregular. In this chapter, we give complete generalizations of these theorems. The results are due to Jha–Johnson [581] and provide criteria by which a translation plane is recognized as being constructed by a Desarguesian plane. It may be noted that there are André planes and generalized André planes that are derivable and when derived produce planes that cannot be generalized André planes as they admit a Baer group B of order $(q^n - 1)/(q - 1)$. These planes also admit a cyclic group Z of order $(q^{2n} - 1)/(q - 1)$ such that $B \times Z$ acts on the plane. Conversely, we may characterize such planes as to their construction.

21.1. The Question of Multiple Hyper-Regulus Replacement.

The fundamental theorem of Jha–Johnson [581] is:

THEOREM 21.1. (Jha–Johnson [581]). Let π be a translation plane of order q^n and kernel $GF(q)$ that admits a cyclic collineation group C of order $q^n - 1$ in the translation complement. If C fixes at least three mutually disjoint n-dimensional $GF(q)$ subspaces, then there exists a Desarguesian plane Σ such that C is a kernel homology group of Σ. Any component L that is in $\pi - \Sigma$ is a subspace of Σ that lies across a set of $(q^n - 1)/(q^s - 1)$, components of Σ, where the 'relative kernel' of L is $GF(q^s)$, and therefore lies in an orbit of length $(q^n - 1)/(q^s - 1)$.

DEFINITION 21.2. A translation plane constructed as in the previous theorem is said to be constructed by 'mixed multiple hyper-regulus replacement'. If all of the degrees of the hyper-reguli are equal, we shall use the term 'multiple hyper-regulus replacement' if there are at least two mutually disjoint hyper-reguli and simple 'hyper-regulus replacement' if there is but one hyper-regulus replacement.

THEOREM 21.3. (Jha–Johnson [581]). Let π be a translation plane of order q^n that admits a cyclic collineation group Z of order $q^n - 1$ in the translation complement. If there is a collineation σ of order p, where $p^r = q$, and p is a prime that commutes with Z, then π may be constructed by multiple hyper-regulus replacement from a Desarguesian affine plane.

21.2. p-Elements.

Now consider if the existence of a p-element σ centralized by a cyclic group Z of order $q^n - 1$ ensures that the plane constructed by multiple hyper-regulus replacement. Let σ be a p-element that is centralized by a cyclic group Z of order $(q^n - 1)$. Let σ fix pointwise a subspace X that is fixed by Z. If X is incident with a component L, then Z fixes L and by order fixes another component M. But σ must move the second fixed component M so Z fixes at least three components. So assume that σ is planar of order p^a, $q = p^r$. So Z fixes Fix σ. Let τ be a p-primitive divisor of $q^n - 1$. The subregular situation is considered in this text and also in Jha and Johnson [**644**], so assume that $n > 2$. Then, τ must fix Fix σ pointwise. Choose any component L that non-trivially intersects Fix σ. Then τ fixes $L \cap$ Fix σ pointwise so then fixes L pointwise, a contradiction. Hence, when there is a p-element, there is an associated Desarguesian affine plane where Z is a kernel homology group. Let L be any component of $\pi - \Sigma$ as in Theorem 21.1. Then whatever set of components L lies across, let $GF(p^a)$ be the minimum dimension of that set. Then there is a group of order exactly $p^a - 1$ that fixes L and is transitive on the vectors on intersection components. This implies that all intersections have cardinality $p^a - 1$ so there is a hyper-regulus of degree $(q^n - 1)/(p^a - 1)$. That is, we have the following result.

THEOREM 21.4. (Jha–Johnson [**581**]). Let π be a translation plane of order q^n that admits a cyclic collineation group Z of order $q^n - 1$ and a p-collineation σ, for $p^r = q$, in the translation complement. If σ commutes with Z, then π may be constructed from a Desarguesian plane by multiple hyper-regulus replacement of hyper-reguli of degrees $(q^n - 1)/(p^a - 1)$, where a divides nr, for a set of divisors a.

It is now possible to construct new planes by the derivation of certain André planes.

21.3. Derivable Generalized André Planes.

Let Σ be the Desarguesian plane of order q^{2n} coordinatized by a field $GF(q^n)$ with spread $x = 0, y = xm; m \in GF(q^n)$. Let π be an André plane of order q^{2n} constructed from Σ of order q^{2n} by the replacement of André nets of degree $(q^{2n} - 1)/(q - 1)$ by nets of components $y = xm$ such that $m^{(q^{2n}-1)/(q-1)} \neq 1$, for $m \in GF(q^n)$. If α is in $GF(q^n)$, then $\alpha^{(q^{2n}-1)/(q-1)} = \delta^{(q^n+1)} = \delta^2$, where $\delta \in GF(q)$. Hence, if we choose André nets $A_\rho^q = \{ y = xm; m^{(q^{2n}-1)/(q-1)} = \rho,\ \rho$ a non-square in $GF(q) \}$, then these André nets will be disjoint from the standard regulus net $GF(q^n) \cup (\infty)$. In this case, there is a homology group of Σ of order $(q^{2n}-1)/(q-1)$ that fixes each André net of degree $(q^{2n}-1)/(q-1)$. The subgroup B of this homology group that fixes $GF(q^n) \cup (\infty)$ has order

$$((q^{2n} - 1)/(q - 1), q^n - 1) = (q^n - 1)/(q - 1)(q^n + 1, q - 1) = 2(q^n - 1)/(q - 1).$$

If we derive any such André plane by replacing $GF(q^n) \cup (\infty)$, we obtain a translation plane of order q^{2n} with kernel $GF(q)$ that admits a Baer homology group of order $2(q^n - 1)/(q - 1)$. This plane is constructed by the replacement of mutually disjoint hyper-reguli of degrees $(q^{2n} - 1)/(q - 1)$ and $(q^{2n} - 1)/(q^n - 1)$. Hence, we have part of the following theorem.

THEOREM 21.5. ((Jha–Johnson [**581**]). (1) For any n and q odd, there are André planes of order q^{2n}, with kernel $GF(q)$, which are derivable and which derive translation planes admitting Baer groups of order $2(q^n - 1)/(q - 1)$. The derived plane is a plane obtained from a Desarguesian plane by the replacement of mutually disjoint hyper-reguli of degrees $(q^{2n} - 1)/(q - 1)$ and $(q^{2n} - 1)/(q^n - 1)$. Such a plane cannot be André or generalized André.

(2) Choose a set of elements δ in $GF(q^2)$, such that δ is not a $q^n + 1^{\text{st}}$ power (either n is odd and δ is not a $q + 1^{\text{st}}$ power or n is even and δ is non-square in $GF(q^2)$, and q is odd in this case). Then choose any André plane by choosing a non-empty subset of André nets A_δ for δ not a $q+1^{\text{st}}$ power, of degree $(q^{2n}-1)/(q^2-1)$. Then by deriving any such plane, we obtain a plane with kernel $GF(q)$, which cannot be a generalized André plane or André plane. This plane admits a Baer collineation group of order $(q^n - 1)/(q-1)$ and is constructed from a Desarguesian plane by the replacement of a set of mutually disjoint hyper-reguli of degrees $(q^{2n} - 1)/(q^2 - 1)$ and $(q^{2n} - 1)/(q^n - 1)$.

(3) Now assume that q is odd and choose two non-empty sets of elements λ_q and λ_{q^2}: λ_q is a subset of the non-squares of $GF(q)$, and λ_{q^2} is a subset of the non-$q + 1^{\text{st}}$ powers of $GF(q^2)$. Note that any element of $GF(q)$ in $GF(q^2)$ is a $q^d + 1^{\text{st}}$ power. Hence, $\lambda_q \cap \lambda_{q^2} = \varnothing$. Choose any set of André nets A_ρ^q of degree $(q^{2n} - 1)/(q - 1)$ where ρ is non-square and choose any set of André nets $A_\delta^{q^2}$ of degree $(q^{2n} - 1)/(q - 1)$. Then the two sets of André nets of different degree form a partial spread disjoint from $GF(q^n) \cup (\infty)$. The plane obtained is a plane derived from a generalized André plane and this plane cannot be a generalized André plane or André plane. This plane admits a Baer collineation group of order $(q^n-1)/(q-1)$ and is constructed from a Desarguesian plane by the replacement of a set of mutually disjoint hyper-reguli of degrees $(q^{2n}-1)/(q^{2n}-1)$, $(q^{2n}-1)/(q^n-1)$ and $(q^{2n} - 1)/(q^n - 1)$.

Although these planes of order q^{2n} are not generalized André or André planes they do admit a group $B \times Z$, where B has order $(q^n - 1)/(q - 1)$ and Z is cyclic of order $(q^{2n} - 1)$. The question is, is it possible to classify translation planes with such groups?

Jha and Johnson [581] actually do characterize translation planes that admit a suitable Baer group using hyper-regulus replacement, up to type of construction.

THEOREM 21.6. (Jha–Johnson [581]). Let π have order q^{2n}, and assume that the kernel contains $GF(q)$. Suppose that π admits a Baer group B of order $(q^n - 1)/(q - 1)$ and a cyclic group of order $q^{2n} - 1$, C, that commutes with B.

(1) Then π may be constructed from a Desarguesian affine plane Σ of order q^{2n} by multiple hyper-regulus replacement of hyper-reguli of degree $(q^{2n} - 1)/(q^a - 1)$ for various divisors a of $2n$, where exactly one hyper-regulus net is a regulus.

(2) If B leaves each hyper-regulus invariant, then π may be constructed from a Desarguesian affine plane of order q^{2n} by multiple hyper-regulus replacement of hyper-reguli of degrees $(q^{2n} - 1)/(q - 1)$, $(q^{2n} - 1)/(q^2 - 1)$ and $(q^{2n} - 1)/(q^n - 1)$.

CHAPTER 22

The Classes of Generalized André Planes.

The following list provides the known classes of generalized André planes and the derived generalized André planes. Many of the planes may be constructed using Ostrom's replacement theorems.

André

Multiple André (Ostrom)

Dickson Nearfield

Johnson (variation on 'all except' method—new replacements of André nets)

Hiramine–Johnson order q^k admit affine homology group order $(q^k - 1)/2$

Draayer order q^k admit affine homology group order $(q^k - 1)/3$

Draayer order q^k admit affine homology groups order $(q^k - 1)/d$,
   where d divides k

Wilke–Zemmer order 3^n, n odd (multiple André)

Narayana Rao and Zemmer

Wilke ('all except' method)

Ostrom 1st and 2nd constructions, coupled with Foulser net replacement

Foulser Q_g planes

Derived André and derived generalized André of Ostrom and Jha–Johnson

Triangle Transitive generalized André of Draayer–Johnson–Pomareda

22.1. Wilke–Zemmer, Order 3^n.

$$
\begin{aligned}
y &= x^3 m, \text{ for } m \text{ non-square,} \\
y &= x^{3^k} m, \text{ for } m^{(3^n - 1)/(3^k - 1)} = 1 \\
y &= xm; \text{ for } m \text{ square but } m^{(3^n - 1)/(3^k - 1)} \neq 1.
\end{aligned}
$$

22.2. Narayana Rao–Zemmer Planes.

Narayana Rao and Zemmer [**1026**] have constructed a set of generalized André planes of order q^d, by multiple André replacement. Let $d = r_1^{\alpha_1} r_2^{\alpha_2} \cdots r_t^{\alpha_t}$ denote the prime decomposition of d (however, we are not ordering the r_i). Choose any set of $k \geq 3$ integers β_i, such that $1 \leq \beta_i \leq \alpha_i$. Let $N = q^d - 1$ and I_N, the integers modulo N. Consider $GF(q^d)^*$, we need to specify a function $\lambda \colon GF(q^d)^* \longmapsto I_N$, which satisfies the conditions for a λ-system in Section 17.1. Let $u = r_1 r_2 \cdots r_k$, $v_i = v/r_i^{\beta_i}$, and let

$$X_i = \{\, j \in I_N; \, j \equiv i \mod (q^{u_i} - 1) \,\}, i = 1, 2, \ldots, k$$

and let $X_{k+1} = I_N - \bigcup_{i=1}^{k} X_i$. Now take the function λ as follows:

$$\lambda(j) = 0 \text{ for } x \in X_{k+1}, \lambda(j) = v_i \text{ for } j \in X_i, \, i = 1, 2, \ldots, k.$$

THEOREM 22.1. (Narayana Rao and Zemmer [**1026**]). Under the above assumptions, λ defines a generalized André plane.

In the Narayana Rao–Zemmer generalized André quasifields, the right and middle associators (nuclei) are equal. By transpose, there are also planes with the left and middle associators equal. These groups are equal to

$$\langle\, g; \, g \text{ has order dividing } LCM(q^{u_i} - 1); \, i = 1, 2, \ldots, k \,\rangle .$$

CHAPTER 23

C-System Nearfields.

In Chapter 16, we have discussed André and generalized André planes, which when the associated semifield is a field, may be defined by spread sets of the following type $x = 0, y = 0, y = x^{p^{\lambda(m)}} m; m \in GF(p^r)$, where λ is a function from $GF(p^r)$ to the set of positive integers modulo r. There is a natural way to generalize spreads of this type.

DEFINITION 23.1. Let $\{F, +, \cdot\}$ define a finite left nearfield so we have $(a + b)c = ab + bc$, for all $a, b \in F$. Let T be an automorphism of $(F, +)$, an additive automorphism such that $T(1) = 1$. Assume that λ is a function from F^* to the integers Z with the following two properties:

(1) $\lambda(1) = 0$ and (2) for all $a, b \in F^*$, there is a unique integer m such that $\lambda(a \cdot (bT^m)^{-1}) = m$. Then $x \circ z = x \cdot zT^{\lambda(x)} \ \forall x, z \in F^*$, $0 \circ z = 0 \ \forall z \in F$ defines a quasifield called a 'C-system'.

PROOF. Dualize both structures to represent the spread using the notation in this text: $x = 0, y = 0, y = xT^{\lambda(m)} \cdot m; m \in D$, D the associated nearfield plane. Since T is an additive automorphism and $\cdot$ distributes on the left, it simply remains to show that the indicated set produces a quasifield. $\qquad\square$

Any generalized André system (λ-system) is a C-system (Wilke–Zemmer [1216]).

This process applied to the irregular nearfields produces several interesting translation planes which are not generalized André planes. These planes are due to M.L. Narayana Rao, D.J. Rodabaugh, F.W. Wilke and J.L. Zemmer [1013].

The main idea is to take an additive automorphism T of an irregular nearfield plane and form the following group $\langle xT \cdot x^{-1} \rangle = G$. Now define a putative λ: $\lambda(x) = 0$ if $x \in G$ and 1 if $x \notin G$. It follows fairly directly using the required conditions to ensure the construction of a plane that such a choice of λ will produce a translation plane; C-system. For example, when G is the full nearfield group, the nearfield is obtained. Hence, what is required is for G to be a proper subgroup. A computer search is employed to determine additive automophisms which generate a proper subgroup, as above. Hence, the constructions will give rise to various translation planes of orders p^2, for $p = 5, 7, 11, 23, 29, 59$. All translation planes of orders 25 and 49 are now known by computer, the planes under consideration were constructed prior to the general computer determinations.

We have discussed the irregular nearfields elsewhere in this text (see Section 8.2), but we list them here for convenience and for notational reference.

Irregular Nearfields;   Group
I. 5^2 ; $M(2,3)$
II. 11^2 ; $M(2,3) \times C_5$
III. 7^2 ; G_3
IV. 23^2 ; $G_3 \times C_{11}$
V. 11^2 ; $M(2,5)$
VI. 29^3 ; $M(2,5) \times C_7$
VII. 59^2 ; $M(2,5) \times C_{29}$

The main results of Narayana Rao–Rodabaugh–Wilke–Zemmer are the determination of the mutually non-isomorphic *C*-system planes obtained. We list therefore only the distinct planes obtained.

Narayana Rao–Rodabaugh–Wilke–Zemmer Planes. The main results of M.L. Narayana Rao, D.J. Rodabaugh, F.W. Wilke and J.L. Zemmer [**1013**] show that there are exactly nine mutually non-isomorphic translation planes obtained from the group systems $\langle xT \cdot x^{-1} \rangle = G$. None of these planes are generalized André planes Letting $T = \begin{bmatrix} 1 & 0 \\ a & b \end{bmatrix}$, the following table provides T as (a,b), a set of generators for G and the order of G. The planes are labeled using an indexing with the number of the irregular nearfield and an integer i. For example, $I - 1$ and $I - 2$ indicate planes obtained from the irregular nearfield of order 25.

The Nine *C*-System Planes of Narayana Rao–Rodabaugh–Wilke–Zemmer

p	System No.	Order of F^*	i	T_i	Generators	Order G
5	I	24	1	$(0,4)$	$(0,1), (0,2)$	8
5	I	24	2	$(1,4)$	$(2,2)$	6
11	II	120	1	$(0,10)$	$(0,1), (1,5)$	24
7	III	48	1	$(0,2)$	$(1,3), (2,0)$	24
7	III	48	3	$(0,6)$	$(3,0), (0,2)$	12
7	III	48	4	$(1,6)$	$(1,2)$	8
11	V	120	1	$(0,10)$	$(1,5), (0,1)$	24
11	V	120	2	$(1,10)$	$(6,3), (8,6)$	20
29	VI	840	1	$(27,28)$	$(0,1), (1,22)$	120

Wilke [**1215**] has constructed what might be called '2nd generation planes' from the planes listed above. In Wilke's construction, certain nets in above planes are found to be replaceable. Indeed, certain planes of orders 5^2 and 7^2 may be obtained by this method. Again, we now recall that all translation planes of orders 25 and 49 are known by computer and may be found in the home page of U. Dempwolff.

In a very general study, Foulser [**370**] developed the notation of 'group replaceable' nets, with the motivation of understanding the planes of Rao–Rodabaugh–Wilke–Zemmer group-theoretically. Foulser points out that the ideas can be applied to the regular or Dickson nearfield planes, but it is not clear if new planes would emerge. Actually, in some sense this is what is done in Chapter 20 on generalized André planes of order q^n admitting affine homology groups of order $(q^n - 1)/2$ by net replacements in the Dickson nearfield planes.

One point that might be made is that the irregular nearfield planes of orders p^2 for $p = 5$ or 7 admit a cyclic homology group of order $p + 1$ and hence correspond to a flock of a quadratic cone. The planes are derivable by any component image of such a group and the derived planes are two of the planes of Rao–Rodabaugh–Wilke–Zemmer.

Five of the irregular nearfield planes are derivable, three of order p^2 having $GF(p)$ in the center of orders 11^2, 23^2, 59^2 (non-solvable cases) and the planes of order 5^2 and 7^2. We may derive any of these. In particular, K. Lueder [**934**] has determined the groups of the derived planes.

CHAPTER 24

Subregular Spreads.

Every translation plane may be obtained by a net replacement applied to a Desarguesian plane, and the simplest type of finite replaceable net corresponds to regulus replacements. Moreover disjoint unions of nets, hence disjoint unions of reguli, are themselves replaceable. So a basic area of study are the planes obtainable by deriving simultaneously the reguli in any collection of pairwise disjoint reguli, of a Desarguesian line spread in $PG(3, q)$: such spreads are *subregular* spreads, obtained by 'multiply deriving' a Desarguesian spread. Thus, the two-dimensional André planes are always subregular, and these are the only generalized André planes that are subregular (since prime-dimensional λ-planes are André planes).

The investigations of subregular spreads was pioneered by Bruck in the 1960s, and has received considerable attention over the years, e.g., [**336, 333**], partly because they often exhibit novel properties, not known to exist in other planes, but also because they display interesting connections with other areas of finite geometries. Moreover, it is known that subregular spreads are not semifield planes.

A common technique for the analysis of finite spreads or finite translation planes is to work combinatorially or by use of the computer with structures of small order that have certain interesting properties. The question then is whether there is an infinite class of translation planes of arbitrarily larger orders that contains the small-order plane. But, however a translation plane is found, the question remains: **Is it always possible to extend the given finite translation plane to an infinite class that contains the original plane and admits similar properties?**

A 'regular spread' is a spread of lines in $PG(3, q)$ such that the unique regulus defined by any three mutually distinct lines is a subset of the spread. It turns out that any regular spread simply defines a Desarguesian spread in $PG(3, q)$. A 'subregular spread' is a spread obtained from a Desarguesian spread by a series of derivations. The theorem of Orr [**1035**] shows that any such subregular spread can be obtained from the original Desarguesian plane by the multiple derivation of a set of mutually disjoint reguli (regulus nets).

In this chapter, we consider the above question for any subregular spread; a spread that may be constructed from a Desarguesian plane by the multiple derivation of a set of mutually disjoint regulus nets. Of course, any André plane of dimension two is a subregular plane and these planes are explicated in detail in Chapter 16 and Section 16.1 in particular. Hence, within the present chapter, we shall be looking for subregular planes that are necessarily not André planes (note that the Hall planes and the regular nearfield planes of dimension two are André planes).

In accordance with our overall approach, we do not distinguish between subregular planes and subregular spreads.

DEFINITION 24.1. Let Δ be any Desarguesian spread in $PG(3, q)$, and suppose $\Theta = \{\, \mathcal{R}_i \, : \, 1 \leq i \leq d \,\}$ is a collection of d pairwise disjoint reguli in Δ. Let $\Theta' = \{\, \mathcal{R}'_i \, : \, 1 \leq i \leq d \,\}$ be the collection of reguli $\mathcal{R}'_i$ that are the opposite of the reguli $\mathcal{R}_i$, for $i = 1 \dots d$.

Then the spread obtained from Δ by replacing each regulus $\mathcal{R}_i$ in Θ by its opposite regulus $\mathcal{R}'_i$:

$$\Delta_\Theta := (\Delta \setminus \Theta) \cup \Theta'$$

is MULTIPLY DERIVED from Δ relative to Θ, and any spread of type Δ_Θ is SUBREGULAR.

The largest subgroup of $\mathrm{Aut}(\Delta)$ that permutes the members of Θ is the INHERITED GROUP of Δ_Θ, relative to Θ.

The definition yields the restriction:

REMARK 24.2. Every subregular spread is a line spread, in some $PG(3, q)$.

By their definition, André planes are multiple derivations of André nets, and since two-dimensional generalized André planes are André planes we have:

REMARK 24.3. A regular spread Σ in $PG(3, q)$ multiply derives to at least $2^{q-1} - 1$ distinct André spreads $\neq \Sigma$. So all the André planes, in $PG(3, q)$, including the Hall and Desarguesian planes, are subregular planes, and a generalized André plane is subregular iff it is an André plane in $PG(3, q)$.

If a regular spread π in $PG(3, q)$ admits Θ pairwise disjoint reguli, then multiply deriving the various subsets of Θ yields $2^{|\Theta|} - 1$ mutually distinct subregular spreads ($\neq \pi$), although there may be isomorphisms among them or with the Desarguesian spread π.

Thus, one expects multiple derivation to provide a potential source of many new line spreads, and the problem of finding multiply-derived Desarguesian spreads reduces to the problem of finding 'non-isomorphic' sets of pairwise disjoint reguli in Desarguesian spreads: here *nets are considered 'isomorphic' if one maps onto the other by a collineation of the underlying plane.*

On the other hand, the following property easily implies that many line spreads are not subregular.

REMARK 24.4. Any subregular spread in $PG(3, q)$ admits a cyclic autotopism group of order $q^2 - 1$: they acts transitively on the $q + 1$ components of each of the derived regulus nets.

PROOF. The kernel homology group of the underlying Desarguesian plane inherits to a group of the claimed type. The group is an autotopism group since at least two components are shared by the subregular and Desarguesian spread.   □

Since subregular spreads, by definition, include derivable nets, any non-derivable line spread cannot be subregular, for instance:

REMARK 24.5. The Lüneburg–Tits planes of even order and the Hering planes of odd order are not subregular.

The main concerns of subregular spread theory might broadly be summarized as follows:

Questions on Subregular Spreads.

(1) Find (maximal) collections of of pairwise disjoint reguli Θ in the regular spread Δ in $PG(3, q)$, and identify techniques for obtaining new Θ from existing ones.

(2) Any subregular $\mathcal{S}_d$ in $PG(3, q)$ involves replacing d reguli in a regular spread by their opposite reguli. Thus $1 \leq d \leq q - 1$. Determine values of d for which all $\mathcal{S}_d$ are known.

(3) Determine when the inherited group of a multiply-derived spread Δ_Θ, from a Desarguesian spread Δ, is the full collineation group of Δ_Θ.

(4) Find subregular spreads with interesting collineation groups or combinatorial properties, including those relating them to other areas.

(5) Identify known classes of line spreads as being subregular (or not—in non-trivial cases!).

(6) Characterize subregular spreads, or classes of such spreads, by their collineation groups or other properties: this is obviously related to (5).

Regarding the final question, we already seen that all subregular spreads are restricted to being line spreads, Remark 24.2, and Remark 24.4 forces subregular spreads of order q^2 to have cyclic autotopism groups of order $q^2 - 1$.

We end by listing the subregular spreads based on extremal values of d, the number of replaced reguli required to the construct them. First consider small d.

REMARK 24.6. Let $\mathcal{S}_d$ be a subregular spread in $OG(3, q)$ obtained by replacing $1 \leq d \leq q-1$ reguli in a Desarguesian lines spread by their opposite reguli. If $d = 1$ then $\mathcal{S}_d$ is a Hall plane, if $d = 2$ then $\mathcal{S}_d$ is an André plane.

PROOF. The case $d = 1$ restates that Desarguesian planes derive to Hall planes. The case $d = 2$ is due to Bruck, who noted that two disjoint reguli may be identified with disjoint André nets. $\qquad\square$

The other extreme, $d = q-1$, is taken care of by a major result due to Orr and Thas.

THEOREM 24.7. (Orr [**1035**], Thas [**1172**]). Any subregular spread of order q^2 obtained from the multiple derivation of $q - 1$ mutually disjoint reguli is a Desarguesian spread.

PROOF. For q odd see Orr [**1035**], and for q even see Thas [**1172**]. $\qquad\square$

Combining Orr–Thas, Theorem 24.7, with Bruck, Remark 24.6, yields the next case:

COROLLARY 24.8. Any subregular spread of order q^2 obtained from the multiple derivation of $q - 2$ mutually disjoint reguli is a Hall spread.

Before returning to subregular spreads, we pause now to provide a brief introduction to 'fans', i.e., component orbits of (partial) spreads under fields of collineations, since subregular spreads and various classes of generalized André planes provide a good source of fans. Some readers might wish to skip the next section on fans altogether.

24.1. Field Orbits and Fans.

Suppose V is a finite vector space over a field F, such that F leaves invariant a spread π on V. If F fixes every component of π, then of course F is merely a (subfield) of the kernel of π. If, however, this is not the case, then F^* is a collineation group of π, with at least one non-trivial orbit of components. In this case we refer to F^* as a *field-group* of π, and the component orbits of F^* (including the trivial orbits) will be called *fans*; we might also refer to F^* or F as a *field of collineations*.

Observe that if F^* is a field-group then the stabilizer of any component is a (sub)field group K^*, hence any fan has size $[F^* : K^*]$: the index of a field group relative to some subfield group.

The reader should note that since V is a vector space over any field-group F^*, it follows *every field-group is fixed-point-free*. On the other hand if a group induces a field-group on a subspace of $W < V$ then F^* need not be a field-group on V, hence not necessarily fixed-point-free. This type of situation occurs, for instance, when we consider homology groups of semifields.

The above conceptual notion of field-groups and fans may of course be stated for partial spreads. Thus, we summarize the above using explicit parameters, as follows.

DEFINITION 24.9. A 'q^e-fan' in a $2ds$-dimensional vector space over $K = GF(q)$ is a set of $(q^w - 1)/(q^e - 1)$ mutually disjoint K-subspaces of dimension ds that are in an orbit under a field group F_w^* of order $(q^d - 1)$, where F_w contains K, and where $w = d$ or $2d$ and in the latter case, by convention, s is odd.

As indicated, fans arise in spreads as follows:

REMARK 24.10. Let π be a translation plane of order q^{ds} and kernel containing K isomorphic to $GF(q)$ that admits a field-group F_w^* of order $(q^w - 1)$ containing K^*, where $w = d$ or $2d$ and s is odd if $w = 2d$. Then, for any component orbit Γ, there is a divisor e_Γ of w such that the orbit length of Γ is $(q^w - 1)/(q^{e_L} - 1)$ so that Γ is a q^{e_Γ}-fan.

In particular, since the kernel homology group of a Desarguesian spread acts on the corresponding André spreads and nets, they give rise to fans. Similarly we have a corresponding result for subregular spreads.

REMARK 24.11. Any André or subregular spread or net of order q^d admits field groups, hence they admit fans of size $(q^d - 1)/(q^e - 1)$, for various e dividing d.

Obviously, this remark applies to a far larger range of spreads, including multiply André planes, based on multiple André planes (based on multiple André partition) and certain other generalized André planes. It appears to be of some interest to compute various fan-sizes.

THEOREM 24.12. Let Σ be a Desarguesian spread of order q^{ds}. For each of the $q - 1$ André nets A_α, choose a divisor e_α of d (these divisors can possibly be equal and/or possibly equal to 1 or d). For each q-André net A_α, there is a corresponding set of k_{e_α} q^{e_α}-fans. Form the corresponding André plane $\Sigma_{(e_\alpha f_\alpha \forall \alpha \in GF(q))}$ obtained with spread:

$$y = x^{q^{e_\alpha f_\alpha}} m \text{ for } m^{(q^{ds}-1)/(q-1)} = \alpha, m \in GF(q^{ds}), (e_\alpha f_\alpha, d) = e_\alpha; x = 0, y = 0.$$

Then $k_{e_\alpha} = \frac{(q^{ds}-1)}{(q-1)} \frac{(q^{e_\alpha}-1)}{(q^d-1)}$, and the spread $\Sigma_{(e_\alpha f_\alpha \forall \alpha \in GF(q))}$ is a union of $\sum_{\alpha=1}^{q-1} k_{e_\alpha}$ q^{e_α}-fans, together with two q^d-fans $x = 0$ and $y = 0$.

24.2. Group-Theoretic Characterization.

We have seen subregular $GF(q)$-spreads of order q^2 admit cyclic groups of order $q^2 - 1$, that fix ≥ 3 pairwise trivial-intersecting 2-dimensional subspaces, Remark 24.4. We establish a converse, ignoring the sporadic 'Zsigmondy' cases.

THEOREM 24.13 (Jha and Johnson). Let π be a spread of order q^2, $q = p^r$, with kernel $\geq K = GF(q)$, defined on $(V, +)$. Then π is a subregular spread if and only if $\mathrm{Aut}(\pi)$ contains an Abelian collineation group $G < GL(V, +)$ of order $q^2 - 1$ that fixes at least three pairwise trivially-intersecting two-dimensional K-subspaces of V.

PROOF. By Remark 24.4, we need only prove the if-case. We only consider the non-exceptional case: when G contains an element θ of order u, a p-primitive divisor of $q^2 - 1$; by Zsigmondy [1219] (see also Theorem 7.1), the remaining cases are so restrictive ($q^2 = p^2$, etc.) that they can be handled by ad hoc methods using q-primitive divisors, roughly following the argument below.

By André, $\theta \in \Gamma L(4, K)$, so by its p-primitivity $\theta \in GL(4, k)$. Since G fixes at least three pairwise trivially intersecting 2-subspaces, X, Y and Z, of V, G acts identically on them. In particular, θ defines a Desarguesian spread Δ_θ (the Ostrom phantom) on $(V, +)$, consisting of all the θ-modules of $(V, +)$ of order p^2, viz., Δ_θ. It remains to show that Δ_θ multiply derives to π.

Step: Δ_θ and π are both K-spreads.
The centralizer of θ in $\Gamma L(V, +)$ (which clearly includes $G \cup K$) permutes the components of Δ_θ hence lies in $\Gamma L(V, F)$, $F \cong GF(q^2)$, where F is the kernel field of Δ_Θ; thus $K < F$ is in the kernel of π and also Δ_θ, so all 2-dimensional K-subspaces of V are lines or components of both spreads.
Step: $G = F^*$ and hence F^* acts on π.
Since G fixes at least three components of Δ_θ it acts isomorphically and irreducibly on every component of Δ_θ that it leaves fixed (since $\theta \in G$ is itself irreducible). On any component ℓ of Δ_θ, Schur's lemma implies that the centralizer of θ is a field, but the centralizer includes $F|\ell$, so $G|\ell = F^*|\ell$ is forced because $|G| = |F^*|$. Since this holds for all components ℓ we have $G = F^*$. Thus F^* is a collineation group of π.
Step: $\mathrm{Orb}_G(\ell)$ is a regulus in π.
Take a component ℓ in $\pi \setminus \Delta_\theta$. So ℓ is Baer subplane of Δ_θ, and $\mathrm{Orb}_{F^*}(\ell)$ consists of $q + 1$ Baer subplanes forming a regulus of Δ_θ and also an orbit of $q + 1$ components of π. Thus $\mathrm{Orb}_{F^*}(\ell)$ is a set of $q + 1$ lines of π that forms a regulus in Δ_θ.
Step: Winding up.
If $\ell, \ell' \in \pi \setminus \pi$, then $\mathrm{Orb}_{F^*}(\ell) \cap \mathrm{Orb}_{F^*}(\ell') = \varnothing$ unless $\mathrm{Orb}_{F^*}(\ell) = \mathrm{Orb}_{F^*}(\ell')$. Hence the component set $\pi \setminus \Delta_\theta$ are partitioned by components. Hence Δ_θ has a collection of pairwise disjoint reguli that are obtained by deriving disjoint reguli in π and all other lines of π and Δ_θ coincide. □

Note that there are two-dimensional spreads (line spreads) that admit cyclic groups of order $q^2 - 1$ which are not subregular spreads. For example, the j-planes, admit such groups but are not subregular.

24.3. Lifting Subregular Spreads.

We have seen that a set of d mutually disjoint reguli $\mathcal{R}$ of a Desarguesian line spread π in $PG(3, q)$ lift uniquely to d mutually disjoint reguli $\mathcal{S} := GF(q^s) \otimes \mathcal{R}$ of a regular spread $\psi = GF(q^s) \otimes \pi$ in $PG(3, q^s)$, for all odd s. Moreover the collineation group of π permuting $\mathcal{R}$—the *inherited group*—extends uniquely to a collineation group of ψ, and this group permutes the d reguli in $\mathcal{S}$.

DEFINITION 24.14. The subregular spread $\psi' := \mathcal{S}'$ of order q^{2s}, obtained by multiply deriving the d reguli of ψ, in $\mathcal{S}$, is the subregular spread LIFTED from the subregular spread based on $\mathcal{R}$.

Recall that Corollary 18.27 demonstrates that lifting provides a recursive construction for an infinite sequence of ascending subregular spreads, starting from any subregular spread. In the language of subregular spreads, Johnson's lifting theorem implies:

THEOREM 24.15. Let π be any subregular spread of order q^2. Then there is an increasing family of subregular spreads $\Pi = (\pi_i)_{i=0}^{\infty}$ such that $\pi_0 = \pi$, and the order of each π_i is q^{2r_i} such that $r_i > 1$ is odd for all $i \geq 1$.

If π is obtained by multiply deriving a set of pairwise disjoint $\mathcal{R} = \{ R_j : 1 \leq j \leq k \}$ from a Desarguesian spread Σ, then the subgroup G of Σ permuting the members of $\mathcal{R}$ acts faithfully on every subregular plane π_i in the chain Π.

Moreover, elations , Baer collineations, etc., when G is veiwed as acting on π, continue to be elations, Baer collineations, etc., when G is viewed as acting on any π_i.

We shall shortly verify, via Walker's inherited group theorem, that all lifted spreads are guaranteed to be non-generalized-André and non-semifield; the latter follows from a much stronger result which is also a consequence of Walker.

24.4. Walker's Inherited Group Theorem.

Walker completely settled the question of when the full collineation group of a subregular spread is the inherited group, cf. Question (3), p. 177.

THEOREM 24.16 (Walker [**1205, 1206**]). Let π be a subregular spread of order q^2 obtained by the multiple derivation of a Desarguesian spread Σ. If π is not an André spread obtained by the replacement of a set S consisting of d mutually disjoint reguli, $d \notin \{1, (q-1)/2, (q-1)\}$, then the full collineation group of π is the subgroup of Σ that permutes the reguli of S.

Walker's theorem implies lifted subregular spreads are guaranteed to be non-André hence also non-generalized-André.

COROLLARY 24.17. Let $\mathcal{S}$ be a subregular spread in $PG(3, q^s)$ obtained by lifting a subregular spread $\mathcal{A}$ in $PG(3, q)$ obtained by replacing d pairwise disjoint reguli in a Desarguesian line spread. Then $\mathcal{S}$ is not a generalized André plane.

PROOF. Let d be the number of derived reguli needed to define $\mathcal{A}$, and hence also $\mathcal{S}$. Thus $d = 1$ yields the Hall plane so assume $1 < d < q - 1$, hence, for $s > 1$, we clearly have $d \notin \{1, (q^s - 1)/2, (q^s - 1)\}$. Hence by Walker, Theorem 24.16, the full group of $\mathcal{S}$ is the inherited group. So $\mathcal{S}$ is not an André plane since it clearly cannot admit admits an affine homology group of order $q^s + 1$ permuting the

d reguli. Hence it is also not generalized André, being two-dimensional, Remark 24.3.   □

COROLLARY 24.18. *If π_k is a subregular plane of order q^2, and is not a (generalized) André plane, then every lifted plane π_k^s of order q^{2s}, for s odd, is also not a (generalized) André plane.*

PROOF. We sketch the proof of this exercise. If the extension is generalized André, then by Remark 24.3 it corresponds to the derivation of a set of André reguli in $PG(3, q^s)$, defined by a pair of common components. It is evidently lifted from a similar collection in $PG(3, q)$, contradicting the assumption that a non-André spread was lifted.   □

COROLLARY 24.19. *Let $q = p^r$, p prime, and suppose π be a non-Desarguesian subregular plane of order q^2 that admits an affine elation group E of order > 2. Then $|E| \leq q/p$.*

PROOF. If π is an André plane, we recall that the full elation group must have order ≤ 2, by Foulser, otherwise Walker's theorem applies and if $|E| > q/p$ then E obviously cannot be inherited.   □

In particular:

COROLLARY 24.20. *A non-Desarguesian subregular spread is not a semifield spread.*

24.5. Inversion Fixed-Subregular Spreads.

An important class of subregular spreads of odd order was found by Ostrom [**1044**]. These planes were explicated by Foulser in [**373**], where the even-order case is also considered. The planes are of order q^4 and the reguli used in the multiple derivation process are all fixed by the collineation $(x, y) \longmapsto (x^q, y^q)$ of an associated Desarguesian affine plane. We consider a generalization of these planes in this section.

DEFINITION 24.21. Let S_k be a set of mutually disjoint reguli in the Desarguesian affine plane Σ_{q^2}. Let R_q be a fixed regulus distinct from any regulus in S_k. S_k will be said to be 'inversion fixed' if and only if the inversion σ of R_q fixes each regulus of S_k and each regulus of S_k non-trivially intersects R_q.

THEOREM 24.22. *Let $\Sigma_{2^{2t}}$ be a Desarguesian affine plane of order 2^{2t}. Let R_{2^t} be a fixed regulus and let S_k be a set of k mutually disjoint reguli each of which share exactly one component with R_{2^t}.*
Then for every Desarguesian affine plane $\Sigma_{2^{2ts}}$ of order 2^{2ts}, where s is odd, there is a fixed regulus $R_{2^{ts}}$ and a corresponding set S_k^s of k mutually disjoint reguli each of which shares exactly one point with $R_{2^{ts}}$.

So, we see that inversion fixed-subregular spreads lift to subregular spreads defined by reguli that are fixed by $\sigma : (x, y) \longmapsto (x^q, y^q)$. We now consider this more generally.

THEOREM 24.23. *Let $\Sigma_{q^{2t}}$ be a Desarguesian affine plane of order q^{2t}, where q is even. Let R_{q^t} be a fixed regulus and coordinatize so that R_{q^t} is coordinatized by $GF(q^t)$. Let R_q denote the subspread coordinatized by $GF(q)$. Let S_k be a set of*

k mutually disjoint reguli each of which share exactly one component with R_q and each is fixed by $\sigma : (x, y) \longmapsto (x^q, y^q)$.

Then for every Desarguesian affine plane $\Sigma_{q^{2ts}}$ of order q^{2ts}, where s is odd, there is a fixed regulus $R_{q^{ts}}$ and a corresponding set S_k^s of k mutually disjoint reguli each of which shares exactly one point with $R_{q^{ts}}$ and these points lie in R_q. Furthermore, the reguli are fixed by σ and such that each regulus contains the corresponding regulus of $\Sigma_{q^{2t}}$.

We now consider the analogous results for odd-order planes. The following lemma is straightforward and the proof is left to the reader.

LEMMA 24.24. Let Σ be a Desarguesian affine plane of odd order $p^{2r} = q^2$ and let R be a fixed regulus in the spread of Σ_{q^2}. Let σ denote the inversion in R_q (the unique Baer involution fixing R linewise). Let $K_q \subseteq K_{q^2}$ denote fields isomorphic to $GF(q)$ and $GF(q^2)$, respectively, that coordinatize Σ_{q^2} and R_q.

Then σ has the following form: $(x, y) \longmapsto (x^q, y^q)$.

THEOREM 24.25. Under the assumptions of the previous lemma, let $S_k = \{ D_i; i = 1, 2, \ldots, k \}$ be any set of mutually disjoint reguli of Σ and distinct from R_q that have the following two properties:
    (i) D in S_k then $D \cap R_q \neq \varnothing$ and
    (ii) D is fixed by σ.
    For D_i in S_k, $D_i \cap R_q = \{P_i, Q_i\}$, $P_i \neq Q_i$.
    Under the assumptions directly above, given any odd integer s, let $K_{q^{2s}}$ be an extension field of K_{q^2} that is isomorphic to $GF(q^{2s})$. Fix a regulus R_{q^s} and coordinatize by a subfield K_{q^s} isomorphic to $GF(q^s)$. Let R_q^+ denote the subspread coordinatized by $GF(q)$. Let $\Sigma_{q^{2s}}$ be the Desarguesian affine plane coordinatized by $K_{q^{2s}}$ and let $\Sigma_{q^2}^+$ denote the subspread coordinatized by K_{q^2} and assume that R_q^+ is a subspread of $\Sigma_{q^2}^+$. Finally, let σ^+ denote the automorphism collineation $(x, y) \longmapsto (x^q, y^q)$, acting in $\Sigma_{q^{2s}}$. If we consider $\Sigma_{q^2}^+$, we may realize Σ_{q^2} as the natural subplane of order q^2. If D_i is a regulus of Σ_{q^2}, let D_i^+ denote the subspread of $\Sigma_{q^{2s}}$ of components defined by D_i. Then there is a corresponding set $S_k^s = \{ R_i^+; i = 1, 2, \ldots, k \}$ of mutually disjoint reguli of $\Sigma_{q^{2s}}$ with the following two properties: (i)^s the reguli of S_k^s are all fixed by σ^+; (ii)^s $D_i^+ \subseteq R_i^+$.

The converse is also valid.

THEOREM 24.26. Consider a Desarguesian affine plane $\Sigma_{q^{2s}}$, for s odd and fix a regulus Z_{q^s}, coordinatize by a field isomorphic to $GF(q^s)$ and let Z_q denote the subspread coordinatized by $GF(q)$. Let $\Sigma_{q^{2s}}$ be coordinatized by a field isomorphic to $GF(q^{2s})$ and let Σ_{q^2} denote the subnet coordinatized by $GF(q^2)$. Let $\Sigma_{q^2}^-$ denote the Desarguesian affine plane coordinatized by $GF(q^2)$. Let $T_k^s = \{ W_i; i = 1, 2, \ldots, k \}$ be a set of k mutually disjoint reguli of $\Sigma_{q^{2s}}$, with the following two properties: (1)$'$ each regulus non-trivially intersects Z_q; (2)$'$ each regulus is fixed by $\sigma^+ : (x, y) \longmapsto (x^q, y^q)$. Then each regulus W_i contains a subspread D_i of Σ_{q^2} that defines a regulus D_i^- of $\Sigma_{q^2}^-$. Hence, there is a corresponding set T_k^- of k mutually disjoint reguli in $\Sigma_{q^2}^-$. Any set of reguli T_k in $\Sigma_{q^2}^-$ that may be constructed from a set of reguli T_k^s is said to be 'contracted' from T_k^s.

The more general version of the first previous theorem is as follows:

THEOREM 24.27. Let $\Sigma_{q^{2t_o}}$ be a Desarguesian affine plane of odd order q^{2t_o} and let $R_{q^{t_o}}$ be a fixed regulus in the spread of $\Sigma_{q^{2t_o}}$. Coordinatize $R_{q^{t_o}}$ by $GF(q^{t_o})$ and let σ denote the collineation $(x, y) \longmapsto (x^q, y^q)$. Let R_q denote the subspread of $R_{q^{t_o}}$ coordinatized by $GF(q)$. Let $S_k = \{ D_i; \ i = 1, 2, \ldots, k \}$ be any set of mutually disjoint reguli of Σ and distinct from R_q that have the following two properties:
(1) D in S_k then $D \cap R_{q^{t_o}}$ has cardinality 2;
(2) D is fixed by σ.
For D_i in S_k, let $D_i \cap R_{q^{t_o}} = \{P_i, Q_i\}$, $P_i \neq Q_i$.
Let $\Sigma_{q^{2t_o s}}$ denote a Desarguesian affine plane containing $\Sigma_{q^{2t_o}}$ as a subplane. If s is odd, then there is a corresponding set $S_k^s = \{ R_i^+; \ i = 1, 2, \ldots, k \}$ of mutually disjoint reguli of $\Sigma_{q^{2t_o s}}$ with the following two properties: $(1)^s$ the reguli of S_k^s are all fixed by σ^+; $(2)^s$ $D_i^+ \subseteq R_i^+$.

24.6. Isomorphism Types.

In this section, we consider the following question: **How does knowledge of the isomorphism type of a subplane affect the isomorphism type of the affine plane containing the subplane?**

There are only a few known classes of subregular planes, the main class of which is the class of André planes. Most of the known classes are discussed in the last section.

Hence, it is always important to determine conditions that ensure that a given subregular plane is not André. We are able to show that if π_k is a subregular plane of order q^2, then any lifted plane π_k^s is André if and only if π_k is André.

THEOREM 24.28. Let π_k be a subregular plane of order q^2 and let π_k^s be the corresponding lifted subregular plane of order q^{2s}, where s is odd.
Then π_k is André if and only if π_k^s is André.

PROOF. If the resulting plane π_k^s is André, then the set S_k^s is linear and we may coordinatize so that we have a set of André nets each of the form

$$\left\{ y = xm; \ m^{q^o+1} = \delta \right\} = A_\delta, \ \text{for} \ \delta \in K_{q^s}.$$

Each of these André nets shares a regulus of $\Sigma_{q^2}^+$, hence there is a subset of elements n in $GF(q^2)$. Then, $n^{q^s+1} = n^{(q+1)(q^s+1)/(q+1)}$, as s is odd, $= \alpha_n^{(q^s+1)/(q+1)} = \delta$, for $\alpha_n \in K_q$. We need to show that $\alpha_n = \alpha_{n^*}$, for n, n^* in $GF(q^2)$.

Since s is odd, and

$$(q^s + 1)/(q + 1) = (1 - q) + (q^2 - 1) + (1 - q^3) + \cdots + (q^{s-1} - 1) + 1,$$

then

$$(q - 1, (q^s + 1)/(q + 1)) = 1.$$

Then $(\alpha_n \alpha_{n^*}^{-1})^{(q^s+1)/(q+1)} = 1$, implies that $\alpha_n \alpha_{n^*}^{-1} = 1$, so that $\alpha_n = \alpha_{n^*}$.

Hence, S_k^+ linear implies that S_k is linear.

Now assume that S_k is linear. Suppose the regulus containing an S_k-André net $y = xn$ such that $n^{q+1} = \alpha$, for $\alpha \in GF(q)$ has components $y = xm$ and $y = xm^*$. Assume that $m^{q^s+1} = \delta$ and $m^{*(q^s+1)} = \delta^*$. There is a unique regulus of $\Sigma_{q^{2s}}$ that contains the S_k^+-André net. Furthermore, this unique regulus must be an André regulus by our previous results. But, consider an André net A_δ of

$\Sigma_{q^{2s}}$ of components $y = xm$ such that $m^{q^s+1} = \delta$. Then, α in $GF(q)$, implies that
$\alpha^{(q^s+1)/(q+1)} = \alpha^{(1+q^2+\cdots+q^{s-1})}\alpha^{-(q+q^3+q^5+\cdots+q^{s-2})} = \alpha^{(s-1)/2+1}\alpha^{-(s-1)/2} = \alpha.$

Thus, $\delta = \delta^* = \alpha$. Hence, there is André net of $\Sigma_{q^{2s}}$ containing the André net of Σ_{q^2} and by uniqueness the regulus in question must be André. This completes the proof of the theorem.   □

Another close cousin of the André planes, which are completely known, is the set of generalized André planes, which are almost completely unknown. Recall that a generalized André plane of order p^t is a translation plane whose spread may be written in the following manner:

Generalized André Spread: $x = 0, y = x^{p^{\lambda(m)}}m; m \in GF(p^t),$

where λ is a function from $GF(p^t)$ to Z_t.

THEOREM 24.29. Let π be a subregular spread. If π is a generalized André plane, then it is an André plane.

PROOF. Let π have order q^2 then the kernel of π contains $GF(q)$. Assume that the plane is not André. Represent π by a generalized André system so that the spread may be written as follows:

$x = 0, y = x^{p^{\lambda(m)}}m; m \in GF(p^{2r})$, where $p^r = q$, and p is a prime,

where $\lambda\colon GF(p^{2r}) \longmapsto \{0, 1, \ldots, 2r-1\}$. Since the plane is not André, the kernel of the plane is a subkernel of Σ and hence may be represented by mappings of the following form:

$(x, y) \longmapsto (xa, ya); a \in GF(q) - \{0\}.$

Each component $y = x^{p^{\lambda(m)}}m$ is fixed if and only if

$a^{p^{\lambda(m)}} = a,$

for all $a \in GF(q) - \{0\}$. Hence, $p^{\lambda(m)} = q$ or q^2, i.e., 1 for all $m \in GF(q^2)$.

The full kernel collineation group

$(x, y) \longmapsto (xa, ya); a \in GF(q^2) - \{0\},$

acts on π and maps $y = x^q m$ onto $y = x^q ma^{1-q}$. However, this means, by definition, that the image of $y = x^q m$ is the set of Baer subplanes of an André net, so the plane is André.   □

COROLLARY 24.30. If π_k is a subregular plane of order q^2 is not a generalized André plane then every lifted plane π_k^s of order q^{2s}, for s odd, is also not a generalized André plane.

PROOF. We have noted previously that π_k is André if and only if π_k^s is André. If π_k^s is generalized André then π_k^s is André, implying the contracted plane π_k is André.   □

Foulser [369] has shown that for any non-Desarguesian, non-Hall generalized André spread, obtained via a spread as noted above, the full collineation group leaves the set $\{x = 0, y = 0\}$ invariant. Since an André spread is also generalized André, then a subregular spread that admits an elation group of order > 2 cannot be generalized André (or André). The question is, of course, how large can an elation group be?

THEOREM 24.31. Let π_k be a non-Desarguesian subregular plane of order q^2. Let E be an elation group of order > 2. Then the order of E is $\leq q/p$, for $p^r = q$, p a prime. In particular, π_k cannot be a semifield plane.

PROOF. First assume that π_k is an André plane. If the full collineation group does not leave invariant a pair $\{x = 0, y = 0\}$ of components then π_k is Hall and does not admit elations of order > 2. If π_k is not Hall and André, then by Foulser [369], the pair of components listed above must be inverted by E, a contradiction.

Hence, assume that π_k is not André. Then the full collineation group of π_k is inherited from that of the associated Desarguesian affine plane Σ of order q^2. Since Σ does not admit Baer collineations of orders larger than 2, it follows that E is an elation group of Σ and must permute a set of fewer than $q - 1$ reguli that must be disjoint from the axis of E since otherwise E would end up being a Baer group of π_k. But then E must permute the set of k reguli semi-regularly. Hence, the order p^t of E must be less than $(q-1)$. Thus, $p^t \leq (q-2)$, which implies that $p^t < q$. □

COROLLARY 24.32. A non-Desarguesian semiregular translation plane of order q^2 that admits an elation group of order > 2 cannot be a semifield plane or a generalized André plane.

24.7. The Known Subregular Spreads and Subregular Lifting.

The main subregular lifting theorem of Johnson then applies to all subregular planes. We shall consider the most interesting of the known planes in this section. Note that when a given subregular plane π_k is new (i.e., not generalized André, etc.) then π_k^s is new. Also, it is conceivable that some of our lifted subregular planes can be or have been realized from the other direction. That is, there may be a subregular plane that admits a contraction to a subregular plane that lifts back to the original plane. Since the lifting process has not been available previously, it is necessary that the original plane would have been initially constructed in a different manner and the connection to the subplane not observed. We shall not give complete constructions to only a few of the following spreads. However, the ideas are straightforward and usually use a group to determine when a set of reguli are mutually disjoint on lines. We shall be interested in 'linear' and 'non-linear' subsets of mutually disjoint reguli.

DEFINITION 24.33. Let π be a Desarguesian affine plane of order q^2 and let S be a set of mutually disjoint reguli in π. Given any regulus R in π, then modulo the kernel homology group, there is a unique involution which fixes each component of R. The set S is 'linear' if and only if the involutions fixing the reguli of S pointwise have a common orbit of length 2 of components under their associated involutions, otherwise, S is said to be 'non-linear'.

24.7.1. Basic Subregular Planes. Before we give a listing of the known subregular planes and how subregular lifting changes the previously known list, we show how easy it is to find non-linear subregular planes. Choose any field F isomorphic to $GF(q^2)$ and a spread for a Desarguesian spread be given as $x = 0, y = xm; m \in F$. Recall that an André regulus A_δ is given by $y = xm; m^{q+1} = \delta$. There are $q - 1$ André partial spreads making up a so-called 'complete linear set'. Now consider the standard regulus R with partial spread $x = 0, y = x\alpha; \alpha \in K \simeq GF(q)$, K a subfield of G. Note that $\beta^{q+1} = \beta^2$, for $\beta \in K$. So, if q is odd take any subset

λ of the subset of $(q-1)/2$ André nets A_δ, for δ a non-square, of cardinality at least two. Each André net is left invariant under a collineation group of order $(q+1)^2$ of the associated Desarguesian plane. Note that an André net A_δ is fixed componentwise by the element $(x, y) \mapsto (y^q, x^q \delta)$. The collineation inverts $x = 0$ and $y = 0$ and maps $y = xn$ onto $y = x\delta n^{-q}$. If ρ is a subset of at least two André nets, say A_δ and A_{δ^*} then ρ is contained in a unique linear set since otherwise, there would be a pair $\{y = xn, y = x\delta n^{-q}\}$ inverted by $(x, y) \rightarrow (y^q, x^q \delta^*)$, which would then force $\delta = \delta^*$, a contradiction. This means that if $\lambda \cup \{R\}$ is linear, then R would be forced to be an André of the form A_γ. Hence, we have proved the following theorem.

THEOREM 24.34. Then $\lambda \cup \{R\}$ is a non-linear set with $3 \le 1 + |\lambda| \le (q+1)/2$ reguli.

Derivation of at least two elements of λ and R produces a non-linear subregular plane. So, we may obtain non-linear subregular planes with $3, 4, \ldots, (q+1)/2$ derived reguli.

24.7.1.1. *The $GL(2,3)$-Spread of Order 5^2.* There is an interesting subregular spread π_3 of order 5^2 obtained by the replacement of three reguli and which is explicated in Johnson and Ostrom [**784**] (see below for a discussion of 'Bruck triple planes') and may be constructed by the previous process. The corresponding translation plane admits a collineation group isomorphic to $GL(2,3)$, where the elements of order 3 are affine homologies. Furthermore, there are exactly 8 centers for homologies. There are $3(5+1)$ centers for involutory affine homologies. Most of the impetus for the original article [**784**] was to try to generalize this plane of order 25, to find translation planes of larger orders that have these properties. Furthermore, any of these planes may be lifted to produce subregular planes with similar properties. Here we are content to consider only the plane of order 25.

Note that any subregular translation plane with spread in $PG(3, q)$ admits a collineation group of order $q^2 - 1$ that acts transitively on the $q + 1$ components of each of the derived regulus nets. Since we may lift central collineations of π_k to central collineations of π_k^s, it follows that any subregular spread of order q^{2s} that is lifted from π_3 will have $3(q^s + 1)$ centers for involutory affine homologies and will have eight centers for affine homologies of order 3. We state this formally. We call the translation plane π_3 of order 25 the '$GL(2,3)$-spread'.

THEOREM 24.35. Let π_3 denote the $GL(2,3)$-spread of order 25. Then any lifted spread π_3^s of order 5^{2s}, for s odd will admit $GL(2,3)$ as a collineation group and have $3(q^s + 1)$ involutory affine homology centers as well as 8 centers for homologies of order 3.

24.7.1.2. *Ostrom, Johnson, Seydel Spread of Order 3^6.* Using the general idea of multiply derivation (see, e.g., [**1063**]), Ostrom gave a construction for an unusual translation plane of order 3^6 that admitted symmetric affine homology groups of order 7. Although there is an error in the calculation, the basic idea of the construction is valid as shown by Johnson and Seydel [**810**]. Any spread with the properties considered can be shown to be a non-generalized André spread. Hence, with subregular lifting and since central collineations also lift to central collineations, there is a spread of order 3^{6s}, for each odd integer s, that admits symmetric affine homology groups of order 7. By the previous section, the lifted planes are also not generalized André planes.

THEOREM 24.36. *Any plane of order 3^{6s}, for s odd, lifted from the Ostrom–Johnson, Seydel plane is a translation plane that is not a semifield or a generalized André plane that admits symmetric homology groups of order 7.*

24.7.1.3. *Ostrom–Foulser Planes of Order q^4.* There are a variety of subregular translation planes of order q^4 for both q even and odd with various interesting properties. For example, there is the Ostrom subregular plane of order q^4 admitting $SL(2,q)$, where $q = p^r$, for p odd, where the p-elements are elations (see Ostrom [**1044**], note that the orders are restricted in this paper but this is not necessary for the construction). In this case, $k = q(q-1)/2$. Hence, there is a class of lifted subregular planes of orders q^{4s}, s odd, admitting $SL(2,q)$, where the p-elements are elations.

Furthermore, Foulser [**373**] has discussed the complete set of subregular spreads of order q^4 whose replacing reguli are fixed by $\sigma : (x,y) \longmapsto (x^q, y^q)$. Furthermore, Foulser has shown that for any Desarguesian plane of order v^{4t}, it is possible to construct subregular planes admitting $SL(2,v)$, where $v = p^z$, and the p-elements are elations. Moreover, Foulser discusses the more complete set of translation planes that may be constructed using what might be called 'inversion constructions'. Any of these planes shall be called Ostrom–Foulser planes.

The question is, if t is odd, does the class of Ostrom–Foulser planes contain those that are lifted from Ostrom–Foulser planes? For example, is any plane lifted from the plane of an Ostrom plane of order v^4, an Ostrom–Foulser plane? Our Theorems 24.23 and 24.27 show that this is, in fact, exactly what occurs.

THEOREM 24.37. *Any Ostrom–Foulser plane of order q^4 admitting $SL(2,q)$ lifts to an Ostrom–Foulser plane of order q^{4s}, for s odd, admitting $SL(2,q)$ as a collineation group, where the p-elements, $q = p^r$, are elations. None of these planes can be generalized André planes or semifield planes.*

Given an Ostrom–Foulser plane of order h^4 admitting $SL(2,h)$. Let v divide h so that $SL(2,v)$ is a subgroup of $SL(2,h)$. If there is an orbit of reguli under $SL(2,v)$ of length $v(v-1)/2$, then there is a translation plane of order $h^4 = v^{4s}$, if $v^s = h$ admitting $SL(2,v)$. What is not completely clear, however, is whether there is always a subplane of order v^4 left invariant by $SL(2,v)$ such that the subregular spread obtained by replacing the orbit of $v(v-1)/2$ reguli contracts to the subplane; contracts to an Ostrom–Foulser plane of order v^4 admitting $SL(2,v)$.

24.7.1.4. *Bruck Triple Planes.* In [**177**], Bruck gives a more or less complete determination of the subregular planes/ spreads that may be obtained by the multiple derivation of three reguli. Subregular lifting shows that whenever π_3 is a subregular plane of order q^2 obtained by the derivation of three mutually disjoint reguli, then there is an infinite class of subregular planes π_k^s of orders q^{2s}, s odd > 1, that contain π_3 as a subplane.

THEOREM 24.38. *Any Bruck triple plane of order q^2 lifts to a Bruck triple plane of order q^{2s}, for s odd > 1. If the original plane is not André, the lifted plane is not André.*

24.7.1.5. *Ebert Quadruple Planes.* Ebert [**333, 336**] has given a theory of subregular spreads obtained from a Desarguesian spread by the replacement of exactly four mutually disjoint reguli. There are a tremendous number of mutually non-isomorphic translation planes of odd order (asymptotically equal to q^6/λ, where

$16 \leq \lambda \leq 3072$ of order q^2). Moreover, Ebert [**333**] has a classification of quadruple planes of odd order, whose collineation group fixing each replaced regulus consists only of the kernel homology group (i.e., acts trivially on the set of "circles" on the line at infinity of the Desarguesian affine plane). The classification is based on the subgroup of S_4 that acts on the set of replaced circles. The main point is that the group K fixing each circle is trivial. There are conditions established by Ebert [**333**, Theorem (2.2)], stating that a set of n mutually disjoint reguli have a trivial group fixing each circle if and only if (i) there is not an additional circle 'orthogonal' to the set and (ii) there is not a circle of the set orthogonal to the remaining circles.

THEOREM 24.39. *Let π be any Ebert quadruple plane of order q^2 admitting a collineation group H such that $H \mid \ell_\infty$ is a subgroup of S_4.*

(1) *Then π may be embedded into a translation plane π^k of order q^{2k}, for k odd, such that π^k is an Ebert quadruple plane with group H.*

(2) *If the subgroup K of π fixing each circle (replaced regulus net) is trivial, then the group K^k of π^k fixing each extended circle is trivial. Hence, a plane of Ebert class (i) (see Ebert [**333**, Theorem (3.7)]) lifts to a plane of Ebert class (i).*

24.7.1.6. *The Orr Spread and Deficiency-One Partial Elliptic Flocks.* It has been proven by Orr [**1035**], for q odd, and Thas [**1171**], for q even, that any subregular spread of order q^2 obtained from the multiple derivation of $q-1$ mutually disjoint reguli is 'linear' and produces a Desarguesian spread. Furthermore, any flock of an elliptic quadric in $PG(3, q)$ corresponds to a subregular spread obtained by the derivation of $q - 1$, so any such flock is necessarily linear in that the set of planes defining the intersections on the quadric share a line. However, it is possible to have partial flocks of elliptic quadrics of deficiency one (i.e., $q-2$ mutually disjoint conics) and these correspond to subregular spreads obtained by the replacement of $q - 2$ mutually disjoint reguli (see Thas [**1171**] and also Johnson [**737**]).

Orr [**1034**] shows that the elliptic quadric of $PG(3, q)$ for $q = 5$ or 9 has a partial flock of deficiency one, which cannot be extended to a flock. In particular, note that the $GL(2, 3)$-plane of order 5^2 is constructed by replacement of $3 = q - 2$ reguli and this plane is non-Desarguesian. Furthermore, Orr constructs a subregular plane of order 9^2 by the replacement of $7 = q-2$ reguli. Indeed, Royle [**1144**] shows by computer that for $q \leq 13$, the above mentioned two partial flocks are the only non-extendible partial flocks of deficiency one.

By our lifting theorem, we obtain the following result (we mentioned the $GL(2, 3)$-spread above):

THEOREM 24.40. *Let π_7 denote the Orr plane of order 9^2, then there is an infinite class of subregular planes π_7^s of orders 9^{2s}, where s is odd, containing π_7 as a subplane and obtained by replacement of 7 mutually disjoint reguli.*

In particular, π_7^s is not a generalized André plane, for any odd integer s.

24.7.1.7. *Jha–Johnson $q/2$-Elation Spreads of Order $q^2 = 8^2$.* It has been mentioned that for a non-Desarguesian subregular plane of order q^2, the maximum-order elation group (with a fixed axis) has order q/p, where $p^r = q$ and p a prime. When $p = 2$, Jha and Johnson [**632**] have constructed a translation plane of order 8^2 that admits a collineation group of order $8(8+1)$ such a Sylow 2-subgroup contains an elation group of order $q/2 = 4$ permuting a set of mutually disjoint reguli in an associated Desarguesian affine plane. Hence, by lifting, we may construct an infinite class of translation planes admitting groups of order $8(8^s + 1)$, where s is any odd integer. Indeed, Jha and Johnson [**644**] have constructed this infinite class directly without lifting.

THEOREM 24.41. The Jha–Johnson $q/2$-plane of order $q^2 = 8^2$ may be lifted to a subclass of Jha–Johnson translation planes of orders q^{2s}, admitting an elation group of order 4 and a collineation group of order $8(q^s + 1)$.

24.7.1.8. *Jha–Johnson 'Moved' Spreads.* In Jha and Johnson [**644**], it is shown that there is a class of Foulser planes of even order q^2 obtained by the replacement of a set $S_{\sqrt{q}}$ of $\sqrt{q}$ mutually disjoint reguli in an orbit under an elation group E of the same order. Furthermore, the axis of E is a line of a regulus R_q that intersects each of the reguli of $S_{\sqrt{q}}$ in a unique component. This is a subclass of the class related to inversion-fixed planes. It is shown in [**644**] that it is possible to 'move' the set $S_{\sqrt{q}}$ so as to obtain a set of reguli in an orbit under E but where the reguli are disjoint from R_q. In this case, it is possible to also derive R_q to obtain a subregular plane of order q^2 admitting a Baer group of order $\sqrt{q}$. The precise manner of construction will be given in the following material.

24.7.1.9. *Prohaska Planes.* Prohaska [**1119**] constructs several classes of subregular translation planes, various of which admit S_4 or A_5 acting on the reguli. Any of these planes can be lifted producing planes admitting similar groups. We shall simply describe two classes of these planes, but see below for the Abatangelo and Larato constructions of planes when $q \equiv -1 \bmod 5$. There is a plane of order 81 that admits $SL(2,5)$, generated by elations. There are $5(5 - 1)/2 = 10$ elation groups of order 3 whose axes define a regulus net R_9. There are six reguli in an orbit under $SL(2,5)$ that are disjoint from R_9. We shall call the subregular plane obtained by the replacement of the six reguli the 'Prohaska plane' π_6 of order 81 and call the subregular plane obtained by the replacement of the seven reguli, the 'derived Prohaska plane' π_7. None of the lifted planes in either situation are generalized André or semifield planes.

THEOREM 24.42. Let π_6 be a Prohaska plane of order 9^2 admitting $SL(2,5)$, generated by elations. Then any lifted plane π_6^s, of order 9^{2s}, for s odd, contains π_6 as a subplane and admits $SL(2,5)$, generated by elations.

THEOREM 24.43. Let π_7 be a derived Prohaska plane of order 9^2 admitting $SL(2,5)$, generated by Baer 3-collineations. Then any lifted plane π_7^s of order 9^{2s}, for s odd, contains π_7 as a subplane and admits $SL(2,5)$, generated by Baer collineations.

24.7.1.10. *Ebert Quintuple Planes.* In [**334**], Ebert provides some examples of planes of orders p^2, for $p = 11, 19, 29, 31$ that may be constructed by replacing five reguli. We call these planes the 'Ebert quintuple planes'. These subregular spreads are of interest in that they do not have the two conditions required for the triviality of the group fixing each circle of the set of five.

THEOREM 24.44. Every Ebert quintuple plane of order p^2, for $p = 11, 19, 29, 31$ lifts to a quintuple plane of order p^{2k}, for k odd.

24.7.1.11. *$SL(2,5)$-Planes.* In the following, we shall mention a variety of subregular planes of order q^2 admitting $SL(2,5)$; the Prohaska, using Miquellian inversive planes; Abatangelo and Larato using more or less standard net replacement; the Bonisoli, Korchmáros, and Szönyi planes, using the ideas of the Italian mathematician E. Ciani (~1906) on the finite discontinuous collineation groups of $PG(3, \mathcal{C})$, for $\mathcal{C}$ the complex numbers, and Charnes–Dempwolff planes, using what they call 'ovoidal' orbits, working in the Klein quadric. There are inevitable interconnections between these constructed planes, but we shall not try to work out the isomorphisms.

24.7.1.12. *Prohaska $SL(2,5)$-Planes.* We have mentioned that there is an interesting class of translation planes due to Prohaska [**1119**] of order 9^2 that admit $SL(2,5)$. More generally, we define a 'subregular $SL(2,5)$ plane' as any subregular plane admitting $SL(2,5)$ as a collineation group. Indeed, if $q = 5^a$ or $q^2 \equiv 1 \bmod 5$, there are Prohaska subregular $SL(2,5)$-planes. Any such plane naturally lifts to a plane of the same type.

24.7.1.13. *Abatangelo and Larato Constructions.* In [**10**], there are constructions of subregular planes of order q^2 admitting $SL(2,5)$ by multiple derivation, for $q \equiv 1 \bmod 4$ and $q \equiv -1 \bmod 5$. All of these planes are the same as the Prohaska planes. All such planes lift to planes of order q^k, k odd, admitting $SL(2,5)$ and we have, by the lifting process, a subregular subplane of order q^2. Furthermore, when $q \equiv 1 \bmod 5$, the same construction may occasionally provide examples of Bruen chains and does provide the example of Pelegrino and Korchmáros when $q = 11$.

Let Σ denote the Desarguesian affine plane coordinatized by $GF(q^2)$ with spread

$$x = 0, y = xm; m \in GF(q^2)$$

and consider the following partial spreads:

$$
\begin{aligned}
C_0 &= \left\{ y = xm;\ m^{q+1} = -1 \right\}, \\
C_1 &= \left\{ y = xm;\ m^{q+1} = 1 \right\}, \\
C_{i+2} &= \left\{ \begin{array}{c} y = xm; (m - 2a^{i(1-q)}b^q c(2b^2 + 1)^{-1})^{q+1} \\ = (2b^2 + 1)^{-2} \end{array} \right\}, i = 0, 1, 2, 3, 4,
\end{aligned}
$$

and where α and β are elements of $GL(2, q^2)$ and are given by

$$
\begin{aligned}
\alpha(x, y) &= (ax, a^q y); a^5 = 1, \\
\beta(x, y) &= (bx + cy, -c^q x + b^q y); b = (a - a^q)^{-1} \text{ and } c \text{ so that } c^{q+1} + b^{q+1} = 1.
\end{aligned}
$$

Then the main results of Abatangelo and Larato show the following:

THEOREM 24.45. The set $\{C_0, C_1, C_2, \ldots, C_6\}$ is a set of mutually disjoint reguli in the associated projective space $PG(3, q)$. The group $\langle \alpha, \beta \rangle \simeq SL(2,5)$ fixes C_0 and acts as $PG(2,5)$ on the partial spreads $\{C_1, C_2, \ldots, C_6\}$. Hence we have the following planes:

(1) By replacing $\{C_1, C_2, \ldots, C_6\}$ but not C_0, there is a translation plane of order q^2, admitting $SL(2,5)$. If q is an even power of 3 then the group is generated by affine elations.

(2) By replacing $\{C_0, C_1, C_2, \ldots, C_6\}$, there is a translation plane of order q^2, admitting $SL(2,5)$. If q is an even power of 3 then the group is generated by Baer 3-collineations.

24.7.1.14. *Bonisoli, Korchmáros, Szönyi Planes.* In [**162**], these three authors construct a class of subregular $SL(2,5)$-planes of odd order q^2, where $q^2 \equiv 4 \bmod 5$. There are three types: (I) five reguli and A_5 acting in its natural representation on the set, for $q \equiv 2 \bmod 3$ (II) ten reguli and A_5 having orbit lengths of $1, 6, 3$,

$q \equiv 1 \bmod 3$ and $q \equiv 1 \bmod 4$ (III) ten reguli and A_5 again having orbit lengths of $1, 6, 3$ but the set of reguli are distinct from that of (II), $q \equiv -1 \bmod 4$, $3 \nmid q$ and $q \equiv 3, 5$ or $6 \bmod 7$ and (IV) fifteen reguli, where A_5 has orbit lengths of $1, 2, 4, 8$, $q \equiv 2 \bmod 3$, $q \equiv 1 \bmod 4$, $q \equiv 1, 2$ or $4 \bmod 7$. However, the construction uses partial flocks of an elliptic quadric $\mathcal{E}$. Assume that we have a set $\mathcal{P}$ of conic intersections. This set consists of mutually disjoint conics if and only for any two planes B and C, containing the conics, the line joining their poles $B^{\perp}$ and $C^{\perp}$ is a chord of $\mathcal{E}$ (intersects $\mathcal{E}$ in two points), for in this case, the intersection of the planes is a line exterior to the quadric constructing the conics. So, a model of $\mathcal{E}$ is given, for which a convenient representation of S_5 may be given. The construction then lists a set of poles and it is then determined that the lines joining any distinct pair for a chord to the elliptic quadric.

The model for the elliptic quadric in $PG(3, q)$, $q^2 \equiv -1 \bmod 5$, making 5 a non-square, is

$$\mathcal{E} : x_1^2 + x_2^2 + x_3^2 + x_4^2 + x_1 x_2 + x_1 x_3 + x_1 x_4 + x_2 x_3 + x_2 x_4 + x_3 x_4 = 0.$$

As for the group, take the $0, 1$ permutation 4×4 matrices and

$$T = \begin{bmatrix} 1 & 0 & 0 & 0 \\ 0 & 1 & 0 & 0 \\ 0 & 0 & 1 & 0 \\ -1 & -1 & -1 & -1 \end{bmatrix}$$

and it turns out that the generated group is isomorphic to S_5.

DEFINITION 24.46. A set of points of $PG(3, q)$ is said to be a 'disjoint set of poles' if and only if the line joining any distinct pair is a chord to $\mathcal{E}$.

The main results of [162] are stated in the following theorem. Note that any subset of a disjoint set of poles produces a set of mutually disjoint conics and hence a set of mutually disjoint reguli within a Desarguesian affine plane. Then derivation of this subset produces the translation plane in question. Hence, there is quite a variety of construction translation planes, some of which lose part of S_5 as a collineation group.

THEOREM 24.47. Under the assumptions above, each of the following sets $\mathcal{P}$, $\mathcal{Q}$, $\mathcal{R}$, $\mathcal{S}$ is a disjoint set of poles.

(I)

$$\mathcal{P} = \{(-4, 1, 1, 1), (1, -4, 1, 1), (1, 1, -4, 1), (1, 1, 1, -4), (1, 1, 1, 1)\}$$

and S_5 permutes $\mathcal{P}$ in an orbit.

(II)

$$
\begin{aligned}
\mathcal{Q} &= \{Q_i; i = 1, 2, \ldots, 10\}, \\
Q_1 &= (1, 0, 0, 0), Q_2 = (0, 1, 0, 0), Q_3 = (0, 0, 1, 0), Q_4 = (0, 0, 0, 1), \\
Q_5 &= (-1, 1, 0, 0), Q_6 = (-1, 0, 1, 0), Q_7 = (-1, 0, 0, 1), Q_8 = (0, -1, 1, 0) \\
Q_9 &= (0, -1, 0, 1), Q_{10} = (0, 0, -1, 1).
\end{aligned}
$$

S_5 acts on $\mathcal{Q}$ as it acts on the set of its dihedral subgroups of order 12; the stabilizer of Q_1 has the following other two orbits:

$$\{Q_2, Q_3, Q_4, Q_5, Q_6, Q_7\}, \{Q_8, Q_9, Q_{10}\}.$$

(III)

$$\begin{aligned}
\mathcal{R} &= \{\, R_i;\ i = 1, 2, \ldots, 10 \,\}, \\
R_1 &= (-3/2, 1, 1, 1), R_2 - (1,\ -3/2, 1, 1),\ R_3 = (1, 1, -3/2, 1), \\
R_4 &= (1, 1, 1, -3/2), R_5 = (3/2, 3/2, -1, -1), R_6 = (3/2, -1, 3/2, -1), \\
R_7 &= (3/2, -1, -1, 3/2), R_8 = (-1, 3/2, 3/2, -1), R_9 = (-1, 3/2, -1, 3/2), \\
R_{10} &= (-1, -1, 3/2, 3/2).
\end{aligned}$$

S_5 acts on $\mathcal{Q}$ as it acts on the set of its dihedral subgroups of order 12; the stabilizer of R_1 has the following other two orbits:

$$\{R_2, R_3, R_4, R_5, R_6, R_7\}, \{R_8, R_9, R_{10}\}.$$

(IV)

$$\begin{aligned}
\mathcal{S} &= \{\, S_i;\ i = 1, 2, \ldots, 15 \,\}, \\
S_1 &= (-1, 1, 1, 0), S_2 = (1, -1, 1, 0), S_3 = (1, 1, -1, 0), S_4 = (-1, 1, 0, 1), \\
S_5 &= (1, -1, 0, 1), S_6 = (1, 1, 0, -1), S_7 = (-1, 0, 1, 1), S_8 = (1, 0, -1, 1), \\
S_9 &= (1, 0, 1, -1), S_{10} = (0, -1, 1, 1), S_{11} = (0, 1, -1, 1), S_{12} = (0, 1, 1, -1), \\
S_{13} &= (1, 1, -1, -1), S_{14} = (1, -1, 1, -1), S_{15} = (1, -1, -1, 1).
\end{aligned}$$

S_5 acts on $\mathcal{S}$ as S_5 acts on its Sylow 5-subgroups.

Since the basic congruences are valid for q^k, k odd, in place of q, we see that type (J) planes lift to type (J) planes, which of course they must due to the action of S_5. However, we now see that there are type (J) subplanes of order q^2 within any type (J) plane of order q^{2k}, for k odd.

24.7.1.15. *Charnes and Dempwolff Planes.* In [**225**, Sections 3 and 4], there are constructions of spreads in $PG(3, q)$ admitting $SL(2, 5) * C$, a subgroup of a $GL(2, q^2)$ of an associated Desarguesian spread, and $*$ denote a central product, C the kernel homology group of order $q^2 - 1$. In particular, we have $(q, 30) = 1 = (q^2 - 1, 5)$. In (4.1) and (4.2), there are constructions of spreads arising from a G-orbit O_i, for $i = 1, 2$ of length $10(q + 1)$ and $5(q + 1)$, respectively, for $q \equiv 1 \bmod 12$ and $q \equiv -1 \bmod 3$. Although determined there in a different manner as sets of points on the Klein quadratic, these correspond to sets of opposite reguli to mutually disjoint reguli of a Desarguesian spread Σ. The construction then finds what are called 'elliptic' orbits which when unioned with O_i produce ovoids and hence, via the Klein quadric, spreads in $PG(3, q)$. The elliptic orbits are pointwise fixed by C, so these become components of Σ. Hence, these spreads are subregular and admit $SL(2, 5)$.

A similar construction occurs for the Dempwolff [**306**] for $q^2 \equiv 1 \bmod 5$.

So, since q^k, for k odd, does not change the congruence statements, we obtain a similar set of subregular spreads in $PG(3, q^k)$ admitting $SL(2, 5)$, now admitting subregular subplanes of order q^2.

24.7.1.16. *Partial Nests of Reguli.* A 'chain' in $PG(3, q)$ is a set of $(q + 3)/2$ reguli with the property that any two distinct reguli have exactly two lines in common and no three mutually distinct reguli share a common line. The concept of a chain was originated by A.A. Bruen, and is sometimes called a 'Bruen chain'. A set of fewer than $(q + 1)/2$ reguli with the above two properties shall be called a 'partial chain'. The interest in chains stems partially from the fact that there is always a replacement set consisting of exactly $(q + 1)/2$ lines from each opposite regulus (see Heden [458]), thus producing a potentially new translation plane. However, it is difficult to find chains and existence is only determined for a few sporadic numbers q ($q = 5, 7, 9, 11, 13, 17, 19, 23, 31$). On the other hand, our lifting theorem shows that it is easy to find partial chains.

THEOREM 24.48. Let Λ be a chain (or partial chain) in $PG(3, q)$ in a Desarguesian affine plane Σ_{q^2}. Then there is a partial chain Λ^s of $(q + 3)/2$ reguli in $PG(3, q^s)$, in a Desarguesian affine plane $\Sigma_{q^{2s}}$, for any odd integer s such that restriction to Σ_{q^2} produces Λ.

The generalizations of chains are due to R.D. Baker and G.L. Ebert and are called t-nests. A 't-nest' of reguli in a Desarguesian affine plane is a set Δ_t of t-reguli with the property that every line of the union of the reguli is common to exactly two distinct reguli of Δ_t. For every known t-nest, with $t \leq q$, there is a replacement set consisting of $(q + 1)/2$ of the lines of each opposite reguli. On the other hand, there are non-replaceable t-nests, for $t = q + 2$ (an infinite class) and $t = q + 1$ and $q + 3$, due to Ebert [344].

Suppose that there is a set of reguli such that each line of the set is in at most two distinct reguli of the set. This might be called a 'partial nest'. The importance of such nests is unclear unless one considers that there may be an additional set of reguli that intersect the lines of the reguli that are 'hit' exactly once. Such constructions occur, for example, in the so-called 'mixed nests' (see Johnson [750]). In any case, partial nests are easily constructed by lifting.

THEOREM 24.49. Let Δ_t be a t-nest of reguli in $PG(3, q)$ in a Desarguesian affine plane Σ_{q^2}. Then there is a partial t-nest Δ_t^s in any $PG(3, q^s)$ in a Desarguesian affine plane $\Sigma_{q^{2s}}$, for s an odd integer, such that restriction to Σ_{q^2} produces Δ_t.

24.8. Characterizing the Foulser–Ostrom Planes.

There are very few translation planes that may be classified according to their collineation groups. If the collineation group is non-solvable, the group often contains $SL(2, q)$ and we might ask what are the translation planes of order q^n that may be classified by a collineation group G containing $SL(2, q)$ in the translation complement? When n divides 2, we have the theorem of Foulser–Johnson.

THEOREM 24.50. (Foulser–Johnson [378, 379]). Let π denote a translation plane of order q^2 that admits a group G isomorphic to $SL(2, q)$ that induces a non-trivial collineation group acting in the translation complement (G need not act faithfully, but does act non-trivially). Then π is one of the following planes: (1) Desarguesian, (2) Hall, (3) Hering, (4) Ott–Schaeffer (5) one of three planes of Walker of order 25 or (6) the Dempwolff plane of order 16.

When $n = 3$, the cyclic semifield planes of Jha and Johnson [611] also produce $SL(2, q)$-planes.

THEOREM 24.51. (Jha and Johnson [607]). Let π be a translation plane of order q^3 admitting $GL(2, q)$, acting canonically. Then π is a generalized Desarguesian plane; π may be constructed using cyclic semifield planes of order q^3.

When $n = 4$, we also have a classification result of a general type:

THEOREM 24.52. (Jha and Johnson [**597**]). Let π be a translation plane of order q^4 admitting $SL(2,q) \times Z_{1+q+q^2}$ in the translation complement. Then, the p-elements, for $q = p^r$, are elations, the kernel is $GF(q)$ and there is a set of $1 + q + q^2$ derivable nets sharing the elation net of degree $q + 1$. Given any elation axis, there is a Desarguesian parallelism of $PG(3, q)$. Conversely, any Desarguesian parallelism admitting a cyclic group of order $1 + q + q^2$ produces a translation plane admitting $SL(2,q) \times Z_{1+q+q^2}$.

There are a variety of very interesting planes constructed from a Desarguesian plane by the multiple derivation of a set of line-disjoint reguli. In particular, in Ostrom [**1044**], there is a class of translation planes of order q^4 admitting $SL(2, q)$, where the p-elements are elations and $q \equiv -1 \bmod 4$. This class was generalized and explicated by Foulser [**373**] who gave a construction based on the Frobenius automorphism of $GF(q^4)$, for arbitrary odd q.

It is furthermore possible to consider translation planes of order q^4 that admit $SL(2, q)Z_{q^4-1}$ and provide to give a very concrete classification result:

24.8.1. Jha–Johnson Theorems on Foulser–Ostrom Planes.

THEOREM 24.53. (Jha, Johnson [**638**]). Let π be a translation plane of order q^4, that admits an Abelian collineation group C of order $q^4 - 1$.

If π admits $SL(2, q)C$, then π is one of the following planes:
(1) Desarguesian,
(2) Hall, or
(3) q is odd and π is the Foulser–Ostrom plane.

24.9. $SL(2, q)Z_{q^n-1}$.

We begin by establishing that any plane admitting such a group may be constructed from a Desarguesian plane.

THEOREM 24.54. Let π be a translation plane of order q^n, for $n > 2$, admitting $SL(2, q)Z_{q^n-1}$. Then the kernel properly contains $GF(q)$ and there is a Desarguesian affine plane Σ such that π may be constructed from Σ by replacing a set of mutually disjoint hyper-regulus nets of degrees $(q^n - 1)/(q^k - 1)$, where k divides n and $n > 1$.

Note that the above result gives a nice characterization of translation planes of order q^n where n is prime.

THEOREM 24.55. Let π be a translation plane of order q^n and kernel $GF(q)$ that admits $SL(2, q)Z_{q^n-1}$. If n is prime then π is Desarguesian.

COROLLARY 24.56. Let π be a translation plane of order q^n that admits $SL(2, q)Z_{q^n-1}$ then the kernel strictly contains $GF(q)$ or $n = 2$ and the plane is Hall.

The class of subregular planes due to Ostrom and Foulser of order q^4 admitting $SL(2, q)$ may be constructed as follows: Let Σ denote the Desarguesian affine plane coordinatized by a field $GF(q^4)$. Let $\sigma : (x, y) \longmapsto (x^q, y^q)$ acting in Σ. Let $GF(q^2) \cup (\infty) = R^{q^2}$ denote the standard regulus;

$$x = 0, y = x\alpha; \alpha \in GF(q^2),$$

and let R^q denote

$$x = 0, y = x\alpha; \alpha \in GF(q).$$

Then σ has orbits of lengths 2 on $R^{q^2} - R^q$ and 4 on $\Sigma - R^{q^2}$. Assume that q is odd. There is a unique regulus $S_{\{P,\sigma(P)\}}$, where P is in $R^{q^2} - R^q$. Foulser [373] shows that $\bigcup S_{\{P,\sigma(P)\}}$ is a set of $q(q-1)/2$ mutually disjoint reguli in Σ which are in an orbit under $SL(2,q)$. The corresponding translation plane is called the 'Foulser–Ostrom' plane π of order q^4. Note that since the plane is obtained by multiple derivation, the kernel homology group Z_{q^4-1} of Σ acts as a collineation group of $\pi\pi$. Hence, the Foulser–Ostrom plane of order q^4 admits and has kernel $GF(q^2)$.

The main result of Jha and Johnson [638] is a complete classification of all translation planes of order q^4 that admit $SL(2,q)Z_{q^4-1}$.

THEOREM 24.57. (Jha and Johnson [638]). Let π be a translation plane of order q^4, that admits an Abelian collineation group C of order $q^4 - 1$. If π admits $SL(2,q)C$, then π is one of the following planes: (1) Desarguesian, (2) Hall, or (3) q is odd and π is the Foulser–Ostrom Plane.

24.9.1. The Derived Foulser–Ostrom Planes. The Foulser–Ostrom Planes are derivable. To see this, we need to represent the spread in a different manner. Let Σ be a Desarguesian affine plane of order q^2, where $q = u^2$, for q odd. Let K denote the coordinatizing field isomorphic to $GF(q^2)$ and let ω be a generator for K^*. We represent the spread for Σ as

$$x = 0, y = xm; m \in K.$$

We note that Baer subplanes which are $GF(q)$-subspaces that disjoint from the component $x = 0$ have the general form $y = x^q n + xt$, where $n \neq 0, t \in K$. Furthermore, if $y = x^q n + xt$ is such a Baer subplane, it defines a regulus net (derivable net) of Σ as the set of $q + 1$ components of intersection with this Baer subplane as a 2-dimensional $GF(q)$-subspace. The opposite regulus then is the image set under the kernel homology group of Σ, whose elements may be represented by

$$(x, y) \rightarrow (xd, yd); d \in K^*.$$

The image set is of the form

$$\left\{ y = x^q d^{1-q} n + xt; d \in K^* \right\} = \langle y = x^q n + xt \rangle.$$

For the Foulser–Ostrom planes, we require $u(u-1)/2$ Baer subplanes defining mutually disjoint derivable nets and these are as follows:

$$N(\beta,\gamma)' = \left\langle y = x^q \gamma \omega^{u+1} + \beta; \text{ for fixed } \gamma \neq 0, \ \beta \in GF(u) \right\rangle.$$

We note that $N(\beta,\gamma)' = N(\beta,-\gamma)'$. Hence, there are $u(u-1)/2$ distinct regulus nets and actually these turn out to be mutually disjoint. These nets do not intersect $\{ x = 0, y = xv; v \in GF(u) \}$ and are permuted transitively by $SL(2,u)$, generated by elations in the Desarguesian affine plane. Hence, the spread for the Foulser–Ostrom planes is

$$\left\{ x = 0, y = xm; m \neq \omega^{i(q-1)}\gamma^* + \beta^*, \ \gamma^* \neq 0, \ \beta^* \in GF(u) \right\}$$
$$\cup \{ N(\beta,\gamma)'; \gamma \neq 0, \ \beta \in GF(u) \} \cup \{ y = xm; \}.$$

Now consider, for fixed i,

$$D_i = \left\{ x = 0, \ y = x^q \gamma \omega^{i(q-1)} + \beta; \ \forall \gamma, \beta \in GF(u) \right\}.$$

This set of $u^2 + 1 = q + 1$ components of the Foulser–Ostrom plane defines a derivable net, which is not a regulus net in the associated $PG(3, q)$. The group $SL(2, u)$ (in fact $GL(2, u)$) is a collineation group of the derived plane, where the p-elements are Baer.

THEOREM 24.58. *The derived Foulser–Ostrom plane of order* u^4 *admits a collineation group isomorphic to* $SL(2, u)$, *where the* p-elements, $u = p^r$, *are Baer. The kernel is* $GF(u)$ *and the translation complement contains the group of order* $u^2 - 1$ *of the Foulser–Ostrom plane, corresponding to* $GF(u^2)^*$.

24.10. When Is a Plane Subregular?

As we have mentioned, the most commonly cited class of finite translation planes is probably one of the least known, the class of translation planes that may be derived or multiply derived by the replacement of a set of mutually disjoint reguli in the spread of a Desarguesian plane.

Since the spread of a Desarguesian affine plane is 'regular' (contains the unique regulus defined any distinct three of its spread lines), the spreads so obtained by multiple derivation are called 'subregular'.

So the question is: Exactly how 'known' is the class of subregular spreads?

Albert [**26**] has shown that derivation of a finite Desarguesian affine plane (replacement of any regulus by its opposite regulus) produces the Hall plane. These are planes that were originally constructed by distorting the multiplication of a field so that the new coordinate structure coordinatized a 'new' translation plane.

Let Σ denote an affine Desarguesian plane of order q^2 and let $PG(1, q^2)$ denote its line at infinity. Then identifying the set of infinite points of reguli with 'circles', there is a natural Miquellian inversive plane whose points are the elements of $PG(1, q^2)$ and whose circles are the reguli of Σ. In this context, a set Γ_k of k mutually disjoint reguli of Σ is a set of k mutually disjoint circles of $PG(1, q^2)$. Given a regulus R_i of Σ, there is a unique Baer involution σ_i of $\Gamma L(2, q^2)$ whose fixed components are the lines of R_i. Such an involution induces a collineation on $PG(1, q^2) - R_i$ (as a 'circle geometry') that interchanges pairs of points.

A set $\Gamma_k = \{ R_i; \ i = 1, 2, \ldots, k \}$ is said to be 'linear' if and only if there is a unique pair of points (P, Q) such that the unique Baer involution σ_i with fixed components those of R_i interchanges P and Q, for $i = 1, 2, \ldots, k$.

Let the Desarguesian plane Σ of order q^2 be coordinatized by a field F isomorphic to $GF(q^2)$ and let its spread be

$$\{ (x = 0), \ y = xm; \ m \in F \}.$$

We define the André partial spreads A_δ, where δ is an arbitrary non-zero element of $GF(q^2)$ as follows:

$$A_\delta = \left\{ y = xm; \ m^{q+1} = \delta \right\}.$$

We note that A_δ is a regulus whose opposite regulus A_δ^* is as follows:

$$A_\delta^* = \left\{ y = x^q m; \ m^{q+1} = \delta \right\}.$$

We consider the following collineation:

$$\sigma_\delta \colon (x,y) \mapsto (y^q d_\delta^{-q}, x^q d_\delta), \ d_\delta^{q+1} = \delta.$$

Then, it follows that

$$(x, xm) \mapsto (x^q m^q d_\delta^{-q}, x^q d_\delta)$$

and since

$$m^q d_\delta^{-q} m = d_\delta, \quad \Longleftrightarrow \quad y = xm \in A_\delta,$$

then σ_δ is the unique Baer involution whose set of fixed components is A_δ.

Since $\{\, \sigma_\delta; \ \delta \in GF(q) \,\}$ is a set of Baer involutions of Σ that interchange (∞) and (0), it follows that any subset of André reguli is a linear set and, conversely, it turns out that linear sets produce a set of André reguli.

The 'André planes' are those obtained from a Desarguesian plane by replacement of a subset of André reguli and these planes are well known. For example, the subregular spreads found in Hirschfeld [**527**, Chapter 17] are all André spreads.

The collineation groups of André planes are determined by Foulser [**369**] who also shows that the full collineation group of any such non-Desarguesian, non-Hall André plane must leave invariant the set $\{(\infty),(0)\}$.

Returning to the question of how much is known about subregular spreads, we note that any two disjoint reguli can always be considered André reguli by Bruck [**177**]. However, there are non-André planes that may be constructed by the replacement of three disjoint reguli, the 'non-linear triple planes', and these are explicated by Bruck [**177**] so they are considered known. Furthermore, the 'non-linear quadruple planes' of odd order are extensively analyzed by Ebert in [**333**] and [**336**]. Furthermore, Ebert has determined non-linear quintuple planes for orders p^2, for $p = 11, 19, 29$, and 31. (Quadruple and quintuple planes are those obtained by replacing four and five mutually disjoint reguli, respectively.)

As mentioned previously, Ostrom [**1044**] has constructed subregular spreads of order u^4 that admit $SL(2, u)$ as a collineation group, where the p-elements are elations for $u = p^r$, and p an odd prime, where $p \equiv 1 \bmod 4$. Here there are $k = u(u-1)/2$ reguli in a non-linear set. Foulser [**373**] has constructed a somewhat less well-known class of subregular planes of even order v^{4t}, for $t > 1$, that admit $SL(2, v)$ as a collineation group and furthermore shows that the construction of Ostrom is valid for any odd prime. These odd-order planes of order u^4 admitting $SL(2, q)$ planes are now called the Foulser–Ostrom planes of order u^4.

There are two other sporadic subregular spreads due to Ostrom [**1048**], and Johnson and Seydel [**810**] for $k = 7$. Furthermore, there are the Prohaska [**1119**] subregular spreads admitting S_4 or A_5.

In a non-Desarguesian, non-Hall André plane of order $q^2 = p^{2r}$, p a prime, there can be affine elations of order p only if $p = 2$, since there are two components that are fixed or interchanged. Indeed, any elation group with a fixed axis cannot be of order larger than 2 in any such André plane. So the existence of elation groups of large order necessarily forces a subregular plane to be non-André. Perhaps one might consider if it is possible to construct semifield planes by multiple derivation. In this regard, we point out the following result.

THEOREM 24.59. *A finite subregular plane of order q^2 is a semifield plane if and only if the plane is Desarguesian.*

PROOF. By the results of Walker [**1205, 1206**], any collineation group of a subregular plane obtained by the replacement of k reguli is either the subgroup of the Desarguesian plane that permutes the k reguli or the set is linear and either $q = 3$ and $k = 1$, or $k = (q-1)/2$, $(q-1)$ or $(q-2)$. When $k = q$ 1, the plane is Desarguesian and when $k = q - 2$, the plane is Hall.

When $k = (q-1)/2$, the plane is an André plane and Foulser's results on the collineation groups of generalized André planes show that there is an invariant pair of points. If the group must permute the reguli and the plane is a semifield plane, there cannot be mutually disjoint reguli permuted by the affine elation group of order q^2.   □

In this section, we concentrate on the construction of subregular planes that admit affine elation groups, thereby essentially guaranteeing that the planes obtained are not André, generalized André or semifield. We also give a variety of constructions of translation planes with higher-dimensional spreads that admit many elations. We also construct a set of quadruply derived planes of order 2^{2r}, for r odd and > 1. These planes are quite unusual in that they admit elation groups of order 4, showing that they cannot be André planes. In fact, these planes constitute the first infinite class of subregular spreads of even order 2^{2r}, for r odd, that admit an elation group of order > 2 (but again also see Ebert [**333, 336**]).

As mentioned, Foulser [**373**] has given a description of the Ostrom planes of order u^4 and has shown that there is a somewhat analogous set of planes of even order. We shall also define a class of translation planes of order u^4, where u is even, admitting an elation group of order u, obtained by the multiple derivation of a set of u mutually disjoint reguli. Our approach shows that it is also possible to find planes of this type admitting a regulus net fixed by the elation group such that the regulus is disjoint from the u reguli in an elation group orbit. So it is also possible to find translation planes of even order u^4 that admit Baer groups of order u. These translation planes have spreads in $PG(3, u^2)$ and are the first planes of this type known.

There are a variety of interesting subregular spreads admitting $SL(2,5)$. We previously mentioned the planes of Prohaska [**1119**]. But there are also the planes of Abatangelo–Larato [**10**], the Bonisoli–Korchmáros–Szőnyi Planes [**162**], and the planes of Charnes–Dempwolff [**227**], constructed by what they call 'ovoidal' orbits, working in the Klein quadric. There are inevitable interconnections between these constructed planes, but we shall not try to work out the isomorphisms. However, all of these planes admit $SL(2,5)C$, where C is an Abelian group of order $q^2 - 1$, which is centralized by $SL(2,5)$. In some of these constructions, it is obvious that we are considering subregular planes but in the Charnes–Dempwolff planes, for example, this is less apparent. But this does raise the question: If a translation plane with spread in $PG(3,q)$ admits $SL(2,5)C$, where C is an Abelian group of order $q^2 - 1$ centralizing $SL(2,5)$, is this enough to ensure that the plane is subregular?

It is therefore of interest to ask for conditions on the collineation group that ensure a spread in $PG(3,q)$ is a subregular spread, thus giving a valuable tool in the identification of this class of translation planes. More generally, if π is a translation plane admitting an Abelian group C of order $q^2 - 1$ that is centralized by a collineation group N, what conditions on NC force the plane to be subregular? We might note that any j-plane of order q^2 admits a cyclic group of order $q^2 - 1$ and none of these are subregular (see Johnson, Pomareda, Wilke [**806**] for the definition of a j-plane). So we would need to add some sort of condition to N.

We are able to show if p divides the order of N, for $p^r = q$, or if N is non-solvable then the plane is subregular. Thus, Theorem 24.66 and Theorem 24.67, we shall show:

THEOREM 24.60. Let π be a translation plane of order q^2 with spread in $PG(3,q)$, for $q = p^r$, p a prime. Then the following hold:

(1) (cf. Theorem 24.66). Suppose that π admits a non-trivial linear p-element σ in the translation complement. Then π is a subregular plane admitting linear p-elements if and only if σ centralizes an Abelian group of order $q^2 - 1$ in the translation complement.

(2) (cf. Theorem 24.67). If the translation complement of π contains an Abelian collineation group C of order $q^2 - 1$ centralized by a non-solvable collineation group N then π is a subregular spread.

The two parts of the theorem will be established in Section 24.10.

Even though subregular spreads are well known, there are really no results that show when a particular spread is subregular. To investigate such matters, we begin with some fundamental lemmas.

LEMMA 24.61. Let π be a spread of order q^2 in $PG(3,q)$ that admits a collineation $\tau_u \in GL(4,q)$ of prime power order $u = v^b$, v a prime, such that u is a q-primitive divisor of $q^2 - 1$. Then the lines of $PG(3,q)$ that are τ_u-invariant are mutually skew.

The elements of type τ_u above arise in $PG(3,q)$ spreads admitting Abelian groups C of order $q^2 - 1$:

LEMMA 24.62. Let π be a spread of order q^2 with spread in $PG(3,q)$ that admits an Abelian collineation group C of order $q^2 - 1$. If $q \neq 3$ then there is always a q-primitive divisor u of $q^2 - 1$ and an element τ_u in $GL(4,q)$ of order u in C.

LEMMA 24.63. Let Σ be a Desarguesian spread of order q^2 and let G be an Abelian collineation group of order $q^2 - 1$ in the collineation group of Σ that fixes at least three components. Then G is the kernel homology group of Σ.

If π is a subregular spread multiply derived from a Desarguesian spread Σ, then the kern homology group C of Σ is an Abelian group of order $q^2 - 1$ inherited by π, and the orbits of C partition the lines of $\pi - \Sigma$. This property characterizes subregular spreads, and is assumed throughout the paper.

LEMMA 24.64. Let π and Σ be spreads in $PG(3,q)$, the components of both spreads consisting of $GF(q)$ subspaces of the same four-dimensional $GF(q)$ vector space V. Suppose Σ is Desarguesian and let C be the group of kernel homologies of Σ of order $q^2 - 1$. Then π is a subregular spread iff it may be expressed as a union of C-orbits: $\ell \in \pi \implies (\ell)C \subset \pi$.

24.10.1. The Fundamental Identification Theorem of Subregular Planes.

THEOREM 24.65. Let π be a translation plane of order q^2 with spread in $PG(3,q)$. Then π is a subregular translation plane if and only if there is an Abelian collineation group G in the translation complement of order $q^2 - 1$ that fixes at least three, mutually disjoint, line size $GF(q)$-subspaces of π.

24.10.2. The Elation Identification Theorem for Subregular Planes. Since we are interested in translation planes admitting elations, we may improve the previous result as follows.

THEOREM 24.66. (cf. Theorem 24.60(1)). Let π be a translation plane of order $q^2, q = p^r$, p a prime, with spread in $PG(3,q)$ that admits a non-trivial collineation σ of order p in $GL(4,q)$, for $q = p^r$.

Then π is a subregular plane admitting linear p-collineations if and only if σ centralizes A, where A is an Abelian collineation group of order $q^2 - 1$ in the translation complement.

24.10.3. The Non-Solvable Identification Theorem for Subregular Planes. Certainly not all translation planes with spread in $PG(3,q)$ admitting cyclic collineation groups of order $q^2 - 1$ are subregular. For example, the j-planes admit such groups but are not subregular. One might ask if a plane admits a non-solvable group centralized by a cyclic group of order $q^2 - 1$, what would occur? In this regard, we can prove that the plane is subregular.

THEOREM 24.67. (cf. Theorem 24.60(2)). Let π be a translation plane of order q^2 with spread in $PG(3,q)$ that admits an Abelian collineation group C of order $q^2 - 1$. Let N be any non-solvable collineation group that commutes with C. Then π is a subregular translation plane.

24.11. Subregular Spreads with Elation Groups of Order 4.

In this section, we construct an infinite class of non-Desarguesian subregular planes that admit elation groups of order 4, so they cannot be (generalized) André planes or semifield planes.

LEMMA 24.68. Let Σ be the Desarguesian spread over $F = GF(q^2)$, where $q = 2^r$, $r > 1$ odd. Let H be the multiplicative subgroup of $GF(q^2)$ of order $q + 1$, and fix $m \in GF(q) - \{0\}$ such that $m \neq h + h^\sigma$, for all $h \in H$; so there are $\geq q/2 - 1$ choices for m.

In the collineation group of Σ, let $\hat{H}$ be the group of kernel homologies of order $q + 1$, and fix the elation group $E = \{(x,y) \mapsto (x, bx + y) : b \in GF(4)\}$ of order 4. Then:

(1) The $GF(q)$-subspace $B_m = (y = x^\sigma m + xm)$ is a Baer subplane of Σ.
(2) The $\hat{H}$-orbit $\mathcal{R}_0 = (B_m)\hat{H}$ subplane B_m is a partial spread that forms a regulus derived from a regulus of components of Σ.
(3) The partial spreads $\mathcal{R}_e = (\mathcal{R}_0)e$ for $e \in E$ are four distinct reguli that form an E-orbit, each $\mathcal{R}_e$ obtained by deriving a regulus of components in Σ.
(4) The E-orbit $\{\mathcal{R}_e : e \in E\}$ consists of four pairwise disjoint reguli, in the sense that no two of the reguli cover a common non-zero point. Hence they define a subregular spread $\Sigma_{m,E}$, that contains the orbit $\mathcal{O} := (y = x^\sigma m + xm)\hat{H}E$ of subspaces and its remaining lines are the components of Σ that do not overlap with the points covered by $\mathcal{O}$, thus:
$$\Sigma_{m,E} = \mathcal{O} \cup \{ S \in \Sigma; S \cap N = \{0\}, N \in \mathcal{O} \}.$$

(5) The spread $\Sigma_{m,E}$ admits $\sigma : x \mapsto x^q$ as a Baer involution that normalizes the Abelian subgroup $\hat{H}E$, hence $\hat{H}E\langle\sigma\rangle$ is a collineation group of order $8(q+1)$ that contains a characteristic elation subgroup of order 4 and also a Baer involution σ.

We interpret our results in terms of translation planes.

THEOREM 24.69. If $q = 2^r$ and r is odd and > 1, then there is an infinite class of subregular translation planes π of orders 2^{2r} and kernel $GF(2^r)$ that admit a collineation group of order $8(q+1)$ containing a normal elation group of order 4 and a Baer 2-group of order 2. These translation planes are obtained by multiply deriving the Desarguesian spread of order q^2.

(1) The planes π may be chosen to be the translation planes specified by the subregular spreads $\Sigma_{m,E}$ specified in Lemma 24.68(4).
(2) The planes are not Desarguesian, Hall, André or generalized André planes.

REMARK 24.70. Actually, there are several interesting partial elliptic flocks found by computer by Royle [**1144**] of which the Orr spread is but one example. There are a wide variety of maximal partial flocks of an elliptic quadric in $PG(3,q)$, for $q \leq 13$. For example, there is a unique deficiency-one non-linear maximal partial flock for each of the orders $q = 4, 5, 7, 8, 9, 11$ and 13, so the number of reguli in the replacement set is also q. Of course, by the sublifting theorem, there are translation planes of order q^{2s}, constructed by the replacement of q mutually disjoint reguli. We call any of the corresponding planes (other than the Orr plane) a 'Royle subregular plane'.

Applying subregular lifting, we have:

THEOREM 24.71. Let π be any subregular plane of Royle mentioned in the previous remark of order q^2. Then, for any odd integer s, there is a subregular translation plane of order q^{2s}, containing a Royle subplane.

We determine the collineation groups of the above planes, and study their isomorphism classes.

THEOREM 24.72. Consider a plane, specified by the subregular spread $\Sigma_{m,E}$, Lemma 24.68(4), hence of order q^2, where $q = 2^r$, and r is odd.

(1) Then the full collineation group of $\Sigma_{m,E}$ is the group permuting the four reguli. In addition, the full collineation group normalizes the elation group E of order 4.
(2) Furthermore, two planes $\Sigma_{m,E}$ and $\Sigma_{n,E}$ are isomorphic if and only if $n = m^\sigma$, where σ is an automorphism of $GF(q)$.
(3) There are at least $(q/2 - 1)/(q/2 - 1, r)$ mutually non-isomorphic planes.

COROLLARY 24.73. There is a unique spread of order 8^2 of elation 4-type.

24.12. Large Elation or Baer Groups in Characteristic 2.

We construct some subregular translation planes of order $GF(q^2)$, q an even square, admitting large elation groups, and we show that some of these subregular spreads may be derived to yield further regular spreads admitting large Baer groups.

LEMMA 24.74. Let Σ_{q^2} be the Desarguesian spread of even order q^2, $q = 2^{2s}$ a square. Then the following hold.

(1) For any $c \in GF(q^2)$, the partial spread $R_c = \{\, y = x(n+c);\ n^{q+1} = 1 \,\} \subset \Sigma$ is a regulus disjoint from $(x = 0)$.

(2) $E = \{\, (x,y) \mapsto (x, xb + y);\ b \in GF(\sqrt{q}) \,\}$ is an elation group of Σ_{q^2}, with axis $(x = 0)$ and order $\sqrt{q}$, such that set of the images of the regulus R_c under E consists of $\sqrt{q}$ mutually disjoint reguli; so they may be multiply derived to yield a subregular spread.

Note that the regulus R_c meets the standard regulus $\{\, y = xf;\ f \in GF(q^2) \,\} \cup \{x = 0\}$ iff $(n + c)^{q-1} \in \{0, 1\}$, equivalently $n = c$ or $n^q + c^q = n + c$. Choose c so that $c + c^q = \beta$ in $GF(q) - \{0\}$ with β^{-1} having trace 0. Since $n^{q+1} = 1$, either $n = 1$ or $n^q + n = \delta$ such that δ^{-1} has trace 1. Hence, with c chosen in this way, the reguli in the orbit $(R_c)E$ are disjoint from the standard regulus. Thus we may derive the subregular spread, of Lemma 24.74, to obtain another on which E acts as a Baer group. We have now established:

THEOREM 24.75. Let Σ_{q^2} be a Desarguesian spread of even order q^2, q a square. The collection of $\sqrt{q}$ reguli specified by $(R_c)E$, in Lemma 24.74, multiply derives to yield a subregular translation plane of even order q^2, admitting an elation group of order $\sqrt{q}$.

This translation plane may be derived, for appropriate c, to produce a subregular translation plane of order q^2 admitting a Baer group of order $\sqrt{q}$.

24.13. Moving.

In this section Σ continues to be the spread of the Desarguesian plane of **even order** q^2 on $F_{q^2} \times F_{q^2}$ and $F_{q^2}^\infty$ the corresponding line at infinity. The reguli in Σ will be identified with the corresponding circles on $F_{q^2}^\infty$. In particular, the regulus defined by the fixed components of the Baer involution $\sigma : (x, y) \mapsto (x^q, y^q)$, the standard regulus of Σ, is identified with the circle $F_q^\infty = \{\, (f);\ f \in F_q \,\} \cup \{(\infty)\} \subset F_{q^2}^\infty$, where (∞) is the point on the circle through the component $(x = 0)$.

REMARK 24.76. A regulus $R \neq F_q^\infty$ of Σ is σ-invariant iff $|R \cap F_q^\infty| = 1$.

LEMMA 24.77. Let $R \subset F_{q^2}^\infty - \{(\infty)\}$ be a σ-invariant regulus. Then there are elations g with axis $(x = 0)$ such that $(R)g \cap F_q^\infty = \varnothing$.

COROLLARY 24.78. Suppose E is a group of elations of Σ with axis $(x = 0)$ that leaves invariant F_q^∞. Assume that S_k is an E-orbit of k mutually disjoint reguli each of which meets F_q^∞ uniquely. Then there exist elations g, with axis $(x = 0)$, such that $(S_k)g$ forms an E-orbit consisting of k mutually disjoint reguli each of which is disjoint from F_q^∞.

The point of the corollary is that in addition to the subregular spread admitting E as an elation group, obtained by multiply deriving S_k, we may 'move away' S_k from F_q^∞, using a suitable elation g, to obtain an additional subregular spread by multiply deriving $S_k g \cup \{F_q^\infty\}$, that admits E as a *Baer* group.

In order to start the process based on Corollary 24.78 we select S_k from among the reguli associated with the Foulser's subregular spreads of even order q^2, $q = u^2$.

Foulser's Construction for Even Order. For the rest of the section, the even integer q is a square, thus the Desarguesian spread Σ has even order $q^2 = u^4$. Regard $\sigma_1 : (x, y) \mapsto (x^u, y^u)$ as a collineation of Σ. So σ_1 is planar and its fixed plane meets $F_{q^2}^\infty$ in the set $F_u^\infty \subseteq F_q^\infty$. Let $R(P)$ be the set of reguli of Σ, distinct from F_q^∞, that intersect F_u^∞ in P and are fixed by σ_1. Then each regulus of $R(P)$ is fixed by $\sigma = \sigma_1^2$ and hence intersects F_q^∞ exactly in the single component P.

LEMMA 24.79. *Let Q and P be distinct members of F_u^∞. Then:*

(1) *A regulus R_Q in $R(Q)$ and a regulus R_P in $R(P)$ must be disjoint.*
(2) *There are exactly $u - 1$ reguli in $R(P)$ in an orbit under $GL(2, u)_{P,Q}$.*

PROOF. The sets of reguli R_Q and R_P are fixed by σ_1, hence by Remark 24.76 each shares exactly one component with F_q^∞. The intersection $R_Q \cap R_P$ is fixed by σ_1 which has orbits of length 4 on $\Sigma - F_q^\infty$. Thus, R_Q and R_P are disjoint since otherwise these two reguli would share > 3 members. See Foulser [**373**, Lemma 5.1] for part (2). □

So if any of $u - 1$ reguli are chosen in each of $k \leq u + 1$ sets $R(P)$, for $P \in F_u^\infty$, then a set S_k of mutually disjoint reguli are obtained. We consider choosing S_k so that it is invariant as a set under some elation subgroup in $SL(2, u)$.

Note that *we always assume the 'standard' embedding of $SL(2, u)$ in $GL(2, q^2)$*, defined by $SL(2, u) < GL(2, q) < GL(2, q^2) = GL(2, q) \otimes GF(q^2)$, so that $SL(2, u)$ leaves invariant the chain of sublines $F_{q^2}^\infty \supset F_q^\infty \supset F_u^\infty$.

Hence E_u, the 2-Sylow subgroup of $SL(2, u)$ fixing (∞), is an elation group of Σ with axis $(x = 0)$ and $F_u^\infty - \{(\infty)\}$ is a regular point orbit of E_u. Let E_v be any subgroup of order v in E_u.

CONSTRUCTION 24.80. Let $(O_i)_{i=1}^{u/v}$ be the family of the u/v distinct E_v-orbits in $F_u^\infty - \{(\infty)\}$. Fix an arbitrary reference point $P_i \in O_i$, for each $i \in [1, u/v]$, and in each case fix one of the $u - 1$ distinct reguli $R_i \in R(P_i)$, Lemma 24.79(2).

Then the E_v-orbit $\mathcal{R}_i = (R_i)E_v, i \in [1, u/v]$, is a collection of v mutually disjoint reguli of Σ, Lemma 24.79(1).

For any choice of non-empty subset of integers $K \subseteq [1.u/v]$ define the subregular spread $\Sigma^* := \Sigma^*_{(R_1,\ldots,R_{\frac{u}{v}};K;E_v)}$ obtained from Σ by multiple replacement of every regulus in the union of reguli $\bigcup_{i \in K} \mathcal{R}_i$.

We may easily count the distinct Σ^* spreads that arise from the construction. With each i there are $u - 1$ possible choices of R_i, Lemma 24.79(2), and each choice of i fixes uniquely the set of reguli $\mathcal{R}_i$ to be multiply derived. However there is a further option of not altering $\mathcal{R}_i$: this is determined by whether $i \in K$, $|K| = u/v$. Thus, there are $u^{\frac{u}{v}} - 1$ possible Σ^*, associated with fixed E_v. Since E_v is evidently a maximal inherited group of Σ, it follows that the subregular spreads associated with different E_v are pairwise disjoint. Finally, note that since E_u is an elementary Abelian 2-group, the number of available choices for E_v is the Gaussian integer $\binom{\log_2 u}{\log_2 v}_2$. Thus, we have established:

THEOREM 24.81. *Let v divide u and let E_v be any of the $\binom{\log_2 u}{\log_2 v}_2$ distinct subgroups E_v of order v in $SL(2, u)$, with axis $(x = 0)$.*

Then for fixed E_v there are $\left(u^{\frac{u}{v}} - 1\right)$ distinct subregular spreads of type $\Sigma^* := \Sigma^*_{(R_1,\ldots,R_{\frac{u}{v}};K;E_v)}$, described in Construction 24.80. Each such spread admits E_v as a maximal inherited elation group. Hence the total number of distinct subregular spreads of type Σ^* is exactly $\binom{\log_2 u}{\log_2 v}_2 \left(u^{\frac{u}{v}} - 1\right)$.

In the theorem choose $K = \{j\}$ to be a singleton, $j \in [1, u/v]$. Then Corollary 24.78 enables the corresponding $\mathcal{R}_j$ to be 'moved away' from F_q^∞ by an elation g centralizing E, so that $(\mathcal{R}_j)g$ is an E_v-orbit of reguli that are mutually disjoint and also disjoint from F_q^∞. Hence we have a collection of pairwise disjoint reguli $(\mathcal{R}_j)g \cup \{F_q^\infty\}$ that multiply derive to a subregular spread and this subregular spread has E_v acting a Baer group. Hence we have established:

THEOREM 24.82. To each of the $\binom{\log_2 u}{\log_2 v}_2$ elation groups E_v, in the Sylow 2-subgroup E_u of $SL(2, u)$, there corresponds a subregular spread of order q^2, $q = u^2$ even, that admits E_v as a Baer group.

If $|E_v| > 2$, the incompatibility of Baer and elation groups, [**123**, Theorem 24.3.9], guarantees that none of the planes in Theorem 24.81 overlap with the planes arising in Theorem 24.82.

24.14. Odd Order.

Let $q = p^r$, for p odd, with ω a primitive element of $GF(q^2)$. Choose a basis $\{e, 1\}$ for $GF(q^2)$ over $GF(q)$, with $e^2 = \gamma$, where γ is any non-square in $GF(q)$. So $e^q = -e$, since e and e^q are the two solutions of $x^2 = \gamma$. Hence for any $a = e\alpha + \beta \in \langle \omega^{q-1} \rangle$, we have $1 = a^{q+1} = (e\alpha + \beta)^{q+1} = \beta^2 - \gamma\alpha^2$. Hence if $e\alpha + \beta \in \langle \omega^{q-1} \rangle$, then so does $e(\pm\alpha) + (\pm\beta)$, where the $\pm$'s are independent. Assume that $\alpha \neq 0$. Then, for every solution for (α, β) there is a second solution $(-\alpha, \beta)$.

LEMMA 24.83. Let $\Lambda = \{\, 2 - (a + a^q); a \in \langle \omega^{q-1} \rangle \,\}$. Then $|\Lambda| \leq (q+1)/2$, hence there exist non-zero $\mu \in GF(q)$ such that $\mu \notin \Lambda$.

LEMMA 24.84. Suppose $GF(q) = GF(3^r)$. Let $g_b \colon (x, y) \mapsto (x, xb + y)$ be the elation, of the Desarguesian plane Σ, associated with some $b \in \langle \omega^{q-1} \rangle$. So $\langle g_b \rangle = \{\, g_{bj} \colon (x, y) \mapsto (x, xbj + y); j = 0, \pm 1 \,\}$ is an elation group of order 3.

Now $\{\, y = x^q ma; a \in \langle \omega^{q-1} \rangle \,\} \langle g_b \rangle$ is a set of three mutually disjoint reguli, each of which forms an orbits under the full kernel homology group of Σ, of order $q^2 - 1$. Assume $m = 1/\mu^{q+1}$, where $\mu \in GF(q) - \Lambda$, Λ as in Lemma 24.83.

Hence, we have sketched the following result:

THEOREM 24.85. For $q = 3^r > 3$, there is a subregular Bruck spread of order q^2 admitting elations of order 3. The translation plane associated with such a spread cannot be an André plane, a generalized André plane or a plane coordinatized by a semifield.

24.15. Translation Planes with Many Elations.

We now provide several constructions in larger dimension that provide an exceptionally large number of previously unknown translation planes with large numbers of elations. We provide only brief sketches of such constructions and it is clear that we have not exhausted all of the possible examples given by our methods. All of the specific examples are due to Jha and Johnson [**644**].

REMARK 24.86. Let π be any derivable translation plane of even order $q^m = 2^{2t}$. If π admits a Baer involution with axis within the derivable net, then the derived plane admits an elation of order 2.

A simple application of this principle yields subregular spreads with elation groups of order 2. However, these spreads are two-dimensional over their kern.

REMARK 24.87. Let Σ be the Desarguesian spread of even order q^2. Suppose S_k is a set of k mutually disjoint reguli in Σ, with $q - 2 > k > 1$ left invariant by a Baer involution σ that fixes every component of a regulus $S \in S_k$. Then the subregular plane Σ', obtained by multiple-derivation using S_k, has maximal elation groups of order 2.

We focus on obtaining higher-dimensional spreads of even order admitting elations.

EXAMPLE 24.88. Suppose where $q^m = 2^{2t}$, m an even integer. Let Σ be the standard Desarguesian spread on $GF(2^{2t})$, with components given by $(x = 0), y = x\mu$ for μ in $GF(2^{2t})$.

Let A_δ be the André net $\{\, y = x\mu;\ \mu^{(q^m-1)/(q-1)} = \delta \,\}$. Hence A_δ intersects the standard derivable net $D = \{\, (x = 0),\ y = x\alpha;\ \alpha \in GF(2^t) = GF(q^{m/2}) \,\}$, if and only if $\mu^{(q^m-1)/(q-1)} = \delta$, for $\mu \in GF(2^t)$.

For example, let $q = 2^r$, so that $rm = 2t$. If $GF(2^t) \cap GF(2^r) = GF(2^{(r,t)})$ and $\delta \in GF(2^r) - GF(2^{(r,t)})$, then A_δ and D are disjoint. Now take the Baer involution $\sigma : (x, y) \mapsto (x^{2^t}, y^{2^t})$, acting on π. Then $y = x\mu$ maps to $y = x\mu^{2^t}$ and A_δ maps to $A_{\delta^{2^t}}$. Choose δ so that A_δ and $A_{\delta^{2^t}}$ are disjoint from D; for example, as above.

Take a replacement $A_\delta^i = \{\, y = x^{q^i}\mu;\ \mu^{(q^m-1)/(q-1)} = \delta \,\}$ and choose the corresponding replacement $A_{\delta^{2^t}}^i$ so that σ interchanges these two replacement sets.

Replacing D by derivation D^*, A_δ by A_δ^i and $A_{\delta^{2^t}}$ by $A_{\delta^{2^t}}^i$, gives a constructed translation plane admitting an elation. Now the question is whether the elation group with fixed axis has order exactly 2.

The kernel homology group of order $(q^m - 1)$ of π acts transitively on the derived net and on both A_δ^i and $A_{\delta^{2^t}}^i$. This means that there is a dihedral group generated by the elations with axis one of the (2^t+1)-Baer subplanes of D (incident with the zero vector) so that this group has order $2(2^t + 1)$.

It turns out that the elation groups must have order 2. Hence there are $2^t + 1$ elation groups each of order exactly 2.

EXAMPLE 24.89. Generalization of the previous example: One could multiply, derive, and/or choose a different q and m and construct different sized André nets, ensure that these nets are disjoint from the standard derivable net. Then the Baer involution permutes the replacement nets and we may construct translation planes of order 2^{2t} admitting $2^t + 1$ elation groups each of order 2.

EXAMPLE 24.90. For a specific example of the above, let $q^m = 2^6$, and let $m = 3$. Then $GF(2^2) \cap GF(2^3) = GF(2)$ so any $\delta \in GF(2^2) - GF(2)$ will work in the above example. Now $\delta^{2^t} = \delta^{2^3} = \delta^2$, so we are replacing two André nets and since i can be chosen as 1 or 2, there are at least two ways to choose the replacement nets. Hence we obtain two distinct translation planes of order 2^6 admitting $2^3 + 1$ elation axes with each elation group of order 2. Note that the kernel of the plane is $GF(2)$ since the kernel homology group of the associated Desarguesian plane of order $2^6 - 1$ has orbits of lengths $(2^3 + 1)$ on the opposite regulus and $(1 + 2^2 + 2^4)$ on each of the replaced André nets.

EXAMPLE 24.91. Generalizing the previous example, let $q^m = 2^{2m}$ for m odd. Then $GF(2^2) \cap GF(2^m) = GF(2)$ and there are again $2^2 - 2$ possible choices for δ, so exactly one choice for $(\delta, \delta^{2^m} = \delta^2)$.

Hence for every m odd there is a translation plane of order 2^{2m} with kernel $GF(2)$ admitting a dihedral group of order $2(2^m + 1)$ generated by elations of order 2.

EXAMPLE 24.92. Generalizing again, let $q^m = 2^{2^a m}$ for m odd. Then $GF(2^{2^a}) \cap GF(2^{2^{a-1}m}) = GF(2^{2^{a-1}})$. Now there are $2^{2^a} - 2^{2^{a-1}} = 2^{2^{a-1}}\left(2^{2^{a-1}} - 1\right)$ possible choices for δ and $2^{2^{a-2}}$ choices for pairs of André nets which are inverted by the Baer involution σ. For each pair inverted by σ, choose an exponent $i = 1, 2, \ldots, m$. If $i = m$ is chosen since $x^{q^m} = x$, this would essentially mean that the André net reappears in the newly constructed plane. Hence, since for each pair the choice of exponent is independent of any other choice, we have $m^{2^{2^{a-2}}}$ possible translation planes constructed each of which admits a dihedral group of order $2(q^{m/2} + 1)$ generated by elation groups of order 2. Furthermore, each such elation group is the full elation group with the given axis.

EXAMPLE 24.93. Let $q^m = 2^{2^a nm}$ where m is odd and relatively prime to $2^a n$. Then

$$GF(2^{2^a n}) \cap GF(2^{2^{a-1}nm}) = GF(2^{2^{a-1}n}).$$

and there are $2^{2^a n} - 2^{2^{a-1}n} = 2^{2^{a-1}n}\left(2^{2^{a-1}n} - 1\right)$ possible choices for δ and $2^{2^{a-2}}n$ choices for pairs of André nets. In this case, there are $m^{2^{2^{a-2}n}}$ possible translation planes.

THEOREM 24.94. For $q = 3^r$, there is a subregular Bruck spread admitting elations of order 3. Any such spread cannot be generalized André.

24.16. Derivable Subregular Planes.

Subregular planes are, of course, planes obtained from Desarguesian affine planes by the replacement of a set of mutually disjoint derivable nets. In this setting the partial spread for a 'derivable net' becomes a regulus in $PG(3,q)$. However, what if there is a subregular plane that admits a derivable net whose partial spread is not a regulus. This section is concerned with derivable translation planes of order q^2 with spread in $PG(3,q)$ and by 'derivable', we intend this to mean the derivable net is not a regulus net. In this setting, when derivation occurs, the constructed translation plane no longer has its spread in a 3-dimensional projective space. Although one might consider that such translation planes are hard to come by, actually they are ubiquitous. Given any translation plane with spread in $PG(3,q)$ and choose any coordinate quasifield Q for the plane. Then by the construction process of 'algebraic lifting', a new plane may be constructed of order q^4 with spread in $PG(3,q^2)$ (see, e.g., Biliotti, Jha, Johnson [123] for details on this construction method). This new plane π of order q^4 is derivable in the sense mentioned above. The spread for the new plane is of the following general form:

$$x = 0, y = x \begin{bmatrix} u & F(t) \\ t & u^q \end{bmatrix} ; u, t \in GF(q^2),$$

for a function $F \colon GF(q^2) \to GF(q^2)$, such that $F(0) = 0$. Now consider the following partial spread D:

$$x = 0, y = x \begin{bmatrix} u & 0 \\ 0 & u^q \end{bmatrix} ; u \in GF(q^2).$$

D is a derivable partial spread which is not a regulus in $PG(3, q^2)$. Derivation constructs a translation plane of order q^4 with spread in $PG(7, q)$.

Now consider an underlying Desarguesian affine plane Σ of order q^4 and identify the points of Σ and π. If the spread for Σ is taken as

$$x = 0, y = xm; m \in GF(q^4)$$

and we consider that $GF(q^4)$ is an extension of the kernel of π, which is isomorphic to $GF(q^2)$, then we may also regard certain components (lines) of π and Σ to be the same. In particular for every algebraically lifted plane π, we see that $\pi \cap \Sigma$ contains the following partial spread:

$$x = 0, y = x \begin{bmatrix} u & 0 \\ 0 & u^q \end{bmatrix} ; u \in GF(q^2) \text{ and } u^q = u,$$

which is isomorphic to $PG(1, q)$.

For a finite derivable partial spread in $PG(3, q)$, of order q^2, it is shown in Johnson [**705**] that coordinates may be chosen so that the derivable partial spread D_σ has the following matrix form:

$$x \;=\; 0, y = x \begin{bmatrix} u & 0 \\ 0 & u^\sigma \end{bmatrix} ; u \in GF(q),$$

and σ a fixed automorphism of $GF(q)$.

D_σ is a regulus partial spread if and only if $\sigma = 1$. Furthermore, derivation of D_σ of a spread π containing it produces a translation plane of order q^2 with kernel the fixed field of σ. Note that each of these derivable partial spreads D_σ shares at least three components with both $\pi - \Sigma$ and $\Sigma - \pi$, where Σ is the underlying Desarguesian plane sharing the points of π.

We are interested in derivable affine planes π with spreads in $PG(3, q)$ whose derivable partial spread D shares at least three components with both $\pi - \Sigma$ and $\Sigma - \pi$, where Σ is a corresponding Desarguesian affine plane. Since this is always the case, whenever the derivable net is not a regulus net, it would seem that not much can be said. However, when one considers derivable subregular planes, in fact, we are able to completely determine all possible planes.

A finite subregular translation plane π of order q^2 is a translation plane which may be constructed from a Desarguesian affine plane Σ of order q^2 by the multiple derivation of a set of mutually disjoint derivable nets. Of course, many of these subregular planes are in themselves derivable by a derivable net which is in the original Desarguesian affine plane and which is disjoint from the derivable nets used in the construction. For example, certain of the André planes may be derived, where the derivable net in question is another derivable net of the associated Desarguesian plane. It is known by the theorem of Orr–Thas [**1035, 1172**] that replacement of $q - 1$ mutually disjoint derivable nets of Σ produces a Desarguesian subregular plane. Since in this case, π and Σ share exactly two components, it is then possible that a subregular plane π can admit a derivable net within $\pi - \Sigma$ or a derivable net that shares two components with $\Sigma - \pi$ and $q - 1$ in $\pi - \Sigma$.

Now the question is whether it is possible to have a subregular plane π that admits a derivable net D with at least three components in the associated Desarguesian affine plane and outside π, so in $\Sigma - \pi$, and at least three components in $\pi - \Sigma$? Since we have seen in the previous paragraph that this is always the case in algebraically lifted translation planes, which are not subregular planes, and we have noted in Chapter 24 that this can occur in subregular planes. Moreover, it is shown in Johnson [**768**] that, in fact, these planes are the only possible 'derivable subregular' planes.

The general structure theory result is due to Johnson:

THEOREM 24.95. (Johnson [**768**]). Let π be a derivable subregular translation plane of order q^2 and kernel K isomorphic to $GF(q)$ that admits a derivable net D that lies over both π and Σ in the sense that D shares at least three components with $\pi - \Sigma$ and with $\Sigma - \pi$ and which is minimal with respect to derivation.

Then there is a non-identity automorphism σ of K such that the spread S_π for π has the following form, where $\{1, e\}$ is a basis for a quadratic extension K^+ of K (let Fix σ denote the fixed field of σ in K):

(1) If q is even,

$$S_\pi = \left\{ \begin{array}{c} y = x(u + e(u^\sigma + u)) + x^q d^{1-q} e(u^\sigma + u); \\ u \in K - \text{Fix}\,\sigma, d \in K^+ - \{0\} \end{array} \right\} \cup (\pi \cap \Sigma).$$

The corresponding regulus nets in Σ are

$$R_u = \left\{ y = x(d^{1-q} e(u^\sigma + u)) + (u + e(u^\sigma + u)); d \in K^+ - \{0\} \right\},$$

for fixed u not in Fix σ.

(2) If q is odd

$$S_\pi = \left\{ y = x\left(\frac{u + u^\sigma}{2}\right) + x^q d^{1-q}\left(\frac{u - u^\sigma}{2}\right); u \in K - \text{Fix}\,\sigma \right\} \cup (\pi \cap \Sigma).$$

The corresponding regulus nets in Σ are

$$R_u = \left\{ \begin{array}{c} y = x\left(d^{1-q}\left(\frac{u-u^\sigma}{2}\right) + \left(\frac{u+u^\sigma}{2}\right)\right); \\ \text{for fixed } u \in K - \text{Fix}\,\sigma; d \in K^+ - \{0\} \end{array} \right\}.$$

(3) The translation plane obtained by derivation of D has order q^2 and kernel Fix σ.

(4) There are exactly $(q - \text{Fix}\,\sigma)/(2, q - 1)$ mutually disjoint regulus nets of Σ that are derived to construct π.

(5) There is a collineation group isomorphic to $SL(2, \text{Fix}\,\sigma)$, generated by elations that act on the plane π.

(6) There is a collineation group isomorphic to $SL(2, \text{Fix}\,\sigma)$, generated by Baer collineations that act on the plane π^* obtained by derivation of D.

(7) The regulus nets R_u in Σ form an exact cover of $PG(1, q) - PG(1, \text{Fix}\,\sigma)$ and each regulus in Σ intersects this set in exactly $2/(q, 2)$ points, namely $y = xu$ and $y = xu^\sigma$ when q is odd and $y = xu$ for q even, $u \neq u^\sigma$. Each regulus net in π intersects $D - \Sigma$ in exactly $2/(q, 2)$ points.

24.16.1. Johnson's Classification of Derivable Subregular Planes.

Using the previous structure theory, Johnson shows that, in fact, the Foulser–Ostrom planes are the only possible derivable subregular planes.

THEOREM 24.96. (Johnson [**768**]). Let π be a derivable subregular plane of order q^2 with derivable net D that lies over both the plane π and the associated Desarguesian plane Σ, in the sense that D shares at least three components with each of $\pi - \Sigma$ and $\Sigma - \pi$. Then q is a square h^2 and π is the Foulser–Ostrom plane of odd order $h^4 = q^2$.

CHAPTER 25

Fano Configurations.

In 1954, Hanna Neumann [**1027**] showed, using coordinates, that in any 'projective' Hall plane of odd order, there is always a projective subplane of order 2—a Fano plane. We recall that, of course, a Hall plane of order q^2 may be characterized as a translation plane obtained from a Desarguesian affine plane of order q^2 by the replacement of a regulus net and so is subregular. Here now the question is which subset of subregular planes of odd order must admit Fano configurations. If one considers the maximal possible cardinality set of $q-1$ disjoint regulus nets, then a multiple derivation of all of these nets leads to another Desarguesian plane. Hence, it is not true that any subregular plane admits Fano configurations. However, we show that if t is less than or equal to roughly a quarter of the possible $(q-1)$, then such planes always admit Fano configurations.

The results of this chapter are due to Fisher and Johnson [**363**] for subregular planes and Johnson [**653**] for larger-dimensional planes.

The main theorem to point out is that Fano configurations exist in many (if not most) subregular translation planes.

THEOREM 25.1. (Fisher and Johnson [**363**]). Let Σ be a Desarguesian affine plane of odd order q^2 and let Λ be a set of $k+1$ disjoint reguli. Let π denote the translation plane obtained by a multiple derivation replacement of the reguli. If $k+1 < \frac{(q+1)}{4}$, then the projective extension of π contains a Fano configuration.

For André planes the above result can be strengthened to conclude that if the number of disjoint reguli is $< 3\frac{(q-1)}{8}$, then again the plane admits a Fano configuration. We also consider the existence of Fano configurations in planes obtained by lifting subregular planes (see Johnson [**656**]).

THEOREM 25.2. (Fisher and Johnson [**363**]). Let π_1 be any subregular plane of odd order q^2. Then there exist subregular planes π_s of order q^{2s}, for any odd integer $s > 1$, containing π_1 as a subplane such that π_s admits Fano configurations.

The idea in both Fisher and Johnson [**363**] and Johnson [**653**] is based on a simple concept. Most of the known finite projective planes are obtained from a Desarguesian plane by designating a line at infinity and a subset Λ of the points on that line. Then the idea is to start from a complete quadrangle $ABCD$ with its six sides and its three diagonal points (points where pairs of opposite sides intersect) P, Q, R and arrange that the set Λ excludes the slopes of the six sides (so that those six lines will also be lines of the derived plane) while the diagonal points lie in the same component of the derived set. In other words, P, Q, R will lie on a line of the new plane.

THEOREM 25.3. (Fisher and Johnson [**363**]). Let Λ be any set of disjoint reguli of a Desarguesian spread Σ. Assume that $q(q-1)/(2(q+1)) \geq |\Lambda|$.

(1) Then we may choose a representation so that (∞) and (0) are parallel classes disjoint from the parallel classes of Λ. Hence, we may choose a representation so that

$$R_1: y = xm; \; m^{q+1} = 1$$

is in Λ, where

$$R_1^* = x^q m; \; m^{q+1} = 1$$

is the partial spread for the opposite regulus.

Let

$$A = (\infty), \; B = \left(0, \frac{c}{c-1}\right), C = (c,0), D = (1,0),$$

where $c^q = -c$.

(2)

$$AD \text{ is } x = 1, \; AB \text{ is } x = 0 \text{ and } AC \text{ is } x = c,$$
$$CD \text{ is } y = 0,$$
$$BD \text{ is } y = -x\left(\frac{1}{1-c^{-1}}\right) + \frac{c}{c-1}, \text{ and}$$
$$BC \text{ is } y = -x\left(\frac{1}{c-1}\right) + \frac{c}{c-1},$$

where $c^{q-1} = -1$.

(3) The cross joins of the quadrangle are

$$
\begin{aligned}
AB \cap CD &= (0,0), \\
AC \cap BD &= (c,-c), \\
AD \cap BD &= (1,1),
\end{aligned}
$$

and are points of of $y = x^q$.

(4) Assume that in the translation plane π obtained by the multiple derivation of Λ, we have some pair of slopes $\left(\frac{1}{c^{-1}-1}\right)$ and $\left(\frac{1}{1-c}\right)$ for some c in $\Sigma - \Lambda$. Then π is a subregular translation plane of odd order admitting a Fano configuration.

Using this fundamental result and sufficient restrictions on the number of reguli used in the multiple derivation process, it is possible to show the existence of Fano configurations in a variety of situations. In particular, a slightly different version of the theorem mentioned previously is given by Fisher and Johnson.

1/4-Theorem.

THEOREM 25.4. (Fisher and Johnson [**363**]). Let Σ be a Desarguesian affine plane of odd order q^2 and let Λ be a set of t disjoint reguli. Let π denote the translation plane obtained by a multiple derivation replacement of the reguli.

If $q \equiv -1 \bmod 4$ assume that $t \le (q+1)/4$, and if $q \equiv 1 \bmod 4$ assume that $t \le (q-1)/4$.

Then the projective extension of π contains a Fano configuration.

Note that for any value of c, if we merely avoid two possible regulus nets, we then obtain a Fano configuration. Hence, the following theorem then follows immediately.

THEOREM 25.5. (Fisher and Johnson [**363**]). Let Σ denote a Desarguesian affine plane of odd order q^2 and let Λ be a set of t disjoint regulus nets. Assume that for any subset Λ^* of Λ, there is a pair of components disjoint from Λ that is a conjugate pair for a regulus net of Λ^* (for example, this condition is valid in André planes). Then of the $2^t - 1$ possible non-empty subsets of Λ that produce subregular planes, at least $2^{t-2} - 1$ contain Fano configurations.

Subregular Lifting and Fano Configurations. We have discussed subregular lifting in Section 24.3 and noted that Johnson [**656**] proved the following theorem.

THEOREM 25.6. (Johnson [**656**]). If S_k is a set of k disjoint reguli in Σ_{q^2} and s is odd, then S_k^s is a set of k disjoint reguli in $\Sigma_{q^{2s}}$.

We then have the following omnibus theorem regarding Fano configurations in subregular planes, even when the initial subregular plane is Desarguesian. The following result then follows immediately.

THEOREM 25.7. Let π_1 be any subregular plane of order q^2. Then there exist subregular planes π_s of order q^{2s}, for any odd integer $s > 1$, containing π_1 as a subplane such that π_s admits Fano configurations.

25.1. André Planes with Fano Configurations.

The André planes are defined as follows: Let Σ denote the Desarguesian affine plane coordinatized by $GF(q^2)$, for q odd. Let the spread for Σ be

$$x = 0, y = xm; \ m \in GF(q^2).$$

We define an 'André net' R_δ as a translation net defined by a partial spread of the following form:

$$R_\delta = \left\{ y = xm; \ m^{q+1} = \delta \right\}.$$

The opposite regulus to an André net is given by

$$R_\delta^* = \left\{ y = x^q m; \ m^{q+1} = \delta \right\}.$$

So we may choose a set of k André nets, which are disjoint on lines, and derived each such net. The planes obtained are called 'André planes' and are then subregular planes. When the plane is an André plane, then the regularity of the set of regulus nets allows a stronger theorem than the 1/4-theorem.

THEOREM 25.8. (Fisher and Johnson [**363**]). Let π be an affine André plane of odd order q^2 where -1 is a non-square in $GF(q)$ (i.e., $q \equiv -1 \mod 4$).

One of the following two possibilities occurs:

(1) The projective extension of π admits a Fano configuration, or

(2) There is a representation of π so that $\tau : (x, y) \longmapsto (x, ty)$, such that $t^{q+1} = 2$, is a collineation of π.

COROLLARY 25.9. In particular, in the above situation, if there is not a Fano configuration in π and if R_α is an André net of the associated Desarguesian plane used in the multiple derivation to produce π, then so is $R_{2\alpha}$.

It is then not difficult to observe the following:

THEOREM 25.10. (Fisher and Johnson [**363**]). Assume that -1 is a non-square, q is prime and the order of 2 is $(p-1)/2$ is $(p-1)$. Then any non-Desarguesian André plane of order q^2 admits a Fano configuration.

COROLLARY 25.11. If p is an odd prime such that $(p-1)/2$ is prime, then any non-Desarguesian André plane of order p^2 admits a Fano configuration.

Then, by a straightforward count, the following may be established.

COROLLARY 25.12. Any non-Desarguesian André plane of order

$$p^2 \in \left\{3^2, 7^2, 11^2, 19^2, 23^2, 43^2, 47^2, 59^2, 67^2, 71^2, 79^2, 83^2\right\}$$

admits a Fano configuration.

REMARK 25.13. Of the 24 odd primes p between 1 and 100, exactly half of these apply to show that any non-Desarguesian André plane of order p^2 admits Fano configurations.

Similarly, one may go through a list of 200 primes with similar results, about $1/2$ of these primes apply for our results.

3/8-Theorem for André Planes. For André planes, it is possible to improve the 1/4-theorem to what might be called the 3/8-theorem.

THEOREM 25.14. (Fisher and Johnson [**363**]). Let π be an André plane of odd order q^2 obtained by the multiple derivation of $\leq \frac{3}{8}(q-1)$ mutually disjoint regulus nets. Then π admits Fano configurations.

In particular, in the following situations π admits Fano configurations:

(1) If 4 divides $(q-1)$ but 8 does not divide $q-1$, and Λ has $\leq 3+3(q-5)/8$ elements, or

(2) If 4 does not divide $(q-1)$ and 8 divides $(q-3)$, and Λ has $\leq 1+3(q-3)/8$ elements, or

(3) If 4 does not divide $(q-1)$ and 8 does not divide $(q-3)$, and Λ has $\leq 3+3(q-7)/8$ elements.

The subregular nearfield plane requires $(q-1)/2$ mutually disjoint regulus nets to be replaced for its constructions. However, the regularity of this plane allows for a slightly different count. Johnson and Fisher (Fisher and Johnson [**363**]) are able to prove:

THEOREM 25.15. The subregular nearfield of odd order q^2 admits Fano configurations.

25.2. Fano Configurations in Known Subregular Planes.

The question now remains which of the subregular spreads admit Fano configurations. At this point, we then ask which of the known classes satisfy the 1/4-condition, which then admit Fano configurations by the 1/4-Theorem. The simple answer is most of these do admit Fano configurations. We offer a few remarks on the known classes.

25.2.1. Fano in Ostrom, Johnson–Seydel Spread Order 3^6, 7 Reguli.
Since $3^3 \equiv -1 \bmod 4$, any subregular plane constructed with $\leq (3^3+1)/4$ reguli admits Fano configurations and since $(3^3+1)/4 = 7$, this translation plane admits Fano configurations.

25.2.2. Fano in Foulser–Ostrom Planes Order v^{4t}, $v(v-1)/2$ Reguli.
There are a variety of subregular translation planes of order q^4 for both q even
and odd with various interesting properties. For example, there is the Ostrom
subregular plane of order q^4 admitting $SL(2,q)$, where $q = p^r$, for p odd, where
the p-elements are elations. In this case, $k = q(q-1)/2$, which does not satisfy the
$(q^2-1)/4$ bound.

Furthermore, Foulser [373] has discussed the complete set of subregular spreads
of order q^4 whose replacing reguli are fixed by $\sigma : (x,y) \longmapsto (x^q, y^q)$. Furthermore,
Foulser has shown that for any Desarguesian plane of order v^{4t}, it is possible to
construct subregular planes admitting $SL(2,v)$, where $v = p^z$, and the p-elements
are elations. In this setting, there are $v(v-1)/2$ reguli and since $v(v-1)/2 \leq$
$(v^{2t}-1)/4$, for $t > 1$, any such translation plane admits Fano configurations.

25.2.3. Fano in Bruck Triples. In [177], Bruck gives a more or less complete determination of the subregular planes/ spreads that may be obtained by the
multiple derivation of three reguli. Since $3 \leq (q-1)/4$ for $q \geq 13$, we see that any
Bruck triple plane of order q^2 for $q \geq 13$ admits Fano configurations.

25.2.4. Fano in Ebert Quadruples. Ebert [333, 336] has given a theory
of subregular spreads obtained from a Desarguesian spread by the replacement
of exactly four disjoint reguli. There are a tremendous number of mutually non-
isomorphic translation planes of odd order (asymptotically equal to q^6/Λ, where
$16 \leq \Lambda \leq 3072$ of order q^2). Since $4 \leq (q-1)/4$ for $q \geq 17$, we see that any Ebert
quadruple plane of order q^2 for $q \geq 17$ will admit Fano configurations.

25.2.5. Fano in Prohaska Planes. Prohaska [1119] constructs several classes
of subregular translation planes, various of which admit S_4 or A_5 acting on the reguli. Any of these planes can be lifted producing planes admitting similar groups.
Many of these planes are constructed using less than or equal to six reguli and since
$6 \leq (q-1)/4$ provided $q \geq 25$, most of these planes admit Fano configurations.

25.2.6. Fano in Ebert Quintuple. In [334], Ebert provides some examples
of planes of orders p^2, for $p = 11, 19, 29, 31$ that may be constructed by replacing
five reguli. We call these planes the 'Ebert quintuple planes'. These subregular
spreads are of interest in that they do not have the two conditions required for the
triviality of the group fixing each circle of the set of five. Clearly the planes of
orders 29^2 and 31^2 will admit Fano configurations.

25.2.7. Fano in Abatangelo–Larato Planes. In [10], there are constructions of subregular planes of order q^2 admitting $SL(2,5)$ by multiple derivation, for
$q \equiv 1 \bmod 4$ and $q \equiv -1 \bmod 5$. All of these planes are similar to the Prohaska
planes and most of these admit Fano configurations.

25.2.8. Fano in Bonisoli, Korchmáros, Szőnyi Planes. In [162], these
three authors construct a class of subregular $SL(2,5)$-planes of odd order q^2, where
$q^2 \equiv 4 \bmod 5$. There are three types: (I) five reguli and A_5 acting in its natural
representation on the set, for $q \equiv 2 \bmod 3$, (II) ten reguli and A_5 having orbit
lengths of $1, 6, 3$, $q \equiv 1 \bmod 3$ and $q \equiv 1 \bmod 4$, (III) ten reguli and A_5 again
having orbit lengths of $1, 6, 3$ but the set of reguli are distinct from that of (II),
$q \equiv -1 \bmod 4$, $3 \nmid q$ and $q \equiv 3, 5$ or $6 \bmod 7$, and (IV) fifteen reguli, where A_5 has
orbit lengths of $1, 2, 4, 8$, $q \equiv 2 \bmod 3$, $q \equiv 1 \bmod 4$, $q \equiv 1, 2$ or $4 \bmod 7$.

In any case, the maximum number of reguli used is 15, so for any translation plane of order q^2 for $q \geq 61$, there are Fano configurations.

25.2.9. Fano in Charnes and Dempwolff Planes. In [**225**], sections 3 and 4, there are constructions of spreads in $PG(3, q)$ admitting $SL(2, 5) * C$, a subgroup of a $GL(2, q^2)$ of an associated Desarguesian spread, and $*$ denoting a central product, C the kernel homology group of order $q^2 - 1$. In particular, we have $(q, 30) = 1 = (q^2 - 1, 5)$. In (4.1) and (4.2), there are constructions of spreads arising from a G-orbit O_i, for $i = 1, 2$ of length $10(q+1)$ and $5(q+1)$, respectively, for $q \equiv 1 \bmod 12$ and $q \equiv -1 \bmod 3$. These spreads turn out to be subregular and constructed by multiple derivation of at most 10 reguli, so for $q \geq 41$, any such plane admits Fano configurations.

25.2.10. Does Fano Exist in Orr Spread? Orr [**1034**] shows that the elliptic quadric of $PG(3, q)$ for $q = 5$ or 9 has a partial flock of deficiency one, which cannot be extended to a flock. What this means is there is a translation plane of order 9^2 constructed by the replacement of 8 reguli from a Desarguesian spread which is not an André plane. Since this number does not satisfy the $1/4$-bound, it is not known if these planes admit a Fano configuration.

We have shown that a wide variety of subregular planes of odd orders submit Fano configurations. In fact, with the exception of the Ostrom–Foulser plane of order q^4 admitting $SL(2, q)$, and a few small-order planes, all known non-André subregular planes admit Fano configurations. We have not tried to submit our analysis to the Ostrom–Foulser plane, but conceivably this plane also admits Fano configurations. For André planes, we have shown that most of these do, in fact, admit Fano configurations. Hence, we offer the following as a conjecture:

CONJECTURE 25.16. *All non-Desarguesian subregular planes of odd order admit Fano configurations.*

CHAPTER 26

Fano Configurations in Generalized André Planes.

The results of this chapter are due to Johnson [**653**].

We have not dealt with other types of translation planes with spreads in $PG(3, q)$, nor have we considered translation planes of larger dimension or of infinite translation planes of characteristic not 2. Hence, we pose a problem.

PROBLEM 26.1. What classes of large dimension translation planes of odd order or of characteristic not 2 admit Fano configurations?

We realize that subregular planes represent only a small part of the known translation planes. In particular, there are a variety of large dimension and, of course, there are many infinite translation planes for which the question of the existence of Fano configurations has never been asked.

So we consider large dimension planes with spreads in $PG(2t - 1, q)$, for $t > 2$ or infinite planes of characteristic not 2 and ask what known planes, if any, admit Fano configurations.

Johnson is able to show that various André and generalized André translation planes of large dimension admit Fano configurations. It is possible to modify the general construction procedure given in Theorem 25.3 to André planes of odd order q^n, n even, and kernel containing $GF(q)$ and various other types of generalized André planes.

In Chapter 16 on generalized André planes, we described a large class, which we called 'mixed generalized André planes' of order q^n and kernel $GF(q)$. The main result of Johnson shows that if n is even and the number of 'big' nets is roughly $(q - 1)/4$, we always obtain Fano configurations. In this case, a 'big' net is simply an André net of degree $(q^n - 1)/(q - 1)$.

1/4-Theorem for Generalized André Planes.

THEOREM 26.2. (Johnson [**653**]). Let π be any mixed generalized André plane of odd order q^n and kernel containing $GF(q)$. Assume that n is even and that the number of big nets is $\leq \left[\frac{(q-1)}{4} / (\frac{q-1}{2}, n/2) \right]$. Also, if $(n/2, (q-1)/2) = 1$, then the above can be replaced as follows: If $q \equiv -1 \bmod 4$ assume that the number of big nets is $\leq (q-1)/4$ and if $q \equiv 1 \bmod 4$ assume that the number of big nets is $\leq [(q+1)/4]$. Then the projective extension of π admits a Fano configuration.

The theorem that establishes this rests on the following general theorem.

THEOREM 26.3. Let K be a field isomorphic to $GF(q^{2r=n})$, for q odd, and let Σ be the affine Desarguesian plane coordinatized by K. Let π be any mixed generalized André plane of kernel $GF(q)$. Then we may assume that one of the André nets is big (of degree $(q^n - 1)/(q - 1)$) and with appropriate coordinates is

$$A_1 = \left\{ y = xm; \ m^{(q^n-1)/(q-1)} = 1 \right\}.$$

Now assume that a set of λ big nets are considered in the replacement procedure. Without making any assumptions on the nature of the replacement, we may assume that $y = x^{q^i}$ is a component of a replacement net A_1^* for A_1, such that $(i, n) = 1$. Choose c such that $c^q = -c$. We may identify λ with a subset of $GF(q)^*$ with each element of $GF(q)^*$ corresponding to a big net. Hence, $GF(q)^* - \lambda$ will then be identified with the remaining big nets. Note that all big nets are disjoint from $x = 0, y = 0$.

Assume that for some c,

$(*)$: $(1 - c)^{-(q^n-1)/(q-1)}$ and $(1 - c^{-1})^{-(q^n-1)/(q-1)}$ are in $GF(q)^* - \lambda$.

We consider the quadrangle $ABCD$ as follows:
Let

$$A = (\infty), \ B = \left(0, \frac{c}{c-1}\right), \ C = (c, 0), \ D = (1, 0).$$

We note that

$$AD \text{ is } x = 1, \ AB \text{ is } x = 0 \text{ and } AC \text{ is } x = c,$$
$$CD \text{ is } y = 0,$$
$$BD \text{ is } y = -x\left(\frac{1}{1-c^{-1}}\right) + \frac{c}{c-1}, \text{ and}$$
$$BC \text{ is } y = -x\left(\frac{1}{c-1}\right) + \frac{c}{c-1}.$$

Furthermore,

$$AC \cap BD = P = (c, -c),$$
$$AD \cap BC = Q = (1, 1),$$
$$AB \cap CD = R = (0, 0).$$

We note that $y = x^{q^i} m_\beta$ contains the points P, Q, R, since $(i, n) = 1$.

Then, the projective extension of the André plane π is a projective plane of odd order that admits a Fano configuration.

PROOF. The points P, Q, R are collinear in π since $y = x^{q^i}$ is a line. We claim that the lines AB, AC, AD, BC, BD, CD of the quadrangle $ABCD$ are still lines in the André plane. Since (∞) and (0) are parallel classes that are not altered in the construction of the mixed generalized André plane π, it remains to check that BC and BD are still lines of π. Hence, we need to show that these lines are not among the ones that are replaced to construct the mixed generalized André plane π. This is true if and only if

$$\left\{\frac{1}{1-d}\right\}^{(q^n-1)/(q-1)} \in GF(q)^* - \lambda, \text{ for } d \in \{c, c^{-1}\}.$$

Since

$$\left\{\frac{1}{1-d}\right\}^{(q^n-1)/(q-1)} = (1 - d)^{-(q^n-1)/(q-1)},$$

we have the proof.   □

CHAPTER 27

Planes with Many Elation Axes.

It is possible to find subregular planes admitting elations. Only rarely however are there more than one elation axis. In this chapter, we show how easy it is to construct translation planes of even order admitting many elation axes. The great number of examples of this chapter illustrates the difficulty in trying to assign appropriate designations to translation planes. Perhaps the best we can do in these instances is to illustrate the various construction techniques. In this setting, we are mostly considering multiple derivation from a Desarguesian affine plane. The specific examples are due to Jha and Johnson [**644**].

REMARK 27.1. Let π be any derivable translation plane of even order $q^m = 2^{2t}$. If π admits a Baer involution with axis within the derivable net, then the derived plane admits an elation of order 2.

EXAMPLE 27.2. Let π be a Desarguesian affine plane of even order q^2 with spread in $PG(3, q)$. Let S be any set of mutually disjoint regulus nets in π. Choose any of the regulus nets and let σ denote a Baer involution that fixes this net. If σ permutes the remaining reguli, then in the multiply-derived plane we have an elation of order 2 whose full elation group has order 2.

Let π be a Desarguesian plane of order 2^{2t} with standard derivable net D coordinatized by $GF(2^t)$. Choose any replacement net M disjoint from D. Let σ be a Baer involution with axis in D and assume that σ leaves M invariant. Replace both M and D to construct a translation plane admitting an elation of order 2. If the full collineation group of the constructed plane is the group inherited from the original, then the elation group with axis $\mathrm{Fix}\,\sigma$ has order exactly 2.

EXAMPLE 27.3. Let the components of the Desarguesian plane π be given by $x = 0, y = xm$ for m in $GF(2^{2t})$. Let A_δ be the André net $\{ y = xm; \ m^{(q^m-1)/(q-1)} = \delta \}$ where $2^{2t} = q^m$. A_δ intersects the standard derivable net $D = \{ x = 0, y = x\alpha; \ \alpha \in GF(2^t) = GF(q^{m/2}) \}$ if and only if

$$m^{(q^m-1)/(q-1)} = \delta, \text{ for } m \in GF(2^t).$$

For example, let $q = 2^r$, so that $rm = 2t$. If $GF(2^t) \cap GF(2^r) = GF(2^{(r,t)})$ and $\delta \in GF(2^r) - GF(2^{(r,t)})$, there A_δ and D are disjoint. Now take the Baer involution $\sigma : (x, y) \longmapsto (x^{2^t}, y^{2^t})$, acting in π. Then $y = xm$ maps to $y = xm^{2^t}$ and A_δ maps to $A_{\delta^{2^t}}$. So, choose δ so that A_δ and $A_{\delta^{2^t}}$ are disjoint from D; for example, as above.

So, take a replacement $A_\delta^i = \{ y = x^{q^i} m; \ m^{(q^m-1)/(q-1)} = \delta \}$ then choose the corresponding replacement $A_{\delta^{2^t}}^i$ so that σ interchanges these two replacement sets.

Now replace D by derivation D^*, A_δ by A_δ^i and $A_{\delta^{2t}}$ by $A^i_{\delta^{2t}}$, there is a constructed translation plane admitting an elation. Now the question is whether the elation group with fixed axis has order exactly 2.

The kernel homology group of order $(q^m - 1)$ of π acts transitively on the derived net and on both A_δ^i and $A^i_{\delta^{2t}}$. This means that there is a dihedral group generated by the elations with axis one of the (2^t+1)-Baer subplanes of D (incident with the zero vector) so that group has order $2(2^t + 1)$.

It follows that the elation groups must have order 2.

(That is, suppose that there is an elation group of order at least 4 with axis one of these $2^t + 1$ components. Then, the group generated by the elations with axes in D^* isomorphic to $SL(2, 2^s)$ for s at least $2^t + 1$ or possible $S_z(2^{2w+1})$. In the first case, we obtain $SL(2, 2^t)$ of $SL(2, 2^{2t})$ and in both cases, this implies that the plane is Desarguesian. In the second case, there is a Lüneburg–Tits subplane of order 2^{2w+1}, and since there are at least $2^t + 1$ elation axes, there is a Lüneburg–Tits subplane of order 2^t or the plane itself is Lüneburg–Tits. Anyway, none of this can occur.)

Hence, there are $2^t + 1$ elation groups each of order exactly 2.

EXAMPLE 27.4. Generalization of the previous example: One could multiply derive and/or choose a different q and m and construct different sized André nets, ensure that these nets are disjoint from the standard derivable net, the Baer involution permutes the replacement nets and construct translation planes of order 2^{2t} admitting $2^t + 1$ elation groups each of order 2.

EXAMPLE 27.5. For a specific example of the above, let $q^m = 2^6$, and let $m = 3$. Then $GF(2^2) \cap GF(2^3) = GF(2)$ so any $\delta \in GF(2^2) - GF(2)$ will work in the above example. Then, $\delta^{2^t} = \delta^{2^3} = \delta^2$, so we are replacing two André nets and since i can be chosen as 1 or 2, there are at least two ways to choose the replacement nets. Hence, we obtain two distinct translation planes of order 2^6 admitting $2^3 + 1$ elation axes with each elation group of order 2. Note that the kernel of the plane is $GF(2)$ since the kernel homology group of the associated Desarguesian plane of order $2^6 - 1$ has orbits of lengths $(2^3 + 1)$ (on the opposite regulus) and $(1 + 2^2 + 2^4)$ on each of the replaced André nets.

EXAMPLE 27.6. Generalizing the previous example, let $q^m = 2^{2m}$, for m odd. Then, $GF(2^2) \cap GF(2^m) = GF(2)$ and there are again $2^2 - 2$ possible choices for δ, so exactly one choice for $(\delta, \delta^{2^m} = \delta^2)$.

Hence, for every m odd, there is a translation plane of order 2^{2m} with kernel $GF(2)$ admitting a dihedral group of order $2(2^m + 1)$ generated by elations of order 2.

EXAMPLE 27.7. Generalizing again, let $q^m = 2^{2^a m}$, for m odd. Then, $GF(2^{2^a}) \cap GF(2^{2^{a-1}m}) = GF(2^{2^{a-1}})$. Then there are $2^{2^a} - 2^{2^{a-1}} = 2^{2^{a-1}}$ possible choices for δ and $2^{2^{a-2}}$ choices for pairs of André nets which are inverted by the Baer involution σ. For each pair inverted by σ, choose an exponent $i = 1, 2, \ldots, m$. If $i = m$ is chosen since $x^{q^m} = x$, this would essentially mean that the André net reappears in the newly constructed plane.

Hence, since for each pair, the choice of exponent is independent of any other choice, we have $m^{2^{2^{a-2}}}$ possible translation planes constructed each of which admits a dihedral group of order $2(q^{m/2}+1)$ generated by elation groups of order 2. Furthermore, each such elation group is the full elation group with the given axis.

EXAMPLE 27.8. Let $q^m = 2^{2^a nm}$ where m is odd and relatively prime to $2^a n$. Then $GF(2^{2^a n}) \cap GF(2^{2^{a-1}nm}) = GF(2^{2^{a-1}n})$. Then, there are $2^{2^a n} - 2^{2^{a-1}n} = 2^{2^{a-1}n}$, possible choices for δ and $2^{2^{a-2}n}$ choices for pairs of André nets. In this case, there are $m^{2^{2^{a-2}n}}$ possible translation planes.

EXAMPLE 27.9. Multiply replaced using possibly different sizes: Do all of this for different sized q's and m's, choose the André nets to be mutually disjoint. Still get an enormous number of different planes admitting an elation group of order 2. Here, conceivably we get generalized André planes that are not André planes that admit such groups.

CHAPTER 28

Klein Quadric.

Let S be a spread in $PG(3,q)$. Let V_4 denote the associated 4-dimensional vector space over a field K isomorphic to $GF(q)$. Choose any three components and choose bases so that V_4 is represented as the set of vectors (x_1, x_2, y_1, y_2); $x_i, y_i \in K$, and for $x = (x_1, x_2)$, $y = (y_1, y_2)$ then the spread has the following representation

$$x = 0, y = x \begin{bmatrix} g(t,u) & f(t,u) \\ t & u \end{bmatrix}; t, u \in K.$$

There is a natural way to construct a 6-dimensional vector space that has a natural quadric defined on it. The formal process of this construction is to take the exterior power $\wedge^2 V_4$, , which is a $\begin{pmatrix} 4 \\ 2 \end{pmatrix}$-dimensional K-vector space. Moreover, since $\wedge^4 V_4$ is a 1-dimensional K-subspace, we may define a quadratic form Q as follows:

$$Q(x,y) = x \wedge y, \text{ where } x, y \in \wedge^2 V_4.$$

Noting that $x \wedge y = -y \wedge x$, it follows that the polar form of Q is given by

$$\begin{bmatrix} 0 & 0 & 0 & 0 & 0 & 1 \\ 0 & 0 & 0 & 0 & -1 & 0 \\ 0 & 0 & 0 & 1 & 0 & 0 \\ 0 & 0 & 1 & 0 & 0 & 0 \\ 0 & -1 & 0 & 0 & 0 & 0 \\ 1 & 0 & 0 & 0 & 0 & 0 \end{bmatrix}.$$

Thus,

$$(1, a, b, c, d, \Delta),$$

where $\Delta = ad - cb$ is a singular point for all a, b, c, d, as if $(0,0,0,0,1)$. Hence, we obtain the following bijection, the 'Klein map' between points of $PG(5,q)$ of the quadric Q, called the Klein quadric, and 2-dimensional K-subspaces with V_4 or equivalently between lines of $PG(3,q)$:

$$(1, a, b, c, d, \Delta) \longleftrightarrow y = x \begin{bmatrix} a & b \\ c & d \end{bmatrix},$$

$$(0,0,0,0,0,1) \longleftrightarrow x = 0.$$

The quadric Q is a hyperbolic quadric and the associated vector space equipped with such a quadric is often denoted by $\Omega^+(6,q)$.

THEOREM 28.1. (1) The Klein map is bijective.

(2) Furthermore, mutually disjoint 2-dimensional K-subspaces map to singular points that are not on a line of the quadric.

(3) Hence, a spread S of $PG(3,q)$ maps bijectively to an 'ovoid' of the quadric, a set of $q^2 + 1$ singular points no two of which are on a line of the quadric.

(4) A Desarguesian spread of $PG(3, q)$ maps bijectively to an elliptic quadric within a subspace isomorphic to V_4 (or projective isomorphic to $PG(3, q)$).

As noted in Chapters 79, 80, and 81, any two hyperbolic quadrics in $PG(5, q)$ are projectively equivalent.

THEOREM 28.2. Hence, via a basis change, any such quadric may be represented in the form listed above and with associated bijection between spreads in $PG(3, q)$ and ovoids on the quadric.

CHAPTER 29

Parallelisms.

In this chapter, we review the connections between ovoids of hyperbolic quadrics in $PG(5, q)$ and spreads on $PG(3, q)$. Most of the necessary background has already been given in our part on symplectic and orthogonal geometry. Here the focus is on two-dimensional translation planes, that is, spreads in $PG(3, q)$ and parallelisms in $PG(3, q)$ that lead to translation planes of larger dimension.

DEFINITION 29.1. A 'parallelism' of $PG(3, q)$ is a partition of the lines of $PG(3, q)$ into spreads. Hence, a parallelism is a set of $1 + q + q^2$ spreads that partition the line set. A 'Desarguesian or regular parallelism' is a parallelism when the associated spreads are all Desarguesian.

Hence, using the Klein mapping, we see that:

THEOREM 29.2. A Desarguesian parallelism in $PG(3, q)$ is equivalent to a partition of the Klein quadric by a set of $1 + q + q^2$ elliptic quadrics.

We shall also be interested in Desarguesian partial parallelisms of deficiency one; a set of $q + q^2$ mutually disjoint Desarguesian spreads or equivalently a set of $q + q^2$ mutually disjoint elliptic quadrics. It is immediate that given a Desarguesian partial parallelism of deficiency one, there is a parallelism extending the partial parallelisms. The question is whether the extension is also Desarguesian. That is, must the spread adjoined necessarily be Desarguesian? From the standpoint of translation planes, both Desarguesian parallelisms and Desarguesian partial parallelisms of deficiency one produce and are equivalent to certain translation planes.

29.1. Direct Products of Affine Planes.

In this section, we shall review sufficient theory of direct products of affine planes so as to consider 'parallelisms' in the next section.

We shall consider the direct product of two affine planes. This may be generalized to more general products of nets and other structures. The reader is referred to Johnson and Ostrom [791] for more complete details. Here we consider only finite structures and the reader may consider K as a finite field. However, all of the results are valid generally when K is a field except that care must be taken when K is not commutative. Furthermore, in the field case, one must deal with left and vector spaces and then the derivable nets become pseudo-regulus nets instead of regulus nets.

DEFINITION 29.3. Let π_1 and π_2 be affine planes equal to $(\mathcal{P}_i, \mathcal{L}_i, \mathcal{C}_i, \mathcal{I}_i)$, for $i = 1, 2$ respectively, where $\mathcal{P}_i$ denotes the set of points, $\mathcal{L}_i$ denotes the set of lines, $\mathcal{C}_i$ denotes the set of parallel classes and $\mathcal{I}_i$ denotes the incidence relation of π_i.

Assume that there exists a 1–1 correspondence σ from the set of parallel classes of π_1 onto the set of parallel classes of π_2.

We denote the following incidence structure $(\mathcal{P}_1 \times \mathcal{P}_2, \mathcal{L}_1 \times_\sigma \mathcal{L}_2, \mathcal{C}_\sigma, \mathcal{I}_\sigma)$ by $\pi_1 \times_\sigma \pi_2$, and called the 'direct product of π_1 and π_2' defined as follows:

The set of 'points' of the incidence structure is the ordinary direct product of the two point sets $\mathcal{P}_1 \times \mathcal{P}_2$.

The 'line set' of the incidence structure $\mathcal{L}_1 \times_\sigma \mathcal{L}_2$ is defined as follows: Let ℓ_1 be any line of π_1 and let ℓ_1 be in the parallel class α. Choose any line ℓ_2 of π_2 of the parallel class $\alpha\sigma$ and form the cross product $\ell_1 \times \ell_2$ of the set of points of the two lines. The 'lines' of $\mathcal{L}_1 \times_\sigma \mathcal{L}_2$ are the sets of the type $\ell_1 \times \ell_2$.

The set of 'parallel classes' $\mathcal{C}_\sigma$ is defined as follows: Two lines $\ell_1 \times \ell_2$ and $\ell_1' \times \ell_2'$ are parallel if and only if ℓ_1 is parallel to ℓ_1'. Note that this implies that ℓ_2 is parallel to ℓ_2' since both lines are in the parallel class $\alpha\sigma$.

Parallelism is an equivalence relation and the set of classes is $\mathcal{C}_\sigma$.

Incidence $\mathcal{I}_\sigma$ is defined by the incidence induced in the above defined sets of points, lines, and parallel classes.

DEFINITION 29.4. A net is a 'vector-space net' if and only if there is a vector space over a field K such that the points of the net are the vectors and the vector translation group makes the net into an Abelian translation net.

THEOREM 29.5. (1) If π_1 and π_2 are affine planes and σ is a 1–1 correspondence from the parallel classes of π_1 onto the parallel classes of π_2, then $\pi_1 \times_\sigma \pi_2$ is a net whose parallel class set is in 1–1 correspondence with the parallel class set of either affine plane. Furthermore, this net admits Baer subplanes isomorphic to π_1 and π_2 on any given point.

(2) If π_1 and π_2 both have order n, then the order of $\pi_1 \times_\sigma \pi_2$ is n^2 and the degree (number of parallel classes) is $n + 1$.

(3) If π_1 and π_2 are both translation planes with translation groups T_i for $i = 1, 2$ respectively, then $\pi_1 \times_\sigma \pi_2$ is an Abelian translation net with translation group $T_1 \times T_2$.

DEFINITION 29.6. A direct product net where σ is induced from an isomorphism of π_1 onto π_2 is called a 'regular direct product' by π_1. Identifying π_1 and π_2 we also refer to $\pi_1 \times \pi_1$ as a '2-fold product'.

COROLLARY 29.7. A regular direct product net contains points incident with at least three mutually isomorphic Baer subplanes.

THEOREM 29.8. Let M be a vector-space net over a field K where M is a regular direct product net of two isomorphic point-Baer subplanes with kernel K_o.

Then M admits $\Gamma \simeq GL(2, K_o)$ as a collineation group that fixes an affine point and fixes each parallel class.

Furthermore, Γ is generated by the groups which fix point-Baer subplanes pointwise.

If $M = \pi_o \times \pi_o$ and K_o is the kernel of π_o as a left K_o-subspace, then the action of an invertible element $\begin{bmatrix} a & b \\ c & d \end{bmatrix}$ on M is

$$\tau_{(a,b,c,d)} \colon (p_o, p_1) \longmapsto (ap_o + cp_1, bp_o + dp_1)$$

for a, b, c, d in K_o and p_o, p_1 points of π_o.

We now turn to Desarguesian products.

29.1.1. Desarguesian Products. We shall want to consider the nature of the lines of 2-fold products (regular direct products) of Desarguesian affine planes π_i, $i = 1, 2$. We identify π_1 and π_2 and consider the corresponding regular direct product or 2-fold product.

THEOREM 29.9. Let π be a Desarguesian affine plane defined by a spread of t-dimensional vector subspaces of a $2t$-dimensional K-vector space where K is a field.

(1) Then, corresponding to π, is a dimension t extension field K_π of K considered as a K-vector space.

(2) A 2-fold product $\pi \times \pi$ produces a derivable net whose components (line incident with the zero vector) are 2-dimensional K_π-subspaces and $2t$-dimensional K-subspaces of a $4t$-dimensional vector space V_{4t} over K.

(3) There is a natural definition of the set of vectors of $\pi \times \pi$ as a 4-dimensional K_π-subspace and as a $4t$-dimensional K-subspace. Furthermore, the Baer subplanes incident with the zero vector are 2-dimensional K_π-subspaces and $2t$-dimensional K-subspaces.

REMARK 29.10. (1) Let $\mathcal{N}_K$ denote the net defined by the 'subplanes' of $\pi \times \pi$, $\pi \times 0$, $0 \times \pi$ and $\ell_\alpha = \{ (P, \alpha P);\ P \text{ a vector of } \rho\ (= \pi) \}$ for all α in K. Then $\mathcal{N}_K$ is a regulus net which may be coordinatized by a $4t$-dimensional K-vector space whose components (the subplanes) are $2t$-dimensional K-subspaces.

In this representation, $\mathcal{N}_K$ has components $x = 0, y = \alpha x$ for all $\alpha \in K$.

(2) Notice that the set of Baer subplanes of the derived net $(\pi \times \pi)^*$ intersects the original Desarguesian affine plane π in the original spread for π.

THEOREM 29.11. A Desarguesian spread in $PG(2t-1, K)$ produces a derivable partial spread of $(4t - 1)$-dimensional K-subspaces in $PG(4t - 1, K)$.

(1) If ρ and π are two Desarguesian affine planes defined by t-spreads over a field K then both $\rho \times \rho$ and $\pi \times \pi$ may be regarded as the same $4t$-dimensional K-subspace whose corresponding Baer subplanes may be considered $2t$-dimensional K-subspaces.

(2) Define the subnet $\mathcal{N}_K$ given by $\rho \times 0 = \pi \times 0$ as $y = 0, 0 \times \rho = 0 \times \pi$ as $x = 0$, and

$$\{ (P, \alpha P);\ P \text{ a vector of } \rho (= \pi) \}$$

as $y = \alpha x$ for all $\alpha \in K$.

COROLLARY 29.12. Then the derived nets of $\rho \times \rho$ and $\pi \times \pi$ both contain $\mathcal{N}_K$.

Now we shall want to consider some partial spreads in $PG(4t - 1, K)$ defined by spreads in $PG(2t - 1, K)$, whether the spread is Desarguesian or not.

When $t = 2$, we may construct some partial spreads in $PG(7, K)$ which may not come from derivable nets as follows:

Let V, U, W be any three mutually disjoint 4-dimensional left subspaces of a 8-dimensional left K-vector space V_8. (Think of V, U, W as $\pi \times 0, 0 \times \pi$ and ℓ_1 of the previous discussion).

Decompose V_8 relative to V and W so that writing vectors of V_8 in the form (x, y), we assume that V, W and U have the equations $x = 0, y = 0, y = x$.

PROPOSITION 29.13. Consider any 2-dimensional K-subspace X of V. Let X^σ denote the image under the mapping $(x, y) \longmapsto (y, x)$ where x, y are 4-vectors over K.

Then $X \oplus X^{\sigma}$ is a 4-dimensional K-subspace of V_8 which intersects $x = 0, y = 0, y = x$ in 2-dimensional K-subspaces. Conversely, any 4-dimensional K-subspace that intersects $x = 0, y = 0, y = x$ in 2-dimensional K-subspaces is of this form.

LEMMA 29.14. Define an 8-dimensional K-space relative to $V \oplus W$ so that V, U, W are K-subspaces.

Let $\mathcal{N}_K$ denote the regulus net with components $x = 0, y = \alpha x$ for all $\alpha \in K$. Note that the set of components contains 4-dimensional K-subspaces and contains V, U, W. Let $\mathcal{P}(\mathcal{N}_K)$ denote the set of vectors which lie on the components of $\mathcal{N}_K$.

If P is any vector not in $\mathcal{P}(\mathcal{N}_K)$ then there exists a unique 2-dimensional K-subspace of V, X, such that P is incident with $X \oplus X^{\sigma}$.

29.1.2. 2-Fold Products.

THEOREM 29.15. (1) Given a spread S of V of 2-dimensional K-vector subspaces, $\{ X \oplus X^{\sigma}; X \in S \}$ is a partial spread of 4-dimensional K-subspaces which contains the regulus partial spread defining $\mathcal{N}_K$ as a set of Baer subplanes.

(2) Let π_S denote the affine translation plane determined by the spread S. Considering σ as an isomorphism, we may form the regular direct product $\pi_S \times_{\sigma} \pi_S$.

This partial spread is the 2-fold product $\pi_S \times \pi_S$ where π_S is the translation plane determined by the spread S.

(3) $\bigcup \{ X \oplus X^{\sigma} - \mathcal{P}(\mathcal{N}_K); \forall 2\text{-dimensional subspaces } X \text{ of } V \}$ is a disjoint cover of the vectors of V_8 which are not in $\mathcal{P}(\mathcal{N}_K)$.

COROLLARY 29.16. The net $\rho \times \rho$ containing the regulus net $\mathcal{N}_K$ (as a set of Baer subplanes) admits a collineation group G isomorphic to $GL(2, K)$ defined as follows:

$$\left\langle (x, y) \longmapsto \left(\begin{bmatrix} \alpha & \beta \\ \delta & \gamma \end{bmatrix} (x, y); \forall \alpha, \beta, \delta, \gamma \in K; \alpha\beta - \delta\gamma \neq 0 \right\rangle \right.$$ (the elements of K act on the left by scalar multiplication).

Furthermore, G leaves invariant each 4-dimensional K-subspace $X \oplus X^{\sigma}$ for X a 2-dimensional K-subspace of V and acts on the points of $X \oplus X^{\sigma} - \mathcal{P}(\mathcal{N}_K)$.

29.2. Desarguesian Partial Parallelisms.

We are now considering a 'parallelism in $PG(3, K)$', where K is a field, as a set of line-disjoint spreads which cover the line set of the projective geometry. Recall, when K is a field, a spread S is said to be 'regular' exactly when the regulus generated by any three distinct lines of S is contained in S. A 'regular parallelism' is a parallelism where the spreads are all regular.

DEFINITION 29.17. Recall that a 'F-regulus net' is any net which may be coordinatized in a natural way by a field F; the components are of the form $x = 0, y = \delta x$ for all $\delta \in F$ where there is an underlying vector space and δ refers to scalar multiplication. In this case, we call the F-regulus net a net. If an F-regulus net contains a K-regulus net, where F is a field extension of K, we shall say that the F-net is an 'extension of the K-net'.

In this section, we shall use the direct product constructions of the previous section to construct translation planes or rather spreads in $PG(7, K)$ from what we call 'Desarguesian parallelisms'.

29.2.1. Desarguesian t-Spreads.

DEFINITION 29.18. Let V be a vector space over a field K. A t-spread $\mathcal{S}_2$ is a cover of the non-zero vectors of V by t-dimensional K-subspaces. We shall say that a t-spread is 'Desarguesian' if and only if there is an extension field K' of K of degree t such that elements of the t-spread are 1-dimensional K'-subspaces.

Note that if K has a t-dimensional extension field K', then there is a natural 1-spread over K' (covering of V by the 1-dimensional K' subspaces) and since each such 1-space over K' is a t-spread over K, it follows that Desarguesian t-spreads are equivalent to t-dimensional field extensions of K (considered as K-spaces). Analogously, we may consider t-spreads.

DEFINITION 29.19. A set of t-spreads of a vector space V over K is a 'partial t-parallelism' if and only if the t-spreads contain no common t-dimensional subspace. A 't-parallelism' is a partial t-parallelism such that every t-dimensional K-space of V is contained in exactly one of the t-spreads. A partial t-parallelism shall be said to be a partial 'Desarguesian t-parallelism' if and only if each t-spread in the partial t-parallelism is Desarguesian. Similarly, we have the natural definition of a 'Desarguesian t-parallelism'.

We first note that from partial spreads of regulus t-dimensional extensions of K-regulus nets, we obtain Desarguesian partial parallelisms.

THEOREM 29.20. Let S be a partial spread in $PG(2tk - 1, K)$ defined by a set of $(tk - 1)$-dimensional K-subspaces where K is a field as follows: There exists a set λ of regulus t-dimensional extensions of the K-regulus net $\mathcal{N}_K$. Then the set of subplanes of each regulus net in λ induces a Desarguesian t-spread on any component of $\mathcal{N}_K$ and the union of the spreads is a Desarguesian partial t-parallelism of $PG(tk - 1, K)$.

We have seen that spreads consisting of t-dimensional extension regulus nets induce Desarguesian partial t-parallelisms.

What is not clear is if Desarguesian partial t-parallelisms in V_{tk} over K produce spreads consisting of t-dimensional extension regulus nets of a K-regulus net.

For simplicity, we consider the situation when $t = k = 2$.

29.2.2. The 4-Dimensional Case.

We now specialize to the case when V_4 over a field is 4-dimensional so that a 2-spread is a spread of 2-dimensional subspaces of V.

In this case, we call partial 2-parallelisms merely 'partial parallelisms' since a 2-spread S determines an affine translation plane π_S.

Furthermore, we also say that the set of spreads as sets of 1-dimensional projective subspaces in $PG(3, K)_{\mathcal{L}}$ is a partial parallelism of the projective space. But, usually, we shall not make a distinction between partial 2-parallelisms of the vector space and the partial parallelisms of the associated projective space.

Note that a Desarguesian 2-spread is now simply a spread which produces a Desarguesian affine plane.

We now show how to obtain a translation plane with spread in $PG(7, K)$ from a parallelism in $PG(3, K)$.

29.2.3. Spreads in $PG(7, K)$ and Parallelisms in $PG(3, K)$.

THEOREM 29.21. Let $\mathcal{P}$ be a Desarguesian parallelism of $PG(3, K)_{\mathcal{L}}$ where K is a field. Then there is a corresponding spread $\pi_{\mathcal{P}}$ in $PG(7, K)_{\mathcal{R}}$ consisting of derivable partial spreads containing the regulus defined by K. Conversely, a spread in $PG(7, K)_{\mathcal{R}}$ which is a union of derivable partial spreads containing the K-regulus produces a Desarguesian parallelism of $PG(3, K)_{\mathcal{L}}$.

We note that the previous theory is a generalization of the work of Walker [**1203**] and Lunardon [**940**].

We now generalize all of this to Desarguesian partial parallelisms.

29.3. Desarguesian Partial Parallelisms with Deficiency.

In this section, we consider partial parallelisms and their possible extensions. The reader is referred to one of the present authors' works [**753**] for additional details.

DEFINITION 29.22. A partial parallelism in $PG(3, K)$ has 'deficiency $t' < \infty$ if and only if, for each point P, there are exactly t 2-dimensional K-subspaces containing P which are not in a spread of the partial parallelism or equivalently, exactly t lines not incident with P. Thus, a partial parallelism is of deficiency t if and only if the number of spreads is $1 + q + q^2 - t$.

Now we note,

THEOREM 29.23. Let $\mathcal{P}_{-1}$ denote a partial parallelism of deficiency one in $PG(3, K)$ for K a field. Then there is a unique extension of $\mathcal{P}_{-1}$ to a parallelism of $PG(3, K)$.

We now consider Desarguesian partial parallelisms that have extensions.

THEOREM 29.24. Let $\mathcal{S}_{-t}$ be a Desarguesian partial parallelism of deficiency $t < \infty$ in $PG(3, K)$, for K a field. Let $\pi_{\mathcal{S}_{-t}}$ denote the corresponding translation net defined by a partial spread in $PG(7, K)$ consisting of derivable partial spreads sharing the K-regulus $\mathcal{N}_K$.

(1) If $\mathcal{S}_{-t}$ can be extended to a partial parallelism of deficiency $t - 1$ by the adjunction of a spread S, let π_S denote the translation plane determined by the spread.

Then a partial spread Σ in $PG(V_8, Z(K))$ may be defined as follows:

$$\Sigma = (\pi_{\mathcal{S}_{-t}} - \mathcal{N}_K) \cup (\pi_S \times \pi_S),$$

where the notation is intended to indicate the 4-dimensional K-subspace components of the net $\pi_{\mathcal{S}_{-t}}$ which are not in $\mathcal{N}_K$ union the 4-dimensional K-subspace components of $\pi_S \times \pi_S$.

(2) Σ admits $GL(2, K)$ as a collineation group where the subgroup $SL(2, K)$ is generated by collineations that fix Baer subplanes pointwise.

29.3.1. Deficiency-One Partial Spreads Produce Translation Planes.

THEOREM 29.25. (1) Let $\mathcal{P}_{-1}$ denote a Desarguesian partial parallelism of deficiency one in $PG(3, K)$, where K is a field. Let S denote the spread extending $\mathcal{P}_{-1}$ and let π_S denote the affine plane given by the spread S.

Then there is a corresponding translation plane Σ with spread in $PG(V_8, Z(K))$ containing a partial spread $\mathcal{D}$ defined by a set of 4-dimensional K-subspaces giving rise to a net $\mathcal{A}(\mathcal{D}) = \pi_S \times \pi_S$ which contains the K-regulus partial spread $\mathcal{N}_K$ as a set of Baer subplanes and the partial spread $\pi_{\mathcal{P}_{-1}} - \mathcal{N}_K$ of the 4-dimensional K-subspace components of derivable partial spreads which are not in $\mathcal{N}_K$. The spread for Σ is $\mathcal{D} \cup \{\pi_{\mathcal{P}_{-1}} - \mathcal{N}_K\}$.

(2) Conversely, a spread with and 4-spaces defined as above produces a Desarguesian partial parallelism of deficiency one.

(3) The unique extension of a Desarguesian partial parallelism of deficiency one to a parallelism may obtain a characterization by a collineation group: The parallelism is Desarguesian if and only if the partial spread $\mathcal{D}$ defined by the extending spread is derivable.

Hence, we have the following result:

THEOREM 29.26. (1) Given a Desarguesian partial parallelism $\mathcal{P}_{-1}$ of deficiency one in $PG(3, q)$, there is a corresponding translation plane Σ with spread in $PG(7, q)$.

(2) The translation plane admits a collineation group isomorphic to $GL(2, q)$ which contains a subgroup isomorphic to $SL(2, q)$ that is generated by Baer collineations.

Furthermore, (a) there are exactly $(1+q)$ Baer axes which belong to a common net $\mathcal{N}^*$ of degree $(1 + q^2)$, (b) $SL(2, q)$ has exactly $q(q + 1)$ component orbits of length $q(q-1)$. These orbits union the $(1+q)$ Baer axes define derivable nets. The $q(q + 1)$ derivable nets induce a Desarguesian parallel parallelism of deficiency one on any Baer axis considered as $PG(3, q)$, (c) The Desarguesian partial parallelism may be extended to a Desarguesian parallelism if and only if the net $\mathcal{N}^*$ is derivable.

(3) Conversely, let π be any translation plane with spread in $PG(7, q)$ that admits $SL(2, q)$ as a collineation group generated by Baer p-elements, where $q = p^r$ and p is a prime. If the component orbits of $SL(2, q)$ have lengths either 1 or $q^2 - q$, then there is a corresponding Desarguesian partial parallelism of deficiency one induced on any Baer axis.

Here we may re-state an open problem in terms of derivable nets.

PROBLEM 29.27. Let π be a translation plane of order q^4 with spread in $PG(7, q)$ that admits $SL(2, q)$ as a collineation group generated by Baer p-collineations where $q = p^r$, p a prime. If the component orbits of $SL(2, q)$ have lengths 1 or $q^2 - q$, show that the net of orbits of length 1 is a derivable net.

29.4. Maximal Partial Spreads.

THEOREM 29.28. If $\mathcal{P}_{-1}$ is a finite Desarguesian partial parallelism of deficiency one whose unique extension to a parallelism $\mathcal{P}_0$ is not Desarguesian, then there is a maximal partial spread in $PG(7, q)$ constructed as follows: Construct the translation plane $\pi_{\mathcal{P}_{-1}}$ admitting $SL(2, q)$ where the p-elements are Baer and $q = p^r$. Let $\mathcal{M}_\pi$ denote the translation net defined by the $1 + q$ Baer subplanes which are pointwise fixed by p-elements in $SL(2, q)$ and the components of $\pi_{\mathcal{P}_{-1}}$ which are not incident with the net $\mathcal{N}^*$ defined by the Baer subplanes.

(1) $\mathcal{M}_\pi$ is a maximal partial spread in $PG(7, q)$ of deficiency $q^2 - q$.

(2) If q is a prime p then there is a transversal-free translation net which is not an affine plane defined by (1) of order p^4 and degree $p^4 - p^2 + p$.

29.4.1. Partial Parallelisms of Deficiency Two. Now assume that we have a Desarguesian partial parallelism $\mathcal{P}_{-2}$ of deficiency two which can be extended to a partial parallelism of deficiency one. Then, there is an associated translation net of degree

$$(q^2 - q)(q + q^2 - 1) + (1 + q)$$

which then produces a translation net of degree

$$(q^2 - q)(q + q^2 - 1) + (1 + q^2) \;=\; q^4 + 1 - (q^2 - q);$$
$$\text{deficiency } q^2 - q.$$

We want to investigate whether this partial spread Σ_1 may be extended to a spread in $PG(7, K)$, for K isomorphic to $GF(q)$.

THEOREM 29.29. Let $\mathcal{P}_{-2}$ be a Desarguesian partial parallelism of deficiency two of $PG(3, K)$, for K isomorphic to $GF(q)$, which can be extended to a partial parallelism $\mathcal{P}_{-1}$. Since $\mathcal{P}_{-1}$ can always be extended to a unique parallelism, let $\mathcal{P}_0$ denote the extended parallelism and assume that the two spreads extending $\mathcal{P}_{-2}$ are both non-Desarguesian.

(1) Then there is a maximal partial spread in $PG(7, q)$ of deficiency $q^2 - q$.

(2) If q is a prime, then there is a transversal-free net of order p^4 with deficiency $p^2 - p$.

29.4.2. Prince Parallelisms in $PG(3, 3)$. Although there are no regular parallelisms in $PG(3, 3)$, Prince has determined the parallelisms in $PG(3, 3)$ that admit a collineation group of order 5.

THEOREM 29.30. (Prince [**1113**]). There are exactly seven mutually non-isomorphic parallelisms in $PG(3, 3)$ invariant under a collineation of order 5.

Since all spreads are either Hall or Desarguesian, the parallelism is said to be of type $(j, 13 - j)$, if j is the number of Hall spreads in the parallelism.

In particular, there are two parallelisms of type $(2, 11)$. Hence, we have partial Desarguesian parallelisms of deficiency two that may be extended to a parallelism. So, we may apply Theorem 29.24 part (1).

29.4.3. Transversal-Free Nets of Order 3^4 and Deficiency $3^2 - 3$. Hence, using the $(2, 11)$ parallelisms, there are some Desarguesian partial parallelisms of deficiency 2 in $PG(3, 3)$ which may be extended to a parallelism by adding two non-Desarguesian spreads.

Hence, we have the situation depicted in the previous theorems. Since we may initially adjoin either of the two non-Desarguesian spreads, we have two transversal-free translation nets of order 3^4 and degree exactly

$$3^4 + 1 - (3^2 - 3) = 3^4 - 5$$

(since of deficiency $3^2 - 3$) which are not affine planes.

29.4.4. Parallelisms in $PG(3, 2)$. There are exactly two parallelisms in $PG(3, 2)$ and since any affine plane of order 4 is Desarguesian, the parallelisms are regular. These are particularly interesting in that the associated translation planes of order 16 with spread in $PG(7, 2)$ have collineation groups one of which acts as 'tangentially transitive group' with respect to a subplane π_o of order 2. That is to say, there is a collineation group G which fixes π_o pointwise and acts transitively on the lines of the plane which are tangent to the plane π_o at a particular affine point. The other plane is the 'transpose' of the tangentially transitive plane. The two planes are the so-called Lorimer–Rahilly and Johnson–Walker planes of order 16.

We have noticed that the planes constructed from parallelisms in $PG(3, q)$ are always composed of a set of $1 + q + q^2$ derivable nets. The derivation of any of these nets thus produces a corresponding translation plane. In the case of the Lorimer–Rahilly and Johnson–Walker planes, any such derived plane is a semifield plane (a plane admitting an affine elation group of order 16 transitive on the components of the plane not equal to the axis of the group).

The two semifield planes used in the derivation process are 'transposes' of each in the sense that their spreads are connected via a dualization (i.e., a polarity). It turns out that the construction processes of 'transpose' and 'derivation' commute which makes the Lorimer–Rahilly and Johnson–Walker planes transposes of each. Moreover, the derived planes are not isomorphic as the two semifield planes when transposed have different affine homology groups.

So, Letting π_{LR} and π_{JW} denote the Lorimer–Rahilly and Johnson–Walker and π_{LF}^* and π_{JW}^* denote the derived planes, we obtain the following lovely pattern:

$$\begin{bmatrix} \pi_{LF}^* & \leftrightarrow \text{(transpose)} & \pi_{JW}^* \\ \updownarrow \text{(derive)} & & \updownarrow \text{(derive)} \\ \pi_{LF} & \leftrightarrow \text{(transpose)} & \pi_{JW} \end{bmatrix}.$$

The reader interested in the particulars of these planes is referred to Johnson [684] and Walker [1207].

REMARK 29.31. These two planes corresponding to the parallelisms in $PG(3, 2)$ admit collineation groups which act transitive on the $1 + 2 + 2^2$ derivable nets, a fact which, in turn, implies that the group acting on the parallelism is two-transitive on the spreads.

CHAPTER 30

Transitive Parallelisms.

Recently, there has been renewed interest in the study of parallelisms in $PG(3, q)$ and their automorphism groups. The parallelism is said to be 'regular' if and only if all spreads in the parallelism are regular spreads.

To facilitate our discussion, we have brought translation planes into this context. As we have seen in Chapter 29, regular parallelisms in $PG(3, q)$ are equivalent to translation planes with spreads in $PG(7, q)$ admitting a collineation group isomorphic to $SL(2, q)$ generated by elations and which acts $1/2$-transitively on the components which are not elation axes. Furthermore, there are $1 + q + q^2$ derivable nets defined by the $SL(2, q)$-orbits which share the $1 + q$ elation axes. The collineation group of the translation plane normalizes $SL(2, q)$ and so permutes the derivable nets. The nets correspond to regular spreads induced on any elation axis considered as $PG(3, q)$. Hence, the collineation group of the translation plane induces a collineation group in $P\Gamma L(4, q)$ that permutes the associated planes of the regular parallelism.

30.1. Definition and Some Examples.

DEFINITION 30.1. A parallelism is 'transitive' if there exists a collineation group of $PG(3, q)$ which acts transitively on the spreads of the parallelism.

30.1.1. Denniston Parallelisms in $PG(3, 8)$. Denniston [**313**] has constructed a transitive regular parallelism in $PG(3, 8)$ again producing a translation plane of order 8^4.

30.1.2. Cyclic Parallelisms of Prince. Prince has found 45 mutually non-isomorphic transitive (and cyclic) parallelisms in $PG(3, 5)$ of which two are regular thus producing associated translation planes of order 5^4 (see Prince [**1114**]).

Since there are an odd number of mutually non-isomorphic transitive parallelisms in $PG(3, 5)$, there must be at least one parallelism which is isomorphic to its dual parallelism.

30.1.3. Penttila and Williams Parallelisms. Penttila and Williams [**1101**] have constructed an infinite class of transitive and regular parallelisms in $PG(3, q)$ for $q \equiv 2 \bmod 3$ which include the regular parallelisms previously mentioned. Furthermore, Penttila and Williams show that the parallelisms are not isomorphic with their dual parallelisms. The idea of the proof of Penttila and Williams is to find a packing of the Klein quadric by mutually disjoint elliptic quadrics and then use the Klein mapping to establish the corresponding parallelism in $PG(3, q)$.

We have noted that there are two other parallelisms with remarkable properties also included in [**1101**]. These parallelisms are in $PG(3, 2)$, are regular since the spreads produce translation planes of order 4, and are duals of each other.

These two parallelisms give rise to the Lorimer–Rahilly and Johnson–Walker translation planes of order 16. The collineation group of the plane contains a group isomorphic to $SL(2,2) \times PSL(2,7)$. Hence, $PSL(2,7) \simeq GL(3,2)$ induces a collineation group of $P\Gamma L(4,2)$ which acts two-transitively on the $1 + 2 + 2^2$ planes of the parallelisms. So, as we have mentioned previously, the two regular parallelisms in $PG(3,2)$ are two-transitive.

In fact, the two regular parallelisms in $PG(3,2)$ are the only two-transitive parallelisms, regular or not.

30.1.4. Two-Transitive Parallelisms; Theorem of Johnson.

THEOREM 30.2. (Johnson [**756**]). Let $\mathcal{P}$ be a parallelism in $PG(3,q)$ which admits an automorphism group acting 2-transitively on the spreads of $\mathcal{P}$.

Then $q = 2$, $\mathcal{P}$ is one of the two regular parallelisms in $PG(3,2)$ and the group is isomorphic to $PSL(2,7)$.

The proof is based fundamentally upon the following two incompatibility results on certain p-collineations in affine translation planes of order p^{2st} with kernel $GF(p^t)$. The collineations considered in affine translation planes are taken to be within the translation complement which is a subgroup of $\Gamma L(4s,p^t)$. The 'linear' translation complement is the collineation subgroup of $GL(4s,p^t)$. When we consider translation planes with spreads in $PG(3,q)$ then $s = 1$.

THEOREM 30.3. (Foulser [**375**]). Let π be an affine translation plane of odd order p^{2r}. Then there cannot be both non-trivial elations and non-trivial Baer p-elements as collineations of π.

THEOREM 30.4. (Jha and Johnson [**587, 588**]). Let π be an affine translation plane of even order $2^{2r} = q^2$.

(1) Assume that there exists an elation group E of order 2^r which normalizes a non-trivial Baer 2-group B. Then the order of B is 2.

(2) Assume that there exists a Baer 2-group B of order $\geq 2\sqrt{q}$ then any elation group has order ≤ 2.

Armed with these results, we may then use the classification theorem of finite simple groups and apply a case-by-case analysis.

As noted, Penttila and Williams [**1101**] have determined two infinite classes of regular parallelisms in $PG(3,q)$ for $q \equiv 2 \bmod 3$ which contains all previous known regular parallelisms. The two classes are duals of each other. Thus, there is a vast number of new translation planes of order q^4 with spreads in $PG(7,q)$. These translation planes admit $SL(2,q)$ as collineation groups in their translation complements generated by elations. All of these examples admit a cyclic group C_{1+q+q^2} which acts transitively on the spreads of the parallelism and which centralizes the $SL(2,q)$ group, so that the planes admit $SL(2,q) \times C_{1+q+q^2}$. Furthermore, without any assumptions on the nature of the spread or on the action of the p-elements where $p^r = q$, Jha and Johnson [**597**] show that the group action on the associated translation plane determines the parallelisms.

30.2. $SL(2,q) \times C_{(1+q+q^2)}$ and Cyclic Parallelisms.

THEOREM 30.5. (Jha and Johnson [**597**]). Let π be any translation plane of order q^4 admitting a collineation group isomorphic to $SL(2,q) \times C_{1+q+q^2}$. Then the kernel is $GF(q)$, the p-elements are elations, where $p^r = q$, and a regular and cyclic parallelism is induced on any elation axis.

Hence, it is of interest to determine classification theorems regarding arbitrary parallelisms in $PG(3, q)$ by virtue of the associated collineation group.

In particular, it is of interest to determine the transitive groups that can act on a parallelism and we here consider such a study. Actually, we may study parallelisms that admit only a collineation of order a p-primitive divisor of $q^3 - 1$, which is a more general study than merely the consideration of transitivity. Furthermore, the same results may be obtained merely by assuming that there is a linear element of order not three whose order divides $1 + q + q^2$.

Certainly, not all transitive parallelisms can be determined at least easily as we have the large number of non-regular but transitive parallelisms in $PG(3, 5)$ due to Prince [**1114**]. It has been noted that the problem of the determination of all 2-transitive parallelisms has been resolved in [**756**], using the classification of all finite simple groups. However, since the groups involved must lie within $P\Gamma L(4, q)$, it is possible to use more elementary results for the study of p-primitive parallelisms, transitive parallelisms, and 2-transitive parallelisms.

In particular, it is possible to provide a new proof for the 2-transitive parallelisms when $q \neq 4$ and furthermore show that non-solvable groups admitting p-primitive elements have very restrictive structure and involve only $PSL(2, 7)$, A_7 or $PSL(3, 4)$. In the transitive case, only $PSL(2, 7)$ can occur and $q = 2$.

30.3. p-Primitive Parallelism Theorem.

The following results are due to Biliotti, Jha and Johnson [**122**].

THEOREM 30.6. Let $\mathcal{P}$ be a parallelism in $PG(3, q)$, which admits a collineation of order a p-primitive divisor u of $q^3 - 1$.

Then, one of the two situations occurs.

(1) The full group G acting on the parallelism is in $\Gamma L(1, q^3)/Z$ where Z denotes the scalar group of order $q - 1$ and fixes a plane and a point.

(2) $u = 7$ and one of the following subcases occurs:

(a) G is reducible, fixes a plane or a point and either

    (i) G is isomorphic to A_7 and $q = 25$, or

    (ii) G is isomorphic to $PSL(2, 7)$ and either $q = 2$ or q is p or p^2 for an odd prime p.

    If $q = p$ then $p \equiv 2, 4 \bmod 7$.

    If $q = p^2$ then $p \equiv 3, 5 \bmod 7$. Furthermore, when $q = 2$ $\mathcal{P}$ is one of the two regular parallelisms in $PG(3, 2)$.

(b) G is primitive and either

    (i) G is isomorphic to $PSL(3, 4)$ and $q = 9$, or

    (ii) G is isomorphic to $PSL(2, 7)$ or A_7 and $q^3 \equiv 1 \bmod 7$ with $q = p$ or p^2 for p an odd prime.

Often we regard G as a subgroup of $\Gamma L(4, q)$ so that G, modulo the scalar subgroup, is the group acting on $PG(3, q)$.

REMARK 30.7. If there is a linear collineation of order not three and dividing $1 + q + q^2$, the same conclusions can be drawn concerning the full collineation group of a parallelism and the nature of non-solvable parallelisms. Hence, if $q = p^r$ and r odd, then any linear element of order dividing $1 + q + q^2$ will suffice.

COROLLARY 30.8. Let $\mathcal{P}$ be a transitive parallelism in $PG(3,q)$ where $q \neq 4$.

Then either $\mathcal{P}$ is one of the two regular parallelisms in $PG(3,2)$ or the full collineation group is a subgroup of $\Gamma L(1,q^3)/Z$, where Z denotes the scalar group of order $q-1$.

(Johnson [**756**]). The 2-transitive parallelisms in $PG(3,q)$, for $q \neq 4$ are exactly the two regular parallelisms in $PG(3,2)$.

As noted above, the only known Desarguesian parallelisms in $PG(3,q)$ admit a cyclic collineation group of order $1+q+q^2$ acting transitively on the spreads. The construction of these parallelisms using the partitions of the Klein quadric of Penttila and Williams [**1101**] will be given in the next section.

30.3.1. The Penttila–Williams Construction.
Let a 6-dimensional $GF(q)$-vector space V_6 be defined as $GF(q^3) \oplus GF(q^3)$. Define a quadric Q by $Q(x,y) = T(xy)$, where $T(w) = w + w^q + w^{q^2}$, for w in $GF(q^3)$. Assume that $q \equiv -1 \bmod 3$, throughout this subsection. The following establishes the basic quadratic form used.

LEMMA 30.9. Q defines a hyperbolic quadric in V_6, so that V_6 is a $O^+(6,q)$-space.

PROOF. Note for example that the polar form of the quadratic form is f such that $f((u_1,u_2),(v_1,v_2)) = T(u_1 v_2) + T(v_1 u_2)$. Fix v_1 and v_2 and assume that for all u_1,u_2, $T(u_1 v_2) + T(v_1 u_2) = 0$. Let $u_2 = 0$, so we obtain $T(u_1 v_2) = 0$ for all u_1, which clearly implies that $v_2 = 0$. Since the argument is symmetric, we have $v_1 = v_2 = 0$. So, we have a non-degenerate quadric. □

Define $\Sigma = \left\{ (y,z); y^{q^2} + z \in GF(q) \right\}$. Then Σ is a 4-dimensional $GF(q)$-space that intersects the hyperbolic quadric in an elliptic quadric; Σ is a $O^-(4,q)$-space.

LEMMA 30.10. Define G as follows:
$$G = \left\{ g_\mu; \; g_\mu(x,y) = (\mu x, \mu^{-1} y), \; \mu \in GF(q^3)^* \right\}, \text{ for } x,y \in GF(q^3).$$
Then G preserves Q.

THEOREM 30.11. The images of the elliptic quadric of Σ under the group G is a partition of the Klein quadric; we have: $G\Sigma \cap Q = Q$, where Q defines a hyperbolic quadric in V_6. Furthermore, the collineation group of the partition is isomorphic to $\Gamma L(1,q^3)/Z$, where Z is the scalar group of order $q-1$.

COROLLARY 30.12. Let σ be any correlation of the association projective space $PG(3,q)$ that normalizes $GL(1,q^3)$. Then form the associated 'dual' partition $G\sigma\Sigma \cap Q$. Then the two partitions are non-isomorphic.

PROOF. If the two partitions are isomorphic, there is an element g of $P\Gamma L(4,q)$ that normalizes $GL(1,q^3)$ and maps one partition to the other. However, $\Gamma L(1,q^3)$ is the full normalizer within $P\Gamma L(4,q)$. Hence, g would leave both partitions invariant, contrary to our assumptions. □

Hence, we obtain:

THEOREM 30.13. Corresponding to the Penttila–Williams partitions of the Klein quadric by elliptic quadrics, there are associated two classes of translation planes of order q^4 and kernel $GF(q)$ admitting $SL(2,q) \times Z_{1+q+q^2}$. The two planes are 'transposes' of each other and are mutually non-isomorphic.

We have noted previously that the two planes associated with the parallelisms of $PG(3,2)$ are the Johnson–Walker and Lorimer–Rahilly planes of order 16, also admitting $PSL(2,7)$. The planes of Denniston of order 8^4, and the planes of Prince of order 5^4 belong to the planes obtained determined by the construction.

CHAPTER 31

Ovoids.

The reader also might consult Chapter 103 on symplectic and orthogonal geometry for additional background information.

An ovoid in a $\Omega^+(2n, q)$-space is a set of $q^{n-1} + 1$ points (1-dimensional subspaces) of the associated hyperbolic quadric Q, no two of which lie on a line of Q. When $n = 3$, we have the ovoid of $q^2 + 1$ points of the Klein quadric which corresponds to a spread of $PG(3, q)$.

Ovoids cannot exist for $n > 4$ (see Moorhouse [**990, 991**] and Hirschfeld and Thas [**529**]). However, there are a variety of such objects when $n = 4$. When $n = 4$, the $q^3 + 1$ points can lie within a subspace of type $\Omega(7, q)$ as a subspace of $\Omega^+(8, q)$. The importance of ovoids is made manifest by the following construction.

THEOREM 31.1. Let $\mathcal{O}$ be an ovoid in a $\Omega^+(8, q)$-space or $\Omega(7, q)$-space. Choose a point P of $Q - \mathcal{O}$ and let $P^\perp$ denote the polar space of P. Then $P^\perp/P$ is an $\Omega^+(6, q)$-space or $\Omega(5, q)$-space, respectively. Then $\{ z + \langle P \rangle ; z \in P^\perp \cap Q \}$ is an ovoid of the Klein quadric, called a 'slice' of Q.

PROOF. We note simply that there are $2(q+1)(q^2+1)$ totally singular 3-spaces of $P^\perp/P$ that project from the $2(q+1)(q^2+1)$ totally singular 4-spaces of $P^\perp$ that contain P. There are exactly $2(q+1)$ totally singular 4-spaces that contain a given 2-space. Hence, it follows that the cardinality of $P^\perp \cap Q$ is $q^2 + 1$. □

So, we obtain the following result:

THEOREM 31.2. Slices of ovoids in $\Omega^+(8, q)$-space or $\Omega(7, q)$-space produce translation planes with spreads in $PG(3, q)$.

Kantor [**861**] determined all slices of the known ovoids in such spaces. These arise from the unitary groups and the Ree–Tits unitals.

31.1. The Slices of the Unitary Ovoids.

We follow the work of Kantor [**861**] as explicated in Johnson [**731**].

We first determine an appropriate $\Omega^+(8, q)$-space, where $q \equiv 0$ or $2 \bmod 3$. Consider

$$V = \left\{ m = \begin{bmatrix} \alpha & \beta & c \\ \gamma & a & \overline{\beta} \\ b & \overline{\gamma} & \overline{\alpha} \end{bmatrix} ; \begin{array}{l} \alpha, \beta, \gamma \in L \simeq GF(q^2); \\ a, b, c \in K \simeq GF(q), \ L \subset K \end{array} \right\}$$

where $T(\alpha) = \alpha + \alpha^q$, and $\overline{\delta} = \delta^q$, and choose $a + T(\alpha) = 0$.

Define

$$Q_8 \colon V \to K \text{ by } Q_8(m) = \alpha^2 + \alpha\overline{\alpha} + \overline{\alpha}^2 + T(\beta\gamma) + bc.$$

Then we know the following:
 (1) V is an $\Omega^+(8,q)$-space if and only if $q \equiv 0$ or 2 mod 3.
 (2) If $q \equiv 0$ mod 3 then $\text{Rad}\, V$ is the set $\langle I \rangle$ of K-scalar matrices.
 (3) Let

$$
J = \begin{bmatrix} 0 & 0 & 1 \\ 0 & 1 & 0 \\ 1 & 0 & 0 \end{bmatrix}.
$$

Let G denote the unitary group $GU(3,q)$ of all invertible matrices A over L such that $J^{-1}AJ = \overline{A}^{-t}$, where if $A = [a_{ij}]$ then $\overline{A} = [a_{ij}^q]$ and t denotes matrix transposition. Then G acts on V of (1) by conjugation and incudes $PGU(3,q)$ and preserves Q_8.
 (4) $\Omega = \{\, \langle Z \rangle \,;\, 0 \neq Z \in V;\, Z^2 = 0 \,\}$ is an ovoid if $q \equiv 2$ mod 3 and projects to an ovoid of $V/\langle I \rangle$ if $q \equiv 0$ mod 3.
 If Y is a singular point with respect to Q_8 and is not in Ω, form $Y^\perp/Y$. If $q \equiv 2$ mod 3, then $Y^\perp/Y$ is a $\Omega^+(6,q)$-space. If $q \equiv 0$ mod 3, then $(Y^\perp/\langle I \rangle)/(Y/\langle I \rangle)$ is a $\Omega(5,q)$-space. In both cases, there are corresponding spread sets in $PG(3,q)$ corresponding to translation planes.

REMARK 31.3. Ovoids in $\Omega(5,q)$-spaces produce symplectic spreads and symmetric spread sets.

We may select representatives from the G-orbits of singular points to find the associated translation planes.

THEOREM 31.4. (See Johnson [**731**] and Kantor [**861**]).
 (1) If $q \equiv 2$ mod 3, then

$$
Y = \begin{bmatrix} 0 & 1 & 0 \\ 0 & 0 & 1 \\ 0 & 0 & 0 \end{bmatrix} \quad \text{and } Y' = \text{Diag}(w,1,w);\, w^3 \neq 1 \neq w
$$

are representatives.
 (a) The associated slices from Y are the spreads of Walker or Betten for q odd or even, respectively.
 (b) The associated slices from Y' are j-planes for $j = 1$.
 (2) If $q \equiv 0$ mod 3, then the associated slice is the following Kantor–Knuth semifield spread

$$
x = 0, y = x \begin{bmatrix} u & t^3\gamma^{-1} \\ t & u \end{bmatrix};
$$

$$
u, t \in GF(q) \text{ and } \gamma \text{ a non-square in } GF(q).
$$

Note that

$$
\begin{bmatrix} u & t^3\gamma^{-1} \\ t & u \end{bmatrix} \begin{bmatrix} 0 & 1 \\ 1 & 0 \end{bmatrix} = \begin{bmatrix} t^2\gamma^{-1} & u \\ u & t \end{bmatrix}
$$

is a symplectic spread of symmetric matrices.

 (3) If $q \equiv 0$ mod 3, then for $N = \text{Diag}(\lambda, 0, \lambda^q)$, a non-singular point for $\lambda \in L^*$ such that $T(\lambda) = 0$, $N/\langle I \rangle$ is a $\Omega^+(6,q)$-space. The associated slice is a j-plane for $j = 1$.

31.2. The Slices of the Ree–Tits Ovoids.

We now take up the slices of the Ree–Tits ovoid.

Let $K \simeq GF(q)$ and let $V = K^7$, where $q = 3^{2e-1}$. Let $a^\sigma = a^{3e}$ for $a \in K$. Let $x = (x_1, x_2, \ldots, x_7)$ and define Q by

$$Q(x) = x_4^2 + x_1 x_7 + x_2 x_6 + x_3 x_5.$$

Let

$$\begin{aligned}
\Omega &= \langle(0,0,0,0,0,0,1)\rangle \cup \{\langle(1,x,y,z,u,v,w)\rangle\,;\, x,y,z \in K\}, \text{ such that} \\
u &= x^2 y - xy + y^\sigma = x^{\sigma+3}, \\
v &= x^\sigma y^\sigma - z^\sigma + xy^2 + yx - x^{2\sigma+3}, \\
w &= xz^\sigma - x^{\sigma+1}y^\sigma - x^{\sigma+3}y + x^2 y^2 - y^{\sigma+1} - z^2 + x^{2\sigma+4}.
\end{aligned}$$

Then the Ree group $R(q)$ leaves Ω invariant (see Gorenstein [414] p. 480 for a description of the Ree groups). It turns out that $R(q)$ has two orbits of singular points not on Ω. Representatives are as follows:

$$T_r = \langle(0,0,0,0,0,1)\rangle \text{ and } T_r' = \langle(0,0,0,0,1,0)\rangle.$$

Since we are now obtaining planes by slicing ovoids in $\Omega(5,q)$-space, we obtain symplectic spreads represented by symmetric matrices.

THEOREM 31.5. (See Johnson [731] and Kantor [861]).
(1) The symplectic spread arising from T_r is a Kantor–Knuth semifield spread.
(2) The symplectic spread arising from T_r' is as follows:

$$x = 0,$$
$$y = x \begin{bmatrix} u & u^\sigma + t^{2\sigma+3} \\ -t & u \end{bmatrix}; u,t \in K \simeq GF(3^{2e-1}) \text{ and } \sigma = 3^3.$$

$$\begin{bmatrix} u & u^\sigma + t^{2\sigma+3} \\ -t & u \end{bmatrix}\begin{bmatrix} 0 & 1 \\ 1 & 0 \end{bmatrix}$$

$$= \begin{bmatrix} u^\sigma + t^{2\sigma+3} & u \\ u & -t \end{bmatrix} \text{ is the symplectic spread of symmetric matrices.}$$

This spread admits a collineation group EK of order $q(q-1)$, which is disjoint from the kernel homology group, of which there is a normal elation group of order q:

$$E = \left\langle \begin{bmatrix} 1 & 0 & b & b^\sigma \\ 0 & 0 & 0 & b \\ 0 & 0 & 1 & 0 \\ 0 & 0 & 0 & 1 \end{bmatrix}; b \in K \right\rangle,$$

$$K = \left\langle \begin{bmatrix} 1 & 0 & 0 & 0 \\ 0 & k^{\sigma+1} & 0 & b \\ 0 & 0 & k^{\sigma+2} & 0 \\ 0 & 0 & 0 & k^{2\sigma+3} \end{bmatrix}; k \in K^* \right\rangle.$$

(3) Take

$$\langle(0,0,0,1,0,0,0)\rangle = n$$

then $n^\perp$ is a $\Omega^+(6,q)$-space. The associated spread may be represented by the matrix spread set in $PG(3,q)$

$$x = 0, y = x \begin{bmatrix} u^2 t + t^\sigma - u^{\sigma+3} & -u^\sigma t^\sigma - ut^2 + u^{2\sigma+3} \\ u & y \end{bmatrix}; u,t \in GF(3^{2e-1}), \sigma = 3^e.$$

REMARK 31.6. All of the planes of the previous two theorems are collectively referred to as the 'slices of the unitary and Ree–Tits ovoids of Kantor'.

CHAPTER 32

Known Ovoids.

There are three types of quadrics, parabolic, hyperbolic and elliptic giving the spaces $\Omega(2n+1,q), \Omega^+(2n,q), \Omega^-(2n,q)$, respectively, and we have seen that ovoids exist in $\Omega^+(8,q)$-spaces and $\Omega(7,q)$-spaces. However, there are a variety of ovoids in the Klein quadric, which we shall list below. Furthermore ovoids in $PG(3,q)$ may be defined without associated quadrics.

DEFINITION 32.1. An ovoid in $PG(3,q)$ is simply a set of q^2+1 points no three of which are collinear.

REMARK 32.2. The only known ovoids in $PG(3,q)$ are elliptic quadrics and the Tits ovoid. For q even, ovoids in $PG(3,q)$ are equivalent to symplectic spreads by the work of Thas [1186], which are, in turn, equivalent to spreads consisting of symmetric matrices.

Of course, this means that the Lüneburg–Tits planes are symplectic.

For $q = 2^{2^{2r+1}}$, and $\sigma : x \longmapsto x^{2^{r+1}}$, the Tits ovoid in $PG(3,q)$ may be represented as

$$\left\{ (1, st + s^{\sigma+2} + t^\sigma, s, t); \; s, t \in GF(q) \right\} \cup \{(0,1,0,0)\}.$$

We have mentioned that ovoids of parabolic spaces $\Omega(5,q)$ produce symplectic spreads in $PG(3,q)$. Furthermore, commutative semifields transpose and dualize to symplectic spreads, which we shall discuss in Chapter 82. There we shall discuss the Thas–Payne and Penttila–Williams spreads, which are symplectic spreads and are the 5th cousins of the Cohen–Ganley and Bader–Lunardon–Pinneri commutative semifield flocks (although the Penttila–Williams spread came first). We have also seen that the Kantor–Knuth spreads and the plane arising from the Ree–Tits ovoid are also symplectic.

We briefly discuss ovoids in larger-dimensional spaces. This is only a glimpse and the reader should consult other works; for example, see Hirschfeld and Thas [529, pp. 345–348] and the references in that text.

However, ovoids in larger-dimensional spaces do not exist by work of Thas [1175] and Gunawardena and Moorhouse [423].

THEOREM 32.3. (Thas [1175] for q even, Gunawardena and Moorhouse [423] for q odd). There is no ovoid in $\Omega(2n+1,q)$, for $n \geq 4$.

No ovoids are known in $\Omega^+(2n,q)$, for $n > 4$ other than the Desarguesian ovoids for q even (see Thas [1175]). The ovoids in $\Omega^+(8,q)$, for q a non-prime prime power are the Desarguesian ovoids for q even, the Dye ovoid for $q = 8$, [332] the unitary ovoids and the Ree–Tits ovoids for $q = 3^r$.

Below, we shall discuss a construction of ovoids in $\Omega^+(8,p)$ by Conway, Kleidman and Wilson [242], and an extensive generalization by Moorhouse [990].

The ovoids in $\Omega^+(8, p)$, for p a prime, may be constructed by the method of Moorhouse (see below) and include the Conway, Kleidman, Wilson binary and ternary ovoids of the first and second kind, the Cooperstein ovoid [245], the Shult ovoid [1161] (which is a binary CKW-ovoid). The slicings of some of the Conway, Kleidman, and Wilson ovoids have been accomplished both by Moorhouse [990] and by Dempwolff and Guthmann [307]. Charnes [222] has sliced a binary ovoid for $p = 17$ and proved that the associated spread has a trivial collineation group. Various slicings of binary ovoids have also been obtained by Biliotti and Korchmáros [142].

32.1. Conway–Kleidman–Wilson; Moorhouse Ovoids.

In this section, we review the Conway–Kleidman–Wilson constructions and the extension by Moorhouse. The reader is directed to Chapter 102 for a short discussion on real orthogonal groups and lattices.

In this section, a quadric Q is defined generally over any field $GF(p)$, for p an arbitrary odd prime, constructing a variety of $\Omega^+(8, p)$-spaces.

Let E denote the E_8 root lattice; E is the collections of vectors of the 8-dimensional real vector space of the form

$$ E = \left\{ \frac{1}{2}(a_1, a_2, \ldots, a_8); \sum_{i=1}^{8} a_i \equiv 0 \bmod 4 \right\}. $$

Since the coefficients of vectors of E are integers, we may form the 8-dimensional $GF(p)$-subspace E^p by reducing the scalars modulo p, that is, in Z/pZ. Denote by $\overline{v}$ reduction modulo p of the scalars of the vector v. There is a natural quadric defined by E by the standard dot product of the real vector space $Q(x) = \frac{1}{2}x \cdot x$. If we take the scalars modulo p, then a quadric is inherited in E^p, say Q^p. This quadric makes E^p into a $\Omega^+(8, p)$-space.

Let p be an odd prime and let r be an arbitrary positive integer. Let r, p be distinct primes, let Z denote the ring of integers and define

$$ \mathcal{O}_{r,p}(x) = \mathcal{O}_{r,p}(Zx + rE) = \bigcup_{1 \le i \le [\frac{r}{2}]} \left\{ \langle \overline{v} \rangle \, ; \, \|v\|^2 = 2i(r - i)p; \, v \in Zx + rE \right\}. $$

THEOREM 32.4. (Moorhouse [991]).
(1) Either $\mathcal{O}_{r,p}(x)$ consists of the zero vector or is an ovoid in E^p as an $\Omega^+(8, p)$-space. If particular, if $r > p$, ovoids are always obtained.
(2) When $r \in \{2, 3\}$, these are the binary and ternary ovoids of Conway, Kleidman, and Wilson [242].

We note that the idea, more or less, is that the ovoids arise from sections of spheres in real vector space. Take $r = 2$ then these ovoids arise from spheres in real 8-dimensional space.

Charnes [222] investigates a plane obtained by slicing a binary ovoid which has as full collineation group the translations and kernel homologies (the spread has the trivial group as collineation group).

32.2. $SL(2, 9)$ and $G/N \simeq A_5$-Planes.

In [1055], Ostrom considers collineation groups G in the translation complement of a translation plane of order q^2, and spread in $PG(3, q)$, for $p^r = q$, p a prime, where $(|G|, p) = 1$. The two maximal groups occurring are as follows:

Type A : $\quad G \simeq SL(2,9)$, where $G/Z(G) \simeq A_6$,

Type B : $\quad G$ has a normal subgroup $Q \simeq D_8 * D_8$ (central product)
and $G/Q \simeq A_5$.

In the Type A situation, Mason [**971**] determines planes of order 49 with the property that the group G has a normal subgroup $D \simeq GF(7)$ such that G/D is either A_6 or S_6. Both cases occur and the corresponding planes are constructed. Also, Nakagawa [**996**] determined planes of orders $5^2, 7^2$ and 11^2 of Type A. Furthermore, Biliotti and Korchmáros [**142**] using the computer determined all examples of prime square order p^2 for $p \leq 19$. For example, these latter planes would then include the Hering plane of order 25, the Mason planes of order 7^2 and the planes of Nakagawa of order 11^2. There is also an example of an $SL(2,9)$-plane of order 7^2 due to Korchmáros found by determining a Bruen chain which appears in the list of planes.

Concerning Type B planes, Mason and Shult [**974**] give a construction using the Klein quadric of planes of prime square order p^2. S. Caboniss, by computer, found all examples up to $p = 47$ and all such planes are determined. Mason and Ostrom [**973**] determined Type B planes of orders 3^2 (the nearfield plane), 5^2 (one of the Walker planes), a plane of order 7^2 and four planes of order 11^2. Actually, the plane of order 7^2 is derivable and the three of the four planes of order 11^2 are derivable.

Jha and Johnson [**644**] have shown that to decide if a translation plane of order q^2 and spread in $PG(3, q)$ is subregular, it is sufficient to have a collineation group G in the translation complement which is non-solvable and contains a cyclic group of order $q^2 - 1$ in its center. Charles and Dempwolff [**225**] and Dempwolff [**306**] use the Klein quadric to determine planes of Type B for $q^2 \equiv -1 \mod 5$ and $q^2 \equiv 1 \mod 5$, respectively. All of these planes admit a cyclic group of order $q^2 - 1$ in the center of a non-solvable group G and hence are subregular planes. The constructions of these planes are given in Chapter 24 on subregular planes.

Concerning subregular planes, we have mentioned above a variety of subregular planes of order q^2 admitting $SL(2,5)$; the Prohaska, using Miquellian inversive planes; Abatangelo–Larato using more or less standard net replacement; the Bonisoli–Korchmáros, Szönyi Planes. Concerning planes of order 81, the combined work of Johnson and Prince [**807**] and Jha and Johnson [**637**] determine all translation planes of order 81 admitting $SL(2,5)$.

The binary ovoids of Conway, Kleidman and Wilson have been sliced in Moorhouse [**992**] and also by Dempwolff and Guthmann [**307**] and several of these constructions produce translation planes with the following properties:

Type A: There is a collineation in the translation complement isomorphic to $SL(2,9)$.

Type B: There is a collineation in the translation complement G containing a normal subgroup N isomorphic to $D_8 * Q_8$ (central product of a dihedral and quaternion groups of orders 8), such that $G/N \simeq A_5$.

32.3. Slices of the Conway–Kleidman–Wilson Binary Ovoids.

Both Moorhouse [**992**] and Dempwolff and Guthmann [**307**] have sliced the binary ovoids of Conway Kleidman and Wilson. The details proving that the following constructions arise are found in Moorhouse [**992**]. We have mentioned previously that Charnes [**222**] has sliced the binary ovoid for $p = 17$.

Here we indicate three constructions of Moorhouse.

(1) Type I. First assume that $p \equiv 1 \bmod 4$ is prime and list all integer solutions of

$$x_1^2 + x_2^2 + \cdots + x_6^2 = 6p, \ x_i \equiv 1 \bmod 4.$$

There are always $p^2 + 1$ solutions and the corresponding points

$$(x_1, x_2, \ldots, x_6)$$

taken homogeneously form an ovoid O_I in $PG(5, p)$ with respect to the standard quadratic form (if $p \equiv -1 \bmod 4$, the standard quadratic form is elliptic rather than hyperbolic and the associated ovoid does not produce a spread for a translation plane). The ovoid is invariant under a group isomorphic to $Z_2 \times S_6$. Furthermore, S_6 will permute the associated spread obtained using the Klein correspondence. The translation plane of order p^2 will admit $SL(2, 9) \langle \sigma \rangle$, where σ is the Frobenius automorphism of $GF(9)$.

   The translation planes are also self-transpose as each is invariant under a correlation of the projective space $PG(3, q)$.

(2) Type II. Again let $p \equiv 1 \bmod 4$ be prime and list all integer solutions of

$$x_1^2 + x_2^2 + \cdots + x_6^2 = p,$$
$$x_1 + 1 \equiv x_2 \equiv x_3 \equiv \cdots \equiv x_6 \bmod 2,$$
$$\sum x_i \equiv 3 \bmod 4.$$

Similarly, there is an associated ovoid O_{II} in $PG(5, p)$ with respect to the standard quadratic from. There is a group of the ovoid $2 \times 2^4 S_5$, inducing a collineation group of the associated translation plane isomorphic to $2^4 S_5$. The associated planes are self-transpose.

(3) Type III. Let $p \equiv 1 \bmod 3$ be prime. The root lattice of type E_6 is taken as

$$L = \left\{ x = (x_1, x_2, \ldots, x_6); \ x_i \text{ an integer, } \sum x_i \equiv 0 \bmod 3 \right\},$$

using the quadratic form

$$Q(x) = \sum x_i^2 - \frac{1}{9} \left(\sum x_i \right)^2.$$

Let $e = (1^6) \in L$. Consider the vectors $v \in e + 2L$ such that $Q(v) = 2p$, omitting $-v$ if v has already been listed. Reducing modulo p provides an ovoid denoted by O_{III}. The ovoid is invariant under a group isomorphic to $2 \times S_6$ producing self-transpose translation planes admitting $SL(2, 9)$ in the translation complement.

CHAPTER 33

Simple T-Extensions of Derivable Nets.

The aim of this chapter is to consider various methods for generating finite spreads π, and hence also translation planes. The distinguishing feature of these methods is that they each involve a partial spread set $\mathcal{F}$ associated with a rational Desarguesian partial spread and another slope matrix 'T': the spread π is then 'generated', in some case-dependent sense, by $\{T\} \cup \mathcal{F}$.

The exact conditions for T to succeed depend on the individual case, but, in each instance, a wide range of planes can be constructed, in the sense that the dimensions over the kernel can be almost arbitrary.

We shall be discussing the construction methods of 'T-extension/transpose' and 'T-distortion', where the underlying Desarguesian partial spread is a derivable partial spread. Using these techniques, we shall be able to construct all semifields which are two-dimensional over one of their nuclei.

Before describing the methods, we need to take a closer look at spread sets containing fields.

33.1. Spread Sets Containing Fields.

Let $\mathcal{S}$ be a finite spread set, and suppose that $\mathcal{F} \subset \mathcal{S}$, $|\mathcal{F}| > 1$. Hence, $\mathcal{F}$ is a field of linear mappings if and only if it is additively and multiplicatively closed. We examine separately the meaning of additive and multiplicative closure of $\mathcal{F}$ using:

HYPOTHESIS 33.1. Let $\mathcal{S}$ be a spread set associated with the additive group of a *finite* vector space V. Assume that $\mathcal{S}$ is coordinatized by any one of the pre-quasifields $Q_e = (V, +, \circ)$, with $\circ$ specified by choosing the identity $e \in V^*$. Let $\mathcal{F} \neq \{\mathbf{O}\}$ be any non-empty subset of $\mathcal{S}$, and let $F \subset V$ be the set of all elements in V whose slope mappings lie in $\mathcal{F}$ relative to the choice of e as the identity, thus:

$$F := \{\, f \in V \mid f = (e)\phi \,\exists\, \phi \in \mathcal{F} \,\}, \qquad \mathcal{F} = \{\, T_f \mid f \in F \,\},$$

where $T_x \in \mathcal{S}$ denotes the slope mapping of $x \in V$, relative to e as the identity, i.e., $T_x \colon y \mapsto y \circ x$, $y \in V$.

First we consider the additive closure of $\mathcal{F}$.

PROPOSITION 33.2. Assume Hypothesis 33.1, in particular, $F = (e)\mathcal{F} \subseteq V$. Then the following are equivalent:
(1) $\forall x \in V$, $f, g \in F : x \circ (f + g) = x \circ f + x \circ g$.
(2) $\mathcal{F}$ is an additive group.
(3) $\mathcal{F}$ is additively closed.

Now we consider the multiplicative closure of $\mathcal{F}$.

PROPOSITION 33.3. Assume Hypothesis 33.1, in particular, $F = (e)\mathcal{F} \subseteq V$. Then the following are equivalent:

249

(1) $\forall x \in V, f, g \in F : x \circ (f \circ g) = (x \circ f) \circ g$.
(2) $\mathcal{F}$ is a multiplicative group.
(3) $\mathcal{F}$ is multiplicatively closed.

Now consider any quasifield $Q = (V, +, \circ)$ such that a subset $F \subset V$ is a field relative to the quasifield operations and that, for $x \in Q$, the following identities hold:

$$x \circ (f + g) = x \circ f + x \circ g, \quad (x \circ f) \circ g = x \circ (f \circ g).$$

On comparing these conditions with Propositions 33.2 and 33.3, we immediately deduce:

PROPOSITION 33.4. Let $\mathcal{S}$ be any finite spread set containing the identity mapping and associated with the additive group $(V, +)$ of some vector space; so $Q_e = (V, +, \circ)$ denotes the quasifield determined by $\mathcal{S}$ and $e \in V^*$. Assign to any $\{\mathbf{O}\} \subset \mathcal{F} \subseteq \mathcal{S}$ the set of images F of e under $\mathcal{F}$, thus:

$$F := \{ f \in V \mid f = (e)\phi, \ \phi \in \mathcal{F} \}.$$

Then the following are equivalent:
(1) $\mathcal{F}$ is a field of linear mappings.
(2) $\mathcal{F}$ is closed under addition and composition.
(3) For some non-zero e: F is a field and Q_e is a vector space over F.
(4) For all non-zero e: F is a field and Q_e is a vector space over F.

Suppose that $Q_e = (V, +, \circ)$ is a finite quasifield, with identity e, such that Q_e is a vector space over a subfield $F = (U, +, \circ)$, for some additive group $(U, +) \leq (V, +)$. Now, to $(V, +)$ may be assigned the structure of a field $K = (V, +, \bullet)$, such that: $\forall v \in V, f \in F : v \circ f = v \bullet f$.

The proof is an exercise in linear-algebra/field-extensions: if V is a k-dimensional vector space over a field $F = GF(q)$, then V can be represented as an F-linear vector space K, where K is a k-dimensional field extension of the field F. For example, view F as the field of scalar $k \times k$ matrices in $\mathrm{Hom}(k, q)$, and then choose as K a field of matrices of order $|F|^n$; this field exists in $\mathrm{Hom}(k, q)$ by Galois theory.

Hence, $y = x \circ f$ and $y = x \bullet f$ define the same subspace of $V \oplus V$, for all $f \in F$. Therefore, all these subspaces are components shared by the spreads $\pi(Q_e)$ and $\pi(K)$, and this clearly means that the rational partial spread associated with $\pi(F)$ is a subpartial spread of both $\pi(Q_e)$ and $\pi(K)$, and since the latter is Desarguesian, we conclude that $\pi(F)$ determines a *rational Desarguesian* partial spread.

We now consider the converse of this assertion. Hence, our goal is to demonstrate that if $\pi(Q_e)$, the spread associated with a finite quasifield $Q_e = (V, +, \circ)$, contains a rational Desarguesian partial spread δ whose components include X, Y and I, then Q_e contains a subfield F such that $(V, +)$ is a vector space over F and the set of components of δ is the partial spread determined by $\pi(F)$, or equivalently, $\pi(F)$ is a spread across δ.

Since δ is Desarguesian and rational, there is a Desarguesian spread $\Delta = \pi(K)$, where $K = (V, +, \bullet)$ is a field that may be chosen so that it contains a subfield F such that $\pi(F)$ is across δ and contains (e, e). It is possible to insist further that e, the identity of $Q_e = (V, +, \circ)$, is also the identity of K, and hence of F. To see this, use the spread set associated with K—it clearly contains the spread set associated with δ—to define $\bullet$ in terms of e.

Since δ is the rational partial spread determined by $\pi(F)$, and lies in both $\pi(K)$ and $\pi(Q_e)$, we have that the subspace $y = x \circ f$, for $f \in F$, may be expressed as $y = x \bullet f'$ for some $f' \in F$, and vice versa. Choosing $x = e$ shows that in every case $f = f'$, since K and Q_e both have the same multiplicative identity e. Thus, we have the identity: $\forall\, x \in V,\ f \in F : x \circ f = x \bullet f$.

Hence, Q_e is a vector space over F, because K has this property. So, we have established that δ is the rational partial spread determine' 'ny $\pi(F)$, where F is a field in Q_e such that the latter is a vector space over F.

Therefore, we have shown that if a finite quasifield Q is a vector space over a field F then $\pi(Q)$, the rational spread determined by $\pi(F)$, is a rational *Desarguesian* partial spread whose components include the standard components X, Y and I of $\pi(Q)$, and, conversely, a rational Desarguesian partial spread δ in $\pi(Q)$ that includes the standard components among its members must be determined by some $\pi(F)$, where F is a subfield of Q over which the latter is an F-vector space. Thus, the above theorem extends to include another equivalence: the statement '$\pi(F)$ determines a rational Desarguesian spread' is equivalent to all the other parts of the theorem.

In the context of finite spread sets $\mathcal{S} \supset I$, associated with a vector space on $(V, +)$, the above has the following interpretation: $\mathcal{F} \subset \mathcal{S}$ is a field of matrices if and only if the components associated with $\mathcal{F}$ in $\pi(Q_e)$ defines a rational Desarguesian partial spread that contains the three standard components X, Y and I of $\pi(Q_e)$.

Thus Proposition 33.3 may be restated in more detail as follows:

THEOREM 33.5. Assume the hypothesis of Proposition 33.3. Let $\mathcal{S}$ be any finite spread set containing the identity mapping and associated with the additive group $(V, +)$ of some vector space; so $Q_e = (V, +, \circ)$ denotes the quasifield determined by $\mathcal{S}$ and $e \in V^*$. Suppose that $\mathcal{F} \subseteq \mathcal{S}$ and let $F = \{\, f \in V : f = (e)\phi,\ \phi \in \mathcal{F}\,\}$. Then the following are equivalent:

(1) $\mathcal{F}$ is closed under addition and composition.

(2) $\mathcal{F}$ is a field of linear mappings.

(3) $(F, +, \circ)$ is a field and V is a vector space over F, for some choice of $e \in V^*$.

(4) $(F, +, \circ)$ is a field and V is a vector space over F, for all choice of $e \in V^*$.

(5) The partial spread $\pi(F)$ in $\pi(\mathcal{S})$ is a rational Desarguesian partial spread. That is, $\pi(Q_e)$ determines a rational Desarguesian partial spread in $\pi(\mathcal{S})$ that includes its three standard components, X, Y and I.

Note that in attempting to state the infinite analogue of the theorem above, care must be taken regarding two points: (1) multiplicative and additive closure will no longer force $\mathcal{F}$ to be a field, and (2) the field F may not be embeddable in a larger field of dimension k, where $k := \dim_F V$.

33.2. *T*-Extension.

If $\mathcal{S}$ is a finite spread set, in some $\overline{GL(V, +)}$, that contains a field $\mathcal{F}$ then the associated spread $\pi_{\mathcal{S}}$ contains the rational Desarguesian partial spread $\pi_{\mathcal{F}}$. In this section, we consider some ways of extending a field of matrices $\mathcal{F}$ to a spread $\mathcal{S}$ so that the latter is, in some sense, 'generated' by $\mathcal{F} \cup \{T\}$, where T is suitably chosen in $GL(V, +) \setminus \{\mathcal{F}\}$. These procedures will yield classes of semifields, and also spreads of order q^3 admitting $GL(2, q)$.

The first method is based on having available a quasifield $Q = (V, +, \circ)$, of square order, that contains a subfield F, such that Q is a two-dimensional vector space over F. Since such situations arise if and only if the spread $\pi(Q)$ is derivable relative to the slopes of $\pi(F)$, we shall refer to the corresponding spreads as being obtained by 'T-extension' or 'T-transpose-extension'. This method yields a range of semifields that are two-dimensional over at least one of their seminuclei, and, in a somewhat vacuous sense yields them 'all': every such semifield 'yields itself' by the procedure to be described. However, the method is also effective in genuinely constructing long chains of two-dimensional semifields when used sensibly.

We also discuss a method of construction somewhat in-between the previous two, that of 'distortion' which constructs the derived versions of one of the previous.

We direct the reader to Chapter 13 for a discussion on a method which is concerned with 'cyclic T-extensions' of a field $\mathcal{F}$ that also yields semifields of non-square order, but this time the field $\mathcal{F}$ lies in at least two seminuclei: N_m and N_r, but these can be changed by dualizing and/or transposing. Thus, neither of the two constructions indicated so far entirely replace the other.

Also in Chapter 13, we discuss a modification of the above indicated method in the three-dimensional case. This yields spreads (*not* semifield spreads) of order q^3 that admit $GL(2, q)$, acting as it does on the Desarguesian spread of order q^3. The dimension of the spread over its kernel can be made arbitrarily large, demonstrating that non-solvable groups can act on spreads of arbitrarily large dimensions: so far this phenomenon is known in surprisingly few cases.

We now describe each of the above indicated constructions.

33.3. T-Extension/Transpose.

We describe here a method of constructing semifields of order q^2 that have $GF(q)$ over their middle nucleus. By transposing and/or dualizing the resultant semifield plane, the $GF(q)$ can be taken to be any of the three seminuclei. Hence, we focus on the middle- or right-nucleus case (as the various treatments are almost identical) and we shall generally ignore the right nucleus (which involves dualizing the nucleus).

Basically, the method begins with a quasifield $Q = (V, +, \circ)$, of arbitrary order q^2, that contains a subfield $(F, +, \circ) \cong GF(q)$ such that $(V, +)$ is a vector space over F. Such quasifields, as we saw earlier, are essentially those obtainable from spread sets $\mathcal{S}$ on $(V, +)$ that contain a subfield $\mathcal{F}$, or equivalently, from spreads of order q^2 that contain rational Desarguesian partial spreads of degree $q + 1$.

The key idea is that for any choice of $T \in \mathcal{S} \setminus \mathcal{F}$, regardless of the Q yielding $\mathcal{S}$, the additive group $\mathcal{F} + \mathcal{F}T$ is an additive spread set. We shall refer to spreads constructed in this manner, as arising by applying a T-*extension* to $\mathcal{S}$:

PROPOSITION 33.6. Let $\mathcal{S}$ be a spread set (or even a partial spread set!) on a finite additive group $(V, +)$ such that $\mathcal{S} \supset \mathcal{F}$, where $\mathcal{F}$ is a field $\cong GF(q)$, and V has order q^2. Then, for any $T \in \mathcal{S} - \mathcal{F}$, the additive set of matrices

$$\Theta_R := \tau(T, \mathcal{F}) = \{\, a + Tb \mid a, b \in \mathcal{F} \,\}$$

and

$$\Theta_L := \tau(T, \mathcal{F}) = \{\, a + bT \mid a, b \in \mathcal{F} \,\}$$

are spread sets, and, hence, so are their transposes:

$$\Theta_R^T = \{\, a + bT^T \mid a, b \in \mathcal{F}^T \,\}$$

and

$$\Theta_L^T = \{\, a + T^T b \mid a, b \in \mathcal{F}^T \,\}.$$

In particular, $\Theta_R \mathcal{F} = \Theta_R \mathcal{F} \Theta_L = \Theta_L$, $\mathcal{F}^T \Theta_R^T = \Theta_R^T$ and $\Theta_L^T \mathcal{F}^T = \Theta_L^T$.

Note that by allowing $\mathcal{S}$ to be a partial spread, the method can be extended even to cartesian groups $Q = (V, +, \circ)$ of order q^2 that are vector spaces over a subfield $F = GF(q)$, provided that some $t \in Q - F$ defines an additive mapping $x \mapsto x \circ t$ on $(V, +)$.

Recall that we have shown (see Section 5.9 and Section 12.1) that, for additive spread sets $\mathcal{S}$, the middle nucleus corresponds to the largest subset $\mathcal{F}$ such that $\mathcal{F}\mathcal{S} = \mathcal{S}$, and the right nucleus corresponds to the transpose situation, viz., the largest $\mathcal{F}$ such that $\mathcal{S} = \mathcal{S}\mathcal{F}$. Hence, for convenience and for its future role, we shall usually only comment on the middle-nucleus situation. We note that any semifield spread set of order q^2 is obtained by applying a T-extension to itself.

If $\mathcal{T}$ is a spread set of order q^2 containing a field $\mathcal{F} \cong GF(q)$ such that $\mathcal{F}\mathcal{T} \subset \mathcal{T}$ then $\mathcal{T} = \mathcal{F} + \mathcal{F}T$, whenever $T \in \mathcal{T} \setminus \mathcal{F}$; in particular, $\mathcal{T}$ coincides with $\tau(T, \mathcal{F})^T$, using the notation of Proposition 33.6.

33.4. *T*-Distortion.

We have previously mentioned and constructed the translation planes of M. Hall who 'distorted' the multiplication of a finite field to construct a quasifield defining the Hall planes. Recall that, it turns out that the Hall planes are those which may be derived from a Desarguesian plane.

The generalization, 'T-distortion', was first considered by Johnson in [**679**]. Here we follow this method to construct quasifields from any finite derivable quasifield. It will turn out that the quasifields constructed are the 'generalized Hall quasifields' coordinatizing the tangentially transitive translation planes of finite order.

The idea of the construction is phrased most conveniently when the derivable net is contained in a translation plane although this is not actually required.

We have previously discussed the extension of a derivable net written on the right or on the left of a particular mapping producing semifield planes of order q^2 with right or middle nucleus isomorphic to $GF(q)$. Hence, the following construction might merely start with a net of degree $q + 2$ and order q^2 which contains a derivable subnet of degree $q + 1$ which then produces a semifield with the 'right-nucleus property'.

Let $(Q, +, \cdot)$ be a semifield (or quasifield) of order q^2 which contains a field K isomorphic to $GF(q)$ such that Q is a right 2-dimensional vector space over K (again, it is pointed out that we may merely start with an 'extra' component as opposed to the semifield itself). Then the net defined by $x = 0, y = x \cdot \alpha$ for all $\alpha \in K$ is a derivable net. Moreover, let $\{1, t\}$ be a basis for Q over K and write vectors in the form $t \cdot \alpha + \beta$ for $\alpha, \beta \in K$. Then we consider the partial multiplication: $(t \cdot \alpha + \beta) \cdot t = t \cdot f(\alpha, \beta) + g(\alpha, \beta)$, for all $\alpha, \beta \in K$ where f and g are functions from $K \times K$ into K.

DEFINITION 33.7. We define a multiplication $*$ as follows:
$$w * \alpha = w \cdot \alpha \ \forall w \in Q \ \forall \alpha \in K,$$
$$(z \cdot \alpha + \beta) * z = z \cdot f(\alpha, \beta) + g(\alpha, \beta)$$
$$\forall z \in Q - K, \ \forall \alpha, \beta \in K.$$

The structure $(Q, +, *)$ is denoted by $Q_t^{\mathcal{D}}$ and is called the 't-distortion of Q'.

33.4.1. The 'New' Multiplication.

THEOREM 33.8. With distortion, we obtain a new general multiplication which is given as follows:
$$(t\alpha + \beta)(t\delta + \gamma)$$
$$= t\delta f(\delta^{-1}\alpha, \beta - \delta^{-1}\alpha\gamma) + g(\delta^{-1}\alpha, \beta - \delta^{-1}\alpha\gamma) + \gamma f(\delta^{-1}\alpha, \beta - \delta^{-1}\alpha\gamma)$$
when δ is not zero and inherited multiplication when δ is 0.

The t-distortion is a derivable quasifield which derives a semifield of order q^2 with middle nucleus $GF(q)$.

Hence, the t-distortion is a generalized Hall quasifield and coordinatizes a tangentially transitive translation plane.

One of the more interesting features of the previous multiplication is that we are 'forcing' a set of $q(q-1)$ automorphisms of a quasifield.

To see this, we begin again with the partial multiplication:
$$(t\alpha + \beta) \cdot t = tf(\alpha, \beta) + g(\alpha, \beta) \forall \alpha, \beta \in GF(q).$$

Demand that we create a coordinate structure such that the following mappings define automorphisms fixing $GF(q)$ pointwise:
$$\sigma_{\delta,\gamma} : t \longmapsto t\delta + \gamma \ \forall \delta \neq 0, \gamma \in GF(q).$$

Then we would obtain:
$$((t\delta + \gamma)\alpha + \beta) \cdot (t\delta + \gamma) = (t\delta + \gamma)f(\alpha, \beta) + g(\alpha, \beta)$$
$$\forall \alpha, \beta \in GF(q)$$
which is exactly our defined multiplication.

That is, we start with a partial multiplication, enforce an automorphism group and see that, using our extension procedures, the new system is, in fact, a quasifield—nothing short of amazing.

33.5. Construction of 2-Dimensional Semifields.

We see that all semifields that are two-dimensional over their middle (or left) nucleus may be realized as T-extensions of derivable nets either by multiplication on the 'right' or of T by elements from a field. Hence, there is an intrinsic connection between such semifields.

By transposition, we may construct the set of semifields with the middle-nucleus property from the set of semifields with the right-nucleus property. However, transposition of a semifield with large kernel can be difficult since we need to write the coordinate structure in matrix form.

On the other hand, the distortion process allows an indirect construction of the middle-nucleus semifields from the right-nucleus semifields by first distorting and then deriving. That is, we have algebraically bypassed the transposition process.

What we have not done is to show that the distorted-derived-transposed semifield, which is then a right-nucleus semifield, is the original semifield.

However, the process of T-extensions can be effectively used to yield a variety of examples of semifields that are two-dimensional over the middle nucleus, and indeed, by transposing and dualizing, any semifield over a seminucleus. To generate such examples, using T-extensions, one can arbitrarily repeat long chains of steps, each step involving one of dualizing-transposing-T-deriving-recoordinatizing and collecting the required spread sets at each stage, for example by adopting the use of a loop such as the following:

Generating Two-Dimensional Semifields.

(1)

(b) Choose a spread with derivable partial spread δ.

(c) Coordinatize by a quasifield Q so that δ is coordinatized by a field F.

(d) Now either form Q' containing field F' such that Q' coordinatizes the transpose spread and Q' is a vector space over F', a field isomorphic to F, or simply choose $Q' = Q$ and $F' = F$.

(e) Obtain a two-dimensional semifield associated with any $t \in Q' - F'$, with middle nucleus F'.

(f) Dualize and/or transpose the semifield and/or derive relative to F'-slopes.

(g) Return to step (b) or stop.

Certainly many non-isomorphic spreads arise in this way, and as indicated above, 'all' finite semifields that are two-dimensional over a seminucleus are of this form, albeit in a somewhat vacuous sense; although T-extensions provide a useful method for generating examples of two-dimensional semifields it is not meaningful to ask if there are 'other' semifields of order q^2, with $GF(q)$ in seminucleus.

CHAPTER 34

Baer Groups on Parabolic Spreads.

This chapter is a prelude to our study of algebraic lifting, Chapter 35. Algebraic lifting is a powerful recursive technique that converts ('lifts') line spreads in $PG(3, q)$ to line spreads in $PG(3, q^2)$. We note that the term 'lifting' is used in this text three ways. In this chapter, 'lifting' shall refer to 'algebraic lifting'. Actually, lifting is essentially an algebraic process that converts spread sets of 2×2 matrices over $GF(q)$ to spread sets of 2×2 matrix over $GF(q^2)$ by a purely algebraic procedure, and by recoordinatizing the same $GF(q)$-line-spread over by different spread sets one expects to lift to a range of possibly non-isomorphic line spreads over $GF(q^2)$. Hence, it becomes desirable to characterize lifted spread sets geometrically.

In this chapter, we characterize geometrically a class of 2×2 spread sets over $GF(q)$ which specializes to yield a classification of the 'standard' spread sets that coordinatize the line spreads in $PG(3, q^2)$ that are obtained by lifting line spreads in $PG(3, q)$, Corollary 34.13.

Actually, we proceed from the opposite point of view: we begin with a class of geometrically defined line spreads and then characterize them by computing a simple form for their associated spread sets. Later, we shall see that the spread sets thus obtained include all the (finite) lifted spreads among them.

Specifically, we define a line spread π in $PG(3, q)$ to be E-**parabolic** if E is an affine elation group of π whose orbits together with its axis define a derivable net (but usually *not* a regulus net). We characterize such 'E-parabolic' spreads π when they admit a linear Baer group B of order > 2 that normalizes but does not centralize E, by describing their (essentially) canonical spread sets.

34.1. Derivable Line Spreads.

In this section, we establish Johnson's fundamental characterization of derivation in the context of line spreads. The principal objective is to show that if any line spread π in $PG(3, q)$ is derivable relative to a derivation net D then D has at least two transversal lines across it, and hence the spread π has matrix representations such that the non-vertical components of D are associated with a diagonal field of matrices $\text{Diag}(k, k^\sigma)$, for $k \in GF(q)$, $\sigma \in \text{Gal}(GF(q))$. This yields some useful standard bases of derivable line spreads π in $PG(3, q)$.

Consider a right quasifield $Q = (V, +, \circ)$, so that Q is a vector space over the kernel $K \cong GF(q)$ operating from the right, and suppose that $F \cong GF(q)$ is a subfield of Q such that V is a vector space over F operating from the right. Thus, $\pi(Q)$ contains $\pi(F)$ as a subplane and the set of slopes of $\pi(F)$ define a rational Desarguesian subnet of π. Moreover, K^* acts as an F-linear group of V regarded as an F-space acting from the right. Hence, the group $K^* \in GL(V, F)$ permutes the dimension-one F spaces among themselves. The following lemma examines the case when V is dimension-two over F and K. Thus, the associated spread π is in

$PG(3, q)$ and contains a derivation net D coordinatized by F. This implies that the derivation net D includes at least two kern planes, and if more than two are included, then D is actually a regulus of the spread.

LEMMA 34.1. Let $Q = (V, +, \circ)$ be a quasifield with a kernel field $K \cong GF(q)$ (hence V is a vector space with K operating from the right) and let $F \cong GF(q)$ be a subfield of Q such that V is a vector space with F operating from the right. Then K leaves invariant at least two distinct rank-one F-subspace of V.

Hence, the following hold:

(1) If the basis of V is chosen to be more from K-invariant subspaces of V, the group K may be represented by the field of matrices: $\{ \operatorname{Diag}(k, k^\sigma) : k \in K \}$, $\sigma \in \operatorname{Gal}(K)$.

(2) If K^* leaves invariant more than two F-subspaces, then K fixes all F-spaces and commutes multiplicatively with Q, and $F = K$.

In the situation of the lemma, let $\pi_t := \{ (t \circ f_1, t \circ f_2) : f_1, f_2 \in F \}$, $\forall t \in V$; so these are the set of subplanes across the derivation net defined by $\pi(F)$. The action of the kernel homologies, associated with $k \in K^*$ maps π_t to π_{kot}. The lemma implies that at least one π_t is invariant under all $k \in K$, viz., the t such that the F-subspace $t \circ F$ is K-invariant, and hence a K-space. Since the spread under consideration is in $PG(3, K)$ we have:[1]

COROLLARY 34.2. A derivable net in a spread $\pi = PG(3, K)$, K a field, contains a transversal line of $PG(3, K)$ across it.

In the case when K is finite, the fact that K^* leaves invariant one dimension-one F-space $t \circ F$ on the unit line $y = x$ forces it to leave another, viz., its Maschke K^*-complement. These notions readily yield Johnson's characterization of derivable line spreads in the finite case:

THEOREM 34.3. A derivable net D in a spread $\pi = PG(3, K)$, $K \cong GF(q)$, contains at least two transversal lines of $PG(3, K)$ across it. In particular, the slope mappings of the kernel field $GF(q)$ may be identified with a field of matrices

$$\left\{ \begin{bmatrix} k & 0 \\ 0 & k^\sigma \end{bmatrix} : k \in K \right\},$$

for some $\sigma \in \operatorname{Gal}(K)$.

The following propositions are minor variants of the above; they draw attention, from slightly different points of view, to the important 'natural' basis of derivable line spreads π in $PG(3, q)$ that enable any derivation net $\Delta \subset \pi$ to be assigned a diagonal field of matrices as its slope set. These natural bases are determined by any three components of Δ and their intersections with any two kernel planes π_1 and π_2 across Δ. In the first proposition, we introduce these bases using a quasifield approach, and the next proposition obtains the same result by adopting a 'basis-change' attitude.

PROPOSITION 34.4. Let $\pi = (V, \tau)$ be a spread of order q^2 with $GF(q)$ in its kernel, i.e., π corresponds to a line spread in $PG(3, q)$. Suppose that $\Delta \subset \pi$ is a

---

[1]The nets $\pi(F)$ in $\pi(K)$ are derivation nets since they are Desarguesian of the correct 'size', and the converse holds by Johnson's classification of infinite derivable nets.

derivation net, thus Δ has at least two kernel planes, π_1 and π_2, that lie across it, Theorem 34.3.

Coordinatize π by a quasifield Q, using the notation $K = GF(q)$ for its kernel subfield (of order q), with identity e, based on choosing $X, Y \in \Delta$, and the unit point $(e, e) \in \pi_1$. Then the slope mappings $T_m : x \mapsto x \circ m$ for $m \in Q$ relative to any K-basis (e, f) of Q with $(f, f) \in \pi_2$ have matrices of the form: $\mathrm{Diag}(k, k^\sigma)$, for $k \in K$, where $\sigma \in \mathrm{Gal}(K)$; thus Δ is a regulus if and only if σ is the identity.

Hence, on the basis $[(e, 0), (f, 0), (0, e), (0, f)]$ of π, the representation of the spread π on $GF(q)^4$ is such that the non-vertical members of Δ consist of the set of subspaces $\{ y = x \, \mathrm{Diag}(k, k^\sigma) : k \in K \}$, for some $\sigma \in \mathrm{Gal}(K)$.

The following proposition is a minor variant of Proposition 34.4, without reference to quasifields; it is essentially the same result presented from the point of view of basis change rather than coordinatization. Thus, it asserts that, by an appropriate change of basis of a derivable spread π in $PG(3, q)$, any chosen derivation net $\Delta \subset \pi$ may be identified with a derivable spread set coordinatized by a field of matrices of the form $\{ \mathrm{Diag}(k, k^\sigma) : k \in K \}$, for $\sigma \in \mathrm{Gal}(K)$. The natural choice of the basis, is determined by three components of the derivation net and two chosen kernel planes across it; these bases will play an important role in characterizing the line spreads obtained by the process of 'lifting'.

PROPOSITION 34.5. (Natural Basis for Derivable Line Spread). Let $\pi = (V, \tau)$ be a spread of order q^2 with kernel containing $K = GF(q)$. Suppose $\Delta \subset \pi$ is a derivation net; so it has at least two distinct kernel planes π_1 and π_2 across it, Theorem 34.3. Let $X, Y, Z \in \pi$ be any three distinct components of Δ.

Then we may choose a K-basis (e_x, e_y) and (f_x, f_y) of the kernel planes π_1 and π_2, respectively, such that (e_x, f_x), (e_y, f_y) and $(e_x + e_y, f_x + f_y)$ are, respectively, bases of X, Y and Z. Thus, the basis (e_x, f_x, e_y, f_y) defines a linear bijection from the ambient K-space V of π to $K^2 \oplus K^2$, and hence this linear bijection defines an isomorphism from π onto a spread $\nabla(\mathcal{M})$, where $\mathcal{M}$ is a spread set of 2×2 matrices over K. This isomorphism, that is a coordinatization of π by $\mathcal{M}$, is such that it induces a coordinatization of Δ by a field of diagonal matrices

$$\mathcal{D} = \{ \mathrm{Diag}[k, k^\sigma] \mid k \in K \}, \text{ for some } \sigma \in \mathrm{Gal}(K).$$

In particular, the coordinatizing isomorphism from π to the spread $\nabla(\mathcal{M})$, on $K^2 \oplus K^2$, maps Y to $\mathbf{0} \oplus K^2$ and the other components of Δ onto the set of components of type $\{ y = xD : D \in \mathcal{D} \}$. Moreover, the kernel subplanes π_1 and π_2 map, respectively, to (i.e., get coordinatized by) $\{ (k_1, 0, k_2, 0) \mid k_1, k_2 \in K \}$ and $\{ (0, k_1, 0, k_2) \mid k_1, k_2 \in K \}$, respectively.

REMARK 34.6. The basis of type (e_x, e_y, f_x, f_y) as described above is the **natural basis** of π relative to the kernel planes π_1 and π_2 across the selected derivation net $\Delta \subset \pi$.

The partial spread Δ above, with slope set defined by $\mathrm{Diag}(k, k^\sigma)$, for all $k \in K$, is called a 'σ-regulus' ; thus a σ-regulus is a regulus if and only if σ is the identity automorphism. Although this definition is in general unstable, in the sense that $\sigma \in \mathrm{Gal}(K)$ might vary with the choice of the coordinatization, it is nonetheless useful.

DEFINITION 34.7 (σ-regulus). Let π be a line spread in $PG(3, K)$, K a field, that contains a derivable net Δ. Then Δ is a 'σ-regulus' , if for some $\sigma \in \mathrm{Gal}(K)$, if π may be coordinatized by a spread set of 2×2 matrices over K such that Δ is coordinatized by a field of matrices of the form $\{\operatorname{Diag}(k, k^\sigma) : k \in K\}$.

As a digression, we sketch the possibilities for (non-identity) σ above, *when Δ is held fixed*. By Johnson's result in the finite case, Theorem 34.3, every derivable net Δ in a line spread π is a σ-regulus for some $\sigma \in \operatorname{Gal}(K)$. However, recoordinatizing π, by interchanging the X and Y axes, shows that if Δ is a σ-regulus then it is also a σ^{-1}-regulus. Generalizing this procedure, by selecting different triples of components $X, Y, Z \in \Delta$, will normally enable Δ to be regarded as a σ-regulus for a range of distinct σ's. In every case, the rational subnet Φ of Δ that is coordinatized by $\operatorname{Fix}(\sigma)$ admits a Foulser-cover of kernel planes of π. Recoordinatizing by choosing, say, two components X, Y in the 'Foulser-net' Φ associated with σ and the third component $Z \in \Delta \setminus \Phi$, we would normally expect Δ to take the form of a ϕ-regulus, where $\phi \in \operatorname{Gal}(K)$ is such that $\operatorname{Fix}(\phi)$ and $\operatorname{Fix}(\sigma)$ have different sizes. However, this last possibility obviously cannot occur if the spread π admits an automorphism mapping the old 'triad' (X, Y, Z) of Δ to the new 'triad' (X', Y', Z') of Δ.

We end with a proposition that implies fairly severe limitations on a Baer group B acting on an *arbitrary* finite spread π and leaving invariant a derivable net $\Delta \subset \pi$ provided B also fixes at least one component of Δ: this occurs automatically if B normalizes an elation group E of an E-parabolic line spread—the topic of Chapter 46. The final part of the proposition specializes to the case when π is a line spread, so it has special relevance to the next section.

PROPOSITION 34.8. (Derivable Baer Groups Are Small). Let B be a Baer group acting on a spread π of order q^2 that leaves invariant a derivable partial spread $\Delta \subset \pi$ that includes a fixed component ℓ of B. If $|B| > 2$, the following hold:

(1) B fixes all the points of $\ell \cap D$ for a unique Baer subplane D across Δ and B leaves invariant at most two Baer subplanes across Δ.

(2) If B leaves invariant a second component $\mu \neq \ell$ then either π_B is across Δ, or B fixes exactly two Baer subplanes, D_1 and D_2, across Δ and B fixes all the points of $\ell \cap D_i$ and of $\mu \cap D_j$, $\{i, j\} = \{1, 2\}$, and acts semiregularly on the remaining $q - 1$ slopes of Δ.

(3) Specialize to the case when the spread π contains the kernel field $K = GF(q)$, thus π is a line spread in $PG(3, q)$ that contains a derivable net Δ. Suppose additionally that the Baer group B, which leaves invariant the derivable net $\Delta \subset \pi$, is K-linear and fixes at least two components $\ell, \mu \in \Delta$. Then either π_B is a kernel plane across Δ, or B leaves invariant exactly two Baer subplanes across the net Δ and both these planes, D_1 and D_2, are kernel planes.

(4) In the latter event, if we choose a basis $[f_x, f_y, e_x, e_y]$, following the procedure of Proposition 34.5, the associated coordinatization of π on $K^2 \oplus K^2$ is such that the Δ gets coordinatized by components of type

$$\left\{ y = x \begin{bmatrix} k^\sigma & 0 \\ 0 & k \end{bmatrix} \,\middle|\, k \in K \right\} \cup \{\mathbf{0} \oplus K^2\},$$

where $\sigma \in \operatorname{Gal}(K)$.

34.2. Parabolic Spreads in $PG(3, q)$.

We begin by recalling two facts that we shall frequently use.

PROPOSITION 34.9. If an affine plane π admits a group of elations E normalized by a planar group B then π_B, the fixed plane of B, is E-invariant if and only if E and B centralize each other.

PROPOSITION 34.10. If one E-orbit union the axis is a derivable partial spread, then all such orbits define derivable partial spreads.

The following definition tacitly assumes Proposition 34.10.

DEFINITION 34.11 (Parabolic Line Spreads). A spread π in $PG(3, K)$, K a field, is 'E-parabolic', relative to an affine elation group E, if some (and hence every) non-trivial E-orbit of components of π along with the E-axis forms a derivation net of π. These derivation nets are 'E-parabolic', or 'E-derivation nets'; the group E is a 'parabolic' elation group of π. If some (hence every) E-derivation net is a regulus, then the E-derivation nets are 'E-regulus' nets.

The spread sets arising in the following lemma are closely related to a recursive process, presented in the next chapter, that *lifts* spreads in *any* $PG(3, q)$ to spreads in $PG(3, q^2)$. The spreads in $PG(3, q^2)$ that are 'lifted' from $PG(3, q)$ have a simple form for their spread sets and form a proper subclass of the type of spread sets described below.

LEMMA 34.12. Let π be a parabolic line spread, Definition 34.11, in $PG(3, q)$ relative to an affine elation group E, with axis $X \in \pi$. Suppose that π admits a K-linear Baer group B that normalizes but does not centralize E. Let Δ be any E-net that includes among its components a component $Y \neq X$ invariant by B and suppose that B leaves invariant at least two kernel planes across Δ. Then there is a basis of the ambient K-vector space associated with π over which the components of π may be expressed as follows:

$$(34.2.1) \qquad y = x \left\{ \begin{bmatrix} u & H(v) \\ v & u^\rho \end{bmatrix} \ \middle| \ u, v \in K \right\},$$

where $\rho \in \mathrm{Gal}(K)$, and H a function on K. Moreover, this basis can be chosen so that every element $b \in B$ has matrix representation $\mathrm{Diag}(1, u, v, 1)$, for $u, v \in K^*$, hence if b is an involution, then its matrix is $\mathrm{Diag}(1, -1, -1, 1)$.

The form of the spread π above was determined by a Baer group B. The full Baer group with fixed plane π_B is larger (in non-pathological cases) than the smallest B needed to define the spread π, provided the field automorphism σ defining the derivation net is an *involution*. As far as the process of lifting is concerned, only this case is of interest.

COROLLARY 34.13. Let $\overline{K} = GF(q^2) \supset GF(q) = K$ and let $\sigma \in \mathrm{Gal}(\overline{K}/K)$ be the involutory automorphism of $\overline{K}$. Suppose π is a line spread in $PG(3, \overline{K})$ with non-vertical components, on the ambient space $\overline{K}^2 \oplus \overline{K}^2$, specified by:

$$y = x \left\{ \begin{bmatrix} u^\sigma & H(v) \\ v & u \end{bmatrix} \ \middle| \ u, v \in \overline{K} \right\}.$$

Then π admits as a collineation group:

$$B(\sigma) := \{ \mathrm{Diag}(1, a, a, 1) \mid a \in \overline{K}, \ a^{q+1} = 1 \},$$

and this is a Baer group of order $q + 1$ that leaves invariant the components of the derivation net Δ with non-vertical components $y = x \, \mathrm{Diag}(k^\sigma, k), k \in K$, and fixes the X and Y-axis (both of which lie in Δ).

Moreover, the commutator $[E, B] \neq 1$.

We may now describe the form of a spread set associated with an arbitrary finite E-parabolic spread (for which the E-derivable nets are not reguli) when the elation group E is normalized but not centralized by a Baer group of order > 2.

THEOREM 34.14. Let π be a parabolic spread of order q^2, associated with an elation group E whose non-trivial component orbits define derivable E-nets at least one of which is not a regulus.

Suppose that π admits a non-trivial K-linear Baer group B of order > 2 that normalizes E, but does not centralize it. Then there is a basis of the ambient space associated with π over which the components of π may be expressed as follows:

$$ y = x \left\{ \begin{bmatrix} u & H(v) \\ v & u^\rho \end{bmatrix} \;\middle|\; u, v \in K \right\}, $$

where $\rho \in \mathrm{Gal}(K)$ is non-trivial.

In particular, when ρ is an involution, the converse holds, in the sense that the above defined spread admits as a Baer group the diagonal matrix group B, defined in Corollary 34.13, and the group B has order $q + 1 > 2$, it normalizes but does not centralize the E-parabolic net (with the obvious choice for E) associated with the displayed spread.

The following corollary is a consequence of the theorem and Corollary 34.13. We state it for emphasis, since this is what we shall really need.

COROLLARY 34.15. Let $K = GF(q^2)$, and let $\sigma \in \mathrm{Gal}(K)$ be the involutory automorphism in K and let π be a line spread in $PG(3, K)$, associated with an E-parabolic elation group whose non-trivial component orbits define σ-derivable E-nets.

Then π admits a non-trivial K-linear Baer group B of order > 2 that normalizes but does **not** centralize E if and only if π may be coordinatized by a spread set of the form

$$ (34.2.2) \qquad \left\{ \begin{bmatrix} u & H(v) \\ v & u^\sigma \end{bmatrix} \;\middle|\; u, v \in K \right\}. $$

In particular, when π is of the above form, the full Baer group with axis π_B contains a Baer subgroup of order $q + 1$ that normalizes but does not centralize E.

When the field K is permitted to be an infinite field, we return to (34.2.2) above, in which case (34.2.2) might be a quasi-spread-set rather than a spread set. Part of the motivation is that this study leads to a large supply of interesting semifields.

The situation excluded in Corollary 34.15 above, when $|B| = 2$, is also of considerable interest. We refer the reader to Johnson [**744**] for many interesting results concerning this situation; we quote just one rather surprising result:

THEOREM 34.16. (Johnson [**744**]). Suppose that π is a parabolic line spread in $PG(3, q^2)$, relative to an elation group E which is normalized but not centralized by a $GF(q^2)$-linear Baer involution β. Suppose the 'full' Baer group fixing π_β elementwise is $\langle \beta \rangle$. Then π is either a Desarguesian spread or a semifield spread of Knuth type.

CHAPTER 35

Algebraic Lifting.

In this chapter we introduce one of the most powerful techniques, 'algebraic lifting', for 'generating new line spreads from old'. It might be regarded as the only known recursive process available for constructing projective planes: in the finite case one 'lifts' any given two-dimensional translation planes π of (square) order n to another two-dimensional translation plane π' of order n^2, and this process may be repeated as long as desired. Moreover, at each stage the lifted plane depends on the choice of coordinates so that even a one-step lift usually leads to a large number of non-isomorphic translation planes.

The lifting process was originated by Hiramine, Matsumoto and Oyama for semifield planes of odd order (see [**525**]). Our treatment follows that of Johnson [**744**], which is valid for arbitrary finite and infinite orders. In this and related papers, a crucial fact concerning lifting was recognized and exploited: that lifting when expressed in terms of spread sets can be made to work on any line spread; in particular by coordinatizing semifields 'badly', by using non-additive spread sets, the range of line spreads generated by lifting can be enormously increased.

Thus 'lifting', in the finite case, is essentially a procedure that lifts an arbitrary 2×2 *spread set* over $GF(q)$ to a spread set of 2×2 matrices over $GF(q^2)$, and hence the procedure may be repeated to yield spread sets and hence line spreads of order q^{2^n}, $n = 1, 2 \ldots$. Note that although the line spread used to initiate the lifting process is *entirely arbitrary*, we shall see that all other line spreads arising in the lifting chain may be characterized geometrically as being parabolic line spreads, Definition 34.11, such that the associated derivation sets are not reguli, along with a couple of other geometric conditions, see Corollary 34.15. Thus, by deriving lifted spreads, one obtains a large variety of mutually non-isomorphic *four*-dimensional planes. Hence, by building arbitrary finite chains of lifted finite line spreads and deriving the last one, it is possible to obtain an enormous range of mutually non-isomorphic four-dimensional translation planes, Theorem 35.11. A more convoluted derivation, involving the dual of the lifted plane leads similarly to a large range of (P, l)-transitive projective planes that are not translation planes or their duals, Theorem 35.22.

35.1. Spread Sets Associated with Line Spreads.

We recall that every spread set $\mathcal{M}$ coordinatizing may be characterized uniquely by a pair of functions as defined below:

REMARK 35.1. Let $\mathcal{M}$ be a spread set of 2×2 matrices over a (finite) field K. Then there is a unique ordered pair of functions, (g, f), from $K \times K$ to K, sending

$(0,0)$ to 0, such that

(35.1.1) $$\mathcal{M} = \left\{ \begin{bmatrix} g(t,u) & f(t,u) \\ t & u \end{bmatrix} \,\middle|\, \forall\, u, t \in K \right\}.$$

The spread set $\mathcal{M}$ is a '(g,f)-spread-set' over K; thus every spread set is a (g,f)-spread-set, and (g,f) is its 'coordinate function pair' and such pairs are uniquely determined by the spread set.

The mapping

$$(t,u) \mapsto (g(t,u), f(t,u))$$

is a bijective mapping from K^2 to K^2 that fixes $(0,0)$.

Note that the bijectivity of the final mapping alone does not guarantee anything!

PROPOSITION 35.2. Let (g,f) be a pair of functions from $K \times K$ to K, K a (finite) field. Assume further that $(g,f) : (t,u) \mapsto (g(t), f(u))$, $t, u \in K$ is bijective, and fixes $(0,0)$. Show that the collection of matrices $\mathcal{M}$ defined by (f,g), as in Equation (35.1.1), will be a spread set if $\mathcal{M}$ is additively closed, but need not be a spread set otherwise. Hint: test differences.

We note that the Knuth–Kantor spreads over $GF(q)$ spread sets are examples of additive spread sets, as considered in Proposition 35.2 above.

REMARK 35.3. Fix a non-square $c \in K = GF(q)^*$, $q = p^r$, $r > 1$ and p an odd prime. Then

$$\tau = \left\{ \begin{bmatrix} u^\sigma & t \\ t & uc \end{bmatrix} \,\middle|\, t, u \in K \right\},$$

is a spread set, called a 'Knuth' spread set or 'Kantor–Knuth' spread set.

The Knuth spread sets above play a special role in a variety of combinatorial structures related to line spreads. Similarly, the following semifield spread set due to Cohen and Ganley is of singular importance.

PROPOSITION 35.4 (Cohen–Ganley Semifields). Cohen and Ganley [**239**]. Let K be a field of characteristic 3, and n a fixed non-square in K. Then the following defines an additive quasifibration in $PG(3, K)$:

$$x = 0, y = x \begin{bmatrix} u + nt^3 & nt^9 + n^3t \\ t & u \end{bmatrix} \quad \forall\, u, t \in K,$$

which is a non-square. It follows that this must always be the case. The result follows.

35.2. Lifting Spread Sets.

We construct certain $PG(3, q^2)$ spreads by 'lifting' certain line spreads over $GF(q)$. The basic technique was discovered by Hiramine, Matsumoto, and Oyama in the context of semifield spreads of odd order, and later generalized by Johnson to arbitrary 2-dimensional spreads over any not necessarily finite field. The HMO–Johnson construction will be considered here only in the context of arbitrary finite line spreads. The construction will be described within the following framework.

REMARK 35.5 (σ-associated spreads). Let $K = GF(q^2) = \overline{K} \supset K = GF(q)$, q arbitrary, and σ the involutory automorphism in $\mathrm{Gal}\,\overline{K}/K$). Let $H : \overline{K} \times \overline{K} \to \overline{K}$ be any map, with $H(0,0) = 0$. Then

$$\mathcal{M}(H) := \left\{ \begin{bmatrix} u^\sigma & H(v,u) \\ v & u \end{bmatrix} : u, v \in \overline{K} \right\},$$

is a spread set if and only if the following condition holds:

(35.2.1) $\qquad \forall u, v, s, t \in \overline{K}, v \neq t : (v - t)\,(H(v,u) - H(t,s)) \notin (\overline{K})^{\sigma+1}.$

The spread sets of type $\mathcal{M}(H)$, i.e., those for which the above condition holds, are called 'σ-associated spread sets'.

We are ready to describe a construction procedure for obtaining from any spread set $\mathcal{M}_{(g,f)}$ of 2×2 matrices, over an arbitrary finite field K, another 2×2 spread set over any quadratic extension field $\overline{K}$ of K, if such a $\overline{K}$ exists: these 'extended' spread sets are always σ-*associated* spread sets, in the sense of Remark 35.5, and are uniquely specified once a $\theta \in \overline{K} \setminus K$ is selected. Thus, in the finite case, the procedure described 'converts' *arbitrary* spreads in $PG(3,q)$ to σ-*associated* spreads in $PG(3, q^2)$.

DEFINITION 35.6 (Lifting). Let $K = GF(q)$ be any finite field and choose $\overline{K} = K(\theta) + GF(q^2)$, defined by some $\theta \in \overline{K} \setminus K$ with minimal polynomial $\theta^2 = \theta\alpha + \beta$, with $\alpha \in K, \beta \in K^*$. Let σ denote the involutory automorphism in $\mathrm{Gal}(\overline{K}/K)$, so $\sigma : \overline{K} \to \overline{K}$ may be expressed as: $\sigma : \theta t + u \mapsto -\theta t + u + \alpha t \,\forall u, t \in K$.

Suppose (g, f) defines a spread set $\mathcal{M}_{(g,f)}$ on K, and define $h(t,u)$ by $f(t,u) = h(t,u) - \alpha g(t,u)$. Hence, the spread set is

$$\begin{bmatrix} g(t,u) & h(t,u) - \alpha g(t,u) \\ t & u \end{bmatrix} \forall u, t \in K.$$

Then the θ-**lifting** of the spread set $\mathcal{M}_{(g,f)}$ is the following set of 2×2 matrices on $\overline{K}$:

$$\mathcal{M}_{(g,f)}(\theta) := \left\{ \begin{bmatrix} v^\sigma & H(u) \\ u & v \end{bmatrix} \,\middle|\, u, v \in \overline{K} \right\},$$

where the function $H : \overline{K} \to \overline{K}$ is defined by

$$H(\theta t + u) = -g(t,u)\theta + h(t,u) := -g(t,u)\theta + (f(t,u) + \alpha g(t,u)) \,\forall\, u, t \in K.$$

The following theorem asserts that spread sets over $GF(q)$ lift to σ-associated spread sets over $GF(q^2)$. Hence, any line spread in $PG(3,q)$ 'lift' to a line spread in $PG(3, q^2)$. The theorem implies a characterization of lifted line spreads.

THEOREM 35.7 (Lifting Spread Sets). Let $\overline{K} = GF(q^2) \supset GF(q) = K$, σ the involution in $\mathrm{Gal}(\overline{K})$, and choose $\theta \in \overline{K} \setminus K$, so $\theta^2 = \theta\alpha + \beta$, for $(\alpha, \beta) \in K^* \times K$.

Then the θ-**lifting** of the spread set $\mathcal{M}_{(g,f)}$, using the notation of Definition 35.6,

(35.2.2) $\qquad \mathcal{M}_{(g,f)}(\theta) = \left\{ \begin{bmatrix} v^\sigma & H(u) \\ u & v \end{bmatrix} \,\middle|\, u, v \in \overline{K} \right\},$

is a spread set, using the notation

$$H(\theta t + u) := -g(t,u)\theta + h(t,u) := -g(t,u)\theta + (f(t,u) + \alpha g(t,u)) \,\forall\, u, t \in K.$$

Conversely, assume that the collection of matrices $\mathcal{M}_{(g,f)}(\theta)$ defined by (35.2.2) turns out to be a spread set. Then the collection of K-matrices $\mathcal{M}_{(g,f)}$, defined as indicated in Definition 35.6, is also a spread set (and hence lifts to the spread set $\mathcal{M}_{(g,f)}(\theta)$) and is a 'retraction' of the spread set $\mathcal{M}_{(g,f)}(\theta)$.

COROLLARY 35.8. A spread set lifts to an additive spread set if and only if it is itself additive.

PROOF. In the notation of Theorem 35.7, the lifted function H is additive if and only if f and g are both additive. Thus the spread set associated with (g, f) is additive if and only if the associated lifted spread set is additive. □

To show that the lifted semifields in Corollary 35.8 are never spreads we note the stronger result that lifted spread sets, in the most general sense, cf. Equation (35.2.2), are never ever Desarguesian.

REMARK 35.9. Let $\mathcal{S}$ be a spread set of type (34.2.2), over any $K = GF(q)$. Then none of the fields determined by $\mathcal{S}$ are multiplicatively associative, a fortiori, the associated spread is non-Desarguesian.

It is worth stressing the geometric aspect of Theorem 35.7. Using it in conjunction with Corollary 34.15, we have:

COROLLARY 35.10 (Lifted Finite Spreads). Let σ be the involutory automorphism of $GF(q^2)$, and suppose π is a line spread in $PG(3, q^2)$.

Then π is lifted from a line spread in $PG(3, q)$ if and only if π admits an E-parabolic elation group whose non-trivial component orbits define σ-derivable E-nets and π admits a non-trivial K-linear Baer group B of order > 2 that normalizes but does **not** centralize E.

When π is of this form, the full Baer group B fixing π_B contains a Baer subgroup of order $q + 1$ that normalizes but does not centralize E.

Lifting line spreads in $PG(3, q)$ lifts to line spreads in $PG(3, q^2)$, which may be lifted again. Moreover, the existence of the Baer group of order $q+1$ shows that all the lifted line spreads in this recursive process are non-Desarguesian, although they are still two-dimensional over their kernels. But deriving the lifted line spread, at any stage of the lifting chain, yields *four*-dimensional planes, and the collineation group of the derived plane is considerably different from the original since the inherited elation and Baer groups become, respectively, 'Baer-elation' and 'Baer-homology groups' (since the axis of the elation becomes a Baer subplane and the axis of the Baer group remains a Baer subplane in the derived structure). We verify all this now.

THEOREM 35.11 (Lifted-Derived Structures). Any lifted and derived spread from $PG(3, q)$ lies in $PG(7, q)$. The associated translation plane admits a Baer group E of order q^2, and a Baer-homology group B of order $q + 1$ such that B normalizes E but does not centralize it.

We note that a class of σ-associated spread sets convert to the lifted class in Theorem 35.7; these may then be 'retracted' before being lifted to 'new directions'.

COROLLARY 35.12 (Retractions). Let K be a field with a quadratic extension field $\overline{K}$, and let k^σ denote the conjugate of $k \in \overline{K}$ relative to the involutory automorphism $\sigma \in \text{Gal}(\overline{K}/K)$. Suppose the mapping $H : \overline{K} \times \overline{K} \to \overline{K}$, maps $(0, 0)$ to

0 such that the following condition holds:

(35.2.3) $\qquad (v - t)\,(H(v, u) - H(t, s)) \notin K \ \forall\, u, v, s, t \in \overline{K}, v \neq t.$

Then

$$\mathcal{M}(H) := \left\{ \begin{bmatrix} u^\sigma & H(v, u) \\ v & u \end{bmatrix} \forall\, u, v \in \overline{K} \right\},$$

is a σ-associated spread set (apply Remark 35.5).

Define f and g to be functions from $K \times K$ to K by the condition:

$$\forall x, y \in K : H(x\theta + y, 0) = -g(x, y)\theta + (f(x, y) + \alpha g(x, , y)).$$

Then

$$\left\{ \begin{bmatrix} g(x, y) & f(x, y) \\ x & y \end{bmatrix} \,\middle|\, x, y \in K \right\}$$

is a spread set on K called the 'retraction' of the given σ-associated spread set.

The reader should note that retraction is a genuinely new process. Thus, in the context of line spreads, if we retract from a lifted spread H arising from (g, f), by changing θ associated with H, the 'retracted' (g', f') might be a line spread different from the one used to lift to the H-spread. Thus, combining and mixing lifts and retractions enables us to move from a line spread to another where neither are restricted to be of lifted type.

35.3. Lifted Semifields.

In this section, we consider the application of lifting to semifields. Basically, we show that semifield line spreads lift (relative to their additive spread sets) to semifield line spreads that are non-Desarguesian semifield spreads. Thus, one gets infinitely long chains of lifted semifields that are non-Desarguesian. In particular, we have this situation when one starts with Desarguesian planes: and we determine their first step extension explicitly. These extensions lead to many interesting semifields in the infinite situation: needless to say, we do not consider them here.

35.3.1. Semifields Lift to Semifields. We have already seen, Corollary 35.8, that semifield spreads lift to semifield spreads. We stress that such lifted spreads are never Desarguesian, even if the original spreads are!

REMARK 35.13. Every finite semifield line spread (including the Desarguesian ones) can be lifted to a *non*-Desarguesian semifield line spread, using any additive spread set (unique for the non-Desarguesian case) and hence these lifted spread sets can be coordinatized by semifields that are not fields. (This process can of course be continued.)

In particular, we have:

COROLLARY 35.14. The spreads of order q^4 obtained by lifting Desarguesian line spreads in $PG(3, q)$ are **non-Desarguesian** semifield spreads of order q^4.

Thus, starting from Desarguesian spreads of *square* order, the lifting process allows us to break out into (*non*-associative) semifield spreads right away. M. Cordero, in a series of papers (see, for example, Cordero [**246**]), has explored the semifields planes of order p^4, p prime, that are obtained by lifting the Desarguesian planes of order p^2.

We next demonstrate, proceeding in more or less the opposite direction, that if we lift starting from a semifield line spread, we can at any stage of the lifting sequence break away from a semifield plane to planes that are not semifield planes. We begin with a proposition.

PROPOSITION 35.15. If π is a non-Desarguesian translation plane of order n, it is always possible to coordinatize it with a quasifield that is not a semifield, and hence by a non-additive spread set.

PROOF. If not, then every component of the associated spread is the axis of a shears group of order n. But this means the plane is Desarguesian, e.g., by Ostrom's characterization of groups generated by elations.   □

THEOREM 35.16. If π is a non-Desarguesian line spread in $PG(3, q)$, then it may be coordinatized by a non-additive spread set and the lifted line spread ψ in $PG(3, q^2)$ obtained by any such spread set cannot be coordinatized by a semifield.

We might sum up by saying that using lifting, on an arbitrary line spread associated with a semifield, we can either go on forever getting new semifield line spreads, or we may at any stage leave the semifield world and still continue lifting forever, but once we have semifield line spreads, we cannot return to them by lifting.

As an illustrative proposition, we determine the line spreads obtained by a *one-step* lift of a Pappian line spread, using a fixed spread set.

35.3.2. Lifted Desarguesian Spreads. We determine the semifield spreads that are obtainable by lifting a Desarguesian line spread in $PG(3, K)$, K any finite field. We have seen already, Corollary 35.14, that the lifting must produce non-Desarguesian semifield line spreads. We compute their explicit form.

THEOREM 35.17. Let δ be a Desarguesian line spread in $PG(3, K)$, $K = GF(q)$ any field. So, fixing an irreducible quadratic $x^2 = xc + d$ δ over K, δ may be chosen to be the line spread defined by the spread set:

$$(35.3.1) \qquad \left\{ \begin{bmatrix} u - ct & dt \\ t & u \end{bmatrix} : \forall\, u, t \in K \right\}.$$

Then any lifting of this K-spread-set to a spread set over any quadratic extension field $\overline{K} = K(\theta) \cong GF(q^2)$, where θ is chosen to satisfy a K-irreducible quadratic (which need not be $x^2 = xc + d$) is a non-Desarguesian semifield spread set specified by:

$$x = 0, y = x \begin{bmatrix} v + as - s\theta & (a - u + ct)\theta + dt \\ t\theta + u & v + s\theta \end{bmatrix} \forall t, u, s, v \in K.$$

The spread π_θ has as it kernel the field $K(\theta)$.

We may summarize much of the material covered so far concerning result as follows. (An infinite, and much subtler, version of the following result see Johnson [**744**]).

THEOREM 35.18. Let $\overline{K} = GF(q^2) \supset GF(q) = K$ and let σ be the involutory automorphism in $\mathrm{Gal}(\overline{K}/K)$. Suppose that S is a spread set with respect to a pair of skew lines (L, M) in $PG(3, \overline{K})$. Then the following are equivalent.

(1) The spread set S may be represented in the form

$$x = 0, y = x \begin{bmatrix} u^\sigma & H(v) \\ v & u \end{bmatrix} \text{ for all } u, v \text{ in } \overline{K},$$

for some function $H : \overline{K} \to \overline{K}$.

(2) The associated spread π_S admits an elation group E with axis M normalized by a $\overline{K}$-linear Baer group B of order > 2 that normalizes but does not centralize E, and some E-orbit union its axis is a σ-regulus.

(3) The spread set S may be lifted from a spread set in $PG(3, K)$.

(4) There exists a retraction of S to a spread set in $PG(3, K)$.

(5) S has the above form, case 1, and satisfies the condition

$$(v - s)(H(v) - H(s)) \in K \implies v = s;$$

furthermore, for any such H, defined by case 1, each $\theta \in \overline{K} \setminus K$, defining $g(t, u)$ and $f(t, u)$ by the condition

$$H(\theta t + u) = -g(t, u)\theta + (f(t, u) + \alpha g(t, u)),$$

yields an associated quasifibration in $PG(3, K)$ given by

$$x = 0, y = x \begin{bmatrix} g(t, u) & f(t, u) \\ t & u \end{bmatrix} \forall t, u \in K.$$

that lifts to the given quasifibration S.

35.4. Chains of Lifts and Contractions.

We have seen how the processes of lifting and retracting line spreads and quasifibrations can be made to yield large numbers of mutually non-isomorphic line spreads. On the other hand, *lifted* finite line spreads are rather restricted since their geometric characterization shows that they form a quite specialized class, in the sense that it is very easy to provide examples of line spreads that are geometrically unqualified to be lifted line spreads (note that lifted line spreads must be derivable and admit prescribed elation and Baer groups). However, since all finite line spreads do lift to line spreads, in fact to infinitely many, it follows that one assign 'classes' to line spreads: two line spreads are in the same class if and only if they can reach each other by a sequence of lifts and/or retractions. We record the definition in the context of quasifibrations.

DEFINITION 35.19 (Lift-Equivalence of Quasifibrations). Let S and R be quasifibrations (possibly line spreads) in $PG(3, L)$ and $PG(3, K)$, respectively, over arbitrary fields L and K. Then S and R are 'lift-equivalent' if and only if there is a chain of quadratic extensions from one field to the other such that S may be constructed from R by a sequence of lifts and retractions.

This definition enables us to end with an intriguing question:

Are All Finite Line Spreads Lift-Equivalent?!

35.5. Lifting Non-Translation Planes.

We use lifting to construct finite projective planes with the following property. We recall the definition of 'point-line' transitivity.

DEFINITION 35.20. Let π be a projective plane with a point P and line ℓ. Then π is '(P, l)-transitive' if it admits a group of central collineations with center P and axis l which is transitive on the non-fixed points of any line incident with P.

We begin considering derivation in the context of certain dual quasifields. Let $Q = (V, +, \circ)$ be a finite quasifield of order q^2 with kernel $K = GF(q)$. So the dual quasifield $Q' = (V, +, *)$, defined by the identity $a * b = b \circ a$, is a dimension-two vector space with K operating from the right. Now the same argument as applied to quasifields shows that the dual affine translation plane $\pi(Q')^{[\infty]}$, associated with the dual of the translation plane $\pi(Q)$, is derivable relative to the slope set of $\pi(K)$. Note also that the Q's just discussed are precisely those associated with line spreads in arbitrary $PG(3, q)$.

The geometric interpretation of all this is the following result (see, for example, Johnson [**753**]).

REMARK 35.21. The dual π' of a translation plane π associated with a line spread in $PG(3, q)$ is derivable relative to the affine restriction of π' obtained by choosing any line ℓ of π' through its translation center L (viz., the translation axis of π).

We need to describe a derivation of the above type in terms of a quasifield Q when Q is *non-additive*, i.e., not a semifield.

Recall that any non-Desarguesian translation plane with spread in $PG(3, q)$ may be coordinatized by a non-additive quasifield Q, Proposition 35.15. So, by Theorem 35.16, the lifted spread, defined by the slope set of Q is not a semifield spread in $PG(3, q^2)$, but being lifted it admits an elation group of order q^2 whose component orbits are derivable nets in which there are two invariant kernel subplanes.

In the setting mentioned above, choose the parallel class (point of the projective extension) containing the elation axis as the line infinity of the dual translation plane (the dual of the projective extension). The plane then derived is a 'semi-translation' plane but not a translation plane or a dual translation plane. In fact, this derived plane admits exactly one point-line transitivity with the point P incident with the line l. The interested reader is referred to Johnson [**731**] for further details. Thus, we have the desired result.

THEOREM 35.22. If π admits a non-additive coordinate quasifield and π is a spread in $PG(3, q)$, then there is a sequence of constructions: lift-dualize-derive, which produces a projective plane with exactly one incident point-line transitivity.

CHAPTER 36

Semifield Planes of Orders q^4, q^6.

In this chapter, we consider semifield spreads in $PG(3, q^2)$, of order q^4, q^6 with center in the semifield of order q. Such semifield planes are quite special in that it is possible to utilize a geometric construct of the 'translation dual,' although for our treatment it is not necessary to mention this concept. However, the reader is directed to Chapter 85 on the translation dual.

LEMMA 36.1. Let S be any semifield spread in $PG(3, q^2)$. Let F_{q^2} denote the subfield isomorphic to F_q^2 in question. Choose coordinates so that $x = 0$ is the elation axis and that $y = 0, y = x$ are components of S. Then S has the following form:

$$S = \left\{ x = 0, y = x \begin{bmatrix} g_1(t) + g_2(u) & f_1(t) + f_2(u) \\ t & u \end{bmatrix}; u, t \in F_{q^2} \right\},$$

where g_i, f_i are additive functions of F_{q^2}, such that $g_2(1) = 1$, $f_2(1) = 0$.

PROOF. Since u and t are independent and the spread is additive, the form given is the general form. The condition on the g_2 and f_2 simply is the condition required to have $y = x$ in the semifield spread. □

LEMMA 36.2. Let V denote the 4-dimensional F_{q^2}-vector space over the field of scalars isomorphic to F_{q^2}. Assume that S contains a subfield K of the scalar field

$$\left\langle \begin{bmatrix} u & 0 \\ 0 & u \end{bmatrix}; u \in F_{q^2} \right\rangle$$

and K is maximal with this condition. Then K commutes with S and S is a K-vector space.

DEFINITION 36.3. Under the conditions of the previous lemma, we shall say that a semifield field spread S in $PG(3, q^2)$ is a 'K-central vector space.' When K is isomorphic to F_q, we shall say that S is a '4-dimensional central vector space over $GF(q)$.'

LEMMA 36.4. Any semifield spread of order q^4 in $PG(3, q^2)$ that is a 4-dimensional central vector space over $GF(q)$ has a representation as follows:

$$S = $$
$$\left\{ x = 0, y = x \begin{bmatrix} a_1 t + a_2 t^q + b_1 u + (1 - b_1) u^q & c_1 t + c_2 t^q + d_1 u - d_1 u^q \\ t & u \end{bmatrix} \right\}.$$
$$; u, t \in F_{q^2}$$

PROOF. Since S is a vector space over

$$\left\langle \begin{bmatrix} v & 0 \\ 0 & v \end{bmatrix}; v \in F_q \right\rangle,$$

271

it follows that g_i and f_i are all F_q-linear transformations. Hence, all such functions Z have the following general form: $Z(u) = z_1 u + z_2 u^q$, for $u \in F_{q^2}$. Furthermore since $\begin{bmatrix} v & 0 \\ 0 & v \end{bmatrix}$ are now spread set elements, we see that $b_1 u + (1 - u_1) u^q = u$ when $u \in F_q$, and $d_1 u - d_0 u^q = 0$ when $u \in F_q$.   □

Let π be any semifield spread in $PG(3, q)$. Then there is a spread set that has the following form:

$$x = 0, y = x \begin{bmatrix} g(t, u) & f(t, u) \\ t & u \end{bmatrix} ; u, t \in GF(q)$$

where g and f are additive functions (i.e., $g(t, u) = g_1(t) + g_2(u)$, $f(u, t) = f_1(t) + f_2(u)$, where g_i, f_i are additive on $GF(q)$).

Let $GF(q^2)$ have basis $\{\theta, 1\}$ over $GF(q)$, where $\theta^2 = \theta \alpha + \beta$.

Then where g and f are additive functions. Hence, we obtain the algebraic lifted spread

$$\pi^L : x = 0, y = x \begin{bmatrix} (\theta s + w)^q & -\theta g(t, u) + (f(t, u) + \alpha g(t, u)) \\ (\theta t + u) & (\theta s + w) \end{bmatrix} ; s, w, t, u \in GF(q),$$

defines a semifield spread of $PG(3, q^2)$, such that if $s = 0 = t = u$, we obtain a $GF(q)$-regulus

$$x = 0, y = x \begin{bmatrix} w & 0 \\ 0 & w \end{bmatrix} ; w \in GF(q),$$

in the semifield spread. This implies that the semifield of order q^4 has center $GF(q)$.

Hence, we have noted the following:

THEOREM 36.5. Any semifield plane of order q^2 with spread in $PG(3, q)$ lifts (in many ways) to a semifield plane of order q^4 with a coordinate semifield of order q^4 that has $GF(q)$ in the center.

Now the question is whether any of these lifted semifields of order q^4 which are 4-dimensional over the center $GF(q)$ are $GF(q)$-central vector spaces.

Recall that the general form for a $GF(q)$-central vector-space semifield spread is

$$\left\{ y = x \begin{bmatrix} a_1 t + a_2 t^q + b_1 u + (1 - b_1) u^q & c_1 t + c_2 t^q + d_1 u - d_1 u^q \\ t & u \end{bmatrix} \begin{matrix} x = 0, \\ \\ ; u, t \in F_{q^2} \end{matrix} \right\}.$$

Now write $u = (\theta s + w)$ and $t = (\theta t^* + u^*)$. So, any of these are lifted from a semifield spread of the form

$$x = 0, y = x \begin{bmatrix} g(t^*, u^*) & f(t^*, u^*) \\ t^* & u^* \end{bmatrix} ; t^*, u^* \in GF(q),$$

we must have $a_1 + a_2 = b_1 = 0$ (these being in $GF(q^2)$) and

$$c_1 t + c_2 t^q + d_1 u - d_1 u^q = -\theta g(t^*, u^*) + (f(t^*, u^*) + \alpha g(t^*, u^*)).$$

Thus, we must have $d_1(u - u^q)$ to be identically zero so that d_1. Note that $t^q = ((\theta t^* + u^*))^q = -\theta t^* + u^* + \alpha t^*$. So, we obtain:

$$c_1(\theta t^* + u^*) + c_2(-\theta t^* + u^* + \alpha t^*) = -\theta g(t^*, u^*) + (f(t^*, u^*) + \alpha g(t^*, u^*)).$$

Suppose $t^* = 0$; then we obtain

$$(c_1 + c_2)u^* = -\theta g(0, u^*) + (f(0, u^*) + \alpha g(0, u^*)).$$

Letting $c_1 + c_2 = \theta w_1 + w_2$, for $w_i \in GF(q)$, we see that $g(0, u^*) = -w_1 u^*$, so

$$f(0, u^*) = (w_2 + \alpha w_1)u^*, \forall u^* \in GF(q).$$

Then letting $u^* = 0$, we obtain:

$$c_1(\theta t^*) + c_2(-\theta t^* + \alpha t^*) = ((c_1 - c_2)\theta + c_2\alpha)t^*$$
$$= -\theta g(t^*, 0) + (f(t^*, 0) + \alpha g(t^*, 0)).$$

Now write $((c_1 - c_2)\theta + c_2\alpha) = \theta z_1 + z_2$, to obtain

$$g(t^*, 0) = -z_1 t^* \text{ and so}$$
$$f(t^*, 0) = (z_2 + \alpha z_1)t^*.$$

Note that $g(t^*, u^*) = g(t^*, 0) + g(0, u^*)$ and $f(t^*, u^*) = f(t^*, 0) + f(0, u^*)$. Hence, $g(t^*, u^*) = -z_1 t^* - w_1 u^*$ and $f(t^*, u^*) = (z_2 + \alpha z_1)t^* + (w_2 + \alpha w_1)u^*$. Hence, we obtain:

THEOREM 36.6. *Any semifield spread of order q^4 with spread in $PG(3, q^2)$ which is a $GF(q)$-central vector space that is lifted from a semifield spread in $PG(3, q)$ is lifted from a Desarguesian spread.*

PROOF. The spread in $PG(3, q)$ has the form

$$x = 0,$$
$$y = x \begin{bmatrix} -z_1 t^* - w_1 u^* & (z_2 + \alpha z_1)t^* + (w_2 + \alpha w_1)u^* \\ t^* & u^* \end{bmatrix} ; t^*, u^* \in GF(q).$$

□

We return to the spread set for the semifields in question:

$$S = \left\{ \begin{bmatrix} a_1 t + a_2 t^q + b_1 u + (1 - b_1)u^q & c_1 t + c_2 t^q + d_1 u - d_1 u^q \\ t & u \end{bmatrix} ; u, t \in F_{q^2} \right\}.$$

There are certain generalized twisted field planes of order q^4 with spreads in $PG(3, q^2)$ of this type and whose original descriptions must be modified to fit into the forms above. Moreover, if we do not require that $y = x$ is a component, the associated matrix forms may be slightly varied from the above.

Cardinali, Polverino and Trombetti [215] completely determine the possible spreads. We have seen examples of spreads of this type as follows:

Kantor–Knuth $\begin{bmatrix} u & \gamma t^q \\ t & u \end{bmatrix}$; $u, t \in GF(q^2), \gamma$ a non-square,

Knuth type (1) $\begin{bmatrix} u^q + t^q f & ft^q \\ t & u \end{bmatrix}$ or type (2) $\begin{bmatrix} u^q + tg & ft^q \\ t & u \end{bmatrix}$;

$u, t \in GF(q^2)$, $x^{q+1} + gx - f \neq 0$ in $GF(q^2)$,

Hughes–Kleinfeld (previous for $g = 0$)

algebraically lifted from a Desarguesian plane of order q

generalized twisted field $x \circ y = xy - cx^{q^2}y^q$, $x, y \in GF(q^2)$.

But, actually, it is possible to construct Kantor–Knuth planes, Knuth and Hughes–Kleinfeld planes by lifting. Furthermore, there are other semifield planes obtained by lifting that are not of these types due to Cordero and Figueroa [251] for odd order and an analogous class due to Cardinali, Polverino and Trombetti [215] for even order. The general form for lifted planes is

$$\begin{bmatrix} u^q & yl + ft^q \\ t & u \end{bmatrix}; u, t \in GF(q^2), u^{q+1} \mid gx - f \neq 0 \text{ in } GF(q^2).$$

If $f = 0$ then note that

$$\begin{bmatrix} u^q & gt \\ t & u \end{bmatrix}\begin{bmatrix} 0 & 1 \\ g^{-1} & 0 \end{bmatrix} = \begin{bmatrix} t & u^q \\ ug^{-1} & t \end{bmatrix},$$

a Kantor–Knuth spread. If $g = 0$ and $x^{q+1} - f \neq 0$, we obtain a Knuth plane of type (2). When $gf \neq 0$, the spread is due to Cordero–Figueroa [251] for q odd and to Cardinali, Polverino and Trombetti [215] for even order.

REMARK 36.7. The Planes of Cordero–Figueroa for q odd are also those of Boerner–Lantz [152], who constructed these in a different manner.

Theorem of Cardinali, Polverino, and Trombetti.

THEOREM 36.8. (Cardinali, Polverino, and Trombetti [215]). If π is a translation plane of order q^4 with spread in $PG(3, q^2)$ and has $GF(q)$ in the center of the semifield then π is either

(1) Lifted from a Desarguesian plane and is either
   Hughes–Kleinfeld,
   Kantor–Knuth,
   Cordero–Figueroa/Boerner–Lantz, for q odd or
   Cardinali, Polverino, and Trombetti for q even,
(2) Knuth of type (1) or
(3) A generalized twisted field plane determined by $x \circ y = xy - cx^{q^2}y^q$, $x, y \in GF(q^2)$.

We note, of course, that there are many mutually non-isomorphic semifield planes that can be algebraically lifted from a Desarguesian plane.

THEOREM 36.9. (Kantor (unpublished)). Given any spread S in $PG(3, q)$, there are at most q^{10} mutually non-isomorphic translation planes of order q^4 constructed from S by algebraic lifting.

36.1. Semifields 6-Dimensional over Their Centers.

The semifields of order q^{2n}, for n odd of Johnson–Marino–Polverino–Trombetti mentioned in Chapter 36 (see Chapter 37, when $n = 3$) are examples of semifields of order q^6, which are 6-dimensional over their centers $GF(q)$. Recalling that the center is contained in the right, middle, and nucleus (kernel) and in this setting we have kernel $GF(q^3)$, with right and middle nuclei equal and isomorphic to $GF(q^2)$. If the right and middle nuclei are both isomorphic to $GF(q^2)$, then by the fusion methods of Jha and Johnson, these nuclei are equal in some isotopism of the semifield. So, we would have the intersection of the right and middle isomorphic to $GF(q)$. If right and middle nuclei are both isomorphic to $GF(q^2)$, then a complete classification of the possible semifields may be given.

Theorem of Johnson, Marino, Polverino, and Trombetti.

THEOREM 36.10. (Johnson, Marino, Polverino, and Trombetti [**781**]). If π is a semifield of order q^6 with kernel K isomorphic to $GF(q^3)$, with right and middle nuclei isomorphic to $GF(q^2)$, and the associated semifield is 6-dimensional over the center, then we may assume that the semifield spread has the following form:

(1) There is a field K^+ isomorphic to $GF(q^6)$ containing K and the fused right and middle nucleus F isomorphic to $GF(q^2)$, such that the spread is isomorphic to one of the following form:

$$x = 0, y = x(\alpha + \beta e) + x^{q^3} v\gamma; \alpha, \beta, \gamma \in F$$

for constants d and e of K^+, where e is not in the fused right and middle nucleus F, and v^{q^3+1} is not of the form $(\alpha^* + \beta^* e)^{q^3+1}$, for all $\alpha^*, \beta^* \in F$.

(2) There is a Desarguesian subspread of degree $1 + q^4$.

(3) The semifield is a cyclic semifield if and only if $v^{q^3+1} = \alpha_0 + \beta_0 e$, for $\alpha_0, \beta_0 \in GF(q)$.

(4) There is a set of $q+1$ reguli R_i, of Σ (the Desarguesian plane coordinatized by K^+), $i = 1, 2, \ldots, q + 1$, that share exactly $q^2 + 1$ components with the spread if e is in K, so this always occurs in the cyclic semifield case.

We have seen that there are a number of translation planes with the conditions given and each such plane may be given a net replacement interpretation.

THEOREM 36.11. (Johnson, Marino, Polverino, and Trombetti [**781**]). Let Σ be a Desarguesian affine plane coordinatized by a field K^+ isomorphic to $GF(q^6)$. Representing the spread of Σ by

$$x = 0, y = xr; r \in K^+,$$

consider the partial spread P:

$$x = 0, y = xr; r \in K^+ - F,$$

where F is the subfield of K^+ isomorphic to $GF(q^2)$.

Define $R_b(P)$:

$$y = x^{q^3} b(\alpha + \beta z) + x\gamma; \alpha, \beta, \gamma \in F, \text{ and } (\alpha, \beta) \neq (0,0),$$

where z is not in F.

(1) Then $R_{b,z}(P)$ is a replacement for P if and only b may be chosen so that b^{q^3+1} is not of the form $(\alpha + \beta z)^{q^3+1}$ for all $\alpha, \beta \in F$, by a set of Baer subplanes whose spreads are disjoint from $x = 0$.

(2) The translation plane $\pi_{b,z}$ obtained by this replacement is a semifield plane $\pi_{b,t}$ as the spread set is additive.

CHAPTER 37

Known Classes of Semifields.

In Chapter 82, we note the connections between commutative semifields and symplectic semifields. From a semifield flock spread, the 5th cousin will be symplectic. Most of these planes have been constructed previously. Here we provide the cross-references to those construction processes and the constructions which have not been previously mentioned.

REMARK 37.1. There is a class of semifield planes due to Denniston [**315**], with spreads in $PG(3, q)$. It is not difficult to show that these spreads are also Knuth semifield spreads (type D to follow).

THEOREM 37.2. The known finite semifields and general construction processes are as follows (the classes are not necessarily disjoint, see C and D):

B: Knuth binary semifields.

F: Flock semifields and their 5th cousins:

$\quad F_1$: Kantor–Knuth.

$\quad F_2$: Cohen–Ganley, 5th cousin: Payne–Thas.

$\quad F_3$: Penttila–Williams symplectic semifield order 3^5, 5th cousin, Bader, Lunardon, Pinneri flock semifield.

C: Commutative semifields/symplectic semifields:

$\quad C_1$: Kantor–Williams Desarguesian scions (symplectic), Kantor–Williams commutative semifields.

$\quad C_2$: Ganley commutative semifields and symplectic cousins.

$\quad C_3$: Coulter–Matthews commutative semifields and symplectic cousins.

D: Generalized Dickson/Knuth/Hughes–Kleinfeld semifields.

S: Sandler semifields.

JJ: Jha–Johnson cyclic semifields (generalizes Sandler, also of type $S(\omega, m, n)$, or p-primitive type 1, or q-primitive type 2).

$JMPT$: Johnson–Marino–Polverino–Trombetti semifields (generalizes Jha–Johnson type $S(\omega, 2, n)$-semifields).

$JMPT(4^5)$: Johnson–Marino–Polverino–Trombetti non-cyclic semifield of order 4^5.

T: Generalized twisted fields.

JH: Johnson–Huang 8 semifields of order 8^2.

CF: Cordero–Figueroa semifield of order 3^6.

General Construction Processes:

L: Algebraically lifted semifields.

The algebraically lifted Desarguesian spreads are completely determined. These are also known as Cordero–Figueroa/Boerner–Lantz semifield planes for q odd or the Cardinali, Polverino, and Trombetti semifield planes for q even.

277

M: The middle-nucleus semifields by distortion-derivation.

C: $GL(2, q) - q^3$ plane construction of Jha–Johnson.

37.1. The Dickson/Knuth/Hughes–Kleinfeld Semifields.

Let $\{e, 1\}$ be a basis for $GF(q^2)$ over $GF(q)$ Let $1 \neq \sigma, \alpha, \beta, \delta$ be automorphisms of $GF(q)$ such that there exists an element k in

$$GF(q) - GF(q)^{\alpha+1}GF(q)^{\beta+1}GF(q)^{\delta-1}.$$

Assume further that $x^{\sigma+1} + gx - f$ has no root in $GF(q)$, for $g, f \in GF(q)$. Consider a binary operation $\circ$ on $GF(q^2)$, which could construct a semifield. Then for $a, b, c, d \in GF(q)$, Dickson commutative semifields are defined by

$$(a + eb) \circ (c + ed) = (ac + b^\sigma d^\sigma) + e(ab + bc).$$

37.2. Knuth Generalized Dickson Semifields.

The Knuth generalized Dickson semifields of type (α, β, δ) are defined using α, β, δ by

$$(a + eb) \circ (c + ed) = (ac + kb^\alpha d^\beta) + e(a^\delta b + bc).$$

37.3. Hughes–Kleinfeld Semifields.

The Hughes–Kleinfeld semifields as defined by

$$(ea + b) \circ (ec + d) = (ac + kb^\delta d) + e(a^\delta b + bc); \ |\delta| = 2$$

are thus Knuth generalized Dickson of type $(\delta, 1, \delta)$, where $|\delta| = 2$. These semifields may be characterized by having two nuclei equal. Hence, fusion methods are possible to characterize these semifields further.

37.4. Knuth Generalized Dickson Semifield of Type (j).

The Knuth generalized Dickson semifields of σ-type (j), using combinations of the automorphism σ, are constructed as follows:

σ-type (1): $(a + eb) \circ (c + ed) = (ac + fb^\sigma d^{\sigma-2}) + e(a^\sigma b + +gb^\sigma d^{\sigma^{-1}} + bc)$,

σ-type (2): $(a + eb) \circ (c + ed) = (ac + fb^\sigma d) + e(a^\sigma b + +gb^\sigma d + bc)$,

σ-type (3): $(a + eb) \circ (c + ed) = (ac + fb^{\sigma^{-1}} d^{\sigma-2}) + e(a^\sigma b + +gbd^{\sigma^{-1}} + bc)$,

and middle nuclei equal to $GF(q)$,

σ-type (4): $(a + eb) \circ (c + ed) = (ac + fb^{\sigma^{-1}} d) + e(a^\sigma b + +gbd + bc)$,

and right nuclei equal to $GF(q)$.

We mention cubical arrays in Chapter 82. There we note the processes of dualization and transposition. Furthermore, we also may e-extend the various semifield planes. The σ-type (j) semifields may be related by these construction processes.

The type (3) and type (4) Knuth semifields are derivable, either by a middle- or right-nucleus net. Johnson [**717**] has shown that the collineation group of derived semifield planes is the full group when the order is > 16. The determination of translation planes admitting many elations is basically unresolved. That is, Rahilly [**1125, 1126**] has shown that if a generalized Hall (a plane derived from a semifield plane of order q^2 with middle nucleus $GF(q)$) admits at least $q+1$ affine involutions, then the plane is Hall and the associated semifield plane is Desarguesian. More generally, Johnson and Rahilly [**809**] show that any derived semifield plane of order q^2 admitting $q + 1$ affine involutions is Desarguesian. However, derivation of the Knuth type (3) and (4) planes often produce many affine involutions. In particular, derivation of Knuth type (3) or (4) of order 2^{4r} planes by the middle- or right-nucleus nets produce derived semifield planes of order 2^{4r} admitting $2^r + 1$ affine elations. These elations generate dihedral groups of order $2(2^r + 1)$.

37.5. Ganley Commutative Semifields.

$(GF(3^{2n}), +, \circ)$, $n > 2$, n odd, is a commutative semifield where, if $\{e, 1\}$ is a basis for $GF(q^2)$ over $GF(q)$, $q = 3^r$, the multiplication is

$$(\alpha + e\beta) \circ (\delta + e\gamma) = (\alpha\delta - \beta^9\gamma - \beta\gamma^9) + e(\alpha\gamma + \beta\delta + \beta^3\gamma^3), \alpha, \beta, \delta, \gamma \in GF(q).$$

The form for the associated transposed-dualized semifield spread (symplectic spread) is

$$(\alpha + e\beta) \circ (\delta + e\gamma) = (\alpha\delta + \beta\delta^{1/3} - \beta^{1/9}\gamma^{1/9} - \beta^9\gamma) + e(\alpha\gamma + \beta\delta), \alpha, \beta, \delta, \gamma \in GF(q).$$

37.6. Knuth Binary Semifields.

Let q be even, $n > 1$, odd, and let T denote the trace map of $GF(q^n)$ over $GF(q)$. Then $(GF(q^n), +, \circ)$ becomes a pre-semifield where

$$x \circ y = xy + (T(x)y + T(y)x)^2.$$

37.7. Kantor–Williams Desarguesian Scions.

In Chapter 79 and Chapter 103, the Kantor–Williams semifield planes are constructed. However, we also give here a description based only on the semifield multiplication. Consider $F \simeq GF(2^{mn})$, and a sequence of subfields

$$F_n \simeq GF(q) \subseteq F_{n-1} \subseteq \cdots \subseteq F_0 = F, \, n \geq 1.$$

Let T_i denote the trace maps from F to F_i. Now choose any sequence

$$(\zeta_1, \zeta_2, \ldots, \zeta_n)$$

of elements $\zeta_i \in F^*$. Then $(GF(2^{mn}), +, \circ)$ is a symplectic pre-semifield, where the defining multiplication is

$$x \circ y = xy^2 + \sum_{i=1}^{n} T_i(\zeta_i x)y + \zeta_i T_i(xy).$$

Dualizing-transpose produces the associated class of commutative semifields with multiplication $\circ$, where

$$x \circ y = xy + \left(x \sum_{i=1}^{n} T_i(\zeta_i y) + y \sum_{i=1}^{n} T_i(\zeta_i x) \right)^2.$$

37.8. Coulter–Matthews Commutative Semifields.

In Chapter 99, we mention the theory of planar functions of Dembowski and Ostrom [292]. The possible planes constructed by planar functions are either translation planes or planes admitting precisely one points-line transitivity (Pierce and Kallaher [1104]). The planes were originally constructed using planar functions.

$(GF(3^e), +, \circ)$, e odd and $e > 3$, defines a commutative pre-semifield where

$$x \circ y = x^9 y + xy^9 + x^3 y^3 - xy.$$

The transpose-dual symplectic semifields are defined by the following multiplication:

$$x \circ y = x^9 y + (xy)^{1/9} + xh,$$
$$xy = x^9 y + (xy)^{1/9} + xy^{1/3} - xy.$$

37.9. Johnson–Huang Semifields of Order 8^2.

The semifields of order 8^2 are of interest relative to the determination of translation planes of orders q^2 with spread in $PG(3, q)$ that admit a linear group of order q^2. In particular, Biliotti, Jha, Johnson, and Menichetti [133] prove the following:

THEOREM 37.3. Let π be a translation plane of order q^2 with spread in $PG(3, q)$, $q = p^r$, for p a prime. If π admits a linear p-group of order at least $q^2 p$, then $p = 2$, the order is $2q^2$ and π is a semifield plane.

We note that if such a semifield plane exists, then there is a Baer involution in the linear translation complement. However, prior to the semifields under discussion, the only known semifields of even order with a linear automorphism of order 2 are fields. It turns out that such semifield planes with spread in $PG(3, q)$ have the following general spread representation

$$x = 0, y = x \begin{bmatrix} u + v + m(v) & f(v) + m(u) \\ v & u \end{bmatrix};$$

$$u, v \in GF(q), \ m, \ f \text{ additive functions on } GF(q).$$

If $m \equiv 0$, we obtain a conical flock spread. However, by Johnson [710], there are no non-Desarguesian possibilities. The following gives the set of all 8 mutually non-isomorphic semifields of order 8^2 with spread in $PG(3, 8)$ that admit a linear Baer involution. If G denotes the full translation complement of the associated plane and K^* is the kernel homology group of order $8 - 1$, we denote by $\overline{G} = G/K^*$.

Class	$f(x)$	$m(x)$	Remarks
I	x	0	Desarguesian
II	x^2	$x^2 + x^4$	$\lvert \overline{G} \rvert = 3 \cdot 2 \cdot 8^2$
III	x^4	$x^2 + x^4$	$\lvert \overline{G} \rvert = 3 \cdot 2 \cdot 8^2$
IV	$\alpha^6 x^4$	$\alpha x + x^2 + \alpha^3 x^4$	$\alpha^3 = \alpha + 1, \lvert \overline{G} \rvert = 2 \cdot 8^3$
V	$\alpha^6 x^2$	$\alpha x + x^2 + \alpha^3 x^4$	$\alpha^3 = \alpha + 1, \lvert \overline{G} \rvert = 2 \cdot 8^3$
VI	x	$\alpha^3 x + x^2 + \alpha x^4$	$\alpha^3 = \alpha + 1, \lvert \overline{G} \rvert = 2 \cdot 8^3$
VII	$x + \alpha^6 x^2 + \alpha x^4$	$\alpha^3 x + x^2 + \alpha x^4$	$\alpha^3 = \alpha + 1, \lvert \overline{G} \rvert = 2 \cdot 8^3$
VIII	$x + x^2 + \alpha^4 x^4$	$\alpha^3 x + x^2 + \alpha x^4$	$\alpha^3 = \alpha + 1, \lvert \overline{G} \rvert = 2 \cdot 8^3$

We note that classes IV and V are transposes of each other.

REMARK 37.4. The determination of the planes of Huang and Johnson [554] is made using the computer. There may be an infinite class of semifield planes containing the order 8^2 planes but this is an open question. However, using algebraic lifting, each plane of order 8^2 may be lifted to a class of semifield planes of orders $(8^2)^{2^n}$, for all n. However, the Baer involution for these planes is lost in the algebraic lifting process.

37.10. Cordero–Figueroa Sporadic Semifield of Order 3^6.

The following multiplication defines a semifield plane of order 3^6 that admits an autotopism of order 7: Let

$$x \circ m = xm + \gamma x^3 y^{27} + \gamma^{13} x^{27} y^3,$$

for all x, m in $GF(3^6)$.

37.11. Johnson–Marino–Polverino–Trombetti Cyclic Semifields.

Johnson, Marino, Polverino, and Trombetti [780] generalize the Jha–Johnson cyclic semifields of type $S(\omega, 2, n)$, n odd as follows:

THEOREM 37.5 (Johnson, Marino, Polverino, and Trombetti [780]). (1) Any cyclic semifield plane of order q^{2n}, n odd, where the right and middle nuclei are isomorphic to $GF(q^2)$ and the nucleus is isomorphic to $GF(q^n)$ may be represented in the following form:

$$x = 0, y = (\alpha_0 + \alpha_1 u + \cdots + \alpha_{\frac{n-1}{2}} u^{\frac{n-1}{2}})x + b(\beta_0 + \beta_1 u \cdots + \beta_{\frac{n-3}{2}} u^{\frac{n-3}{2}})x^{q^n},$$

where α_i, β_i vary in $\mathbb{F}_{q^2}$, u is a fixed element of $\mathbb{F}_{q^n}$ such that $\{1, u, \ldots, u^{n-1}\}$ is an $\mathbb{F}_{q^2}$-basis of $\mathbb{F}_{q^{2n}}$ and b is a fixed element of $\mathbb{F}_{q^{2n}}^*$ such that $b^{q^n+1} = u$.

(2) Conversely, if $\{1, u, \ldots, u^{n-1}\}$ is an $\mathbb{F}_{q^2}$-basis of $\mathbb{F}_{q^{2n}}$ and b is a fixed element of $\mathbb{F}_{q^{2n}}^*$ such that $b^{q^n+1} = u$ then define $y = xT$ by $y = x^{q^n}b$, and

$$x = 0, y = (\alpha_0 + \alpha_1 u + \cdots + \alpha_{\frac{n-1}{2}} u^{\frac{n-1}{2}})x + b(\beta_0 + \beta_1 u \cdots + \beta_{\frac{n-3}{2}} u^{\frac{n-3}{2}})x^{q^n}.$$

THEOREM 37.6 (Johnson, Marino, Polverino, and Trombetti [780]). Consider the cyclic semifields of order q^{2n}, for odd $n > 3$ with middle nucleus and right nucleus equal to $GF(q^2)$ and left nucleus $GF(q^n)$. Let U denote the union of the proper subfields of $GF(q^n)$. Then the number of mutually non-isomorphic semifield planes is exactly $N(\mathcal{O}_2)$, which is the number of orbits of $\mathrm{Gal}\, GF(q^n)$ on $\mathcal{O}_2$, where $\mathcal{O}_2$ is the set of $(q^n - |U|)/2$ elements

$$\{ u, u^{-1}; u \in GF(q^n) - |U| \text{ modulo } GF(q) \}.$$

Hence, the number of non-isomorphic cyclic semifield planes is at least

$$(q^n - |U|)/2ne,$$

where $q = p^e$, for p a prime.

REMARK 37.7. Now when $n = 3$, each cyclic semifield produces a generalized Desarguesian plane of order q^6 (i.e., $GL(2, h)$ planes of order h^3, for $h = q^2$), by the method of Jha–Johnson. Non-isomorphic cyclic semifield planes of cubic order will produce non-isomorphic generalized Desarguesian planes. When $n = 3$ then

U is simply $GF(q)$. Then the number of mutually non-isomorphic cyclic semifields and hence mutually non-isomorphic generalized Desarguesian planes is at least

$$(q^3 - q)/6e.$$

Kantor [864] gave a construction for the generalized Desarguesian planes of order q^6 and also calculated this same number of mutually non-isomorphic generalized Desarguesian planes.

REMARK 37.8. Let $\ell = \mathrm{lcm}(m, n)$, where $m, n > 1$ are integers, and $\ell > 1$. Consider the mapping T:

$$T : x \rightarrow \omega x^{q^n},$$

where ω is a primitive element of $GF(q^\ell)$. Then (see [606, Theorem 2]) Jha and Johnson show that there is an associated cyclic semifield plane, called the 'Jha–Johnson cyclic semifield (q, m, n).' Here the middle and right nuclei contain $GF(q^m)$ and the left nucleus (kernel) is $GF(q^n)$ (actually this is the dual semifield to the semifield constructed by Jha and Johnson). When $m = 2$ and n is odd, we obtain a cyclic semifield of order q^{2n} with kernel $GF(q^2)$ and middle and right nuclei $GF(q^2)$. Hence, some of the semifields constructed by Johnson, Marino, Polverino and Trombetti [780] are Jha–Johnson cyclic semifields. The constructions here require a basis $\{1, u, u^2, \ldots, u^{n-1}\}$ for $GF(q^n)$ over $GF(q)$, and an element b such that $b^{q^n+1} = u$, so any element u in $GF(q^n) - U$ will suffice. The order of b is then $(q^n + 1) |u|$, so that b is primitive if and only if $|u| = q^n - 1$. The mapping $(x, y) \rightarrow (xz, yz)$ replaces b by bz^{1-q^n}, so if b is primitive so is bz^{1-q^n}, as is b^σ, where σ is an automorphism of $GF(q^n)$. Hence, the number of non-isomorphic Jha–Johnson cyclic semifields of this type is $\leq \varphi(q^{2n} - 1)/(q^n - 1)$. In any case, the question of isomorphism is not considered in Jha–Johnson [606].

REMARK 37.9. It is possible to construct semifield planes that are not isotopic to cyclic semifield planes using the same basic setup where the putative spread is

$$\left\{ x = 0, \ y = \left(\alpha_0 + \alpha_1 u + \cdots + \alpha_{\frac{n-1}{2}} u^{\frac{n-1}{2}} \right) x \right.$$
$$\left. + b \left(\beta_0 + \beta_1 u + \cdots + \beta_{\frac{n-3}{2}} u^{\frac{n-3}{2}} \right) x^{q^n} ; \ \alpha_i, \beta_i \in GF\left(q^2\right) \right\}.$$

In Johnson, Marino, Polverino, and Trombetti [780], the semifields isotopic to cyclic semifields of this type are completely determined. In particular, there is a sporadic semifield plane of order 4^5 constructed in this manner which is not isotopic to a cyclic semifield plane. In this case, $b^{q^5+1} = b^{33} = \omega^{21}$, where ω is a primitive element of $GF(2^5)$.

Methods of Oyama, and the Planes of Suetake.

In this chapter, we describe a model for spreads based on the work of Oyama [**1079**]. Let V_{2n} be a $2n$-dimensional vector space over $GF(q)$. Then a spread of V_{2n} is, of course, a set of $q^n + 1$ mutually disjoint n-dimensional $GF(q)$-subspaces. Choose a basis and represent vectors in V_{2n} as $2n$-tuples $(x_1, x_2, \ldots, x_n, y_1, y_2, \ldots, y_n)$, $x_i, y_i \in GF(q)$. The idea is to consider $GF(q^n) \times GF(q^n)$ as V_{2n} and find a suitable representation for spreads. Basically, this is set up so that for $x \in GF(q^n)$, we consider the n-tuple

$$(x, x^q, x^{q^2}, x^{q^3}, \ldots, x^{q^{n-1}}) = [x],$$

so that for $x, y \in GF(q^n)$, $([x], [y])$ will represent our $2n$-vector over $GF(q)$. Let

$$\omega = \begin{bmatrix} 0 & . & . & . & . & 0 & 1 \\ 1 & 0 & . & . & . & . & 0 \\ 0 & 1 & . & . & . & . & 0 \\ . & 0 & 1 & . & . & . & 0 \\ . & . & 0 & 1 & . & . & . \\ . & . & . & . & . & . & . \\ . & . & . & . & . & . & . \\ 0 & . & . & . & . & 1 & 0 \end{bmatrix}.$$

Define $\mathcal{A} = \{\, \alpha \in GL(n, q^n); \alpha^q = \alpha \omega \,\}$, where $\alpha^q = [a_{i,j}^q]$, for $\alpha = [\alpha_{i,j}]$. Choose any element ρ of $GL(n, q)$, where $\rho = [r_{i,j}]$ as a matrix is an element of $GL(n, q^n)$, for $r_{i,j} \in GF(q)$. If $\beta \in \mathcal{A}$ then $\rho\beta \in \mathcal{A}$, since $(\rho\beta)^q = \rho^q\beta^q = \rho\beta^q = \rho(\beta\omega) = (\rho\beta)\omega$. If β and $\beta^* \in \mathcal{A}$, then $(\beta\beta^{*^{-1}})^q = (\beta\omega)(\beta^*\omega)^{-1} = \beta\beta^{*-1}$. Hence, $\beta\beta^{*-1} \in GL(n, q)$. Hence, $\mathcal{A} = GL(n, q)\beta_0$, for any $\beta_0 \in \mathcal{A}$. Furthermore, if $\alpha = [a_{ij}] \in GL(n, q^n)$, then $\alpha \in \mathcal{A}$, if and only if

$$\alpha = \begin{bmatrix} a_{11} & a_{11}^q & . & . & . & a_{11}^{q^{n-2}} & a_{11}^{q^{n-1}} \\ a_{21} & a_{21}^q & . & . & . & a_{21}^{q^{n-2}} & a_{21}^{q^{n-1}} \\ a_{31} & a_{31}^q & . & . & . & & a_{31}^{q^{n-1}} \\ . & . & . & . & . & & . \\ . & . & . & . & . & & . \\ . & . & . & . & . & & . \\ a_{n1} & a_{n1}^q & . & . & . & a_{n1}^{q^{n-2}} & a_{n1}^{q^{n-1}} \end{bmatrix}.$$

That is, simply note that since $\alpha^q = \alpha\omega$, then

$$
\begin{bmatrix}
a_{11}^q & a_{12}^q & \cdots & & a_{1n}^q \; 1 & a_{1n}^q \\
a_{21}^q & a_{22}^q & \cdots & & & a_{2n}^q \\
a_{31}^q & a_{32}^q & \cdots & & & a_{3n}^q \\
\cdot & \cdot & \cdot & \cdot & \cdot & \cdot \\
\cdot & \cdot & \cdot & \cdot & \cdot & \cdot \\
\cdot & \cdot & \cdot & \cdot & \cdot & \cdot \\
a_{n1}^q & \cdot & \cdots & & a_{nn-1}^q & a_{nn}^q
\end{bmatrix}
$$

$$
=
\begin{bmatrix}
a_{11} & a_{12} & \cdots & & a_{1n-1} & a_{1n} \\
a_{21} & a_{22} & \cdots & & & a_{2n} \\
a_{31} & a_{32} & \cdots & & & a_{3n} \\
\cdot & \cdot & \cdot & \cdot & \cdot & \cdot \\
\cdot & \cdot & \cdot & \cdot & \cdot & \cdot \\
\cdot & \cdot & \cdot & \cdot & \cdot & \cdot \\
a_{n1} & \cdot & \cdots & & a_{nn-1} & a_{nn}
\end{bmatrix}
\begin{bmatrix}
0 & \cdot & \cdot & \cdot & 0 & 1 \\
1 & 0 & \cdot & \cdot & \cdot & 0 \\
0 & 1 & \cdot & \cdot & \cdot & 0 \\
\cdot & 0 & 1 & \cdot & \cdot & 0 \\
\cdot & \cdot & 0 & 1 & \cdot & \cdot \\
\cdot & \cdot & \cdot & \cdot & \cdot & \cdot \\
0 & \cdot & \cdot & \cdot & 1 & 0
\end{bmatrix}
$$

$$
=
\begin{bmatrix}
a_{12} & a_{13} & \cdots & & a_{1n} & a_{11} \\
a_{22} & a_{23} & \cdots & & \cdot & a_{21} \\
a_{32} & a_{33} & \cdots & & \cdot & a_{31} \\
\cdot & \cdot & \cdot & \cdot & \cdot & \cdot \\
\cdot & \cdot & \cdot & \cdot & \cdot & \cdot \\
\cdot & \cdot & \cdot & \cdot & \cdot & \cdot \\
a_{n2} & \cdot & \cdots & & a_{nn} & a_{n1}
\end{bmatrix}.
$$

Hence, $a_{11}^q = a_{12}$, $a_{12}^q = a_{13}$ so that $a_{11}^{q^2} = a_{13}$, so clearly $a_{1k} = a_{11}^{q^{k-1}}$. Similarly, $a_{j1}^{q^{k-1}} = a_{jk}$.

Noting that $\alpha^{-q} = \omega^{-1}\alpha^{-1}$, we see that for $\beta \in \mathcal{A}$, the general form for $\beta^{-1} = [b_{ij}]$ is

$$
[b_{ij}]^{-1} =
\begin{bmatrix}
a_{11} & a_{12} & \cdots & & a_{1n-1} & a_{1n} \\
a_{11}^q & a_{12}^q & \cdots & & \cdot & a_{1n}^q \\
a_{11}^{q^2} & a_{12}^{q^2} & \cdots & & \cdot & a_{1n}^{q^2} \\
\cdot & \cdot & \cdot & \cdot & \cdot & \cdot \\
\cdot & \cdot & \cdot & \cdot & \cdot & \cdot \\
\cdot & \cdot & \cdot & \cdot & \cdot & \cdot \\
a_{11}^{q^{n-1}} & a_{12}^{q^{n-1}} & \cdots & & a_{1n-1}^{q^{n-1}} & a_{1n}^{q^{n-1}}
\end{bmatrix}.
$$

We now connect $GL(n, q)$ to $GL(n, q^n)$, as follows:

CLAIM 38.1. For $\alpha \in \mathcal{A}$, then

$$
GL(n, q)^\alpha = \left\{ \beta \in GL(n, q^n); \; \beta^q = \beta^\omega = \omega^{-1}\beta\omega \right\}.
$$

Then

$$\beta = \begin{bmatrix} a_1 & a_n^q & a_{n-1}^{q^2} & \cdot & \cdot & a_3^{q^{n-2}} & a_2^{q^{n-1}} \\ a_2 & a_1^q & a_n^{q^2} & \cdot & \cdot & & a_3^{q^{n-1}} \\ a_3 & a_2^q & a_1^{q^2} & \cdot & \cdot & \cdot & a_{1n}^{q^2} \\ \cdot & \cdot & \cdot & & \cdot & \cdot & \cdot \\ \cdot & \cdot & \cdot & & \cdot & \cdot & \cdot \\ \cdot & \cdot & \cdot & \cdot & \cdot & \cdot & \cdot \\ a_n & a_{n-1}^q & a_{n-2}^{q^2} & \cdot & \cdot & a_2^{q^{n-2}} & a_1^{q^{n-1}} \end{bmatrix}$$

PROOF. To see this, just note that $(\alpha^{-1}\delta\alpha)^q = \alpha^{-q}\delta\alpha^q = \omega^{-1}\alpha^{-1}\delta\alpha\omega$. Now if $\beta \in GL(n, q^n)$, such that $\beta^q = \omega^{-1}\beta\omega$, then $(\alpha\beta\alpha^{-1})^q = \alpha\omega\beta^q\omega^{-1}\alpha^{-1} = \alpha\omega\omega^{-1}\beta\omega\omega^{-1}\alpha^{-1}$, so that $\alpha\beta\alpha^{-1} \in GL(n, q)$. □

Therefore, we have shown

PROPOSITION 38.2.

$$GL(n, q)^\alpha =$$

$$\left\{ \begin{bmatrix} a_1 & a_n^q & a_{n-1}^{q^2} & \cdot & \cdot & a_3^{q^{n-2}} & a_2^{q^{n-1}} \\ a_2 & a_1^q & a_n^{q^2} & \cdot & \cdot & & a_3^{q^{n-1}} \\ a_3 & a_2^q & a_1^{q^2} & \cdot & \cdot & \cdot & a_{1n}^{q^2} \\ \cdot & \cdot & \cdot & & \cdot & \cdot & \cdot \\ \cdot & \cdot & \cdot & & \cdot & \cdot & \cdot \\ \cdot & \cdot & \cdot & \cdot & \cdot & \cdot & \cdot \\ a_n & a_{n-1}^q & a_{n-2}^{q^2} & \cdot & \cdot & a_2^{q^{n-2}} & a_1^{q^{n-1}} \end{bmatrix} \in GL(n, q^n); a_i \in GF(q^n) \right\}.$$

Note that for $(x_1, x_2, \ldots, x_n)$ in $V_4/GF(q)$, the $(1, 1)$ element of

$$(x_1, x_2, \ldots, x_n) \begin{bmatrix} a_{11} & a_{11}^q & \cdot & \cdot & \cdot & a_{11}^{q^{n-2}} & a_{11}^{q^{n-1}} \\ a_{21} & a_{21}^q & \cdot & \cdot & \cdot & a_{21}^{q^{n-2}} & a_{21}^{q^{n-1}} \\ a_{31} & a_{31}^q & \cdot & \cdot & & & a_{31}^{q^{n-1}} \\ \cdot & \cdot & \cdot & \cdot & \cdot & & \cdot \\ \cdot & \cdot & \cdot & \cdot & \cdot & & \cdot \\ \cdot & \cdot & \cdot & \cdot & \cdot & & \cdot \\ a_{n1} & a_{n1}^q & \cdot & \cdot & \cdot & a_{n1}^{q^{n-2}} & a_{n1}^{q^{n-1}} \end{bmatrix}$$

is $\sum_{i=1}^{n} x_i a_i = x$. Since the set $\{ a_i; i = 1, 2, \ldots, n \}$ is linearly independent over $GF(q)$, then $x \in GF(q^n)$ and the mapping

$$(x_1, x_2, \ldots, x_n) \longmapsto \sum x_i a_i = x$$

is an injective map from $V_4/GF(q)$ to $GF(q^n)$. Conversely, if $x \in GF(q^n)$, then there exist $x_i \in GF(q)$, $i = 1, 2, \ldots, n$ such that

$$\sum_{i=1}^{n} x_i a_i = x.$$

Then, clearly

$$(x_1, x_2, \ldots, x_n) \begin{bmatrix} a_{11} & a_{11}^q & \cdot & \cdot & \cdot & a_{11}^{q^{n-2}} & a_{11}^{q^{n-1}} \\ a_{21} & a_{21}^q & \cdot & \cdot & \cdot & a_{21}^{q^{n-2}} & a_{21}^{q^{n-1}} \\ a_{31} & a_{31}^q & \cdot & \cdot & \cdot & & a_{31}^{q^{n-1}} \\ \cdot & \cdot & \cdot & & & & \cdot \\ \cdot & \cdot & & \cdot & & & \cdot \\ \cdot & \cdot & & & \cdot & & \cdot \\ a_{n1} & a_{n1}^q & \cdot & \cdot & \cdot & a_{n1}^{q^{n-2}} & a_{n1}^{q^{n-1}} \end{bmatrix}$$

$$= (x, x^q, x^{q^2}, \ldots, x^{q^{n-1}}).$$

Now consider $(x, x^q, x^{q^2}, \ldots, x^{q^{n-1}})$, for $x \in GF(q^n)$. For

$$\alpha = \begin{bmatrix} a_{11} & a_{11}^q & \cdot & \cdot & \cdot & a_{11}^{q^{n-2}} & a_{11}^{q^{n-1}} \\ a_{21} & a_{21}^q & \cdot & \cdot & \cdot & a_{21}^{q^{n-2}} & a_{21}^{q^{n-1}} \\ a_{31} & a_{31}^q & \cdot & \cdot & \cdot & & a_{31}^{q^{n-1}} \\ \cdot & \cdot & \cdot & \cdot & \cdot & \cdot & \cdot \\ \cdot & \cdot & & \cdot & \cdot & \cdot & \cdot \\ \cdot & \cdot & & & \cdot & & \cdot \\ a_{n1} & a_{n1}^q & \cdot & \cdot & \cdot & a_{n1}^{q^{n-2}} & a_{n1}^{q^{n-1}} \end{bmatrix},$$

with the second subscript on the entries ignored, we see that

$$\{\, a_i;\ i = 1, 2, \ldots, n \,\}$$

is linearly independent over $GF(q)$. Hence, there exist $x_i \in GF(q)$, $i = 1, 2, \ldots, n$, such that $x = \sum_{i=1}^{n} x_i a_i$. Therefore

$$(x, x^q, x^{q^2}, \ldots, x^{q^{n-1}})\alpha^{-1} = (x_1, x_2, x_3, \ldots, x_n).$$

NOTATION 38.3. $V_n\alpha$ is denoted by V_n^* and is an isomorphic n-dimensional vector space with vectors $[x] = (x, x^q, x^{q^2}, \ldots, x^{q^{n-1}})$. We denote $V_n^* \oplus V_n^*$ by V_{2n}^* with vectors $([x], [y])$ where

$$(x_1, x_2, \ldots, x_n, y_1, y_2, \ldots, y_n)\alpha = ([x], [y]) \text{ if and only if}$$

$$x = \sum_{i=1}^{n} x_i a_i, y = \sum_{i=1}^{n} y_i a_i.$$

A matrix $B = [b_{ij}]$ in $GL(n, q)$ then becomes

$$\alpha^{-1}B\alpha = B^* = \begin{bmatrix} a_1 & a_n^q & a_{n-1}^{q^2} & \cdot & \cdot & a_3^{q^{n-2}} & a_2^{q^{n-1}} \\ a_2 & a_1^q & a_n^{q^2} & \cdot & \cdot & & a_3^{q^{n-1}} \\ a_3 & a_2^q & a_1^{q^2} & \cdot & \cdot & & a_{1n}^{q^2} \\ \cdot & \cdot & \cdot & \cdot & \cdot & & \cdot \\ \cdot & \cdot & \cdot & & \cdot & \cdot & \cdot \\ \cdot & \cdot & \cdot & & & \cdot & \cdot \\ a_n & a_{n-1}^q & a_{n-2}^{q^2} & \cdot & \cdot & a_2^{q^{n-2}} & a_1^{q^{n-1}} \end{bmatrix} \in GL(n, q^n),$$

for some $GF(q)$-linearly independent set $\{\, a_i;\ i = 1, 2, \ldots, n \,\}$.

Since any matrix is then uniquely determined by its first column, we may choose

$$\begin{bmatrix} a_1 & a_2 & a_3 & \ldots & a_n \end{bmatrix}^t$$

to represent the associated linear transformation B^*.

Hence, we have another model to present spreads, and hence translation planes, of order q^n with kernel containing $GF(q)$.

Now, for x in $GF(q^n)$, let $x^{(i)} = x^{q^i}$.

Let $[a] = (a, a^{(1)}, a^{(2)}, \ldots, a^{(n-1)})$. Let $N(x) = \prod_{i=0}^{n-1} x^{(i)}$, so $N(x) \in GF(q)$. Let $V_n = \{ [a]; a \in GF(q^n) \}$, and note that V_n is an n-dimensional $GF(q)$-vector space. Set

$$[a_1, a_2, a_2, \ldots, a_n] = \begin{bmatrix} a_1 & (a_{n-1})^{(1)} & \ldots & (a_2)^{(n-1)} \\ a_2 & (a_1)^{(1)} & \ldots & (a_3)^{(n-1)} \\ \vdots & \vdots & \vdots & \vdots \\ a_n & (a_{n-1})^{(1)} & \ldots & (a_1)^{(n-1)} \end{bmatrix},$$

for $a_i \in GF(q^n)$, $i = 1, 2, \ldots, n$.

38.1. Suetake's Planes.

Set $I(x) = [x, 0, 0, \ldots, 0]$ and $J = [0, 1, 0, \ldots, 0]$. Furthermore, let

$$P(x) = I(x)(J^i I(a) + J^{-i} I(b)) I(x)$$

$$\text{for } (i, n) = 1, \ 1 \le i \le n - 1,$$

$$\text{for } N(a) + N(b) \neq 0, \ a, b \text{ fixed in } GF(q^n) - \{0\}.$$

THEOREM 38.4. (Suetake [**1168**, (2.7)]). Let q and n be odd and using the above notation, the following set $\pi_{i,a,b}$ is a spread:

$$\{ [x] = [0], [y] = [0] \} \cup \{ [y] = [x]P(z); z \in GF(q^n)^* \}$$

$$\cup \{ [y] = [x]uP(z^{-1})^{-1}; x \in GF(q^n)^* \},$$

where u is a non-square in $GF(q)$.

We note that these planes of order q^n are generalizations of planes of orders q^3 of Suetake [**1169**]. In that paper, the construction is restricted to cases where 2 is a non-square in $GF(q)$. Translation planes of order q^3 with similar group properties are constructed by Hiramine [**506**] q^3, where this restriction is not required. About this time (early 1990s) in unpublished notes, R. Pomareda constructed a class of translation planes of order q^5. The planes of Pomareda are those constructed later by Suetake when $n = 5$. Indeed, Suetake's original constructions were for $n = 3$ and then for $n = 5$. However, the construction was eventually generalized to any odd n using suggestions of Y. Hiramine. In any case, certain of the constructed planes are also due to Hiramine for $n = 3$ and Pomareda for $n = 5$.

REMARK 38.5.

$$\{ uP(z^{-1})^{-1}; z \in GF(q^n)^* \}^{-1} = \{ uP(z); z \in GF(q^n)^* \}.$$

To see this, we note that

$$((uP(z^{-1}))^{-1})^{-1} = uu^{-2}P(z^{-1}) = uu^{-2}I(z^{-1})P(1)I(z^{-1})$$

$$= uI((uz)^{-1})P(1)I((uz)^{-1}) = uP((uz)^{-1}).$$

THEOREM 38.6. (Suetake [**1168**, (3.1)]). The following is the spread of a generalized twisted field plane, denoted by $\pi_{i,a,b}^{GT}$:

$$\{[x] = [0], [y] = [0]\} \cup \{ [y] - [r]P(z); \ z \in GF(q^n) \}$$

$$\cup \{ [y] = [x]uP(z); \ z \in G\Gamma(q^n) \}.$$

REMARK 38.7.

$$G = \left\langle \left[\begin{array}{cc} I(x)^{-1} & 0 \\ 0 & I(x) \end{array} \right] ; x \in GF(q^n)^* \right\rangle$$

is a collineation group of $\pi_{i,a,b}$ and since $uP(x^{-1})^{-1} = uI(x)(J^iI(a)+J^{-i}I(b))^{-1}I(x)$, G is also a collineation group of $\pi_{i,a,b}^{GT}$.

CHAPTER 39

Coupled Planes.

We have noted the Suetake planes [**1168**], a class of translation planes of order q^n and kernel $GF(q)$, that admit two orbits of components, of lengths 2 and $(q^n - 1)$. Furthermore, by net replacement in these planes, we have noted that there is an associated class of generalized twisted field planes. Thus, it is possible to construct the Suetake planes from generalized twisted field planes by reversing the construction process. However, the form that Suetake uses to represent the spreads is not particularly compatible with the usual representation of the twisted field planes. Suetake also leaves open the question of whether his new planes are generalized André planes.

In this chapter, we settle both the question of the type of generalized twisted field plane that constructs the Suetake plane and also the problem of whether the Suetake planes are generalized André planes (they are not!).

DEFINITION 39.1. A pair of translation planes of order q^n are said to be 'weakly coupled' if and only if the two planes admit the following collineation group: Planes π_1 and π_2 both admit

$$ G = \left\langle \begin{bmatrix} T^{-1} & 0 \\ 0 & T \end{bmatrix} ; T \in MatGF(q^n)^* \right\rangle, $$

where the notation is meant to indicate that the element T is taken from a matrix field of $n \times n$ matrices over $GF(q)$ containing the $GF(q)^*$-scalar group of order $q - 1$. Furthermore, we must have $x = 0, y = 0$ are components of both translation planes but $y = x$ is not necessarily a component of either. Consider two translation planes that are weakly coupled and their spreads are related by the inversing map $(x, y) \longrightarrow (y, x)$, in the sense that there is a matrix subspread $\{ y = xM; M \in \lambda \}$ of one spread such that $\{ y = xM^{-1}; M \in \lambda \}$ is a subspread of the other and the two spreads agree on the other components. Then we shall say that the translation planes are 'coupled.'

DEFINITION 39.2. A translation plane is said to be 'triangle transitive' if and only if there is a collineation group G of the translation complement, which leaves invariant a pair $\{x = 0, y = 0\}$ of components and acts transitively on the non-fixed components of the three projective lines $x = 0, y = 0, \ell_\infty$.

Each Suetake plane is coupled with a generalized twisted field plane and both planes are triangle transitive. The corresponding generalized twisted field planes of order q^n admit a group G fixing both of the affine axes $x = 0, y = 0$ and acting transitively on $x = 0, y = 0, \ell_\infty$, whereas the Suetake planes are triangle transitive but the stabilizer of $x = 0$ and $y = 0$ (the autotopism group) has two orbits on ℓ_∞ of lengths $(q^n - 1)/2$. A major question is if there are other translation planes that are coupled with semifield planes. It is possible to form a characterization of the Suetake planes by showing that any translation plane coupled with a non-Desarguesian semifield plane is a Suetake plane. In particular, Jha and Johnson show the following.

THEOREM 39.3. (Jha–Johnson [583]). (1) If a non-Desarguesian generalized twisted field plane of order q^n is one plane of a coupling, then q is odd, n is odd, and the second plane of the coupling is a Suetake plane.

(2) The generalized twisted field planes of order q^n corresponding to the Suetake planes are as follows:

$$x = 0, y = 0, y = xm - cx^{q^{2b^*}} m^{q^{b^*}},$$

where $(2b^*, n) = (b^*, n) = 1$ and q is any odd prime power, n an odd integer.

(3) If a finite translation plane π is coupled with a non-Desarguesian semifield plane, then π is a Suetake plane.

THEOREM 39.4. (Jha–Johnson [583]). Let the following spread denote the generalized twisted field plane spread from which the Suetake planes may be constructed:

$$x = 0, y = 0, y = xm - cx^{q^{2b^*}} m^{q^{b^*}}, \qquad \text{for } m \in GF(q^n) - \{0\},$$

where $(2b^*, n) = (b^*, n) = 1$ and q is any odd prime power.

Let G denote the collineation group

$$\left\langle \begin{bmatrix} t^{-1} & 0 \\ 0 & t^{q^*} \end{bmatrix} ; t \in GF(q^n)^* \right\rangle.$$

Let $f(x) = x^{q^{-2b^*}} - c^{q^{-b^*}} x$ and note that f is injective on $GF(q^n)$. Let the two component orbits of length $(q^n - 1)/2$ be denoted by Γ_1 and Γ_2, where $y = x - cx^{q^{2b^*}}$ is in Γ_1 and $y = xm_0 - cx^{q^{2b^*}} m_0^{q^{b^*}}$ is in Γ_2.

Then the Suetake planes have the following spread:

$$x = 0, y = 0, (y = (xm_0 - cx^{q^{2b^*}} m_0^{q^{b^*}})G) \cup (y = (f^{-1}(x))G).$$

Since a Desarguesian plane is also a semifield plane, it is also possible to show:

THEOREM 39.5. (Jha–Johnson [583]). A translation plane π of order p^{rn} is coupled with a Desarguesian plane if and only if π is an André plane with one of the following spreads:

$$x = 0, y = 0, y = x^{p^{2s}} wt^2 xz^2, y = xz^2$$

for w a fixed non-square, for all $t, z \in GF(q^n)^*$, where $(2s, rn) = r, x^{p^{2s}} \neq x, \forall x$.

COROLLARY 39.6. (Jha–Johnson [583]).

(1) A plane coupled with a Desarguesian plane admits a collineation group generated by $\sigma : (x, y) \longmapsto (y^{p^{-s}}, x^{p^s} w)$, where w is the non-square in the above theorem.

(2) The plane above admits two cyclic homology groups with affine axes of orders $(p^{rn} - 1)/2$.

(3) The group G has two component orbits of length $(p^{rn} - 1)/2$ if and only if $(p^s + 1, p^{rn} - 1) = 2$.

(4) There is always an orbit of length 2, $x = 0$ and $y = 0$ are inverted by a collineation

$$\sigma : (x, y) \longmapsto (y^{p^{-s}}, x^{p^{-s}} w).$$

Thus, these planes are always triangle transitive under the group $\langle G, \sigma \rangle$.

(5) There is a homology group of order $p^m - 1$ if and only if the order of the plane is $p^{rn} = p^{4s}$.

Suetake's Planes Are Not Generalized André. In Suetake's article, the problem was left open as to whether any of the constructed planes could actually be generalized André planes. However, using the ideas of coupled planes and generalized twisted field planes that, in fact, these planes are not generalized André planes.

We begin with a general theorem on generalized André planes. First we clarify what we mean by the 'inversing map'.

REMARK 39.7. (i) If two spreads admit the group

$$G = \left\langle \begin{bmatrix} T^{-1} & 0 \\ 0 & T \end{bmatrix}; T \in MatGF(q^n)^* \right\rangle,$$

then it is possible that $y = x$ is not a component of either spread, although we assume that $x = 0, y = 0$ represent components to both spreads.

(ii) Two spreads S_1 and S_2 are related by the inversing mapping $(x, y) \longmapsto (y, x)$ in the sense that there is a matrix subspread set $\mathcal{M} = \{ y = xM; M \in \lambda \}$ of one spread, say S_1, such that $\mathcal{M}^{-1} = \{ y = xM^{-1}; M \in \lambda \}$ is a matrix subspread of the other spread S_2, and $S_2 = (S_1 - \mathcal{M}) \cup \mathcal{M}^{-1}$.

(iii) If G is represented in another form, say by

$$\left\langle \begin{bmatrix} A^{-1}T^{-1}A & 0 \\ 0 & B^{-1}TB \end{bmatrix}; T \in MatGF(q^n)^* \right\rangle,$$

and we require the spreads to be coupled, then the corresponding inversing mapping is given by

$$(x, y) \longmapsto (yB^{-1}A, x(B^{-1}A)^{-1}).$$

In the following, we shall discuss a more general inversing mapping: $(x, y) \longmapsto (y^{p^{-s}}, x^{p^s})$, for generalized André planes of order $p^w = q^n$, of kernel $GF(q)$. In this setting, note that $y = x^{q^{\lambda(m)}} m$ will map onto $y = x^{q^{-\lambda(m)}p^{2s}} m^{-q^{\lambda(m)}p^s}$, so it is essentially immediate if this is the sort of inversing map that we are using then a generalized André plane will produce a generalized André plane under such a mapping.

THEOREM 39.8. (Jha–Johnson [**583**]). Let Σ be a finite non-Desarguesian generalized André plane with kernel $GF(q)$ and order $q^n > 9$. There are two special lines being axes admitting affine homologies. Representing these lines as $x = 0$, $y = 0$, the set S of remaining lines of the spread has the following general form:

$$S = \{ y = x^{q^{\lambda(m)}} m; \ m \in GF(q^n) \},$$

where $1 \leq \lambda(m) \leq n$. Further, denote $y = x^{q^{\lambda(m)}} m = xM_m$ where M_m is a non-singular $n \times n$ matrix over $GF(q)$, depending on m.

Choose any subset Φ of $GF(q^n)^*$ and assume that

$$\{ y = x^{q^{-\lambda(m)}p^{2s}}; \ m \in \Phi \} = \Phi^{*(-p^s)}$$

is a replacement for the net

$$\{ y = xM_m; \ m \in \Phi \} = \{ y = x\Phi \},$$

using the inversing map $(x, y) \longmapsto (y^{p^{-s}}, x^{p^s})$.

Then the translation plane with spread

$$(S - \{y = x\Phi\}) \cup \{y = x\Phi^{*(-p^{-s})}\}$$

is also a generalized André plane.

LEMMA 39.9. Suppose that π is a non-Desarguesian generalized André plane of order q^n, with n odd and $GF(q)$ the kernel of π, $q = p^r$, p a prime, and such with some coordinatization we have

$$G = \left\langle \begin{bmatrix} T^{-1} & 0 \\ 0 & T \end{bmatrix} ; T \in MatGF(q^n)^* \right\rangle,$$

as a collineation group. Then, we may choose a generalized André system of kernel $GF(q)$ and order q^n to represent π of the form

$$x = 0, y = 0, y = x^{q^{\lambda(m)}} m; m \in GF(q^n),$$

where $y = x$ is a component and G becomes

$$G = \left\langle \begin{bmatrix} t^{-1} & 0 \\ 0 & t^{p^s} \end{bmatrix} ; t \in GF(q^n)^* \right\rangle.$$

DEFINITION 39.10. Let π be a Desarguesian plane of odd order q^n, for $p^r = q$. If all non-square André nets are replaced relative to $GF(p^{2s})$, and we choose the components $y = x^{p^{2s}} m$, for all replacements, we obtain the spread:

$$
\begin{aligned}
x &= 0, y = 0, y = x^{p^{2s}} wt^2; w \text{ is a non-square,} \\
y &= xz^2, \text{ for all } t, z \in GF(q^n)^*, \\
\text{where } (2s, rn) &= r \text{ and } p^{2s} \neq 1.
\end{aligned}
$$

Such an André plane admits the collineation group

$$G = \left\langle \begin{bmatrix} t^{-1} & 0 \\ 0 & t^{p^s} \end{bmatrix} ; T \in MatGF(q^n)^* \right\rangle,$$

and the André spread is obtained from the Desarguesian spread using the inversing map $(x, y) \longmapsto (y^{p^{-s}}, x^{p^s})$. That is, the two spreads are coupled.

THEOREM 39.11. A translation plane π of order p^{rn} is coupled with a Desarguesian plane if and only if π is an André plane with one of the following spreads:

$$
\begin{aligned}
x &= 0, y = 0, y = x^{p^{2s}} wt^2; w \text{ is a fixed non-square, } y = xz^2, \\
\text{for all } t, z &\in GF(q^n)^*, \text{ where } (2s, rn) = r.
\end{aligned}
$$

COROLLARY 39.12. (1) A plane coupled with a Desarguesian plane admits a collineation group

$$\sigma : (x, y) \longmapsto (y^{p^{-s}}, x^{p^s} w), \text{ where } w \text{ is the non-square in the above theorem.}$$

(2) The plane above admits two cyclic homology groups with affine axes of orders $(p^{rn} - 1)/2$.

(3) The group G has two component orbits of length $(p^{rn} - 1)/2$ if and only if $(p^s + 1, p^{rn} - 1) = 2$.

(4) There is always an orbit of length 2, $x = 0$ and $y = 0$ are inverted by a collineation

$$\sigma : (x, y) \longmapsto (y^{p^{-s}}, x^{p^{-s}} w).$$

Thus, these planes are always triangle transitive.

(5) There is a homology group of order $p^m - 1$ if and only if the order of the plane is $p^{rn} = p^{4s}$.

DEFINITION 39.13. An André plane that is coupled to a Desarguesian plane shall be called a 'simple full non-square' André plane, to indicate that only one type of replacement component is considered and that the non-square André nets are all replaced.

Thus, we see from results of Jha and Johnson [583] that the following remarks hold.

REMARKS 39.14. (1) The Suetake planes are coupled to a class of generalized twisted field planes by the group G and the inversing map $(x, y) \longmapsto (y, x)$.

REMARK 39.15. (2) The Suetake planes of order q^n do not admit affine homology groups of orders $(q^n - 1)/2$ with axis $[y] = [0]$ and coaxis $[x] = [0]$.

THEOREM 39.16. (3) The Suetake planes are not generalized André planes.

(4) Two Suetake planes are isomorphic if and only if the corresponding generalized twisted field planes are isomorphic.

REMARK 39.17. (5) Any two planes coupled to generalized twisted field planes are isomorphic if and only if the generalized twisted field planes are isomorphic.

39.1. Translation Planes Coupled with Semifield Planes.

If we have coupled planes, where one of the planes is a generalized twisted field plane, then we may use the analysis of Biliotti, Jha and Johnson [121] for this study since all such planes are known, to a certain extent. This theory will be explicated more completely in Chapter 11. The group listed above as G in the previous material on coupled planes has the form that we called C under an appropriate basis change. In this case, we must have a field group induced on the axis $x = 0$. We then have the following results.

THEOREM 39.18. (1) If a non-Desarguesian generalized twisted field plane of order q^n is one plane of a coupling then q is odd, n is odd and the second plane of the coupling is a Suetake plane.

(2) The generalized twisted field planes of order q^n corresponding to Suetake planes are as follows:

$$x = 0, y = 0, y = xm - cx^{q^{2b^*}} m^{q^{b^*}},$$

where $(2b^*, n) = (b^*, n) = 1$ and q is any odd prime power.

Finally, using the Ganley–Jha, Cordero–Figueroa theorem (see [392] and [253]); that any finite semifield plane of order $p^n \neq 2^6$, p a prime and $n \geq 3$, that admits an autotopism group transitive on one side of the autotopism triangle is a generalized twisted field plane, Jha and Johnson [583] improve the previous theorem as follows:

THEOREM 39.19. (Jha and Johnson [583]). Let π be a finite translation plane that is coupled with a semifield plane. Then π is a Suetake plane or a simple full non-square André plane.

We now have that the group

$$\left\langle \begin{bmatrix} t^{-1} & 0 \\ 0 & t^{q^{b^*}} \end{bmatrix} ; t \in GF(q^n)^* \right\rangle$$

acts on both the generalized twisted field plane and the Suetake plane and we have by our previous analysis the form of the inversing map is

$$(x, y) \longmapsto (y^{q^{-b^*}}, x^{q^{b^*}}).$$

Hence, we may obtain the Suetake planes from the generalized twisted field planes listed and construct an alternative spread set for these planes.

THEOREM 39.20. (Jha and Johnson [583]). Let the following spread denote the generalized twisted field plane spread from which the Suetake planes may be constructed:

(39.1.1) $x = 0, y = 0, y = xm - cx^{q^{2b^*}} m^{q^{b^*}}$,   for $m \in GF(q^n) - \{0\}$,

where $(2b^*, n) = (b^*, n) = 1$ and q is any odd prime power.

Let G denote the collineation group

$$\left\langle \begin{bmatrix} t^{-1} & 0 \\ 0 & t^{q^*} \end{bmatrix} ; t \in GF(q^n)^* \right\rangle.$$

Let $f(x) = x^{q^{-2b^*}} - c^{q^{-b^*}} x$ and note that f is injective on $GF(q^n)$. Let the two component orbits of length $(q^n - 1)/2$ be denoted by Γ_1 and Γ_2, where $y = x - cx^{q^{2b^*}}$ is in Γ_1 and $y = xm_0 - cx^{q^{2b^*}} m_0^{q^{b^*}}$ is in Γ_2.

Then the Suetake planes have the following spread:

(39.1.2) $x = 0, y = 0, (y = (xm_0 - cx^{q^{2b^*}} m_0^{q^{b^*}})G) \cup (y = (f^{-1}(x))G)$.

This is $y = (f^{-1}(x))$.

39.2. Generalized André and Inversing Replacements.

We now raise the question of what generalized André planes admit replacements of the type introduced in the previous section.

First consider the generalized André field planes π of order q^2 and kernel $GF(q)$ that admit replacements of inverse type by the map $(x, y) \longmapsto (y^{p^{-s}}, x^{p^s})$.

Then the components of π are of the form:

$$y = x^{q^{\lambda(m)}} m; m \in GF(q^2)$$

where $\lambda(m) = 0$ or 1. The inverse map takes $y = x^q m$ onto

$$y = x^{q^{-1} p^{2s}} m^{-qp^s} = x^{qp^{2s}} m^{-qp^s}.$$

Furthermore,

$$y = x \longmapsto y = x^{p^{2s}}.$$

Choose $p^s = q$. Then replacement using the inverse map takes

$$y = x^q m \longmapsto y = xm^{-1} \text{ and}$$
$$y = xn \longmapsto y = x^q m^{-q}.$$

This is a replacement if and only if $\lambda(m) = 1$ if and only if $\lambda(m^{-1}) = 0$. For more general choices of p^s, there will be large number of possible planes and where the kernels of the two planes may be quite different.

Similarly, there are a variety of non-trivial replacements when $n > 2$. If $p^s = 1$, for example, $y = x^{q^{\lambda(m)}}m$ maps to $y = x^{q^{n-\lambda(m)}}m^{-q^{\lambda(m)}}$. Suppose we have a multiply-replaced André plane where $\lambda(m) = \lambda(n)$ if and only if $m^{(q^{\lambda(m)}-1)/(q-1)} = n^{(q^{\lambda(n)}-1)/(q-1)}$. Note that, in this case, $m^{(q^{\lambda(m)}-1)/(q-1)} = \alpha \in GF(q)$ if and only if $m^{-q^{\lambda(m)}} = \alpha^{-1}$. Let $\lambda(m) = \lambda_m$, a constant throughout the André net $\mathcal{A}_\alpha^{\lambda_m}$, then the André net $\mathcal{A}_\alpha^{\lambda_m}$ maps to the André net $\mathcal{A}_{\alpha^{-1}}^{n-\lambda_m}$ under the inversing mapping. This means that if we agree to replace an α-André net if and only if we replace an α^{-1}-André net, we would then obtain a replacement net using the inversing replacement procedure.

Hence, we obtain:

THEOREM 39.21. (Jha and Johnson [**583**]). Let π be a generalized André plane of order q^n and kernel $GF(q)$ that admits a subnet A_π which is a union of André nets such that an α-André net is in A_π if and only if a corresponding α^{-1}-André net is in A_π then π admits replacement by inversing.

Of course, it is possible to obtain more general replacements when $p^s \neq 1$. Now suppose that we demand that the two generalized André planes are coupled and that one of the coupled planes in the manner in which the Suetake planes are coupled to a class of generalized twisted field planes.

THEOREM 39.22. (Jha and Johnson [**583**]). Assume two generalized André planes of order q^n are coupled by G and that G has two orbits on the line at infinity of lengths $(q^n - 1)/2$ on both of the planes. Then either

(1) One of the planes is Desarguesian and the other is a simple full non-square André plane, or

(2) Both of the planes are simple full non-square André planes.

CHAPTER 40

Hyper-Reguli.

In Chapters 11 and 16 through 18, we see that there are hyper-reguli which are not André hyper-reguli but nevertheless are built from André nets; the replacements not being André replacement.

So, the question arises whether there are hyper-reguli of order q^n and degree $(q^n-1)/(q-1)$, which are not in fact André hyper-reguli, even though we know that not all replacements of André hyper-reguli need to lead to André planes. When $n = 3$, it is known by Bruck [178] that every hyper-regulus is an André hyper-regulus. However, it still is possible that a set of André hyper-reguli might not correspond to an André set of hyper-reguli; that is, might not determine an André plane (in the latter case, the set is said to be 'linear'). In fact, this is done in Dover [323, 325], where sets of two and three André hyper-reguli are constructed that are not linear, thus constructing non-André translation planes. Furthermore, Culbert and Ebert [267] have determined several classes of sets of hyper-reguli of order q^3, of large order. In addition, Basile and Brutti [103, 104] analyze sets of hyper-reguli of order q^3 that produce either André or generalized André planes.

We consider the class of translation planes obtained by the replacement of such a hyper-regulus and show that we obtain a new class of translation planes, which are not André or generalized André. These planes are interesting in that they have a very small affine homology group, whereas in the André or generalized André planes the affine homology group is rather larger (order $(q^n-1)/(q-1)$ for André planes). In fact, when n and q are both > 3, and n is odd, the affine homology group of these planes is trivial.

We repeat here for emphasis some of the material on André nets.

40.1. André Nets and Planes.

It is, of course, well known that a regulus R in $PG(3, q)$ is a set of $q + 1$ lines that are covered by another set of $q+1$ lines, the opposite regulus R^*. Considering this in the associated 4-dimensional $GF(q)$-vector space, we have a set of $q + 1$ 2-dimensional $GF(q)$-subspaces that are covered by a set of $q + 1$ Baer subplanes that share the zero vector. If we have a regulus R in V_4, considered within a Desarguesian affine plane with spread

$$x = 0, y = xm; m \in K^2 \simeq GF(q^2),$$

we consider the kernel homology group

$$Kern^2 \colon \left\{ \sigma_d : (x,y) \longmapsto (xd, yd); d \in K^2 - \{0\} \right\}.$$

Then if we select a Baer subplane of R, π_o (line of R^*), it follows that

$$R^* = \pi_o Kern^2.$$

For example, if we consider the sets

$$A_\delta = \left\{ y = xm; \; m^{q+1} = \delta \right\}, \text{ for } \delta \in F \simeq GF(q),$$

we have a union of $q-1$ reguli that are disjoint from $x = 0, y = 0$. Any of these may be selected to be derived as above. In this setting the opposite reguli $A_\delta^* = \{ y = x^q m; \; m^{q+1} = \delta \}$.

More generally, if $K^n \simeq GF(q^n)$, and we consider the Desarguesian affine plane with spread

$$x = 0, y = xm; m \in K^n \simeq GF(q^n),$$

the sets

$$A_\delta = \left\{ y = xm; \; m^{(q^n-1)/(q-1)} = \delta \right\}, \text{ for } \delta \in F \simeq GF(q),$$

are called the 'André' nets of degree $(q^n - 1)/(q-1)$. Here, there are corresponding sets that cover these André nets, defined as follows:

$$A_\delta^{q^k} = \left\{ y = x^{q^k} m; \; m^{(q^n-1)/(q-1)} = \delta \right\}, \text{ for } \delta \in F \simeq GF(q), \; k = 1, 2, \ldots, n-1.$$

The analogous construction process of derivation then may be more generally considered as follows: If $(k, n) = 1$ and π_o is any subspace $y = x^{q^k} m_o$ such that $m_o^{(q^n-1)/(q-1)} = \delta$, then

$$A_\delta^{q^k} = \pi_o Kern^n.$$

If $(k, n) \neq 1$ we may also construct a covering set of degree $(q^n - 1)/(q^{(k,n)} - 1)$ of

$$A_{\delta,k} = \left\{ y = xm; \; m^{(q^n-1)/(q^{(k,n)}-1)} = \delta \right\}, \text{ for } \delta \in F^{(k,n)} \simeq GF(q^{(k,n)}),$$

by taking

$$A_\delta^{q^j} = \pi_o Kern^n.$$

DEFINITION 40.1. A 'hyper-regulus' in a vector-space of dimension $2n$ over $GF(q)$ is a set of $(q^n - 1)/(q - 1)$ n-dimensional $GF(q)$-subspaces that have a replacement set of $(q^n - 1)/(q - 1)$ n-dimensional $GF(q)$-subspaces, so that each subspace of the replacement set 'lies over' each of the original subspaces in a 1-dimensional $GF(q)$-subspace.

Hence, we see that the André nets defined above are hyper-reguli and if the degree is q^n, there are at least $n-1$ replacement sets. These replacement sets are called the 'André' replacements, which are also hyper-reguli. We note that any André replacement hyper-regulus is also an André hyper-regulus using a different Desarguesian affine plane.

Moreover, each André net in a Desarguesian affine plane Σ admits an affine homology group of Σ that is transitive on the components:

$$\left\langle (x, y) \longmapsto (x, ya^{(q-1)}); \; a \in GF(q^n) \right\rangle$$

acts transitively on A_1. We note the following for André nets of order q^n and degree $(q^n - 1)/(q - 1)$, for $n > 2$.

André planes may be determined by the existence of certain affine homologies. But, also, we note that

THEOREM 40.2. (See, for example, Jha–Johnson [**582**]). Given an André net $\mathcal{A}$ in a Desarguesian plane Σ of order q^n, $n > 2$, there is a unique pair of components L and M external to $\mathcal{A}$ such that $\mathcal{A}$ admits non-trivial affine homologies with axis L and coaxis M, and with axis M and coaxis L.

The André nets of a Desarguesian plane Σ are in an orbit under $GL(2, q^n)$.

We isolate the following two problems.

PROBLEM 40.3. Is any hyper-regulus in a $2n$-dimensional $GF(q)$-vector space always an André hyper-regulus?

PROBLEM 40.4. If $\mathcal{A}$ is an André hyper-regulus, is every replacement hyper-regulus also an André hyper-regulus (an André replacement hyper-regulus)?

We note that strictly as an abstract net of degree $(q^n - 1)/(q - 1)$, and order q^n, an André net may be embedded in a Desarguesian affine plane, and if $n > 2$, this Desarguesian affine plane is unique. We define the more general term 'André net' in a Desarguesian affine plane of order q^n by means of the following theorem.

THEOREM 40.5. Let $\mathcal{A}$ be any André net of degree $(q^n - 1)/(q - 1)$ and order q^n, where $n > 2$. Then there is a unique Desarguesian affine plane Σ containing $\mathcal{A}$. Now choose any two components L and M of Σ. Then there is a set of $(q - 1)$ mutually disjoint André nets of degree $(q^n - 1)/(q - 1)$ such that the union of these nets with L and M is the spread for Σ. Any such net shall be called an André net.

Therefore, we obtain the following corollaries.

COROLLARY 40.6. Within any Desarguesian affine plane, there are exactly

$$(q^n + 1)q^n(q - 1)/2$$

André nets in Σ. And, there is a corresponding set of

$$(n - 1)(q^n + 1)q^n(q - 1)/2$$

replacement sets for André nets of Σ.

THEOREM 40.7. If $n > 2$, there are exactly

$$(|\Gamma L(2n, q)| / |\Gamma L(2, q^n)|)(q^n + 1)q^n(q - 1)/2$$

André nets in a $2n$-dimensional $GF(q)$-vector space V_{2n}.

COROLLARY 40.8. Let Σ be an affine Desarguesian plane of order q^n, $n > 2$, and let $\mathcal{A}$ be a hyper-regulus whose components lie in Σ. If $\mathcal{A}$ is not an André hyper-regulus of Σ, then $\mathcal{A}$ is not an André hyper-regulus.

When $n = 3$, Bruck [**179**] has shown that every hyper-regulus that sits in a Desarguesian affine plane is an André hyper-regulus. Furthermore, Pomareda [**1108**] proved that each André hyper-regulus of order q^3 can have exactly two possible André replacements.

So, the question is whether there could be hyper-reguli that are not André of order q^n, for $n > 3$.

40.2. Non-André Hyper-Reguli.

We have noted above that certain subspaces $y = x^{q^k}$ 'lie over' sets of $(q^n - 1)/(q^{(k,n)} - 1)$ components in the Desarguesian affine plane. When we count André hyper-reguli of degree $(q^n - 1)/(q - 1)$, then we can count these by counting the number of subspaces that lie over $(q^n - 1)/(q - 1)$ components and are in an orbit of some $y = x^{q^k}$. Thus, we may determine André hyper-reguli by consideration of the number of lying-over subspaces in orbits under $GL(2, q^n)$. In particular, we have:

THEOREM 40.9. Let Σ be a Desarguesian affine plane of order q^n.
There are exactly

$$q^n(q^{2n} - 1)\varphi(n)/2$$

lying-over subspaces defining André hyper-reguli and each such subspace is in an orbit of $GL(2, q^n)$ under a subspace of the form $y = x^{q^k}$, where $(k, n) = 1$.

COROLLARY 40.10. Let Σ be a Desarguesian affine plane of order q^n. If L is a lying-over subspace which is not an image of $y = x^{q^k}$ under $GL(2, q^n)$, for $(k, n) = 1$, then the set $\mathcal{A}$ of components that L lies over is a hyper-regulus that is not an André hyper-regulus. Furthermore, $LKern^n$ is a replacement set for $\mathcal{A}$.

40.2.1. The Fundamental Hyper-Reguli of Jha–Johnson.
In a series of papers by Jha and Johnson [**582, 584, 642, 643**], a large class of new hyper-reguli are constructed. The number of different isomorphism classes is extraordinarily large and none of these planes are generalized André planes or André planes. There are few collineations and fewer central collineations. The basic idea is to find a subspace that 'lies over' a Desarguesian net in the sense that the components of intersection are all of the same dimension. As basic as this seems, this is a very difficult problem. However, to begin, we note that it is possible to determine a hyper-regulus of order q^n that is not an André hyper-regulus for $n > 3$.

THEOREM 40.11. (Jha–Johnson [**582**]). Let Σ denote the Desarguesian affine plane of order q^n. Consider the following set of subspaces

$$\left\{ y = x^{q^k} d^{1-q^k} + x^{q^{n-k}} d^{1-q^{n-k}} b; \ d \in GF(q^n) - \{0\} \right\},$$

where $b^{(q^n-1)/(q^{(k,n)}-1)} \neq 1$.
Then this set is the replacement set for a hyper-regulus in Σ of order q^d and order $(q^n - 1)/(q^{(k,n)} - 1)$; $y = x^{q^k} + x^{q^{n-k}} b$ is a $GF(q^{(k,n)})$-subspace and lies over Σ in the sense that the subspace intersects exactly $(q^n - 1)/(q^{(k,n)} - 1)$ components in 1-dimensional $GF(q^{(k,n)})$-subspaces. If $n/(n,k) > 3$, the hyper-regulus of components that this subspace lies over is not an André hyper-regulus.

DEFINITION 40.12. Any hyper-regulus of order q^n and degree $(q^n - 1)/(q^{(k,n)} - 1)$ of a Desarguesian affine plane that has a replacement set of the form

$$\left\{ y = x^{q^k} ad^{1-q^k} + x^{q^{n-k}} d^{1-q^{n-k}} c; \ d \in GF(q^n) - \{0\} \right\}, \text{ for } a, c \in GF(q^n),$$

is called a 'fundamental hyper-regulus'.

REMARK 40.13. A fundamental hyper-regulus for $c = 0$, and $a \neq 0$, is an André hyper-regulus.

COROLLARY 40.14. Any fundamental hyper-regulus for $ac \neq 0$ is isomorphic to

$$\left\{ y = x^{q^k} d^{1-q^k} + x^{q^{n-k}} d^{1-q^{n-k}} b; \ d \in GF(q^n) - \{0\} \right\},$$

for some b.

PROOF. Simply take the mapping $(x, y) \longmapsto (x, ya^{-1})$ as the isomorphism and let $a^{-1}c = b$. $\square$

So there are hyper-reguli that are not André hyper-reguli (in the strictest sense of this concept), but the question is whether the associated translation planes may be considered new. The derivation of any regulus net in an affine Desarguesian produces the Hall plane regardless of the particular regulus net. This is far from the case when replacing a particular fundamental hyper-regulus net of Jha and Johnson, as will be seen.

PROPOSITION 40.15. When $n/(k,n) > 2$, each hyper-regulus of degree $(q^n - 1)/(q^{(k,n)} - 1)$ and order q^n may be embedded in at most one Desarguesian spread of order q^n.

THEOREM 40.16. (Jha–Johnson [582]). Let Σ be a Desarguesian affine plane of order q^n. Form the translation plane π of order q^n, $n/(k,n) > 2$, with spread

$$\left\{ y = x^{q^k} d^{1-q^k} + x^{q^{n-k}} d^{1-q^{n-k}} b; \ d \in GF(q^n) - \{0\} \right\} \cup M,$$

where M is the subspread of the associated Desarguesian plane Σ whose components do not intersect the hyper-regulus. Note that M contains $x = 0$.

If $q > 3$, then the full collineation group of π is the group inherited from the associated Desarguesian plane Σ.

Generalized André planes always have fairly large affine homology groups. That the planes of Jha–Johnson do not have such groups shows how different they are from generalized André planes.

THEOREM 40.17. Assume that $q > 3$ and that π is a Jha–Johnson plane as in the previous theorem. Then the following hold:

(1) If $n/(k,n)$ is odd > 3, then the plane π admits no affine central collineation group and the full collineation group in $GL(2, q^n)$ has order $(q^n - 1)$ and is the kernel homology group of the associated Desarguesian plane.

(2) If $n/(k,n)$ is even > 3, then the plane π admits symmetric affine homology groups of order $q^{(k,n)} + 1$ but admits no elation group. The full collineation group in $GL(2, q^n)$ has order $(q^{(k,n)} + 1)(q^n - 1)$, and is the direct product of the kernel homology group of order $(q^n - 1)$ by a homology group of order $q^k + 1$.

COROLLARY 40.18. The translation plane of order q^n, $n/(k,n) > 3$, $q > 3$, has kernel $GF(q^{(n,k)})$ and cannot be a generalized André or André plane.

40.2.2. Replacements. Since there are $n - 1$ different replacement nets for a given André net of degree $(q^n - 1)/(q - 1)$ and order q^n, one could ask if there are more than one replacement for the non-André hyper-regulus that we have determined previously. When n is odd, we may show that there are at least two such replacements.

THEOREM 40.19. If n is odd and $\alpha \in GF(q) - \{0\}$ and $(k, n) = 1$, then

$$\{x^{q^k} \alpha d^{1-q} + x^{q^{n-k}} \alpha^{-1} b d^{1-q^{n-1}}\}$$

and

$$\{x^{q^k} \alpha^{-1} b^{q^k} d^{1-q} + x^{q^{n-k}} \alpha d^{1-q^{n-1}}\}$$

are replacements for each other, provided

$$b^{(q^n-1)/(q-1)} \neq \alpha^{\pm 2n}.$$

Hence, in each situation above, we may replace any subset of the set of hyper-reguli in at least two ways, when n is odd.

40.2.3. Isomorphism. We have seen that any fundamental hyper-regulus is isomorphic to one of the following form

$$\mathcal{H}_{b,k} = \left\{ y = x^{q^k} d^{1-q^k} + x^{q^{n-k}} d^{1-q^{n-k}} b; \; d \in GF(q^n) - \{0\} \right\},$$

for $b^{(q^n-1)/(q-1)} \neq 1$, where $n/(n, k) > 3$.

The question is: when are $\mathcal{H}_{b,k}$ and $\mathcal{H}_{b^*,k^*}$ isomorphic?

THEOREM 40.20. (Jha–Johnson [**582**]). For hyper-reguli given by

$$\mathcal{H}_{b,k} = \left\{ y = x^{q^k} d^{1-q^k} + x^{q^{n-k}} d^{1-q^{n-k}} b; \; d \in GF(q^n) - \{0\} \right\} \cup M,$$

we have the following isomorphisms.

(1) $\mathcal{H}_{b,k}$ and $\mathcal{H}_{b^*,k}$ are isomorphic if and only if

$$b^* = b^{p^s} a^{-q^{n-k}+q^k} = b^{p^s} a^{-q^k(q^{n-2k}-1)}$$

for some element a of $GF(q^n) - \{0\}$.

(a) When $(n, 2k) = (n, k)$ there are at least

$$(q^{(n,k)} - 2)/(|\mathrm{Gal}\, GF(q^n)|, (q^{(n,k)} - 2))$$

mutually non-isomorphic hyper-reguli.

(b) When $(n, 2k) = 2(n, k)$, there are at least

$$(q^{(n,k)} + 1)(q^{(n,k)} - 2)/(|\mathrm{Gal}\, GF(q^n)|, (q^{(n,k)} + 1)(q^{(n,k)} - 2))$$

mutually non-isomorphic hyper-reguli.

(2) $\mathcal{H}_{b,k}$ and $\mathcal{H}_{b,k^*}$ are isomorphic if and only if

$$k^* \in (k, n - k) \text{ modulo } n.$$

Thus, we may work out the isomorphisms as follows: From the above theorem, we may choose k in any of $(n - 1)/2$ ways if n is odd and in any of $n/2$ ways if n is even. Note that

$$\Delta_k = (q^{(n,k)} + 1)^{((n,2k)-(n,k))/(n,k)}$$

is 1 or $(q^{(n,k)} + 1)$ exactly when $(n, 2k) = 2(n, k)$ or (n, k), respectively. However, these choices are restricted by the assumption that $q > 3$ and $n/(k, n) > 3$. Let

$$\Gamma_k = |\mathrm{Gal}\, GF(q^d)|, (q^{(n,k)} + 1)^{((n,2k)-(n,k))/(n,k)} (q^{(n,k)} - 2).$$

COROLLARY 40.21. For $q > 3$ and for all k such that $n/(k,n) > 3$, there are at least

$$\sum_{k=1; n/(k,n)>3}^{[n/2]} \Delta_k(q^{(n,k)} - 2)/\Gamma_k$$

mutually non-isomorphic translation planes of order q^n and kernels $GF(q^{(j,n)})$ that may be obtained from a Desarguesian plane by replacement of a hyper-regulus of the type $\mathcal{H}_{b,j}$.

To get a sense of the size of the preceding number, we note that when n is highly composite, there are a wide variety of planes of various size kernels $GF(q^{(n,j)})$. When n is prime we obtain only planes of kernel $GF(q)$. For example, when n is a prime > 3, we obtain at least $(n-1)/2$ possible k's so we obtain, for $q = p^t$,

$$\sum_{k=1}^{(n-1)/2} (q-2)/(|\operatorname{Gal} GF(q^n)|, (q-2)) = (n-1)(q-2)/2(nt, q-2)$$

possible mutually non-isomorphic translation planes of order q^n and kernel $GF(q)$. To compare this with the number of non-isomorphic André replacements, we see that there are exactly $n - 1$ replacements of which there are exactly $[n/2]$ that are non-isomorphic. So, roughly speaking, the number of isomorphisms of these new planes is approximately a factor of $(q - 2)$ times the number of isomorphisms of André planes obtained by the replacement of exactly one André hyper-regulus.

REMARK 40.22. If $n = q$ is an odd prime > 3, then $(n, n-2) = 1$, so we obtain exactly $(q - 1)(q - 2)/2$ mutually non-isomorphic planes. Hence, the number of isomorphism classes grows with the order of the plane.

40.3. Multiple Hyper-Regulus Planes.

In the previous section, there have been constructed hyper-reguli in a Desarguesian affine plane of order q^n. The construction is valid for any prime power q and any integer $n > 2$. The nice feature here is that these new hyper-reguli are never André when $n > 3$. In fact, $n > 3$, replacement of such a hyper-regulus produces a plane which is not André or generalized André. When there are sets of at least two hyper-reguli and $n = 3$, it is possible that there is some overlap with certain of the sets of hyper-reguli of Ebert and Culbert (see Section 43.1). In any case, in general, it is of great interest to find sets of mutually disjoint hyper-reguli.

It is actually possible to find sets of mutually disjoint hyper-reguli of various cardinalities. Replacement of any subset of a set of these mutually disjoint hyper-reguli will produce a new translation plane, a plane that is not André or generalized André.

THEOREM 40.23. (Jha–Johnson [**584**]). Let Σ denote a Desarguesian affine plane of order q^n, for $n > 2$, coordinatized by a field isomorphic to $GF(q^n)$. Assume that k is an integer less than n such that $n/(n, k) > 2$. Let ω be a primitive element of $GF(q^n)^*$, then for ω^i, attach an element $f(i)$ of the cyclic subgroup of $GF(q^n)^*$ of order $(q^n - 1)/(q^{(n,k)} - 1)$, $C_{(q^n-1)/(q^{(n,k)}-1)}$, and for $\omega^{-iq^{n-k}}$, attach an element $f(i)^{-q^{n-k}}$. Hence, we have a coset representative set

$$\{\omega f(1), \omega^2 f(2), \dots, \omega^{(q^{(n,k)}-1)} f(q^{(n,k)} - 1)\}$$

for $C_{(q^n-1)/(q^{(n,k)}-1)}$. Let

$$\mathcal{H}^* = \left\{ \begin{array}{c} y = x^{q^k}\omega^{i_j}f(i_j)cd^{1-q^k} + +x^{q^{n-k}}\omega^{-i_j q^{n-k}}f(i_j)^{-q^{n-k}}bd^{1-q^{n-k}} \\ ; \, d \in GF(q^n)^*, \\ \text{for } i_j \in \lambda \subseteq \{1, 2, \ldots, q^{(n,k)} - 1\}, \\ \text{assume } \left(\frac{b}{c}\right)^{(q^n-1)/(q^{(n,k)}-1)} \notin (\omega^{i_j+i_z})^{(q^n-1)/(q^{(n,k)}-1)}, \\ \text{for all } i_j, i_z \in \lambda \end{array} \right\}.$$

Then $\mathcal{H}^*$ is a set of $|\lambda|$ mutually disjoint hyper-reguli of order q^n and degree $(q^n - 1)/(q^{(n,k)} - 1)$.

Using this general theorem, it is possible to apply it to construct a variety of different sets of mutually disjoint hyper-reguli, from which come a large number of translation planes of order q^n and kernel $GF(q^{(n,k)})$. When n is odd, each hyper-regulus has a replacement hyper-regulus of the same general form.

One of the nicest applications of this main result involves what we are calling 'group-constructed' sets of hyper-reguli and is as follows.

THEOREM 40.24. (Jha–Johnson [584]). Let Σ denote a Desarguesian affine plane of order q^n, for $n > 2$, coordinatized by a field isomorphic to $GF(q^n)$. Assume that k is an integer less than n such that $n/(n,k) > 2$. Let ω be a primitive element of $GF(q^n)^*$, then for ω^i, attach an element $f(i)$ of the cyclic subgroup of $GF(q^n)^*$ of order $(q^n - 1)/(q^{(n,k)} - 1)$, $C_{(q^n-1)/(q^{(n,k)}-1)}$, and for $\omega^{-iq^{n-k}}$, attach an element $f(i)^{-q^{n-k}}$. Hence, we have a coset representative set

$$\{\omega f(1), \omega^2 f(2), \ldots, \omega^{(q^{(n,k)}-1)} f(q^{(n,k)} - 1)\}$$

for $C_{(q^n-1)/(q^{(n,k)}-1)}$. Let $e \neq 1$ be a divisor of $q^{(n,k)} - 1$ and let

$$\mathcal{H}^*_{e,(n,k)} = \left\{ \begin{array}{c} y = x^{q^k}\omega^{i_j}f(i_j)cd^{1-q^k} + +x^{q^{n-k}}\omega^{-i_j q^{n-k}}f(i_j)^{-q^{n-k}}bd^{1-q^{n-k}} \\ ; \, d \in GF(q^n)^*, \\ \text{for } i_j \in \lambda = \{\, ei; \, i = 1, 2, \ldots, (q^{(n,k)} - 1)/e\,\}, \\ \text{choose } \left(\frac{b}{c}\right)^{(q^n-1)/(q^{(n,k)}-1)} \notin C_{(q^{(n,k)}-1)/e} \end{array} \right\}.$$

Then $\mathcal{H}^*_{e,(n,k)}$ is a set of $(q^{(n,k)} - 1)/e$ mutually disjoint hyper-reguli of order q^n and degree $(q^n - 1)/(q^{(n,k)} - 1)$.

Although it might appear that sets of mutually disjoint hyper-reguli using groups might imply that the group acts transitively on the set of hyper-reguli, this is far from what actually occurs as it turns out that ordinarily only the group isomorphic to $GF(q^n)^*$ acts as a collineation group of any of the constructed translation planes, which further distinguishes this class from the class of André planes.

Actually, the constructions that we established are somewhat ubiquitous in that they work in all Pappian spreads with minimum assumptions. For example, Jha and Johnson prove the following theorem and note the difference between the finite and infinite cases can result in the construction of translation Sperner spaces.

THEOREM 40.25. (Jha–Johnson [584]). Let L be any field and σ be any automorphism of L. Assume that $L^{\sigma-1} \neq L$. Then $(L - \{0\})^{\sigma-1} = L^{*(\sigma-1)}$ is a proper subgroup of $L - \{0\} = L^*$. Let $\mathcal{B}$ be a coset representative set for $L^{*(\sigma-1)}$.

Let λ be a subset of $\mathcal{B}$ so that $\bigcup \omega_i \omega_j L^{*(\sigma-1)} \neq L^*$, for all ω_i, ω_j in λ. If $b \in L^* - \bigcup \omega_i \omega_j L^{*(\sigma-1)}$ then

$$\left\{ y = x^\sigma \omega_i d^{1-\sigma} + x^{\sigma^{-1}} \omega_i^{-\sigma^{-1}} d^{1-\sigma^{-1}} b; \; d \in L^*, \omega_i \in \lambda \right\}$$

is a set of mutually disjoint hyper-reguli, where $y = xm; m \in L, x = 0$, defines the corresponding Pappian spread Σ coordinatized by L.

Choose any subset λ^* of λ. Then there is a corresponding Sperner space constructed using these subspaces together with the components of Σ not intersecting the components of λ^* and forming a spread. If the spread is a congruence partition (for example, in the finite-dimensional or finite case), any such translation plane will have kernel containing the fixed field of σ.

In all of our examples, any associated translation plane (say in the finite case) admits the cyclic kernel homology group of order $q^n - 1$ of the associated Desarguesian affine plane. When this occurs, it is possible to construct subgeometry and quasi-subgeometry partitions, which then are new whenever the associated translation planes are new. In fact, our constructions produce the first sets of non-André hyper-reguli whose associated translation planes produce a variety of non-André subgeometry partitions. For true subgeometry partitions to occur, the key feature is that there is an associated field $GF(q^2)$ whose multiplicative group contains a scalar (kernel homology group) of order $q - 1$ but yet is not itself a kernel homology group. So, in particular, when n is even and larger than or equal to 4, new subgeometry partitions may be constructed. For example, we provide the following construction.

THEOREM 40.26. (Jha–Johnson [584]). If n is even and > 2, there are sets of hyper-reguli in a Desarguesian plane of order q^n that produce non-generalized-André partitions of $PG(n-1, q^2)$ consisting of $PG(n/2-1, q^2)$'s and $PG(n-1, q)$'s admitting a collineation group of order $(q^n - 1)/(q^2 - 1)$ fixing each $PG(n/2-1, q^2)$ with orbits of length $(q^n - 1)/(q^2 - 1)$ on the set of $PG(n-1, q)$'s.

In terms of subgeometry partitions, we are also interested in determining something of a classification of such partitions. We are able to show the following, which provides a classification of sorts.

THEOREM 40.27. (Jha–Johnson [584]). Assume a partition of $PG(n-1, q^2)$ consisting of $PG(n/2 - 1, q^2)$'s and $PG(n-1, q)$'s admits a collineation group of order $(q^n - 1)/(q^2 - 1)$ fixing each $PG(n/2 - 1, q^2)$ with orbits of length $(q^n - 1)/(q^2 - 1)$ on the set of $PG(n-1, q)$'s and there are at least three $PG(n/2-1, q^2)$ and $n > 2$. Then the associated translation plane π may be constructed from a Desarguesian affine plane by multiple hyper-regulus replacement.

40.3.1. General Multiplicative Theorem. There are a number of ways to obtain sets of mutually disjoint reguli based on what we call 'multiplicative sets'. Again, these constructions are not necessarily of the group-constructed of maximal type sets that we have previously constructed.

THEOREM 40.28. (Jha–Johnson [584]). Suppose that $\{ \alpha_i; \alpha_i \in GF(q) - \{0\}, i = 1, 2, \ldots, t \}$ is a set of elements of $GF(q)$ such that

$$\left(\frac{\alpha_i}{\alpha_j} \right)^{(n,q-1)} \neq 1, \; \alpha_j \neq \alpha_i,$$

and there exists an element b in $GF(q^n) - \{0\}$ for which the following conditions hold:

$$b^{(q^n-1)/(q-1)} \neq \left(\frac{\alpha_i}{\alpha_1 \cdots \widehat{\alpha}_i \cdots \alpha_t} \right)^n,$$

$$b^{(q^n-1)/(q-1)} \neq \left(\frac{1}{\alpha_1 \cdots \widehat{\alpha}_i \cdots \widehat{\alpha}_j \cdots \alpha_t} \right)^n,$$

for $\alpha_i \neq \alpha_j$, where $\widehat{\alpha}_i$ indicates that the element α_i is not in the product. Then

$$\mathcal{R}_t = \Big\{ y = x^{q^k} \alpha_i d^{1-q} + x^{q^{n-k}} \alpha_1 \alpha_2 \cdots \widehat{\alpha}_i \cdots \alpha_t b d^{1-q^{n-1}}; \ i = 1, 2, \ldots, t;$$

$$d \in GF(q^n) - \{0\} \Big\}, \ \text{for } (k, n) = 1,$$

is a partial spread of degree $t(q^n - 1)/(q-1)$ that lies over a set of $t(q^n - 1)/(q-1)$ components in a Desarguesian affine plane Σ which is defined by a set of t hyper-reguli. If $\mathcal{M}$ denotes the set of components of $\Sigma - \mathcal{R}_t$, then

$$\mathcal{R}_t \cup \mathcal{M}$$

is a spread with kernel $GF(q)$.

DEFINITION 40.29. Any set of mutually disjoint hyper-reguli obtained from a set λ as above shall be called a 'multiplicative' set of hyper-reguli. More precisely, we call this a 'multiplicative set of degree 1'.

For more applications of the multiplication theorem and other related concepts, the reader is directed to the article by Jha and Johnson [584].

40.4. Classifying the Planes.

In the previous sections of this chapter, we have given a variety of constructions of sets of mutually disjoint hyper-reguli in Desarguesian affine planes of order q^n. If $n > 3$, none of these hyper-reguli can be André nets by an argument similar to the one showing that

$$\{ y = x^q d^{1-q} + x^{q^{n-1}} b d^{1-q^{n-1}}; \ d \in GF(q^n)^* \}$$

defines a non-André hyper-regulus given in Jha and Johnson [582]. Furthermore, it is shown that if one replaces exactly one of these hyper-reguli, the corresponding translation plane can never be André or generalized André, and therefore is essentially new.

Now if we take a set of mutually disjoint hyper-reguli of the type that we have constructed and replace a subset of at least one such hyper-regulus (in at least two ways if $n/(n, k)$ is odd), the question is: are these planes also not André or generalized André? It is noted in Jha and Johnson [582] that the group that leaves a hyper-regulus invariant has a very small affine homology group, while the kernel is at least $GF(q)$. However, since André and generalized André planes of kernel $GF(q)$ have rather larger affine homology groups (i.e., of order $(q^n - 1)/(q - 1)$ for André planes), we see that none of these planes can be André or generalized André.

CHAPTER 41

Subgeometry Partitions.

Concerning the possibility of partitions of projective geometries, A.A. Bruen and J.A. Thas [198] determined that it is possible to construct partitions of the points of $PG(2s-1, q^2)$ by projective subgeometries isomorphic to $PG(s-1, q^2)$'s and $PG(2s-1, q)$'s. Since there are at least two possible types of geometries such partitions perhaps could be called 'mixed subgeometry partitions'. Bruen and Thas showed that there is an associated translation plane of order q^{2s} and kernel containing $GF(q)$, to contrast with a construction using Segre varieties. That is, a more general construction using Segre varieties is given in Hirschfeld and Thas [529], which generalizes the previously mentioned results of Bruen and Thas and includes Baer subgeometry partitions of $PG(2s, q^2)$ by $PG(2s, q)$'s. In this latter case, there is an associated translation plane of order q^{2s+1} and kernel containing $GF(q)$.

Actually, it is possible to consider subgeometry partitions of $PG(n-1, q^2)$ by subgeometries isomorphic to $PG(n/2-1, q)$ and $PG(n-1, q)$, for n even. Also, analogously, translation planes may be constructed. The associated translation planes are said to be 'geometrically lifted' by the subgeometry partition. Conversely, Johnson [755] introduced a procedure called 'retraction' by which certain translation planes may be retracted to a subgeometry partition. This procedure may be considered a reconstruction in the sense that applying the geometric lifting procedure to the plane produces the original planes. The fundamental property of importance in the construction is the existence of a collineation group of order $q^2 - 1$ containing a kernel homology group of order $q - 1$ and whose union with the zero mapping becomes a field isomorphic to $GF(q^2)$. Therefore, a subgeometry partition may be obtained from a translation plane of order q^{2n}, if it is known that an associated field group of order $q^2 - 1$ exists.

Let Σ be a Desarguesian affine plane of order q^{2n}. Assume that there is a subspace S of dimension n over $GF(q)$ that has the property that the non-zero intersections with components are all 1-dimensional $GF(q)$-subspaces. Let K^{2n} denote the kernel homology group of Σ of order $q^{2n}-1$. Then SK^{2n} is a replacement for the net of components of non-trivial intersection. Any such net of degree $(q^{2n}-1)/(q-1)$ and order q^{2n} that admits a replacement of this sort is called a 'hyper-regulus' net. Until very recently, the only known hyper-reguli were André hyper-reguli. In fact, when $n = 2$ or 3, all hyper-reguli are André. In any case, if $\mathcal{H}$ is a set of mutually disjoint hyper-reguli with replacements given as indicated, then by replacing any subset, there is a corresponding translation plane of order q^{2n} that admits a collineation group of order $q^{2n}-1$, with the required properties to produce a subgeometry partition using the subgroup of order $q^2 - 1$. It might be noted that there is a generalization of subgeometry partitions, to which field groups of order

307

$q^k - 1$ correspond. Such structures are called 'quasi-subgeometry partitions' (see Johnson [**762**]).

Hence, subgeometry partitions may be constructed provided sets of mutually disjoint hyper-reguli may be found in Desarguesian planes. Therefore, there are a variety of mutually disjoint André hyper-reguli that produce subgeometry partitions and these are explicated in Johnson [**760, 762**], André planes of order q^m and kernel $GF(q)$ admit symmetric affine homology groups of order $(q^m - 1)/(q - 1)$ and are constructed by the replacement of a set of mutually disjoint hyper-regulus nets. Conversely, a set of mutually disjoint hyper-regulus sets of a Desarguesian affine plane is said to be 'linear' if and only if there is an affine homology group of order $(q^m - 1)/(q - 1)$ that leaves each net invariant, where the axis and coaxis of the homology group lie external to the set of hyper-reguli.

Thus, we are looking for non-linear sets of mutually disjoint hyper-reguli, even if the hyper-reguli individually may be André. Another approach is to find new replacements for existing André hyper-reguli, which will then produce non-André subgeometry partitions. This has been done in Johnson [**760**].

The construction results of Jha and Johnson construct a wide variety of new translation planes of order q^m by multiple hyper-regulus replacement, where, for $m > 3$, are never André hyper-reguli. It is further shown that there are a number of ways of choosing mutually disjoint hyper-reguli and when $n > 3$, none of these hyper-reguli are André. For example, for $n > 2$, it is possible to have sets of $(q-1)/2$, $(q-3)/2$ and $q/2 - 1$ mutually disjoint hyper-reguli for $q \equiv 1, 3, 2 \bmod 4$, respectively. For $n > 3$, the replacement of any subset of these sets produces translation planes that are not generalized André. For $n > 2$, the replacement of any subset of cardinality at least 2 produces non-generalized-André planes. All such hyper-reguli originate in Desarguesian affine planes and have unusual replacements but are of the type mentioned above, where the resulting translation planes of order q^m admit a cyclic collineation group of order $q^m - 1$. Hence, we may then construct new quasi-subgeometry partitions (here the partitioning is more complex than indicated above) and when m is even, $m = 2n$, we may construct new subgeometry partitions of $PG(n - 1, q^2)$ by subgeometries isomorphic to $PG(n/2 - 1, q^2)$ and $PG(n - 1, q)$.

For example, to stick strictly with subgeometry partitions, the following provides a variety of new partitions.

THEOREM 41.1. *If n is even and > 2, there are sets of hyper-reguli in a Desarguesian plane of order q^n that produce non-generalized-André partitions of $PG(n - 1, q^2)$ consisting of $PG(n/2 - 1, q^2)$'s and $PG(n - 1, q)$'s admitting a collineation group of order $(q^n - 1)/(q^2 - 1)$ fixing each $PG(n/2 - 1, q^2)$ with orbits of length $(q^n - 1)/(q^2 - 1)$ on the set of $PG(n - 1, q)$'s.*

In particular, every subset of any class of mutually disjoint sets of hyper-reguli of Jha and Johnson produces non-generalized-André partitions.

Thus, since these constructions involve such a vast array of new subgeometry partitions, it might appear that there is no hope of formulating a classification. However, it still is possible to give something of a characterization of the type of partitions obtained in terms of the collineation group

41.1. Group Characterization.

We have a group C of order $q^n - 1$ acting on a translation plane of order q^n. This group arises from a Desarguesian affine plane of order q^n, and if we replace hyper-regulus nets of degree $(q^n - 1)/(q - 1)$, then the kernel homology group of order $q - 1$ is contained in the group. If n is even, there is a subgroup of order $q^2 - 1$ such that union 0 is a field and contains the kernel $GF(q)$. Hence, applying Johnson [**755**], since we have a spread in $PG(2n - 1, q)$, let $n = 2n^*$. Then there is a corresponding subgeometry partition of $PG(2n^* - 1, q^2)$ consisting of $PG(n^* - 1, q^2)$'s and $PG(2n^* - 1, q)$'s.

For example, if we replace exactly one hyper-regulus, there are exactly $(q^n - 1)/(q^2 - 1)$ subspaces isomorphic to $PG(n - 1, q)$ and $q^n + 1 - (q^n - 1)/(q^2 - 1)$ subspaces isomorphic to $PG(n/2 - 1, q^2)$. In this case, there is a cyclic collineation of order $(q^n - 1)/(q^2 - 1)$ acting transitively on the $PG(n - 1, q)$'s and fixing each of the remaining $PG(n/2 - 1, q^2)$'s. The question is whether such subgeometry partitions can be characterized by such a collineation group.

The $PG(n - 1, q)$ subspaces lift to reguli in the associated vector space of dimension $2n$ over $GF(q)$. These regulus nets are each fixed by a collineation group H of order $q^2 - 1$ such that $H \cup \{0\}$ is isomorphic to $GF(q^2)$ and H contains the kernel homology group of order $q - 1$ and the $q + 1$ components of each are in an orbit under H. Hence, there is a collineation group G containing H of order $(q^n - 1)$ that fixes each of the components of the translation plane corresponding to the $PG(n/2 - 1, q^2)$ subspaces. Assume that $n > 2$ so there is a collineation τ in G of order a prime p-primitive divisor of $q^n - 1$. By Johnson [**700**], there is a corresponding Desarguesian affine plane Σ of order q^n whose components are the τ-invariant subspaces of line size. Since τ fixes exactly $(q^n + 1) - (q^n - 1)/(q - 1)$ components, it follows that the plane in question π and the Desarguesian plane Σ share exactly these components. We know from the construction that H fixes each of the components of $\pi \cap \Sigma$ and the action of G/H is such that it follows that G fixes each of the components of $\pi \cap \Sigma$. By cardinality, it follows easily that G is a collineation group of Σ, so is in $\Gamma L(2, q^n)$. Since the number of fixed components is strictly larger than the order of a subfield of $GF(q^n) + 1$, then G is the kernel homology group of Σ.

Now consider the more general situation, that there is a set λ of $PG(n - 1, q)$ subspaces in orbits of length $(q^n - 1)/(q^2 - 1)$ under a cyclic group of order $(q^n - 1)/(q^2 - 1)$ that fixes the remaining $PG(n/2 - 1, q^2)$'s and assume that $\lambda < (q - 1)$. Then G fixes at least $2 + (q^n - 1)/(q - 1)$ $PG(n/2 - 1, q^2)$'s. The previous argument then applies to show that the corresponding translation plane π shares at least $2 + (q^n - 1)/(q - 1)$ components of an associated Desarguesian affine plane Σ of order q^n. Furthermore, it also follows that G is a subgroup of $GL(2, q^n)$ that is the kernel homology group of Σ of order $q^n - 1$. This means that the remaining components are in orbits of length $(q^n - 1)/(q - 1)$ since G contains the kernel homology group of order $q - 1$. This forces the remaining components to be in a set of mutually disjoint hyper-reguli and entails that the plane is obtained from a Desarguesian plane by multiple hyper-regulus replacement.

THEOREM 41.2. Assume a partition of $PG(n - 1, q^2)$ consisting of $PG(n/2 - 1, q^2)$'s and $PG(n - 1, q)$'s admits a collineation group of order $(q^n - 1)/(q^2 - 1)$ fixing each $PG(n/2 - 1, q^2)$ with orbits of length $(q^n - 1)/(q^2 - 1)$ on the set of $PG(n - 1, q)$'s and there are at least three $PG(n/2 - 1, q^2)$ and $n > 2$. Then the associated translation plane π may be constructed from a Desarguesian affine plane by multiple hyper-regulus replacement.

PROOF. The assumptions imply that there are at least $2 + (q^n - 1)/(q - 1)$ components fixed by the collineation group acting on the associated translation plane. Our previous argument then applies.   □

So, we have a tremendous number of new translation planes, but what we have not determined is the isomorphism classes. For every integer $k < n$ and divisor $e \neq 1$, of $q^{(n,k)} - 1$, we have constructed at least

$$\left((q^{(n,k)} - 1)/e \right)^{(q^n - 1)/(q^{(n,k)} - 1)}$$

sets of $(q^{(n,k)} - 1)/e$ mutually disjoint hyper-reguli. For different values of k and e it is not clear if the two classes produce isomorphic planes. Furthermore, there are a variety of ways of choosing these sets by varying the constants b and c, the number of which is not counted in the above number. Moreover, we have given sets of $(q^{(n,k)} - 3)/2$ and $q^{(n,k)}/2 - 1$ hyper-reguli. Furthermore the method of 'avoidance' applies to connect group-constructed sets of hyper-reguli and multiplicative sets of hyper-reguli. What has not really been considered yet is the determination of the corresponding isomorphism classes. Since this requires the determination of the collineation groups of each plane.

All of the examples constructed have the property that the number of hyper-reguli is $\leq (q^{(n,k)} - 1)/2$ or $q^{(n,k)}/2 - 1$, for q odd or even, respectively. In this case, Jha and Johnson [642] show how to determine the full collineation groups of the constructed translation planes, which provides a means to determine the isomorphism classes. In fact, it is then shown that there are extraordinarily large numbers of mutually non-isomorphic planes. The reader is directed to [642] for the complete description of the classes.

CHAPTER 42

Groups on Multiple Hyper-Reguli.

Since we now have non-André hyper-reguli, there is the problem of providing criteria to show that a given hyper-regulus cannot be André. In this regard, we prove the following characterization theorem.

THEOREM 42.1. Suppose a hyper-regulus of order q^n and degree $(q^n - 1)/(q - 1)$ in a Desarguesian affine plane Σ of order q^n for $n > 3$ has a homology-type replacement which has a Desarguesian subset D of cardinality $> ((q^n - 1)/(q - 1) + (q^{[n]} + q))/2$. Then the hyper-regulus is André.

Concerning the collineation group of the translation planes, Jha and Johnson [642] provide the following results.

THEOREM 42.2. Let $\mathcal{H}$ be any set of hyper-reguli of degrees $(q^n - 1)/(q - 1)$ and order q^n in a Desarguesian affine plane Σ of order q^n, $q > 3$, each of which has a homology-type replacement.
(1) If the cardinality of $\mathcal{H}$ is $< (q - 1)/2$, then the full collineation group of a translation plane obtained by the homology-type replacement of any subset is the group inherited from the Desarguesian plane Σ.
(2) Assume the cardinality of $\mathcal{H}$ is $(q - 1)/2$ and the union of the replacement hyper-reguli is not Desarguesian. Then the full collineation group of a translation plane obtained by the homology-type replacement of any subset is the group inherited from the Desarguesian plane Σ.

We have given various constructions of mutually disjoint hyper-reguli previously considered in the main construction theorem of Jha and Johnson, of which the group-constructed sets are most easily described.
The collineation group of any translation plane obtained by the replacement of a subset of a group-constructed set of hyper-reguli is the group inherited from the associated Desarguesian plane and is usually quite small, containing essentially only the cyclic kernel homology group of order $(q^n - 1)$ that fixes each hyper-regulus.

THEOREM 42.3. Let π be a translation plane of order q^n, and kernel $GF(q^{(n,k)})$, for $n/(n, k) > 3$, obtained by the replacement of a subset of hyper-reguli of type (b, k, t) of order q^n and degrees $(q^n - 1)/(q^{(n,k)} - 1)$ from a Desarguesian affine plane Σ.
Then the full collineation group of π is a subgroup of $\Gamma L(2, q^n)$ and the intersection with $GL(2, q^n)$ is either the kernel homology group of Σ or $(q^{(k,n)} - 1)/t$ is even and there is an affine homology of order 2 permuting the subset of hyper-reguli used in the replacement in orbits of length 2.
In particular, the full group of affine homologies is either trivial or of order 2.

Normally isomorphism results are difficult to come by, but in contrast the above results show that there are vast numbers of mutually non-isomorphic planes that may be constructed in this manner.

THEOREM 42.4. Let π be a translation plane of order $q^u = p^{nd}$, for p a prime, and kernel $GF(q^{(n,k)})$, for $n/(n,k) > 3$, obtained by the replacement of a subset of a group-constructed set of hyper-reguli of type (b, k, t) of order q^n and degrees $(q^n - 1)/(q^{(n,k)} - 1)$ from a Desarguesian affine plane Σ (note that we require $((q^{(n,k)} - 1)/t, n/(n,k)) = 1$).

(1) Then there are at least

$$\frac{\sum_{i=1}^{(q^{(n,k)}-1)/t} \binom{(q^{(n,k)}-1)/t}{i}}{((q^{(n,k)} - 1)/t, 2) \, |\mathrm{Gal}\, GF(q^n)|}$$

mutually non-isomorphic translation planes constructed.

(2) If π is constructed using a subset of (b, k, t) and ρ is constructed using a subset of (b^*, k^*, t), where $(n, k) = (n, k^*)$, then

$$k^* \in \{k, n - k\}, \mod n,$$

and

(a) When $(n, 2k) = (n, k)$ there are at least

$$(q^{(n,k)} - 2)/(|\mathrm{Gal}\, GF(q^n)|, (q^{(n,k)} - 2))$$

ways of choosing the term b or b^* so that no hyper-regulus of one set is isomorphic to any hyper-regulus of the second set, while

(b) When $(n, 2k) = 2(n, k)$, there are at least

$$(q^{(n,k)} + 1)(q^{(n,k)} - 2)/(|\mathrm{Gal}\, GF(q^n)|, (q^{(n,k)} + 1)(q^{(n,k)} - 2))$$

ways of choosing the term b or b^* so that no hyper-regulus of one set is isomorphic to any hyper-regulus of the second set.

Let $\theta(k) = \{ j; 1 \le j \le n; (k, n) = (j, n) \}$.

(3) Hence, when $(n, 2k) = (n, k)$, there are at least

$$\left(\frac{\sum_{i=1}^{(q^{(n,k)}-1)/t} \binom{(q^{(n,k)}-1)/t}{i}}{((q^{(n,k)} - 1)/t, 2) \, |\mathrm{Gal}\, GF(q^n)|} \right) \cdot \left(\frac{(q^{(n,k)} - 2)}{(|\mathrm{Gal}\, GF(q^n)|, (q^{(n,k)} - 2)) \, |\theta(k)|/2} \right)$$

mutually non-isomorphic translation planes obtained using the same group, i.e., the same t, and when $(n, 2k) = 2(n, k)$, there are at least

$$\left(\frac{\sum_{i=1}^{(q^{(n,k)}-1)/t} \binom{(q^{(n,k)}-1)/t}{i}}{((q^{(n,k)} - 1)/t, 2) \, |\mathrm{Gal}\, GF(q^n)|} \right) (q^{(n,k)} + 1)(q^{(n,k)} - 2)/\Delta,$$

where

$$\Delta = \left(|\mathrm{Gal}\, GF(q^n)|, (q^{(n,k)} + 1)(q^{(n,k)} - 2) \right) |\theta(k)|/2,$$

mutually non-isomorphic translation planes, all with kernel $GF(q^{(n,k)})$.

(4) If $(n, k) \ne (n, k^*)$ a plane obtained using the group $C_{(q^{(n,k)}-1)/t}$ and a plane obtained using the group $C_{(q^{(n,k^*)}-1)/t^*}$ cannot be isomorphic. Let $\delta(n)$ denote the number of divisors of n not including the integer n. For each $\delta(n)$, we may choose a corresponding t_n. Furthermore, let $\delta_2(n)$ denote the divisors z of n so that $2z$ is also a divisor of n and let $\delta_{2'}(n)$ denote the complement of $\delta_2(n)$ in $\delta(n)$. In

the following summations, we assume that (n,k) is appropriate to either $\delta_2(n)$ or $\delta_{2'}(n)$.

Then there are at least

$$\sum_{(n,k)=1}^{\delta_2(n)} \left(\frac{\sum_{i=1}^{(q^{(n,k)}-1)/t_n} \binom{(q^{(n,k)}-1)/t_n}{i}}{\left((q^{(n,k)}-1)/t_n, 2 \right) |\operatorname{Gal} GF(q^n)|} \right) (q^{(n,k)}+1)(q^{(n,k)}-2)/\Delta$$

$$+ \sum_{(n,k)=1}^{\delta_{2'}(n)} \left(\frac{\sum_{i=1}^{(q^{(n,k)}-1)/t} \binom{(q^{(n,k)}-1)/t}{i}}{\left((q^{(n,k)}-1)/t, 2 \right) |\operatorname{Gal} GF(q^n)|} \right)$$

$$\cdot \left(\frac{(q^{(n,k)}-2)}{\left(|\operatorname{Gal} GF(q^n)|, (q^{(n,k)}-2) \right) |\theta(k)|/2} \right),$$

where

$$\Delta = \left(|\operatorname{Gal} GF(q^n)|, (q^{(n,k)}+1)(q^{(n,k)}-2) \right) |\theta(k)|/2,$$

mutually disjoint translation planes of order q^n and kernel containing $GF(q)$ that may be obtained from a Desarguesian affine plane by the replacement of a subset of a group-constructed set of hyper-reguli.

We now consider the collineation group of a translation plane constructed from a Desarguesian plane by multiple hyper-regulus replacement.

THEOREM 42.5. If the set of hyper-reguli of order q^n, for $n > 3$ and $q > 3$, has cardinality $\leq (q-1)/2$ and contains a non-André regulus, then the full collineation group is the group inherited from the Desarguesian affine plane Σ.

THEOREM 42.6. Assume that π has order q^3, for $q > 3$. If a set of hyper-reguli of cardinality $\leq (q-1)/2$ is not linear, then the full collineation group is the inherited group.

Actually, if π has order q^n, for $q > 3$, $n \geq 3$, constructed by homology-type replacement of a non-Desarguesian set of hyper-reguli of cardinality $\leq (q-1)/2$, then the full group is the inherited group.

COROLLARY 42.7. Assume that π has order q^3, for $q > 3$. If a set of hyper-reguli of cardinality $\leq (q-1)/2$ is not linear, then π is not a generalized André plane.

42.1. Isomorphisms of the Fundamental Planes.

We have noted the wide variety of constructions of mutually disjoint hyper-reguli. Choosing any subset of such produces a translation plane. For simplicity, any such plane constructed in this manner shall be called a 'Jha–Johnson hyper-regulus plane'.

The fundamental hyper-reguli are of the following general form:

$$\left\{ y = x^{q^k} \alpha d^{1-q^k} + x^{q^{n-k}} \alpha^{-1} \rho d^{1-q^{n-k}} b; \ d \in GF(q^n) - \{0\} \right\},$$

where α, ρ are in $GF(q^{(n,k)}) - \{0\}$, and where the condition for the existence of this hyper-regulus is $b^{(q^n-1)/(q^{(k,n)}-1)} \neq \rho^n \alpha^{-n/(k,n)}$.

The following isomorphism theorem provides when two hyper-reguli of this general type could be isomorphic.

THEOREM 42.8. (Jha–Johnson [**582**]). Let

$$\mathcal{H}_{b,k} = \left\{ y = x^{q^k} d^{1-q^k} + x^{q^{n-k}} d^{1-q^{n-k}} b; \ d \in GF(q^n) - \{0\} \right\} \cup M.$$

(1) $\mathcal{H}_{b,k}$ and $\mathcal{H}_{b^*,k}$ are isomorphic if and only if

$$b^* = b^{p^s} a^{-q^{n-k}+q^k} = b^{p^s} a^{-q^k(q^{n-2k}-1)}$$

for some element a of $GF(q^n) - \{0\}$.

(a) When $(n, 2k) = (n, k)$, there are at least

$$(q^{(n,k)} - 2)/(|\mathrm{Gal}\,GF(q^n)|, (q^{(n,k)} - 2))$$

mutually non-isomorphic hyper-reguli.

(b) When $(n, 2k) = 2(n, k)$, there are at least

$$(q^{(n,k)} + 1)(q^{(n,k)} - 2)/(|\mathrm{Gal}\,GF(q^n)|, (q^{(n,k)} + 1)(q^{(n,k)} - 2))$$

mutually non-isomorphic hyper-reguli.

(2) $\mathcal{H}_{b,k}$ and $\mathcal{H}_{b,k^*}$ are isomorphic if and only if

$$k^* \in \{k, n - k\}, \ \mathrm{mod}\ n.$$

CHAPTER 43

Hyper-Reguli of Dimension 3.

Here we restrict attention to hyper-reguli in translation planes of order q^n for $n = 3$. As mentioned, when $n = 3$, Bruck [179] has shown that every hyper-regulus that sits in a Desarguesian affine plane is an André hyper-regulus. Furthermore, Pomareda [1108] proved that each André hyper-regulus of order q^3 can have exactly two possible André replacements.

Furthermore, we have given constructions of hyper-reguli that are not André due to Jha and Johnson of order q^n, $n > 3$, and noted a general construction procedure to construct sets of mutually disjoint hyper-reguli. The collineation groups of the translation planes are relatively well behaved when $n > 3$.

Now restricting the main construction theorem of Jha–Johnson listed in Section 40.3 for $n = 3$, we have:

THEOREM 43.1. (Jha–Johnson [584]). Let Σ be a Desarguesian affine plane of order q^3. Let ω be a primitive element of $GF(q^3)$, then for ω^i, attach an element $f(i)$ of the cyclic subgroup of $GF(q^3)^*$ of order $(q^3 - 1)/(q - 1)$, $C_{(q^3-1)/(q-1)}$. Let

$$\mathcal{H}^* =$$

$$\left\{ \begin{array}{c} y = x^q \omega^{i_j} f(i_j) d^{1-q} + x^{q^2} (\omega^{i_j})^{-q^2} g(i_j) b d^{1-q^2}; d \in GF(q^3)^*, \\ \text{for } i_j \in \lambda \subseteq \{1, 2, \ldots, q-1\}, \text{ assume } b^{(q^3-1)/(q-1)} \notin (\omega^{i_j+i_k})^{(q^3-1)/(q-1)} \end{array} \right\}.$$

Then $\mathcal{H}^*$ is a set of $|\lambda|$ mutually disjoint hyper-reguli.

In particular, (1) then $\mathcal{H}^*_{i_j}$, for i_j fixed, defines a replacement set for a hyper-regulus $\mathcal{H}_{i_j}$ of Σ.

(2) Furthermore,

$$\left\{ y = x^q \omega^{i_j} f(i_j) d^{1-q} + x^{q^2} (\omega^{i_j} f(i_j))^{-q^2} b d^{1-q^2} \right\} \text{ and}$$

$$\left\{ y = x^q (\omega^{i_j} f(i_j))^{-1} b^q d^{1-q} + x^{q^2} (\omega^{i_j} f(i_j))^{q^2} d^{1-q^2} \right\}$$

are replacements for each other.

DEFINITION 43.2. A set of hyper-reguli in $PG(5, q)$ is said to be 'linear' if and only if the set belongs to a standard set of André hyper-reguli invariant under the same affine homology group of order $(q^3 - 1)/(q - 1)$.

The question is if there are any linear subsets of cardinality > 1. In fact, we show that there are no such sets. That is, there are no two hyper-reguli in a linear set. If there is a set of $t > 1$ hyper-reguli of the type represented above, we show that every subset of $j \leq t$, $j > 1$, will produce a non-André plane and hence a non-generalized-André plane.

Culbert and Ebert [267] have constructed several classes of sets of hyper-reguli in $PG(5, q)$, and from the construction it turns out that no two hyper-reguli fall into a linear set. This construction uses the analysis of cubic functions by Sherk [**1159**] and constructs the hyper-reguli directly from the associated Desarguesian affine plane (more precisely from the associated line at infinity), where our construction finds first the replacement hyper-reguli and then only indirectly can one construct the hyper-reguli in the associated Desarguesian plane. Although there may be some overlap between these two classes, this is completely undetermined.

Using algebraic methods concerning linear sets, Jha and Johnson [**584, 643**] prove the following general result:

THEOREM 43.3. Let Σ be a Desarguesian affine plane of order q^3. If we have a set of mutually disjoint hyper-reguli of the type $\{y = x^q \alpha d^{1-q} + x^{q^2} \alpha^{-q^2} b d^{1-q^2}\}$, where α is in $\lambda \subseteq GF(q^3)^*$, then no two of these hyper-reguli are in a linear set.

COROLLARY 43.4. For any subset of at least two mutually disjoint hyper-reguli of the type stated in the previous theorem, for each hyper-regulus, choose one of the two possible replacements. Then a translation plane of order q^3 and kernel $GF(q)$ is constructed that admits a cyclic collineation group of order $q^3 - 1$ but is not an André or generalized André plane.

43.1. A Few Examples of Cubic Planes.

We mention just a few of the many examples of sets of mutually disjoint hyper-reguli producing Jha–Johnson planes of order q^3. The following examples refer to Theorem 43.1.

EXAMPLE 43.5. (1) Let q odd be and $\lambda = \{2, 4, \ldots, (q-1)/2\}$ and $i_j = 2j$. Then $(\omega^{2j+2k})^{(q^3-1)/(q-1)} \in C_{(q-1)/2}$. Choose b so that $b^{(q^3-1)/(q-1)} \notin C_{(q-1)/2}$. Then, we obtain a set of $(q-1)/2$ mutually disjoint hyper-reguli.

(2) Similarly, for q odd, if we take $\lambda = \{2, 4, \ldots, (q-1)/2\} + 1$ and choose b so that $b^{(q^3-1)/(q-1)} \notin \omega C_{(q-1)/2}$, we obtain a set of $(q-1)/2$ mutually disjoint hyper-reguli.

(3) For q odd, take $\lambda = \{1, 2, 3, 4, \ldots, (q-3)/2\}$. Then note that since for i, j in λ we have $i + j \leq q - 3$, we may take $b^{(q^3-1)/(q-1)} = 1$ to construct a set of $(q-3)/2$ mutually disjoint hyper-reguli.

(4) Actually, take any set of $(q-3)/2$ elements of $\{1, 2, \ldots, q-1\}$ such that $i + j$ is not congruent to 0 mod $q - 1$. Then for $b^{(q^3-1)/(q-1)} = 1$, we may construct a set of $(q-3)/2$ mutually disjoint hyper-reguli. More generally, choose any i_0 from $\{1, 2, 3, \ldots, q-1\}$. Then choose any set of $(q-3)/2$ elements so that $i + j$ is not congruent to i_0 mod $(q-1)$. Letting $b^{(q^3-1)/(q-1)} = \omega^{i_0(q^3-1)/(q-1)}$ produces a set of $(q-3)/2$ mutually disjoint hyper-reguli.

For example take $q = 9$. Taking $\lambda = \{2, 4, 6, 8\}$ produces sets of four disjoint hyper-reguli. Taking $\lambda = \{1, 2, 3\}$ produces a set with three disjoint hyper-reguli.

(5) Let q be even and let $\lambda = \{1, 2, 3, \ldots, q/2 - 1\}$ and let $b^{(q^3-1)/(q-1)} = 1$. Then for i, j in $\{1, 2, \ldots, q/2 - 1\}$, we have $i + j \leq q - 2$, so this will produce a set of $q/2 - 1$ mutually disjoint hyper-reguli.

(6) Let k divide $q - 1$. Take $\lambda = \{k, 2k, \ldots, (q-1)/k\}$. Then $i + j = kz$ and $\omega^{kz(q^3-1)/(q-1)} \in C_{(q-1)/k}$. Take b so that $b^{(q^3-1)/(q-1)} \notin C_{(q-1)/k}$ to produce a set of $(q-1)/k$ mutually disjoint hyper-reguli.

(7) In the previous example, let i_0 be any element of

$$\{1, 2, \ldots, q-1\} - \{3, 6, 9, \ldots, (q-1)/k\}$$

and consider $\lambda = \{ i_0 + ktk; t = 1, 2, \ldots, (q-1)/k \} \cup \{k, 2k, \ldots, (q-1)/k\}$. Then $(\omega^{i+j})^{(q^3-1)/(q-1)} \in C_{(q-1)/k} \cup \omega^{i_0} C_{(q-1)/k} \cup \omega^{2i_0} C_{(q-1)/k}$. So, if $k > 3$, we may add on to obtain $2(q-1)/k$ mutually disjoint hyper-reguli.

Note when we obtain a set of t mutually disjoint hyper-reguli, we obtain $(t)^{(q^3-1)/(q-1)}$ different sets of t mutually disjoint hyper-reguli. So, if q is odd, we may obtain $((q-1)/2)^{(q^3-1)/(q-1)}$ sets of $(q-1)/2$ mutually disjoint hyper-reguli and when q is even, we may obtain $(q/2 - 1)^{(q^3-1)/(q-1)}$ sets of $q/2 - 1$ hyper-reguli. Furthermore, all of these sets correspond to the same choice for b. Other choices for b that have the same avoidance condition also work. For example, if b works, so does b^* so that $b^{*(q^3-1)/(q-1)} = b^{(q^3-1)/(q-1)}$. In the group-type sets, if $t = (q-1)/2$, for example, then there are actually $(q^3-1)/2$ choices for b. Hence, there are then at least $((q-1)/2)^{(q^3-1)/(q-1)}(q^3-1)/2$ possible sets of $(q-1)/2$ mutually disjoint hyper-reguli. Moreover, for each coset representative defining a hyper-regulus, we may replace the hyper-regulus with the replacement mentioned above and still produce a set of mutually disjoint hyper-reguli.

We ask if the plane obtained could be an André plane (or generalized André plane). If so, there is an affine homology group fixing each André replacement set which is induced from a homology group of the associated Desarguesian affine plane. So, if this is a set of André nets with carrying lines L and M of Σ, assume first that $x = 0$ and $y = xm_0$ are the axis and coaxis.

Hence, the following theorem shows that none of the possible constructed planes are generalized André.

THEOREM 43.6. (Jha, Johnson [**584**]). Let $\mathcal{H}$ be a set of hyper-reguli of the form $\mathcal{H}^*_{\alpha, \alpha^{-q^2}, b}$ where $\alpha \in \lambda \subseteq GF(q^3)^*$. Choose any subset of hyper-reguli of cardinality at least two. Then the corresponding translation plane is not generalized André or André.

43.2. The Cubic Planes of Culbert and Ebert.

Culbert and Ebert [**267**] have constructed various sets of hyper-reguli of order q^3 and degree $1+q+q^2$. Bruck [**178**] shows that any hyper-regulus in a Desarguesian affine plane of degree $1+q+q^2$ and order q^3 is actually an André hyper-regulus and Pomareda [**1108**] showed there are actually two possible replacements, namely the André replacements. Recall that in the standard setting there are $q-1$ mutually disjoint André hyper-regulus with two components $x = 0, y = 0$ and a corresponding affine homology group H of order $1 + q + q^2$ leaving invariant each André hyper-regulus. A subset of hyper-regular is said to be 'linear' if and only if the set is a subset of an André set of $q-1$ hyper-reguli which are orbits under a group isomorphic to H.

Culbert and Ebert [**267**] actually found sets of mutually disjoint hyper-reguli (so each is necessarily André) with the property that no subset of at least two hyper-reguli is linear. Any such subset then can be replaced in two ways producing a translation plane which cannot be André or indeed generalized André.

The construction of the sets of mutually disjoint hyper-reguli operates on the line at infinity of a Desarguesian affine plane of order q^3. In this context, consider any hyper-regulus $\mathcal{H}$ then there is a collineation of order 3 isomorphic to the Frobenius automorphism of $GF(q^3)$ over $GF(q)$ which fixes $\mathcal{H}$ pointwise. The group generated by such a collineation is called the 'stability group' of the hyper-regulus.

Choose two points on ℓ_∞, P and Q, and denote by $\phi(P,Q)$ the group generated by the stability groups of all hyper-reguli containing P and Q. An orbit under $\phi(P,Q)$ of a point R not equal to P or Q is called a 'cover' ('Bruck cover') and Bruck [**179**] shows that covers are hyper-reguli (the infinite points of hyper-reguli in the associated affine Desarguesian plane).

Let N denote the norm function from $GF(q^3)$ to $GF(q)$, so that $N(x) = x^{1+q+q^2}$. Then any hyper-regulus has the following form:

(i) $\left\{ x \in GF(q^3); \; N(x-a) = f \right\}$, for some $a \in GF(q^3)$, $f \in GF(q)^*$, or

(ii) $\left\{ x \in GF(q^3) \cup \{\infty\}; \; N(\frac{x-a}{x-b}) = f \right\}$, for some $a,b \in GF(q^3)$, $f \in GF(q)^*$.

The main construction device depends on results on 'Sherk surfaces':

DEFINITION 43.7. Let N and T denote the norm and trace, respectively, of $GF(q^3)$ over $GF(q)$. Let $f,g \in GF(q)$, $\alpha, \delta \in GF(q^3)$. Then the 'Sherk surface' is defined by:

$$S(f,\alpha,\delta,g) = \left\{ z \in GF(q^3) \cup \{\infty\}; \; fN(z) + T(\alpha^{q^2} z^{q+1}) + T(\delta z) + g = 0 \right\}.$$

The set of Sherk surfaces can be partitioned into four orbits under $P\Gamma L(2,q^3)$. Two surfaces of the same cardinality are in the same orbit. The cardinalities of the surfaces are 1, $q^2 - q + 1$, $q^2 + 1$, $q^2 + q + 1$. The Sherk surfaces of cardinality $q^2 + q + 1$ are covers or rather hyper-reguli.

Now take a $GF(q)$-linear combination of two Sherk surfaces that have no intersection. Then this linear combination will define a set of Sherk surfaces which are mutually disjoint. The subset of covers of this linear combination consists then of mutually disjoint hyper-reguli. Then by various choices of generating Sherk surfaces of the linear combination, certain subsets of covers arise. For example,

THEOREM 43.8. (Part of Lemma 1 and Theorems 2 and 3, Culbert and Ebert [**267**]). Assume that q is an odd prime.

(1) For $u \in GF(q)$, let $S = S(0,1,0,u)$. Then S is a cover precisely when u is a square in $GF(q)$.

(2) Let B denote the bitrace, $B(x) = T(x^{q+1})$. Let $S = S(0,\alpha,\delta,g)$, for $\alpha, \delta \in GF(q^3)$, not both zero, $g \in GF(q)$. Define

$$\Delta = 4N(\alpha)g - B((\alpha\delta)^q + (\alpha\delta)^{q^2} - \alpha\delta).$$

Then S is a cover precisely when Δ is a non-square.

(3) Let $S = S(1,\alpha,\delta,g)$, a Sherk surface which is not a simple point. Define $\Delta' = 4N(\delta') + (g')^2$, where $\delta' = \delta - \alpha^{q^2+q}$ and $g' = g + 2N(\alpha) - T(\alpha\delta)$. Then S is a cover precisely when Δ' is a square.

Similarly for q even, the main results identifying covers is the following theorem.

THEOREM 43.9. (Part of Lemma 4 and Theorems 5 and 6, Culbert and Ebert [**267**]). Let q be even. Let T_0 denote the trace function from $GF(q)$ to $GF(2)$.

(1) Let $u \in GF(q)^*$. If $S = S(0,1,1,u)$ then S is a cover precisely when $T_0(u+1) = 0$.

(2) Let $\alpha, \delta \in GF(q^3)$, not both zero and let $g \in GF(q)$. Let $S = S(0,\alpha,\delta,g)$. Then S is a cover if $T(\alpha\delta) \neq 0$ and $T_0(c) = 0$, where $c = (gN(\alpha) + B(\alpha\sigma))/T(\alpha\delta)^2$.

(3) Let $S = S(1,\alpha,\delta,g)$ be a Sherk surface that is not a single point. Define $\delta' = \delta + \alpha^{q^2+q}$ and $g' = g + T(\alpha\delta)$. Then S is a cover if $g \neq T(\alpha\delta)$ and $T_0(c') = 0$ where $c' = N(\delta')/(g')^2$.

The main result of Culbert and Ebert for odd order is the following theorem.

THEOREM 43.10. (part of Lemma 1 and Theorems 2,3 Culbert and Ebert [**267**]). Assume that q is an odd prime.
(1) For $u \in GF(q)$, let $S = S(0, 1, 0, u)$. Then S is a cover precisely when u is a square in $GF(q)$.
(2) Let B denote the bitrace, $B(x) = T(x^{q+1})$. Let $S = S(0, \alpha, \delta, g)$, for $\alpha, \delta \in GF(q^3)$, not both zero, $g \in GF(q)$. Define

$$\Delta = 4N(\alpha)g - B((\alpha\delta)^q + (\alpha\delta)^{q^2} - \alpha\delta).$$

Then S is a cover precisely when Δ is a non-zero square.
(3) Let $S = S(1, \alpha, \delta, g)$, a Sherk surface which is not a simple point. Define $\Delta' = 4N(\delta') + (g')^2$, where $\delta' = \delta - \alpha^{q^2 + q}$ and $g' = g + 2N(\alpha) - T(\alpha\delta)$. Then S is a cover precisely when Δ' is a non-zero square.

Similarly for q even, the main results identifying covers is the following theorem.

THEOREM 43.11. (Part of Lemma 4, Theorems 5,6 Culbert and Ebert [**267**]) Let q be even. Let T_0 denote the trace function from $GF(q)$ to $GF(2)$.
(1) Let $u \in GF(q)^*$. If $S = S(0, 1, 1, u)$ then S is a cover precisely when $T_0(u + 1) = 0$.
(2) Let $\alpha, \delta \in GF(q^3)$, not both zero and let $g \in GF(q)$. Let $S = S(0, \alpha, \delta, g)$. Then S is a cover if $T(\alpha\delta) \neq 0$ and $T_0(c) = 0$, where $c = (gN(\alpha) + B(\alpha\delta))/T(\alpha\delta)^2$.
(3) Let $S = S(1, \alpha, \delta, g)$ be a Sherk surface that is not a single point. Define $\delta' = \delta + \alpha^{q^2 + q}$ and $g' = g + T(\alpha\delta)$. then S is a cover if $g \neq T(\alpha\delta)$ and $T_0(c') = 0$ where $c' = N(\delta')/(g')^2$.

Equipped with these theorems, it is then possible to determine the set of mutually disjoint covers within a linear combination of two appropriately selected Sherk surfaces.

The main results of Culbert and Ebert on pencils of Sherk surfaces are given in the following theorem.

THEOREM 43.12. (part of Theorems 11,12,13,14 Culbert and Ebert [**267**])
(i) Let q be an odd prime power ≥ 7. Consider the F-linear combination

$$fS(1, 0, -1, 0) + g(S(0, 0, 0, 1); \ f, g \in GF(q).$$

Then the subset of (mutually disjoint) covers has cardinality $\frac{(q-3)}{2}$. Furthermore, this set is

$$\left\{ (S(1, 0, -1, g); \ -4 + g^2 \text{ is a non-zero square in } GF(q), g \in GF(q) \right\}.$$

(ii) Let q be an odd prime power ≥ 5 and let u be a fixed non-square. Consider the F-linear combination

$$fS(1, 0, -u, 0) + g(S(0, 0, 0, 1); \ f, g \in GF(q).$$

Then the subset of (mutually disjoint) covers has cardinality $\frac{(q-1)}{2}$. This subset of covers is

$$\left\{ S(1, 0, -u, g); \ -4u^3 + g^2 \text{ a non-zero square in } GF(q) \right\}.$$

(iii) Let $q = 2^m$, with $m \geq 3$. Let $v \in GF(q) - \{0, 1\}$.

$$fS(0, 1, 1, v) + g(S(1, 0, 0, 0); \ f, g \in GF(q).$$

Then the subset of (mutually disjoint) covers contains a subset of cardinality $\frac{(q-2)}{2}$.
Furthermore,

(a) if $T_0(v+1) = 0$ then the subset of covers is

$$\left\{ S(1, f^{-1}, f^{-1}, vf^{-1}); vf^{-1} \neq 1; f \in GF(q) - \{0\}; T_0(c') - 0 \right\}$$
$$\cup \left\{ S(0, 1, 1, v); T_0(v+1) = 0 \right\},$$

(b) if $T_0(v+1) = 1$ then the subset of covers is

$$\left\{ S(1, f^{-1}, f^{-1}, vf^{-1}); vf^{-1} \neq 1; f \in GF(q) - \{0\}; T_0(c') = 0 \right\},$$

where T_0 is the trace function from $GF(q)$ to $GF(2)$ and $c' = \frac{N(f^{-1} + f^{-q^2-q})}{vf^{-1} + T(f^{-2})}$, for $vf^{-1} \neq 1$ in both cases.

Since these sets of mutually disjoint hyper-reguli arise from a linear combination of Sherk surfaces, any two hyper-reguli will then linearly generate the same set. Since a linear set of $q - 1$ covers is not generated in any of these these cases, it follows that no two hyper-reguli belong to a linear set. In other words, any subset of at least two hyper-reguli from either of these three situations will produce a 'non-linear' set. Furthermore, each hyper-regulus has two independent replacements and the corresponding translation planes can never be André or generalized André.

We have seen similar types of hyper-regulus replacements of Jha and Johnson of order q^3. However, since the methods of Culbert and Ebert first find the hyper-reguli as images on the Desarguesian line, and Jha and Johnson find the replacements for the hyper-reguli and not explicitly the hyper-reguli, it is not clear how these two sets of translation planes intersect, if they do at all.

We therefore list the following open problem.

PROBLEM 43.13. Determine if any of the classes of Culbert and Ebert listed above are isomorphic to classes of Jha and Johnson in the examples using Theorem 43.1, with the same properties.

REMARK 43.14. Basile and Brutti [**106, 107**] give necessary and sufficient conditions for a subspace disjoint from $x = 0$ to determine a hyper-regulus of order q^3 and degree $1 + q + q^2$. All of these hyper-reguli determine André planes.

43.3. The Hyper-Regulus Planes of Dover.

Dover in [**323, 325**] found pairs and triples of non-linear sets of hyper-reguli of order q^3 and degree $1 + q + q^2$.

The constructions are as follows:

THEOREM 43.15. (Dover [**325**]). Let q be an even power. For $d \in GF(q^3)$, let N denote the norm of $GF(q^3)$ over $GF(q)$ and assume that $N(d) \neq 0, 1$. Assume that $m \in GF(q) - \{0, 1\}$ such that

$$T_0 \left(\frac{mN(d)}{1 + (N(d))^2} \right) = 0,$$

where T_0 is the trace function of $GF(q)$ over $GF(q)$. Let $b \in GF(q^3)$ such that $N(b)$ satisfies:

$$N(b)^2 + m(N(d) + 1)N(b) + m^3 N(d) = 0.$$

(1) Suppose

$$x\varphi = \frac{x+b}{(md/b)x+d}.$$

Then the André partial spread $\{\, y = xm;\ N(m) = 1\,\} = N_1$ and $N_1\varphi$ are disjoint on components.

(2) The set $\{N_1, N_1\varphi\}$ is non-linear.

Similarly, for odd order, there is a corresponding theorem.

THEOREM 43.16. (Dover [**323**]). Let $d \in GF(q^3)$, q odd, and $q \geq 5$, such that $N(d) = d^{1+q+q^2} \neq 0, 1$. Let $m \in GF(q) - \{0, 1\}$ such that $N(d)m + \frac{1}{4}(N(d) - 1)^2$ is a square or 0 in $GF(q)$. Assume that b is an element of norm

$$\frac{1}{2}(N(d) - 1)m + n\left\{ N(d)m + \frac{1}{4}(N(d) - 1)^2 \right\}^{\frac{1}{2}}.$$

If $N_1 = \{\, x \in GF(q^3);\ N(x) = 1 \,\}$ and $g \in PGL(1, q^3)$ such that

$$xg = \frac{x+b}{(md/b)x+d}.$$

Then N_1 and N_1g define line-disjoint hyper-reguli, which form a non-linear set.

Dover also constructs non-linear triples.

THEOREM 43.17. (Dover [**323**]).

(1) Let $q = 3^{2r}$ and let i be a square root of -1 in $GF(q)$. Consider $N_1 = \{\, x \in GF(q^3);\ N(x) = 1 \,\}$, $xf = \frac{x+(i+1)}{(i-1)x-1}$, $xg = \frac{x+(1-i)}{(-1-i)x-1}$. Then $\{N_1, N_1f, N_1g\}$ is a non-linear set of mutually disjoint hyper-reguli.

(2) Assume that $q \equiv 1 \bmod 4$ and $q \neq 0 \bmod 5$. Define $xh = \frac{x+\delta}{-ix+\delta}$, $xf = \frac{x-(i+1)}{(i-1)x-1}$, where $\delta \in GF(q^3)$, such that $N(\delta) = 1 + i$. Then $\{N_1, N_1f, N_1h\}$ is a non-linear set of mutually disjoint hyper-reguli.

Hence, we may choose one of two replacements for each of the two hyper-regulus nets above and construct translation planes which are not André or generalized André.

CHAPTER 44

Elation-Baer Incompatibility.

The theory of groups generated by elations in translation planes, and Baer incompatibility is considered in the *Foundations* book [**123**], and the reader is referred to this text for details. Here we merely list the main points.

44.1. Generalized Elations.

DEFINITION 44.1. Let V denote the additive group of some vector space. Then $\sigma \in GL(V, +)$ is a 'generalized elation' if σ induces the identity on $V/\operatorname{Fix}(\sigma)$. The subspace $\operatorname{Fix}(\sigma)$ is the 'axis' of the generalized elation σ and $C(\sigma) = (\sigma - 1)V$ is the 'center' of σ.

If $\sigma \in GL(V, K)$, for some field K, then σ has 'type t', relative to K, if $\operatorname{Fix}(\sigma)$ has K-dimension t.

Note that although the type t of a generalized elation $\sigma \in GL(V, K)$ obviously depends on the chosen field $K = GF(q)$, over which σ is regarded as a K-linear map of V, the cardinality of $|\operatorname{Fix}(\sigma)| = q^t$ is independent of K.

Elations are generalized elations whose axis coincides with a component.

REMARK 44.2. Suppose that $\sigma \in GL(V, +)$ is an automorphism of an arbitrary spread $\pi = (V, \mathfrak{F})$. Then σ is an elation if and only if it is a generalized elation whose axis is a component.

If V is a vector space of dimension $2n$ over a field K and $\sigma \in GL(V, K)$ is an elation, then it is a generalized elation of type n.

PROOF. Let σ be an elation of π with component F. Then σ must fix the parallel class associated with F. Hence, for $v \in V$, the parallel class $v + F$ is σ invariant, so $\sigma(v) - v \in F$, as desired. The statement on type-size follows from the fact that since a spread is a direct sum of any two isomorphic subspaces, each component has half the dimension of the whole space. $\qquad\square$

However, elations are not the only collineations that are generalized elations. As we shall see, Baer p-elements, of spreads of order p^{2r}, are also generalized elations that, moreover, have the same type as elations. The notion of a generalized elation thus enables results concerning elation groups to be converted to corresponding results concerning Baer groups, and also leads to striking results concerning the interaction of Baer p-groups and elation groups. To discuss all this, we consider further characterizations of generalized elations.

REMARK 44.3. A mapping $\sigma \in GL(V, K)$ is a generalized elation if and only if $(\sigma - 1)^2 = 0$.

PROOF. If σ is a generalized elation, then $(\sigma - 1)V \subset V$, and this implies that $(\sigma - 1)^2 = 0$, since $\sigma - 1$ maps $\operatorname{Fix}(\sigma)$ to zero. The converse is equally trivial. $\qquad\square$

In dealing with generalized elations σ, of a finite vector space V of order p^n, p prime, it will be convenient to regard σ as being a generalized elation relative to $GF(p)$; thus if σ has type t, then the axis has cardinality p^t. It is clear from the foregoing that all this involves no loss in generality.

PROPOSITION 44.4. Let σ be a generalized elation of type t of a vector space V of order p^n. Then the following hold.
(1) The order of σ is p;
(2) $\dim C(\sigma) + \dim \mathrm{Fix}(\sigma) = n$;
(3) $t \geq n/2$;
(4) Relative to any ordered basis of V whose first t elements form a basis of $\mathrm{Fix}(\sigma)$, the generalized elation σ has the following matrix representation

$$\begin{pmatrix} \mathbf{I}_t & A \\ \mathbf{O} & \mathbf{I}_{n-t} \end{pmatrix}$$

where A is a $t \times (n-t)$ matrix.

PROOF. For $x \in V$ we have $\sigma(x) = x + f$, for $f \in \mathrm{Fix}(\sigma)$. Thus, $\sigma^p(x) = x + pf = x$. Case (1) follows. Case (2) holds, because $\ker(\sigma - 1) = \mathrm{Fix}(\sigma)$ and the image of $(\sigma - 1)$ is, by definition, $C(\sigma)$. Now case (3) follows, since $C(\sigma) \subseteq \mathrm{Fix}(\sigma)$. The case (4) is verified by noting that the basis elements e_i, that are not fixed, map to elements of type $e_i + f$, for $f = \sum_{j=1}^{t} f_i a_{ij}$, where $(f_i)_{i=1}^{t}$ is the chosen basis of $\mathrm{Fix}(\sigma)$ and a_{ij} are in the field. $\square$

COROLLARY 44.5. Let V be a vector space of dimension $2n$ over $GF(p)$. Suppose α and β are generalized elations of V with a common axis F of rank n. Then the commutator $[\alpha, \beta] = 1$, so $\alpha\beta$ has order p.

PROOF. Apply Proposition 44.4 (4): obtain upper triangular matrices for both mappings and observe that these must commute. $\square$

However, the set of generalized elations with the same axis (plus the identity) does not always form a group, as the axis of a product $\alpha\beta$ might be larger than the common axis of the generalized elations α and β. Thus, some caution is needed in dealing with generalized elations, as the discussion below indicates.

COROLLARY 44.6. Let G be a subgroup of $GL(V, +)$, V an elementary Abelian p-group of order p^{2n}. Suppose that all of the non-trivial generalized elations in G are of type n: that is, they are of minimal type. Then the generalized elations in G fixing elementwise a subspace F of size p^n form an elementary Abelian group G_F and the non-trivial elements of G are the generalized elations in G with axis F.

We summarize below some equivalent conditions for an automorphism of a *finite* spread to be a generalized elation.

THEOREM 44.7. Let $\sigma \in GL(V, K)$, $K = GF(q)$, be an automorphism of a K-spread $\pi = (V, \mathcal{S})$ of order $n = q^r$ and characteristic p. Regard σ as an element of $GL(V, +)$. Then the following are equivalent.
(1) σ is a generalized elation.
(2) σ is a minimal generalized elation: thus $|\mathrm{Fix}(\sigma)| = \sqrt{|V|}$ (or equivalently σ has type r over $GF(p)$).
(3) σ is an elation or a Baer p-element of π.
(4) σ is a quadratic mapping; thus, $(\sigma - 1)^2 = 0$.

PROOF. By Proposition 44.4, case (3), a generalized elation σ cannot act on a spread of order n unless $|\operatorname{Fix}(\sigma)| = n$; thus the type $t = r$ is forced. However, in a translation plane of order n only Baer collineations and affine central collineations may fix $n = q^r$ affine points. Since $|\sigma|$ is also a p-element, it thus must be a Baer p-element or an elation. (Note that for elations, we could, alternatively, have applied Remark 44.2—to the above and to the remainder of the proof).

Conversely, since elations and Baer p-elements both have matrix representation of form

$$\begin{pmatrix} \mathbf{I} & \mathbf{I} \\ \mathbf{O} & \mathbf{I} \end{pmatrix},$$

they are generalized elations of minimal type. All the desired equivalences follow (if we recall Remark 44.3). □

As indicated in the above proof, by Remark 44.2, elations are easily seen to be generalized elations, even in the infinite case (by a matrix-free argument). However, in attempting to extend the above to the infinite Baer case, we would need to define infinite Baer subplanes—leading to complications that we wish to avoid.

The above suggests how the concept of an elation might be extended to a partial spread.

DEFINITION 44.8. An 'elation of a partial spread' $\pi := (V, \mathcal{F})$ is a mapping $\sigma \in GL(V, +)$ in $\operatorname{Aut}(\pi)$ such that
(1) $\operatorname{Fix}(\sigma)$ is a component of π;
(2) σ induces the identity on $V/\operatorname{Fix}(\sigma)$.

We turn to the study of elations of partial spreads, generalizing the connection between elations of spreads and generalized elations. This approach displays some striking parallels between elations and Baer p-elements that gives rise to some of the most powerful techniques for the study of finite translation planes.

Elations of finite spreads generalize readily to elations of partial spreads. However, we go over the details, partly because this provides an opportunity to fix some notation.

NOTATION 44.9. If K is a field then, for any $n \times n$ matrix T over K, the associated 'elation matrix' is the $2n \times 2n$ matrix over K specified as follows:

$$\rho_T = \begin{pmatrix} \mathbf{I} & T \\ \mathbf{O} & \mathbf{I} \end{pmatrix}.$$

We begin with some generalizations of elations acting on spreads to *partial* spreads. The proofs are the same as for elations on spreads and hence will not be repeated.

REMARK 44.10. Let the dimension-$2n$ vector space $V = K^n \oplus K^n$, for K a finite field, admit a partial spread $\mathfrak{F}$ whose components include $X = K^n \oplus \mathbf{O}$ and $Y = \mathbf{O} \oplus K^n$. Then the following hold:
(1) There is a partial spread set $\mathcal{F}$ of $n \times n$ matrices over K, that includes zero, such that

$$\mathfrak{F} = \{'y = xF'|F \in \mathcal{F}\} \cup \{Y\}.$$

(2) The subset of $\mathcal{F}$ given by

$$\mathcal{E} := \{F \in \mathcal{F}|F + \mathcal{F} \subseteq \mathcal{F}\},$$

is an additive group of matrices contained in $\mathcal{F}$. Hence, $\mathcal{E} \subseteq \mathcal{F}$ is a partial subspread and $\mathcal{F}$ is a union of some cosets of $\mathcal{E}$.

(3) The elations of $\mathfrak{F}$ with axis Y, or the 'Y-shears group' is the subgroup of $GL(V, +)$:

$$\mathfrak{E} := \{ 'y = x\rho_T' | T \in \mathcal{E} \} ,$$

where ρ_T is the elation matrix associated with T, as specified in Notation 44.9.

Therefore, if $y = x$ is a member of $\mathfrak{F}$ then an elation η of $\mathfrak{F}$, with axis the y-axis, maps the x-axis onto the unit line $y = x$ if and only if $\eta = \rho_1$.

(4) The orbit of the component $y = xF$, $F \in \mathcal{F}$, under the elation group $\mathfrak{E}$, is the set $\{ 'y = x(F + E)' | E \in \mathcal{E} \}$.

Hence, the set of slopes of the $\mathfrak{E}$-orbit of X is the set of components with slope set in $\mathcal{E}$ and, more generally, the non-trivial orbits of components of the shears group are the cosets of $\mathcal{E}$ that lie in $\mathcal{F}$.

Later, we consider partial spreads admitting non-trivial elations that do not share their axes. This essentially reduces to the following situation.

REMARK 44.11. Let the dimension-$2n$ vector space $V = K^n \oplus K^n$, for K a field, admit a partial spread $\mathfrak{F}$ whose components include the x-axis, the y-axis and the component '$y = xA$'. The partial spread $\mathfrak{F}$ includes an elation σ_A with axis the x-axis that maps the y-axis to the component $y = xA$ if and only if $\sigma_A \in GL(V, +)$ is an automorphism of $\mathfrak{F}$ of the form

$$\sigma_A = \begin{pmatrix} \mathbf{I} & \mathbf{O} \\ A & \mathbf{I} \end{pmatrix}.$$

Hence, an elation σ with axis the x-axis maps the y-axis to the unit line if and only if $\sigma := \sigma_1$.

PROOF. An automorphism σ of the given form clearly is an elation of the desired type. Conversely, consider an elation σ with axis the x-axis and mapping the y-axis to the unit line. Since σ is the identity on the x-axis, and since it fixes the factor space pointwise, being a generalized elation, it has the form

$$\sigma = \begin{pmatrix} \mathbf{I} & \mathbf{O} \\ A & \mathbf{I} \end{pmatrix},$$

and this maps the component $y = 0$ to $y = xA$, yielding the desired result. □

44.2. Properties of $SL(2, q)$.

In this section, we collect some standard properties of $SL(2, q)$ that we shall use, subsequently. We first note the size of the conjugacy classes of its Sylow subgroups of order q.

REMARK 44.12. Let P be a p-Sylow subgroup of $SL(2, q)$, q a power of the prime p. Let $N(P)$ denote the normalizer of P in $SL(2, q)$. Then the following hold:

(1) For q odd, P splits into two non-trivial conjugacy classes in $N(P)$, each of size $(q - 1)/2$.

(2) For q even, all non-trivial elements of P are conjugate in $N(P)$.

PROOF. The identity

$$\begin{pmatrix} x & 0 \\ 0 & x^{-1} \end{pmatrix} \begin{pmatrix} 1 & a \\ 0 & 1 \end{pmatrix} \begin{pmatrix} x^{-1} & 0 \\ 0 & x \end{pmatrix} = \begin{pmatrix} 1 & ax^2 \\ 0 & 1 \end{pmatrix},$$

implies the desired result.　　　　□

We shall make essential use of the following fundamental theorem of Dickson:

THEOREM 44.13 (Dickson). Let $F = GF(p^r)$, $p > 2$ prime, and let λ be a generator of $F = GF(p^r)$ over the prime subfield Z_p of F, thus, $F = Z_p(\lambda)$. Then the subgroup of $GL(2, p^r)$ specified by

$$L := \left\langle \begin{pmatrix} 1 & 0 \\ 1 & 1 \end{pmatrix}, \begin{pmatrix} 1 & \lambda \\ 0 & 1 \end{pmatrix} \right\rangle$$

is either
  (1) $SL(2, p^r)$, or
  (2) $p^r = 9$ and $L \cong SL(2, 5)$.

PROOF. Gorenstein [**414**, (8.4), pp. 44–55].　　　　□

We shall frequently need to consider the standard representations associated with $GL(2, q)$ and $SL(2, q)$. Our goal is to describe these in concrete terms, in a form suitable for our purposes. We begin with a couple of generalities.

The groups $GL(2, L)$ and $SL(2, L)$, for L any field, have standard actions on $V_2(L)$: these yield their degree-2 representations over L, and hence of all their subgroups. In particular, noting the chain of subgroups $SL(2, K) \subset GL(2, K) \subset GL(2, L)$, for K a subfield of L, we have degree-2 representations, over L, of each of the groups in the chain. Moreover, viewing $V_2(L)$ as a K-vector space of dimension $2n$, where $n = [L : K]$, we see that the representations just considered may be regarded as *standard* K-representations of degree $2n$, of $GL(2, K)$, $SL(2, K)$ (and $GL(2, L)$).

In the reverse situation one might ask: when does $GL(2, K)$, and hence $SL(2, K)$, admit such 'standard' representations of degree $2n$ over K? In the case of interest, $K = GF(q)$, the answer is—always: since K admits a unique extension field $L = GF(q^n)$.

It will prove desirable to work with concrete matrix versions of these representations by choosing $L = GF(q^n) \supset K$ to be a *matrix* field in $GL(n, K)$. Thus, we elaborate on the above as follows.

REMARK 44.14 (Standard Degree $2n$ Representations). The 'standard representations of $GL(2, K)$ and $SL(2, K)$' of degree $2n$ are defined to be those representations obtained by the following procedure:

  (1) Choose any matrix field $L \subseteq \overline{GL(n, K)}$, containing the scalar field K, such that $L \cong GF(q^n)$. Let K^n be regarded as an L-module, based on the standard action of matrices in $M_n(K)$, the ring of $n \times n$ matrices over K, on K^n.

  (2) Thus, $GL(2, L)$ consists of non-singular matrices that partition into 2×2 blocks chosen from the field of matrices L; each of these blocks is a matrix in $M_n(K)$.

  (3) The 'canonical representation' of $GL(2, L)$ is the action of $GL(2, L)$, viewed as a subring of $M_{2n}(K)$, on $V = K^n \oplus K^n$.

  (4) Let $\mathcal{G} = GL(2, K) \subseteq GL(2, L)$, and let $\mathcal{S}$ be the unimodular subgroup of $\mathcal{G}$, viewed as a group of 2×2 matrices over the matrix field L.

(5) Then the representations of $\mathcal{G}$ and $\mathcal{S}$ obtained by restricting the canonical representation of $GL(2, L)$ to these groups are, respectively, the 'standard degree-$2n$ K-linear representations' of $GL(2, K)$ and $SL(2, K)$.

CHAPTER 45

Hering–Ostrom Elation Theorem.

We now turn to the main topic of interest: the investigation of a partial spread $\pi := (V, \mathfrak{F})$, over a prime field $F = GF(p)$, $p > 3$, when π admits non-trivial elations α and β with distinct axes. More precisely, our goal is to determine the associated group generated by the two elations $G := \langle \alpha, \beta \rangle$, and, furthermore, to determine the action of G on V.

The method of investigation will be based on considering G_K the action of G on the vector space $V_K := V \otimes K$, where $K \supset GF(p)$ is an appropriately chosen finite field. It turns out that K may be chosen so that G leaves invariant a dimension-two K-space that may be regarded as a Desarguesian spread on which G induces *faithfully* $SL(2, K)$; this implies $G \cong SL(2, K)$ has *standard* action on V.

All of this essentially involves considering the following situation:

HYPOTHESIS 45.1. $\mathcal{N} \supset \{X, Y, \mathbf{I}\}$ is a partial spread on $V = F^n \oplus F^n$, for F a finite field, admitting an automorphism group $G = \langle \sigma, \rho_A \rangle$, for some $A \neq \mathbf{O}$, where:

$$\sigma := \begin{pmatrix} \mathbf{I} & \mathbf{O} \\ \mathbf{I} & \mathbf{I} \end{pmatrix}; \quad \rho_A := \begin{pmatrix} \mathbf{I} & A \\ \mathbf{O} & \mathbf{I} \end{pmatrix}.$$

The following result analyzes the above situation over the field $F = GF(p)$, mainly for $p > 3$. It shows that if a partial spread admits non-trivial elations ρ and σ with distinct axes then the group they generate $G := \langle \rho, \sigma \rangle \cong SL(2, K)$, for some field $K = GF(q)$, and equally important, the components corresponding to non-trivial elation axes in G form a rational Desarguesian subnet of the given partial spread. Thus knowing that a net is invariant under elations with distinct axes provides considerable information about the structure of the net, as well as its collineation group. The case $p = 3$ creates some special but minor problems, because of the need to apply Dickson's theorem; we shall not treat this exceptional situation in full detail here.

THEOREM 45.2. Assume Hypothesis 45.1, taking $F = GF(p)$, for some odd prime p. Let λ be any eigenvalue of the F-matrix A, as in Hypothesis 45.1, in the algebraic closure of F, and let $K = F(\lambda)$, the cyclic extension field of F by λ. Assume[1] $|K| > 9$ if $p = 3$.

Then the following hold.

(1) The group $G = \langle \sigma, \rho_A \rangle$ leaves invariant a dimension-two K-subspace U of $V \otimes K$, such that the restriction $G^U = SL(2, K)$.

(2) The subgroup of Y-shears in G is the following multiplicative matrix group

$$\Theta := \left\{ \rho_M := \begin{pmatrix} \mathbf{I} & M \\ \mathbf{O} & \mathbf{I} \end{pmatrix} \mid M \in \mathcal{A} \right\},$$

---

[1]To avoid considering the exceptional situation in Dickson's theorem.

where $\mathcal{A} \cong (K, +)$ is an additive group of matrices.

(3) The identity $\mathbf{I} \in \mathcal{A}$ and $\mathcal{A}^*$ is closed under (multiplicative) inversion.

(4) The polynomial ring $F[A]$ coincides with $\mathcal{A}$, and, in fact, the ring $F[A]$ is a field $\cong K$.

(5) The group $G = SL(2, F[A]) \cong SL(2, K)$, and G leaves invariant the rational Desarguesian partial spread:

$$\mathcal{T} := \{y = xT | T \in F[A]\} \cup \{\mathbf{O} \oplus F^n\} \subseteq \mathcal{N}.$$

Hence, the $\mathcal{N}$-elation axes in G consist of the components of the rational Desarguesian spread $\mathcal{T}$ and these form a G-orbit.

(6) The eigenvalues of the matrix A are all conjugates to λ and they lie in the field $K = F(\lambda)$.

(7) The $GF(p)$-space $V = F^n \oplus F^n$ is a vector space, of rank 2, over the field $F[A] \cong K = GF(q)$, under the standard action:

$$(x \oplus y)\phi = (x)\phi \oplus (y)\phi \; \forall x, y \in F^n, \phi \in F[A].$$

In addition, the components $F^n \oplus \mathbf{O}$, $\mathbf{O} \oplus F^n$ and the unit line $y = x$ of $\mathcal{N}$ are all $F[A]$ subspaces of V.

Moreover, $G = SL(2, F[A]) \cong SL(2, K)$, has the standard action on V. Thus, G is the full unimodular subgroup of $GL(V, F[A])$ and G leaves invariant the net $\mathcal{N}$. The Sylow p-subgroups of G are elation groups of $\mathcal{N}$; so each such group fixes a different component of $\mathcal{N}$.

COROLLARY 45.3. Let $\mathcal{N} \supset \{X, Y, \mathbf{I}\}$ be a partial spread on $V = F^n \oplus F^n$, $F = Z_p$, p a prime larger than 2. Let F_n be the ring of $n \times n$ matrices over F, and define the corresponding upper triangular Sylow p-subgroup in $GL(n, F)$ by

$$\Theta(F) := \left\{ \rho_M := \begin{pmatrix} \mathbf{I} & M \\ \mathbf{O} & \mathbf{I} \end{pmatrix} | M \in F_n \right\}.$$

Suppose the partial spread $\mathcal{N}$ admits an automorphism group $G = \langle \sigma, T \rangle$, for some subgroup $T \leq \Theta(F)$, where

$$\sigma := \begin{pmatrix} \mathbf{I} & \mathbf{O} \\ \mathbf{I} & \mathbf{I} \end{pmatrix}.$$

Assume that $|T| > 9$ if $p = 3$. Then, $G \cong SL(2, q)$ such that the p-elements in G are the affine elations in G, and the axes of these elations form a rational Desarguesian partial spread in $\mathcal{N}$ on which G has the standard action.

In particular, the subgroup of Y-shears in G is the following multiplicative matrix group:

$$\Theta := \left\{ \rho_M := \begin{pmatrix} \mathbf{I} & M \\ \mathbf{O} & \mathbf{I} \end{pmatrix} | M \in \mathcal{A} \right\},$$

Thus, we have listed Ostrom's original theorem involving the group generated by two elations with distinct axes. What we have not considered is the group generated by the set of all affine elations in a finite translation plane, nor have we completed the case when the characteristic is three. This has been done by combining the work of C. Hering and T.G. Ostrom and shall be listed below. The interested reader is referred to Hering [476] and Ostrom [1047, 1052].

Hering–Ostrom Theorem.

THEOREM 45.4 (Hering–Ostrom theorem). Let π be a finite translation plane of order p^r, for p a prime. Let E denote the group that is generated by all affine elations in the translation complement (that fix the zero vector). Then one of the following situations occurs:

(1) E is elementary Abelian, and there is a unique elation axis.

(2) E is isomorphic to $SL(2, p^t)$, and there are exactly $p^t + 1$ elation axes, and $p^t + 1$ elementary Abelian elation groups of order p^t.

(3) E is isomorphic to $SL(2, 5)$, $p = 3$, and the number of elation axes is 10.

(4) E is isomorphic to $S_z(2^{2t+1})$ (a Suzuki group), and the number of elation axes is $2^{2(2t+1)} + 1$.

(5) The order of E is $2d$ where d is an odd number and $p = 2$.

CHAPTER 46

Baer Elation Theory.

Here we demonstrate the high degree of incompatibility between Baer p-elements and affine elations, acting on a translation plane π of order p^{2r}. Among the most startling of such results is Foulser's theorem, asserting that non-trivial Baer p-elements and non-trivial affine elations cannot simultaneously act on π if p is odd. The first section of this chapter establishes striking constraints of this type, all due to Foulser, that apply to translation planes of odd order. The second section, due to Jha and Johnson, is concerned with the even order versions of an analogous theory: here affine elations and Baer 2-elements *are* compatible, but they constrain each other quite severely.

46.1. Odd Order.

We consider distinct Baer p-elements, σ and τ, acting on a spread of order p^{2r}.

THEOREM 46.1. Let π be a translation plane of order p^{2k}, for $p > 3$. Suppose that π admits Baer p-elements σ and τ in the translation complement with fixed Baer subplanes π_σ and π_τ, containing the origin **O**.

Then either the axes π_σ and π_τ coincide, or the only affine point shared by π_σ and π_τ is the origin **O**.

Furthermore, Foulser shows that all Baer axes of p-collineations share their parallel classes, for $p > 3$. This might be regarded as a stronger version of the previous theorem.

THEOREM 46.2. Let π^ℓ be a finite affine translation plane of odd order p^{2k}, for $p > 3$, and suppose that $\mathcal{B}$ denotes the set of axes of the Baer p-collineations of π. Then each Baer subplane $B \in \mathcal{B}$ has the same slope set. Thus, $\ell \supset \delta$ such that $|\delta| = p^k + 1$ and δ is the slope set of every $B \in \mathcal{B}$.

It has been pointed out that Ostrom's theorem can be stated for $p = 3$ also and in this case, it is possible that $SL(2,5)$ is generated. This also applies to the result of Foulser that ultimately uses Ostrom's results on generalized elations.

We have mentioned above that an adaptation of the proof of case (3) will show that it is not possible to have both Baer p-collineations and elations acting on a translation plane of odd order. We state this formally. We note that this is valid for 'any' odd prime p.

THEOREM 46.3. Let π be a finite translation plane of odd order p^r.

Then the collineation group of π does not contain both Baer p-collineations and elations.

46.2. Incompatibility: Even Order.

We have seen in the previous section that, when p is odd, it is not possible that elations and Baer p-collineations can coexist in translation planes of order p^r. This is definitely not the case in planes that are not translation planes. For example, there exist 'semi-translation planes' of order q^2 derived from dual translation planes for which there is a Baer group of order q and an elation group of order q as well. Furthermore, it is possible that Baer involutions and elations exist even in Desarguesian affine planes of even order. If π is Desarguesian of order q^2 coordinatized by $GF(q^2)$, then the field automorphism of order 2 that fixes $GF(q)$ pointwise induces a Baer involution.

Note that, in a semifield plane of even order q^2, if there exists a Baer subplane sharing the special point on the line at infinity, then there exists an elation group of order q that leaves the subplane invariant. Here we consider this more generally. That is, we consider translation planes of even order q^2 that admits a Baer 2-group $\mathcal{B}$ and group of elations $\mathcal{E}$ that normalizes $\mathcal{B}$.

46.2.1. Maximal Elation Groups and Baer Involutions.

THEOREM 46.4. Let π be a translation plane of even order q^2 for $q = 2^r$. We assume that all groups are subgroups considered over the prime subfield $GF(2)$. Let π_o be a Baer subplane of π that is fixed pointwise by a Baer 2-group $\mathcal{B}$.
(1) If π admits an elation group $\mathcal{E}$ of order q that normalizes $\mathcal{B}$ then $|\mathcal{B}| \leq 2$.
(2) If $|\mathcal{B}| = 2$ then the full collineation group that fixes π_o pointwise has order 2.

46.2.2. Large Baer Groups and Elations.
Considering possible incompatibility relations, we consider the coexistence of a 'large' Baer group and an elation group of order > 2. Recall that it follows from the previous subsection that the existence of a Baer group of order $> \sqrt{q}$ shows that the Baer axis is a Desarguesian subplane. In this subsection, we consider the possible incompatibility with Baer groups of order $> \sqrt{q}$ and elation groups of order > 4.

REMARK 46.5. Let π be a translation plane of order 2^r that admits a Baer group $\mathcal{B}$ of order $> \sqrt{q}$. Let E be any affine elation group. Let S_2 be a Sylow 2-subgroup of the translation complement containing the full elation group E^* with axis E. Show that there exists a Baer group $\mathcal{B}^*$ of order $|\mathcal{B}|$ contained in S_2. Show that $\mathcal{B}^*$ normalizes the full group E^*.

We may assume, by the above exercise, that $\mathcal{B}$ leaves invariant the axis L for the elation group E. Let E^+ denote the full elation group with axis L. Clearly, we may arrange the representation so that L is $x = 0$ and the representation for the Baer group is as previously presented. Therefore, $\mathcal{B}$ normalizes E^+. Now if E normalizes $\mathcal{B}$, then E centralizes $\mathcal{B}$. We assume throughout this subsection that $|\mathcal{B}| > \sqrt{q}$, implying that $\operatorname{Fix} \mathcal{B} = \pi_o$ is Desarguesian.

LEMMA 46.6. An element $g \in E^+$ normalizes $\mathcal{B}$ if and only if, for

$$g = \begin{pmatrix} I & 0 & E_1 & E_2 \\ 0 & I & E_3 & E_4 \\ 0 & 0 & I & 0 \\ 0 & 0 & 0 & I \end{pmatrix},$$

then $E_3 = 0$, $E_1 = E_4$, and $E_2 = f(E_1)$.

COROLLARY 46.7. *If there exists an element $g \in E$ that does not normalize $\mathcal{B}$, then there is an elation subgroup E_o of E^+ of order $> \sqrt{q}$ that does normalize $\mathcal{B}$.*

Therefore, in the following, we may assume that either E normalizes $\mathcal{B}$ or, if not, there exists a subgroup E_o of E^+ of order $> \sqrt{q}$ that does normalize $\mathcal{B}$. In either case, we shall prove that any normalizing subgroup has order at most 2. Hence, in the latter case, we have $2 > \sqrt{q}$ so that $4 > q$, implying $q = 2$, so that the translation plane is Desarguesian of order 4 implying that the order of $\mathcal{B}$ is 2.

Therefore, we have an elation group E_1 normalizing $\mathcal{B}$ which is either E or E_o of order $> \sqrt{q}$.

THEOREM 46.8. *Let π be a translation plane of order $q^2 = 2^{2r}$ that admits a Baer group $\mathcal{B}$ of order $> \sqrt{q}$. If E is any elation group of π then either $|E| \leq 2$, or π is the Desarguesian plane of order 4, and $\mathcal{B}$ has order 2.*

We note that in Jha–Johnson [602], the proof of the above result required a group-theoretic result due to Dempwolff [293]. Here, our analysis was more combinatorial, thus eliminating the need for this result.

46.2.3. Baer Groups with non-Normalizing Elation Groups.

We have shown that if E is an elation group of order q 'all' of whose elements normalize a Baer group $\mathcal{B}$, then the order of $\mathcal{B}$ is at most 2. In this subsection, we consider the opposite situation: when E is an elation group such that 'no' element of $E - \{1\}$ normalizes $\mathcal{B}$.

Our main result of this subsection is:

THEOREM 46.9. *Let π be a translation plane of even order q^2 admitting a non-trivial Baer 2-group $\mathcal{B}$ and an elation group E of order q such that no element of $E - \{1\}$ normalizes $\mathcal{B}$. Then, π is a semifield plane and $|\mathcal{B}| = 2$.*

Conversely,

THEOREM 46.10. *Let π be a semifield plane of even order q^2 that admits a non-trivial Baer 2-group $\mathcal{B}$. Then, there exists an elation subgroup E of order q such that no element of $E - \{1\}$ normalizes $\mathcal{B}$.*

Note that we have shown that the existence of 'large' Baer groups (i.e., order $> \sqrt{q}$) implies that any elation group has order at most 2, or $q = 2$ and the plane is Desarguesian. What we have not discussed is whether any such elation group must necessarily centralize $\mathcal{B}$, when $q > 2$. However, this is valid as well.

CHAPTER 47

Spreads Admitting Unimodular Sections—Foulser–Johnson Theorem.

We have seen previously that there are very few types of collineation groups that are known to act on finite translation planes. We mentioned the Hering–Ostrom theorem that describes the groups generated by the set of affine elations in the translation complement. In the non-solvable case, if the order of the translation plane is p^n, the possible groups are isomorphic to $SL(2, p^t)$, $S_z(2^t)$, for $p = 2$ and $SL(2, 5)$ for $p = 3$ and all of these groups can occur. In particular, there are a variety of translation planes of order p^n admitting $SL(2, p^t)$, where the p-elements are elations. However, it is also possible that a translation plane of order p^n could admit a collineation group in the translation complement isomorphic to $SL(2, p^t)$, where the p-elements are say Baer (fix a Baer subplane pointwise). There are important classes of translation planes of order q^2 that admit a collineation group $SL(2, q)$ in the translation complement. So, it is of fundamental importance to determine the complete set of translation planes that admit such groups.

Here we discuss the translation planes of order q^2 that admit $SL(2, q)$. One of the most applicable results in the study of the collineation groups of translation planes is the classification of such planes. This result has a history of sorts beginning with the classification of spreads in $PG(3, q)$ admitting $SL(2, q)$ first for the odd-order-q case by Walker [**1203**] and then the even-order-q case by Schaeffer [**1150**]. These results use strongly the fact that the translation plane exists within an ambient 4-dimensional vector space over $GF(q)$. A nice treatment of the work of Walker and Schaeffer appears in Lüneburg's book [**961**].

In Section 101.1, we mention that there are exactly 8 translation planes of order 16, all of which may be derived from the three semifield planes, which we mention formally.

THEOREM 47.1. (See Johnson [**684**]).
(1) There are exactly three semifield planes of order 16, one each with kernel $GF(2)$, $GF(4)$, $GF(16)$.
(2) All other translation planes of order 16 may be obtained by derivation of a semifield plane.
(3) The Dempwolff plane of order 16 admits $SL(2, 4)$ as a collineation group and may be derived from the semifield plane with kernel $GF(2)$.

Hence, any determination of the translation planes of order q^2 admitting $SL(2, q)$ in the translation complement would necessarily include the Dempwolff plane with kernel $GF(2)$, which would not then be included in the work of Walker and Schaeffer.

Using representation of groups, Foulser and Johnson [**378, 379**] completed the study without any hypothesis on the kernel. A sketch of this result will follow, but basically, except when $q = 4$, the idea is to argue using group theory that only the kernel $GF(q)$ is possible, where the associated vector space is forced to be 4-dimensional.

We first consider the examples that arise in the classification.

47.1. The Examples.

47.1.1. Desarguesian Planes. A Desarguesian plane of order q^2 may be coordinated by a field $F \simeq GF(q^2)$ and admits $\Gamma L(2, q^2)$ in the translation complement, where the p-elements are elations, for $p^r = q$. In particular, there is a regulus net R that is left invariant by a subgroup isomorphic to $GL(2, q)$.

47.1.2. Hall Planes. If the net R is derived, the group $GL(2, q)$ is inherited as a collineation group of the derived plane. Hence, the Hall planes admit $GL(2, q)$ where the p-elements are Baer p-collineations.

47.1.3. Hering and Ott–Schaeffer Planes.

DEFINITION 47.2. Let $\mathcal{Q}$ be any set of $q + 1$ points in $PG(3, q)$ such that no four of the points are coplanar. Then $\mathcal{Q}$ is called a $(q + 1)$-arc.

The $(q + 1)$-arcs are all determined as follows:

THEOREM 47.3. Let $\mathcal{Q}$ be a $(q + 1)$-arc then $\mathcal{Q}$ may be represented as follows:
(1) (Segre [**1154**]). If q is odd, then the representation is $\{(s^3, s^2 t, st^2, t^3); s, t$ in $GF(q)$, $(s, t) \neq (0, 0)\}$. If q is even or odd and if an arc has this representation, we call this a 'twisted cubic' $\mathcal{Q}^3$.
(2) (Casse and Glynn [**216**]). If q is even then the representation is $\mathcal{Q}^\alpha = \{s^{\alpha+1}, s^\alpha t, st^\alpha, t^{\alpha+1}); s, t$ in $GF(q)$, $(s, t) \neq (0, 0)\}$, where α is an automorphism of $GF(q)$ that is a generator of the Galois group.

THEOREM 47.4. Let V_4 denote a 4-dimensional vector space over $K \simeq GF(q)$. Consider the following matrix group:

$$
S^\beta = \left\langle \begin{bmatrix} a^{\beta+1} & ba^\beta & ab^\beta & b^{\beta+1} \\ ca^\beta & da^\beta & cd^\beta & db^\beta \\ ac^\beta & bc^\beta & ad^\beta & bd^\beta \\ cc^\beta & dc^\beta & cd^\beta & d^{\beta+1} \end{bmatrix} ; a, b, c, d \in K \text{ and } ad - bc \neq 0 \right\rangle.
$$

(1) If q is not 3^r or 2 and $\beta = 2$, then $S^{\beta=2}$ is isomorphic to $GL(2, q)$ and acts triply transitive on the points of the twisted cubic $\mathcal{Q}^3$. Furthermore, S^2 acts irreducibly on V_4.
(2) If $q = 2^r$ and β is an automorphism α of K, then $S^{\beta=\alpha}$ is isomorphic to $GL(2, q)$ and acts triply transitive on the points of the $(q+1)$-arc $\mathcal{Q}^\alpha$. Furthermore, S^α acts irreducibly on V_4.

THEOREM 47.5. Let Σ be $PG(3, q)$ and consider the plane $x_4 = 0$ where the points are given homogeneously by (x_1, x_2, x_3, x_4), for x_i in $GF(q)$, $i = 1, 2, 3, 4$.
(1) Then $x_1 x_3 = x_2^\beta$ for $\beta \in \{2, \alpha\}$ defines an 'oval cone' C_β with vertex $(0, 0, 0, 1)$ and oval $\mathcal{O}_\beta = \{(1, t, t^\beta, 0), (0, 0, 1, 0); t \in GF(q)\}$ in $x_4 = 0$.
(2) The $(q + 1)$-arc $\mathcal{Q}^\beta = \{(1, t, t^\beta, t^{\beta+1}), (0, 0, 0, 1); t \in GF(q)\}$ is contained in C_β and the q lines $L_t = \langle (0, 0, 0, 1), (1, t, t^\beta, t^{\beta+1}) \rangle$ intersect $\mathcal{O}_\beta$ in $(1, t, t^\beta, 0)$. Hence, there is a unique line $L_\infty = \langle (0, 0, 0, 1), (0, 0, 1, 0) \rangle$ of the oval cone that does not contain a point of $\mathcal{Q}_\beta$.

We shall call L_∞ the 'tangent' line to $(0, 0, 0, 1)$. More generally, any image of L_∞ under an element of the group S^β is called the tangent line at the corresponding image point.

(3) Consider the plane $x_1 = 0$ that intersects $\mathcal{Q}^\beta$ in exactly the point $(0, 0, 0, 1)$. We shall call $x_1 = 0$ the 'osculating' plane at $(0, 0, 0, 1)$. Each image of $x_1 = 0$ under an element of S^β is also called an osculating plane and the corresponding image point.

THEOREM 47.6. If $\mathcal{Q}^\beta$ is a twisted cubic then the set of $q + 1$-tangents form a partial spread $\mathcal{T}$.

THEOREM 47.7. Assume q is even and $\beta = \alpha$, for some automorphism of $GF(q)$. Let S_2 denote a Sylow 2-subgroup of S^α.

(1) Then S_2 fixes a unique point P of $\mathcal{Q}^\alpha$ and fixes the tangent plane $T(P)$.

(2) Choose any point Q of $\mathcal{Q}^\alpha - \{P\}$ and form the lines XQ and then the intersection points $I = T(P) \cap XQ$ and then the lines PI of $T(P)$ incident with P. Let $N_i(P)$ denote the two remaining lines of $T(P)$ incident with P for $i = 1, 2$.

Then $\mathcal{R}_i = N_i(P)S^\alpha$ is a regulus and $\mathcal{R}_j$ is the opposite regulus to $\mathcal{R}_i$ for $i \neq j$.

To construct the Hering and Ott–Schaeffer planes, we require that $q \equiv -1$ (mod 3).

THEOREM 47.8. When $q \equiv -1$ (mod 3) any element ρ of order 3 in S^β fixes a 2-dimensional subspace M pointwise.

(1) There is a unique Maschke complement L for ρ such that $V_4 = L \oplus M$.

(2) If $\beta = 2$ and q is odd then $\mathcal{T} \cup LS^2 \cup MS^2$ is the unique S-invariant spread of V_4.

The corresponding translation plane is called the 'Hering plane' of order q^2.

(3) If $\beta = \alpha$ and q is even then $\mathcal{R}_i \cup LS^\alpha \cup MS^\alpha$ is an S-invariant spread of V_4, for $i = 1$ or 2, and, for any automorphism α of $GF(q)$.

The corresponding translation planes are called the 'Ott–Schaeffer planes'.

REMARK 47.9. (1) The Hering and Ott–Schaeffer planes admit affine homologies of order 3 with $q(q - 1)$ distinct axes.

(2) Schaeffer determined the planes when α is the Frobenius automorphism and Ott generalized this to arbitrary automorphisms. (See Hering [473], Schaeffer [1150] and Ott [1073].)

(3) Each Ott–Schaeffer plane is derivable. If α is an automorphism for a given Ott–Schaeffer plane, then α^{-1} is the automorphism for its corresponding derived plane (see, e.g., Johnson [714]). If $q = 2^r$, it turns out that the number of mutually non-isomorphic planes is $\varphi(r)$, as the automorphisms used in the construction are generators of the cyclic group of order r.

47.1.4. The Three Walker Planes of Order 25. Let

$$
\tau_s = \begin{bmatrix} 1 & 0 & 0 & 0 \\ s & 0 & 0 & 0 \\ 3s^2 & s & 1 & 0 \\ s^3 & 3s^2 & s & 1 \end{bmatrix} ; s \in GF(5)
$$

and

$$\rho = \begin{bmatrix} 0 & 1 & 0 & 0 \\ -1 & 0 & 0 & 0 \\ 0 & 0 & 0 & 1 \\ 0 & 0 & -1 & 0 \end{bmatrix}.$$

Then $\langle \tau_s, \rho \rangle = S \simeq SL(2,5)$.

Furthermore, let the subgroup H of S be

$$H = \left\langle \begin{bmatrix} t & 0 & 0 & 0 \\ 0 & t^{-1} & 0 & 0 \\ 0 & 0 & t & 0 \\ 0 & 0 & 0 & t^{-1} \end{bmatrix} ; t \in GF(5) - \{0\} \right\rangle.$$

Then, there are exactly three mutually non-isomorphic spreads π_2, π_4, π_6 of order 25 that admit S such that H fixes exactly 6 components of each plane and ρ fixes either 2, 4, or 6 of these components, respectively. These planes are determined by Walker in [1203].

47.1.5. The Dempwolff Plane of Order 16.

Of course, all translation planes of order 16 are determined by computer (Dempwolff and Reifart [309]), but actually Johnson [699] shows that it is possible to determine all translation planes of order 16 that admit a non-solvable collineation group without the use of computer. In any case, how does one construct a translation plane of order 16 that admits $SL(2,4)$ as a collineation group in the translation complement? It has been mentioned that every translation plane of order 16 may be constructed by the derivation of a semifield plane, of which there are three, one each with kernel $GF(16), GF(4)$, and $GF(2)$. The Dempwolff plane may be constructed by derivation of the semifield with kernel $GF(2)$. Johnson [693] considers the translation planes admitting $SL(2,4)$ in the translation complement by analysis of the fixed-point spaces of Sylow 2-subgroups. In the Dempwolff case, these fixed-point spaces are Baer subplanes, since the planes are derived from a semifield plane and the elation group of order 4 fixing the derivable net becomes a Baer subgroup of order 4. Johnson [717] has shown that when the order of a dual translation plane of order q^2 is larger than 16, the collineation group of the derived plane is inherited from the group of the plane from which one is deriving. In the order-16 case, we have the very special situation when the derived net is moved by a collineation in the derived plane. In the Dempwolff-plane case, the group $SL(2,4)$ leaves invariant precisely two components. So, it is then possible to consider the spread for the Dempwolff plane directly by use of the group. Let $SL(2,4) = \langle Q_0, Q_1 \rangle$, it is possible to determine the form of these groups as follows

$$Q_0 = \left\langle \begin{bmatrix} I & 0 & 0 & 0 \\ A & I & 0 & 0 \\ 0 & 0 & I & 0 \\ 0 & 0 & A & I \end{bmatrix} \right\rangle,$$

$$Q_1 = \left\langle \begin{bmatrix} I & 0 & 0 & 0 \\ C & I & 0 & 0 \\ 0 & 0 & I & 0 \\ 0 & 0 & 0 & I \end{bmatrix} \begin{bmatrix} I & A & 0 & 0 \\ 0 & I & 0 & 0 \\ 0 & 0 & I & A \\ 0 & 0 & 0 & I \end{bmatrix} \begin{bmatrix} I & 0 & 0 & 0 \\ C & I & 0 & 0 \\ 0 & 0 & I & 0 \\ 0 & 0 & 0 & I \end{bmatrix} \right\rangle,$$

$\forall A \in K$, K a field of 2×2 matrices isomorphic to $GF(4)$, $C = \begin{bmatrix} 0 & 1 \\ 1 & 0 \end{bmatrix}$.

The two common components are $x = 0, y = 0$. The next step determines the components fixed by the two Baer groups and their images. The three other components fixed by Q_0 have the following form:

$$y = x \begin{bmatrix} A^i & 0 \\ 0 & A \end{bmatrix}, i = 1, 2, 3, A = \begin{bmatrix} 0 & 1 \\ 1 & 1 \end{bmatrix}$$

and the three other components fixed by Q_1 turn out to have the following form:

$$y = x \begin{bmatrix} A^i & CA^j \\ CA^j & 0 \end{bmatrix}, j = 1, 2, 3; A = \begin{bmatrix} 0 & 1 \\ 1 & 1 \end{bmatrix}$$

and the images of such components under $Q_0 - \{1\}$ have the form:

$$y = x \begin{bmatrix} A^i + CA^{j+k} & CA^j \\ CA^{j+k} & A^{k-j}C \end{bmatrix}, j, k = 1, 2, 3; A = \begin{bmatrix} 0 & 1 \\ 1 & 1 \end{bmatrix}.$$

These subspaces then form a spread admitting $SL(2,4)$. It follows that the Dempwolff plane has order lengths of $1, 1, 15$.

47.1.6. Spreads in $PG(3, q)$ Admitting $SL(2, q)$. The translation planes of order q^2 with kernels containing $GF(q)$ and admitting $SL(2, q)$ as a collineation group are completely determined by Walker and Schaeffer. This work is explicated in Lüneburg [**961**].

THEOREM 47.10. Let π be a translation plane of order q^2 with spread in $PG(3, q)$ that admits $SL(2, q)$ as a collineation group.
Then π is one of the following types of planes:
(1) Desarguesian,
(2) Hall,
(3) Hering and q is odd,
(4) Ott–Schaeffer and q is even, or
(5) One of three planes of order 25 of Walker.

47.1.7. Arbitrary Dimension. Using methods of combinatorial group theory and linear algebra, Foulser and Johnson were able to prove that the only translation plane of order q^2 that admits $SL(2, q)$ as a collineation group and whose spread is not in $PG(3, q)$ is, in fact, the Dempwolff plane.

THEOREM 47.11. (Foulser–Johnson [**378, 379**]). Let π be a translation plane of order q^2 that admits a collineation group isomorphic to $SL(2, q)$ in its translation complement.

Then either the plane has its spread in $PG(3,q)$ (and, is Desarguesian, Hall, Hering, Ott–Schaeffer, or Walker of order 25), or is the Dempwolff plane of order 16.

Actually, the way that the proof was given, it was not necessary to assume that $SL(2,q)$ acts faithfully on the translation plane. That is, it is possible that $PSL(2,q)$ acts on the plane. In fact, this essentially never occurs.

COROLLARY 47.12. Let π be a translation plane of order q^2 that admits a collineation group isomorphic to $PSL(2,q)$ then q is even.

PROOF. The proof given by Foulser–Johnson actually shows that either q is even, or possibly the plane is Desarguesian. However, the collineation group of the 'affine' Desarguesian plane of order q^2 is $\Gamma L(2,q)$ which does not contain $PSL(2,q)$ unless q is even.   □

Actually, there are two instances where knowledge of the fixed-point spaces over the Sylow p-subgroups determines the plane.

THEOREM 47.13. (Foulser–Johnson–Ostrom [380]). If a translation plane of order q^2 admits $SL(2,q)$ in the translation complement and the Sylow p-subgroups, $p^r = q$ fix components pointwise, then the plane is Desarguesian.

Also, when q is even the Sylow 2-subgroups fix q points (i.e., a Baer subline) pointwise, Johnson [690] determines the plane as an Ott–Schaeffer plane (i.e., each two Baer collineations share exactly q fixed points).

THEOREM 47.14. (Johnson [690]). If a translation plane of even order q^2 admits $SL(2,q)$ in the translation complement and the Sylow 2-subgroups, $p^r = q$ fix Baer sublines pointwise, then the plane is Ott–Schaeffer.

So, armed with these results, the sketch of the proof is as follows:

47.1.8. Sketch of the Proof of Foulser–Johnson Theorem. (1) Using Brauer and Nesbitt [166], determine the representation modules of $SL(2,q)$ over $GF(q)$, $q = p^r$, p a prime.
(2) If N is an irreducible $GF(q)SL(2,q)$-module, let K be the smallest field over which the matrices for the group may be written. Determine the irreducible $GF(p)SL(2,q)$-modules.
A few of the details of this initial reduction are as follows: Let $\theta : x \to x^p$. Let N_i denote the representation module of $SL(2,q)$ of homogeneous polynomials in X and Y of degree i. The dimension of N_i is $i+1$. The $GF(q)SL(2,q)$-module $N_i^{\theta^j}$ is obtained from the module N_i by entries β in the matrices by β^{p^j}. Brauer and Nesbitt [166] then determine the irreducible $GF(q)$-modules N as follows:

$$N = N_{i_0} \oplus N_{i_1}^\theta \otimes N_{i_2}^{\theta^2} \otimes N_{i_1}^\theta \otimes \cdots \otimes N_{i_{r-1}}^{\theta^{r-1}}; 0 \le i_j \le p-1.$$

This means, of course, that $\dim N = \prod_{j=0}^{r-1}(i_j+1)$.

We note that if s is the minimum positive integer such that N^{θ^s} is isomorphic to N, then K is $GF(p^s)$. Let B be a $GF(q)$-basis for N. Define $N_K = KB$, the d-dimensional vector space over K generated by the basis B. Then N_K is a $KSL(2,q)$-module and N is isomorphic to $N_K \otimes_K GF(q)$.

Now the elements of K may be thought of as $s \times s$ matrices over $GF(p)$, and replacing each element of K in a d-tuple over $GF(q)$ produces a ds-tuple over $GF(p)$.

In this way, N_K becomes an irreducible $GF(p)SL(2,q)$-module and furthermore, each irreducible module may be obtained in this manner. In fact, if M is an irreducible $GF(p)SL(2,q)$-module then

$$M \otimes_{GF(p)} GF(q) = R \oplus R^\theta \oplus R^{\theta^2} \oplus \cdots \oplus R^{\theta^{s-1}} \text{ and}$$

$$R = N_{i_0} \oplus N_{i_1}^\theta \otimes N_{i_2}^{\theta^2} \otimes N_{i_1}^\theta \otimes \cdots \otimes N_{i_{r-1}}^{\theta^{r-1}} ; 0 \le i_j \le p-1.$$

In this way, it is possible to consider all irreducible $GF(p)SL(2,q)$-modules of any dimension d. Now if d is bounded by $4r$, where the group acts on a translation plane of order p^{2r}, the irreducible modules may be constructed as follows:

Case	R	K	$\dim_{GF(p)} M$	Comments
(a)	N_1	$GF(p^r)$	$2r$	Canonical
(b)	$N_1 \otimes N_1^{\theta^{r/3}} \otimes N_1^{\theta^{2r/3}}$	$GF(p^{r/3})$	$8r/3$	$3 \mid r$
(c)	N_3	$GF(p^r)$	$4r$	$p \ne 2,3$
(d)	$N_1 \otimes N_1^{\theta^{r/2}}$	$GF(p^{r/2})$	$2r$	$2 \mid r$
(e)	N_2	$GF(p^r)$	$3r$	$p \ne 2$
(f)	$N_1 \otimes N_1^{\theta^i} ; 1 \le i < r$	$GF(p^r)$	$4r$	—
(g)	$N_1 \oplus N_1^{\theta^{r/4}} \otimes N_1^{\theta^{2r/4}} \otimes N_1^{\theta^{3r/4}}$	$GF(p^{r/4})$	$4r$	$4 \mid r$

(3) Consider the cases:

(i) $SL(2,q)$ acts irreducibly on π.

(ii) $SL(2,q)$ is completely reducible and decomposable on π.

(iii) $SL(2,q)$ is reducible but indecomposable on π.

Since we know the irreducible modules, we have a method of attacking this problem. In case (i), it is possible to show that only the modules of type (a) for q odd or type (f) for q even occur, leading to the Hering and Ott–Schaeffer planes, respectively, as it may be shown that the components are K-subspaces in these cases.

In case (ii), the p-groups either fix components or Baer subplanes pointwise, showing that only the Desarguesian and Hall planes of Dempwolff planes occur, respectively.

When we have situation (iii), only the three Walker planes of order 25 occur. One of these planes is also the Hering plane of order 25 with a reducible and indecomposable group. In this case, it is possible to show that an irreducible $GF(p)SL(2,q)$-module W_1 is such that $\pi/W_1 = W_2$ is also irreducible of the same degree, which eventually leads to the case when $W_1 \simeq W_2 \simeq N_1$ or $W_1 \simeq W_2 \simeq N_1 \otimes N_1^{\theta^{r/2}}$. When q is odd a combinatorial matrix argument may be used to reduce to the Walker order-25 planes.

So, let q be even and let L be an algebraically closed field containing $GF(2^r)$. We form the $LSL(2,q)$-module $N_1 \otimes_{GF(q)} L = V_1$ and let $V_1^{\theta^i} = V_{i+1}$. Since N_1 is an absolutely irreducible $GL(q)SL(2,q)$-module, then V_{i+1} is an irreducible $LSL(2,q)$-module. Let $T = \{1,2,\ldots,r\}$ and for $I \subseteq T$, form $V_I = \otimes_{i \in I} V_i$, where V_ϕ is the zero module. V_I are also irreducible $LSL(2,q)$-modules.

Now let $\mathrm{Ext}^1_{LG}(V_I, V_J)$ denote the space of extensions of V_I by V_J, where R is an $LSL(2,q)$ module such that $V_I \subseteq R$ and $R/V_I \simeq V_J$. A theorem of Alperin [**30**, Theorem 3, pp. 221, 229] applies to show that normally $\mathrm{Ext}^1_{LG}(V_I, V_J) = 0$, that is, the extensions all split. However, in our situation $\mathrm{Ext}_{GF(p)SL(2,q)}(W_1, W_2) \ne 0$. Moreover, there is the complication of writing our modules over $GF(2)$ over L, and when this is done, we need to retain the non-splitting situation. Indeed, when this rewriting process is accomplished, we obtain irreducible L-modules $U_i^{t_i}$, where τ_i is an automophism of K_i, arising from W_i, $i = 1, 2$, by the method mentioned in steps (1) and (2), and it is then shown that $\mathrm{Ext}^i_{LSL(2,q)}(U_i^{t_i}, U_i^{t_i}) \neq 0$. However, the conditions imposed by Alperin's theorem show that when W_1 and W_2 are either both N_1 or both $N_1 \otimes N_1^{\theta^{r/2}}$, a contradiction is obtained.

This completes the sketch of the proof.

COROLLARY 47.15. *Let π be a translation plane of order q^2 that admits a collineation group isomorphic to $PSL(2,q)$ then q is even.*

PROOF. The proof given by Foulser–Johnson actually shows that either q is even, or possibly the plane is Desarguesian. However, the collineation group of the 'affine' Desarguesian plane of order q^2 is $\Gamma L(2,q)$, which does not contain $PSL(2,q)$ unless q is even.   □

47.2. Applications.

Let π be a translation plane of odd order $q^2 = p^{2r}$ that admits at least two Baer p-groups B_1 and B_2 in the translation complement with distinct Baer axes. Assume that $|B_i| > \sqrt{p^r} \geq 3$. Then, by Foulser's work (which works in the characteristic-3 case in this situation), it follows that the Baer axes lie in the same net of degree $p^r + 1$. The Baer groups generate a group G isomorphic to $SL(2,p^s)$ for $p^s > p^{r/2}$. From here, it follows that the group G must be $SL(2,q)$. Applying the previous theorem of Foulser–Johnson, we have:

THEOREM 47.16. (Jha and Johnson [**602**]). *Let π be a translation plane of odd order $q^2 = p^{2r}$ that admits at least two Baer p-groups of order $> \sqrt{p^r} \geq 3$. Then π is the Hall plane of order q^2.*

Recall that Foulser's result is not necessarily valid in translation planes of even order but there is considerable incompatibility between elation and Baer 2-groups.

Dempwolff analyzed the groups generated by two Baer 2-groups with distinct axes and orders $\sqrt{2^r}$ if the translation plane is of order 2^{2r}.

THEOREM 47.17. (Dempwolff [**293**]). *Let π be a translation plane of even order q^2 and let G be a collineation group in the translation complement that contains at least two Baer 2-groups of orders $> \sqrt{q}$ with distinct axes. Let N denote the subgroup of G generated by affine elations.*

Then one of the following situations occurs:

(1) $q^2 = 16, G \simeq SL(3,2) \simeq PSL(2,7)$ and π is either the Lorimer–Rahilly or Johnson–Walker plane, or

(2) $G/N \simeq SL(2,2^z)$ where $2^z > \sqrt{q}$ and $N \subseteq Z(G)$.

Using the incompatibility results mentioned in Chapter 46, we know that any elation group centralizing a Baer 2-group must have order ≤ 2. If the order is 1 then we argue that, in fact, we obtain $SL(2,q)$ so that the results of Foulser and Johnson apply. If the order is 2 then some group representation theory shows that $G \simeq SL(2,2^z) \oplus N$ and we argue that $SL(2,2^z)$ contains a Baer group of order $> \sqrt{q}$ which again shows that $SL(2,q)$ is a collineation group. We note that the Dempwolff plane of order 16 does not occur here, since there are no large Baer 2-groups in this plane.

Hence, Jha and Johnson show that the existence of two large Baer groups is enough to classify the planes containing them.

THEOREM 47.18. (Jha and Johnson [**598**]). Let π be a translation plane of even order q^2 that admits at least two Baer groups with distinct axes and orders $\geq 2\sqrt{q}$ in the translation complement. Then, either π is Lorimer–Rahilly or Johnson–Walker of order 16 or π is a Hall plane.

PROOF. Note that if we had assumed that the order is merely $> \sqrt{q}$, we could have included the Desarguesian plane of order 4 in the set of planes satisfying the hypothesis. □

47.2.1. Elations and Large Baer Groups; Jha–Johnson Incompatibility. Finally, we return to the question of elations and large Baer groups, first mentioned in Chapter 46.

The following result uses a variety of different theorems that we have presented, so it is a suitable conclusion to this text.

THEOREM 47.19. Let π be a translation plane of order 2^{2r} that admits a Baer 2-group B of order $\geq 2\sqrt{q}$. Then any elation axis lies within $\operatorname{Fix} B = \pi_o$.

PROOF. Sketch: Deny! Then there exist at least two distinct Baer groups of order $\geq 2\sqrt{q}$, and we may apply the previous theorem to conclude that the plane is a Hall plane or Lorimer–Rahilly or Johnson–Walker of order 16.

Now, it turns out that in a Hall plane of order other than 9, the net contain a Baer 2-group axis of order at least $2\sqrt{q}$ is invariant under the full group.

In the L–R or J–W planes, there are exactly three elation axes. For each axis, there are seven Baer subplanes containing this line (i.e., these Baer axes overlap). Hence, all elation axes are in $\operatorname{Fix} B = \pi_o$. However, this is a contradiction in either case, as exterior elations must move the net of degree $q + 1$ containing a Baer axis. □

We now consider the extremal situation; $q + 1$ elation axes with a 'large' Baer 2-group B. We have noticed that all axes are within $\operatorname{Fix} B$.

THEOREM 47.20. If π is a translation plane of order 2^{2r} that admits at least $q + 1$ non-trivial elation axes and a Baer 2-group B of order at least $2\sqrt{q}$, then the following statements hold:
(1) π is derivable.
(2) There is a Desarguesian affine plane Σ of order q^2 such that the group G generated by B and the elations is the product of an elation group of Σ, a Baer involution of Σ and the cyclic kernel homology group C_{q+1} of Σ; $G = B \langle \sigma \rangle C_{q+1}$.

PROOF. The group H generated by the elations must fix $\pi_o = \operatorname{Fix} B$. We assert that it acts faithfully. We know, by the theorem of Hering–Ostrom, that H has order $2d$ where d is odd. Hence, if there is a group element g of H that fixes π_o pointwise then the order of g divides $q - 1$. However, this means that there exist at least two and, hence, at least three 'Desarguesian' Baer subplanes of the net N containing π_o (why?). But, H must fix each of these subplanes, since H is generated by elations that necessarily must fix each Baer subplane of N.

Hence, H acts faithfully on π_o.

Since π_o is Desarguesian, it follows that H is a subgroup of $\Gamma L(2, q)$, and, from this, it follows that H is a dihedral group of order $2(q + 1)$ and, hence, has a cyclic stem C_{q+1} of order $q + 1$.

Now, it turns out that, if q is not 8, we can guarantee that there is a prime 2-primitive divisor u of $q^2 - 1$ that must, necessarily, divide $q + 1$. Hence, there is an element in C_{q+1}, say g, which has order u. But, even for $q = 8$, we see that $q + 1 = 9$ is a 2-primitive divisor.

Note that, in either situation, g must fix at least two components of $\pi - N$ and fixes π_o. Hence, g fixes at least three mutually disjoint linesize subspaces which implies, by the p-primitive collineation theorem, that there is a Desarguesian affine plane of order q^2, whose spread consists of g-invariant linesize subspaces.

Now since B centralizes all elations then the group generated, $G = \langle B, elations \rangle$, is a subgroup of $\Gamma L(2, q^2)$, the translation complement of Σ. Moreover, g is a kernel homology of Σ. The cyclic stem C_{q+1} of H fixes a component $(\pi_o = \text{Fix } B)$ of Σ, and C_{q+1} fixes at least two components of $\pi - N$ that consequently are also components of Σ. Thus, C_{q+1} is a subgroup of $GL(2, q^2)$, as it commutes with g. However, a cyclic group of $GL(2, q^2)$ that fixes at least three components of Σ must, necessarily, be a subgroup of the kernel homology group.

Now, a component that is an elation axis of π, is then, a Baer subplane of Σ, as it is fixed pointwise by an involution σ acting in $\Gamma L(2, q^2)$. Hence, the kernel subgroup C_{q+1} acts transitively on a set of $q + 1$ Baer subplanes of Σ that are, in fact, the $q + 1$ components of the net N. Hence, the net N is an 'opposite' regulus net in Σ, so the plane π is derivable.

The rest of the theorem now follows directly. $\square$

47.3. $G/N \simeq PSL(2, q)$.

As mentioned previously, the theorem of Foulser and Johnson, which classifies the translation planes of order q^2 that admit $SL(2, q)$ or $PSL(2, q)$ as a collineation group in the translation complement, has important applications. This theorem invariably arises when considering the action of a group on a finite spread or corresponding geometry. For example, applications arise in the analysis of doubly transitive groups acting non-solvably on a subset of points on a finite spread or acting on a set of spreads forming a parallelism or partial parallelism. Other examples involve the action of groups on flocks of quadratic cones or on partial flocks of elliptic or hyperbolic quadrics.

The problem is that in certain applications the group $SL(2, q)$ is not presented as directly acting in the translation complement but there is a group G of the translation complement that induces $SL(2, q)$ or $PSL(2, q)$ on a set L of points, or more generally, there is a normal subgroup N of G so that G/N is isomorphic to $PSL(2, q)$. So, here we see that it is possible to resolve this technicality and show that even in this case, the complete classification of such translation planes is exactly as in the Foulser–Johnson theorem. Furthermore, we note that it is not necessary to assume that G is in the translation complement.

However, this may be overcome using a combination of group-theoretic and spread-theoretic arguments. This result is due to Biliotti–Jha–Johnson [131].

THEOREM 47.21. Let π denote an affine translation plane of order q^2. Assume that π admits a collineation group G containing a normal subgroup N, such that G/N is isomorphic to $PSL(2, q)$.

Then π is one of the following planes:
(1) Desarguesian,
(2) Hall,
(3) Hering,
(4) Ott–Schaeffer
(5) One of three planes of Walker of order 25, or
(6) The Dempwolff plane of order 16.

We shall give a sketch of these results:

47.3.1. Sketch of the Proof of Biliotti–Jha–Johnson Theorem. The idea of the proof is to establish that the collineation group G may be considered a subgroup of the translation complement and then to show that G contains $SL(2,q)$ or $PSL(2,q)$, whereupon one may apply the theorem of Foulser–Ostrom. So, if G is a subgroup of the translation complement, it becomes a subgroup of $GL(4r,p)$, if $p^r = q$.

For the group-theoretic part, we may begin by assuming only that G is a subgroup of $GL(4r,p)$, where p is a prime integer, containing a subgroup N such that G/N is isomorphic to $PSL(2,q)$, for $q = p^r$.

Then the following may be proved.

THEOREM 47.22. Let V be a $4r$-dimensional vector space over $GF(p)$, for p a prime, and let $q = p^r$. Let G be a subgroup of $GL(4r,p)$ containing a normal subgroup N such that G/N is isomorphic to $PSL(2,q)$.

Then one of the following occurs:
(1) $q = 2, 3, 4, 8, 9$.
(2) $q = 5, 7, 17, 5^2, 3^3$ and the following hold:
   (a) $O_{p'}(G)$ is a 2-group S_2, or when $q = 7$, $O_{p'}(G)$ is the product of a 2-group S_2 and a 3-group S_3.
   (b) G/S_2 is isomorphic to $PSL(2,q)$, or when $q = 7$, $G/S_2 \times S_3$ is isomorphic to $PSL(2,q)$.
(3) $O_p(G)$ has order $\geq q^2$.
(4) G contains $SL(2,q)$ or $PSL(2,q)$.

REMARK 47.23. By Zsigmondy [1219] (see also Theorem 7.1), either there is a p-primitive divisor of $q^2 - 1$, or $q = p$ and $p + 1 = 2^b$, for some positive integer b, or $q = 8$.

We assume that $q > 3$ so that $PSL(2,q)$ is simple.

47.3.2. N is the Frattini Group. It is first established that N may be considered the Frattini subgroup of G, G is perfect; $G = G'$ and that no proper subgroup Z of G has the property that ZN/N is isomorphic to $PSL(2,q)$. It is then not difficult to establish that

LEMMA 47.24. N contains all proper normal subgroups of G. Hence,

$$N = O_p(G) \times O_{p'}(G),$$

and:
   (1) Any two Sylow p-subgroups S_p^1 and S_p^2 such that $S_p^1 \nsubseteq S_p^2 N$ generate G.
   (2) If p-primitive elements u exist, then any two distinct groups T_u^1 and T_u^2 such that $T_u^1 \nsubseteq T_u^2 N$ of order u generate G.

(3) If p-primitive elements do not exist and p is odd, then any two Sylow 2-subgroups S_2^1 and S_2^2 such that $S_2^1 \nsubseteq S_2^1 N$ generate G.

(4) If $q = 2^6$, then any two Sylow 7-subgroups S_7^1 and S_7^2 such that $S_7^1 \nsubseteq S_7^2 N$ generate G.

The main use of p-primitive divisors provides the following.

LEMMA 47.25. Assume that q is not 8. Then $O_p(G) \le Z(G)$ or $|O_p(G)| \ge q^2$.

PROOF. Assume that $O_p(G)$ has order $< q^2$. Assume that u is a p-primitive divisor of $q^2 - 1$. Then every u-element will commute with $O_p(G)$, implying that $O_p(G)$ is centralized by G. Now assume that $q = p$ and $p + 1 = 2^a$. Then if $O_p(G)$ has order $< p^2$, the order of $O_p(G)$ is p. Then all p-elements acting on $O_p(G)$ will necessarily commute with $O_p(G)$, so that $O_p(G)$ is again central.   □

Then the use of central extensions establishes:

LEMMA 47.26. Assume that q is not 4 or 9. If $O_p(G)$ is central, then we may assume that $O_p(G) = \langle 1 \rangle$ and $G/O_{p'}(G)$ is isomorphic to $PSL(2, q)$.

PROOF. Since $N/O_{p'}(G)$ is isomorphic to $O_p(G)$, $G/O_{p'}(G)$ is perfect, and $N/O_{p'}(G)$ is central in $G/O_{p'}(G)$. Hence, $G/O_{p'}(G)$ is a representation group for $PSL(2, q)$ or is isomorphic to $PSL(2, q)$. In the former case, $p = 2$ but then $PSL(2, q)$ and $SL(2, q)$ are isomorphic. Thus, N is $O_{p'}(G)$.   □

At this point, it is possible that $PSL(2, q)$ can act on an elementary Abelian v-subgroup, where v is not p. Hence, the results of Harris and Hering [**438**] or Landazuri and Seitz [**906**] may be used, which provide bounds on the degrees of the actions. To give an idea of how this might work, we include the following lemma.

LEMMA 47.27. Assume q is not 4 or 9. Assume that $O_p(G)$ is trivial. If p is odd, assume that no v-subgroup of $O_{p'}(G)$ is of order at least $v^{(q-1)/2}$ and if $p = 2$ assume that no v-subgroup of $O_{2'}(G)$ is of order at least v^{q-1}. Then G contains a group isomorphic to $PSL(2, q)$ or $SL(2, q)$.

PROOF. Consider $G/N' = \overline{G}$, where N' is the commutator subgroup of N. Then, $\overline{G}$ induces an automorphism group on $N/N' = \overline{N}$, with kernel the centralizer $\overline{M}$ of $\overline{N}$ in $\overline{G}$. Since $\overline{N}$ is Abelian, $\overline{N}$ is contained in $\overline{M}$. Hence, $\overline{M}/\overline{N}$ is a normal subgroup of $\overline{G}/\overline{N}$, which is isomorphic to $PSL(2, q)$. So, $\overline{M}$ is either $\overline{G}$ or $\overline{N}$. Assume that $\overline{M}$ is $\overline{N}$. Then $PSL(2, q)$ acts on an Abelian subgroup $\overline{N}$ and further acts on the product of the elementary Abelian subgroups of $\overline{N}$ and, in fact, acts on each elementary Abelian v-subgroup. By Landazuri and Seitz [**906**], the degree is at least $(q - 1)/2$, a contradiction, to our assumptions. Hence $PSL(2, q)$ commutes with the product of the elementary Abelian subgroups of $\overline{N}$. Let $\overline{N}^*$ denote this subgroup. Then $PSL(2, q)$ acts on $\overline{N}/\overline{N}^* = \overline{N}_2$ as an automorphism group. However, $PSL(2, q)$ then centralizes the product of the maximal elementary v-subgroups of $\overline{N}_2$. An obvious induction argument shows that $PSL(2, q)$ commutes with $\overline{N}$. Now we have $\overline{G}/\overline{N}$ isomorphic to $PSL(2, q)$ and $\overline{N}$ is central in $\overline{G}$, which is a perfect group. By Schur multipliers, either $\overline{G}$ is $PSL(2, q)$ or $SL(2, q)$. By induction either N centralizes G or at some point in the derived series of N, we have $\overline{G}/\overline{P}$ isomorphic to $SL(2, q)$, where $\overline{P}$ is an Abelian group. The group induced on $\overline{P}$ has a kernel which is either trivial, a group of order 2, or the full group. If the kernel is a subgroup of order 2, we are back into the $PSL(2,q)$ situation. In any case, we have a projective representation of an elementary Abelian v-subgroup of $\overline{P}$, which again by Landazuri and Seitz [**906**] is a contradiction. Hence, by induction, we see that G contains either $SL(2,q)$ or $PSL(2,q)$. $\qquad\square$

47.3.3. When the Landazuri–Seitz Bound Holds. At this point, combinatorial number theory and elementary use of the computer show that a few cases remain: $q = 2,3,4,5,7,8,3^2,17,5^2,3^3$. So, we then consider the group G as a collineation group of an associated translation plane. To illustrate the connections and arguments with spreads of translation planes, we include a few more details in the translation-plane case.

47.4. The Translation-Plane Case.

Now assume that G leaves invariant a spread for a translation plane π of order q^2, $q = p^r$, and that contains a subgroup N such that G/N is isomorphic to $PSL(2,q)$. We show that G may be considered a subgroup of the translation complement.

LEMMA 47.28. Assume that $q > 3$. There is a subgroup G^* of the translation complement such that G^*/N^* is isomorphic to $PSL(2,q)$, where N^* is a normal subgroup of G^*.

PROOF. Let T denote the subgroup of translations of π. Then GT/T is isomorphic to a subgroup of the translation complement. TN is a normal subgroup of GT and $(GT/N)/(TN/N)$ is isomorphic to $PSL(2,q)$. Since $PSL(2,q)$ is simple, it follows that $PSL(2,q)$ is in the composition series for GT/T. However, this means that $PSL(2,q)$ is in the composition series for a subgroup of the translation complement. Thus, there is a subgroup G^* and a normal subgroup N^* of G^* such that G^*/N^* and G^* is a subgroup of the translation complement. $\qquad\square$

Henceforth, since we may assume that $q > 3$, we may assume that G is a subgroup of the translation complement and hence is a subgroup of $GL(4r,p)$, where $p^r = q$.

We shall utilize our main structure theorem, Theorem 47.22 given in the previous section, to show that G contains a group isomorphic to $SL(2,q)$ or $PSL(2,q)$ so that the Foulser–Johnson theorem may be applied.

There are essentially two obstacles: First it is possible that $|O_p(G)| \geq q^2$ and second, the Landazuri bound holds when $q = 2,3,4,5,7,8,3^2,17,5^2,3^3$.

REMARK 47.29. All translation planes of orders $4,9,16,25,49$ are known and the only planes admitting a group G as in the previous statement are the Desarguesian, Hall planes, Hering, Ott–Schaeffer, Walker of order 25 and Dempwolff of order 16. For example, see Johnson [**699**] for the translation planes of order 16 that admit non-solvable groups, see Czerwinski [**269**] to check the planes of order 25, and see Dempwolff [**304, 305**] to check that only the planes of order 49 indicated have the given group.

Hence, we may assume that $q > 7$. Let $q = p^r$, where p is a prime and r a positive integer. We begin by showing that the possibility that $|O_p(G)| \geq q^2$ cannot hold in translation planes. We actually shall require a stronger version of this when $q = 8$.

LEMMA 47.30. *If q is not 8, and the order of $O_p(G)$ is at least q^2, then $O_p(G)$ is an elation group.*

PROOF. Assume that there is a prime p-primitive divisor u of $q^2 - 1$ and let y_u be an element of order u. Let X be fixed pointwise by $O_p(G)$. If X does not have order q^2, then y_u will fix X pointwise and since G is generated by its p-primitive elements, then G fixes X pointwise. If L is a component of π which non-trivially intersects X, then L is fixed pointwise by the same argument, implying that G is a central collineation group, a contradiction, by order if nothing else. Hence, X is a component. Hence the order of X is q^2 and X cannot be Baer since the p-group fixing a Baer subplane pointwise has order at most q. Thus, X is a component.

Now assume that $p + 1 = 2^a$ and $q = p^2$. Since X must now be a 1 or 2-dimensional $GF(p)$-subspace, we are finished by the above argument unless possible X is a 1-dimensional subspace. But, the largest p-subgroup fixing a 1-dimensional subspace and acting faithfully on the component L containing X has order q. Furthermore, we know that there is a Sylow p-subgroup of order p^3 which must leave X invariant. But, this implies that there is an elation group of order p^2 and a Baer group of order p, a contradiction by Foulser [**375**]. □

The remaining part of the proof involves combinatorial arguments on translation planes and group theory. In fact, this part of the proof proves to be the most difficult, and in particular when $q = 8$, p-primitive arguments do not apply and when $q = 9$, Schur multiplier arguments tend not to work.

CHAPTER 48

Spreads of Order q^2—Groups of Order q^2.

The idea of isolating translation planes of order q^2 admitting collineation groups in the translation complement of order q^2 probably originated with Bartolone [**100**], who actually looked at such planes admitting larger groups of order $q^2(q-1)$. Of course, any semifield plane of order q^2 with kernel $GF(q)$ admits such a group. But, there are also the interesting planes of Walker and Betten that also admit groups of these orders. The Lüneburg–Tits translation planes of order q^2 admit $Sz(q)$ as a collineation group in the translation complement, which acts doubly transitively on the infinite points and the stabilizer of an infinite point (∞) has the required order. Bartolone's theorem states that under certain assumptions, these are exactly the possible translation planes admitting such groups.

THEOREM 48.1. (Bartolone [**100**]). Let π be a translation plane of order q^2 with spread in $PG(3, q)$. Let G be a collineation group within $GL(4, q)$ (the linear translation complement) of order $q^2(q-1)$. Assume further that
(1) There is a normal subgroup H of order q^2 that fixes an infinite point and acts transitively on the remaining infinite points.
(2) $G = HN$, where N has order $q-1$ and
(3) There is a subgroup H^* of HN which is a Frobenius group with complement N.
Then π is a Lüneburg–Tits, Walker, or Betten plane.

Considering what can be said of translation planes of order q^2 that admit only a group of order q^2, Kantor [**860**] developed the ideas of 'likeable' planes and gave important examples, which turn out to also produce flocks of quadratic cones. The main criterion that a spread in $PG(3, q)$ must have to connect and produce a corresponding flock of a quadratic cone is that there is a set of q reguli that mutually intersect in a line (component) and when such reguli exist, it is possible to show that that there is an associated elation group E of order q, which is 'regulus-inducing' in the sense that the axis and any component orbit define the q reguli. So, often it will turn out that the existence of a group of order q^2 will force the existence of a regulus-inducing elation group. Hence, these planes structurally are important to the overall theory of translation planes.

Assume that π is a translation plane of order q^2 that admits a linear collineation group of order $q^2 = p^{2r}$, for p a prime. The question is then, why should the group of order q^2 fix a component and act transitively on the remaining components? If the group is not transitive, then there is a planar p-subgroup in the group, which is Baer if the spread is in $PG(3, q)$ and the group is linear. But any group of order q^2 must fix a component L and induce upon that component a subgroup of $GL(2, q)$. This says that the group must contain an elation group of order q, even though the elation group may not be regulus-inducing. However, there is an incompatibility of elations and Baer p-groups due to Foulser [**375**] so any odd-order group of order q^2 must fix a component and act transitively on the remaining components. There is an analogous theory of incompatibility of elations and Baer involutions for even-order translation planes due to Jha–Johnson, as we have mentioned previously, and while elations and Baer involutions are not incompatible, there are strong restrictions on the relative orders of elation groups and Baer 2-groups. This incompatibility leads to severe restrictions on the group of order q^2 as we have noticed in Chapter 46. So, we see that knowledge of groups of order q^2 in translation planes of order q^2 proves to a valuable tool in the analysis of related geometries.

48.1. Regular Groups.

We begin by listing the most general result on translation planes of order q^2 admitting 'regular' groups of order q^2, that is the group fixes a component and acts transitively on the remaining components. We note that if G is the group fixing the component $x = 0$, then the spread is uniquely determined as

$$x = 0, (y = 0)g; g \in G.$$

THEOREM 48.2. (See Johnson [**705**]). Let π be a translation plane of order $q^2 = p^{2r}$, for p a prime, with spread in $PG(3, q)$ that admits a collineation group G of order q^2 in the linear translation complement (i.e., in $GL(4, q)$). Assume that G fixes a component $x = 0$ and acts transitively on the remaining components. Then G may be represented as follows:

$$G = \langle M(b, u); b, u \in GF(q) \rangle,$$

where

$$M(b, u) = \begin{bmatrix} 1 & T(b) & u + L(b) & m(u) + uT(b) + R(b) \\ 0 & 1 & b & u \\ 0 & 0 & 1 & T(b) \\ 0 & 0 & 0 & 1 \end{bmatrix},$$

where

(1) m and T are additive functions on $GF(q)$ and $m(a) = 0$ for all $a \in GF(p)$.
(2) $|\{T(b); b \in GF(q)\}| = p^{r-t}$ for a particular integer t, $0 \leq t \leq r$.
(3) The group E of elations with axis $x = 0$ has order p^{r+t} and

$$E = \langle M(b, u); T(b) = 0 \rangle.$$

(4) $L(a) + L(b) + T(b)a = L(a + b) + bT(a)$ for all $a, b \in GF(q)$.
(5) $R(a) + R(b) + L(b)T(a) = m(bT(a)) + bT(a)T(a + b) + R(a + b)$ for all $a, b \in GF(q)$.

Using this general result, it is possible to show

COROLLARY 48.3. For q even then $L(x) = xT(x) + \ell(x)$, where ℓ is an additive function on $GF(q)$. Hence,

$$G = \langle M(b, u); b, u \in GF(q) \rangle,$$

where

$$M(b, u) = \begin{bmatrix} 1 & T(b) & u + bT(b) + \ell(b) & m(u) + uT(b) + R(b) \\ 0 & 1 & b & u \\ 0 & 0 & 1 & T(b) \\ 0 & 0 & 0 & 1 \end{bmatrix},$$

where T, m and ℓ are additive functions on $GF(q)$ and $m(1) = 0$. Furthermore, $R(a) + R(b) + \ell(b)T(a) = m(bT(a)) + bT(a)^2 + R(a + b)$ for all $a, b \in GF(q)$.

While the result for even order does not particularly specify the group, the odd-order case is much more tractable.

COROLLARY 48.4. For q odd then E has order q or q^2 and is a semifield plane in the latter case. Then $p \neq 3$ and

$$G = \langle M(b, u); b, u \in GF(q) \rangle ,$$

where

$$M(b, u) = \begin{bmatrix} 1 & b & u & -\frac{1}{3}b^3 + J(a) + m(u) \\ 0 & 1 & b & u \\ 0 & 0 & 1 & b \\ 0 & 0 & 0 & 1 \end{bmatrix},$$

where m is additive, $m(1) = 0$ and $J(a+b)+m(ab) = J(a)+J(b)$ for all $a, b \in GF(q)$.

DEFINITION 48.5. For either even or odd order, if the order of E is q, then T is bijective on $GF(q)$. In this case, the plane is said to be 'elusive' when q is even and 'desirable' when q is odd.

If m is identically zero, the plane is said to be 'likeable'.

DEFINITION 48.6. When m and J both are identically zero, and q is odd, the plane is called the 'likeable plane of Walker', and when q is even, the plane is referred to as the 'likeable plane of Betten'.

DEFINITION 48.7. For $q = 5^r$, m identically zero and $J(a) = k^{-1}a + ka^5$, for k a non-square in $GF(q)$, the plane is called the 'likeable plane of Kantor'.

REMARK 48.8. The Lüneburg–Tits planes are elusive. Let $q = 2^{2k+1}$, with $\alpha : x \to x^{2^{k+1}}$. Then taking $T(b) = b^{\alpha^{-1}}$, $m(u) = u + u^\sigma$, $R(b) = b + b^\alpha + b^{\alpha+1}$ and $\ell(b) = b$, the group is then completely determined as

$$G = \left\langle M(b^{\alpha^{-1}}, u); b, u \in GF(q) \right\rangle ,$$

where

$$M(b, u) = \begin{bmatrix} 1 & b^{\sigma^{-1}} & u + b^{1+\alpha^{-1}} & u + u^\alpha + ub^{\alpha^{-1}} + b + b^\alpha + b^{\alpha+1} \\ 0 & 1 & b & u \\ 0 & 0 & 1 & b^{\alpha^{-1}} \\ 0 & 0 & 0 & 1 \end{bmatrix}.$$

There are two sporadic planes of order 64 which are due to Biliotti–Menichetti [144, 145] and Jha–Johnson [585].

DEFINITION 48.9. Let $q = 8$ and G defined as follows:

$$G = \langle M(b, u); b, u \in GF(8) \rangle ,$$

where

$$M(b, u) = \begin{bmatrix} 1 & b^2 & u + b^3 + b^4 & u^2 + u^4 + ub^2 + b + b^2 + b^4 + b^5 + b^6 \\ 0 & 1 & b & u \\ 0 & 0 & 1 & b^2 \\ 0 & 0 & 0 & 1 \end{bmatrix}.$$

The associated translation plane is called the 'Biliotti–Menichetti plane of order 64'.

DEFINITION 48.10. Let q be even. Consider planes obtained from the group G defined as follows:

$$G = \langle M(b, u); b, u \in GF(q) \rangle,$$

where

$$M(b, u) = \begin{bmatrix} 1 & t_0(b+b^2) & l(u,b) & k(u,b) \\ 0 & 1 & b & u \\ 0 & 0 & 1 & t_0(b+b^2) \\ 0 & 0 & 0 & 1 \end{bmatrix},$$

$$l(u, b) = u + bt_0(b+b^2) + b + t_0(b^2+b^4),$$
$$k(u, b) = u + u^2 + ut_0(b+b^2) + t_0^2 b^3(1+b)^3 + S(b),$$

where S is an additive function and t_0 is a constant. For $q = 64$ there is a unique plane when $q = 64$ (where we may take $t_0 = 1$ and S identically zero). The associated translation plane is called the 'Jha–Johnson plane of order 64'. Existence of a plane when q is > 8 is open. The interesting feature of these planes is that there is an elation group of order $2q$.

48.2. Non-Regular Groups.

We have noted that when q is even and we have a group of order q^2 in $GL(4, q)$, then the group is always regular. When q is even, using the incompatibility results of Jha–Johnson mentioned in Chapter 46, it is possible to show that if the group is not regular, then there are two orbits of lengths $q^2/2$. Furthermore, in this setting, the center of the group has order $\leq q$. That is, Abelian groups are always regular and indeed can be proven to always be linear (Jha–Johnson [586]).

THEOREM 48.11. (Johnson [705]). Let π be a translation plane of even order $q^2 = q^{2r}$ that admits a linear group G of order q^2. Assume that G is not regular. Then there are two orbits of lengths $q^2/2$ and there exists an additive subgroup Σ of $GF(q)$ of order $q/2$ and functions m and f on $GF(q)$ such that

$$G = \langle \tau_b, \sigma_u, \tau; b \in \Sigma, u \in GF(q) \rangle,$$

where $\langle \sigma_u \rangle = Z(G)$ (center of G) and

$$\tau_b = \begin{bmatrix} 1 & T(b) & b+m(b)+bT(b) & f(b) \\ 0 & 1 & b & 0 \\ 0 & 0 & 1 & T(b) \\ 0 & 0 & 0 & 1 \end{bmatrix}, b \in \Sigma,$$

$$\sigma_u = \begin{bmatrix} 1 & 0 & u & m(u) \\ 0 & 1 & 0 & u \\ 0 & 0 & 1 & 0 \\ 0 & 0 & 0 & 1 \end{bmatrix}; u \in GF(q), \text{ so } Z(G) \text{ has order } q,$$

$$\tau = \begin{bmatrix} 1 & 1 & 0 & 0 \\ 0 & 1 & 0 & 0 \\ 0 & 0 & 1 & 1 \\ 0 & 0 & 0 & 1 \end{bmatrix}.$$

where f is a bijective function on $GF(q)$, m is additive and $m(0) = m(1) = 0$ such that

$$f(a+b) = f(a) + f(b) + m(aT(b)) + aT(b)^2 + T(b)(a + m(a)), \forall a \in GF(q), \forall b \in \Sigma.$$

Furthermore, the spread for π is

$$x = 0, y = x \begin{bmatrix} u + v + m(v) & f(v) + m(u) \\ v & u \end{bmatrix} ; u, v \in GF(q)$$

THEOREM 48.12. (Biliotti, Jha, Johnson, Menichetti [**133**]).

Let π be a translation plane of even order q^2 with spread in $PG(3,q)$ that admits a linear group of order q^2 that fixes a component and acts regularly on the remaining components. Then coordinates may be chosen so that the group G may be represented as follows:

$$\left\langle \sigma_{b,u} = \begin{bmatrix} 1 & T(b) & u + bT(b) + l(b) & uT(b) + R(b) + m(u) \\ 0 & 1 & b & u \\ 0 & 0 & 1 & T(b) \\ 0 & 0 & 0 & 1 \end{bmatrix} ; b, u \in GF(q) \right\rangle$$

where T, m, and l are additive functions on $GF(q)$ and $m(1) = 0$ and such that

$$R(a+b) + R(a) + R(b) + l(a)T(b) + m(aT(b)) + a(T(b))^2 = 0, \forall a, b \in GF(q).$$

Using these results, the following general classification theorems may be proven.

THEOREM 48.13. (Biliotti–Menichetti [**145**], Jha–Johnson–Wilke [**647**]). Let π be a translation plane of even order q^2 with spread in $PG(3,q)$ that admits a linear collineation group of order q^2 with the property that either the elation subgroup has order q or q^2. Then the plane is one of the following types:
(1) A semifield plane,
(2) Lüneburg–Tits,
(3) Betten or
(4) The Biliotti–Menichetti plane of order 64.

The following extension of the theorem of Bartolone is also proved with these methods.

THEOREM 48.14. (Jha–Johnson–Wilke [**647**]). Let π be a translation plane of order q^2 that admits a linear collineation group G such that if K^* is the kernel homology group of order $q - 1$, then GK/K has order $q^2(q - 1)$. Then π is one of the following planes:
(1) A semifield plane,
(2) Lüneburg–Tits,
(3) Betten for q even or Walker for q odd.

For even order the previous result may be improved as follows:

THEOREM 48.15. (Jha–Johnson [**590**]). Let π be a translation plane of even order q^2 with spread in $PG(3,q)$ that admits a linear collineation group G where uq^2 divides $|G|$ and u is a 2-primitive divisor of $q - 1$. Then the plane is
(1) A semifield plane,
(2) Lüneburg–Tits, or
(3) Betten.

CHAPTER 49

Transversal Extensions.

Johnson's work on derivable nets shows that every derivable net is combinatorially equivalent to a three-dimensional projective space over a skewfield K. More precisely, the points and lines of the net become the lines and points skew to a fixed line N. Furthermore, Knarr [**895**] proved that, for any derivable affine plane π, every line ℓ not belonging to the derivable net embeds to a set of lines in the projective space such this set union N becomes a spread $S(\ell)$ of $PG(3, K)$. Each such spread obtained in this way is also a dual spread.

DEFINITION 49.1. A set of lines S in $PG(3, K)$, for K a skewfield, is a dual spread if and only if each plane of $PG(3, K)$ contains a unique line of S; S covers the set of planes.

Since in the finite case, every spread is a dual spread, given any spread, there is an associated spread obtained by taking the dualizing of the associated projective space wherein lives the original spread. The associated duality in terms of matrix spread sets has a particularly nice representation. In particular, if

$$x = 0, y = xM; M \in \mathcal{M}$$

is a matrix spread set for a translation plane, the

$$x = 0, y = xM^{-t}; M \in \mathcal{M},$$

where A^t is the transpose of the matrix A, is a matrix spread set of the dual spread. If we recoordinatize by $(x, y) \to (y, x)$, then there is an isomorphic spread

$$x = 0, y = xM^t; M \in \mathcal{M}$$

which may not be isomorphic to the original spread.

DEFINITION 49.2. Given any matrix spread set S, the matrix spread set S^t, obtained by taking the transposes of the associated matrices of S is called the 'transposed spread' and the corresponding translation planes are said to be 'transposes of each other'.

Note that every finite spread is automatically a dual spread, and conversely. Hence, it is of interest to ask what sorts of spreads arise from a given derivable affine plane. We first point out that it is not actually the existence of the derivable plane that provides the spread nor that of the stronger condition that there is an affine plane containing a derivable net, but simply the existence of a 'transversal' to a derivable net. That is, given a transversal to a derivable net, there is a corresponding spread of the projective space associated combinatorially or 'geometrically' by the embedding process. The nature of the transversal determines whether the spread constructed is also a dual spread.

The reader may wish to consult Chapters 97, 98, and 99 on affine planes that are not translation planes and the definition of a semi-translation plane. The following discussion is considered for arbitrary derivable nets; that is, finite or infinite since the composition is essentially the same for either case. However, there are some differences if the associated skewfield K is not commutative. So, the reader interested solely in the finite case, can easily reduce the material for K isomorphic to $GF(q)$.

DEFINITION 49.3. (1) A spread of a three-dimensional projective space over a skewfield shall be said to be a 'transversal spread' if and only if it arises from a transversal to a derivable net by the embedding process mentioned above.

(2) A transversal spread shall be said to be a 'planar transversal spread' if and only if there is an affine plane π containing the derivable net such that the transversal to the derivable net is a line of π.

We show that every spread is a transversal spread and, in fact, every spread is a planar transversal spread. But more can be said concerning transversal spreads by applying some ideas of Ostrom regarding extension of derivable nets. We show how the extension techniques of Ostrom can be related to the question of transversal spreads and show how to interconnect these ideas with the geometric embedding process to determine that all spreads are planar transversal spreads.

Using these ideas, we have a framework to be able to discuss the structure of derivable affine planes which admit various collineation groups. In particular, we show that a finite derivable affine plane of order q^2 which admits a (linear) collineation group leaving the derivable net $\mathcal{D}$ invariant and acting flag-transitively on flags on lines not in $\mathcal{D}$ must always be a semi-translation plane which admits either an elation group of order q or a Baer group of order q. When the semi-translation plane is a translation plane, the structure of the spread is more or less determined in this situation.

The assumption on partial flag-transitivity may be relaxed to assumptions on orders of certain groups. In this case, we are able to provide some structure theory for finite affine planes as follows:

THEOREM 49.4. (Johnson [**758**]). (1) If a derivable affine plane π of order q^2 admits a linear p-group of order q^5 if q is odd or $2q^5$ if q is even that fixes the derivable net $\mathcal{D}$, then π contains a group that acts transitively on the affine points.

(2) Furthermore, the group contains either an elation group of order q or a Baer group of order q with axis a subplane of $\mathcal{D}$. If the order of the stabilizer of a point H is at least $2q$, then the order is $2q$ and H is generated by an elation group of order q and a Baer involution with axis in $\mathcal{D}$ or by a Baer group of order q and an elation.

(3) π is a non-strict semi-translation plane of order q^2 admitting a translation group of order q^3p^γ. Furthermore, either π is a translation plane or there is a unique $((\infty), \ell_\infty)$-transitivity and the remaining infinite points are centers for translation group of orders qp^γ.

(4) If π is non-Desarguesian in the elation case above, then π admits a set of q derivable nets sharing the axis of the elation group of order q.

49.1. Extensions of Derivable Nets.

As is well known, the concept of the derivation of a finite affine plane was conceived by Ostrom in the 1960s. During this period, one of the associated problems that Ostrom considered concerned the extension of the so-called derivable nets to either a supernet or to an affine plane. At that time, coordinate geometry was the primary model in which to consider extension questions. With a particular vector-space structure assumed for a derivable net, Ostrom [**1039**] was able to show that any transversal to such a finite derivable net allowed its embedding into a dual translation plane. One of the present authors (Johnson) was able to extend this to the arbitrary or infinite case. Before we proceed with other models, we review some of the definitions and recall some of the results of the coordinate or algebraic method.

DEFINITION 49.5. Let K be a skewfield and V a right two-dimensional vector space over F. A 'vector-space derivable net' $\mathcal{D}$ is a set of 'points' (x, y) $\forall x, y \in V$ and a set of 'lines' given by the following equations:

$$x = c, \ y = x\alpha + b \ \forall c, b \in V, \ \forall \alpha \in F.$$

DEFINITION 49.6. A 'transversal' T to a net $\mathcal{N}$ is a set of net points with the property that each line of the net intersects T in a unique point and each point of T lies on a line of each parallel class of $\mathcal{N}$.

A 'transversal function' f to a vector-space derivable net is a bijective function on V with the following properties:

(i) $\forall c, d, \ c \neq d$ of V, $f(c) - f(d)$ and $c - d$ are linearly independent,
(ii) $\forall \alpha \in F$ and $\forall b \in V$, there exists a $c \in V$ such that $f(c) = c\alpha + b$.

It follows from Johnson [**666**] that transversals and transversal functions to vector-space derivable nets are equivalent, each giving rise to the other. It should be noted that everything can be phrased over the 'left' side as well. That is, a 'right vector-space net' over a skewfield F is naturally a 'left vector-space net' over the associated skewfield F^{opp} where multiplication $\bullet$ in F^{opp} is defined by $a \bullet b = ba$ where juxtaposition denotes multiplication in F.

THEOREM 49.7. (Johnson [**666**, (1.7)]). Let $\mathcal{D}$ be a vector-space derivable net and let T be a transversal. Then there is a transversal function f on the associated vector space V such that $\mathcal{D}$ may be extended to a dual translation plane with lines given as follows:

$$x = c, y = f(x)\alpha + x\beta + b \ \forall \alpha, \beta \in K \text{ and } \forall b, c \in V.$$

Conversely, any dual translation plane whose associated translation plane has its spread in $PG(3, K)$ may be constructed from a transversal function as above.

We will be recalling a more geometric approach shortly and to distinguish between the two, we formulate the following definition:

DEFINITION 49.8. Let $\mathcal{D}$ be a (right) vector-space derivable net with transversal function f then the dual translation plane with lines given by

$$x = c, y = f(x)\alpha + x\beta + b \ \forall \alpha, \beta \in K \text{ and } \forall b, c \in V$$

shall be called the 'algebraic extension' of $\mathcal{D}$ by f and the set of such shall be termed the set of 'algebraic extensions of $\mathcal{D}$'.

49.2. Geometric Extension.

Historically, perhaps the most important question left open by the coordinate approach was whether 'derivation' could be considered a geometric construction, and the text of one of the present authors *Subplane Covered Nets* [**753**] examines this question in detail. In particular, the structure of a derivable net $\mathcal{D}$ has been determined in Johnson [**723**] and [**724**]. The points and lines can be embedded as the lines and points of a 3-dimensional projective space Σ isomorphic to $PG(3, K)$, K a skewfield, which are skew or non-incident with a fixed line N. The parallel classes of the net become the planes of Σ containing N and the Baer subplanes of the net become the planes of Σ that do not contain N. It then turns out that every derivable net is a 'right' vector-space net over a skewfield F and, of course, may be considered a 'left vector-space net' over F^{opp}. In particular, when the net is considered a 'left net' the embedding into the projective space is determined by a left 4-dimensional vector space over the associated skewfield.

When one has a derivable affine plane π, Knarr [**895**] asked of the general nature of the lines of an affine plane π containing $\mathcal{D}$ in terms of the embedding. Knarr showed that every line of $\pi - \mathcal{D}$ produces a spread of lines of Σ that contain N. Also, this spread is a dual spread.

Since we are interested in the more general situation, we assume only that there is a transversal T to the derivable net $\mathcal{D}$ which defines a simple net extension $\mathcal{D}^{+T}$. In Johnson [**749**], it is pointed out that it is possible to embed any derivable net into an affine plane where the affine plane may not be derivable itself. Hence, we distinguish between having a net extension and having a 'derivable extension', by which we mean that each Baer subplane of the net remains Baer when considered within the extension net; each point is on a line of each subplane, taken projectively (the subplane structure is 'point-Baer') and each line is incident with a point of each subplane, taken projectively (the subplane structure is 'line-Baer'). In essence, we would merely require that T intersect each Baer subplane.

THEOREM 49.9. (See Knarr [**895**]). Let $\mathcal{D}$ be a derivable net and assume that T is a transversal to $\mathcal{D}$ defining an extension net $\mathcal{D}^{+T}$.

(1) Then the points of T determine a spread $S(T)$ of lines in the projective space Σ associated with $\mathcal{D}$ that contains the special line N.

(2) If the net extension is a derivable extension, then $S(T)$ is a dual spread.

(3) Conversely, if $S(T)$ is a dual spread, for each line of $T - \mathcal{D}$, then the net extension is a derivable extension.

Now we see that any derivable net is, in fact, a vector-space derivable net so the two approaches merge.

DEFINITION 49.10. Let $\mathcal{D}$ be any derivable net. Then $\mathcal{D}$ may be considered a 'left' vector-space net over a skewfield K. Let Σ denote the three-dimensional projective space $PG(3, K)$ for K a skewfield, with special line N defined combinatorially by $\mathcal{D}$ and so that $\mathcal{D}$ may be embedded in Σ. Let T be any transversal to $\mathcal{D}$ and let $S(T)$ denote the spread of Σ defined by the net-points of T as lines of Σ together with the line N. Let $\pi_{S(T)}$ denote the associated translation plane and let $\pi^D_{S(T)}$ denote any affine dual translation plane whose projective extension dualizes to $\pi_{S(T)}$, taken projectively. Then $\pi^D_{S(T)}$ contains a derivable net isomorphic to $\mathcal{D}$ but considered as a 'right' vector-space net over the skewfield K^{opp}.

We shall call $\pi^D_{S(T)}$ a 'geometric extension of $\mathcal{D}$' by $S(T)$.

Hence, given a derivable net $\mathcal{D}$ with transversal T, we may consider two possible situations. First of all, we know that $\mathcal{D}$ may be considered a right vector-space net over a skewfield F and there is an associated transversal function which we may use to extend $\mathcal{D}$ to a dual translation plane π^D_f (the algebraic extension). On the other hand, we may consider $\mathcal{D}$ as a left vector-space net over $F^{\text{opp}} = K$, embed the net combinatorially into a (left) three-dimensional projective space Σ isomorphic to $PG(3, K)$, with distinguished line N and then realize that the transversal T, as a set of points of $\mathcal{D}$, is a set of lines whose union with N, is a spread of Σ which defines a translation plane with an associated dual translation plane $\pi^D_{S(T)}$ (the geometric extension).

Hence, we arrive at the following fundamental question:

Given a derivable net D with transversal T and associated transversal function f, is the algebraic extension of D by f isomorphic to a geometric extension of D by $S(T)$?

Before we consider this question, we note the following connection with spreads in three-dimensional projective space and transversals to derivable nets. However, the reader might note that there are possible left and right transversal functions depending on whether the vector-space derivable net is taken as a left or a right vector space and this will play a part in our discussions.

THEOREM 49.11. The set of spreads in three-dimensional projective spaces is equivalent to the set of transversals to the set of derivable nets; every spread is a transversal spread.

49.3. Planar Transversal Extensions and Dual Spreads.

THEOREM 49.12. (See Knarr [**895**, (2.7)]). Let $\mathcal{P}$ be any spread in $PG(3, K)$, for K a skewfield. Let $\mathcal{D}$ be a derivable net and T a transversal to it which geometrically constructs $\mathcal{P}$ by the embedding process.

Then there is a dual translation plane π^T_D constructed by the algebraic extension process. For any line of $\pi^T_D - \mathcal{D}$, the spread in the associated three-dimensional projective geometry obtained by the geometric embedding process produces a translation plane with spread $S(T)$ isomorphic to $\mathcal{P}$ and whose dual is isomorphic to π^T_D.

Hence, all spreads are planar transversal spreads.

We have not yet dealt with the possibility that a planar transversal spread may not actually arise from a derivable affine plane, that it may be possible that the spread is not a dual spread (recall that a 'dual spread' in $PG(3, K)$, for K a skewfield, is a set S of mutually skew lines such that given any plane π, there is a unique line of S contained in π).

In Johnson [**749**], similar constructions to the following are given and the reader is referred to this article for additional details.

THEOREM 49.13. Let $\mathcal{D}$ be a derivable net and let Σ be isomorphic to $PG(3, K)$ and correspond to $\mathcal{D}$ with special line N. If K is infinite, then there exists a dual translation plane π extending $\mathcal{D}$ and a line T of $\pi - \mathcal{D}$ such that $S(T)$ is a spread of $PG(3, K)$ which is not a dual spread. In particular, if K is a field, then $S(T)$ is non-Pappian.

There exist planar transversal spreads that are not dual spreads.

49.4. Translation Extension Nets.

Suppose that $\mathcal{D}$ is a derivable net and there exists a transversal T and construct the spread $S(T)$. In Knarr [**895**], the question was raised when $S(T)$ is Pappian or what happens when $\mathcal{D}$ is contained in a translation plane. By Sections 49.1 and 49.2, we may apply the algebraic extension process to consider such questions. In particular, there are lines $x = c$, $y = f(x)\alpha + x\beta + b$ which define any affine plane containing the derivable net $\mathcal{D}$. Hence, if $\mathcal{D}$ is contained in a translation plane, then the dual translation plane containing $\mathcal{D}$ is a translation plane, which implies that $S(T)$ is a semifield spread. Knarr observes this fact by noting that the 'translation' collineation group of the derivable net would then act on the net extended by the transversal implying a collineation group fixing a component N and transitive on the remaining components of the spread (points of T). Hence, we obtain:

THEOREM 49.14. (See Knarr [**895**], also see Johnson [**758**]). Let $\mathcal{D}$ be a derivable net and let T be a transversal. If the extension net $\langle\mathcal{D}\cup\{T\}\rangle$ defined by $\mathcal{D}\cup\{T\}$ is a translation net, then the spread $S(T)$ defines a semifield plane.

Furthermore, all semifield spreads in $PG(3, K)$, for K a skewfield, are 'semifield planar transversal spreads' (arising from semifield planes).

Let $\mathcal{P}$ be any non-Desarguesian semifield spread in $PG(3, K)$, for K a skewfield. If we choose the axis of the affine elation group to be N and view the spread as a transversal to a derivable net with the embedding in $PG(3, K) - N$, the corresponding dual translation plane will be a semifield plane. On the other hand, if any other line of $\mathcal{P}$ is chosen as N in the embedding, the affine dual translation plane will not be an affine semifield plane. So, a semifield spread in $PG(3, K)$ could arise as a planar transversal spread without the affine plane containing the derivable net being an affine semifield plane.

With the remarks of the previous paragraph in mind, we now examine the semifield spreads that can be obtained when $\langle\mathcal{D}\cup\{T\}\rangle$ is a translation extension-net.

By Sections 49.1 and 49.2, the affine plane containing the derivable net will correspond to the dual translation plane side where the components are left subspaces over the skewfield K^{opp}, provided the geometric embedding is in the left projective space $PG(3, K)$. By the arguments of Johnson [**724**], there is a vector space V over a prime field $\mathcal{P}$ of the form $W \oplus W$ such that points are the vectors (x, y) for $x, y \in W$ and we may choose a basis so that $x = 0, y = 0, y = x$ belong to the derivable net $\mathcal{D}$. We want to consider the derivable net as a right vector-space net over a skewfield at the same time we are considering the vector space and the components of the derivable net as left spaces over the same skewfield. Furthermore, there is a skewfield K^{opp} such that $W = K^{\text{opp}} \oplus K^{\text{opp}}$ as a left K^{opp}-vector space and components of $\mathcal{D}$ may be represented as follows:

$$x = 0, y = x \begin{bmatrix} B & 0 \\ 0 & B \end{bmatrix} ; B \in K^{\text{opp}}.$$

We again note that the components of $\mathcal{D}$ are not necessarily all right K^{opp}-subspaces, although we say that $\mathcal{D}$ is a 'right' vector-space net over K^{opp}. We note that following the ideas in Sections 49.1 and 49.2, we are working in the dual translation plane side which contains the derivable net. The translation plane obtained by dualization has its spread in $PG(3, K)$. Recall that this means that the so-called 'right nucleus' of the semifield in question is K^{opp}.

It also follows that any translation net has components of the general form $y = xT$ where T is a $\mathcal{P}$-linear bijection of W. If W is decomposed as $K^{\mathrm{opp}} \oplus K^{\mathrm{opp}}$ over the prime field $\mathcal{P}$, choose any basis $\mathcal{B}$ for K^{opp} over $\mathcal{P}$. Then, we may regard V as (x_1, x_2, y_1, y_2) where x_i, y_i are in K for $i = 1, 2$ and also may be represented as vectors over $\mathcal{B}$. That is, for example, $x_j = (x_{j,i}; i \in \lambda)$, for $j = 1, 2$, with respect to $\mathcal{B}$ for $x_{j,i} \in \mathcal{P}$ for some index set λ. With this choice of basis, we may represent T as follows:

$$ T = \left(y = x \begin{bmatrix} T_1 & T_2 \\ T_3 & T_4 \end{bmatrix} \right), $$

where T_i are linear transformations over $\mathcal{P}$ represented in the basis $\mathcal{B}$. Note that we are not trying to claim that the T_i's are K^{opp}-linear transformations, merely $\mathcal{P}$-linear.

The action is then

$$ (y_1, y_2) = (x_1, x_2) \begin{bmatrix} T_1 & T_2 \\ T_3 & T_4 \end{bmatrix} = (x_1 T_1 + x_2 T_3, x_1 T_2 + x_2 T_4). $$

where the x_i and y_j terms are considered as $\mathcal{P}$-vectors.

Hence, the T_i's are merely additive mappings on K^{opp} but not necessarily K^{opp}-linear.

In reading the next theorem, it might be kept in mind that a derivable net may always be considered algebraically a pseudo-regulus net with spread in $PG(3, K^{\mathrm{opp}})$ when the geometric embedding is in $PG(3, K)$. When K is a field, this is not to say that these two projective spaces are the same as $\mathcal{D}$ can be a regulus in a three-dimensional projective space while being embedded in another and both projective spaces are isomorphic.

THEOREM 49.15. (Johnson [**758**]). If $\mathcal{D}$ is a derivable net and $\langle \mathcal{D} \cup \{T\} \rangle$ a translation net regarded as a left vector-space net over the associated prime field $\mathcal{P}$, and $\mathcal{D}$ regarded as a right vector-space net over K^{opp} then the geometric embedding of $\mathcal{D}$ into Σ isomorphic to $PG(3, K)$ is considered as a 'left' space embedding.

Representing $\mathcal{D}$ as

$$ x = 0, y = x \begin{bmatrix} B & 0 \\ 0 & B \end{bmatrix} ; B \in K^{\mathrm{opp}}, $$

we may represent T as

$$ \left(y = x \begin{bmatrix} T_1 & T_2 \\ T_3 & T_4 \end{bmatrix} \right) $$

where the T_i are additive mappings of K^{opp} and $\mathcal{P}$-linear transformations.

(1) The line $y = x \begin{bmatrix} T_1 & T_2 \\ T_3 & T_4 \end{bmatrix}$ determines a semifield spread admitting an affine homology group with axis $y = 0$ and coaxis $x = 0$ isomorphic to $K^{\mathrm{opp}} - \{0\}$ (the dual semifield plane has its spread in $PG(3, K)$ and is $S(T)$).

The semifield spread has the following form:

$$ x = 0, y = x \left(\begin{bmatrix} T_1 & T_2 \\ T_3 & T_4 \end{bmatrix} \begin{bmatrix} A & 0 \\ 0 & A \end{bmatrix} + \begin{bmatrix} B & 0 \\ 0 & B \end{bmatrix} \right) \ \forall A, B \in K^{\mathrm{opp}}. $$

(2) The semifield spread with spread in $PG(3, K)$ (the $S(T)$) has the following form (where here the T_i's are considered additive mappings of K):

$$ x = 0, y = x \begin{bmatrix} sT_1 + vT_3 & sT_2 + vT_4 \\ s & v \end{bmatrix} \ \forall v, s \in K. $$

(3) The semifield spread in $PG(3, K)$ is a skew-Desarguesian spread if and only if the T_i's are all K-linear transformations (i.e., multiplication by elements of K) if and only if $\mathcal{D}^{+T}$ is a partial spread in $PG(3, K^{\text{opp}})$; considering $\mathcal{D}$ as a pseudo-regulus in $PG(3, K^{\text{opp}})$, T is a subspace in the projective spread $PG(3, K^{\text{opp}})$.

(4) See Knarr [**895**]. If K is a field, then the semifield spread in $PG(3, K)$ is Pappian if and only if the T_i's are all K-linear transformations (i.e., multiplication by elements of K) if and only if $\langle \mathcal{D} \cup \{T\} \rangle$ is a partial spread in $PG(3, K)$; T is a subspace in the projective spread $PG(3, K)$ wherein $\mathcal{D}$ is considered a regulus.

We noted in part (4) that the associated spreads in $PG(3, K)$ are Pappian if a derivable net is a regulus net in $PG(3, K)$ and the transversal is a subspace within the same $PG(3, K)$. We might inquire as to the nature of the semifield spreads if we assume initially that the transversal is a subspace in $PG(3, K)$ but the derivable net is not necessarily a K-regulus.

Derivable nets in $PG(3, K)$, for K a field may be given a particularly nice form (see Johnson [**705**] for the finite case and Jha–Johnson [**619**] for the infinite case). A derivable net $\mathcal{D}$ with partial spread in $PG(3, K)$, for K a field, may be represented in the following form:

$$x = 0, y = x \begin{bmatrix} u & A(u) \\ 0 & u^\sigma \end{bmatrix} \forall u \in K$$

and where σ is an automorphism of K, and where x and y are 2-vectors over K and A is a function on K such that

$$\begin{bmatrix} u & A(u) \\ 0 & u^\sigma \end{bmatrix} \forall u \in K$$

is a field isomorphic to K.

LEMMA 49.16. *Furthermore, $A \equiv 0$ in the finite case, or when there are at least two Baer subplanes incident with the zero vector that are K-subspaces. When there is exactly one K-subspace Baer subplane, the characteristic is two, $\sigma = 1$ and $A(u) = Wu + uW$ for some linear transformation W of K over the prime field.*

Hence, we obtain:

THEOREM 49.17. (Johnson [**758**]). *A derivable net $\mathcal{D}$ with transversal extension T giving a partial spread $\langle \mathcal{D} \cup \{T\} \rangle$ that is in $PG(3, K)$, for K a field, and such that there are at least two Baer subplanes that are K-subspaces, constructs a semifield spread in $PG(3, K)$ of the following form:*

$$x = 0, y = x \begin{bmatrix} v^\sigma + sk & s^{\sigma^{-1}}l \\ s & v \end{bmatrix} \forall v, s \in K,$$

for σ an automorphism of K, and constants $k, l \in K \in K$.

The spreads mentioned above are considered in Johnson [**744**] and are a generalization of spreads originally defined by Knuth and hence, perhaps, should be called 'generalized Knuth spreads'.

THEOREM 49.18. (Johnson [**758**]). *Let a derivable net $\mathcal{D}$ with transversal extension T giving a partial spread $\langle \mathcal{D} \cup \{T\} \rangle$ that is in $PG(3, K)$, for K a field, such that there is exactly one Baer subplane that is a K-subspace. Then there is* an associated semifield spread in $PG(3, K)$ of the following form:

$$x = 0, y = x \begin{bmatrix} s(a + Wc) + vc & s((a + Wc)W + b + Wd) + v(cW + d) \\ s & v \end{bmatrix}$$

$$\forall v, s \in K$$

where W is some prime field linear transformation of K, a, b, c, d constants in K.

49.5. Vector-Space Transversals.

Now assume that we have a derivable net and a vector-space transversal. We represent the derivable net as in the previous section as

$$x = 0, y = x \begin{bmatrix} B & 0 \\ 0 & B \end{bmatrix} \forall B \in K^{\mathrm{opp}}$$

where $K^{\mathrm{opp}} = F$ is a skewfield and we represent the transversal in the form

$$\left(y = x \begin{bmatrix} T_1 & T_2 \\ T_3 & T_4 \end{bmatrix} \right) = xT$$

where the T_i's are prime field linear transformations and additive F-mappings.

First let F be isomorphic to $GF(q)$ for $q = p^r$, p a prime. Let $H_{p,f,g,u}$ be a group of order $q - 1$ whose elements are defined as follows:

$$\tau_u = \begin{bmatrix} p(u) & 0 & 0 & 0 \\ 0 & f(u) & 0 & 0 \\ 0 & 0 & p(u)\lambda(u) & 0 \\ 0 & 0 & 0 & f(u)\lambda(u) \end{bmatrix} \text{ for } u \in F - \{0\}$$

and p, f, λ are functions on F. We require that the derivable net is left invariant. For this, we must have

$$\begin{bmatrix} p(u)^{-1}p(u)\lambda(u) = \lambda(u) & 0 \\ 0 & f(u)^{-1}f(u)\lambda(u) = \lambda(u) \end{bmatrix}$$

for some function v of u.

We consider situations under which

$$x = 0, y = x \begin{bmatrix} p(u)^{-1} & 0 \\ 0 & f(u)^{-1} \end{bmatrix} T \begin{bmatrix} p(u)\lambda(u) & 0 \\ 0 & f(u)\lambda(u) \end{bmatrix} + wI$$

for all $u \neq 0, w \in F$ defines a spread. We note that there is an associated elation group E of order q and the spread fixes $x = 0$.

DEFINITION 49.19. When the above set defines a spread, we denote the spread by $\pi_{T, H_{p,f,p\lambda,f\lambda}}$ and call the spread a '$(T, EH_{p,f,p\lambda,f\lambda})$-spread'.

More generally, it might be possible to have a group containing an elation group E of order q and a group H such that EH acts transitively on the components of the spread not in the derivable net but H may not be diagonal. In the more general case, we refer to the spread as a 'partially transitive elation group spread'.

Any $(T, H_{p,f,p\lambda,f\lambda})$-spread is derivable and the derived plane admits a collineation group fixing the spread and acting transitively on the components not in the derivable net. The elation group is turned into a Baer group B and the group $H_{p,f,p\lambda,f\lambda}$ is turned into the group $H_{p,p\lambda,f,f\lambda}$. We call such a spread a '$(T^*, BH_{p,p\lambda,f,f\lambda})$-spread'. Also, more generally, if we have a partially transitive elation group spread, it derives to a so-called 'partially transitive Baer group spread'.

EXAMPLE 49.20. Examples of partially transitive elation group spreads are as follows:

(1) Any semifield plane of order q^2 whose semifield is of dimension two over its middle nucleus. This plane is a $(T, EH_{\lambda^{-1},\lambda^{-1},1,1})$-spread which derives to a $(T^*, BH_{\lambda^{-1},1,\lambda^{-1},1})$-spread that is also known as a 'generalized Hall spread' (i.e., of type 1).

Note that if $y = xT$ is a vector-space transversal to a finite derivable net, then we may realize the derivable net as a left vector-space net and automatically use the 'left' extension process to construct a dual translation plane which then becomes a semifield plane with spread:

$$x = 0, y = \alpha T + \beta I, \forall \alpha, \beta \in F.$$

(2) Any semifield plane of order q^2 whose semifield is of dimension two over its right nucleus. This plane provides a $(T, EH_{1,1,\lambda,\lambda})$-spread which derives a $(T^*, BH_{1,\lambda,1,\lambda})$-spread that is also known as a 'generalized Hall spread of type 2'.

This is merely the situation with which we began, realizing the derivable net as a right vector space and constructing the dual translation plane which is then a semifield plane with spread:

$$x = 0, y = T\alpha + \beta I, \forall \alpha, \beta \in F.$$

(3) Other known examples all correspond to situations where the $y = xT$ is a line in the projective space wherein the derivable net is a regulus. The partially transitive elation spreads correspond to flocks of quadratic cones. When the order q^2 is odd, there is a classification of such partially transitive elation spreads in Hiramine and Johnson [523] and the possibility that the group H may not be diagonal is included in this study. However, the known examples are $(T, EH_{p,f,p\lambda,f\lambda})$-spreads. When $q = 2$, any partially transitive elation spread turns out to be a $(T, EH_{p,f,p\lambda,f\lambda})$-spread. Furthermore, combining this work with a study of Penttila and Storme [1100] shows that the known examples are the only possible examples. For details and additional information, the reader is referred to the survey paper of Johnson and Payne [792].

To list but one of these examples in the odd-order case, we consider the spread:

$$x = 0, y = x \begin{bmatrix} u & \gamma t^\sigma \\ t & u \end{bmatrix}; \forall u, t \in GF(q)$$

where γ is a non-square, q is odd and σ is an automorphism of $GF(q)$.

We note that the elements of the elation group E have the following form:

$$\begin{bmatrix} 1 & 0 & u & 0 \\ 0 & 1 & 0 & u \\ 0 & 0 & 1 & 0 \\ 0 & 0 & 0 & 1 \end{bmatrix}; \forall u \in GF(q).$$

Furthermore, the group is $H_{1,f,\lambda,f\lambda}$ where $f(u) = u^\sigma, \lambda(u) = u^{\sigma+1}$.

One of the early results using group theory to determine the structure of finite affine planes is the result of Wagner [1202] who proved that finite affine planes that admit a collineation group acting transitively on the flags of the plane are always translation planes.

A related problem of interest is

PROBLEM 49.21. Let π be a finite derivable affine plane and let $\mathcal{D}$ denote the derivable net.

(1) If π admits a collineation group G leaving $\mathcal{D}$ invariant that acts transitively on the flags of π on lines not in $\mathcal{D}$, is π a translation plane?

(2) If π is a translation plane, is it always either a partially transitive elation plane or a partially transitive Baer plane?

DEFINITION 49.22. We shall call a derivable affine plane that admits a collineation group leaving the derivable net invariant and acting flag-transitively on the flags on lines not in the derivable net a 'partially flag-transitive plane'.

49.5.1. Dual Translation Planes. Let π be a translation plane with spread in $PG(3, q)$. Choose any component $x = 0$ and let (∞) denote the parallel class containing the component. Take any 1-dimensional $GF(q)$-subspace X on any component $y = 0$. Then the lines $x(\infty)$ union (∞), for all x in X, form a derivable net in the dual translation plane obtained by taking (∞) as the line at infinity in the dual plane. Notice that the dual translation plane admits a collineation group of order $q^3(q - 1)$ which leaves the derivable net invariant. This group consists of the translation group of order q^3 generated by the subgroups with center (∞) of order q^2 and the group of order q with center (0) which leaves X invariant. The kernel homology group H of order $q - 1$ leaves X invariant. Hence, the product of these two groups is a group W of the dual translation plane π^*. Furthermore, there is an infinite point $(\infty)^*$ of the derivable net $\mathcal{D}^*$ which is fixed by W. The group of order q mentioned above becomes an elation group E of order q. Note that HE acts transitively on the lines of (∞) not in $(\infty)X$ since E fixes no lines and has q orbits of length q and H fixes exactly one line and has orbits of length $q - 1$. Note that E is normal in EH so that H can fix exactly one orbit of E. Hence, HE has orbits of length q and $q(q - 1)$.

Thus, EH acting on the dual translation plane is transitive on the infinite points of the plane which are not in the derivable net $\mathcal{D}^*$.

Now assume that the associated translation plane admits a collineation group of order q^2 that is in the linear translation complement (i.e., in $GL(4, q)$), fixes (∞) and acts transitively on the components other than $x = 0$.

It follows from the work of Johnson and Wilke [**814**] that the group of order q^2 may be represented so that acting on $x = 0$, the group is a subgroup of

$$\left\langle \begin{bmatrix} 1 & a \\ 0 & 1 \end{bmatrix} ; a \in GF(q) \right\rangle.$$

Hence, it follows that, representing the vector space by 4-vectors (x_1, x_2, y_1, y_2), the lines $x = (0, \alpha) \ \forall \ \alpha \in GF(q)$ are fixed. Thus, if we take

$$X = \{(0, \alpha, 0, 0); \alpha \in GF(q)\}$$

as defining the derivable net $\mathcal{D}^*$, we have a group S of order q^2 acting on the dual translation plane. Furthermore, we then obtain a collineation group of order $q^5(q - 1)$ which fixes the derivable net and acts partially flag-transitively on the affine dual translation plane. In order to see this, we note that the group of order $q(q - 1)$ mentioned previously is transitive on the parallel classes. The translation group with center (∞) becomes a translation group of the dual translation plane with center $(\infty)^*$ and this group acts transitively on the affine lines of any parallel class not equal to $(\infty)^*$. Take any point P of the translation plane such that $P(\infty)$

is not a line of $(\infty)X$. Then, using the translation group, there is a group conjugate to S by a translation which acts transitively on the lines incident with P other than $P(\infty)$.

Therefore:

THEOREM 49.23. (Johnson [**758**]). Let π be any translation plane of order q^2 with spread in $PG(3, q)$. Assume that there exists a collineation group in the linear translation complement of order q^2 that fixes a component and acts transitively on the remaining components of π.

Then the dual translation plane is a partially flag-transitive derivable affine plane admitting a collineation group of order $q^5(q-1)$.

The dual translation plane is a translation plane (and hence a semifield plane) if and only if the group of order q^2 mentioned above is an elation group of the associated translation plane.

COROLLARY 49.24. Any semi-translation plane obtained by the derivation of a partially flag-transitive dual translation plane is also partially flag-transitive. The dual translation plane is of 'elation' type whereas the semi-translation plane is of 'Baer' type (the dual translation plane admits an elation group of order q and the semi-translation plane admits a Baer group of order q).

The previous examples all involve solvable groups. Are there 'non-solvable' partially flag-transitive affine planes?

The Hall and Desarguesian planes are the only translation planes of order q^2 that admit $SL(2, q)$ and are partially flag-transitive affine planes.

49.6. Transposed Spreads and Reconstruction.

In this section, we ask the following question:

Let $\mathcal{D}$ be a derivable net and T a transversal such that $\mathcal{D} \cup \{T\}$ is a derivable extension. Let $\widehat{\mathcal{D}} \cup \{T\}$ denote the corresponding derivable extension, where $\widehat{\mathcal{D}}$ is the derived net of $\mathcal{D}$. When is the spread $S(T)$ with respect to $\mathcal{D} \cup \{T\}$, $S(T)_{\mathcal{D}+T}$, isomorphic to the spread $S(T)_{\widehat{\mathcal{D}}+T}$ with respect to $\widehat{\mathcal{D}} \cup \{T\}$?

First of all, we note that if the corresponding projective space is $PG(3, K)$ of $\mathcal{D}^{+T}$, then $S(T)_{\mathcal{D}+T}$ is in $PG(3, K)$, whereas $S(T)_{\widehat{\mathcal{D}}+T}$ is in $PG(3, K^{\text{opp}})$. However, the points and (Baer) subplanes of $PG(3, K)$ are subplanes and points of $PG(3, K^{\text{opp}})$, respectively. Hence, the question only makes genuine sense when K is a field.

THEOREM 49.25. Let $\mathcal{D}$ be a derivable net embedded in $PG(3, K)$, where K is a field, and let T be a transversal to $\mathcal{D}$ which is also a transversal to the derived net $\widehat{\mathcal{D}}$. Then the spread $S(T)_{\mathcal{D}+T}$ corresponding to $\mathcal{D}^{+T}$ is isomorphic to the spread $S(T)_{\widehat{\mathcal{D}}+T}$ corresponding to $\widehat{\mathcal{D}}^{+T}$ if and only if there is a duality of $PG(3, K)$ which maps one spread to the other.

49.6.1. Reconstruction. We have noted that transversals to derivable nets are basically equivalent to spreads in $PG(3, K)$. However, although transversals are then used to construct dual translation planes, there are other affine planes containing a derivable net which are not dual translation planes. The question is whether there is a way to use various sets of spreads to construct or reconstruct an affine plane containing a derivable net.

DEFINITION 49.26. A 'skew parallelism' of $PG(3, K) - N$ is a set of spreads each containing N which forms a disjoint cover of the lines of $PG(3, K)$ skew to N.

Let $\mathcal{S}$ be a skew parallelism of $PG(3, K) - N$ and let $\mathcal{P}$ be any spread containing N. We shall say that $\mathcal{P}$ is 'orthogonal' to $\mathcal{S}$ if and only if $\mathcal{P}$ intersects each spread of $\mathcal{S}$ in a unique line $\neq N$.

A set of skew parallelisms of $PG(3, K) - N$ is 'orthogonal' if and only if each spread of any one skew parallelism is orthogonal to each of the remaining skew parallelisms.

A set $\mathcal{A}$ of skew parallelisms of $PG(3, K) - N$ is said to be 'planar' if and only if, given any two lines ℓ_1 and ℓ_2 of $PG(3, K)$ which are skew to N, there is a skew parallelism of $\mathcal{A}$ containing a spread sharing ℓ_1 and ℓ_2.

If the spreads of a set of skew parallelisms are all dual spreads, we shall say that the set is a 'derivable' set of skew parallelisms.

THEOREM 49.27. (Johnson [**758**]). Given an orthogonal and planar set $\mathcal{A}$ of skew parallelisms of $PG(3, K) - N$, then there is a unique affine plane $\pi_{\mathcal{A}}$ containing a derivable net such that the set of transversals to the derivable net are the spreads of the set $\mathcal{A}$.

Conversely, any affine plane containing a derivable net corresponds to a uniquely defined orthogonal and planar set of skew parallelisms.

Hence, the set of derivable affine planes is equivalent to the set of derivable, orthogonal and planar sets of skew parallelisms.

Now we ask the nature of a 'transitive' skew parallelism.

DEFINITION 49.28. A skew parallelism of $PG(3, K) - N$ is 'transitive' if and only if there exists a subgroup of $P\Gamma L(4, K)_N$ that acts transitively on the spreads of the skew parallelism.

A planar and orthogonal set of skew parallelisms is 'transitive' if and only if there exists a subgroup of $P\Gamma L(4, K)_N$ which acts transitively on the set.

We shall say that the set is 'line-transitive' if and only if the stabilizer of a spread is transitive on the lines not equal to N of the spread, for each spread of the skew parallelism.

We have seen the following in Section 49.1; we reintroduce the ideas again here. Let π^D be a dual translation plane with transversal function $f(x)$ to a right vector-space derivable net so that lines have the equations:

$$x = 0, y = f(x)\alpha + x\beta + b$$

for all $\alpha, \beta \in K$ and for all $b \in V$ (see the notation of Section 49.1). Then, there is a collineation group of π^D which leaves invariant the derivable net and acts transitively on the lines not in the derivable net and of the form $y = f(x)\alpha + x\beta + b$ where $\alpha \neq 0$. The 'translation group' T is transitive on the lines of each such parallel class and represented by the mappings: $(x, y) \longmapsto (x, y + b) \; \forall b \in V$. The affine elation group E is represented by mappings of the form $(x, y) \longmapsto (x, x\beta + y)$ $\forall \beta \in K$ and the affine homology group H is represented by mappings of the form: $(x, y) \longmapsto (x, x\alpha) \; \forall \alpha \in K - \{0\}$. Notice that T and E correspond to certain translation subgroups with fixed centers of the corresponding translation plane and the group H corresponds to the kernel homology group of the associated translation plane.

(1) It also follows that any derivable affine plane coordinatized by a cartesian group will admit a group isomorphic to T and hence corresponds to a transitive skew parallelism.

(2) Any such dual translation plane will produce a transitive planar and orthogonal set of transitive skew parallelisms.

(3) Any semifield spread which contains a derivable net as above will admit a translation group with center $(f(x))$ which fixes $f(x)$ and acts transitively on the points of $f(x)$, which implies that the transversal spread is transitive.

REMARK 49.29. Hence, any semifield spread produces a line-transitive planar and orthogonal set of skew parallelisms.

So, we would ask whether a line-transitive planar and orthogonal set of skew parallelisms corresponds either to a translation plane, a dual translation plane or a semi-translation plane. Hence, not all such sets of skew parallelisms in three-dimensional projective spaces can force the affine plane containing the derivable net to be a semifield plane or even a translation plane. We formulate a fundamental question?

Is a finite derivable partially flag-transitive plane of order q^2 a translation plane, a dual translation plane or a semi-translation plane and if so, is there either a Baer group or elation group of order q?

49.7. Partially Flag-Transitive Planes.

Assume that π is a finite partially flag-transitive affine plane. The derivable net D is combinatorially equivalent to a projective space $PG(3, K)$, where K is isomorphic to $GF(q)$, relative to a fixed line N of $PG(2, K)$. Furthermore, the full collineation group of the net D is $P\Gamma L(4, K)_N$. Assume now that the given collineation group G of π is linear; i.e., in $PGL(4, K)_N$. It follows that the linear subgroup which fixes an affine point and the derivable net (which is now a regulus net) is a subgroup of $GL(2, q)GL(2, q)$, where the product is a central product with common group the center of order $q - 1$. Note we are not trying to say that the derivable net corresponds to a regulus in the particular $PG(3, q)$ wherein lives the skew parallelisms, merely that the derivable net can be realized as arising from a regulus in some three-dimensional projective space.

Let $p^r = q$. Then, there must be a group of order divisible by $q^2(q^2)(q^2 - q)$ by the assumed transitive action. Hence, the p-groups have orders divisible by q^5 and note that the full linear p-group of the derivable net has order q^6. Any such p-group S_p must leave invariant an infinite point (∞) of the derivable net $\mathcal{D}$ as the derivable net consists of $q + 1$ parallel classes.

So, again let G denote the full collineation group of the associated affine plane π, under the assumption that the group is 'linear' with respect to the derivable net and let T denote the translation group with center (∞) of S_p. We note that T is normal in S_p. Let ℓ be any transversal line to the derivable net. Then there exists a collineation group G_ℓ which acts transitively on the points of ℓ.

Hence, we obtain:

THEOREM 49.30. (Johnson [**758**]). (1) A finite derivable partially flag-transitive affine plane of order q^2 with linear group is a non-strict semitranslation plane with a translation group of order $q^3 p^\gamma$.

(2) The plane admits either an elation group or a Baer group of order q.

(3) Furthermore, in the elation case, either the plane is Desarguesian or the plane admits a $((\infty), \ell_\infty)$-transitivity, the point (∞) is invariant and the infinite points not equal to (∞) are centers for translation groups of order qp^γ.

COROLLARY 49.31. A finite derivable partially flag-transitive affine plane of order q^2 where $(p, r) = 1$ for $p^r = q$, is a non-strict semitranslation plane with a translation group of order $q^3 p^\gamma$.

We have seen that dual translation planes arising from translation planes with spreads in $PG(3, q)$ that admit a collineation group of order q^2 in the translation complement and transitive on the components other than a fixed component admit collineation groups of order $q^5(q - 1)$ fixing a derivable net $\mathcal{D}$. Since the plane is a dual translation plane, there is an elation group of order q^2. Hence, there is a collineation group of order $q^6(q - 1)$. We may ask if partially flag-transitive planes admitting the larger group must be dual translation planes. The following properties are established by Johnson [**758**].

THEOREM 49.32. Let π be a finite derivable affine plane of order q^2 admitting an elation group H of order q leaving invariant a derivable net. Then the H-orbits of infinite points union the center of H define a set of q derivable nets of π.

Hence, we see that

THEOREM 49.33. Let π be a non-Desarguesian partially flag-transitive affine plane of order q^2 with linear group and of elation type. Then the corresponding group G fixes one derivable net containing the axis of (∞) and acts transitively on the remaining $q - 1$ derivable nets sharing (∞).

THEOREM 49.34. (Johnson [**758**]). Let π be a non-Desarguesian partially flag-transitive affine plane of order q^2 with linear group and of elation type. If $q = p^r$, assume that $(p, r) = 1$.
(1) If π admits a p-group S of order q^6, or $2q^6$ if $q = 2$, then π admits a collineation group which fixes an infinite point (∞) and is transitive on the remaining infinite points.
(2) Furthermore, either the plane is a dual translation plane or the full group acts two-transitively on a set of q derivable nets sharing the infinite point (∞).

COROLLARY 49.35. Let π be a finite derivable partially flag-transitive affine plane with linear group which is a translation plane and assume that the order is even. Then π is either a $(T, EH_{p,f,p\lambda,f\lambda})$-plane or a $(T^*, BH_{p,p\lambda,f,f\lambda})$-plane.

Hence, in both cases, we obtain a group of order q^4 or $2q^4$ that acts transitively on the affine points of the derivable affine plane.

THEOREM 49.36. (Johnson [**758**]). (1) If a derivable affine plane π of order q^2 admits a linear p-group of order q^5 if q is odd or $2q^5$ if q is even, then π contains a group that acts transitively on the affine points.
(2) Furthermore, the group contains either an elation group of order q or a Baer group of order q with axis a subplane of $\mathcal{D}$. If the order of the stabilizer of a point H is at least $2q$, then the order is $2q$ and H is generated by an elation group of order q and a Baer involution with axis in $\mathcal{D}$ or by a Baer group of order q and an elation.
(3) π is a non-strict semi-translation plane of order q^2 admitting a translation group of order $q^3 p^\gamma$. Furthermore, either π is a translation plane or there is a unique

$((\infty), \ell_\infty)$-transitivity and the remaining infinite points are centers for translation group of orders qp^γ.

(4) If π is non-Desarguesian in the elation case above, then π admits a set of q derivable nets sharing the axis of the elation group of order q.

COROLLARY 49.37. Let π be a derivable affine plane of order q^2 that admits a p-group of order q^6 if q is odd or $2q^6$ if q is even containing a linear subgroup of order q^5 or $2q^5$ leaving invariant the derivable net $\mathcal{D}$ invariant. Assume that when $q = p^r$ then $(r, p) = 1$ and assume that π admits an elation group H of order q.

(1) Then either the plane is Desarguesian or the center of H is invariant.

(2) Then π admits a collineation group fixing an affine point of order q^2 or $2q^2$ that fixes an infinite point (∞) of $\mathcal{D}$ and acts transitively on the remaining infinite points.

(3) Either the plane is a dual translation plane or the group acts transitively on a set of q derivable nets sharing (∞).

49.8. Subplane Covered Nets.

All of these ideas can be generalized by replacing the word 'derivable net' by 'subplane covered net' in any of the definitions. Given a subplane covered net $\mathcal{S}$, there is a projective geometry Σ and a codimension two subspace N of Σ such that the points, lines, parallel classes, subplanes of $\mathcal{S}$ are, respectively, the lines skew to N, points of $\Sigma - N$, hyperplanes of Σ containing N and planes of Σ each of which intersects N in a point (see Johnson [**753**]).

THEOREM 49.38. Let $\mathcal{S}$ be a subplane covered net and let T be a transversal to $\mathcal{S}$. Then, using the embedding of $\mathcal{S}$ into the projective space Σ with distinguished codimension two subspace N, it follows that T, as a set of lines of Σ, is a partial line spread of Σ which covers $\Sigma - N$.

We consider the generalization of the problem on partially flag-transitive affine planes, this time with respect to a subplane covered net. We leave the following as an open problem.

PROBLEM 49.39. Let π be a finite affine plane containing a subplane covered net $\mathcal{S}$. If there exists a collineation group G which leaves $\mathcal{S}$ invariant and acts flag-transitively on the flags on lines not in $\mathcal{S}$, can π be determined?

CHAPTER 50

Indicator Sets.

Indicator sets provide an alternative manner of determining spreads and were developed initially by R.H. Bruck. We begin with indicator sets which produce spreads of $PG(3, q)$.

Consider a 4-dimensional vector space V over a field K isomorphic to $GF(q)$. Form the tensor product of V with respect to a quadratic field extension F of K, F isomorphic to $GF(q^2)$, $V \otimes_K F$. If we form the corresponding lattices of subspaces to construct $PG(3, F)$, we will have a $PG(3, K)$ contained in $PG(3, F)$, such that, with respect to some basis for V over K, (x_1, x_2, y_1, y_2), for $x_1, x_2, y_1, y_2 \in K$ represents a point homogeneously in both $PG(3, K)$ and $PG(3, F)$. The Frobenius automorphism mapping defined by

$$\rho_q : (x_1, x_2, y_1, y_2) \longmapsto (x_1^q, x_2^q, y_1^q, y_2^q)$$

is a semi-linear collineation of $PG(3, F)$ with set of fixed points exactly $PG(3, K)$. We use the notation $Z^q = (x_1, x_2, y_1, y_2)^q = (x_1^q, x_2^q, y_1^q, y_2^q)$. Finally, choose a line $PG(1, K)$ within the given $PG(2, K)$ in the analogous manner so that there is a corresponding $PG(1, F)$. Hence, we have

$$PG(1, K) \subseteq PG(3, K),$$
$$PG(1, K) \subseteq PG(1, F) \subseteq PG(2, F).$$

Now choose a $PG(2, F)$ such that $PG(2, F) \cap PG(3, K) = PG(1, K)$. For example, take $PG(2, F)$ as the lattice arising from the 3-dimensional vector space

$$\langle (1, 0, 0, 0), (0, 1, 0, 0), (0, 0, 1, e) \rangle ; e \in F - K.$$

We note that given any point Z of $PG(2, F) - PG(1, F)$ then $\langle Z, Z^q \rangle$ is a 2-dimensional F-vector subspace, since if $Z^q = \lambda Z$, for $\lambda \in F$, then projectively $Z = Z^q$ and $Z \in PG(3, K)$. Note if $e^q = e\alpha_0 + \beta_0$, for $\alpha_0 \neq 0$ and $\beta_0 \in K$, then $(0, 0, 1, e)^q = (0, 0, 1, e\alpha_0 + \beta_0)$. Since $e^q + e \in K$, it follows that $\alpha_0 = -1$. Now $\langle (0, 0, 1, e), (0, 0, 1, e)^q = (0, 0, 1, -e + \beta_0) \rangle$. Hence, within this subspace is $(0, 0, 2, \beta_0)$. For example, if q is odd, we may take $\beta_0 = 0$, implying that $(0, 0, 1, 0)$ is in the subspace. But, this then implies that $(0, 0, 0, e)$ and hence $(0, 0, 0, 1)$ is in the subspace. This means that $\langle (0, 0, 1, e), (0, 0, 1, e)^q \rangle \cap V/K = \langle (0, 0, 1, 0), (0, 0, 0, 1) \rangle$ is a 2-dimensional K-subspace which has trivial intersection with $\langle (1, 0, 0, 0), (0, 1, 0, 0) \rangle$.

More generally, $\langle Z, Z^q \rangle \cap V/K$ is a 2-dimensional K-vector subspace for all $Z \in PG(2, F) - PG(1, F)$.

DEFINITION 50.1. A space $PG(2, F)$ with the above properties (i.e., contains $PG(1, K) \subseteq PG(1, F)$, $PG(2, F) \cap PG(3, K) = PG(1, K)$ (so $\langle Z, Z^q \rangle \cap PG(3, K)$ is a line skew to $PG(1, F)$, for $Z \in PG(2, F) - PG(1, F)$) is called an 'indicator space'.

DEFINITION 50.2. Let $PG(2, F)$ be an indicator space within $PG(3, F)$. An 'indicator set' I of $PG(2, F)$ is a set of q^2 points in $PG(2, F) - PG(1, F)$ such that the line AB, for all $A, \neq B \in S$ intersects $PG(1, F) - PG(1, K)$. Note that $\langle A, A^q \rangle$ now becomes a line of $PG(2, F)$, which intersects $PG(3, K)$ in a line skew to $PG(1, F)$.

RESULT 50.3. If $PG(2, F)$ is an indicator space, then $PG(2, F)^q$ is an indicator space such that $PG(2, F)^q \cap PG(2, F) = PG(1, F)$. Hence, if I is an indicator set, then for $A \neq B$ of I then $\langle A, B \rangle \cap \langle A, B \rangle^q$ is a point of $PG(1, F) - PG(1, K)$.

THEOREM 50.4. If I is an indicator set then

$$\{\langle A, A^q \rangle \cap PG(3, K); A \in I\} \cup PG(1, K)$$

is a spread of $PG(3, K)$.

PROOF. There are $q^2 + 1$ lines, it is only then necessary to show these lines are skew or rather that the associated 2-dimensional vector subspaces are mutually disjoint as subspaces. Suppose that $C = \alpha A + \beta A^q = \alpha^* B + \beta^* B^q$, for $\alpha, \beta, \alpha^*, \beta^* \in K$. If C is (or represents) a point of $PG(3, K)$ then $C^q = C$, implying that $\alpha = \beta^q$ and $\alpha^* = \beta^{*q}$. But then $(\beta^q A - \beta^{*q} B) = -(\beta A^q - \beta^* B^q) = -(\beta^q A - \beta^{*q} B)^q$, which implies that this intersection of $\langle A, B \rangle$ and $\langle A, B \rangle^q$ is a point of $PG(1, K)$, which then implies that all scalars are zero. Hence, the indicated subspaces are disjoint on V/K; the lines are skew in $PG(3, K)$. □

The strength of the use of indicator sets is in their intrinsic simplicity but the weakness of such sets is the projectively equivalent indicator sets could produce isomorphic translation planes.

50.1. Affine Version.

50.1.1. Transversals to Derivable Nets. Now we consider the affine version of the above result with a different visual image. Consider a finite derivable net. By the work of De Clerck and Johnson [**274**] and of Johnson [**753**], every finite derivable net is a regulus net. Hence, there is an ambient 4-dimensional vector space V over a field K isomorphic to $GF(q)$ such that the derivable net, when affinely presented, corresponds to $PG(1, q)$ on a Desarguesian line $PG(1, q^2)$ and may be given the following partial spread representation:

$$N : x = 0, y = x\alpha; \alpha \in K \simeq GF(q)$$

and we may consider the Desarguesian affine plane of order q^2, coordinatized by a quadratic field extension $F \supseteq K$ and isomorphic to $GF(q^2)$. Now take an indicator set I of q^2 "affine" points. Without loss of generality, take I to contain the zero vector of the associated vector space. Take any line of the derivable net (regulus net) N, either $y = x\alpha + b$, for $\alpha \in K$ and $b \in F$ or $x = c$, for $c \in F$. Consider the points A_i, $i = 1, 2, \ldots, q^2$ of I fix a point A_1 and consider the Desarguesian lines $A_1 A_i$, $i = 2, 3, \ldots, A_{q^2}$. We know that these Desarguesian lines do not lie on a parallel class of N and each line must intersect each line of parallel class of N in a unique point. Let λ be a parallel class of N and let ℓ be a line of λ. There are q^2 lines of λ each of which can contain at most one point of I, since I is an indicator set. However, on a given Desarguesian line $A_1 A_i$, there are at least two lines of λ, say ℓ_1 and ℓ_i, each of which shares exactly one point of I. Now consider $A_1 A_j$, for $i \neq j$, and let ℓ_j share A_j. If $\ell_i = \ell_j$, we have the obvious contradiction. Hence, there are $q^2 - 1 + 1$ different lines of λ that intersect I in exactly one point. Recall the following definition:

DEFINITION 50.5. A 'transversal' T to a net $\mathcal{N}$ is a set of net points with the property that each line of the net intersects T in a unique point and each point of T lies on a line of each parallel class of $\mathcal{N}$.

Clearly, an indicator set provides a transversal to a derivable net and hence produces a dual translation plane π whose dual translation plane has a spread in $PG(3, K)$. We now consider the converse. Assume that T is a transversal to a finite derivable net. Think of the scenario depicted previously, so we may consider T to be a set of q^2 points of a Desarguesian affine plane coordinatized by a field $F \supseteq K$, where K is a field isomorphic to $GF(q)$, which coordinatizes the derivable net as a regulus net. We have a natural associated 4-dimensional K-vector space V in which the derivable net N lives and a natural 3-dimensional projective space $PG(3, K)$. Form $V \otimes_K F$ and construct the natural 3-dimensional projective space $PG(3, F)$ containing $PG(3, K)$, think of the Desarguesian affine plane projectively as a $PG(2, K)$ and embed this within $PG(2, F)$ contained in $PG(3, F)$ so that $PG(2, F) - PG(1, F)$ (the original line at infinity) contains no points of $PG(3, K)$. Consider two distinct points A and B of T. Suppose that the Desarguesian line AB is in a parallel class of the derivable net N. Then there is a line of the net N which contains two points of the transversal, a contradiction. Hence, AB intersects 'outside' and hence in $PG(1, F) - PG(1, K)$. That is, a transversal to a derivable net produces an indicator set. Hence, we have proved the following theorem.

THEOREM 50.6. (Also see Bruen [191]). Finite indicator sets of $PG(2, q^2)$ are equivalent to transversals to derivable nets.

From Section 49.1, we know that there is a spread arising geometrically from the embedding of the derivable net combinatorially in $PG(3, K)$. We know that the corresponding translation plane is dual to the one we obtained considering the extension theory using a transversal. However, if we use the transversal as an indicator set, there is another spread of $PG(3, K)$. The question is: Is this spread dual to the associated dual translation plane? Normally the answer to this question is no! What actually occurs is that the 'transversal spread', the spread arising from the transversal extension theory is 'dual' to the spread arising from the indicator spread, say the 'indicator spread'. We now consider the coordinate matrices, following ideas of Bruen [191].

50.1.2. Spread Set for the Transversal Spread. Choose a basis $\{e, 1\}$ for F over K and let T be a transversal to a derivable net. Then, there is an associated transversal function f such that the lines of the associated dual translation plane are $y = f(x)\alpha + x\beta + b$, $x = c$, for all $\alpha, \beta \in GF(q)$ and for all $b, c \in GF(q^2)$. Then the associated multiplication in the right quasifield is

$$x * (e\alpha + \beta) = f(x)\alpha + x\beta.$$

So, we obtain a left quasifield which will define the transversal spread as follows:

$$(e\alpha + \beta) \circ x = f(x)\alpha + x\beta.$$

Let $x = e\rho + \gamma$. Let $f(x) = F(\rho, \gamma)e + G(\rho, \gamma)$. So,

$$f(x)\alpha + x\beta = F(\rho, \gamma)e\alpha + G(\rho, \gamma)\alpha + e\rho\beta + \gamma\beta = (F(\rho, \gamma)\alpha + \rho\beta)e + (G(\rho, \gamma)\alpha + \gamma\beta)$$

Also, denoting $(ec + d) = (d, c)$, we have:

$$(\beta, \alpha) \circ (\gamma, \rho) = (\beta, \alpha) \begin{bmatrix} \gamma & \rho \\ G(\mu, \gamma) & F(\rho, \gamma) \end{bmatrix}.$$

Hence, as x varies over $GF(q^2)$, we obtain the 'transversal spread'

$$\left\{ x = 0, y = x \begin{bmatrix} \gamma & \rho \\ G(\rho, \gamma) & F(\rho, \gamma) \end{bmatrix}; \gamma, \rho \in GF(q) \right\}.$$

We consider points $(x, y) = (x_1 e + x_2, y_1 + y_2 e) = (x_1, x_2, y_1, y_2)$ over K.

50.1.3. Spread Set for the Indicator Spread. To consider $V \otimes_K F$ over F and write points (x, y) over F. We represent $V \otimes_K F$ by elements (x_1, x_2, y_1, y_2) over F.

We now take the indicator space $PG(2, F)$ as the lattice arising from the 3-dimensional vector space

$$\{(x_1, x_2, y_1 e, y_1); x_i, y_1 \in F, \ i = 1, 2\}.$$

If we have a transversal function $f(x)$, we have an associated indicator set. Furthermore, the line $\langle A, A^q \rangle \cap PG(3, K)$ corresponding to a point A of the indicator set is the unique line of $PG(3, q)$ which lies on A. For example, if $A = (ea_1 + a_2, eb_1 + b_2, ec_1 + c_2, ed_1 + d_2)$, for $a_i, b_i, c_i, d_i \in K$, is in $PG(3, F) - PG(3, K)$, then (a_1, b_1, c_1, d_1) and (a_2, b_2, c_2, d_2) are points of $PG(3, K)$. But, $\langle (a_1, b_1, c_1, d_1), (a_2, b_2, c_2, d_2) \rangle_F$ contains A. This must be the line $\langle A, A^q \rangle \cap PG(3, K)$. The (affine) points of $y = f(x)$ have the form

$$(x, f(x)) = x(1, 0, 0, 0) + f(x)(0, 1, 0, 0) + e(0, 0, 1, 0) + (0, 0, 0, 1) = (x, f(x), e, 1)$$

in terms of our representation, noting that these points are not on the line at infinity, which we designate as $y_1 = 0$. Our previous representation $x = e\rho + \gamma$, and $f(x) = F(\rho, \gamma)e + G(\rho, \gamma)$, produces the form

$$(e\rho + \gamma, F(\rho, \gamma)e + G(\rho, \gamma), e, 1)$$

and the unique line of $PG(3, K)$ incident with this point is

$$\langle (\rho, F(\rho, \gamma), 1, 0), (\gamma, G(\rho, \gamma), 0, 1) \rangle_K.$$

Take the scalar multiples

$$(\gamma, G(\rho, \gamma), 0, 1)x_1 + (\rho, F(\rho, \gamma), 1, 0)x_2 = \left((x_1, x_2) \begin{bmatrix} \gamma & G(\rho, \gamma) \\ \rho & F(\rho, \gamma) \end{bmatrix}, (x_1, x_2) \right).$$

Choose another basis as $(y = 0, x = 0)$ (that is, the new points are (y, x) if the original are (x, y)). Then we obtain

$$\left((x_1, x_2), (x_1, x_2) \begin{bmatrix} \gamma & G(\rho, \gamma) \\ \rho & F(\rho, \gamma) \end{bmatrix} \right)$$

Hence, we obtain the 'indicator spread:

$$\left\{ x = 0, y = x \begin{bmatrix} \gamma & G(\rho, \gamma) \\ \rho & F(\rho, \gamma) \end{bmatrix}; \gamma, \rho \in GF(q) \right\},$$

which is the dual spread of the transversal spread. Therefore, we obtain:

THEOREM 50.7 (Bruen's indicator spread set theorem). (Bruen [**191**]). The indicator spread and transversal spreads are dual to each other.

Now again consider a derivable net D and a transversal T to D. It has been pointed out in Johnson [**753**], that the way that the derived net D^* is related to the derivable net D using the geometric embedding in a 3-dimensional projective space $PG(3, q)$ is that the structure is determined by a polarity of the original $PG(3, q)$; to abuse the language 'derivation is a polarity'. If we derive the net D to D^*, then T now becomes a transversal to D^*, as can be seen by reconsidering D^* in the standard form and embedding in a different $PG(3, K^*)$. That is, T remains a set of q^2 points such that the new Desarguesian line AB, for $A, B \in T$ is not in the set of parallel classes defining the new derivable net N^*. To see this, we note that a line of D^* is a Baer subplane of D and in the original Desarguesian plane $AG(2, F)$, a Desarguesian line AB which is not in the parallel classes of D must intersect a Baer subplane in a unique point. Now the counting argument establishing that we have the appropriate intersection for an indicator set to be a transversal works here as well and shows that T is a transversal to D^*. Now consider the geometric spread obtained by considering T as a set of lines of the new $PG(3, K^*)$ so that $T \cup N^*$ (the adjoined line) becomes a spread of the dual space to $PG(3, K)$.

In order to see this algebraically, we note that the so-called 'Albert switch' of coordinates will re-coordinatize D^* is standard form and map a point (x_1, x_2, y_1, y_2) of $AG(2, F)$ into the point (x_1, y_1, x_2, y_2). We are not, however, deriving. We want to see the new representation of the point set within $PG(3, F)$:

$$(e\rho + \gamma, F(\rho, \gamma)e + G(\rho, \gamma), e, 1).$$

Making this switch over F instead of over K, produces the point:

$$(e\rho + \gamma, e, F(\rho, \gamma)e + G(\rho, \gamma), 1)$$

and the corresponding line in $PG(2, K)$: The corresponding line of the spread is now

$$\langle (\rho, 1, F(\rho, \gamma), 0), (\gamma, 0, G(\rho, \gamma), 1) \rangle_K.$$

In order to re-represent this in matrix form, we note that taking scalar multiples

$$x_1(\gamma, 0, G(\rho, \gamma), 1) + x_2(\rho, 1, F(\rho, \gamma), 0)$$
$$= (x_1\gamma + x_2\gamma, x_2, x_1 G(\rho, \gamma) + x_2 F(\rho, \gamma), x_1),$$

and then mapping this point

$$(x_1\gamma + x_2\gamma, x_2, x_1 G(\rho, \gamma) + x_2 F(\rho, \gamma), x_1)$$
$$\longmapsto (x_1, x_2, x_1\gamma + x_2\gamma, x_1 G(\rho, \gamma) + x_2 F(\rho, \gamma))$$

reproduces the transversal-spread set

$$\left\{ x = 0, y = x \begin{bmatrix} \gamma & \rho \\ G(\rho, \gamma) & F(\rho, \gamma) \end{bmatrix}; \gamma, \rho \in GF(q) \right\}.$$

Hence, we have the following connections:

THEOREM 50.8. Let D be a finite derivable net and let T be a transversal to D.

(1) Form the transversal spread π_D^T (i.e., form the dual translation plane, dualize and find the spread within $PG(3, K)$, for K isomorphic to $GF(q)$).

(2) Realize T as an indicator set and form the indicator spread π_D^I.

(3) Derive D to D^* and realizing that T is a transversal to D^*, form the transversal spread $\pi_{D^*}^T$.

(4) Realize T as an indicator set relative to D^* and form the indicator spread $\pi_{D^*}^I$.

(5) Form from D and the transversal T the geometric extension spread π_D^{GT} obtained by realizing D combinatorially within $PG(3, K)$.

(6) Form from D^* and the transversal T the geometric extension spread $\pi_{D^*}^{GT}$ obtained by realizing D^* combinatorially within $PG(3, K)$ using a polarity of $PG(3, K)$.

Then the transversal spreads are isomorphic to the geometric spreads, respectively, and are dual to each other.

The indicator spreads are dual to the transversal spreads, respectively.

Hence, a given indicator spread is isomorphic to the 'derived' version of the original transversal spread (by 'derived' we mean derive the net and use the original transversal).

All of the statements in the above theorem are more generally true, with appropriate modifications, for constructions using derivable nets and transversal over infinite skewfields. However, the derivation process will reverse 'right' and 'left' so that the 3-dimensional projective spaces are over K and K^{opp}. In any case, the indicator spreads still can be realized as the geometric extension spreads of the derived derivable net, using the original transversal function.

COROLLARY 50.9. *The dual translation plane obtained from an indicator set I of $PG(2, q^2)$ using I as a transversal function to the natural derivable net (regulus net) defined using $PG(1, q)$, is the dual plane to the translation plane constructed from the spread in $PG(3, q)$ using I as in Theorem 50.4.*

50.1.4. Blocking Sets and Transversals.

DEFINITION 50.10. *Let Π be a finite projective plane of order n. A 'blocking set' B is a set of points of Π such that each line of Π non-trivially intersects B. If $|B| = n + t$ and some line contains t points, the blocking set is said to be of 'Rédei type'.*

There is a voluminous theory of blocking sets, inspired by the work of Bruen [**187**] who noted that any Baer subplane B of a projective plane of order q^2 is a blocking set. However, it is not our intention to delve much into this theory. However, we are interested in how this might connect with transversal extension theory. In particular, the recent paper by Lunardon [**948**] discusses connections with semifield planes and certain blocking sets of Rédei type. Here we note the following connection and generalization.

THEOREM 50.11. *Let N be a finite derivable net and let T be a transversal to N. Consider the associated Desarguesian affine plane $AG(3, q^2)$ as defining the ambient 4-dimensional vector space V containing N. Define $T^* = T \cup \{\ell_\infty - N_\infty\}$ (the infinite points of $AG(3, q^2)$ which are not in the derivable net).*

Then T^* is a blocking set of $PG(2, q^2)$ of Rédei type.

PROOF. Every affine line of $PG(2, q^2)$ is either in the derivable net and hence non-trivially intersects T and hence T^*, or it is a line that naturally intersects $\{\ell_\infty - N_\infty\}$. Finally, the infinite line ℓ_∞ shares exactly $q^2 - q$ points of the set T^* of cardinality $q^2 + q^2 - q$ and hence we have a blocking set of Rédei type. $\qquad\square$

In Section 49.1, we discussed 'vector-space' transversals and showed that semifield planes with spreads in $PG(3, K)$ are equivalent to vector-space transversals of derivable nets. Since transversals are basically indicator sets, we define the indicator set arising from a vector-space transversal to be a 'linear indicator set'. Hence, we obtain:

THEOREM 50.12. (See Lunardon [948, Corollary 1]). Semifield spreads in $PG(3, q)$ are equivalent to linear indicator sets of $PG(2, q^2)$.

Indicator sets may be generalized to produce both the arbitrary t-spreads in $PG(2t - 1, q)$ and also to connect to infinite translation planes and their spreads. Furthermore, the concept of a transversal does not require finiteness nor does the concept of a 'blocking set'. So, the results presented in this subsection are not stated in their most general form.

50.2. The Hermitian Sequences.

In this section, we connect spreads in $PG(3, q)$ with certain ovoids ('locally Hermitian') of the Hermitian surface $H(3, q^2)$, and spreads (locally Hermitian) of the elliptic quadric $Q^-(5, q)$. Furthermore, the linear representation of $H(3, q^2)$ provides additional spreads. The connection with ovoids of $H(3, q^2)$ is work due to Shult [1163] and Lunardon [948]. The theory of spreads arising from the linear representation is the work of Lunardon [948], which is adapted from the theory of J.A. Thas [1176], which also relies on work of Cossidente, Ebert, Marino and Siciliano [261]. We review only the part of Hermitian varieties required for our constructions. The text of Hirschfeld and Thas [529] provides all of the details.

DEFINITION 50.13. Let V_{2k} be a $2k$-dimensional vector space over a field L isomorphic to $GF(q^2)$. A 'Hermitian form' is a mapping s with the following properties:

$$s \ : \ V_{2k} \oplus V_{2k} \longmapsto L,$$
$$s(x + w, y + z) = s(x, y) + s(w, y) + s(x, z) + s(w, z)$$
$$s(cx, dy) = cd^q s(x, y) \text{ and } s(x, y) = s(y, x)^q$$
$$s(x_0, V_{2k}) = 0, \text{ implies } x_0 = 0 \text{ (i.e., a 'non-Degenerate Hermitian form')}$$

Now assume that $k = 2$. Given a Hermitian form (non-degenerate), given a vector subspace S, form S^δ as follows:

$$S^\delta = \{v \in V; s(v, S) = 0\}.$$

Then the mapping

$$S \longmapsto S^\sigma$$

is a polarity of the associated projective 3-space $PG(3, q^2)$, which is said to be a 'Hermitian polarity', or 'unitary polarity'. The subgroup of $\Gamma L(4, q^2)$ that preserves the Hermitian form is called the 'unitary group'. This group is denoted by $\Gamma U(4, q^2)$. The associated group $\Gamma U(4, q^2)/Z(\Gamma U(4, q^2))$ is called the 'projective unitary group'.

A subspace S is said to be 'totally isotropic' if and only if

$$S \cap S^\delta = S.$$

In $PG(3, q^2)$, the set of totally isotropic points and totally isotropic lines form the point-line geometry $H(3, q^2)$, the 'Hermitian surface'. Projectively there is a canonical form for $H(3, q^2)$:

$$\{(x_1, x_2, x_3, x_4); x_1 x_4^q + x_2 x_3^q + x_3 x_2^q + x_4 x_1^q = 0\}.$$

An 'ovoid' in this setting is a set of $q^3 + 1$ points of $H(3, q^2)$ which forms a cover of the set of totally isotropic lines. For any two points A and B of $H(3, q^2)$ then the line AB contains $q + 1$ or $q^2 + 1$ points of $H(3, q^2)$. A 'tangent line' to a point C of $H(3, q^2)$ is a line containing exactly one point of $H(3, q^2)$.

DEFINITION 50.14. A 'tangent plane' to $H(3, q^2)$ at a point P of $H(3, q^2)$ is the image plane of a point under the associated polarity. This plane will intersect $H(3, q^2)$ in exactly $q + 1$ lines of $H(3, q^2)$ incident with P.

Given a plane Π of $PG(3, q^2)$, then Π intersects $H(3, q^2)$ either at a point, a line, a unital or is a tangent plane. Hence, if a plane intersects in a line at points not on that line, then the plane is a tangent plane.

By Johnson [**728**], we consider the combinatorial structure of 'P-points' as lines of N and 'P-lines' as points of D considered as the set of intersecting lines. Furthermore, we call the parallel classes of N the 'P-hyperplanes.' We embed this structurally in a projective 3-space $PG(3, q)$ by adjoining a line N to all P-hyperplanes. In this way, a derivable net of parallel classes, lines, and points becomes the set of hyperplanes of $PG(3, q)$ incident with a particular line N, points of $PG(3, q) - N$, and lines of $PG(3, q)$ which are skew to N. In this model, T becomes a set of lines of $PG(3, q)$ whose union with N is a spread. We know that this spread is isomorphic to the transversal spread and is the dual spread of the associated indicator spread.

If we consider the corresponding $PG(2, F)$ and consider T as an indicator set, form the dual plane again isomorphic to $PG(2, F)$ so that T is now a set of q^2 lines with the property that the join of any two distinct 'lines' A and B does not intersect the dual of $PG(1, K)$. Note that we may consider the $PG(1, K)$ as a 'Hermitian line'. This means that if we dualize $\Pi = PG(2, F)$, we find that $\Pi \cap H(3, q^2) = \Delta$ is a set of $q + 1$ lines and the dual of T, T^D, is a set of q^2 lines that do not contain the point Q, which is the $PG(1, F)$, with the property that no two intersect on a line of Δ. Such a set of lines is said to be a 'Shult set' [**1163**]. Shult sets are equivalent to certain ovoids of $H(3, q^2)$.

DEFINITION 50.15. An ovoid Φ of $H(3, q^2)$ is a set of $q^3 + 1$ points that cover the set of totally isotropic lines of $PG(3, q^2)$ (each totally isotropic line is incident with exactly one point of Φ). Note that there will be exactly $q + 1$ totally isotropic lines incident with each point. Choose any two distinct points Q and Z of Φ, then there are $q + 1$ points of $H(3, q^2)$ on the line QZ. If for a fixed point Q and for all $Z \in \Phi$, the points on QZ are in Φ, we call Φ a 'locally Hermitian' ovoid with respect to Q.

DEFINITION 50.16. If a locally Hermitian ovoid with fixed point Q admits a group leaving $H(3, q^2)$ invariant, fixing all lines of $H(3, q^2)$ incident with Q and which acts transitively on the remaining points of the ovoid, we say that the ovoid is a 'translation ovoid' (i.e., a 'locally Hermitian translation ovoid).

Form the Shult set T^D arising from the indicator set T, take the set of polar lines $T^{D\delta}$, with respect to the Hermitian polarity δ. Shult [**948**] shows that $T^{D\delta}$ is a set of q^2 lines incident with Q, such that there are $q+1$ points of $H(3, q^2)$ on each such line and the union of this set of points of $H(3, q^2)$ forms a locally Hermitian ovoid. Conversely, any locally Hermitian ovoid of $H(3, q^2)$ forms a Shult set, which dualizes to an indicator set, which constructs an indicator spread. Furthermore, if the original indicator set is a vector-space transversal then the constructed locally Hermitian ovoid admits a collineation group which fixes all lines of Q and acts transitively on the remaining point of the ovoid, that is, the locally Hermitian ovoid becomes a translation ovoid.

Hence, the work of Shult and Lunardon [**948**], together with our interpretation of transversals to derivable nets, produces the following construction:

THEOREM 50.17 (Shult–Lunardon Hermitian ovoids theorem). (1) Locally Hermitian ovoids of $H(3, q^2)$ are equivalent to transversals of finite derivable nets; one constructs the other.

(2) The associated indicator spread is a semifield spread if and only if the locally Hermitian ovoid is a translation ovoid, which is the case if and only if the associated transversal is a vector-space transversal.

The Dual Structure. We dualize $H(3, q^2)$ combinatorially to $Q^-(5, q)$, embed this within $PG(5, q)$ and realize that a locally Hermitian ovoid dualizes to a 'spread' of $Q^-(5, q)$, that is, a set S of mutually disjoint maximal singular subspaces (i.e., lines) that partition the point set, with the property that if A^Q arises from Q of the previous notation, and A^Q and B are distinct lines in S, then $\langle A, B \rangle$ is 3-dimensional K-space such that $\langle A, B \rangle \cap Q^-(5, q)$ is a regulus containing A and B, say $R(A, B)$, that is also contained in S. Note that we are using the notations $Q^-(5, q)$ and $\Omega^-(6, q)$ interchangeably.

DEFINITION 50.18. A spread of $Q^-(5, q)$ is a set S of singular lines with respect to the elliptic polarity, which partitions the point set. Fix a point A of S. If A, B in S generate a 3-dimensional projective space $\langle A, B \rangle$ such that the regulus $R(A, B) = \langle A, B \rangle \cap Q^-(5, q) \subset S$, for all $B \in S - \{A\}$, then S is said to be 'locally Hermitian' with respect to A.

If there is a group leaving $Q^-(5, q)$ invariant, which fixes each point of A and acts transitively on the lines of $S - \{A\}$, the spread is said to be a 'locally Hermitian translation spread'.

THEOREM 50.19. (1) The dual structure of $H(3, q^2)$ is isomorphic to $Q^-(5, q)$, the elliptic quadric of $PG(5, q)$.

(2) A locally Hermitian ovoid of $H(3, q^2)$ dualizes to a locally Hermitian spread of $Q^-(5, q)$.

(3) A locally Hermitian translation ovoid dualizes to a locally Hermitian translation spread of $Q^-(5, q)$.

Let S be a locally Hermitian spread of $Q^-(5, q)$ with respect to a line A. Let δ_{Q^-} denote the associated polarity of $PG(5, q)$ and let $A^{\delta_{Q^-}} = \Lambda$ be the corresponding 3-dimensional projective subspace of $PG(3, q)$ and note that $\Lambda \cap Q^-(5, q) = A$. For $B \in S - \{A\}$, then $\langle A, B \rangle^{\delta_{Q^-}}$ is a line, which is disjoint from $\langle A, B \rangle$. But for $C \in S - \{A, B\}$, $\langle A, C \rangle^{\delta_{Q^-}}$ and $\langle A, B \rangle^{\delta_{Q^-}}$ are lines disjoint from each other since $\langle A, B \rangle \cap \langle A, C \rangle = A$. Hence, from a transversal of q^2 points, we arrive at a set of q^2 lines of Λ that are mutually disjoint and disjoint from A. Hence,

$$A \cup \left\{ \langle A, B \rangle^{\delta_{Q^-}} ; B \in S - \{A\} \right\}$$

is a spread of Λ. This construction is due to Thas [**1176**], adapted by Lunardon [**948**].

THEOREM 50.20. (1) (Thas [**1176**]). A locally Hermitian spread S of $Q^-(5,q)$ with respect to a line A constructs a spread

$$A \cup \left\{ \langle A, B \rangle^{\delta_{Q^-}} ; B \in S - \{A\} \right\},$$

of $A^{\delta_{Q^-}}$, where δ_{Q^-} is the associated polarity of $PG(5,q)$.

(2) (Lunardon [**948**]). A locally Hermitian translation spread constructs a semifield spread.

(3) A locally Hermitian ovoid of $H(3,q^2)$ dualizes to a locally Hermitian spread of $Q^-(5,q)$, which constructs a spread of $PG(3,q)$. We call the constructed spread the associated 'elliptic spread-spread'.

50.2.1. Hermitian Sequence of Elliptic Type. We now form the following set of spreads of $PG(3,q)$:

THEOREM 50.21. Let D be a finite derivable net and let T be a transversal to D.

(1) Then we obtain the following sequence of spreads in $PG(3,q)$, which we call the 'short Hermitian sequence':

$$\left[\begin{array}{c} \pi_D^I; \text{ the indicator spread} \rightleftarrows (\text{dualize}) \\ \text{Form the Shult set and associated locally Hermitian ovoid of } H(3,q^2) \\ \rightleftarrows (\text{dualize}) \\ \text{Form the locally Hermitian spread of } Q^-(5,q) \longmapsto \text{form associated} \\ \pi_{D(E)}^I; \text{ the elliptic spread-spread.} \end{array} \right].$$

(2) If T is a vector-space transversal, then all of the spreads in the short Hermitian sequence are semifield spreads, the ovoid of $H(3,q^2)$ is a translation ovoid and the spread of $O^-(5,q)$ is a translation spread.

(3) We also obtain the transversal spread, the derived transversal spread and the two geometric extension spreads.

THEOREM 50.22. (1) If the transversal is a vector-space transversal, then all of the planes of the Hermitian sequence of elliptic type are semifield planes. Actually, if any of the planes of this sequence is a semifield plane, then they all are semifield planes.

(2) The associated locally Hermitian ovoids with tangent point P become 'translation ovoids', i.e., there is a collineation group of $H(3,q^2)$ fixing all lines of $H(3,q^2)$ incident with the point of tangency P and acting transitively on the remaining points of the ovoid.

(3) The associated locally Hermitian spread with line L of $Q^-(5,q)$ is a 'translation spread' if and only if there is a collineation group of $Q^-(5,q)$, fixing all points of L and transitive on the remaining lines of the spread.

PROOF. The indicator set produces a semifield plane if and only if the transversal is a vector-space transversal. This implies that the associated locally Hermitian ovoid of $H(3,q^2)$ is a translation ovoid, which implies that spread of $Q^-(5,q)$ is a translation spread, which implies that the constructed spread produces a semifield plane (see, for example, Lunardon [**948**]). Then the new transversal of the associated dual semifield plane (which admits a regulus in its spread) will have a vector-space transversal and the same argument applies to the second round in the sequence. Similarly, if any of the spreads in the sequence are semifield spreads then they all are semifield spreads.   □

REMARK 50.23. Let π be a translation plane which admits a regulus in its spread, let the spread be in $PG(3, q)$ and let T be any component external to the regulus D. The 'T-extension' process creates a semifield plane π, using T as a transversal to D. The associated indicator spread produces a semifield plane which is the 'transpose' of π, π^t (the associated dual spread of the spread of π). Hence, there is an associated elliptic spread-spread, which is a semifield spread.

REMARK 50.24. Given a finite derivable net D in the spread of any translation plane, if T_1 and T_2 are components exterior to D, then the T_i-extensions of the derivable net can produce non-isomorphic semifield spreads and then produce mutually non-isomorphic Hermitian sequences of elliptic type.

50.2.2. The Hermitian Sequence of Hyperbolic Type. In this subsection, we make the 'short Hermitian sequence' a bit longer by considering another representation of the Hermitian surface and then reconnecting locally Hermitian ovoids arising via Shult sets from indicator sets or rather from transversals to derivable nets. We first show how to construct copies of projective spaces with others using 'normal line spreads'. The material presented in this subsection more or less mirrors Lunardon [**948**], which uses, in particular, the work of Thas [**1176**].

DEFINITION 50.25. A 'line spread' S of a projective space is a covering of the points by a set of mutually skew lines. The line spread is said to be 'normal' if and only if the line spread induces a line spread in each of the 3-dimensional subspaces generated by a distinct pairs of lines of S; $A, B \in S$, implies that $\langle A, B \rangle \cap S$ is a spread of $\langle A, B \rangle$ (as a $PG(3, K)$).

The main use of normal line spreads in our treatment will involve the following proposition.

PROPOSITION 50.26. Let N be a normal line spread of a hyperplane H of a projective space $H^+ \simeq PG(2t, q)$. Define the following incidence geometry: $\rho(H^+, H, N)$:

> 'Points' are the points of $H^+ - H$ and the elements of N.
>
> 'Lines' are the planes of H^+ that intersect H in a line of N
>
> or the spreads $S_{\langle A,B \rangle}$, $A, \neq B \in N$ of $\langle A, B \rangle \cap N$.
>
> 'Incidence' is that inherited within H^+.

Then $\rho(H^+, H, N)$ is isomorphic to $PG(t, q^2)$. Furthermore, the collineation group of H^+ leaving N invariant is isomorphic to the collineation group of $\rho(H^+, H, N)$ as $PG(t, q^2)$ which fixes a hyperplane.

PROOF. Note that vectorially, $H^+ - H$ has $q^{2t+1} - q^{2t} = q^{2t}(q-1)$ vectors and projectively, we have q^{2t} points. In order that a vector-space normal line spread cover $q^{2t} - 1$ non-zero vectors, it follows that N consists of $(q^{2t} - 1)/(q^2 - 1)$ 2-dimensional vector subspaces, so that N has this same number of lines, considered projectively. Since $(q^{2t} - 1)/(q^2 - 1) + q^{2t} = (q^{2(t+1)} - 1)/(q^2 - 1)$, we see that combinatorially, we had in $\rho(H^+, H, N)$, exactly the required number of points in a $PG(t, q^2)$. Choose two elements A, B of N as 'points'. Then $\langle A, B \rangle$ intersects N in a spread of $q^2 + 1$ lines of N. Hence, two points of the N-type are incident with a unique 'line' of $q^2 + 1$ 'points'. The remaining isomorphism and group connections are straightforward to establish. □

Now the idea is to establish another way to consider $H(3, q^2)$ within $PG(3, q^2)$ as a geometry $\rho(H^+, H, N) \simeq PG(3, q^2)$. Therefore, we require H^+ to be isomorphic to $PG(6, q)$, H a hyperplane isomorphic then to $PG(5, q)$, and we need to find an appropriate normal line spread of $PG(5, q)$. This is done using $PG(7, q^2)$ as follows: Let V_8 be an 8-dimensional $F \simeq GF(q^2)$ vector space and let $(x, y) \in V_8$, where x and y are 4-vectors, relative to a fixed basis. Define the following involution σ:

$$\sigma : (x, y) \longmapsto (y^q, x^q), \text{ where } z^q = (z_1^q, z_2^q, z_3^q, z_4^q), \text{ for a 4-vector } z.$$

Then there are three subgeometries of importance: The fixed-point set of σ, taken homogeneously,

$$\text{Fix} \, \sigma = \{ (x, x^q); \, x \text{ is a 4-vector} \} \simeq PG(7, q),$$

and the two $PG(3, q^2)$'s, $x = 0$ and $y = 0$ (and note that $(x = 0)^\sigma = (y = 0)$).

LEMMA 50.27. Define $N = \{ L(x) = \langle (x, 0), (0, x^q) \rangle \cap \text{Fix} \, \sigma; \, \forall \, x \neq 0 \}$. For m a line of $(y = 0)$, define $N(m) = \{ L(x); \, (x, 0) \in m \}$.
   (1) $N(m)$ is a regular (i.e., Desarguesian) spread of

$$\langle (x, 0), (0, x^q); \, (x, 0) \in m \rangle \cap \text{Fix} \, \sigma \simeq PG(3, q).$$

   (2) N is a normal line spread of $\text{Fix} \, \sigma \simeq PG(7, q)$.
   (3) The mapping $(x, 0) \longmapsto L(x)$ establishes an isomorphism from $y = 0$ to N, where the associated 'lines' are the $N(m)$. Hence, N may be considered isomorphic to $PG(3, q^2)$. When N is considered a $PG(3, q^2)$, we shall use the notation $P(N)$.

PROOF. Note that $(cx, c^q x^q)$ for c in F, define projectively the $q + 1$ points $L(x)$ in $\text{Fix} \, \sigma$. Since $(x = 0)$ and $(y = 0)$ are disjoint, it follows that N is a line spread. Now consider $N(m)$, vectorially, so that a component as a 2-dimensional $GF(q)$-subspace has the basic form, for x fixed as $\{ (cx, c^q x); \, c \in GF(q^2) \}$, which is clearly a 2-dimensional $GF(q)$-space. However, we may also make this subspace into a 1-dimensional $GF(q^2)$-space as well by defining $c \cdot (x, x^q) = (cx, c^q x^q)$. Since we may consider m as a $PG(1, q^2)$, then vectorially, m is the line at infinity of a Desarguesian affine plane coordinatized by $GF(q^2)$; that is $N(m)$ is a regular spread (Desarguesian spread).
   Choose two lines $L(w)$ and $L(v)$ of N and consider $\langle L(w), L(v) \rangle$. Let $m = (w, w^q)(v, v^q) = \langle (w, w^q), (v, v^q) \rangle$, the line joining the two points in $\text{Fix} \, \sigma$. Then in $\langle L(w), L(v) \rangle$ admits an $N(m)$-spread, implying that N is a normal line spread. □

We now determine a quadric whose intersection becomes a quadric associated with the Hermitian surface $H(3, q^2)$:

$$\{ (x_1, x_2, x_3, x_4); \, x_1 x_4^q + x_2 x_3^q + x_3 x_2^q + x_4 x_1^q = 0 \}.$$

LEMMA 50.28. (1) $x = 0$ and $y = 0$ are subspaces of the hyperbolic quadric $Q^+(7, q^2)$, which has the following equation:

$$x_1 y_4 + x_2 y_3 + x_3 y_2 + x_4 y_1 = 0.$$

   (2) $Q^+(7, q^2) \cap \text{Fix} \, \sigma = Q^+(7, q)$ (an associated hyperbolic quadric), with the equation:

$$x_1 x_4^q + x_2 x_3^q + x_3 x_2^q + x_4 x_1^q = 0.$$

   (3) $\{ L(x); \, L(x) \cap Q^+(7, q) \neq \phi \} = S^+$ is a line spread of $Q^+(7, q)$ (i.e., a set of lines that partitions the quadric) and $\{ (x, 0); \, L(x) \in S^+ \}$ is a Hermitian surface $H(3, q^2)$ of $(y = 0) \simeq PG(3, q^2)$.

PROOF. If $L(x)$ non-trivially intersects $Q^+(7, q^2) \cap \mathrm{Fix}\,\sigma = Q^+(7, q)$, then $L(x)$ contains a point $(cx, c^q x^q)$ on $Q^+(7, q)$ if and only if for $x = (x_1, x_2, x_3, x_4)$ we have

$$c^{1+q}(x_1 x_4^q + x_2 x_3^q + x_3 x_2^q + x_4 x_1^q) = 0.$$

Hence, if there is one point of intersection then $L(x)$ lies in $Q^+(7, q)$. The remaining statements now follow. □

Returning to the connection with the spaces $\rho(H^+, H, N)$, we require an appropriate 6-dimensional projective space and look for a natural normal spread of some hyperplane. For this, take any two $L(w), L(v)$ and form the 3-dimensional projective subspace $\langle L(w), L(v) \rangle$, which admits a spread induced from N. Take any line $L(u)$ which is not a line of this spread of $\langle L(w), L(v) \rangle$ and take $\langle L(u), L(w), L(v) \rangle$, which is a 5-dimensional projective subspace of $\mathrm{Fix}\,\sigma$. There are $q^2 + 1$ lines of N in $\langle L(w), L(v) \rangle$. For each such line $L(x)$, form $\langle L(x), L(u) \rangle$ to construct a set of $(q^2 + 1)^2 + 1$ lines of N within $\langle L(u), L(w), L(v) \rangle$. Since a line spread of a 5-dimensional subspace requires $(q^6 - 1)/(q^2 - 1) = 1 + q^2 + q^4$ lines, it follows we have a line spread of $\langle L(u), L(w), L(v) \rangle$, which is, of course, a normal line spread. We have proved the following lemma:

LEMMA 50.29. (1) Choose $L(u)$ not in $\langle L(w), L(v) \rangle$. Then there is an induced normal line spread N_H in $\langle L(u), L(w), L(v) \rangle = H$.

(2) If H^+ is any hyperplane of $\mathrm{Fix}\,\sigma$ containing H, there is a corresponding geometry $\rho(H^+, H, N_H) \simeq PG(3, q^2)$.

Now every line $L(x)$ of N either lies in H^+ or intersects H^+ in a unique point. Consider the mapping which maps $L(x)$ to $L(x) \cap H^+$, recalling that the points of $\rho(H^+, H, N_H)$ are the points of $H^+ - H$ and the lines of N_H. All lines of N within N_H then mapped to lines of N_H. Every line of $N - N_H$ cannot intersect H, but must intersect $H^+ - H$, since H is a hyperplane of H^+. From here, it follows that this mapping is an isomorphism from $P(N)$ to $\rho(H^+, H, N_H)$. If we now compose the mapping $(x, 0) \longmapsto L(x)$, to form an isomorphism γ from $y = 0$ to $\rho(H^+, H, N_H)$, we have the following connection.

PROPOSITION 50.30. First $y = 0$ is isomorphic to $\rho(H^+, H, N_H)$ by $\gamma : (x, 0) \longmapsto L(x) \longmapsto L(x) \cap H^+$. Then $\gamma(\{ (x, 0); L(x) \in S^+ \})$ is a Hermitian surface $H(3, q^2)$ of $(y = 0) \simeq PG(3, q^2)$, which is a Hermitian surface of $\rho(H^+, H, N_H)$.

Let $\Omega_H = \{ (x, 0); L(x) \in H \}$; then Ω_H is a 2-dimensional projective subspace of $y = 0$. Then

$$\langle (x, 0), (0, x^q); L(x) \in H \rangle \cap \mathrm{Fix}\,\sigma = \Omega_H^+$$

is a 5-dimensional subspace, and as such it must be H. Consider $\Omega_H \cap H(3, q^2)$. Since Ω_H is a 2-dimensional projective subspace, if the intersection is non-singular, we have a conic intersection of $q + 1$ points. This would then imply that $H \cap Q^+(7, q) = Q^-(5, q)$. On the other hand, if Ω_H is tangent to $H(3, q^2)$ at $(w_0, 0)$ then $H \cap Q^+(7, q)$ will contain the line $L(w_0)$ and the set of points of $H \cap Q^+(7, q)$ be a 'cone' with vertex $L(w_0)$. Furthermore, H^+ will be a tangent hyperplane (or tangent prime) on a point Z incident with $L(w_0)$ (recall that a tangent hyperplane at Z consists of the points on tangents to the quadric and incident with Z union the set of points of the quadric on lines of the quadric through Z (see Chapter 103 for details on quadrics). By proper choice of generator lines, we may assume that $H \cap Q^+(7,q)$ is a cone with vertex $L(w_0)$, so that we may assume that Ω_H is tangent to $H(3,q^2)$ at $(w_0,0)$.

Hence, we obtain:

PROPOSITION 50.31. *The Hermitian surface $H_\rho(3,q^2)$ of $\rho(H^+, H, N_H)$ is the set $Q^+(7,q) \cap H^+$. Hence, the points of $H_\rho(3,q^2)$ are the points of $Q^+(7,q) \cap H^+ - H$ and the set of lines of N_H within $Q^+(7,q)$. $H \cap Q^+(7,q)$ is a quadratic cone with vertex $L(w_0)$.*

PROOF. That is, if $L(x)$ shares two points with H^+ then $L(x)$ lies in $Q^+(7,q)$ and since H^+ is tangent at Z, $L(x)$ lies in $H \cap Q^+(7,q)$. Since $\rho(H^+, H, N_H) \simeq PG(3,q^2)$, and $H \cap Q^+(7,q)$ is a cone with vertex $L(w_0)$, then the mapping under γ becomes a cone with vertex $L(w_0)$, a point of $\rho(H^+, H, N_H)$, i.e., a quadratic cone in $\rho(H^+, H, N_H)$. $\square$

Our primary objective is to create another spread in $PG(3,q)$ from an indicator spread in $PG(3,q)$. So, given a transversal T to a finite derivable net, we may form an indicator spread then dualize to create a Shult set, which forms a locally Hermitian ovoid Θ of $H(3,q^2)$ within $PG(3,q^2)$. Thinking of this ovoid as tangent at a point $W = (w_0,0)$ with the subspace $y = 0$, we have a set of $q^3 + 1$ points, no two on a line of $H(3,q^2)$, with the property that these points lie on q^2 lines of $PG(3,q^2)$ incident with W. Since W maps to $L(w_0)$, we need to determine the image of the set of q^2 lines of $PG(3,q^2)$ incident with W, none of which are in $H(3,q^2)$. The lines incident with W but not in $H(3,q^2)$ map under γ to the planes of H^+ which intersect H in the line $L(w_0)$. Let Π denote such a plane and consider $\Pi \cap (Q^+(7,q) \cap H^+)$. The q points of $H(3,q^2)$ on a line π such that $\gamma\pi = \Pi$, map to q points of $Q^+(3,7) \cap H^+$ which must be on a line ℓ_Π of Π necessarily incident with Z, as H^+ is tangent at Z. Hence, we have a set of q^2 lines distinct from $L(w_0)$ incident with Z.

PROPOSITION 50.32. *The $q^3 + 1$ points of Θ tangent at $W = (w_0,0)$ determine q^2 lines incident with $(w_0,0)$, none of which are in $H(3,q^2)$, and hence determine a set of q^2 lines incident with Z and distinct from $L(w_0)$. Hence, there is a corresponding set S_Θ of $q^2 + 1$ lines of H^+ concurrent at Z.*

Now take a hyperplane H_1 of H^+ which does not contain Z. Then $H_1 \cap Q^+(7,q) = Q^+(5,q)$. Furthermore, $H_1 \cap H$ is the tangent hyperplane to $Q^+(5,q)$ at $L(w_0) \cap H_1 = K$. The intersection with the set S_Θ is a set of $q^2 + 1$ points containing $L(w_0) \cap H_1 = K$ of $Q^+(5,q)$, Φ_Θ. Note that $\Phi_\Theta = \{ R; RK \in S_\Theta \}$.

THEOREM 50.33. (Lunardon [**948**, Theorem 6]). *Φ_Θ is an ovoid of $Q^+(5,q)$.*

PROOF. Let A and B be two distinct points of Φ_Θ. Assume that the line AB is a line of $Q^+(5,q)$. Consider $AB \cap H = W$. Since W is a point of $Q^+(7,q)$, it follows that the unique line $L(w_1)$ of N_H which is incident with W contains a point of $Q^+(7,q)$ and hence lies within $Q^+(7,q)$. Assume first that $(w_0,0) = (w_1,0)$ in $y = 0$. Every line of $y = 0$, which is incident with $(w_0,0)$ and does not belong to $Q^+(7,q)$ maps under γ to a plane of H^+ which intersects H in the line $L(w_0)$. Let m_A denote the line of $y = 0$, which constructs the plane containing the line $\langle ZA, L(w_0) \rangle$. If AB intersects $L(w_0)$, then since $\langle ZA, L(w_0) \rangle = \langle ZB, L(w_0) \rangle$, in this case, it follows that $ZA = ZB$, since there is a unique line in $Q^+(7,q) - \{L(w_0)\}$ in $Q^+(7,q)$, a contradiction. We claim that the line $\langle (w_0,0), (w_1,0) \rangle = m_\infty$ is in

$H(3, q^2)$. The corresponding distinct lines $L(w_0)$ and $L(w_1)$ are in H^+, and hence the line spread of $\langle L(w_0), L(w_1) \rangle$ belongs to H^+. Then under γ^{-1}, this line spread maps to a line of $H(3, q^2)$.

Consider the plane of $y = 0$, $\langle \langle (w_0, 0), (w_1, 0) \rangle, m_A \rangle = \pi$. Since m_A has exactly $q + 1$ points of $H(3, q^2)$ and m_∞ is a line of $H(3, q^2)$, this plane π must be tangent to $H(3, q^2)$ at a point $(w_2, 0)$ which necessarily lies on m_∞. Consider $\langle \pi, \pi^q \rangle \cap$ Fix $\sigma = H_1$. This is a 5-dimensional projective of Fix σ. So, $\pi\gamma = H_1$. Note that $\langle L(w_0), L(w_1) \rangle$ and ZA are in H_1. Therefore, B and hence ZB are also in H_1. This means if m_B is the line incident with $(w_0, 0)$ which maps to ZB, then m_B is in the plane π. If $(w_0, 0) \neq (w_2, 0)$, then a line of $H(3, q^2)$ incident with $(w_2, 0)$ will meet both m_A and m_B in a point of $H(3, q^2)$; however, since we started with an ovoid, there is a unique line of $H(3, q^2)$ on each point of the ovoid, so we have a contradiction.

Hence, $(w_0, 0) = (w_2, 0)$. In this case, there are $q + 1$ lines of $H(3, q^2)$ incident with $(w_0, 0)$. However, since we have an ovoid, these are the $q + 1$ lines of the plane π_H, which maps to H. That is, this will force $\pi = \pi_H$, and in turn, form $H = H_1$. But, A is a point of $H_1 - H$, a contradiction. This completes the proof. □

Actually, when the transversal to a finite derivable net is a vector-space transversal, so the associated indicator spread is a semifield spread, it turns out that the corresponding ovoids are translation ovoids.

THEOREM 50.34. (See Lunardon [**948**]). Corresponding to a transversal T to a finite derivable net, there is an associated ovoid of $Q^+(5, q)$. This ovoid produces a spread of $PG(3, q)$, via the Klein quadric mapping. We call the associated spread a 'hyperbolic spread'.

T is a vector-space transversal if and only if the indicator spread is a semifield spread if and only if the associated ovoid of $Q^+(5, q)$ is a translation ovoid, which is valid if and only if the hyperbolic spread is a semifield spread.

So starting with an indicator spread π_D^I, there is a corresponding hyperbolic spread $\pi_{D(H)}^I$ via the previous sequence. Starting with the hyperbolic spread, we may repeat the process using an indicator set from the new translation plane and construct another hyperbolic spread. We call this a 'Hermitian sequence of hyperbolic type'.

$$\pi_D^{I_1} \longmapsto \pi_{D(H)}^{I_1} \longmapsto \pi_{D^2(H)}^{I_2} \longmapsto \pi_{D^3(H)}^{I_3} \longmapsto \cdots \longmapsto \pi_{D^t(H)}^{I_t} \longmapsto \cdots.$$

Hermitian sequences producing spreads in $PG(3, q^2)$.

However, by the use of coordinates of indicator sets, Bader, Marino, Polverino and Trombetti [**55**] prove that all Hermitian sequences consist of isomorphic spreads. We state this result in terms of transversals to derivable nets.

50.2.2.1. *Bader, Marino, Polverino, Trombetti Theorem on the Hermitian Sequences.*

THEOREM 50.35. (Bader, Marino, Polverino and Trombetti [**55**]). Given a transversal to a finite derivable net the corresponding translation planes constructed using either the elliptic or hyperbolic Hermitian sequences are all isomorphic (or transpose).

As pointed out, non-equivalent indicator sets can produce isomorphic planes, but the question is: do these produce equivalent locally Hermitian ovoids? This question is considered and solved in Cossidente, Lunardon, Marino, and Polverino [**263**].

50.2.2.2. *Cossidente, Lunardon, Marino, Polverino Hermitian Isomorphism Theorem.*

THEOREM 50.36. (Cossidente, Lunardon, Marino, and Polverino [**263**]). Two locally Hermitian ovoids of the Hermitian surface are isomorphic if and only if the corresponding indicator sets are equivalent.

So, we have the interesting question of determining for certain interesting spreads in $PG(3, q)$ the non-equivalent indicator sets that construct them, for these produce the set of mutually non-isomorphic locally Hermitian ovoids. One such spread is the flock spread of Kantor–Knuth, probably most interesting for the fact that the planes containing the conics of the corresponding flock contain a point and the flocks may be characterized by this property. Actually it is shown that there are at least three mutually non-isomorphic locally Hermitian ovoids corresponding to the Kantor–Knuth spreads.

Cossidente, Ebert, Marino and Siciliano [**262**] construct two classes of interesting translation ovoids of the Hermitian surface directly by determining certain particular Shult sets. Hence, there are corresponding classes of semifield spreads. One of these classes of spreads is shown to be a semifield flock spread and is naturally of interest. The flock semifield spreads are shown to be Kantor–Knuth by Cossidente, Lunardon, Marino, and Polverino [**263**]. This result is also proved by Johnson [**654**] and in that work it is also shown that the other class of semifields of Cossidente, Ebert, Marino and Siciliano is isomorphic to the dual transpose to a class of Hughes–Kleinfeld semifields.

50.2.2.3. *The Pabst–Sherk Planes.* In this subsection, we construct the 'Pabst–Sherk Planes' [**1160**] plane by indicator sets. We connect spreads in $PG(3, q)$ with the set of points of $AG(2, q^2) - \{L\}$, where L is a fixed line. Choose coordinates for L, so it may be represented by $x = 0$, where points of $AG(2, q^2)$ are represented by pairs (x, y), for $x, y \in GF(q^2)$. Let $\{1, e\}$ be a basis for $GF(q^2)$ over $GF(q)$, and let $e^2 = eg + f$. The connection between 2-dimensional vector subspaces, now representable in the form $y = xM$, where $M = \begin{bmatrix} a & b \\ c & d \end{bmatrix}$ and elements (x, y), such that x is not zero is

$$\begin{bmatrix} a & b \\ c & d \end{bmatrix} \iff (a + ce, b + de).$$

Recalling that an indicator set is a set of q^2 points of $AG(2, q^2)$, such that the join of any pair is in the slope set determined by $GF(q^2) - GF(q)$. For example, consider the set of q^2 points on the line $y = xe$ of $AG(2, q^2)$. This would be an indicator set and would 'indicate' the Desarguesian spread. In particular, the set of $q + 1$ points of the form $R_\lambda = \{(x, xe) = x^{q+1} = \lambda\}$, may be seen to indicate a regulus of $PG(3, q)$. Hence, the indicator set $y = xe$ will determine a Desarguesian spread which is the union of a set of $q - 1$ mutually disjoint reguli union two carrying lines $x = 0, y = 0$ corresponding to an elliptic flock. Now take q odd and $e^2 = f$, where f is a non-square in $GF(q)$. Let $m \in GF(q^2) - \{GF(q), \pm e\}$, such that $m^2 - f$ is non-square in K. Then let $S_\sigma = \{(x, xm); m^{q+1} = \sigma\}$, and note that this set indicates a regulus in $PG(3, q)$. There is a set of $(q - 3)/2$ reguli indicated by R_λ, which are componentwise disjoint from the regulus indicated by S_σ. Now take the collineation group

$$H = \left\langle (x,y) \longmapsto (x,y) \begin{bmatrix} u & tf \\ t & u \end{bmatrix}; u^2 - t^2 f = 1 \right\rangle$$

of $AG(2,q^2)$. Pabst and Sherk [**1160**] show that the indicated reguli, which are componentwise disjoint from the regulus indicated by S_σ, are also componentwise disjoint from the regulus indicated by any image set of S_σ under H. Furthermore, these $(q+1)$-images are mutually disjoint. We call the sets indicating the regulus 'compatible' when the corresponding reguli are mutually disjoint of lines. Now take the associated kernel homology group of order $(q+1)$, K^+:

$$K^+ : \left\langle (x,y) \longmapsto (ax, ay); |a| \text{ divides } q+1 \right\rangle.$$

It is easy to see that $H \cap K^+$ has order 2, so that there are $(q+1)^2/2$ points in an order of $S_\sigma HK^+$. This means that there is a corresponding spread of $PG(3,q)$. In the following theorem, there is constructed an indicator set for a class of planes that we shall call the 'Pabst–Sherk planes'.

THEOREM 50.37. (Pabst and Sherk [**1160**]). Let $AG(3,q^2)$ be the indicator space for $PG(3,q)$. If $\{1,e\}$, such that $e^2 = f$, for f non-square in $GF(q)$, is a basis for $GF(q^2)$ over $GF(q)$ then for any $m \in GF(q^2) - \{GF(q), \pm e\}$ such that $m^2 - f$ is non-square in $GF(q)$, let $S_\sigma = \left\{ (x, xm); m^{q+1} = \sigma \right\}$. Let

$$H = \left\langle (x,y) \longmapsto (x,y) \begin{bmatrix} u & tf \\ t & u \end{bmatrix}; u^2 - t^2 f = 1 \right\rangle$$

and

$$K^+ : \left\langle (x,y) \longmapsto (ax, ay); |a| \text{ divides } q+1 \right\rangle.$$

and form $HK^+ S_\sigma$. Then this is a compatible set (the corresponding reguli are mutually line-disjoint). Then there is a set Ω of $(q-3)/2$, sets

$$R_\lambda = \left\{ (x, xe) = x^{q+1} = \lambda \right\},$$

each of which is compatible with $HK^+ S_\sigma$ such that

$$\Omega \cup HK^+ S_\sigma$$

in an indicator set producing a corresponding translation plane (i.e., the Pabst–Sherk plane) with spread in $PG(3,q)$.

Now the resulting translation plane may be seen to be constructed via $(q+1)$-nest replacement as well. Hence, we note the following theorem.

THEOREM 50.38. The Pabst–Sherk planes of order q^2 with spread in $PG(3,q)$ may be constructed from a Desarguesian affine plane of order q^2 by $(q+1)$-nest replacement. These planes admit an affine homology group of order $q+1$, represented by H in the above theorem and there is a unique orbit of length $(q+1)^2/2$ of components, representing the $(q+1)$-nest.

50.3. General Indicator Sets.

It is possible to generalize the concept of an indicator set producing spreads in $PG(3, q)$ to sets indicating spreads in arbitrary dimensional projective spaces. This extension is due to Sherk [**1155**]. For a projective geometry $PG(2t - 1, q)$, a 't-spread' is a set of mutually disjoint $t - 1$-dimensional projective spaces that partition the point set.

We consider a t-spread in $PG(2t - 1, q)$, try to determine an 'indicator' set of points of the affine geometry $AG(t, q^t)$. Choose a basis $\{1, e, e^2, \ldots, e^{t-1}\}$ for $GF(q^t)$ over $GF(q)$, where $e^t = \sum_{i=0}^{t-1} e^i f_i$, for $f_i \in GF(q)$. We now consider a point of $AG(t, q^2)$ as a vector $(z_1, z_2, \ldots, z_t)$, such that $z_i \in GF(q^t)$ and hence may be represented in by $z_i = \sum_{i=0}^{t-1} e^i g_i$, for $g_i \in GF(q)$. As in the $PG(3, q)$ case, we choose a fixed t-space of $AG(2t, q)$ and choose a basis for the underlying $2t$-dimensional vector space over $GF(q)$ so that the fixed space has the form $x = 0$, where (x, y) is a point of $AG(2t, q)$, x and y are t-vectors over $GF(q)$. We note that any t-dimensional subspace disjoint from $x = 0$, will have the form $y = xM$, where M is a $t \times t$ matrix over $GF(q)$. We form the following bijection:

$$M = [\alpha_{1,i}^T, \alpha_{2,i}^T, \ldots, \alpha_{t,i}^T] \Longleftrightarrow \left(\sum_{i=0}^{t-1} e^i \alpha_{1,i}, \sum_{i=0}^{t-1} e^i \alpha_{2,i}, \ldots, \sum_{i=0}^{t-1} e^i \alpha_{t,i} \right),$$

where

$$\alpha_{z,i}^T = \begin{bmatrix} \alpha_{z,1} \\ \alpha_{z,2} \\ \vdots \\ \alpha_{z,t} \end{bmatrix}.$$

Now the condition that two points $z = (z_1, z_2, \ldots, z_t)$ and $w = (w_1, w_2, \ldots, w_t)$ of $AG(t, q^t)$ indicate two mutually disjoint t-subspaces is that z and w are not $GF(q)$-linearly dependent. Any two such points are said to be 'compatible'.

LEMMA 50.39. A set of points of $AG(t, q^t)$ corresponds to a set of mutually disjoint t-subspaces of $V_{2t}/GF(q)$, exactly when the set is $GF(q)$-linearly independent.

DEFINITION 50.40. An 'indicator' set of $AG(t, q^t)$ is a set of q^t points which are $GF(q)$-linearly independent.

THEOREM 50.41. (Sherk [**1155**]). The set of translation planes of order q^t with kernel containing $GF(q)$ is equivalent to the set of indicator sets of $AG(t, q^t)$.

There is more to be said about indicator sets and the effect of isomorphic planes on the associated indicator sets.

We end this section by showing how two interesting planes of order 27 may be visualized using indicator sets in $AG(3, 3^3)$. We follow Sherk [**1155**] in the next subsection.

50.3.1. Hering's Plane of Order 27.
Hering's translation plane of order 27 admits the collineation group $SL(2, 13)$ and is flag-transitive.

Let

$$E = \begin{bmatrix} 0 & -1 & 0 \\ -1 & 0 & -1 \\ 1 & 0 & 0 \end{bmatrix}, \; A = \begin{bmatrix} -1 & 0 & 1 \\ -1 & 0 & 0 \\ 0 & -1 & 0 \end{bmatrix}, B = \begin{bmatrix} -1 & 0 & 0 \\ -1 & 1 & 0 \\ 1 & -1 & -1 \end{bmatrix},$$

$$C = \begin{bmatrix} -1 & 0 & 0 \\ -1 & 1 & 0 \\ -1 & -1 & -1 \end{bmatrix}, D = \begin{bmatrix} 1 & 0 & 0 \\ 1 & -1 & 0 \\ -1 & 1 & 1 \end{bmatrix}.$$

Consider the matrix elements g and h:

$$g = \begin{bmatrix} E & 0 \\ 0 & A \end{bmatrix}, h = \begin{bmatrix} B & B \\ C & D \end{bmatrix}.$$

Then $\langle g, h \rangle \simeq SL(2,13)$. Furthermore,

THEOREM 50.42. *Under the above assumptions,*

$$\{x = 0, y = 0\} \langle g, h \rangle$$

is a spread (the Hering spread of order 27), which is

$$\left\{ x = 0, y = xM; \; M \in \left\{ 0, E^{-i}A^i, E^{-i}C^{-1}DA^i; \; i = 0, 1, \ldots, 12 \right\} \right\}.$$

We note that the set of points obtained from the bijection

$$M = [\alpha_{1,i}^T, \alpha_{2,i}^T, \ldots, \alpha_{t,i}^T] \iff \left(\sum_{i=0}^{t-1} e^i \alpha_{1,i}, \sum_{i=0}^{t-1} e^i \alpha_{2,i}, \ldots, \sum_{i=0}^{t-1} e^i \alpha_{t,i} \right),$$

where

$$\alpha_{z,i}^T = \begin{bmatrix} \alpha_{z,1} \\ \alpha_{z,2} \\ \vdots \\ \alpha_{z,t} \end{bmatrix},$$

produces the associated indicator set (note that $M = 0$ maps to the points $(0,0,0,\ldots,0)$).

In Chapter 38 on the method of Oyama, we construct the infinite class of translation planes known as the Suetake planes, which contains the Hering plane of order 27 as the smallest-order plane of this set. Suetake proves that these planes correspond to a class of semifield planes called generalized twisted field planes by a certain net replacement procedure. This fact was also proved by Sherk [1155] for the Hering planes.

Apply to the Hering spread the mapping

$$M \to \begin{bmatrix} 0 & 0 & 1 \\ 0 & 1 & 0 \\ 1 & 0 & 0 \end{bmatrix}^{-1} M \begin{bmatrix} 0 & 0 & 1 \\ 0 & 1 & 1 \\ 1 & -1 & -1 \end{bmatrix}.$$

This equivalent spread set for the associated translation plane has the form:

$$\left\{ x = 0, y = xM; \; M \in \left\{ 0, U^{3i}P_0U^i, U^{3i}QU^i; \; i = 0, 1, \ldots, 12 \right\} \right\},$$

where

$$U = \begin{bmatrix} 0 & -1 & 1 \\ -1 & 1 & 1 \\ 1 & -1 & -1 \end{bmatrix}, \quad P_0 - R^{-1}S - \begin{bmatrix} 1 & -1 & -1 \\ 0 & 1 & 1 \\ 0 & 0 & 1 \end{bmatrix},$$

$$Q = R^{-1}C^{-1}DS = \begin{bmatrix} -1 & 1 & 0 \\ 0 & -1 & -1 \\ 0 & 0 & -1 \end{bmatrix}.$$

Now

$$\left\{ x = 0, y = xM; \; M \in \left\{ \alpha U^3 P_0 U + \beta U^6 P_0 U^2 + \gamma P_0; \; \alpha, \beta, \gamma \in GF(3) \right\} \right\},$$

is an additive spread and hence is a semifield spread of order 27, which actually corresponds to a twisted field plane. Therefore, we see that the set

$$\left\{ y = xM; \; M \in \left\{ U^{3i} Q U^i; \; i = 0, 1, 2, \ldots, 12 \right\} \right\}$$

replaces the set

$$\left\{ y = xN; \; N \in \left\{ U^{3i}(-P_0)U^i; \; i = 0, 1, 2, \ldots, 12 \right\} \right\}.$$

It turns out that this semifield spread is the unique non-Desarguesian semifield spread of order 27. We note that all spreads of order 27 have been determined by computer by Dempwolff [**304, 305**].

THEOREM 50.43. *The Hering plane of order 27 may be constructed by net replacement from the unique semifield plane of order 27 with kernel* $GF(3)$.

Both the Hering plane of order 27 and the semifield plane of order 27 are 'symplectic' planes in that their components are totally isotropic subspaces of an ambient symplectic form. More generally, the Suetake planes may be recently shown to be symplectic by Ball, Bamberg, Lavrauw and Penttila [**81**] by noting the associated generalized twisted field plane admits a symplectic form inherited under a new replacement procedure, which makes both of these classes of planes symplectic.

CHAPTER 51

Geometries and Partitions.

In this chapter, we touch on some of the many varieties of designs and geometries that may be connected to or constructed from translation planes. There are also many connections to partial and semi-partial geometries, gamma spaces, and so on, so our treatment will only scratch the surface of connections with translation planes. The reader is directed to Beth–Jungnickel–Lenz [**111**] for a rigorous general treatment of designs. We are content here merely to point to a few connections with designs and translation planes and with classical geometries and translation planes. For example, we shall mention a few connections with unitals, inversive planes and divisible designs.

DEFINITION 51.1. A 't-(v, k, λ)-design' is a set of v points and b blocks such that each line is incident with k points and every set of t distinct points is contained in λ blocks.

DEFINITION 51.2. A 2-$(q^3 + 1, q + 1, 1)$-design is called a 'unital', and a 3-$(n^2 + 1, n + 1, 1)$-design is called an 'inversive plane.'

51.1. Divisible Designs.

DEFINITION 51.3. Let V be a finite set of order which permits an equivalence relation R, where the classes all have cardinality s. A subset S of V is said to be an R-transversal if S intersects each equivalence class in at most one element. Let B be a family of R-transversal subsets of V and t, k, λ positive integers with $2 \leq k \leq v$. The pair $D = (V, B)$ is called a t-(s, k, λ) divisible design if (i) $|b| = k$ for each $b \in B$ and (ii) for each transversal subset Z of V of size t, there are exactly λ elements of B counting S.

Divisible designs may be constructed from translation planes admitting a doubly transitive group on a subset of slopes on the line at infinity. In Chapter 69, we consider a variety of problems involving groups acting doubly transitive on a subset of slopes on the line at infinity. For example, the associated planes include the Lüneburg–Tits planes of order q^2 admitting $S_z(q)$, the Hering and Ott–Schaeffer planes of order q^2 admitting $SL(2, q)$ (permuting $q+1$ points on the line at infinity), any generalized Desarguesian of Jha and Johnson of order q^3 admitting $SL(2, q)$, generated by elations (and hence permuting $q+1$ on the line at infinity), any translation plane of order q^4 corresponding to a Desarguesian parallelism in $PG(3, q)$ admitting $SL(2, q)$ (permuting $q + 1$ points on the line at infinity) and generated by elations.

The following ideas are due to Spera [**1164**] and Schulz and Spera [**1152**].

Choose a translation plane of order q^n admitting such a group G acting doubly transitive on the set of $q+1$ points on the line at infinity. Take any subset of slopes of the $q+1$ and take a line from each parallel class of this set. This subset becomes the transversal set B, since element is a line. Take all lines on this $q+1$ set on the line at infinity, call this set $\widetilde{O}$. Then $(\widetilde{O}, B)$ becomes a divisible design with $q^n(q+1)$ 'points'. The 'blocks' become the images of a particular transversal set—so each block has k points on it. The equivalent classes all have size q^n. Take two 'points', i.e., two lines of the translation plane on the $q+1$ orbit Γ in different parallel classes. This is a 'transversal' set of size $t = 2$. Take the intersecting point P. Then incident with P is exactly one line of each parallel class, and hence the question is how many elements of the B^g's set contain the transversal set? This is equivalent to asking questions about the stabilizer subgroup. If the group is $SL(2,q)$, then the global stabilizer of two lines has order $(q(q^2-1))/(q(q+1)) = (q-1)$ or possibly $(q-1)/2$. So, we need the orbit length of the global stabilizer of two components of the set B. This constant becomes λ.

Spera [1164] and Schulz and Spera [1152] consider various of the planes listed above to construct divisible designs.

THEOREM 51.4. (Spera [1164] and Schulz and Spera [1152]). Any translation plane of order q^n admitting a collineation group doubly transitive on a set of $q+1$ points on the line at infinity constructs a divisible design with $q^n(q-1)$ points.

So, we get a 2-(v, k, λ)-design with the added feature that the blocks arise as 'transversal' subsets from an equivalence relation on the points.

REMARK 51.5. Using results of GJDLCF (see index under 'theorems') (the classification of spreads fixing one component and doubly transitive on the rest are generalized twisted field planes) and the results of Biliotti–Jha–Johnson [121] on generalized twisted fields, it is possible to give considerable generalizations of these constructions on divisible designs.

There are a variety of other constructions, and indeed there are connections to semifield spreads by Cerroni [220] and Cerroni and Spera [221] and with t-spreads by O'Keefe and Rahilly [1030] and Rahilly [1127].

51.2. Inversive, Minkowski and Laguerre Planes.

In this section, we note some of the basic connections with classical geometries and affine translation planes (or more probably with affine planes).

DEFINITION 51.6. A 'Laguerre plane (P, C, E) is a set of points P, circles C and an equivalence relation E called 'parallelism on points' with the following properties:

(i) For any three mutually non-parallel points, there is a unique circle incident with these points.

(ii) Given a flag (x, c), where x is a point incident with a circle c, then for any point y which is not parallel to x, there is a unique circle incident with x and y which is tangent to c (intersects c in exactly one point z).

(iii) Each circle is incident with at least three points and not all points are on the same circle.

As we are mainly concerned with finite structures in this handbook, we now look at the so-called 'oval Laguerre planes', which include all known finite examples.

DEFINITION 51.7. Let K be a field isomorphic to $GF(q)$ and consider an oval O in a Desarguesian plane Π within $PG(3, K)$. Choose a point v_0 (the 'vertex') exterior to the plane Π and form the lines $v_0 P$ for all points $P \in O$ to form the 'oval cone' $\mathcal{C}$. These lines are called the 'generators' of the cone. Define 'points' to be the elements of the set of non-vertex points on the generators and 'circles' to be the plane intersections not including the vertex by planes of $PG(3, K)$. Define two points to be parallel if and only if they lie on a generator.

This set of points, circles and parallelism forms a Laguerre plane called an 'oval Laguerre plane.'

If the oval O is a conic, we obtained the 'classical Laguerre plane' and the oval cone is said to be a 'quadratic cone'.

DEFINITION 51.8. A 'partial flock of an oval Laguerre plane' is a partial cover of the non-vertex points by a set of mutually disjoint circles. A 'flock' is a partial flock which forms a cover of the non-vertex points.

DEFINITION 51.9. A 'residue of a point p' of a Laguerre plane is defined by the set of points which are not parallel to p and the set of circles which are incident with p. The residue becomes an affine plane and in the finite oval case, the residue is a Desarguesian affine plane of order q if K is isomorphic to $GF(q)$.

There are many different ovals of a Desarguesian, so there are a great many different oval Laguerre planes. Flocks and partial flocks of oval Laguerre planes are of great interest and, in particular, flocks and certain partial flocks of quadratic cones may be shown to produce and be equivalent to certain translation planes. We shall be discussing these connections in the chapters on flocks of quadratic sets (see Chapters 54 through 61). Concerning 'flocks' of other objects, we consider objects such as hyperbolic and elliptic quadrics and the more general 'ovoids' all of which lie in $PG(3, q)$.

DEFINITION 51.10. An 'ovoid O' in $PG(3, q)$ is a set of $q^2 + 1$ points such that plane intersections are either tangent (intersect uniquely) or are ovals in the plane. Furthermore, tangent planes at a point P are uniquely determined by the set of lines of $PG(3, q)$ that intersect O exactly at P.

DEFINITION 51.11. As noted above, an 'inversive plane' is a 3-$(q^2 + 1, q + 1, 1)$-design. In this case, call the blocks 'circles'. Let O be an ovoid in $PG(3, q)$. Define points to be points of O and circles to be the non-tangent and non-trivial plane intersections. This structure forms a inversive plane called an 'ovoidal inversive plane'. If the ovoid is an elliptic quadric, the inversive plane is said to be 'classical'.

DEFINITION 51.12. Let I be a finite inversive plane and let p be a point. The 'residue of p' is the set of points not equal to p and the set of circles incident with p. The residue then becomes a finite affine plane of order q. When the inversive plane is classical, the residues are Desarguesian planes.

DEFINITION 51.13. A 'partial flock of an ovoidal inversive plane' is a partial covering of the points by a set of mutually disjoint, non-trivial and non-tangent, plane intersections. A 'flock of an ovoidal inversive plane' is a partial flock which covers all but one point.

There are strong connections to translation planes with partial flocks and partial flocks of classical inversive planes, that is, when the ovoid is an elliptic quadric, which we shall mention shortly.

DEFINITION 51.14. A 'Minkowski plane' is a set of points and circles, equipped with two parallelism relations on points, called 'plus parallelism' and 'minus parallelism' with the following properties:

(i) Given any three distinct points no two of which are plus or minus parallel, there is a unique circle incident with these points.

(ii) Given a circle C and a point p not incident with C, there are unique points a, b incident with C such that a is plus parallel to p and b is minus parallel to b.

(iii) Given a flag (x, C), where x is a point and C is a circle and any point y not incident with C which is neither plus nor minus parallel to x, there is a unique circle incident with y which is tangent to C.

(iv) Every circle is incident with at least three points and not all points are incident with the same circle.

DEFINITION 51.15. A 'partial flock of a Minkowski plane' is a partial cover of the points by mutually disjoint circles. A 'flock' of a Minkowski plane is a partial flock which covers the points.

REMARK 51.16. Let $\mathcal{H}$ be a finite hyperbolic quadric in $PG(3, q)$. Then the points and plane intersections which do not contain a line define a finite Minkowski plane, where the two sets of reguli R_+ and R_- are used to define the plus and minus parallelisms; two points p and q are plus or minus parallel, respectively as they are incident with a line of R_+ or R_-. This plane is called the 'classical Minkowski plane'.

The main connections with flocks and partial flocks of inversive, Laguerre and Minkowski planes arise in the classical cases. Hence, we consider flocks and partial flocks of elliptic quadrics, quadratic cones, and hyperbolic quadrics, respectively. All of these flocks live in $PG(3, q)$.

We formalize this as follows:

DEFINITION 51.17. We define a 'quadric set' as either an elliptic quadric, quadratic cone or hyperbolic quadric in $PG(3, q)$. A 'partial flock of a quadric set' is a set of mutually disjoint plane intersections of conics of the set. The three cases are as follows:

(1) P_E^t, for $1 \leq t \leq q + 1$, a partial flock of an elliptic quadric by t conics; P_E^t is 'conical'.

(2) P_C^t, for $1 \leq t \leq q$, a partial flock of a quadratic cone by t conics; P_C^t is 'elliptic'.

(3) P_H^t, for $1 \leq t \leq q - 1$, a partial flock of a hyperbolic quadric by t conics; P_H^t is 'hyperbolic'.

DEFINITION 51.18. The 'deficiency of a partial flock P^t of t conics of a quadric set' is the cardinality of the sets of conics of a flock minus t. Hence, if P^t is elliptic, the deficiency is $q - 1 - t$, if P^t is conic, the deficiency is $q - t$, and if P^t is hyperbolic, the deficiency is $q + 1 - t$.

The direct connections with translation planes arise using the Klein quadric.

51.2.1. The Thas–Walker Construction. The following construction is due independently to Walker [**1203**] and Thas (see related work [**1172, 1173, 1178**]). The form given here is for partial flocks as opposed to the original construction using flocks.

Let P^t be a partial flock of a quadric set in $PG(3,q)$. Let Σ^t denote the set of t planes which contain the t conics of P^t. Embed the projective space $PG(3,q)$ in a 5-dimensional projective space $PG(5,q)$ in such a way so that the quadric set lies in the Klein (hyperbolic quadric) quadric. Let $(\Sigma^t)^*$ denote the set of planes polar to the planes of Σ^t. Then it turns out that the planes of $(\Sigma^t)^*$ share $0,1,2$ if and only if P^t is elliptic, conical or hyperbolic.

There is a bijection between points of the Klein quadric in $PG(5,q)$ and lines of $PG(3,q)$ so that mutually skew lines of $PG(3,q)$ are projected to points of the Klein quadric which are not collinear with a generator line. We note that these mutually skew lines are mutually disjoint (as subspaces) 2-dimensional $GF(q)$-vectors spaces in the corresponding 4-dimensional $GF(q)$-vector space.

The original partial flocks comprise t mutually disjoint conics; the polarity changes this configuration as noted. Hence, in the elliptic case, there are t mutually disjoint reguli in the associated partial spread. In the conical case, there are t reguli, but these share a common component (line) and in the hyperbolic case, there are t reguli that mutually share two components. Hence, we obtain partial spreads of t components (2-dimensional $GF(q)$-subspaces) from partial flocks of quadric sets with t conics with certain configurations of reguli in the spreads.

Conversely, Gevaert, Johnson and Thas [**397**] have essentially shown that partial spreads with such configurations of reguli reproduce at least in the conical case. The other cases are consequences of work of Thas [**1177**], Gevaert, Johnson and Thas [**397**], and Johnson [**736**]. Hence, we have:

51.2.2. Conical, Elliptic, Hyperbolic Partial Spreads.

THEOREM 51.19. (i) A conical partial flock of t conics in $PG(3,q)$ is equivalent to a partial spread of t reguli in $PG(3,q)$, which mutually share a component (line). We call such a partial spread a 'conical partial spread' if it corresponds to a conical partial flock.

(ii) An elliptic partial flock of t conics in $PG(3,q)$ is equivalent to a partial spread of t mutually disjoint reguli in $PG(3,q)$. We use the term 'elliptic partial spread' to denote the corresponding partial spread.

(iii) A hyperbolic partial flock of t conics in $PG(3,q)$ is equivalent to a partial spread of t reguli, which mutually share two components (two lines). In this situation, we call the partial spread 'a hyperbolic partial spread'.

In general, a partial flock of one of the above types is said to be a 'partial spread of quadratic type'.

We note that we may identify any partial spread with the translation net it defines. Hence, a spread defines a translation plane. Notationally, if P is a partial spread then π_P shall denote the associated translation net and if π is a translation net then P^π shall denote the associated partial spread. Note also in this section that all of our translation nets have elementary Abelian translation groups and all partial spreads are 'congruence partial spreads' in the sense that any two components generate the entire vector space.

The work of Gevaert and Johnson [**395**] and Johnson [**737**] establishes group connections with conical, elliptic and hyperbolic partial flocks.

THEOREM 51.20. *Let P^t be a partial spread of quadric type in $PG(3, q)$.*

(i) P^t *is conical if and only if there exists an affine elation group of order q fixing the common component, whose t orbits of length q are reguli.*

(ii) P^t *is elliptic if and only if there exists a group of order $q^2 - 1$ containing the scalar group of order $q - 1$ such that there are t non-trivial orbits consisting of reguli.*

(iii) P^t *is hyperbolic if and only if there exists an affine homology group H of order $q - 1$ fixing the two common components such that the t non-trivial orbits consist of reguli.*

Actually, more can be said of the elliptic case.

THEOREM 51.21. *Let P^t be an elliptic partial flock of t conics. Then there is a Desarguesian affine plane Σ and t identified mutually disjoint regulus nets such that multiple derivation by these t regulus nets correspond to P^t.*

Hence, every elliptic partial flock of t conics may be embedded in a translation plane obtained from a Desarguesian affine plane by the multiple derivation of a set of mutually disjoint regulus nets. Such translation planes are said to be 'subregular' and we consider subregular translation planes in Chapter 24. If $t = q - 1$, it turns out that the associated translation plane is always Desarguesian.

THEOREM 51.22. (Thas [**1172**] and Orr [**1035**]). *A subregular translation plane arising from an elliptic flock is Desarguesian.*

51.3. Deficiency-One Partial Flocks and Baer Groups.

However, we note that it is possible to find translation planes which are not Desarguesian from a partial elliptic flock of $q - 2$ conics in $PG(3, q)$. That is, this would be a partial elliptic flock of deficiency one. In fact, there is such a partial elliptic flock due to Orr [**1034**] in $PG(3, 7)$.

Although partial elliptic flocks correspond to translation planes, it is not always the case that partial flocks of general quadric sets also correspond to set of translation planes. However, deficiency-one partial flocks of quadric sets do correspond. The way this is seen is through the use of Baer groups.

DEFINITION 51.23. *Let π be an affine plane of order q^2. A 'Baer group' is a group of collineations that fix an affine Baer subplane of π pointwise.*

It has been mentioned that partial conical flocks correspond to conical partial spreads which are equipped with an elation group E of order q with $PG(3, q)$ and that the conical partial spread consists of reguli that share a component. The elation group E leaves each regulus invariant. Hence, if one further derives one of these reguli, the corresponding translation plane still admits the collineation group E but now the group fixes a Baer subplane pointwise. The Baer subplane fixed pointwise by E together with each orbit of components of E form a set of $q - 1$ reguli that share a 'component'; we have a conical partial spread of deficiency one. It follows from the work of Johnson [**726**] that one can reverse this idea. That is, suppose there is a translation plane of order q^2 with spread in $PG(3, q)$ which admits a Baer group of order q. Then the Baer group is necessarily structured so that any 2-dimensional subspace which is disjoint from the net containing the Baer axis union with the Baer axis forms a regulus in $PG(3, q)$. Hence, we have a set of $q - 1$ reguli that share the Baer axis. That is, we have a conical partial flock of deficiency one.

THEOREM 51.24. (Johnson [**726**]). Partial conical flocks of deficiency one in $PG(3, q)$ are equivalent to translation planes with spreads in $PG(3, q)$ that admit a Baer group of order q. Furthermore, the partial flock may be extended to a flock if and only if the net of degree $q + 1$ defined by the Baer axis is derivable.

It turns out that Baer groups of order relatively prime to q of translation planes have the property that each Baer axis (pointwise fixed subplane) has a unique companion Baer subplane in the same net of degree $q + 1$ as the axis, which is fixed by the Baer group. Now suppose we have a hyperbolic flock in $PG(3, q)$, which then corresponds to a translation plane with spread in $PG(3, q)$, which consists of $q + 1$ reguli that mutually share two components. Again, there is a connection with a central collineation group. In this case, there is an affine homology group H of order $q - 1$ which leaves each of these $q + 1$ reguli invariant. If we form the translation plane by derivation of one of these reguli, we obtain a translation plane with spread in $PG(3, q)$ with the property that H fixes two Baer subplanes, one pointwise. Now we reverse this concept. Suppose that there is a translation plane with spread in $PG(3, q)$, which admits a Baer group H of order $q - 1$. Now we have two Baer subplanes in the same net of degree $q + 1$ fixed by H. There are q other orbits of length $q - 1$ and the nature of the Baer group forces these orbits together with the two Baer subplanes to form reguli. Hence, we have q reguli that mutually share the two Baer subplanes (as lines).

THEOREM 51.25. (Johnson [**726**]). Partial hyperbolic flocks of deficiency one in $PG(3, q)$ are equivalent to translation planes with spreads in $PG(3, q)$ that admit a Baer group of order $q - 1$. Furthermore, a partial flock of deficiency one may be extended to a flock if and only if the net of degree $q + 1$ defined by the Baer axis is a regulus net.

There are many more connections and results regarding conical flocks, which we consider in other chapters (see Chapters 54 through 61). We end this chapter with the classification of flocks of hyperbolic quadrics. Note that although there is such a classification of flocks, there is no such classification of deficiency-one hyperbolic flocks, since there are non-trivial examples, which we shall list subsequently.

DEFINITION 51.26. An affine translation of odd order is said to be a 'Bol' plane if and only if there are two components L and M such that given any other component N, there is an affine collineation of order 2 with axis N that interchanges L and M.

REMARK 51.27. We note that any nearfield plane with spread $x = 0, y = 0, y = xM$, where $x = 0$ and $y = 0$ are axes of affine homologies admits a collineation group $(x, y) \longrightarrow (y, x)$. This collineation has order 2 and fixes $y = x$ pointwise (which we assume is a component) and also leaves invariant $y = -x$ (which is $y = x$ if the characteristic is two).

LEMMA 51.28. If π is a finite Bol translation plane with components L and M that are interchanged by the set of central involutions, then the group generated by the involutions has an index two subgroup which acts transitive on the non-zero vectors of L.

We shall discuss more fully the finite Bol planes in Chapter 72, where we point out that all finite Bol translation planes are actually nearfield planes.

51.4. Tubes in $PG(3, q)$.

DEFINITION 51.29. A 'partial tube' in $PG(3, q)$ is a pair $\mathcal{T} = \{L, \mathcal{L}\}$, where $\{L\} \cup \mathcal{L}$ is a partial spread of lines in $PG(3, q)$ such that any plane π containing L intersects $\mathcal{L}$ in an arc. If $|\mathcal{L}| = q + 1$, the partial tube is said to be an 'oval tube'. If $|\mathcal{L}| = q + 2$, when q is even then the arc is a hyperoval and the partial tube is called a 'tube' in this case (when q is odd 'tubes' are also partial tubes with associated complete arcs).

DEFINITION 51.30. An oval tube in the even-order case may not produce a hyperoval tube. However, if $\mathcal{L}$ is a set of $q + 1$ lines of a regulus and L is a line exterior to $\mathcal{L}$, then $\{L, \mathcal{L}\}$ is an oval tube, since the regulus is defined by a hyperbolic quadric. In this case, we call the oval tube a 'quadratic tube'. Furthermore, for quadratic tubes, we may adjoin $L^\perp$ to $\mathcal{L}$, where $\perp$ is the polarity associated with the hyperbolic quadric, to form a hyperoval tube.

Tubes are studied by Cameron and Ghinelli [211] and Cameron and Knarr [212]. Let $\mathcal{R}(L, M, N)$ denote the unique regulus in $PG(3, q)$ generated by three mutually disjoint lines L, M and N. The following lemma is due to Cameron and Knarr.

LEMMA 51.31. (Cameron and Knarr [212, (2.1)]). If $\mathcal{T} = \{L, \mathcal{L}\}$ is a partial tube, then

$$\mathcal{P}_k = \bigcup_{i \neq k} \mathcal{R}(L, M_i, M_k); \mathcal{L} = \{M_1, \dots, M_m\}, \ k = 1, 2, \dots, m,$$

is a partial spread of order q^2 and degree $2 + (m - 1)(q - 1)$.

When one has an even-order tube, then $\mathcal{P}_k$ has degree $2 + (q + 1)(q - 1)$ so is a spread which is the union of hyperbolic reguli sharing two components. Hence, we have an associated flock of a hyperbolic quadric, which means that associated spreads correspond to Desarguesian planes.

In a finite Desarguesian spread S in $PG(3, q)$ of odd order, let L be a line, take an oval Ω (hyperoval) of a plane π in $\pi - \{L\}$ of $PG(3, q)$ containing L and let $\mathcal{L}$ denote the set of all lines of S intersecting Ω in a point. Then $\{L, \mathcal{L}\}$ is an oval tube (tube, respectively). Cameron and Knarr [212] show that all oval tubes of odd order may be constructed in this manner. We call a tube or oval tube so constructed a 'regular tube' or 'regular oval tube'.

THEOREM 51.32. (Cameron and Knarr [212]). Tubes (i.e., hyperoval tubes) for q even and oval tubes for q odd are regular.

The proof for the even and odd orders are quite different and are sketched below. Actually, as mentioned, we could start with oval tubes for even order to produce a hyperoval tube or 'tube' in the even-order case. In both cases, we choose L as $x = 0 = (x_1, x_2)$, where we regard $PG(3, q)$ as a 4-dimensional vector space V_4 with points (x_1, x_2, y_1, y_2). When q is even, in this case, the above partial spread is a spread corresponding to a hyperbolic flock of a quadratic cone in $PG(3, q)$ and hence is Desarguesian (since it has even order). Hence, tubes in the even order case are regular.

When q is odd, it is possible to transform the tube into a set of matrices $\mathcal{L}_M = \{M_i; i = 1, 2, \ldots, q + 1\}$, with the property that $-M_i \in \mathcal{L}_M$, whenever $M_i \in L_M$. Then the following shows that

$$x = 0, y = x(\alpha M_i + \beta M_k); \alpha, \beta \in GF(q), \, k \neq \pm i$$

are Desarguesian spreads $\Sigma_{i,k}$. Note that any such spread shares $y = x(\pm M_i)$ and $y = x(\pm M_k)$ of $\mathcal{L}$. Since a conic is uniquely determined by five points, it follows that if any other $y = xM_j$ lies within $\Sigma_{i,k}$ then the oval tube is regular. Cameron and Knarr [212] use Pascal's theorem on conics (every oval in a Desarguesian plane is a conic in the odd-order case) to show that all other $y = xM_j$ lie in any $\Sigma_{i,k}$, thus proving that every oval tube of odd order is regular.

Considering that for q odd, oval tubes are sets of $q + 1$ components in a Desarguesian affine plane with spread in $PG(3, q)$, Cameron and Knarr [212] determine the possibilities. Let q be odd and let the Desarguesian spread be $x = 0, y = xm; m \in GF(q^2)$. Then every oval tube (relative to $x = 0$) is projectively equivalent to a set $\mathcal{L}_b = \{y = x(w + bw^q); b \in GF(q^2); w^{q+1} = 1\}$. When $b = 0$, we have a quadratic tube.

Below, we discuss partial flocks of hyperbolic quadrics of deficiency t. When $t = 1$, there are associated spreads in $PG(3, q)$ and this concept arises in the concept of an oval tube.

Although we have seen that oval tubes of odd order are regular, it is not clear that oval tubes of even order are, in fact, always regular; however, if the oval is a conic, the proof of Cameron and Knarr [212] may possibly apply. Note also that for oval tubes of orders even or odd,

$$\mathcal{P}_k = \bigcup_{i \neq k} \mathcal{R}(L, M_i, M_k); \mathcal{L} = \{M_1, \ldots, M_{q+1}\}, \, k = 1, 2, \ldots, q+1,$$

may be extended to a spread π_k by the work of Johnson [726] and corresponds to a partial flock of a hyperbolic quadric of deficiency one. Hence there are $q + 1$ spreads in $PG(3, q)$ that correspond to an oval tube.

PROBLEM 51.33. Show that oval tubes for q even are regular. The translation planes corresponding to π_k admit a Baer group of order $q - 1$ fixing L pointwise and leaving L_k invariant, where L and L_k are Baer subplanes of the translation plane. The partial flock of the hyperbolic quadric is a flock precisely when the net defined by L or L_k is a regulus net.

Suppose on the contrary that there is an oval tube which is not regular. Then none of the $q + 1$ translation planes π_k can be derivable by a regulus. Hence, we would obtain $q + 1$ non-extendable partial flocks of deficiency one of a hyperbolic quadric.

51.5. Sharply Transitive Sets.

In Chapter 8, we have seen connections to sharply transitive and sharply doubly transitive sets of permutations with translation planes. Consider now a set of permutations S which acts sharply three transitive on a finite set X. Call the elements of $X \times X$ 'points' and the 'circles' the elements of S. This structure then becomes a Minkowski plane, if we define plus and minus parallelism by (a, b) and (c, d) are plus parallel if and only if $a = c$ and minus parallel if and only if $b = d$. Furthermore, it turns out that:

REMARK 51.34. Finite Minkowski planes are equivalent to sharply three transitive subsets acting on finite sets.

The hyperbolic quadric as a Minkowski plane has the group $PGL(2, q)$ acting sharply three transitive on $PG(1, q) \times PG(1, q)$. In this sense, a flock of a hyperbolic quadric becomes a sharply transitive subset of the group. We note that sharply transitive subsets of $PGL(2, q)$ then produce flocks of hyperbolic quadrics and hence associated translation planes.

51.5.1. Flocks of the Minkowski Planes $M(q, p^t)$. The remaining finite Minkowski planes have the following construction: consider the following set S_σ of sharply 3-transitive permutations on $PG(1, q)$, for q odd: $x \longmapsto (x^\sigma a + b)/(x^\sigma c + d); \sigma = 1$ if $ab - bc$ is square, σ is fixed: $x \longmapsto x^{p^t}$ if $ab - cd$ is non-square.

We denote the associated Minkowski plane by $M(q, p^t)$. A 'flock' now is simply a sharply 1-transitive subset of S_σ, which would consist of $(q + 1)/2$ elements.

Bonisoli [**155**] has shown that if E is a cyclic semi-regular subgroup of $PSL(2, q)$ of order $(q + 1)/2$, then for particular values of q, there exists an element $g \in P\Gamma L(2, q)$, with the property that $E \cup Eg$ is a sharply 1-transitive subset of S_σ. The significance for us here in this text is the result of Knarr [**895**] extending flocks of hyperbolic quadrics.

THEOREM 51.35. (Knarr [**895**]). Given a sharply 1-transitive subset of $P\Gamma L(2, q)$, there is a corresponding translation plane of order q^2 whose spread is the union of $q + 1$ derivable nets that share two components $x = 0, y = 0$.

It is pointed out that the spread constructed above from a sharply 1-transitive set is not necessarily in $PG(3, q)$, and in fact, is only within $PG(3, q)$ if g is in $PGL(2, q)$.

Recalling that a translation plane of order q^2 is equivalent to a sharply 1-transitive subset of matrices in $GL(2n, p)$, where $p^n = q$, we need only create a sharply 1-transitive subset in $\Gamma L(2, q)$ from a sharply 1-transitive subset S of $P\Gamma L(2, q)$. This is accomplished as follows: If $x \longmapsto x^{\sigma_M} M$ is an element of S then $x \longmapsto (ux)^{\sigma_M} M; u \in GF(q)^*$ is in $\Gamma L(2, q)$, implying that a set of $(q + 1)$ elements of S produces a set of $(q + 1)(q - 1)$ of $S^+ \subset \Gamma L(2, q)$, and, it turns out that S^+ is sharply transitive. Hence, we obtain a translation plane of order q^2. Furthermore, there is an associate group of affine homologies generated by $(x, y) \longmapsto (xu, y); u \in GF(q)^*$ with axis $x = 0$ and coaxis $y = 0$ of order $q - 1$. Any orbit of components of this group union the axis and coaxis becomes a derivable net. Actually, there are two homology groups, the other generated by elements of the form $(x, y) \longmapsto (x, yu); u \in GF(q)^*$.

The structure of Bonisoli's flocks involves the group E of order $(q + 1)/2$. This group, coupled with the affine homology groups of order $q - 1$ adjoined form affine homology groups of order $(q^2 - 1)/2$. Recall that a nearfield plane is one which has affine homology groups of order $(q^2 - 1)$.

Hence, we have:

THEOREM 51.36. Bonisoli's flocks of the Minkowski planes $M(q, p^s)$ produce translation planes of order q^2 that admit $q + 1$ derivable nets sharing two components. Furthermore, there are symmetric affine homology groups of orders $(q^2 - 1)/2$ with axis and coaxis the two shared components.

Is was initially thought that the planes of Bonisoli were not planes known at that time. In fact, they turn out to be André planes.

51.5.1.1. *The Bonisoli Planes.*

THEOREM 51.37. (Johnson [738]). The Bonisoli planes are André planes.

Now there are two natural ways to generalize this discussion:
(1) Determine the translation planes of order n that admit two symmetric affine homology groups of order $(n - 1)/2$.
(2) Generalize the ideas of sharply 1-transitive subsets of $P\Gamma L(2, q)$ to sharply 1-transitive subsets of $PGL(2n, p)$.
The first problem was solved by Hiramine and Johnson [521].

51.5.2. Two Homology Groups of Index Two.

THEOREM 51.38. (Hiramine and Johnson [521]). Let π be a translation plane of order $q^n = p^{nr}$, for p a prime, that admits two homology groups of order $(q^n - 1)/2$. Then one of the following occurs:
(1) π is Desarguesian,
(2) $q^n = 7^2$ and the plane is the irregular nearfield plane,
(3) $q^n = 7^2$ and the plane is the exceptional Lüneburg plane,
(4) $q^n = 23^2$ and the plane is the irregular nearfield plane,
(5) π is a generalized André plane and the associated right and middle nuclei of the (sub)homology groups are equal.

The second area of research is developed in Johnson [738, 739].

51.5.3. Baer Groups of Order $q - 1$ in Planes of Order q^2. But, there
is another, less obvious, way to consider this material. Recall that Baer groups of order $q - 1$ acting on translation planes of order q^2 with spreads in $PG(3, q)$ turn out to define partial flocks of hyperbolic quadrics of deficiency one. Actually, there are two Baer groups of order $q - 1$, whose orbits on the line at infinity are the same. So, the question would be can there be such a theory of Baer groups equivalent to partial flocks of Minkowski planes of deficiency one? Actually, one can simply bypass the Minkowski plane and ask the same question of partial sharply 1-transitive subsets of cardinality q, instead of $q + 1$. Furthermore, there are a variety of generalizations of such ideas in Johnson [739] of which we mention exactly one result. When a Baer group of order relatively prime to the order of the translation plane acts, it fixes two Baer subplanes that share the same parallel classes, the fixed pointspace, which we call the 'Baer axis' and another uniquely defined Baer subplane, which is called the 'Baer coaxis'.

THEOREM 51.39. (Johnson [739, (3.6)]) There is a 1–1 correspondence between partial sharp subsets of $P\Gamma L(2, q)$ of deficiency one and translation planes of order q^2 that admit two distinct Baer groups of order $q-1$ which have the same component orbits. If B is one of the Baer groups of order $q-1$ then any component orbit union the subspaces Baer axis and Baer coaxis of B is a derivable partial spread. The net containing the Baer axis is derivable if and only if the partial sharp subset of $P\Gamma L(2, q)$ can be extended to a sharply transitive set.

CHAPTER 52

Maximal Partial Spreads.

There is, of course, another related area of interest, partial and maximal partial spreads, of which there is a voluminous amount of constructions and theory. For example, the text by Beth, Jungnickel and Lenz [111] gives a nice presentation of many of the maximal partial spreads. We do not attempt any sort of treatment of this area other than to point out some of the nice consequences of ideas of Bruen [190]. Take any spread in $PG(3,q)$, form the associated translation plane and choose any 2-dimensional $GF(q)$-subspace π_o which is not a component (line). Then the net of degree $q + 1$ and order q^2 defined by the subspace π_o may or may not be a regulus net (corresponds to a regulus in $PG(3,q)$). If it is not a regulus, then it turns out that it is either 1 or 2 Baer subplanes of this net that can arise from 2-dimensional $GF(q)$-subspaces. Define a new translation net by taking the Baer subplanes of a non-regulus net constructed in the above manner together with the components of the spread not in the partial spread defining the net. This will be forced to be a maximal partial spread of either $q^2 - q + 1$ or $q^2 - q + 2$ subspaces. The way to see this is to note that any other 2-dimensional subspace will be forced to define a Baer subplane of the net in question. However, this will then force the net to be a regulus net. Jungnickel [815] calls such maximal partial spreads 'of small deficiency' and there is a corresponding theory of such structures.

THEOREM 52.1. (Bruen [190]). Given any non-Desarguesian spread in $PG(3,q)$. Then there is a corresponding maximal partial spread of degree $q^2 - q + i$, where $i = 1$ or 2.

We point out that there are maximal partial flocks of hyperbolic quadrics of deficiency one, which cannot be extended to a flock (where the theorem of Thas [1179] and Bader–Lunardon [49]) would classify them. However, each of these defines a translation plane of order q^2 with spread in $PG(3,q)$ with the property that there is a Baer group of order $q-1$ fixing two Baer subplanes of a net of degree $q + 1$ and order q^2. Choosing these two Baer subplanes, together with the $q^2 - q$ components exterior to this net, forms an example of the type of maximal partial spread obtainable from the ideas of Bruen.

THEOREM 52.2. (See Johnson [726]). Any non-extendible partial flock in $PG(3,q)$ of a hyperbolic quadric of deficiency one produces a maximal partial spread in $PG(3,q)$ of degree $q^2 - q + 2$.

We discuss Desarguesian parallelisms in this text and the translation planes with spread in $PG(7,q)$ associated. If there is a parallelism in $PG(3,q)$ with all but one of its spreads Desarguesian, then there is still an associated translation plane. For Desarguesian parallelisms, the translation planes of order q^4 admit $SL(2,q)$ as a collineation group where the p-elements for $q = p^r$, p a prime, are elations. For partial Desarguesian partial parallelisms of deficiency one in $PG(3,q)$, the associated translation plane of order q^4 also admits $SL(2,q)$ as a collineation group but where the p-elements fix Baer subplanes pointwise. If N is the net of degree $q^2 + 1$ defined by the Baer axes, there are at least $q + 1$ Baer subplanes sharing the zero vector, which are $GF(q)$-subspaces. If there is another such Baer subplane, then the net is a derivable net and the derived plane corresponds to a Desarguesian parallelism.

It is an open question whether Desarguesian partial parallelisms of deficiency one exist. If they do not, there are associated maximal partial spreads.

THEOREM 52.3. (Johnson [753]). Let $\mathcal{P}$ be a Desarguesian partial parallelism of deficiency one in $PG(3,q)$, which cannot be extended to a Desarguesian partial parallelism.

(1) Then there is an associated translation plane of order q^4 admitting $SL(2,q)$ as a collineation group, where the $q + 1$ Sylow p-subgroups, for $p^r = q$ and p a prime, are Baer groups.

(2) Define a point-line geometry by taking the points to be the vectors and the spread elements to be the $q+1$ Baer subplanes together with the $q^4 - q^2$ components of the associated translation plane. This forces a maximal partial spread of degree $q^4 - q^2 + q + 1$.

A related structure of interest are the so-called 'transversal-free' translation nets, which are defined by maximal partial spreads that cannot even be extended by one 'line', i.e., 'transversal.' It turns out that the maximal partial spreads constructed in the previous theorem are also transversal-free, but more can be said of any spread in $PG(3,q)$, which is not Desarguesian.

52.1. Ubiquity of Maximal Partial Spreads.

THEOREM 52.4. (Jha and Johnson [630]). Every non-Desarguesian spread in $PG(3,q)$ produces a maximal partial spread of degree $q^2 - q + i$, for $i = 1$ or 2, which remains maximal when considered a partial spread in $PG(4s - 1, p)$, where $q = p^r$. Furthermore, this maximal partial spread is transversal-free.

CHAPTER 53

Sperner Spaces.

There is a more or less classical geometry called a Sperner space, the theory of which essentially parallels the theory of translation planes. We have discussed spreads for translation planes. In this setting, we have a $2n$ dimension vector space over $GF(q)$ and a set of mutually disjoint (as subspaces) n-dimensional subspaces forming a partition of the non-zero vectors. If we relax the condition that the partition of subspaces be n-dimensional, and define the resulting geometry by taking points as vectors and lines as translates of the 'spread' subspaces, we would obtain a Sperner space.

DEFINITION 53.1. A 'congruence spread' is a partition of a $2n$-dimensional vector space by mutually disjoint n-dimensional subspaces. A 'spread' or 'general spread' of a k-dimensional vector space is a partition of the vector spaces by mutually disjoint m-dimensional subspaces (hence, m divides k). A 'generalized spread' is a partition of a vector space by mutually disjoint subspaces (of perhaps different dimensions).

DEFINITION 53.2. A 'translation Sperner spread' is a spread of a finite-dimensional vector space. A 'translation Sperner space' is a geometry obtained from a Sperner spread by taking the points as vectors and the lines as translates of the general spread components.

A general Sperner space may be defined as follows:

DEFINITION 53.3. A geometry of points and lines satisfying the axioms of an affine plane without the assumption that two lines of different parallel classes must intersect is called a 'Sperner space'.

For example, since general spreads are fairly easy to come by, we may construct a variety of translation Sperner spaces. Furthermore, infinite Sperner spaces are easy to construct from spreads that are not dual spreads.

53.1. Sperner Spaces by Derivation.

DEFINITION 53.4. Let V be a vector space of dimension $2n$ over a field K. A covering of the hyperplanes of $PG(V, K)$ by a set of mutually disjoint n-dimensional K-subspaces is called a 'hyperplane spread'. Given a congruence spread S in the projective space $PG(V, K)$, (the lattice of subspaces of an associated vector space V over a field K). Form the natural dual of V, $\widehat{V}$. Then in $PG(\widehat{V}, K)$ there is an associated hyperplane spread formed by taking the images of the spread components under the natural polarity. If the hyperplane spread is a spread, it is called the 'spread dual to S.' Furthermore, a spread is said be a 'dual spread' if the associated hyperplane spread is also a spread.

It is an easy counting argument to show that finite spreads are always dual spreads. However, their duals may not be isomorphic to the original spread. However, in the infinite case, for every field K, there are spreads in $PG(3, K)$ that are not dual spreads. Take a spread in $PG(3, K)$ which is not a dual spread. We have seen that spreads in $PG(3, q)$ construct translation planes and hence dual translation planes that are derivable. This is still true here but what actually occurs is that there is an embedded derivable 'net' in the associated dual translation plane which is actually not derivable in the entire plane, in the sense the subplanes of the corresponding net are not actually Baer subplanes of the planes, merely Baer subplanes of the net. However, one can still derive this structure to obtain a Sperner space.

THEOREM 53.5. (Johnson [**749**], and Cameron [**210**] for the existence of spreads that are not dual spreads). For every skewfield K, there are spreads in $PG(3, K)$ that are not dual spreads. Furthermore, there are derivable nets in each associated dual translation plane. There exist dual translation planes admitting derivable nets, chosen so that the translation plane itself is not derivable. The corresponding derived structure is a Sperner space.

The reader is referred to Johnson's text [**753**] for more details about derivable nets and the corresponding Sperner spaces.

CHAPTER 54

Conical Flocks.

In this chapter and the following, where flocks of quadric sets are discussed, there were essentially no flocks or corresponding translation planes listed or mentioned. In this chapter, we give a description of the known infinite classes of flocks of quadratic cones. Actually, it is possible to consider infinite flocks of quadratic cones, the variety of examples is extremely large. To illustrate, let K be an arbitrary field that admits a quadratic extension K^+. In $PG(3, K)$, choose a plane containing a conic and from a point v_0 exterior to the plane, form the natural cone. Just as in the finite case, a 'flock' is a partition of the points of the cone other than v_0 by plane intersections.

From the standpoint of translation planes, we are interested in methods that produce examples, classification results and interconnecting theories.

A flock of a quadratic cone in $PG(3, q)$ is equivalent to a translation spread with spread in $PG(3, q)$, covered by a set of q reguli that share a line. Considering the associated translation planes, there are then q possible ways to derive the plane to produce a possibly new plane. Hence, when we consider flocks of quadratic cones and their associated translation planes, we can also consider a variety of translation planes obtained by derivation.

We provide here some of the many connections with flocks of quadratic cones and other interesting geometric structures.

54.1. Spread Sets.

In Chapter 5, we have shown that any translation plane can be given by a spread of an associated vector space V over a skewfield K. Furthermore, we have noted that when V is of dimension two over K (K-dimension one), there is, up to isomorphism, exactly one translation plane, the Desarguesian plane over K. Later in the text, we shall produce a variety of spreads and translation planes with various dimensions over K. Here we begin with a particular type of spreads associated with flocks of quadratic cones, since they provide an enormous variety of interesting examples of translation planes with spreads in $PG(3, q)$.

Many of the important examples of translation planes have the property that V is of dimension 4 over K or K-dimension two, so we pause in our general theory to give an introduction to spreads in $PG(3, K)$.

As the following 'matrix spread set' construction is basic to many of our constructions, we give the following in some detail and include a series of exercises. For simplicity, we restrict ourselves to the dimension-4 vector space situation, although we repeat some of these ideas more generally later and the reader might note that such restrictions are not required.

409

THEOREM 54.1. Let $\mathcal{S}$ be a spread in $PG(3, K)$, where K is a field and let V denote the associated 4-dimensional K-vector space such that the components of $\mathcal{S}$ are 2-dimensional K-subspaces. Choose two components say L and M. Let $\mathcal{B}_L$ and $\mathcal{B}_M$ be bases for L and M, respectively, so that $\{\mathcal{B}_L, \mathcal{B}_M\} = \mathcal{B}$ is a basis for V and represent $V = L \oplus M$.

Since L and M are K-isomorphic, identify L and M and rewrite $V = L \oplus L$ where then $L \oplus 0$ is L and $0 \oplus L$ represents M. Let $\mathcal{B}_L = \{e_1, e_2\}$ and $\mathcal{B}_M = \{f_1, f_2\}$.

(1) Let $x = x_1 e_1 + x_2 e_2$ and $y = y_1 f_1 + y_2 f_2$ and write elements of V over $\mathcal{B}$ as $(x, y) = (x_1, x_2, y_1, y_2)$. In this context, the component M has the 'equation' $x = 0$ and the component L has the 'equation' $y = 0$.

Then any other component N distinct from L and M has an equation of the form $y = xT$ where T is a non-singular 2×2 matrix with entries from K.

(2) Given any three distinct components L, M, and N of the spread $\mathcal{S}$, there is a basis for V such that 'points' have the form (x, y) and L, M and N have the equations $y = 0, x = 0, y = x$, respectively. Furthermore, any other component of $\mathcal{S}$ has the following general form:

$$y = x \begin{pmatrix} g(u,t) & f(u,t) \\ t & u \end{pmatrix} \forall t, u \in K \text{ for } (t, u) \neq (0, 0)$$

and where g and f are functions: $K \times K \longmapsto K$ and where the associated matrices are non-singular and $g(1, 0) = 1$, $f(1, 0) = 0$.

We now assume that K is finite and establish one of the most important techniques for the construction of translation planes.

THEOREM 54.2. Let $\mathcal{M}$ be any set of $q^2 - 1$ non-singular 2×2 matrices with elements from a field $K \simeq GF(q)$ such that $\mathcal{M}$ contains the identity matrix.

If the difference of any two distinct matrices of $\mathcal{M}$ is non-singular, then

$$x = 0, y = 0, y = xM \ \forall M \in \mathcal{M}$$

defines a spread in $PG(3, q)$.

In this setting, $\mathcal{M}$ is said to be a 'matrix spread set'.

Notice how this theorem opens up the theory to questions about functions on finite fields.

Since we do not yet have any examples, suppose we simplify matters and consider spreads of the following form:

DEFINITION 54.3. Any spread in $PG(3, K)$, for K finite or infinite, which has a 'matrix spread set' as in the above theorem with defining functions g and f such that

$$g(u,t) = u + g(t) \text{ and } f(u,t) = f(t)$$

for all $u, t \in K$ shall be called a 'conical spread'.

There are connections between covers of quadratic cones in $PG(3, K)$ by plane intersections (conics) and conical spreads whereby these spreads receive their designation. The interested reader is referred to Johnson [**753**] for more details and background.

There are a large variety of such conical spreads and we shall be content here to study only a few that are easily constructed.

EXAMPLE 54.4. Let $K \simeq GF(q)$, q odd, where we choose $g(t) = 0$ and $f(t) = \gamma t^\sigma$ where γ is a non-square in K and σ is an automorphism of K. That is, the spread is given by:

$$x = 0, y = x \begin{pmatrix} u & \gamma t^\sigma \\ t & u \end{pmatrix} \forall t, u \in K.$$

Hence, by the above exercise, we obtain a class of spreads called the set of 'Kantor Knuth conical spreads'.

54.1.1. The Regulus Partial Spread. In various spreads in $PG(3, K)$, for K a field, we notice that there is a subset of the spread (called a 'partial spread' and discussed in Chapter 5) with the following form:

$$x = 0, y = x \begin{pmatrix} u & 0 \\ 0 & u \end{pmatrix} \forall u \in K.$$

Consider the following sets:

$$\pi_{a,b} = \{(a\alpha, b\alpha, a\beta, b\beta); \alpha, \beta \in K\}$$

where a and b are fixed elements of K not both of which are zero. We note that $\pi_{1,0}$ and $\pi_{0,1}$ are Desarguesian planes. Then it is not difficult to show that all such substructures $\pi_{a,b}$ are Desarguesian planes. When $K \simeq GF(q)$, show that there are exactly $q + 1$ distinct such 'subplanes'.

When we consider the set projectively, it is a set of $q + 1$ lines in $PG(3, q)$. But, the 'subplanes' are also 2-dimensional K-subspaces so also become 'lines' in $PG(3, q)$ that form a cover of the original set.

DEFINITION 54.5. A set of $q + 1$ lines of $PG(3, q)$ that is covered pointwise by a second set of $q + 1$ lines is said to be a 'regulus'. The covering set is called the 'opposite regulus' and it is also a regulus that is covered pointwise by the original set.

It is straightforward to show that the form we have chosen to represent a regulus is 'canonical'. That is, any regulus in $PG(3, q)$ can be brought into the form:

$$x = 0, y = x \begin{pmatrix} u & 0 \\ 0 & u \end{pmatrix} \forall u \in K \simeq GF(q).$$

We shall refer to this as the 'standard form' of a regulus.

The following theorem shows the importance of the existence of a regulus in a spread in $PG(3, q)$.

54.1.2. Derivation of Spreads in $PG(3, q)$. This material also appears in the foundations part (Chapters 2–7, see also [**123**]), but we repeat this here to illustrate the technique.

THEOREM 54.6. Let S be a spread represented in the form:

$$x = 0, y = x \begin{pmatrix} g(u,t) & f(u,t) \\ t & u \end{pmatrix} \forall t, u \in K.$$

Assume that S contains a regulus $\mathcal{R}$ in standard form

$$x = 0, y = x \begin{pmatrix} u & 0 \\ 0 & u \end{pmatrix} \forall u \in K \simeq GF(q).$$

Hence, $g(u, 0) = u$ and $f(u, 0) = 0$ for all $u \in K$.

Then,

$$x = 0, y = x \begin{pmatrix} u & 0 \\ 0 & u \end{pmatrix} \forall u \in K \simeq GF(q)$$

together with

$$y = x \begin{pmatrix} -g(u,t)t^{-1} & f(u,t) - g(u,t)t^{-1}u \\ t^{-1} & ut^{-1} \end{pmatrix} \forall t \neq 0, u \in K$$

is also a spread in $PG(3,q)$.

This spread is called the 'derived spread' of the original. (See also the foundations part, Chapters 2–7 and also [123], for material on derivation.)

REMARK 54.7. In general a derivable net within a spread in $PG(3,q)$ may be given the following representation

$$x = 0, y = x \begin{pmatrix} u & 0 \\ 0 & u^\sigma \end{pmatrix} \forall u \in K \simeq GF(q), \sigma \in \mathrm{Aut}\, GF(q).$$

When this occurs and σ is not 1, then the derived translation plane does not have its spread in $PG(3,q)$, so this is a simple way of creating large dimensional spreads.

So, from the Knuth spreads, we obtain:

COROLLARY 54.8. The following represents a spread in $PG(3,q)$:

$$x = 0, y = x \begin{pmatrix} v & 0 \\ 0 & v \end{pmatrix}, y = x \begin{pmatrix} -ut^{-1} & \gamma t^\sigma - u^2 t^{-1} \\ t^{-1} & ut^{-1} \end{pmatrix}$$

$\forall t \neq 0, v, u \in GF(q)$ for q odd, γ a non-square and $\sigma \in \mathrm{Gal}\, GF(q)$.

54.2. s-Inversion and s-Square.

We noted above that the conical spreads contain a regulus partial spread and that there is then a so-called 'derived spread' whenever a spread contains such a partial spread.

In the special case when the spread is a conical spread, there is another construction technique that might be of interest to us. These constructions originated not in the theory of translation planes but in connections to other geometries. The interested reader might look at the survey article by Johnson and Payne [792] for further information on from where such constructions can be traced.

DEFINITION 54.9. Let $\mathcal{S}$ be any conical spread represented as follows:

$$x = 0, y = x \begin{pmatrix} u + g(t) & f(t) \\ t & u \end{pmatrix} \forall t, u \in K \simeq GF(q).$$

Assume that q is odd.

For a fixed $s \in K$, we define the 's-inversion' $\mathcal{S}^{-s}$, a set of 2-dimensional subspaces, as follows:

$\mathcal{S}^{-s}$: $x = 0$ and

$$y = x \left(\begin{pmatrix} -(g(t) - g(s))/2 & f(t) - f(s) \\ t - s & (g(t) - g(s))/2 \end{pmatrix}^{-1} + \begin{pmatrix} u & 0 \\ 0 & u \end{pmatrix} \right)$$

for all $t, u \in GF(q)$.

DEFINITION 54.10. We now consider a similar situation when q is even. Let $\mathcal{S}$ be a conical spread with defining spread

$$x = 0, y = x \begin{pmatrix} u + g(t) & f(t) \\ t & u \end{pmatrix} \forall u, t \in K \simeq GF(q).$$

Then, for each $s \in K$, we define the structure $\mathcal{S}_2^s$, called the 's-square'

$$x = 0, y = x \left(1/(g(t) + g(s))^2 \begin{pmatrix} g(t) + g(s) & f(t) + f(s) \\ t + s & 0 \end{pmatrix} + u I_2 \right),$$

$$y = x \begin{pmatrix} v & 0 \\ 0 & v \end{pmatrix}$$

$\forall t, t \neq s, \forall u, v \in K.$

Again, the origin of this structure comes not from the theory of translation planes necessarily, but from related geometries called 'generalized quadrangles'. These will not concern us in this text except to note that there are many connections with the types of structures that we are encountering and other types of geometries and we will provide the connections in the part in this text on flocks.

THEOREM 54.11. If q is even then an s-square conical spread is a conical spread in $PG(3, q)$.

Before we take up the odd-order situation, we take up some examples in the even-order case.

Consider the following 2-dimensional K-subspaces:

$$x = 0, y = x \begin{pmatrix} u + t^2 & t^3 \\ t & u \end{pmatrix} \forall u, t \in K \simeq GF(2^{2r+1}).$$

It is not difficult to show that the set is a conical spread in $PG(3, K)$. We simply note that $\text{trace}((u/t^2)^2 + (u/t^2) + 1) = 1$.

Thus, we obtain several other conical spreads, the s-square conical spreads and, of course, each of these may be derived:

We now turn back to the odd-order situation.

THEOREM 54.12. If $\mathcal{S}$ is a conical spread in $PG(3, q)$ and q is odd, then an s-inverted conical spread is a spread in $PG(3, q)$.

We note that it is a simple matter to now determine the matrix spread sets for the s-inverted conical spreads arising from a Knuth conical spread mentioned above. However, it is also direct to determine the form for the derivation of any s-inverted conical spread in general and then use this to determine the form for the derivation of any s-inverted Knuth conical spread.

Hence, the translation planes of order q^2 with spread in $PG(3, q)$ that we shall be considering admit an affine elation group E of order q, one of whose component orbits (and hence all component orbits) is a regulus in $PG(3, q)$. In the 4-dimensional vector space V_4 over $GF(q)$, represent components by the 4-tuple, (x_1, x_2, y_1, y_2); $x_i, y_i \in GF(q)$, where a basis has been chosen so as to represent the axis of E as $x = 0 = (x_1, x_2) = (0, 0)$. In this case, we represent the vectors in the form (x, y), for $y = (y_1, y_2)$. We may also represent a given regulus that has $x = 0, y = 0, y = x$ as lines (2-dimensional $GF(q)$-subspaces) in the following form:

$$x = 0, y = x \begin{bmatrix} u & 0 \\ 0 & u \end{bmatrix}; u \in GF(q).$$

We call this the 'standard regulus'. Then an elation group E with axis $x = 0$, which fixes the standard regulus and which acts transitive on the remaining q components not equal to $x = 0$, will have the following form:

$$E = \left\langle \begin{bmatrix} 1 & 0 & u & 0 \\ 0 & 1 & 0 & u \\ 0 & 0 & 1 & 0 \\ 0 & 0 & 0 & 1 \end{bmatrix} ; u \in GF(q) \right\rangle.$$

In general, a spread for a translation plane with spread in $PG(3, q)$ has the following form:

$$x = 0, y = x \begin{bmatrix} g(t, u) & f(t, u) \\ t & u \end{bmatrix} ; t, u \in GF(q),$$

where g and f are functions from $GF(q) \times GF(q)$ to $GF(q)$. Therefore, in this situation, $g(0, u) = u$ and $f(0, u) = 0$. Let $g(t, 0) = g(t)$ and $f(t, 0) = f(t)$. Hence, we have components in the form

$$y = x \begin{bmatrix} g(t) & f(t) \\ t & 0 \end{bmatrix} ; t \in GF(q).$$

If we do not apply E to this set of components, we will fill out the spread of $q^2 + 1$ components as:

DEFINITION 54.13. Consider a spread of the form

$$x = 0, y = x \begin{bmatrix} u + g(t) & f(t) \\ t & u \end{bmatrix} ; u, t \in GF(q), g(0) = f(0) = 0.$$

Since these are the type of spreads that correspond to flocks of quadratic cones, we call any such spread a 'conical flock spread' and the translation plane a 'conical flock plane'.

Note for functions g and f of $GF(q)$, the above putative spread is a spread if and only if the differences of the matrices are identically zero or non-singular. It is immediate that this is equivalent to the quadratic

$$(*) : x^2 + x(g(t) - g(s)) - (t - s)(f(t) - f(s))$$

having no non-zero solutions for all $t, s \in GF(q)$.

54.2.1. Flocks. A flock of a quadratic cone in $PG(3, q)$ is a set of q conics that partitions the non-vertex points of the cone. It is then convenient to establish coordinates. In $PG(3, q)$, represent coordinates as above (x_1, x_2, y_1, y_2), $x_i, y_i \in GF(q)$, taken homogeneously. Any quadratic cone $\mathcal{C}$ may be represented as follows: Choose $(0, 0, 0, 1)$ as the vertex v_0 and use the form of a conic $x_1 x_2 = y_1^2$, in the plane $y_2 = 0$. We represent a general plane π_t in $PG(3, q)$, which does not contain v_0 as follows:

$$\pi_t : tx_1 - f(t)x_2 + g(t)y_1 + y_2,$$

where t is in $GF(q)$ and f and g are functions of t. To obtain a flock, we require q conics on the cone, which are mutually disjoint. Hence, we would require that the intersection of π_t and π_s, for $t, s \in GF(q)$, intersects $\mathcal{C}$ trivially. Using the form of the quadratic cone, this is equivalent to $(*)$. Hence, there is a simple and convenient device to construct a flock from a conical flock plane or conversely a conical flock plane from a flock. Although there are more synthetic methods to establish these connections, it is this algebraic device that probably has proven the more productive. Hence, we have:

THEOREM 54.14. (Gevaert–Johnson–Thas [**397**]). The set of flocks of quadratic cones and the set of conical flock spreads are equivalent.

54.2.2. Baer Groups. Note that since there are q reguli in a conical flock spread, any such regulus net may be derived and since the elation group fixes any of these regulus nets, the constructed spread in $PG(3, q)$ admits the group E as a Baer group (fixes a Baer subplane pointwise). Indeed, it was probably Kantor [**868**], who first noticed the connection between a set of q translation planes admitting Baer groups and flocks of quadratic cones. Actually, it is possible to reconstruct a conical flock plane directly from any such Baer group (Johnson [**722**], Payne–Thas [**1094**]). Put another way, if a translation plane with spread in $PG(3, q)$ admits a Baer group of order q, it turns out that the Baer subplane pointwise fixed by the group defines a regulus net, which when defined, produces a conical flock spread, since the Baer group now becomes an elation group of order q of the correct type, which we call 'regulus-inducing'. Hence, we see:

THEOREM 54.15. (Johnson [**722**] and Payne–Thas [**1094**]). The set of translation planes with spreads in $PG(3, q)$ admitting Baer groups of order q are equivalent to flocks of quadratic cones.

54.2.3. The Known Deficiency-One Partial Hyperbolic Flocks. It is shown in Johnson [**722**] that the semifield of order 16 with kernel $GF(4)$ admits an automorphism group of order 3. This means that there is a Baer group of order 3, that is we have a translation plane of order q^2, with spread in $PG(3, q)$, that admits a Baer group of order $q - 1$. Hence, there is an associated partial flock of a hyperbolic quadric of deficiency one. There is another such example in Johnson and Pomareda [**801**] of order 9^2, as well as an example of order 5^2 in Biliotti and Johnson [**138**]. Actually, all such deficiency-one partial hyperbolic flocks in $PG(3, q)$, for $q \leq 13$ are determined by Royle [**1144**]. It turns out that for $q = 4$ and 9, there is a unique such partial hyperbolic flock and for $q = 5$ and 7, there are exactly two deficiency-one partial hyperbolic flocks. There are no deficiency-one partial flocks for $q = 8, 11$ or 13.

54.2.4. Generalized Quadrangles. The theory of generalized quadrangles as connected to flocks of quadratic cones is due primarily to J.A. Thas and S.E. Payne. Furthermore, many of the connecting ideas and constructions are due to W.M. Kantor. In terms of flocks of quadratic cones, spreads that are unions of reguli sharing a line, and generalized quadrangles, the generalized quadrangle probably plays the more unifying role. In this subsection, we certainly do not do justice to the intrinsic character of generalized quadrangles and only intend this to be a brief sketch of the ideas.

DEFINITION 54.16. A point-line geometry is a 'generalized quadrangle of order (s, t)', for s, t both at least 1, if and only if the following properties hold:

(1) Each line contains $1 + s$ points, and two distinct lines are incident with at most one point,

(2) Each point contains $1 + t$ points, and two distinct points are incident with at most one line,

(3) If A is a point and ℓ is a line not incident with A, there is a unique point B incident with ℓ such that A and B are incident with a line AB.

Generalized quadrangles of order (s,t) have $(s+1)(ts+1)$ points and $(t+1)(ts+1)$ lines.

There are certain classical generalized quadrangles, based on polarities, to which we shall return when we consider the so-called Hermitian sequences of spreads.

Here we are interested in generalized quadrangles of order (q^2, q). Note that a generalized quadrangle of this order will have $(q^2+1)(q^3+1)$ points and $(q+1)(q^3+1)$ lines. The idea for the following construction is due to Kantor, and constructs a generalized quadrangle using cosets of a particular group, under certain conditions.

Let
$$G = \{(\alpha, c, \beta); \alpha, \beta \in GF(q) \times GF(q), c \in GF(q)\}.$$

Let $u \cdot v$ denote the ordinary dot product in $GF(q) \times GF(q)$ and define an operation in G by
$$(\alpha, c, \beta) \cdot (\overline{\alpha}, \overline{c}, \overline{\beta}) = (\alpha + \overline{\alpha}, c + \overline{c} + \beta \cdot \overline{\alpha}, \beta + \overline{\beta}).$$

Then G becomes a group of order q^5. Let
$$A(\infty) = \{(0, 0, \beta); \beta \in GF(q) \times GF(q)\},$$

which is an Abelian subgroup of order q^2. Let
$$A_t = \begin{bmatrix} t & g(t) \\ 0 & -f(t) \end{bmatrix}, t \in GF(q), g, f \text{ functions on } GF(q).$$

Define
$$A(t) = \{\alpha, \alpha A_t \alpha^T, \alpha(A_t + A_t^T); \alpha \in GF(q) \times GF(q)\},$$

where α^T denotes the transpose of α. Let
$$C = \{(0, c, 0); c \in GF(q)\},$$

and define
$$A^*(t) = A(t)C, \text{ for } t \in GF(q) \cup \{\infty\}.$$

Form the following incidence structure I_G: Points are of three types:

(1) The elements of G (number q^5),

(2) The right cosets $A^*(t)g$ for $g \in G$, $A^*(t) \in J^* = \{A^*(t); t \in GF(q)\} \cup \{\infty\}$ (number q^2 for each, $q^2(q+1)$ in total),

(3) The symbol (∞).

Note that we obtain $q^5 + q^3 + q^2 + 1 = (q^2+1)(q \cdot q^2+1)$ points, the correct number for a generalized quadrangle of order (q^2, q).

Lines are of two types:

(1) The right cosets $A(t)g$, for $g \in G$, for $t \in GF(q) \cup \{\infty\}$ (number q^3 for each, $(q+1)q^3$ in total).

(2) The symbols $[A(t)]$, for $t \in GF(q) \cup \{\infty\}$ (number $q+1$).

Hence, we obtain $q^4 + q^3 + 1 = (q+1)(q \cdot q^2 + 1)$, the number required for a generalized quadrangle of order (q^2, q).

Incidence is defined as follows: The point h of G of type (1) is incident with each line $A(t)h$ of type (1) for $t \in GF(q) \cup \{\infty\}$, the point $A^*(t)h$ of type (2) is incident with a line of type (2), $[A(t)]$. The point (∞) is incident with each line $[A(t)]$. Further, the point $A^*(t)h$ is incident with a line $A(t)g$ of type (1) if and only if $A(t)g \subseteq A^*(t)h$.

We also may refer to this structure with the shorthand notation:

$$\begin{bmatrix} t & g(t) \\ 0 & -f(t) \end{bmatrix}; t \in GF(q).$$

Thus, we have:

THEOREM 54.17.

$$\begin{bmatrix} t & g(t) \\ 0 & -f(t) \end{bmatrix}; t \in GF(q)$$

defines a generalized quadrangle of order (q^2, q) if and only if

$$(*) : x^2 + x(g(t) - g(s)) - (t - s)(f(t) - f(s))$$

has no non-zero solutions for all $s, t \in GF(q)$.

DEFINITION 54.18. Any generalized quadrangle of order (q^2, q) of type

$$\begin{bmatrix} t & g(t) \\ 0 & -f(t) \end{bmatrix}; t \in GF(q)$$

is said to be a 'flock generalized quadrangle'.

Hence, we obtain:

THEOREM 54.19. (Thas [1177], Gevaert, Johnson, and Thas [397]). The set of flocks of quadratic cones in $PG(3, q)$ is equivalent to the set of conical flock spreads in $PG(3, q)$,which is equivalent to the set of flock generalized quadrangles.

REMARK 54.20. The set of matrices listed above

$$\left\{ A_t = \begin{bmatrix} t & g(t) \\ 0 & -f(t) \end{bmatrix}; t \in GF(q) \right\}$$

can be viewed abstractly to construct a generalized quadrangle of flock type. In this setting, a matrix B is said to be 'anisotropic' if and only if $\alpha B \alpha^T = 0$ if and only if $\alpha = 0$. If the differences $A_t - A_s$ for $s \neq t$ are anisotropic, then a flock generalized quadrangle is constructed. This set is called a 'q-clan'. Hence, the set of all q-clans is equivalent to flocks of quadratic cones.

54.2.5. *BLT*-Sets. Assume that q is odd. Let $Q(4, q)$ be a non-singular quadric in $PG(4, q)$. A set $\mathcal{B}$ of $q+1$ points of $Q(4, q)$ such that no three are collinear in $Q(4, q)$ is called a 'BLT-set' (after Bader, Lunardon and Thas [54]). Assume that $\mathcal{B} = \{P_i; i = 0, 1, 2, \ldots, q\}$. Then, for a fixed P_k, the lines of $Q(4, q)$ incident with P_k form a quadratic cone $\mathcal{C}_k$. Furthermore, the set of conics $C_{k,j} = P_k^\perp \cap P_j^\perp \cap Q(4, q)$, for $0 \leq j \leq q$ is a flock of $\mathcal{C}_k$, where $P_k^\perp$ refers to the polarity defined by $Q(4, q)$.

Furthermore, from a flock of a quadratic of odd order, it is possible to construct an associated BLT-set.

Hence:

THEOREM 54.21. (Bader, Lunardon, and Thas [54]). The set of BLT-sets in $Q(4, q)$ is equivalent to the set of flocks of quadratic cones of odd order.

54.2.6. Herds of Hyperovals. When q is even, flocks of quadratic cones turn out to be equivalent to certain sets of hyperovals ($q+2$ arcs in $PG(2,q)$). We shall use the notation $(t, -f(t), g(t))$ to designate the flock with planes

$$\pi_t : tx_1 - f(t)x_2 + g(t)y_1 + y_2.$$

THEOREM 54.22. (Payne [**1082**], Cherowitzo, Penttila, Pinneri, and Royle [**232**], Storme and Thas [**1165**] (attributed to Payne and Penttila)). The set $(t, -f(t), g(t))$ corresponds to a flock of a quadratic cone if and only if

$$K_{(a_1,a_2)} = \left\{ \left(1, \sqrt{a_1^2 t - a_1 a_2 f(t) + a_2^2 g(t)}, \ t \in GF(q) \right) \right\} \cup \{(0,1,0),(0,0,1)\}$$

is a hyperoval for each pair $(a_1, a_2) \neq (0,0)$.

DEFINITION 54.23. A set of $q^2 - 1$ hyperovals of type listed above is called a 'herd of hyperovals'.

THEOREM 54.24. For q even, the set of flocks of quadratic cones and the set of herds of hyperovals are equivalent.

54.2.7. Hyperbolic Fibrations. A 'hyperbolic fibration' is a set $\mathcal{Q}$ of $q-1$ hyperbolic quadrics and two carrying lines L and M such that the union $L \cup M \cup \mathcal{Q}$ is a cover of the points of $PG(3,q)$. (More generally, one could consider a hyperbolic fibration of $PG(3, K)$, for K an arbitrary field, as a disjoint covering of the points by a set of hyperbolic quadrics union two carrying lines.) The term 'regular hyperbolic fibration' is used to describe a hyperbolic fibration such that for each of its $q-1$ quadrics, the induced polarity interchanges L and M. When this occurs, and (x_1, x_2, y_1, y_2) represent points homogeneously, the hyperbolic quadrics have the form

$$V(x_1^2 a_i + x_1 x_2 b_i + x_2^2 c_i + y_1^2 e_i + y_1 y_2 f_i + y_2^2 g_i)$$

for $i = 1, 2, \ldots, q-1$ (the variety defined by the quadrics). When $(e_i, f_i, g_i) = (e, f, g)$ for all $i = 1, 2, \ldots, q-1$, the regular hyperbolic quadric is said to have 'constant back half'.

The main theorem of Baker, Ebert and Penttila [**78**] is equivalent to the following.

THEOREM 54.25. (Baker, Ebert, and Penttila [**78**]).
(1) Let $\mathcal{H} := V(x_1^2 a_i + x_1 x_2 b_i + x_2^2 c_i + y_1^2 e + y_1 y_2 f + y_2^2 g)$ for $i = 1, 2, \ldots, q-1$ be a regular hyperbolic fibration with constant back half.

Consider $PG(3,q)$ as (x_1, x_2, x_3, x_4) and let C denote the quadratic cone with equation $x_1 x_2 = x_3^2$.

Define

$$\pi_0 : x_4 = 0, \ \pi_i : x_1 a_i + x_2 c_i + x_3 b_i + x_4 = 0 \quad \text{for } 1, 2, \ldots, q-1.$$

Then

$$\{\pi_j, j = 0, 1, 2, \ldots, q-1\}$$

is a flock of the quadratic cone C.

(2) Conversely, if $\mathcal{F}$ is a flock of a quadratic cone, choose a representation as $\{\pi_j, j = 0, 1, 2, \ldots, q-1\}$ above. Choose any convenient constant back half (e, f, g), and define $\mathcal{H}$ as $V(x_1^2 a_i + x_1 x_2 b_i + x_2^2 c_i + y_1^2 e + y_1 y_2 f + y_2^2 g)$ for $i = 1, 2, \ldots, q-1$. Then $\mathcal{H}$ is a regular hyperbolic fibration with constant back half.

Now for each of the $q-1$ reguli, choose one of the two reguli of totally isotropic lines. Such a choice will produce a spread and a translation plane. Hence, there are potentially 2^{q-1} possible translation planes obtained in this way.

Now it is possible to reconstruct a hyperbolic fibration with constant back half from any translation plane of order q^2 with spread in $PG(3, q)$ that admits a cyclic homology group of order $q + 1$. We consider this more generally in Chapter 61. Here we simply list the main connecting result.

THEOREM 54.26. (Johnson [**769**]). Translation planes with spreads in $PG(3, q)$ admitting cyclic affine homology groups of order $q + 1$ are equivalent to flocks of quadratic cones.

CHAPTER 55

Ostrom and Flock Derivation.

For the following discussion of Ostrom derivation, Ostrom-derivates and flock-derivates, we follow the work of Jha and Johnson [**639**], although most of the proofs are omitted.

A spread in $PG(3, q)$ associated with a flock of a quadratic cone consists of a set of q reguli that mutually share exactly one line. These reguli, called the 'base' reguli, may be derived producing 'new' spreads.

Hence, we point out that there are q possible distinct spreads obtained by the replacement of the q reguli and the number of isomorphism classes is exactly the number of orbits of base reguli of the full collineation group of the original translation plane.

In particular, the q reguli share a line that is the axis of an elation group E of order q fixing each of these and the planes that admit a linear group acting transitively on the reguli are essentially determined by Johnson–Wilke [**814**], Biliotti, Jha, Johnson, and Menichetti [**133**] and Jha, Johnson, and Wilke [**647**] and more recently by Jha and Johnson for even-order planes in [**641**]. Hence, modulo a small class of planes, and by restriction to orders forcing a group to be linear, it would normally be possible to claim that any conical flock spread produces at least two non-isomorphic derived flock spreads.

Is it true that any conical flock spread produces at least two mutually non-isomorphic translation planes obtained from derivation of the reguli of the spread?

Actually, the answer is no since this is not true in Desarguesian, Fisher–Thas–Walker (Walker or Betten) or Knuth semifield flock spreads. However, it turns out that these are exactly the exceptions.

55.1. Flock-Derivations and Ostrom-Derivates.

Given a conical flock spread, there are $q + 1$ conical flock spreads (containing the given one) in what is called the 'skeleton' of the spread. When q is odd, these correspond to the flocks obtained via the derivation procedure of Bader, Lunardon and Thas [**54**] and when q is even, there is an algebraic process of transforming the spread set corresponding to a given spread into q others. We call such transformed spreads 'flock-derivates' and the set of all such spreads the 'skeleton' $S(\pi)$ of a given spread π. We shall distinguish between the spreads derived by replacement or derivation of a base regulus and the flock-derivates by calling the former 'Ostrom-derivates'.

We would like to claim that normally there are at least two non-isomorphic Ostrom-derivates arising from a skeleton. An automorphism of a skeleton is an element of $\Gamma L(4, q)$ that permutes the members of the skeleton. We first show that any collineation group of a flock spread must permute the remaining spreads of the associated skeleton and thus act as an automorphism of the skeleton. Then if all Ostrom-derivates are isomorphic, it will force the existence of a group that acts doubly transitively on the flock-derivates. In this regard, Jha and Johnson [**618**] prove:

THEOREM 55.1. (Jha and Johnson [**618**]). If π is a flock spread of a quadratic cone in $PG(3, q)$, then there is an automorphism group acting doubly transitively on the skeleton of π if and only if π is one of the following types of spreads:
  (1) Desarguesian,
  (2) Walker when q is odd or Betten when q is even, or
  (3) Knuth semifield and q is odd.

Hence, we obtain:

THEOREM 55.2. If π is a flock spread that is not Desarguesian, Walker, Betten or Knuth semifield, then there are at least two non-isomorphic Ostrom-derivates.

Generally speaking, there are many mutually non-isomorphic Ostrom-derivates and, in particular, we prove:

THEOREM 55.3. If the skeleton of π, $S(\pi)$, is not transitive, and $q = 2^r$, for r odd, then there exist at least four non-isomorphic Ostrom-derivates.

On the other end of the spectrum are flock spreads in $PG(3, q)$ that provide $q(q + 1)$ Ostrom-derivates. Do there exist such flock spreads?

In order to see how all of these fit together, we must deal with various geometric structures equivalent to each other and equivalent to a flock of a quadratic cone. Some of the following repeats and emphasizes material in the previous sections.

The main question when deriving (i.e., in the sense of Ostrom) an affine plane is the nature of the collineation group of the derived plane. If a regulus net which is invariant under an elation group of order q then derivation of this regulus net produces a corresponding translation plane of order q^2 and spread in $PG(3, q)$ admitting a Baer group of order q. Hence, results on Baer groups provide the connection. There are two types of Baer-elation incompatibility, one each for odd- and even-order planes. These are as follows:

55.2. Baer Groups in Translation Planes.

THEOREM 55.4. (Foulser [**375**]). Let π be a translation plane of order $q^2 = p^r$, where p is an odd prime.
  (1) Then π cannot admit both elations and Baer p-collineations.
  (2) If $p > 3$, and π admits Baer p-collineations then the net of degree $q + 1$ defined by any pointwise fixed Baer subplane is left invariant under the full collineation group of π. Furthermore, the group generated by all Baer p-collineations satisfies the conditions of the Hering–Ostrom theorem for groups generated by elations in translation planes.
  (3) If $p = 3$ and there exists a Baer 3-group of order at least 9, then the statements of part (2) hold.

THEOREM 55.5. (Jha and Johnson [**598**]). Let π be a translation plane of even order q^2 admitting at least two Baer groups of order $2\sqrt{q}$ in the translation complement. Then either π is Lorimer–Rahilly or Johnson–Walker of order 16 or π is Hall.

Thus, using such results, it is possible to note that the group inherited from a conical flock plane in a derived conical flock plane is, in fact, the full collineation group. We have noted previously that flocks of quadratic cones are equivalent to spreads in $PG(3, q)$ with regulus-inducing elation groups. Specifically, the structure is as follows:

55.3. Flocks of Quadratic Cones.

THEOREM 55.6. (Gevaert and Johnson [**395**]). Let π denote a translation plane of order q^2 that corresponds to a flock of a quadratic cone.

(1) Then π admits an elation group E of order q such that the component orbits union the axis of E are reguli.

(2) Coordinates may be chosen so that E has the following form:

$$\left\langle \begin{bmatrix} 1 & 0 & u & 0 \\ 0 & 1 & 0 & u \\ 0 & 0 & 1 & 0 \\ 0 & 0 & 0 & 1 \end{bmatrix} ; u \in GF(q) \right\rangle.$$

THEOREM 55.7. (Gevaert and Johnson [**395**]). Let π be a translation plane of order q^2 with spread in $PG(3, q)$ that admits an affine elation group E of order q such that there is at least one orbit of components union the axis of E that is a regulus in $PG(3, q)$. Then π corresponds to a flock of a quadratic cone in $PG(3, q)$.

In the case above, the elation group E is said to be 'regulus-inducing' as each orbit of a 2-dimensional $GF(q)$-vector space disjoint from the axis of E will produce a regulus. The reguli of the spread are called the 'base reguli'.

The relationship between the collineation group of a conical flock and that of a conical flock plane is as follows:

THEOREM 55.8. (Gevaert, Johnson, and Thas [**397**] and Payne and Thas [**1094**]). Let π be a finite conical flock translation plane with spread in $PG(3, q)$.

(1) If π is not Desarguesian, then the full collineation group of π permutes the base reguli.

(2) Let $\mathcal{F}$ be the collineation group of the associated flock, $\mathcal{G}$ the full translation complement of π. Let E be the regulus-inducing elation group of π and K^* the scalar group of order $q - 1$.

Then $\mathcal{F} \simeq \mathcal{G}/EK^*$.

55.4. Generalized Quadrangles, *BLT*-Sets, Payne–Rogers Recoordinatization.

The concept of *BLT*-sets appears previously; we recall this again here to emphasize the 'even-order' case.

DEFINITION 55.9. Assume that q is odd. Let $Q(4, q)$ be a non-singular quadric in $PG(4, q)$. A set $\mathcal{B}$ of $q + 1$ points of $Q(4, q)$ such that no three are collinear in $Q(4, q)$ is called a '*BLT*-set'. Assume that $\mathcal{B} = \{P_i; i = 0, 1, 2, \ldots, q\}$. Then, for a fixed P_k, the lines of $Q(4, q)$ incident with P_k form a quadratic cone $\mathcal{C}_k$. Furthermore, the set of conics $C_{k,j} = P_k^\perp \cap P_j^\perp \cap Q(4, q)$, for $0 \le j \le q$ is a flock of $\mathcal{C}_k$, where $P_k^\perp$ refers to the polarity defined by $Q(4, q)$.

Furthermore, a flock of a quadratic cone induces a BLT-set so that, for q odd, a BLT-set is equivalent to a flock of a quadratic cone.

Note that from a BLT-set there is a set of $q+1$ flocks and $q+1$ conical spreads. These are the flock derivates when q is odd and the associated spreads form the skeleton on any one of them. To see how this works in the even-order case, we recall the following theorem of Payne and Rogers.

THEOREM 55.10. (Payne and Rogers [**1093**]). Let Q be a generalized quadrangle corresponding to a conical flock spread π. Let $S(\pi)$ denote the skeleton of π. In Q, let (∞) denote the special point and let $[A(j)]$, $j = \infty, 1, \ldots, q$, denote the lines incident with (∞). In addition, let 0 denote the point corresponding to the identity in the group defining the points and lines of Q.

Then each conical flock spread of $S(\pi)$ corresponds to a recoordinatization of Q by a line $[A(i)]$ taken as $[A(\infty)]$; each spread of the skeleton of π corresponds to the same generalized quadrangle.

55.5. Ostrom-Derivations and Skeletons.

DEFINITION 55.11. Any spread in $PG(3, K)$, for K a finite or infinite field, which has a 'matrix spread set' as in the following theorem with defining functions g and f such that

$$g(u, t) = u + g(t) \text{ and } f(u, t) = f(t)$$

for all $u, t \in K$ shall be called a 'conical spread' and the corresponding translation plane is called a 'conical translation plane'.

We have called an Ostrom-derivate a translation plane that may be derived from a conical translation plane. We now make specific the exact nature of the spreads.

We shall normally not use different notation to distinguish between a spread and the associated translation plane. We developed the algebraic relationships between a spread in $PG(3, q)$ and its Ostrom-derivation, in Chapter 54. This spread is called the 'derived spread' of the original. In particular, we noted above that the conical spreads contain a regulus partial spread and that there is then a so-called 'derived spread' whenever a spread contains such a partial spread. We may recoordinatize by taking any regulus defined by $x = 0$ and an image set under the regulus-inducing elation group E to obtain another spread of the above form. That is, if

$$x = 0, y = x \begin{pmatrix} u + g(t) & f(t) \\ t & u \end{pmatrix} \forall t, u \in K$$

is a conical flock spread then the q-reguli are

$$R_t : x = 0, y = x \begin{pmatrix} u + g(t) & f(t) \\ t & u \end{pmatrix} ; \forall u \in K, \text{ for } t \text{ fixed in } K.$$

Fixing t as t_0, we apply the mapping $(x, y) \longmapsto (x, y) \begin{pmatrix} -g(t_0) & -f(t_0) \\ t_0 & 0 \end{pmatrix}$ to translate the spread into the form

$$x = 0, y = x \begin{pmatrix} u + g(t) - g(t_0) & f(t) - f(t_0) \\ t - t_0 & u \end{pmatrix} \forall t, u \in K.$$

Note that the net R_{t_0} becomes R_0 in the new representation.

Hence, there are potentially q distinct Ostrom-derivates from a given conical translation plane.

In the special case when the spread is a conical spread, there is another construction technique that might be of interest to us. The construction may be given from the associated generalized quadrangle by recoordinatization.

DEFINITION 55.12. Let $\mathcal{S}$ be any conical spread in $PG(3, q)$ represented as follows:

$$x = 0, y = x \begin{pmatrix} u + g(t) & f(t) \\ t & u \end{pmatrix} \forall t, u \in K \simeq GF(q).$$

Assume that q is odd.

For a fixed $s \in K$, we define the 's-inversion' $\mathcal{S}^{-s}$, a set of 2-dimensional subspaces, as follows:

$\mathcal{S}^{-s}$: $x = 0$ and

$$y = x \left(\begin{pmatrix} -(g(t) - g(s))/2 & f(t) - f(s) \\ t - s & (g(t) - g(s))/2 \end{pmatrix}^{-1} + \begin{pmatrix} u & 0 \\ 0 & u \end{pmatrix} \right)$$

for all $t, u \in GF(q)$.

THEOREM 55.13. If $\mathcal{S}$ is a conical spread in $PG(3, q)$ and q is odd, then an s-inverted conical spread is a spread in $PG(3, q)$.

DEFINITION 55.14. We now consider a similar situation when q is even. Let $\mathcal{S}$ be a conical spread with defining spread

$$x = 0, y = x \begin{pmatrix} u + g(t) & f(t) \\ t & u \end{pmatrix} \forall u, t \in K \simeq GF(q).$$

Then, for each $s \in K$, we define the structure $\mathcal{S}_2^s$, called the 's-square'

$$x = 0, y = x \left(1/(g(t) + g(s))^2 \begin{pmatrix} g(t) + g(s) & f(t) + f(s) \\ t + s & 0 \end{pmatrix} + uI_2 \right),$$

$$y = x \begin{pmatrix} v & 0 \\ 0 & v \end{pmatrix}$$

$\forall t, t \neq s, \forall u, v \in K.$

THEOREM 55.15. If q is even then an s-square spread is a conical spread in $PG(3, q)$.

DEFINITION 55.16. We shall call the set of s-inverted spreads or s-square spreads of a conical flock spread π, the 'skeleton $S(\pi)$' of π. Although it is possible to prove directly that an s-inverted spread or an s-square spread is a conical spread, these are simply the forms that arise from the associated generalized quadrangle by recoordinatization.

THEOREM 55.17. (Johnson [736]). Assume that π is a non-Desarguesian conical flock plane of odd order q^2.

Then the translation complement acts as a permutation group on the planes of the skeleton $S(\pi) - \{\pi\}$. The number of orbits of this group acting on the base reguli is the same as the number of orbits of the planes not equal to π of the skeleton.

THEOREM 55.18. (Jha and Johnson [639]). The full group of a derived conical flock plane of finite order by one of the base reguli is the group inherited from the conical flock plane or the conical plane is Desarguesian of order 4 or 9.

We have seen that flocks of quadratic cones correspond to spreads in $PG(3, q)$ admitting regulus-inducing elation groups of order q. We are interested in whether there are collineation groups that are transitive on the skeleton of the flock. This implies that the q reguli containing a component in the spread of the associated conical flock are in a transitive class. Hence, there is an associated collineation group of order q^2 in the plane of order q^2. There is an accompanying theory for such planes.

55.6. q^2-q^2 Structure Theory—Even Order.

THEOREM 55.19. (Johnson [710]). Let π denote a translation plane of even order q^2 that corresponds to a flock of a quadratic cone.

Then π is a semifield plane if and only if π is Desarguesian.

THEOREM 55.20. (See Biliotti and Menichetti [145], Jha and Johnson [586] and Johnson and Wilke [814]). Let π denote a translation plane of even order q^2 with spread in $PG(3, q)$ that admits a linear collineation group G of order q^2. If the elation subgroup of G has order q, then π is one of the following planes:
  (1) Betten,
  (2) Lüneburg–Tits,
  (3) Desarguesian, or
  (4) Biliotti–Menichetti of order 64.

THEOREM 55.21. (Jha and Johnson [617]). If a translation plane of order 2^t admits an affine elation group of order $2^t/2$, then the plane is a semifield plane.

THEOREM 55.22. (Biliotti, Jha, Johnson, and Menichetti [133]). Let π be a translation plane of even order q^2 with spread in $PG(3, q)$ that admits a linear group of order q^2 that fixes a component and acts regularly on the remaining components. Then coordinates may be chosen so that the group G may be represented as follows:

$$\left\langle \sigma_{b,u} = \begin{bmatrix} 1 & T(b) & u + bT(b) + l(b) & uT(b) + R(b) + m(u) \\ 0 & 1 & b & u \\ 0 & 0 & 1 & T(b) \\ 0 & 0 & 0 & 1 \end{bmatrix} ; b, u \in GF(q) \right\rangle$$

where T, m, and l are additive functions on $GF(q)$ and $m(1) = 0$ and such that

$$R(a + b) + R(a) + R(b) + l(a)T(b) + m(aT(b)) + a(T(b))^2 = 0, \forall a, b \in GF(q).$$

It may occur that there is a collineation group properly containing the regulus-inducing elation group E and the associated (scalar) kernel homology group K^* (of order $q - 1$) that leaves each regulus invariant. In this case, the flock is said to be 'rigid' and there is a corresponding theory for such spreads due to Jha and Johnson [617]).

THEOREM 55.23. (Jha and Johnson [617]). Let π be a translation plane of even order q^2 that corresponds to a flock of a quadratic cone. Let $\mathcal{B}$ denote the set of reguli that are defined via the regulus-inducing elation group E. Let K^* denote the kernel homology group of order $q - 1$.
  (1) If there exists a collineation group properly containing EK^* that fixes each regulus of $\mathcal{B}$, then π is Desarguesian.
  (2) If π is non-Desarguesian, linear Baer involutions cannot exist.

Putting such ideas together for even, Jha and Johnson [641] are able to prove that such transitive skeletons are quite rare:

THEOREM 55.24. (Jha and Johnson [641]). Let π be a translation plane of even order that corresponds to a flock of a quadratic cone. If π admits a linear collineation group G (i.e., in $GL(4,q)$) that acts transitively on the components not equal to a fixed component $x = 0$, then π is one of the following:
   (1) Betten or
   (2) Desarguesian.

We have noted that there are perhaps mutually non-isomorphic flocks arising from flock-derivation. The question arises if non-isomorphic flocks produce non-isomorphic translation planes.

THEOREM 55.25. (Gevaert and Johnson [395]). Two conical flocks in $PG(3,q)$ are isomorphic if and only if the corresponding translation planes are isomorphic.

Transitive skeletons for odd order are considered by Johnson, Lunardon and Wilke, who prove:

THEOREM 55.26. (Johnson, Lunardon, and Wilke [779]). Let π be a conical flock plane and let $S(\pi)$ denote the skeleton of π. Assume that the collineation group of π acts transitively on the base reguli.
   (1) If q is odd and all of the planes of the skeleton are isomorphic, then π is one of the following types of planes:
      (a) Desarguesian,
      (b) Walker, or
      (c) Knuth semifield of conical plane type.
   (2) If $S(\pi)$ is a doubly transitive skeleton, then π is one of the planes of part (1).

Again, part of the problem deals with general translation planes of odd order q^2 that admit collineation groups of order q^2. The accompanying q^2-q^2-theory for odd order is due to Johnson and Wilke.

THEOREM 55.27. (See Johnson and Wilke [814]). Let π denote a conical flock plane of odd order q^2 that admits a collineation group of order q^2 in the linear translation complement. Then either π is a semifield plane or the spread for π may be represented in the following form:

$$x = 0, y = x \begin{bmatrix} u - a^2 & \frac{-a^3}{3} + J(a) \\ a & u \end{bmatrix} ; u, a \in GF(q), J \text{ a function}$$

on $GF(q)$ such that the following group G is a collineation group:

$$G = \left\{ \begin{bmatrix} 1 & a & u & ua - \frac{a^3}{3} + J(a) \\ 0 & 1 & a & u \\ 0 & 0 & 1 & a \\ 0 & 0 & 0 & 1 \end{bmatrix} ; u, a \in GF(q) \right\}.$$

55.7. Fundamental Theorems on Generalized Quadrangles.

To piece these previous ideas together, we begin with the fundamental theorem regarding derived conical flock planes.

THEOREM 55.28. (Also see Johnson and Payne [**792**, part of (10.1) and (10.7)]). Let π_1 and π_2 be non-Desarguesian conical flock spreads in $PG(3, q)$ and let Q_1 and Q_2 denote the corresponding generalized quadrangles with special points $(\infty)_i$, 0_i and lines $[A_i(j)]$, $j = \infty, 1, \ldots, q$, for $i = 1, 2$.

(1) If π_1 and π_2 are represented in the conical flock form listed above, and the generalized quadrangles are obtained via the associated q-clans then π_1 and π_2 are isomorphic if and only if there is an isomorphism of Q_1 onto Q_2 that maps $(\infty)_1$ onto $(\infty)_2$, maps 0_1 onto 0_2 and maps $[A_1(\infty)]$ onto $[A_2(\infty)]$.

(2) If π_1 and π_2 belong to the same skeleton let π_i correspond to the line $[A(j_i)]$ in $Q_1 = Q_2$.

(a) Then π_1 is isomorphic to π_2 if and only if there is a collineation of $Q_1 = Q_2$ fixing $(\infty)_1$ and 0_1 and mapping $[A(j_1)]$ onto $[A(j_2)]$.

(b) In particular, let σ be a collineation of π_1 in the translation complement. Denote $\pi_2\sigma$ by π_3. Then there is an induced collineation σ^* of Q_1 that fixes $[A(j_1)]$ and maps $[A(j_2)]$ onto $[A(j_2)]\sigma^*$. The conical flock spread of $S(\pi_1)$ corresponding to $[A(j_3)] = [A(j_2)]\sigma^*$ is π_3.

(c) The collineation group of π_1 permutes the elements of $S(\pi_1) - \{\pi_1\}$ in the sense that an image is an isomorphic translation plane to the preimage. If π_1 is not Desarguesian, then the collineation group permutes the base reguli and the group acting on the base reguli is permutation isomorphic to the group acting on the planes $\neq \pi_1$ of the skeleton.

(d) π_2 is in the skeleton of π_1 $\iff$ $S(\pi_2) = S(\pi_1)$.

THEOREM 55.29. If π is a conical flock plane of order q^2, for $q > 3$, let R_i for $i = 1, 2, \ldots, q$ denote the set of 'base' reguli and let π_i^* denote the translation plane obtained by the replacement of R_i by its opposite regulus (net) R_i^*.

Then π_i^* is isomorphic to π_j^* if and only if there is a collineation in π mapping R_i to R_j.

Of course there is some confusion with the use of the term 'derivation' as applied to 'Ostrom-derivation' of a net corresponding to a regulus and the term 'derivation' as applied to the process of Bader, Lunardon and Thas [**54**], 'BLT-derivation' obtaining exactly q 'other' flocks and hence conical flock spreads from a given one. This derivation process works geometrically for q odd using BLT-sets but there is an algebraic process for q even due to Payne and Rogers [**1093**], by recoordinatizations of the corresponding generalized quadrangle, as we noted in the previous section.

The algebraic process also works just as well when q is odd, and we have presented the form of the constructed spreads in Chapter 6 and reviewed it in Section 54.1.2. The main point is that what we have called the s-inverted constructions define the spreads that correspond to the BLT-derivation in the odd-order case. The s-square construction is then the analogue in the even-order case. Hence, our use of the term 'skeleton' just refers to the $q + 1$ conical flock spreads that may be obtained from a given one algebraically as noted.

We shall use the term 'flock-derivation' to describe the associated spreads in the skeleton of a given one. In Chapter 6 and Section 54.1.2, we have included a note on the derivation process using a 'standard' regulus. Since we may derive using any of the q-reguli in any of the conical flock spreads, we have $q(q+1)$ distinct spreads arising from a given conical flock spread by the standard derivation of the reguli in the spreads of the skeleton of a given spread.

Here we note the fundamental connection. There is a corresponding but stronger fundamental theorem.

We recall that the set of flock-derivates of a conical flock plane π, $S(\pi)$, is called the 'skeleton' of π. Furthermore, $S(\pi) = S(\rho)$ for $\rho \in S(\pi)$.

THEOREM 55.30. *If π is a conical flock plane of order q^2, for $q > 3$, let $\rho \in S(\pi)$ and let R_i^{ρ} for $i = 1, 2, \ldots, q$, denote the set of 'base' reguli for ρ and let ρ_i^* denote the translation plane obtained by the replacement of R_i^{ρ} by its opposite regulus (net).*

Then π_i^* is isomorphic to ρ_j^* if and only if there is an element of $\Gamma L(4, q)$ mapping π to ρ that also maps R_i^{π} to R_j^{ρ}.

DEFINITION 55.31. The set of Ostrom-derivations of planes of the skeleton $S(\pi)$ of a conical flock spread π is called the 'Baer skeleton'. The subset of the Baer skeleton of Ostrom-derivations of $\rho \in S(\pi)$ is called the 'ρ-Baer skeleton'.

THEOREM 55.32. *Let π be a non-Desarguesian conical flock translation plane of order q^2 with spread in $PG(3, q)$.*

Then the number of mutually non-isomorphic Ostrom-derivations of π, the π-Baer skeleton, is precisely the number of orbits of the set of flock-derivations in $S(\pi) - \pi$ under the collineation group of π.

CHAPTER 56

Transitive Skeletons.

This chapter refers directly to Chapter 55.

DEFINITION 56.1. We shall say that a skeleton is 'cyclic' if and only if there is a cyclic collineation group in $P\Gamma L(4, q)$ that acts transitively on the $q + 1$ conical flock planes of the skeleton of any of them. More generally, a skeleton is 'transitive' if and only if there is a transitive permutation group acting on the skeleton.

REMARK 56.2. If π is a conical flock spread and $S(\pi)$ is a transitive skeleton, then either there are at least two non-isomorphic planes of the Baer skeleton corresponding to π or there is a group that acts doubly transitive on the skeleton $S(\pi)$.

56.1. Double Transitivity.

By Theorem 55.26, we have a classification of all doubly transitive skeletons, when q is odd.

When q is even, we may consider the s-squares of a given flock plane. Hence, we have a set of $q + 1$ spreads algebraically constructed from one of them π; the skeleton $S(\pi)$ of π.

One question that could be asked is if two of the spreads are isomorphic by an element σ of $P\Gamma L(4, q)$, does σ act on the remaining skeleton? If $S(\pi)$ is the skeleton of π and $\pi\sigma \in S(\pi)$, is $S(\pi)\sigma = S(\pi)$? In fact, Theorem 55.28 says exactly that.

Hence, to classify the doubly transitive skeletons, we may use the classification theorem of finite simply groups and apply the action of the stabilizer of π as a quotient of a collineation group of the associated translation plane.

THEOREM 56.3. A doubly transitive skeleton where q is even exists if and only if the skeleton is Betten or Desarguesian.

The proof depends on the following lemma and the classification theorem of finite simple groups.

LEMMA 56.4. Let π be a non-Desarguesian conical flock plane with spread in $PG(3, q)$, q even, and let E denote the regulus inducing elation group of order q and let K^* denote the kernel homology group of order $q - 1$. Let G denote the full collineation of π. Then G normalizes E and therefore permutes the base reguli. Then G/EK^* is the group induced on the flock skeleton $S(\pi) - \{\pi\}$.

Hence, this theorem coupled with the odd-order theorem of Johnson–Wilke–Lunardon mentioned previously shows the following corollary to be true.

COROLLARY 56.5. Let π be a conical flock of a quadratic cone that is not Desarguesian, Walker–Betten or Knuth semifield. Then there are at least two mutually non-isomorphic Ostrom derived planes from the skeleton of π.

COROLLARY 56.6. Let π be a conical flock plane and let $Q(\pi)$ be the generalized quadrangle corresponding to π. Let (∞) denote the special point of $Q(\pi)$. Assume that π is not one of the spreads mentioned above.

Then the number of orbits of lines incident with (∞) of $Q(\pi)$ is strictly less than the number of mutually non-isomorphic Ostrom-derivates.

56.1.1. Even-Order Transitive Skeletons.
If the skeleton is transitive, we still obtain at least two non-isomorphic Ostrom-derivates. Hence, we consider the situation when the skeleton is not transitive and there are at least two orbits of planes. Within each other, we ask if there are at least two Ostrom-derivates. If not, there is a collineation group that acts transitively on the base reguli of a given flock plane.

As mentioned in the previous remarks, Jha and Johnson [**641**] have completely determined the conical flock planes of even order that admit a linear collineation group acting transitively on the base reguli of the flock plane.

THEOREM 56.7. (See Jha and Johnson [**641**]). Let π be a conical flock translation plane of even order 2^{2r} admitting a linear collineation group in the translation complement that acts transitively on the base reguli.

Then π is one of the following planes:

(1) Betten, or

(2) Desarguesian.

Note, that we cannot be certain that a group that acts transitively must, in fact, be linear unless $q = 2^r$ and r is odd. So, we make this assumption when considering Ostrom-derivates. Note that in both of the above types of planes the skeleton is transitive. Hence, we then have the following:

THEOREM 56.8. Let π be a conical flock plane of even order 2^{2r}, where r is odd. Assume that the skeleton $S(\pi)$ is not transitive. Assume that there are $k \geq 2$ orbits of planes within the skeleton with representatives say π_i ; $i = 1, \ldots, k$.

(1) Corresponding to each plane π_i, there are at least two mutually non-isomorphic Ostrom-derivates.

(2) Hence, there are at least $2k$ mutually non-isomorphic Ostrom-derivates.

PROOF. If the skeleton is not transitive, assume that there are k orbits in the skeleton. Consider a representative plane π_i. Assume that there is a unique Ostrom-derivate arising from π_i. Then there is a group acting transitively on the base reguli. However, this implies that the plane is Betten or Desarguesian. □

COROLLARY 56.9. Any even-order conical flock plane of order 2^{2r}, for r odd, which does not belong to a transitive skeleton, produces at least four mutually non-isomorphic Ostrom-derivates.

CHAPTER 57

BLT-Set Examples.

For many of the known conical flock planes, the full collineation group is also known. This means that using this group, we may determine how many mutually non-isomorphic Ostrom-derivates actually are produced per skeleton. We have shown that normally, there are at least two non-isomorphic Ostrom-derivates when the skeleton is transitive. Indeed, for the known transitive skeletons, such as arising from the Penttila flocks or the Adelaide flocks, there are an enormous number of 'missing' spreads that may be obtained by derivation of the base reguli.

57.1. The Penttila Flocks/Spreads.

The Penttila flocks/spreads are most easily described using *BLT*-sets: For example, let K be isomorphic to $GF(q)$ and let F extend K and be isomorphic to $GF(q^2)$. Let q be odd and ≥ 5. Let α be a primitive element for E and let $\beta = \alpha^{q-1}$ of order $q+1$. Let $V = \{ (x, y, a); x, y \in F \text{ and } a \in K \}$ considered as a 5-dimensional K-space. Define

$$Q \text{ by } Q(x, y, a) = x^{q+1} + y^{q+1} - a^2.$$

Then Q is a quadratic form on V. Furthermore,

$$\{(2\beta^{2j}, \beta^{3j}, \sqrt{5}) \in V; 0 \leq j \leq q\}, q \equiv \pm 1 \bmod 10$$

is a *BLT*-set (see Chapters 54 and 55).

We note that

$$\tau_\beta : (x, y, b) \longmapsto (x\beta, y\beta, b)$$

defines an isometry, since $Q(x, y, a) = Q(x\beta, y\beta, a)$. Indeed, $\langle \tau_\beta \rangle$ is a group that acts transitively on the *BLT*-set. We notice that a group acting transitively on the *BLT*-set implies that there is a cyclic group induced on the skeleton of planes; we have a 'cyclic skeleton'.

In order to determine how many Ostrom-derivates exist, we need to determine the stabilizer of any point of the *BLT*-set for this induces the collineation group of the associated translation plane, which must permute the q base reguli. The number of orbits of the reguli is the number of mutually non-isomorphic Ostrom-derivates. The number of orbits of the reguli is the number of orbits minus 1 of the points of the *BLT*-set of the stabilizer of any given point.

Here we list the following interesting problem:

PROBLEM 57.1. Determine the stabilizer of a point of the *BLT*-set producing the Penttila flocks. Use this to enumerate the number of mutually non-isomorphic Ostrom-derivates.

57.2. The Fisher Flocks/Spreads.

A BLT-set for the Fisher flocks is

$$\{\, (\beta^{2j}, 0, 1), (0, \beta^{2j}, 1); \ 1 \leq j \leq (q+1)/2 \,\},$$

(see Payne [**1089**], form determined by T. Penttila). Consider the following group:

$$\langle \sigma, g_\beta \rangle \,;\, \sigma : (x, y, a) \longmapsto (y, x, a) \text{ and } g_\beta : (x, y, a) \longmapsto (x\beta^2, y\beta^2, a),$$

a group isomorphic to $Z_2 \times Z_{(q+1)/2}$, where the notation of the previous section is utilized. Hence, we again have a transitive, but not necessarily cyclic skeleton.

The stabilizer of a point induces a collineation group of the Fisher planes.

THEOREM 57.2. The full collineation group of the Fisher planes is the group inherited from the associated Desarguesian affine plane.

For a nice representation of the Fisher planes of arbitrary order, the reader is referred to the article of Jha and Johnson [**628**].

THEOREM 57.3. (Jha and Johnson [**628**]). Let K be a full field of odd or 0 characteristic and let $K[\theta]$ be a quadratic extension of K that is also a full field. Let Σ be the Pappian affine plane coordinatized by $K[\theta]$ and let H be the kernel homology group of squares in Σ.

Let s be any element of $K[\theta]$ such that $s^{\sigma+1}$ is non-square in K if -1 is a non-square in K, and $s^{\sigma+1}$ is square in K if -1 is a square in K. Let E denote the regulus-inducing group and H is the homology group of squares of kernel homologies in Σ. Then

$$EH(y = x^\sigma s) \cup \{\, y = xm; \ (m+\beta)^{\sigma+1} \neq s^{\sigma+1} \ \forall \beta \in K \,\}$$

is a generalized Fisher conical spread in $PG(3, K)$.

In the finite case, the spread is a union of q reguli that share a component. There are $(q+1)/2$ reguli in an orbit under a group EH, of order $q(q+1)$, where the group H of order $q+1$ is a kernel homology subgroup of the associated Desarguesian affine plane Σ, and the group E is a regulus-inducing elation group of order q. There are $(q-1)/2$ reguli from Σ, each of which is left invariant by EH.

Hence, we may consider a collineation g of Σ that fixes $y = x^q s$ and fixes $x = 0$ and is in $\Gamma L(2, q^2)$. Let

$$g : (x, y) \longmapsto (x^\tau a, x^\tau b + y^\tau c),$$

acting on Σ represented in the standard manner. Then,

$$y = x^q s \longmapsto y = x^q a^{-\tau q} s^\tau c + x a^{-\tau} b = x^q s,$$

if and only if

$$b = 0 \text{ and } c = s^{1-\tau} a^q.$$

Hence, g fixes $x = 0, y = 0$ and must normalize E, implying that g must fix the standard regulus net so that

$$s^{1-\tau} a^{q-1} \in GF(q) - \{0\}.$$

Moreover, we may assume, using the kernel homology group of order $q - 1$, that g fixes pointwise a 1-dimensional K-subspace of $y = x^q s$, say generated by $(x_o, x_o^q s)$.
So,

$$a = x_o^{1-\tau}.$$

Therefore, the collineation g has the form:

$$g : (x,y) \longmapsto (x^\tau x_o^{1-\tau}, y^\tau x_o^{(1-\tau)q} s^{1-\tau}).$$

So, if $\tau = 1$ then $g = 1$. It follows that there is a cyclic group of order $2r$, where $q^2 = p^{2r}$, and p is a prime, acting on the Fisher spreads. Thus, we have:

THEOREM 57.4. Let π be a Fisher plane of odd order p^{2r}, where p is a prime. Then, there are at least $1 + \frac{(q-1)}{2([2r])}$ mutually non-isomorphic Ostrom-derivates.

If $(r, q-1) = 1$ then there are: (a) $1 + (q-1)/2$ mutually non-isomorphic Ostrom-derivates if $q \equiv -1 \bmod 4$ and (b) $1 + (q-1)/4$ mutually non-isomorphic Ostrom-derivates if $q \equiv 1 \bmod 4$.

CHAPTER 58

Many Ostrom-Derivates.

In this chapter, we continue with the ideas developed in the two preceding chapters, Chapters 56 and 57, and determine a few conical flock spreads that have a number of mutually non-isomorphic Ostrom-derivates.

58.1. The Adelaide Flocks/Spreads.

We shall provide the construction of these flock planes in a chapter on the set of all infinite classes (see Chapter 59). The Adelaide flocks in $PG(3, 2^e)$ are cyclic admitting a group acting on the skeleton isomorphic to $C_{q+1} \times C_{2e}$. This means that the stabilizer of a given flock has order $2e$ in the latter case. Hence, the group acting on the q base reguli of the associated spread also has order $2e$. Thus, there are at least $q/2e$ orbits of base reguli providing at least this number of Ostrom-derivates from a given Adelaide flock.

THEOREM 58.1. For any Adelaide conical flock plane of order 2^{2e}, there are at least $[2^{e-1}/e]$ mutually non-isomorphic Ostrom-derivates.

For general background on the above remarks, the reader is referred to the survey article by Payne [1089].

58.2. Law–Penttila Flocks/Spreads.

When the skeleton is not transitive, we have given one general result for conical flocks of even order but again this is a very rough lower bound for the number of mutually non-isomorphic planes obtained.

We mention again the following question raised previously: **Can there exist a conical flock plane in $PG(3, q)$ that produces $(q + 1)q$ mutually non-isomorphic Ostrom-derivates?**

For example, the infinite class of Law–Penttila flocks of order/spreads of order q^2 and characteristic three come fairly close to providing such numbers of Ostrom-derivates. The spreads for these conical planes are as follows:

$$x \;=\; 0, y = x \begin{bmatrix} u - t^4 - nt^2 & n^{-1}t^9 - t^7 - n^2t^3 + n^3t \\ t & u \end{bmatrix}; u, t \in GF(3^e),$$

and n is a fixed non-square in $GF(3^e)$.

For example, the Law–Penttila flocks of order 3^{p^h} have $3^{p^{h-1}} \left(\frac{3^{\phi(p^h)}-1}{2p^h} \right)$ orbits of flocks in the skeleton which have trivial automorphism groups. This means that each such flock contributes $q = 3^{p^h}$ mutually non-isomorphic Ostrom-derivate planes. There are at least $3^{p^{h-1}} \left(\frac{3^{\phi(p^h)}-1}{2p^h} \right) 3^{p^h}$ mutually non-isomorphic Ostrom-derivates from these orbits. The flocks in other orbits contribute at least $[q/2p^h]$ Ostrom-derivates (see Payne [1089] for details on the orbit structure of the group).

THEOREM 58.2. Let π denote a Law–Penttila conical plane of order 3^{2p^h}. Then, there are at least $3^{p^{h-1}}\left(\frac{3^{\phi(p^h)}-1}{2p^h}\right)3^{p^h}+[3^{p^h}/2p^h]$ mutually non-isomorphic Ostrom-derivates.

Finally, we mention a very general problem, which is of general interest.

PROBLEM 58.3. Study the known conical flock planes and determine the number of mutually non-isomorphic Ostrom-derivates obtained from each plane.

CHAPTER 59

Infinite Classes of Flocks.

Since there are so many connections with flocks of quadratic cones, there are a variety of different methods of construction and hence there are competing designations for many of the infinite classes. We begin with a rough list of the infinite classes, after which, we shall add comments indicating the complexity of the situation. To add to the confusion, there are the derived flocks (the s-square and s-inverted spreads), many of which were determined after the initial discovery of the flock. Hence, we prefer to list the infinite classes in terms of their skeletons. We formulate a very rough classification of the skeletons to include 'semifield skeletons', where at least one member of the skeleton is a semifield flock, 'monomial skeletons', where at least one member of the skeleton has defining functions $(t, -f(t), g(t))$, where both f and g are monomials, 'likeable skeletons', where the conical flock spread is defined using a likeable function, 'transitive skeletons', where there is a transitive action on the members of the skeleton, 'Rigid skeletons', where at least one flock member of the skeleton admits only the identify collineation group, and the 'Adelaide super skeleton', containing skeletons corresponding to the Adelaide flocks and the Subiaco flocks. Finally, the 'linear skeleton' consisting only of the linear flock or Desarguesian conical flock spread is not listed; however, the 'almost linear skeletons', where at least one member of the skeleton contains a linear subset of $(q-1)/2$ conics, consists of precisely the Fisher skeleton. The explicit constructions and comments are directly below. We provide the conical flock spread representation or the BLT-set representation, except for the Adelaide super skeleton, which is provided directly after the statement of the theorem.

THEOREM 59.1. The infinite classes of flock skeletons are as follows:

S: *Semifield Skeletons*:

    S_1: The Kantor–Knuth skeleton,

    S_2: The Cohen–Ganley skeleton.

M: *Monomial Skeletons*:

    M_1: The Walker for q odd and Betten for q-even Skeletons (Narayana Rao and Satyanarayana for characteristic 5),

    M_2: The Barriga/Cohen–Ganley/Kantor/Payne skeleton.

L: *Likeable Skeletons*:

    L_1: The Kantor likeable skeleton.

AL: *Almost Linear Skeleton*:

    AL_1: The Fisher skeleton.

T: *Transitive Skeletons*:

    T_1: The Penttila skeleton,

    T_2: The Adelaide skeleton.

R: *Rigid Skeletons*:

    R_1: The Law–Penttila skeleton.

A: *The Adelaide Super Skeletons*:

    A_1: The Betten skeleton,

    A_2: The Adelaide skeleton,

    A_3: The Subiaco skeleton.

Whenever possible, we list the examples in translation plane, spread set form

$$x = 0, y = x \begin{bmatrix} u + g(t) & f(t) \\ t & u \end{bmatrix} ; u, t \in GF(q).$$

However, there are other ways that such translation planes might occur, in particular, they could have arisen from associate generalized quadrangles or from translation planes admitting Baer groups of order q or from infinite analogues, as for example, from topological translation planes. Whenever possible, we indicate the individuals connected with the various translation planes, either directly or indirectly. Otherwise, we give the BLT-set representation, if not already previously given. The designation of the flock derived spreads is not given explicitly.

59.1. Semifield Skeletons.

S_1: Knuth–Kantor Semifield Flocks ($q = p^e$, p odd, $e > 1$)

$$\begin{bmatrix} u & t^\sigma \gamma \\ t & u \end{bmatrix}, \ \sigma \in \operatorname{Aut} GF(q), \ \gamma \text{ a non-square in } GF(q).$$

REMARK 59.2. The translation plane is due to Knuth [**898**], and the associated generalized quadrangle was discovered by Kantor [**859**].

S_2: Cohen–Ganley Semifield Flocks ($q = 3^e$, $e > 1$)

$$\begin{bmatrix} u + nt + n^{-1}t^9 & f(t) \\ t & u \end{bmatrix} ; \ n \text{ a non-square in } GF(q)$$

Derived flock planes: Payne–Thas–Johnson–Lunardon–Wilke.

REMARK 59.3. This class of flocks arise from semifields of Cohen and Ganley [**238, 239**], and are commonly referred to as 'Ganley' flocks, but the semifields are implicit in Cohen and Ganley's work. Cohen and Ganley studied semifields of order q^2 that commute over a middle nucleus $GF(q)$, which forces the semifields to be [**395**] commutative. We note from Kantor [**877**] that they are associated with symplectic semifield planes, which are considered in Chapter 82. The spreads given above are obtained from the transpose and dualization of the middle nucleus semifields, producing semifields with left nucleus $GF(q)$, with spread in $PG(3, q)$. The form given is due to Gevaert and Johnson obtained by this method.

The derived flock planes are determined independently in Payne and Thas [**1094**] and Johnson, Lunardon and Wilke [**779**].

There are no non-linear semifield flocks of even order, by Johnson [**710**].

REMARK 59.4. The translation planes obtained by derivation are also due to Narayana Rao and his co-authors, around 1987 ([**1023, 1024**]. The Cohen–Ganley planes date from 1984.

59.2. Monomial Skeletons.

M_1: Walker/Betten/Denniston

$$\begin{bmatrix} u - t^2 & -\frac{t^3}{3} \\ t & u \end{bmatrix}; q \equiv -1 \bmod 3$$

REMARK 59.5. The flock quadrangles are determined by Kantor [859]. These flocks are also sometimes referred to as the Fisher–Thas–Walker flocks. Betten [112] found the even-order analogue from topological considerations. When $q = 5^r$, r odd, these planes arising from the derived side are also due to Narayana Rao and Satyanarayana [1014]. We shall be discussing hyperbolic fibrations which are usually equivalent to flocks of quadratic cones. Denniston [315] constructs a class of regulus-free spreads in $PG(3, q)$ by a partition of $PG(3, q)$, $q = 2^{2r+1}$, by a set of $q - 1$ quadrics union two lines, a so-called 'hyperbolic fibration'. The particular hyperbolic fibration constructs the flock planes.

M_2: Barriga/Cohen–Ganley/Kantor/Payne

$$\begin{bmatrix} u - \beta t^3 & -\gamma t^5 \\ t & u \end{bmatrix}; q = p^r,\ r \text{ odd},\ p \equiv \pm 2 \bmod 5, \beta^2 = 5\gamma$$

Derived flock: Johnson–Payne–Lunardon.

REMARK 59.6. Barriga [95] determined the spreads in the Baer or derived version using Chebyshev polynomials. The Baer or derived version are also implicit in Cohen and Ganley [239]. The quadrangles are due to Kantor [870] for q odd and Payne [1082] for q even. The derived flocks are determined using group theory in Johnson [736], and geometrically in Payne [1084]. Furthermore, these derived flocks are implicit in Lunardon [943].

59.3. Likeable Skeletons.

L_1: Kantor

$$\begin{bmatrix} u - t^2 & -\frac{1}{3}t^3 + kt^5 + k^{-1}t \\ t & u \end{bmatrix}; q = 5^r$$

Derived flocks: Bader–Lunardon–Thas.

REMARK 59.7. The Walker and Betten flock spreads are also likeable. The translation planes are due to Kantor [876]. Bader–Lunardon–Thas [54] used the classification theorem to determine the derived flock, whereas there is a geometric argument in Payne and Rogers [1093].

59.4. Almost Linear Skeletons.

Fisher Flocks (q odd, $q > 3$)

BLT-set: $\{ (\beta^{2j}, 0, 1), (0, \beta^{2j}, 1);\ 1 \le j \le (q+1)/2 \}$, β of order $q + 1$.

REMARK 59.8. The flocks originally discovered by Fisher produce associated generalized quadrangles found by Thas [**1177**] and further analyzed in Payne and Thas [**1094**]. In Payne and Thas [**1094**], it is shown that the Fisher flocks may be characterized as exactly the non-linear flocks (the Desarguesian spreads correspond to 'linear flocks' where the planes of $PG(3, q)$ of intersection share a line) that admit a linear subset of $(q-1)/2$ conics. This means that in the associated translation plane, of the q reguli that share a line, there is a subset of $(q-1)/2$ reguli whose associated translation net is Desarguesian. This characterization may be generalized in two ways. First, for odd order, Johnson [**743**] shows that any non-linear partial flock containing a linear subset of $(q-1)/2$ conics may be uniquely extended to a flock (the Fisher flock). In Johnson [**710**], it is shown that there are no (nonlinear) semifield flocks of even order. Recall that semifield spreads correspond to additive sets of matrices. However, for even order, Jha and Johnson [**627**] show that it is possible to extend a linear subset of $q/2 - 1$ flocks to a non-linear maximal partial flock. We note that the flock derivation process is valid for partial flocks (Johnson [**736**]) and the derived partial flocks are additive and non-linear.

59.5. Adelaide Super Skeleton.

In the Adelaide super skeleton, there are actually, more or less, three skeletons: The Subiaco, Betten and Adelaide skeletons. The Subiaco flocks are due to Cherowitzo, Penttila, Pinneri and Royle [**232**]. There is a unified construction of the Subiaco and Adelaide flocks due to Cherowitzo, O'Keefe and Penttila [**230**]. We shall give this construction. Let $q = 2^e$ and assume that $\beta \in GF(q^2)$ such that $\beta^{q+1} = 1$. Let T denote the trace function over $GF(q)$: $T(x) = x + x^q$ for all $x \in GF(q^2)$.

Let

$$a = \frac{T(\beta^m)}{T(\beta)} + \frac{1}{T(\beta^m)} + 1.$$

Define functions f and g on $GF(q)$ as follows:

$$f(t) = f_{m,\beta}(t) = \frac{T(\beta^m)(t+1)}{T(\beta)} + \frac{T((\beta t + \beta^q)^m)}{T(\beta)(t + T(\beta)^{1/2} + 1)^{m-1}} + t^{1/2}$$

and

$$ag(t) = ag_{m,\beta}(t) = \frac{T(\beta^m)}{T(\beta)}t + \frac{T((\beta^2 t + 1)^m)}{T(\beta)T(\beta^m)(t + T(\beta)t^{1/2} + 1)^{m-1}} + \frac{1}{T(\beta^m)}t^{1/2}.$$

We consider the putative conical flock spread

$$x = 0, y = x \begin{bmatrix} u + t^{1/2} & ag(t) \\ f(t) & u \end{bmatrix}; u, t \in GF(q).$$

The following theorem is stated in terms of translation planes and our designations:

THEOREM 59.9. (Cherowitzo, O'Keefe, and Penttila [**230**]). The above set defines a conical flock plane of the indicated type in the following situations:

(1) If $m \equiv \pm 1 \mod (q+1)$, the Desarguesian plane is obtained.

(2) If $q = 2^e$ and e is odd and $m \equiv \pm\frac{q}{2} \mod (q+1)$ the associated translation plane is Betten.

(3) If $q = 2^e$ and $m = \pm 5$, assume that β is such that for $\beta = \lambda^{k(q-1)}$ and λ a primitive element of $GF(q^2)$. Then if $q + 1$ does not divide km, a Subiaco plane is constructed.

(4) If $q = 4^e > 4$ and $m \equiv \frac{q-1}{3} \bmod (q+1)$ then for all β, an Adelaide plane is constructed.

It turns out that any Adelaide skeleton admits a group transitive on the skeleton. More generally, we consider transitive skeletons.

59.6. Transitive Skeletons.

T_1: Penttila

BLT-set: $\{\, (2\beta^{2j}, \beta^{3j}, \sqrt{5}) \in V; \, 0 \le j \le q \,\}$, $q \equiv \pm 1 \bmod 10$,

β of order $q + 1$.

59.7. Rigid Skeletons.

R_1: Penttila–Law

$$x = 0, y = x \begin{bmatrix} u - t^4 - nt^2 & n^{-1}t^9 - t^7 - n^2t^3 + n^3t \\ t & u \end{bmatrix} ; u, t \in GF(3^e),$$

n a fixed non-square.

CHAPTER 60

Sporadic Flocks.

Apart from the infinite classes, there are a number of conical flocks of small order. Actually, all flocks of prime-power order ≤ 29 are known, mostly by the use of the computer, and the flocks of order 32 are completely known. The use of the computer is probably pushed about as far as it can be, at least with current memory and speed. For example, it is estimated by Law and Penttila [**913**] that it might require eight years of computer time to determine the flocks of order 31. Still it is important to know flocks of small order, as these conceivably point to infinite classes. For example, there is a flock of order 11 (i.e., in $PG(3, 11)$) whose associated translation plane q^2 admits a collineation group of order $q(q+1)$ in the translation complement. Jha and Johnson [**632, 633**] completely determine the translation planes of order q^2 with spreads in $PG(3, q)$ that admit linear groups of order $q(q+1)$. When q is odd, all of these planes are either conical flock planes or Ostrom derivates of conical flock planes. The classification method is one of structure, in that in all but two sporadic cases, such planes are shown to be related to a Desarguesian plane by a single or multiple nest replacement procedure. For example, the Fisher conical flock planes are constructed from a Desarguesian affine plane by q-nest replacement. The more general concept of multiple nest replacements is developed in Jha and Johnson [**632, 633**] to analyze the planes obtained. In particular, the De Clerck–Herssens–Thas conical flock plane of order 11, originally found by computer, admits a collineation group of order $q(q+1)$, when $q = 11$, and may be shown to be constructed from a Desarguesian plane by double-nest replacement.

In the following table, we show the list of the conical flocks of orders q, for $q \leq 29$ and $q = 32$. This list is compiled from the work of Law and Penttila [**913**]. In that work, the term 'Kantor semifield flock' replaces our designation of 'Kantor–Knuth' flock, 'Thas–Fisher–Walker' replaces our designation of Walker for q odd or Betten for q even. The term 'Mondello' family refers to flocks determined by Penttila [**1098**] and the designation PRj refers to flocks found by Penttila and Royle [**1099**] where j refers to a group order. The group orders refer to groups acting on the associated BLT-sets. What is called a 'Kantor' monomial flock, we used the designation Barriga/Cohen–Ganley/Kantor/Payne flock(s), including both the even or odd orders. We use the term BCGKP for these flocks in both odd and even cases.

The flocks of orders $2, 3, 4$ are determined by Thas [**1177**], of orders $5, 7, 8$ by De Clerck, Gevaert and Thas [**272**], order 9 by Mylle [**995**] of order 11 and 16 by De Clerck and Herssens [**273**] (there is a class of order-11 partial flocks determined synthetically by Thas [**1187**], so when $q = 11$, the flock is called the De Clerck–Herssens–Thas flock). When $q = 13$ or 17, the flocks are determined by Penttila and Royle [**1099**]. For $q = 19, 23, 25, 27, 29$ Law and Penttila [**913**] find all flocks.

Finally, when $q = 32$, the possible flocks are listed in Brown, O'Keefe, Payne, Penttila, and Royle [**170**].

In the following, we list the order q, then the possible flocks, followed by comments, if any.

$q = 2, 3, 4$, linear.

$q = 5, 7, 8$, linear or Fisher.

$q = 9$, linear, Fisher, Kantor–Knuth.

$q = 11$, linear, Fisher, Walker, Mondello.

$q = 13$, linear, Fisher, two in BCGKP-skeleton.

$q = 16$, linear, Subiaco.

$q = 17$, linear, Fisher, Walker,

      two BCGKP-skeleton, two in De Clerck–Herssens-skeleton,

      two in Penttila–Royle-skeleton.

$q = 19$, linear, Fisher, Mondello, PR20, four from PR16-skeleton.

$q = 23$, linear, Fisher, Walker, PR1152, PR24,

      two in DCH72-skeleton, two in BCGKP-skeleton, four in PR16-skeleton,

      five in PR6-skeleton.

$q = 25$, linear, Fisher, Kantor–Knuth, two in Kantor Likeable-skeleton,

      three in PR16-skeleton, four in PR8-skeleton.

$q = 27$, linear, Fisher, Kantor–Knuth, two in Cohen–Ganley-skeleton,

      two in BCGKP-skeleton, seven in Law–Penttila-skeleton.

$q = 29$, linear, Fisher, Walker, Mondello, LP720, two in LP48-skeleton,

      five in LP8-skeleton, seven in LP6-skeleton, ten in LP3-skeleton.

$q = 32$, linear, Walker, BCGKP, Subiaco.

As mentioned, going further than 29 by computer, except for 'nice' orders, demands extraordinary computer memory. Actually, there are two special orders, $q = 23$ and 47, that merit mention. The classification results of Jha and Johnson [**633**] show that any translation plane with spread in $PG(3, q)$ admitting a linear collineation group of order $q(q+1)$ is either a conical flock plane or a derived conical flock plane. Further, with the exceptions of orders 23 and 47, all of the associated translation planes may be constructed from a Desarguesian affine plane of order q^2 by a described net replacement procedure. For $q = 23$ the planes of Penttila–Royle admit a group of order $23(24)$, PR24. Moreover, there are some sporadic flocks of order 47, one of which is due to Penttila–De Clerck, which admits a group of order $47(48)$. These exceptional cases are part of the general classification results of Jha and Johnson [**633**].

60.1. Law–Penttila $27 \leq q \leq 125$.

There are several other examples of BLT-sets found by computer. For example, Law and Penttila [**1099**], who find new examples of order q various orders. Let N denote the number of new mutually non-isomorphic BLT-sets determined. Then we have

q	N
27	1 (member of an infinite family)
29	5
31	5
37	3
41	3
43	1
47	2
49	2
53	2
59	3
125	1

The reader is directed to Law and Penttila [**1099**] for the description of the BLT-sets and other details.

60.2. Law $q = 83, 89$.

In addition to the BLT-sets listed in Chapter 57, there are other examples, one each for $q = 83$ and 89 in Law [**911**].

60.3. Lavrauw–Law $q = 167$.

In Lavrauw and Law [**910**], it is shown that there are two conjugacy classes of subgroups isomorphic to S_4 in $P\Gamma LO(5, q)$, for $q \equiv 5 \bmod 6$. Then a method for searching for BLT-sets admitting S_4 is given. A variety of BLT-sets are constructed, for $q = 23, 47, 71$ and a new BLT-set for $q = 167$. This method allows the possibility of continuing with the computation of new examples with the parameters given.

CHAPTER 61

Hyperbolic Fibrations.

In this chapter, we discuss another geometry associated with flocks of quadratic cones. Some of the material is taken from Johnson [**769**] in abbreviated form.

There is tremendous interest in what might be called the geometry of flocks of quadratic cones in $PG(3, q)$. This geometry includes certain translation planes whose corresponding spreads are unions of q reguli sharing a common line (see Gevaert, Johnson, and Thas [**397**]), certain generalized quadrangles of type (q^2, q) (see Thas [**1177**]), and translation planes with spreads in $PG(3, q)$ admitting Baer groups of order q (see Johnson [**726**] and Payne and Thas [**1094**]). In the last situation, there is a deficiency-one partial flock of a quadratic cone due to the work of Johnson [**726**]. Furthermore, partial flocks of deficiency one may be extended uniquely to flocks by Payne and Thas [**1094**]. There are also connections to sets of ovals, called 'herds' (see, e.g., [**1094**]), the existence of which provides a more general extension theory when q is even (see Storme and Thas [**1165**]). The reader interested in these and other connections is referred to the survey article by Johnson and Payne [**792**].

Another connection has emerged due to the work of Baker, Ebert and Penttila [**78**]. This is, that flocks of quadratic cones in $PG(3, q)$ are equivalent to the so-called 'regular hyperbolic fibrations with constant back half.'

A 'hyperbolic fibration' is a set $\mathcal{Q}$ of $q - 1$ hyperbolic quadrics and two carrying lines L and M such that the union $L \cup M \cup \mathcal{Q}$ is a cover of the points of $PG(3, q)$. (More generally, one could consider a hyperbolic fibration of $PG(3, K)$, for K an arbitrary field, as a disjoint covering of the points by a set of hyperbolic quadrics union two carrying lines.) The term 'regular hyperbolic fibration' is used to describe a hyperbolic fibration such that for each of its $q - 1$ quadrics, the induced polarity interchanges L and M. When this occurs, and (x_1, x_2, y_1, y_2) represent points homogeneously, the hyperbolic quadrics have the form

$$V(x_1^2 a_i + x_1 x_2 b_i + x_2^2 c_i + y_1^2 e_i + y_1 y_2 f_i + y_2^2 g_i)$$

for $i = 1, 2, \ldots, q - 1$ (the variety defined by the quadrics). When $(e_i, f_i, g_i) = (e, f, g)$ for all $i = 1, 2, \ldots, q - 1$, the regular hyperbolic quadric is said to have 'constant back half'.

The main theorem of Baker, Ebert and Penttila [**78**] is equivalent to the following.

THEOREM 61.1. (Baker, Ebert, Penttila [**78**]).
(1) Let $\mathcal{H} := V(x_1^2 a_i + x_1 x_2 b_i + x_2^2 c_i + y_1^2 e + y_1 y_2 f + y_2^2 g)$ for $i = 1, 2, \ldots, q - 1$ be a regular hyperbolic fibration with constant back half.

Consider $PG(3, q)$ as (x_1, x_2, x_3, x_4) and let C denote the quadratic cone with equation $x_1 x_2 = x_3^2$.

Define

$$\pi_0 : x_4 = 0, \ \pi_i : x_1 a_i + x_2 c_i + x_3 b_i + x_4 = 0 \quad \text{for } 1, 2, \ldots, q-1.$$

Then $\{\pi_j, j = 0, 1, 2, \ldots, q-1\}$ is a flock of the quadratic cone C.

(2) Conversely, if $\mathcal{F}$ is a flock of a quadratic cone, choose a representation as $\{\pi_j, j = 0, 1, 2, \ldots, q-1\}$ above. Choose any convenient constant back half (e, f, g), and define $\mathcal{H}$ as $V(x_1^2 a_i + x_1 x_2 b_i + x_2^2 c_i + y_1^2 e + y_1 y_2 f + y_2^2 g)$ for $i = 1, 2, \ldots, q-1$. Then $\mathcal{H}$ is a regular hyperbolic fibration with constant back half.

Now for each of the $q-1$ reguli, choose one of the two reguli of totally isotropic lines. Such a choice will produce a spread and a translation plane. Hence, there are potentially 2^{q-1} possible translation planes obtained in this way.

Using only affine homology groups Johnson [**769**] shows that it is possible to get to a flock using the ideas of hyperbolic fibrations.

61.1. Homology Groups and Hyperbolic Fibrations.

The following lemmas, given without proof, indicate how to connect translation planes admitting regulus-inducing affine homology groups to hyperbolic fibrations. All of this material may be given without finiteness, which allows for other interesting questions to emerge.

LEMMA 61.2. (1) Let $\mathcal{H}$ be a hyperbolic fibration of $PG(3, K)$, for K a field (a covering of the points by a set λ of hyperbolic quadrics union two disjoint carrying lines). For each quadric in λ, choose one of the two reguli (a regulus or its opposite). The union of these reguli and the carrying lines form a spread in $PG(3, q)$.

(2) Conversely, any spread in $PG(3, K)$ that is a union of hyperbolic quadrics union two disjoint carrying lines produces a hyperbolic fibration.

The important connection is that there is an associated Pappian plane and a corresponding model for a regulus using the Galois group of the associated field.

LEMMA 61.3. Let π be a translation plane with spread in $PG(3, K)$, for K a field, that admits a homology group H, such that some orbit of components is a regulus in $PG(3, K)$. Let Γ be any H-orbit of components.

Then there is a unique Pappian spread Σ containing Γ and the axis and coaxis of H.

LEMMA 61.4. Under the assumptions of the previous lemma, we may represent the coaxis, axis and Γ as follows:

$$x = 0, y = 0, y = xm; \ m^{\sigma+1} = 1; m \in K^+,$$

where m is in the field K^+, a 2-dimensional quadratic extension of K, and σ is the unique involution in $\mathrm{Gal}_K K^+$.

A basis may be chosen so that Σ may be coordinatized by K^+ as $\begin{bmatrix} u & t \\ ft & u+gt \end{bmatrix}$ for all u, t in K, for suitable constants f and g.

LEMMA 61.5. Under the previous assumptions, if $\{1, e\}$ is a basis for K^+ over K then $e^2 = eg + f$, and $e^\sigma = -e + g$, $e^{\sigma+1} = -f$. Furthermore, $(et + u)^{\sigma+1} = 1$ if and only in matrix form $et + u = \begin{bmatrix} u & t \\ ft & u+gt \end{bmatrix}$, such that $u(u+gt) - ft^2 = 1$.

The opposite regulus

$$y = x^\sigma m; m^{\sigma+1} = 1$$

may be written in the form

$$y = x \begin{bmatrix} 1 & 0 \\ g & -1 \end{bmatrix} \begin{bmatrix} u & t \\ ft & u+gt \end{bmatrix}; u(u+gt) - ft^2 = 1.$$

All of this provides an algebraic description of the associated translation planes.

LEMMA 61.6. If π is a translation plane of order q^2 with spread in $PG(3, K)$ admitting a homology group H such that one component is a regulus in $PG(3, K)$, then, choosing the axis of H as $y = 0$ and the coaxis as $x = 0$, we have the following form for the elements of H:

$$\begin{bmatrix} I & 0 \\ 0 & T \end{bmatrix}; T^{\sigma+1} = I.$$

Furthermore, we may realize the matrices T in the form $\begin{bmatrix} u & t \\ tf & u+gt \end{bmatrix}$ such that $u(u+gt) - t^2 f = 1$.

We then obtain the following representation of the regulus corresponding to the orbit of $y = x$ under H.

LEMMA 61.7. The associated hyperbolic quadric corresponding to $(y = x)H$ has the following form: $V[1, g, -f, -1, -g, f]$.

Using this homology group, we obtain:

LEMMA 61.8. It is possible to represent the spread as follows:

$$x = 0, y = 0, y = xM_iT \quad \text{for } i \in \rho, T^{\sigma+1} = I,$$

where $i \in \rho$, some index set. We thus have a set of ρ reguli that have the property that $x = 0$ and $y = 0$ are interchanged by the polarity induced by the associated hyperbolic quadric.

LEMMA 61.9. The spread for π has the following form:

$$x = 0, y = 0, y = xM_i \begin{bmatrix} u & t \\ ft & u+gft \end{bmatrix}; u(u+gt) - ft^2 = 1$$

and M_i a set of 2×2 matrices over K, where $i \in \rho$, some index set. Let

$$R_i = \left\{ y = xM_iT; T^{\sigma+1} = 1 \right\} \quad \text{for } i \in \rho, \text{ where } M_1 = I.$$

Then R_i is a regulus in $PG(3, K)$.

The opposite regulus to R_i may be represented as

$$R_i^* = \left\{ y = xM_i \begin{bmatrix} 1 & 0 \\ g & -1 \end{bmatrix} T; T^{\sigma+1} = I \right\}.$$

Using these ideas, it is possible to determine the associated quadratic forms.

LEMMA 61.10. The quadratic form for R_i is

$$V \left(xM_i \begin{bmatrix} 1 & g \\ 0 & -f \end{bmatrix} M_i^t x^t - y \begin{bmatrix} 1 & g \\ 0 & -f \end{bmatrix} y^t \right).$$

Also, recalling that it is really the set of hyperbolic quadrics that is of interest, the derived planes may be easily described.

LEMMA 61.11. The regulus

$$R_i = \left\{ y = x M_i T; \ T^{\sigma+1} = 1 \right\} \quad \text{for } i \in \rho, \text{ where } M_1 = I$$

and its opposite regulus

$$R_i^* = \left\{ y = x M_i \begin{bmatrix} 1 & 0 \\ g & -1 \end{bmatrix} T; T^{\sigma+1} = 1 \right\}$$

are interchanged by the mapping

$$\rho = \begin{bmatrix} I & 0 \\ 0 & P \end{bmatrix}, \text{where } P = \begin{bmatrix} 1 & 0 \\ g & -1 \end{bmatrix}; \ (x, y) \longmapsto (x, y^\sigma).$$

61.1.1. Johnson's Homology Theorem for Hyperbolic Fibrations.

THEOREM 61.12. (Johnson [**769**]). Let π be a translation plane with spread in $PG(3, K)$, for K a field. Assume that π admits an affine homology group H so that some orbit of components is a regulus in $PG(3, K)$.

(1) Then π produces a regular hyperbolic fibration with constant back half.

(2) Conversely, each translation plane obtained from a regular hyperbolic fibration with constant back half admits an affine homology group H, one orbit of which is a regulus in $PG(2, K)$.

H is isomorphic to a subgroup of the collineation group of a Pappian spread Σ, coordinatized by a quadratic extension field K^+, $H \simeq \langle g^{\sigma+1}; g \in K^+ - \{0\} \rangle$, where σ is the unique involution in $\mathrm{Gal}_K K^+$.

(3) Let $\mathcal{H}$ be a regular hyperbolic fibration with constant back half of $PG(3, K)$, for K a field. The subgroup of $\Gamma L(4, K)$ that fixes each hyperbolic quadric of a regular hyperbolic fibration $\mathcal{H}$ and acts trivially on the front half is isomorphic to $\langle \rho, \langle g^{\sigma+1}; g \in K^+ - \{0\} \rangle \rangle$, where ρ is defined as follows: If $e^2 = ef + g$, f, g in K and $\langle e, 1 \rangle_K = K^+$ then ρ is $\begin{bmatrix} I & 0 \\ 0 & P \end{bmatrix}$, where $P = \begin{bmatrix} 1 & 0 \\ g & -1 \end{bmatrix}$.

In particular, $\langle g^{\sigma+1}; g \in K^+ - \{0\} \rangle$ fixes each regulus and opposite regulus of each hyperbolic quadric of $\mathcal{H}$ and ρ inverts each regulus and opposite regulus of each hyperbolic quadric.

In the finite case, we may improve this as follows:

THEOREM 61.13. Let π be a translation plane with spread in $PG(3, q)$. Assume that π admits a cyclic affine homology group of order $q + 1$.

Then π produces a regular hyperbolic fibration with constant back half.

COROLLARY 61.14. Finite regular hyperbolic fibrations with constant back half are equivalent to translation planes with spreads in $PG(3, q)$ that admit cyclic homology groups of order $q + 1$.

61.2. K^+-Partial Flocks.

We now consider more completely the forms of the quadrics associated with translation planes admitting a homology group as above.

61.2.1. The Forms. Let π be a translation plane with spread in $PG(3, K)$ that produces a hyperbolic fibration with carriers $x = 0, y = 0$. Represent the spread of π as follows:

$$x = 0, y = 0, y = x \begin{bmatrix} u & t \\ F(u,t) & G(u,t) \end{bmatrix}; u, t \in K,$$

for functions F and G on $K \times K$ to K.

Let $\delta_{u,t} = \det \begin{bmatrix} u & t \\ F(u,t) & G(u,t) \end{bmatrix}$. We may assume that when $u(u+gt) - ft^2 = 1$ then $F(u,t) = ft$ and $G(u,t) = u + gt$. We now compute

$$V\left(xM_i \begin{bmatrix} 1 & g \\ 0 & -f \end{bmatrix} M_i^t x^t - y \begin{bmatrix} 1 & g \\ 0 & -f \end{bmatrix} y^t \right).$$

The operand

$$xM_i \begin{bmatrix} 1 & g \\ 0 & -f \end{bmatrix} M_i^t x^t - y \begin{bmatrix} 1 & g \\ 0 & -f \end{bmatrix} y^t$$

for $M_i = \begin{bmatrix} u & t \\ F(u,t) & G(u,t) \end{bmatrix}$ is easily calculated as follows:

$$x \begin{bmatrix} u^2 + (ug - tf)t = \delta_{u,t} & uF(u,t) + (ug - tf)G(u,t) \\ F(u,t)u + (gF(u,t) - fG(u,t))t & F(u,t)^2 + (gF(u,t) - fG(u,t))G(u,t) \end{bmatrix} x^t$$

$$- y \begin{bmatrix} 1 & g \\ 0 & -f \end{bmatrix} y^t.$$

We first show that the $(2, 2)$-entry in the front half of the quadric is a function of $\delta_{u,t}$. If not, consider $y = xM_i$, corresponding to $\delta_{u,t}$, and $y = xM_j$, corresponding to δ_{u^*,t^*}, and assume that the corresponding $(2, 2)$-elements are equal in both front halves. Then let $x = (0, x_2)$, and realize that then $(0, x_2, (0, x_2)M_i)$ is on the $\delta_{u,t}$-quadric and $(0, x_2, (0, x_2)M_j)$ is on the δ_{u^*,t^*}-quadric. This implies that

$$((0, x_2)M_i) \begin{bmatrix} 1 & g \\ 0 & -f \end{bmatrix} ((0, x_2)M_i)^t = ((0, x_2)M_j) \begin{bmatrix} 1 & g \\ 0 & -f \end{bmatrix} ((0, x_2)M_j)^t.$$

But this means that $(0, x_2, (0, x_2)M_i)$ is on both quadrics, a contradiction. Hence, this says that $\delta_{u,t}$ is a function of the $(2, 2)$-entry, and since the argument is essentially symmetric, we have that the $(2, 2)$-entry is a function of the $(1, 1)$-entry, say $f(\delta_{u,t})$. Similarly, if the sum of the $(1, 2)$- and the $(2, 1)$-entries of the front half is not a function of $\delta_{u,t}$ then there would be two distinct sums for a given $\delta_{u,t}$. But this again would say that $(x_1, 0, (x_1, 0)M_i)$ would be in two quadrics. Hence, the sum of the $(1, 2)$- and $(2, 1)$-elements is a function of $\delta_{u,t}$, say $g(\delta_{u,t})$. Consider a corresponding translation plane with components $y = xM$ and $y = xN$. Notice that the $(2, 2)$-element of the previous matrix for the front half is

$$\det \begin{bmatrix} F(u,t) & G(u,t) \\ fG(u,t) & F(u,t) + gG(u,t) \end{bmatrix}.$$

Hence, we obtain the next result, noting that the mapping from K^+ to K, mapping k in K^+ to $\det k$ in K, may not be onto.

THEOREM 61.15. (Johnson [**769**]). A regular hyperbolic fibration with constant back half in $PG(3, K)$, K a field, with carrier lines $x = 0, y = 0$, may be represented as follows:

$$V\left(x\begin{bmatrix} \delta & g(\delta) \\ 0 & -f(\delta) \end{bmatrix} x^t - y\begin{bmatrix} 1 & g \\ 0 & -f \end{bmatrix} y^t\right)$$

$$\text{for all } \delta \text{ in } \left\{\begin{bmatrix} u & t \\ ft & u+gt \end{bmatrix}; u, t \in K, (u,t) \neq (0,0)\right\}^{\sigma+1},$$

where

$$\left\{\begin{bmatrix} \delta & g(\delta) \\ 0 & -f(\delta) \end{bmatrix}; \delta \in \left\{\begin{bmatrix} u & t \\ ft & u+gt \end{bmatrix}^{\sigma+1}; u, t \in K, (u,t) \neq (0,0)\right\}\right\} \cup \left\{\begin{bmatrix} 0 & 0 \\ 0 & 0 \end{bmatrix}\right\}$$

corresponds to a partial flock of a quadratic cone in $PG(3, K)$, and where f and g are functions on $\det K^+$.

Johnson [**769**] also shows the following connection:

THEOREM 61.16. (Johnson [**769**]). The correspondence between any spread π in $PG(3, K)$ corresponding to the hyperbolic fibration and the partial flock of a quadratic cone in $PG(3, K)$ is as follows:

If π is

$$x = 0, y = 0, y = x\begin{bmatrix} u & t \\ F(u,t) & G(u,t) \end{bmatrix},$$

then the partial flock is given by $\begin{bmatrix} \delta_{u,t} & g(\delta_{u,t}) \\ 0 & -f(\delta_{u,t}) \end{bmatrix}$ with

$$\delta_{u,t} = \det\begin{bmatrix} u & t \\ ft & u+gt \end{bmatrix},$$
$$g(\delta_{u,t}) = g(uG(u,t) + tF(u,t)) + 2(uF(u,t) - tfG(u,t)),$$
$$-f(\delta_{u,t}) = \delta_{F(u,t),G(u,t)},$$

where

$$\delta_{M_i} = \det M_i$$

and

$$\delta_{F(u,t),G(u,t)} = \det\begin{bmatrix} F(u,t) & G(u,t) \\ fG(u,t) & F(u,t) + gG(u,t) \end{bmatrix} \in \det K^+.$$

THEOREM 61.17. If we have a hyperbolic fibration in $PG(3, K)$, there are corresponding functions given in the previous theorem such that the corresponding functions

$$\phi_s(t) = s^2 t + sg(t) - f(t)$$

are injective for all s in K and for all $t \in \det K^+$.

Indeed, the functions restricted to $\det K^+$ are surjective on $\det K^+$.

It is also possible to prove the converse.

THEOREM 61.18. (Johnson [**769**]). Any partial flock of a quadratic cone in $PG(3, K)$, with defining set λ (i.e., so t ranges over λ and planes of the partial flock are defined via functions in t) equal to $\det K^+$, whose associated functions on $\det K^+$, as above, are surjective on $\det K^+$ (K^+ some quadratic extension of K), produces a regular hyperbolic fibration in $PG(3, K)$ with constant back half.

61.2.2. Homology Groups in $GL(4, q)$. Suppose that H is a homology group of order $q + 1$ of a translation plane with spread in $PG(3, q)$. Let K^* denote the kernel homology group of the plane. Then HK^* fixes the axis and coaxis of the plane. Knowledge of the subgroups of $\Gamma L(2, q)$ implies the following:

LEMMA 61.19. If q is even then H is cyclic.

LEMMA 61.20. Assume that q is odd and $q \equiv 1 \bmod 4$. Then H is cyclic.

LEMMA 61.21. Let π be a translation plane of order q^2 admitting an Abelian homology group H. Then H is cyclic.

Combining the above three lemmas, then flocks of quadratic cones and translation planes of order q^2 with spreads in $PG(3, q)$ admitting affine homology groups of order $q + 1$ are interconnected as follows:

COROLLARY 61.22. Let π be a translation plane with spread in $PG(3, q)$ that admits an affine homology group H of order $q + 1$ in the translation complement. If any of the following conditions hold, π constructs a regular hyperbolic fibration with constant back half and hence a corresponding flock of a quadratic cone.
  (1) q is even,
  (2) q is odd and $q \equiv 1 \bmod 4$,
  (3) H is Abelian,
  (4) H is cyclic.

REMARK 61.23. There are translation planes of order $q^2 = 7^2$ with spread in $PG(3, 7)$ that admit quaternion homology groups of order $q + 1 = 8$, due to Heimbeck [**467**].

We have shown that it is possible to construct partial flocks of quadratic cones from regular hyperbolic fibrations in $PG(3, K)$, where K is an arbitrary field. Furthermore, we have shown that translation planes with spreads in $PG(3, K)$, admitting 'regulus-inducing' homology groups produce regular hyperbolic fibrations. Given two functions on K, $f(t)$, $g(t)$, form the functions $\phi_s \colon \phi_s(t) = s^2 t + sg(t) - f(t)$. A flock of a quadratic cone in $PG(3, K)$ is obtained if and only if the function ϕ_s is bijective for each s in K. Let K^+ be a quadratic extension of K (required for the construction of a flock) and let $\det K^+$ denote the set of determinants of elements of K^+, K^+ written as a 2×2 matrix field over K. We have seen that there is a corresponding regular hyperbolic fibration if and only if $\phi_s \mid \det K^+$ is surjective on $\det K^+$. Of course, this is trivially true in the finite case. We list this as an open problem, in the general case.

PROBLEM 61.24. Does every flock of a quadratic cone in $PG(3, K)$, for K an infinite field, produce a regular hyperbolic fibration?

We have not considered the collineation group of the translation planes corresponding to regular hyperbolic fibrations. In particular, for finite translation planes, we have a variety of open problems.

PROBLEM 61.25. Let π be a translation plane with spread in $PG(3, q)$ that admits a cyclic affine homology group H of order $q + 1$.
  (1) Is H normal in the full collineation group of π?
  (2) Is the full collineation group of π a subgroup of the group of the corresponding hyperbolic fibration?

We notice that there are Heimbeck planes of order $7^2 = q^2$ that admit quaternion homology groups H of order $8 = q + 1$, where H is not normal in the full collineation group of the plane. Hence, we may generalize the previous problem as follows:

PROBLEM 61.26. Let π be a translation plane with spread in $PG(3, q)$ that admits at least three affine homology groups of order $q + 1$. Completely classify the possible planes.

61.3. Geometric Constructions.

Brown, Ebert and Luyckx [169] give geometric constructions connecting regular hyperbolic fibrations and flocks of quadratic cones. Starting from a regular hyperbolic fibration, the construction does not require that the regular hyperbolic fibration have constant back half (or front half) and produces flocks. Since any flock produces, in turn, a regular hyperbolic fibration with constant back half, there is the possibility that from a translation plane covered by reguli, there is a construction which produces a similar plane that admits a cyclic affine homology group of order $q + 1$. For q even, the constructions of Luyckx [963] are more direct; however, here we give the construction that works for both q even and odd.

61.3.1. Translation Planes Covered by Reguli. Let π_1 be a translation plane of order q^2 with spread in $PG(3, q)$ whose spread is covered by a set of $q - 1$ mutually disjoint reguli together with two components L and M. Brown, Ebert and Luyckx [169] show that if q is even then the associated hyperbolic fibration is regular. Assume for either even or odd order that the associated hyperbolic fibration is regular. The construction of Brown, Ebert and Luyckx produces an associated flock of a quadratic cone. We call such flocks 'BEL-FLOCKS'. Conversely, using the flock, there is a construction of a set of associated regular hyperbolic fibrations with constant back half. Choose one. Hence, there is a set of translation planes admitting a cyclic homology group of order $q + 1$. Choose one of these, say π_2. Hence, we have a construction from any even-order translation plane whose spread is covered by a set of $q - 1$ mutually disjoint reguli together with two components to a translation plane (actually a set of such planes) that admit affine cyclic homology groups of order $q + 1$. What is the connection between these two planes? We formulate the following definition for even-order translation planes, but the definition may be applied for generally for such planes that induce regular hyperbolic fibrations.

DEFINITION 61.27. Let π be a translation plane of even order q^2 with spread in $PG(3, q)$ whose spread is covered by a set of $q - 1$ mutually disjoint reguli together with two components L and M. Consider the following sequence:

$\pi \to$ Hyperbolic Fibration $H_1 \to$ BEL-Flock

$\qquad \to$ Regular Hyperbolic Fibration $\to \Pi$-Cyclic Homology Plane.

This sequence is explained in the previous remarks. Any translation plane Π, admitting a cyclic affine homology group of order $q + 1$ constructed from π be such a sequence shall be called 'cyclic homology plane associated with π'.

PROBLEM 61.28. Given a translation plane of even order π with spread in $PG(3, q)$, which is covered by a set of $q - 1$ mutually disjoint reguli. Completely determined the cyclic homology planes associated with π.

61.3.1.1. *The Constructions of Brown, Ebert and Luyckx.* Let π be a translation plane of order q^2 with spread in $PG(3,q)$ whose spread is a union of $q-1$ mutually disjoint reguli union two lines ('carrying lines') $x = 0, y = 0$. Assume that the associated hyperbolic fibration is regular (it is necessarily regular, for example, when q is even).

Choose one of the carrying lines L. In the associated projective space $PG(3,q)$, let π_1 be a plane containing L. Let $n = \pi_0 \cap M$, where M is the remaining carrying line. The key point is that each of the reguli becomes in this setting a hyperbolic quadric $Q_i^+(3,q)$ that intersects π_0 in a non-degenerate quadric C_i, for $i = 1, 2, \ldots, q-1$. Furthermore,

$$\bigcup C_i \cup \{L, n\} = \pi_1, \text{ (is a covering)}.$$

When q is odd, n is the pole of L with respect to each C_i and is the nuclei of each C_i, when q is even.

LEMMA 61.29. (Brown, Ebert and Luyckx [**169**]). Let p and s be points on $M - \{n\}$. Then there is a unique elliptic quadric $Q_i^-(3,q)$ through p, s and C_i such that if ρ_i is the associated polarity of $Q_i^-(3,q)$ then $L^{\rho_i} = M$. Thus, each of the elliptic quadrics has common tangent planes $\langle L, p \rangle$ and $\langle L, s \rangle$ at p and s, respectively.

It turns out that the $q-1$ elliptic quadrics pairwise intersect precisely in the points p and s. Take a plane $\pi_1^* \neq \langle L, p \rangle$ on p but not on s, and set $\pi_1^* \cap \langle L, p \rangle = T$, and $\pi_1^* \cap \langle L, s \rangle = N_1$. Now to find the required cone, take any plane ρ_0 on N_1 distinct from π_1^* and note that $T \cap N_1 \cap L = \{x\}$. In ρ_0 choose any non-degenerate conic C_0' disjoint from x such that there is a tangent line ℓ to C_0' through x and form the corresponding quadratic cone $\mathcal{K}$, using the conic C_0', with x as vertex. Let $\ell = xy$, where y is not on N_1 and $y \in C_0'$.

REMARK 61.30. $\pi_1^* \cap Q_i^-(3,q) = C_i^*$, for $i = 1, 2, \ldots, q-1$ are conics that intersect pairwise in the point p and have T as common tangent line at p. Form the quadratic cones $\mathcal{K}_i$, with vertex y and base conic C_i^* and note that the cones $\mathcal{K}_i$ pairwise mutually share precisely the points on the line yp. Furthermore, the cones have a common tangent plane at yp, the plane $\langle y, T \rangle$ and $\mathcal{K} \cap \mathcal{K}_i$ consists of yp and q other points.

Brown, Ebert and Luyckx [**169**] show that the points of $\mathcal{K} \cap \mathcal{K}_i - \{yp\}$ lie in a plane ρ_i, for $i = 1, 2, \ldots, q-1$.

The main result then is:

THEOREM 61.31. (Brown, Ebert and Luyckx [**169**]). $\{\rho_0, \rho_1, \ldots, \rho_{q-1}\}$ is a flock of the quadratic cone $\mathcal{K}$.

Brown, Ebert and Luyckx also provide a geometric construction of a hyperbolic fibration from a flock, which actually constructs a regular hyperbolic fibration with constant back half and the reader is directed to their article for this construction.

Assume that we do have a regular hyperbolic fibration and the question is whether there is a constant back half. For any of the associated translation planes of order q^2, if there exists a cyclic affine homology group of order $q+1$, then the regular hyperbolic fibration has constant back half. To see how this works algebraically, we note the following lemma.

LEMMA 61.32. Let π be a translation plane of order q^2 with spread in $PG(3, q)$, which is covered by a set of $q - 1$ mutually disjoint reguli R_i, $i = 1, 2, \ldots, q - 1$, together with two lines L and M. The associated hyperbolic fibration is regular if and only if each regulus net defined by R_i together with $\{L, M\}$ can be embedded into a unique Desarguesian affine plane Σ_i such that there is an affine homology group H_i with axis M and coaxis L that act regularly on the regulus R_i.

PROOF. The idea of the proof is that each regulus net union L and M becomes a natural André net with carrying lines L and M. $\qquad\square$

So, we may begin by selecting any regulus R_1 and assume that L is $x = 0$, M is $y = 0$ and R_1 contains $y = x$ in the Desarguesian affine plane Σ_1. Then the partial spread is then represented in the form

$$\left\{ \begin{array}{c} x = 0, y = 0, y = x \begin{bmatrix} u & t \\ tf_1 & u + tg_1 \end{bmatrix}; \\ u, t \in GF(q); u(u + tg_1) - t^2 f_1 = 1 \end{array} \right\},$$

where $x^2 + xg_1 - f_1$ is $GF(q)$-irreducible.

Now in π, assume that components $y = xM_i$ are in R_i, for $i = 1, 2, \ldots, q - 1$, where M_i are 2×2 matrices over $GF(q)$ and $M_1 = I_2$. Embed R_i and $x = 0, y = 0$ into a unique Desarguesian affine plane Σ_i and change bases by $(x, y) \to (x, yM_i^{-1})$ to realizing $y = x$ as a component of Σ_i. Hence, we obtain for R_i as representation

$$\left\{ \begin{array}{c} y = xM_i \begin{bmatrix} u & t \\ tf_i & u + tg_i \end{bmatrix}; \\ u, t \in GF(q); u(u + tg_i) - t^2 f_i = 1 \end{array} \right\},$$

where $x^2 + xg_i - f_i$ is $GF(q)$-irreducible for all $i = 1, 2, \ldots, q - 1$. Thus, the spread for a translation plane, whose spread is covered by $q - 1$ mutually disjoint reguli in this way has the following form:

$$\left\{ \begin{array}{c} x = 0, y = 0, y = xM_i \begin{bmatrix} u & t \\ tf_i & u + tg_i \end{bmatrix}; \\ u, t \in GF(q); u(u + tg_i) - t^2 f_i = 1, i = 1, 2, \ldots, q - 1, M_1 = I_2 \end{array} \right\}.$$

Now define the following fields

$$K_i = \begin{bmatrix} u & t \\ tf_i & u + tg_i \end{bmatrix}; u, t \in GF(q), i = 1, 2, \ldots, q - 1.$$

By the results of Johnson [**769**], there is an affine homology if and only if all of the fields K_i are equal to K_1. Hence, we note the following problem:

PROBLEM 61.33. A regular hyperbolic fibration is equivalent to a spread of the following form:

$$\left\{ \begin{array}{c} x = 0, y = 0, y = xM_i \begin{bmatrix} u & t \\ tf_i & u + tg_i \end{bmatrix} \\ u, t \in GF(q); u(u + tg_i) - t^2 f_i = 1, i = 1, 2, \ldots, q - 1, M_1 = I_2 \end{array} \right\}.$$

Showing that all fields $K_i = \begin{bmatrix} u & t \\ tf_i & u + tg_i \end{bmatrix}; u, t \in GF(q), i = 1, 2, \ldots, q - 1$ are the same is equivalent to showing that the regular hyperbolic fibration has constant back half.

Therefore, the problem is to show that the fields all must be equal.

REMARK 61.34. As mentioned all hyperbolic fibrations for q even has been shown to be regular by Brown, Ebert and Luyckx [169]. For q even, it turns out that

$$M_i = \begin{bmatrix} a_i & 0 \\ G_i(a_i) & F(a_i) \end{bmatrix}; \{a_i; i = 1, 2, \ldots, q - 1\} = GF(q)^*.$$

Hence, a hyperbolic fibration for q even is equivalent to a spread of the following form:

$$\left\{ \begin{array}{c} x = 0, y = 0, y = x \begin{bmatrix} a_i & 0 \\ G_i(a_i) & F(a_i) \end{bmatrix} \begin{bmatrix} u & t \\ tf_i & u + tg_i \end{bmatrix} \\ u, t \in GF(q); u(u + tg_i) - t^2 f_i = 1, i = 1, 2, \ldots, q - 1 \end{array} \right\},$$

where $a_1 = 1$ and $G_1(a_1) = 0$ and $F(a_1) = 1$.

CHAPTER 62

Spreads with 'Many' Homologies.

In this chapter we consider the possible flocks of quadratic cones that produce associated translation planes admitting at least two cyclic regulus-inducing affine homology groups. Intrinsic to this study are two independent studies of translation planes admitting homologies. First we may consider translation planes admitting 'many' homologies and translation planes admitting 'several' large homology groups.

Assume that π is a translation plane of order q^2 with spread in $PG(3,q)$ that admits an affine homology group H of order $q+1$ in the translation complement. In two papers ([**796, 798**]), Johnson and Pomareda completely classified the translation planes with spreads in $PG(3,q)$ that admit 'many' homology axes of groups of prime odd order u. The term 'many' is defined to be $> q + 1$ axes. The main theorem is as follows:

THEOREM 62.1. (Johnson and Pomareda [**796**, Theorem 2]). Let π be a translation plane of order q^2 with spread in $PG(3,q)$ that admits $> q+1$ axes (or $> q+1$ coaxes) of homologies of odd order $u \neq 1$.

Then one of the following situations occurs:

(i) All of the homology groups have the same axis or all have the same coaxis and there is an elation group of order $> q + 1$ with affine axis,

(ii) π is Desarguesian,

(iii) π is Hall,

(iv) π is Ott–Schaeffer of order 2^{2r} where r is odd and the order $u = 3$,

(v) π is a Hering plane of order p^{2r} where r and p are both odd and $u = 3$,

(vi) G is $SL(2,9)$, $u = 3$ and $q = 7, 11, 13, 17$ and the planes are enumerated in Biliotti–Korchmáros [**142**], or

(vii) G is $GL(2,3)$, $u = 3$ and $q = 5$, and the plane is determined as in Johnson and Ostrom [**784**].

In Johnson and Pomareda [**796**], there is also an analysis of translation planes admitting two cyclic homology groups of order $q + 1$.

THEOREM 62.2. (Johnson and Pomareda [**796**, Theorem 18]). Let π be a translation plane with spread in $PG(3,q)$ that admits at least two cyclic homology groups of order $q + 1$. Then π is one of the following types of planes:

(1) André,

(2) q is odd and π is constructed from a Desarguesian spread by $(q + 1)$-nest replacement, or

(3) q is odd and π is constructed from a Desarguesian spread by a combination of $(q+1)$-nest and André net-replacement.

However, the case when q is prime p and $p+1 = 2^a$ was considered in Johnson and Pomareda [**796**] only in the case when there might be p-elements in the group generated by the affine homology groups. However, this case may be proved in more general situations and is mentioned in the next section.

It has been previously pointed out in Johnson and Pomareda [**796**] that the irregular nearfield planes of order 5^2 and 7^2 may be obtained from a Desarguesian by $q + 1$-nest replacement.

62.1. Two $(q + 1)$-Homology Groups.

In Chapter 61, flocks of quadratic cones and hyperbolic fibrations are intrinsically connected. In particular, flocks of quadratic cones in $PG(3, q)$ are equivalent to translation planes of order q^2 with spread in $PG(3, q)$ that admit affine cyclic homology groups of order $q + 1$. A major question concerning the relationship with the flock and the translation plane is the determination of the collineation group of the associated translation plane admitting a cyclic homology group as related to the group of the flock. In particular, if there is a cyclic homology group of order $q+1$, is the set $\{axis, coaxis\}$ invariant by the full collineation group? Furthermore, if the set $\{axis, coaxis\}$ is invariant under the full collineation group is the cyclic homology group normal in the full collineation group? For example, could there be two distinct cyclic homology groups with the same axis and coaxis?

We consider the following general problem in this section. **Determine the translation planes of order q^2 with spread in $PG(3, q)$ that admit at least three affine homology groups of order $q + 1$.** In the second section, we shall review the various examples. That these previous examples of translation planes are exceptionally rare, is made manifest, as this paper demonstrates. The principal result of Johnson [**765**] is the following:

THEOREM 62.3. (Johnson [**765**]). Let π denote a translation plane of order $q^2 = p^{2r}$, for p a prime, with spread in $PG(3, q)$ that admits an affine homology group H of order $q + 1$, in the translation complement.

(1) Then either the set $\{axis, coaxis\}$ of H is invariant under the full collineation group or π is one of the following planes:

    (a) Desarguesian,

    (b) The Hall plane,

    (c) The Heimbeck plane of order 49 of type III.

(2) If there exists at least three mutually distinct affine homology groups of order $q + 1$ but one or two axes, we have one of the following situations:

The plane is either

    (a) The irregular nearfield plane of order 25,

    (b) The irregular nearfield plane of order 49,

    (c) $q = 11$ or 19 and admits $SL(2, 5)$ as an affine homology group, (the irregular nearfield plane of orders 11^2 and 19^2 and the two exceptional Lüneburg planes of order 19^2 are examples),

    (d) $q \equiv -1 \bmod 4$ and there are two homology groups of order $q + 1$, with the same axis M and same coaxis L, exactly one of which is cyclic. Furthermore, the group generated by the two groups $H_1 H_2$ has order $2(q + 1)$. If K^* denotes the kernel homology group, then $H_1 H_2 K^*$ induces the regular nearfield group of dimension two on the coaxis L.

Also, there is a corresponding flock of a quadratic cone admitting a collineation g fixing one regulus and permuting the remaining $q - 1$ reguli in $(q - 1)/2$ pairs. In this case, there is an affine homology group of order 2 acting on the flock plane.

Indeed, the conical flock spread is either Desarguesian, Fisher or constructed from a Desarguesian spread by $3q$-double-nest construction and the conical flock plane is described in Theorem 62.11.

More generally, Johnson proves the following result when the assumption is that there are but two homology groups of order $q + 1$.

THEOREM 62.4. (Johnson [**765**]). Let π be a translation plane with spread in $PG(3, q)$ that admits at least two homology groups of order $q + 1$. Then one of the following occurs:

(1) $q \in \{5, 7, 11, 19, 23\}$, (the irregular nearfield planes and the exceptional Lüneburg planes are examples),

(2) π is André,

(3) q is odd and π is constructed from a Desarguesian spread by $(q + 1)$-nest replacement (actually $q = 5$ or 7 for the irregular nearfield planes also occur here),

(4) q is odd and π is constructed from a Desarguesian spread by a combination of $(q + 1)$-nest and André net-replacement,

(5) q is odd and $q \equiv -1 \bmod 4$ and the axis/coaxis pair is invariant under the full collineation group (furthermore, there is a non-cyclic homology group of order $q + 1$), or

(6) $q = 7$ and the plane is the Heimbeck plane of type III with 10 homology axes of quaternion groups of order 8.

62.2. Applications.

These general results applied to cyclic homology groups then provide the following theorem connecting the collineation groups of the translation planes associated with a regular hyperbolic fibration with constant back half and the collineation groups of the associated translation planes arising from flocks of quadratic cones. This material is taken from Johnson [**765**] with slight modification.

THEOREM 62.5. Let π_H be a translation plane of order q^2 with spread in $PG(3, q)$ admitting a cyclic affine homology group H of order $q + 1$. Let $\mathcal{H}$ denote the regular hyperbolic fibration obtained from π_H and let π_E be the corresponding conical flock spread. Then one of the following occurs:

(1) H is normal in the collineation group F_H of π_H. In this case, the collineation group of π_H is a subgroup of the group of the regular hyperbolic fibration and induces a permutation group F_H/H on the associated q reguli sharing a component, fixing one, of the corresponding conical flock spread π_E.

If Ker is the subgroup of π_E that fixes each regulus, then π_H is either

(a) Desarguesian or

(b) Kantor–Knuth of odd order or

(c) $Ker = K^*$, the kernel homology group of order $q - 1$.

(2) H is not normal in F_H and there is a collineation inverting the axis and coaxis of H. In this case, π_H is one of the following planes:

Then π is one of the following types of planes:

(a) André,

(b) q is odd and π is constructed from a Desarguesian spread by $(q+1)$-nest replacement, or

(c) q is odd and π is constructed from a Desarguesian spread by a combination of $(q + 1)$-nest and André net-replacement.

REMARK 62.6. According to Theorem 62.2, every translation plane of order q^2 that admits at least two cyclic homology groups of order $q + 1$ is one of the following:

(1) André,

(2) q is odd and π is constructed from a Desarguesian spread by $(q + 1)$-nest replacement, or

(3) q is odd and π is constructed from a Desarguesian spread by a combination of $(q + 1)$-nest and André net-replacement.

62.2.1. The Exceptional Lüneburg Planes Admitting $SL(2,3) \times SL(2,3)$.

There are eleven translation planes of order p^2 admitting $SL(2,3) \times SL(2,3)$ generated by affine homologies, admitting a collineation group G, such that G_L is doubly transitive on L for any component L, and such that the component orbits of the two homology groups isomorphic to $SL(2,3)$ are identical. These collectively are known as the exceptional Lüneburg planes of type $F * p$ (see section 19 of Lüneburg [**961**]). In this case, $p \in \{5, 7, 11, 23\}$.

We are interested in when $SL(2,3)$ contains two distinct groups of order $p + 1$.

Consider A_4 and note that there is a unique Sylow 2-subgroup S_2 such that any 3-group acts transitively on $S_2 - \{1\}$ by conjugation. Hence, there are no subgroups of A_4 of order 6. Since there is a unique involutory affine homology with a given axis in a translation plane, it follows that there are not two groups of order 12 in $SL(2,3)$. Hence, p cannot be 11. We know that the affine homology group induced on the coaxis is a subgroup of $SL(2,3)Z_{p-1}$, implying that there cannot be two homology subgroups of order 24 with a given axis. Hence, when $p = 23$, there are two homology groups of order $23 + 1$ but not three.

However, when $p = 5$, there are four Sylow 3-subgroups and hence four groups of order 6 in $SL(2,3)$, implying that the irregular nearfield plane of order 5^2 has at least three affine homology groups of order $5 + 1$.

However, when $p = 7$, there are two distinct quaternion groups of order 8 in $SL(2,3) \times SL(2,3)$, and two cyclic groups of order 8 in $GL(2,3) \times GL(2,3)$. Hence, in the irregular nearfield plane there are at least four homology groups of order 8. In the exceptional Lüneburg plane, there are two distinct quaternion groups of order 8.

In summary, we have:

SUMMARY 62.7. (1) When $p = 5$ or 7 the irregular nearfield planes of order p^2 admit at least three affine homology groups of order $p + 1$.

(2) When $p = 7$, the exceptional Lüneburg plane of order 7^2 admits at least two affine homology groups of order $7 + 1$.

(3) When $p = 23$, the irregular nearfield plane admits at least two affine homology groups of order $23 + 1$.

62.2.2. The Exceptional Lüneburg Planes Admitting $SL(2,5) \times SL(2,5)$.

There are fourteen translation planes of order p^2 admitting $SL(2,5) \times SL(2,5)$, generated by affine homologies, where there is a group G such that the stabilizer G_L of a non-axis, non-coaxis component L, is transitive on the non-zero points of L and the orbits of the two homology groups isomorphic to $SL(2,5)$ are identical (see Lüneburg [**961**, Section 18]). Among these planes are the irregular nearfield planes. When $p = 11^2$, there is only the irregular nearfield plane and when $p = 19^2$, there is the irregular nearfield plane and three others. The question is are there three affine homology groups of order $p + 1$?

Note that in $SL(2,5)$, there are six Sylow 5-subgroups of order 5, normalized by a group of order 4. Hence, there are at least two affine homology groups of order $20 = 19 + 1$ with the same axis. Hence, the irregular nearfield plane and the three exceptional Lüneburg planes of order 19^2 admit at least three affine homology groups of order $19 + 1$.

In this setting $p \in \{11, 19, 29, 59\}$.

Consider A_5, and the subgroup $\langle (123)(45)(23) \rangle$ and note that

$$(45)(23)(123)(45)(23) = (132).$$

Hence, A_5 has subgroups of order 6 (i.e., the normalizer of a Sylow 3-subgroup), implying that $SL(2,5)$ has at least two subgroups of order 11+1. Thus, the irregular nearfield plane of order 11^2 admits at least three homology groups of order $11 + 1$.

SUMMARY 62.8. (1) The irregular nearfield plane of order 11^2 with non-solvable homology groups admits at least three homology groups of order $11 + 1$.

(2) The irregular nearfield plane and the three exceptional Lüneburg planes of order 19^2 admit at least three homology groups of order $19 + 1$.

62.2.3. The Regular Nearfield Planes of Dimension 2. Since a homology group is generated from the Frobenius automorphism of $GF(q)$ and the cyclic group of order $(q^2 - 1)/2$ of $\Gamma L(1, q^2)$, and is sharply transitive on the non-zero vectors of the coaxis, the regular nearfield planes of odd order q^2, for $q \equiv -1 \mod 4$, admit homology subgroups of order $2(q+1)$, containing a cyclic subgroup of order $(q+1)$ and a non-Abelian subgroup of order $(q+1)$, containing a cyclic subgroup of order $(q+1)/2$.

Hence, there are at least three homology groups of order $q + 1$.

62.2.4. Conical Flock Planes; Two Groups. Now consider a conical flock plane; a translation plane corresponding to a flock of a quadratic cone, when q is odd and $q \equiv -1 \mod 4$. Assume that there is an affine homology of order 2 acting on the conical flock plane. If the corresponding spread is

$$x = 0, y = x \begin{bmatrix} u + g(t) & f(t) \\ t & u \end{bmatrix} ; u, t \in GF(q),$$

where f and g are functions on $GF(q)$, then the collineation takes the form:

$$(x, y) \longmapsto (x, -y).$$

From the calculations of Baker, Ebert and Penttila [78] (see p. 6), it follows that there is an associated collineation of the corresponding hyperbolic fibration of the form

$$(x, y) \longmapsto (x, yA),$$

where A is the matrix of an associated field involutory automorphism of an associated field times an element of the field:

$$\begin{bmatrix} u & t \\ ft & u + gt \end{bmatrix} ; u, t \in K.$$

Note that this mapping normalizes the cyclic homology group of order $(q + 1)$ arising from the construction of the hyperbolic fibration. So, the orbits of this homology group of order $(q+1)$ are permuted by the collineation in question. By a proper choice of reguli, there are translation planes admitting both groups as affine homology groups. Hence,

$$A = \begin{bmatrix} 1 & 0 \\ g & -1 \end{bmatrix} \begin{bmatrix} u & t \\ ft & u + gt \end{bmatrix}.$$

For example, if $g = 0$, and $u^2 + t^2 = -1$, then $A^2 = -I_2$. Note that $A = \begin{bmatrix} u & t \\ t & -u \end{bmatrix}$, does not have an eigenvalue of 1 and hence is fixed-point-free on the coaxis. Then the group generated by A, the field elements M of determinant 1 and the scalar group K^* of order $q - 1$ will induce a group on the coaxis $x = 0$ which is regular and corresponds to the regular nearfield group of dimension 2. We note that $AM = M^{-1}A$, for field elements of determinant 1, as $M^q = M^{-1}$. Hence, we obtain an affine homology group of order $2(q + 1)$ admitting two subgroups of order $q + 1$, $\langle M \rangle$, and $\langle A, M^2 \rangle$, exactly one of which is cyclic (which is, of course, normal).

We have proved the following result:

THEOREM 62.9. Let ρ be a finite conical flock plane of odd order q^2, $q \equiv -1 \bmod 4$ admitting an affine homology group of order 2; the spread has the following form

$$x = 0, y = x \begin{bmatrix} u + g(t) & f(t) \\ t & u \end{bmatrix}; u, t \in GF(q),$$

$$f(-t) = -f(t), g(-t) = -g(t).$$

Then there are associated translation planes of order q^2 admitting a cyclic homology group of order $q + 1$ that also admit an affine homology group of order $2(q + 1)$, with the following presentation

$$\left\langle g, h; h^{q+1} = g^4 = 1, g^2 = h^{(q+1)/2}, gh = h^{-1}g \right\rangle,$$

admitting two distinct affine homology groups of order $q + 1$;

$$\langle h \rangle, \left\langle g, h^2 \right\rangle.$$

62.2.5. Conical Flock Planes; Three Groups. In this section, we shall be considering three types of translation planes: translation planes π_H admitting cyclic affine homology groups of order $q + 1$, conical flock planes π_E, corresponding to these previous planes and Desarguesian planes Σ that are related as noted below.

Concerning translation planes π_H of order q^2 admitting cyclic affine homology groups of order $q + 1$ and their associated conical flock translation planes π_E, we see that if G is the group of a plane π_H and H is a cyclic homology group of order $q + 1$, then G/H is a group that acts on the set of reguli of the associated conical flock spread π_E. This group fixes one regulus and if π_H is not an André plane, then π_E is not Desarguesian so that G/H arises from a group G^+ of the translation plane π_E such that action on the set of reguli is G/H. It is possible that there is a collineation in G^+ that is not in the kernel homology group K^* of order $q - 1$ but fixes each regulus. However, the possible planes can only be the Kantor–Knuth of Desarguesian planes by the theory of 'rigidity' of Jha and Johnson [617]. The Kantor–Knuth spread is monomial and all monomial conical flock spread correspond to j-planes (see Johnson [655]). Hence, either the plane π_E is Kantor–Knuth or the group G^+ may be assumed to contain K^* and G^+/K^*

is isomorphic to G/H. Furthermore, note that the Kantor–Knuth planes do not admit groups of order$(q + 1)$.

Now assume that there are at least three distinct affine homology groups with affine axes acting in π_H. The axes could be distinct also, but assume that they are not. Then there must be two groups with the same axis and coaxis. It turns out that normally (and it is the object of this chapter to prove this) the third group must have axis and coaxis equal to the coaxis and axis, respectively, of the preceding two homology groups. In this case, except for a few sporadic cases, we are able to show that there is a cyclic homology group of order $q + 1$ on the two-group axis, where the group generated by the two homology groups has order $2(q + 1)$. In this case, $q \equiv -1 \bmod 4$. Furthermore, one of the groups of order $q + 1$ is cyclic and there is an associated flock of a quadratic cone by Theorem 61.13 and Corollary 61.14.

If there are two cyclic homology groups of order $(q + 1)$, then the translation plane admitting the homology groups is determined as either an André plane, a plane constructed by $(q + 1)$-nest replacement or a combination of André replacement and $(q+1)$-nest replacement. Let H denote the cyclic homology group of order $q + 1$. Then, $N_G(H)/H$ induces a collineation group on the associated flock and this group leaves one of the conics invariant. Considering what this says about the corresponding translation plane, we have a group of order $2(q + 1)$ that normalizes the 'regulus-inducing' elation group E of order q (assuming that the associated flock is not linear or equivalently that the translation plane is not Desarguesian). Hence, we obtain a collineation group acting on the translation plane of order $q(2(q + 1))t$, where there is a subgroup of order $2t$ that fixes each of the q reguli of the spread, so by the above remarks, this is a subgroup of the kernel homology group K^* of order $q - 1$. Since this group arises from a subgroup of $GL(4, q)$, it follows from the action on the conical flock that the corresponding group is a subgroup of $GL(4, q)$ that fixes a regulus and normalizes E. Hence, there is a subgroup of order $2(q+1)t$ that fixes the axis $x = 0$ of E and a second component $y = 0$. Since this group is a subgroup of $GL(4, q)$ that leaves invariant a regulus, it is also in $GL(2, q)*GL(2, q)$, where the $*$ denotes a central product by the center of either group. There is then a subgroup of $GL(2, q) * GL(2, q)$ that fixes two components of the regulus net and order $2(q + 1)t$. We know that in the affine homology translation plane, any subgroup of odd prime-power order of the homology group of order $(q + 1)$ (that is not cyclic) is cyclic. Moreover, the product of the Sylow t-subgroups of odd order forms a normal and cyclic subgroup. Assume that $q + 1$ has an odd prime factor u. Then there is an element τ_u of order u that fixes at least three components. Note that τ_u must centralize E, implying that τ_u fixes the regulus linewise. Hence, there is an associated Desarguesian affine plane Σ admitting a group G of order $2(q+1)t$, there a subgroup of order $(q + 1)_{2'}$ is a cyclic homology group (since the cyclic group is in G/K^*, and $(q + 1, q - 1) = 2$). Therefore, the intersection with $GL(2, q^2)$, the linear collineation group of Σ, has order divisible by $(q + 1)t$ and this group fixes two components of the conical flock plane which are then components of Σ as well. Moreover, the regulus of the conical flock plane also becomes a regulus of Σ fixed by G, so the stabilizer in $G \cap GL(2, q^2)$ of three components is a kernel homology group. Hence, we have a kernel homology group of order divisible by $(q + 1)/2$ of the Desarguesian plane Σ.

Now assume that q is odd and $q+1 = 2^a$, so that $q = p$ is prime. In the original affine homology group plane π_H, there is a homology group of order $(q+1)$ and since we are assuming that we do not have two cyclic homology groups of order $(q + 1)$, this group must be generalized quaternion. Therefore, there is a cyclic subhomology group of order $(q + 1)/2$. So, there is a cyclic subgroup acting on the conical flock, so there is a group G of order $2(q+1)t$, normalizing the elation group E, such that the group induced on the set of q reguli contains a cyclic group of order $(q + 1)/2$. Thus, it follows that G contains a cyclic group C of order divisible by $(q+1)/2$. If $q > 3$, the same ideas show that there is an associated Desarguesian affine plane Σ admitting the normalizer of C as a collineation group. Since the groups originate from a direct product of affine homology groups, it follows that the cyclic affine homology group of order $(q + 1)/2$ is normalized by the group of order $2(q + 1)^2$ of the affine homology group plane. This means that the group of order $2(q + 1)$ induced (not necessarily faithfully) on the set of reguli of the conical flock plane also normalizes the corresponding cyclic group. Since $q = p$, the conical flock plane cannot be Kantor–Knuth without being Desarguesian. Hence, G^+/K^* contains a normal cyclic subgroup of order $(q + 1)/2$, so let N be a normal subgroup of G^+ containing K^* such that N/K^* is cyclic of order $(q + 1)/2$. Let $N/K^* = \langle gK^* \rangle$, such that $g^{(q+1)/2} \in K^*$. Since we may assume that $q \equiv -1 \bmod 4$, (as otherwise all affine homology groups of order $(q + 1)$ are cyclic), it follows that there is an element g of order either $(q + 1)/2$ or $(q + 1)$ such that gK^* generates N/K^*.

Since all of these groups arise from subgroups of $GL(4, q)$, it follows that $\langle g \rangle$ is normal in N and N is Abelian. Therefore, there exists a unique Sylow 2-subgroup of N, implying that $\langle g \rangle$ is characteristic in N of order $(q + 1)$, so that we have a normal subgroup of order divisible by $(q+1)/2$ in a subgroup of order $2(q + 1)t$. Furthermore, since we have a cyclic group of order $(q+1)/2$ and if $q > 3$, the same ideas show that there is an associated Desarguesian affine plane Σ admitting as a collineation group the normalizer of this cyclic group of order $(q + 1)/2$. Hence, we again have a subgroup of $\Gamma L(2, q^2)$ of order $2(q + 1)t$, that fixes two components and normalizes an elation group E of order q. So, this group intersects $GL(2, q^2)$ is a group of order divisible by $(q + 1)t$, implying there is a kernel homology group of order at least $(q + 1)/2$.

Hence, we have shown the following:

THEOREM 62.10. Let π_H be a translation plane of order q^2 with spread in $PG(3, q)$ that admits at least three affine homology groups, where one is assumed cyclic, so q is odd. Then there is an associated Desarguesian affine plane Σ such that the conical flock plane π_E associated with π_H admits a Desarguesian collineation group of order $q(q+1)/2$, which is a product of a regulus-inducing elation group of order q and a kernel homology group of order $(q + 1)/2$ of Σ.

Such planes have been classified by Jha and Johnson [**632, 633, 636**]:

62.2.6. Jha–Johnson's $q(q+1)$-Theorem.

THEOREM 62.11. (Jha and Johnson [**636**]). Let π be a translation plane of order q^2 with spread in $PG(3, q)$ that admits a linear group G with the following properties:

(i) G has order $q(q+1)/2$,

(ii) There is an associated Desarguesian affine plane Σ of order q^2 such that $G = EZ$ where E is a normal, regulus-inducing elation group of Σ and Z is a kernel homology group of order $(q+1)/2$ of Σ.

Then

(1) π is either a conical flock plane or a derived conical flock plane.

(2) π is either Desarguesian or Hall, or

(3) If 4 does not divide $(q+1)$ then π is Fisher or derived Fisher.

(4) If π is of odd order q^2, $4 \mid (q+1)$, then either π is one of the planes of part (2) or (3) or π may be constructed from a Desarguesian plane by either

    (a) Double-nest replacement of a $3q$-double-nest, or

    (b) Derived from a plane which may be so constructed, by a base regulus net fixed by the group of order $q(q+1)/2$.

(5) If π is constructed by $3q$-double-nest replacement, the replacement net consists of a set of exactly $3(q+1)/4$ base reguli (E-orbits of components of Σ). This set is replaced by $\{\pi_i EZ; i = 1, 2, 3\}$ where π_i are Baer subplanes of Σ that intersect exactly $(q+1)/2$ base reguli in two components each.

The sets $\mathcal{B}_i$ of $(q+1)/2$ base reguli of intersection pairwise have the property that $|\mathcal{B}_i \cap \mathcal{B}_j| = (q+1)/4$ for $i \neq j$, $i, j = 1, 2, 3$ and $\mathcal{B}_1 \cap \mathcal{B}_2 \cap \mathcal{B}_3 = \varnothing$.

Hence, the type of associated conical flock plane that is obtained from a translation plane of order q^2 admitting at least three affine homology groups of order $(q+1)$ is determined. We shall see later that if there are three affine homology groups of order $q + 1$, then either the order is in $\{5, 7, 11, 19\}$ or admit homology groups of order $q + 1$ with symmetric axes (the axis and coaxis of one group is the coaxis and axis of the second group, respectively), at least one of which is cyclic. Hence, we have the following theorem.

THEOREM 62.12. Let π be a translation plane of order q^2, with spread in $PG(3, q)$, for $q \notin \{5, 7, 11, 19\}$ that admits at least three affine homology groups of order $(q+1)$. Then π corresponds to a flock of a quadratic cone and the corresponding conical flock spread is either

(1) Desarguesian,

(2) Fisher, or

(3) $q \equiv -1 \bmod 4$ and the plane may be constructed from a Desarguesian plane by $3q$-double nest replacement.

62.2.7. The Heimbeck Planes. Heimbeck [467] classifies all translation planes of order 7^2 that admit a quaternion affine homology group of order 8. There are exactly ten planes, of which there is a unique plane, type III, that admits at least three quaternion homology groups. The set of orbit lengths of the full collineation group on the components is $[10, 40]$.

62.2.8. The Hall Planes. Of course, the very exceptional Hall plane of order 9 admits 10 homology axes of groups of order 4. The Hall planes of order q^2, $q > 3$ admit $(q^2 - q)$ homology axes of cyclic groups of order $q + 1$.

Hence, we have examples in each of the possibilities listed in the main theorem when there are at least three affine homology groups of order $q + 1$. The specific examples are the irregular nearfield planes of orders $5^2, 7^2, 11^2, 19^2$ and the exceptional Lüneburg planes of order 19^2, the Heimbeck planes of order 7^2, the Hall and Desarguesian planes and planes arising from flocks of quadratic cones admitting involutory affine homologies and arising from the Desarguesian, Fisher, or $3q$-double nest replacements. Furthermore, the following result may be obtained partially by use of the computer program listing translation planes of order 49.

THEOREM 62.13. If π has order 7^2 and $\{\text{axis}, \text{coaxis}\}$ is not invariant, then π is Desarguesian, Hall or the Heimbeck plane of class III.

CHAPTER 63

Nests of Reguli.

In this chapter, we give a discussion of 'nests of reguli'. This concept originated with 'Bruen chains', which is generalized by Baker and Ebert. Although this text is concerned with finite translation planes, the discussion given here is valid for arbitrary orders and follows ideas of Johnson [**750**], who developed the general cases again following ideas of Baker and Ebert in the finite case.

DEFINITION 63.1. A 'chain' (or 'Bruen chain') of reguli in $PG(3, q)$ is a set of $(q + 3)/2$ reguli in a Desarguesian spread such that each line of the set is in exactly two reguli of the set.

Chains of reguli were originated by Bruen in 1978 [**195**] for the construction of translation planes. Bruen determined a chain in $PG(3, q)$, $q = 5, 7$, with the property that there is a corresponding replaceable translation net of degree $(q + 1)(q + 3)/4$ and order q^2, $q = 5$, in an associated Desarguesian affine plane. The replacement for the translation net is constructed from $(q + 1)/2$ components (lines) from the opposite regulus net of each of the $(q + 3)/2$ regulus nets (reguli). Replacement of this net produces a 'new' translation plane. We shall give a complete list of the known Bruen chains later in the section. In general, the question is whether all Bruen chains determine a replaceable translation net as in the case of $q = 5$ and 7, and in fact, as in all the known examples. The theorem of Heden [**458**] establishes this fact.

THEOREM 63.2. (Heden [**458**]). Finite Bruen chains in $PG(3, q)$ determine replaceable translation nets, where the replacement partial spread consists of exactly $(q + 1)/2$ components of the opposite regulus of each of the $(q + 3)/2$ reguli.

Actually, it is possible to give a definition of a 'chain' of reguli more generally as follows:

DEFINITION 63.3. A 'chain' of reguli in $PG(3, K)$, for K, a field that permits a quadratic extension field K^+ is a set of reguli in an associated Pappian spread such every two of the reguli share exactly two lines and no three reguli share a line.

A counting argument shows that the two definitions are the same for finite fields. There are no infinite examples of chains and there are no infinite classes of finite chains.

Recalling that the main interest in chains arises from the fact that from them interesting translation planes may be constructed, we isolate on the following generalization of chains due to Baker and Ebert (see, e.g., [**69, 70, 73, 74, 80**]). Because the arbitrary case may be easily carried out, we give the definition over arbitrary fields K.

For finite structures, as all chains determine replaceable translation nets, we may consider a model connecting reguli in $PG(3, q)$ with circles in the corresponding inversive plane obtained from $PG(1, q^2)$, by taking the 'points' as the points of $PG(1, q^2)$ and the 'circles' as the points at infinity of regulus nets. Since this may also be accomplished for any field K that admits a quadratic field extension K^+, we consider this more generally in $PG(3, K^+)$. The fields that are appropriate are the so-called 'full fields'.

DEFINITION 63.4. A 'full field' L of characteristic 0 or odd is a field such that the product of any two non-squares is a square that there exist non-squares. A field of characteristic not 2 if full if and only if the multiplicative subgroup of non-zero squares in an index 2 subgroup.

NOTATION 63.5. Let K be a full field of characteristic not 2, which has a full field quadratic extension $K(\sqrt{\gamma})$, where γ is a non-square in K. Let $\{1, e\}$ be a basis for $K(\sqrt{\gamma})$ such that $e^2 = \gamma$.

Let σ denote the involutory automorphism such that $(te + u)^\sigma = (-te + u)$, for $t, u \in K$. Define $a^{\sigma+i} = a^\sigma a^i$, for i an integer and $a \in K(\sqrt{\gamma})$.

The Pappian plane Σ coordinatized by $K(\sqrt{\gamma})$ has a spread representation as as follows:

$$x = 0, y = x \begin{bmatrix} u & \gamma t \\ t & u \end{bmatrix}; u, t \in K,$$

where the underlying vector space is 4-dimensional over K and x and y are 2-vectors over K.

DEFINITION 63.6. Let K be a full field of characteristic 0 or odd and let $K(\sqrt{\gamma})$ be a full field quadratic extension of K. For the corresponding Pappian plane Σ. A 't-nest of reguli' is a set of t-reguli in the spread for Σ with the property that given any component L of the union of the reguli then L is contained in exactly two reguli of the set.

Hence, for finite fields K isomorphic to $GF(q)$, a $(q + 3)/2$-nest of reguli is a Bruen chain.

There are infinite classes of t-nests of reguli, which define replaceable translation nets exactly as in the Bruen chain situation, where $t = q - 1, q, q + 1$ and $2(q - 1)$, constructed by Baker and Ebert (see, e.g., [**69, 70, 73, 74, 80**]). Furthermore, there are various sporadic t-nests. All such t-nests shall be discussed in due course. Actually, not all t-nests are replaceable.

THEOREM 63.7. (Ebert [**344**]). If C is a circle in $PG(1, q^2)$, q odd, which is tangent to linear flock of circles (q circles which mutually share exactly one point) there are two base circles of tangency R_1 and R_2.

(1) If E is the corresponding group of order q which leaves each flock circle invariant, then $EC \cup \{R_i; i = 1, 2\}$ is a $(q + 2)$-nest which is not replaceable.

(2) There are various sporadic nests which are not replaceable, all either $(q+1)$-nests or $(q + 3)$-nests, for $q = 5, 7, 11, 13$ or $(q + 4)$ for $q = 11$ or $(q + 7)$ for $q = 17$.

REMARK 63.8. For a t-nest to exist, we have $\frac{(q+3)}{2} \le t \le 2(q-1)$; the conjecture is if $t \le q$, then the t-nest is replaceable.

All of the infinite classes of t-nests in the finite case may be given by group-theoretic constructions and, as such, may be more generally considered in the arbitrary case.

DEFINITION 63.9. Let Σ be a Pappian plane coordinatized by a full field $K(\sqrt{\gamma})$. The kernel homology group H of Σ is the collineation group which fixes each component of the spread. Denote by $H^2 = \{h^2; h \in H\}$. Let G be a collineation group of Σ within $GL(4, K)$, which fixes $x = 0, y = 0$ (or merely $x = 0$ in some cases), which contains H^2.

Specify a set $\mathcal{B}$ of reguli within the spread of Σ such that G fixes each regulus of $\mathcal{B}$. Let $I = \cap \mathcal{B}$ denote the set of common components of the reguli in $\mathcal{B}$. These reguli are called the 'base reguli' for G provided the following properties hold:

(i) G acts transitively on the set of components of each base regulus disjoint from I,

(ii) The kernel subgroup H^2 within G has two orbits on the 1-dimensional K-subspaces of any component, so

(iii) The 1-dimensional subspaces of the base reguli not in I are in two orbits under G.

(iv) If $I = \{x = 0, y = 0\}$, then G acts transitively on the components of the opposite regulus of each base regulus (automatic in the finite case).

If G satisfies the conditions listed above, G is said to be a 'nest group'.

DEFINITION 63.10. Let G be a nest group. Assume that there is a Baer subplane π_0 of Σ which is a 2-dimensional K-subspace such that

(i) π_0 shares 0 or 2 components with each base regulus and

(ii) The two 1-dimensional K-subspaces of intersection on the components of a given bases regulus of intersection are in distinct G-orbits.

Let N_{π_0} denote the net in Σ defined by the components of the Baer subplane π_0. Assume that the partial spread corresponding to $N_{\pi_0}G$ is a nest of reguli.

(1) If $\pi_0 G$ is a replaceable partial spread for the partial spread of $N_{\pi_0}G$, we call the corresponding translation plane with spread determined by

$$\{\Sigma - N_{\pi_0}G\} \cup \pi_0 G,$$

a 'group replaceable translation plane'.

(2) If $N_{\pi_0}G$ is a nest of reguli, we shall call $N_{\pi_0}G$ a G/H^2-nest of reguli.

A classification result of Johnson [**750**] determines all of the G/H^2-nests of reguli.

THEOREM 63.11. (Johnson [**750**]). Let Σ denote the Pappian plane coordinatized by $K(\sqrt{\gamma})$, where K is a full field of zero or odd characteristic and where $K(\sqrt{\gamma})$ is a full field. Let σ denote the involutory automorphism of $\mathrm{Gal}_K K(\sqrt{\gamma})$ and assume that the mapping $m \longmapsto m^{\sigma+1}$ is surjective onto K, and that $|K| > 3$.

Let π be a group replaceable translation plane by use of a group G containing the index 2 kernel homology group H^2. Let π_0 be an associated Baer subplane and let N_{π_0} denote the regulus net in Σ defined by the components of the subplane π_0. Then $\pi_0 G$ is a replacement for the partial spread of $N_{\pi_0}G$ and one of the following situations must occur:

(1) $N_{\pi_0}G$ is a $\langle a^{\sigma+1}; a \in K(\sqrt{\gamma})^* \rangle$-nest of reguli and G is

$$\langle (x, y) \longmapsto (xa^{\sigma+2}, ya); a \in K(\sqrt{\gamma})^* \rangle H^2 \simeq \langle a^{\sigma+1}; a \in K(\sqrt{\gamma})^* \rangle H^2.$$

In the finite case, this is a $(q - 1)$-nest. (There is an associated affine homology group of order $(q - 1)/2$).

(2) $N_{\pi_0} G$ is a $\langle a^{\sigma-1}; a \in K(\sqrt{\gamma})^* \rangle$-nest of reguli and G is

$$\langle (x,y) \longmapsto (x, ya^{\sigma-1}); a \in K(\sqrt{\gamma})^* \rangle H^2 \simeq \langle a^{\sigma-1}; a \in K(\sqrt{\gamma})^* \rangle H^2.$$

In the finite case, we have a $(q+1)$-nest. (The affine homology group has component orbits which define reguli. This implies that there is an associated partial flock of a quadratic cone by the homology group theorem of Johnson [**726**]. This partial flock is a flock in the infinite case.)

(3) $N_{\pi_0} G$ is a $\left\langle \begin{bmatrix} 1 & u \\ 0 & 1 \end{bmatrix}; u \in K \right\rangle$-nest of reguli and G is

$$\left\langle (x,y) \longmapsto (x, xu+y); u \in K \right\rangle H^2 \simeq \left\langle \begin{bmatrix} 1 & u \\ 0 & 1 \end{bmatrix}; u \in K \right\rangle H^2.$$

There is a corresponding flock of a quadratic cone in $PG(3,K)$ (a generalization due to Jha and Johnson [**615**]) of work in the finite case due to Payne and Baker–Ebert (Payne [**1083**] for arbitrary square orders, generalizing Baker–Ebert [**70**] for prime square orders). In the finite case, we have a q-nest.

This nest is also called an E-nest (where E denotes the associated affine elation group).

(4) $N_{\pi_0} G$ is the set of all André nets either all of square type or all of non-square type. And, in either case, replacement of this partial spread leads to the full square André nearfield plane. (In a Pappian spread $x=0, y=0, y=xm$; $m \in K(\sqrt{\gamma})^*$, an André net determined by the partial spread $\{y=xm; m^{\sigma+1} = \delta\}$ is said to be 'square' or 'non-square', respectively as δ is square or non-square.)

(5) $N_{\pi_0} G$ is a $\langle a^{\sigma+1}; a \in K(\sqrt{\gamma})^* \rangle$-nest of reguli and G is

$$\langle (x,y) \longmapsto (x, ya^{\sigma+1}); a \in K(\sqrt{\gamma})^* \rangle H^2 \simeq \langle a^{\sigma+1}; a \in K(\sqrt{\gamma})^* \rangle.$$

In this situation, there is a corresponding flock of a hyperbolic quadric in $PG(3,K)$. This case does not occur in the finite case due to the theorem of Thas [**1178, 1179**] and Bader–Lunardon [**49**].

The corresponding constructions are as follows and are generalizations of work of Baker and Ebert (see references to t-nests in bibliography) in the finite case. Using the groups listed above in the cases (1), (2), and (3), we have putative examples of nests of reguli in the various cases but not a necessary Baer subplane. It turns out that Johnson [**750**] shows the existence, in each of these cases, of an appropriate Baer subplane. The arguments given in Johnson [**750**] are blends of what might be called 'circle geometry' and 'André methods'. Before the construction statements are given, we give a short sketch of how these two methods connect.

63.1. Circle Geometry over a Full Field.

We repeat for convenience our notation and setup.

NOTATION 63.12. Let K be a full field of characteristic not 2, which has a full field quadratic extension $K(\sqrt{\gamma})$, where γ is a non-square in K. Let $\{1, e\}$ be a basis for $K(\sqrt{\gamma})$ such that $e^2 = \gamma$.

Let σ denote the involutory automorphism such that $(te+u)^\sigma = (-te+u)$, for $t, u \in K$. Define $a^{\sigma+i} = a^\sigma a^i$, for i an integer and $a \in K(\sqrt{\gamma})$.

The Pappian plane Σ coordinatized by $K(\sqrt{\gamma})$ has a spread representation as follows:

$$x = 0, y = x \begin{bmatrix} u & \gamma t \\ t & u \end{bmatrix}; u, t \in K,$$

where the underlying vector space is 4-dimensional over K and x and y are 2-vectors over K.

We define the 'points' of the circle geometry to be the elements of $PG(1, K(\sqrt{\gamma}))$ and the 'circles' to be the set of 2×2 matrices of the following form:

$$\left\{ \begin{bmatrix} a & \alpha \\ \beta & -a^\sigma \end{bmatrix} \ \forall\, a \in K(\sqrt{\gamma}) \text{ and } \forall\, \alpha, \beta \in K; \ a^{\sigma+1} + \alpha\beta \neq 0 \right\}.$$

We identify the points of the geometry with the elements of $K(\sqrt{\gamma}) \cup \{(\infty)\}$. The set of points on a circle $\begin{bmatrix} a & \alpha \\ \beta & -a^\sigma \end{bmatrix}$ are the elements z that satisfy:

$$z = (az^\sigma + \alpha)/(\beta z^\sigma - a^\sigma),$$

noting that (∞) is a point of the circle if and only if $\beta = 0$.

The function Q which maps a circle to its determinant is a quadratic form which has the following associated bilinear form h:

$$h : h(C, D) = \mathrm{Det}(C + D) - \mathrm{Det}\, C - \mathrm{Det}\, D, \text{ for circles } C, D.$$

Then the condition of tangency, disjoint and secant of circles is determined as follows:

THEOREM 63.13. (See Johnson [**750**, (4.1)]). Two circles C and C are tangent, disjoint or secant if and only if

$(h(C, D)/2)^2 - \mathrm{Det}\, C \, \mathrm{Det}\, D$ is 0, a non-zero square, or non-square, respectively.

Circles are the infinite points of regulus nets.

THEOREM 63.14. (Johnson [**750**, (4.2)]). Consider a circle represented by

$$\begin{bmatrix} a & \alpha \\ \beta & -a^\sigma \end{bmatrix}; a^{\sigma+1} + \alpha\beta \neq 0.$$

Let $a = ea_1 + a_2$, for $a_i \in K$, $i = 1, 2$. Let $z = et + u$ be a point of the circle. Then the corresponding regulus net in the associated Pappian affine plane Σ coordinatized by $K(\sqrt{\gamma})$ has the following form:

$$\left\{ \begin{array}{c} y = x \begin{bmatrix} u & \gamma t \\ t & u \end{bmatrix}; \ (u^2 - \gamma t^2)\beta + 2a_1\gamma t - 2a_2 u - \alpha = 0; \\ \text{and } x = 0 \iff \beta = 0 \end{array} \right\}.$$

DEFINITION 63.15. In the Pappian plane Σ, an 'André net' A_δ is defined by the partial spread

$$\left\{ y = xm; m^{\sigma+1} = \delta \right\}.$$

An André net is a regulus net in Σ.

Considered as a circle, an André net is obtained if and only if $a = 0$.

Since we are interested in when Baer subplanes defined by 2-dimensional K-subspaces in the associated vector space are appropriate intersections with regulus nets, we require an associated connection with the circle geometry.

THEOREM 63.16. (Johnson [**750**, (4.4)]). Under the above assumptions and notations, we consider a particular 'base regulus'

$$\left\{ (x = 0), \; y = x \begin{bmatrix} u_0 & \gamma t_0 \\ t_0 & u_0 \end{bmatrix} \delta I; \; \delta \in K \right\} = R_{(u_0,t_0)},$$

for fixed $u_0, t_0 \in K$, not both zero.

Then a regulus net

$$\left\{ y = x \begin{bmatrix} u & \gamma t \\ t & u \end{bmatrix} ; \; (u^2 - \gamma t^2)\beta + 2a_1\gamma t - 2a_2 u - \alpha = 0; \\ \text{and } x = 0 \iff \beta = 0 \right\}$$

has $1, 0,$ or 2 components of intersection with $R_{(u_0,t_0)}$ if and only if

$$(a_1\gamma t_0 - a_2 u_0)^2 + \alpha(u_0^2 - \gamma t_0^2)\beta \text{ is zero, a non-zero square,}$$

$$\text{or non-square, respectively.}$$

Since our interest is when a Baer subplane intersects in two components, when it intersects we note that this occurs when

$$(a_1\gamma t_0 - a_2 u_0)^2 + \alpha(u_0^2 - \gamma t_0^2)\beta = 0$$

has no roots for any u, t in K, not both zero.

DEFINITION 63.17. Given a Baer subplane π_0, and associated regulus net N_{π_0} defined by the components of π_0, a base regulus is said to be a 'hit' of N_{π_0} if the number of component intersections with N_{π_0} is 2 and a 'miss' if there are no component intersections.

So, hits and misses refer only to the case when there are no tangent intersections.

Given two reguli R and R^*, if the hits of R are the misses of R^*, we shall call the two reguli 'companions'.

DEFINITION 63.18. The base regulus net with partial spread

$$\left\{ (x = 0), \; y = x \begin{bmatrix} 1 & 0 \\ 0 & 1 \end{bmatrix} \delta I; \; \delta \in K \right\} = R_{(1,0)},$$

for fixed $u_0, t_0 \in K$, not both zero,

is called the 'standard regulus net'. The corresponding circle is called the 'standard circle'.

THEOREM 63.19. (Johnson [**750**, (4.7)]).

(1) Any regulus net (respectively, circle) is the image of the standard regulus net (respectively, standard circle) under $GL(2, K(\sqrt{\gamma}))$ (respectively, $PGL(2, K(\sqrt{\gamma}))$).

(2) Any regulus net that does not contain $x = 0$ may be realized as an image of an André net under $GL(2, K(\sqrt{\gamma}))$. Hence, any regulus that does not contain $x = 0$ may be represented by

$$\left\{ y = xm; \; (m - b)^{\sigma+1} = \alpha \right\}$$

(3) The corresponding circle not containing (∞) has the form

$$\begin{bmatrix} b & \alpha \\ 1 & b^\sigma \end{bmatrix}$$

such that

$$m^{\sigma+1} - (mb^{\sigma} + m^{\sigma}b) - \alpha = 0,$$

represents points m on the circle.

(4) The Baer subplanes of the regulus net with the above form have the equations:

$$y = x^{\sigma}n + xb; n^{\sigma+1} = \beta.$$

Now it turns out that given a secant regulus to a base regulus, then companions always exist. That is,

THEOREM 63.20. (Johnson [**750**, (4.8)]). If R is a secant regulus with form

$$\left\{ y = x \begin{bmatrix} u & \gamma t \\ t & u \end{bmatrix}; (u^2 - \gamma t^2)\beta + 2a_1\gamma t - 2a_2 u - \alpha = 0 \text{ for } \beta \neq 0 \right\}$$

the companion R^* has the form

$$\left\{ \begin{array}{c} y = x \begin{bmatrix} u & \gamma t \\ t & u \end{bmatrix}; \\ (u^2 - \gamma t^2)\beta + 2a_2\gamma t - 2a_1\gamma u - \gamma(\alpha + (a_2^2 - \gamma a_1^2)/\beta = 0, \text{ for } \beta \neq 0 \end{array} \right\}.$$

Armed with this method and results, the problem of finding appropriate Baer subplanes may be resolved.

63.2. E-Nests of Reguli; q-Nests.

THEOREM 63.21. (Jha and Johnson [**615**]). (See Johnson [**750**, (3.1)]). Let K be a full field of 0 or odd characteristic and let $K(\sqrt{\gamma})$ be a full field quadratic extension of K. Let Σ denote the Pappian affine plane coordinatized by $K(\sqrt{\gamma})$. Let E denote the elation group with axis L which acts regularly on the components not equal to L of a K-regulus containing L. Let H^2 denote the group of squares of kernel homologies of Σ. Define the nets with defining partition spreads the component orbits of E union L as 'base regulus nets'.

Choose any Baer subplane π_0 which is a 2-dimensional K-subspace, let N_{π_0} denote the regulus net defined by the partial spread of π_0 and assume that N_{π_0} hits a base regulus R (i.e., two components of intersection). Then there are two Baer subplanes (incident with the zero vector) of the associated base regulus net such that π_0 intersects these Baer subplanes non-trivially. Assume these two Baer subplanes of intersection are in different H^2-orbits.

(1) Then, wherever π_0 intersects a component of a base regulus net, there are two components of intersection of this base regulus net (i.e., a 'hit') and, for each such base regulus net, the two Baer subplanes of intersection are in distinct H^2-orbits.

(2) There is a translation plane Σ_{π_0} obtained from the Desarguesian affine plane Σ by nest replacement of an E-nest of regulus.

The E-nest is the set of images of the regulus net N_{π_0} under E. If N is the set of components of the base regulus nets other than the axis of E, then $N_{\pi_0}E = N$. Furthermore, the replacement partial spread defining the replacement translation net is $\pi_0 E H^2$.

(3) In the finite case, for odd order, there is an associated replaceable q-nest.

REMARK 63.22. For specific examples, merely take a 1-dimensional K-space in a component of a base regulus and a second 1-dimensional K-space in a component of a second base regulus, with the property that the two 1-dimensional K-spaces are in different H^2-orbits and such that the generated 2-dimensional K-subspace is disjoint from $x = 0$. The natural spread on the 2-dimensional vector space produces a desired Baer subplane.

63.3. $\langle a^{\sigma-1} \rangle$-Nests of Reguli; $(q+1)$-Nests.

THEOREM 63.23. (Johnson [**750**, (5.2)]). Let K be a full field of 0 or odd characteristic and let $K(\sqrt{\gamma})$ be a full field quadratic extension of K. Let Σ denote the Pappian affine plane coordinatized by $K(\sqrt{\gamma})$, such that $|K| > 3$. Let σ denote the involution in $\mathrm{Gal}_K K(\sqrt{\gamma})$. Assume that the mapping from $K(\sqrt{\gamma})$ to K defined by $b \longmapsto b^{\sigma+1}$ is surjective. Let G denote the collineation group of the Pappian plane Σ coordinatized by $K(\sqrt{\gamma})$ defined by the elements

$$(x, y) \longmapsto (xa^{\sigma-1}, yb^{\sigma-1}); a, b \in K(\sqrt{\gamma})^*,$$

(contains the group H^2 of squares of kernel homologies). Define the 'base regulus nets' to be the André nets A_δ.

(1) Then there are Baer subplanes π_0 that intersect the base regulus nets only in 0 or two components (there are only 'hits').

Let λ denote the set of base reguli which are hits.

Let N_{π_0} denote the regulus net defined by the components of π_0.

(2) Then $N_{\pi_0}G$ defines a $\langle a^{\sigma-1} \rangle$-nest of reguli.

(3) There is a nest replacement $\pi_0 G = \bigcup \{ R_\beta; \ \beta \in \lambda \}$ constructed.

(4) In the finite case, and odd order, there is a corresponding replaceable $(q+1)$-nest.

REMARK 63.24. For specific examples, take the Baer subplane π_0 as follows:

$$\left\langle (1,1), (e, -ed); \ d^\sigma = d^{-1} \neq \pm 1, \text{ if } -1 \text{ is a square in } K \right\rangle,$$

$$\left\langle (1,1), (c, c^\sigma d); \ d^\sigma = d^{-1} \neq \pm 1, d^{-1} \neq \pm c^{\sigma-1}, \text{ if } -1 \text{ is a non-square in } K \right\rangle.$$

63.4. $\langle a^{\sigma+1} \rangle$-Nests of Reguli; $(q-1)$-Nests.

THEOREM 63.25. (Johnson [**750**, (7.3)]). Let K be a full field of 0 or odd characteristic and let $K(\sqrt{\gamma})$ be a full field quadratic extension of K. Let Σ denote the Pappian affine plane coordinatized by $K(\sqrt{\gamma})$, such that $|K| > 5$. Let σ denote the involution in $\mathrm{Gal}_K K(\sqrt{\gamma})$. Assume that the mapping from $K(\sqrt{\gamma})$ to K defined by $b \longmapsto b^{\sigma+1}$ is surjective. Let G denote the collineation group of the Pappian plane Σ coordinatized by $K(\sqrt{\gamma})$ defined by the elements

$$(x, y) \longmapsto (xa^{\sigma+2}b^2, yab^2); a, b \in K(\sqrt{\gamma})^*$$

(so contains H^2). Consider the subgroup G^-:

$$\left\langle (x, y) \longmapsto (xa^{\sigma+2}, ya) \right\rangle$$

and define 'base regulus nets' as defined by orbits of components under G^- union $x = 0, y = 0$ (these turn out to be reguli).

(1) Then there are Baer subplanes π_0 which intersect the base regulus nets in only 0 or 2 components (all 'hits') and whose non-trivial G-images are mutually

disjoint. Let λ denote the set of base regulus nets that are hits (intersect in two components). Let N_{π_0} denote the regulus net defined by the components of π_0.

(2) Then $N_{\pi_0} G$ defines a $\langle a^{\sigma+1} \rangle$-nest of reguli.

(3) There is a nest replacement $\pi_0 G = \bigcup \{ R_\beta; \, \beta \in \lambda \}$, which constructs a translation plane via $\langle a^{\sigma+1} \rangle$-replacement.

(4) In the finite case, there is a corresponding replaceable $(q-1)$-nest of reguli and an associated translation plane.

REMARK 63.26. For specific examples, take the Baer subplane π_0 as follows:

Write the vector space 4-dimensionally over K, using the basis $\{1, e\}$, write $t + ue$ as (t, u) and take

$$\left\langle \begin{array}{c} (1, 0, 1, 0), (0, 1, 0, \rho_o); \\ \rho_0 \text{ is a non-square} \neq \pm 1, \text{ if } -1 \text{ is a square in } K \end{array} \right\rangle ,$$

$$\left\langle \begin{array}{c} (1, 0, 1, 0), (a_2, 1, 0, -1); \\ a_2^2 - \gamma \text{ is non-zero square if } -1 \text{ is a non-square in } K \end{array} \right\rangle$$

(for example, if -1 is a non-square, $a_2 = \gamma + 1/4$ will work, if $\gamma \neq -1/4$).

63.5. Double $\langle a^{\sigma+1} \rangle$-Nests of Reguli; $2(q-1)$-Nests—Companion Constructions.

Baker and Ebert [**69**] have constructed replaceable $2(q-1)$-nests; these also have a generalization.

We state this theorem jointly with the theorem of the previous subsection. We shall see that there are potentially four translation planes constructed from a suitable subplane choice.

THEOREM 63.27. (Johnson [**750**, (7.3)]). Let K be a full field of 0 or odd characteristic and let $K(\sqrt{\gamma})$ be a full field quadratic extension of K. Let Σ denote the Pappian affine plane coordinatized by $K(\sqrt{\gamma})$, such that $|K| > 5$. Let σ denote the involution in $\mathrm{Gal}_K K(\sqrt{\gamma})$. Assume that the mapping from $K(\sqrt{\gamma})$ to K defined by $b \longmapsto b^{\sigma+1}$ is surjective. Let G denote the collineation group of the Pappian plane Σ coordinatized by $K(\sqrt{\gamma})$ defined by the elements

$$(x, y) \longmapsto (xa^{\sigma+2}b^2, yab^2); a, b \in K(\sqrt{\gamma})^*$$

(so contains H^2). Consider the subgroup G^-:

$$\langle (x, y) \longmapsto (xa^{\sigma+2}, ya) \rangle$$

and define 'base regulus nets' as defined by orbits of components under G^- union $x = 0, y = 0$ (these turn out to be reguli).

(1) (a) Then there are Baer subplanes π_0 which intersect the base regulus nets in only 0 or 2 components (all 'hits') and whose non-trivial G-images are mutually disjoint. Let λ denote the set of base regulus nets that are hits (intersect in two components). Let N_{π_0} denote the regulus net defined by the components of π_0.

(b) Then $N_{\pi_0} G$ defines a $\langle a^{\sigma+1} \rangle$-nest of reguli.

(c) There is a nest replacement $\pi_0 G = \bigcup \{ R_\beta; \, \beta \in \lambda \}$, which constructs a translation plane via $\langle a^{\sigma+1} \rangle$-replacement.

(d) In the finite case, there is a corresponding replaceable $(q-1)$-nest of reguli and an associated translation plane.

(2) (a) There is a Baer subplane π_0 satisfying part (1), a corresponding regulus net N_{π_0}, which has an associated Baer subplane π_0^*, such that the corresponding regulus net $N_{\pi_0^*} = N_{\pi_0}^*$ is a companion to N_{π_0} (the hits of N_{π_0} are the misses of $N_{\pi_0}^*$), such that hits of $N_{\pi_0}^*$ are in the set $\Lambda - \lambda$, where Λ denotes the complete set of base regulus nets.

(b) $N_{\pi_0}^* G$ defines a $\langle a^{\sigma+1} \rangle$-nest of reguli.

(c) There is a nest replacement $\pi_0^* G = \bigcup \{ R_\beta; \, \beta \in \Lambda - \lambda \}$, which constructs a translation plane via $\langle a^{\sigma+1} \rangle$-replacement.

(d) In the finite case, there is a second corresponding replaceable $(q-1)$-nest of reguli and another associated translation plane.

(e) The union of the two nests of reguli $N_{\pi_0} G$ and $N_{\pi_0}^* G$ covers the components of $\Sigma - \{x = 0, y = 0\}$ and using both replacements produces a double $\langle a^{\sigma+1} \rangle$-nest of reguli. In the finite case, there is a $2(q-1)$-nest of reguli.

63.6. Double-Doubles.

It is not difficult to show that the collineation $\tau \colon (x, y) \longmapsto (xk, y); k$ is a non-square, in not in G, but τ^2 is in G. But, τ will permute the set of base reguli and we note that π_0 and $\pi_0 \tau$ will turn out to be in different G-orbits. If we are just considering one $\langle a^{\sigma+1} \rangle$-nest of reguli, we obtain another nest of reguli and the corresponding translation planes will, of course, be isomorphic. However, when considering a double $\langle a^{\sigma+1} \rangle$-nest of reguli. We may combine this second nest with the original companion nest to produce another translation plane and as we may reverse the procedure, there are potentially four translation planes constructed in this manner, noting that π_0^* is NOT an image of π_0.

COROLLARY 63.28. Under the hypothesis of the above theorem, assume that π_0 leads to a secant regulus net N_{π_0} such that $N_{\pi_0} G$ defines a $\langle a^{\sigma+1} \rangle$-nest of reguli with replacement $\pi_0 G$. Assume further that a companion secant regulus net denoted by $N_{\pi_0^*}$ is such that $N_{\pi_0^*} G$ is a $\langle a^{\sigma+1} \rangle$-nest of reguli with replacement $\pi_0^* G$, where the partial spreads defining N_{π_0} and $N_{\pi_0^*}$ are disjoint. Then there is a translation plane with spread

$$\{x = 0, y = 0, \pi_0 \tau G, \pi_0^* G\}.$$

Hence, for a given Baer subplane which produces a secant regulus net with respect to the base reguli, there are potentially four mutually non-isomorphic translation planes that can be constructed by a combination of a replacement of either one or two $\langle a^{\sigma+1} \rangle$-nests of reguli originating from that subplane.

63.7. Mixed Nests.

Again, we develop a generalization of an idea of a 'mixed nest' due to Baker and Ebert [74], although Johnson and Pomareda [797] also considered such nests in the finite case as well. This is considered over an arbitrary full field K, which admits a full field quadratic extension field $K(\sqrt{\gamma})$, in Johnson [750]. The main idea of a mixed nest comes from consideration of André nets in an associated Pappian affine plane with spread in $PG(2, K)$, where we are considering a $\langle a^{\sigma-1} \rangle$-nest. The setup breaks apart the group used. If

$$G = \left\langle (x, y) \longmapsto x, yb^{(\sigma-1)}; \, b \in K \right\rangle H^2,$$

the group used in a $\langle a^{\sigma-1} \rangle$-nest construction, let G^- be

$$G^- = \Big\langle (x,y) \longmapsto x, yb^{2(\sigma-1)}; \, b \in K \Big\rangle H^2.$$

Assume that we have a set of base regulus nets λ with respect to a Baer subplane π_0, which produces the required replacement for the $\langle a^{\sigma-1} \rangle$-nest.

DEFINITION 63.29. We shall say that an André net R of λ is a 'D/S-net (different/same net)' (respectively, 'S/D-net (same/different net)') if and only if the two components of intersection with N_{π_0} are in different (respectively, the same) G^--orbit(s) and the two Baer subplanes of R of intersection are in the same (respectively, different) G^--orbit(s).

THEOREM 63.30. (Johnson [**750**, discussion on pp. 236–239]). Every André net of intersection in λ is either a D/S-net or an S/D-net.

NOTATION 63.31. $\lambda(D/S)$ shall denote the set of all D/S-nets of λ. Hence, $\lambda - \lambda(D/S)$ is the set $\lambda(S/D)$ of S/D-nets.

D/S-nets and S/D-nets are connected as follows:

THEOREM 63.32. (Johnson [**750**, (6.3)]). Let Σ denote the Pappian plane coordinatized by $K(\sqrt{\gamma})$. Let π_0 denote a Baer subplane which produces a $\langle a^{\sigma-1} \rangle$-nest of reguli with group G as above. Derive each André net, producing another Pappian plane Σ^*, where the components of Σ^* are $y = x^\sigma m \iff y = xm$ is a component of Σ, together with $x = 0, y = 0$, where σ is the involution in $\mathrm{Gal}_K K(\sqrt{\gamma})$.

(1) G is a collineation group of Σ^*.

(2) Every André net of intersection of Σ (respectively, Σ^*) which is a D/S-net (respectively, S/D-net) in λ (respectively, λ^* (the corresponding base regulus set of intersection)) is an S/D-net (respectively, D/S-net) in λ^* (respectively, λ). Simply put, the multiple derivation process makes a D/S-net into an S/D-net, and conversely.

THEOREM 63.33. (Johnson [**750**, (6.1), (6.2), and (6.4)]).

(1) For every André net in $\lambda(D/S)$, the 1-dimensional K-subspaces within the net are in exactly four G^--orbits.

(2) The stabilizer subgroup of any André Baer subplane of an André net in $\lambda(D/S)$ has two orbits of 1-dimensional K-subspaces in different G^--orbits.

(3) For every André net in $\lambda(D/S)$, the 1-dimensional subspaces of a Baer subplane of intersection of the André net are covered by $\pi_0 G^-$.

(4) If $R_\beta \in \lambda(D/S)$, there exists a set B_β of Baer subplanes of R_β such that the set of 1-dimensional K-subspaces on Baer subplanes of B_β is covered by $\pi_0 G^-$ and, for any element g in $G - G^-$, $B_\beta g \cup B_\beta$ is a disjoint union of the set of all Baer subplanes of R_β, which are incident with the zero vector (and hence are 2-dimensional K-subspaces).

(5) Let R be an André net in $\lambda(S/D) = \lambda - \lambda(D/S)$. Then the 1-dimensional K-subspaces of each line in the G^--orbit containing the components of intersection are in the same G^--orbit.

(6) The set of lines in the same orbit as the components of intersection are covered by $\pi_0 G^-$ and the set of lines in the other G^--orbit of lines are completely uncovered by $\pi_0 G^-$.

With these omnibus results on André nets, we now are in a position to define a 'mixed nest'.

DEFINITION 63.34. Consider the regulus net N_{π_0}, which is defined by the spread for π_0. Then the set of regulus nets in $N_{\pi_0}G^-$ is called the 'set of group-regulus nets'. This set turns out to be disjoint from the set of D/S-André nets. The union of these two sets of reguli is called a 'mixed nest' of reguli.

THEOREM 63.35. (Johnson [**750**, (6.6), (6.7) and the last part of (6.8)]).
(1) A mixed nest of reguli is a nest of reguli.
(2) Let M be a mixed nest of regulus net. Then the Baer subplanes of the base regulus nets $R_\beta \in \lambda(D/S)$ may be partitioned into two disjoint sets B_β and $B_\beta g; g \in G - G^-$, such that $\bigcup B_\beta$ is covered by $\pi_0 G^-$.
(3) For each André net in $\lambda(D/S)$, let B_β^* denote the set of Baer subplanes of $B_\beta g$ for $g \in G - G^-$. Then

$$M = N_{\pi_0}G^- \cup \{ R_\beta; R_\beta \in \lambda(D/S) \}$$

has a nest replacement by the following set:

$$\pi_0 G^- = \{ B_\beta^*; R_\beta \in \lambda(D/S) \}.$$

(4) There is a corresponding translation plane obtained from this mixed nest replacement, which we denote by Σ_{π_0}.
(5) Multiply derive the Pappian plane Σ by deriving all of the base regulus (André) nets to construct the Pappian plane Σ^*. Then there is another translation plane obtained by mixed nest replacement. Let the subplane be denoted by π_0^* is the corresponding Pappian plane Σ^* and let $N_{\pi_0^*}$ denote the regulus which contains π_o^*. Then the set

$$M = N_{\pi_{0^*}}G^- \cup \{ R_\beta^*; R_\beta \in \lambda(S/D) = \lambda^*(D/S) \}$$

is a mixed nest which permits nest replacement analogously as in part (3).

Note in the finite case in $PG(3,q)$, there are $q-1$ André nets, so if there are d_s D/S-André nets, there are $q-1-d_s = s_d$ S/D-André nets. The construction implies that we may use a set of either d_s nets in Σ or s_d nets in Σ^* to construct translation planes.
(6) In the finite case, mixed nests become $(\frac{(q+1)}{2}+i)$-nests, where i is the number of D/S-André nets. Hence, the above construction produces $(\frac{(q+1)}{2} + (q-1) - i)$-nests from a mixed nest.

Furthermore, the calculation of the possible numbers i has not been completely determined.

Weida [**1209**] shows that the possible replacements of t-nests are the expected type and also determines the collineation group.

THEOREM 63.36. (Weida [**1209**]). Let N be a finite replaceable t-nest and let V be a replacement.
(1) V consists of $(q+1)/2$ lines from each of the reguli of N.
(2) The collineation group of the translation plane π obtained from the Desarguesian plane Σ by replacement of V is inherited form Σ, provided $q > 7$.

REMARK 63.37. Several of the finite translation planes obtained from t-nest replacement may be constructed in different ways. For example:

(1) Payne [**1084**], for arbitrary q, and Baker and Ebert [**70**], for prime q, have shown that translation planes obtained from q-nest replacement correspond to the Fisher flock of a quadratic cone in $PG(3, q)$.

(2) The planes obtained from $(q+1)$-nests are those of Sherk and Pabst [**1160**], defined using indicator sets.

(3) Translation planes of order q^2 with spread in $PG(3, q)$ admitting cyclic affine homology groups are equivalent to flocks of quadratic cones by Johnson [**769**]. The $(q + 1)$-nest planes may then be constructed from the Fisher flocks.

REMARK 63.38. For mixed nests, t-nests of type $(\frac{(q+1)}{2} - i)$, there have been only partial determination of the possible values for i. However, there is a bound for i,

$$\frac{(q+1)}{4} - \frac{\sqrt{q}}{2} \le i \le \frac{(q+1)}{4} + \frac{\sqrt{q}}{2},$$

which is due to Baker and Ebert [**74**]. When $q = 37$, we have $26 \le t \le 31$, and there are t-nests for each of the possible values.

63.8. Classification Theorems for t-Nests.

REMARK 63.39. Most translation planes obtained from t-nest replacement also permit derivation or multiple derivation. As such, it might be difficult to recognize the newly constructed translation plane. On the other hand, certain groups will be inherited under the derivation process and it might be possible to reconstruct the translation plane using the group.

Now given a translation plane π of order q^2 with spread in $PG(3, q)$, when can it be determined that π has been constructed from a t-nest of a Desarguesian affine plane? We consider, $(q - 1)$-nests, $(q + 1)$-nests, and mixed nests. Other results of Jha and Johnson [**632**, **633**] on translation planes of order q^2 admitting groups of order $q(q + 1)$ will be given in Chapter 65.

THEOREM 63.40. (Johnson [**720**]). Let π be a translation plane of order q^2 with spread in $PG(3, q)$. If π admits a cyclic collineation group of order $q^2 - 1$ in the linear translation complement (i.e., in $GL(4, q)$), which has an orbit of components of length $q - 1$, then π is one of the following planes:

(1) A Desarguesian plane,

(2) A Hall plane,

(3) A regular nearfield plane or a derived regular nearfield plane,

(4) A plane constructed from a Desarguesian plane by replacing a $(q - 1)$-nest,

or

(5) A plane derived from a $(q - 1)$-nest plane.

For $(q + 1)$-nests, we have:

THEOREM 63.41. (Johnson and Pomareda [**794**]). Let π be a translation plane of order q^2 with spread in $PG(3, q)$. If π admits two commuting cyclic affine homology groups of order $(q + 1)$, then one of the following holds:

(1) π is an André plane,

(2) q is odd and π is obtained from a Desarguesian plane by replacing a $(q+1)$-nest, or

(3) q is odd and π is obtained from a Desarguesian plane by replacing a $(q+1)$-nest and a set of mutually disjoint reguli.

Finally, for mixed nests, there is a characterization as follows:

THEOREM 63.42. (Johnson and Pomareda [**797**]). Let π be a translation plane of odd order q^2, $q \neq 3, 7$, with spread in $PG(3, q)$. Assume that π admits two commuting affine homology groups H_1 and H_2 of order $(q+1)/2$. Let $e = (2, (q-1)/2)$ and assume that there is an Abelian group G of order $\frac{1}{2e}(q+1)^2$ in the linear translation complement which contains $H_1 \times H_2$.

If there is exactly one component orbit of length $> \frac{(q+1)}{e}$ under G, then π is one of the following:

(1) A plane constructed from a Desarguesian plane by replacing a mixed nest $(\frac{(q+1)}{2} + i)$-nest, or

(2) A plane constructed from a Desarguesian plane by replacing a mixed nest and a set of André nets whose carriers are the axes of H_1 and H_2.

CHAPTER 64

Chains.

We now list the known finite chains of reguli. The possible collineation groups are determined by L.M. Abatangelo [**4**].

THEOREM 64.1. (L.M. Abatangelo [**4**]). Let π be a translation plane obtained from a Desarguesian plane by $\frac{(q+3)}{2}$-nest replacement. Let G denote the full translation complement and K^* the kernel homology group of order $q-1$. Then G/K^* contains a subgroup N of index ≤ 4 with the following properties:

(1) N is isomorphic either to A_4, S_4 or A_5, or

(2) N is dihedral or cyclic with order bounded by a function of q (the complete bound is also given).

Here we develop our connections with chains of reguli and circle geometry with the following notation changes: Recall that a circle not containing (∞) has the form $\begin{bmatrix} b & \alpha \\ 1 & -b^\sigma \end{bmatrix}$ with points m such that $(m-b)^{\sigma+1} = \beta$. Hence, it follows that $m = b + ah$, where $h^{\sigma+1} = 1$ and $a^{\sigma+1} = \alpha$. Now assume that $\sigma = q$, for the finite case. Hence, in general if

$$H = \left\langle\, h \in GF(q^2)^*; \, h^{\sigma+1} = h^{q+1} = 1 \,\right\rangle,$$

we may represent

$$\begin{bmatrix} b & \alpha \\ 1 & -b^\sigma \end{bmatrix} \;=\; \left\{\, b + aH; \, a,b \text{ fixed in } GF(q^2) \,\right\}, \text{ or merely by}$$

$$b + aH.$$

Heden [**458**] determines Bruen chains by computer, using a set of what he calls 'even' circles.

DEFINITION 64.2. Given a circle $b + aH$, let Ω denote a fixed set of coset representatives of the group $\langle -1 \rangle$ and let $H = \langle \eta \rangle$. Then it follows that $a = rh$ or $a = rh\eta$ for $h \in H$ and $r \in \Omega$, and the representation in terms of elements of Ω, H and η is unique. Then $b + rH$ is said to be an 'even circle' and $b + rH\eta$ is called an 'odd circle'. In general, for a circle $b + aH$, b is said to be the 'center' of the circle and $|a| = \sqrt{a^{q+1}}$ is said to be the 'radius' of the circle.

It turns out that a Bruen chain of circles has either all even or all odd circles and one may transform from one set to the other; an 'even' Bruen chain is equivalent to an 'odd' Bruen chain.

Heden's algebraic characterization of Bruen chains follows:

THEOREM 64.3. (Heden [**458**, Theorem (6.1)]). Define $R_i = a_i + b_iH$, where $b_i \in \Omega$, for $i = 1, \ldots, (q+3)/2$ be an even circle, for each i. Let $\mathcal{B}$ denote the set of these circles. Define the functions i and s on $\mathcal{B} \times \mathcal{B}$ as follows:

$$i(R_i, R_j) = (a_i - a_j)^{q+1} - (b_i^{q+1} + b_j^{q+1})^2 - 4(a_i - a_j)^{q+1} b_j^{q+1},$$

$$s(R_i, R_j) = (a_i - a_j)^{q+1} - (b_i - b_j)^{q+1}.$$

Now define functions p on $\mathcal{B} \times \mathcal{B}$ as follows:

$$p(R_i, R_j) = \left\{ \begin{array}{c} 1 \text{ if } - s(R_i, R_j)/b_i b_j \text{ is a non-zero square of } GF(q), \\ 0 \text{ if } - s(R_i, R_j)/b_i b_j = 0, \\ -1 \text{ if } - s(R_i, R_j)/b_i b_j \text{ is a non-square of } GF(q) \end{array} \right\}.$$

Define par by

$$par(R_i) = p(R_i, R_1).$$

Then $\mathcal{B}$ is a Bruen chain if and only if the following two conditions are satisfied:
(1) $i(R_i, R_j)$ is a non-square of $GF(q)$ for any two i and j, $i \neq j$,
(2) $par(R_i) = p(R_i, R_j)par(R_j)$, for any two i and j, $i \neq j$.

Heden uses the above theorem to prove that Bruen chains are always nest replaceable. Also, this theorem provides a convenient algorithm for the construction of chains of small orders.

NOTATION 64.4. For a circle $a + bH$, let

$$\alpha = \left\{ \begin{array}{c} -1 \text{ if } a = 0 \\ t \text{ if } a = \eta^t, \text{ for } 0 \leq t \leq q^2 - 1 \end{array} \right\}.$$

We use the notation

$$(\alpha, \sqrt{b^{q+1}}, par(a + bH))$$

to represent the circle.

Hence, a Bruen chain may be represented by an appropriate set of $(q+3)/2$ 3-tuples.

REMARK 64.5. (1) Heden [458] determined by computer all Bruen chains in $PG(3, p)$ for odd prime $p \in \{5, 7, 11, 13, 17, 19, 23\}$.

(2) Heden [459] determined by computer all Bruen chains for $p = 29, 31$ and 37.

(3) Heden and Saggese [464] determined computer all Bruen chains in $PG(3, q)$, for $q = 9 = 25, 27$.

(4) Cardinali, Durante, Penttila, and Trombetti [214] have shown, also by computer, that the chains of the previous three parts are all of the chains in $PG(3, q)$, for q an odd prime power ≤ 49 and conjecture that this is a complete list of Bruen chains.

THEOREM 64.6. The following list of Bruen chains constitutes a complete list of the chains in $PG(3, q)$ for $q \leq 49$. The chains shall be denoted by a set of $(q+3)/2$ 3-tuples, where the notation is explained previously. We shall list the chain and usually the group Aut permuting the circles. We designate the chains by the mathematician originally constructing the chain, together with the size q. For example, (Bruen(5)) denotes a chain constructed by Bruen in $PG(3, 5)$. There are exactly nineteen chains.

(1)(Bruen(5)) : $q = 5$: $\{(-1, 1, 1), (0, 1, 1), (4, 1, 1), (17, 2, -1)\}$,

Aut $/ \langle e \rangle \simeq S_4$, e central involution; Bruen [195].

$$(2)(\text{Bruen}(7)) : q = 7 : \left\{ \begin{array}{c} (-1,1,1), (1,1,1), (3,1,1), \\ (15,1,1), (17,3,-1) \end{array} \right\},$$

Aut $/ \langle e \rangle \simeq S_4$, e central involution; Bruen [195].

$(3)(\text{Korchmáros}(7)) : q = 7 :$

$$\{(-1,1,1), (1,1,1), (25,1,1), (21,2,1), (45,2,1)\},$$

Aut $\simeq S_5$; Korchmáros [903].

$(4)(\text{V. Abatangelo}(9)) : q = 9 :$

$$\left\{ \begin{array}{c} (-1,0,1), (1,0,-1), (3,0,-1), \\ (18,0,1), (58,0,1), (65,0,1) \end{array} \right\},$$

Abatangelo [7].

$(5)(\text{Heden–Saggese}(9)) : q = 9 :$

$$\left\{ \begin{array}{c} (-1,0,1), (1,0,-1), (6,0,1), \\ (27,0,-1), (46,0,1), (75,0,-1) \end{array} \right\},$$

Heden and Saggese [464].

$(6)(\text{Capursi}(11)) : q = 11 :$

$$\left\{ \begin{array}{c} (-1,1,1), (0,1,-1), (7,1,1), \\ (50,1,-1), (115,3,1), (79,4,1), (103,4,1) \end{array} \right\},$$

Aut $\simeq D_{10} \times C_2$; Capursi [213].

$(7)(\text{Korchmáros}(11)) : q = 11 :$

$$\left\{ \begin{array}{c} (-1,1,1), (1,1,1), (4,2,1), \\ (34,2,1), (85,3,1), (45,5,1), (92,5,1) \end{array} \right\},$$

Aut $\simeq PSL(2,5)$; Korchmáros [902].

$(8)(\text{Raguso}(13)) : q = 13 :$

$$\left\{ \begin{array}{c} (-1,1,1), (3,1,-1), (10,1,1), \\ (165,1,-1), (158,5,-1), \\ (165,5,1), (131,6,1), (160,6,1) \end{array} \right\},$$

Aut $\simeq D_{12} \times C_2$; Raguso [1124].

$(9)(\text{Heden}(13)) : q = 13 :$

$$\left\{ \begin{array}{c} (-1,1,1), (3,1-1), (10,1,1), \\ (111,2,1), (73,3,-1), \\ (154,4,1), (0,6,1), (160,6,-1) \end{array} \right\},$$

Aut $\simeq S_4$; Heden [458].

$(10)(\text{Heden}(17)) : q = 17 :$

$$\left\{ \begin{array}{c} (-1,1,0), (0,1,1), (16,1,1), \\ (53,2,-1), (215,2,1), (26,3,1), \\ (170), (3,5,-1), (67,5,-1), (251,7,1) \end{array} \right\},$$

Aut $\simeq D_{12}$; Heden and Saggese [464].

(11)(Baker–Ebert–Weida(17)) : $q = 17$:

$$\left\{ \begin{array}{c} (-1,1,1), (0,1,1), (27,1,-1), \\ (37,2,-1), (105,2,-1), (200,2,1), \\ (188,3,-1), (83,4,-1), (98,6,1), (225,8,1) \end{array} \right\},$$

Aut $\simeq D_{16}$; Baker, Ebert and Weida [**80**].

(12)(Heden(19)$_1$) : $q = 19$:

$$\left\{ \begin{array}{c} (-1,1,1), (1,1,1), (21,1,1), \\ (83,1,1), (125,1,1), (131,1,1), \\ (145,1,1), (4,2,-1), (119,2,-1), (92,4,1), (229,4,1) \end{array} \right\},$$

Aut $\simeq D_{12}$; Heden [**458**].

(13)(Heden(19)$_2$) : $q = 19$:

$$\left\{ \begin{array}{c} (-1,1,1), (1,1,1), (64,1,-1), \\ (226,1,-1), (247,4,1), (45,5,1), \\ (155,5,1), (293,6,-1), (355,7,-1), (170,9,1), (194,9,1) \end{array} \right\},$$

Aut $\simeq S_4$; Heden [**458**].

(14)(Heden(23)) : $q = 23$:

$$\left\{ \begin{array}{c} (-1,1,1), (0,1,-1), (4,1,-1), \\ (15,1,1), (147,1,1), (148,2,1), (192,3,-1), \\ (404,5,1), (369,6,1), (212,8,1), \\ (493,9,-1), (279,11,1), (299,11,-1) \end{array} \right\},$$

Aut $\simeq D_8$; Heden [**458**].

(15)(Heden–Saggese)(25)$_1$) : $q = 25$:

$$\left\{ \begin{array}{c} (-1,1,0, (2,0,1), (10,0,1), (470,0,1), \\ (591,26,1), (199,52,1), (342,130,1), \\ (121,156,1), (464,156,-1), \\ (572,156,1), (604,156,1), (81,182,1), \\ (205,208,-1), (610,260,1) \end{array} \right\},$$

Heden and Saggese [**464**].

(16)(Heden–Saggese)(25)$_2$) : $q = 25$:

$$\left\{ \begin{array}{c} (-1,0,1), (2,0,1), (13,0,-1), \\ (257,0,-1), (440,26,-1), (118,78,1), \\ (289,78,-1), (91,104,1), (557,104,-1), \\ (592,130,-1), (372,208,-1), \\ (513,208,-1), (278,260,-1), (1,286,1) \end{array} \right\},$$

Heden and Saggese [**458**].

(17)(Heden–Saggese)(27)) : $q = 27$:

$$\left\{ \begin{array}{c} (-1, 0, 1), (2, 0, -1), (18, 0, -1), \\ (200, 0, -1), (260, 0, -1), (283, 28, -1), \\ (377, 56, -1), (151, 84, 1), (420, 112, -1), \\ (611, 196, -1), (470, 224, -1), \\ (644, 224, 1), (308, 252, -1), \\ (58, 308, -1), (350, 308, -1) \end{array} \right\},$$

Heden and Saggese [**458**].

(18)(Heden)(31)) : $q = 31$:

$$\left\{ \begin{array}{c} (-1, 1, 1), (1, 1, 1), (113, 1, 1), \\ (746, 2, 1), (626, 4, -1), (119, 5, 1), \\ (255, 6, 1), (402, 6, 1), (73, 7, 1), \\ (586, 7, 1), (84, 8, 1), (408, 8, 1), (72, 9, 1), \\ (285, 10, 1), (897, 11, 1), (707, 14, 1), (498, 15, 1) \end{array} \right\},$$

Heden [**458**].

(19)(Heden)(37)) : $q = 37$:

$$\left\{ \begin{array}{c} (-1, 1, 1), (3, 1, -1), (9, 1, -1), \\ (625, 1, -1), (1057, 1, -1), (1085, 1, -1), \\ (258, 2, -1), (406, 2, 1), (534, 2, -1), \\ (183, 3, 1), (891, 3, -1), (507, 4, 1), \\ (898, 4, -1), (201, 6, -1), (694, 11, 1), \\ (746, 11, -1), (962, 12, 1), \\ (922, 17, 1), (424, 18, 1), (976, 18, 1) \end{array} \right\},$$

Heden [**459**].

CHAPTER 65

Multiple Nests.

In Chapter 63, it has been pointed out that the Fisher planes, corresponding to the Fisher flocks of a quadratic cone in $PG(3,q)$ correspond to the q-nest planes. The Fisher planes admit a collineation group of order $q(q+1)$ and indeed, any translation plane of order q^2 with spread in $PG(3,q)$ that admits an affine elation group of order q is potentially a conical flock plane, and is one, provided the elation group has orbits defining reguli. However, in general a translation plane of order q^2 admitting a collineation group of order $q(q+1)$ may not be Fisher, but there may be a more general connection with Desarguesian planes and construction methods by use of sets of reguli. The general connection involves what are called 'multiple nests', the existence of which is not completely clear, although certain of these structures do occur in a natural manner.

In the following two definitions for Desarguesian planes, we note that any Baer subplane incident with the zero vector is always a line in $PG(3,q)$.

DEFINITION 65.1. Let Σ be a Desarguesian spread of order q^2 in $PG(3,q)$. A '$(k;t)$-nest' of reguli is a set of t reguli of Σ such that for each line L of the union, there are exactly $2k$ reguli of Σ that contain L and the cardinality of lines of the union is $t(q+1)/2k$ (where $2k$ divides $q+1$). When $k=1$, 2, 3 we use the terms 't-nest', 'double t-nest' and 'triple t-nest' of reguli, respectively.

If $\mathcal{P}$ is a $(k;t)$-nest of reguli, assume that for each regulus R of $\mathcal{P}$, there are exactly $(q+1)/2k$ Baer subplanes of the associated affine plane incident with the zero vector such that the union of these $t(q+1)/2k$ subplanes covers $\mathcal{P}$. We call such a set $\mathcal{P}^*$ of Baer subplanes a '$(k;t)$-nest replacement' for $\mathcal{P}$ and say that $\mathcal{P}$ is '$(k;t)$-nest replaceable'.

DEFINITION 65.2. A set $\mathcal{P}$ of t reguli in a Desarguesian affine plane of order q^2 with spread in $PG(3,q)$ is said to be a 'multiple-nest of type $(k_1,\ldots,k_z;t)$' or a '$(k_1,k_2,\ldots,k_z;t)$-nest', where the k_j are distinct integers, $k_i \le k_{i+1}$, such that the t reguli can be partitioned into z sets $\mathcal{S}_1,\mathcal{S}_2,\ldots,\mathcal{S}_z$ with the following properties:

(i) $\bigcup_{j=1}^{z}\mathcal{S}_j = \mathcal{P}$,

(ii) Lines of $\mathcal{S}_1$ are lines of exactly $2k_1$ reguli of $\mathcal{P}$, lines of $\mathcal{S}_2 - \mathcal{S}_1$ are lines of exactly $2k_2$ reguli of $\mathcal{P}$, lines of $\mathcal{S}_3 - \bigcup_{j=1}^{2}\mathcal{S}_j$ are lines of exactly $2k_3$ reguli of $\mathcal{P}$, $\ldots$, lines of $\mathcal{S}_w - \bigcup_{j=1}^{w-1}\mathcal{S}_j$ are lines of exactly $2k_w$ reguli of $\mathcal{P}$, for $w=2,3,\ldots,z$.

Note that if $k_i = k_{i+1}$, we may take $\mathcal{S}_i \cup \mathcal{S}_{i+1}$ as one of the sets of the partition. However, when we consider replacements for this set of reguli, we might lose information as to the replacement set (see below).

If $\mathcal{P}$ is a $(k_1,k_2,\ldots,k_z;t)$-nest of reguli, assume that for each regulus R of $\mathcal{S}_j$, there are exactly $w_j(q+1)/2k_z$ Baer subplanes of the associated affine plane incident with the zero vector such that the union of these subplanes covers $\mathcal{P}$. We call such a set $\mathcal{P}^*$ of Baer subplanes a '$(k_1, k_2, \ldots, k_z; t)$-nest replacement' for $\mathcal{P}$ and say that $\mathcal{P}$ is '$(k_1, k_2, \ldots, k_z; t)$-nest replaceable'.

Note that two reguli in different sets can nevertheless share lines.

Let $|\mathcal{S}_j| = t_j$ so that $\sum_{j=1}^{z} t_w = t$ then the cardinality of the set of lines of the union is

$$\sum_{j=1}^{z} t_j w_j (q+1)/2k_z.$$

We shall also use the notation

$$(k_1, k_2, \ldots, k_z; \{t_1, w_1\}, \{t_2, w_2\}, \ldots, \{t_z, w_z\})$$

for a $(k_1, k_2, \ldots, k_z; t)$ nest where $\sum_{j=1}^{z} t_w = t$ and $\sum_{j=1}^{z} t_j w_j (q+1)/2k_z$ is the degree of the net in the Desarguesian plane defined by the set of reguli.

REMARK 65.3. Let $\mathcal{P}$ be a $(k_1, k_2, \ldots, k_z; t)$-nest that is a nest-replaceable multiple nest of reguli in a Desarguesian spread Σ, with $(k_1, k_2, \ldots, k_z; t)$-nest replacement $\mathcal{P}^*$, then $(\Sigma - \mathcal{P}) \cup \mathcal{P}^*$ is a spread in $PG(3, q)$.

In the q-nest setting, there is a q-nest of reguli in a Desarguesian affine plane of odd order and a Baer subplane π_o such that H^- is the kernel homology subgroup of squares and E is a regulus-inducing elation group of order q then $\pi_o E H^2$ is a replaceable partial spread of $q(q+1)/2$ 2-dimensional K-subspaces. Replacement of the corresponding q-nest of reguli in the Desarguesian plane Σ, produces the Fisher planes. Note that we merely consider that H^- has order $(q+1)$. Then, since 2 divides $(q+1)$, there is always a $GF(q)$-group of order 2 fixing all Baer subplanes (that are $GF(q)$-subspaces).

We consider a more general construction. Now assume that 4 divides $q+1$, H is the homology group of order $(q+1)$ and we take the subgroup H^2 of order $(q+1)/2$ and consider three different q-reguli as follows:

Consider three distinct Baer subplanes of Σ, π_1, π_2, π_3 such that the reguli R_i of Σ containing π_i, $i = 1, 2, 3$, respectively, share exactly two components of $(q+1)/2$ base reguli (E-orbits of components). Let the set of such base reguli be denoted by B_i, $i = 1, 2, 3$. Furthermore, assume that B_i and B_j share components of exactly $(q+1)/4$ base reguli for all $i \neq j$, $i, j = 1, 2, 3$ and the union of the three sets of base reguli has line cardinality $3(q+1)/4$ and $B_1 \cap B_2 \cap B_3 = \varnothing$.

Then $\pi_i E H^2$ is a partial spread of $q(q+1)/4$ 2-dimensional K-subspaces involving q reguli and each 'line' of this set is in exactly four of the q-reguli.

To be specific concerning double and triple nests, we repeat part of the previous definition.

DEFINITION 65.4. If H^2 is the homology group of order $(q+1)/2$, for $(q+1)/2$ even, of the Desarguesian affine plane Σ then $\bigcup_{i=1}^{3} \pi_i E H^2$ is a set of $3(q(q+1)/4)$ 2-dimensional K-subspaces, and there are exactly $3q$ reguli and each 'line' lies on exactly 4 of the reguli. We shall call this a 'double-nest' of $3q$-reguli.

If $\bigcup_{i=1}^{3} \pi_i E H^2$ is a partial spread, we shall call the double-nest of $3q$-reguli 'double-nest replaceable'.

Hence, in this case, there is a corresponding translation plane admitting the collineation group $E H^2$ of order $q(q+1)/2$ constructed from a Desarguesian affine plane. Since E is regulus-inducing, it follows that the constructed translation plane is a conical flock plane.

DEFINITION 65.5. Suppose we have a set of Nq reguli of a Desarguesian affine plane of order q^2 covering exactly $N(q+1)/8$ lines and such that each line lies on exactly 8 reguli. We shall call such a set of reguli a 'triple-nest of Nq reguli.

Assume that there is a regulus-inducing group E of order q and a kernel homology group of H^4, of order $(q+1)/4$ of the kernel homology group. Assume that there are N subplanes π_i such that $\pi_i E H^4$ are partial spreads and that the associated sets of base reguli B_i and B_j corresponding to π_i and π_j share exactly $(q+1)/8$ base reguli, the intersection of any three distinct sets of B_k is empty and $\bigcup_{i=1}^{N} \pi_i E H^4$ involves $Nq(q+1)/8$ lines such that each line is incident with exactly 8 reguli. If this set is a partial spread, we call the triple-nest of Nq reguli 'triple-nest replaceable.'

Hence, we obtain a translation plane admitting a collineation group EH^4 of order $q(q+1)/4$. Since E is regulus-inducing, the plane is a conical flock plane.

REMARK 65.6. In the more general situation, that of a $(k_1, k_2, \ldots, k_z; t)$-nest and corresponding replacement denoted by

$$(k_1, k_2, \ldots, k_z; \{t_1, w_1\}, \{t_2, w_2\}, \ldots, \{t_z, w_z\})\text{-nest},$$

Jha and Johnson [636] have completely determined the possibilities when there is a group EH^4 as above. Apart from the double-nest and triple-nest situations, there are exactly the following types: a $(2, 4; 4q)$ or $(2, 4; 5q)$-nest of degree $Nq(q+1)/8$ for $N = 6$ or 7 or a $(3, 4; 6q)$-nest of degree $7q(q+1)/8$.

The main result of Jha and Johnson [636] is as follows:

THEOREM 65.7. (Jha and Johnson [636]). Let π be a translation plane of order q^2 with spread in $PG(3, q)$ that admits a linear group G with the following properties:

(i) G has order $q(q+1)$,

(ii) There is an associated Desarguesian affine plane Σ of order q^2 such that $G = EH$ where E is a normal, regulus-inducing elation group of Σ, and H a subgroup of $\Gamma L(2, q^2)$.

Then

(1) π is either a conical flock plane or a derived conical flock plane.

(2) If 8 does not divide $q+1$ then π is either

    (a) Desarguesian,

    (b) Hall,

    (c) Fisher,

    (d) Derived Fisher,

    (e) Constructed by double-nest replacement of a $3q$-double nest, or

    (f) Derived from a plane which may be so constructed, by a base regulus net fixed by the group of order $q(q+1)/4$.

(3) If 8 divides $q+1$ and is not one of the planes of part (2), then π may be constructed from the Desarguesian plane Σ by one of the following construction procedures:

    (a) By triple-nest replacement of a Nq-triple nest where $N = 5, 6$ or 7,

    (b) Derived from a plane which may be so constructed as in (a), by a base regulus net fixed by the group of order $q(q+1)$,

    (c) Constructed from a Desarguesian plane by replacing a $(2, 4; 4q)$ or $(2, 4; 5q)$-nest of degree $Nq(q+1)/8$ for $N = 6$ or 7,

(d) Derived from a plane which may be so constructed as in (c), by a base regulus net fixed by the group of order $q(q+1)$,

(e) Constructed from a Desarguesian plane by replacing a $(3, 4; 6q)$-nest of degree $7q(q+1)/8$, or

(f) Derived from a plane which may be so constructed as in (e), by a base regulus net fixed by the group of order $q(q+1)$.

PROBLEM 65.8. Find an infinite class of $3q$-double nests in Desarguesian planes of order q^2, show they are double-nest replaceable and show that there is a corresponding infinite class of conical flock planes.

PROBLEM 65.9. Determine if it is possible to construct Nq-triple nests. Determine the exact integers $N = 5, 6, 7$ that are possible. If there are Nq-triple nests, find a corresponding triple-nest replacement and show that there are infinite classes of conical flock planes corresponding to these triple-nests.

PROBLEM 65.10. Determine the full linear translation complement of either a double or triple nest plane.

PROBLEM 65.11. If π is a translation plane of order q^2 and spread in $PG(3, q)$ that may be obtained by q-nest replacement using a group EH of order $q(q+1)$, show that E must be regulus-inducing so that π is a Fisher plane.

PROBLEM 65.12. Show that there cannot exist a replaceable double-nest of q-reguli in a Desarguesian plane of order q^2.

PROBLEM 65.13. Determine the 'spectrum' of $(k_1, k_2, \ldots, k_z; t)$-nests for any of t reguli in a Desarguesian affine plane; find all possible partitions of t into subsets of reguli satisfying the multiple-nest definitions.

Actually, it is possible to completely determine the spreads in $PG(3, q)$ that admit a linear collineation group of order $q(q+1)$. Note that there is particular interest in the case when all subgroups S_q of order q fix a line of $PG(3, q)$ and act non-trivially on it, for in this setting, there are connections to flocks of quadratic cones. In this latter case, we provide a complete classification of all such spreads.

It is now well known that corresponding to flocks of quadratic cones in $PG(3, q)$, there is a corresponding spread in $PG(3, q)$ consisting of a set of q reguli that share a line, and conversely such a spread produces a flock of a quadratic cone. By the work of Gevaert and Johnson [395] and Gevaert, Johnson and Thas [397], it turns out that with the collineation group of the translation plane associated with a so-called 'conical flock spread' is an elation group E of order q with axis the shared line of the set of reguli of the spread whose 'component' orbits union the axis are reguli. Conversely, the existence of such an elation group in an arbitrary translation plane with spread in $PG(3, q)$ is sufficient to guarantee an associated conical flock. We call such elation groups 'regulus-inducing'.

Furthermore, if there is a linear Baer group of order q of a translation plane with spread in $PG(3, q)$, it turns out that these spreads are simply those obtained from conical flock spreads by derivation of any of its reguli sharing the axis of the regulus-inducing group. In terms of our assumption that a group of order q must fix a line of $PG(3, q)$ and act non-trivially on it, we note that a Baer subplane fixed pointwise by a Baer group of order q is a 2-dimensional $GF(q)$-subspace and hence a line in $PG(3, q)$. But, every Baer group of order q fixes each of the components of the underlying translation plane that non-trivially intersect the fixed point space. Hence, any Baer group of order q automatically satisfies our assumption that a group of order q must fix a line of $PG(3, q)$ and act non-trivially on it.

Actually, considering translation planes of order q^2 admitting Baer groups of order q, it is shown in Johnson [726] that there is a corresponding 'partial flock of deficiency one' of a quadratic cone and conversely such partial flocks produce translation planes admitting Baer groups of order q. The partial flock may be extended to a flock if and only if the net of degree $q + 1$ containing the Baer subplane fixed pointwise by the Baer group is derivable. However, it is known by Payne and Thas [1094] that every partial flock of deficiency one may, in fact, be uniquely extended to a flock. This means that Baer groups of order q in translation planes of order q^2 are essentially equivalent to flocks of quadratic cones.

THEOREM 65.14. (Johnson, Payne–Thas). If π is a translation plane of order q^2 with spread in $PG(3, q)$ that admits a Baer group of order q then π is a derived conical flock plane.

So, if there is a Baer group of order q or there is a regulus-inducing elation group of order q, there is a corresponding flock of a quadratic cone. Hence, assuming that there is a linear group (the group is in $GL(4, q)$) of order $q(q + 1)$, there will be at least some of the associated spreads admitting such groups that correspond to flocks of quadratic cones.

There has been considerable interest in flocks of quadratic cones, as besides being of intrinsic importance in finite geometry, produce and correspond to certain generalized quadrangles, and, as we have mentioned, spreads of translation planes. A great deal of the focus has been on the construction of these objects as opposed to understanding situations giving rise to these structures.

Previously, Jha and Johnson [608] have been able to develop results for odd-order translation planes with spreads in $PG(3, q)$ that might correspond to flocks of quadratic cones. One particular result when $q \equiv 1 \bmod 4$ is of interest as it gives abstract conditions when certain spreads correspond to the Fisher flocks (see Jha and Johnson [608]). We shall call spreads and translation planes corresponding to flocks of quadratic cones 'conical flock spreads' and 'conical flock planes', respectively.

The Fisher flocks are of particular interest due to their unique configuration of having exactly $(q - 1)/2$ conics whose corresponding planes in $PG(3, q)$ share a line (a linear subset of $(q - 1)/2$). However, the spread sets describing the flocks are given by somewhat awkward functions on $GF(q)$. Hence, it is of interest to describe the associated translation planes abstractly, perhaps group-theoretically.

There is now a complete theory of deficiency-one flocks of quadratic cones and realize that these always correspond to translation planes with spreads in $PG(3, q)$ that admit a Baer collineation group B of order q. The set of $(q - 1)$ orbits of B union Fix B form reguli that correspond to the partial flock of $q - 1$ conics of a quadratic cone.

The results of Johnson on Baer groups, Payne and Thas on extensions, and Jha and Johnson on generalized Fisher planes provide the foundation for a new approach to the problem of determining all spreads in $PG(3, q)$ that admit linear collineation groups of order $q(q + 1)$ whose associated p-groups fix a 2-dimensional subspace (line in the projective space) and act non-trivially on it.

Hence, it is possible to give a complete characterization of spreads in $PG(3,q)$, $q = p^r$, whose associated translation planes admit linear collineation groups of order $q(q+1)$. Either there is an elation group E of order q or a Sylow p-subgroup leaves invariant a projective line and acts non-trivially on it. We have mentioned previously that the Fisher planes admit such groups. These planes may be constructed by replacement in an associated Desarguesian affine plane of a set of q reguli the union of which forms a partial spread of degree $q(q+1)/2$. This set of reguli is called a q-nest of reguli. If a translation plane admits a linear group of order $q(q+1)$ and is not Fisher, then unless $q = 23$ or 47, there is always an associated Desarguesian affine plane which produces the translation plane by a generalization of the idea of nest replacement of a q-nest that we call a 'multiple-nest' replacement, of which 'double-nest' replacement is of particular importance.

We have discussed the notion of 'multiple-nests' and specifically 'double-nests' and have noted, in particular, that the Thas, Herssens, De Clerck conical plane of order 11^2 [**273**], obtained originally by computer, may be constructed from a Desarguesian plane by 'double-nest replacement'. Furthermore, we have seen that Jha and Johnson completely determine the translation planes of order q^2 that admit a linear collineation group EH and have an associated Desarguesian affine plane Σ admitting EH such that E is a normal regulus-inducing elation group acting on Σ (see Theorem 65.7).

When $q = 23$ or 47, it is possible that there is not an associated Desarguesian affine plane that constructs the plane by multiple nest replacement. In this situation, there is a collineation group G in $GL(4,q)$ containing a normal subgroup E of order q that fixes a 2-dimensional $GF(q)$-subspace L pointwise. $(G \mid L)/Z(G \mid L)$ is isomorphic to A_4 or S_4 when $q = 23$ or 47. When $q = 23$, there is a unique conical flock of this type due to Penttila and Royle (see Chapter 60). The group of order $q + 1 = 24$ is an extension of A_4. The geometric construction of the BLT in shows that the BLT-group is $S_4 wr \, S_2$ of order $2^7 3^2$. When $q = 47$, there is at least one flock of the indicated type admitting S_4 due to De Clerck and Penttila where the BLT-group has order $2^8 3^2$. The intersection with the group of order $q + 1$ shows that our group is either an extension of S_4 by the kernel involution or there is a second possible situation, when the group induces A_4 on the elation axis. In this setting, the group induced on the spread is $A_4 \times Z_2$. Thus, there are two possible exceptional situations when $q = 47$, when the group induced on the spread is S_4 or $A_4 \times Z_2$, and one for $q = 23$, where the group induced on the spread is A_4.

The results give a complete classification of the spreads in $PG(3,q)$ satisfying the above conditions in so far as we determine exactly how such spreads are constructed.

The main theorem of Jha and Johnson [**633**] is as follows:

THEOREM 65.15 (Jha–Johnson $q(q + 1)$-Classification Theorem; Odd Order). (Jha and Johnson [**633**]). Let π be a translation plane of odd order q^2 with spread in $PG(3,K)$, K isomorphic to $GF(q)$, $q = p^r$.

If π admits a linear collineation group G of order $q(q+1)$, assume that a Sylow p-subgroup leaves invariant a 2-dimensional K-subspace and acts non-trivially on it. Then, the following are valid:

(i) There is a G-invariant 2-dimensional K-subspace fixed pointwise by a Sylow p-subgroup or π is Desarguesian or Hall of order 9.

(ii) π is either:

(a) A conical flock spread, or

(b) Derived from a conical flock spread.

(iii) If $q \equiv 1 \bmod 4$, then π is one of the following planes:

(a) Desarguesian,

(b) Hall,

(c) A Fisher conical plane, or

(d) A plane derived from a Fisher conical plane.

(iv) If $4 \mid (q+1)$ but $8 \nmid (q+1)$, then either π is one of the planes of part (iii) or π may be constructed from a Desarguesian plane by either

(a) Double-nest replacement of a $3q$-double nest, or

(b) Derived from a plane which may be so constructed, by a base regulus net fixed by the group of order $q(q+1)$.

(v) If π has order 23^2 or 47^2, then one of the following possibilities occurs:

(a) π has order 23^2 and is the Penttila–Royle conical flock plane, whose spread admits A_4 as an automorphism group,

(b) π has order 23^2 and is derived from the Penttila–Royle flock spread,

(c) π has order 47^2 and is a conical flock plane, whose spread admits S_4 as an automorphism group; the Penttila–De Clerck spread is an example,

(d) π has order 47^2 and is derived from a conical flock plane, whose spread admits S_4 as an automorphism group,

(e) π has order 47^2 and is a conical flock spread and admits $SL(2,3)\langle \rho \rangle$ as a collineation group, when ρ is an affine homology of order 2 and $SL(2,3)$ fixes a base regulus linewise (the group induced on the spread is $A_4 \times Z_2$ in this case), or

(f) π has order 47^2 and is a derived conical flock spread and admits $SL(2,3)\langle \rho \rangle$ as a collineation group, when ρ is a Baer involution and $SL(2,3)$ has four orbits of length 12 on the derived regulus (the group induced on the spread is $A_4 \times Z_2$ in this case).

(vi) If $8 \mid (q+1)$ then either π is one of the planes of parts (iv) or (v) (e.g., it is possible that π could be constructed using a double-nest of $3q$ reguli) or π may be constructed from a Desarguesian plane by one of the following construction procedures:

(a) By triple-nest replacement of a Nq-triple nest where $N = 5, 6$ or 7,

(b) Derived from a plane which may be so constructed as in (a), by a base regulus net fixed by the group of order $q(q+1)$,

(c) Constructed from a Desarguesian plane by replacing a $(2,4;4q)$- or $(2,4;5q)$-nest of degree $Nq(q+1)/8$ for $N = 6$ or 7,

(d) Derived from a plane which may be so constructed as in (c), by a base regulus net fixed by the group of order $q(q+1)$,

(e) By replacing a $(3,4;6q)$-nest of degree $7q(q+1)/8$, or

(f) Derived from a plane constructed as in (e), by a base regulus net fixed by the group of order $q(q+1)$.

65.1. Planes of Prime-Square Order.

Now let $p = q$ and note that any elation group of order p must be regulus-inducing as follows: Let σ be an elation and let the axis of σ be $x = 0$ and let σ map $y = 0$ to $y = x$ by choice of basis. Then, σ must have the form $(x, y) \longmapsto (x, x + y)$

and the generated group is

$$E : \langle (x,y) \longmapsto (x, x\alpha + y); \, \alpha \in GF(p) \rangle .$$

Hence, each orbit of components union $x = 0$ is a regulus so that E fixes each of a set of Baer subplanes incident with the zero vector of each such regulus net. So, we have satisfied our hypothesis, obtaining the following theorem:

THEOREM 65.16. Let π be a translation plane of order p^2, p a prime, that admits a linear collineation group G of order $p(p + 1)$. Then, the following are valid:

(i) There is a G-invariant 2-dimensional $GF(p)$-subspace fixed pointwise by a Sylow p-subgroup.

(ii) π is either a conical flock spread or derived from a conical flock spread.

(iii) If $p \equiv 1 \bmod 4$, then π is one of the following planes:

(a) Desarguesian,

(b) Hall,

(c) A Fisher conical plane, or

(d) A plane derived from a Fisher conical plane.

(iv) If $4 \mid (p+1)$ but $8 \nmid (p+1)$, then either π is one of the planes of part (iii) or π may be constructed from a Desarguesian plane by

(a) Double-nest replacement of a $3p$-double nest, or

(b) Derived from a plane which may be so constructed, by a base regulus net fixed by the group of order $p(p+1)$.

(v) If π has order 23^2 or 47^2, then one of the following possibilities occurs:

(a) π has order 23^2 and is the Penttila–Royle conical flock plane, whose spread admits A_4 as an automorphism group,

(b) π has order 23^2 and is derived from the Penttila–Royle flock spread,

(c) π has order 47^2 and is a conical flock plane, whose spread admits S_4 as an automorphism group (the Penttila–De Clerck spread is an example),

(d) π has order 47^2 and is derived from a conical flock plane, whose spread admits S_4 as an automorphism group,

(e) π has order 47^2 is a conical flock spread and admits $SL(2,3) \langle \rho \rangle$ as a collineation group, when ρ is an affine homology of order 2 and $SL(2,3)$ fixes a base regulus linewise (the group induced on the spread is $A_4 \times Z_2$ in this case), or

(f) π has order 47^2 is a derived conical flock spread and admits $SL(2,3) \langle \rho \rangle$ as a collineation group, when ρ is a Baer involution and $SL(2,3)$ has four orbits of length 12 on the derived regulus (the group induced on the spread is $A_4 \times Z_2$ in this case).

(vi) If π is of odd order and $8 \mid (p+1)$, then either π is one of the planes of parts (iv) or (v) or π may be constructed from a Desarguesian plane by one of the following construction procedures:

(a) Triple-nest replacement of a Np-triple nest where $N = 5$, 6 or 7,

(b) Derived from a plane which may be so constructed, by a base regulus net fixed by the group of order $p(p+1)$,

(c) Constructed from a Desarguesian plane by replacing a $(2,4;4p)$- or $(2,4;5p)$-nest of degree $Np(p+1)/8$ for $N = 6$ or 7,

(d) Derived from a plane which may be so constructed as in (c), by a base regulus net fixed by the group of order $p(p+1)$,

(e) Constructed from a Desarguesian plane by replacing a $(3, 4; 6p)$-nest of degree $7p(p+1)/8$, or

(f) Derived from a plane which may be so constructed as in (e), by a base regulus net fixed by the group of order $p(p+1)$.

65.2. Even Order.

Previously was mentioned the odd-order characterization of translation planes of order q^2 admitting collineation groups of order $q(q+1)$. It is also possible to completely classify the even-order spreads in $PG(3, q)$ that admit a linear collineation group of order $q(q+1)$. However for odd-order spreads there is an added hypothesis that a subgroup of order q must fix a line and act non-trivially on it. In the odd-order case, the planes are all related to flocks of quadratic cones in $PG(3, q)$. In the even-order case, this may be proved to occur.

The main theorem is as follows:

THEOREM 65.17 (Jha–Johnson $q(q+1)$-Classification Theorem; Even Order). (Jha and Johnson [**632**]). Let π be a translation plane of even order q^2 with spread in $PG(3, K)$, K isomorphic to $GF(q)$, that admits a linear collineation group G of order $q(q+1)$ (i.e., in $GL(4, q)$).

Then π is one of the following types of planes:
(i) Desarguesian,
(ii) Hall, or
(iii) A translation plane obtained from a Desarguesian plane by multiple derivation of a set of $q/2$ mutually disjoint regulus nets that are in an orbit under an elation group of order $q/2$.

65.3. The De Clerck–Herssens–Thas Double-Nest Plane of Order 11^2.

.

Actually, it is possible to show that certain multiple nests occur.

THEOREM 65.18. (Jha and Johnson [**636**]). The Thas–Herssens–De Clerck conical plane of order 11^2 [**273**] may be constructed from a Desarguesian plane of order 11^2 by double-nest replacement of a $3 \cdot 11$-double nest.

We shall illustrate the T–H–DC plane of order 11^2 in the form

$$EH^-\{y = x^q s \pm xt, y = x^q \alpha_s s\} \cup \mathcal{B},$$

and provide the explicit elements s and t and scalar element α_s. These matrices were found by analyzing the group EH isomorphic to $Z_2 \times S_3$ on the line at infinity of the associated translation plane.

We define the field coordinatizing the associated Desarguesian affine plane Σ as

$$\left\langle \begin{bmatrix} u & -t \\ t & u \end{bmatrix} ; u, t \in GF(11) \right\rangle.$$

We let

$$s = \begin{bmatrix} 3 & -2 \\ 2 & 3 \end{bmatrix}, \, t = \begin{bmatrix} 1 & 1 \\ -1 & 1 \end{bmatrix}, \, \alpha_s = 4.$$

Let

$$\theta = \begin{bmatrix} 5 & -3 & 0 & 0 \\ 3 & 5 & 0 & 0 \\ 0 & 0 & 5 & -3 \\ 0 & 0 & 3 & 5 \end{bmatrix}, \text{ and note that } \theta^3 = I.$$

We also obtain the following characterization theorem when $p = 11$.

THEOREM 65.19. (Jha and Johnson [**633, 636**]). Let π be a translation plane of order 11^2 that admits a linear collineation group of order $11(12)$; then π is one of the following six planes:

(1) Desarguesian,
(2) Hall,
(3) Fisher,
(4) Derived Fisher,
(5) Thas–Herssens–De Clerck, or
(6) Derived Thas–Herssens–De Clerck.

CHAPTER 66

A Few Remarks on Isomorphisms.

66.1. Fingerprint.

If π_1 and π_2 are finite translation planes, in order that π_1 and π_2 are said to be 'isomorphic', both must have the same order p^r, where p is a prime and r a positive integer and there is an element g of $GL(2r, p)$ which is a bijection from the spread S_1 for π_1 to the spread S_2 for π_2. If the kernels are not known explicitly, then the task of determining numbers of translation planes isomorphic to a given translation plane is quite difficult. Indeed, even the planes are of order q^2 and the kernels isomorphic to $GF(q)$, the method of determining bijections in $\Gamma L(4, q)$ is often not a particularly efficient means of determining isomorphisms. For computer searches of mutually non-isomorphic planes of a particular order, the 'fingerprint' of the plane is sometimes employed. We follow Charnes and Dempwolff [**226**] for the discussion.

DEFINITION 66.1. Let π denote a translation plane of order p^n and let S denote the spread in $PG(2n - 1, p)$. Choose any matrix spread set

$$\{\, x = 0, y = xA; \; A \in S_M \,\},$$

where S_M is a set of p^n+1 $n{\times}n$ matrices over $GF(p)$, one of which is the zero matrix, such that the matrices and the differences of pairs of matrices are non-singular or identically zero. Let x_0 be the zero matrix and let

$$S_M = \{\, x_i; \; 0 \le i \le p^n - 1 \,\}.$$

We may also choose one of the matrices to be the identity matrix. Let x be an $n \times n$ matrix over $GF(p)$ and define

$$[x] = \det x/p \text{ for } \det x \neq 0 \text{ and } 0 \text{ otherwise.}$$

Define an $(p^n - 1 + 2, p^n - 1 + 2)$-matrix $Q = (q_{ij})$ such that

$$q_{ij} = \left| \sum_{k=0}^{p^n-1} [x_i - x_k][x_j - x_k] + 1 \right|, \; 0 \le i, j \le p^n - 1,$$

$$q_{i\infty} = q_{\infty i} = \left| \sum_{k=0}^{p^n-1} [x_i - x_k] \right|, \; 0 \le i \le p^n - 1,$$

$$q_{\infty\infty} = p^n.$$

Then the multiset of entries of $Q = Q(S)$ is an invariant of isomorphism. Q is said to be the 'fingerprint' of the matrix spread set S_M.

We note that two non-isomorphic planes may have two spread sets with the same fingerprint but the multiset Q is an isomorphism invariant with respect to recoordinatization of matrix spread sets.

REMARK 66.2. Specifically, recoordinatization of a given matrix spread set is possible only under a finite set of applications of the following mappings:

$$O_1(x) \quad : \quad S_M \to x S_M x^{-1}; \ x \text{ a non-singular matrix},$$
$$O_2(x) \quad : \quad S_M \to t_i^{-1} S_M; \ 1 \leq i \leq p^n - 1,$$
$$O_3(x) \quad : \quad \to S_M \to S_M^{-1} \ (\text{where } 0^{-1} = 0),$$
$$O_4 \quad : \quad S_M \to 1 - S_M.$$

66.2. Leitzahl.

There are two other invariants of isomorphism which are useful.

DEFINITION 66.3. For an $n \times n$ matrix over $GF(q)$, define $\langle\langle x \rangle\rangle = 1$ if $\det x = 1$ and 0 otherwise. Define

$$\ell(S_M) = \sum_{0 \leq i < j \leq p^n - 1} \ \sum_{k \neq i,j, k=0}^{p^n - 1} \left\langle \left\langle (x_i - x_k)(x_k - x_j)^{-1} \right\rangle \right\rangle.$$

The value $\ell(S_M)$ is called the 'Leitzahl' and is invariant under the above operations with the possible exception of type O_3.

66.3. Kennzahl.

DEFINITION 66.4. For an $n \times n$ matrix over $GF(q)$, define $((x)) = 1$ if $\det x = 0$ and 0 otherwise.
Let

$$k(S_M) = \sum_{1 \leq i < j \leq p^n - 1} ((x_i + x_j)).$$

The value $k(S_M)$ is invariant under the above operations with the possible exception of type O_4. This value is called the 'Kennzahl' of the matrix spread set.

So, the fingerprint, the Leitzahl, and the Kennzahl are computationally expedient tools for translation planes of small orders but would not establish isomorphism criteria for general classes of translation planes, where we are reduced to considerations of $\Gamma L(2n, p)$.

CHAPTER 67

Flag-Transitive Geometries.

If we were to deal completely with all geometries admitting transitive actions, we would need to double the size of this book. For example, one could consider group actions on affine planes, projective planes, parallelisms of projective spaces, partial parallelisms of projective spaces, actions on projective spaces or affine spaces, on Steiner triple systems, on general designs and so on. We necessarily need to restrict our remarks here to include results that impinge most directly on translation planes, even if this becomes quite subjective. So, we shall mainly focus directly on affine and projective planes and their interconnections when considering the myriad problems considering transitive actions on geometries.

67.1. Flag-Transitive Translation Planes.

In this section, we list the set of known finite flag-transitive translation planes. We first note Wagner's theorem.

THEOREM 67.1. (Wagner [**1202**]). Finite flag-transitive affine planes are translation planes.

With a flag-transitive plane, the associated group G can be non-solvable or solvable. Furthermore, the non-solvable finite flag-transitive affine planes are completely determined. The following theorem is a special case of a more general result on flag-transitive designs.

THEOREM 67.2. (BDDKLS-Theorem).(Buekenhout, Delandtsheer, Doyen, Kleidman, Liebeck, and Saxl [**201**]). Let π be a non-solvable flag-transitive finite affine plane.

Then π is one of the following types of affine planes: (1) Desarguesian, (2) Lüneburg–Tits, (3) Hering of order 27, or (4) Hall of order 9.

For completeness, we give a concrete representation for the Lüneburg–Tits and Hering order-27 spreads.

The Lüneburg–Tits planes are defined in Chapter 48 on translation planes of order q^2 admitting collineation groups of order q^2 and the Hering plane of order 27 is listed in Section 50.3 on indicator sets. However, here is another construction of the Hering plane.

67.1.1. The Hering Plane of Order 27. Hering's translation plane of order 27 admits the collineation group $SL(2,13)$ and is flag-transitive. Let

$$E = \begin{bmatrix} 0 & -1 & 0 \\ -1 & 0 & -1 \\ 1 & 0 & 0 \end{bmatrix}, \quad A = \begin{bmatrix} -1 & 0 & 1 \\ -1 & 0 & 0 \\ 0 & -1 & 0 \end{bmatrix}, \quad B = \begin{bmatrix} -1 & 0 & 0 \\ -1 & 1 & 0 \\ 1 & -1 & -1 \end{bmatrix},$$

$$C = \begin{bmatrix} -1 & 0 & 0 \\ -1 & 1 & 0 \\ -1 & -1 & -1 \end{bmatrix}, \quad D = \begin{bmatrix} 1 & 0 & 0 \\ 1 & -1 & 0 \\ -1 & 1 & 1 \end{bmatrix}.$$

Consider the matrix elements g and h:

$$g = \begin{bmatrix} E & 0 \\ 0 & A \end{bmatrix}, h = \begin{bmatrix} B & B \\ C & D \end{bmatrix}.$$

Then $\langle g, h \rangle \simeq SL(2,13)$. And,

THEOREM 67.3. Under the above assumptions, $\{x = 0, y = 0\} \langle g, h \rangle$ is a spread (the Hering spread of order 27), which is

$$\left\{ x = 0, y = xM; \ M \in \left\{ 0, E^{-i}A^i, E^{-i}C^{-1}DA^i; \ i = 0, 1, \ldots, 12 \right\} \right\}.$$

We fist discuss flag-transitive translation planes with spreads in $PG(3, q)$ and sketch a construction technique of Baker and Ebert [**68**]. There is an important geometric connection with such translation planes and 'caps' in $PG(3, q)$, which are sets of points no three points of which are collinear.

67.2. Flag-Transitive Planes with Spreads in $PG(3, q)$.

It is possible (Ebert [**349**]) to partition the points of $PG(3, q)$ into caps of size $q^2 + 1$, which are actually elliptic quadrics. Let $\mathcal{C}$ denote the partition. The caps arise as orbits under a particular 'Singer' cycle subgroup as follows.

We represent the 4-dimensional vector space over $GF(q)$ by $GF(q^4)$ over $GF(q)$. Hence, the elements of the vector space and the elements of the field $GF(q^4)$. Ultimately, we wish to represent $PG(3, q)$ as the lattice of subspaces of V_4 and hence points will be represented by 1-dimensional vector subspaces over $GF(q)$. Let e denote an element of order $q^4 - 1$ and let θ_e denote the $GF(q)$-linear transformation:

$$\theta_e : \theta_e(x) = xe, \text{ for all } x \in V_4.$$

Then $\langle \theta_e \rangle$ has order $q^4 - 1$ and acts transitively and regularly on the non-zero vectors. We call such a collineation group (or orbit) a 'Singer cycle'. If we take the group $\langle \theta_e^{q+1} \rangle$, then each orbit will have length $(q^2 + 1)(q - 1)$. Taking the associated set of 1-dimensional $GF(q)$-subspaces, we will obtain a set of $q^2 + 1$ points, no three of which are collinear. That is, if three are collinear, they will generate a 3-dimensional vector space V_3, containing $q^3 - 1$ non-zero vectors. The group $\langle \theta_e^{q+1} \rangle$ is fixed-point-free (no non-identity element fixes a non-zero point), it follows that $q^3 - 1$ must divide $q^3 - 1$, a contradiction. Hence, the orbits are caps and are actually elliptic quadrics in $PG(3, q)$. There are $(q^4 - 1)/(q-1) = (q^2+1)(q+1)$, Hence, we have a partition $\mathcal{C}$ of the points of $PG(3, q)$ into $q + 1$ caps, which are elliptic quadrics. Take an arbitrary line ℓ of $PG(3, q)$ so that ℓ intersects each cap in 1 or 2 points. If we wish to construct a flag-transitive translation plane, one method would be to find a spread S of the form $\ell \left\langle \theta_e^{q^2-1} \right\rangle$. This would work provided ℓ intersects each cap in exactly one point. In the odd-order case, unfortunately, this method never works, as it turns out (Ebert [**349**]) that any line ℓ either has exactly 2 intersections with $(q+1)/2$ caps (the 'secant' type lines) or has exactly 2 intersections with $(q-1)/2$ caps and is tangent to exactly two caps (the 'tangent' type). Hence, using the full group $\left\langle \theta^{q^2-1} \right\rangle$ will never work. However, it is possible to use the group $\left\langle \theta^{q^2-1} \right\rangle^2$ and a secant type line to construct a partial spread of $(q^2+1)/2$ lines.

So, the question is when must a group such as $\left\langle \theta^{q^2-1} \right\rangle^2$ be used in the construction of flag-transitive spreads, by finding two secant type lines ℓ_i, $i = 1, 2$ such that the spread is

$$\{\, \ell_i; \, i = 1, 2 \,\} \left\langle \theta^{q^2-1} \right\rangle^2 ?$$

In this setting the group is non-solvable. Since the Lüneburg–Tits spread in $PG(3,q)$, for q even, are flag-transitive, this procedure will not give all even-order flag-transitive spreads in $PG(3,q)$. However, if the group G of a flag-transitive spread is solvable, then Foulser [**365**] has determined all possibilities for G:

$$G \subseteq \langle\, \theta_e, \, x \longmapsto x^p; \, q = p^r, \, p \text{ a prime} \,\rangle.$$

What is required for the construction is to ensure that G of order $(q^2+1)/2$ is necessarily linear. We call this simply the 'linear condition'. There is then a combinatorial condition for the existence of such linear groups. We shall provide constructions by a different and more general method to follow, but list the following theorem of Ebert [**349**] and Baker and Ebert [**68**]. Also, Narayana Rao [**999**] has similar results by slightly different methods.

Flag-Transitive Construction Theorem of Baker–Ebert and Narayana Rao.

THEOREM 67.4. (Ebert [**349**], Baker and Ebert [**68**], and Narayana Rao [**999**]). A flag-transitive translation plane of odd order q^2 and spread in $PG(3,q)$, whose group G satisfies the linear condition may be constructed by finding pairs of secant lines ℓ_i, so that the spread is

$$\{\, \ell_i; \, i = 1, 2 \,\} \left\langle \theta^{q^2-1} \right\rangle^2.$$

67.3. The Known Flag-Transitive Planes with Solvable Group.

67.3.1. The Flag-Transitive Planes with Dimension 2.

67.3.1.1. *Planes of M.L. Narayana Rao.*

67.3.1.2. *Planes of Baker–Ebert (Same As Narayana Rao).*

67.3.1.3. *Plane of Order 49 of M.L. Narayana Rao and K. Kuppuswamy Rao.*

67.3.1.4. *Planes of Order 25 of Foulser.* Here we concentrate on finite flag-transitive affine planes that of dimension > 2 although the constructions provide the dimension 2 (i.e., spreads in $PG(3,q)$ as well).

REMARK 67.5. (1) Again, we note that all planes of orders $16, 25, 27, 49$ are known by computer. These planes may be found in the home page of Ulrich Dempwolff: http://www.mathematik.uni-kl.de/~dempw/ (or 'google' home page of Ulrich Dempwolff). The flag-transitive planes of order 27 are also known by computer work of Prince [**1111**].

(2) There is no non-Desarguesian flag-transitive plane of order 16.

(3) There are exactly two flag-transitive planes of order 25, the two Foulser planes.

(4) There are exactly four flag-transitive planes of order 27, the Desarguesian, Hering and two others due to Narayana Rao, Kuppuswamy Rao, and by Narayana Rao, Kuppuswamy Rao and Satyanarayana. Furthermore,

(5) The planes of order 125 are known by computer work of Prince [**1112**] and are given by constructions that we describe below.

67.3.2. The Flag-Transitive Planes of Large Dimension.

67.3.2.1. *Planes of Order* 27 *of M.L. Narayana Rao and K. Kuppuswamy Rao* [**1002**].

67.3.2.2. *Planes of Order* 27 *of M.L. Narayana Rao, K. Kuppuswamy Rao and K. Satyanarayana* [**1012**]. Again these planes may be viewed in the home page of Ulrich Dempwolff.

67.3.2.3. *Plane of Order* 125 *of M.L. Narayana Rao and K. Kuppuswamy Rao.*

67.3.2.4. *The Desarguesian Scions of Kantor and Williams.* These planes are constructed in Chapter 79 and the reader is directed to that chapter. Furthermore, these planes are included below in the general constructions.

67.3.2.5. *Planes of Suetake* [**1167**].

67.3.2.6. *Generalizations of Suetake by Kantor and Suetake* [**879**].

67.3.2.7. *Planes of Kantor* [**874**].

67.3.2.8. *Planes of Narayana Rao and Kuppuswamy Rao* [**1005**].

67.3.2.9. *Planes of Narayana Rao, Kuppuswamy Rao and Durga Prasad Order* 81 [**1004**].

67.3.2.10. *Planes of Durga Prasad Order* 81 [**331**].

The planes of Narayana Rao, Kuppuswamy Rao and Durga Prasad of order 81 [**1004**], of 1988 and the general class of Narayana Rao and Kuppuswamy Rao constructed in 1989 [**1005**] predate those of Suetake [**1167**] 1991, Suetake and Kantor [**879**] 1994 and Kantor 1992, 1993 [**874**], [**875**], respectively. However, the general constructions of Suetake and Kantor, and Kantor include the planes of Narayana Rao, Kuppuswamy Rao and of Narayana Rao [**1005**] order, of Kuppuswamy Rao and Durga Prasad [**1004**], and Durga Prasad [**331**] of order 81, so we give, in the following, the general constructions. Similarly the planes of Durga Prasad [**331**] of order 81 are included in the constructions.

We now provide the constructions which give all known finite flag-transitive affine planes.

67.3.2.11. *The General Constructions of Kantor and Suetake.* To construct a flag-transitive plane of order q^n, for q a prime power p^r, we consider a field $GF(q^{2n})$ as a 2n-dimensional $GF(q)$-vector space. Let G be a cyclic subgroup of $GF(q^{2n})$ of order $(q^n + 1)(q - 1)$.

We begin with the Desarguesian scion flag-transitive planes of even order. These planes constitute the complete set of known even-order flag-transitive planes admitting solvable groups. Note that these planes are symplectic planes.

THEOREM 67.6. (Kantor [**862**]). Assume that q is even and n is odd > 1. Let T denote the trace map from $GF(q^n)$ to $GF(q)$ and let $h(x) = T(x) + rx$, where $r \in GF(q^2) - GF(q)$. Then for $L = GF(q^n)$,

(1) $S_r = \{ s^i h(L); 0 \le i \le q^n; \langle s \rangle \}$ defines a flag-transitive affine plane.

(2) The flag-transitive group acting on the affine plane (that is, transitive on the affine points and transitive on the infinite points) is

$$\left\langle \, z \longmapsto s^j + w; \ s^j \in \langle s \rangle, \ w \in GF(q^{2n}) \, \right\rangle.$$

(3) S_r and $S_{r'}$ are isomorphic if and only if $r' + 1 = k(r + q)^\sigma$, for $\sigma \in \operatorname{Aut} GF(q^2)$, $k \in GF(q)^*$.

The following flag-transitive planes generalize those of Suetake [1167].

THEOREM 67.7. (Kantor [875], Suetake [1167]). Let q be odd, n odd and let $b \in GF(q^{2n})$, such that $b^q = -1$. Let $h(x) = x + bx^\sigma$, where $1 \neq \sigma \in \operatorname{Gal}_{GF(q)} GF(q^n)$. Then

$$S_{b,\sigma} = \left\{ \, s^i h(L); \ 0 \leq i \leq q^n, \langle s \rangle = G \, \right\}$$

defines a flag-transitive affine plane $\pi_{S_{b,\sigma}}$. The associated group is given by

$$z \longmapsto s'z + w; s' \in \langle s \rangle, w \in GF(q^{2n}).$$

Furthermore, $\pi_{S_{b,\sigma}}$ is isomorphic to $\pi_{S_{b^*,\sigma^*}}$ then $\sigma^* = \sigma^{\pm 1}$ and indeed, $b^* = b^{-1}$ and $\sigma^* = \sigma^{-1}$ produces isomorphic planes. When $\sigma^* = \sigma$ then the affine planes are isomorphic if and only if $b^* = \alpha^{1-\sigma} b^\rho$, where $\alpha \in GF(q^n)^*$ and $\rho \in \operatorname{Gal} GF(q^{2n})$.

REMARK 67.8. The previous two sets of examples admit cyclic collineation groups acting transitively on the components of the associated translation planes. We shall see in Section 67.4 that there is a connection with certain flag-transitive planes and Baer subgeometry partitions, when there is a cyclic and transitive action.

In the following two constructions, the following flag-transitive planes are made up of two sets of $(q^n + 1)/2$ components in an orbit under a cyclic group. This first type of planes are generalizations of the two-dimensional planes of Baker–Ebert [68], Narayana Rao [999] and Foulser and also generalizes a construction of Suetake [1167].

THEOREM 67.9. (Kantor [874], Suetake [1167]). Let $q^n \equiv 1 \bmod 4$ and b and h are as in the previous theorem. Choose $\sigma \in \operatorname{Gal}_{GL(q)} GF(q^{2n}) - \operatorname{Gal}_{GF(q)} GL(q^n)$; then

$$S'_{b,\sigma} = \left\{ \, s^i h(L); \ 0 \leq i \leq (q^n - 1)/2, \ \langle s \rangle \, \right\} \cup \left\{ \, s^{2i} h(L); \ 0 \leq i \leq (q^n - 1)/2, \ \langle s \rangle \, \right\}$$

is a flag-transitive spread giving an affine plane $\pi_{S'_{b,\sigma}}$. Furthermore, $\pi_{S'_{b,\sigma}}$ is isomorphic to $\pi_{S'_{b^*,\sigma^*}}$ then $\sigma^* = \sigma^{\pm 1}$ and indeed, $b^* = b^{-1}$ and $\sigma^* = \sigma^{-1}$ produces isomorphic planes. When $\sigma^* = \sigma$ then the affine planes are isomorphic if and only if $b^* = \alpha^{1-\sigma} b^\rho$, where $\alpha \in GF(q^n)^*$ and $\rho \in \operatorname{Gal} GF(q^{2n})$. Indeed, if $\operatorname{Fix} \sigma \subset GF(q^n)$ then there is no cyclic group acting transitively on the spread (Suetake [1167]).

The final set of flag-transitive planes are variations of the previous class. In the following, when σ is the identity, the planes of Narayana Rao and Kuppuswamy Rao [1005] are included in the class, as are the planes of Kuppuswamy Rao and Durga Prasad [1004], and Durga Prasad [331] of order 81.

THEOREM 67.10. (Kantor [874]). Let q and n be odd and s and h as before. Let $\sigma \in \operatorname{Gal}_{GL(q)} GF(q^{2n}) - \operatorname{Gal}_{GF(q)} GL(q^n)$ and let $\mu \in GF(q^{2n})^*$ such that μ^{1+q^n}

is fixed by σ but is not a square in $GF(q^n)$. Then

$$S''_{b,\sigma} = \left\{ s^{2i} h(L); 0 \le i \le (q^n - 1)/2, \langle s \rangle \right\}$$

$$\cup \left\{ \mu s^{2i} (h(L))^{q^n}; 0 \le i \le (q^n - 1)/2, \langle s \rangle \right\}.$$

Also, if $\mathrm{Fix}\,\sigma \subset GF(q^n)$ then there is no cyclic group transitive on the spread.

For flag-transitive planes of order q^3 with kernel $GF(q)$, we have examples of cyclic groups acting transitively on the spread and examples where there is a cyclic action of only 'half' of the spread components in two orbits. Also, Baker, Culpert, Ebert and Mellinger [**61**] show that if there is such a cyclic group action with one or two orbits as mentioned, then the known examples are the only possibilities. Furthermore, when $q = p^r$, for p a prime, and if $((q^3 + 1)/2, 3r) = 1$, then only the cyclic group action mentioned is possible. Hence, under such conditions all 3-dimensional flag-transitive affine planes are completely determined.

67.4. Flag-Transitive Planes and Subgeometry Partitions.

Let π be a flag-transitive finite affine plane. Then π is a translation plane of order q^n and kernel $GF(q)$, where q is a prime power. There is a collineation group G in $GL(2n, q)$ which acts transitively on the components. The non-solvable planes (where G is non-solvable) are known as the Desarguesian, Lüneburg–Tits, Hering of order 27 and Hall of order 9. Suppose that G is solvable. Then by Foulser [**365**], we know the form of G as a subgroup of the Frobenius automorphism $z \longmapsto z^p$, for $q = p^r$ and a cyclic field group of order $q^n - 1$. We are interested in when π might correspond directly to a subgeometry partition (mixed or not). Such an investigation has been considered by Ebert [**349**] for various examples by determining the associated subgeometries directly. Furthermore, Baker, Dover, Ebert, and Wantz [**63**], show that any collineation group G admitting a cyclic collineation group of order $q + 1$, acting semi-regularly on the spread components will produce a Baer subgeometry partition of $PG(2n - 1, q^2)$. Here we consider only the question of whether there is a field group of order $q^2 - 1$. Hence, we are interested simply in asking if G contains a cyclic group of order $q^2 - 1$. Actually, G will have order divisible by $(q^n + 1)$ and we let K^* denote the associated kernel homology group of order $q - 1$, then GK^* will have order $(q^n + 1)(q - 1)$. Hence, there is a Baer subgeometry partition precisely when n is odd and there is a cyclic subgroup of order $q+1$ acting subregularly on the line at infinity, (see Baker, Dover, Ebert, and Wantz [**63**]).

We consider the group

$$\mathcal{F} = \left\langle \theta_e, \alpha : \theta_e(x) = xe; e \text{ has order } q^{2n} - 1, \alpha(z) = z^p \right\rangle, \text{ for } x, z \in GF(q^n)^*.$$

If we consider a vector space of dimension $2n$ over $GF(q)$ as $GF(q^{2n})$, a flag-transitive plane which admits a solvable group G may be constructed by taking an n-dimensional $GF(q)$-subspace S, so that SG is a set of $(q^n + 1)$ mutually disjoint subspaces. So, the question is when $((q^n + 1), nr) = 1$ with $q = p^r$. Whenever this occurs, there is an associated Baer subgeometry partition.

CHAPTER 68

Quartic Groups in Translation Planes.

In the theory of finite translation planes of order p^r, there are various important results classifying the collineation subgroups generated by collineations g of order p when the fixed point nature of the collineation is specified.

For example, if Fix g is a component of the plane, the Hering–Ostrom theorem, see Chapter 45 or Hering [476], Ostrom [1047], provides a complete classification of the possible groups generated.

If Fix g is a Baer subplane and p is strictly larger than 3, the Foulser theorem (see Foulser [375]) on collineation groups generated by Baer p-collineations shows that the generated groups fall into the same general classes as do the groups generated by elations.

When the spread for the translation plane is within $PG(3, q)$, any collineation g of order p that lies within $GL(4, q)$ (the linear translation complement) is either an elation, a Baer p-element or has minimal polynomial $(x - 1)^4$ and Fix g is a 1-dimensional $GF(q)$-subspace. In particular, we will be interested in groups generated by 'quartic groups,' so to be clear, we provide an explicit definition.

DEFINITION 68.1. Let π be a translation plane of order $q^2, q = p^r$, p a prime, with spread in $PG(3, q)$. A 'quartic group' T is an elementary Abelian p-group all of whose non-identity elements are quartic (i.e., have minimal polynomials $(x-1)^4$) and which fix the same 1-dimensional $GF(q)$-subspace pointwise. The fixed-point space is called the 'quartic center' of the group and the unique component of π containing the center is called the 'quartic axis'. If T and S are quartic groups, we shall say that T and S are 'trivially intersecting' if and only if their quartic centers are distinct (i.e., if and only if when $t \in T - \{1\}$ and $s \in S - \{1\}$ then Fix $t \cap$ Fix $s = \langle 0 \rangle$).

On the other hand, if a translation plane of order q^2, $p^r = q$, p a prime, admits a collineation group in the translation complement isomorphic to $SL(2, q)$, the Foulser–Johnson Theorem, see Chapter 47 (see [378]) does provide a complete classification of the possible planes and within this class are the Hering planes of odd order for which the p-elements are quartic and the Sylow p-subgroups of $SL(2, q)$ are 'quartic groups' of order q. Indeed, also the Walker planes of order 25 admit a collineation group isomorphic to $SL(2, 5)$ where the 5-elements are quartic.

However, utilization of the Foulser–Johnson theorem is often obtained via an indirect approach and to resolve certain of such technicalities, the authors (see Biliotti, Jha, Johnson [131]) extended the Foulser–Johnson theorem to determine the possible translation planes of order q^2 that admit a collineation group G and a set of points L such that G induces $SL(2, q)$ on L (the reader is directed to Chapter 47 for discussion of groups G such that $G/N \simeq PSL(2, q)$ on translation planes of order q^2).

Previously, Jha and Johnson [**598, 602**]) have studied what might be called 'large Baer group' planes and asked the following question: What are the finite translation planes admitting two trivially intersecting large Baer p-groups? In planes of order q^2, in this context, 'large' means of order $> \sqrt{q}$, for q odd and $\geq 2\sqrt{q}$ when q is even. Such planes were completely determined as either the Hall planes or the Dempwolff plane of order 16.

In this chapter, we ask the same sort of question regarding quartic groups: What are the finite translation planes with spreads in $PG(3, q)$ that admit two trivially intersecting 'large' quartic groups? It is possible to then attack this problem using the Foulser–Johnson Theorem and its generalization by Biliotti–Jha–Johnson.

To reiterate, a 'large' quartic group of odd order is a quartic p-group of order $> \sqrt{q}$. The main result of Biliotti, Jha and Johnson on quartic groups is as follows:

THEOREM 68.2. (Biliotti, Jha, and Johnson [**126**]). Let π be a translation plane with spread in $PG(3, q)$, where $q = p^r$, p a prime. If π admits at least two trivially intersecting quartic p-groups of orders $> \sqrt{q}$, then

(1) The group generated by the quartic p-elements is isomorphic to $SL(2, q)$, and

(2) The plane is one of the following planes: (a) a Hering plane or (b) $q = 5$ and the plane is one of the three exceptional Walker planes.

There are three exceptional Walker planes of order 5^2 admitting $SL(2, 5)$. In this situation, the group $SL(2, 5)$ leaves invariant a component of the spread and induces $SL(2, 5)$ faithfully on that component. The group is reducible but not completely reducible. One of these planes is the Hering plane of order 5^2 and admits a second group isomorphic to $SL(2, 5)$ that acts irreducibly. When we refer to the Walker planes, we shall use the group that acts reducibly and reference to the Hering plane shall indicate that we are considering the irreducible $SL(2, 5)$ group.

68.1. Even-Order Quartic Groups.

Although there are studies of translation planes of even order admitting large Baer 2-groups due to Jha and Johnson [**598**], (a Baer group is a group that fixes a Baer subplane pointwise), there is no existing general theory for groups generated by Baer involutions, where not all involutions fix the same Baer subplane, unless it is forbidden that such groups contain elations (e.g., Johnson and Ostrom [**786**]) and the dimension is 2 (the spread is in $PG(3, q)$). Furthermore, we note that the studies of translation planes of even order admitting large Baer 2-groups due to Jha and Johnson [**598**] may be proved by methods of quartic groups. Whereas the group structure is not generally well understood in the odd-order case, it is possible to make exceptional progress when q is even.

Hence, we consider what might be called a 'quartic group' generated by Baer involutions. We shall see that, when q is even, we may obtain a reasonable classification of groups generated by quartic groups with no assumption on the dimension of the plane.

We shall also adopt a definition of the 'index' of a quartic group. As mentioned above, we shall require that our group be an elementary Abelian 2-group. The term 'quartic' will refer to the fact that there are q^4 points and the group fixes q of these. For purposes of composition, we also allow that Baer groups be called quartic in some cases as seen below in the definition.

DEFINITION 68.3. Let π be a translation plane of even order q^2 and let Q be an elementary Abelian 2-group consisting of Baer involutions whose axes share a common line (subline). For example, Q could be a Baer group, otherwise, there is a unique common line, called the 'center' of the group and the unique component containing the center is called the axis. Any such group Q shall be called a 'quartic group'. Any quartic group that is not Baer shall be called a 'proper quartic group'.

Now defining 'index'.

DEFINITION 68.4. Let Q be a quartic group. If there is a partition of Q into Baer groups of order 2^i (i.e., the subgroups of the partition of order 2^i fix Baer subplanes pointwise), we shall say that Q is a quartic group of index i. We further assume that there is no subgroup of Q of order strictly larger than 2^i that fixes a Baer subplane pointwise. If the partition is trivial in the sense that there is a unique element of the partition, we shall say that the group of order 2^a has 'maximal index a' or equivalently that the group is 'Baer'.

(1) The Ott–Schaeffer planes admit quartic groups of index 1. What this means is that there is a partition of the elementary Abelian group by Baer involutions of order 2. (2) The Hall planes admit quartic groups of maximal index; the 2-groups are Baer. (3) Using the above definition, it is possible to show that two quartic groups of index i will normally generate non-solvable groups, where the centers of the Sylow 2-subgroups are of index $j \geq i$. This means that index i groups might generate higher-dimensional quartic groups. However, the only examples that are known involve dimension 1 or maximal dimension.

The most general result for arbitrary quartic groups is due to Biliotti, Jha and Johnson [130] and is as follows:

THEOREM 68.5. (Biliotti, Jha and Johnson [130]). Let Q be a quartic group in a collineation group G of the translation complement of a translation plane π of even order q^2. Then one of the following occurs:

(1) Q contains a Baer group of index 1 or 2.

(2) Q is contained in an elementary Abelian normal 2-subgroup of G.

(3) Q contains a Baer group Q^- of index 2 and there exists an element x of $G - N_G(Q)$, such that $\langle Q, Q^x \rangle$ fixes a Baer subplane π_o that is fixed pointwise by Q^- and induces a solvable group of order $2d$, for d odd, on π_o.

(4) $|Q| = 8$ and there is a conjugate Q^x, for $x \in G - N_G(Q)$ such that $\langle Q, Q^x \rangle$ is the universal covering group of $SL(2,4)$.

(5) There is an elementary Abelian 2-subgroup $\overline{Q}$ containing Q and fixing a Baer subline pointwise and a solvable subgroup W of order $2d$, $d > 1$ generated by elations and a subgroup S of G such that S/W is isomorphic to $SL(2, |\overline{Q}|)$, $S_z(|\overline{Q}|)$, $SU(3, |\overline{Q}|)$ or $PSU(3, |\overline{Q}|)$. In all cases, $|\overline{Q}W/W|$ is the order of the center of a Sylow 2-subgroup of S/W.

Similarly, Biliotti, Jha and Johnson consider arbitrary quartic groups with index as follows:

THEOREM 68.6. (Biliotti, Jha and Johnson [130]). Let Q be a quartic group of index i in a collineation group G of the translation complement of a translation plane π of even order q^2. Then one of the following occurs:

(1) Q is contained in an elementary Abelian normal 2-subgroup of G.

(2) $|Q| \leq 4$ and $i = 1$.

(3) $|Q| = 8$ and there is a conjugate Q^x, for $x \in G - N_G(Q)$ such that $\langle Q, Q^x \rangle$ is the universal covering group of $SL(2,4)$. In this setting, i must be 1.

(4) There is an elementary Abelian 2-subgroup $\overline{Q}$ containing Q and fixing a Baer subline pointwise and a solvable subgroup W of order $2d$, $d > 1$ generated by elations and a subgroup S of G such that S/W is isomorphic to $SL(2, |\overline{Q}|)$, $S_z(|\overline{Q}|)$, $SU(3, |\overline{Q}|)$ or $PSU(3, |\overline{Q}|)$. In all cases, $|\overline{Q}W/W|$ is the order of the center of a Sylow 2-subgroup of S/W.

Applying these results for 'large' quartic groups, Biliotti, Jha and Johnson obtain some strong classification results.

THEOREM 68.7. (Biliotti, Jha and Johnson [**130**]). Let π be a translation plane of even order q^2 and let Q be a quartic group of order at least $\max(16, 4\sqrt{q})$ of a collineation group G of a translation plane π of order q^2. Then one of the following occurs:

(1) Q is contained in a normal elementary Abelian 2-group of G,
(2) π is Hall, or
(3) π is Ott–Schaeffer.

When the spread is contained in $PG(3, q)$, it is actually possible to go further to obtain the following result:

THEOREM 68.8. (Biliotti, Jha and Johnson [**130**]). Let π be a translation plane of even order q^2 that contains an elementary Abelian 2-group Q in the linear translation complement G and which contains no elations.

(1) If the order of Q is at least $\max(16, 4\sqrt{q})$, then one of the following occurs: (a) Q is contained in an elementary Abelian normal subgroup of G, (b) π is Hall, or (c) π is Ott–Schaeffer.

(2) Assume that the order of Q is at least $\max(16, 2\sqrt{q})$. Assume that Q is Baer, or is not Baer and does not contain a Baer group of index 2. For example, assume that Q has index i. Then one of the following occurs: (a) Q is contained in an elementary Abelian normal subgroup of G, (b) π is Hall, or (c) π is Ott–Schaeffer.

If the order is at least 16, then

THEOREM 68.9. (Biliotti, Jha and Johnson [**130**]). Let π be a translation plane of even order q^2 that admits a quartic group Q of order at least 16 in a subgroup G of the translation complement.

Then one of the following holds: (1) Q contains a Baer group of index 1 or 2, (2) Q is contained in an elementary Abelian normal subgroup of G or (3) Let Q_1 be a maximal elementary Abelian 2-group containing Q and fixing a Baer subline pointwise. Then G contains a subgroup S^* isomorphic to $SL(2, 2^b)$ or $S_z(2^b)$, generated by quartic groups, where the order of Q_1 is either 2^b or 2^{b+1}.

COROLLARY 68.10. (Biliotti, Jha and Johnson [**130**]). Assume that Q is a quartic group of order at least 16, does not contain a Baer group of index 1 or 2 and is not contained in an elementary Abelian normal subgroup of G. If the order of Q is at least $2\sqrt[4]{q}$, then there is a collineation subgroup of G isomorphic to $SL(2, 2^t \geq \sqrt[4]{q})$, generated by quartic groups.

68.1.1. Characterization of Planes. Now assume that Q is a quartic group of order at least $\max(2\sqrt{q}, 16)$ and assume that Q is not contained in a normal elementary Abelian subgroup of G. Then there is a subgroup S^* isomorphic to $SL(2, 2^b)$ or $S_z(2^b)$, generated by quartic groups. Since $2\sqrt{q} > 2\sqrt[4]{q}$, applying Corollary 68.10, we may obtain the following of Biliotti, Jha and Johnson.

THEOREM 68.11. (Biliotti, Jha and Johnson [130]). Let π be a translation plane of even order q^2 admitting a quartic group Q of order at least $\max(16, 2\sqrt{q})$. Assume that Q is not contained in an elementary Abelian normal subgroup of the translation complement.

Then either Q contains a Baer group of index 1 or 2 or π is Ott–Schaeffer.

68.2. Baer Groups.

We have seen in the previous section that if we have a 'large' quartic group which does not contain a Baer subgroup of index 1 or 2, then the planes admitting such groups may be determined, provided the quartic group is not contained in a normal elementary Abelian normal subgroup of a group G containing Q. Planes admitting at least two large Baer groups have been completely determined as Hall planes or the Dempwolff plane of order 16 by Jha and Johnson [598], using a classification of Dempwolff [304].

Here we note that it is possible to give a different proof of the result of Jha and Johnson, which is more in keeping with the ideas presented here, using the following device: Suppose that Q is a Baer group of order at least 4 in a collineation group G of the translation complement. Consider $Q \cap Q^x$, for $x \in G - N_G(Q)$. If $g \neq 1$ is in $Q \cap Q^x$, it follows that $\langle Q, Q^x \rangle$ is elementary Abelian since both Q and Q^x will then fix the same Baer subplane pointwise. Hence, assume that Q_1 is a maximal Baer group of G containing Q. Then, Q_1 has trivial intersection with its conjugates and we may apply Hering's non-trivial intersection theorem (see Hering [477] for the statement). Using the general set-up, the following results may be obtained:

THEOREM 68.12. (Biliotti, Jha and Johnson [130]). Let π be a translation plane of even order q^2 admitting a Baer group Q of order at least 4 in a collineation group G of the translation complement. Let Q_1 be a maximal Baer group of G containing Q. Then either Q_1 is normal or the normal closure of Q_1 is isomorphic to $SL(2, 2^b)$, $S_z(2^b)$, $SU(3, 2^b)$ or $PSU(3, 2^b)$.

COROLLARY 68.13. (Biliotti, Jha and Johnson [130]). If the order of a Baer group Q is at least $2\sqrt{q}$ and is not normal in the translation complement, then there is a subgroup isomorphic to $SL(2, 2^b)$ generated by Baer groups and the plane is either Hall or Dempwolff of order 16.

Thus, combining the results on quartic groups and Baer groups, Biliotti, Jha and Johnson prove the following result:

THEOREM 68.14. (Biliotti, Jha and Johnson [130]). Let π be a translation plane of even order q^2 admitting a quartic group Q of order at least $(16, 4\sqrt{q})$. Assume that Q is not contained in an elementary Abelian normal subgroup of the translation complement. Then π is one of the following planes: (1) Hall, or (2) Ott–Schaeffer.

68.2.1. Dimension Two Quartic Groups.
Now assume that the spread for a translation plane π of order q^2 is in $PG(3, q)$, and assume that G is a subgroup of the linear translation of π. Then any 2-group of G fixes a 1-dimensional K-subspace, where $K \simeq GF(q)$ and K is contained in the kernel of π. Now let Q be

any elementary Abelian 2-subgroup of G. Any elementary Abelian 2-subgroup Q of G that contains no elations is a quartic group.

We therefore have the following result:

THEOREM 68.15. (Biliotti, Jha and Johnson [**130**]). Let π be a translation plane of even order q^2 that contains an elementary Abelian 2-group Q in the linear translation complement G and which contains no elations.

(1) If the order of Q is at least $\max(16, 4\sqrt{q})$, then one of the following occurs: (a) Q is contained in an elementary Abelian normal subgroup of G, (b) π is Hall or (c) π is Ott–Schaeffer.

(2) Assume that the order of Q is at least $\max(16, 2\sqrt{q})$. Assume that Q is Baer, or is not Baer and does not contains a Baer group of index 2. For example, assume that Q has index i. Then one of the following occurs: (a) Q is contained in an elementary Abelian normal subgroup of G, (b) π is Hall, or (c) π is Ott–Schaeffer.

We have discussed the classification of translation planes with spreads in $PG(3, q)$ that admit 'large quartic groups', where there are two quartic groups but here we allow that such groups are 'skew' but still have the quartic axis.

DEFINITION 68.16. If T and S are quartic groups, we shall say that T and S are 'skew' if and only if their quartic centers are distinct (i.e., if and only if $t \in T - \{1\}$ and $s \in S - \{1\}$). Let T and S be quartic 2-groups. We shall say that T and S are 'skew' if and only if the quartic center of T and the quartic center of S are disjoint subspaces.

The main result concerning planes of even order admitting two large trivially intersecting quartic groups is as follows and used the theory of collineation groups of order q^2 in translation planes of order q^2, see Chapter 48.

THEOREM 68.17. (Biliotti, Jha and Johnson [**130**]). Let π be a translation plane of even order with spread in $PG(3, q)$ that admits at least two skew quartic 2-groups of orders at least $2\sqrt{q}$ then π is an Ott–Schaeffer plane.

For odd order, we have a very different sort of result, the reader is directed to Chapter 48 for the definition of a 'desirable' plane.

THEOREM 68.18. (Biliotti, Jha and Johnson [**127**]). Let π be a translation plane of odd order q^2 and spread in $PG(3, q)$ that admits two trivially intersecting quartic groups with orders q that share their quartic centers. (1) Then π is a 'desirable' plane. (2) If π is a desirable plane, then there is a group G of order q^2 that contains two trivially intersecting quartic groups T and S with the same quartic axis such that $\langle T, S \rangle = G$.

CHAPTER 69

Double Transitivity.

In this chapter, we discuss the major results involving translation planes admitting collineation groups which act doubly transitively on a set of points, or on a set of parallel classes (infinite points). For reasons of space, we give only a few details on these results. We will pay particular attention to collineation groups with doubly transitive group actions on line size sets.

In the affine situation, if a collineation group G fixes a set S of affine points and acts doubly transitively on S, then we may assume that S contains the zero vector and that G_0 acts transitively on $S - \{0\}$. Furthermore, if S is a subspace to begin with and G_0 acts transitively on $S - \{0\}$, then adjoining the translation subgroup T_S that acts transitively on S, we have that $G_0 T_S$ is a group acting doubly transitively on S. So, doubly transitive groups acting on subspaces amount to transitive actions on the non-zero vectors of the subspace.

A collineation group can fix a component and act transitively on the non-zero vectors or could fix a subplane containing the zero vector and act transitively on the non-zero points. If the subplane is Baer, then the number of points being permuted is the same as the order of the plane, the number of points on a line. On the other hand, if a group acts doubly transitively on a set of line size, the set may not be a subspace. For example, the set could be the set of affine points of a parabolic oval or the set could be the set of affine points of a parabolic unital. In the case that the set S is a set of parallel classes, S could be the entire set, in which case the group that we obtain is 'flag-transitive'. Or S could be the set of points on a line of a hyperbolic unital. To bring into focus what might be considered, we first mention results of Cofman and Kantor [855].

THEOREM 69.1. (Cofman [236, 237]). Let π be a finite affine or projective plane and Γ a set of affine points upon which a collineation group G induces a two-transitive group. Then Γ is either (1) a collinear set, (2) a $(q+1)$-arc or (3) a 2-$(v, k, 1)$-design.

Case (3) is covered by Kantor classification of the doubly transitive designs as follows:

THEOREM 69.2. (Kantor [855]). Let D be a 2-$(v, k, 1)$-design that admits a collineation group doubly transitive on its points. Then D is one of the following: (1) $PG(a, h)$, (2) $AG(d, h)$, (3) the points and secant lines of a unital where $q+1 = h^2 + 1$, or (4) D is the affine Hall plane of order 9 or (5) the affine Hering plane of order 27.

69.1. Doubly Transitive on Line-Size Sets.

There are a variety of situations in this case, but we first consider the general problems.

PROBLEM 69.3. Assume that π is a finite translation plane. If there is a collineation group in the translation complement which fixes a component and acts transitively on the non-zero points of a component L, determine the group and classify the translation planes.

For example, consider a Bol plane.

DEFINITION 69.4. An affine translation of odd order is said to be a 'Bol' plane if and only if there are two components L and M such that given any other component N, there is an affine collineation of order 2 with axis N that interchanges L and M.

Any nearfield plane with spread $x = 0, y = 0, y = xM$, where $x = 0$ and $y = 0$ are axes of affine homologies, admits a collineation group $(x, y) \longrightarrow (y, x)$. This collineation has order 2 and fixes $y = x$ pointwise (which we assume is a component) and also leaves invariant $y = -x$ (which is $y = x$ if the characteristic is two). What this means is every nearfield is a Bol plane. The question is, are there other Bol planes?

Consider a Bol plane and assume that G is the group generated by the set of central involutions. Then $G_L = G_{L,M}$ is a subgroup of index 2. In general, let $x = 0, y = 0, y = xA; A \in S$ denote a spread where $\{L, M\} = \{x = 0, y = 0\}$. Then $\sigma_A : (x, y) \longrightarrow (yA^{-1}, xA)$ fixes $y = xA$ pointwise and leave invariant $y = -xA$. Then $\sigma_I \sigma_A : (x, y) \longrightarrow (xA, yA^{-1})$. Since the group $\langle x \longrightarrow xA; A \in S \rangle$ is sharply transitive, we have the following:

THEOREM 69.5. If π is a finite Bol plane, then the stabilizer of the two special components L and M is transitive on the non-zero vectors of each of them, L and M.

Note that in even-order Bol planes, the group is clearly transitive on the infinite points not in $\{(\infty), (0)\}$, since the remaining components are elation axes.

DEFINITION 69.6. A finite translation plane that admits an autotopism collineation group (i.e., fixes a triangle in the projective plane and fixes the vertices of the triangle) that acts transitively on the non-vertex points of the autotopism triangle is said to be 'triangle transitive'.

We consider Bol planes and triangle transitive planes in Chapters 72 and 70, respectively.

The standard approach of any such problem is based on the now widely used classification of finite 2-transitive groups, see Chapter 104 for the statement of the theorem, and particularly Hering's classification of transitive linear groups in $GL(n, p)$ in terms of the maximal fields over which the groups are semilinear. Solvable doubly transitive groups are known by Huppert [565]. The main theorem of Hering [479] incorporates this work.

THEOREM 69.7. (Hering [479]). Let G be a 2-transitive affine permutation group of degree p^d, with socle $V \simeq (Z_p)^d$ for some prime p, and let G_0 be the stabilizer of the zero vector in V. Then G_0 belongs to one of the following classes (and conversely, each of the classes below does give a 2-transitive affine group).

(A) Infinite classes:
(1) $G_0 \leq \Gamma L(1, p^d)$;
(2) $G_0 \rhd SL(a, q)$ and $p^d = q^a$;

(3) $G_0 \triangleright Sp(2a, q)$ and $p^d = q^{2a}$;

(4) $G_0 \triangleright G_2(q)'$, $p^d = q^6$ and $p = 2$.

(B) Extraspecial classes: with the notation of part (B) of the theorem of Liebeck (see Chapter 104 for Liebeck's theorem on primitive rank three groups):

(i) $p^d = 5^2, 7^2, 11^2, 23^2, r = 2$ and $R = Q_8$, or

(ii) $r = 2$, $p^d = 3^4$ and $R = R_2^2 \simeq D_8 \circ Q_8$ and $G_0/R \le S_5$.

(C) Exceptional classes:

(i) $G_0 \triangleright SL(2, 5)$ and $p^d = 9^2, 11^2, 19^2, 29^2, 59^2$ and $SL(2, 5) < SL(2, p^{d/2})$,

(ii) G_0 is A_6, $p^d = 2^4$ and $A_6 \simeq S_p(4, 2)'$,

(iii) G_0 is A_7, $p^d = 2^4$ and $A_7 < A_8 \simeq SL(4, 2)$, or

(iv) G_0 is $SL(2, 13)$, $p^d = 3^6$ and $SL(2, 13) < Sp(6, 4)$.

In general, there are a number of translation planes that admit solvable collineation groups that act transitively on the non-zero vectors of a component.

69.2. Solvable Doubly Transitive Groups.

There are a variety of finite translation planes that admit solvable collineation groups acting doubly transitively on an affine line or a Baer subplane. For example, let Σ be a finite Desarguesian affine plane. Take any subspace H of line size, which is not a line of Σ of the associated vector space and such that H intersects components of Σ in subspaces of the same dimension, when the intersection is non-trivial and write the order as q^n, making the intersections 1-dimensional $GF(q)$-subspaces. There is a kernel homology group of Σ, K^* of order $q^n - 1$ such that K^*H then forms a cover of the set of components of non-trivial intersection. We may form a new translation plane Σ_H by taking new components as elements of K^*H together with the components of non-intersection of Σ. We discuss such constructions elsewhere in this text as 'hyper-regulus' construction. Since K^* acts transitively on all components of Σ, and the components of intersection are now subspaces of Σ_H, we have a collineation group of Σ_H which acts doubly transitively on a line-size set. Of course, in this setting the doubly transitive group is solvable. This situation occurs often, and there is probably not a satisfactory way to classify such translation planes. The idea of constructing translation planes using the kernel homology group of an associated Desarguesian translation plane is due to Ostrom and there are a tremendous variety of translation planes so constructed in Ostrom's seminal monograph [**1046**]. On the other hand, one can make some progress if we ask what makes a translation plane into one which can be so constructed.

DEFINITION 69.8. Let Σ be a finite Desarguesian affine plane of order p^n. Assume that a translation plane π may be constructed from Σ from a set S of subspaces H, each of which is of line size such that the nets of intersection N_H are mutually disjoint on lines. That is, if F^* is the kernel homology group of Σ, then the components of N_H form the set HF^*. Then, we shall call π a 'cyclic Ostrom plane'.

DEFINITION 69.9. A finite group G acting as a transitive group on a set X is said to be '3/2-transitive' if and only if orbits of the stabilizer G_0 not equal to $\{0\}$ have the same length. A finite transitive group G has 'rank $k + 1$' if and only if the number of orbits of G_0 is $k + 1$ (including the trivial orbit $\{0\}$). A transitive group is said to be 'stable' if every non-trivial orbit B of G_0 is such that some element of $G - G_0$ leaves invariant $B \cup \{0\}$.

Consider a cyclic Ostrom plane of order n. Then there is an associated translation group T of order n^2 and if we compose the kernel homology group K^* of order $n - 1$ with T, we have a collineation group of order $n^2(n - 1)$. This group is transitive and the stabilizer of 0 (the zero vector) has $n + 1$ orbits of length $n - 1$. Hence, this group G is 3/2-transitive of the rank $n + 2$ and stable since the translations of $G - G_0$ satisfy the condition required.

When Ostrom [**1046**] considered such methods of construction as we are discussing, he also considered a classification of such translation planes using groups. In our terminology, Ostrom's theorem is as follows:

THEOREM 69.10. (Ostrom [**1046**]). Let γ be a spread of order $n = p^r$. Then γ is a cyclic Ostrom spread if and only if it admits a cyclic automorphism group K^* of order $n - 1$ such that the associated vector space V contains $n + 1$ distinct additive subspaces $\{W_i; 1 \leq i \leq n + 1\}$, each of order n such that K^* acts transitively on the non-zero points of each W_i.

But, more can be said of cyclic Ostrom spreads, even without initially assuming that one is dealing with a spread, that is, with a translation plane. The following result of Jha and Johnson [**624**] shows the necessary conditions are those of 3/2-transitivity, rank and stability.

69.2.1. Jha–Johnson Classification of Cyclic Ostrom Spreads.

THEOREM 69.11. (Jha and Johnson [**624**]). If a finite affine plane π of order $n \geq 2$ admits a collineation group G that is 3/2-transitive and stable of rank $n + 2$ then π is a translation plane producing a cyclic Ostrom spread.

We point out that we did not initially assume that the affine plane is a translation plane and were able to prove this using our assumptions. Actually, if G is a 3/2-transitive collineation group, then it turns out that G is either primitive or a Frobenius group. The question of whether a finite affine plane admitting a transitive and primitive collineation group is quite an old problem.

DEFINITION 69.12. A 'primitive affine plane' is a finite affine plane that admits a transitive and primitive collineation group on the points.

This problem was first considered by Keiser [**884**] who proved the following:

THEOREM 69.13. (Keiser [**884**]). Let π be a primitive affine plane of order n. If n is either a non-square or $n = m^2$, for $m \equiv 0, 2, 3 \bmod 4$, then π is a translation plane.

This problem was completely resolved by Hiramine [**509**].

69.2.2. Hiramine/Keiser Theorem on Primitive Affine Planes.

THEOREM 69.14. (Hiramine [**509**]). Primitive affine planes are translation planes.

We note that many results on solvable doubly transitive groups involve doubly transitive subgroups of $A\Gamma L(1, p^r)$. These groups are known by Foulser [**365**].

69.3. Non-Solvable Doubly Transitive Groups.

Now apart from a few sporadic orders, it is possible to give a classification of all finite translation planes of order n that admit a collineation group G that induces a non-solvable doubly transitive collineation group on a set Ω of n affine points.

Furthermore, it is possible that a collineation group G could act doubly transitively on a parabolic oval and certain generalized twisted field planes admit such groups. In this case, the group G leaves invariant an infinite point and acts doubly transitively on Ω, the remaining n points of the oval. However, in this case, the group is always solvable. The reader interested in 2-transitive parabolic ovals is referred to the papers of Enea and Korchmáros [353] and Biliotti, Jha and Johnson [120, 124, 125]. In addition, certain generalized twisted field planes admit collineation groups which act doubly transitively on one or more lines, and the reader is referred to Biliotti, Jha and Johnson [121] for a complete discussion of such planes, but again we note that all groups are solvable in these planes.

From a group-theoretic standpoint, it might be possible to determine the collineation groups G that can act on a finite affine plane and induce a non-solvable doubly transitive group on a line-sized set Ω.

Actually, Hiramine has, more or less, considered this problem when the line-sized set Ω is a line.

Under these conditions, the possibilities for G are severely restricted, but without some further hypothesis, not much can be deduced about the isomorphism types of the associated planes.

THEOREM 69.15. (Hiramine [513]). Let π be a finite affine plane of order n that admits a collineation group G which fixes an infinite point (∞) and acts doubly transitively on the remaining n infinite points $\ell_\infty - \{(\infty)\}$. Then n is a power of a prime and the doubly transitive group induced by G on the line at infinity contains a regular normal subgroup. Let H denote the subgroup of G which fixes a second point on the infinite line and let $\overline{H}$ denote the group induced on the set of infinite points. Then one of the following conditions holds:

(1) $\overline{H} \leq \Gamma L(1, p^r)$ and $n = p^r$,
(2) $SL(2, q) \leq \overline{H} \leq \Gamma L(2, q)$ and $n = q^2$,
(3) $n \in \{2^4, 3^2, 3^4, 3^6, 5^2, 7^2, 11^2, 19^2, 23^2, 29^2, 59^2\}$.

So, we may utilize the theorem of Hiramine when the set Ω is a line. Furthermore, if π is a translation plane, then case (2) of Hiramine's result implies that $H \cong SL(2, q)$ acts faithfully on π of order q^2. In this situation, the theorem of Foulser and Johnson completely determines the translation planes.

THEOREM 69.16. (Foulser and Johnson [378, 379]). Let π denote a translation plane of order q^2 that admits a collineation group isomorphic to $SL(2, q)$. Then π is one of the following planes: (1) Desarguesian, (2) Hall, (3) Hering, (4) Ott–Schaeffer, (5) one of three Walker planes of order 25, (6) the Dempwolff plane of order 16.

It is noted therefore that the argument of the following results of Ganley–Jha–Johnson is not strictly a theorem involving group theory. In fact, geometric arguments in translation planes are essential for the determination of the structure of the set Ω being permuted and for the determination of the isotopism classes of the translation planes.

69.3.1. Ganley–Jha–Johnson.

THEOREM 69.17. Let π be a translation plane of order $n = p^r$, where $n \neq 3^4, 3^6, 11^2, 19^2, 29^2, 59^2$, which admits a collineation group that induces a non-solvable doubly transitive group on a set Ω of n affine points. Then π is one of the following types of planes: (I) Desarguesian, (II) Hall, (III) one of three Walker planes of order 25, (IV) the Dempwolff plane of order 16.

In each of the cases (I)–(IV) in the above theorem, Ω is invariant under a collineation group $G = G_O T$ that acts faithfully and 2-transitively on Ω, where T is a translation group of order $n = q^2$ and $G_O \cong SL(2, q)$. In the case of the Hall plane Ω is Baer subplane, and in the other cases Ω is a line.

The excluded n, from Theorem 69.17, correspond to certain sporadic cases which are yet to be resolved. It is conceivable that a computer program for these cases might then provide a solution to the general problem but we have not attempted such a study. It will become clear from the proof that dealing with sporadic cases would require considerable combinatorial argument. However, there are at least some non-Desarguesian examples, satisfying the hypothesis of Theorem 69.17, associated with each sporadic order.

REMARK 69.18. (I) The possibilities of sporadic planes of order p^2 where p is a prime include not only exceptional nearfield planes, but also the planes of Lüneburg of type $R * p$: these are not always nearfield planes for $p \neq 11, 59$.

(II) The only known possibilities of order 3^6 are the Hall plane and the Desarguesian planes; these admit $G = SL(2, 3^3)$ acting transitively on the non-zero points of a component and a subspace, respectively. Thus $G \geq SL(2, 13)$, as mentioned in the proof of the theorem.

(III) The Prohaska planes, of order 3^4 admit a group G isomorphic to the central product $SL(2, 5)K^*$ where K^* is a cyclic group of order 8. The axes of the elations in G define a derivable net D and G acts transitively on the non-zero points of each Baer subplane in D that passes through $\mathbf{0}$.

(IV) The derived Prohaska planes, obtained by deriving relative to D in case (III) above, also admits G. In this situation G acts 2-transitively on each line through $\mathbf{0}$ that corresponds to the Baer subplanes in D of the original Prohaska plane.

This provides an example of the exceptional case listed in Hiramine's theorem (case (3)).

69.3.2. Sketch of the Proof.

The general approach to proving Theorem 69.17 is first of all to reduce to the situations in the Hering theorem. Once this is done, we eliminate many of the groups arising in Hering's list to reduce to the Foulser–Johnson theorem using geometric methods. Specifically, it is possible to eliminate many of the possibilities in Hering's list by repeatedly appealing to two facts: (1) large Baer groups force their fixed planes to be Desarguesian, and (2) large planar p-groups P, acting on translation planes of order p^r, satisfy $|P| \leq p^{r-1}$, and this bound is technically sharp. The fact that these two results, and also the Foulser–Johnson classification, have only been established (to duality) in *translation* planes is one of the main reasons why the main theorem is restricted to translation planes.

CHAPTER 70

Triangle Transitive Planes.

An 'autotopism group' of an affine translation plane π is a group of collineations G that fixes two points on the line of infinity of the projective extension π^+ of π and fixes an affine point P. The triangle so formed in π^+ is called the 'autotopism' triangle of G. Since the collineation group of π is a semidirect product of the stabilizer of a point by the translation group, the affine point fixed by an autotopism group may always be taken arbitrarily. When autotopism groups become important for the analysis of a translation plane, there is a natural choice for the two infinite points fixed by the group. For example, if π is a finite non-Desarguesian semifield plane of order p^u, every collineation fixes the center of the affine elation group E of order p^u, which we denote by (∞). Furthermore, the second fixed infinite point may be chosen arbitrarily since E is transitive on the set of remaining infinite points. More generally, we may coordinatize so that the two infinite points are (∞) and (0) and the affine fixed point is the zero vector $(0,0)$ of the associated underlying vector space.

As the title might suggest, we are interested in 'triangle transitive translation planes'. Since we will be dealing with autotopism groups, such planes are defined as translation planes that admit an autotopism group that acts transitively on the set of non-vertex points of each side of the autotopism triangle. (In other considerations, a triangle transitive plane may be required only to have a group leaving invariant the triangle and acting transitively on the set of non-vertex points of each side of the triangle.) When this occurs, there is essentially a fixed triangle to consider since Jha and Johnson [621] have shown that the plane may be completely determined if either of the vertices on the line at infinity are moved outside of vertex set of the given triangle.

But, of course, another definition might be that a 'triangle transitive translation plane' is a translation plane that admits a collineation group that acts transitively on all affine triangles. However, it is clear that the only possible finite translation planes of this type are Desarguesian, since there is a corresponding group acting doubly transitively on the affine points. So, we are content with our definition of triangle transitivity as it coincides with the autotopism groups with a certain maximal level of transitivity. A major problem in translation planes is the following:

Completely determine the class of triangle transitive translation planes.

70.1. Known Triangle Transitive Translation Planes.

Finite generalized twisted field planes are triangle transitive precisely when the right, middle and left nuclei of the associated semifield plane coincide (see Biliotti, Jha, and Johnson [121]). If the order is p^u, there is a cyclic autotopism group of order $p^u - 1$ that acts on the plane. Furthermore, the Suetake planes [1168] are triangle transitive under the less restrictive definition.

Perhaps the most standard examples of triangle transitive planes are the near-field planes of order p^u. In this case, there is an autotopism group of order $(p^u - 1)^2$ that is triangle transitive, the group being the direct product of two affine homology groups of order $p^u - 1$ where the axis and co-axis of one group is the co-axis and axis of the second group, respectively.

There are a great variety of triangle transitive planes of order p^u due to M.E. Williams that admit a cyclic autotopism group of order $p^u - 1$ arising from orthogonal spreads. In particular, the number of non-isomorphic planes is not bounded by any polynomial in $q = p^u$.

In this section, we shall construct a class of triangle transitive planes of order p^u that admit an autotopism group of order $(p^u - 1)^2/2$. These planes are generalized André planes and also admit symmetric affine homology groups of orders $(p^u - 1)/2$ but fail to be nearfield planes.

If a triangle transitive group has order $(p^u - 1)$, it may be quite difficult to determine the general nature of the plane as noted above, given the great variety of triangle transitive planes with 'small' groups. So, the question is whether the existence of a 'large' triangle transitive autotopism group necessarily forces the plane to be a nearfield plane or at least belong to a known class of planes. We define 'large' to any group of order divisible by $(p^u - 1)_{2'}^2$. We consider this question in Section 71.3 and show that, with the exception of a few sporadic orders, such a plane is at least forced to be a generalized André plane. We summarize the triangle transitive planes discussed above as follows:

- Finite generalized twisted field planes are triangle transitive precisely when the right, middle and left nuclei of the associated semifield plane coincide (see Biliotti, Jha, and Johnson [121]).
- The Suetake planes.
- The planes of Kantor and Williams.
- Certain generalized André planes.
- The nearfield planes.
- The non-solvable triangle planes consist of the irregular nearfield planes with non-solvable groups, orders 11^2, 29^2 or 59^2.

CHAPTER 71

Hiramine–Johnson–Draayer Theory.

To elaborate on the comments of Chapter 70 and to digress slightly, a nearfield plane is a translation plane of order k that admits a group of affine homologies of order $k - 1$. In this case, it follows directly that there are two affine homology groups of order $k - 1$. The basic question is how large must a homology group be before it can be concluded that one obtains, in fact, the full group? For example, suppose there is a homology group of order $(k - 1)/2$ or perhaps two such groups. Must then the plane be a nearfield plane? Actually, the answer is 'no'; there are translation planes of order k that admit one or even two affine homology groups of order $(k-1)/2$. The complete answer when there are two groups is due to Hiramine and Johnson.

71.1. When There Are Two Groups.

THEOREM 71.1. (Hiramine–Johnson [**521**]). Let π be a translation plane of order k that admits two distinct homology groups of order $(k - 1)/2$ with distinct axes. Then one of the following occurs: (1) π is Desarguesian or (2) the axis and coaxis of one homology group is the coaxis and axis, respectively, of the remaining homology group. Furthermore, either the translation plane is (a) a generalized André plane or (b) the order is 7^2 and the plane is the irregular nearfield plane, (c) the order is 7^2 and the plane is the exceptional Lüneburg plane, or (d) the order is 23^2 and the plane is the irregular nearfield plane.

So, it might be said that such translation planes are 'known'. However, the question then remains to determine all generalized André planes of order k that admit two affine homology groups of order $(k - 1)/2$. This problem was considered by Hiramine and Johnson also. However, it was improperly argued that when $k = q^n$ then $\{q, n\}$ is always a Dickson pair. In fact, this need not be the case, although it is true that the prime divisors of n will divide $q - 1$, the problem is when this occurs and $n \equiv 0 \bmod 4$, as was pointed out and corrected in Draayer [**326**]. The following theorem combines Hiramine and Johnson and the work of Draayer to completely determine the generalized André planes of order $k = q^n$ admitting symmetric homology groups of order $(q^n - 1)/2$, where it is assumed that the right and middle nuclei of the corresponding quasifield coordinatizing π are equal. We state the result below although, it tuns out that one of the cases (case (c)) does not occur.

In the statement of the following theorem, we shall assume that the order of the plane p^w is q^n, there all prime divisors of n divide $q - 1$ and either $\{q, n\}$ is a Dickson pair or $q \equiv -1 \bmod 4$ and $n \equiv 0 \bmod 4$.

THEOREM 71.2. (See Hiramine–Johnson [**520**] and Draayer [**326**, (4.2) and (5.2)]). Let π be a generalized André plane of order q^n that admits symmetric homology groups of order $(q^n - 1)/2$ (the axis and coaxis of one group are the coaxis and axis, respectively, of the second group). Assume that the right and middle nuclei of the corresponding quasifield coordinatizing π are equal. Then the spread may be represented as follows:

$$x = 0, y = x^{q^i} w^{s(q^i-1)/(q-1)} \alpha, y = x^{p^t q^j} w^{v q^j} w^{s(q^j-1)/(q-1)} \beta$$

where $\alpha, \beta \in \mathcal{A}$ a subgroup of $GF(q^n)^*$ of order $(q^n-1)/2n$, where $i, j = 0, 1, \ldots, n-1$ and where t and v have the following restrictions:

for $q \equiv 1 \bmod 4$ or n odd, we may choose $s = 2$, $v = 1$, $t \geq 0$,

for $q \equiv -1 \bmod 4$ and $n \equiv 2 \bmod 4$, we may choose

$s = 1$, $v = n$, $t \geq 0$ and even, or

for $q \equiv 3 \bmod 8$ and $n \equiv 0 \bmod 4$, we may choose

$s = 1$, $v = 2[n]_{2'}$, $t \geq 0$ and even.

Furthermore, we have the congruence:

$$s(p^t - 1) \equiv v(q-1) + 2nz \bmod (q^n - 1), \text{ for some } z \text{ so that}$$
$$s(p^t - 1) \equiv v(q-1) \bmod 2n.$$

Conversely, generalized André planes with the above conditions have two symmetric homology groups corresponding to equal right and middle nuclei in the associated quasifield.

This brings up an interesting question: **What is the subclass of translation planes of order q^n, where $\{q, n\}$ is not a Dickson pair that admit two homology groups of order $(q^n - 1)/2$, where the associated middle and right nuclei are equal? And, what is the subclass when the associated middle and right nuclei are not equal?**

Moreover, it is of general interest to completely determine the generalized André planes of order q^n admitting two affine homology groups of order $(q^n - 1)/2$, where the associated nuclei are not equal. We have previously called planes admitting two 'half-order' homology groups, 'near nearfield planes'. When the associated nuclei are not equal, Hiramine and Johnson show that such translation planes admit an autotopism group that acts transitively on the non-vertex points on the 'infinite line' and completely classify such planes. Furthermore, it is possible to show that any near nearfield plane always admits an autotopism group that acts transitively on the non-vertex points of each affine side of the autotopism triangle, thus providing additional examples of triangle transitivity planes.

We first note the following theorem of Hiramine and Johnson [**520**].

THEOREM 71.3. (Hiramine and Johnson [**520**]). Let π be a generalized André plane of order q^n admitting symmetric homology groups of order $(q^n-1)/2$. Assume that the right and middle nuclei of the corresponding quasifield coordinatizing π are equal. Then $\{q, n\}$ is a Dickson pair.

Draayer considers the more general situation when there is but one such nucleus.

THEOREM 71.4. (Draayer [**326**]). Let π be a generalized André plane of order q^n that admits an affine homology group H_y of order $(q^n - 1)/2$ with axis $y = 0$ and co-axis $x = 0$. Then the spread for π may be represented as follows:

$$x = 0, y = x^{q^i} w^{s(q^i-1)/(q-1)} \alpha, \quad y = x^{p^t q^j} w^{v q^j} w^{s(q^j-1)/(q-1)} \beta$$

where $\alpha, \beta \in \mathcal{A}$ a subgroup of $GF(q^n)^*$ of order $(q^n-1)/2n$, where $i, j = 0, 1, \ldots, n-1$ and where t and v have the following restrictions:

for $q \equiv 1 \bmod 4$ or n odd, let $s = 2$, $v = 1$, $t \geq 0$,

for $q \equiv -1 \bmod 4$ and $n \equiv 2 \bmod 4$, let $s = 1, v = n, t \geq 0$ and even, or

for $q \equiv 3 \bmod 8$ and $n \equiv 0 \bmod 4$, let $s = 1, v = 2[n]_{2'}$, $t \geq 0$ and even.

Furthermore, the group H_y has the following form:

$$\left\langle (x, y) \longmapsto (x, y^{q^i} w^{s(q^i-1)/(q-1)} \alpha); \ \alpha \in \mathcal{A}, \text{ where } i = 0, 1, \ldots, n-1 \right\rangle.$$

REMARK 71.5. In Hiramine and Johnson [520], there is an analysis of generalized André systems of order q^n that admit right sub-nuclei of order $(q^n-1)/2$. Such right sub-nuclei correspond to the affine homology group with axis $y = 0$ and coaxis $x = 0$. This analysis applies equally, in theory, to such systems that admit middle sub-nuclei of order $(q^n-1)/2$. Such middle sub-nuclei correspond to the affine homology group with axis $x = 0$ and coaxis $y = 0$. In particular, the set of generalized André systems of order q^n with right sub-nuclei of order $(q^n-1)/2$ are in 1–1 correspondence with the set of all generalized André systems of order q^n with middle sub-nuclei of order $(q^n-1)/2$ under the mapping $(x, y) \longmapsto (y, x)$.

71.1.1. Hiramine, Johnson / Draayer Theorem Two Large Homology Groups.
When there are two large homology groups, Hiramine and Johnson [520, 521] and Draayer [326] completely resolve the problem.

THEOREM 71.6. (Hiramine and Johnson [520, 521], Draayer [326]). Let π be a translation plane of order $q^n = p^{nr}$, for p a prime, that admits two homology groups of order $(q^n-1)/2$. Then one of the following occurs: (1) π is Desarguesian, (2) $q^n = 7^2$ and the plane is the irregular nearfield plane, (3) $q^n = 7^2$ and the plane is the exceptional Lüneburg plane, (4) $q^n = 23^2$ and the plane is the irregular nearfield plane, (5) π is a generalized André plane and the associated right and middle nuclei of the (sub) homology groups are equal. Then the spread may be represented as follows:

$$x = 0, y = x^{q^i} w^{s(q^i-1)/(q-1)} \alpha, y = x^{p^t q^j} w^{vq^j} w^{s(q^j-1)/(q-1)} \beta$$

where $\alpha, \beta \in \mathcal{A}$ a subgroup of $GF(q^n)^*$ of order $(q^n-1)/2n$, where $i, j = 0, 1, \ldots, n-1$, and where t and v have the following restrictions:

for $q \equiv 1 \bmod 4$ or n odd, let $s = 2$, $v = 1$, $t \geq 0$, or

for $q \equiv -1 \bmod 4$ and $n \equiv 2 \bmod 4$, let $s = 1, v = n, t \geq 0$ and even.

Furthermore, we have the congruence:

$$s(p^t - 1) \equiv v(q - 1) + 2nz \bmod (q^n - 1), \text{ for some } z.$$

(6) π is a generalized André plane such that there are subnuclei of the associated (sub) homology groups of index two and are not equal. Then one of two situations occurs: (a) π is a Dickson nearfield plane and the orbit lengths of components under the full group is 2, $q^n - 1$ or (b) π is not a nearfield plane but nevertheless, there is a transitive autotopism group and the spread may be represented as follows:

$$x = 0, y = 0, y = x^{q^i} w^{s(q^i-1)/(q-1)} \alpha,$$
$$y = x^{p^t q^j} w^{vq^j} w^{s(q^j-1)/(q-1)} \beta,$$

where $\alpha, \beta \in Z^-$, the cyclic subgroup of $GF(q^n)^*$ of order $(q^n - 1)/2n$, $i, j = 1, \ldots, n$, w a primitive element of $GF(q^n)^*$ and

$$q \equiv 3 \bmod 8 \text{ and } [n]_2 = 4, \; s = 1, \text{ and } v = n/2 \text{ and } t \in (0, nr/2).$$

DEFINITION 71.7. Translation planes of order q^2 with spread in $PG(3, q)$ admitting two symmetric homology groups of order $(q^2 - 1)/2$ are said to be 'near nearfield' planes. (There are connections with nearfield planes which we develop in due course.)

COROLLARY 71.8. Any near nearfield plane is affine transitive (the autotopism group is transitive on the non-zero affine points of each affine axis of the autotopism triangle).

COROLLARY 71.9. A near nearfield plane of order q^n has an affine transitive autotopism group G of order $(q^n - 1)^2/2$. When $\{q, n\}$ is not a Dickson and, where $\mathcal{A}$ is a cyclic group of order $(q^n - 1)/2n$ in $GF(q^n)^*$.

71.2. Triangle Transitive Planes with Large Homology Groups.

In the previous sections, we have considered the set of translation planes of orders q^n that admit two homology groups of orders $(q^n - 1)/2$, say H_y and H_x such that $\langle H_y, H_x \rangle$ is a transitive autotopism group (fixes two infinite points, and an affine point and acts transitively on the remaining points of the infinite side of the autotopism triangle). However, it might be possible that there is a transitive autotopism group and the two homology groups have the same two orbits on the line at infinity. In this case, there would be an element g that interchanges the two orbits of H_y and fixes $x = 0$ and $y = 0$.

Specifically, suppose that we have a non-Desarguesian translation plane of order $q^n = p^{nr}$ with two homology groups of order $(q^n-1)/2$. Since we have a generalized André plane (except for the indicated few special orders), there are two components $x = 0$ and $y = 0$ that are fixed or interchanged. Suppose that there is a group G containing the homology groups such that G is a transitive autotopism group (i.e., fixes $x = 0$ and $y = 0$ and is transitive on the remaining components). If the two homology groups lead to different right and middle nuclei, we have a complete classification. However, it may be possible such transitive autotopism groups G (infinite side transitive) where the two right and middle nuclei are actually equal. We explore this situation in the current section.

Recall that we have the spreads

$$x = 0, y = x^{q^i} w^{s(q^i-1)/(q-1)}\alpha, y = x^{p^t q^j} w^{vq^j} w^{s(q^j-1)/(q-1)}\beta$$

where $\alpha, \beta \in \mathcal{A}$ a subgroup of $GF(q^n)^*$ of order $(q^n-1)/2n$, where $i, j = 0, 1, \ldots, n-1$ and where t and v have the following restrictions:

for $q \equiv 1 \bmod 4$ or n odd, let $s = 2$, $v = 1$, $t \geq 0$, or

for $q \equiv -1 \bmod 4$ and $n \equiv 2 \bmod 4$, let $s = 1, v = n, t \geq 0$ and even.

Furthermore, we have the congruence:

$$s(p^t - 1) \equiv v(q - 1) + 2nz \bmod (q^n - 1), \text{ for some } z, \text{ I} \iff$$
$$s(p^t - 1) \equiv v(q - 1) \bmod 2n.$$

If we assume that $q \equiv -1 \bmod 4$, so that $s = 1$ and $v = n$, then n divides $p^t - 1$.

The other possibility is to assume $q \equiv 1 \bmod 4$ (n may be even or odd), and let $s = 2$, $v = 1$, $t \geq 0$,

$$2(p^t - 1) \equiv (q - 1) \bmod 2n.$$

In any case, since G is an autotopism group, by our previous results, either the plane is a Dickson nearfield plane or the homology group H_y is normal in G. Hence, we may assume that there exists an element g in G that interchanges the two H_y-orbits of components. Furthermore, we know that there is an affine transitive autotopism group acting transitively on each of the affine axes of the autotopism triangle. Since the group element $g_n : (x, y) \longmapsto (xw^n, yw^n)$ normalizes both H_y and H_x, and fixes $y = x$, we may assume that g_n is in G. Let $G^- = \langle g_n, H_y, H_x \rangle$ and note that G^- normalizes H_y. Let g interchange the two H_y-orbits of components, we claim that g may be assumed to fix some point $(1, z)$. Assume that g maps 1 to s_0. Then, there exists an element of G^- that maps $(s_0, g(z))$ to $(1, d)$ for some element d. So, we may assume, without loss of generality, that g fixes 1 on $y = 0$.

We recall from Chapter 17 that it was mentioned (Section 17.2) that Foulser [369] determines the full collineation group of a generalized André plane. In particular, if the plane is not Desarguesian or a Hall plane, then the collineation group fixes the autotopism triangle and fixes or interchanges the two infinite vertices. Furthermore, the collineation group may be represented using the group $\Gamma L(2, q^n)$ and the automorphism group of $GF(q^n)$ that fixes the kernel elementwise.

71.2.1. Theorem of Draayer–Johnson–Pomareda Triangle Transitive.
Using a variety of results on generalized André planes and affine homology groups, Draayer, Johnson and Pomareda [328] prove the following result.

THEOREM 71.10. (Draayer, Johnson and Pomareda [328]). Let π be a non-Desarguesian affine plane of order p^w that admits two affine homology groups H_x and H_y of order $(p^w - 1)/2$ with axis (respectively, co-axis) $x = 0$ and co-axis (respectively, axis) $y = 0$. Then there is an autotopism group that is triangle transitive if and only if either (1) the plane is a Dickson nearfield plane, (2) $p^w = 7^2$ or 23^2 and the plane is an irregular nearfield plane, (3) $p^w = q^n$ and the plane is a generalized André plane of order q^n, where

$$q \equiv 3 \bmod 8, \ n \equiv 0 \bmod 4, \ n/4 \text{ is odd.}$$

Furthermore, the spread may be represented as follows:

$$\left\{ x = 0, y = 0, \{y = x, y = w^{n/2}\} H_y \right\}.$$

71.3. Non-Solvable Triangle Transitive Planes.

We remind the reader that all the non-Desarguesian triangle transitive planes known to date have solvable triangle transitive groups, apart from a few famous counterexamples, namely certain of the irregular nearfields. In particular, the irregular nearfield planes, one of order 11^2 and the two of order 29^2 and 59^2 are non-solvable triangle transitive planes since the direct product of the two homology groups is an autotopism group transitive on the non-vertex points on each side of the autotopism triangle.

Using the theorem of Ganley–Jha–Johnson [393], and the classification theorem of finite doubly transitive groups (see Chapter 104 for a statement), it is possible to show that these three planes are exactly the non-solvable triangle transitive planes. For example, the initial results show that the order $q^n \in \{11^2, 19^2, 29^2, 59^2, 3^4, 3^6\}$.

THEOREM 71.11. (Johnson [**764**]). Let π be a finite non-solvable triangle transitive plane then π is an irregular nearfield.

71.3.1. Large Triangle Transitive Groups. We have seen that triangle transitive groups which are not generalized André planes have relatively small-order groups. The question is how large an autotopism group must be in order to ensure that the plane is a generalized André plane?

THEOREM 71.12. (Draayer, Johnson, and Pomareda [**328**]). Let π be a triangle transitive translation plane of order p^r.

(1) If π is non-solvable (admits a non-solvable group), then π is the irregular nearfield plane of order $11^2, 29^2$ or 59^2 with non-solvable group.

(2) If π is the Hall plane of order 9, then π is a triangle transitive translation plane.

(3) If π is solvable assume that q^2 is not in $\left\{5^2, 7^2, 11^2, 23^2, 3^4\right\}$. (a) If $p^r - 1$ admits a p-primitive divisor u and if π admits a group of order divisible by $(p^r - 1)\frac{2}{2}$, then π is a generalized André plane. (b) If $p^r - 1$ does not admit a p-primitive divisor and $r = 2$, assume that the order of the group is divisible by $(p^2 - 1)^2/2$. Then π is a generalized André plane.

PROBLEM 71.13. Determine the triangle transitive generalized André planes that admit 'large' symmetric homology groups.

CHAPTER 72

Bol Planes.

A Bol translation plane is a translation plane with two distinguished components L and M such that any other component N there is an involutory collineation σ_N which fixes N pointwise and interchanges L and M. For example, every near-field plane is a Bol translation plane and it has been conjectured that these are the only possibilities. On the other hand, Burn [202] has shown that there are infinite Bol planes that are not nearfield planes.

Bol planes are connected to flocks of hyperbolic quadrics in the following way. Given a flock of a hyperbolic quadric in $PG(3, q)$, Thas [1178] shows there is a set of involutions that act on the flock. Using the Thas–Walker construction, Bader and Lunardon [49] realized that there are associated Bol translation planes. Of course, these Bol spreads are in $PG(3, q)$, so one does not require complete knowledge of all Bol spreads to classify spreads arising from flocks of hyperbolic quadrics, but it would be nice to have such a result.

The group G generated by the involutory collineations can either be solvable or non-sovable, of course, and Kallaher and Ostrom [844, 845] have shown the solvable case can almost be resolved. The group G_L stabilizing L and M is transitive on the non-zero vectors and if the group is solvable either G_L is a subgroup of $\Gamma L(1, q^n)$, if the plane has order q^n or the order is in $\{3^2, 5^2, 7^2, 11^2, 23^2, 3^4\}$. The structure of the Bol group shows that, except for these sporadic orders, the Bol plane must be a generalized André plane.

THEOREM 72.1. (Kallaher and Ostrom [844, 845]). A finite generalized André plane which is a Bol plane is a nearfield plane; a Bol plane with associated group in $A\Gamma L(1, q^n)$ is a nearfield plane.

THEOREM 72.2. (Kallaher and Ostrom [844, 845])). A finite solvable Bol plane is a nearfield plane or the order is in $\{3^2, 5^2, 7^2, 11^2, 23^2, 3^4\}$.

In general, Kallaher was able to resolve most of the non-solvable cases and proved:

THEOREM 72.3. (Kallaher [822, 826]). A finite Bol translation plane is either a nearfield or the order is in $\{5^2, 7^2, 11^2, 19^2, 23^2, 59^2, 3^4, 3^6\}$.

More recently, Hanson and Kallaher have resolved most of these remaining cases.

THEOREM 72.4. (Hanson and Kallaher [437]). A finite Bol plane of order in $\{5^2, 7^2, 11^2, 19^2, 23^2, 59^2\}$ is a nearfield plane.

Hence, the remaining two orders 3^4 and 3^6 remain in question.

Recently, the work of Johnson and Prince, and Jha and Johnson proves the following:

THEOREM 72.5. (Johnson–Prince [**807**] and Jha–Johnson [**637**]). There are exactly 14 translation planes of order 81 admitting $SL(2,5)$ as a collineation group in the translation complement.

It turns out that this result may be utilized to study Bol planes of order 3^4. Recently, the non-solvable triangle transitive translation planes have been classified (a triangle transitive translation plane is a translation plane admitting an autotopism collineation group which acts transitively on the non-vertex points of each leg of the triangle). As mentioned we have the non-solvable triangle transitive theorem:

THEOREM 72.6. (Johnson [**764**]). Let π be a finite non-solvable triangle transitive plane then π is an irregular nearfield plane of order 11^2, 29^2 or 59^2.

Indeed, the cases of order 3^4 and 3^6 must be analyzed in the triangle transitive planes in a manner similar to that of Bol planes, as Bol planes admit a group that acts transitively on two legs of an autotopism triangle. These analyses of planes of order 3^4 and 3^6 of Jha, Johnson and Prince may be applied for the study of Bol planes of these orders. Using these ideas, and the classification theorem for doubly transitive groups, it is shown in Johnson [**767**] that it is possible to complete the analysis of the two remaining orders to finish the study of Bol translation planes.

THEOREM 72.7. (Johnson [**767**]). If π is a Bol translation plane of order 3^4 or 3^6 then π is a nearfield plane.

Hence, we obtain:

THEOREM 72.8 (The Classification Theorem of Bol Planes; Kallaher–Ostrom, Hanson, Johnson). Finite Bol translation planes are nearfield planes.

CHAPTER 73

2/3-Transitive Axial Groups.

Bol planes are translation planes that admit collineation groups that act transitively on two components. Hence, we have an autotopism group that fixes two components and acts transitively on the components.

73.1. Acting Transitively on the Autotopism Triangle.

DEFINITION 73.1. If a finite translation plane π admits an autotopism group G that acts transitively on i sides of non-vertex points of the autotopism triangle, for $i \geq 1$, we refer to such a plane as an '$i/3$-triangle transitive plane'.

By the theorem of Ganley–Jha–Johnson [**393**], see Chapter 104, we know that non-solvable $i/3$-transitive planes are Hall unless the order is a member of the set $\{3^4, 3^6, 11^2, 19^2, 29^2, 59^2\}$. So, the question is which of these orders admit non-solvable $i/3$-triangle transitive planes, which are not Hall planes. We have noted in the triangle transitive chapter, Chapter 70, that if an autotopism group induces a non-solvable group on the non-vertex points of the infinite line of the autotopism triangle, then the group must induce a non-solvable group on at least one of the affine axes. Using the Foulser–Johnson theorem of Chapter 47 and the p-planar bound of Jha (see, e.g., the *Foundations* text of Biliotti, Jha, and Johnson [**123**], it then follows that the non-solvable groups involve $SL(2,5)$ or $SL(2,13)$ and the order is 3^6.

In the latter case, the 3-elements are planar and since $SL(2,13)$ must act faithfully on both affine axes (since the only possible non-solvable central collineation group involves $SL(2,5)$ as it is a Frobenius complement). The central involution must be the kernel involution, so any such plane cannot also be triangular transitive. The 3-groups are forced to fix subplanes of order 3 pointwise. Then the group is transitive on the two affine axes, but has two orbits of length $13 \cdot 7 \cdot 4$ on the infinite line. Hence, we mentioned the following open problem:

PROBLEM 73.2. Let π be a translation plane of order 3^6 admitting $SL(2,13)$ in the translation complement and fixing two components L and M. Then $SL(2,13)$ is transitive on the non-zero vectors of L and of M and has orbit lengths on the non-fixed infinite points of size $13 \cdot 7 \cdot 4$. Determine the possible planes π.

Turning to the prime square orders, it turns out that there are a variety of planes of these types. The known planes of such orders admit $SL(2,5) \times SL(2,5)$ as a collineation group generated by affine homology groups isomorphic to $SL(2,5)$. More generally, there are planes where the group is solvable and contains $SL(2,3) \times SL(2,3)$ as a collineation generated by affine homology groups isomorphic to $SL(2,3)$. The planes include the exceptional Lüneburg planes and the Hiramine $SL(2,5) \times SL(2,5)$ planes; see Chapter 9.

In the exceptional Lüneburg planes, the stabilizer of a component N is doubly transitive on N. We visit this idea in the following.

73.1.1. Generalized André Planes and Doubly Transitive Groups. If π is a generalized André plane that is neither Desarguesian nor Hall, Foulser [**369**] proves that the full collineation group leaves invariant a pair $\{L, M\}$ of components. Indeed, Lüneburg (see, e.g., [**961**]) characterizes generalized André planes by their collineation groups as follows:

THEOREM 73.3. (Lüneburg [**961**]). Let π be an affine plane that admits an Abelian autotopism group G which fixes two components L and M such that for any component not equal L and M the stabilizer subgroup G_N is irreducible acting on N. Then π is a generalized André plane.

We have discussed the 'cyclic Ostrom planes' and have noticed that there is a corresponding group characterization theorem. For example, all André planes are cyclic Ostrom planes as are all of the hyper-regulus multiple replacement planes of Jha and Johnson (see Chapter 40 on hyper-regulus replacements) and, in fact, all subregular planes. Furthermore, there are semifield planes that admit groups G that fix one component and that stabilizer of any component G_N is transitive on the non-zero vectors.

When G is an autotopism collineation group such that the stabilizer of any component G_N is doubly transitive, we would expect, at least in the odd-order case, that the corresponding translation plane would become a generalized André plane and there are various studies that point to this always being the case (see, for example, Kallaher [**830**] and Kallaher–Ostrom [**846**]).

THEOREM 73.4. (Kallaher [**830**]). Let π be a finite translation plane that admits a collineation group G in the linear translation complement such that the stabilizer of each component of N is transitive on the non-zero vectors. If π has odd non-square order and the characteristic is not 3, then π is either a generalized André plane or a semifield plane.

73.1.2. Doubly Transitive on a Subset. If we consider problems of collineation groups of translation planes acting doubly transitively not on a set of affine points but on a set of infinite points, there are several important problems that arise.

73.1.2.1. *Doubly Transitive on the Line at Infinity.* We first consider a collineation group that acts doubly transitively on the line at infinity. Hence, the plane is flag-transitive and the non-solvable planes are determined. Of these, only the Desarguesian and Lüneburg–Tits admit doubly transitive groups. Actually, once the group is determined as to contain $SL(2, n)$ if the plane is of order n or $S_z(q)$ if the plane has order q^2, the planes are then determined by Lüneburg [**952**] and Liebler [**918**].

THEOREM 73.5. (Lüneburg [**952**]). An affine plane of order n admitting $SL(2, n)$ is Desarguesian.

THEOREM 73.6. (Liebler [**918**]). An affine translation plane of order q^2 admitting $S_z(q)$ is Lüneburg–Tits.

However, previously Czerwinski [**268**] had essentially resolved this result except in the case when the order is m^2 and $m \equiv 1 \bmod 4$, and there are Baer involutions. Similarly, Schulz [1151] for m not congruent to 3 mod 4, required no Baer involutions, but otherwise determined the possible groups. This result may be obtained more generally using the classification theorem of all doubly transitive groups, which was done by M.J. Kallaher in unpublished notes. Note, for example, that the group must be solvable and of degree $n + 1$, where n is a prime power, which almost never occurs and when it does, ad hoc arguments suffice to resolve the problem.

THEOREM 73.7. (KCLLS theorem (Kallaher, Czerwinski, Liebler, Lüneburg, and Schulz)). (Czerwinski [268], Schulz [1151], Lüneburg [952], Liebler [918], Kallaher). A finite translation plane admitting a doubly transitive group on the line at infinity is Desarguesian or Lüneburg–Tits.

73.2. Doubly Transitive with a Fixed Component.

When a group leaves invariant a component and acts doubly transitively on remaining components (i.e., doubly transitively on a set of all but one parallel class), the translation plane is shown to be a semifield plane by Ganley and Jha [392].

THEOREM 73.8. (Ganley and Jha [392]). A finite translation plane admitting a collineation group that fixes one parallel class and acting doubly transitively on the remaining parallel classes is a semifield plane.

We note that this says that the plane is a semifield plane with an autotopism group that acts transitively on the non-vertex points of the infinite side of the autotopism triangle. For example, considering generalized twisted field planes, it is generally known when a group is transitive in this manner. Using the model of Oyama, Dempwolff [301], under the assumption that the autotopism group is cyclic, showed that, in fact, the plane must be a generalized twisted field plane. There is a strong connection between distance transitive graphs and triangle transitive semifield planes, in particular the very question of when we have such planes, and work by R.A. Liebler, who showed that such planes are generalized twisted field planes, is important in this context (see work of Liebler in [411, 567, 923]). More generally, Cordero and Figueroa, in a series of papers, showed that except possibly for order 64, the following result holds.

THEOREM 73.9. (Cordero–Figueroa [253]). Let π be a semifield plane of order $n \neq 64$. If there is an autotopism group which acts transitively on the non-vertex points of any side of the autotopism triangle, then the plane is a generalized twisted field plane.

Hence, we obtain the following classification theorem:

Theorem of Ganley–Jha/Dempwolff/Liebler/Cordero–Figueroa.

THEOREM 73.10. (GJDLCF theorem). (Ganley and Jha [392], Dempwolff [301], Liebler [411, 567], Cordero and Figueroa [253]). Let π be a finite translation plane that admits a collineation group fixing a component and acting doubly transitively on the remaining components. Then π is a generalized twisted field plane.

CHAPTER 74

Doubly Transitive Ovals and Unitals.

DEFINITION 74.1. Let O be an oval in a projective plane Π. Let ℓ be a tangent line and form the associated affine plane Π_ℓ by removing ℓ. The O is said to be a 'parabolic oval' of Π_ℓ. The point of O on the line at infinity is called the 'parabolic point'. We say that O is a 'doubly transitive parabolic oval' if and only if there is a collineation group of the affine plane that leaves O invariant and induces a doubly transitive group on the affine points of O.

In a study of doubly transitive parabolic ovals, many of the results on doubly transitive group actions are brought into consideration, in particular, the Ganley–Jha, Dempwolff, Liebler, Cordero–Figueroa result that a translation plane admitting a collineation group fixing a component and acting doubly transitively on the remaining components is a generalized André plane. Enea and Korchmáros [**353**] have analyzed doubly transitive parabolic ovals in odd-order translation planes and Biliotti–Jha–Johnson [**121**] have generalized studied collineation groups of generalized André planes. These results culminate in the classification of doubly transitive parabolic ovals in the odd-order case.

74.1. Classification of Odd-Order Planes with Doubly Transitive Parabolic Ovals; Enea–Korchmáros.

THEOREM 74.2. (Enea–Korchmáros [**353**]). A translation plane of odd order q^2 that admits a doubly transitive parabolic oval is Desarguesian or a commutative twisted field plane and the oval consists of the absolute points of an orthogonal polarity.

74.2. Doubly Transitive Parabolic Ovals in Even-Order Planes; Biliotti–Jha–Johnson.

In the even-order case, there are few results on parabolic ovals, particularly since no non-Desarguesian translation planes are known to admit any such ovals. Moreover, the arguments in the even-order case are apparently more difficult for a variety of reasons. However, Foulser's result on the doubly transitive groups of $A\Gamma L(1, 2^r)$ play a pivotal role is any such analysis.

THEOREM 74.3. (Foulser [**365**]).
Let the elements of $A\Gamma L(1, 2^r)$ be represented by the mappings $x \longmapsto gx^\mu + b$ over $GF(2^r)$ with $b \in GF(2^r)$, $g \in GF(2^r)^*$ and $\nu \in \text{Aut}\, GF(2^r)$. Let w denote a primitive root of $GF(2^r)$. Let w, α, τ_a denote the mappings

$$x \longmapsto wx, x \longmapsto x^2, x \longmapsto x + a,$$

respectively. Let $T = \langle \tau_a; a \in K = GF(2^r) \rangle$.
Let F be a doubly transitive subgroup of $A\Gamma L(1, 2^r)$. Then

(1) $T \subset F$ and $F = T \cdot H$ is the split extension of T by $H = F_0$.

(2) H may be represented in the form $\langle w^d, w^e \alpha^s \rangle$, where

    (i) $d > 0$ and $d \mid 2^r - 1$,

    (ii) $s > 0$ and $r \equiv 0 \pmod{sd}$,

    (iii) $0 \leq e < d$ and $(d, e) = 1$,

    (iv) The prime divisors of d divide $2^s - 1$,

    and $H_1 = F_{0,1} = \langle \alpha^{sd} \rangle$.

(3) F contains a minimal sharply 2-transitive subgroup. Furthermore, $r = sd$ in a representation of a minimal 2-transitive group.

(4) Two doubly transitive solvable subgroups $\overline{G} = T \cdot \overline{H}$ and $\widetilde{G} = T \cdot \widetilde{H}$ of $A\Gamma L(1, 2^r) = T \cdot \langle w, \alpha \rangle$ are isomorphic if and only if they are conjugate under an element of $\langle w, \alpha \rangle$.

To see how Foulser's theorem comes into question, we mention the following general result on parabolic ovals.

THEOREM 74.4. (Biliotti–Jha–Johnson [**124**]). Let π be a finite affine plane of even order n that admits a 2-transitive parabolic oval O with group G. Then $n = 2^r$. Furthermore,

(1) O is a translation oval (the affine part admits a transitive translation subgroup) and

(2) $G \leq A\Gamma L(1, 2^r)$.

It might be asked in even order if generalized twisted field planes can admit doubly transitive parabolic ovals and this question is resolved in Biliotti–Jha–Johnson [**125**].

THEOREM 74.5. (Biliotti–Jha–Johnson [**125**]). A generalized twisted field of even order admitting a doubly transitive parabolic oval is Desarguesian.

Similar to cyclic Ostrom planes, it is interesting to consider a plane covered by two transitive ovals.

THEOREM 74.6. (Biliotti–Jha–Johnson [**124**]). Let π be a finite affine plane of even order $n \neq 2^6$ that admits two distinct 2-transitive parabolic ovals that share at least two points. Then π is Desarguesian.

74.3. Doubly Transitive Hyperbolic Unitals; Biliotti–Jha–Johnson–Montinaro.

We will be discussing the construction of unitals in affine translation planes of order q^2 in Chapter 86. These come in two types; parabolic and hyperbolic, depending on whether the unital shares 1 or $q + 1$ points with the line at infinity. The question for us here is what could be said about 'doubly transitive hyperbolic lines', that is about translation planes of order q^2 admitting a hyperbolic unital and collineation group fixing the unital and acting doubly transitively on the points on the line at infinity. Actually, such planes are completely determined.

THEOREM 74.7. (Biliotti, Jha, Johnson, and Montinaro [**134, 135, 136, 137**]). A translation plane of order q^2 admitting a doubly transitive hyperbolic line is Desarguesian and the unital is classical.

74.4. Related Group Actions; Parallelisms and Skeletons.

What is meant by related doubly transitive group actions would include such actions as doubly transitive groups on the spreads of a parallelism in $PG(3, q)$ or by doubly transitive groups on the skeleton of a conical flock and a variety of other situations which are probably beyond the scope of this text. We end by mentioning what can be said about parallelisms and skeletons of conical flocks.

THEOREM 74.8. (Johnson [**756**]). Let $\mathcal{P}$ be a parallelism in $PG(3, q)$ (that is a set of $1 + q + q^2$ spreads such that each line of $PG(3, q)$ lines in exactly one of the spreads) that admits a collineation group acting doubly transitively on the spreads.

Then $q = 2$ and $\mathcal{P}$ is one of two Desarguesian parallelisms dual to each other under a polarity of the projective space.

THEOREM 74.9. (Johnson–Lunardon–Wilke [**779**], Jha and Johnson [**639**]). If π is a flock spread of a quadratic cone in $PG(3, q)$, then there is an automorphism group acting doubly transitively on the skeleton of π if and only if π is one of the following types of spreads:

(1) Desarguesian,

(2) Walker when q is odd, or Betten when q is even, or

(3) Knuth semifield and q is odd.

CHAPTER 75

Rank 3 Affine Planes.

75.1. Rank k Affine Planes.

A finite affine plane π is said to be a 'rank k affine plane' if and only if there is a collineation group G that acts transitively on the affine points of π and for an affine point 0, the stabilizer subgroup G_0 has exactly k affine point orbits one of which is $\{0\}$. If G is non-solvable, the plane is said to be a 'non-solvable rank k affine plane', otherwise the term used is 'solvable rank k affine plane.'

For rank k affine planes, if G_0 has exactly k affine point orbits distinct from $\{0\}$ and k infinite point orbits, the affine plane is determined as follows:

75.1.1. Lüneburg, Kallaher–Johnson (k, k)-Theorem.

THEOREM 75.1. (Lüneburg [**959**], Kallaher and Johnson [**771**]). Let π be a rank k affine plane with group G such that G_0 has k non-trivial point orbits and k infinite point orbits. Then π is a translation plane and G contains the translation group.

However, the impetus for the study of rank k affine planes certainly arose from the consideration of flag-transitive affine planes, doubly transitive affine planes and their generalizations. By fundamental work of A. Wagner in 1964–1965 [**1202**], any flag-transitive affine plane is a translation plane. Furthermore, Wagner showed that a finite affine plane that admits a group transitive on the affine points and transitive on the infinite points is flag-transitive. Also, in 1964, it was proved by Foulser [**366**] that for solvable rank 2 affine planes (doubly transitive) only the Desarguesian planes and the Hall plane of order 9 are possible.

In the later 1960s and early 1970s, fundamental work of Kallaher and Liebler established the basic structure of rank 3 affine planes.

75.1.2. Kallaher and Liebler Rank 3 Theorem.

THEOREM 75.2. (Kallaher [**817**] and Liebler [**917**]). Let π be a rank 3 affine plane then π is a translation plane.

In 1985 Kantor [**877**] working more generally with doubly transitive designs showed, in particular, that the only non-solvable rank 2 affine planes are the Desarguesian, Hall of order 9 or Hering of order 27. This result also follows from the work of Pfaff who worked on affine Designs [**1103**]. Furthermore, there is a short proof of this in Jha and Johnson [**620**] using results particular to translation planes.

Hence, we obtain:

THEOREM 75.3. (Foulser [**365**], Kantor [**877**], Pfaff [**1103**] and Jha and Johnson [**620**]). The rank 2 affine planes are: (1) Desarguesian, (2) Hall of order 9, (3) Hering of order 27.

There has been considerable interest in the construction of flag-transitive planes and there are a variety of such techniques of finding solvable flag-transitive planes, and all of these are in Chapter 67.

For non-solvable flag-transitive planes, it is possible to provide a complete classification. The later study has been completed by Buekenhout, Delandtsheer, Doyen, Kleidman, Liebeck, and Saxl who worked more generally on flag-transitive designs.

75.1.3. Buekenhout, Delandtsheer, Doyen, Kleidman, Liebeck, Saxl Theorem.

THEOREM 75.4. (Buekenhout, Delandtsheer, Doyen, Kleidman, Liebeck, and Saxl [**201**]). Let π be a non-solvable flag-transitive affine plane. Then π is one of the following types of affine planes: (1) Desarguesian, (2) Lüneburg–Tits, (3) Hering of order 27 or (4) Hall of order 9.

Turning to rank 3 affine planes, we see that any finite nearfield plane is an example of a rank 3 affine plane. Furthermore, various generalized André planes and generalized twisted field planes are also rank 3 planes. However, most of these examples are solvable rank 3 planes.

In fact, the rank 3 semifield planes are completely determined by Biliotti, Jha and Johnson as follows:

THEOREM 75.5. (Biliotti, Jha and Johnson [**121**, (7.3)]). Let π be a semifield plane of order $p^r \neq 2^6$. Then π is of rank 3 if and only if π is a generalized twisted field plane such that the left, middle and right nuclei of the corresponding semifield are all equal.

Similar to the analysis of non-solvable flag-transitive planes, we shall be interested primarily in non-solvable rank 3 planes. Biliotti and Johnson are [**141**] able to provide a complete classification.

The primary tool in this study is Liebeck's work on the affine permutation groups of rank 3 with the assumption that the group acts primitively, which is listed in Chapter 104. Keeping this in mind, the following result of Kallaher is fundamental to the approach.

THEOREM 75.6. (Kallaher [**817**]). Let π be a rank 3 affine plane with corresponding group G. Then one of the following holds: (1) G has no fixed infinite point and acts primitively on the affine points of π, (2) G has a fixed infinite point and acts imprimitively on the set of points of π.

Hence, prior to the application of Liebeck's results, situation (2) must be considered.

For existing examples, we note that the irregular nearfield planes of orders 11^2, 29^2 and 59^2 with non-solvable groups are examples of non-solvable rank 3 planes admitting $SL(2,5)$. The Lüneburg–Tits planes of order q^2 for $q = 2^r$ are non-solvable rank 3 planes admitting the Suzuki group $Sz(q)$. Furthermore, Mason and Ostrom (see, e.g., [**973**]) have shown that there are non-solvable rank 3 planes of orders 5^2 and 7^2 whose infinite orbits have lengths $\{10, 16\}$ and $\{10, 40\}$, respectively. In addition, there is an unusual plane of Korchmáros [**903**] of order 7^2 which is a non-solvable rank 3 plane with infinite orbits of length $\{20, 30\}$, again admitting $SL(2,5)$. The reader is directed to the Models part of the index for information locating these examples with the text.

As remarked above, T.G. Ostrom has shown that one of the exceptional Walker planes of order 25 is non-solvable rank 3. It actually turns out that all three of the exceptional Walker planes of order 25 are non-solvable rank 3 planes. Using various combinatorial arguments, the classification of non-solvable rank 3 planes is completely determined, showing that the planes mentioned above together with the Hall and Desarguesian planes form a complete list.

We give below our main result which indicates the planes, the group and the orbit lengths on the line at infinity. However, the non-solvable rank 3 groups of the Desarguesian plane are given by Foulser and Kallaher [**381**, Theorem (5.1)] and shall not be completely listed, although we offer a few remarks below.

It can be seen in the proofs that there are two main non-solvable rank 3 actions on a Desarguesian plane. When the plane has order q^2, the group is $SL(2,q)Z_{q^2-1}$ and the orbits lengths are $(q+1, q^2-q)$. When the order of the plane is q^3, the group is $SL(2,q)Z_{q^3-1}$ and the orbit lengths are $(q+1, q^3-q)$. However, for various small orders, it is possible that the group involves A_5 as a quotient. Furthermore, Foulser and Kallaher [**381**, Theorem (5.3), p. 128] provide precisely the possible orders and orbit lengths. The orders and orbit lengths that are not prime are as follows: $\{(3^2, (10)), (2^4, (6, 12)), (5^2, (6, 20)), (7^2, (20, 30)), (2^6, (5, 60)), (5^3, (6, 120))\}$.

However, we note that $SL(2,4) \simeq PSL(2,5) \simeq A_5$ so that the cases when the orders are $4^2, 4^3, 5^2, 5^3$ are included in the previous group actions. Hence, apart from the classical actions of $SL(2,q)$, the only other Desarguesian non-solvable rank 3 groups occur for orders 9 and 49.

75.1.4. The Non-Primitive Case. Let π be a non-solvable rank 3 plane with non-solvable collineation group G. Since π is a translation plane of order p^r, we may assume that the subgroup which fixes the zero vector G_0 is in $\Gamma L(2r, p)$. Moreover, since there are exactly two remaining affine point orbits distinct from $\{0\}$, we see that either the group G is primitive or there is a fixed component by G_0. It is then possible to employ the Theorem of Ganley–Jha–Johnson on doubly transitive groups on sets of line size.

THEOREM 75.7. (Ganley, Jha and Johnson [**393**]). Let π be a finite translation plane of order p^r which admits a collineation group inducing a non-solvable doubly transitive group on a set of size p^r. Then either $p^r \in \{3^4, 3^6, 11^2, 19^2, 29^2, 59^2\}$ or π is one of the following types of planes: (1) Desarguesian, (2) Hall, (3) the Dempwolff plane of order 16, or (4) one of the three exceptional Walker planes of order 25.

When G_0 induces a non-solvable group on a fixed component, we have the following two results of Biliotti and Johnson.

THEOREM 75.8. (Biliotti and Johnson [**141**]). If π is a rank 3 plane with a non-solvable group G that fixes a component ℓ, then π induces a solvable group on ℓ or π is one of the exceptional Walker planes of order 25. Hence, there is a non-trivial elation group with axis ℓ.

THEOREM 75.9. (Biliotti and Johnson [**141**]). If π is a non-solvable rank 3 affine plane and G the corresponding collineation group, then G cannot fix a component unless π is an exceptional Walker plane of order 25. Hence, if the order of the plane is not 25, then G is a primitive group acting on the affine points of the plane and is a non-solvable affine rank 3 group since G contains the translation subgroup of

π. If π is a non-solvable rank 3 affine plane and G the corresponding collineation group, then G cannot fix a component unless π is an exceptional Walker plane of order 25. Furthermore, if the order of the plane is not 25, then G is a primitive group acting on the affine points of the plane and is a non-solvable affine rank 3 group since G contains the translation subgroup of π.

Using analysis of the primitive and non-primitive cases allows a complete classification. There are a few remarks on the proof given in Chapter 104.

75.2. Classification of Non-Solvable Rank 3 Affine Planes.

The main result of Biliotti and Johnson provides the classification of rank 3 affine planes with non-solvable collineation groups.

THEOREM 75.10. (Biliotti and Johnson [141]). Let π be a non-solvable rank 3 affine plane. Then π is one of the following types of planes: (1) Desarguesian, (2) Hall, (3) Lüneburg–Tits (4) irregular nearfield plane of order 11^2, 29^2 or 59^2, (5) Korchmáros of order 49, (6) Mason–Ostrom of order 49, or (7) one of the three exceptional Walker planes of order 25.

As mentioned above, the groups and orbits for non-solvable Desarguesian rank 3 planes are given in Foulser and Kallaher [381](section 5). For non-Desarguesian non-solvable rank 3 planes, Biliotti and Johnson prove the following corollary:

COROLLARY 75.11. (Biliotti and Johnson [141]). Let π be a non-Desarguesian, non-solvable rank 3 affine plane. Then one of the following three situations occurs: (1) π is a Hall plane of order $q^2 \neq 9$, containing the group $SL(2,q)Z_{q^2-1}$ with orbits on the line at infinity of lengths $q + 1$, $q^2 - q$, (2) π is a Lüneburg–Tits planes of order q^2 containing the group $S_z(q)Z_{q-1}$ which is transitive on the infinite points and the affine point orbits are the non-zero points fixed pointwise by Sylow 2-subgroups of length $(q - 1)(q^2 + 1)$ and the remaining non-zero affine points of length $(q^2 - q)(q^2 + 1)$, or (3) the plane, minimal group and orbit lengths on the line at infinity are as in the following table:

$$
\begin{bmatrix}
\text{Type:} & \text{Group:} & \ell_\infty\text{-orbit lengths:} \\
\text{Irregular nearfield,} & \begin{matrix} SL(2,5) \\ \text{as a homology} \\ \text{group} \end{matrix} & (2, p^2 - 1) \\
\text{order } p^2,\ p = 11, 29, 59 & & \\
\begin{matrix} \text{Korchmáros} \\ \text{order 49} \end{matrix} & SL(2,5) & (20, 30) \\
\begin{matrix} \text{Mason–Ostrom} \\ \text{order 49} \end{matrix} & \begin{matrix} \text{Subgroup } H_0 \\ H_0/(D_8 \circ Q_8) \simeq A_5 \end{matrix} & (10, 40) \\
\begin{matrix} \text{Likeable Walker, I} \\ \text{order 25} \end{matrix} & \begin{matrix} \text{Non-primitive group} \\ SL(2,5) \cdot 5 \end{matrix} & (1, 25) \\
\begin{matrix} \text{Exceptional Walker, II} \\ \text{order 25} \end{matrix} & \begin{matrix} \text{Subgroup } H_0 \\ H_0/(D_8 \circ Q_8) \simeq A_5 \end{matrix} & (10, 16) \\
\begin{matrix} \text{Exceptional Walker, III} \\ \text{order 25/Hering} \end{matrix} & G_0/Z(G_0) \simeq S_6 & (6, 20)
\end{bmatrix}
\ .
$$

CHAPTER 76

Transitive Extensions.

There are a number of important translation planes of order q^n that admit a collineation group isomorphic to $SL(2, q)$. In fact, when $n = 2$, we have the classification theorem of Foulser–Johnson [**378**, **379**]; see Chapter 47 on actions of $PSL(2, q)$, which details the possible planes, and these are the Desarguesian, Hall, Hering, Ott–Schaeffer, Dempwolff of order 16 and three Walker planes of order 25.

When $n = 3$, we noted in Chapters 13 and 14 on semifield planes and related planes that the cyclic semifield planes of order q^3 are equivalent to the so-called generalized Desarguesian planes of order q^3 admitting $GL(2, q)$ acting in a canonical manner.

When $n = 4$, a result of Jha and Johnson [**597**] shows that the set of translation planes of order q^4 admitting $SL(2, q) \times Z_{1+q+q^2}$ is equivalent to the set of Desarguesian parallelisms in $PG(3, q)$ (i.e., sets of $1 + q + q^2$ Desarguesian spreads that partition the line set).

In various of the possible translation planes mentioned above, there is a subplane of order q admitting $SL(2, q)$ contained in a collineation group G of an extension translation plane of order q^n with the property that G fixes the subplane and induces a collineation group transitive on the infinite points of the affine subplane and also transitive on the infinite points outside of the subplane.

We recall that a 'flag-transitive affine plane' is an affine plane admitting a collineation group that admits transitive on the flag (incidence point-line pairs) of the affine plane. We recall that Wagner's theorem [**1202**] shows that finite flag-transitive affine planes are translation planes. In this case, flag-transitive translation planes are simply translation planes admitting a collineation group in the translation complement that is transitive on the infinite points. Suppose a flag-transitive plane is contained as a subplane in a translation plane that retains the flag-transitive group and also admitting a group acting transitive on the infinite points outside those of the subplane, we call the superplane an 'extension of the flag-transitive subplane'. Specifically, we consider planes of the following type.

DEFINITION 76.1. An affine plane π is said to be an 'extension of a flag-transitive plane' if and only if π contains a subplane π_o and a collineation group G which leaves π_o invariant and acts transitively on the sets of affine and infinite points of π_o and on the infinite points of $\pi - \pi_o$. Note that in the finite case this makes the subplane π_o into a translation plane.

DEFINITION 76.2. If an affine plane π of order q^n admits a collineation group G which has an infinite point orbit of length $q + 1$ and i infinite point orbits of length $(q^n - q)/i$ for $i = 1$ or 2, we shall call π a '$(q+1, (q^n - q)/i)$-transitive plane' and G a '$(q+1, (q^n - q)/i)$-transitive group'. If G leaves a subplane π_o of order q invariant within the net of length $q + 1$ and there is a collineation group transitive

543

on the sets of affine and infinite points of π_o and the infinite points of $\pi - \pi_o$ then π_o is a flag-transitive affine plane and we shall call π an 'extension of π_o'. If the group of an extension is solvable, we shall call the plane a 'solvable extension of a flag-transitive plane'.

Of course, we are concerned in this text with finite translation planes. For reasons that will become clear, we are most interested in quadratic, cubic and quartic extensions.

DEFINITION 76.3. In the situation listed above, a translation plane of order q^n admitting a subplane of order q an 'n-dimensional extension of a flag-transitive plane'. When $n = 2$, the extension is said to be 'quadratic', for $n = 3$, the extension is 'cubic' and for $n = 4$, the extension is 'quartic'.

The Hall, Desarguesian and Hering planes are translation planes with spreads in $PG(3,q)$ that admit collineation groups G in the linear translation complement that have two or three point orbits at infinity. In particular, there is always one orbit of length $q + 1$ and i orbits of length $(q^2 - q)/i$ for $i = 1, 2$.

It is possible to actually determine the spreads that admit such groups. Hiramine, Jha, and Johnson completely characterized translation planes of odd order with spreads in $PG(3,q)$ that have such collineation groups. Indeed all of the work related to extensions of flag-transitive planes is due to Hiramine, Jha, and Johnson [**514, 515, 516, 517, 518, 519**].

One important situation occurs when there is a subgroup isomorphic to $SL(2,q)$ acting on a subplane of order q. When this occurs, and the order is not too large, for example, for order q^n, and $n = 2, 3, 4$ various classifications may be obtained and certain well-known classes of planes occur. Hence, in the following, we indicate briefly what can be said for quadratic, cubic and quartic extensions. Furthermore, parallelisms in higher dimensional projective spaces also involve $SL(2,q)$. If the order of a subplane is small in relation to the superplane it is increasingly difficult for a group $SL(2,q)$ to have much impact. An example of this are the planes of Foulser–Ostrom of order v^{2r}, admitting $SL(2,v)$, which exist for essentially any r. For example, there are the Foulser–Ostrom derivable and subregular planes of order q^4 that admit $SL(2,q)$. By the process of geometric lifting of Johnson, it is possible to construct subregular planes of order q^{4s} admitting $SL(2,q)$, for any odd integer s. Conversely, however, it is a very challenging problem to provide characterization results for translation planes of such orders admitting $SL(2,v)$ as a collineation group.

On the other hand, if one assumes that there is a solvable extension, the group possibilities often indicated the existence of an accompanying Desarguesian plane, called an Ostrom phantom, which exist whenever there is a collineation τ_u of order a prime p-primitive divisor u of $q^n - 1$, where the order of the plane is q^n. Solvability often is sufficient to show the full group is a Desarguesian group acting on the Ostrom phantom. So, for solvable extensions, one can essentially give somewhat of a complete theory and we sketch what is known in the solvable case. We have mentioned the importance of quadratic, cubic and quartic extensions of flag-transitive plane. However, for planes of order q^n, $n > 5$, the assumption of solvability basically resolves the problems and it is possible to show that the plane is normally a Hall plane, the existence of a subplane along with orbits provides sufficient tools to accomplish the classification.

76.1. Quadratic Extension Planes.

Let Σ be a Desarguesian affine plane of order q^2 and let N denote a regulus subnet of Σ. The collineation group G generated by the affine elations with axes the components of N is isomorphic to $SL(2,q)$. Furthermore, all such central collineations with center in N must leave invariant every Baer subplane incident with the zero vector 0 and G acts transitively on the components of $\Sigma - N$. Letting T denote the translation subgroup which acts regularly on a given subplane π_o of N, we have that GT is a collineation group which is flag-transitive on π_o and acts transitively on the components of $\Sigma - N$. Hence, the Desarguesian plane of order q^2 is a quadratic extension of a flag-transitive plane of order π_o and GT is non-solvable unless $q = 2$ or 3.

Now derive N to construct the Hall plane π. But, note that G does not leave invariant a Baer subplane of the derived net N^*. So, let BC denote the central collineation group of Σ of order $q(q - 1)$ with fixed axis ℓ, a component of N, that acts 2-transitively on the components different from ℓ of N. Let K denote the kernel homology group of order $q^2 - 1$ of Σ and note that K fixes all components of Σ and acts transitively on the Baer subplanes of N incident with the zero vector.

Now acting in π, BCK leaves the Baer subplane ℓ invariant, acts transitively on the components of N^* (the Baer subplanes of N) and acts transitively on the components of $\Sigma - N$ which are also components of $\pi - N^*$.

Hence, if T is the translation subgroup of π acting regularly on ℓ as a Baer subplane then $BCKT$ is flag-transitive on ℓ and transitive on the components of $\pi - N^*$.

There are times, however, that it is possible to relax the assumption of the existence of a subplane and assume merely that there is a group with orbits on the line at infinity of lengths $q + 1$ and $q^2 - q$. For example, for spreads of odd order with spread in $PG(3,q)$, the following theorem due to Hiramine, Jha, and Johnson is applicable.

THEOREM 76.4. (Hiramine, Jha, and Johnson [**515**]). Let π be a translation plane of odd order q^2 with spread in $PG(3,q)$ that admits a linear collineation group G with point orbits at infinity one of length $q + 1$ and i of length $(q^2 - q)/i$ for $i = 1, 2$. Then π is one of the following types of planes: (i) Desarguesian, the group G is reducible and there exists an elation—in this case, $i = 1$; (ii) Hall, the group G is reducible and there exists a Baer p-element, $q = p^r$—here, $i = 1$; (iii) Hering, the group G is irreducible and $q = p^r$ for r odd and, in this case, $i = 2$.

Notice that the group is reducible in both the Desarguesian and Hall cases and, in either case, there is an invariant subplane of order q. In fact, it is possible to completely classify, without any further conditions, translation planes that are quadratic extensions of a flag-transitive plane.

The main result of Hiramine, Jha, and Johnson on quadratic extensions is as follows.

THEOREM 76.5. (Hiramine, Jha, and Johnson [**516**]). Let π be a finite translation plane which is a quadratic extension of a flag-transitive plane π_o. Then π is either Desarguesian or Hall or the derived likeable Walker plane of order 25. In particular, (1) if the associated collineation group is non-solvable, then π is Desarguesian or the derived likeable Walker plane of order 25, and (2) if the associated collineation group is solvable, then π is Hall or Desarguesian of order 4 or 9.

The proof is possible due to the many theorems available for translation planes of order q^2. In particular, since there is a group acting transitively on a subplane of order q, the group acting on the subplane often involves $SL(2, q)$, where it is possible to apply the theory of Foulser and Johnson, see Chapter 47 or 104.

76.2. Cubic Extensions.

In a more general setting, an analysis of translation planes of order q^n that are n-dimensional extensions of a flag-transitive plane π_o might be considered. If $n = 3$, for example, in Jha–Johnson [609, 611], , the classification of generalized Desarguesian planes of order q^3 is given. These are translation planes admitting $GL(2, q)$ and are cubic extensions of a flag-transitive plane of order q. However, there is a tremendous number of mutually non-isomorphic planes of this type. In such planes, the associated vector space is a standard $GF(q)GL(2, q)$ module. This means that the group $SL(2, q)$ is generated by elation groups and that $GL(2, q)$ leaves invariant each subplane of order q incident with the zero vector in the associated $GF(q)$-regulus net defined by the elation axes of $SL(2, q)$. Furthermore, there are always infinite point orbits of lengths $q+1$ and $q^3 - q$, which implies that we have a very large assortment of cubic extensions of a flag-transitive plane. All of these planes are related to cyclic semifields and to the theory of cone representations of Glynn (see Chapters 14 and 15).

Hence, we see that the problem even for cubic extensions becomes exponentially more difficult than for quadratic extensions.

76.2.1. Cubic Extensions and the Lüneburg–Tits Planes. In particular, for order q^3 cubic extensions of flag-transitive planes of order q and when the order is even but without any further assumptions, it is always possible to show that there is a collineation group isomorphic to either $SL(2, q)$ or $S_z(\sqrt{q})$ which is generated by elation groups.

We have noted that there are a variety of translation planes of orders q^3 admitting a collineation group isomorphic to $SL(2, q)$ which are cubic extensions of a Desarguesian affine plane of order q. We show that it is possible to have extensions of non-Desarguesian planes of order q. We first note a result on the structure of nets containing sufficiently many subplanes.

THEOREM 76.6. (See, e.g., Hiramine, Jha, and Johnson [519]). Let π be a translation plane of order q^n admitting a subplane π_o of order q and kernel D isomorphic to $GF(p^{t_o})$ where $q = p^r$. Let $\mathcal{N}$ denote the net of degree $q+1$ determined by the components of π_o. If there exist $n + 1$ subplanes of $\mathcal{N}$ such that any n of them direct sum to π then all subplanes are isomorphic and there are exactly $(p^{t_o n} - 1)/(p^{t_o} - 1)$ subplanes of $\mathcal{N}$ incident with the zero vector.

In this section, we consider whether there are Lüneburg–Tits planes of order q^3 which are cubic extensions of a flag-transitive plane π_o. The subplane π_o is Desarguesian or Lüneburg–Tits. In order to better consider the action of the collineation group, we mention some background on these planes and their representation.

PROPOSITION 76.7. Let π be a Lüneburg–Tits plane of order $2^{2(2r+1)}$ with spread in $PG(3, K \simeq GF(2^{2r+1}))$. Denote the points of π by (x_1, x_2, y_1, y_2) for all x_i, $y_i \in K$, $i = 1, 2$. Let $x = (x_1, x_2)$, $y = (y_1, y_2)$. Let $\sigma : x_1 \longmapsto x_1^{2^{r+1}}$ so that

$x_1^{\sigma^2} = x_1^2$ for all $x_1 \in K$. Then the spread has the following representation:

$$\{(x = 0)\} \cup \left\{ y = x \left[\begin{array}{cc} b^\sigma & b + a^{\sigma+1} \\ b + a^{\sigma+1} & a^\sigma \end{array} \right] ; \, a, b \in K \right\}.$$

It is generally known that the Lüneburg–Tits spreads are regulus-free and this is easily verified using the above matrix representation.

THEOREM 76.8. (Hiramine, Jha, and Johnson [**519**]). The spreads for the Lüneburg–Tits planes are regulus-free. Let π be a Lüneburg–Tits plane of order $2^{2(2r+1)}$.

(1) Then the spread for π is a union of Desarguesian partial spreads of degree 5 that share a line. Hence, there are Desarguesian subplanes π_o of order 4.

(2) If $2r_1 + 1$ divides $2r + 1$ and $2r_1 + 1 > 1$ then there is a Lüneburg–Tits subplane π_{r_1} of order $2^{2(2r_1+1)}$.

(3) Let G denote the full translation complement of π. Let K_{r_1} denote the subfield of K isomorphic to $GF(2^{2r_1+1})$. The stabilizer of π_{r_1}, i.e., $G_{\pi_{r_1}}$, is

$$\langle \omega, \tau(a, b), \eta(k), \mathrm{aut}(\rho_z), s(\alpha) \rangle$$

$$\forall \, a, b \in K_{r_1}, \, \forall \, k, \alpha \in K_{r_1} - \{0\} \text{ and } \forall \, \rho_z \in \mathrm{Aut} \, K.$$

(4) Given any subplane π_{r_1} and $r_1 > 0$, there exist exactly $(2^{2r+1} - 1)/(2^{2r_1+1} - 1)$ Lüneburg–Tits subplanes that share the same components as π_{r_1}.

We now consider if there are Lüneburg–Tits planes that are n-dimensional extensions of either a Desarguesian or a Lüneburg–Tits subplane.

THEOREM 76.9. (Hiramine, Jha, and Johnson [**519**]). A Lüneburg–Tits plane of order h^{2n} is an n-dimensional extension of a subplane of order h^2 if and only if one of the following occurs:

(1) $h = 2$ and $n = 3$, or

(2) $h = 2^3$ and $n = 3$.

Hence, there is a 'chain' of 3-dimensional extensions $\pi_o \subseteq \pi_1 \subseteq \pi$ where π_o is Desarguesian of order 2^2, π_1 is a Lüneburg–Tits plane of order 2^6 and π is a Lüneburg–Tits plane of order 2^{18}.

COROLLARY 76.10. (Hiramine, Jha, and Johnson [**519**]). The Lüneburg–Tits plane π of order 2^{18} is a cubic extension of a Lüneburg–Tits subplane π_1 of order 2^6. Furthermore, this is the unique cubic extension of a translation plane of order 2^6 with kernel $GF(2^9)$ that admits $S_z(2^3)$. The net of degree $2^6 + 1$ defined by π_1 admits exactly $1 + 2^3 + 2^6$ Lüneburg–Tits subplanes incident with the zero vector.

We now point out that only the groups $SL(2, q)$ or $S_z(\sqrt{q})$ are possible for even-order cubic extensions.

THEOREM 76.11. (Hiramine, Jha, and Johnson [**519**]). Let π be a cubic extension translation plane of even order q^3 with subplane π_o of order q. Then π_o is Desarguesian or Lüneburg–Tits.

It turns out that the involutions are always elations so that the group always contains a more or less standard normal subgroup generated by elations. This is accomplished in two theorems. First the nature of the abstract group actually can be established.

THEOREM 76.12. (Hiramine, Jha, and Johnson [**515, 519**]). Let π be a translation plane which is a cubic extension of a subplane π_o of order q.

(1) If $q \neq 4$, then the full collineation group G contains a group H acting on π_o isomorphic to $SL(2, q)$ or $S_z(\sqrt{q})$, respectively as the subplane π_o is Desarguesian or Lüneburg–Tits.

(2) If $q = 4$ and the group is non-solvable, then there is a subgroup acting on π_o isomorphic to $SL(2, 4)$. If the group is solvable, then the Sylow 2-subgroups are cyclic of order 4.

We now eliminate the possibility that the involutions are Baer.

THEOREM 76.13. (Hiramine, Jha, and Johnson [**519**]). Let π be a translation plane which is a cubic extension of a subplane π_o of even order q.

(1) Then the involutions are elations.

(2) If $q = 2^r$ and r is odd, then π_o is Desarguesian and the group generated by elations is isomorphic to $SL(2, q)$.

When q is even, there is a class of translation planes of order q^3 that admit two groups isomorphic to $GL(2, q)$ both of which contain a group $\mathcal{S}$ isomorphic to $SL(2, q)$ where the involutions are elations. One of these groups is $\mathcal{S} \times K^*$ where K^* is the kernel homology group of order $q - 1$ and K is the kernel of order q that commutes with $\mathcal{S}$. The other group is defined as follows:

$$\left\langle \begin{bmatrix} \alpha & \beta \\ \delta & \gamma \end{bmatrix} \forall \alpha, \beta, \delta, \gamma \in F \ni \alpha\gamma - \beta\delta \neq 0 \right\rangle$$

where F is a field isomorphic to K. The components are $x = 0, y = x\alpha$ for all α in F and the images of $y = xT$ under the standard action of the above group where $\alpha T = T\alpha^\sigma$ where σ is an automorphism of F. Note that the group elements when $\beta = \sigma = 0$ and $\gamma = \alpha^\sigma$ define the stabilizer of $y = xT$. We note that the group K^* does not leave the subplanes of the elation net invariant whereas the group defined above does leave every subplane invariant.

Hence, it is possible to have a group which is a $(q + 1, q^3 - q)$-transitive group that leaves a subplane of order q invariant and also a $(q+1, q^3 - q)$-transitive group that does not leave a subplane of order q invariant. However, in either case, there is a subplane of order q within the orbit of length $q + 1$ and there is a subgroup that leaves the subplane invariant and induces $SL(2, q)$ on the subplane.

Also, when $SL(2, q)$ is generated by elations, the elation net is a regulus net and hence there are $1 + q + q^2$ Desarguesian subplanes incident with the zero vector. In general, the subplane structure is not known when $S_z(\sqrt{q})$ acts.

THEOREM 76.14. (Hiramine, Jha, and Johnson [**519**]). Let π be a translation plane of order q^3 that admits $S_z(\sqrt{q})$ generated by elations. Then there is a net $\mathcal{N}$ of degree $q + 1$ containing either $1, 2, 3, \sqrt{q} + 1$ or $1 + \sqrt{q} + q$ Lüneburg–Tits subplanes incident with the zero vector.

THEOREM 76.15. (Hiramine, Jha, and Johnson [**518, 519**]). Let π be a cubic extension translation plane of even order q^3 for $q > 4$. Assume that there is a $(q + 1, q^3 - q)$-transitive group G that does not leave invariant a subplane of order q. If there is a subplane π_o of the net of degree $q + 1$ such that some subgroup of G leaves π_o invariant and induces either $SL(2, q)$ or $S_z(\sqrt{q})$ on the subplane, then there is a collineation group isomorphic to $SL(2, q)$ or $S_z(\sqrt{q})$ where the involutions are elations. Furthermore, when $T \simeq SL(2, q)$ is a collineation group, there are exactly $1 + q + q^2$ subplanes incident with the zero vector that are left invariant by the group T and when $T \simeq S_z(\sqrt{q})$ is a collineation group, there are either $1 + \sqrt{q}$ or $1 + \sqrt{q} + q$ subplanes which are invariant under T.

76.2.2. Cubic Chains. We have noticed that there are chains of cubic extensions. We indicate the extent of such chains as follows.

THEOREM 76.16. (Hiramine, Jha, and Johnson [**519**]). Let $\pi_o \subseteq \pi_1 \subseteq \pi$ be a set of finite translation planes such that π_1 is a cubic extension of π_o and π is a cubic extension of π_1. Assume that the order of π_o is q so that the orders of π_1 and π are q^3 and q^9, respectively, where q is even. Then one of the following occurs:

(1) The extensions are non-solvable–non-solvable and one of the following occurs: (a) both π_o and π_1 are Desarguesian, (b) π_o and π_1 are both Lüneburg–Tits planes of orders $q = 2^6$ and 2^{18}, respectively.

(2) The extensions are solvable–non-solvable, π_o is Desarguesian of order $q = 4$, and π_1 is Lüneburg–Tits or Desarguesian of order 4^3.

76.2.3. Cubic Extensions for Spreads in $PG(5, q)$. The investigation of whether there exist solvable cubic extensions may be seen to present problems. However, one may study cubic extensions with spreads in $PG(5, q)$.

THEOREM 76.17. (Hiramine, Jha, and Johnson [**515, 519**]). Let π be a cubic extension translation plane of order q^3 and kernel containing $K \simeq GF(q)$ which contains a subplane π_o of odd order $q > 3$. If G is a collineation group which leaves π_o invariant and acts as a $(q + 1, q^3 - q)$-transitive group G, then G is non-solvable. Furthermore, either (a) π_o is Desarguesian and induces a non-solvable group on π_o containing $SL(2, q)$ or (b) $q = 9$ and $G \mid \pi_o \cap \ell_\infty \simeq A_5$.

76.2.4. Cubic Extensions of Orders q^3 for $q \equiv -1 \bmod 4$. In the indicated situation, there is the theorem of Hiramine, Jha, and Johnson:

THEOREM 76.18. (Hiramine, Jha, and Johnson [**515, 519**]). Let π be a translation plane of order q^3 which admits a collineation group G that fixes a subplane π_o of order q and is a $(q + 1, q^3 - q)$-transitive group.

If $q \equiv -1 \bmod 4$ and q is not 3, then G is non-solvable.

Furthermore, the subplane is Desarguesian and $G \mid \pi_o$ contains $SL(2, q)$.

76.2.5. Even-Order Cubic Extensions. Indeed, for even order the is also the following result:

THEOREM 76.19. (Hiramine, Jha, and Johnson [**515, 519**]). Let π be a cubic extension of even order q^3 with subplane π_o of order q. Then π_o is Desarguesian or Lüneburg–Tits and $G \mid \pi_o$ contains a group isomorphic to $SL(2, q)$ or $S_z(\sqrt{q})$ acting transitively on the infinite points of π_o and $\pi - \pi_o$ where the involutions are elations on π_o. In particular, if $q = 2^r$ and r is odd, then π_o is Desarguesian.

Actually, Hiramine, Jha, and Johnson are able to show that the translation plane admits $SL(2, q)$ or $S_z(\sqrt{q})$.

THEOREM 76.20. (Hiramine, Jha, and Johnson [**519**]). Let π be a cubic extension translation plane of even order $q^3 > 8$ which admits a $(q + 1, q^3 - q)$-transitive group G that leaves invariant a Desarguesian subplane π_o and that is such that $G \mid \pi_o$ contains a subgroup isomorphic to $SL(2, q)$ or $S_z(\sqrt{q})$. Then there is a collineation group of π isomorphic to $SL(2, q)$ or $S_z(\sqrt{q})$ whose involutions are elations.

THEOREM 76.21. (Hiramine, Jha, and Johnson [519]). Let π be a cubic extension translation plane of even order $q^3 > 8$. If there is a $(q+1, q^3-q)$-transitive group G and there is a subplane π_o of the net of degree $q+1$ such that some subgroup of G leaves π_o invariant and induces either $SL(2, q)$ or $S_z(\sqrt{q})$ on the subplane, then there is a collineation group isomorphic to $SL(2, q)$ or $S_z(\sqrt{q})$ where the involutions are elations. Furthermore, when $T \simeq SL(2, q)$ is a collineation group, there are exactly $1 + q + q^2$ subplanes incident with the zero vector which are left invariant by the group T and when $T \simeq S_z(\sqrt{q})$ is a collineation group, there are either 1 or $1 + \sqrt{q} + q$ subplanes which are invariant under T.

COROLLARY 76.22. Let π be a solvable extension of order q^n, $q > 5$ of a proper flag-transitive plane of order q. If the spread for π is in $PG(2n - 1, q)$, then π is the Hall plane of order q^2.

COROLLARY 76.23. Let π be a solvable extension of order $q^n, q > 5$ of a proper flag-transitive plane of order q. If either q is even or $q \equiv -1 \bmod 4$, then π is the Hall plane of order q^2.

76.3. Quartic Extensions.

For extensions of dimension 4, the problem turns somewhat on the construction of regular packings (regular parallelisms). In Jha–Johnson [596] and [597], regular parallelisms and associated translation planes are considered. Furthermore, Penttila and Williams [1102] have constructed an infinite class of cyclic regular parallelisms in $PG(2, q)$ for $q \equiv 2 \bmod 3$. The previously known regular parallelisms are also cyclic and lie in $PG(3, 2)$, $PG(3, 5)$ and $PG(3, 8)$.

In general, there are corresponding translation planes of order q^4 admitting a collineation group isomorphic to $SL(2, q) \times Z_{1+q+q^2}$ when the parallelism lies in $PG(3, q)$. In this case, there are $(q^4-q)/(q^2-q) = 1+q+q^2$ derivable nets containing a net R of degree $1 + q$ and the group Z_{1+q+q^2} acts regularly on the set of these derivable nets. The group $SL(2, q)$ fixes a derivable net and acts transitively on the $q^2 - q$ components not in R and transitively on the components of R. Moreover, R is a K-regulus net for some field K isomorphic to $GF(q)$ so there are $1 + q + q^2 + q^3$ subplanes. It follows that Z_{1+q+q^2} leaves invariant a subplane and $SL(2, q)$ leaves invariant all of the translation planes π of order q^4 that contain a subplane π_o of order q such that π is a transitive extension of a flag-transitive plane.

So, the Johnson–Walker and Lorimer–Rahilly translation planes of order 2^4 admit collineation groups with orbits of length $2 + 1$ and $2^4 - 2$ and also there is an invariant subplane of order 2 within the net of degree 3. The planes of Prince [1114] of order 5^4, the planes of Denniston [313] of order 8^4 as well as all of the translation planes arising from the parallelisms of Penttila and Williams [1102] then are examples of translation planes of order q^4 that admit a collineation group with infinite point orbits of lengths $q + 1$ and $q^4 - q$ that leave invariant a subplane of order q; examples of dimension 4 extensions of a flag-transitive plane.

Thus, we see that to solve a more general problem, the translation planes of order q^n admitting $SL(2, q)$ probably need to be determined. When $n = 2$, the classification is given by Foulser [378, 379]. Furthermore, the analysis of Baer groups allows a description of the subplane when the group is solvable, which ultimately leads to the use of nets of critical deficiency. The reader is directed to Chapter 29 for details about how transitive Desarguesian parallelisms are connected to translation planes admitting the following types of groups.

THEOREM 76.24. (Jha and Johnson [597]). Let π be any translation plane of order q^{2r} for $r > 1$ admitting a collineation group in the translation complement isomorphic to $SL(2, q) \times Z_{(q^{2r-1}-1)/(q-1)}$. (1) Then the kernel is isomorphic to $GF(q)$, the p-elements are elations where $q = p^t$ and there is a regular partial 2-parallelism induced on any elation axis. (2) π is an extension of a flag-transitive plane.

So, we see that the complete determination of the translation planes of order q^n which are extensions of flag-transitive planes of order q is probably not possible when n is 3 due to the existence of the generalized Desarguesian planes and is made further difficult by the possibility of planes which may be constructed from regular 2-parallelisms. However, when the group is solvable much more can be said.

76.4. Solvable Extensions.

In the above known situations for extensions that are at least cubic, the collineation group is always non-solvable unless $q = 2$ or 3. Hence, if we assume that we have a 'solvable extension of a flag-transitive plane'; that is, if we assume that the group is solvable then there may be a chance to obtain a complete classification.

To see that a solvable group occurs in the Hall planes, consider the Hall plane π of order q constructible from a Desarguesian plane Σ of order q^2 by the replacement of a regulus net R. There exists a central collineation group C of order $q(q - 1)$ which fixes a component L of R pointwise and which leaves R invariant and acts transitively on the points at infinity not in R. So, C fixes all Baer subplanes of R incident with the zero vector. Let H^* denote the homology group of order $q^2 - 1$ of Σ. H^* acts transitively on the set of $q + 1$ Baer subplanes of R which are incident with the zero vector. Let π^* denote the Hall plane obtained by the replacement of the net R and let R^* denote the derived net and $\pi_o^* = L$ be a Baer subplane of R^*. Hence, CH^* acts transitively on the infinite points of R^* and transitively on the remaining infinite points and leaves $L = \pi_o^*$ invariant. That is, CH^* is a solvable group which fixes a subplane of order q and has two points orbits at infinity of lengths $q + 1$ and $q^2 - q$.

Hence, we shall find the Hall planes in any concluding statement on the classification of solvable extensions of flag-transitive planes. Are there other solvable extensions?

The main general result on solvable extensions, due to Hiramine, Jha and Johnson, is as follows:

THEOREM 76.25. (Hiramine, Jha, and Johnson [517]). Let π be a finite translation plane of order q^n which is a solvable extension of a proper flag-transitive plane π_o of order q. Let G denote the corresponding group. Then one of the following occurs:

(1) π is Desarguesian and (q, n) is in $\{(2, 2), (2, 3), (3, 2), (3, 3), (2, 5)\}$. (a) For $(2, 2), (2, 3)$, the group $SL(2, 2)$ is a $(3, 2)$- or $(3, 6)$-transitive group, respectively. (b) For $(3, 2), (3, 3)$, the group $SL(2, 3)$ is a $(4, 6)$- or $(4, 24)$-transitive group, respectively. (c) For $(2, 5)$, the group $SL(2, 2) \times Z_5$ is a $(3, 30)$-transitive group.

(2) π is Hall and $n = 2$.

(3) $n = 3$.

(4) $n > 3$ and one of the following occurs: (a) $q = 2$ and there is a normal sub-group generated by elations isomorphic to $SL(2, 2)$ which acts doubly transitively on the infinite points of π_o. Also, the Sylow 2-subgroups have order 2 and the full group $G_{[\pi_o]}$ which fixes π_o pointwise has index 6 so that $SL(2, 2)G_{[\pi_o]}$ is the full translation complement. In addition, if n is even then the spread is a union of De-sarguesian nets of degree 5 containing π_o and there is a regular partial 2-parallelism of $2^{n-1} - 1$ 2-spreads in $PG(2n - 1, 2)$, (b) $q = 3$ and n is even. Furthermore, there is a normal subgroup generated by 3-elements such that the restriction to π_o is iso-morphic to $SL(2, 3)$ and which acts doubly transitively on the infinite points of π_o. The Sylow 3-subgroups are non-planar groups of order 3 and the full group $G_{[\pi_o]}$ which fixes π_o pointwise has index 24 so $SL(2, 3)G_{[\pi_o]}$ is the full translation comple-ment. If the 3-elements are elations, the spread is a union of Desarguesian nets of degree 10 containing π_o and there is a regular partial 2-parallelism of $(3^{n-1} - 1)/2$ 2-spreads in $PG(2n - 1, 3)$. Furthermore, if the 3-elements are not elations, then $n \geq 20$. (c) $q = 4$ and $n = 4$. (d) $q = 4$ and $n > 4$. Then all involutions are elations and there is a normal subgroup generated by elations that acts doubly transitively on the infinite points of π_o. Furthermore, the Sylow 2-subgroups are cyclic of order 4 and there is a normal 2-complement. If τ is a collineation of order 4, then π may be decomposed into a direct sum of n cyclic $\tau GF(2)$-submodules of dimension 4 and each Sylow 2-group pointwise fixed subspace has cardinality 2^n.

We note that there are examples in case 4(a) of the Lorimer–Rahilly and Johnson–Walker planes of order 16. For the other possibilities, it is not known whether non-Desarguesian, non-Hall examples exist.

Thus, we have:

COROLLARY 76.26. (Hiramine, Jha, and Johnson [**517**]). Let π be a finite translation plane of order q^n which is a solvable extension of a proper flag-transitive plane π_o of order q. If $q > 5$ and $n \neq 3$, then π is the Hall plane of order q^2.

The main result on solvable extensions does not provide any information on what occurs when $n = 3$. As we have seen, there exist infinite families of such planes admitting $SL(2, q)$ which are cubic extensions of flag-transitive planes and conceivably cubic extension $\geq q^{n-1}$.

76.4.1. Solvable Extensions of Even Order. Hiramine, Jha, and Johnson complete the problem on solvable extensions for even order as follows:

COROLLARY 76.27. (Hiramine, Jha, and Johnson [**517**]). Let π be a translation plane of even order q^n which is a solvable extension of a flag-transitive plane of order q. Then we have one of the following: (1) $q = 2$ or 4, or (2) π is Hall.

We note that there are examples of solvable n-dimensional extensions that are not Hall when $(q, n) \in \{(2, 2), (2, 3), (2, 4), (2, 5), (4, 3)\}$.

The main result on solvable extensions does not provide any information on what occurs when $n = 3$. As we have seen, there exists infinite families of such planes admitting $SL(2, q)$ which are cubic extensions of flag-transitive planes and conceivably cubic extension planes always admit non-solvable groups when $q > 3$. It turns out that cubic extensions almost never admit solvable groups acting as in the manner under discussion. The reader is referred back to the sections on cubic extensions. We list only the results relative to solvability.

COROLLARY 76.28. (Hiramine, Jha, and Johnson [**517**]). Let π be a solvable extension of order q^n, $q > 5$ of a proper flag-transitive plane of order q. If the spread for π is in $PG(2n - 1, q)$, then π is the Hall plane of order q^2.

COROLLARY 76.29. (Hiramine, Jha, and Johnson [**517**]). Let π be a solvable extension of order q^n, $q > 5$, of a proper flag-transitive plane of order q. If either q is even or $q \equiv -1 \bmod 4$, then π is the Hall plane of order q^2.

CHAPTER 77

Higher-Dimensional Flocks.

77.1. Connections and Definitions.

It is fairly well known that there are connections to the theories of flocks of hyperbolic quadrics in $PG(3,q)$ and translation planes with spreads in $PG(3,q)$. In particular, a spread in $PG(3,q)$ that is a union of reguli sharing two common lines produces a flock of a hyperbolic quadric and conversely. Using a blend of these theories, a complete characterization of hyperbolic flocks is given by the beautiful theorem of Thas [1178] and Bader–Lunardon [49]. The flocks corresponding to the regular nearfield planes were constructed by J.A. Thas by geometric methods. That there are flocks corresponding to certain irregular nearfields was independently determined by Bader [38], Baker and Ebert [67] for $p = 11$ and 23, and Johnson [718].

THEOREM 77.1. (Thas [1178] and Bader–Lunardon [49]). A flock of a hyperbolic quadric in $PG(3,q)$ is one of the following types:
(i) Linear,
(ii) A Thas flock, or
(iii) A Bader/Baker–Ebert/Johnson flock of order p^2 for $p = 11, 23, 59$.

Recently, Bader, Cossidente, and Lunardon [41, 42] generalized the notions of hyperbolic flocks in a way that connects with a theory of spreads or translation planes depending on the notion of an (A, B)-regular spread.

DEFINITION 77.2. An (A, B)-regular spread in $PG(2n+1, q)$ is a spread S such that for C in $S - \{A, B\}$, the q-regulus generated by $\{A, B, C\}$ is in S.

In terms of the associated translation plane π_S corresponding to a spread S, a coordinate structure Q for the translation plane (called a 'quasifield') may be taken so that any given q-regulus corresponds to a subfield K isomorphic to $GF(q)$ of Q that multiplicatively commutes with Q. It is important to note that in this case, the corresponding spread may be realized in $PG(2n - 1, q)$ and Q is a K-vector space. In this latter case, we say that K is in the 'kernel' of Q. To emphasize the importance of the kernel, for example there may be a subfield F of Q distinct from the subkernel K that multiplicatively commutes with Q but Q may not be an F-vector space. In the case that K commutes with Q and Q is a vector space over K, the associated translation plane then admits a Desarguesian net of degree $q + 1$ that may be coordinatized by K. In attempting to determine situations when a given spread may be (A, B)-regular, it is convenient to consider a more general situation when the corresponding translation plane contains nets that may be coordinatized by a subfield of an associated coordinate quasifield, whether that subfield commutes with the quasifield, whether the quasifield is a vector space over that field, or whether the associated subfield is in the kernel. Furthermore, we are then concerned with spreads that are unions of such nets.

DEFINITION 77.3. Let R be a net of degree $1 + q$ and order q^n corresponding to a partial spread in $PG(2n - 1, K)$, where K is a field isomorphic to $GF(q)$.

(i) If R contains a Desarguesian subplane of order q, R is said to be a 'rational net'. The associated partial spread is called a 'rational partial spread'.

(ii) If R is a rational net that may be embedded into a Desarguesian affine plane, the partial spread is called a 'rational Desarguesian net'. The associated partial spread is called a 'rational Desarguesian partial spread'.

(iii) A 'component' of a spread S in $PG(2n-1, K)$ is one of the $n-1$-dimensional K-subspaces of S.

(iv) A 'hyperbolic cover of order q' of a spread in $PG(2n - 1, q)$ is a set of $(q^n - 1)/(q - 1)$ rational Desarguesian partial spreads each of degree $q + 1$ that share two components of S and whose union is S.

(v) If the rational Desarguesian partial spreads are all K-reguli, we call the hyperbolic cover a 'regulus hyperbolic cover'.

DEFINITION 77.4. A 'semifield' is a non-associative division ring without zero divisors. In particular, a translation plane is a 'semifield plane' if and only if it has a coordinate system (quasifield) that is a semifield. Let $Q = (Q, +, \cdot)$ be a semifield. Then, the 'seminuclei', 'center' and 'nucleus' are defined as follows:

(i) The 'left seminucleus' N_L is $\{ k \in Q; \ k \cdot (x \cdot y) = (k \cdot x) \cdot y; \ \forall x, y \in Q \}$,

(ii) The 'right seminucleus' N_R is $\{ k \in Q; \ (x \cdot y) \cdot k = x \cdot (y \cdot k); \ \forall x, y \in Q \}$, and

(iii) The 'middle seminucleus' N_M is $\{ k \in Q; \ x \cdot (k \cdot y) = (x \cdot k) \cdot y; \ \forall x, y \in Q \}$.

(iv) We shall say that a seminucleus is 'central' if and only if it commutes with Q.

(v) The 'nucleus' is the intersection of the left, right and middle seminuclei.

(vi) The 'center' is the set of elements in the nucleus that commute with Q.

REMARK 77.5. The seminuclei of any quasifield may be defined exactly as for a semifield. However, without both distributive laws, the existence of a seminucleus may not imply the simultaneous existence of an associated vector space over that seminucleus, even if the seminucleus commutes with the quasifield.

The importance of semifields now becomes apparent in the following remark.

REMARK 77.6. (1) The seminuclei of a finite semifield are fields and the semifield is a vector space over any of the seminuclei.

(2) If Q is a semifield with left nucleus N_L isomorphic to $GF(q)$ and Q has order q^n then form $Q \times Q$ and call the elements of this set 'points'. The 'lines' of the corresponding semifield plane then have the following equations:

$$x = c, \ y = x \cdot t + b, \ \forall t, b, c \in Q.$$

Furthermore, the set

$$\{ x = 0, \ y = x \cdot t, \ \forall t \in Q \},$$

is a spread in $PG(2n - 1, N_L)$.

(3) Let N_R, and N_M have order q_R, and q_M, respectively. Then we obtain the following rational Desarguesian partial spreads in $PG(2n - 1, N_L)$:

$$\{ x = 0, \ y = x \cdot k, \ \forall k \in N_M \},$$

$$\{\, x = 0, \ y = x \cdot t, \, \forall t \in N_R \,\},$$

called, respectively, a 'middle seminucleus net' of degree $q_M + 1$ and a 'right seminucleus net' of degree $q_R + 1$, respectively. Note that the order of the net depends on the order of the seminucleus in question.

(4) Any semifield plane admits an elation group (shears group) that fixes a component pointwise (the 'shears axis') and acts regularly on the remaining components. If the plane is finite and non-Desarguesian, there is a unique shears axis Y.

(a) Let X be any component not equal to the shears axis Y. The right seminucleus corresponds to an affine homology group of the semifield plane that fixes X pointwise and leaves the shears axis Y invariant.

(b) The middle seminucleus corresponds to an affine homology group of the semifield plane that fixes the shears axis Y pointwise and leaves X invariant.

(c) The left seminucleus corresponds to the homology group of the semifield plane that fixes each point at infinity and has an affine point as center.

(d) Coordinatizing with Y as $x = 0$, X as $x = 0$, the nucleus corresponds to the intersection of the right and middle homology groups defined by scalar elements. That is, the subgroups are $\left\langle \begin{bmatrix} 1 & 0 \\ 0 & \lambda \end{bmatrix} \right\rangle, \left\langle \begin{bmatrix} \lambda & 0 \\ 0 & 1 \end{bmatrix} \right\rangle$, respectively, so that $\left\langle \begin{bmatrix} \lambda & 0 \\ 0 & \lambda \end{bmatrix} \right\rangle$ is a subgroup of the homology group corresponding to the left seminucleus.

DEFINITION 77.7. (1) A subplane of a semifield plane that is invariant under a left-seminucleus homology group is called a 'kernel' subplane.

(2) A subplane that is invariant under a right- or middle-seminucleus homology group is called a 'right seminuclear' or 'middle seminuclear' subplane, respectively.

(3) A subplane that is left invariant under subgroups corresponding to the nucleus is called a 'nuclear' subplane.

(4) A subplane that may be coordinatized by the center is called a 'central' subplane. This occurs, for example, when the seminuclei all coincide and one is central. The central plane is then also a nuclear subplane.

The main reason that we are discussing semifield planes in this section is that we always obtain hyperbolic covers.

THEOREM 77.8. (Jha and Johnson [**634**, see Proposition 7]).
Let π be a finite semifield plane with shears axis Y and a second component X not equal to Y. Let the left, middle, and right seminuclei of a coordinate semifield (with $x = 0$ as Y) have orders q_L, q_M, and q_R, respectively, where the order of Q is $q_L^{n_L}$, $q_M^{n_M}$, and $q_R^{n_R}$, respectively.
Then, considering the associated spread S in $PG(2n_L - 1, q_L)$, there is a unique hyperbolic cover $\mathcal{E}_M$ of order q_M and a unique hyperbolic cover $\mathcal{E}_R$ of order q_R.

REMARK 77.9. If a subright and submiddle seminucleus can be identified with a subfield of the left seminucleus and is central, then, considering the spread over a smaller subfield if necessary, the right or middle seminucleus partial spread is a regulus.

Furthermore, if the orders of the seminuclei are all equal, the question is when such seminuclei are actually equal becomes of utmost importance in trying to determine if any given hyperbolic cover is actually a regulus hyperbolic cover. In [**634**] Jha and Johnson study the so-called 'fusion' of the seminuclei in finite semifield planes. In particular, given a semifield of order q^r with left, right and middle seminuclei isomorphic to $GF(q)$, it can be shown that the three subnuclei can be 'fused' (identified). With the fusion considerations, there is a development of hyperbolic covers corresponding to the right or middle seminuclei isomorphic to $GF(q)$ just as we have outlined above; these are covers of the components by a set of rational Desarguesian partial spreads sharing two components. When the left, right and middle subnuclei commute with the associated quasifield, the hyperbolic cover is, in fact, a 'regulus hyperbolic cover'; the rational Desarguesian partial spreads are reguli in an associated projective space over the left nucleus (the left nucleus of a semifield plane is also called the 'kernel').

THEOREM 77.10. (Jha and Johnson [**634**, Theorem 16]). Let π be a finite semifield spread, with shears axis Y and let $X \in \pi \setminus \{Y\}$. For any semifield D coordinatizing π, use the notation $N_1 = N_L(D)$, $N_2 = N_M(D)$, and $N_3 = N_R(D)$. Then the following hold.

(1) If two fields N_i and N_j, for $i, j \in \{1, 2, 3\}$, each contain a subfield isomorphic to $GF(q)$, then π may be recoordinatized by a semifield E such that

$$N_i(E) \cap N_j(E) \supseteq F \cong GF(q).$$

(2) Suppose each N_i, $i = 1, 2, 3$, contains a subfield isomorphic to $GF(q)$, but some semifield coordinatizing π does *not* have $GF(q)$ in its center.

   (a) Then π has order-q hyperbolic covers $\mathcal{E}_M$ and $\mathcal{E}_R$, associated, respectively, with the middle and right seminucleus and with carriers $\{Y, X\}$, and there is a non-trivial partition of $r = \sum_{i=1}^{s} d_i$, for $s \neq r$ and the hyperbolic covers share a set of exactly

$$\left(\sum_{j=1}^{s} (q^{d_j} - 1)/(q-1) \right)$$

rational Desarguesian nets $\mathcal{R}$.

   (b) Furthermore, there is another non-trivial partition $\sum_{i=1}^{t} k_i$, $t \neq r$ and each of these nets contain exactly

$$\left(\sum_{j=1}^{t} (q^{k_j} - 1)/(q-1) \right)$$

nuclear planes of order q (and these constitute the set of all the nuclear planes of order q that contain $\{Y, X\}$).

(3) Suppose each N_i, $i = 1, 2, 3$, contains a subfield isomorphic to $GF(q)$. Then either all the seminuclei of order q in every coordinatizing semifield are in the center or π has no central planes of order q and there are two non-trivial partitions of $r = \sum_{j=1}^{s} d_j$, $s \neq r$ and $r = \sum_{i=1}^{t} k_i$, $t \neq r$ and

there are exactly

$$\left(\sum_{j=1}^{s}(q^{d_j}-1)/(q-1)\right)\left(\sum_{j=1}^{t}(q^{k_j}-1)/(q-1)\right)$$

nuclear planes of order q passing through the axes $\{X, Y\}$.
(4) If each of the three seminuclei, N_1, N_2, and N_3, contains a subfield isomorphic to $GF(q)$, then π may be coordinatized by a semifield E such
that the nucleus of E contains a field F isomorphic to $GF(q)$, that is

$$GF(q) \cong F \subseteq N(E) = N_L(E) \cap N_M(E) \cap N_R(E).$$

Now as mentioned previously, Bader, Cossidente and Lunardon ([**41, 42**]) have
extended the notions of hyperbolic quadrics and the connecting concepts of (A, B)-
regular spreads.

The hyperbolic quadrics correspond to covers of the Segre variety $S_{1,1}$ by conics
(caps of size $q + 1$). Bader, Cossidente, and Lunardon [**42**] consider the partition of
Segre varieties $S_{n,n}$ by caps of size $(q^{n+1} - 1)/(q - 1)$ as a 'flock' of $S_{n,n}$. However, in
order to have a corresponding translation plane, something more must be assumed:
It must be that the caps are Veronese varieties obtained as sections of $S_{n,n}$ by linear
subspaces of the projective space in which $S_{n,n}$ resides. These flocks are said to be
'flat flocks'.

In Bader, Cossidente, and Lunardon [**41**], some new flat flocks are given that
correspond to the Dickson nearfield translation planes. In this case, the flat flocks
are 'transitive': they admit a collineation group that acts transitively on the caps
of the flock.

Now the transitive flat flocks correspond naturally to translation planes admitting regulus hyperbolic covers that further admit a collineation group transitive on
the reguli of the cover.

In terms of translation planes, we are then interested in the more general concept of translation planes admitting arbitrary hyperbolic covers and, in particular,
those that admit collineation groups that act transitively on these hyperbolic covers.

These considerations then lead us to consider finite translation planes that
admit a collineation group fixing two components and acting transitively on the
remaining components, whether a hyperbolic cover is present or not. So, an important problem in the theory of translation planes is as follows:

PROBLEM 77.11. *Determine all finite translation planes that admit a collineation group that fixes two components and acts transitively on the remaining
components.*

In this class, there are the so-called 'j-planes' as well as the generalized twisted
field (semifield) planes (see below). We shall recall the definitions of j-planes and
their generalizations $j \ldots j$-planes.

DEFINITION 77.12. Let K be a finite field of order q^n, $q = p^r$, for p a prime.
Write K as an $n \times n$ matrix field and for $M \in K - \{0\}$, let δ_M denote the determinant
of M.

Let J_M^J denote an $n \times n$ diagonal matrix group with (i, i)-entries $\delta_M^{j_i}$, where
$j_1 = 0$, and j_j are integers between 0 and $r - 1$ and independent of M. Consider

the following $2n \times 2n$ matrix group:

$$J^{J^J} = \left\langle \begin{bmatrix} J_M^J & 0 \\ 0 & M \end{bmatrix} ; M \in K - \{0\} \right\rangle.$$

Then, if

$$\left\{ x = 0, y = 0, (y = x)g; \, g \in J^{J^J} \right\}$$

is a partial spread in $PG(2n-1, q)$, we obtain a spread and an associated translation plane called a '$j \ldots j$-plane'.

If $n = 2$, the plane is called a j-plane.

To give an idea of the importance of $j \ldots j$-planes, as regards the ideas of hyperbolic covers, we mention the following result, which is immediate.

THEOREM 77.13. Any $j \ldots j$-plane of order q^{q-1} and kernel containing $GF(q)$ produces a $GF(q)$-regulus hyperbolic cover. Furthermore, there is a cyclic group acting transitively on the regulus cover.

Hence, there is a corresponding cyclic and transitive flat flock of $S_{q-2, q-2}$ over $GF(q)$.

We now define the generalized twisted field planes.

DEFINITION 77.14. Let K be a field isomorphic to $GF(q^r)$. Define points as elements of $K \times K$ and spread components as follows:

$$x = 0, y = xm - cx^\alpha m^\beta \ \forall m \in K \simeq GF(p^r),$$

where α, β are automorphisms of K and c is a constant such that $x^{1-\alpha} m^{1-\beta} \neq c$ for all $x, m \in K$.

Then this defines a spread for a semifield plane called a 'generalized twisted field plane'.

Generalized twisted field planes are considered prominently in this text and here we are particularly interested in determining the class of semifield planes that admit regulus hyperbolic covers, and those that admit collineation groups that act transitively on the covers thus producing flat flocks and transitive flat flocks, respectively.

Using fusion and results on generalized twisted field planes, we are able to substantially add to the class of flat flocks and completely classify the transitive semifield flat flocks (those corresponding to semifield planes admitting such transitive groups).

DEFINITION 77.15. A 'flat flock' is a partition of a Segre variety that corresponds to a translation plane. If a translation plane is of type T, then we shall say that we have a 'type T flat flock'.

In particular, if the translation plane is a semifield plane, we shall use the term 'semifield flat flock' to designate the corresponding flat flock.

Furthermore, if the translation plane associated with a flat flock admits a collineation group transitive on the regulus hyperbolic cover, we shall call the associated flock a 'transitive flat flock'. If the transitive flat flock is also a semifield flat flock, we use the term 'transitive semifield flat flock'.

77.1.1. Semifield Flat Flocks. The main results of Jha and Johnson [**640**] are actually corollaries to more general results on semifield planes.

COROLLARY 77.16. *Let K be an arbitrary subfield of a finite field F, where F is isomorphic to $GF(q)$, for $p^s = q$, p a prime, and s a positive integer. Let t be a divisor of s so that K is isomorphic to $GF(p^{s/t})$. Then, there exist a $K = GF(p^{s/t})$-regulus hyperbolic cover of order q^n which is non-classical (non-Desarguesian), thus constructing a semifield flat flock.*

The main results of Jha and Johnson [**640**] on transitive semifield flat flocks are as follows.

COROLLARY 77.17. *Let $\mathcal{F}$ be a transitive semifield flat flock in $S_{n,n}$ over $GF(q)$, $q \neq 2^6$.*

(1) Then $\mathcal{F}$ is a generalized twisted field flat flock. Furthermore, it also follows that $\mathcal{F}$ has a collineation group of order $(q^{n+1} - 1)/(q - 1)$ acting transitively on the caps.

(2) For any subfield $GF(q')$ of $GF(q)$, let $q = p^s$ and $q' = p^{s/t}$. Then, there is an associated transitive semifield flat flock $\mathcal{F}^{s/t}$ in $S_{nt+t-1,nt+t-1}$ over $GF(q')$.

(3) In either situation (1) or (2), whenever there is a transitive group, there is an associated cyclic group of order $(q'^{(nt+t)} - 1)/(q' - 1)$ acting transitively on the caps.

For semifield planes, Jha and Johnson [**634**] show that in semifield planes, right-seminucleus and middle-seminucleus homology groups provide the spread with hyperbolic covers, and furthermore show that every right and/or middle seminucleus of a semifield plane produce such hyperbolic covers.

The problem becomes whether, in any translation plane, the existence of a hyperbolic cover implies the existence of homology groups. If the hyperbolic cover is a 'regulus' cover, then it turns out that there are always associated homology groups for both the right and middle seminuclei. However, it is an open question whether this is true for arbitrary hyperbolic covers.

THEOREM 77.18. *Let π be a translation plane of order q^r which admits a hyperbolic cover of degree-$(q+1)$ rational Desarguesian partial spreads sharing the components $x = 0$ and $y = 0$.*

(1) Then, there are exactly $(q^r - 1)/(q - 1)$ such rational Desarguesian partial spreads, $\mathcal{D}_i$, for $i = 1, 2, \ldots, (q^r - 1)/(q - 1)$:

$$\mathcal{D}_i : x = 0, y = 0, y = xA_{ij} \text{ for } j = 1, 2, \ldots, q - 1,$$

where

$$\{\, A_{ij};\ i = 1, 2, \ldots, (q^r - 1)/(q - 1),\ j = 1, 2, \ldots, q - 1 \,\} \cup \{0\}$$

is a matrix spread set, where we may let $A_{11} = I$.

Hence,

$$\begin{aligned} \mathcal{M}_i &= \{A_{ij}A_{i1}^{-1}, j = 1, 2, \ldots, q - 1\} \cup \{0\} \text{ and} \\ \mathcal{R}_i &= \{A_{i1}^{-1}A_{ij}, j = 1, 2, \ldots, q - 1\} \cup \{0\} \end{aligned}$$

are fields isomorphic to $GF(q)$, for each $i = 1, 2, \ldots, (q^r - 1)/(q - 1)$.

(2) (a) A right homology group of order $q - 1$ leaving invariant each rational Desarguesian partial spread is obtained if and only if $\mathcal{R}_i = \mathcal{R}_j$ for all $i, j = 1, 2, \ldots, (q^r - 1)/(q - 1)$.

(b) A middle homology group of order $q-1$ leaving invariant each rational
Desarguesian partial spread is obtained if and only if $\mathcal{M}_i = \mathcal{M}_j$ for
all $i, j = 1, 2, \ldots, (q^r - 1)/(q - 1)$.

COROLLARY 77.19. A regulus cover (regulus hyperbolic cover) is obtained in
either of the following two situations:

(1) If and only if there are right and middle homology groups of order $q - 1$
whose associated right and middle seminuclei are in the center of the associated
quasifield, or

(2) If and only if the associated fields $\mathcal{M}_i = \mathcal{M}_j = \mathcal{R}_i = \mathcal{R}_j$ for all $i, j =
1, 2, \ldots, (q^r - 1)/(q - 1)$.

Jha and Johnson [634] have discussed the Hughes–Kleinfeld semifields of order
h^2 that admit right, left and middle seminuclei isomorphic to $GF(h)$ and such that
the center is not $GF(h)$. A related question is whether a semifield plane of order
h^t, for t odd, could have right, left and middle nuclei all isomorphic to $GF(h)$ and
such that the center is not $GF(h)$. That is, in such a situation must we always
obtain regulus hyperbolic covers?

THEOREM 77.20. Let π be a semifield plane of order p^{2t}, p a prime and t odd,
that admits left, right and middle seminuclei isomorphic to $GF(p^2)$. Then, the
hyperbolic covers are always regulus hyperbolic covers.

COROLLARY 77.21. Every finite semifield of order p^{2t}, with t odd that admits
left, right and middle seminuclei isomorphic to $GF(p^2)$ produces a semifield flat
flock.

Now in order to produce a large set of examples of the above type, we turn to
'cyclic' semifields.

77.2. Cyclic Semifields and Hyperbolic Covers.

We now recall the class of cyclic semifields. From such semifields, we may
construct also a variety of regulus hyperbolic covers.

DEFINITION 77.22. Let F be a field isomorphic to $GF(q)$, let V be an n-
dimensional F-vector space and let T be a strictly semilinear transformation of V
which is irreducible. Then let $\alpha T = T \alpha^\sigma$ for all $\alpha \in F$, where σ is an non-identify
automorphism of F. Consider x and y are n-vectors and define the following spread
in $PG(2n - 1, F)$:

$$x = 0, \; y = x(\alpha_1 I + \alpha_2 T + \alpha_3 T^2 + \cdots + \alpha_{n-1} T^{n-1})$$
$$\forall \, \alpha_i \in F, \; i = 1, 2, \ldots, n - 1.$$

Then, this is a semifield spread called the 'cyclic semifield defined by (T, F, σ)'.

From the generator T over the field F, we take the intersection of the kernel
(left seminucleus) with F. We note that F is contained in both the right middle
seminuclei. For example, we also know that $\alpha T = T \alpha^\sigma$ for some automorphism σ
of F. Hence, in particular the fixed field of σ in F may be fused into right, middle,
and left subnuclei also producing a regulus hyperbolic cover.

THEOREM 77.23. (1) Let π be any cyclic semifield plane of order q^n defined by the F-irreducible and strictly semilinear element T and the field F; components are

$$x = 0, \; y = x(\alpha_1 I + \alpha_2 T + \alpha_3 T^2 + \cdots + \alpha_{n-1} T^{n-1})$$
$$\forall \, \alpha_i \in F, \; i = 1, 2, \ldots, n-1,$$

where $\alpha T = T\alpha^\sigma$ for some non-identity automorphism σ of F.

Then, there exists a subfield $K = \mathrm{Fix}\,\sigma$ of F which is in the center and thus produces a K-regulus hyperbolic cover.

(2) Let F be isomorphic to $GF(p^t)$ for p a prime. Then, for any t' between 1 and $t-1$, there exists a cyclic semifield $(T, F, p^{t'})$ producing a $GF(p^{t'})$-regulus hyperbolic cover.

REMARK 77.24. For the construction of a cyclic semifield, let S be any linear cyclic Singer cycle and let σ be any automorphism of F then $T = S\sigma$ will produce the required cyclic semifield.

The ubiquity of regulus hyperbolic covers corresponding to cyclic semifields is now apparent. The reader is referred an article by two of the authors [606] for more details on cyclic semifields.

COROLLARY 77.25. Let F be a finite field isomorphic to $GF(q)$, for $q = p^s$, p a prime and s a positive integer. For any integer n and for any proper subfield K of F, isomorphic to $GF(p^{s/t})$, there exist a $K = GF(p^{s/t})$-regulus hyperbolic cover of order q^n which is non-classical (non-Desarguesian).

77.2.1. Segre Varieties. As mentioned, in Bader, Cossidente, and Lunardon [41, 42], there are connections between what are called (A, B)-regular spreads and flocks of Segre varieties.

We remind the reader that our definition of a spread admitting a 'hyperbolic cover' is clearly equivalent to an (A, B)-regular spread if and only if the hyperbolic cover is a regulus cover.

DEFINITION 77.26. The 'Segre variety' of $PG(m, q)$ and $PG(k, q)$ is the variety $S_{m,k}$ of $PG(N, q)$ consisting of the points given by the vectors $v \otimes u$ as v and u vary over $PG(m, q)$ and $PG(k, q)$, respectively, (the decomposable tensors in the associated tensor product of the vector spaces).

When $m = k = n$, a 'flock' of $S_{n,n}$ is a partition of the point-set into caps (a set of points no three of which are collinear) of size $(q^{n+1} - 1)/(q - 1)$. Note that there would also be $(q^{n+1} - 1)/(q - 1)$ such caps.

A flock is 'flat' if its caps are Veronese varieties $\mathcal{V}_n$ obtained as sections of $S_{n,n}$ by dimension $n(n+3)/2$ projective subspaces of $PG(N, q)$, for $N = (n+1)^2 - 1$.

REMARK 77.27. The reader is referred to Hirschfeld and Thas [529, Chapter 25] for background on Veronese and Segre varieties. In particular, the variety considered in the previous definition is the Veronesian of quadrics of $PG(n, q)$, and sits naturally in $PG(n(n+3)/2, q)$. There is a bijection from the Veronesian of quadrics $\mathcal{V}_n$ onto $PG(n, q)$ so that $|\mathcal{V}_n| = (q^{n+1} - 1)/(q - 1)$. Using [529], Theorems (25.1.8), (25.5.4) and (25.5.8), we see that (i) $\mathcal{V}_n$ is a cap of $PG(n(n+3)/2, q)$, (ii) $|S_{n,n}| = ((q^{n+1} - 1)/(q - 1))^2$, and (iii) the intersection of appropriate projective subspaces of dimension $n(n+3)/2$ of $PG((n+1)^2 - 1, q)$ with $S_{n,n}$ are Veronesians of quadrics of type $\mathcal{V}_n$.

When $n = 1$, $S_{1,1}$ is a hyperbolic quadric in $PG(3,q)$ and the Veronesian is a conic of $q+1$ points in $PG(2,q)$ considered as a section of $S_{1,1}$ by a $PG(2,q)$. Hence, the concept of a flock of Segre varieties $S_{n,n}$ by caps that are Veronese varieties of type $\mathcal{V}_n$ is a direct generalization of flocks of hyperbolic quadrics in $PG(3,q)$.

77.2.2. Theorem of Bader, Cossidente, and Lunardon on (A, B)-Regular Spreads. Bader, Cossidente, and Lunardon show that the Veronese varieties correspond to $GF(q)$-reguli and, furthermore, show that there is a corresponding spread.

THEOREM 77.28. (Bader, Cossidente, and Lunardon [**41**, Theorem 3.4]).
Flat flocks of $S_{n,n}$ and (A, B)-regular spreads in $PG(2n+1, q)$ are equivalent.

Hence, we obtain:

COROLLARY 77.29. Flat flocks of $S_{n,n}$ are equivalent to regulus hyperbolic q-covers in translation planes of order q^{n+1}.

Hence, we have shown the following theorem.

THEOREM 77.30. Consider the Segre variety $S_{n,n}$ in $PG((n+1)^2 - 1, q)$. If $n + 1$ is not prime, then there exist non-classical cyclic semifield flat flocks of $S_{n,n}$ corresponding to non-Desarguesian semifield planes of order q^{n+1}.

We now turn to a discussion of transitive flat flocks.

77.3. Transitive Semifield Flat Flocks.

DEFINITION 77.31. Let π be a semifield plane with shears axis Y and choose another component X. An 'autotopism group' of π is a collineation group of π that fixes Y and X. The triangle determined by Y, X, and the line at infinity ℓ_∞ is called the 'autotopism triangle' with 'vertex points' $Y \cap X$, $Y \cap \ell_\infty$, and $X \cap \ell_\infty$ (considering the components projectively). A 'transitive autotopism group' is an autotopism group that acts transitively on the components of the spread for π different from Y and X.

REMARK 77.32. The finite generalized twisted field planes of order $p^n \neq 2^6$ admitting an autotopism group that acts transitively on the non-vertex points of one side of the autotopism triangle (Y, or X or the line at infinity) are completely determined by Biliotti–Jha–Johnson [**121**]. The reader is referred to this article for additional details.

Let π be a generalized twisted field plane of order q^{n+1} with center $GF(q)$ admitting a $GF(q)$-linear autotopism group acting transitively on the non-vertex infinite points. Now consider a hyperbolic cover. We see that any right- or middle-seminucleus homology group of order $q - 1$ will be normalized by the indicated group. Hence, it follows that the hyperbolic cover will be preserved under the group. Thus, we obtain:

THEOREM 77.33. Any generalized twisted field plane admitting a transitive autotopism group $\mathcal{A}$ may be written as of order q^{n+1} with center containing $GF(q)$ where $\mathcal{A}$ is a $GF(q)$-linear group.

(1) Furthermore, the group $\mathcal{A}$ acts transitively on the reguli in the regulus hyperbolic cover.

(2) This implies that we have a non-Desarguesian generalized twisted field flat flock of $S_{n,n}$ admitting a collineation group acting transitively on its caps.

Moreover, we provide a characterization of transitive semifield flat flocks as follows.

THEOREM 77.34. *Let π be a finite semifield plane of order $p^r = q^{n+1} \neq 2^6$ with center containing $GF(q)$ and p a prime. Let $\mathcal{R}$ denote the corresponding regulus hyperbolic cover.*

(1) If π admits a collineation group that is transitive on the reguli of the cover $\mathcal{R}$, then π is a generalized twisted field plane.

(2) Furthermore, let the components of π of (1) be written as

$$x = 0, y = xm - cx^\alpha m^\beta, \ \forall m \in GF(p^r), \ \alpha, \beta \in \text{Gal } GF(p^r).$$

Let $\alpha = p^a$ and $\beta = p^b$ then $(r, a, b) = (r, a)$ and $GF(q) \subseteq \text{Fix } \alpha \cap \text{Fix } \beta = K$.

(3) Conversely, any generalized twisted field plane with spread given as in (2) where $(r, a, b) = (r, a)$ and $GF(q')$ any subfield of $K = \text{Fix } \alpha \cap \text{Fix } \beta$ admits a q'-regulus hyperbolic cover and a collineation group acting transitively on the reguli of the cover.

Hence, we obtain the following characterization of transitive semifield flat flocks mentioned in the introduction.

COROLLARY 77.35. *Let $\mathcal{F}$ be a transitive semifield flat flock in $S_{n,n}$ over $GF(q)$, $q \neq 2^6$.*

(1) Then $\mathcal{F}$ is a generalized twisted field flat flock. Furthermore, it also follows that $\mathcal{F}$ has a collineation group of order $(q^{n+1} - 1)/(q - 1)$ acting transitively on the caps.

(2) For any subfield $GF(q')$ of $GF(q)$, let $q = p^s$ and $q' = p^{s/t}$. Then, there is an associated transitive semifield flat flock $\mathcal{F}^{s/t}$ in $S_{nt+t-1,nt+t-1}$ over $GF(q')$.

(3) In either situation (1) or (2), whenever there is a transitive group, there is an associated cyclic group of order $(q'^{(nt+t)} - 1)/(q' - 1)$ acting transitively on the caps.

The study of finite semifield planes of order q^n reveals that, contrary to the case of $n = 2$, such planes provide and correspond to a rich and varied set of semifield flat flocks of Segre varieties. We have seen that the cyclic semifields provide a wealth of examples. Also, there are a variety of other regulus hyperbolic covers in semifield planes that we have not particularly considered. For example, Jha and Johnson [634] have shown that there are Hughes–Kleinfeld semifield planes of order q^4 with center $GF(q)$ that admit such regulus hyperbolic covers. Thus, there are corresponding semifield flat flocks in $S_{3,3}$ over $GF(q)$. Furthermore, Jha and Johnson prove that any semifield of order p^{2t}, t odd, with left, right and middle seminuclei all of order p^2 produce semifield flat flocks. In addition, certain generalized twisted field planes produce the interesting 'transitive semifield flat flocks'.

But, perhaps the most dramatic statement that can be made is the following which shows the complete ubiquity of semifield flat flocks located by associated semifields.

THEOREM 77.36. *Any non-Desarguesian semifield plane of order p^{n+1} has subright, submiddle and subleft seminuclei isomorphic to $GF(p)$ in its center. Hence, there are 'semifield' flat flocks of $S_{n,n}$ over $GF(p)$.*

So, we must conclude that although it is of general interest to connect somewhat disparate geometries, we see that clearly it would be next to impossible to attempt a complete classification of the flat flocks analogous to the classification of the flocks of hyperbolic quadrics in $PG(3,q)$. These flocks, however, can be recognized within the projective space, as Bader, Cossidente, and Lunardon [**41**, Theorem (4.3)] have characterized semifield flat flocks by a configuration property in the ambient space. On the other hand, we have made some progress in classifying these objects by showing that any transitive semifield flat flock can be completely determined.

Considering transitive flat flocks, we have noticed that, in addition to semifield flat flocks, there are certain 'transitive' flat flocks corresponding to $j \ldots j$-planes. In recent work on $j \ldots j$-planes, there is some evidence to indicate that at least for Abelian transitive groups, there is some hope of a complete classification. In general, it may be possible to completely determine the transitive flat flocks by analysis of translation planes. We mention the following open problem which speaks to this.

PROBLEM 77.37. Show that any finite translation plane admitting a collineation group fixing two components and transitive on the remaining components is either a $j \ldots j$-plane or admits a regulus hyperbolic cover when the spread is considered over a specified field.

77.4. André Flat Flocks.

In this section, a class of André planes are shown to produce flat flocks. This section follows the work of Jha and Johnson [**635**]. The basic question then is:

Is there a class of translation planes of order q^n and kernel containing $GF(q)$ such that each plane of this class corresponds to and produces a flat flock of a Segre variety $S_{n-1,n-1}$?

Since we will be guided by such considerations in André planes, we ask if there is a fixed order q^n such that every André of this order produces a flat flock of a Segre variety?

Jha and Johnson [**635**] show that flat flocks are ubiquitous in André planes of order $q^{t(q-1)}$. That is, every such André plane in this class produces a flat flock.

THEOREM 77.38. Any André spread of order $q^{t(q-1)}$ and kernel containing $GF(q)$ admits a regulus hyperbolic cover.

COROLLARY 77.39. Any André spread of order $q^{t(q-1)}$ and kernel containing $GF(q)$ produces a flat flock of a Segre variety $S_{t(q-1)-1,t(q-1)-1}$ over $GF(q)$.

We now consider the converse. We first establish that the 'carriers' of a regulus hyperbolic cover are normally uniquely defined in generalized André planes.

THEOREM 77.40. Let π be a non-Desarguesian and non-Hall generalized André plane of finite order q^n. If $q > 2$ and π admits a regulus hyperbolic cover then the set of two common components is uniquely defined.

Now assume that an André spread of order q^n and kernel containing $GF(q)$ corresponds to a flat flock of a Segre variety. Assume that the plane is not Desarguesian or Hall. Then, there is a homology group of order $q - 1$ as above, and $q - 1 > 2$ by assumption. Furthermore, there is a homology group of order $(q^n - 1)/(q - 1)$. By the previous theorem, we may assume that the axes and centers of both homology groups are the same (note there are 'symmetric' homology groups of both orders).

Assume that every André plane π of this order has a regulus-inducing homology group. Then, there exists an André spread where exactly one André partial spread is replaced. Since the plane is not Hall, then $n > 2$. There are $q^n + 1 - (q^n - 1)/(q - 1)$ remaining components of π. Suppose that this replaced partial spread is moved by an element σ of the regulus-inducing homology group of order $q - 1$. Since the components of π are either of the form $y = x^{q^i} m$ or $y = xn$, it follows that the regulus-inducing group must leave invariant the set of all components of the general form $y = x^{q^i} m$; that is this group must leave the replaced net invariant. Since this affine homology group of order $q - 1$ must leave this net invariant then $q - 1$ divides $(q^n - 1)/(q - 1) = 1 + n - 1 + \sum_{j=1}^{n-1}(q^j - 1)$, implying that $q - 1$ divides n. Hence, we obtain the following result:

THEOREM 77.41. *Every André plane of the class of André planes of order* q^n *and kernel containing* $GF(q)$ *corresponds to a flat flock of a Segre variety in* $S_{n-1,n-1}$ *over* $GF(q)$ *if and only if* $q - 1$ *divides* n.

REMARK 77.42. It is possible to determine exactly the number of isomorphic planes within the class of André planes of order q^n and kernel containing $GF(q)$. Roughly speaking, we have $(q - 1)$ choices for δ and n choices for the nets A_δ^i. Hence, potentially, there are $(q - 1)^n$ possible planes. Thus, as a rough count, there are approximately $(q - 1)^{t(q-1)}$ corresponding flat flocks of Segre varieties $S_{t(q-1)-1,t(q-1)-1}$.

77.5. Generalized André Planes and Flat Flocks.

We now vary the subfield required in the construction of André nets and construct a tremendous number of generalized André planes of order q^{q-1} and kernel containing $GF(q)$ that produce flocks of Segre varieties.

The idea is to construct André nets of degree $(q^{t(q-1)} - 1)/(q^z - 1)$ with kernel containing $GF(q)$ for a set of subfields $GF(q^z)$ of $GF(q^{q-1})$ that admit the homology group of order $q - 1$. We require that $(q - 1)$ divides $(q^{t(q-1)} - 1)/(q^z - 1)$. Since $(q - 1)$ divides $(q^{t(q-1)} - 1)/(q - 1)$. But, this requirement is satisfied provided z divides t.

Hence, take a subset of fields $\{ GF(q^{t_i}); t_i \mid t \text{ for } i = 1, 2, \ldots, k \}$. For each such field, consider an André net A_{δ_j} where $\delta_j \in GF(q^{t_j})$ and choose a replacement net $A_{\delta_i}^{i(\delta_j)}$. Assume that the André nets chosen are mutually disjoint. Now if we replace by $\bigcup_{j=1}^{k} A_{\delta_j}^{i(\delta_j)}$, we obtain a generalized André plane that admits the homology group of order $q - 1$ mentioned previously; the regulus-inducing homology group. Hence, we obtain generalized André planes that are not necessarily André planes whose spreads are unions of $GF(q)$-reguli sharing two components. We formalize this as follows.

THEOREM 77.43. *Let* π *be an André plane of order* $q^{t(q-1)}$ *and kernel containing* $GF(q)$. *Using a set* $\{ GF(q^{t_i}); t_i \mid t \text{ for } i = 1, 2, \ldots, k \}$, *choose* k *mutually disjoint André nets* A_{δ_j} *corresponding to the subfields* $GF(q^{t_i})$ *where* $\delta_i \in GF(q^{t_i})$. *Then, choose any replacement net* $A_{\delta_j}^{i(\delta_j)}$ *where* $i(\delta_j)$ *is an integer between* 0 *and* $t_i - 1$. *Let* $\mathcal{N}$ *denote the net defined by the components of* π *not in one of the chosen André nets.*

Then $\bigcup_{j=1}^{k} A_{\delta_j}^{i(\delta_j)} \cup \mathcal{N}$ *is a generalized André plane and corresponds to a flat flock of a Segre variety in* $S_{q-2,q-2}$ *over* $GF(q)$.

REMARK 77.44. Johnson [**760**] constructs non-André replacements of André nets, thereby constructing generalized André planes. Any of these with the restrictions of the previous theorem will produce another class of generalized André planes that correspond to flat flocks.

CHAPTER 78

$j \ldots j$-Planes.

The 'geometry' of flocks of quadratic cones has now reached into many diverse areas of incidence geometry. For example, it is know that if there is a translation plane of order q^2 with spread in $PG(3, q)$ that admits a Baer group of order q (fixes a Baer subplane pointwise) there is a corresponding flock of a quadratic cone. In Johnson [**726**], it was shown that the $q - 1$ orbits of length q of the Baer group on the components of the spread define reguli that share the pointwise fixed subspace, which, in turn, defines a partial flock of deficiency one of a quadratic cone. Payne and Thas [**1094**] then show that any deficiency-one partial flock may always be extended to a flock of a quadratic cone. This means that the net of degree $q + 1$ defined by the components of the Baer subplane is a regulus net and by derivation of this net, there is an associated translation plane with spread in $PG(3, q)$ where the Baer group now becomes an affine elation group. We call such elation groups 'regulus-inducing'.

Hence, translation planes admitting regulus-inducing elation groups are equivalent to flocks of quadratic cones. However, using the fundamental analysis of Baker, Ebert and Penttila [**78**], it has now been shown in Johnson [**769**] that it is always possible to connect flocks of quadratic cones with translation planes admitting cyclic homology groups.

THEOREM 78.1. (Johnson [**769**]). The set of translation planes of order q^2 with spread in $PG(3, q)$ that admit cyclic affine homology groups of order $q + 1$ are equivalent to the set of flocks of a quadratic cone.

So, there are intrinsic connections and a complete equivalence between the set of translation planes q^2 admitting regulus-inducing elation groups and with the set of translation planes of order q^2 admitting cyclic homology groups of order $q + 1$.

An important class of translation planes of order q^2 admitting cyclic homology groups of order $q + 1$ are the j-planes that cyclic collineation groups of order $q^2 - 1$, within which there is an affine cyclic homology group of order $q + 1$. Another equally important subset is the class of planes obtained from a Desarguesian plane by the replacement of a $(q + 1)$-nest of reguli. These planes admit a collineation group of order $(q + 1)(q^2 - 1)/2$ that contains two distinct cyclic affine homology groups of order $q + 1$.

One of the questions that we are concerned with in this section is how the j-planes and $(q + 1)$-nest planes are related to flocks of quadratic cones and, more generally, how the collineation group of such translation planes relates to the collineation group of the associated flock or flocks.

In Baker, Ebert, and Penttila [**78**], the connection with a flock is not particularly with a given translation plane but with a set of translation planes connection between what are called 'regular hyperbolic fibrations with constant back half.'

A 'hyperbolic fibration' is a set $\mathcal{Q}$ of $q-1$ hyperbolic quadrics and two carrying lines L and M such that the union $L \cup M \cup \mathcal{Q}$ is a cover of the points of $PG(3,q)$. The term 'regular hyperbolic fibration' is used to describe hyperbolic fibrations such that for each of its $q-1$ quadrics, the induced polarity interchanges L and M. When this occurs, and (x_1, x_2, y_1, y_2) represent points homogeneously, the hyperbolic quadrics have the form $V(x_1^2 a_i + x_1 x_2 b_i + x_2^2 c_i + y_1^2 c_i \mid y_1 y_2 f_i + y_2 g_i)$, for $i = 1, 2, \ldots, q-1$ (the variety defined by the quadrics). When $(e_i, f_i, g_i) = (e, f, g)$ for all $i = 1, 2, \ldots, q-1$. The regular hyperbolic quadric is said to have 'constant back half'.

We recall the principal theorem of Baker, Ebert, and Penttila [**78**].

THEOREM 78.2. (Baker, Ebert, and Penttila [**78**]).

(1) Let $\mathcal{H} := V(x_1^2 a_i + x_1 x_2 b_i + x_2^2 c_i + y_1^2 e + y_1 y_2 f + y_2 g)$, for $i = 1, 2, \ldots, q-1$ be a regular hyperbolic fibration with constant back half.

Consider $PG(3,q)$ as (x_1, x_2, x_3, x_4) and let C denote the quadratic cone with equation $x_1 x_2 = x_3^2$.

Define

$$\pi_0 : x_4 = 0, \ \pi_i : x_1 a_i + x_2 c_i + x_3 b_i + x_4 = 0, \text{ for } 1, 2, \ldots, q-1.$$

Then

$$\{\pi_j, j = 0, 1, 2, \ldots, q-1\},$$

is a flock of the quadratic cone C.

(2) Conversely, if $\mathcal{F}$ is a flock of a quadratic cone, choose a representation as $\{\pi_j, j = 0, 1, 2, \ldots, q-1\}$ above. Then, choosing any convenient constant back half (e, f, g), and defining $\mathcal{H}$ as $V(x_1^2 a_i + x_1 x_2 b_i + x_2^2 c_i + y_1^2 e + y_1 y_2 f + y_2 g)$, for $i = 1, 2, \ldots, q-1$, then $\mathcal{H}$ is a regular hyperbolic fibration with constant back half.

Now for each of the $q-1$ reguli, it is then possible to select one of the two reguli of totally isotropic lines. Such a choice will produce a spread and a translation plane. Hence, there are perhaps 2^{q-1} possible translation planes obtained in this way.

Here, we show how to sift such translation planes into isomorphism classes.

But all of this works for infinite translation planes as well. In particular, it is possible to construct infinite flocks of quadratic cones and translation planes admitting regulus-inducing elation groups. Moreover, there are infinite versions of j-planes and 'nest' planes.

78.1. Hyperbolic Fibrations Revisited.

In order to understand the method of going back and forth between flocks of quadratic cones and translation planes admitting affine homology groups, we remind the reader of the constructions of Johnson [**769**] by revisiting the various results.

LEMMA 78.3. (1) Let $\mathcal{H}$ be a hyperbolic fibration of $PG(3,q)$ (a covering of the points by a set λ of hyperbolic quadrics union two disjoint carrying lines. For each quadric in λ, choose one of the two reguli (a regulus or its opposite). The union of these reguli and the carrying lines form a spread in $PG(3,q)$.

(2) Conversely, any spread in $PG(3,q)$ that is a union of hyperbolic quadrics union two disjoint carrying lines produces a hyperbolic fibration.

LEMMA 78.4. *Let π be a translation plane with spread in $PG(3, K)$, for K a field isomorphic to $GF(q)$, that admits a cyclic affine homology group H. Let Γ be any H-orbit of components.*

(1) *Then there is a unique Desarguesian spread Σ containing Γ and the axis and coaxis of H.*

(2) *Furthermore, we may represent the coaxis, axis and Γ as follows:*

$$x = 0, y = 0, y = xm; m^{q+1} = 1; m \in K^+$$

where m is in the field K^+, a 2-dimensional quadratic extension of K, so K^+ is isomorphic to $GF(q^2)$.

(3) *A basis may be chosen so that Σ may be coordinatized by K^+ in the form*
$\begin{bmatrix} u & t \\ ft & u + gt \end{bmatrix}$, *for all u, t in K, for suitable constants f and g.*

(4) *If $\{1, e\}$ is a basis for K^+ over K then $e^2 = eg + f$, and $e^\sigma = -e + g$, $e^{\sigma+1} = -f$. Furthermore, $(et + u)^{q+1} = 1$ if and only in matrix form $et + u = \begin{bmatrix} u & t \\ ft & u + gt \end{bmatrix}$, such that $u(u + gt) - ft^2 = 1$.*

(5) *The opposite regulus*

$$y = x^q m; m^{q+1} = 1,$$

may be written in the form

$$y = x \begin{bmatrix} 1 & 0 \\ g & -1 \end{bmatrix} \begin{bmatrix} u & t \\ ft & u + gt \end{bmatrix}; u(u + gt) - ft^2 = 1.$$

LEMMA 78.5. *The spread for π has the following form:*

$$x = 0, y = 0, y = xM_i \begin{bmatrix} u & t \\ ft & u + gft \end{bmatrix}; u(u + gt) - ft^2 = 1$$

and M_i a set of 2×2 matrices over K, where $i \in \rho$, some index set. Let

$$R_i = \left\{ y = xM_iT; T^{q+1} = 1 \right\}, \text{ for } i \in \rho.$$

Then R_i is a regulus in $PG(3, K)$.

LEMMA 78.6. *The quadratic form for R_i is*

$$V \left(xM_i \begin{bmatrix} 1 & g \\ 0 & -f \end{bmatrix} M_i^t x^t - y \begin{bmatrix} 1 & g \\ 0 & -f \end{bmatrix} y^t \right).$$

The term $xM_i \begin{bmatrix} 1 & g \\ 0 & -f \end{bmatrix} M_i^t x^t$ is self-transpose and

thus equal to $xM_i \begin{bmatrix} 1 & g \\ 0 & -f \end{bmatrix}^t M_i^t x^t$.

THEOREM 78.7. (Johnson [**769**]). *Let π be a translation plane with spread in $PG(3, K)$, for K a field. Assume that π admits an affine homology group H so that some orbit of components is a regulus in $PG(3, K)$.*

(1) *Then π produces a regular hyperbolic fibration with constant back half.*

(2) *Conversely, each translation plane obtained from a regular hyperbolic fibration with constant back half admits an affine homology group H, one orbit of which is a regulus in $PG(2, K)$.*

H is isomorphic to a subgroup of the collineation group of a Pappian spread Σ, coordinatized by a quadratic extension field K^+,

$$H \simeq \left\langle g^{\sigma+1}; g \in K^+ - \{0\} \right\rangle,$$

where σ is the unique involution in $\mathrm{Gal}_K K^+$.

(3) Let $\mathcal{H}$ be a regular hyperbolic fibration with constant back half of $PG(3, K)$. The subgroup of $\Gamma L(4, K)$ that fixes each hyperbolic quadric of a regular hyperbolic fibration $\mathcal{H}$ and acts trivially on the front half is isomorphic to

$$\left\langle \rho, \left\langle g^{\sigma+1}; g \in K^+ - \{0\} \right\rangle \right\rangle,$$

where ρ is defined as follows: If $e^2 = ef + g$, f, g in K and $\langle e, 1 \rangle_K = K^+$ then ρ is $\begin{bmatrix} I & 0 \\ 0 & P \end{bmatrix}$, where $P = \begin{bmatrix} 1 & 0 \\ g & -1 \end{bmatrix}$.

In particular, $\left\langle g^{\sigma+1}; g \in K^+ - \{0\} \right\rangle$ fixes each regulus and opposite regulus of each hyperbolic quadric of $\mathcal{H}$ and ρ inverts each regulus and opposite regulus of each hyperbolic quadric.

78.1.1. General Matrix Forms. If we consider the arbitrary case of spreads in $PG(3, K)$, the connection between translation planes admitting appropriate affine homology groups and flocks of quadratic cones is not direct when K is infinite. We recall the pertinent theorem of Johnson [**769**].

THEOREM 78.8. (Johnson [**769**]). A regular hyperbolic fibration with constant back half in $PG(3, K)$, K a field, with carrier lines $x = 0, y = 0$, may be represented as follows:

$$V\left(x \begin{bmatrix} \delta & g(\delta) \\ 0 & -f(\delta) \end{bmatrix} x^t - y \begin{bmatrix} 1 & g \\ 0 & -f \end{bmatrix} y^t\right)$$

$$\text{for all } \delta \text{ in } \left\{ \begin{bmatrix} u & t \\ ft & u+gt \end{bmatrix} ; u, t \in K, (u,t) \neq (0,0) \right]^{\sigma+1} \right\},$$

where

$$\left\{ \begin{bmatrix} \delta & g(\delta) \\ 0 & -f(\delta) \end{bmatrix} ; \delta \in \left\{ \begin{bmatrix} u & t \\ ft & u+gt \end{bmatrix}^{\sigma+1} ; u, t \in K, (u,t) \neq (0,0) \right\} \right\}$$

$$\cup \left\{ \begin{bmatrix} 0 & 0 \\ 0 & 0 \end{bmatrix} \right\}$$

corresponds to a partial flock of a quadratic cone in $PG(3, K)$, and where f and g are functions on $\det K^+$.

THEOREM 78.9. (Johnson [**769**]). The correspondence between any spread π in $PG(3, K)$ corresponding to the hyperbolic fibration and the partial flock of a quadratic cone in $PG(3, K)$ is as follows:

If π is

$$x = 0, y = 0, y = x \begin{bmatrix} u & t \\ F(u,t) & G(u,t) \end{bmatrix}$$

then the partial flock is given by $\begin{bmatrix} \delta_{u,t} & g(\delta_{u,t}) \\ 0 & -f(\delta_{u,t}) \end{bmatrix}$ with

$$\delta_{u,t} = \det \begin{bmatrix} u & t \\ ft & u+gt \end{bmatrix},$$

$$g(\delta_{u,t}) = g(uG(u,t) + tF(u,t)) + 2(uF(u,t) - tfG(u,t)),$$

$$-f(\delta_{u,t}) = \delta_{F(u,t),G(u,t)},$$

where

$$\delta_{M_i} = \det M_i$$

and

$$\delta_{F(u,t),G(u,t)} = \det \begin{bmatrix} F(u,t) & G(u,t) \\ fG(u,t) & F(u,t) + gG(u,t) \end{bmatrix} \in \det K^+.$$

THEOREM 78.10. (Johnson [**769**]). If we have a hyperbolic fibration in $PG(3,K)$, there are corresponding functions given in the previous theorem such that the corresponding functions

$$\phi_s(t) = s^2 t + sg(t) - f(t)$$

are injective for all s in K and for all $t \in \det K^+$.

Indeed, the functions restricted to $\det K^+$ are surjective on $\det K^+$.

Furthermore, we obtain:

THEOREM 78.11. (Johnson [**769**]). Any partial flock of a quadratic cone in $PG(3,K)$, with defining set λ (i.e., so t ranges over λ and planes of the partial flock are defined via functions in t) equal to $\det K^+$, whose associated functions on $\det K^+$, as above, are surjective on $\det K^+$ (K^+ some quadratic extension of K), produces a regular hyperbolic fibration in $PG(3,K)$ with constant back half.

78.2. j-Planes.

The concept of a j-plane grew out of the work of Figueroa [**356**] and the realization that certain slices of ovoids produce such j-planes.

In Johnson, Pomareda, and Wilke [**806**], j-planes are constructed and developed in the finite case. Here we wish to consider this in a more general manner and over any $PG(4,K)$, where K is a field.

DEFINITION 78.12. Let K be a field and K^+ a quadratic field extension of K represented as follows:

$$K^+ = \left\{ \begin{bmatrix} u & t \\ ft & u+gt \end{bmatrix} ; u,t \in K \right\}.$$

Consider the following group:

$$G_{K^+,j} = \left\{ \begin{bmatrix} 1 & 0 & 0 & 0 \\ 0 & \delta_{u,t}^{-j} & 0 & 0 \\ 0 & 0 & u & t \\ 0 & 0 & ft & u+gt \end{bmatrix} ; u,t \in K, (u,t) \neq (0,0) \right\},$$

where j is a fixed integer and $\delta_{u,t} = \det \begin{bmatrix} u & t \\ ft & u+gt \end{bmatrix}$.

A 'j-plane' is any translation plane π containing $x=0, y=0$ and $y=x$ that admits $G_{K,j}^+$ as a collineation group acting transitively on the components of $\pi - \{x=0, y=0\}$.

PROPOSITION 78.13. A *j*-plane produces a regular hyperbolic fibration with constant back half and hence a corresponding det K^+-partial flock of a quadratic cone in $PG(3, K)$.

THEOREM 78.14. (Johnson [655]). A *j*-plane with spread set

$$\left\{ x = 0, y = 0, y = x \begin{bmatrix} 1 & 0 \\ 0 & \delta_{u,t}^j \end{bmatrix} \begin{bmatrix} u & t \\ ft & u + gt \end{bmatrix} \right\}; u, t \in K,$$

$$(u, t) \neq (0, 0), K \text{ a field} \Bigg\},$$

produces a monomial det K^+ partial flocks of a quadratic cone with monomial functions

$$f(\delta_{u,t}) = f\delta_{u,t}^{2j+1}, \; g(\delta_{u,t}) = g\delta_{u,t}^{j+1}.$$

Now all even-order monomial flocks have been determined by Penttila and Storme [1100]. Hence, we have the following corollaries.

78.2.1. Penttila, Storme / Johnson Classification *j*-Planes.

THEOREM 78.15. (Johnson [655]). The set of *j*-planes with spread in $PG(3, q)$ is equivalent to the set of monomial conical flocks in $PG(3, q)$.

When q is even the *j*-planes are completely arising from the monomial flocks with functions

$$f(z) = fz^{2j+1}, \; g(z) = gz^{j+1}, \text{ for } j = 0, 1, 2.$$

More than twenty years ago, T.G. Ostrom asked N.L. Johnson whether there exist non-André translation planes of order q^n that admit an affine homology group of order $(q^n - 1)/(q - 1)$. The André planes of order q^n admit two affine homology groups of order $(q^n - 1)/(q - 1)$ that fix a pair of components L and M such that one group has axis L and coaxis M and the remaining group has axis M and coaxis L. If the groups are cyclic, planes with two homology groups of such order may be characterized. We call such homology groups as above to be symmetric to each other. When $n = 2$, Johnson [765] recently described the planes with spread in $PG(3, q)$ that admit such groups.

THEOREM 78.16. (Johnson [765]). Let π be a translation plane with spread in $PG(3, q)$ that admits at least two homology groups of order $q + 1$. Then one of the following occurs:

(1) $q \in \{5, 7, 11, 19, 23\}$, (the irregular nearfield planes and the exceptional Lüneburg planes are examples),

(2) π is André,

(3) q is odd and π is constructed from a Desarguesian spread by $(q + 1)$-nest replacement (actually $q = 5$ or 7 for the irregular nearfield planes also occur here),

(4) q is odd and π is constructed from a Desarguesian spread by a combination of $(q + 1)$-nest and André net-replacement,

(5) $q \equiv -1 \bmod 4$ and the axis/coaxis pair is invariant under the full collineation group. Furthermore, there is a non-cyclic homology group of order $q + 1$,

(6) $q = 7$ and the plane is the Heimbeck plane of type III with 10 homology axes of quaternion groups of order 8.

Furthermore, Johnson has shown that every translation plane of order q^2 with spread in $PG(3, q)$ that admits a cyclic homology group produces a flock of a quadratic cone.

THEOREM 78.17. (Johnson [**769**]). The set of translation planes of order q^2 with spread in $PG(3, q)$ that admit cyclic affine homology groups of order $q + 1$ is equivalent to the set of flocks of a quadratic cone.

Therefore, we may really consider Ostrom's question for planes of order q^n, $n > 2$. When there are two cyclic symmetric affine homology groups of order $(q^n - 1)/(q - 1)$, the following result shows that we really are dealing only with the André planes.

THEOREM 78.18. (Johnson and Pomareda [**794**]). Let π be a translation plane of order q^n that admits symmetric cyclic affine homology groups of orders $(q^n - 1)/(q - 1)$, $n > 2$.
Then the plane π is André.

So, the problem posed above can be reduced to asking if one affine homology group of order $(q^n - 1)/(q - 1)$ is sufficient to classify such planes.

Again, when $n = 2$, since every flock of a quadratic cone gives at least one translation plane with the required homology group, there are tremendous varieties of such translation planes. In particular, the so-called '*j*-planes' of order q^2 admit a cyclic collineation group of order $q^2 - 1$, of which there is an affine homology subgroup of order $q + 1$. Hence, in particular, *j*-planes correspond to flocks of quadratic cones (in fact, *j*-planes correspond to monomial flocks).

So, the question now turns to whether there are non-André planes of order q^n admitting an affine homology group of order $(q^n - 1)/(q - 1)$, for $n > 2$. The connection with *j*-planes and cyclic homology groups of order $q + 1$ in translation planes of order q^2 and then with corresponding flocks of quadratic cones ultimately depends upon the partition of $PG(3, q)$ into a set of $(q - 1)$ mutually disjoint hyperbolic quadrics unioned with two carrying lines. When considering whether there are non-André planes admitting such large groups, we note the following:

THEOREM 78.19. Let π be a translation plane of order q^n that admits a cyclic affine homology group H of order $(q^n - 1)/(q - 1)$. Then any component orbit union the axis and coaxis of the group is a Desarguesian partial spread.

COROLLARY 78.20. Let π be a translation of order q^n and kernel containing $GF(q)$ that admits a cyclic affine homology group H of order $(q^n - 1)/(q - 1)$ and let S^π denote the spread for π. Then S^π is the union of a set of $(q - 1)$ André nets A_i in Desarguesian spreads Σ_i, $i = 1, 2, \ldots, q - 1$ union the axis and coaxis of H.
Furthermore, any such André net has at least $n - 1$ André replacements.

When $n = 2$, André nets in this context are reguli. The connection with flocks of quadratic cones and translation planes of order q^2 with spreads in $PG(3, q)$ admitting cyclic homology groups of order $q + 1$ is made due to 'hyperbolic fibrations'; a covering of $PG(3, q)$ by a set of $(q - 1)$ hyperbolic quadrics union two carrying lines. Noticing the similarity with the content of the above corollary, we formulate the following definition. Although there is a projective definition, we prefer the vector-space version.

DEFINITION 78.21. Let V_{2n} be a $2n$-dimensional $GF(q)$-vector space. A 'hyper-regulus' is a partial spread of order q^n and degree $(q^n - 1)/(q - 1)$ of n-dimensional $GF(q)$-subspaces that has a replacement partial spread of the same degree such that each component of the replacement set intersects each component of the original partial spread in a 1-dimensional $GF(q)$-subspace.

DEFINITION 78.22. A 'hyperbolic fibration of dimension n' is a partition of V_{2n} into $(q - 1)$ hyper-reguli union two carrying lines.

Hence, we see that any translation of order q^n and kernel $GF(q)$ admitting a cyclic homology group of order $(q^n - 1)/(q - 1)$ produces a hyperbolic fibration of dimension n. Recently, Culbert and Ebert [**267**], point out the possibility of extending the nature of hyperbolic fibrations to correspond to a situation such as described in the above corollary and notes that so far there are no known examples of such generalizations.

About 1992, in unpublished work, N.L. Johnson constructed an affine plane of order 4^3 and kernel $GF(4)$ admitting an affine homology group of order $(4^3 - 1)/(4-1)$, which, in fact, is not André. F.W. Wilke was able to use the computer to construct a large set of such planes; however, no classification was attempted at that time. The present work extends both of the previously mentioned constructions and is part of O. Vega's Ph.D. thesis at the University of Iowa [**1194**].

The idea of the basic constructions involves extending the definition of j-planes for planes of order q^2, to $j \ldots j$-planes for planes of order q^n. When $n = 3$, we call these 'jj-planes', or 'j, k-planes', to fix the notation.

DEFINITION 78.23. Let K be a matrix field over $GF(q) \simeq GF(q^n)$, for $a = p^r$, p a prime. Define a cyclic group as follows:

$$\left\langle \begin{bmatrix} D_M & 0 \\ 0 & M \end{bmatrix}; M \in K - \{0\} \right\rangle,$$

where D is a diagonal matrix with (i, i) entries δ^{ji}, for $i = 1, 2, \ldots, n$, and $\delta = \det M$. If

$$x = 0, y = 0, (y = x) \left\langle \begin{bmatrix} D_M & 0 \\ 0 & M \end{bmatrix}; M \in K - \{0\} \right\rangle$$

$$= (y = x D_M^{-1} M; M \in K^*)$$

is a spread then the associated translation plane with kernel containing $GF(q)$ is called a '$j \ldots j$-plane'.

It fact, Vega [**1194**] has shown that there are infinite classes of André $j \ldots j$-planes of various orders (the reader is directed to Vega [**1194**] for more details on $j \ldots j$-planes).

Using the computer, it is possible to completely determine all jj planes of order 4^3. As noted in the above corollary, we now have three André nets, each of which admit two André replacements. Hence, we obtain additional translation planes, each of which admits affine homology groups of order $(4^3 - 1)/(4 - 1)$. Of course, there is the unique nearfield plane of order 4^3 with kernel $GF(4)$, which also admits a cyclic homology group of order $(4^3 - 1)/(4 - 1)$.

Let K^* denote the kernel homology group of order $4 - 1 = 3$. Within CK^*, it is possible to find so-called 'regulus-inducing' homology groups of order 3; the axis and coaxis together with any component orbit of length 3 define a regulus in

$PG(5,4)$. Hence, we have a covering of the spread by a set of reguli that share two components. We note that Jha and Johnson [**640**] have shown that such spreads correspond to and produce flat flocks of Segre varieties. Hence, any such plane constructed that admits this homology group of order 3 produces a flat flock.

We also note that any homology group of order $(4^3-1)/(4-1)$ contains a cyclic affine homology group of order $8-1$, where the plane has order 8^2. Since any such orbit union the axis and coaxis is contained in a unique Desarguesian spread, it follows that a cyclic affine homology group of order $8-1$ corresponds to the cyclic homology group arising from $GF(8)$ in $GF(8^2)$. It then follows that any such orbit union the axis and coaxis defines a derivable net (this is a regulus in the $PG(3,8)$ wherein the unique Desarguesian spread lives). If any such net is derived we obtain a new translation plane that retains the group of order 7 but loses the group of order 3 and, in fact, the kernel of these derived planes becomes $GF(2)$.

78.2.2. The j,k-Planes of Johnson, Vega, and Wilke. The main results of Johnson, Vega, and Wilke [**813**] are as follows:

THEOREM 78.24. There are three isomorphism classes of j,k-planes of order 4^3.

One of these planes is a nearfield plane and the other two are new planes. All such planes have kernel $GF(4)$ and spreads in $PG(5,4)$.

Each j,k-plane admits a collineation group G of order 4^3-1 fixing two components and transitive on the remaining components.

Within G, there is an affine homology group of order $(4^3-1)/(4-1)$ producing three nets (André nets) of the same size that are replaceable by two distinct replacements.

THEOREM 78.25. Using the replacements listed in the theorem above, there are exactly two mutually non-isomorphic planes obtained by multiple André replacement, which are not j,k-planes. These two planes admit affine homology groups of order $(4^3-1)/(4-1)$ but not the larger group of order 4^3-1. These planes also have kernel $GF(4)$ and spreads in $PG(5,4)$.

THEOREM 78.26. Each of the four new translation planes listed in the two previous theorems admits a cyclic affine homology group of order 7. The component orbits union the axis and coaxis of the group define derivable nets.

Deriving such planes provides 'nine' mutually non-isomorphic and new translation planes. Each derived plane has kernel $GF(2)$ and spread in $PG(11,2)$.

THEOREM 78.27. Each j,k-plane and replaced j,k-plane whose spreads have kernel $GF(4)$ admit affine homology groups of order 3. The component orbits union the axis and coaxis are $GF(4)$-reguli. Hence, we have a 'regulus hyperbolic cover'. Each such regulus hyperbolic cover produces a flat flock of the Segre variety $S_{2,2}$ by Veronesians.

COROLLARY 78.28. Every j,k-plane and replaced j,k-plane of order 4^3 induces a flat flock.

REMARK 78.29. Note that the non-André j,k-planes induce new flat flocks. The André j,k-plane is one of the regular nearfield planes.

78.2.3. Vega's Planes of Small Orders. O. Vega [**1194**] has determined several other $j \ldots j$-planes of small orders. Vega has also shown that there are infinite classes of $j \ldots j$-planes corresponding to certain André planes, so it is probable that infinite classes of $j \ldots j$-planes that are not André might be found. In fact, Vega has determined various $j \ldots j$-translation planes of small orders $7^3, 3^4, 4^4, 5^4$. Various of these planes admit net replacement, so there are other constructible planes of these orders. Here we describe only the $j \ldots j$-planes. For the planes of order 7^3, we use a $GF(7)$-irreducible polynomial

$$x^3 - ax^2 - bx - c$$

and for the other planes we use an irreducible polynomial

$$x^4 - ax^3 - bx^2 - cx - d.$$

So, in order to designate the planes, we need an ordered set of integers, which represents the exponents j on the determinants δ^{-j} to the matrices M in the associated field, where the ordering is from the $(2,2)$-element j_2 to the $(3,3)$-element j_3, or $(4,4)$-element j_4 in the p^4-case.

THEOREM 78.30. (Vega [**1194**]). The following table shows the constructed $j \ldots j$-planes of orders 7^3:

(a, b, c)	(j_2, j_3)
$(0, 0, 5)$	$(0, 4)$
$(0, 0, 5)$	$(2, 4)$

Net replacement procedures produce many other new translation planes of order 7^3.

THEOREM 78.31. (Vega [**1194**]). The following table shows the constructed $j \ldots j$-planes of orders 3^4:

(a, b, c, d)	(j_2, j_3, j_4)
$(0, 2, 0, 1)$	$(1, 0, 1)$
$(0, 1, 0, 1)$	$(1, 0, 1)$

Since there are always homology groups in $j \ldots j$-planes of order q^d, we see that there is an affine homology group of order $(3^4 - 1)/2$.

THEOREM 78.32. (Vega [**1194**]). The following table shows the constructed $j \ldots j$-planes of orders 4^4: Let $\alpha^2 = \alpha + 1$ in the subfield of order 4:

(a, b, c, d)	(j_2, j_3, j_4)
$(1, 1, 0, \alpha)$	$(2, 0, 1)$
$(\alpha + 1, \alpha, 0, \alpha)$	$(2, 0, 1)$

Again, replacements produce a variety of new planes of order 4^4.

THEOREM 78.33. (Vega [**1194**]). The following table shows the constructed $j \ldots j$-planes of orders 5^4. One of the planes found by computer is André (the second plane listed), whereas the others are all new planes as are the replacements. The planes are as follows:

(a,b,c,d)	(j_2,j_3,j_4)
$(0,0,0,3)$	$(2,0,2)$
$(0,0,0,3)$	$(1,2,3)$
$(0,2,0,2)$	$(2,0,2)$
$(0,2,0,2)$	$(1,0,1)$
$(0,2,0,2)$	$(3,0,3)$
$(0,1,0,3)$	$(2,0,2)$
$(0,1,0,3)$	$(1,0,1)$
$(0,1,0,3)$	$(3,0,3)$

In addition to standard net replacement these planes are also derivable, so there are many new interesting planes of order 5^4.

78.3. One-Half j-Planes.

Consider a translation plane π of odd order q^2 with spread in $PG(3,q)$ that admits a collineation group of order q^2-1 in the linear translation complement that trivially intersects the kernel group of order $q-1$. Figueroa [356] shows that when G is Abelian then π is either Desarguesian, a nearfield plane or a j-plane. Furthermore, if G is not Abelian but a Sylow 2-subgroup S has a non-cyclic center, there is an interesting class of translation planes that might be called 'one half j-planes' in that there is a partial spread of degree $2+(q^2-1)/2$, which is defined exactly as in j-planes. In this setting, there is a cyclic collineation group C of order $(q^2-1)/2$ and a Baer collineation σ in G, which does not commute with C and $\langle C,\sigma\rangle = G$. We formulate Figueroa's theorem slightly differently than stated in [356].

THEOREM 78.34. (Figueroa [356]). Let π be a translation plane of order q^2 with spread in $PG(3,q)$ that admits a linear collineation group G of order q^2-1 that trivially intersects the kernel homology group K^* of order $q-1$.

(1) If G is Abelian and q is odd, then π is Desarguesian, a nearfield plane or a j-plane.

(2) If G is Abelian, q is even and G fixes two components, then π is Desarguesian or a j-plane.

(3) If a Sylow 2-subgroup of G has a non-cyclic center, then $q \equiv 1 \bmod 4$, G contains a Baer involution σ and a cyclic subgroup C of order $(q^2-1)/2$ such that $G = \langle C,\sigma\rangle$. There is a field

$$\mathcal{K} = \left\langle \begin{bmatrix} u & t \\ tf & u+tg \end{bmatrix} ; u,t \in GF(q) \right\rangle,$$

such that

$$C = \left\langle \begin{bmatrix} 1 & 0 & 0 & 0 \\ 0 & \delta^{-j} & 0 & 0 \\ 0 & 0 & u & t \\ 0 & 0 & tf & u+gt \end{bmatrix} ; \right.$$

$$\delta = \det \begin{bmatrix} u & t \\ tf & u+gt \end{bmatrix} ; \ u,t \in GF(q), \ (u,t) \neq (0,0),$$

$$\left. \begin{bmatrix} u & t \\ tf & u+gt \end{bmatrix} \text{ of order dividing } (q^2-1)/2 \right\rangle .$$

Then either π is Desarguesian or a nearfield or $i \neq (q-1)/2$. Then

$$\sigma \text{ is } (x,y) \rightarrow \left(x \begin{bmatrix} 1 & 0 \\ 0 & -1 \end{bmatrix}, y \begin{bmatrix} 1 & 0 \\ 0 & -1 \end{bmatrix} \right).$$

Let τ denote the affine homology $(x,y) \rightarrow (x,-y)$ and then the $q+1$ components

$$y = x \begin{bmatrix} 0 & u \\ u^{2j+1}w & 0 \end{bmatrix} ; \ u \in GF(q), \ x=0,$$

$$v_0 \text{ a non-zero constant} \neq t^{2(j+1)} \ \forall t \in GF(q)^*,$$

are fixed by the Baer involution $\tau\sigma$. Hence, the spread is

$$x = 0, y=0, y=x \begin{bmatrix} 1 & 0 \\ 0 & \delta^j \end{bmatrix} \begin{bmatrix} 0 & 1 \\ w & 0 \end{bmatrix}^z \begin{bmatrix} u & t \\ tf & u+gt \end{bmatrix} ;$$

$$\delta = \det \begin{bmatrix} u & t \\ tf & u+gt \end{bmatrix}, \forall \begin{bmatrix} u & t \\ tf & u+gt \end{bmatrix} \text{ of order dividing } (q^2-1)/2$$

$$z = 0 \text{ or } 1, w \text{ a constant}, x^2+gx-f \ GF(q)\text{-irreducible}.$$

DEFINITION 78.35. Any spread as defined in the previous theorem and is not a j-spread is called a 'one-half j-spread' (even possibly when q is not congruent to 1 mod 4). We note that one half j-spreads admit cyclic affine homology groups of orders $q-1$ and $(q+1)$ in GK^*, where K^* is the kernel homology group of π, which generate a cyclic group of order (q^2-1), one group has axis $y=0$ and coaxis $x=0$ and the other group has axis $x=0$ and coaxis $y=0$. This group, however, will contain a kernel involution. Note that it is possible that a one-half j-plane could also be a j^*-plane.

Note that since any one-half j-plane admits a cyclic affine homology group of order $q+1$ so there is an associated regular hyperbolic fibration and consequently associated conical flock planes. Furthermore, the associated flock will have $(q-1)/2$ $+1$ of its regulus nets determined by monomial functions. It is known that there cannot be two flocks that share $(q+1)/2$ of their conics, but it is possible that a flock can be 1/2-monomial without being monomial.

We recall that from Johnson [655] j-planes produce monomial flocks with functions $f(t) = ft^{2j+1}$, $g(t) = gt^{j+1}$, where $f \neq 0, g$ are constants in $GF(q)$. The known monomial flocks exist for $j = 0, 1, 2$ and for $j = (p^s-1)/2$, $p^r = q$, p odd (and $g = 0$). Also, we know that j-planes and $j + (q-1)/2$-planes (see below) will actually produce the same monomial flocks.

PROBLEM 78.36. If a flock of a quadratic cone is $1/2$-monomial, show that it is monomial.

The following constructions arising from work of Johnson, Pomareda, and Wilke [806] illustrate how one-half j-planes may be obtained from j-planes.

THEOREM 78.37. (Johnson, Pomareda, and Wilke [806]). Let π be a j-plane of odd order q^2, where the associated field corresponds to the irreducible $x^2 - f$, where f is a non-square in $GF(q)$. Then there is an associated cyclic affine homology group H whose component orbits define reguli in $PG(3, q)$. Let G denote the associated cyclic collineation group whose intersection with the kernel homology group K^* of order $q - 1$ is trivial. Let

$$G = \left\langle \begin{bmatrix} 1 & 0 & 0 & 0 \\ 0 & \delta^{-j} & 0 & 0 \\ 0 & 0 & u & t \\ 0 & 0 & tf & u \end{bmatrix}; u, t \in GF(q), \text{ not both zero} \right\rangle, \delta = \det \begin{bmatrix} u & t \\ tf & u \end{bmatrix}.$$

(1) Then there are two component orbits of length $(q^2 - 1)/2$ under the group G^2. Each orbit is a set of exactly $(q - 1)/2$ reguli.

(2) The Frobenius automorphism induces a collineation $\sigma : (x, y) \longmapsto (x^q, y^q) = \left(x \begin{bmatrix} 1 & 0 \\ 0 & -1 \end{bmatrix}, y \begin{bmatrix} 1 & 0 \\ 0 & -1 \end{bmatrix} \right)$ that fixes each regulus net.

(3) Multiply derive the $(q - 1)/2$ reguli in the orbit not containing $y = x$. This produces a one-half j-plane π^*admitting the collineation group

$$\langle \sigma, G^2 \rangle$$

of order $q^2 - 1$ that trivially intersects the kernel homology group K^*.

The spread for π^* is

$$\left\{ x = 0, y = x \begin{bmatrix} 1 & 0 \\ 0 & \delta^j \end{bmatrix} \begin{bmatrix} 1 & 0 \\ 0 & -1 \end{bmatrix}^\rho \begin{bmatrix} u & t \\ ft & u \end{bmatrix} \right\};$$

$$\rho = 1 \text{ or } 0, \text{ respectively as } \det \begin{bmatrix} u & t \\ tf & u \end{bmatrix} \text{ is non-square or square.}$$

Then $\delta^{(q-1)/2} = \pm 1$ and

$$\begin{bmatrix} 1 & 0 \\ 0 & \delta^j \end{bmatrix} \begin{bmatrix} 1 & 0 \\ 0 & -1 \end{bmatrix}^\rho \begin{bmatrix} u & t \\ ft & u \end{bmatrix} = \begin{bmatrix} 1 & 0 \\ 0 & \delta^{j+(q-1)/2} \end{bmatrix} \begin{bmatrix} u & t \\ ft & u \end{bmatrix},$$

for all $u, t \in GF(q)$, not both zero.

Hence, this one-half j-plane is also a $j + (q-1)/2$-plane.

PROOF. Note that $\begin{bmatrix} 1 & 0 \\ 0 & -1 \end{bmatrix} \begin{bmatrix} 0 & 1 \\ f & 0 \end{bmatrix} = \begin{bmatrix} 1 & 0 \\ -f & 0 \end{bmatrix}$, hence take $w = -f$ in Figueroa's theorem to see that we obtain a one-half j-plane. $\square$

78.3.1. Figueroa's One-Half j-Planes. Figueroa [356] constructed several one-half j-planes, some of which may be identified by the previous construction process. However, it is an open problem whether all such planes are $j + (q-1)/2$ planes.

In the previous representation, take $g = -2a$, so that $a^2 + f$ is a non-square n. Thus, the above note shows that if n^* is taken to be $-f$, it is possible to construct one-half j-planes from existing j-planes. There are known j-planes for $j = 0, 1, 2$, some of which produce one-half j-planes. Figueroa constructs one-half j-planes corresponding to the following table:

q	n	j	a	w
5	2	1	2	4
5	2	1	0	± 2
13	2	1	6	5
13	2	3	0	± 2
17	3	1	5	8
17	3	2	7	11
17	3	4	0	± 3
17	3	5	5	2
17	3	6	7	7
29	2	1	9	6
29	2	7	0	± 2
29	2	9	9	4
37	2	2	11	31
37	2	9	0	± 2
37	2	14	11	31
41	3	1	14	21
41	3	10	0	± 3
41	3	13	14	33

PROBLEM 78.38. Determine the flocks of quadratic cones corresponding to Figueroa's one-half j-planes. Noting that all flocks of order ≤ 29 are known, the question is if the associated flocks of orders 37 and 41 are known.

PROBLEM 78.39. Show that every one-half j-plane arises from a j-plane by multiple derivation.

CHAPTER 79

Orthogonal Spreads.

In this chapter, we give a reasonably complete synthesis of the methods of orthogonal geometry. These methods becomes increasingly important as ovoids in projective spaces and a variety of important construction techniques become available producing translation planes and spreads. Our treatment here provides more proofs than in other chapters in the text as we consider this fundamental and indispensable material. The reader is directed to Chapter 103 on Aspects of Symplectic and Orthogonal Geometry for background.

We shall consider (V, Q) quadratic forms defined on a vector space V over a field L. This means that the map: $(x, y) = Q(x + y) - Q(x) - Q(y)$ is a bilinear form on V. We shall work mainly with a concrete vector space equipped with a specific form; later we shall see that this does not lead to any loss of generality.

Let L be a finite field of characteristic 2 and let L^{4m} be the vector space of $4m$-tuples over L; we take the ordered basis of L^{4m} to be $(e_1, e_2, \ldots e_{2m}, f_1, f_2, \ldots f_{2m})$, and we usually write the general element $v \in L^{4m}$ in the form:

$$v = (x_1, x_2, \ldots, x_{2m}, y_1, y_2, \ldots, y_{2m}),$$

thus x_i is the coordinate associated with the basis element e_i and y_j is the coordinate associated with the basis element f_j.

Assign to L^{4m} the quadratic form $q \colon L^{4m} \to L$ given by:

$$q(x_1, x_2, \ldots, x_{2m}, y_1, y_2, \ldots, y_{2m}) = \sum_{i=1}^{2m} x_i y_i \qquad (= \text{'}x \cdot y\text{'}).$$

To verify that q is a quadratic form is equivalent to checking that its putative bilinear form

$$(z, z') = q(z + z') - q(z) - q(z'), \forall z, z' \in L^{4m}$$

is a genuine bilinear form. This is easily verified by writing the left-hand side in explicit form:

$$((x, y), (x', y')) = (x + x') \cdot (y + y') - x \cdot y - x' \cdot y'$$
$$= x \cdot y' + x' \cdot y,$$

which is bilinear.

The following consequences follow trivially.

REMARK 79.1. For $1 \leq i \leq j \leq 2m$:

(1) $q(e_i) = q(e_j) = 0$,
(2) $(e_i, e_j) = (f_i, f_j) = 0$,
(3) $(e_i, f_j) = \delta_{ij}$.

It follows that the matrix for the bilinear form $(\,\cdot\,,\,\cdot\,)$, associated with the quadratic form q, is

$$\begin{pmatrix} 0 & 1 \\ 1 & 0 \end{pmatrix}.$$

A vector $x \in V$ is called singular if $q(x) = 0$, and a subspace W of V is said to be totally singular if $q(W) = 0$. We now consider totally singular subspaces of q of rank $2m$; such subspaces are totally singular subspaces of maximum possible rank—the *generators* of the form q.

Using our customary notation, let

$$L^{4m} = L^{2m} \oplus L^{2m} = X \oplus Y,$$

where X and Y are the subspaces of V with bases $(e_i)_{i=1}^{2m}$ and $(f_i)_{i=1}^{2m}$, respectively. Thus, X and Y themselves are totally singular subspaces of rank $2m$, and we consider the general form of totally singular subspaces G of rank $2m$ that are disjoint from Y. A fortiori, G is a subspace of $X \oplus Y$ of form $y = xM$, where M is a non-singular matrix. So consider when $y = xM$ is totally singular relative to q. Recalling that $q(x \oplus y) := x \cdot y$, we see that $y = xM$ is totally singular iff the dot product $x \cdot (xM)$ vanishes identically for $x \in L^{2m}$ and this is equivalent to:

$$x(xM)^T = 0 \,\forall\, x \in L^{2m}$$
$$\Leftrightarrow \sum_{1 \le i,j \le 2m} x_i m_{ij} x_j = 0$$

and this is equivalent to the matrix M being alternating or *skew symmetric*—skew symmetric with all diagonal entries zero—in the terminology of [**862**]. Moreover, $y = xM$ is disjoint from X if and only if M is non-singular. Thus we have shown:

REMARK 79.2. The totally singular subspaces Z of $V = X \oplus Y$ that have the same rank as X and Y but are disjoint form Y are just the spaces of V of form $y = xM$, where M is either zero (yielding X) or a non-singular skew symmetric matrix.[1]

Thus, if M is skew symmetric, then $y = xM$ is a totally singular subspace of rank $2m$ consisting of $|L|^{2m} - 1$ non-zero singular vectors. It can be shown that altogether there are $(|L|^{2m-1}+1)(|L|^{2m}-1)$ non-zero singular vectors. Hence, there is a putative partial spread consisting of $|L|^{2m-1}+1$ components that partitions the totally singular subspaces of q. Such a partial spread will be called an *orthogonal spread*.

In the language of generators, an orthogonal spread is a partial spread consisting of pairwise skew generators that partition the totally singular points of the quadratic form q.

Thus, by Remark 79.2, a partial spread set consisting of $|L|^{2m-1}$ skew symmetric matrices (including the zero matrix) defines an orthogonal spread that includes X and Y among its components, and conversely any orthogonal spread that includes X and Y among its components is necessarily of this type. Partial spread sets of this type are called 'Kerdock sets'.

---

[1]We regard this as a synonym for an alternating matrix.

79.1. The Connection.

We now show that orthogonal spreads on an orthogonal space V yield 'ordinary' spreads on an ambient vector space Y, in the André sense, with a distinguishing feature: the components of the spread are maximal totally isotropic subspaces of the ambient space Y, when Y is equipped with some non-singular alternating bilinear form. Such spreads, and the associated translation planes, are called *symplectic*.

In this section, we demonstrate an essential equivalence between symplectic and orthogonal spreads, via a process called slicing. Starting from any symplectic or orthogonal spread, slicing and related operations tend to yield large quantities of non-isomorphic symplectic translation planes and orthogonal spreads.

Let y be a non-singular vector in V, so Ly is a rank one space whose non-zero vectors are all non-singular. Then the vector space $V_y = y^\perp/Ly$, whose rank is $4m - 2$, is a symplectic space relative to the induced alternating form:

$$(u + \lambda, v + \mu) = (u, v), \quad u, v \in y^\perp, \lambda, \mu \in Ly,$$

and Σ 'induces' a spread on V_y in the sense that

$$\Sigma_y := \left\{ \langle y^\perp \cap X, y \rangle / Ly \mid X \in \Sigma \right\}$$

is a spread on V_y. This spread is actually a *symplectic* spread in the sense that its components are totally isotropic subspaces, of rank $2m - 1$.

79.2. Slicing.

It has been obtained by *slicing* the orthogonal spread at y.

The procedure can actually be reversed: symplectic spreads can always be expressed in the form Σ_y for some orthogonal spread Σ, and Σ is essentially unique.

THEOREM 79.3 (Extension Theorem). Let Σ be a symplectic spread in a vector space of rank $4m-2$ vector space over a $GF(q)$ of even order. Then up to equivalence there is a unique orthogonal spread Σ^* in a vector space of rank $4m$ over $GF(q)$ such that Σ is equivalent to a slice of Σ^*.

PROOF. Identify the symplectic spread with a symplectic spread on some V_y. This identifies the components of the spread with totally singular subspaces of V_y that are pairwise disjoint and of rank $2m - 1$. Each of these subspaces extends to a totally singular subspace of rank $2m$ in one of two ways: all those extensions in the same ruling class are pairwise disjoint since they have rank at most one, and one is impossible for members of the same class. So we obtain an orthogonal spread in two ways. These are equivalent because the orthogonal transvection with axis $b^\perp$ interchanges the two classes of totally singular $2n$-spaces. [This will show the 'uniqueness'.]   □

79.3. Spreading.

The process of going from a symplectic spread to an orthogonal spread as in the above proof is called *spreading*.

We note that any totally isotropic subspace W, associated with a quadratic form, meets the associated quadric in at least a hyperplane of W.

LEMMA 79.4. Let W be a totally isotropic subspace associated with a finite $GF(q)$ vector space V, equipped with a quadratic form $Q: V \to GF(q)$. Then the set of totally singular elements in W:

$$S = \{\, w \in W \mid Q(w) = \mathbf{0} \,\},$$

is either a hyperplane of W or coincides with W.

PROOF. It is sufficient to show that every rank two subspace $T \le W$ meets S non-trivially. This holds because $Q: T \to GF(q)$ cannot be injective (by the size of T) and yet Q is additive on T, since T is totally isotropic.   □

We may thus speak of the *singular part* of any totally isotropic space T, and of the *singular hyperplane* of T when T itself is not totally isotropic.

The following proposition is the basis of the fundamental connection between symplectic and orthogonal spreads. It demonstrates a natural lattice-like connection between certain maximal totally singular subspaces of an orthogonal space (V, Q) and the maximal totally isotropic spaces of any symplectic space $y^\perp/y$, defined by a non-singular point y of V.

PROPOSITION 79.5. Let V be a vector space of rank $2n$ over $GF(q)$, q even, equipped with a non-singular quadratic form Q, thus:

$$Q(x + y) - Q(x) - Q(y) = \Lambda(x, y),$$

where Λ is a non-singular alternating form on V.

Let $Y = y^\perp$ be the hyperplane annihilating a non-singular vector y, and regard the quotient space $y^\perp/\langle y\rangle$ as a symplectic space, induced by Λ (and hence by Q) in the standard way. Let

$$\pi_y: y^\perp \to y^\perp/y$$

denote the quotient linear from y to $y^\perp$. Then the following hold.

(1) $y^\perp/\langle y\rangle$ is non-singular as a symplectic space.
(2) If F is a totally singular subspace of $y^\perp$ of rank $n-1$, then $(\langle y\rangle + F)/\langle y\rangle$ is a maximal totally isotropic subspace of $y^\perp/\langle y\rangle$.

   Conversely, every maximal totally isotropic subspace F' of $y^\perp/\langle y\rangle$ is of form $(\langle y\rangle + F)/\langle y\rangle$, where F is the singular hyperplane (of rank $n-1$) in $\pi_y{}^{-1}(F')$, the full preimage of F' relative to $\pi_y: y^\perp \to y^\perp/\langle y\rangle$. Thus F is the unique totally singular complement of $\langle y\rangle$ in $\pi_y{}^{-1}(F')$; this means F is the totally singular part of (the clearly totally isotropic space) $\pi_y{}^{-1}(F')$.
(3) Let F' and G' be maximal totally isotropic spaces of $y^\perp/y$. So $F' = \langle y\rangle + F$ and $G' = \langle y\rangle + G$, where F and G are the singular parts of F' and G': thus F and G are the unique totally singular spaces that are the complements of $\langle y\rangle$ in F and G, respectively. Then F' and G' are mutually skew iff F and G are mutually skew.
(4) Let F' be a maximal totally isotropic subspace of $y^\perp/\langle y\rangle$ and F the totally singular complement of $\langle y\rangle$ in $\pi_y{}^{-1}(F')$. Assume that $n = 2m$ is even. Then F lies in exactly two maximal totally singular subspaces, F_1 and F_2, of V. Thus F_1 and F_2 are rank n subspaces that share F as a hyperplane, and are in one of the two families of totally singular subspaces of Q.

PROOF. (1) holds because otherwise a rank-two space $\langle y, z\rangle$ of V has annihilator with rank $2n - 1 > 2n - 2$. So consider (2).

Since V has rank $2n$, the maximum-rank totally singular subspaces of V have rank n. Since $y^\perp/\langle y\rangle$ has rank $2n-2$ its maximum-rank totally isotropic subspaces have rank $n-1$. Hence the rank of the maximal totally singular subspaces of $y^\perp/\langle y\rangle$ is $n-1$. Thus if $F \le V$ is a totally singular subspace of rank $n-1$ in $y^\perp$ then $\langle y\rangle + F/\langle y\rangle$, which is clearly totally isotropic in $y^\perp/\langle y\rangle$, has the correct rank to be a maximal totally isotropic subspace of $y^\perp/\langle y\rangle$.

Conversely, let F' be any maximal totally isotropic subspace of the non-singular symplectic space $y^\perp/\langle y\rangle$. Hence the preimage $\pi_y-1(F') \le y^\perp$ has rank n and is a totally isotropic subspace of $y^\perp$ containing $\langle y\rangle$. So by Lemma 79.4, and the non-singularity of $\langle y\rangle$, we have: $F' = \langle y\rangle + F$, where F is the rank $n-1$ subspace of $y^\perp$ consisting of all its totally singular points. Now (2) follows.

Consider (3). If F and G overlap non-trivially then do so outside $\langle y\rangle$, hence F' and G' also overlap. The converse is equally obvious.

Consider (4). So $n = 2m$ and the rank $n-1$, of each maximal totally singular subspace F of $y^\perp$, is odd. Hence any two distinct totally singular subspaces of V, say F_1 and F_2, that contain F are subspaces of rank n that contain the odd-rank subspace F as a hyperplane. Thus F_1 and F_2 cannot lie in the same ruling class: as there are exactly two ruling classes, the result follows. $\square$

By counting the sizes of putative symplectic spreads and putative orthogonal spreads it is now clear that every symplectic spread yields an orthogonal spread and vice versa.

THEOREM 79.6. Denote by $\mathcal{A}$ either one of the two ruling classes associated with the non-singular quadratic form Q on the vector space V of rank $4n$ over $GF(q)$. Let $\mathcal{F}'$ be a symplectic spread in $y^\perp/\langle y\rangle$, and let $\mathcal{F}$ be the associated set of singular y-complements in $y^\perp$. Then $\mathcal{G}$, the set of all maximal singular spaces of V each of which contains at least one element of $\mathcal{F}$, is an orthogonal spread.

Conversely the components of any orthogonal spread $\mathcal{G}$ of V define a set

$$\mathcal{F} := \left\{\, G \cap y^\perp \mid G \in \mathcal{G} \,\right\},$$

of totally singular subspaces of rank $n-1$ in $y^\perp$, such that in the symplectic space $y^\perp/\langle y\rangle$ the set

$$\mathcal{F}' := \{\, (F + \langle y\rangle)/\langle y\rangle \mid F \in \mathcal{F} \,\},$$

constitutes a symplectic spread.

CHAPTER 80

Symplectic Groups—The Basics.

A vector space V, equipped with a non-singular alternating form, has an associated semilinear *symplectic* group that preserves the associated inner product *projectively*:

DEFINITION 80.1. Let $\Gamma\mathrm{L}(2m, K)$ be the group of invertible semilinear transformations of the vector space V of rank $2m$ over a field K. Then relative to a given non-singular alternating form $(\,\cdot\,,\,\cdot\,)$ on V, the associated symplectic semilinear group $\Gamma\mathrm{Sp}(2m, K)$ is the subgroup of $\Gamma\mathrm{L}(2m, K)$ consisting of all

$$\{\, g \in \Gamma\mathrm{L}(2m, K) \mid \exists\, \sigma \in \mathrm{Gal}(K),\ c \in K^* \ni (x^g, y^g) = c(x, y)^\sigma \ \forall x, y \in K \,\}$$

The POLARITY determined by the inner product maps each subspace $W \leq V$ to its annihilator $W^\perp$.

When dealing with two isomorphic vector spaces V_1 and V_2 we may extend the above concept in the obvious way.

DEFINITION 80.2. Let $g\colon V_1 \to V_2$ be a K-linear isomorphism, K a finite field, between the K-vector spaces V_1 and V_2. Suppose V_1 and V_2 are equipped with non-singular inner products $(\,\cdot\,,\,\cdot\,)_1$ and $(\,\cdot\,,\,\cdot\,)_2$, respectively. Then $g\colon V_1 \to V_2$ is a semilinear SYMPLECTIC isomorphism, relative to the given inner products, if there are $a \in K^*$ and $\sigma \in \mathrm{Gal}(K)$ such that $(x^g, y^g)_2 = a(x, y)_1^\sigma \ \forall x, y \in V_1$.

THEOREM 80.3. [**319**, 6, chapter III, §3]. Let V_i, $i = 1, 2$, be alternating spaces over the same field K and let θ_i, $i = 1, 2$, be the associated standard polarities. Then any bijective semilinear isomorphism $g\colon V_1 \to V_2$, such that $\theta_1{}^g = \theta_2$ is a symplectic semilinear isomorphism.

80.1. The Fundamental Theorem of Symplectic Spreads.

We verify a simple fact concerning arbitrary permutations.

LEMMA 80.4. Let θ and ϕ denote involutory permutations of the sets X and Y, respectively, and suppose $g\colon X \to Y$ is a bijection; thus the map

$$h = \phi\theta^g = g^{-1}\theta g\colon Y \to Y,$$

is a bijection of Y. Suppose h has odd order $2j+1$ and put $f := h^j$. Then $(\theta^g)^f = \phi$.

PROOF. This amounts to verifying that in any associative system $(ab)^k = a(ba)^{k-1}b$ is a generic identity.

$$(\theta^g)^f = (\phi\theta^g)^{-j}\theta^g(\phi\theta^g)^j$$
$$= (\theta^g\phi)^j\theta^g(\phi\theta^g)^j, \text{ since } \phi \text{ and } \theta \text{ are both involutions,}$$
$$= \theta^g \underbrace{(\phi\theta^g)(\phi\theta^g)\dots(\phi\theta^g)}_{j \text{ terms}}(\phi\theta^g)^j$$
$$= \theta^g(\phi\theta^g)^{2j}$$
$$= \theta^g h^{-1}$$

and since ϕ, θ^g are both involutions

$$= \theta^g\theta^g\phi$$
$$= \phi, \text{ as } \theta^g \text{ is an involution.} \qquad \square$$

We can now establish the fundamental theorem for symplectic spreads on a vector space V, of rank $2m$ over $GF(q)$, q even. This asserts that S_1 and S_2 are isomorphic spreads of V (or equivalently, they lie in the same orbit of $GL(V,+)$) if *and only if* they are in the same orbit of $\text{Symp}(2m, q)$.

THEOREM 80.5 (Fundamental Theorem for Symplectic Spreads). Let V_i, for $i = 1, 2$, be non-singular symplectic vector spaces of rank $2m$ over $GF(q)$, q even. Suppose Σ_i are symplectic spreads on V_i, for $i = 1, 2$, such that the associated translation planes are isomorphic. Then there is a symplectic isomorphism $s\colon V_1 \to V_2$ such that $s(\Sigma_1) = \Sigma_2$.

PROOF. Since the two spreads are isomorphic as ordinary spreads, there is a semilinear bijection $g\colon V_1 \to V_2$ such that $(\Sigma_1)g = \Sigma_2$. Let θ and ϕ be the associated standard polarities of V_1 and V_2, relative to their given symplectic form. Since θ is the standard polarity, it maps every $S \in \Sigma_1$ onto $S^\perp$, and since Σ_1 is also *symplectic* it follows that $S \le S^\perp$, but since both spaces are rank m, we now have $S = S^\perp$ for $S \in \Sigma_1$. Hence the polarity θ fixes every member of Σ_1 and, similarly, ϕ fixes every member of Σ_2.

It follows that the conjugate $g^{-1}\theta g\colon V_2 \to V_2$ fixes every $S \in \Sigma_2$ is also a polarity. Hence $h = g^{-1}\theta g\phi$ is a collineation of Σ_2, viewed as $PG(2m-1, q)$, that fixes every the components of Σ_2. Hence h is in the kern of Σ_2 and is thus a scalar map of V_2 as a $GF(q)$-space. Since q is even h can be taken to have odd order $2j+1$. So by Lemma 80.4 above $(\theta^g)^f = \phi$, where $f = h^j$. Now $gf\colon V_1 \to V_2$ is a linear bijection such that $\theta^{gf}\colon V_2 \to V_2$ coincides with ϕ. Hence gf projectively[1] preserves the inner product associated with ϕ; hence it maps Σ_1 onto Σ_2. Hence by Theorem 80.3, the desired result follows, because ϕ leaves Σ_2 invariant since f does. $\qquad \square$

80.2. Desarguesian Symplectic Spreads.

The goal here is to obtain the orthogonal spreads associated with a Desarguesian spread of even order, when viewed as a symplectic spread. Thus we may

---

[1]It might be helpful to recall that g is *semi*linear and f is scalar.

assume:

(80.2.1) $$F = GF(q^d) \supset GF(q) = K \supseteq GF(2),$$

and $T\colon F \to K$ is the trace map from F to the proper subfield K. Next we define on $V = F \times K \times F \times K$, viewed as a K-space, the map

$$Q\colon F \times K \times F \times K \to K$$
$$(\alpha, a, \beta, b) \mapsto T(\alpha\beta) + ab,$$

which we claim is a quadratic form from V to K. To verify this we check:

LEMMA 80.6. *The K-valued map:*

$$\beta\colon V \times V \to K$$
$$(x, y) \mapsto Q(x + y) - Q(x) - Q(y)$$

is a K-bilinear form on $V \times V$ that may be written:

(80.2.2) $$\beta(x, y) = T(\alpha\beta' + \beta\alpha') + ab' + ba',$$

and hence $Q\colon V \to K$ is a quadratic form that polarizes to β.

PROOF.

$$Q(\alpha + \alpha', a + a', \beta + \beta', b + b') + Q(\alpha, a, \beta, b) + Q(\alpha', a', \beta', b') =$$
$$T((\alpha + \alpha')(\beta + \beta')) + (a + a')(b + b') + T(\alpha\beta) + ab + T(\alpha'\beta') + a'b'$$

and since the K-linear map T induces the identity on K, the right-hand side, viewed as a map $\beta\colon V \times V \to K$, simplifies to:

$$\beta[(\alpha, a, \beta, b), (\alpha', a', \beta', b')] = T(\alpha\beta' + \beta\alpha') + ab' + ba'.$$

So β is clearly K-bilinear on V and the lemma follows. $\square$

Thus (V, Q) is a quadratic form that polarizes to β. However, we need the form to be *non-singular* and *hyperbolic*. The latter cannot hold unless V has *odd* projective dimension, so this will be our standard hypothesis from now.

HYPOTHESIS 80.7. *The rank $d > 1$ of the field $F = GF(q^d)$ over the subfield $K = GF(q)$ is an odd integer.*

A trivial but repeatedly used consequence of this assumption is:

REMARK 80.8 (Trace-Remark). *The (K-linear) trace $T\colon F \to K$ fixes K elementwise, hence also:*

$$F = \ker T \oplus K$$

PROOF. The K-linear trace maps K to zero, unless it fixes K identically, but $T(1) = 1$ because d is odd and F has even characteristic 2. Hence also $\ker T \cap K = 0$, and hence the second assertion holds since K is also the image of T. $\square$

We can establish that Q is non-singular and corresponds to a *hyperbolic* quadric.

PROPOSITION 80.9 (Trace Form for Hyperbolic Quadric). *Suppose $[F : K] = d$ is an odd integer > 1. Thus $V = F \times K \times F \times K$ is a rank $d + 1$ vector space over K and $Q\colon V \to K$ is a quadratic form that polarizes to β. Let $0 \times 0 \times F \times K$ be a totally singular subspace $\mathcal{Y}$. Then the following hold:*

(1) *The symplectic form $\beta\colon V \times V \to K$ is non-singular.*

(2) The quadratic form Q is non-singular and the associated quadric is HY-PERBOLIC.

PROOF. Suppose (α, a, β, b) is in the radical of (V, β). Then for $x, y \in F$ and $u, v \in K$:

$$\beta[(\alpha, a, \beta, b), (x, u, y, v)] = T(\alpha y + \beta x) + av + bu = 0$$

and the latter identity immediately yields $a = b = 0$ so for all $x, y \in F$:

$$T(\alpha y + \beta x) = 0$$

and hence $T(K) = 0$, contrary to the trace-remark above.

Thus Q is a non-singular quadric on V that polarizes to the non-singular symplectic form β. It remains to show that the quadric is hyperbolic, i.e., has maximum Witt (or equivalently projective) index. But this is evident since the subspace $0 \times 0 \times F \times K$ is totally singular, by the definition of Q, and the size of space corresponds to subspace of maximum Witt index. □

The generator $0 \times 0 \times F \times K$ of the hyperbolic quadric associated with Q will arise frequently, so it will be convenient to write: $\mathcal{Y} := 0 \times 0 \times F \times K$.

80.2.1. An Orthogonal Spread. We now show that V admits an orthogonal spread whose other components, besides $\mathcal{Y}$, consist of the following subspaces:

$$(80.2.3) \qquad \forall s \in F \colon V_s := \left\{ \left(\alpha, a, s^2 \alpha + s T(s\alpha) + sa, T(s\alpha) \right) \mid \alpha \in F, a \in K \right\}.$$

We first verify that each of the above V_s's are totally singular, relative to our standard hyperbolic form.

LEMMA 80.10. The K-space V_s is a totally singular subspace of rank $\dim_K F + 1$ relative to the standard quadratic from $Q(\alpha, a, \beta, b) = T(\alpha\beta) + ab$ on $V = F \times K \times F \times K$.

PROOF. It is evident that V_s is a K-subspace of the stated dimension. We verify that it is totally singular.

$$Q\left(\alpha, a, s^2\alpha + sT(s\alpha) + sa, T(s\alpha)\right) = T\left(\alpha\left(s^2\alpha + sT(s\alpha) + sa\right)\right) + aT(s\alpha)$$
$$= T\left((s\alpha)^2\right) + T(s\alpha)T(s\alpha) + T(sa\alpha) + aT(s\alpha)$$

and since T commutes with the Frobenius map $x \mapsto x^2$ and is linear over K

$$= T(s\alpha)^2 + T(s\alpha)^2 + aT(s\alpha) + aT(s\alpha)$$
$$= 0. \qquad \square$$

Next we verify that distinct V_s's have trivial intersection.

LEMMA 80.11. For $s, t \in F$: $V_s \cap V_t = \mathbf{O} \implies s = t$.

PROOF. The condition

$$\left(\alpha, a, s^2\alpha + sT(s\alpha) + sa, T(s\alpha)\right) \in V_s = \left(\beta, b, t^2\beta + tT(t\beta) + tb, T(t\beta)\right) \in V_t$$

cannot hold unless $\alpha = \beta$, $a = b$. Hence a non-trivial intersection cannot arise unless:

$$(80.2.4) \qquad\qquad s^2\alpha + sT(s\alpha) + sa = t^2\alpha + tT(t\alpha) + ta$$

and moreover

$$T(s\alpha) = T(t\alpha) \implies T((s + t)\alpha) = 0.$$

But Equation (80.2.4) yields:

(80.2.5) $(s+t)^2\alpha + (s+t)T(s\alpha) + (s+t)a = 0,$

hence for $s \neq t$:

(80.2.6) $(s+t)\alpha + T(s\alpha) + a = 0.$

But since $T((s+t)\alpha) = 0$, and T^K is the identity, applying T to Equation (80.2.6) converts it to $T(s\alpha) + a = 0$ and now Equation (80.2.6) yields $(s+t)\alpha = 0$ and since $s \neq t$ this means $\alpha = 0$, and hence Equation (80.2.5) further forces $a = 0$, but this cannot happen unless $V_s \cap V_t = \mathbf{O}$, as claimed.   □

Similarly, it immediately follows that none of the V_s's above meet $\mathcal{Y}$. Thus we have established:

THEOREM 80.12. The collection of subspaces of a quadratic space (V, Q) given by: $\{\, V_s \mid s \in F \,\} \cup \{\mathcal{Y}\}$, is an orthogonal spread called the DESARGUESIAN orthogonal spread.

The choice of the terminology arises from the fact that *some* of the symplectic spreads that it defines are Desarguesian spreads in the André sense.

PROPOSITION 80.13. We continue with the notation: $V = F \times K \times F \times K$ and

$$V_s := \left\{\, (\alpha, a, s^2\alpha + sT(s\alpha) + sa, T(s\alpha)) \mid \alpha \in F, k \in K \,\right\}.$$

Let $y = (0, 1, 0, 1) \in F \times K \times F \times K$. $Q(y) = 1$; thus y is non-singular, and we may project the Desarguesian spread to a symplectic spread at y, on $y^\perp/y$.
   (1) The component V_s, of the orthogonal Desarguesian spread, meets $y^\perp$ in the hyperplane of V_s given by:

$$y^\perp \cap V_s = \left\{\, (\alpha, a, s^2\alpha, a) \mid \alpha \in F, a \in K \,\right\}.$$

   (2) If W is a subspace of maximum dimension on the quadric associated with Q, thus W is a GENERATOR of the quadric, then $y^\perp$ is a hyperplane of V and $y^\perp \cap W$ is a hyperplane of W. (See Remark 103.43.)
   (3) $(0 \times 0 \times F \times K) \cap y^\perp = 0 \times 0 \times F \times 0$.

PROOF. $Q(y) = T(0 \times 0) + 1 \times 1 = 1$. The elements of V_s in $y^\perp$ must satisfy:

$$y \cdot (\alpha, a, s^2\alpha + sT(s\alpha) + sa, T(s\alpha)) = 0,$$

and this holds iff $T(s\alpha) = a$, and this means $s^2\alpha + sT(s\alpha) + sa$ collapses to $s^2\alpha$, as required. The other cases are equally straightforward.   □

Bearing in mind the above proposition, we immediately obtain:

THEOREM 80.14. The orthogonal Desarguesian spread when symplectically projected at $y = (0, 1, 0, 1)$ yields the symplectic K-spread on $y^\perp/y$ with components:

$$\left\{\, (W_s \oplus \langle y \rangle) / \langle y \rangle \mid s \in F \,\right\} \cup \{\, (0 \times 0 \times F \times 0 \oplus \langle y \rangle) / \langle y \rangle \},$$

and this spread is isomorphic to the standard Desarguesian F-spread.

Now comes the interesting thing: the *Desarguesian* orthogonal spread also yields *non-Desarguesian* spreads, by projecting onto symplectic spreads arising from non-singular projective points $\langle z \rangle$ not in the orbit of $\langle y \rangle$.

An example of this phenomenon is Kantor's 'third cousin'. This is based on the projection at $z = (0, k + 1, 0, 1)$, where $k \in K - GF(2)$; if $k = 1$ then the point is singular and hence not permitted, and if $k = 0$ we are back to the previous $y = (0, 1, 0, 1)$—corresponding to the Desarguesian situation.

LEMMA 80.15. For $s \in F$, let

$$V_s := \left\{ \left(\alpha, a, s^2\alpha + sT(s\alpha) + sa, T(s\alpha)\right) \mid \alpha \in F, a \in K \right\}.$$

Relative to the inner product $\beta(\cdot, \cdot)$, Equation (80.2.2), put $y := \langle(0, k + 1, 0, 1)\rangle$, and let $W_s := V_s \cap y^\perp$, for $s \in F$. Then

$$\forall s \in F \colon W_s := \left\{ \left(\alpha, (k + 1)T(s\alpha), s^2\alpha + sT(s\alpha) + s(k + 1)T(s\alpha), T(s\alpha)\right) \mid \alpha \in F \right\}.$$

Moreover, $W_s \cap W_u = \mathbf{o}$ whenever $s, u \in F$, $s \neq u$.

PROOF. Since

$$\beta\left(\left(\alpha, a, s^2\alpha + sT(s\alpha) + sa, T(s\alpha)\right), (0, k + 1, 0, 1)\right) = 0$$

$$\implies a = (k + 1)T(s\alpha),$$

and substituting for a in V_s yields the required form for W_s.

If $s \neq t$ then by Lemma 80.11, $V_s \cap V_t = \mathbf{O}$, a fortiori $W_s \cap W_t = \mathbf{O}$. $\square$

It follows that in the factor space $y^\perp/y$ the images of W_s form a system of pairwise disjoint totally isotropic subspaces any two of which direct sum to $y^\perp/y$, and hence define a symplectic spread on the symplectic space $y^\perp/y$.

In order to find a convenient representation of the slope maps of this spread, we choose a simplified representation of $y^\perp/y$ in the obvious way. Since $y = (0, k + 1, 0, 1)$ we identify $y^\perp/y$ with the K-space whose elements are form $(x, z(k + 1), y, z) \to (x, y)$. We consider the surjective linear map $y^\perp/y \to F \times F$, induced by

$$(x, z(k + 1), y, z) \mapsto (x, y) \forall x, y, z \in F.$$

This map sends the subspace W_s onto $\left\{ \left(\alpha, s^2\alpha + skT(s\alpha)\right) \mid \alpha \in F \right\}$. Hence these W_s-images, along with the image of $\mathcal{Y}$, $\mathbf{O} \oplus Y$, form a symplectic spread on $F \oplus F$.

It is evident that the slope set of this spread is the set of maps:

(80.2.7) $\left\{ \alpha \mapsto \left(\alpha, s^2\alpha + skT(s\alpha)\right) \mid s \in F \right\}.$

It is desirable to verify directly that these maps do actually define a spread set—in fact we see they correspond to a pre-semifield.

THEOREM 80.16. Let $F = GF(q^d) > GF(q) = K \geq GF(2)$, with $d > 1$ odd, and suppose $k \in K \setminus \{1\}$. Let T be the trace map for $\mathrm{Gal}(F : K)$. Then $(F, +, \circ)$ is a pre-semifield where:

$$x \circ y = x\alpha + xy^2\alpha + xyk \; \forall \, x, y \in F.$$

The associated spread is Desarguesian if $k = 1$.

PROOF. If $k = 0$ then the situation has been considered earlier. So assume $k \notin GF(2)$; so the slopes correspond to the maps as in Equation (80.2.7). This is a set of additive maps that is additively closed. So a spread arises iff all the non-zero maps in Equation (80.2.7) are injective. But since these maps are additive, it is sufficient to show

$$s^2\alpha + skT(s\alpha) = 0,$$
$$\text{so } s\alpha = kT(s\alpha).$$

Taking traces and noting the trace fixes all $k \in K$ hence the right-hand side:

$$T(s\alpha) = kT(s\alpha),$$
$$(1 - k)T(s\alpha) = 0$$

and since $k \neq 1$

$$T(s\alpha) = 0$$

so by the first equation

$$s^2\alpha = 0.$$

Hence s or α is zero, as required. $\square$

CHAPTER 81

Symplectic Flag-Transitive Spreads.

Our aim here is to study symplectic spreads of even order that are flag transitive relative to what might be called the *rotation group*, associated with the *unit circle*, of an appropriate field. We begin by introducing these and related terms.

DEFINITION 81.1. (Unit Circle). Let $F^{(2)}$ be a quadratic extension of a field F and let α be the involutory automorphism with fixed field F. Then $\bar{x} := \alpha(x)$, for $x \in F^{(2)}$, is the CONJUGATE of x. The UNIT CIRCLE of $F^{(2)}$, relative to α is the multiplicative group

$$C := \{\, z \in F^{(2)} \mid z\bar{z} = 1, z\, inF \,\}.$$

The SCALAR GROUP of $F^{(2)}$ consists of all bijections of $F^{(2)}$ of form $\tilde{s}\colon x \mapsto xs$, for $s \in F^{(2)*}$, and the subgroup

$$\tilde{C} := \{\, \tilde{c}\colon x \mapsto xz \mid z \in C \,\}$$

is the ROTATION GROUP.

So the scalar group coincides with the kern homology group, when $F^{(2)}$ is regarded as a Desarguesian spread, and the rotation group may be identified with the unit circle.

We summarize below some notation that will be used here quite frequently. Reminders regarding their meaning will be issued whenever it seems appropriate.

(1) The field $F^{(2)}$ is a finite field of characteristic two (and contains as a subfield all the other fields of interest).
(2) If G is any subfield of $F^{(2)}$ then $G^{(2)}$ is the quadratic extension field of G that lies in $F^{(2)}$.
(3) C is the unit circle of $F^{(2)}$.
(4) If K is a subfield of F then $T_K\colon F \to K$ is the K-linear trace map of F. And if L is a subfield of K then

$$T_{K,L} := \{\, k \in K \mid T_L(k) = 0 \,\}.$$

A major reason for imposing characteristic 2 is that it forces the multiplicative group of $F^{(2)}$ to be a *direct* product of F^* and C:

REMARK 81.2. Suppose the field $F^{(2)}$ is a quadratic extension of any field F and let C be the unit circle. Then

$$F^{(2)*} = F^* \times C.$$

PROOF. If $x \in C \cap F^*$ then

$$x\bar{x} = 1 \implies x^2 = 1 \implies x = 1.$$

Let $z = f/\bar{f}$ and $r = f\bar{f}$. Then $z\bar{z} = 1$, $r \in F$ and $zr = f^2$ □

PROBLEM 81.3. Suppose $F^{(2)}$ is a quadratic extension of a general field. Explore the extent to which the remark remains valid.

It follows immediately, from Remark 81.2, that $F^{(2)}$ partitions into a spread whose components constitute the $\tilde{C}$-orbit of F: this spread is clearly the natural Desarguesian spread, and C acts as a regular flag-transitive group

In order to obtain other flag-transitive spreads, one might try and replace F by other additive subspaces S of $F^{(2)}$, where $|S| = |F|$; so W will be a subspace over some subfield $K \leq F$, and the extremal case, $F = K$, corresponds to the Desarguesian situation just discussed.

The main theme here is to consider the situation when S is chosen to be symplectic, relative to an alternating form under which the rotation group is a group of *isometries*. This will ensure, for well-chosen S, that the resulting spread—the S-orbit under the rotation group $\tilde{C}$—will be symplectic and flag-transitive. Naturally, the symplectic property will lead to further spreads: via orthogonality, slicing and expansion.

Thus, we need first of all an alternating form on $F^{(2)}$, regarded as a vector space over a subfield K of F, *such that the rotation group of $F^{(2)}$ is a group of isometries* relative to the form.

This can be arranged when the alternating form is trace-defined as follows.

DEFINITION 81.4. Suppose K is a subfield of F such that $[F : K] \geq 1$ is odd. Then the STANDARD INNER PRODUCT $(\,\cdot\,,\,\cdot\,)_K$ on $F^{(2)}$, associated with the trace map $T_K \colon F \to K$, is the bilinear form on $F^{(2)}$:

$$\text{For } x, y \in F^{(2)} : (x, y)_K := T_K \left(x\overline{y} + y\overline{x} \right).$$

Note that $(x, y)_K$ makes sense since $x\overline{y} + y\overline{x}$ is self conjugate and hence lies in F. We now verify that this bilinear form is preserved by the rotation group and that the form is *non-singular*: this uses our hypothesis that $[F : K]$ is odd.

REMARK 81.5. The group

$$\tilde{C} := \left\{ \tilde{\theta} \colon x \mapsto \theta x \mid x \in F^{(2)} \right\} \cong C^*$$

is a group of isometries of $F^{(2)}$, relative to the form $(\,\cdot\,,\,\cdot\,)_K$.

PROOF. To verify $c \in C$ defines an isometry, note that:

$$(cx, cy)_K = T_K \left(cx\overline{cy} + cy\overline{cx} \right)$$
$$= T_K \left(c\overline{c} \left(x\overline{y} + y\overline{x} \right) \right)$$

and by $c\overline{c} = 1$

$$= T_K \left(\left(x\overline{y} + y\overline{x} \right) \right) = (x, y)_K.$$

To verify that the form $(\,\cdot\,,\,\cdot\,)_K$ is non-singular, first note that $T_K(1) = 1$, since $[F : K]$ is odd. To get a contradiction assume that $x \in F^{(2)}$ is in the radical. Then:

$$T_K \left(x\overline{y} + y\overline{x} \right) = 0 \; \forall \, y \in F^{(2)} \implies$$
$$T_K \left((x + \overline{x}) f \right) = 0 \; \forall \, f \in F,$$

but $x + \overline{x} \in F$ now forces $T_K(F) = 0$, contrary to $T_K(1) = 1$, unless $x = \overline{x}$. But in the latter event $x = f \in F^*$, and we similarly have $T_K\left(f\left(y + \overline{y}\right)\right) = 0$, for all $y \in F^{(2)}$, and hence again $T_K(1) = 0$. $\square$

We may now return to the question of when the rotation orbit, of a K-subspace $S \leq F^{(2)}$, actually is a spread. It will be convenient to refer to such spreads as being *circular*.

DEFINITION 81.6. Let $(\,\cdot\,,\,\cdot\,)_K$ be the alternating form on $F^{(2)}$ associated with a subfield K of F, where $[F : K] \geq 1$ is odd. Then a spread on $F^{(2)}$ whose components form an orbit of the rotational group and consist of totally isotropic subspaces, relative to $(\,\cdot\,,\,\cdot\,)_K$, is a CIRCULAR SPREAD, or a K-CIRCULAR SPREAD, of $F^{(2)}$.

Since some component, of any circular spread, must contain 1, we may assume without loss of generality that some component $S \supset K$, and hence $S = W \oplus K$, where W is a K-subspace of $F^{(2)}$, of rank $[F : K] - 1$. We may now specify the form of any circular spread.

REMARK 81.7. A spread ζ on $F^{(2)}$ is circular relative to $K \leq F$ iff it can be expressed as follows

$$\zeta := \{\,\theta(W + K) \mid \theta \in C\,\},$$

where W is a totally isotropic K-space such that the K-rank of W is $[F : K] - 1$.

PROOF. Any circular spread is of the specified form by the comments preceding the remark. The converse, that spreads which may be written as above are symplectic, follows upon noting that $W + K$ must be totally isotropic as K is also totally isotropic—it is a one-space of an alternating form. $\square$

Thus, the above characterizes all circular spreads in terms of W—any complement of the component containing K. The following definition yields W meeting the requirements of the Remark 81.7, and hence such W, when they exist, will always yield circular spreads.

DEFINITION 81.8 (KW-Space). Let $[F : K] \geq 1$ be odd. Suppose W is a K-subspace of $F^{(2)}$ and $\gamma \in C$, the unit circle of $F^{(2)}$. Then (W, γ, K) is a KW-TRIPLE, if the following conditions hold:

KW1 $W \cap K = \{0\}$ and $W \oplus K\gamma$, as a K-subspace, has rank $[F : W]$, i.e., is 'linesized'.
KW2 $T_K(w\overline{w}) = 0 \forall w \in W$;
KW3 $(W, K\gamma)_K = 0$;
KW4 The $\tilde{C}$-orbit of $W \oplus K\gamma$ is a spread ζ of $F^{(2)}$.

81.1. Kantor–Williams Flag-Transitive Spreads.

DEFINITION 81.9. The spread ζ will be called the KW-spread, or the KANTOR–WILLIAMS spread, associated with the given KW-triple. The KW-triple, and the associated Kantor–Williams spread, are STRONGLY ORTHOGONAL if the following condition holds:

(KW3a) $\left(W, K^{(2)}\gamma\right) = 0.$

Note that the terminology implies that axiom KW3a automatically forces KW3. The strong-orthogonality condition KW3a is not necessary to force the KW-spread to be a symplectic circular spread; its importance stems from the fact that it is essentially a recursive condition: this will lead to many new strongly orthogonal KW systems, and hence spreads, starting from a given one. Similarly, $[F : K] = 1$ is permitted so as to start the induction.

We first verify that the axiom KW2 forces any KW-spread to be a circular symplectic spread. The class of circular symplectic spreads is probably larger than the class of KW-spreads.

REMARK 81.10. If W is a KW-subspace of $F^{(2)}$, then the associated Kantor–Williams spread will be a circular spread. In particular, the spread admits K as a group of kern homologies (under its standard action) and $\tilde{C}$ acts as a regular K-linear flag-transitive group on the spread.

PROOF. For $w_1, w_2 \in W$:

$$T_K\left((w_1 + w_2)\overline{(w_1 + w_2)}\right) = 0$$
$$\implies T_K\left((w_1\overline{w_2}) + (w_2\overline{w_1}) + (w_1\overline{w_1}) + (w_2\overline{w_2})\right) = 0$$
$$\implies T_K\left((w_1\overline{w_2}) + (w_2\overline{w_1})\right) = 0$$
$$\implies (w_1, w_2)_K = 0,$$

thus W is totally isotropic.

Since by hypothesis $W \perp K\gamma$, it follows that $W \oplus K\gamma$ is totally isotropic ($K\gamma$ itself is totally isotropic because any one-space of an alternating space is isotropic). Since $\tilde{C}$ is a group of isometries, the set of images of $W \oplus K\gamma$ under $\tilde{C}$ are also totally isotropic. Hence the Kantor–Williams spreads are symplectic. The result follows.   □

We now turn to the existence of KW-triples and strongly orthogonal KW-triples. As already indicated, *strong* orthogonality permits a recursive construction of strongly orthogonal symplectic spreads.

To start the recursion we first note that Desarguesian spreads correspond to *strongly orthogonal* KW-triples, and they also provide our first examples of KW-triples.

PROPOSITION 81.11. If $\gamma = 1$ and $W = \ker T_K$, then (γ, W) corresponds to a KW-triple and hence the associated symplectic KW-spread, the orbit of $F = W \oplus K$ under the rotation group $\tilde{C}$, is a Desarguesian spread—the SYMPLECTIC DESARGUESIAN spread. Moreover, the spread is strongly orthogonal, Definition 81.8[KW3a].

PROOF. By convention $[F : K]$ is odd, so $F = \ker T_K \oplus K$. But since, by definition, F is the fixed field of the conjugating involution $\alpha\colon x \mapsto \overline{x}$, we have

$$T_K(f\overline{f}) = \sum(f\overline{f})^\rho = \sum(f^2)^\rho = \left(\sum(f^\rho)\right)^2 = T_K(f)^2,$$

and hence $T_K(f\overline{f}) = T_K(f)^2 = 0$, whenever $f \in \ker T_K$.

We now verify the strong orthogonality condition (thus implying condition KW3):

(81.1.1) $$\left(W, K^{(2)}\right)_K = 0.$$

Remembering that $W = \ker T \subset F = \mathrm{Fix}(\alpha)$ yields $\overline{w} = w$, for $w \in W$, and that $t + \overline{t} \in K$, even when $t \in K^{(2)}$, we have:

$$
\begin{aligned}
(w, t)_K &= T_K(w\overline{t} + t\overline{w}) \\
&= T_K(w\overline{t} + tw) \\
&= (\overline{t} + t)\, T_K(w) = 0,
\end{aligned}
$$

since $w \in \ker T_K$. Thus the strong orthogonality condition holds, and hence also KW3.

Finally, the orbits of F^* under $\tilde{C}$ are clearly disjoint—they correspond to a system of multiplicative cosets of F^* in $(F^{(2)})^*$. Hence the $\tilde{C}$ orbit of F is a spread: the usual Desarguesian spread.

Thus, all the KW-conditions are met for *strong* orthogonality, so $\tilde{C}$ is a strongly symplectic spread. It is Desarguesian because it coincides with the standard Desarguesian spread on $F^{(2)}$, since this spread has as its components the members of the orbit of F under the action of the group C. $\qquad\square$

In Proposition 81.11 above, the reason why $T_K \oplus K\gamma$ yields a *spread*, for $\gamma = 1$, is trivial: in this case $T_K \oplus K\gamma = T_K \oplus K$, which is just F. However, it appears less easy to determine when $T_K \oplus K\gamma$ is a spread for more general γ. We shall see, in Corollary 81.28, that if $\gamma \in K^{(2)}$ then we still get a (circular) spread: an essential step in the recursion indicated above.

To establish this and other aspects of the desired recursion, we make systematic use of the connections between symplectic and orthogonal spreads, based on expanding the K-vector spaces of interest, by viewing them as vector spaces over appropriate subfields L of K.

Before going into details, we give a vague sketch of the inductive procedure to be developed. Starting with a *strongly orthogonal* KW-system, associated with $F \supset K$, the associated Kantor Williams spread is symplectic and hence can be extended to an orthogonal spread in $F^{(2)} \times K^{(2)}$, and this spread may be expanded to an orthogonal spread over a subfield $L \subset K$: the associated quadratic form Q_L is based on $F^{(2)} \times L^{(2)}$. Now suitable symplectic slices of this quadric, in the initial symplectic spread $F^{(2)}$, viewed as L-space, yields new *strongly orthogonal* KW-triples relative to L and hence new circular spreads with L in the kern—in many cases this is the full kern.

Thus, for this type of recursive construction to work, the only conditions that need to be imposed are that $[K : L] > 1$ is odd, and that the initial K corresponds to a *strongly orthogonal* KW-triple.

81.2. Expanding the Quadratic Form.

We first introduce the quadric associated with $L \subset K \subseteq F$. This will be represented as a trace-defined quadratic form Q_L that ensures that the unit circle C may be identified with an isometry group of Q_L.

DEFINITION 81.12. Let L be a proper subfield of K such that the degree $[K : L]$ is odd. On the vector space $V^{(2)} = F^{(2)} \oplus L^{(2)}$ define the map

$$
Q_L((x, \lambda)) := T_L(x\overline{x}) + \lambda\overline{\lambda}.
$$

REMARK 81.13. The map $Q_L \colon F^{(2)} \oplus L^{(2)} \to L$ is a quadratic form that polarizes to a non-singular L-symplectic form, on $F^{(2)} \oplus L^{(2)} \times F^{(2)} \oplus L^{(2)}$, given by:

$$\beta((x, \lambda) \cdot (y, \lambda)) := T_L(x\overline{y} + y\overline{x}) + \lambda\overline{\mu} + \mu\overline{\lambda}.$$

PROOF. By a routine computation, remembering the characteristic is 2, it can be verified that

$$Q_L((x, \lambda) + (y, \mu)) - Q_L((x, \lambda)) - Q_L((y, \mu)) = T_L(x\overline{x}) + \lambda\overline{\lambda},$$

thus establishing that Q_L is a quadratic form that polarizes to β.

It remains to verify that β is non-singular. So assume (x, λ) is in the rad(β). Thus

$$\forall y \in F^{(2)}, \mu \in L^{(2)} \colon \beta((x, \lambda)(y, \mu)) = 0$$

$$\implies \forall y \in F^{(2)}, \mu \in L^{(2)} \colon \beta((x, \lambda)(y, \mu)) = 0$$

$$\implies \forall y \in F^{(2)}, \mu \in L^{(2)} \colon T_L(x\overline{y} + y\overline{x}) + \lambda\overline{\mu} + \mu\overline{\lambda} = 0,$$

and now using judiciously chosen values for y and μ force $x = \lambda = 0$, as required. For example, $y = 0$ yields $\lambda\overline{\mu} + \mu\overline{\lambda} = 0$, and now $\mu = 1$ forces $\lambda = \overline{\lambda}$ and hence also that $\overline{mu} = \mu$ for all $\mu \in L^{(2)}$, unless $\lambda = 0$. But the involution α acts faithfully on $L^{(2)}$, so $\overline{mu} = \mu$ is a contradiction unless $\lambda = 0$, so we are left with:

$$\forall y \in F^{(2)}, T_L(x\overline{y} + y\overline{x}) = 0,$$

and $y = 1$ forces $x = \overline{x}$ and if this is non-zero then as above we have the contradiction that $\overline{y} = y$ for all $y \in F^{(2)}$. Hence (x, λ) cannot be in the radical of β unless it is zero. $\qquad\square$

Thus the quadratic form Q_L polarizing to β corresponds to either a hyperbolic quadric or an elliptic quadric. However, it will emerge, cf. Corollary 81.19, that Q_L is always hyperbolic.

81.2.1. An Orthogonal Spread for Q_L. The following remark applies to the traces of Galois fields in general.

REMARK 81.14. Let $A < B < C$ be a chain of Galois fields, and let Θ be a set of coset representatives of $\mathrm{Gal}(C; A)$ over the subgroup $\mathrm{Gal}(C; B)$, thus $|\Theta| = |\mathrm{Gal}(C; A)|/|\mathrm{Gal}(C; B)|$. Then the trace maps T_B and T_A, onto B and A respectively, are related by:

$$\forall x \in C \colon T_A(x) = \sum_{\theta \in \Theta} T_B(x)^\theta.$$

Hence $\ker T_C \leq \ker T_B$.

PROOF. Writing T_B in full, shows that the right-hand side of the summation ranges over all elements of $\mathrm{Gal}(C : A)$, and thus defines T_A. $\qquad\square$

LEMMA 81.15.

$$\forall x, y \in F^{(2)} \colon (x, y)_K = 0 \implies T_L(x\overline{y} + y\overline{x}) = 0.$$

PROOF. By definition, $T_K \colon F^{(2)} \to K$ satisfies

$$(x, y)_K = T_K(x\overline{y} + y\overline{x}),$$

and hence Remark 81.14 applies. $\qquad\square$

We note that the kernel of the restriction of T_L to K is a hyperplane:

LEMMA 81.16. *The L-subspace of K:*

$$T_{K,L} := \{\, k \in K \mid T_L(k) = 0 \,\}$$

is an L-hyperplane of K.

PROOF. Since $[F : L]$ is odd, $T_L|L$ is the identity map; but L is a proper subfield of K, so $T_L|K$ is not the zero map. $\qquad\square$

We introduce a type of L-subspaces of $F^{(2)} \times L^{(2)}$ that is intended to be a component of an orthogonal spread, based on our earlier definition of γ and W. We first note that the size is correct.

LEMMA 81.17. *For any $\gamma \in C$, the unit circle of $F^{(2)}$, where $W + K\gamma = W \oplus K\gamma = F$, the L-spaces in $F^{(2)} \times L^{(2)}$ of form*

$$(W + T_{K,L}\gamma, 0) + L^{(2)}\,(1, \gamma) := \left\{\, (w + t\gamma + \lambda\gamma, \lambda) \mid w \in W, t \in T_{K,L}, \lambda \in L^{(2)} \,\right\}$$

have L-rank $[F : L] + 1$.

PROOF. By the direct sum hypothesis, $W \oplus K\gamma = F$, along with the fact that $T_{K,L}$ has L-rank $[K : L] - 1$, we conclude that $W + T_{K,L}\gamma = W \oplus T_{K,L}\gamma$ is an L-subspace of F of L-rank $[F : L] - 1$. Identifying this subspace with $(W + T_{K,L}\gamma, 0)$ and, noting that this subspace is disjoint from the rank 2 L-space $L^{(2)}\,(1, \gamma)$, we see that the sum of these two spaces is $[F : K] - 1 + 2$ as required. $\qquad\square$

We now verify that the spaces, $(W + T_{K,L}\gamma, 0) + L^{(2)}\,(1, \gamma)$, of the lemma above, are totally singular, when associated with a *strongly orthogonal* KW-triple. Thus, strongly orthogonal triples yield totally singular subspaces of Q_L—with Witt index corresponding to hyperbolic quadrics.

In particular, see Corollary 81.19, Q_L is always hyperbolic.

THEOREM 81.18. *Let $F > K > L$ is a chain of Galois fields of even order, and let C be the unit circle of $F^{(2)}$. Assume $\gamma \in C$, and W is a K-subspace of $F^{(2)}$ such that $(W, K\gamma)$ is a strongly orthogonal KW-triple, cf. Definition 81.8. Then the subspace of $F^{(2)} \times L^{(2)}$:*

$$(W + T_{K,L}\gamma, 0) + L^{(2)}\,(1, \gamma) := \{(w + t\gamma + \lambda\gamma, \lambda) \mid w \in W, t \in T_{K,L}, \lambda \in L\}$$

is a totally singular subspace for Q_L, and its Witt index is $[F : L] + 1$.

PROOF. By Lemma 81.17, it is sufficient to verify that the subspace above is totally singular:

$$Q_L\left((w + t\gamma + \lambda\gamma, \lambda)\right) = T_L\left((w + (t + \lambda)\gamma)\,\overline{(w + (t + \lambda)\gamma)}\right) + \lambda\bar{\lambda}$$

$$= T_L\left(w\overline{w}\right) + T_L\left((t + \lambda)\gamma\overline{(t + \lambda)\gamma}\right) + T_L\left(\overline{w}(t + \lambda)\gamma + w\overline{(t + \lambda)\gamma}\right) + \lambda\bar{\lambda}$$

However, by Definition 81.8 [KW2], $T_L(w\overline{w}) = 0$ and $\gamma\bar{\gamma} = 1$, for $\gamma \in C$ the above reduces to:

$$Q_L\left((w + t\gamma + \lambda\gamma, \lambda)\right) = T_L\left((t + \lambda)\overline{(t + \lambda)}\right) + T_L\left(\overline{w}(t + \lambda)\gamma + w\overline{(t + \lambda)\gamma}\right) + \lambda\bar{\lambda}$$

Since $t \in K$, we have $(t + \lambda)\gamma \in K^{(2)}\gamma$, hence $(w, (t + \lambda)\gamma)_K = 0$, and, by Lemma 81.15, this forces

$$T_L\left(\overline{w}(t + \lambda)\gamma + w\overline{(t + \lambda)\gamma}\right) = 0,$$

so we are left with

$$Q_L\left((w + t\gamma + \lambda\gamma, \lambda)\right) = T_L\left((t + \lambda)\overline{(t + \lambda)}\right) + \lambda\overline{\lambda}$$

and since $t \in K < F$ is self-conjugate:

$$= T_L\left(t^2 + t(\overline{\lambda} + \lambda) + \lambda\overline{\lambda}\right) + \lambda\overline{\lambda}$$

and since both $\lambda + \overline{\lambda}$ and $\lambda\overline{\lambda}$ are in L:

$$= T_L\left(t^2\right) + (\overline{\lambda} + \lambda)T_L\left(t\right) + \lambda\overline{\lambda}T_L(1) + \lambda\overline{\lambda}$$

but since T_L is the identity on L and $t \in \ker T_L$:

$$= T_L\left(t\right)^2 + \lambda\overline{\lambda} + \lambda\overline{\lambda} = 0,$$

as required.   □

COROLLARY 81.19. *The quadric Q_L is hyperbolic.*

PROOF. By Remark 81.13, Q_L is always non-singular. So it is sufficient to check that it is associated with a totally singular L-subspace of rank $[F : L] + 1$, i.e., of the correct Witt index. By Proposition 81.11, $(W = \ker : T_K : F^{(2)} \to K, \gamma = 1, K)$ corresponds to a strongly orthogonal KW-triple. Hence, by Theorem 81.18, Q_L has the correct Witt index to be hyperbolic.   □

REMARK 81.20. Let g be a collineation of a projective space $PG(n, q)$ of odd order. If g leaves invariant a hyperbolic quadric $\mathcal{H}$, then for every totally singular subspace $H \in \mathcal{H}$, both H and $g(H)$ have the same type.

In particular, g fixes each of the two ruling classes of $\mathcal{H}$.

PROOF. The collineation g preserves the lattice of subspaces of $PG(n, q)$ and hence the type classes. The result follows, because there are two type-classes and g has odd order.   □

DEFINITION 81.21. Let C be the unit circle of $F^{(2)}$, and define the action of $c \in C$ on $F^{(2)} \oplus L^{(2)}$ by

$$\tilde{c}\colon F^{(2)} \oplus L^{(2)} \to F^{(2)} \oplus L^{(2)}$$
$$(f, \lambda) \mapsto (fc, \lambda),$$

and let

$$\tilde{C} = \{\tilde{c} \mid c \in C\}.$$

We note that $\tilde{C}$ is a group of type-fixing isometries associated with Q_L:

REMARK 81.22. The group $\tilde{C}$ preserves the hyperbolic quadric associated with the quadratic form Q_L and maps each totally singular subspace onto a subspace of the same type.

PROOF. It is straightforward to verify that $\tilde{C}$ is an isometry group, and its order is odd because the unit circle for $F^{(2)}$ has odd order. Hence it preserves type-classes by Remark 81.20.   □

We now return to the generator of the hyperbolic quadric Q_L, constructed in Theorem 81.18. We show that under the group $\tilde{C}$, of circular Q_L-isometries $\tilde{C}$ of $F^{(2)}$, the orbit of our generator forms an orthogonal spread.

THEOREM 81.23. Assume the hypothesis of Theorem 81.18. Thus, $\gamma \in C$, the unit circle of $F^{(2)}$, and W is a K-subspace of $F^{(2)}$ such that $(W, K\gamma)$ is a strongly orthogonal KW-triple, Definition 81.8; in particular, the C-orbit of $W \oplus K\gamma$ is a spread $\zeta[\gamma]$ of $F^{(2)}$.

Then the $\tilde{C}$-orbit of the subspace $F^{(2)} \times L^{(2)}$ of

$$(W + T_{K,L}\gamma, 0) + L^{(2)}(1, \gamma) := \left\{ (w + t\gamma + \lambda\gamma, \lambda) \mid w \in W, t \in T_{K,L}, \lambda \in L^{(2)} \right\}$$

is an orthogonal spread $\Sigma_\zeta[\theta]$, of Q_L, and hence the components of $\Sigma_\zeta[\theta]$ are all subspaces of $F^{(2)} \times L^{(2)}$ of the form:

$$(*) \qquad \Sigma_\zeta[\theta] := \left\{ ((w + t\gamma + \lambda\gamma)\,\theta, \lambda) \mid w \in W, t \in T_{K,L}, \lambda \in L^{(2)} \right\}, \text{ for } \theta \in C.$$

PROOF. Since the putative components consist of an isometry-group orbit of the generator $\Sigma_\zeta[1]$, of Q_L, and this orbit has the correct number of components to be a spread, it is sufficient to verify that:

$$\theta \in C \setminus \{1\} \implies \Sigma_\zeta[1] \cap \Sigma_\zeta[\theta] = \mathbf{O}.$$

However, since $\Sigma_\zeta[1]$ and $\Sigma_\zeta[\theta]$ are distinct generators of Q_L that lie in the same ruling class they are either disjoint or meet in at least a rank two L-subspace, in which case they meet the L-hyperplane $F^{(2)} \times L$, of the L-space $F^{(2)} \times L^{(2)}$, non-trivially. Thus if $\Sigma_\zeta[1]$ and $\Sigma_\zeta[\theta]$ have non-trivial intersection, then the intersection includes non-trivial vectors whose final components lie in the *subfield* $L < L^{(2)}$, and hence we have:

$$((w + (t + \lambda)\gamma), \lambda) = ((w'\theta + (t' + \lambda')\gamma\theta), \lambda')$$

for some $w, w' \in W$, $t, t' \in \ker T_L \to T_L | K$ and $\lambda, \lambda' \in L$. Our goal is to show that all of these parameters are forced to be zero. Noting that $\lambda = \lambda' \in L$ is forced, we have $t + \lambda \in K$ and hence, since there cannot be a non-zero overlap of the subspaces $W + K\gamma$ and $(W + K\gamma)\theta$—they are *distinct* components of the symplectic spread $\zeta[\gamma]$—the only possibility is for $w = w' = 0$ and $t + \lambda = 0 = t' + \lambda'$. However, since $t, t' \in \ker T_L \to T_L | K$ and $\text{Fix}(T_L)$ includes L we have:

$$T_L(\lambda) = \lambda = 0 = T_L(\lambda') = \lambda'$$

and hence also $t = 0 = t'$. $\qquad\square$

LEMMA 81.24. Suppose $(0, z) \in F^{(2)} \times L^{(2)}$, where $z \in L^{(2)} \cap C$. Then, relative to the quadratic form Q_L, $(0, z)$ is non-singular and $(0, z)^\perp = F^{(2)} \times zL$. Hence, in the notation of $(*)$ in Theorem 81.23:

$$\forall \theta \in C \colon \Sigma_\zeta[\theta] \cap (0, z)^\perp = \{((w + t\gamma + z\lambda\gamma)\,\theta, z\lambda) \mid w \in W, t \in T_{K,L}, \lambda \in L\}.$$

PROOF. Since $Q_L((0, z)) = 0\bar{0} + z\bar{z} = 1$, $(0, z)$ is non-singular. Since Q_L polarizes to the symplectic form

$$(x, \lambda) \cdot (y, \mu) = T(x\bar{y} + \bar{x}y) + \lambda\bar{\mu} + \mu\bar{\lambda},$$

we have

$$(f, \lambda) \in (0, z)^\perp \iff \lambda\bar{z} + z\bar{\lambda} = 0 \iff \overline{\lambda\bar{z}} = \lambda\bar{z} \iff \lambda\bar{z} \in L \iff \lambda \in \frac{1}{\bar{z}}L = zL$$

as required. $\qquad\square$

We may identify the quotient symplectic L-space $F^{(2)} \times zL/\{0\} \times zL$ with $F^{(2)}$, using $(f, zl)/\{0\} \times zL \mapsto f$; note that this simply corresponds to truncating (f, zl) to f: $(f, zl) \equiv (f', zl') \pmod{zL}$ iff $(f - f', z(l - l')) \in (0, zL)$ iff $f = f'$; so the alternating form inherited is simply

$$x \cdot y = T_L\left(x\overline{y} + y\overline{x}\right), \forall x, y \in F^{(2)}.$$

Thus, we may treat the map:

$$\pi_z \colon F^{(2)} \times zL \to F^{(2)}$$

$$(f, z\lambda) \mapsto f$$

as the projection of $(0, z)^\perp$ onto a symplectic space that maps each orthogonal spread of Q_L onto its symplectic slice at $(0, z)$.

In particular, the slice at $(0, z)$, corresponding to the given orthogonal spread, is the symplectic spread on $F^{(2)}$ whose components are all subspaces of from

$$\left\{ ((w + t\gamma + z\lambda\gamma)\,\theta) \mid w \in W, t \in T_{K,L}, \lambda \in L \right\}, \ \theta \in C.$$

Thus we have established:

THEOREM 81.25. *The orthogonal spread* $\{\Sigma_\zeta[\theta] \mid \theta \in C\}$, *of Theorem 81.23, when sliced at* $(0, z)L$, *yields the symplectic spread* Slice(F, W, K, γ, L, z) *which may be identified with the symplectic spread on* $F^{(2)}$ *given by:*

$$\left\{ (W + T_{K,L}\gamma + zL\gamma)\theta \mid \theta \in C \right\},$$

when the symplectic form on $F^{(2)}$ *is given by:*

$$x \cdot y = T_L(x\overline{y} + y\overline{x}), \forall x, y \in F^{(2)}.$$

We note in passing that the above splitting of components are *direct* sums.

REMARK 81.26. *Assume the hypothesis of 81.25. For* $z \in L^{(2)} \cap C$:

$$W + T_{K,L}\gamma + zL\gamma = W \oplus T_{K,L}\gamma \oplus zL\gamma$$

PROOF. Since the left-hand side is a component of a L-spread of order $|F|$, its L-rank is $[F : L]$. Hence, recalling that $W + T_{K,L}\gamma$ is an L-hyperplane of $W + K\gamma = W \oplus K\gamma + F$, Lz must be disjoint from $W \oplus K\gamma$. The result follows. □

We now show that the spreads sliced by Theorem 81.25 may be viewed as strongly orthogonal KW-triples, *relative to* L. This has the implication that Theorem 81.25 may actually be applied inductively to produce large quantities of spreads.

THEOREM 81.27. *Let* $W' = W + T_{K,L}\gamma$. *The triple* $(W', L, \gamma z)$ *is a strongly orthogonal* KW-triple *on* $F^{(2)}$ *with the standard* T_L-alternating form, and the associated spread consists of the L-spaces:*

$$\left\{ (W + T_{K,L}\gamma + zL\gamma)\theta \mid \theta \in C \right\}$$

is the spread associated with a strongly orthogonal KW-triple.

PROOF. We have just seen that the given subspaces do form a symplectic spread. We check $T_L\left(w'\overline{w'}\right) = 0$, for $w' \in W'$. We may write $w' = w + t\gamma$, $t \in T_{K,L}$, so

$$T_L\left(w'\overline{w'}\right) = T_L\left(w\overline{w}\right) + T_L\left(t\gamma\overline{t\gamma}\right)$$

$$= 0 + T_L\left(t\overline{t}\gamma\overline{\gamma}\right)$$

and since $\gamma\bar{\gamma} = 1$:

$$= T_L\left(t^2\right)$$
$$= T_L\left(t\right)^2$$
$$= 0,$$

since $t \in \ker T_L$ by definition.

We verify the strong orthogonality condition:

$$(w' + t\gamma, \gamma z\lambda)_L := T_L\left(w'\overline{\gamma z\lambda} + \gamma z\lambda\overline{w'}\right)$$
$$= T_L\left((w + t\gamma)\overline{\gamma z\lambda} + \gamma z\lambda\overline{(w + t\gamma)}\right)$$

applying distributivity to the inner product

$$= T_L\left(w\overline{\gamma z\lambda} + \gamma z\lambda\overline{w}\right) + T_L\left(t\gamma\overline{\gamma z\lambda} + \gamma z\lambda\overline{t\gamma}\right).$$

Now it is sufficient to show that this vanishes when L is replaced by the containing field K. But then the term becomes

$$(w, \gamma z\lambda)_K + (t, \gamma z\lambda)_K$$

and since $z\lambda \in K^{(2)}$, recall that by definition $z \in L^{(2)}$, the first term vanishes—by the strong orthogonality hypothesis on the triple (W, γ, K)—so we are left with

$$(w', \gamma z\lambda)_L = T_L\left(t\gamma\overline{\gamma z\lambda} + \gamma z\lambda\overline{t\gamma}\right)$$

and by $\gamma\bar{\gamma} = 1$

$$= T_L\left(t\left(\overline{\gamma z\lambda} + z\lambda\right)\right)$$

and as $z \in L$ by self-conjugacy we have $\left(\overline{\gamma z\lambda} + z\lambda\right) \in L$, hence

$$= \left(\overline{\gamma z\lambda} + z\lambda\right) T_L(t) = 0$$

as $t \in \ker T_L$. Hence $(w', \gamma z\lambda)_L = 0$ and the strong orthogonality follows. $\qquad\square$

In view of the importance of the construction in Theorem 81.27, let us review it explicitly and illustrate how it can be applied recursively.

We begin with a strongly orthogonal KW-triple, (W, K, γ) where $K \subset F$ and $\gamma \in C$: the strong orthogonality condition might impose constraints on γ vis-a-vis K, for example, if $W = \ker T_K : F \to K$, then it appears necessary to insist that $\gamma \in C \cap K^{(2)}$, to guarantee strong orthogonality.

Theorem 81.27 is concerned with how to distort $W \oplus K\gamma$ so that the rotation orbit of the 'distortion' of $W \oplus K\gamma$ continues to be a spread of $F^{(2)}$—which is *strongly orthogonal relative to subfield* $L \subset K$, where $[K : L] > 1$ is odd. Specifically we seek to distort the $K\gamma$-part of $W \oplus K\gamma$, in terms of a subfield L, while leaving W intact. Writing $T_{K,L}\gamma \oplus L\gamma = K\gamma$, the proposed distortion simply involves 'rotating' the second part $L\gamma$ by an element $z \in L^{(2)} \cap C$, yielding:

$$W \oplus K\gamma = W \oplus T_{K,L}\gamma \oplus L\gamma \to W \oplus T_{K,L}\gamma \oplus zL\gamma;$$

now the rotation orbit of the left-hand side is in general a different spread from the right-hand side; we shall soon see that the kern of the right-hand side is often just L, and hence different from the kern of the left-hand side which, of course, must include K.

To enable smooth induction, we explicitly record an important case of the above discussion—for *starting* the induction hinted at in Theorem 81.27.

COROLLARY 81.28. *For $z \in K^{(2)} \cap C$, the orbit of $T_K + zK$, under the rotation group $\tilde{C}$, is a strongly orthogonal symplectic K-spread. But if $z \in K \cap C$, then the spread consisting of the orbit of $T_K + zK$ is simply the circular Desarguesian spread in $F^{(2)}$, with F as one of its components.*

PROOF. We apply the notation of Theorem 81.25, using 1-subscripted variables to avoid confusion. Thus, taking $\gamma_1 = 1$, $K_1 := F$ and $L_1 := K$. So $W_1 := 0$, $T_{K_1,L_1} := T_K$, the kernel of the trace $T_K \colon F \to K$, and $z_1 \in K^{(2)} \cap C$. Thus by Theorem 81.25, applied to the 1-subscripted variables, we have $T_K \oplus z_1 K$ defines a strongly orthogonal spread consisting of the $\tilde{C}$-orbit of $z_1 \in K^{(2)} \cap C$ is a strongly orthogonal symplectic spread. Moreover, for $z_1 = 1$ we have the Desarguesian case: the spread is simply the orbit of F under the rotation group. The final sentence follows because if $z \in K$ then we have:

$$T_K + zK = T_K + K = F,$$

on recalling that $[F : K]$ odd forces $T_K + K = F$. $\qquad\square$

REMARK 81.29. Let K be any subfield of $F^{(2)}$. Suppose $z \in K^{(2)} \cap C$. Then $z \in K$ if and only if $z = 1$.

The final sentence of the corollary, means no new spread is obtained if $z \in K$ (and this only happens if $z = 1$): so we might say that the 'chain' $F \supset K$ can be 'collapsed' to the 'simpler chain' F, unless $z \in K^{(2)} \setminus K$. This notion becomes helpful in the analysis of more complicated towers of fields that lead to spreads.

The above corollary may be viewed as the first non-trivial step in a recursive procedure on a tower of subfields F to obtain new circular spreads on $F^{(2)}$ from given (strongly orthogonal) spreads, particularly when the initial spread is the circular Desarguesian spread.

We apply the recursive procedure of Corollary 81.28 once more, partly as a simple exercise, and partly to display a more generic-looking example of a recursively generated circular spread. The presence of collapsing is again indicated.

Thus, starting from the situation in Corollary 81.28, viz., $W \oplus K = T_K \oplus K z_1$, for $z_1 \in K^{(2)} \cap C$, we 'distort' $K z_1$ by first writing it as $K z_1 = (T_{K,L} \oplus L) z_1$ and then rotate the second part $L z_1$ by $z_2 \in K^{(2)} \cap C$ to yield the L-subspace $T_K \oplus T_{K,L} z_1 \oplus L z_1 z_2$ which corresponds to a strongly orthogonal KW-triple $(T_K \oplus T_{K,L} z_1, L, z_2)$.

It is also obvious that if $z_2 \in L$ then $z_2 = 1$, by Remark 81.29, so we are back to the previous situation (collapsing $F \supset K \supset L$ to $F \supset L$), which means the 'new' spread is the same as the rotation orbit of $T_K + zK$. Thus we might sum up as follows:

COROLLARY 81.30. *Suppose the chain of Galois fields of even order $F \supset K \supset L$, satisfies the condition that $[F : K]$ and $[K : L]$ are both odd. Choose $z_1 \in K^{(2)} \cap C$ and $z_2 \in L^{(2)} \cap C$. Then the orbit Θ of the L-subspace of*

$$T_K + z_1 T_{K,L} + z_1 z_2 L = T_K \oplus z_1 T_{K,L} \oplus z_1 z_2 L,$$

under the rotation group $\tilde{C}$, is a strongly orthogonal symplectic L-spread, or equivalently $(T_K + z_1 T_{K,L} + z_1 z_2 L, L, z_2)$ is a strongly orthogonal KW-triple.

If $z_2 \in L$ (or equivalently $z_2 = 1$, Remark 81.29) the spread Θ becomes the same as the spread as in the previous corollary, viz., the rotation orbit of $T_K \oplus z_1 K$, and if $z_1 \in K$ (which means $z_1 = 1$), the spread Θ becomes the rotation orbit of $T_L \oplus z_2 L$ (with L playing the role of K).

Thus, we might say that if the distorting rotation z_i lies in the field F_i (i.e., $z_i = 1$ by Remark 81.29), then we can eliminate the field F_i from the tower, without losing any new spreads.

We now turn to the general version of the two corollaries above.

We shall consider the inductive procedure, indicated in Corollaries 81.28 and 81.30, in more detail later on.

Returning to orthogonal spreads in the context of Theorem 81.25, one might say that the symplectic slice *at* $(0, z)$, arising from the orthogonal spread $\Sigma_\zeta[\theta]$, is obtained by restricting $L^{(2)}$-valued parameters to L values and completely slicing off the final $L^{(2)}$-position, from $F^{(2)} \times L^{(2)}$.

We now show that L can sometimes be forced to be the *full* kern. Thus, by varying L we can produce a range of flag-transitive spreads, of given order $2^{2n+1} > 8$, that are non-isomorphic and even have kerns of different size: much more can be achieved, but we set ourselves modest goals for now.

We first review the normalizer-version of Schur's lemma, for finite fields.

REMARK 81.31. (Schur Lemma for Normalizers). Let $F = GF(p^r)$ be any finite field, and U a non-trivial u-subgroup of $GL(F, +)$ where u is a p-primitive divisor of $p^r - 1$. For $f \in F$ let

$$\tilde{f} \colon F \to F$$

$$\tilde{f} \colon f \mapsto fx, \quad \text{for } f \in F$$

and let

$$\tilde{F}^* = \{\tilde{f} \mid f \in F^*\}.$$

Then the normalizer of U in $GL(F, +)$ is $\Gamma L(1, F)$.

PROOF. Since the Sylow u-subgroups of $GL(F, +)$ are irreducible and hence semiregular, they must be conjugate to the Sylow u-subgroups of the cyclic group $\tilde{F}$. Hence, applying Schur's lemma, the centralizer of U must correspond to the multiplicative group of a field containing U and hence must be $\tilde{F}$: if it were smaller, U would fail to be irreducible as it would fix the subfield. So the normalizer of U must normalize $\tilde{F}$ and hence must be a group of semilinear maps of F as an F-space: this means $U \leq \Gamma L(1, F)$, as required. $\qquad\square$

We can now demonstrate that we have considerable control on the kern-size of circular spreads obtained by a single 'perturbation' of a Desarguesian spread. Later, Theorem 81.46, we shall state an inductive version of this result.

THEOREM 81.32. Choose $W = \ker T_K \colon F \to K$. Then kern of the sliced spread Slice(F, W, K, γ, L, z), of the L-space $F^{(2)}$, is precisely L, unless $z \in L$: in which case the sliced spread is simply the standard Desarguesian spread whose components in $F^{(2)}$ are simply the non-zero multiplicative cosets of F—and F^* is the kern group.

PROOF. Let u be any 2-primitive divisor of $q^2 - 1$, where $q = |F|$. Since C is transitive on the components of the spread C contains a non-trivial u-Sylow subgroup, say U. U is semilinear over the kern Λ of the given spread. However,

$|\Lambda| - 1$ must divide $|F| - 1 = q - 1$, since F is an L-space, so U must centralize, rather than just normalize, the multiplicative group of Λ. But by Schur's lemma the centralizer of U can only be $F^{(2)}$. Hence Λ must act on $F^{(2)}$ by an action of type $h_a \colon x \mapsto xa$, for some $a \in F^{(2)}$.

Writing $W = \ker T_K$, $T = T_{K,L}$, we must have h_a fixing each component in the orbit of the subspace $W + T + Lz$ and, in particular, we have:

$$a(W + T + Lz) = W + T + Lz.$$

Thus aW, a subspace of the left-hand side, lies in the right-hand side. However W and aW are subspaces of the right-hand side, each with $GF(2)$-rank $> \frac{1}{2}\operatorname{rank}(F)$. Since the right-hand side has size $|F|$, this forces $aW \cap W \neq \{0\}$. So $aw_1 = w_2 \neq 0$, for some $w_1, w_2 \in W$, hence $a = w_1/w_2 \in W < F$. In particular $\bar{a} = a$.

Similarly, choosing $l_0 \in L^*$, we have

$$al_0 z = w + t + lz \ \exists\, w \in W,\ t \in T,\ l \in L,$$

and adding this to its conjugate, while noting that $a, l, l_0, t \in F$ must be self-conjugate, we have:

$$al_0 (z + \bar{z}) = l (z + \bar{z})$$

and this forces $a = l/l_0 \in L$ unless $z = \bar{z}$. Hence the result follows: in the exceptional case when $z \in F$ (recall $z \in L^{(2)}$) our notation implies that the component being considered is simply F and its C-orbit is clearly the Desarguesian spread on $F^{(2)}$ consisting of the non-zero multiplicative cosets of F.   □

We now obtain a powerful tool for analyzing isomorphisms among circular spreads, given in Kantor and Williams [**880**]. This applies to more general representations of alternating forms than we have so far considered.

Any non-zero additive map $\phi \colon F \to GF(2)$ defines a non-singular alternating form $\phi(x\bar{y} + y\bar{x})$ on $F^{(2)}$, that includes all of the types of alternating forms on $F^{(2)}$ that we have considered so far. In particular, the rotation group is a group of isometries for any such form, and in fact any scalar isometry must correspond to a rotation. Thus, we may discuss the isomorphisms among symplectic spreads defined in this broader context, rather than considering only the trace-defined symplectic forms discussed above.

We begin by recording and verifying the assertions just indicated.

PROPOSITION 81.33. Let $F^{(2)}$ be regarded as a $GF(2)$ space and let $\phi \colon F \to GF(2)$ be any non-zero additive map. Then

(a) The $GF(2)$-bilinear form on $F^{(2)}$ defined by:

$$(x, y)_\phi = \phi(\bar{x}y + \bar{y}x)$$

is a non-singular alternating $GF(2)$-bilinear form on $F^{(2)}$, and the rotation group $\tilde{C}$ preserves the form.

(b) The rotation group, associated with C, is a group of isometries for the alternating form $(\cdot, \cdot)_\phi$ on $F^{(2)}$, and a scalar map of form $x \mapsto xf$, for $f \in F^{(2)}$, is an isometry iff $f \in C$.

PROOF. (a) If $x \neq 0$ then $\bar{x}y$ ranges over $F^{(2)}$ as y ranges over $F^{(2)}$, hence $\bar{x}y + \bar{y}x$ ranges over

$$\left\{ f + \bar{f} \mid f \in F^{(2)} \right\} = F,$$

thus the defined alternating form has x in its radical iff $x = 0$.

(b) The map $x \mapsto ax$ is an isometry iff:

$$(x, y)_\phi = \phi(ax, ay)_\phi$$
$$\Longleftrightarrow \phi(x\overline{y} + y\overline{x}) = \phi(ax\overline{ay} + ay\overline{ax})$$
$$\Longleftrightarrow \phi(x\overline{y} + y\overline{x}) = a\overline{a}\phi(x\overline{y} + y\overline{x})$$
$$\Longleftrightarrow (x, y)_\phi = a\overline{a}(x, y)_\phi,$$

and since the inner product is not identically zero, $a\overline{a} = 1$, yielding the desired result. $\qquad\square$

We return to the circular spreads in $F^{(2)}$; the spreads are permitted to symplectic relative to any fixed $(\,\cdot\,,\,\cdot\,)_\phi$, of the proposition above. It is rather surprising that if the *translation planes* associated with any two such symplectic spreads, S and S', are isomorphic then the isomorphism is also induced by a field isomorphism of $F^{(2)}$, that is, $S^\sigma = S'$, for $\sigma \in \text{Gal}\left(F^{(2)}\right)$. We now establish this.

THEOREM 81.34. Let $\phi \colon F \to GF(2)$ be any non-zero additive map and regard $F^{(2)}$ as a $GF(2)$ space equipped with the non-singular alternating bilinear form

$$(x, y)_\phi = \phi\left(\overline{x}y + \overline{y}x\right).$$

Suppose S and S' are rotation-invariant symplectic spreads of $F^{(2)}$, relative to the alternating form $(x, y)_\phi$.

Then the translation planes associated with S and S' are isomorphic only if there is a $\sigma \in \text{Gal}\left(F^{(2)}\right)$ such that $S^\sigma = S'$.

PROOF. Recall that symplectic spreads yield isomorphic planes iff there is an isometry mapping one onto the other. Hence an isometry $g \colon F^{(2)} \to F^{(2)}$ satisfies $g(S) = S'$.

Let $|F^{(2)}| = 2^n - 1$ and suppose $u^a \| 2^n - 1$, where u is any 2-primitive divisor of $2^n - 1$: by Zsigmondy this must exist since, by hypothesis, the odd integer $n > 2$ (see Theorem 7.1).

Hence, the u-Sylow subgroups of any transitive subgroup of $Sp(F^{(2)}, +)$, including $\tilde{C}$, are irreducible groups of order u^a. Thus, $\tilde{C}$ contains an irreducible u-Sylow subgroup U, and U leaves invariant all the circular spreads in $F^{(2)}$, including S and S'. It follows that U^g lies in $\text{Aut}(S') \cap Sp(F^{(2)}, +)$ and hence, by the conjugacy of U and U^g in the symplectic group, we may choose $h \in \text{Aut}(S') \cap Sp(F^{(2)}, +)$ such that $U^g h = U$. Thus gh normalizes the irreducible group U and hence by the normalizer-version of Schur's lemma, Remark 81.31,

$$gh \in \Gamma L(1, F^{(2)}) \cap Sp(F^{(2)}, +) = \{x \mapsto ax^\sigma \mid a \in C, \sigma \in \text{Gal}(F^{(2)})\},$$

since $x \mapsto ax^\sigma$ is an isometry if and only if $a \in C$, by Proposition 81.33. So

$$S = (S)gh = (S')h = a(S')^\sigma,$$

yielding

$$a^{-1}(S) = (S')^\sigma,$$

and since $a \in \tilde{C}$ it fixes all circular symplectic spreads, so

$$a^{-1}(S) = (S) = (S')^\sigma,$$

as required. $\qquad\square$

REMARK 81.35. Assume the hypothesis of Theorem 81.34. By considering the Vaughan polynomial of the additive map ϕ, it is not difficult to show that the converse of the theorem may fail, i.e., not all σ are isometries: in fact σ commutes with ϕ, or equivalently, σ is an isometry, iff the coefficients of the Vaughan polynomial are in the fixed field of σ.

We have shown in this section how to construct large numbers of *symplectic* flag-transitive spreads that are odd-dimensional over their kerns. Not only are the planes obtained non-isomorphic, but their dimensions over their kerns can be varied almost at will.

We now show that certain *orthogonal* spreads are non-isomorphic, following approximately the proof of Theorem 81.34 above.

LEMMA 81.36. The normalizer of the group $C \times 1$ leaves invariant the subspace $F^{(2)} \times 0$, of the L-space $F^{(2)} \times L^{(2)}$.

PROOF. Since $C \times 1$, by definition, leaves $F^{(2)} \times 0$ it is sufficient to show that it has a unique Maschke complement of $0 \times L^{(2)}$. Since, by its size, it acts irreducibly on its Maschke complements they must be disjoint. Hence, if $n + 2$ denotes the L-rank of $F^{(2)} \times L^{(2)}$, we have $n + 2 \geq 2n$, contradicting our hypothesis that $n > 1$ is odd.   □

In the following, the ambient group Θ is the full semilinear orthogonal group of $F^{(2)} \times L^{(2)}$ relative to the standard quadratic form associated with Q_L.

LEMMA 81.37. The normalizer H of the group $C \times 1$ is a map of form:

$$h \colon F^{(2)} \times L^{(2)} \to F^{(2)} \times L^{(2)}$$
$$(x, \lambda) \mapsto (ax^\sigma, \nu(\lambda)),$$

where ν is some non-singular additive map of $L^{(2)}$. Moreover, if $h \in \Theta$ then $a \in C$ and if further h leaves invariant a non-zero subset of $F \times 0$ then $a = 1$.

PROOF. By the lemma above, the normalizer must fix $F^{(2)} \times 0$, and h is of the desired form by applying to the restriction $C \times 1 || L^{(2)}$ the normalized version of Schur's lemma, Remark 81.31.

If $h \in \Theta$ then $Q_L(h((x, 0))) = Q_L((x, 0))$ for $x \in F^{(2)}$, and this means

$$T_L(ax^\sigma, \overline{ax^\sigma}) = T_L(x, \overline{x})$$
$$T_L(a\bar{a}x^\sigma, \overline{x^\sigma}) = T_L(x, \overline{x})$$
$$T_L((a\bar{a} + 1)x^\sigma, \overline{x^\sigma}) = 0,$$

and since $x^\sigma, \overline{x^\sigma}$ runs over all F we have $(a\bar{a} + 1)x^\sigma, \overline{x^\sigma}$ running over all F unless $a\bar{a} = 1$, and this must hold since $T_L(1) = 1$ prevents T_L from vanishing on any field. Hence $a \in C$ is the only possibility. If now h were to fix a non-zero subset of F then we would have $a \times f = f_2$, $f_1, f_2 \in F$ forcing $a \in F$. But $C \cap F = 1$.   □

LEMMA 81.38. Let $W = \ker T_K$ and $T = T_{K,L}$. Suppose that $\mathcal{Q}_i$, for $i = 1, 2$, are isomorphic circular spreads associated with the $C \times 1$ orbits respectively of the totally singular subspaces Q_i, $i = 1, 2$ given by:

$$Q_i = (W \oplus T) + (z_i, 1) L + (z_i, 1) \iota L.$$

Then the isomorphism h can be chosen so that it maps Q_1 onto Q_2 and has the form:

$$h \colon F^{(2)} \times L^{(2)} \to F^{(2)} \times L^{(2)}$$
$$(x, \lambda) \mapsto (x^{\sigma}, \nu(\lambda)),$$

and the isomorphism cannot exist unless $z_2 = z_1^{\sigma}$.

PROOF. Let g, in the group of orthomorphisms Θ, be such that $g \colon \Theta_1 \to \Theta_2$. Hence Θ_2 admits C and C_g. Let U be a u-Sylow subgroup of $C \times \{1\}$ and so this group and U^g are in $\mathrm{Aut}(\Theta_2)$, and must be the u-Sylow subgroup of this group. Hence there is an $h \in \mathrm{Aut}(\Theta_2)$ such that $U^{gh} = U$, so gh normalizes U and hence also $C \times \{1\}$. Thus gh must, by the previous lemma, be of the form

$$h \colon F^{(2)} \times L^{(2)} \to F^{(2)} \times L^{(2)}$$
$$(x, \lambda) \mapsto (ax^{\sigma}, \nu(\lambda)),$$

where ν is some non-singular additive map of $L^{(2)}$, and $a \in C$. Since the Θ_i's admit $C1$ as a flag transitive group, we may apply the automorphism $(x, l) \to (xA, l)$, for some $A \in C$ to Θ_2 so that Q_1 maps onto Q_2 as well: that is without loss of generality h maps Q_1 onto Q_2.

Noting that $Q_i \cap F \times 0 = (W \oplus T) \times 0$ and h leaves invariant $F \times 0$ and both Q_i's, we conclude that h leaves invariant $W \oplus T \subset F$, and hence $a(W \oplus T)^{\sigma} = W \oplus T \subset F$, and now the previous lemma forces $a = 1$.

Now apply the map h to the generic element of Q_1:

$$(w + t, 0) \oplus x(z_1, 1) \oplus y\imath(z_1, 1),$$

means that the image is

$$(w' + t', 0) \oplus x(z_1^{\sigma}, 1) \oplus y\imath^{\sigma}(z_1^{\sigma}, 1) = (w + t, 0) \oplus x(z_1^{\sigma}, 1) \oplus y\imath^{\sigma}(z_1^{\sigma}, 1),$$

since σ is the identity on F, and this coincides with some element of Q_2

$$(w + t, 0) \oplus x'(z_2, 1) \oplus y'\imath(z_2, 1),$$

and in particular $w = t = 0$ yields:

$$x(z_1^{\sigma}, 1) \oplus y\imath^{\sigma}(z_1^{\sigma}, 1) = x'(z_2, 1) \oplus y'\imath(z_2, 1).$$

Now comparing the components:

$$x z_1^{\sigma} + y\imath^{\sigma} z_1^{\sigma} = x' z_2 + y'\imath z_2$$
$$x + y\imath^{\sigma} = x' + y'\imath$$

and now substituting for $y'\imath$ in the first equation yields the condition:

$$x z_1^{\sigma} + y\imath^{\sigma} z_1^{\sigma} = x' z_2 + z_2(x + y\imath^{\sigma} + x')$$

which reduces to

$$x z_1^{\sigma} + y\imath^{\sigma} z_1^{\sigma} = z_2(x + y\imath^{\sigma}),$$

and equating the real and imaginary parts yields:

$$x z_1^{\sigma} = x z_2 \quad \text{and} \quad y z_1^{\sigma} = y z_2,$$

and these force $z_1^{\sigma} = z_2$. $\qquad \square$

This brings us to the main conclusion:

THEOREM 81.39. The isomorphism classes of the orthogonal spreads associated with the Q_i correspond to the number of orbits of the rotation group of $L^{(2)}$ under the field automorphism group of $F^{(2)}$.

So we have now seen examples of non-isomorphic *orthogonal* spreads. Later on an inductive version of this result will be presented, Theorem 81.49

81.3. Flag-Transitive Scions of Desarguesian Spreads.

We now survey the classification, due to Kantor and Williams, of all the flag-transitive scions of Desarguesian spreads. Speaking loosely, the *scions* of any symplectic spread are all the spreads obtainable from it by applying a sequence of up-and-down moves, i.e.,

symplectic → orthogonal, or orthogonal → symplectic(slices),

possibly interspersed with field changes (expansion). We stress that the classification we consider deals only with the *flag-transitive* scions of Desarguesian spreads: other types of scions of these spreads are not considered here.

We have already seen that strongly orthogonal KW-spreads could be recursively defined, essentially by taking appropriate scions of some pre-selected strongly-orthogonal KW-spreads, and since any KW-spread is flag-transitive there is a recursive method for obtaining flag-transitive scions of any given strongly-orthogonal KW-spread.

Below we develop explicitly a class of such scions for Desarguesian spreads, by relating them to certain towers of fields: this correspondence turns out to yield all the flag-transitive scions of the Desarguesian spreads; the complete solution of the isomorphism problem—for two such scions to yield isomorphic flag-transitive spreads—is also mentioned.

Let

$$F = F_0 \supset \cdots \supset F_i \supset \cdots \supset F_n$$

denote a tower of finite fields, of even order, so each of the quadratic extensions $F_i^{(2)} \supset F_i$ lies in $F^{(2)}$ and $F_i^{(2)} \backslash F \neq \varnothing$. We shall assume further that, for $0 \leq i < n$, each degree $[F_i : F_{i+1}]$ is *odd* and > 1.

Let $T_i \colon F \to F_i$ denote the trace map from the top field F to the subfield F_i, for $i = 1, \ldots, F_n$, and let W_i denote the kernel of the trace map T_i restricted to T_{i+1}, thus for $i \geq 0$:

$$W_i := \ker : T_{i+1} | F_i \to F_{i+1}$$
$$x \mapsto T_{i+1}(x).$$

Using the condition that $[F_i : F_{i+1}]$ is odd, we have:

REMARK 81.40. For $0 \leq i < n$: $F_i = W_i \oplus F_{i+1}$.

PROOF. Since $[F : F_{i+1}]$ is clearly odd, and the characteristic is 2, T_{i+1} must fix identically all the elements of F_{i+1}: recall that T_{i+1} is sum of all the elements of $\mathrm{Gal}\,(F : F_{i+1})$. In particular $\ker T_{i+1}$ and F_{i+1} have trivial intersection. Hence $T_{i+1} | F_{i+1}$ maps F_i onto its subfield F_{i+1}, and induces the identity on it. The result follows on recalling that W_i denotes the kernel of the restriction $T_{i+1} | F_{i+1}$. □

Hence by an easy induction we obtain:

(81.3.1) $F = W_0 \oplus \cdots \oplus W_{n-1} \oplus F_n$

Recall that we have already encountered this situation earlier for $n = 1$ and $n = 2$, where W_0 represents ker : T_K, from our earlier notation.

$$F = \ker T_K \oplus z_1 K, \exists z_1 \in K^{(2)}$$

$$F = \ker T_K \oplus T_{K,L} z_1 \oplus L z_1 z_2 \,\exists\, z_2 \in L^{(2)}$$

We noted that if $z_1 \in K$, respectively $z_2 \in L$, then we have $z_1 = 1$, respectively $z_2 = 1$, inducing appropriate 'collapses' in the two towers, in the sense that the collapsed towers lead to the same strongly orthogonal spreads as the uncollapsed version.

One can of course easily generalize all this to arbitrary towers of fields where each field is an odd dimensional extension of its successor. This results in towers that correspond to strongly orthogonal spreads on $F^{(2)}$, and that repeated collapsing of the tower eventually leads to a unique 'irreducible' tower that yields the same spread as the initial tower.

To describe the spreads associated with any such tower, we need to define the permitted 'rotation' γ_n, intended to distort the final field of the given tower by:

$$\zeta_0 = 1$$

$$\zeta_i \in F_i^{(2)} \cap C \,\forall\, 1 \leq n$$

$$\gamma_n := \prod_0^n \zeta_i \,\forall\, 1 \leq i \leq n,$$

and this yields the permitted distortion:

$$W_0 \gamma_0 + \cdots + W_{n-1} \gamma_{n-1} + F_n \gamma_n \to W_0 \gamma_0 \oplus \cdots \oplus W_{n-1} \gamma_{n-1} \oplus F_n \gamma_n,$$

where the directness of the sum on right-hand side (or the left-hand side) follows, for example, by the fact that the right-hand side is spread-sized, since its rotation orbit is a spread (by an easy induction):

THEOREM 81.41. (Desarguesian Scions). The $\hat{C}$-orbit

$$\mathcal{Z}\left((F_i)_0^n, (\zeta_i)_0^n\right) := \left\{ \theta \left(\sum_{i=0}^{n1} W_i \gamma_i + F_n \gamma_n \right) \,\middle|\, \theta \in C \right\}$$

is a strongly orthogonal spread on $F^{(2)}$ over the field F_n.

To describe the isomorphism classes among the spreads associated with the theorem above, we observe that collapsing a tower does not change the associated spread, and every tower collapses to a unique 'irreducible tower'. Hence we introduce the following:

DEFINITION 81.42. The pair

$$\mathcal{Z} = \left((F_i)_0^n, (\zeta_i)_0^n\right),$$

is a TOWER, and if $\zeta_i \neq 1$, for all $i > 1$, then the tower is REDUCED. A REDUCTION of any tower is obtained by deleting from it, for some $I \geq 1$, the field F_I in the tower $(F_i)_0^n$, and also the corresponding ζ_I in the ζ-sequence $(\zeta_i)_0^n$, but leaving unchanged the order of the remaining n fields in the tower sequence and the remaining n terms in the ζ-sequence.

Thus, starting from a given defining pair one can apply a series of reductions to obtain a unique reduced tower. So all towers define strongly orthogonal spreads, and reducing any tower leads to the same spread. Thus:

THEOREM 81.43. (Reduced Pairs). [**862**, Corollary 4.5]. The symplectic spread associated with a tower is the same as the symplectic spread obtained by reducing it, and hence every tower defines the same symplectic spread ζ as a reduced tower. Towers based on any all-one ζ-sequence always yield the (circular) Desarguesian spread.

Thus the isomorphism problem for the scions of Desarguesian spreads that are associated with towers (and hence are flag-transitive) is equivalent to the isomorphism problem for spreads corresponding to reduced towers. The complete answer to the isomorphism of spreads associated with towers is due to Kantor and Williams: two towers define isomorphic translation plane only in the obvious case—when the towers may be obtained from the other by applying a field automorphism of $F^{(2)}$:

THEOREM 81.44. (Classification of Desarguesian Scions). Given reduced Desarguesian towers $((F_i)_0^n, (\zeta_i)_0^n)$ and $\left((F_i')_0^{n'}, (\zeta_i)_0^{n'}\right)$ corresponding to the field $F^{(2)}$, of order > 8, the corresponding translation planes are isomorphic if and only if for some $\sigma \in \operatorname{Gal} F^{(2)}$:

$$\forall i: \quad n = n', \ F_i = F_i', \quad \text{and} \quad \zeta_i' = \zeta_i \sigma.$$

We shall omit the proof of the theorem which, although fairly short, [**862**, Corollary 4.5], is rather intricate. Instead we mention an obvious, but important, consequence: [**862**, Corollary 4.5].

COROLLARY 81.45. For $|F^{(2)}| > 8$, two defining ζ-sequences, $(\zeta_i)_0^n$ and $(\zeta_i')_0^n$, based on the *same* tower of fields $(F_i)_0^n$, produce isomorphic planes if and only if for some $\sigma \in F^{(2)}$

$$\zeta_i' = \sigma(\zeta_i) \quad \forall i.$$

It is also worth observing that the kern of a reduced tower is the smallest field of the tower: this is an inductive version of Theorem 81.32, where 'towers' corresponding to $F \supset K \supset L$ were considered.

THEOREM 81.46. The kernel of the spread of a reduced tower, associated with a field chain $F \supset F_1 \supset F_2 \supset \cdots \supset F_n$, is the field F_n.

Thus, generally speaking, there exist many non-isomorphic flag-transitive affine translation planes of a given order n, whenever $\log_2 n > 3$ is an integer with an odd divisor > 1, but there is a range of possible kern sizes for a fixed n. We have shown how to construct such examples by taking tower-based scions of Desarguesian spreads.

The question still remains whether there are any further Desarguesian *flag-transitive* scions of Desarguesian spreads. This is not the case, as stated in the following theorem, so the theorem implies that all *flag-transitive* scions of Desarguesian spreads are completely classified.

THEOREM 81.47. (Desarguesian Scions). The $\hat{C}$-orbit

$$\mathcal{Z}\left((F_i)_0^n, (\zeta_i)_0^n\right) := \left\{ \theta \left(\sum_{i=0}^{n1} W_i \gamma_i + F_n \gamma_n \right) \ \middle| \ \theta \in C \right\}$$

is a strongly orthogonal spread on $F^{(2)}$ over the field F_n, for all $\zeta_i \in F_i$, $\zeta_i \neq 1$ for $i \geq 1$ and $\zeta_0 = 1$.

Moreover, every flag-transitive translation plane that is associated with a scion of a Desarguesian spread must be isomorphic to one of these spreads.

PROOF. We only provide a vague sketch. To see that the spreads mentioned include all the flag-transitive scions, consider their associated orthogonal spreads. By the flag-transitive hypothesis, note that they admit cyclic groups of order $|C|$ and that two such groups in $\Gamma O^+ \left(F^{(2)} \times L^{(2)} \right)$ are mutually conjugate, and hence they are all isomorphic to $\tilde{C} \times 1$. The flag-transitive spreads may thus be assumed to admit this group, and the only permitted flag-transitive slices are the standard ones, leading to one of the spreads of the tower. $\qquad \square$

The classifications above also point to a limitation of Desarguesian scions.

REMARK 81.48. If p is a prime then all flag-transitive scions of Desarguesian spreads of order 2^p are Desarguesian spreads.

Thus, it appears to be an open question whether there are any symplectic flag-transitive spreads of order 2^p. More generally, the reader will have noticed that the hardest place to find new translation planes of non-prime order n occur when $n = 2^p$, p a prime.

We end by observing that the natural *orthogonal* spreads associated with flag-transitive scions, of Desarguesian symplectic spreads, are isomorphic as *orthogonal* spreads—in an appropriate sense—iff their associated towers correspond to isomorphic affine translation planes. This is an inductive version of Theorem 81.39 proved earlier.

THEOREM 81.49. (Orthogonal Spreads of Desarguesian Scions). Let L be a subfield of F, $|F| > 8$, such that $[F : L]$ is odd. Suppose $\zeta = ((F_i)_0^n, (\zeta_i)_0^n)$ and $\zeta' = \left((F_i')_0^{n'}, (\zeta_i)_0^{n'} \right)$ are Desarguesian towers of F such that L is a proper subfield of the field $F_n \cap F_n'$.

Then the orthogonal spreads Σ_ζ and $\Sigma_{\zeta'}$ of the standard L-space $F^{(2)} \times L^{(2)}$ are equivalent under $\Gamma O^+ \left(F^{(2)} \times L^{(2)} \right)$ if and only if the associated affine planes A_ζ and A_ζ' are isomorphic.

Note that the theorem implies that many non-isomorphic orthogonal L-spreads exist, at least in the technical sense that are not equivalent under $\Gamma O^+ \left(F^{(2)} \times L^{(2)} \right)$.

Symplectic Spreads.

Let π be a semifield flock of a quadratic cone translation plane with spread in $PG(3, q)$. In 1987, Johnson [710] showed that semifield flocks of quadratic cones, semifields of order q^2 that commute over a left nucleus $GF(q)$, are equivalent to semifields of order q^2 that commute over a right or middle nucleus isomorphic to $GF(q)$. Since semifields of order q^2 that commute over a middle nucleus $GF(q)$ are commutative, there is an implicit connection with commutative semifields of order q^2 with middle nucleus $GF(q)$ and semifield flocks.

In 1994, Thas and Payne [1094] have shown geometrically that given a (Cohen and Ganley [238]) semifield flock, there is a corresponding semifield spread in $PG(3, q)$, whose dual lies in $Q(4, q)$; the semifield spread is symplectic with symplectic dimension 2. Even though Thas and Payne worked solely with the Cohen–Ganley flock, this result is also more generally valid by work of several authors Lunardon [945], Bloemen [147], Thas [1185] (see also Johnson and Payne [792] (section 14)). The construction connects the associated generalized quadrangle of order (q^2, q), forms the translation dual of order (q, q^2), within which it is realized that there are subquadrangles isomorphic to $Q(4, q)$, and subsequently determines ovoids of isotropic points of a symplectic polarity within $PG(4, q)$. The dual of $Q(4, q)$ is $M(3, q)$, the set of isotropic points and lines of a symplectic polarity of $PG(3, q)$. Thus, the ovoids of $Q(4, q)$ determine symplectic spreads of $PG(3, q)$ that turn out to be semifield spreads, although this is not altogether completely obvious from this construction. In Chapter 49 on transversal spreads, we discuss the more general ideal of a 'translation dual' to translation ovoids corresponding to semifield spreads in $PG(3, q)$.

Recently, in 2003, relying on what we will call the Thas–Payne construction, Ball and Brown [83] have shown that there are six semifield spreads associated with a semifield flock spread, two of which are spreads in $PG(3, q)$. Furthermore, these two spreads in $PG(3, q)$ are isomorphic if and only if the semifield flock spread is Kantor–Knuth. Also, in 2003, Kantor [877] has now connected symplectic semifield spreads with commutative semifields by the iterative construction process of transpose and dualization (from the commutative side). We begin with this construction.

82.1. Cubical Arrays.

Let $(GF(q^n), +, \circ)$ define a pre-semifield, where the binary operation $\circ$ is $GF(q)$-linear in the sense that the functions R_y, and L_y such that $R_y(x) = x \circ y$ and $L_y(x) = y \circ x$ are $GF(q)$-linear. Choose a basis $\{v_i; i = 1, 2, \ldots, n\}$ of $GF(q^n)$ over

$GF(q)$. Then any multiplication

$$\sum_{i=1}^{n} t_i v_i \circ \sum_{i=1}^{n} z_i v_i,$$

for t_i, $z_i \in GF(q)$, $i = 1, 2, \ldots, n$ is determined uniquely by the elements $v_i \circ v_j$. Hence,

$$v_i \circ v_j = \sum_{k=1}^{n} a_{ijk} v_k$$

for $a_{ijk} \in GF(q)$. The set $\{a_{ijk}\}$ is called a 'cubical array' (Knuth [**898**]). We may then define the 'dual' and 'transpose' of a semifield by this process. If $x * y = y \circ x$, then we obtain a semifield $(GF(q^n), +, *)$, which is said to be 'dual' to the original. In this case, the dualization process in terms of cubical arrays is $\{a_{ijk}\} \longmapsto \{a_{jik}\}$. If we write the corresponding spread using a set of matrices M, then there is a corresponding spread with matrices M^t (the transpose). The spread is called the 'transposed spread'. This is done easily by writing v_j as a matrix as

$$[a_{ijk}]_{i,k=1,2,\ldots,n}.$$

If we transpose these matrices we obtain the corresponding matrix spread set of

$$[a_{kji}]_{i,k=1,2,\ldots,n}.$$

The corresponding cubical array then is obtained from the original as follows

$$\{a_{ijk}\} \longmapsto \{a_{kji}\}.$$

REMARK 82.1. These processes, dualization and transposition may be iterated producing six possible semifields, which is equivalent to the permutations of the subscripts on the cubical arrays. The connection with dualization and transposition is considered and is implicit in Johnson [**666**] and is further explicated in Ball et al. [**81**]. We note that the transposed spread corresponds to a duality of the corresponding projective space and we obtain a spread in the dual vector space. Such a spread "a dual spread" always occurs and is then a spread in particular when considering finite spreads. The dual spread to the spread is then obtained directly from the matrix spread set by transposing the matrices.

The polarity used may be considered a symplectic polarity defined by

$$\begin{bmatrix} 0 & I \\ -I & 0 \end{bmatrix}.$$

Considering transposition, it is possible that the transposed spread is identical to the spread. If the matrices are self-transpose, i.e., symmetric, we obtain a 'symplectic spread'. Conversely, a spread is symplectic provided there is a symplectic form leaving invariant the spread members and may be coordinatized so that the above form applies.

We note that a semifield is commutative if applying dualization to the semifield, the original semifield is obtained. That is, if

$$\{a_{ijk}\} \longmapsto \{a_{jik}\} = \{a_{ijk}\}.$$

Similarly, the semifield is symplectic if applying transposition to the semifield, the original semifield is obtained. That is,

$$\{a_{ijk}\} \longmapsto \{a_{kji}\} = \{a_{ijk}\}.$$

Use $\{a_{ijk}\}^d$, to denote the dualization process. Similarly, use $\{a_{ijk}\}^t$, to denote the transposition process. Assume that we begin with a commutative semifield, so that the dualization process produces the same semifield, then apply transposition and then dualization. That is, we consider

$$\{a_{ijk}\}^{dtd},$$

when $\{a_{ijk}\}^d = \{a_{ijk}\}$. But, note that dualization is a (12) permutation and transposition is a (13) permutation and $(12)(13)(12)(13) = (13)(12)$. This says the (12) permutation produces the same cubical array, then the semifield plane obtained by transposing and then dualizing has the property that it is self-transpose individually. That is,

$$\left\{\{a_{ijk}\}^{dtd}\right\}^t = \{a_{ijk}\}^{td}.$$

Hence, we have:

PROPOSITION 82.2. (Kantor [**877**]). The transposition and dualization of a commutative semifield is a symplectic semifield. Conversely, the dualization and transposition of a symplectic semifield is a commutative semifield.

82.2. Symplectic Flocks.

As noted in Section 82.1 on cubical arrays, a commutative semifield of order q^2 with middle nucleus $GF(q)$ will construct a symplectic spread in $PG(3, q)$, with symplectic dimension two. Actually, we show that any symplectic semifield spread in $PG(3, q)$ must have symplectic dimension two.

In this setting, if one tries to construct semifield spreads by an iteration of the construction processes of transpose and dualization, it turns out that there are not six but three possible spreads (see also Theorem 82.22). Using the Thas–Payne construction, there are three additional semifield spreads. But, also using the constructions of distortion, t-extension and derivation, it is possible to construct the same semifield spreads completely algebraically.

Hence, we see that there is now a complete equivalence of symplectic semifield spreads in $PG(3, q)$ and semifield flocks of quadratic cones in $PG(3, q)$ (see section 82.3), using only the algebraic methods of construction of various semifields of transpose, dualization, t-extension, derivation and t-distortion. In this way, Biliotti, Jha and Johnson [**128**] obtain the results of Ball and Brown [**83**], independent of the Thas–Payne construction of ovoids of $Q(4, q)$. However, we also obtain generalized Hall planes in our sequence of spreads as well. It might be noted that Jha [**570**] showed that generalized Hall planes are precisely those that are derived from semifield planes of order q^2 that have a middle nucleus isomorphic to $GF(q)$. Any coordinate semifield derived relative to the middle nucleus derives a generalized Hall plane and conversely. Also, the construction process of 'distortion' takes a derivable net D and any transversal T to it and distorts the multiplication relative to the pair (D, T), creating a generalized Hall plane. Hence, a two-step process from a semifield of order q^2 containing a derivable net containing the elation axis of distortion and derivation will produce a semifield of order q^2 with middle nucleus $GF(q)$.

There are no semifields with spreads in $PG(3, q)$ that are symplectic but not of dimension two. However, this certainly raises the question whether any symplectic spread in $PG(3, q)$ must have symplectic dimension two. In particular, are there flock spreads in $PG(3,q)$ that are of symplectic dimension two? We show that the only such flocks are Desarguesian or Kantor–Knuth.

As noted above, there is an interesting connection between commutative semifields and symplectic spreads, shown by Kantor [877]; every semifield plane that has a coordinate commutative semifield produces, by transpose and dualization, a symplectic spread. Furthermore, Maschietti [967] points out that 2×2 matrix spread sets consisting of symmetric matrices produce and are equivalent to symplectic spreads, considered over a 4-dimensional vector space.

We note subsequently that there is a conical flock spread, the Kantor–Knuth flock spread that admits a representation of symmetric 2×2 matrices. In general, any flock spread has the following standard form:

$$(*) : x = 0, y = x \left[\begin{array}{cc} u + G(t) & F(t) \\ t & u \end{array} \right], \forall t, u \in GF(q).$$

Clearly, we are not considering a representation where the 2×2 matrices are symmetric, unless $F(t) = t$ and we note below that this implies that the conical flock spread is Desarguesian when q is even or Kantor–Knuth or Desarguesian when q is odd.

The question then becomes: What are the flock spreads that are symplectic and can also be written as symmetric spread sets of 2×2 matrices? We show that the only possibilities are the Desarguesian and Kantor–Knuth flock spreads.

The following is easily proved (also see Maschietti [967]).

THEOREM 82.3. Let π be a finite translation plane. Then, there is a matrix spread set $\mathcal{S}$ for π that is symplectic if and only if there is a set of matrices such that if M is in $\mathcal{S}$ then $M = M^t$, (transpose).

DEFINITION 82.4. A symplectic spread S will be said to have 'symplectic dimension k' if and only if some matrix spread set representing S is elementwise self-transpose, the matrices are $k \times k$ and k is minimum in any self-transpose representation. If S has order p^t then $1 \le k \le t$.

Note the symplectic spreads of symplectic dimension 1 are exactly the Desarguesian spreads.

In the above representation, it may not be the case that the matrix spread set of symmetric matrices contains the identity matrix. If this is required, we would need to adjust the symplectic form more generally to be: $\left[\begin{array}{cc} 0 & A \\ -A & 0 \end{array} \right]$ where $A = A^t$.

EXAMPLE 82.5. The Kantor–Knuth conical flock spreads are symplectic spreads of symplectic dimension 2.

In fact, the following may be proved:

THEOREM 82.6. (Biliotti, Jha, and Johnson [128]). A flock spread is symplectic of dimension 1 or 2 if and only if it is either Desarguesian or Kantor–Knuth.

82.3. Symplectic and Flock Spreads.

The connection between semifield flock spreads and symplectic spreads has an elegant algebraic component.

THEOREM 82.7. (Biliotti, Jha, and Johnson [**128**]). Let S be a semifield flock spread in $PG(3, q)$.

(1) Then applying the following sequence of construction operations produces a symplectic semifield spread S_{sym} in $PG(3, q)$:

$$S(flock) \longmapsto \text{dualize} \longmapsto \text{distort} \longmapsto \text{derive}$$
$$\longmapsto \text{transpose} \longmapsto \text{dualize} \longmapsto S_{\text{sym}}(\text{symplectic}).$$

(2) Let S_{sym} be a symplectic semifield whose spread is in $PG(3, q)$. Then applying the following sequence of constructions produces a semifield flock spread in $PG(3, q)$:

$$S_{\text{sym}}(\text{symplectic}) \longmapsto \text{dualize} \longmapsto \text{transpose} \longmapsto \text{derive}$$
$$\longmapsto \text{extend} \longmapsto \text{dualize} \longmapsto S(\text{flock}).$$

DEFINITION 82.8. Let S be a semifield flock spread and let S_{sym} denote the associated symplectic semifield spread constructed and connected as in the above theorem. In either case, the remaining semifield is the '5th cousin' of the former. Thus, the 5th cousin of a symplectic semifield spread in $PG(3, q)$ is a semifield flock spread and the 5th cousin of a semifield flock spread is a symplectic semifield spread. More generally, any two spreads constructed from one another by i-iterations of the construction techniques of dualization, transpose, derivation, extension, dualization are said to be 'ith cousins'.

82.4. The Symplectic Semifield Flock Spreads.

When we have a semifield flock spread in $PG(3, q)$, there is a corresponding symplectic flock spread in $PG(3, q)$, the 5th cousin. We intend to show that a semifield flock spread is symplectic if and only if it is Kantor–Knuth or Desarguesian.

Our proof is based on the fact that if the semifield flock is symplectic then the dual transpose is a commutative semifield spread. We are not assuming anything regarding the symplectic dimension. So, there is a problem dealing with transposed spreads, since in the standard model the transposed spreads may be obtained via the transpose of matrix spreads, which might be large-dimensional.

We choose a different symplectic form for which a more convenient model is possible when the original spread in $PG(3, q)$. We begin with a more general analysis of the dual transpose of a semifield spread in $PG(3, q)$.

The following lemmas, given without proofs, indicate how the proof proceeds.

LEMMA 82.9. Let

$$x = 0, y = x \left[\begin{array}{cc} g_1(t) + g_2(u) = g(t, u) & f_1(t) + f_2(u) = f(t, u) \\ t & u \end{array} \right] ; t, u \in GF(q),$$

be a semifield spread in $PG(3, q)$, where g_i and f_i are additive functions from $GF(q)$ to $GF(q)$, $i = 1, 2$.

Define the associated pre-semifield $(S = GF(q) \times GF(q), +, \cdot)$ as follows:

$$(c, d) \cdot (t, u) = (c(g_1(t) + g_2(u)) + dt, c(f_1(t) + f_2(u)) + du).$$

Assume that the spread is non-Desarguesian, then this defines a pre-semifield with left nucleus $\{ (0, \alpha); \alpha \in GF(q) \}$.

LEMMA 82.10. Define the dual pre-semifield $(S, +, *)$ by taking
$$(a, b) * (c, d) = (c, d) \cdot (a, b) = (cg(a, b) + da, cf(a, b) + db),$$
which defines a pre-semifield with right nucleus $\{ (0, \alpha); \alpha \in GF(q) \}$.

LEMMA 82.11. Let $p^r = q$, for p a prime, so that $g_i(x) = \sum_{j=0}^{r-1} g_{i,j} x^{p^j}$ and $f_i(x) = \sum_{j=0}^{r-1} f_{i,j} x^{p^j}$. Then $(S, +, \diamond)$, with

$$(s, t) \diamond (c, d) = \left(\sum_{j=0}^{r-1} g_{2,j}^{1/p^j} (ct)^{1/p^j} + \sum_{j=0}^{r-1} f_{2,j}^{1/p^j} (cs)^{1/p^j} + ds, \right.$$
$$\left. \sum_{j=0}^{r-1} g_{1,j}^{1/p^j} (ct)^{1/p^j} + \sum_{j=0}^{r-1} f_{1,j}^{1/p^j} (cs)^{1/p^j} + dt \right),$$

defines a pre-semifield which coordinatizes the dual transpose of the original semifield plane.

LEMMA 82.12. Under the assumptions of Lemma 82.11, $(0, 1) \diamond (0, 1) = (0, 1)$ and defining a multiplication $\odot$ by
$$((s, t) \diamond (0, 1)) \odot ((0, 1) \diamond (c, d)) = (s, t) \diamond (c, d)$$
will produce the dual transpose semifield which has middle nucleus $\{ (0, \alpha); \alpha \in GF(q) \}$. Let $\tilde{g}_i(x) = \sum_{j=0}^{r-1} g_{i,j}^{1/p^j} x^{1/p^j}$ and $\tilde{f}_i(x) = \sum_{j=0}^{r-1} f_{i,j}^{1/p^j} x^{1/p^j}$, for $i = 1, 2$.
Then
$$(s, t) \odot (\tilde{g}_2(c), \tilde{g}_1(c) + d) = \left(\tilde{g}_2(ct) + \tilde{f}_2(cs) + ds, \tilde{g}_1(ct) + \tilde{f}_1(cs) + dt \right)$$
is the semifield $(S, +, \odot)$, with middle nucleus $\{ (0, \alpha); \alpha \in GF(q) \}$ and identity $(0, 1)$.

LEMMA 82.13. Any semifield $(S, +, \odot)$ of order q^2 and middle nucleus $GF(q)$ that commutes over $\{(0, \alpha); \alpha \in GF(q)\}$ may be written in the form
$$(s, t) \odot (c, d) = (H(sc) + ct + sd, M(sc) + dt)$$
and constructs by transpose and dualization a symplectic spread of dimension two
$$x = 0, y = x \begin{bmatrix} \widetilde{M}(t) + \widetilde{H}(u) & u \\ u & t \end{bmatrix} ; t, u \in GF(q).$$

Assume an isotope is commutative. Then the original semifield may be re-coordinatized by functions g_i and f_i such that $g_1(t) = 0$, and $g_2(u) = u$, for all $t, u \in GF(q)$. In this case, the original semifield spread is symplectic of dimension two defined by
$$x = 0, y = x \begin{bmatrix} u & f_1(t) + f_2(u) \\ t & u \end{bmatrix} \begin{bmatrix} 0 & 1 \\ 1 & 0 \end{bmatrix} = \begin{bmatrix} f_1(t) + f_2(u) & u \\ u & t \end{bmatrix} ; t, u \in GF(q).$$

We adopt the notation developed in the previous lemmas. The previous results produce specific functions when we have semifield flock spreads. First to construct the associated commutative semifield:

PROPOSITION 82.14. When $(S, +, \cdot)$ defines a semifield flock spread, we may assume that $g_2(x) = x$ and $f_2(x) = 0$. Hence, $\tilde{g}_2(x) = x$ and $\tilde{f}_2(x) = 0$. Then
$$(s, t) \odot (c, d) = \left(ct + (d - \tilde{g}_1(c))s, \tilde{g}_1(ct) + \tilde{f}_1(cs) + (d - \tilde{g}_1(c))t \right)$$

is the dual transposed semifield with middle nucleus $\{\,(0,\alpha);\ \alpha \in GF(q)\,\}$.

When $(S,+,\cdot)$ defines a semifield flock spread with additive functions $g(t)$ and $f(t)$ so that

$$x = 0, y = x\left[\begin{array}{cc} u + g(t) & f(t) \\ t & u \end{array}\right]; t, u \in GF(q)$$

is the associated spread, then the commutative semifield $(S,+,\odot)$ obtained by

$$S(\text{flock}) \longmapsto \text{dualize} \longmapsto \text{distort} \longmapsto \text{derive} = (\text{commutativesemifield})$$

may be defined by

$$(s,t) \odot (c,d) = (-g(sc) + ct + sd, f(st) + dt).$$

We now produce the form for the symplectic semifield spread associated with the semifield flock spread.

PROPOSITION 82.15. When $(S,+,\cdot)$ defines a semifield flock spread with additive functions $g(t)$ and $f(t)$ so that

$$x = 0, y = x\left[\begin{array}{cc} u + g(t) & f(t) \\ t & u \end{array}\right]; t, u \in GF(q),$$

then the symplectic spread in $PG(3,q)$ constructed using

$$S(\text{flock}) \longmapsto \quad \text{dualize} \longmapsto \text{distort} \longmapsto \text{derivation}$$
$$\longmapsto \quad \text{transpose} \longmapsto \text{dualize} \longmapsto S_{\text{sym}}(\text{symplectic})$$

is

$$x = 0, y = x\left[\begin{array}{cc} -\widetilde{g}(u) + \widetilde{f}(t) & u \\ u & t \end{array}\right]; t, u \in GF(q).$$

PROPOSITION 82.16. Any semifield spread in $PG(3,q)$ which is symplectic is symplectic of dimension two.

As corollaries to our analysis, we have:

COROLLARY 82.17. Let π be a semifield flock spread in $PG(3,q)$. Then π is symplectic if and only if π is a Kantor–Knuth plane or Desarguesian.

COROLLARY 82.18. The symplectic 5th cousin of a semifield flock spread is a flock spread if and only if the semifield flock is Kantor–Knuth.

82.4.1. The Roman Son. In Payne and Thas [1094], there is a construction of a semifield spread in $PG(3,q)$, using the Cohen–Ganley semifield flock spread. Basically, one considers the associated generalized quadrangle $Q(CG)$, of type (q^2,q) and forms the translation dual of type (q,q^2), the 'Roman generalized quadrangle'. This generalized quadrangle is shown to have subquadrangles isomorphic to $Q(4,q)$, the set of totally isotropic points and totally isotropic lines with respect to a quadric in $PG(4,q)$. Payne and Thas [1094] show that there are ovoids in $Q(4,q)$. Since the dual of $Q(4,q)$ is $M(3,q)$, the set of totally isotropic points and totally isotropic lines of a symplectic polarity of $PG(3,q)$, we obtain from any such ovoid a symplectic spread in $PG(3,q)$ (given by a symplectic form over the associated 4-dimensional vector space V_4, 'symplectic dimension 2'). It is furthermore determined by Lunardon [945] that these symplectic spreads are semifield spreads. We call any such semifield spread a 'Roman Son'. In fact, it is noted in Payne and Thas [1094] that all Roman Sons are isomorphic.

Hence, we obtain:

THEOREM 82.19. The Roman Son (Thas–Payne spread) and the Cohen–Ganley flock spread are non-isomorphic 5th cousins.

82.4.2. Penttila–Williams' 5th Cousin. There is a sporadic ovoid in $Q(4, 3^5)$ and hence a symplectic spread of symplectic dimension 2 in $PG(3, 3^5)$, due to Penttila and Williams [**1102**]. This turns out to be a semifield spread, so the 5th cousin is a semifield flock spread in $PG(3, 3^5)$. This is the semifield flock spread constructed in Bader, Lunardon, and Pinneri [**53**].

THEOREM 82.20. The symplectic Penttila–Williams semifield spread is not isomorphic to its 5th cousin, the associated semifield flock spread of Bader, Lunardon, and Pinneri.

REMARK 82.21. (1) Ball and Brown [**83**] show, using the Thas–Payne construction, that there are six semifield spreads corresponding to a semifield flock. Furthermore, they show that the symplectic semifield spread and semifield flock spread are isomorphic if and only if the semifield flock spread is Desarguesian or Kantor–Knuth (our version of this did not require that the symplectic semifield spread had symplectic dimension two).

In our sequence of constructions

$$S(\text{flock}) \quad \longmapsto \quad \text{dualize} \longmapsto \text{distort} \longmapsto \text{derive}$$
$$\longmapsto \quad \text{transpose} \longmapsto \text{dualize} \longmapsto S_{\text{sym}}(\text{symplectic})$$

and

$$S_{\text{sym}}(\text{symplectic}) \quad \longmapsto \quad \text{dualize} \longmapsto \text{transpose} \longmapsto \text{derive}$$
$$\longmapsto \quad \text{extend} \longmapsto \text{dualize} \longmapsto S(\text{flock}),$$

we see that there are also five possible semifields and a generalized Hall quasifield and these are mutually non-isomorphic provided the flock is not Desarguesian or Kantor–Knuth; two each with left, right nucleus isomorphic to $GF(q)$ and one with middle nucleus isomorphic to $GF(q)$. Furthermore, if one considers the sequence

$$S(\text{flock}) \longmapsto \text{dualize} \longmapsto \text{transpose},$$

another semifield with middle nucleus $GF(q)$ is constructed. The six semifield spreads in our sequences are isomorphic to the six semifield spreads of Ball and Brown, who also included dual and transpose versions of the various semifield spreads.

(2) If one considers an iteration of transpose-dual-transpose, etc., it is potentially possible to construct twelve semifield spreads. To see that there are, in fact, only six, Ball and Brown [**83**] show that any transpose of a semifield flock spread is isomorphic to the original, using the Klein quadric.

More generally, one could ask if the transpose of a flock spread is isomorphic to itself. And, in fact, this is valid as well, and an algebraic proof using transposed matrices is all that is required.

THEOREM 82.22. The dual spread of a flock spread is isomorphic to the flock spread.

PROOF. Represent the flock spread as follows:

$$x = 0, y = x \begin{bmatrix} u + g(t) & f(t) \\ t & u \end{bmatrix}; u, t \in GF(q),$$

where g and f are functions on $GF(q)$ (not necessarily additive). The dual spread is

$$x = 0, y = x \begin{bmatrix} u + g(t) & f(t) \\ t & u \end{bmatrix}^T = \begin{bmatrix} u + g(t) & t \\ f(t) & u \end{bmatrix}; u, t \in GF(q).$$

Now note that

$$\begin{bmatrix} 0 & 1 \\ -1 & 0 \end{bmatrix} \begin{bmatrix} u + g(t) & t \\ f(t) & u \end{bmatrix} \begin{bmatrix} 0 & 1 \\ -1 & 0 \end{bmatrix} = \begin{bmatrix} -u & f(t) \\ t & -u - g(t) \end{bmatrix},$$

and letting $-u - g(t) = v$, the previous matrix is

$$\begin{bmatrix} v + g(t) & f(t) \\ t & v \end{bmatrix}.$$

Since the operations listed merely correspond to a basis change, we have the proof. $\square$

REMARK 82.23. It also is now trivial to note that there are no non-Desarguesian symplectic semifield spreads in $PG(3, q)$, when q is even. That is, the symplectic dimension must be two so there is an associated flock semifield plane of even order q^2, which is necessarily Desarguesian by Johnson [**710**].

82.5. The Omnibus Flock Theorem.

We list here the various connections that have been established between flocks of quadratic cones and related geometries.

THEOREM 82.24. Each of the following sets are equivalent to the set of flocks of quadratic cones in $PG(3, q)$:

(1) Conical flock spreads in $PG(3, q)$ (translation planes with spreads in $PG(3, q)$ admitting regulus-inducing elation groups of order q).

(2) Translation planes with spreads in $PG(3, q)$ admitting Baer groups of order q,

(3) Flock generalized quadrangles of order (q^2, q),

(4) When q is odd, the set of BLT sets of $Q(4, q)$,

(5) When q is even, the set of herds of hyperovals in $PG(2, q)$,

(6) Translation planes of order q^2 with spreads in $PG(3, q)$ that admit cyclic homology groups of order $q + 1$.

(7) Furthermore, the set of semifield flocks is equivalent to the set of symplectic semifield spreads in $PG(3, q)$.

82.6. General Semifield 5th Cousins.

It is possible to start with any semifield spread S_1 in $PG(3, q)$ and construct the associated 5th cousins in two ways. First the kernel $GF(q)$ of S_1 corresponds to the left nucleus so the dual of S_1, S_2 has right nucleus $GF(q)$, which makes this a derivable semifield plane. However, since there is now a derivable net, we may also apply the construction processes of t-extension or t-distortion or transposition of the associated matrices. If we apply dualization followed by distortion, we construct a generalized Hall plane G_3. Any such generalized Hall plane is also derivable and when derived constructs a semifield plane S_4 of order q^2 with middle nucleus isomorphic to $GF(q)$. Any such semifield S_4 when transposed constructs a semifield S_5 of order q^2 and right nucleus isomorphic to $GF(q)$. The dualization of S_5

produces a semifield S_6 of order q^2 and kernel $GF(q)$. We may call S_6 the '5th cousin' of S_1. The sequence may be abbreviated as:

$$S_1 \longmapsto \quad \text{dualize} = S_2 \longmapsto \text{distort} = G_3 \longmapsto \text{derive} = S_4$$
$$\longmapsto \quad \text{transpose} = S_5 \longmapsto \text{dualize} = S_6.$$

Similarly, we may start with a semifield S_1' with spread in $PG(3, q)$ and dualize to construct a semifield with right nucleus $GF(q)$, S_2'. At this point, transpose to construct a semifield S_3' with middle nucleus $GF(q)$. Any semifield has a middle-nucleus derivable net and when derived produces a generalized Hall plane G_4'. We may also apply t-extension to G_4' to construct a semifield plane S_5' with right nucleus $GF(q)$. Then dualization of S_5' produces a semifield S_6' with left nucleus $GF(q)$. This sequence may be abbreviated as:

$$S_1' \longmapsto \quad \text{dualize} = S_2' \longmapsto \text{transpose} = S_3' \longmapsto \text{derive} = G_4'$$
$$\longmapsto \quad \text{extend} = S_5' \longmapsto \text{dualize} = S_6'.$$

The second sequence can usually be viewed as the reverse of the first sequence. When S_1 is a flock semifield then S_6 is a symplectic semifield and when S_1' is a symplectic semifield then S_6' is a flock semifield.

REMARK 82.25. It is possible to construct 5th cousins geometrically and some of the above sequences have a geometric counterpart called the 'translation dual', originated by Lunardon [**947**] as adapted from work of Thas on the translation dual of a generalized quadrangle. We shall return to this concept in Chapter 85.

82.6.1. Lifting and More Cousins. Given any semifield spread in $PG(3, q)$, the plane defined by dualizing the semifield may be derived to obtain a non-semifield spread corresponding to the semifield spread. Furthermore, any semifield flock spread may be derived using one of the base reguli, to obtain a spread of order q^2 admitting a Baer group of order q. The reader is referred to the paper of Jha and Johnson [**639**], where these derived planes are discussed in greater detail. Any of these might be called a 'cousin'. If cousins of semifields may only be semifields, then there is the matter of 'lifting'.

Given any spread in $PG(3, q)$, there is an associated spread in $PG(3, q^2)$, obtained by a process called 'lifting'. In fact, the construction is not spread dependent but quasifield dependent, in that it is possible to construct two different lifted spreads from the same spread.

The following is from the authors' text [**123**, 29.5, p. 450]. Let K be isomorphic to $GF(q)$ and let K^+ be a quadratic extension of K, by θ, where $\theta^2 = \theta\alpha + \beta$, for $\alpha, \beta \in K$. Let

$$\pi : x = 0, y = x \begin{bmatrix} g(t, u) & f(t, u) \\ t & u \end{bmatrix}; u, t \in K,$$

then

$$\pi^L : x = 0, y = x \begin{bmatrix} (\theta s + w)^q & -\theta g(t, u) + (f(t, u) + \alpha g(t, u)) \\ (\theta t + u) & (\theta s + w) \end{bmatrix}; s, w, t, u \in K$$

is a spread in $PG(3, K^2)$, called a spread 'lifted' from π. In turn, π is called a 'contraction' of π^L. Of course, this process may be repeated indefinitely to obtain an infinite chain of lifted planes.

It is clear from the construction that if the original spread set is additive, then the lifted spread set is additive; semifields lift to semifields. In particular, this means that a semifield flock or a symplectic semifield spread constructs an infinite chain of lifted planes. But, also note that this means that a semifield flock spread in $PG(3, q)$ could be the 'nth cousin' of a semifield spread in $PG(3, q^{2^n})$, obtained by a series of n contractions. To complicate matters, we note that we may derive the net

$$x = 0, y = x \begin{bmatrix} (\theta s + w)^q & 0 \\ 0 & (\theta s + w) \end{bmatrix}; s, w \in K,$$

to produce spreads in $PG(7, q)$. Are these 'cousins'?

82.6.1.1. *Lifting Semifield Flocks and Their 5th Cousins.* As an example of the enormous variety of semifield spreads associated with a given one, we consider a semifield flock spread

$$x = 0, y = x \begin{bmatrix} u + g(t) & f(t) \\ t & u \end{bmatrix}; u, t \in GF(q),$$

where g and f are additive functions. Hence, we obtain the lifted spread

$$\pi^L : x = 0, y = x \begin{bmatrix} (\theta s + w)^q & -\theta(g(t) + u) + (f(t) + \alpha g(t)) \\ (\theta t + u) & (\theta s + w) \end{bmatrix}; s, w, t, u \in K.$$

Consider the 5th cousin of the semifield spread

$$x = 0, y = x \begin{bmatrix} -\widetilde{g}(u) + \widetilde{f}(t) & t \\ t & u \end{bmatrix}; u, t \in GF(q).$$

This spread lifts to

$$x = 0, y = x \begin{bmatrix} (\theta s + w)^q & -\theta(-\widetilde{g}(u) + \widetilde{f}(t)) + (t + \alpha(-\widetilde{g}(u) + \widetilde{f}(t))) \\ (\theta t + u) & (\theta s + w) \end{bmatrix}; s, w, t, u \in K.$$

Obviously, we have a longer chain connecting a semifield spread and its 5th cousin, if we begin with one of the lifted versions. But, what is a direct connection between the two lifted 5th cousins?

So, the main question seems to be: **Are all spreads in $PG(3, q)$ cousins?**

82.6.1.2. *The Middle-Nucleus Semifield.* Thus, we find that the transpose of the middle-nucleus semifield $(S, +, \odot)$ is the right-nucleus counterpart $(S, +, \diamond)$, so

$$(d_1, d_2, \ldots, d_n) \odot (c_1, c_2, \ldots, c_n)$$

$$= \left(\left(\sum_{i=1}^{n} \sum_{j=1}^{n-1} \widetilde{g}_{j,i,n-e+1}(d_{n-i+1}c_j) + d_e c_n \right) \right)_{1 \times n}, \text{ for } e = 1, 2, \ldots, n,$$

where

$$(d_1, d_2, \ldots, d_n) \diamond (c_1, c_2, \ldots, c_n) = (c_1, c_2, \ldots, c_n) \cdot (d_1, d_2, \ldots, d_n)$$

$$= \left(\left(\left(\sum_{i=1}^{n-1} c_i \sum_{k=1}^{n} g_{z,i,k}(s_k) \right) + c_n s_z \right) \right)_{n \times 1}, \text{ for } z = 1, 2, \ldots, n.$$

Notice that we are connecting pre-semifields, not necessarily semifields.

82.6.1.3. *Commutative Semifields.* Now if we have a commutative pre-semifield S of dimension n over the middle nucleus M, then S(transpose-dual) defines a symplectic spread given by a symplectic pre-semifield of dimension n over the left nucleus M. We point out that if the transpose-dual is re-coordinatized by the backward identity permutation matrix, we obtain a matrix spread set of symmetric matrices. Of course, Kantor [**877**] showed that if a commutative semifield is written over its prime field then using the standard symplectic form, the transpose-dual spreads consist of symmetric matrices. It is sometimes more convenient to write the semifield as a vector space over its middle nucleus. Thus, one nice feature of this procedure is that the commutative pre-semifields written over their middle nuclei give rise directly to symmetric matrix spread sets.

THEOREM 82.26. *If a pre-semifield $(S, +, \odot)$ is written over its middle nucleus, then the pre-semifield is commutative if and only if the transpose-dual pre-semifield $(S, +, \cdot)$, followed by the backward identity permutation matrix, is a symplectic pre-semifield, whose spread in matrix form consists of symmetric matrices.*

Using these ideas, the following may be proved:

COROLLARY 82.27. *Any finite symplectic semifield of dimension n over its left nucleus has symplectic dimension n.*

REMARK 82.28. The previous corollary is valid for any symplectic spread as proved by Kantor [**849**].

82.6.1.4. *The Knuth Symplectic Spreads.* There are a variety of symplectic semifield planes and hence commutative semifield planes of even order in existence due to the 'slicing' and 'up and down' methods of Kantor discussed more fully in Chapter 79 on orthogonal spreads. Certain generalized twisted field planes are symplectic and these may be net replaced to construct the Suetake planes. Furthermore, the Suetake planes (and the Hering planes) are the only known planes of odd order and odd dimension over their kernel that are symplectic, but which are not semifield planes (see Ball et al. [**81**]). For dimension 2 over the kernel, the known planes correspond to ovoids in $O(5, q)$-spaces. The corresponding translation planes are quite rare, being the Lüneburg–Tits planes, The Thas–Payne spread (the 5th cousin of the Cohen–Ganley semifield spreads), the Kantor–Knuth spreads, and the Penttila–Williams spreads (5th cousin of the Bader–Lunardon–Pinneri spread of order 3^{10}) , and a slice of the Ree–Tits ovoids in $O(7, 3^k)$ (see, e.g., Johnson [**731**] and Penttila and Williams [**1102**]).

Hence, the known symplectic spreads are:
    Suetake (includes Hering 27)
    (only known symplectic, non-semifield, odd-order and odd-dimension),
    The symplectic generalized twisted field spread
    Lüneburg–Tits
    Payne–Thas (5th cousin of Cohen–Ganley semifield flock)
    Kantor–Knuth flock spreads
    Penttila–Williams
    (5th cousin of Bader–Lunardon–Pinneri semifield flocks order 3^{10})
    Symplectic Slice of Ree–Tits ovoids,
    Kantor and Kantor–Williams symplectic semifield planes

CHAPTER 83

When Is a Spread Not Symplectic?

This chapter follows more or less the work of Johnson and Vega [**812**] with fewer details.

Let V be a $2d$-dimensional vector space over $GF(q)$, which admits a symplectic form. Then there is a basis for V so that the form is $\begin{bmatrix} 0 & I \\ -I & 0 \end{bmatrix}$, since all forms are conjugate. If V admits a spread of totally isotropic subspaces, we call the spread a 'symplectic spread' and say that we also have a 'symplectic translation plane'. By noting that the symplectic group acts doubly transitive on totally isotropic subspaces, we see that we may choose

$$x = 0, y = 0, \ y = xM; M^t = M$$

as a matrix spread set (see also Biliotti, Jha, and Johnson [**128**]). More generally, we may allow any two components to be called $x = 0, y = 0$, but we shall revisit this below. We call this a 'symmetric representation'.

Indeed, W.M. Kantor, has recently shown that any symplectic spread remains symplectic when considered over the kernel of the translation plane. In other words, we may always assume that $K \simeq GF(q)$ is the kernel of the plane and that M is a set of $d \times d$ matrices over K such that $M^t = M$.

Hence, the known semifield flock spreads provide us also with symplectic semifields. It is known also that the Lüneburg–Tits planes and the Hering planes of order 27 are symplectic. There is an infinite class of planes due to Suetake [**1168**] that contain the Hering planes of order 27. In fact, the Suetake planes are known to be net replaceable from a certain class of generalized twisted field planes, which are, in fact, symplectic. That the replacement procedure preserves the symplectic form is a recent result of Ball, Bamberg, Lavrauw, and Penttila [**81**].

In general trying to determine if any given translation plane is also a symplectic plane is difficult without knowledge of a given symplectic form for then the question boils down to asking when there is a matrix spread set of symmetric matrices. For example, given an ovoid in $PG(3, q)$, for q even, the work of Thas [**1171**] shows that there is a corresponding symplectic spread in $PG(3, q)$, hence providing a matrix spread set of symmetric 2×2 matrices. That is there is a symmetric representation of the following form:

$$x = 0, y = x \begin{bmatrix} f(t, u) & t \\ t & u \end{bmatrix} ; t, u \in GF(q).$$

Brown [**168**] shows that if an ovoid in $PG(3, q)$, for q even, has at least one hyperplane intersection that is a conic, then the ovoid is an elliptic quadric. Maschietti [**967**] points out that this is equivalent to having a regulus in the symmetric matrix representation. This means that coordinates may be chosen so that $f(0, u)$. So, it would potentially be possible to show that if $f(0, u) = u$ for all $u \in GF(q)$, then the spread is Desarguesian, which would be an algebraic proof of the theorem of Brown. This formulation of a deep result into a problem involving symmetric representations of symplectic spreads illustrates how potentially difficult it would be to use only the symmetric representations to attack problems on symplectic spreads.

Hence, it would be nice to have if not a criterion for the existence of a symplectic spread then perhaps a criterion for the non-existence of a symplectic spread. In other words, when can it be guaranteed that a spread is not symplectic? By again noting the connection with commutative semifield spreads and symplectic semifield spreads, this would speak to the problem of asking when a given semifield spread has an isotopic image which is commutative.

83.1. Symplectic Planes and Affine Homologies.

If we have a symplectic plane, we may regard the kernel as the prime subfield and choose the symplectic form as $\begin{bmatrix} 0 & I \\ -I & 0 \end{bmatrix}$, which means if we consider the spread written over the prime field then we have a set of symmetric matrices forming a spread.

Let π be a symplectic plane. Then there is a matrix spread set so that the spread is $x = 0, y = 0, y = xM$, such that $M^t = M$. If we change bases by the mapping $(x, y) \longrightarrow (x, x(-Z) + y)$, where $Z^t = Z$ then the new spread has the form $x = 0, y = (N - Z)$, where N is either 0 or M such that $M^t = M$. Since $(N - Z)^t = N^t - Z^t = N - Z$, we still have a symplectic representation. We call this 'elation sliding'.

Similarly, if we change bases by $(x, y) \longrightarrow (y, x)$, then the new representation is $x = 0, y = xM^{-1}$, such that $M^t = M$. But $(M^{-1})^t = (M^t)^{-1} = (M)^{-1} = M^{-1}$, so the new representation is still symplectic. We call this 'inverting'. Hence, we have:

THEOREM 83.1. If we have a symmetric representation, then any iteration of processes obtained by elation sliding and inverting preserves a symmetric representation. So, in particular, choose any two distinct components L and M of a symplectic spread in a symmetric representation. Then there is a basis change so that L may be represented by $x = 0$ and the matrix representation is still symmetric.

PROOF. By elation sliding and inverting, we can assume that L is $x = 0$ and retain a symmetric representation. In this representation, we may apply elation sliding which fixes $x = 0$ and takes the new representation of M into $y = 0$. More precisely, if L is $y = xZ$ use the elation slide of $(x, y) \to (x, -xZ + y)$ to rewrite L as $y = 0$, while retaining the symmetric representation. Now use inverting to represent L as $x = 0$, while still maintaining a symmetric representation. Now M has the form $y = xT$ and use elation sliding to represent M as $y = 0$ while fixing $x = 0$ and maintaining a symmetric representation.   □

Now assume that π is a symplectic plane. Assume that we have an affine homology group. By the previous theorem, there is a symmetric representation so that $y = 0$ is the axis and $x = 0$ is the center. In such a plane, there is a representation so that components $x = 0, y = 0$ are chosen so that there is a homology group with axis $y = 0$ and co-axis $x = 0$. Considering that we are working over the kernel then the affine homology has the form $(x, y) \longrightarrow (x, yM)$, where we regard all of the matrices over the primitive field. That is, by the previous theorem, we may assume that we have a symplectic representation $x = 0, y = 0, y = xT$, such that $T^t = T$. Hence, we have the following conditions:

$$(TM)^t = TM = M^t T^t = M^t T,$$

for all T in S (of cardinality q^n) and for all M in a cyclic group G of order $(q^n - 1)/(q - 1)$. Thus, we have

$$T - M^t T M^{-1},$$

for all T in S. Now consider the mapping $(x, y) \to (xM^{-t}, yM^{-1})$, which maps $y = xT$ onto $y = M^t T M^{-1} = T$. This then is a kernel homology of the plane π of order equal to the order of the matrix M, which in turn, is the order of the affine homology. That is, note that the square of the previous mapping is $(x, y) \longrightarrow (xM^{-2t}, yM^{-2})$. So we obtain a group of order equal to the order of M, which acts as a kernel homology of the translation plane. Hence, we have proved the following theorem.

THEOREM 83.2. (Johnson and Vega [812]). Let π be a symplectic translation plane of order q^n and kernel isomorphic to $GF(q)$. Then any affine homology of π must have order dividing $q - 1$.

THEOREM 83.3. (Johnson and Vega [812]). Let π be a non-Desarguesian symplectic plane of order $q^n - 1$. Then there cannot exist an affine homology of order a prime p-primitive divisor of $q^n - 1$.

PROOF. If there is such an affine homology, then the kernel of π must be $GF(q^n)$, so that the plane is Desarguesian.

We now have an affine homology group of order $(q^n - 1)/(q - 1)$ which must divide $(q - 1)$, impossible unless $n = 1$. $\qquad \square$

Now assume that we have a symplectic plane of order q^n and kernel $GF(q)$. If $\tau_\alpha : (x, y) \to (x, y\alpha)$ is a collineation, then so is $(x, y) \to (x\alpha^t, y\alpha)$, implying that $\rho_{\alpha^t} : (x, y) \to (x\alpha^t, y)$ is a collineation. Similarly, if $(x, y) \to (x\beta, y)$ is a collineation, so is $(x, y) \to (x\beta, y\beta^t)$, implying that $(x, y) \to (x, y\beta^t)$ is a collineation. Mapping τ_α to ρ_{α^t} is a monomorphism and mapping ρ_β to τ_{β^t} is a monomorphism. Hence, the two affine groups are isomorphic. Furthermore, since the groups define isomophic subgroups of the kernel homology group, they are isomorphic. Hence, in symplectic planes the right and middle nuclei are isomorphic.

Now we may choose the kernel so that the kernel homology groups are represented as diagonal matrices. For example, we can certainly do this if there is an element $y = x$ in the matrix spread set. In our situation, a basis change by $(x, y) \to (x, yT^{-1})$ produces such a matrix spread set (although we lose potentially the symmetric representation). By Kantor [849], we may assume that the matrix spread set are matrices over the kernel. When we have $y = x$ in the matrix spread set, then the corresponding affine colllineations $(x, y) \to (x, yM)$ force M to be in the matrix spread set. In other words, in the symmetric representation $M = T_2^{-1} T_1$, where T_2 and T_1 are in the symmetric set S. In any case, M is a matrix over the kernel. Hence, this shows that as kernel homologies, the mappings are of the form $(x, y) \to (xG, yG)$, where G are diagonal matrices. So, there is a representation so that the right and middle associator groups (right and middle nuclei) are identical and contained in the kernel.

Hence, we have proved:

THEOREM 83.4. (Johnson and Vega [**812**]). Let π be a symplectic translation plane. Then there is a coordinatization so that the right and middle associated groups (right and middle nuclei) are equal and contained in the left nucleus (coordinate kernel) of the translation plane.

COROLLARY 83.5. (Johnson and Vega [**812**]). Let π be a symplectic semifield plane. Then there is a coordinatization so that the right and middle nuclei are equal fields and contained in the left nucleus (kernel) of the semifield.

Now considered a symplectic semifield plane with semifield S. If we dualize S then the left nucleus and right nucleus are interchanged and the middle nucleus remains the same. Hence, we now have a semifield plane with left and middle nucleus equal and contained in the right nucleus. When we transpose, the middle and right nuclei are interchanged and the left nucleus remains the same. So, we have a semifield with left and right nuclei equal and contained in the middle nucleus. Since this set isotopically defines a commutative semifield plane by Kantor [**877**], we have the following corollary.

COROLLARY 83.6. Let Σ be a finite commutative semifield plane. Then there is a coordinate semifield so that the left and right nuclei are equal and contained in the middle nucleus.

COROLLARY 83.7. Let Σ be a finite commutative and symplectic semifield plane. Then there is a coordinate semifield so that left, right and middle nuclei are equal.

The idea that one can 'fuse' (sub)nuclei of the same size is considered generally in Jha and Johnson [**634**] where it is shown that any semifield plane that has two or three subnuclei of the same order has a coordinate semifield when the two or three subnuclei are fused; that is, we may assume they are all equal in that coordinate semifield. For the example, the Hughes–Kleinfeld semifields of order q^n have right and middle nuclei equal and isomorphic to $GF(q)$. If it is asked if these semifields might be symplectic, this would force the left nucleus to have order at least q and there is a fusion result that identifies the right and middle nuclei with a sub-left-nucleus isomorphic to $GF(q)$. For example, the above result shows that the Hughes–Kleinfeld semifields of order q^2 cannot be symplectic.

83.2. Self-Transpose.

A translation plane is said to be 'self-transpose' if it is isomorphic to its dual plane. Hence, a translation plane is self-transpose if and only if it is invariant under a correlation of the projective space within which lives the corresponding spread. But, a symplectic spread is invariant under a symplectic polarity of the space and there is a matrix spread set of symmetric matrices. It would be interesting to find translation planes that are invariant under a polarity and ask how far away from symplectic could be such associated spreads.

DEFINITION 83.8. A finite translation plane is said to be 'set-transpose' if and only if the associated spread is invariant under a polarity of the associated projective space. Hence, there is an associated spread set $\mathcal{M}$, such that

$$M \in \mathcal{M} \implies M^t \in \mathcal{M},$$

where M^t denotes the transpose of M.

We note that our arguments for symplectic spreads and affine homologies will not quite work for set-transpose planes, but note the following connection with affine homology groups.

THEOREM 83.9. Let π set a set-transpose plane with spread

$$x = 0, y = 0, y = xM, \ M \in \mathcal{M}, \text{ such that } M^t \in \mathcal{M}.$$

(1) Then the homology group with axis $y = 0$ and coaxis $x = 0$ is isomorphic to the homology group with axis $x = 0$ and coaxis $y = 0$.

PROOF. If $(x, y) \to (x, yB)$ is a collineation of π then $MB \in \mathcal{M}$, implies $B^t M^t \in \mathcal{M}$, for all $M^t \in \mathcal{M}$, which implies that $(x, y) \to (xB^{-t}, y)$ is a collineation of π. $\qquad\square$

Since symplectic spreads are difficult to construct, one wonders if this would be also true of set-transpose spreads. Actually, these are easy to construct from derivable symplectic spreads as follows.

THEOREM 83.10. Let π be a derivable symplectic translation plane with derivable net D.

(1) Then derivation of π produces a set-transpose plane.

(2) More generally, any multiply derived plane from a set of mutually disjoint derivable nets produces a set-transpose plane.

(3) Any subregular translation plane is set-transpose.

PROOF. Choose a matrix spread set with $x = 0$ and $y = 0$ not components of D and such that

$$x = 0, y = 0, y = xM, \ M \in \mathcal{M}, \text{ such that } M^t = M.$$

Now any Baer subplane of D incident with the zero vector will now have the general form $y = xN$, where N is a matrix over the prime field. If we choose the matrix spread set over the prime field, we may still have the properties mentioned. Consider $y = xN$. This subspace will now non-trivially intersect the components $y = xM$ of D non-trivially and hence $y = xN^t$ will non-trivially intersect the components $y = xM^t$ of D^t. Since this set is D again, we see that $y = xN^t$ becomes a Baer subplane of the net D. This means that upon derivation we have a set-transpose spread but not necessarily a symplectic spread. In the multiply derived case, we need only choose two components not in the set being replaced, which we may always do, to be called $x = 0, y = 0$.

This proves (1) and (2). Finally, any Desarguesian plane is symplectic so any subregular translation plane then is set-transpose. $\qquad\square$

The subregular spreads that we observe are set-transpose might better be called 'semi-symplectic of type i', where i is the number of components that are totally isotropic under the symplectic polarity. In the case of subregular planes, it is not at all clear when the type i might be. Still another variation of self-transposed spreads are what are called 'non-singular pairs' in Weintraub [1210] (we shall call such pairs 'symplectic pairs'). These would be our type 0, or perhaps 'symplectically paired planes' and symplectic planes of order q^n would be of type $(q^n + 1)$. These are defined in our context as follows:

DEFINITION 83.11. Let π be a finite translation plane whose spread is invariant under a symplectic polarity that permutes the components in orbits of length 2. Then the orbits are called 'symplectic pairs'. We shall call the plane a 'symplectically paired plane.'

Weintraub [**1210**] shows that for any vector space V of dimension $2n$, n even over a field F isomorphic to $GF(q)$, q odd (or over a field F of characteristic not 2 which admits a cyclic Galois field extension of degree n), there is always a set of symplectic pairs that partition $V - \{0\}$. In the context of translation planes, this would mean that the translation plane is anti-symplectic. It turns out that the infinite class that Weintraub found always determines the Hall planes of odd order q^2.

THEOREM 83.12. (Johnson and Vega [**812**]). Let π_σ denote a Desarguesian affine translation plane of odd order q^2 with spread determined as follows:

$$x = 0, y = x \begin{bmatrix} u & t \\ \gamma t & u \end{bmatrix} ; u, t \in GF(q),$$

where γ is a non-square. Define a symplectic form for the associated 4-dimensional $GF(q)$-vector space as follows:

$$\langle (x_1, x_2, y_1, y_2), (z_1, z_2, w_1, w_2) \rangle = (x_1, x_2, y_1, y_2) \begin{bmatrix} 0 & 0 & 2 & 0 \\ 0 & 0 & 0 & -2\gamma \\ -2 & 0 & 0 & 0 \\ 0 & 2\gamma & 0 & 0 \end{bmatrix} \begin{bmatrix} z_1 \\ z_2 \\ w_1 \\ w_2 \end{bmatrix},$$

$\forall x_i, y_i, z_i, w_i \in GF(q)$, where $i = 1, 2$.

Then

$$y = x \begin{bmatrix} u & t \\ \gamma t & u \end{bmatrix} \text{ and } y = \begin{bmatrix} u & -t \\ -\gamma t & u \end{bmatrix} \text{ for } t \neq 0$$

are orthogonal under the symplectic form and mutually disjoint, thus forming exactly $(q^2-1)/2$ symplectic pairs of mutually disjoint 2-dimensional $GF(q)$-subspaces.

Consider the regulus net

$$x = 0, y = x \begin{bmatrix} u & 0 \\ 0 & u \end{bmatrix} ; u \in GF(q).$$

The opposite regulus has Baer subplanes of the following form:

$$\pi_\beta = \{(x_1, \beta x_1, x_2, \beta x_2); x_i \in GF(q), i = 1, 2\},$$
$$\pi_\infty = \{(0, y_1, 0, y_2); y_i \in GF(q), i = 1, 2\} \; \forall \beta \in GF(q).$$

Then

$$\{\pi_0, \pi_\infty\} \text{ and } \{\pi_\beta, \pi_{1/\gamma\beta}\} \text{ for } \beta \neq 0$$

are symplectic $(q+1)/2$ pairs of mutually orthogonal 2-dimensional $GF(q)$-subspaces.

(1) Then the union of these pairs forms an anti-symplectic spread that is clearly isomorphic to the spread obtained by the derivation of the Desarguesian spread π by the replacement of a single regulus. Hence, the resulting infinite class of symplectically paired spreads corresponds to the Hall spreads.

(2) In the infinite case, note that all of this may be accomplished for fields K of characteristic not 2 that admit a quadratic extension—obtain the infinite Hall spreads relative to K and the field extension.

PROOF. Relative to the symplectic form given, it is a straightforward calculation to see that $y = x \begin{bmatrix} a & b \\ c & d \end{bmatrix}$ will map to $y = x \begin{bmatrix} a & -c/\gamma \\ -b\gamma & d \end{bmatrix}$ under the symplectic form. If these two components belong to the same spread and if $-c \neq b\gamma$ we would have a symplectic pair. In our situation, consider

$$y = x \begin{bmatrix} u & t \\ \gamma t & u \end{bmatrix} \text{ and } y = \begin{bmatrix} u & -t \\ -\gamma t & u \end{bmatrix} \text{ for } t \neq 0.$$

Hence, we have symplectic pairs when t is non-zero. It is straightforward to see that

$$\{\pi_0, \pi_\infty\} \text{ and } \{\pi_\beta, \pi_{1/\gamma\beta}\} \text{ for } \beta \neq 0$$

are symplectic $(q+1)/2$ pairs of mutually orthogonal 2-dimensional $GF(q)$-subspaces. Hence, the Hall plane is anti-symplectic. $\square$

Since the Hall planes are André, it would be natural to ask if there are other André planes which are anti-symplectic. In fact, almost all André planes may be shown to anti-symplectic. Specifically, Johnson and Vega [812] prove the following theorem.

THEOREM 83.13. (Johnson and Vega [812]). Let π be a Desarguesian affine plane of odd order q^2. Choose a standard coordinatization and let R denote the regulus net coordinatized by $GF(q)$. Let K^* denote the kernel homology group of order $q^2 - 1$.

(1) If $y = x^q m + xn$, for $m \neq 0$ is a Baer subplane of π then $K^*(y = x^q m + xn)$ is the opposite regulus of a regulus of π.

(2) Define a symplectic form as follows:

$$\langle (x, y), (x^*, y^*) \rangle = \text{trace}_{GF(q)} \left(\begin{bmatrix} x & y \end{bmatrix} \begin{bmatrix} 0 & 1 \\ -1 & 0 \end{bmatrix} \begin{bmatrix} x^{*q} \\ y^{*q} \end{bmatrix} \right).$$

Then any subspace of the form $y = x^q m$ is totally isotropic and any subspace of the form $y = xn$ maps to $y = xn^q$ under the symplectic form. Hence

$$y = x^q m + xn \text{ and } y = x^q m + xn^q$$

are symplectic pairs if and only if $n \notin GF(q)$.

(3) $K^*(y = x^q m + xn) = \{ y = x^q md^{1-q} + xn; d \in GF(q^2)^* \}$.

(4) If $\mathcal{A}_R$ is an André set of $q - 1$ mutually disjoint reguli containing R then the Frobenius mapping $(x, y) \longrightarrow (x^q, y^q)$ leaves $\mathcal{A}_R$ invariant. Furthermore, this mapping maps $K^*(y = x^q m + xn)$ to $K^*(y = x^q m + xn^q)$. If $n^q \neq n$, we have a set of $(q + 1)/2$ symplectic pairs.

(5) If $\mathcal{A}_R$ is an André set of $q - 1$ mutually disjoint reguli containing R then there is a unique opposite regulus $K^*(y = x^q m + xn)$ such that $n \in GF(q)$. The set of remaining $(q - 3)$ reguli corresponds to the paired set of opposite reguli of the following form:

$$\mathcal{P}_R = \{ (K^*(y = x^q m_i + xn_i), K^*(y = x^q m_i + xn_i^q)) \},$$

where $n_i^q \neq n_i$, for $1, 2, \ldots, (q - 3)/2$, $m_i \neq 0$.

(6) Choose any subset λ of $\mathcal{P}_R$ of symplectic pairs. Form the translation plane $\pi_{R,\lambda}$ by multiple derivation of π by replacement of the regulus nets corresponding to λ and also derive R.

Then $\pi_{R,\lambda}$ is an anti-symplectic translation plane.

Hence, there are at least $2^{(q-3)/2}$ possible symplectically paired translation planes constructed (not all of these are necessarily non-isomorphic).

(7) Specifically, assume that $K^*(y = xm_0; m_0^{q+1} = \alpha)$. Choose a basis $\{1, t\}$ for $GF(q^2)$ over $GF(q)$ so that $t^2 = \alpha$, α a non-square. Then $\mathcal{A}_R$ may be chosen to have the following form:

$$\mathcal{A}_R = \left\{ K^*(y = xm_0; m_0^{q+1} = \alpha), K^*(y = x^q m_\gamma \pm x\sqrt{\alpha\gamma^2}; m_\gamma^{q+1} = \alpha\gamma^2) \right\},$$

where $\gamma \neq 0$ or 1.

Note that

$$(y = x^q m_\gamma d^{1-q} + x\sqrt{\alpha\gamma^2}; m_\gamma^{q+1} = \alpha\gamma^2)$$
$$\leftrightarrows (y = x^q m_\gamma d^{1-q} - x\sqrt{\alpha\gamma^2}; m_\gamma^{q+1} = \alpha\gamma^2)$$

under the symplectic form, for all $d \in GF(q)^{2*}$.

There are many variations on this theme and many symplectic spreads of type i, and notice that the same plane can be symplectic of a variety of types. The symplectic group is doubly transitive on totally isotropic subspaces and transitive on pairs (Weintraub [1210]). Hence, there is a symplectic form such that with a choice of basis, we may assume that $x = 0, y = 0$ are components that are either both totally isotropic or form a pair (if $x = 0, y = xT$ is a pair, a change of basis shows that $x = 0, y = 0$ is a pair for some symplectic form). Note in the examples given in the previous result a basis change by $\begin{bmatrix} 1 & 0 & 0 & 0 \\ 0 & 0 & 1 & 0 \\ 0 & 1 & 0 & 0 \\ 0 & 0 & 0 & 0 \end{bmatrix}$ will change π_0 and π_∞ to $x = (x_1, x_2) = 0$ and $y = (y_1, y_2) = 0$, respectively. Therefore, if L and M are components such that there is an affine homology group $H_{[L],M}$ with axis L and coaxis M as well as an affine homology group $H_{[M],L}$ with axis M and coaxis L then the only way that the spread could be symplectic of any type is if $H_{[L],M}$ is isomorphic to $H_{[M],L}$.

We note, for example, that all of the anti-symplectic translation planes of order q^2 that we have found have isomorphic homology groups of order $(q + 1)$.

Since there are a variety of translation planes that admit homology groups quite unrelated to the kernel of the translation plane, we have then a criterion to decide when a translation plane cannot be symplectic. For example, no André plane, generalized André plane, j-plane, $j \ldots j$-plane, nearfield plane, of order q^2 with affine homology groups of order not dividing $q - 1$ such as those admitting cyclic homology groups of order $q+1$ obtained from a flock of a quadratic cone (see for example, Johnson [655] for this connection with flocks of quadratic cones). Also, note that there are translation planes whose affine homology groups are not cyclic, such as the Heimbeck planes, and so forth, which therefore cannot be symplectic.

Furthermore, these results give criteria to decide if a given semifield has a commutative isotopic version. If the right and left nuclei are not the same size or one is larger than the middle nucleus, this cannot occur.

The only known commutative semifield planes that are also symplectic are generalized twisted field planes of order p^n with spreads given as follows:

$$x = 0, y = xm - cx^{p^a} m^{p^b},$$

where $c = -1$, $2a = b$, n and p of odd order and $z \to z^{p^a}$ of order 3 (Biliotti, Jha, and Johnson [121] have pointed out that the kernel (left nucleus) is isomorphic to $GF(p^{(n,a)})$, the middle nucleus is isomorphic to $GF(p^{(n,a-b)})$ and the right nucleus is isomorphic to $GF(p^{(n,b)})$). So, for such a generalized twisted field plane to be symplectic the necessary condition would be that $(n,b) = (n,a-b) \leq (n,a)$. And for commutative and symplectic semifields, we would require $(n,b) = (n,a-b) = (n,a)$. In this context, we recall the following results:

THEOREM 83.14. (Jha and Johnson [583]). (1) If a non-Desarguesian generalized twisted field plane of order q^n is one plane of a coupling then q is odd, n is odd and the second plane of the coupling is a Suetake plane.

(2) The generalized twisted field planes of order q^n corresponding to the Suetake planes are as follows:

$$x = 0, y = 0, y = xm - cx^{q^{2b^*}} m^{q^{b^*}},$$

where $(2b^*, n) = (b^*, n) = 1$ and q is any odd prime power, n an odd integer.

(3) If a finite translation plane π is coupled with a non-Desarguesian semifield plane then π is a Suetake plane.

THEOREM 83.15. (Jha and Johnson [583]). Let the following spread denote the generalized twisted field plane spread from which the Suetake planes may be constructed:

$$(83.2.1) \qquad x = 0, y = 0, y = xm - cx^{q^{2b^*}} m^{q^{b^*}}, \qquad \text{for } m \in GF(q^n) - \{0\},$$

where $(2b^*, n) - (b^*, n) = 1$ and q is any odd prime power.

Let G denote the collineation group

$$\left\langle \begin{bmatrix} t^{-1} & 0 \\ 0 & t^{q^*} \end{bmatrix} ; t \in GF(q^n)^* \right\rangle.$$

Let $f(x) = x^{q^{-2b^*}} - c^{q^{-b^*}} x$ and note that f is injective on $GF(q^n)$. Let the two component orbits of length $(q^n - 1)/2$ be denoted by Γ_1 and Γ_2, where $y = x - cx^{q^{2b^*}}$ is in Γ_1 and $y = xm_0 - cx^{q^{2b^*}} m_0^{q^{b^*}}$ is in Γ_2.

Then the Suetake planes have the following spread:

$$(83.2.2) \qquad x = 0, y = 0, (y = (xm_0 - cx^{q^{2b^*}} m_0^{q^{b^*}})G) \cup (y = (f^{-1}(x))G).$$

So, the generalized twisted field planes that may be net replaced to construct the Suetake planes are such that n and p are odd, $a = 2b$. In this setting, $(n,b) = (n,a-b) = (n,a) = r$. As noted previously, both the generalized twisted field planes and the Suetake planes are symplectic but not all of the semifield planes are commutative. In any case, in the generalized twisted field case, we have fusion among the nuclei.

For symplectic spreads and their variation, there are many variations on this theme and many symplectic spreads of type i. Notice that the same plane can be symplectic of a variety of types. The symplectic group is doubly transitive on totally isotropic subspaces and transitive on pairs (Weintraub [1210]). Hence, there is a symplectic form such that with a choice of basis, we may assume that $x = 0, y = 0$ are components that are either both totally isotropic or form a pair (if $x = 0, y = xT$ is a pair, a change of basis shows that $x = 0, y = 0$ is a pair for some symplectic form). Note in the examples given in the previous result a basis change by $\begin{bmatrix} 1 & 0 & 0 & 0 \\ 0 & 0 & 1 & 0 \\ 0 & 1 & 0 & 0 \\ 0 & 0 & 0 & 0 \end{bmatrix}$ will change π_0 and π_∞ to $x = (x_1, x_2) = 0$ and

$y = (y_1, y_2) = 0$, respectively.

We have also given various infinite classes of anti-symplectic spreads from which it is very easy to construct various other classes of set-transpose spreads.

Again, perhaps it is of importance to ask when such constructions of set-transpose, symplectic, symplectically paired spreads are never possible. Therefore, if L and M are components such that there is an affine homology group $H_{[L],M}$ with axis L and coaxis M as well as an affine homology group $H_{[M],L}$ with axis M and coaxis L then the only way that the spread could be symplectic of any type is if $H_{[L],M}$ is isomorphic to $H_{[M],L}$.

83.3. Symplectically Paired Spreads and Hemisystems.

Let σ be a symplectic form of a 4-dimensional vector space V_4 over a field K isomorphic to $GF(q)$. We have noted in the previous section that the André planes that can be so are symplectically paired planes. Let $\mathcal{W}_3(q)$ denote the associated point-line geometry (generalized quadrangle) of absolute points and lines of σ. An 'm-ovoid' of $\mathcal{W}_3(q)O$ is a set of points such that each line contains exactly m points. The order of $\mathcal{W}_3(q)$ is (q, q).

DEFINITION 83.16. A 'hemisystem' of the dual of $\mathcal{W}_3(q)$ is the dual of an $\frac{(q+1)}{2}$-ovoid.

In general, hemisystems of generalized quadrangles may be defined and are connected to graphs which are 'strongly regular'. Although we make no attempt at connecting the various geometries associated with strongly regular graphs, the definition is as follows.

DEFINITION 83.17. A 'strongly regular graph' on n vertices has parameters $(n, k, l.m)$, so that for any vertex, there are exactly k vertices adjacent, for any pair of adjacent vertices, there are exactly l vertices adjacent to each and for any pair of non-adjacent vertices, there are exactly m vertices adjacent to each.

The connection of symplectically paired spreads and ovoids is as follows (we are using different terminology than in the work described in the next theorem).

THEOREM 83.18. (Cossidente, Culbert, Ebert, and Marino [**258**]). A symplectically paired spread of $\mathcal{W}_3(q)$ gives rise to $2^{(q^2+1)/2} \frac{(q+1)}{2}$-ovoids and hence to associated hemisystems of the dual of $\mathcal{W}_3(q)$, $Q(4, q)$.

Hence, whenever possible, the André spreads produce hemisystems of $Q(4, q)$.

PROBLEM 83.19. Determine non-André symplectically paired spreads in $PG(3, q)$.

CHAPTER 84

When Is a Spread Symplectic?

We mentioned in Chapter 51 on classical geometries that we cannot do justice to the many connections with designs and translation planes. When considering symplectic spreads of even order, there are intrinsic connections with certain designs constructed in the symplectic group, which were introduced by Kantor [856]. Furthermore, the question of finding an internal condition to force a spread to be symplectic is elusive. For arbitrary order, there is a group criterion, which when satisfied shows that a spread is *not* symplectic. For even order, Maschietti [968] has found the existence of a certain type of line oval to provide a necessary and sufficient condition for a spread to be symplectic. Furthermore, Maschietti also shows how to connect the designs of Kantor to any symplectic spread.

84.1. Dual Line Ovals.

We first point out how to connect dual line ovals.

DEFINITION 84.1. An 'oval' in a projective plane of order n is a set of $n+1$ points no three of which are collinear. If n is even, the lines tangent to the points of the oval are concurrent at a point called the 'knot' (or 'nucleus') of the oval. The set of these $n+2$ points, oval and knot, is also a set of points no three of which are collinear and is called a 'hyper-oval'. A 'dual oval' ('dual hyper-oval') is a set of lines in a projective plane which is an oval (hyper-oval) in the corresponding dual projective plane. We call the line dual to the knot of the associated hyper-oval the 'nucleus' of the line oval.

Let $V(2n, q)$ be a $2n$-dimensional vector space over $GF(q)$ and let $\mathcal{S}$ denote a symplectic spread;

$$\mathcal{S} = \{S_0, S_1, \ldots, S_{q^n}\},$$

where S_i for $i = 0, 1, \ldots, q^n$ are totally isotropic subspaces with respect to a nondegenerate alternating bilinear form (symplectic form). Then the following fundamental theorem may be proved.

THEOREM 84.2. (Maschietti [968, part of Theorem 7]). Assume the conditions above with $q^n \geq 8$.

$$\{ S_0, S_1, S_{i+1} + v_i; \ S_0 - \{0\} = \{ v_i; \ i = 1, \ldots, q^n - 1 \} \}$$

is a line oval with nucleus at the line at infinity.

The line oval that exists in all even-order symplectic planes satisfies a strong combinatorial condition called 'complete regularity'. We shall give the definition after introducing some notation.

NOTATION 84.3. Let $\mathcal{O}$ be a line oval in a projective plane of even order and let N denote the nucleus of the line oval. Form the affine plane by removing N and let $B(\mathcal{O})$ to denote the affine points Q such that Q is on a line of $\mathcal{O}$.

DEFINITION 84.4. Let $\mathcal{O}$ be a line oval with nucleus N and P a point on N (a point on the line at infinity). Then $\mathcal{O}$ is said to be 'P-regular' if and only if for any pair of distinct lines L and M incident with P and disjoint from $\mathcal{O}$, there is a third line Z incident with P and disjoint from $\mathcal{O}$ with the property that any line ℓ which is not incident with P intersects L, M or Z in a point of $B(\mathcal{O})$. The line oval $\mathcal{O}$ is said to be 'completely regular' if and only if the oval is P-regular for every point of the nucleus.

84.1.1. Maschietti's Theorem on Line Ovals. The main theorem of Maschietti [**968**] characterizes the line ovals associated with symplectic spreads as completely regular and conversely establishes that such ovals in an affine translation plane imply that the plane is symplectic.

THEOREM 84.5. (Maschietti [**968**, Theorem 3]). An affine translation plane π of even order 2^d, for $d \geq 3$ is symplectic if and only if π admits a completely regular line oval with nucleus the line at infinity.

84.2. The Associated Designs.

Continuing with the previous notation, note that for every translation g of an affine translation plane π of even order q admitting a line oval $\mathcal{O}$ with nucleus the line at infinity then $\mathcal{O}g$ is also a line oval with nucleus the line at infinity and $B(\mathcal{O}g)$ is the set of affine points that line on lines of $\mathcal{O}g$. Form the point-line incidence structure of 'points' as affine points and 'blocks' the sets $B(\mathcal{O}g)$, for all translations g of π. Then this structure $\mathcal{D}(\mathcal{O})$ is a symmetric design (same number of points and blocks) (v, k, λ)-design with parameters $v = q^2$, $k = q(q+1)/2$ and $\lambda = \frac{q^2}{4} + \frac{q}{2}$ (Theorem 4 Maschietti [**968**]). When the affine translation plane is symplectic, we know that the line oval is completely regular. When this occurs, the symmetric design takes on a particular structure.

Let V_{2d} be a $2d$-dimensional vector space over $GF(2)$ admitting the quadratic form $O^+(2d, 2)$ of Witt index d and let $Sp(2d, 2)$ denote the associated symplectic group. Let

$$H(2) = \begin{bmatrix} -1 & 1 & 1 & 1 \\ 1 & -1 & 1 & 1 \\ 1 & 1 & -1 & 1 \\ 1 & 1 & 1 & -1 \end{bmatrix},$$

and let $H(2d)$ denote the tensor product of d copies of $H(2)$. Form a symmetric design as follows: the rows of $H(2d)$ are the 'points' and the columns of $H(2d)$ are the 'blocks' and a point is incident with a block if and only if the corresponding entry is 1. Then the structure is a symmetric design $\vartheta_1(2d)$ with the same parameters as $\mathcal{D}(\mathcal{O})$. In 1973 Kantor [**856**] proved that the Desarguesian and Lüneburg–Tits symplectic planes of even order 2^d produce symmetric designs $\mathcal{D}(\mathcal{O})$ isomorphic to $\vartheta_1(2d)$. In 2003 Maschietti proved that this is always the case.

THEOREM 84.6. (Maschietti [**968**, Theorem 9]). Let π be a symplectic translation plane of even order 2^d. Then the completely regular line oval $\mathcal{O}$ constructs a symmetric design $\mathcal{D}(\mathcal{O})$ isomorphic to $\vartheta_1(2d)$.

CHAPTER 85

The Translation Dual of a Semifield.

In this section, we construct a semifield plane $S^\perp$ with spread in $PG(3,q)$ from a spread S with spread in $PG(3,q)$. $S^\perp$ is called the 'translation dual'. This concept is explicated in Lunardon [**947**], and this concept grows out the similar idea of a translation dual of a generalized quadrangle due to J.A. Thas, wherein symplectic semifield spreads and semifield flock spreads are connected. In Chapter 82 on symplectic spreads, we discussed the general construction of semifield 5th cousins and pointed out that semifield flock spreads are equivalent to symplectic semifield spreads in $PG(3,q)$. It is also true that the translation dual of a semifield flock spread is symplectic (see Lunardon [**947**]). However, there have been no concrete connections between the translation dual of a semifield spread in $PG(3,q)$ and its 5th cousin.

For this section, we more or less follow Lunardon [**947**].

We shall detail the construction in the following, but here is the general idea: First realize a semifield 2×2 matrix spread set of order q^n as a $2n$-dimensional vector space over the center $GF(q)$ of the semifield and embedded as a subspace within the associated 4-dimensional vector space V over $GF(q^n)$, regarded as a $4n$-dimensional vector space over $GF(q)$. Then considering the hyperbolic quadric of the associated $PG(3,q^n)$ followed by the $GF(q)$-trace map, there is an associated non-degenerate quadric α in $PG(4n-1,q)$. Finally, realizing a matrix semifield spread S set as a translation ovoid $O(S)$ in $\Omega^+(6,q^n)$, there is a way to pass from $O(S)$ to $O(S^\alpha)$ another translation ovoid in $\Omega^+(6,q^n)$, producing a corresponding semifield spread in $PG(3,q)$, the 'translation dual'. We note that ultimately many of these ideas can be traced back to Barlotti, Cofman, Knarr, and Thas.

First we formalize the idea of a 'translation ovoid' on $\Omega^+(6,h)$; the Klein quadric.

DEFINITION 85.1. An ovoid of $\Omega^+(6,h)$ is a set of h^2+1 points no two of which lie on a line of the quadric. Any such ovoid is equivalent to a matrix spread set of an associated 4-dimensional vector space V_4 over $GF(h)$, of the following form:

$$x = 0, y = x \begin{bmatrix} g(u,t) & f(u,t) \\ u & t \end{bmatrix} ; u, t \in GF(h),$$

where g, f are functions from $GF(q) \times GF(q)$ to $GF(q)$. If the spread set is a semifield spread set then a representation may be chosen so that $x = 0$ is the axis of an elation group E of order h^2, which acts transitively on the components not equal to $x = 0$. Considered as an ovoid, E fixes one point P, fixes all lines of $\Omega^+(6,h)$ incident with P and acts transitively on remaining points of the ovoid. Any such ovoid of $\Omega^+(6,h)$ which has this property on groups is said to be a 'translation ovoid'.

We note that any ovoid of a hyperplane of $PG(5, h)$ produces a 'symplectic translation ovoid'.

Since translation ovoids of $\Omega^+(6, h)$ and semifield spreads of $PG(3, h)$ are equivalent objects, the focus is now on such translation ovoids.

We begin the so-called linear representation.

85.1. $GF(q)$-**Linear Representation of** $PG(3, q^n)$.

DEFINITION 85.2. Let $\Omega = PG(r - 1, q^n) = PG(V, q^n)$, $q = p^h$, p prime, and let L be a set of points of Ω. The set L is said to be a $GF(q)$-*linear* set of Ω if it is defined by the non-zero vectors of a $GF(q)$-subspace U of V, i.e., $L = L_U = \{\langle \mathbf{u} \rangle_{\mathbb{F}_{q^n}} : \mathbf{u} \in U \setminus \{\mathbf{0}\}\}$.

If $\dim_{\mathbb{F}_q} U = t$, we shall say that L has *rank* t. If $\Lambda = PG(W, q)$ is a subspace of Ω and L_U is an $GF(q)$-linear set of Ω, then $\Lambda \cap L_U$ is an $GF(q)$-linear set of Λ defined by the $GF(q)$-linear vector subspace $U \cap W$.

DEFINITION 85.3. If $L = L_U$ is an $GF(q)$-linear set of Ω of rank t, we shall say that a point $P = \langle \mathbf{u} \rangle_{GF(q^n)}$, $\mathbf{u} \in U$, of L has '*weight* i' in L_U if P as the associated $n - 1$-dimensional subspace over $GF(q)$, intersects L in an $(i - 1)$-dimensional projective subspace and we write $\omega(P) = i$.

It can be seen that if L_U contains a point of weight n, then L_U is a union of lines through such a point. Also a line r is contained in L_U if and only if the rank of the $GF(q)$-linear set $L_U \cap r$ is at least $n + 1$.

Now consider a semifield spread in $PG(3, q^n)$, written as

$$x = 0, y = x \left[\begin{array}{cc} g(t, u) & f(t, u) \\ t & u \end{array} \right] ; u, t \in GF(q^n),$$

where g and f are functions additive in each variable. The set of matrices, since additive, forms a vector space $V_{\mathcal{M}}$ of cardinality q^{2n}. If we allow that $GF(q)$ is the center of the associated semifield then this set of vectors is a vector space over $GF(q)$. Hence, the set of matrices is a $2n$-dimensional vector space over $GF(q)$.

We then may consider the vector space V of dimension 4 over $GF(q^n)$, also a vector space of dimension $4n$ over $GF(q)$. In this sense, we have that $V_{\mathcal{M}}$ is a vector subspace of dimension $2n$ of the $4n$-dimensional vector space V over $GF(q)$. This means that projectively, the semifield spread becomes a $GF(q)$-linear set of $PG(3, q^n)$, considered also as a $PG(4n - 1, q)$ projective space.

PROPOSITION 85.4. Any semifield spread in $PG(3, q^n)$, with center $GF(q)$, becomes, in a natural way, a linear set in $PG(3, q^n)$, considered as a $PG(4n - 1, q)$.

DEFINITION 85.5. For any point x of $PG(3, q^n)$, considered as a $PG(4n - 1, q)$, we let $P(x)$ denote the associated $(n - 1)$-dimensional projective subspace. The set $\mathcal{S}$ of all such subspaces $P(x)$ for $x \in PG(3, q^n)$, is a 'normal $n - 1$-spread' of $PG(4n - 1, q)$, in the following sense: If U is a $2n$-dimensional vector subspace of V over $GF(q)$ containing two elements $P(x_1)$ and $P(x_2)$ as n-dimensional vector spaces, then an n-spread is induced in U by $\mathcal{S}$, or projectively an $(n - 1)$-projective spread is induced by $\mathcal{S}$. This normal $(n - 1)$-spread $\mathcal{S}$ is called the '$GF(q)$-linear representation of $PG(3, q^n)$'.

The reader is referred to the foundations part of this text (Chapters 2–7, see also [**123**]) for a reminder of how the Bruck–Bose and André methods fit together and for a better understanding of the phrase 'Bruck–Bose is André at infinity'. Consider now the affine space $AG(4, q^n)$, and by adjunction of S, we may form a space isomorphic to $PG(4, q^n)$, denoted by $\mathcal{P}$ (the 'points' of $\mathcal{P}$ are the affine points and the $P(x)$'s for $x \in PG(3, q^n)$). Hence, we also arrive at a projective space isomorphic to $PG(4n, q)$.

Another word of what we are trying to do is appropriate here: From a semifield with spread in $PG(3, q^n)$ and center $GF(q)$, we construct from its matrix spread set a vector space of dimension $2n$ over $GF(q)$ and hence a corresponding projective subspace U of dimension $2n - 1$. Taking the standard hyperbolic quadric of $PG(3, q^n)$, we follow this mapping by the trace mapping over $GF(q)$ to produce a non-singular quadric of $PG(4n-1, q)$. The associated polarity α maps U to U^α, also a subspace of projective dimension $2n - 1$. However, we now do not know that U^α necessarily corresponds to a semifield spread or equivalently to a translation ovoid in the associated $\Omega^+(6, q^n)$. That is, we need a model of $\Omega^+(6, q^n)$ that understands ovoids using such $(2n-1)$-dimensional subspaces. The model used will show that a matrix spread set for the semifield plane may be chosen so that the matrices as certain points are disjoint from the standard hyperbolic quadric. In our terminology, this means that U (the associated $(2n-1)$-dimensional projective space) is disjoint from $P(x)$, where $x \in \Omega^+(4, q)$. If $PG(4n - 1, q)$ is regarded as a hyperplane of $PG(4n, q)$ above and U, the $(2n - 1)$-dimensional subspace is embedded into the unique subspace U' such that $U = U' \cap PG(4n - 1, q)$ then the associated ovoid is $O(U) = \{(\infty), U' - U\}$ (note the matrix spread set omits $x = 0$, which is now essentially adjoined). The beautiful part of this is that since U^α will also be disjoint from $P(x)$ for all $x \in \Omega^+(4, q)$, it then follows that $O(U^\alpha)$ is a translation ovoid and hence defines a corresponding semifield spread, the 'translation ovoid'. Now when $O(U)$ is a translation ovoid corresponding to a semifield flock spread then $O(U^\alpha)$ is a translation ovoid of $\Omega(5, q^n) = \Omega^+(6, q^n) \cap PG(4, q^n)$, which means that the associated matrices as subspaces are totally isotropic subspaces of an underlying symplectic polarity. Thus, the translation dual of a semifield flock spread is symplectic. Since the 5th cousin of a semifield flock spread also is a symplectic spread, we note the following problems.

PROBLEM 85.6. Show that the 5th cousin of a semifield flock spread is the associated translation dual, which is symplectic.

Actually, the above problem may be solved using the techniques of Biliotti, Jha, and Johnson [**128**] and Ball and Brown [**83**]. The following problem is, however, open.

More generally:

PROBLEM 85.7. Show that the 5th cousin of a semifield spread in $PG(3, q)$ is the associated translation dual.

Actually, it follows directly as in Biliotti, Jha, and Johnson that a 5th cousin of a semifield flock spread is isomorphic to the original spread if and only if the semifield spread is Desarguesian or Kantor–Knuth.

PROBLEM 85.8. Show that the 5th cousin of a semifield spread in $PG(3, q)$ is isomorphic to the original spread if and only if the spread is Desarguesian or a Kantor–Knuth spread. Similarly, show that the translation dual of a semifield spread in $PG(3, q)$ is isomorphic to the original if and only if the spread is Desarguesian or a Kantor–Knuth spread.

We begin with the model of $\Omega^+(6, q^n)$, mentioned in the previous paragraph.

85.2. A Model for $\Omega^+(6, q^n)$ and Its Linear Representation.

The goal now is to represent a semifield spread as a set of matrices considered a set of points disjoint from a hyperbolic quadric in $PG(3, q^n)$. Let $\Omega^+(6, q^n)$ (or $Q^+(5, q)$) denote the Klein quadric in $PG(3, q^n)$, with homogeneous coordinates $(x_0, x_1, x_2, x_3, x_4, x_5)$ and equation $x_0 x_5 + x_1 x_4 - x_2 x_3$.

Take Σ' as the $PG(4, q^n)$, with equation $x_5 = 0$ and $\Sigma \subset \Sigma'$ with equation $x_0 = 0 = x_5$. Then $\Sigma \cap \Omega^+(6, q^n)$ is the quadric with equation $x_1 x_4 - x_2 x_3$ and coordinates (omitting x_0 and x_5) (x_1, x_2, x_3, x_4), the associated $\Omega^+(4, q^n)$. Define the following point-line geometry $\mathcal{H}$:

$$\text{Points: } \{(\infty)\} \cup \left\{ \begin{array}{c} \text{3-spaces of } \Sigma' \\ \text{intersecting } \Sigma \text{ in a plane tangent to } \Omega^+(4, q^n) \end{array} \right\}$$
$$\cup \{\text{points of } \Sigma' - \Sigma\},$$

$$\text{Lines: } \{\text{points of } \Omega^+(4, q^n)\}$$
$$\cup \{\text{lines of } \Sigma' \text{ that intersect } \Sigma \text{ in a point of } \Omega^+(4, q^n)\}$$
$$\cup \{\text{planes of } \Sigma' \text{ that intersect } \Sigma \text{ in a line of } \Omega^+(4, q^n)\},$$

where all incidences are inherited, except that (∞) is incident exactly with the 'lines' that are points of $\Omega^+(4, q^n)$. We recall that under the standard mapping the points of $\Omega^+(6, q^n)$ and components of the associated 4-dimensional vector space are connected as follows:

$$(1, a, b, c, d, \Delta) \quad \longleftrightarrow \quad y = x \begin{bmatrix} a & b \\ c & d \end{bmatrix},$$
$$(0, 0, 0, 0, 0, 1) \quad \longleftrightarrow \quad x = 0,$$

where $\Delta = ad - bc$.

REMARK 85.9. Then $\mathcal{H}$ is isomorphic to $\Omega^+(6, q^n)$ by a mapping σ that, in particular, maps points as follows:

$$(0, 0, 0, 0, 0, 1) \quad \rightarrow \quad (\infty),$$
$$(1, a, b, c, d, \Delta) \quad \rightarrow \quad (1, a, b, c, 0).$$

REMARK 85.10. Let O be an ovoid of $\Omega^+(6, q^n)$ and let π denote the projection of O^σ from $(1, 0, 0, 0, 0, 0)$ to $\Sigma \simeq PG(3, q)$.

Assume that $y = x \begin{bmatrix} a & b \\ c & d \end{bmatrix}$ maps to $(1, a, b, c, d, \Delta)$ under the Klein map, so that $(1, a, b, c, d, 0)$ then maps to $(0, a, b, c, d, 0)$ in Σ. Then $(0, a, b, c, d, 0) \equiv (a, b, c, d)$ is not on the associated quadric with equation $x_1 x_4 - x_2 x_3$, since otherwise $ad - bc = 0$, but the matrix is non-singular.

So, a point $(0, a, b, c, d, 0)$ identified as (a, b, c, d) cannot satisfy the equation

$$x_1 x_4 - x_2 x_3 = 0,$$

and hence we may think of a semifield with spread in $PG(3, q)$ as a set of points disjoint from a hyperbolic quadric. Considering a semifield spread in $PG(3, q^n)$ as a $(2n-1)$-dimensional subspace U in $PG(4n-1, q)$, which cannot be incident with a point $P(x)$, as an $(n-1)$-dimensional $GF(q)$-subspace, for $x \in \Omega^+(4, q^n)$.

REMARK 85.11. In considering the original $PG(3, q^n)$ within a particular $PG(4, q^n)$, form the natural tensor product of the associated 4-dimensional vector space V_4 by a 1-dimensional $GF(q^n)$-subspace $\langle e \rangle$. If we take V_4 as a $4n$-dimensional $GF(q)$ space and then a $4n$-basis of V_4 over $GF(q)$ and adjoin e, we obtain a natural $(4n + 1)$-dimensional $GF(q)$-space and hence an associated projective space isomorphic to $PG(4n, q)$ containing the original $PG(4n - 1, q)$ as a hyperplane.

Under this convention, a projective 3-space of Σ' intersecting Σ in a plane tangent to $\Omega^+(4, q^n)$ becomes a $3n + 1$-$GF(q)$-vector space, that is, a $3n$-projective space.

85.3. The Linear Representation of $\mathcal{H}$.

We now consider a semifield spread in $PG(3, q^n)$ thought of as a $PG(4n-1, q)$, Σ, embedded into a $PG(4n, q)$, Σ' in a natural manner.

Let $\perp$ denote the hyperbolic polarity of $PG(3, q^n)$ and let

$$\mathcal{Q} = \left\{ P(x); x \in \Omega^+(4, q^n) \right\}.$$

We note that $x^\perp$ is a 3-dimensional vector $GF(q^n)$-subspace, a plane tangent, so a $3n$-dimensional $GF(q)$-vector subspace and hence a $(3n - 1)$-dimensional projective subspace, with notation $P(x^\perp)$. Hence,

$$\widehat{\mathcal{Q}} = \left\{ P(x^\perp); x \in \Omega^+(4, q^n) \right\}$$

is a set of $(3n-1)$-dimensional projective subspaces. Considering Σ' as the $PG(4, q^n)$ above, also regarded as a $PG(4n, q)$ with the convention noted, then there are certainly $3n$-dimensional projective subspaces containing the $P(x^\perp)$. Recalling that the 3-dimensional projective subspaces of Σ' that intersect Σ in a plane tangent to $\Omega^+(4, q^n)$ are points of $\mathcal{H}$, become $3n$-dimensional projective $GF(q)$-spaces, we form the following point-line geometry:

$P(\mathcal{Q})$:

Points: $\{(\infty)\} \cup \left\{ \begin{array}{c} 3n\text{-spaces of } \Sigma', \text{ considered as } PG(4n, q), \\ \text{intersecting } \Sigma \text{ in a plane tangent to } \Omega^+(4, q^n), \\ \text{considered as a } 3n - 1\text{-space of } \widehat{\mathcal{Q}} \end{array} \right\}$

$\cup \{\text{points of } \Sigma' - \Sigma\}$, considered as $PG(4n, q) - PG(4n - 1, q)$.

Lines: $\left\{ P(x) \text{ spaces for } x \text{ points of } \Omega^+(4, q^n) \right\}$

$\cup \left\{ \begin{array}{c} \text{lines of } \Sigma' \text{ that intersect } \Sigma \text{ in a point of } \Omega^+(4, q^n), \\ \text{considered as } n\text{-projective spaces} \end{array} \right\}$

$\cup \left\{ \begin{array}{c} \text{planes of } \Sigma' \text{ that intersect } \Sigma \text{ in a line of } \Omega^+(4, q^n), \\ \text{considered as } 2n\text{-projective spaces} \end{array} \right\}$,

where incidence is as in $\mathcal{H}$, as regarding (∞), and considered in $PG(4n, q)$.

PROPOSITION 85.12. Under the convention given, $\mathcal{H}$ is isomorphic to $P(\mathcal{Q})$ so that $P(\mathcal{Q})$ is isomorphic to $\Omega^+(6, q^n)$ $(Q^+(5, q^n))$.

Now under the above isomorphism, a semifield spread in $PG(3, q)$ becomes a translation ovoid and hence a $(2n - 1)$-dimensional projective subspace U that is disjoint from $P(x)$ for $x \in \Omega^+(4, q^n)$ $(Q^+(3, q^n))$. To retract a semifield spread from such a $(2n - 1)$-dimensional projective subspace, we note the following important theorem.

THEOREM 85.13. (Lunardon [**947**, Theorems 8 and 10]). Let U be a subspace of dimension $2n - 1$ in a $PG(4n - 1, q)$ such that U is disjoint from $P(x)$, for $x \in \Omega^+(4, q^n)$. Embed U into a subspace of dimension $2n$ U' in an associated $PG(4n, q)$. Then

$$O(U) = \{(\infty)\} \cup \{U' - U\}$$

is a translation ovoid of $P(\mathcal{Q})$ (isomorphic to $\Omega^+(6, q^n)$).

Furthermore, any translation ovoid of $P(\mathcal{Q})$ may be assumed to have this form.

PROOF. We note that $O(U)$ does have $q^{2n} + 1$ points. Assume that two points are collinear with a line ℓ of the isomorphic copy of the Klein quadric $P(\mathcal{Q})$ then ℓ is in U' but intersects $PG(4n - 1, q)$ is a point which must be of the from $P(x)$ for $x \in \Omega^+(4, q^n)$, a contradiction, implying that $O(U)$ is an ovoid. To see that it is a translation ovoid, we note that the group of elations with axis $PG(4n - 1, q)$ and center in U defines a subgroup of the group of $P(\mathcal{Q})$, which fixes the point (∞) and all of the lines of $P(\mathcal{Q})$ incident with it and acts regularly on $O(U) - \{(\infty)\}$.

The second part of this theorem is essentially immediate from the general setup but also see Lunardon [**947**, Theorem 10]. $\qquad\square$

85.4. The Translation Dual.

Let π be a semifield spread in $PG(3, q^n)$ and form the associated subspace U of dimension $2n - 1$ in $PG(4n - 1, q)$ as in the preceding section. Let $\perp$ denote the hyperbolic polarity of $PG(3, q^n)$ in the form agreed upon in our convention; follow this by the trace mapping over $GF(q)$. This then defines a non-singular quadric in $PG(4n - 1, q)$, which we denote by α. Then we note that U^α is also a $(2n - 1)$-dimensional subspace. Furthermore, it is not difficult to show that given that U and $P(x)$ for $x \in \Omega^+(4, q^n)$ are disjoint, implies the same condition for U^α. Hence, using the previous theorem, we see that $O(U^\alpha)$ is a translation ovoid of $P(\mathcal{Q})$. Hence, we have an associated semifield plane π^α with spread in $PG(3, q)$.

DEFINITION 85.14. Given any semifield spread π in $PG(3, q^n)$, using the previous construction, form the associated semifield spread π^α. π^α is said to be the 'translation dual' of π.

REMARK 85.15. If π is a semifield flock spread in $PG(3, q^n)$, then it turns out that there is a point y in $PG(3, q^n)$, such that $P(y)$ is contained in the associated $(2n - 1)$-dimensional space U, implying then that y is $PG(3, q^n) - \Omega^+(4, q^n)$ so that $P(y^\perp)$ is a non-singular plane. Embed this plane into a hyperplane of Σ', a $PG(4, q^n)$. Then $\mathcal{H}_{P(y^\perp)}$, the point-line geometry of intersection subspaces turns out to be isomorphic to $\Omega(5, q^n)$ $(Q(4, q^n)$ (see section 6, Lunardon [**947**]). Since U^α then belongs to $P(y^\perp)$, it follows that π^α consists of totally isotropic subspaces of a symplectic polarity; π^α is a semifield symplectic spread.

Again, π^α is the 5th cousin of π using the algebraic approach.

CHAPTER 86

Unitals in Translation Planes.

In 1972, Ganley [**385**] showed the existence of unitary polarities in certain finite semifield planes of order q^2; certain of the Hughes–Kleinfeld semifield planes, certain of the Knuth semifield planes, as well as the Dickson commutative semifield planes. The unitals then turn out to be parabolic in the sense that the line at infinity is tangent to the unitals. More recently, Abatangelo, Korchmáros, and Larato [**9**] have studied 'transitive parabolic unitals in semifield planes' (by this we mean that there is a collineation group of the associated semifield plane that fixes the parabolic unital and acts transitively on the affine points). If a translation plane of order q^2 admits a collineation group G of order q^3 that acts transitively on the affine points of a parabolic unital, the nature of G has not been generally determined. However, Abatangelo, Enea, Korchmáros, and Larato [**8**] show that when the plane is a semifield plane coordinatized by a commutative Dickson semifield, then G is never Abelian.

So, the unital recently the only known finite translation planes that admit transitive parabolic unitals are the commutative twisted field planes, the commutative Dickson semifield planes and, of course, the Desarguesian planes. It has been an open question to determine those translation planes that admit transitive parabolic unitals and particularly to determine those that are semifield planes.

A unital U in a projective plane of order q^2 is a set of $q^3 + 1$ points such that the lines of the plane are either 'tangent', that is are incident with exactly one point of U or are 'secant', incident with exactly $q + 1$ points of U. If π is an affine plane restriction of the projective plane, then the unital is said to be 'parabolic' or 'hyperbolic', respectively as the line at infinity is a tangent line or a secant line. Buekenhout [**199**] shows for any spread in $PG(3, q)$, there is a corresponding parabolic unital in the associated affine translation plane arising from the classical unital in $PG(2, q^2)$ of the absolute points of a unitary polarity. This point set then becomes a unital in any such translation plane. Buekenhout points that the results do not depend upon finiteness and that if there were to be a suitable definition of 'unital' in the infinite case, this result would be valid for spreads in $PG(3, K)$, for K a field. So, we see that classical unitals as the absolute points of unitary polarities, we would need that the field K admits a quadratic field extension K^+. However, initially, we do not necessarily require such an assumption, and we formulate a definition of unitals in translation planes with spreads in $PG(3, K)$.

DEFINITION 86.1. Let π be a translation plane with spread in $PG(3, K)$, where K is a field. Let π^+ denote the projective extension of π. A set of points U in $PG(4, K)$ is said to be a 'unital' if and only if it has the following properties:
  (1) Every point of U has a unique tangent line of π^+,
  (2) For each point S of U the secant lines of π^+ incident with S (i.e., the non-tangent lines on points on U) non-trivially intersect U and cover U,

(3) For each point S of U and secant line L incident with S, then $\mathrm{Card}(L - \{S\}) = \mathrm{Card}\,K$,

(4) Every line of π^+ is either a tangent line or a secant line.

The unital is said to be 'parabolic' or 'hyperbolic' if and only if the line at infinity of π is a tangent line or secant line, respectively.

REMARK 86.2. Let π be a translation plane with spread in $PG(3, q)$. Let U be a point set satisfying the previous definition of 'unital'. Let P be a point of U and by (1) let T_P denote the unique tangent line of the projective extension π^+ to U. Then the q^2 remaining lines of π^+ each share $q = \mathrm{Card}\,K$, when $K \simeq GF(q)$. Hence, there are exactly $q^2 q + 1 = q^3 + 1$ points. Hence, this shows that the more general definition of unital fits the standard situation in the finite case.

It is a natural question whether there are affine translation planes that admit transitive parabolic unitals. It turns out this is the most natural situation, as is shown in Johnson [657].

THEOREM 86.3. (Johnson [657]). Let π be any semifield plane with spread in $PG(3, K)$, where K is a field that has a quadratic extension K^2.

(1) Then π admits a transitive parabolic unital $\mathcal{U} - (\infty)$.

(2) More precisely, if the spread for the associated Pappian plane Σ is

$$x = 0, y = x \begin{bmatrix} u + gt & ft \\ t & u \end{bmatrix}, u, t \in K,$$

and the spread for π is represented in the form:

$$x = 0, y = x \begin{bmatrix} u & F(t, u) \\ t & G(t, u) \end{bmatrix}, u, t \in K,$$

assume that F and G are additive functions from $K \times K$ to K, where $x = 0$ is identified in both spreads. That is,

$$F(t + t^*, u + u^*) = F(t, u) + F(t^*, u^*) \text{ and}$$
$$G(t + t^*, u + u^*) = G(t, u) + G(t^*, u^*).$$

Then the following set $\mathcal{G}$ is a collineation group of π that acts sharply transitively on $\mathcal{U} - (\infty)$: Let

$$\sigma_{u,t,\beta} \colon (x, y) \longmapsto \left(x + \left(\frac{2u - gt}{4f + g^2}, \frac{-gu - 2ft}{4f + g^2} \right), x \begin{bmatrix} u & F(t, u) \\ t & G(t, u) \end{bmatrix} + y \right.$$
$$\left. + \left(\left(\frac{-gu - 2ft}{4f + g^2} \right) \left(\left(\frac{-gu - 2ft}{4f + g^2} \right) + g \left(\frac{2u - gt}{4f + g^2} \right) \right) - f \left(\frac{2u - gt}{4f + g^2} \right)^2, \beta \right) \right).$$

Then

$$\mathcal{G} = \{ \sigma_{u,t,\beta} \}, \ u, t, \beta \in K.$$

We provide a sketch of this proof since it basically also provides an alternative proof of Buekenhout's construction. In order to determine whether any of the collineation group of the classical unital in a Desarguesian spread is inherited, in some sense, in a translation plane, we require an algebraic approach. Hence, we provide an alternative proof of the theorem of Buekenhout which is also valid for partial spreads and their associated affine nets. For example, it is possible for a maximal partial spread to admit a parabolic unital. Armed with this technique, we are able to show that any semifield plane with spread in $PG(3, K)$, where K admits a quadratic extension K^+, always admits a transitive parabolic unital.

86.1. Buekenhout's Parabolic Theorem.

As mentioned Buekenhout [**199**] discusses unitals in translation planes and in particular shows that any translation plane of order q^2 with spread in $PG(3,q)$ admits a parabolic unital. In this context, a 'parabolic unital' is a unital in the projective extension of the translation plane such that the line at infinity contains exactly one point of the unital. This is done by considering the classical unital in a Desarguesian plane and then showing that this set becomes a unital in any translation plane with spread in $PG(3,q)$. However, the method of proof does not consider whether any of the collineation group of the unital that fixes the point (∞) is inherited as a collineation group of the translation plane. When the plane is Desarguesian, there is a considerable amount of work done in determining the nature of a unital admitting a large group. In particular, if there is a linear group of order $q^3(q-1)$ acting on a Desarguesian plane and leaving invariant a unital, then it is known (see Abatangelo and Larato [**11**], and Ebert and Wantz [**352**]) that the unital is a Buekenhout–Metz unital.

So, we provide an algebraic proof of Buekenhout's parabolic theorem, which will point the way to consideration of a possible corresponding group action. We do not require finiteness for our results (Buekenhout also points out that his construction does not require finiteness).

The following is immediate.

LEMMA 86.4. Let K be a field that admits a quadratic extension K^+ and let σ denote the unique involution in $\mathrm{Gal}_K K^+$. Consider the affine Pappian plane Σ coordinatized by K^+ and let $\mathcal{U}$ denote the classical unital in the associated projective plane Σ^+. Choose a tangent line as the line at infinity of the affine restriction, which we take as Σ. Then $\mathcal{U}$ may be represented in the form

$$\left\{ (c, bc^{\sigma+1} + \beta); c \in K^+, \beta \in K \right\}, \; b \notin K.$$

Here we are considering the translation plane Σ to have the following spread:

$$x = 0, y = xm; m \in K^+.$$

LEMMA 86.5. Let K be a field. A basis may be chosen so that Σ may be coordinatized by K^+ as $\begin{bmatrix} u & t \\ ft & u+gt \end{bmatrix}$ for all $u, t \in K$, for suitable constants f and g. Under the previous assumptions, if $\{1, e\}$ is a basis for K^+ over K then $e^2 = eg + f$, and $e^\sigma = -e + g$, $e^{\sigma+1} = -f$. Furthermore,

$$(et + u)^{\sigma+1} = u(u+gt) - ft^2$$

if and only if, in matrix form, $et + u = \begin{bmatrix} u & t \\ ft & u+gt \end{bmatrix}$, $u(u+gt) - ft^2$ is the determinant of the matrix.

PROOF. This is proved in Johnson [**657**]. □

REMARK 86.6. In the above lemma, the basis $(\{1,e\},\{1,e\})$ is used for the 4-dimensional K-vector space. If we use the basis $(\{e,1\},\{e,1\})$ then the matrix representing $(et+u)$ is given by $\begin{bmatrix} u+gt & ft \\ t & u \end{bmatrix}$; $u(u+gt)-ft^2$ is still $(et+u)^{\sigma+1}$. In the following, we shall use the basis $(\{e,1\},\{e,1\})$ and furthermore let $et+u = (t,u)$.

LEMMA 86.7. Let π be a translation plane with spread in $PG(3, K)$, where K^+ is a quadratic field extension.

(1) If the spread for the associated Pappian plane Σ is

$$x = 0, y = x \begin{bmatrix} u + gt & ft \\ t & u \end{bmatrix}, u, t \in K,$$

then the spread for π may be represented in the form:

$$x = 0, y = x \begin{bmatrix} u & F(t, u) \\ t & G(t, u) \end{bmatrix}, u, t \in K,$$

where F and G are functions from $K \times K$ to K, and where $x = 0$ may be identified in both spreads.

Furthermore, the sets of points of each affine plane are identified as the elements $(\alpha e + \beta, \delta e + \rho)$, where $\alpha, \beta, \delta, \rho$ are in K. We further denote this by $(\alpha, \beta, \delta, \rho)$.

(2) Note, in fact, that any partial spread of degree at least three may be represented in the following form:

$$x = 0, y = 0, y = xM_i, \text{ for } i \in \lambda,$$

where M_i is a non-singular 2×2 matrix with entries in K such that $M_i - M_j$ is non-singular.

PROOF. Notice that the first column of an associated matrix spread set for π varies over $K \times K$ so may be chosen as indicated. Since both subspaces represented by $x = 0$ are simply 2-dimensional K-subspaces, we lose no generality in making this identification.

For part (2), note that the first column of M_i may not vary completely over $K \times K$, but does uniquely determine the second column of the matrix M_i. □

LEMMA 86.8. Under the assumptions above, the classical unital $\mathcal{U} - (\infty)$ has the following form:

$$\left\{ (c_1, c_2, k(c_2(c_2 + gc_1) - fc_1^2), \beta); c_1, c_2, \beta \in K \right\},$$

where k is a constant in K.

PROOF. If $c = (c_1, c_2)$ is in K^+, then

$$\left\{ (bc^{\sigma+1} + \beta); \beta \in K \right\} = \left\{ k(c_2(c_2 + gc_1) - fc_1^2, \beta); \beta \in K \right\},$$

using the previous lemmas. If $bc^{\sigma+1} = (b_1, b_2)(0, c^{\sigma+1}) = (b_1 c^{\sigma+1}, b_2 c^{\sigma+1})$. Since b is not in K, then $b_1 \neq 0$. □

REMARK 86.9. In the following, we shall take $b = (1, 0)$, so that $k = 1$ in the previous lemma.

LEMMA 86.10. Let $\mathcal{P}$ be any partial spread of degree at least three:

$$x = 0, y = 0, y = xM_i, \text{ for } i \in \lambda,$$

where M_i is a non-singular 2×2 matrix with entries in K such that $M_i - M_j$ is non-singular. Let $\mathcal{U}$ denote the set in the form

$$(\infty) \cup \left\{ (c_1, c_2, k(c_1(c_1 + gc_2) - fc_2^2), \beta); c_1, c_2, \beta \in K \right\}.$$

Then $y = 0$ intersects $\mathcal{U}$ in exactly the point $(0,0,0,0)$ (note the two uses of zero) and $y = xM_i = x \begin{bmatrix} u & b \\ ft & d \end{bmatrix}$; $ud - bf \neq 0$, intersects $\mathcal{U}$ in a non-degenerate conic, for u and t fixed in K.

Furthermore, $y = c$ and $y = xM_i + d$, for $c, d \in K^+$ intersect $\mathcal{U}$ in a unique point or a non-degenerate conic.

Since the previous lemma now applies more generally for spreads or partial spreads, we now prove Buekenhout's parabolic theorem, stated more generally for partial spreads.

THEOREM 86.11. Let $\mathcal{A}$ be an affine net with partial spread of degree at least three and in $PG(3, K)$, where K is a field admitting a quadratic extension K^+.

Then the classical parabolic unital $\mathcal{U}$ remains a unital in $\mathcal{A}$.

In particular, if $\mathcal{A}$ is a translation plane with spread in $PG(3, K)$, then $\mathcal{U}$ is a parabolic unital in $\mathcal{A}$.

PROOF. The lines of $\mathcal{A}$ are of the form $x = c, y = d, y = xM_i + e$, where c, d, e in K^+. Every line $x = c$ is a line of the associated Pappian plane and inherits from Σ the intersection property, using the previous lemma.   □

86.2. Johnson's Theorem on Transitive Parabolic Unitals.

In this section, we ask if the embedding of the classical unital in a translation plane π with spread in $PG(3, K)$ allows the determination of the collineation group of the unital in π. We are interested in determining whether there is a collineation group of the affine plane π that acts transitively on $\mathcal{U} - (\infty)$.

We choose the coordinate structure as in the previous section. So, we have a Pappian affine plane Σ with spread

$$x = 0, y = x \begin{bmatrix} u + gt & ft \\ t & u \end{bmatrix}, u, t \in K,$$

and the spread for π may be represented in the form:

$$x = 0, y = x \begin{bmatrix} u & F(t, u) \\ t & G(t, u) \end{bmatrix}, u, t \in K.$$

When π is a semifield plane if $F(t, u)$ and $G(t, u)$ are additive functions on the sets (t, u). (Note that $F(t, u) = F_1(t) + F_2(u)$, where both F_1 and F_2 are additive on K).

We have the collineation group E whose elements are given as follows:

$$(x, y) \longmapsto \left(x, x \begin{bmatrix} u & F(t, u) \\ t & G(t, u) \end{bmatrix} + y \right); u, t \in K.$$

Furthermore, we have the translation group T whose elements are given by

$$(x, y) \longmapsto (x + a, y + b); a, b \in K^+.$$

We are looking for a collineation group G, which is a subgroup of ET that acts transitively and regular on the affine points of the unital

$$\mathcal{U} = (\infty) \cup \left\{ (c_1, c_2, (c_2(c_2 + gc_1) - fc_1^2, \beta); c_1, c_2, \beta \in K \right\}.$$

Consider typical elements

$$(x, y) \longmapsto \left(x + (c_1, c_2), x \begin{bmatrix} u & F(t, u) \\ t & G(t, u) \end{bmatrix} + y + (c_2(c_2 + gc_1) - fc_1^2, 0) \right)$$

86. UNITALS IN TRANSLATION PLANES.

and

$$(x,y) \longmapsto \left(x + (c_1^*, c_2^*), x \begin{bmatrix} u^* & F(t^*, u^*) \\ t^* & G(t^*, u^*) \end{bmatrix} + y + (c_2^*(c_2^* + gc_1^*) - fc_1^{*2}, 0) \right)$$

and consider the composition in the order given. We then obtain:

$$(x,y) \longmapsto \left(x + (c_1, c_2) + (c_1^*, c_2^*), x \left(\begin{bmatrix} u^* & F(t^*, u^*) \\ t^* & G(t^*, u^*) \end{bmatrix} + \begin{bmatrix} u & F(t, u) \\ t & G(t, u) \end{bmatrix} \right) \right.$$

$$+ (c_1, c_2) \begin{bmatrix} u^* & F(t^*, u^*) \\ t^* & G(t^*, u^*) \end{bmatrix}$$

$$\left. + (c_2(c_2 + gc_1) - fc_1^2, 0) + y + (c_2^*(c_2^* + gc_1^*) - fc_1^{*2}, 0) \right).$$

Since $(c_1, c_2) + (c_1^*, c_2^*) = (c_1 + c_1^*, c_2 + c_2^*)$, we would like

$$(c_1, c_2) \begin{bmatrix} u^* & F(t^*, u^*) \\ t^* & G(t^*, u^*) \end{bmatrix} + (c_2(c_2 + gc_1) - fc_1^2, 0) + y + (c_2^*(c_2^* + gc_1^*) - fc_1^{*2}, 0)$$

$$= ((c_2 + c_2^*)((c_2 + c_2^*) + g(c_1 + c_1^*)) - f(c_1 + c_1^*)^2, \beta),$$

for some element $\beta \in K$. Hence, we would require

$$(c_2 + c_2^*)((c_2 + c_2^*) + g(c_1 + c_1^*)) - f(c_1 + c_1^*)^2$$
$$= (c_1 u^* + c_2 t^*) + (c_2(c_2 + gc_1) - fc_1^2) + y + (c_2^*(c_2^* + gc_1^*) - fc_1^{*2})$$

and

$$(c_1 F(t^*, u^*) + c_2 G(t^*, u^*) = \beta.$$

Since the second equation is obviously possible as β may range over K, we turn to the first equation. By cancellation of like terms, we obtain the following requirement:

$$c_2^* gc_1 + 2c_2 c_2^* + c_2 gc_1^* - 2fc_1 c_1^* = c_1 u^* + c_2 t^*.$$

This is valid if and only if

$$c_1(c_2^* g - 2fc_1^* - u^*) = c_2(t^* - gc_1^* - 2c_2^*).$$

Take

$$(c_2^* g - 2fc_1^* - u^*) = 0 = (t^* - gc_1^* - 2c_2^*).$$

Hence, we obtain the following system of linear equations:

$$-2fc_1^* + gc_2 = u^*$$
$$gc_1^* + 2c_2^* = t^*.$$

It is easy to check that $4f + g^2 \neq 0$, since we have a field defined by the basis $\{1, e\}$ and $e^2 = eg + f$. Hence, we obtain

$$c_1^* = \frac{\det \begin{bmatrix} u^* & g \\ t^* & 2 \end{bmatrix}}{4f + g^2}, \qquad\qquad c_2^* = \frac{\det \begin{bmatrix} -2f & u^* \\ g & t^* \end{bmatrix}}{4f + g^2},$$

$$c_1^* = \frac{2u^* - gt^*}{4f + g^2}, \qquad\qquad c_2^* = \frac{-gu^* - 2ft^*}{4f + g^2}.$$

Hence, we consider the elements of a putative group as

$$\sigma_{u,t,\beta}\colon (x,y) \longmapsto \left(x + \left(\frac{2u-gt}{4f+g^2}, \frac{-gu-2ft}{4f+g^2}\right), x\begin{bmatrix} u & F(t,u) \\ t & G(t,u) \end{bmatrix} + y\right.$$

$$\left. + \left(\left(\frac{-gu-2ft}{4f+g^2}\right)\left(\left(\frac{-gu-2ft}{4f+g^2}\right) + g\left(\frac{2u-gt}{4f+g^2}\right)\right) - f\left(\frac{2u-gt}{4f+g^2}\right)^2, \beta\right)\right).$$

Our previous argument implies that

$$\sigma_{u,t,\beta}\sigma_{u^*,t^*\beta^*} = \sigma_{(u+u^*),(t+t^*),\rho} \text{ (read left to right)}$$

where

$$\rho = (\beta + \beta^*) + \left(\left(\frac{2u-gt}{4f+g^2}\right)F(t^*,u^*) + \left(\frac{-gu-2ft}{4f+g^2}\right)G(t^*,u^*)\right).$$

Note that since $u(u+gt) - ft^2 \neq 0$, for $(u,t) \neq (0,0)$, taking $t = 1$, the discriminant of this quadratic is $g^2 + 4f$, so $g^2 + 4f \neq 0$.

Also note that $\begin{bmatrix} v+gs & fs \\ s & v \end{bmatrix} = \begin{bmatrix} u & -- \\ t & -- \end{bmatrix}$, by choosing $s = t$ and $v+gs = u$, so this again fits into our scheme of choosing coordinates.

THEOREM 86.12. Let π be any semifield plane with spread in $PG(3,K)$, where K is a field that has a quadratic extension K^2.

(1) Then π admits a transitive parabolic unital $\mathcal{U} - (\infty)$.

(2) More precisely, if the spread for the associated Pappian plane Σ is

$$x = 0, y = x\begin{bmatrix} u+gt & ft \\ t & u \end{bmatrix}, u,t \in K,$$

and the spread for π is represented in the form:

$$x = 0, y = x\begin{bmatrix} u & F(t,u) \\ t & G(t,u) \end{bmatrix}, u,t \in K,$$

assume that F and G are additive functions from $K \times K$ to K, where $x = 0$ is identified in both spreads. That is,

$$F(t+t^*, u+u^*) = F(t,u) + F(t^*,u^*) \text{ and}$$
$$G(t+t^*, u+u^*) = G(t,u) + G(t^*,u^*).$$

Then the following set $\mathcal{G}$ is a collineation group of π that acts sharply transitively on $\mathcal{U} - (\infty)$: Let

$$\sigma_{u,t,\beta}\colon (x,y) \longmapsto \left(x + \left(\frac{2u-gt}{4f+g^2}, \frac{-gu-2ft}{4f+g^2}\right), x\begin{bmatrix} u & F(t,u) \\ t & G(t,u) \end{bmatrix} + y\right.$$

$$\left. + \left(\left(\frac{-gu-2ft}{4f+g^2}\right)\left(\left(\frac{-gu-2ft}{4f+g^2}\right) + g\left(\frac{2u-gt}{4f+g^2}\right)\right) - f\left(\frac{2u-gt}{4f+g^2}\right)^2, \beta\right)\right).$$

Then

$$\mathcal{G} = \{\sigma_{u,t,\beta}\}, \ u,t,\beta \in K.$$

Thus, we have that any semifield plane with spread in $PG(3,K)$, where K is a field that admits a quadratic field extension K^+, admits a transitive parabolic unital. In particular, the point set of the unital is the classical unital in a Pappian projective plane coordinatized by the field K^+, and the lines are modified relative to the lines of the semifield plane. This construction of the associated group uses an algebraic technique or variation of the construction of Buekenhout [199], whereby the possible collineation group "inherited" from the group of the classical unital is made more accessible. Of course, the classical unital is not the only unital that may be constructed in a translation plane with spread in $PG(3, K)$. The main ingredient that is essential to the Buekenhout construction is that the classical unital, when considered parabolically in an associated Desarguesian affine plane, becomes a cone over an ovoid. Thus, as pointed out by Metz [**981**], it is not necessary to begin with a classical unital to accomplish this.

Hence, for example, when one considers an elliptic quadric Ω in $PG(4, q)$, it is possible to form an appropriate associated cone over Ω and construct an associated parabolic unital in any translation plane with spread lying in $PG(3, q)$. Now assume that one tries this over arbitrary fields K. In order to have an elliptic quadric in $PG(3, K)$, it would be necessary and sufficient to have a quadratic field extension K^+. So, it might (and, in fact, 'is') be possible to construct what might be called infinite or arbitrary versions of Buekenhout–Metz unitals. More generally, let Φ be an ovoid in $PG(3, K)$, for K a field. In the finite case, and q is odd, it is known that Φ must be an elliptic quadric. When q is even, except for the elliptic quadrics, the only known ovoids are the Tits' ovoids. However, for arbitrary fields, there are perhaps a more extensive variety of ovoids. For any ovoid, it may be possible to consider a cone over an ovoid to construct parabolic unitals in $PG(3, K)$. For example, certainly the Tits' ovoid construction does not require finiteness, but does require a particular restriction on the field K. Are there corresponding parabolic unitals in any such translation plane with spread in such a $PG(3, K)$? For example, we do know that for every field (or skewfield) there are spreads in $PG(3, K)$. So, we arrive at several questions.

Assume that K is any field admitting a quadratic field extension. Is it possible to construct parabolic unitals from elliptic quadrics in any translation plane with spread in $PG(3, K)$, just as in the finite case? If so, choose any semifield with spread lying in $PG(3, K)$, is there an associated collineation group leaving invariant the unital and acting transitively on the affine points?

More generally, let Φ be an ovoid in $PG(3, K)$, where K is a field. Is it possible to construct parabolic unitals from Φ (as cones over Φ) to construct a parabolic unital in any translation plane with spread in $PG(3, K)$?

Buekenhout [**199**] also constructs hyperbolic unitals in translation planes with spreads in $PG(3, q)$ that contain reguli by constructing a quadric in $PG(4, q)$ that intersects a hyperplane $PG(3, q)$ in the $(q + 1)^2$ points of a regulus. There is a unique non-degenerate quadric in $PG(4, q)$, up to isomorphism.

Let K be a field and assume that Q_4 is a quadric in $PG(4, K)$ that intersects a hyperplane $PG(3, K)$ in the points of a regulus in $PG(3, K)$. Let π be a translation plane with spread in $PG(3, K)$ that contains a regulus, is it possible to construct hyperbolic unitals from Q_4 in π?

Now consider what sorts of groups can exist in the putative hyperbolic unitals.

Assume that it is possible to construct hyperbolic unitals as suggested above in any translation plane π with spread in $PG(3, K)$, that contains a regulus R. Suppose that G is any collineation group of $AGL(4, K)$ of π that leaves R invariant. Is it possible that G leaves invariant the constructed hyperbolic unital?

It is possible to show that for any spread S in $PG(3, K)$, for K a field that admits a quadratic extension field K^+ then for every component L of the associated translation plane π_S, π_S admits a Buekenhout–Metz parabolic unital U^+ that contains the infinite point of L (the associated parallel class) (L), so that U^+ is affine. And, indeed, when the plane is a semifield plane, there is a collineation group fixing the parabolic point and acting transitively on the unital. More generally, in the finite case, Johnson and Pomareda prove the following theorem characterizing those planes admitting such groups.

THEOREM 86.13. (Johnson and Pomareda [**804**]). Given a Buekenhout–Metz unital U^+, a translation plane π with spread in $PG(3,q)$ admits a transitive and regular parabolic group in $AGL(4,q)$ acting on U^+ if and only if π is a semifield plane.

If a finite Buekenhout–Metz unital is embedded in a Desarguesian plane, it admits a collineation group in $AGL(4,q)$, it admits a collineation group of order $q^3(q-1)$ containing a normal transitive and regular parabolic group (see Ebert and Wantz [**352**]). We may consider the following more general problem:

Determine the finite translation planes with spreads in $PG(3,q)$ that admit a collineation group in $AGL(4,q)$ of order $q^3(q-1)$ containing a normal regular parabolic subgroup.

In this regard, Johnson and Pomareda [**804**] prove the following theorem.

THEOREM 86.14. (Johnson and Pomareda [**804**]). Let π be a translation plane of odd order q^2 with spread in $PG(3,q)$ that admits a transitive and regular parabolic unital. If there is a collineation group of order $q^3(q-1)$ that leaves the unital invariant then π is one of the following types of planes:
(1) A semifield plane,
(2) Betten of even order of Walker of odd order q^2, where $q = p^r$, and r is odd,
(3) Lüneburg–Tits of even order q^2.

We also consider Abelian parabolic groups acting on Buekenhout–Metz unitals. The main result that obtained in Johnson and Pomareda is the following.

THEOREM 86.15. (1) Let π be any semifield plane with spread in $PG(3,K)$, where K admits a quadratic extension K^+. Assume that G is a transitive and regular parabolic group acting on a Buekenhout–Metz unital. If G is Abelian then π is Pappian.
(2) Let π be any translation plane with spread in $PG(3,q)$. Assume that G is a transitive and regular parabolic group acting on a Buekenhout–Metz unital. If G is Abelian then π is Desarguesian, q is odd and the unital is partial pencil of conics.

Concerning hyperbolic unitals, when a translation plane π_S with spread in $PG(3,K)$ contains a regulus R_S in $PG(3,K)$, then both π_S and the derived plane π_S^* admit what we are calling hyperbolic Buekenhout unitals U^+ and U^{+*} such that $U^+ - U^{+*}$ is the set of infinite points of R_S and $U^{+*} - U^+$ is the set of infinite points of R_S^*, the associated opposite regulus of R_S. Our analysis allows us to consider the associated collineation group of regulus as the stabilizer of a point of the unital and from there we are able to associate various classes of translation planes admitting groups fixing a regulus and acting on an associated hyperbolic Buekenout unital.

86.3. Buekenhout–Metz Unitals.

In Hirschfeld [**527**], a general construction of parabolic unitals in $PG(4,q)$ from cones over elliptic quadrics is given. The question is if this same analysis is possible over projective geometries $PG(4, K)$, where K admits a quadratic field extension K^+. Since in Hirschfeld, the main ingredient is simple an irreducible quadric over $GF(q)$, we begin by ensuring that unitals are obtained for any such field K, in essentially the same manner.

THEOREM 86.16. (Johnson and Pomareda [804]). Let K be a field that admits a quadratic extension K^+, so there exists a Pappian spread Σ in $PG(3, K)$. Let $g(x_1, x_2)$ be an irreducible quadric over K;

$$g(x_1, x_2) = x_1^2 \alpha_1 + x_2^2 \alpha_2 + x_1 x_2 \alpha_3, \text{ where} \alpha_i \in K, \ i = 1, 2, 3.$$

Consider the 4-dimensional vector space V_4 over K. Choose any spread in $PG(3, K)$, choose any component L and coordinatize so that L becomes $x = 0 = (x_1, x_2)$, where the associated vector space is represented in the form

$$\{ (x_1, x_2, y_1, y_2); \ x_i, y_i \in K \}.$$

Then the spread may be represented as follows:

$$x = 0, y = x \begin{bmatrix} u & F(t, u) \\ t & G(t, u) \end{bmatrix}; u, t \in K,$$

where F and G are functions from $K \times K$ to K. Now consider the following point set:

$$U = \{(c_1, c_2, g(c_1, c_2), \beta); c_1, c_2, \beta \in K\}.$$

Then $U \cup (\infty) = U^+$ is a unital (a 'parabolic unital') in the projective extension π_S^+, where (∞) is the infinite point of $x = 0$. We call such a unital a 'Buekenhout–Metz' unital.

As pointed out, Johnson has shown that the classical unital as a point set becomes a transitive parabolic unital in any semifield plane, where the associated group is in $AGL(4, K)$. The question then becomes whether this is still true for Buekenhout–Metz unitals in $PG(3, K)$, for K an arbitrary field. Furthermore, it is an open question whether any other translation planes with spreads in $PG(3, K)$ can admit transitive parabolic groups in $AGL(4, K)$ for any or a particular Buekenhout–Metz embedded unital. For arbitrary, the following theorem shows that for any semifield plane with spread in $PG(3, K)$ and any Buekenhout–Metz unital, there is an associated transitive parabolic collineation group acting on the unital.

THEOREM 86.17. (Johnson and Pomareda [804]). Given a parabolic Buekenhout–Metz unital U^+ and given any semifield plane π with spread in $PG(3, K)$, where K is a field admitting a quadratic field extension K^+. If the irreducible quadric is $g(c_1, c_2) = c_1^2 \alpha_1 + c_2^2 \alpha_2 + c_1 c_2 \alpha_3$, assume that $4\alpha_1 \alpha_2 - \alpha_3^2 \neq 0$ and note that $\alpha_1 \alpha_2 \neq 0$.

Then π admits a collineation group G acting transitively and regularly on the affine points of U^+.

If the unital is

$$U^+ = \{ (c_1, c_2, g(c_1, c_2), \beta); c_1, c_2, \beta \in K \} \cup \{(\infty)\}$$

in particular, the collineation group

$$x = 0, y = x \begin{bmatrix} u & F(t, u) \\ t & G(t, u) \end{bmatrix}; u, t \in K,$$

where F and G are additive functions from $K \times K$ to K. Then the following defines a transitive collineation group:

$$\langle\, \sigma_{c_1,c_2,\beta};\, c_1, c_2, \beta \in K \,\rangle,$$

$$\sigma_{c_1,c_2,\beta} : (x, y) \longmapsto$$

$$\left(x + (c_1, c_2),\, x \begin{bmatrix} c_2\alpha_3 + 2c_1\alpha_1 & F(2c_2\alpha_2 + c_1\alpha_3, c_2\alpha_3 + 2c_1\alpha_1) \\ 2c_2\alpha_2 + c_1\alpha_3 & G(2c_2\alpha_2 + c_1\alpha_3, c_2\alpha_3 + 2c_1\alpha_1) \end{bmatrix} \right.$$

$$\left. +\, y + (g(c_1, c_2), \beta) \right).$$

The proof of this theorem follows along the same lines as the theorem of Johnson [657] on transitive parabolic unitals.

86.4. Buekenhout Hyperbolic Unitals.

The next two sections follow Johnson and Pomareda [804].

Consider $PG(4, K)$, for K a field admitting a quadratic extension field K^+ and let

$$Q_4 : Q_4(x_1, x_2, y_1, y_2, z) = x_1 y_2 - x_2 y_1 - \gamma z^2,$$

for γ a constant, be a non-degenerate quadric in $PG(4, K)$. Note that if $\Sigma_3 \simeq PG(3, K)$ is given by $z = 0$, then $Q_4 \cap \Sigma_3$ in a regulus in Σ_3. Consider the mapping

$$e_\alpha : (x_1, x_2, y_1, y_2, z) \longmapsto (x_1, x_2, x_1\alpha + y_1, x_2\alpha + y_2, z), \text{ for } \alpha \in K.$$

Since

$$x_1(x_2\alpha + y_2) - x_2(x_1\alpha + y_1) - \gamma z^2 = 0$$

if

$$x_1 y_2 - x_2 y_1 - \gamma z^2 = 0,$$

we see that there is a group

$$\langle\, e_\alpha;\, \alpha \in K \,\rangle$$

acting on Q_4 and leaving invariant Σ_3. Consider a Pappian spread S of the general form

$$x = 0, y = xm; m \in K^+,$$

written in Σ_3 and realized affinely in $AG(4, K)$, i.e., over the associated 4-dimensional K-vector space V_4. Noting that

$$x = 0, y = x\alpha; \alpha \in K$$

is a regulus R, and writing $x = (x_1, x_2)$, $y = (y_1, y_2)$ and $x\alpha = (x_1\alpha, x_2\alpha)$, we see that S admits an affine elation group in the associate Pappian affine plane Σ_S corresponding to $\langle e_\alpha \rangle$ that fixes one component $x = 0$ pointwise of R and acts transitively on $R - \{x = 0\}$. Similarly, consider the mapping

$$g_\alpha : (x_1, x_2, y_1, y_2, z) \longmapsto (x_1 + y_1\alpha, x_2 + y_2\alpha, z)$$

and similarly note that g_α leaves Q_4 and Σ_3 invariant and the associated group induces an affine elation group fixing the component $y = 0$ pointwise of R and acting transitively on $R - \{y = 0\}$. Hence, we observe the following results.

THEOREM 86.18. Let $PG(4, K)$ be a 4-dimensional projective space over a field K admitting a quadratic field extension K^+. Let

$$Q_4 : Q_4(x_1, x_2, y_1, y_2, z) = x_1 y_2 - x_2 y_1 - \gamma z^2,$$

be a non-degenerate quadric of $PG(4, K)$. Then there is a Pappian spread S in $z = 0$; Σ_3 and an associated affine Pappian translation plane in $AG(4, K)$ that admits a collineation group isomorphic to $SL(2, K)$ that fixes Q_4 and acts doubly transitively on the set of lines of a regulus R of $Q_4 \cap \Sigma_3$.

THEOREM 86.19. Let K be any field admitting a quadratic field extension K^+. Let Q_4 be the irreducible $PG(4, K)$ that intersects some $\Sigma_0 \simeq PG(3, K)$ in a hyperbolic quadric H contained in some $PG(2, K)$. H as a set of points may be partitioned into two reguli, R_H and its opposite regulus R_H^*.

Choose either of these two reguli R and assume that a spread S of Σ_0 contains R. Let R_∞ denote the points on the line at infinity of the associated translation plane π_S.

(1) Then

$$(Q_4 - H) \cup R_\infty$$

is a unital of the projective extension π_S^+.

(2) If R^* is the opposite regulus of R, then π_S^*, the derived plane of π_S, contains R^* and if R_∞^* is the set of points of R^* on the line at infinity of π_S^* then

$$(Q_4 - H) \cup R_\infty^*$$

is a unital (a 'hyperbolic unital') of the projective extension π_S^{*+}.

CHAPTER 87

Hyperbolic Unital Groups.

Consider $PG(4, K)$, for K a field admitting a quadratic extension field K^+ and let

$$Q_4 : Q_4(x_1, x_2, y_1, y_2, z) = x_1 y_2 - x_2 y_1 - \gamma z^2,$$

for γ a constant, be a non-degenerate quadric in $PG(4, K)$. Note that if $\Sigma_3 \simeq PG(3, K)$ is given by $z = 0$, then $Q_4 \cap \Sigma_3$ in a regulus in Σ_3. Consider the mapping

$$e_\alpha : (x_1, x_2, y_1, y_2, z) \longmapsto (x_1, x_2, x_1\alpha + y_1, x_2\alpha + y_2, z), \text{ for } \alpha \in K.$$

Since

$$x_1(x_2\alpha + y_2) - x_2(x_1\alpha + y_1) - \gamma z^2 = 0,$$

if

$$x_1 y_2 - x_2 y_1 - \gamma z^2 = 0,$$

we see that there is a group

$$\langle e_\alpha; \alpha \in K \rangle$$

acting on Q_4 and leaving invariant Σ_3. Consider a Pappian spread S of the general form

$$x = 0, y = xm; m \in K^+,$$

written in Σ_3 and realized affinely in $AG(4, K)$, i.e., over the associated 4-dimensional K-vector space V_4. Noting that

$$x = 0, y = x\alpha; \alpha \in K$$

is a regulus R, and writing $x = (x_1, x_2)$, $y = (y_1, y_2)$ and $x\alpha = (x_1\alpha, x_2\alpha)$, we see that S admits an affine elation group in the associate Pappian affine plane Σ_S corresponding to $\langle e_\alpha \rangle$ that fixes one component $x = 0$ pointwise of R and acts transitively on $R - \{x = 0\}$. Similarly, consider the mapping

$$g_\alpha : (x_1, x_2, y_1, y_2, z) \longmapsto (x_1 + y_1\alpha, x_2 + y_2\alpha, z)$$

and similarly note that g_α leaves Q_4 and Σ_3 invariant and the associated group induces an affine elation group fixing the component $y = 0$ pointwise of R and acting transitively on $R - \{y = 0\}$. Hence, we observe the following result:

THEOREM 87.1. Let $PG(4, K)$ be a 4-dimensional projective space over a field K admitting a quadratic field extension K^+. Let

$$Q_4 : Q_4(x_1, x_2, y_1, y_2, z) = x_1 y_2 - x_2 y_1 - \gamma z^2,$$

be a non-degenerate quadric of $PG(4, K)$. Then there is a Pappian spread in the 3-dimensional projective space S with equation $z = 0$; Σ_3 and an associated affine Pappian translation plane in $AG(4, K)$ that admits a collineation group isomorphic to $SL(2, K)$ that fixes Q_4 and acts doubly transitively on the set of lines of a regulus R of $Q_4 \cap \Sigma_3$.

PROOF. It is clear that $\langle e_\alpha, g_\beta; \alpha, \beta \in K \rangle \simeq SL(2, K)$.

Let $x = c = (c_1, c_2)$. If $c_1 = c_2 = 0$ then $x = 0$ (extended projectively) intersects

$$(Q_4 - H) \cup R_\infty$$

in a unique point. Otherwise, we have the solution set (y_1, y_2, z) to

$$c_1 y_2 + c_2 y_1 + z^2 = 0.$$

When $(c_1, c_2) \neq (0, 0)$, and $z = 0$, assume that $c_1 \neq 0$. Then, we obtain the homogeneous point $(1, -c_2/c_1, 0)$. Hence, the cardinality of the intersection on a point of

$$(Q_4 - H) \cup R_\infty$$

is $1 + \operatorname{card} K$. By double transitivity on the components and the fact that the group $SL(2, K)$ preserves the unital, it follows that we obtain $SL(2, K)$ as maintained. □

The following then is immediate from our previous results.

THEOREM 87.2. In the above theorem, let R^* denote the opposite regulus. Then the derived plane of the Pappian plane, the associated Hall plane, admits a hyperbolic unital Q^* fixed by a collineation group isomorphic to $SL(2, K)$ generated by Baer collineations.

THEOREM 87.3. Let $PG(4, K)$ be a 4-dimensional projective space over a field K admitting a quadratic field extension K^+. Let

$$Q_4 : Q_4(x_1, x_2, y_1, y_2, z) = x_1 y_2 - x_2 y_1 - \gamma z^2,$$

be a non-degenerate quadric of $PG(4, K)$.

(1) Let S be any spread in $PG(3, K)$ that contains a regulus R. If R is identified as $Q_4 \cap (z = 0)$, and the associated translation plane π_S admits a group E that fixes a component $x = 0$ and acts transitively on $R - \{x = 0\}$ then E is an elation group and S is a conical flock spread (corresponding to a flock of a quadratic cone in $PG(3, K)$).

(2) The derived conical flock plane π_S^* admits a Baer group B fixing a regulus R^*, that fixes one Baer subplane and acts transitively on the remaining Baer subplanes incident with the zero vector that preserves a hyperbolic unital.

Note that it follows that any linear group of a regulus, i.e., a subgroup of $GL(2, q) \circ GL(2, q)$, where the product is a central product by a group of order $q - 1$ acting affinely on a regulus net then becomes a group of Q_4 fixing $z = 0$ and the affine space $AG(4, K)$. Hence, we obtain immediately the following result.

THEOREM 87.4. Let π be a translation plane with spread in $PG(3, K)$ that admits a regulus R in its spread. Assume that G is a collineation group of $GL(4, K) \cap GL(2, q) \circ GL(2, q)$ that leaves invariant R. Then there is a Buekenhout hyperbolic unital Q that is left invariant by G.

PROOF. G acting on R is a subgroup of $GL(2, q) \circ GL(2, q)$ and this group preserves the quadric Q. If G is not faithful on R then there exists an element g of G fixing each 1-dimensional $GF(q)$-subspace and hence it can only be a kernel homology relative to K. But, the kernel homology also leaves Q_4 invariant, since a point $(x_1^*, x_2^*, y_1^*, y_2^*, z^*)$ satisfying

$$x_1 y_2 - x_2 y_1 + \gamma z^2 = 0$$

maps to a point $(x_1^*k, x_2^*k, y_1^*k, y_2^*k, z^*) \equiv (x_1^*, x_2^*, y_1^*, y_2^*, z^*/k)$ satisfying the same equation. □

COROLLARY 87.5. *The Desarguesian and Hall planes of order q^2 with spread in $PG(3, q)$ admit collineation groups isomorphic to $SL(2, q)$ preserving a hyperbolic unital.*

PROOF. Noting that each of these planes admits a regulus and admits groups isomorphic to $SL(2, q)$ with the required central product preserving the regulus, the corollary then follows from the previous theorem. □

COROLLARY 87.6. *The translation planes of order 81 of Johnson and Prince [807], with spreads in $PG(3, 9)$ admit $SL(2, 5)$ fixing a regulus and preserving a Buekenhout hyperbolic unital. In these planes, the group is either generated by affine elations or by Baer 3-collineations. The group is either transitive on the components of the regulus or transitive on the Baer subplanes incident with the zero vector of the regulus.*

PROOF. See Johnson and Prince [807] for the details of these planes. There are 10 groups of order 3 in $SL(2, 5)$ and in the elation case, each is an elation group. □

Similarly, we obtain:

COROLLARY 87.7. *(1) Every translation plane π corresponding to a flock of a quadratic cone with spread in $PG(3, K)$ admits a collineation group E that fixes a regulus, fixes one component of the regulus and acts transitively on the remaining components. There is a hyperbolic unital in π preserved by E.*

(2) Every derived conical flock plane admits a group B that fixes a regulus, fixes one Baer subplane of the regulus net and acts transitively on the remaining Baer subplanes incident with the zero vector of the regulus net. There is a hyperbolic unital in π that is preserved by B.

PROOF. By Gevaert and Johnson [395] and Jha and Johnson [615], every flock of a quadratic cone in $PG(3, K)$, for K a field, produces a translation plane with associated elation group E with the properties indicated. The spread is a union of reguli that mutually share one component. Replacement of any of these regulus nets produces a Baer group B and applying the previous theorem, we see that there is a hyperbolic unital in each plane (conical flock plane or derived conical flock plane) preserved by the group E or B, respectively. □

COROLLARY 87.8. *Any translation plane of order q^2 with spread in $PG(3, q)$ that admits a cyclic homology group H of order q^2 admits a set of $q - 1$ regulus nets, which are mutually disjoint on lines, the components of which are orbits under H.*

(1) For each of these reguli, there is a hyperbolic unital with hyperbolic line the set of infinite points of this regulus which is preserved by the homology group H.

(2) For each of these reguli, we may derive the corresponding regulus to obtain a corresponding translation plane containing a hyperbolic unital.

(3) If there is a collineation group fixing the axis and coaxis of the homology group H that acts transitively on the set of $q - 1$ regulus nets, then there are $q - 1$

distinct hyperbolic unitals in the translation plane and a group of order $q^2 - 1$ that acts transitively on this set of unitals.

PROOF. By Jha and Johnson [**593**], such translation planes admit reguli as indicated. We now may apply our previous results.   □

COROLLARY 87.9. Any j-plane of order q^2 admits a set of $q - 1$ hyperbolic reguli in an orbit under a collineation group of order $q^2 - 1$.

87.1. Theorem of Johnson–Pomareda on Parabolic Unitals.

If a parabolic unital is Buekenhout–Metz embedded in a translation plane with spread in $PG(3, q)$ admitting a transitive group then it is possible to show that the plane must be a semifield plane. This work is due to Johnson and Pomareda [**804**].

Let U^+ be any parabolic unital in a translation plane π with spread in $PG(3, q)$. Assume that there is a transitive and regular parabolic collineation group G of π in $AGL(4, q)$. Let T denote the translation group then assume that 0 is a point of U^+. Then there is a unique component L of π incident with 0, which is tangent to U^+. Let P_∞ denote the parabolic point on the line at infinity and L_∞ the infinite point of L, assume that some collineation g of G fixes L_∞.

Note that G is in TF_0, where F_0 is the translation complement of the translation plane. Note the line at infinity is the unique tangent line to P_∞ of U^+. Thus, the group G is transitive on the affine lines of π in the parallel class P_∞. Consider $h \in G \cap T$. If the center of h is not P_∞ then a fixed line that intersects the unital does so in $q+1$ points, permuted by h. But, h cannot fix an affine point, a contradiction. Hence, $G \cap T$ is a subgroup of the translation group T_∞ with center P_∞ that fixes U^+, so has order at most q. Then, $(GT)_0$ has order at least q^3/q. Hence, we have a subgroup of the "linear" translation complement of order at least q^2. Any such group has an elation group of order at least q. If q is odd, then there can be no Baer p-elements. Therefore, if g above fixes L_∞ and P_∞ and is not 1, then g cannot be a translation, so in $(GT)_0$ there would be a Baer p-collineation.

Therefore, assume that q is even and there is an element of order 2 fixing L_∞ and P_∞. By Baer's theorem on involutions, there must be a Baer involution σ fixing U^+. Furthermore, since the group G is regular on U^+, the affine Baer subplane Fix σ is disjoint from U^+. However, P_∞ is an infinite point of Fix σ and all affine lines incident with P_∞ are secant lines. Choose any affine point Q of Fix σ. Consider the set of q fixed lines of σ incident with P_∞ other than QP_∞. If all of these lines are tangent lines to Q, then there is exactly one additional tangent line incident with Q, since Q cannot be a point of U^+. But then this last tangent line would be fixed by σ, a contradiction. Hence, at least one of the fixed lines incident with Q are secant lines. But the line at infinity is also a tangent line, so there are exactly $q + 1$ affine points of U^+ on a fixed line incident with Q, permuted by σ. But $q + 1$ is odd, σ must fix a point of U^+, a contradiction.

We have proved the following result.

THEOREM 87.10. Let π be a translation plane with spread in $PG(3, q)$ that admits a parabolic unital U^+. Let P_∞ denote the parabolic point on the line at infinity.

If there exists a collineation group G in $AGL(4, q)$ that acts transitively and regularly on the q^3 affine points of U^+ then

(1) G contains a translation group of order q with center P_∞ and G contains no other translations,

(2) G acts transitively on the affine lines incident with P_∞, and

(3) G acts transitively on the infinite points of ℓ_∞.

If T is the translation group then G acting on ℓ_∞ is equal to $(GT)_0$ acting on ℓ_∞.

THEOREM 87.11. Let π be a translation plane with spread in $PG(3, q)$ that admits a parabolic unital U^+.

(1) If there exists a collineation group G in $AGL(4, q)$ that acts transitively and regularly on the q^3 affine points of U^+ then in $(GT)_0$, there is an affine elation group of order at least q.

(2) Moreover, If q is odd, either π is a semifield plane so that $(GT)_0$ is an elation group, or the affine elation subgroup of $(GT)_0$ has order q.

PROOF. Once we know that we have a collineation group in $(GT)_0$ and in $GL(4, q)$ of order q^2 that acts transitively on $\ell_\infty - \{\infty\}$, then the theorem follows from Jha–Johnson–Wilke [**647**], Biliotti–Menichetti–Jha–Johnson [**133**], and Johnson [**705**].   □

Hence, we may now determine a rough form for the parabolic unital. Choose P_∞ as (∞). There is a translation group T_∞ of order q, normalized by G. Notice that T_∞ is normal in GT. The elements of T_∞ will have the form: $(x, y) \longrightarrow (x, y + (b_{1i}, b_{2i}))$, for $b_{1i}, b_{2i} \in GF(q)$, for any choice of basis of the associated 4-dimensional vector space, and where

$$\{ (b_{1i}, b_{2i}); \; i = 1, 2, \ldots, q \}$$

is an additive group of order q. If we choose $(0, 0, 0, 0)$ to belong to U^+, then $(0, 0, b_{1i}, b_{2i})$ also belongs to U^+. There is a unique tangent line on $(0, 0, 0, 0) = 0$, which we always choose to be $y = 0$.

THEOREM 87.12. Under the assumptions of the previous theorem and the previous note, that is, G is a 'transitive parabolic group', choosing any basis for the 4-dimensional $GF(q)$-vector space, and $P_\infty = (\infty)$ then U^+ has the following form: Assume that $(0, 0, 0, 0)$ is a point of U^+ then

$$U^+ = (\infty) \cup \left\{ \begin{array}{l} (c_1, c_2, g(c_1, c_2) + b_{1i}, f(c_1, c_2) + b_{2i}); \; c_1, c_2 \in GF(q), \\ \{ (b_{1i}, b_{2i}); \; i = 1, 2, \ldots, q \} \text{ an additive group of order } q \end{array} \right\},$$

where g and f are functions from $GF(q) \times GF(q)$ to $GF(q)$, where

$$(g(c_1, c_2) + b_{1i}, f(c_1, c_2) + b_{2i}) = (0, 0)$$

if and only if $c_1 = c_2 = b_{1i} = b_{2i} = 0$.

Furthermore, if there is to be a transitive and regular parabolic group G, then since there are elations axis $0P_\infty$ in $(GT)_0$, it follows immediately that if there are two such axes, then $SL(2, q)$ is contained in the group generated by the elation groups, forcing the plane to be Desarguesian by Foulser–Johnson–Ostrom [**380**]. Note the main idea of the following is that given the group $(GT)_0$ of order q^2, there is a corresponding set of q^2 translations such that for each element g of $(GT)_0$, there is a unique translation τ_g in T modulo T_∞ such that $g\tau_g \in G$. Putting this another way, if one considers the translation

$$\tau_{(c_1, c_2)} \colon (x, y) \mapsto (x + (c_1, c_2), y + (g(c_1, c_2), f(c_1, c_2))),$$

then there is a unique element $g_{(c_1,c_2)}$of $(GT)_0$ such that $\tau_{(c_1,c_2)}g_{(c_1,c_2)}$ is an element of G taken modulo the normal subgroup T_∞. Since we are free to choose the representation for $(GT)_0$, we may employ the results of Johnson [**705**] so that $(GT)_0$ has the following form, recalling the following theorem, which is stated for even order in the original, but this is not essential. In the odd-order case, there is a more convenient form. We shall provide both forms. Note that the choice of representation for the group might affect any given initial representation of the unital.

THEOREM 87.13. (Biliotti, Jha, Johnson, and Menichetti [**133**]). Let π be a translation plane of even order q^2 with spread in $PG(3, q)$ that admits a linear group $(GT)_0$ of order q^2 that fixes a component and acts regularly on the remaining components. Then coordinates may be chosen so that the group $(GT)_0$ may be represented as follows:

$$\left\langle \sigma_{b,u} = \begin{bmatrix} 1 & T(b) & u+bT(b)+l(b) & uT(b)+R(b)+m(u) \\ 0 & 1 & b & u \\ 0 & 0 & 1 & T(b) \\ 0 & 0 & 0 & 1 \end{bmatrix} ; b,u \in GF(q) \right\rangle$$

where T, m, and l are additive functions on $GF(q)$ and $m(1) = 0$ and such that

$$R(a+b) + R(a) + R(b) + l(a)T(b) + m(aT(b)) + a(T(b))^2 = 0, \; \forall\, a, b \in GF(q).$$

THEOREM 87.14. (See Johnson [**705**]). Let π be a translation plane of odd order q^2 with spread in $PG(3, q)$ that admits a linear group $(GT)_0$ of order q^2 that fixes a component and acts regularly on the remaining components. Then either π is a semifield plane or the elation subgroup of $(GT)_0$ has order q. Then coordinates may be chosen so that the group $(GT)_0$ may be represented as follows:

$$\left\langle \sigma_{b,u} = \begin{bmatrix} 1 & b & u & -\frac{1}{3}b^3 + J(b) + m(u) + ub \\ 0 & 1 & b & u \\ 0 & 0 & 1 & b \\ 0 & 0 & 0 & 1 \end{bmatrix} ; b,u \in GF(q) \right\rangle$$

where J, m are functions from $GF(q)$ to $GF(q)$, such that m is additive and $m(GF(p)) = 0$, and

$$J(a+b) + m(ab) = J(a) + J(b).$$

We may use the same general setup notationally for both q odd or even provided in the odd-order case: we merely take

$$T(b) = b, \; l(b) = -b^2, \; R(b) = -\frac{1}{3}b^3 + J(b).$$

Hence, given $\sigma_{b,u}$, there is a unique translation (modulo T_∞),

$$\tau_{(c_1,c_2)} \colon (x,y) \mapsto (x + (c_{1,b,u}, c_{2,b,u}), y + (g(c_{1,b,u}, c_{2,b,u}), f(c_{1,b,u}, c_{2,b,u})))$$

such that

$$(x, y) \mapsto \left(x \begin{bmatrix} 1 & T(b) \\ 0 & 1 \end{bmatrix} + (c_{1,b,u}, c_{2,b,u}), \right.$$

$$x \begin{bmatrix} u + bT(b) + l(b) & uT(b) + R(b) + m(u) \\ b & u \end{bmatrix} + y \begin{bmatrix} 1 & T(b) \\ 0 & 1 \end{bmatrix}$$

$$\left. + (g(c_{1,b,u}, c_{2,b,u}), f(c_{1,b,u}, c_{2,b,u})) \right),$$

$$(x, y) \mapsto (x, y + (b_{1i}, b_{2i})),$$

the latter forming the normal translation group T_∞ of order q. Let G denote the group whose elements are these mappings. This group G of order q^3 acts regularly on the affine points of the unital

$$U^+ = (\infty) \cup \left\{ \begin{array}{l} (c_1, c_2, g(c_1, c_2) + b_{1i}, f(c_1, c_2) + b_{2i}); \ c_1, c_2 \in GF(q), \\ \{(b_{1i}, b_{2i}); \ i = 1, 2, \ldots, q\} \ \text{an additive group of order } q \end{array} \right\}.$$

We shall be interested in general transitive parabolic unitals as well as those of Buekenhout–Metz. Ebert and Wantz [**352**] showed that in a Desarguesian spread the collineation group of a Buekenhout–Metz unital which is not classical has a collineation group of order $q^3(q-1)$, with a normal subgroup of order q^3. We consider a group of this size here, but first we recall the main theorem of Jha–Johnson–Wilke [**647**]. We recall that a Walker plane of odd order q^2, $p^r = q$, r odd has the form given above where $J \equiv m \equiv 0$, identically. If q is even and $q = 2^r$, r odd, then this same form also provides a translation plane called the Betten plane. The Lüneburg–Tits plane of even order q^2 admits a group of order $q^2(q^2 + 1)(q - 1)$, the Suzuki group $Sz(q)$ as a group of collineations within the translation complement. In this case, the representation of the group of order q^2 is as follows:

$$\left\langle \sigma_{b,u} = \begin{bmatrix} 1 & b^{\alpha^{-1}} & u + b^{1+\alpha^{-1}} & u + u^\alpha + ub^{\alpha^{-1}} + b + b^\alpha + b^{\alpha+1} \\ 0 & 1 & b & u \\ 0 & 0 & 1 & b^{\alpha^{-1}} \\ 0 & 0 & 0 & 1 \end{bmatrix} ; \right.$$

$$\left. b, u \in GF(q) \right\rangle,$$

where for $q = 2^{2k+1}$, $\alpha = 2^{k+1}$.

THEOREM 87.15. (Jha–Johnson–Wilke [**647**]). Let π denote a translation plane of order q^2 with spread in $PG(3, q)$. If π admits a collineation group of order $q^2(q-1)$ in the linear translation complement then π is one of the following types of planes:
   (1) A semifield plane,
   (2) Betten of even order of Walker of odd order q^2, where $q = p^r$, and r is odd,
   (3) Lüneburg–Tits of even order q^2.

Hence, we obtain:

THEOREM 87.16. Let π be a translation plane of odd order q^2 with spread in $PG(3, q)$ that admits a transitive and regular parabolic unital. If there is a

collineation group of order $q^3(q-1)$ that leaves the unital invariant, then π is one of the following types of planes:
(1) A semifield plane,
(2) Betten of even order of Walker of odd order q^2, where $q = p^r$, and r is odd,
(3) Lüneburg–Tits of even order q^2.

87.2. Buekenhout–Metz Forms.

Now we specialize to the situation that we are considering. Assume that U^+ is a Buekenhout–Metz unital in the form that we have established. In this case, in the above representation, we would have $g(c_1, c_2) = c_1^2\alpha_1 + c_2^2\alpha_2 + c_1 c_2 \alpha_3$, an irreducible quadric, $f(c_1, c_2) = 0$, $b_{1i} = 0$ and b_{2i} taking on all possible elements of $GF(q)$. If we start with a Buekenhout–Metz unital in this form, this is a unital in any translation plane with spread

$$x = 0, y = x \begin{bmatrix} u & F(t,u) \\ t & G(t,u) \end{bmatrix}; u, t \in GF(q).$$

The problem now is that by choosing this form, we may have altered the form of the group. However, the form is not altered significantly. For example, we know that we have in $(GT)_0$ an elation in the center of the group of order q^2. Suppose that this elation σ has the following form $(x, y) \to (x, xT + y)$, mapping $y = 0$ to $y = xT$. If we choose a new basis by $\begin{bmatrix} I & 0 \\ 0 & T^{-1} \end{bmatrix}$, then since we have any group element in the general form $\begin{bmatrix} A & B \\ 0 & C \end{bmatrix}$, then choosing the element σ as $(x, y) \mapsto (x, x + y)$, it now follows, since the element is in the center, that $A = C$. Now since our group $(GT)_0$ has order q^2, and is in $GL(4, q)$, it will fix pointwise a 1-dimensional $GF(q)$-subspace on $x = 0$. We may pre-choose a basis so that $C = \begin{bmatrix} 1 & a \\ 0 & 1 \end{bmatrix}$, for some $a \in GF(q)$. This basis change could have been accomplished by a mapping of the form $\begin{bmatrix} I & 0 \\ 0 & D \end{bmatrix}$.

This means that if we wish to have the form of the group listed above but start with a Buekenhout–Metz unital in the form

$$U^+ = (\infty) \cup \{ (c_1, c_2, g(c_1, c_2), \beta); c_1, c_2, \beta \in GF(q) \},$$

for $g(c_1, c_2)$ an irreducible quadric, then after the basis changes, we obtain the form

$$U^+ = (\infty) \cup \{ (c_1, c_2, g(c_1, c_2), \beta)E; c_1, c_2, \beta \in GF(q) \},$$

where E is a non-singular matrix $\begin{bmatrix} e_1 & e_2 \\ e_3 & e_4 \end{bmatrix}$.

We now show that when this occurs, it can only be that $T(b) = 0$; that we obtain a semifield plane.

THEOREM 87.17. Given a Buekenhout–Metz unital U^+, a translation plane π with spread in $PG(3, q)$ admits a transitive and regular parabolic group in $AGL(4, q)$ acting on U^+ if and only if π is a semifield plane.

Hence, we have proved the following theorem.

THEOREM 87.18. (Johnson and Pomareda [**804**]). (1) Let π be any semifield plane with spread in $PG(3, K)$, where K admits a quadratic extension K^+. Assume that G is a transitive and regular parabolic group acting on a Buekenhout–Metz unital. If G is Abelian then π is Pappian.

(2) Let π be any translation plane with spread in $PG(3, q)$. Assume that G is a transitive and regular parabolic group acting on a Buekenhout–Metz unital. If G is Abelian then π is Desarguesian, q is odd and the unital is a partial pencil of conics.

CHAPTER 88

Transitive Parabolic Groups.

This chapter is concerned with translation planes of order q^2 admitting collineation groups preserving a parabolic unital, or more precisely with translation planes of order q^2 admitting collineation groups G of order q^3u, where u is a prime p-primitive divisor of $q^2 - 1$. The work follows Jha and Johnson [645].

The situation we wish to consider arises when G is a subgroup of $PSU(3, q)$ acting on a projective Desarguesian plane and G preserves a classical unital. Since $PSU(3, q)$ has order $(q^3 + 1)q^3(q^2 - 1)/(3, q+1)$, acting on the Desarguesian plane, it follows that the stabilizer of a point of the associated unital has order $q^3(q^2 - 1)/(3, q + 1)$. Taking the tangent line as the line at infinity, it follows that there is an affine group of order $q^3(q^2 - 1)/(3, q + 1)$ acting on the affine Desarguesian plane of order q^2.

We note that since there is a subgroup E of order q^3 acting on a unital, it then follows that the translation subgroup has order at most q. Furthermore, E is a normal subgroup in this case. We note that there is a subgroup of order q^3u, unless $q = 8$ or q is a Mersenne prime or when $q + 1 = 3 \cdot 2^a$.

Jha and Johnson have classified the translation planes of order q^2 with spreads in $PG(3, q)$ that admit collineation subgroups G of $GL(4, q)$, of order $q(q + 1)$ (see [632, 633]). In this setting, there are various conical flock planes that admit such a group. When this occurs, there is a unique point (∞) on the line at infinity that is fixed by G. Let T denote the translation subgroup of order q^2 whose elements have center (∞). Then T is normalized by the collineation group of order $q(q+1)$. Hence, GT has order $q^3(q + 1)$. When $q + 1 \neq 2^a$, there is a collineation group of order q^3u, where u is a prime p-primitive divisor of $q^2 - 1$. Therefore, a complete classification of all translation planes of order q^2 admitting collineation groups of order q^3u would necessarily involve an analysis of conical flock planes. On the other hand, if it is assumed that when there is a translation subgroup of a group of order q^3u, the translation subgroup with fixed center does not have order q^2, it turns out that the problem can be solved.

Hoffer [553] showed that a projective plane π of order q^2 admitting $PSU(3, q)$ is necessarily Desarguesian and, as we have noted, when $q^2 - 1$ has a p-primitive divisor u then there is an associated subgroup of order q^3u of $PGU(3, q)$ as E is normal in the stabilizer.

In the following, we mention some extensions of Hofer's result due to Jha and Johnson.

THEOREM 88.1. (Jha and Johnson [645]). Assume a translation plane of order q^2 admits a collineation group G of order q^3u, where u is a p-primitive divisor of $q^2 - 1$. If the plane admits an affine homology τ of order u. Then there is always a translation subgroup of order q^2p^b, normalized by τ, for some divisor p^b of q^2.

Conversely, if there is a proper translation subgroup of order $> q^2$, normalized by an element τ of order u then τ is an affine homology and there exists a translation subgroup of order q^2 with fixed center the center of τ.

Jha–Johnson Theorem on Parabolic Unitals.

THEOREM 88.2. (Jha and Johnson [**645**]). Let π be a translation plane of order q^2 that admits a collineation group G of order $q^3 u$, where u is a prime p-primitive divisor of $q^2 - 1$. Assume that G does not contain a translation subgroup of order a multiple of q^2. Then the plane π is Desarguesian.

COROLLARY 88.3. Let π be a translation plane of order q^2 that admits a collineation group G of order $q^3 u$ that preserves a parabolic unital, where u is a prime p-primitive divisor of $q^2 - 1$.

Then π is Desarguesian.

We have pointed out that if a translation plane of order q^2 admits a collineation group of order $q^3 u$ preserving a parabolic unital, then the translation plane is Desarguesian. What has not been determined is the type of unital that is preserved. However, by the collineation group, the unital cannot be of Buekenhout–Metz type, unless the unital is the classical unital.

PROBLEM 88.4. Show that the parabolic unital preserved by a collineation group of order $q^3 u$ is classical.

However, Jha and Johnson show the following:

THEOREM 88.5. (Jha and Johnson [**645**]). Let π be a translation plane of order q^2 that admits a collineation group G of order $q^3 u(q - 1)$, where u is a p-primitive divisor of $q^2 - 1$, admitting a normal subgroup of order q^3.

If G preserves a parabolic unital, then π is Desarguesian. If the group then is in $AGL(2, q^2)$, the unital is classical.

COROLLARY 88.6. Assume that π is a translation plane of order q^2, $p^r = q$, such that $(2r, q - 1) = (2, q - 1)$. Let π admit a collineation group of order $q^3 2u(q - 1)$, containing a normal subgroup of order q^3, where u is a prime p-primitive divisor of $q^2 - 1$.

Then π is Desarguesian and the unital is classical.

CHAPTER 89

Doubly Transitive Hyperbolic Unital Groups.

Let U be a hyperbolic unital of Buekenhout type embedded in a translation plane π of order q^2 with spread in $PG(3, q)$. That is to say that U arises from a quadric in $PG(4, q)$ and in this setting, U has $q + 1$ points on the line at infinity that form the infinite points of a regulus net with the translation plane. It is shown in Johnson and Pomareda [**804**] that any collineation subgroup of $SL(2, q) \circ SL(2, q)$ of π that leaves invariant the regulus net can be arranged to leave invariant the hyperbolic unital.

For example, any conical flock plane admits an elation group of order q that leaves a regulus invariant, that is, fixes one component and acts regularly on the remaining components and hence the elation group E acts also on the hyperbolic unital. Similarly, the planes of Johnson and Prince [**807**] of order 81 admitting $SL(2, 5)$ leaving invariant a regulus are of this type as well; the group leaves invariant a hyperbolic unital of Buekenhout type. In the first situation there is a group E fixing a point and transitive on the remaining q of the $q + 1$ points and in the second case, various of the planes of order 81 admit groups transitive on the $q + 1$ points but not two-transitive. Of course, the Desarguesian plane admits $SL(2, q)$ acting two-transitively on $q + 1$ points of a regulus and the question is: are there other translation planes admitting groups acting two-transitively on a set of $q + 1$ points on the line at infinity that preserve a unital?

PROBLEM 89.1. Determine the translation planes of order q^2 and spread in $PG(3, q)$ that admits a hyperbolic unital with a secant line at infinity such that the translation plane admits a doubly transitive group on the secant line. Furthermore, once the planes are determined, determine the unital.

Actually, we may generalize this problem and consider translation planes of order q^2 without reference to the kernel of the plane and without reference to a unital that admit doubly transitive groups on a set of $q + 1$ points on the line at infinity.

PROBLEM 89.2. Classify the translation planes of order q^2 admitting a collineation group acting two-transitively on a set of $q + 1$ points on the line at infinity.

Of course, included in such a putative classification there would be the Ott–Schaeffer planes and the Hering planes as well as the Desarguesian since these admit $SL(2, q)$ acting transitively on $q + 1$ points, even though these points do not always belong to a regulus. But, note again that we are not making any assumptions on the dimension of the associated vector space; we do not require the spread to initially be in $PG(3, q)$. But if we are to use the nature of a translation plane, it might be asked why the set of $q + 1$ points should be required to lie on the line at infinity. Could there be translation planes π, whose projective extensions π^+ admit such groups? Also, must the $q + 1$ points necessarily be restricted to a line? For example, could the set Γ of $q + 1$ points be an arc or a unital design?

Let ρ^+ be a projective Desarguesian plane of order h^2. Then ρ^+ contains a classical unital U_h of absolute points and non-absolute lines of a unitary polarity that admits a collineation group G isomorphic to $PSU(3, h)$. In this setting U has $h^3 + 1$ points. ρ^+ is isomorphic to $PGL(3, h^2)$ and may be coordinatized by a field isomorphic to $GF(h^2)$. Choose any cubic extension $GF(h^6)$ of this field and consider the Desarguesian plane π^+ isomorphic to $PGL(3, h^6)$ coordinatized by $GF(h^6)$. In this case, there is an over-group $PSU(3, q)$ of $PSU(3, h)$, $h^3 = q$ and a unital U_q of $q^3 + 1$ points containing U_h. So, there is a projective extension of an affine translation plane of order q^2 that admits a set of $q + 1$ points and a set of $h^4 + h^2 + 1 - (1 + h^3) = h^4 - h^3 + h^2$ lines that form a unital design Γ admitting a collineation group isomorphic to $PSU(3, h)$ acting two-transitively on Γ, namely the Desarguesian plane of order q^2. One of the questions that might be considered is whether there are others.

Similarly, if γ^+ is a projective Desarguesian plane of order h^3, let C_h be a conic of $h^3 + 1$ points of γ^+. Then there exists a collineation group isomorphic to $PSL(2, h^3)$ acting two-transitively on the points of C_h. Let π^+ be a projective Desarguesian plane of order h^6 and C_q be a conic of $h^6 + 1$ points admitting $PSL(2, h^6)$, for $h^6 = q^2$. Then C_h is an arc of $q + 1$ points admitting a collineation group of π^+ isomorphic to $PSL(2, q)$.

Consider a projective plane of order q^2 that contains a set Γ of $q + 1$ points admitting a collineation group G inducing a two-transitive group on Γ. Is it possible to determine the projective plane and the group G? Biliotti, Jha, Johnson and Montinaro [**134, 135, 136, 137**] consider this question under the hypothesis that the projective plane is the projective extension of an affine translation plane.

PROBLEM 89.3. Let π be a translation plane of order q^2 and let π^+ be the projective extension of π. Let Γ be a set of $q + 1$ points of π^+. Determine the projective translation planes π^+ that admit doubly transitive collineation groups acting on a set Γ of $q + 1$ points of π^+.

Recently, A. Montinaro [**983**] studied the projective planes of order n admitting a two-transitive orbit of large cardinality (i.e., at least $n/2$), a problem whose origins go back to Cofman [**236**], who asked what could be said of finite projective planes and the groups acting two-transitively on a subset of points. So, our study requires that the projective plane is the projective extension of a translation plane and we have a smaller than $q^2/2$ orbit but nevertheless, various ideas from the work of Montinaro [**983**] may be employed for the solution of this problem.

To consider the general problem listed above, it is then possible and appropriate to use the classification theorem of finite groups. In particular, the groups with non-solvable socles are well known (see classification theorem listed below) and it is possible to show that the collineation group G induces $PSL(2, q)$ on the set Γ, and, in some sense, the motivation for this work came from this connection and translation planes.

Recall that the theorem of Biliotti–Jha–Johnson [**131**] shows the planes to be Desarguesian, Hall, Hering, Ott–Schaeffer, Walker order 25 or Dempwolff order 16.

Of the planes listed, the Hall planes and the planes of orders 16 and 25 do not admit groups acting two-transitively on $q + 1$ points on the line at infinity. So, it would be expected that at least the Desarguesian, Hering and Ott–Schaeffer planes would be included in the classification. In fact, it turns out that these are exactly the set of translation planes admitting two-transitive groups on an infinite subline. Furthermore, it is possible to show that if the set of points is not on the line at infinity, the plane must be Desarguesian.

The main theorems in this regard are due to Biliotti–Jha–Johnson–Montinaro [**134, 135, 136, 137**].

Biliotti–Jha–Johnson–Montinaro Theorems on Unitals.

THEOREM 89.4. Let π be a translation plane of order q^2 and let π^+ denote the associate projective extension. Let π^+ admit a collineation group G inducing a two-transitive group on a set Γ of $q + 1$ points of π^+. Then π is one of the following types of planes:

(1) Desarguesian,
(2) Hering, or
(3) Ott–Schaeffer.

THEOREM 89.5. Let π^+ be a projective Desarguesian plane of order q^2 containing a set Γ of $q + 1$ points upon which there is a collineation group of π^+ inducing a two-transitive group on Γ. Then one of the following occurs:

(1) Γ is a set of $q + 1$ points on a line L. If π is the affine restriction by L then there is an affine group isomorphic to $SL(2, q)$ of π inducing $PSL(2, q)$ on Γ.

(2) Γ is a unital design of $q+1$ points and there is a collineation group isomorphic to $PSU(3, \sqrt[3]{q})$ of π^+ induced on Γ.

(3) Γ is an arc contained in a conic and there is a collineation group isomorphic to $PSL(2, q)$ of π^+ induced on Γ.

Rethinking the main initial impetus for this problem, one could now ask which of these classes of translation planes admit hyperbolic unitals U equipped with a group G leaving U invariant and acting two-transitively on the points of the infinite secant line of U. Since the Desarguesian plane does admit such groups, acting for example on the Buekenhout hyperbolic unital, the question remains whether the Hering or Ott–Schaeffer planes can admit such unitals and group actions. Biliotti et al. are able to show that the action of the group $SL(2, q)$ acting on the Hering or Ott–Schaeffer planes prohibits such group actions on unitals. Now, the question of the nature of the unital then becomes of interest. However, it is also possible to revolve this question as well, proving the unital is classical (Buekenhout). Indeed, one may formulate the problem in the projective extension of the affine translation plane for a stronger theorem. The main results therefore related to unitals are as follows:

THEOREM 89.6. (Biliotti, Jha, Johnson, and Montinaro [**134, 135, 136, 137**]). Let π be a translation plane of order q^2 and let π^+ denote the projective extension. Assume that π^+ admits a collineation group G inducing a two-transitive group on a set S of $q + 1$ points contained in a line.

Then π is one of the following types of planes:

(1) Desarguesian,
(2) Hering, or
(3) Ott–Schaeffer.

THEOREM 89.7. (Biliotti, Jha, Johnson, and Montinaro [**134, 135, 136, 137**]). Let π be a translation plane of order q^2 and let π^+ denote the projective extension.

Assume that π^+ contains a unital U and admits a collineation group G that leaves U and a secant line L invariant and acts two-transitively on $L \cap U$.

Then π is Desarguesian and the unital is classical.

CHAPTER 90

Retraction.

A 'subgeometry partition' is a partition of a projective space by subgeometries. For example, when Σ is isomorphic to $PG(2m, q^2)$, the partition components are Baer subgeometries isomorphic to $PG(2m, q)$ and we say that the partition is a 'Baer subgeometry partition'. When Σ is isomorphic to $PG(2n-1, q^2)$, it is possible to have partition of β $PG(n-1, q^2)$'s and α $PG(2n-1, q)$'s. The configuration is such that $\alpha(q+1) + \beta = q^{2n} + 1$ and the partition is called a 'mixed subgeometry partition'.

The interest in such partitions lies in the fact that they may be used to construct spreads and hence translation planes. Baer subgeometry partitions produce translation planes of order q^{2m+1} where mixed partitions produce translation planes of order q^{2n}. These constructions are applications of the theory of Segré varieties and are given in Hirschfeld and Thas [529], p. 206. In particular, all Baer subgeometries produce $GF(q)$-reguli in the associated spread. When the partition is a Baer subgeometry partition, the spread is a union of mutually disjoint $GF(q)$-reguli. Furthermore, mixed partitions of $PG(2m-1, q^2)$ by $PG(m-1, q^2)$'s and $PG(2m-1, q)$'s produce spreads in $PG(4m-1, q)$ which contain d $GF(q)$-reguli provided there are d $PG(2m-1, q)$'s in the mixed partition.

Recently, there has been considerable work by Baker, Dover, Ebert and Wantz (see [63] and [77]) on flag-transitive translation planes using Baer subgeometry partitions of $PG(2m, q^2)$; partitions of $PG(2m, q^2)$ by $PG(2m, q)$'s. Such Baer subgeometry partitions produce spreads in $PG(4m+1, q)$ which are unions of $GF(q)$-reguli. One of their main results shows that any flag-transitive translation plane of order q^{2m+1} with spread in $PG(4m+1, q)$ admitting a collineation group G such that acting on the spread is cyclic and regular must come from a Baer subgeometry partition of $PG(2m, q^2)$.

One comes then naturally to the general problem of deciding abstractly what translation planes subgeometry partitions can produce. It turns out that the key property required is the existence of the 'field' group; that is, a collineation group of order $q^2 - 1$ such that adjoint the zero mapping, one obtains a field. So, it is possible to give a reverse procedure called 'spread retraction' which produces partitions of projective geometries. This idea is due to Johnson [755], who shows that this retraction procedure is equivalent to the construction method of Hirschfeld and Thas.

THEOREM 90.1. (Johnson's Spread-Retraction Theorem). Let π be a translation plane with spread in $PG(4m-1, q)$. Suppose the associated vector space may be written over a field K isomorphic to $GF(q^2)$ which extends the indicated field $GF(q)$ as a $2m$-dimensional K-vector space. If the scalar mappings with respect to K over V_{2m}/K act as collineations of π, assume that the orbit lengths of components are either 1 or $q + 1$ under the scalar group of order $q^2 - 1$. Let δ denote the number of components of orbit length 1 and let $(q + 1)d$ denote the number of components of orbit length $q+1$. Then there is a mixed partition of $PG(2m-1, q^2)$ of δ $PG(m - 1, q^2)$'s and d $PG(2m - 1, q)$'s.

DEFINITION 90.2. Under the above conditions, we shall say that the mixed partition of $PG(2m-1, q^2)$ is a 'retraction' of the spread of π or a 'spread retraction'.

THEOREM 90.3. Let π be a translation plane of order q^{2m+1} with kernel containing $GF(q)$, with spread in $PG(4m + 1, q)$, whose underlying vector space is a $GF(q^2)$-space and which admits as a collineation group the scalar group of order $q^2 - 1$. If all orbits of components have length $q+1$ corresponding to $K - \{0\}$, then a Baer subgeometry partition of $PG(2m, q^2)$ may be constructed.

DEFINITION 90.4. A Baer subgeometry partition produced from a spread as above is called a 'spread retraction'.

PROOF. The main idea of the proof is the following two points: (1) Any orbit of size $q + 1$ of $GF(q)$-components under the scalar group forms a $GF(q)$-regulus whose subplanes of order q incident with the zero vector are $GF(q^2)$-1-dimensional vector subspaces.

To see why (1) is valid, notice that we have subspaces of dimension q^k (for $k = 2m + 1$) that are actually covered by the point orbits under $GF(q^2)^*$. Each point orbit union the zero vector is isomorphic to $GF(q^2)$ which is a 2-dimensional $GF(q)$-vector space. Since this vector space is decomposed naturally into $q + 1$ $GF(q)$-spaces lying on components of the orbit of length $q + 1$, we obtain a spread of $GF(q^2)$. Hence, each point orbit within an orbit of components of length $q + 1$ is a Desarguesian affine subplane. Hence, we have a subplane covered net by Desarguesian subplanes of order q; a $GF(q)$-regulus net. (Note that this argument is independent of the value for k).

(2) If R is a regulus net of order q^{2m} and degree $q + 1$, then taking the subplanes incident with the zero vector as 'points', the structure becomes a projective geometry $PG(2m - 1, q)$. Then starting from a vector space V_{2m} over $GF(q^2)$, to form the projective space by taking the lattice of $GF(q^2)$-subspaces, it follows that the orbits of $GF(q)$-subspaces under the group $GF(q^2)^*$ become $PG(2m - 1, q)$'s and the components of the spread which are $GF(q^2)$-subspaces become natural $PG(m - 1, q^2)$'s.

Assume that we have a spread S of order q^{2m+1} with kernel containing $GF(q)$ and a corresponding translation plane π. Then π is a $2(2m + 1)$-vector space over $GF(q)$. Assume that there is a field K isomorphic to $GF(q^2)$ such that π is also a $(2m + 1)$-K-vector space and assume that the scalar group of order $q^2 - 1$ relative to K acts as a collineation group of S. Assume that each orbit of components has length $q + 1$.

We again note: Each orbit of length $q + 1$ is a $GF(q)$-regulus and the subplanes incident with the zero vector of the regulus net are K-1-dimensional vector subspaces and each orbit of components corresponds in the projective geometry $PG(2m, K)$ to a $PG(2m, q)$. This completes the proof.   □

90.1. Geometric Lifting.

In this section, we give a vector-based version of geometric lifting which ultimately will show that spread retraction and geometric lifting are equivalent.

We begin with the embedding fundamentals associated with the construction process as developed in Hirschfeld and Thas [**529**, p. 206].

Consider Π isomorphic to $PG(4n - 1, q)$ or $PG(4m + 1, q)$, respectively, and embed Π in Ω and isomorphic to $PG(4n - 1, q^2)$ or $PG(4m + 1, q^2)$, respectively. Choose a projective subspace Σ disjoint from Π in Ω where Σ is isomorphic to either $PG(2n - 1, q^2)$ or $PG(2m, q^2)$, respectively as Ω is isomorphic to $PG(4n - 1, q^2)$ or $PG(4m + 1, q^2)$. For v a point in Ω, for any vector basis and $v = (x_i)$ where $x_i \in GF(q^2)$, define $v^q = (x_i^q)$.

We consider the same setup with Π the set of points of Ω each of which is fixed under the q-mapping above. For more details, the reader is referred to Mellinger [**978**].

LEMMA 90.5. (1) For any point v in Σ, $v^q \notin \Sigma$ and $\langle v, v^q \rangle \cap \Sigma$ is a Baer subline of $q + 1$ points which may be taken in the form:

$$\left\{ kv + v^q;\ k \in GF(q^2),\ |k| \text{ dividing } q + 1 \right\}.$$

(2) The set of points of Π is

$$\left\{ sv + v^q;\ v \in \Sigma,\ s \text{ of order dividing } q + 1 \right\}.$$

PROOF. Part (1), in the mixed case when Σ is $PG(2n - 1, q^2)$ is in Mellinger [**978**]. The proof to the general case is virtually identical completing part (1).

To see that (2) is valid, note that $sv + v^q = tw + w^q$ if and only if $s = t$ and $v = w$ and then $(q+1)(q^{2(2n)} - 1)/(q^2 - 1) = (q^{4n} - 1)/(q - 1)$ and $(q+1)(q^{2(2m+1)} - 1)/(q^2 - 1) = (q^{4m+2} - 1)/(q - 1)$ which are the number of points in $PG(4n - 1, q)$ and $PG(4m + 1, q)$, respectively. $\square$

LEMMA 90.6. let s, t be in the group C_{q+1} of order $q + 1$ of $GF(q^2)^*$. Let $\theta_{s,t}$ be an element such that $\theta_{s,t}^q(s/t) = \theta_{s,t}$. Without loss of generality, $\theta_{s,t} = \theta_{ks,kt}$ for $k \in C_{q+1}$.

PROOF. There exist exactly $q - 1$ non-zero elements m such that $m^{q-1} = s/t$ and there exist exactly $q - 1$ non-zero elements n such that $n^{q-1} = kt/ks = t/s$. $\square$

LEMMA 90.7. Let $sv + v^q$ and $tw + w^q$ be distinct points of Π and let $L_{(s,v),(t,w)}$ denote the unique line of Π incident with these two points. Then

$$L_{(s,v),(t,w)} = \left\{ s(v + \theta_{s,t}^q w) + (v + \theta_{s,t}^q w)^q;\ \forall \theta_{s,t} \right\} \cup \left\{ sv + v^q, tw + w^q \right\}.$$

PROOF. The $GF(q^2)$-vector space generated by the two 'vectors' is

$$\langle sv + v^q, tw + w^q \rangle = l(sv + v^q) + m(tw + w^q)\ \forall\ l, m \in GF(q^2).$$

Note that

$$s(v + \theta_{s,t}^q w) + (v + \theta_{s,t}^q w)^q = sv + v^q + \theta_{s,t}(tw + w^q)$$

since $\theta_{s,t}^q = \theta_{s,t} t/s$. Hence, the indicated points are $q + 1$ points of Π and are on the line generated by $sv + v^q$ and $tw + w^q$. $\square$

THEOREM 90.8. Define $\sigma_k\colon sv + v^q \longmapsto ksv + v^q$ for $k \in C_{q+1}$. Then σ_k is a collineation of Π.

PROOF. It suffices to check that σ_k is a bijection on the points which maps lines to lines and preserves incidence. It is claimed that $L_{(s,v),(t,w)}$ is mapped onto $L_{(ks,v),(kt,w)}$ by σ_k. We note that

$$L_{(ks,v),(kt,w)} = \left\{ ks(v + \theta_{ks,kt}^q w) + (v + \theta_{ks,kt}^q w)^q;\ \forall \theta_{ks,kt} \right\} \cup \{ ksv + v^q, ktw + w^q \}.$$

Since $\theta_{ks,kt} = \theta_{s,t}$, it follows that $L_{(ks,v),(kt,w)} = L_{(s,v),(t,w)}\sigma_k$. $\square$

LEMMA 90.9. Let B be a $PG(2n-1,q)$ or a $PG(2m,q)$ of Σ, respectively as Σ is $PG(2n-1,q^2)$ or $PG(2m,q^2)$.

(1) Then for each $s \in C_{q+1}$, $B_s^+ = \{\, sv + v^q;\ v \in B\,\}$ is a subspace of Π isomorphic to $PG(2n-1,q)$ or $PG(2m+1,q)$ respectively. Furthermore, $B_s^+\sigma_k = B_{ks}^+$. Thus, $\{\, B_t^+;\ t \in C_{q+1}\,\}$ is an orbit under $\langle\, \sigma_k;\ k \in C_{q+1}\,\rangle$.

(2) For a subgeometry S isomorphic to $PG(n-1,q^2)$ when Σ is $PG(2n-1,q^2)$, then $S^+ = \{\, sv + v^q;\ v \in S,\ \text{for all } s \in C_{q+1}\,\}$ is a subspace of Π is isomorphic to $PG(2n-1,q)$. Furthermore, S^+ is invariant under $\langle\, \sigma_k;\ k \in C_{q+1}\,\rangle$.

PROOF. Note that the lines $L_{(s,v),(s,w)}$ have points of Π of the form:

$$L_{(s,v),(s,w)} = \{\, s(v + \theta_{s,s}^q w) + (v + \theta_{s,s}^q w)^q;\ \forall \theta_{s,s}\,\} \cup \{ sv + v^q, sw + w^q \}.$$

Thus, we see that

$$s(v + \theta_{s,s}^q w) + (v + \theta_{s,s}^q w)^q = sv + v^q + \theta_{s,s}(sw + w^q)$$

which is $s(v + \theta_{s,s}w) + (v + \theta_{s,s}w)^q$. Also, v and $w \in B$ implies that $v + \theta_{s,s}w \in B$. Hence, all points of the line generated by two points of B_s^+ are points of B_s^+ so that it follows that B_s^+ is a subspace. Since the number of points is the number of points of $PG(2n-1,q)$ or $PG(2m+1,q)$, we have the proof. This part also may be deduced using Segré varieties as in Hirschfeld and Thas. The proof of part (2) is similar and left to the reader. $\square$

The following lemma is now immediate from our previous lemmas.

LEMMA 90.10. For any partition of the projective geometry, the cyclic group $\langle \sigma_k; k \in C_{q+1}\rangle$ is a collineation group of the projective spread in Π.

We now show the converse and prove:

THEOREM 90.11. (1) Any mixed partition $\mathcal{M}$ of $PG(2n-1,q^2)$ or Baer partition $\mathcal{B}$ of $PG(2m,q^2)$ gives rise to a spread S in $PG(4n-1,q)$ or, respectively, in $PG(4m+1,q)$ defining a translation plane $\pi_{\mathcal{M}}$ or $\pi_{\mathcal{B}}$, respectively. In either case, the translation plane is a $GF(q^2)$-vector space that admits the scalar group $GF(q^2)^*$ of order $q^2 - 1$ as a group of collineations.

(2) For mixed partitions with α $PG(2n-1,q)$'s and β $PG(m-1,q^2)$'s, there exist α orbits of components under $GF(q^2)^*$ of length $q+1$ each forming a $GF(q)$-regulus and there exist β components that are fixed by $GF(q^2)^*$.

PROOF. By the Fundamental Theorem of Projective Geometry, the preimage of $\langle \sigma_k;\ k \in C_{q+1}\rangle$ acts on the $GF(q)$-vector space V and permutes the vector space spread. Hence, the preimage G is a subgroup of $\Gamma L(V,q)$. First assume that $\alpha = 1$. So, we have a cyclic group acting on the line at infinity of order $q+1$ that fixes each component of the spread. Thus, we have a kernel homology group of the translation plane so that the translation plane has kernel containing $GF(q^2)$. Now assume that $\alpha \neq 1$.

Since we know that each $sv + v^q$ for all $s \in C_{q+1}$ is a Baer subline; i.e., isomorphic to $PG(1,q)$. The preimage space is then a Desarguesian plane isomorphic to $GF(q^2)$. In other words, the group G must be $GF(q^2)^*$. In the orbits of length $q+1$, this Desarguesian plane becomes a natural subplane of order q which lies on the orbit of components. That is, we obtain a $GF(q)$-regulus net from each $q+1$

orbit. Since we now have a group isomorphic to $GF(q^2)^*$, we may define a vector space on V over $GF(q^2)$ by taking $v \cdot g = vg$ for all $g \in GF(q^2)$. Since $GF(q^2)^*$ is fixed-point-free and the group action is the scalar action, we obtain a proper vector space admitting the group as maintained with the component orbits of lengths 1 or $q = 1$. This completes the proof.   □

90.2. The Fundamental Theorem on Geometric Lifting.

THEOREM 90.12. Geometric lifting and spread retraction are equivalent. Furthermore, the partition obtained by spread retraction geometrically lifts back to the original spread.

PROOF. Given a spread which permits retraction, we obtain a partition of the projective space. Conversely, given a partition of the projective space, we have constructed by geometric lifting a spread which permits retraction. The remaining question is whether the translation plane obtained by geometric lifting is isomorphic to the original translation plane.

Suppose there is a spread admitting retraction. To fix the situation, suppose that we have a translation plane with spread in $PG(4n-1, q)$ admitting $GF(q^2)^*$ as a collineation group. Pull back to the vector space V_{2n}/q^2 and consider this as $PG(2n-1, q^2)$. We consider the mapping $v \longmapsto kv + v^q$ for all $k \in C_{q+1}$. We have seen that we can make this set into the set of 1-dimensional $GF(q)$-subspaces of a $GF(q^2)$-subspace. As a 1-dimensional $GF(q^2)$-subspace, call the image space W_{2n}/K where K is isomorphic to $GF(q^2)$ and denote the 1-dimensional K-subspace $kv + v^q$ for all $k \in C_{q+1}$ as $\tilde{v}$. Hence, we have a mapping from V_{2n}/q^2 onto W_{2n}/K by $\Gamma: v \longmapsto \tilde{v}$. This is clearly an isomorphism of $GF(q^2)$-vector spaces. Now think of V_{2n}/q^2 as V_{4n}/q with distinguished subspaces as components of a spread which are either fixed by $GF(q^2)^*$ or in orbits of length $q+1$ under $GF(q^2)^*$. It is clear that Γ will map orbits of length 1 to orbits of length 1 and orbits of length $q+1$ to orbits of length $q+1$. Hence, the spread $(S^-)^+$ obtained by geometric lifting from the mixed partition obtained by spread retraction by S is clearly isomorphic to S. The proof for the Baer subgeometry partition is virtually identical.   □

Using the fundamental theorem, we may also discuss the associated collineation groups and isomorphisms.

DEFINITION 90.13. Let π be a translation plane with spread which permits spread retraction. The collineation group of π that normalizes the scalar group of order $q^2 - 1$ is called the 'inherited group', which we denote by $\mathcal{I}(\pi)$. Note that the inherited group permutes the components fixed by the scalar group (if there are any) and permutes the $GF(q)$-regulus nets.

THEOREM 90.14. Let π be a translation plane with spread which permits spread retraction. Let $\mathcal{F}_{\mathrm{Proj}}$ denote the full automorphism group in the associated projective group of the constructed partition of the associated projective geometry. Then $\mathcal{I}(\pi)/GF(q^2)^* \simeq \mathcal{F}_{\mathrm{Proj}}$.

The following result now follows essentially immediately. We note the second part of the corollary was proved by Baker et al. by different methods.

COROLLARY 90.15. (1) Two mixed partitions of $PG(2m-1, q^2)$ are projectively equivalent if and only if their geometric lifts are isomorphic.

(2) (See also Baker et al. [**63**, (4.2)]). Two Baer subgeometry partitions of $PG(2m, q^2)$ are projectively equivalent if and only if their geometric lifts are isomorphic.

PROOF. Clearly two isomorphic translation planes have spreads that retract to projectively equivalent partitions of the associated projective geometry, which geometrically lift back to the original or mutually isomorphic spreads. □

Now we are able to use the fundamental theorem to provide various new mixed partitions. Before we do this, we show an easy way to construct spreads satisfying the retraction requirement.

THEOREM 90.16. (Net Replacement Version). Let π be any spread of order q^{2m} with spread in $PG(2m - 1, q^2)$. Let K denote the kernel homology group of order $q^2 - 1$. Suppose we take any replaceable net N by $GF(q)$-subspaces such that $NK = K$ and such that the replaceable net N^* contains exactly δ $GF(q^2)$-subspaces and the remaining subspaces are in orbits of length $(q + 1)$ under K. Suppose that the degree of N is $\delta + d(q + 1)$.

Then there exists a mixed partition of $PG(2m - 1, q^2)$ of d $PG(2m - 1, q)$'s and $q^{2m} + 1 - d(q + 1)$ $PG(m - 1, q^2)$'s.

90.2.1. André Constructions. Let π be a Desarguesian spread of order q^4 considered as a spread in $PG(3, q^2)$. Hence, the components are lines of $PG(3, q^2)$. Consider an André net A of degree $(q^4 - 1)/(q - 1)$. This net has a replacement net of components $y = x^{q^i} m$ where $y = xm$ is in the original André net. For i odd, the components $y = x^{q^i} m$ are not $GF(q^2)$-subspaces, i.e., not lines of $PG(3, q^2)$. Now take the André net projectively as a lattice of vector $GF(q^2)$-subspaces. The question is what happens to the components? The Desarguesian affine plane admits a kernel homology group H of order $q^4 - 1$ that acts transitively on the non-zero points of each component. Furthermore, there are exactly $q^2 + 1$ $GF(q^2)$-1-subspaces on each component in an orbit under the group and each fixed by the group H^- of order $q^2 - 1$. The group H acts transitively on the $(q^4 - 1)/(q - 1)$ components of the form $y = x^{q^i} m$ and the group H^- permutes these components in orbits of length $(q + 1)$. Hence, when we go to the projective version, the $(q + 1)$ vector subspaces of dimension 4 over $GF(q)$ basically collapse to one such projective space $PG(3, q)$. So, the collapsing or retraction process identifies one $PG(3, q)$ for a set of $q + 1$ of these. The geometric lifting process will construct all of these back by constructing the subplanes (of dimension two) sitting on them.

LEMMA 90.17. Let A be the André net $y = xm$ such that $m^{(q^4-1)/(q-1)} = 1$. And let $y = x^q m$ for all such m's denote the replacement set. Let the group generated by $(x, y) \longmapsto (ex, ey)$ for $|e|$ of order $q^2 - 1$. Then the image sets have the form $\{y = x^q m e^{i(1-q)}\}$, for m fixed, each of which is a set of $q + 1$ replacement components.

(1) Each image set forms a $GF(q)$-regulus.
(2) Modulo $GF(q^2)$, each such $GF(q)$-regulus is a $PG(3, q)$.

We note that we could have accomplished the same result by subspaces $y = x^{q^3} m$ instead of $y = x^q m$ since the images of $y = x^{q^i}$ under the indicated group are of the form $y = x^{q^i} e^{j(1-q^i)}$. Hence, if $(1 - q^i, 1 - q^2) = 1 - q^{(i,2)} = q - 1$, the previous proof applies. We call such André replacements the 'odd' André replacements. Hence, we obtain the following result:

THEOREM 90.18. (Johnson [**755**]). Every Desarguesian plane of order q^4 produces, from each of its 'odd' André replacements, a mixed partition of $(q^2 + 1)$ $PG(3, q)$'s of $PG(3, q^2)$ and the remaining $q^4 + 1 - (q^2 + 1)(q + 1)$ lines of $PG(3, q^2)$.

Since we may multiply André replacements, replacing any subset of $q - 1$ André nets each in any of three possible ways, we obtain a variety of mixed partitions. In particular, replacing say $\lambda \leq q - 1$ André nets, we obtain:

COROLLARY 90.19. Replacement of λ André nets by 'odd' replacement produces a mixed partition of $\lambda(q^2 + 1)$ $PG(3, q)$'s and $q^4 + 1 - \lambda(q^2 + 1)(q + 1)$ lines of $PG(3, q^2)$.

This generalizes as follows:

THEOREM 90.20. (Johnson [**755**]). Let π be a Desarguesian plane of order q^{2n} with spread in $PG(2n - 1, q^2)$. Then any of the n odd André replacements lead to a mixed partition of $(q^{2n} - 1)/(q^2 - 1)$ $PG(2n - 1, q)$'s and the remaining $q^{2n} + 1 - (q^{2n} - 1)/(q^2 - 1)$ $PG(n - 1, q^2)$'s of $PG(2n - 1, q^2)$.

COROLLARY 90.21. Replacement of λ André nets all by odd André replacement produces a mixed partition of $\lambda(q^{2n} - 1)/(q^2 - 1)$ $PG(2n - 1, q)$'s and $q^{2n} + 1 - \lambda(q^{2n} - 1)/(q^2 - 1)$ $PG(n - 1, q^2)$'s.

Now there was nothing particular about a Desarguesian plane, merely that when the order is q^4 we have a spread in $PG(3, q^2)$ and we have a replacement partial spread of $GF(q)$-subspaces admitting the kernel homology group H^- of order $q^2 - 1$. That is, we have a replacement net N such that $NH^- = N$. However, in our construction, we required that $q + 1$ divides $|N|$. Actually, this was merely to be able to count the component parts of the mixed partition. It is possible that some of the replacement components are $GF(q^2)$-spaces; 'lines'. This still produces a mixed partition.

REMARK 90.22. Moreover, for Desarguesian spreads of order q^{2n} also for 'even' André replacements, as long as one of the André replacements is odd, we obtain a truly 'mixed' partition. Hence, for each André net we may choose any of the $2n - 1$ possible replacements and as long as one of the replacements is odd, we still obtain a non-trivial mixed partition of $PG(2n - 1, q^2)$.

90.2.1.1. *Retractions Containing Reguli.* Suppose that we have a spread of order q^4 with spread in $PG(3, K \simeq q^2)$ admitting two symmetric groups of affine homologies which fix a derivable net and both groups act transitively on the non-axis/co-axis components. Assume that the derivable net is not a K-regulus.

Hence, we may represent the two groups with axis and coaxis (coaxis and axis) $x = 0$ and $y = 0$ as follows:

$$H_y = \left\langle \begin{bmatrix} 1 & 0 & 0 & 0 \\ 0 & 1 & 0 & 0 \\ 0 & 0 & u & 0 \\ 0 & 0 & 0 & u^\sigma \end{bmatrix} ; u \in K - \{0\} \right\rangle,$$

for some non-trivial automorphism $\sigma \in \mathrm{Gal}(K)$, and

$$H_x = \left\langle \begin{bmatrix} u & 0 & 0 & 0 \\ 0 & u^\sigma & 0 & 0 \\ 0 & 0 & 1 & 0 \\ 0 & 0 & 0 & 1 \end{bmatrix} ; u \in K - \{0\} \right\rangle,$$

for some non-trivial automorphism σ.

So, the two homology groups are groups of order $q^2 - 1$ and both correspond to the multiplicative group of a field. We note the derivable partial spread D with component set

$$x = 0, y = x \begin{bmatrix} u & 0 \\ 0 & u^\upsilon \end{bmatrix} ; u \in K$$

is a 'regulus' in the associated projective geometry $PG\left(3, \begin{bmatrix} u & 0 \\ 0 & u^\sigma \end{bmatrix}\right)$. We note that the group:

$$H = \left\langle \begin{bmatrix} u & 0 & 0 & 0 \\ 0 & u^\sigma & 0 & 0 \\ 0 & 0 & u & 0 \\ 0 & 0 & 0 & u^\sigma \end{bmatrix} ; u \in K - \{0\} \right\rangle$$

acts as a collineation group of the translation plane. Moreover, this group is fixed-point-free and decomposes the vector space into a 4-dimensional $\begin{bmatrix} u & 0 \\ 0 & u^\sigma \end{bmatrix}$-vector space.

Now assume that $\sigma = q$ and multiply each term in H by $u^{-1}I_4$ to obtain a Baer group

$$B = \left\langle \begin{bmatrix} 1 & 0 & 0 & 0 \\ 0 & u^{q-1} & 0 & 0 \\ 0 & 0 & 1 & 0 \\ 0 & 0 & 0 & u^{q-1} \end{bmatrix} ; u \in K - \{0\} \right\rangle.$$

Note that this Baer group fixes the derivable net componentwise and has remaining component orbits of lengths $q + 1$. It then follows immediately that the group H fixes exactly $q^2 + 1$ components of the derivable net and has $(q^4 - q^2)/(q + 1) = q^2(q - 1)$ orbits of length $q + 1$. We notice that we are now considering the situation that we have previously, we have a vector space over $\begin{bmatrix} u & 0 \\ 0 & u^q \end{bmatrix}$ with exactly $q^2 + 1$ fixed components. The remaining components are not $\begin{bmatrix} u & 0 \\ 0 & u^q \end{bmatrix}$-subspaces but are $GF(q)$-subspaces (i.e., $\begin{bmatrix} v & 0 \\ 0 & v \end{bmatrix}$-subspaces for $v \in$ the subfield of K isomorphic to $GF(q)$).

Moreover, it is almost immediate that the component orbits of length $q + 1$ are $GF(q)$-reguli and the point orbits are the subplanes of the regulus net incident with the zero vector which are, of course, $\begin{bmatrix} u & 0 \\ 0 & u^q \end{bmatrix}$-1-dimensional vector subspaces.

Now, actually we do not require the individual homology groups, only the group H. Hence, we obtain:

THEOREM 90.23. (Johnson [**755**]). Let π be a translation plane of order q^4 with spread in $PG(3, K \simeq GF(q^2))$ admitting a derivable net of the form:

$$D : x = 0, y = x \begin{bmatrix} u & 0 \\ 0 & u^q \end{bmatrix} ; u \in K$$

and a collineation group of order $q^2 - 1$ of the form

$$H = \left\langle \begin{bmatrix} u & 0 & 0 & 0 \\ 0 & u^q & 0 & 0 \\ 0 & 0 & u & 0 \\ 0 & 0 & 0 & u^q \end{bmatrix} ; u \in K - \{0\} \right\rangle.$$

(1) Then there is an associated mixed partition of $PG\left(3, \begin{bmatrix} u & 0 \\ 0 & u^q \end{bmatrix} \simeq GF(q^2)\right)$ consisting of a regulus of $q^2 + 1$ lines and $q^2(q-1)$ $PG(3,q)$'s.

(2) The translation plane obtained by derivation also retracts to a regulus of $q^2 + 1$ lines and $q^2(q-1)$ $PG(3,q)$'s.

COROLLARY 90.24. Let π be a translation plane of order q^4 with spread in $PG(3, q^2)$ of the following form:

$$x = 0, y = x \begin{bmatrix} u & t^q f \\ t & u^q \end{bmatrix} ; u, t \in GF(q^2) \text{ and } f \in GF(q^2) - GF(q).$$

Then there are two associated mixed partitions of some $PG(3, q^2)$ consisting of a regulus and $q^2(q-1)$ $PG(3,q)$'s obtained from the plane and its derived plane.

REMARK 90.25. (1) The form of the spread listed above forces the associated translation plane to be a semifield plane coordinatizable by a semifield of order q^4 where all nuclei are isomorphic to $GF(q^2)$ and the right and middle nuclei are identical. Hence, the translation plane is a Hughes–Kleinfeld plane.

(2) In Johnson [744], there is a condition given which describes the spreads in $PG(3, q^2)$ which may be algebraically retracted to spreads in $PG(3, q)$. The condition is that there is an elation group of order q^2, E, and a Baer group of order $q + 1$, B such that $[E, B] \neq 1$. Hence, it follows that the above Hughes–Kleinfeld plane can be algebraically retracted. It is not difficult to check that the algebraic retraction is a Desarguesian plane.

Hence, the Hughes–Kleinfeld planes may be algebraically lifted from Desarguesian planes.

90.3. Baer Subgeometry Partitions.

The non-solvable flag-transitive planes have been completely determined by Buekenhout et al. [201]. These are the Desarguesian, Lüneburg–Tits, Hering of order 27 and Hall of order 9.

In Baker, Dover, Ebert, and Wantz [63], it is shown that cyclic and regular groups acting flag-transitively on odd order planes force the spreads to arise from Baer subgeometry partitions. Here, we provide a variation of this result based upon the spread-retraction ideas of the fundamental theorem. Basically, for such spreads, we need only show that the vector space is a $GF(q^2)$-space and that the component orbit lengths are all $q + 1$.

Hence, consider solvable flag-transitive planes of order q^n where n is odd > 1 and kernel containing $GF(q)$. The collineation group of any such plane is a subgroup of $\Gamma L(1, q^{2n})$ by Foulser (see [365] and [366]). Basically, the points of the plane are identified with the points of the Desarguesian affine plane (or of $GF(q^{2n})$). That is, in this instance, the vector space may be considered over $GF(q^2)$. Now assume that the collineation group contains $GL(1, q^2)$ so that there is a linear and cyclic subgroup of order $q^2 - 1$ that contains the kernel homology group of order $q - 1$. Let L be a component and consider $GL(1, q^2)_L$. Since the full group is transitive and normalizes any cyclic subgroup of $GL(1, q^2)$, we see that $GL(1, q^2)_L$ must fix each component of the translation plane and hence indices a kernel homology subgroup. If $GF(q)^*$ is properly contained in $GL(1, q^2)_L$, then the kernel of the translation plane contains the ring generated by $\{GL(1, q^2)_L, GF(q)\}$, which is clearly $GL(1, q^2)$. Hence, the kernel of the translation plane contains $GF(q^2)$. However, the translation plane is of order q^{2m+1} and is a $2k$-dimensional vector space over the kernel $GF(q^z)$ so is $2kz$-dimensional over $GF(q)$. Thus $2kz = 2(2m + 1)$, so that kz is odd, implying that both k and z are odd, a contradiction. Thus, $GL(1, q^2)_L = GF(q)^*$ for all components L.

Thus, we obtain the following theorem:

THEOREM 90.26. (Johnson [755]). Let π be a non-Desarguesian flag-transitive translation plane of order q^{2m+1} and kernel containing $GF(q)$ with group G.

(1) If G is non-solvable then π is the Hering plane of order 27.

(2) If G is solvable then G is a subgroup of $\Gamma L(1, q^{2(2m+1)})$. If $GL(1, q^2) \subseteq G$ then π corresponds to a Baer subgeometry partition of $PG(2m, q^2)$.

COROLLARY 90.27. Let π be a non-Desarguesian flag-transitive plane of order q^{2m+1}, $m > 0$ which contains $GF(q)$ in the kernel and let $q = 2^r$.

(1) If $(r(2m + 1), q + 1) = 1$ then π arises from a Baer subgeometry partition of $PG(2m, q^2)$.

(2) Hence, any non-Desarguesian flag-transitive plane of order 2^{2m+1} for m odd and not divisible by 3 arises from a Baer subgeometry partition of $PG(2m, 4)$.

As an application to the above corollary, since $(2nr, q^n + 1) = 1$ implies that $(nr, q + 1) = 1$, we obtain a result of G.L. Ebert.

COROLLARY 90.28. Let π be a non-Desarguesian flag-transitive plane of order q^n, $n > 1$ odd which contains $GF(q)$ in the kernel and let $q = 2^r$.

If $(2rn, q^n + 1) = 1$ then π arises from a Baer subgeometry partition.

As a further application of this result, now assume that we have a cyclic and regular group on the line at infinity and the kernel contains $GF(q)$. Assume that $n > 1$ and q is not 2: then there must be a p-primitive divisor of $q^{2n} - 1$ that is linear, i.e., is in $GL(1, q^{2n})$. The centralizer of an element of this order is $GL(1, q^{2n})$. Since we do have a cyclic group of order $q^{2n} - 1$ which centralizes the indicated element, we have $GL(1, q^{2n}) \supseteq GL(1, q^2)$.

Hence, we have:

COROLLARY 90.29. (Baker, Dover, Ebert, and Wantz [63, see Theorem (2.2)]).

Any flag-transitive translation plane of order q^n and kernel containing $GF(q)$ for n odd > 1 and $(n, q) \neq (3, 2)$ that admits a collineation group inducing a cyclic and regular group on the line at infinity may be obtained by geometric lifting from a Baer subgeometry partition.

It might be noted that any three components of a translation plane of even order automatically form a $GF(2)$-regulus. Hence,

THEOREM 90.30. Let π be a translation plane of order 2^n for n odd.

(1) Then the spread for π is a union of $GF(2)$-reguli.

(2) π arises from a Baer subgeometry partition if and only if π admits a non-planar collineation of order 3.

In [**63**], the following two questions are asked? Does a spread of order q^n for n odd and > 1, which is a disjoint union of $GF(q)$-reguli, arise from a Baer subgeometry partition? And, if there are more than $(q^n + 1)/(q + 1)$ $GF(q)$-reguli, must the spread be regular (Desarguesian)? It is also pointed out that the Hering spread of order 27 shows that the general answer to these question is no. However, the question was raised as to whether the Hering spread is the only exception.

Note that, in particular, any semifield plane of order 2^n for n odd cannot admit a collineation group of order 3 which does not leave invariant a component and hence is planar. Hence any such spread cannot arise from a Baer subgeometry partition. Furthermore, even when there is a non-planar element of order 3, there are $\binom{2^n+1}{3} \binom{2^n+1}{3}$ $GF(2)$-reguli.

Hence, we see that the $GF(2)$-reguli trivially answer these questions negatively. So, these questions more properly need to exclude the case when $q = 2$.

90.3.1. Transitive Baer Partitions. Some easy consequences of the fundamental theorem regarding transitivity from the other direction are as follows.

DEFINITION 90.31. A Baer subgeometry partition of $PG(2m, q^2)$ is said to be 'transitive' or 'two-transitive' if and only if there is a collineation group in $P\Gamma G(2m + 1, q^2)$ that acts transitively or two-transitively, respectively, on the sets of $PG(2m, q)$'s on the partition.

THEOREM 90.32. A transitive Baer subgeometry partition of $PG(2m, q^2)$ produces a flag-transitive translation plane of order q^{2m+1} with spread in $PG(4m+1, q)$ admitting $GF(q^2)^*$ as a collineation group.

Conversely a flag-transitive spread permitting spread retraction (whose group is in $\Gamma L(2m + 1, q^2)$) produces a transitive Baer subgeometry partition.

THEOREM 90.33. A two-transitive Baer subgeometry partition of $PG(2m, q^2)$, for $(q, 2m + 1) \neq (2, 5)$ produces a Desarguesian spread of order q^{2m+1} with spread in $PG(4m + 1, q)$.

An open problem of some importance and interest is whether all flag-transitive planes of order q^n with kernel containing $GF(q)$ and n odd correspond to Baer subgeometry partitions. The only known plane that may not be so constructed is the Hering plane of order 27.

90.3.2. The Hering Plane of Order 27. Mathon and Hamilton [**975**] have shown that the Hering plane of order 27 does not correspond to a Baer subgeometry partition of $PG(2, 9)$ by exhaustive computer search.

We may use the fundamental theorem to give an easy proof that the Hering plane of order 27 does not correspond to a Baer subgeometry partition of $PG(2, 9)$. We need to ask if there is a collineation group isomorphic to $GF(9)^*$ with point orbits of length 8 and component orbits of length 4. However, the full collineation group (full translation complement) of the Hering planes is $SL(2, 13)$ (see Dempwolff [**304**]). Moreover, in $SL(2, 13)$, the 2-groups are generalized quaternion of order 8. Hence, no such cyclic group exists. Hence, by the fundamental theorem:

THEOREM 90.34. The Hering plane of order 27 does not arise from a Baer subgeometry partition of $PG(2, 9)$.

90.4. Maximal Partial Projective Partitions.

Suppose we have a maximal partial spread in $PG(4n-1, q)$ that admits $GF(q^2)^*$ as a collineation group that has components of lengths 1 or $q+1$. Then clearly, we may form a 'retraction' of the partial spread.

THEOREM 90.35. (Johnson [755]). (1) Let $\mathcal{P}$ be a translation net with partial spread in $PG(4n-1, q)$ such that the associated vector space is a $GF(q^2)$-space for some quadratic extension field of the underlying field isomorphic to $GF(q)$ and $GF(q^2)^*$ acts as a collineation group of $\mathcal{P}$. Assume that the partial spread components are in orbits of lengths 1 or $q+1$ under $GF(q^2)^*$ and that there are β orbits of length 1 and α orbits of length $q+1$.

Then there is a mixed partial partition of $PG(n-1, q^2)$ of β $PG(n-1, q^2)$'s and α $PG(2n-1, q)$'s.

Note that it is possible to have $\beta = 0$ in this case producing a partial Baer subgeometry partition of $PG(2n-1, q^2)$.

(2) Let $\mathcal{P}$ be a translation net with partial spread in $PG(4m+1, q)$ such that the associated vector space is a $GF(q^2)$-space for some quadratic extension field of the underlying field isomorphic to $GF(q)$ and $GF(q^2)^*$ acts as a collineation group of $\mathcal{P}$. Assume that the partial spread components are in orbits of lengths $q+1$ under $GF(q^2)^*$ and that there are α orbits of length $q+1$.

Then there is a partial Baer subgeometry partition of $PG(2m, q^2)$ by α $PG(m-1, q^2)$'s.

(3) Define the partial spread $\mathcal{P}$ to be '$GF(q^2)$-maximal' if and only if there is no extension partial spread in $PG(4n-1, q^2)$ or, respectively, $PG(4m+1, q)$ admitting $GF(q^2)^*$ as a collineation group.

Then $\mathcal{P}$ is $GF(q^2)$-maximal if and only if the associated partial mixed or partial Baer subgeometry partition is maximal in $PG(2n-1, q^2)$ or $PG(2m, q^2)$, respectively.

THEOREM 90.36. A translation plane with spread in $PG(3, q^2)$ admitting two mutually disjoint derivable nets that are not regulus nets and exactly one of which has all transversal Baer subplanes as $GF(q)$-subspaces produces a maximal mixed partial partition of $PG(3, q^2)$ of $q-1$ $PG(3, q)$'s and $q^4 - 2q^2 + 3$ lines.

90.4.1. Fixed-Point-Free Groups.
The previously mentioned group retraction theorem of Johnson is generalized by Johnson and Mellinger as follows:

THEOREM 90.37. (Johnson and Mellinger [782]). Let π be a translation plane of order q^t and kernel containing K isomorphic to $GF(q)$. Let G be a collineation group of π such that GK^* is fixed-point-free (where K^* denotes the kernel homology group of order $q-1$).

If all point orbits union the zero vector are 2-dimensional K-subspaces, then GK^* union the zero mapping is a field isomorphic to $GF(q^2)$. Furthermore,

(1) Every component orbit under GK^* has length 1 or $q+1$.

(2) Furthermore, all component orbits have length $q+1$ if and only if t is odd.

(3) All component orbits have length 1 if and only if GK^* is a kernel subgroup of the translation plane.

(4) The plane π permits spread retraction and produces a partition of a corresponding projective space.

There is another geometric interpretation for the last theorem. Consider the group G as acting on the spread S associated with the translation plane π. Then, the orbits of G form lines of the space Π containing S. Because of the action of G, these lines must form a geometric 1-spread of Π. This geometric 1-spread can then be used to retract S to a mixed partition just as in [**755**]. See Mellinger [**978**] for a complete geometric description.

Note that if we assume that GK^* is cyclic of order $q^2 - 1$ and $GK^* \cup \{0\}$ is a field isomorphic to $GF(q^2)$, it then follows that GK^* is generated by an element Z and

$$GK^* = \{ Z\alpha + \beta; \ \alpha, \beta \in K; (\alpha, \beta) \neq (0,0) \}.$$

Let w be any non-zero vector. Then, the 2-dimensional K-subspace generated by w and wZ, $\langle wZ, w \rangle_K$, is $\{ w(Z\alpha + \beta); \ \alpha, \beta \in K \}$; that is, by similar arguments as above, every point-orbit is a 2-dimensional K-subspace, and hence we may apply the above theorem to obtain the following.

THEOREM 90.38. (Johnson and Mellinger [**782**]). Let π be a translation plane of order q^t and kernel containing K isomorphic to $GF(q)$ that admits collineation group G such that GK^* is fixed-point-free and $GK^* \cup \{0\}$ is a field of order q^2. Then the spread permits spread retraction.

We now assume that GK^* is cyclic.

THEOREM 90.39. (Johnson and Mellinger [**782**]). Let π be a translation plane of order q^t and kernel containing K isomorphic to $GF(q)$. Let G be a collineation group of π such that GK^* is cyclic and fixed-point-free (where K^* denotes the kernel homology group of order $q - 1$).

If there exists a set S of $t+1$ point-orbits which, together with the zero vector, are 2-dimensional K-subspaces and any t of these direct sum to π, then GK^* union the zero vector is a field and π permits spread retraction.

EXAMPLE 90.40. Let π be a translation plane of order q^{2r} with kernel containing K^+ isomorphic to $GF(q^2)$ and assume that there are both a right nucleus and a middle nucleus of a coordinatizing quasifield equal to the kernel K^+. Furthermore, assume that the quasifield is a vector space over the right and/or middle nucleus (for example, this would be the case when the quasifield is a semifield). Furthermore assume that K isomorphic to $GF(q)$ is in the center of the nuclei.

Then, the plane admits retraction by three distinct groups whose generators are given as follows:

$$(x, y) \longmapsto (xa, ya^q),$$
$$(x, y) \longmapsto (xa, ya),$$
$$(x, y) \longmapsto (ax, ay),$$

where a is a primitive element of K^+. The groups are of the form GK^*, the union with the zero mapping produces fields and the groups are fixed-point-free and hence each produces retractions.

In the following sections, we shall be interested in whether the various partitions induced from a given translation plane are isomorphic. Clearly, for mixed partitions, if the partition numbers in the partitions are distinct, then the partitions are non-isomorphic. However, potentially, two partitions can be non-isomorphic and have identical partition numbers.

THEOREM 90.41. Two partitions of a projective space that produce the same translation plane are isomorphic if and only if the corresponding collineation groups in the translation plane are conjugate within the full collineation group of the plane.

90.4.1.1. *Partitions of $PG(V,4)$ and Planes of Order* 16. This study of retraction actually originated from analysis of the mixed partitions of $PG(3,4)$. We note that all translation planes of order 16 may be obtained from one of the three semifield planes by a derivation (see, e.g., Johnson [684]). Considering these implications, the following general result may be proved.

THEOREM 90.42. (Johnson and Mellinger [782]). Let π be a finite translation plane of order 2^n and suppose π admits a fixed-point-free collineation of order 3.

(1) If n is odd, then there is a corresponding Baer subgeometry partition of $PG(n-1,4)$.

(2) If n is even, then the partition is a mixed partition of $PG(n-1,4)$. Also, when $n = 2m$, the fixed components correspond to $PG(m-1,4)$'s and the orbits of length 3 correspond to $PG(2m-1,2)$'s.

Mellinger [978] enumerates all mixed partitions in $PG(3,4)$ using Magma. In the following sections, we shall consider general situations for translation planes producing several mixed partitions. First, we revisit the conclusions in light of fixed-point-free collineations for translation planes of order 16.

90.4.1.2. *The Planes of Order* 16. For purposes of identification, we shall denote the eight translation planes of order 16 as the Desarguesian plane, the semifield plane with kernel $GF(4)$, the semifield plane with kernel $GF(2)$ and the five derived planes. The plane derived from the Desarguesian plane is the Hall plane. There are three planes derived from the semifield plane with kernel $GF(4)$, called the Lorimer–Rahilly plane, the Johnson–Walker plane and another plane which we shall simply call the $GF(4)$-derived plane. The plane derived from the semifield plane with kernel $GF(2)$ is the Dempwolff plane.

In this setting, K is always isomorphic to $GF(2)$.

THEOREM 90.43. (Johnson and Mellinger [782], also see Mellinger [978]). All translation planes of order 16 admit fixed-point-free groups of order 3 that fix a vector subspace of dimension two over $GF(2)$, and hence correspond to mixed partitions of $PG(3,4)$. The ℓ fixed components correspond to $PG(1,4)$'s and the b orbits of length 3 correspond to $PG(3,2)$'s; type (ℓ, b).

More specifically,

(1) If π is Desarguesian with group $\langle (x,y) \longmapsto (xa, xa^2) \rangle$, there is an associated $(2,5)$-partition.

(2) Derivation of π using a derivable net invariant under the group of (1) produces a $(5,4)$-partition in the Hall plane.

Let the semifield of order 16 with the kernel $GF(4) =$ right nucleus $=$ left nucleus be denoted by $(S, +, *)$. Consider the group

$$G_j K^* = \left\langle (x,y) \longmapsto (x * a, x * a^{2^j}) \right\rangle,$$

for j fixed.

(3) $j = 0$.

   (a) Then the semifield plane produces a $(5,4)$-partition.

(b) In part (a), choose a derivable net as a right = middle nucleus net to produce a $(5,4)$-partition in the derived semifield plane.

(c) There is another derivation that produces a $(8,3)$-partition in the same derived semifield plane using the same group G_oK^* as above.

(4) Choose $j = 1$.

(a) Then the semifield plane produces a $(5,4)$-partition but from the group G_1K^*.

(b) Using the same net as in (3)(b), the derived plane produces another $(5,4)$ using the group G_1K^*.

(c) There is another derivation that produces a $(8,3)$-partition in the derived semifield plane using the group G_1K^*.

(5) Consider the kernel homology group G_3K^*: $\langle (x,y) \longmapsto (a*x, a*y) \rangle$. This produces a $(14,1)$-partition of the derived plane.

Hence, the semifield plane with kernel $GF(4)$ produces two partitions of type $(5,4)$ and the derived plane from the semifield plane with kernel $GF(4)$ corresponds to two partitions each of types $(5,4), (8,3)$ and a partition of type $(14,1)$. The partitions of the same type induce isomorphic planes.

(6) The Lorimer–Rahilly and Johnson planes derived from the semifield plane with kernel $GF(4)$ also produce partitions of types $(14,1)$.

(7) The Dempwolff plane admits a fixed-point-free group within $SL(2,4)$ producing a partition of type $(8,3)$.

Hence, we see that a variety of mixed partitions of different types can correspond to the same translation plane that admit different fixed-point-free groups.

CHAPTER 91

Multiple Spread Retraction.

In this chapter, we consider whether two subgeometry partitions may be obtained from the same translation plane.

DEFINITION 91.1. Let π be a translation plane admitting spread retraction and let $PG(\pi)$ denote the associated partitions. Two partitions $PG(\pi_1)$ and $PG(\pi_2)$ are said to be 'paired' if and only if the associated translation planes π_1 and π_2 are isomorphic.

We identify the translation planes π_1 and π_2 and continue to use the term 'paired' for partitions of the projective space.

If, in this setting, we allow that the two associated cyclic groups $G_i K^*$, respectively, for π_i, $i = 1, 2$, of order $q^2 - 1$ of the translation plane normalize each other, we shall say that the partitions are 'normalizing, paired partitions'.

Furthermore, we shall use the term 'double retraction' to describe the construction of a normalizing paired partition. In general, 'triple retraction' refers to the construction of three partitions which are mutually normalizing paired partitions, and 'multiple retraction' (or 'k-retraction') shall refer to the construction of a number of (respectively k) partitions which are mutually normalizing paired partitions.

LEMMA 91.2. Given a translation plane of order q^t with kernel containing K, two cyclic groups $G_i K^*$, $i = 1, 2$, of order $q^2 - 1$ corresponding to spread retractions centralize each other if they normalize each other.

LEMMA 91.3. Let P_1 and P_2 be two distinct paired partitions and let $G_1 K^*$ and $G_2 K^*$ denote the corresponding groups in the translation plane π. Then $G_1 K^* \cap G_2 K^* = K^*$.

We begin with a general consideration of normalizing, paired partitions corresponding to translation planes of order q^t. We have seen that there are several mixed partitions that a given translation plane can produce, so this occurs when t is even. The question is, can this happen when t is odd?

THEOREM 91.4. (Johnson and Mellinger [782]). Let π be a translation plane of order q^t that admits spread retraction and produces two distinct normalizing, paired partitions of projective spaces; π permits double retraction. Then for t odd, one of the following occurs:
    (a) q odd, $(q + 1)/2$ is odd, $(q + 1)/2$ divides t,
    (b) q even, $(q + 1)$ divides t.

91.1. Spreads Admitting Triple Retraction.

First assume that we have a translation plane of order q^4 with spread in $PG(3, q^2)$ such that the middle and right nuclei correspond to fields of order q^2 with respect to a fixed coordinate system and choice of base axes, $x = 0, y = 0, y = x$. Assume that the nuclei are equal to each other and equal to the kernel, and that the subkernel group K of order $q - 1$ commutes within the underlying quasifield $Q = (Q, +, *)$.

Then, we obtain a group of the following form:

$$\left\langle (x, y) \longmapsto ((a * x) * b, (a * y) * c);\ a, b, c \in K^+ \simeq GF(q^2) - \{0\} \right\rangle.$$

We consider the following three groups: H_i for $i = 1, 2, 3$:

$$H_1 = \left\langle (x, y) \longmapsto (a * x, a * y);\ a \in K^+ \simeq GF(q^2) - \{0\} \right\rangle,$$
$$H_2 = \left\langle (x, y) \longmapsto (x * b, y * b^q);\ b \in K^+ \simeq GF(q^2) - \{0\} \right\rangle,$$
$$H_3 = \left\langle (x, y) \longmapsto (x * c, y * c);\ c \in K^+ \simeq GF(q^2) - \{0\} \right\rangle.$$

We know that all groups union the zero mapping produce fields isomorphic to $GF(q^2)$, all groups are fixed-point-free and each group contains the K^*-kernel homology group due to the assumption on the centrality of K in the quasifield.

Hence, each group necessarily produces a mixed partition of $PG(2, q^2)$ of the given translation plane but we shall be more interested in the derived planes.

Assume that the middle nucleus defines derivable nets and we shall require that these are not K^+-reguli. Then, the kernel homology group H_1 must fix exactly two Baer subplanes of any such derivable net and permute the remaining Baer subplanes into $q - 1$ sets of size $q + 1$.

So, we obtain:

THEOREM 91.5. (Johnson and Mellinger [**782**]). The derived plane produces a mixed partition in $PG(3, q^2)$ of type $(q^4 - q^2 + 2, q - 1)$ using the group H_1.

Now consider the group H_3. It is clear that the plane admits a mixed partition. Furthermore, $y = x * m$ is fixed if and only if

$$(x * b) * m = (x * m) * b$$

for all x and for all $b \in K^{+*}$. Since b is in the middle and right nucleus, this is equivalent to

$$b * m = m * b.$$

We know that the orbits have length 1 or $q + 1$. Let $\{1, t\}$ be a basis for Q over K^+ and write $m = m_1 * t + m_2$ for $m_i \in K^+$. Then, since K^+ is in the kernel of the translation plane, and, as we are assuming that the middle nucleus defines a derivable net, and since all such derivable nets are defined from 'right' 2-dimensional vector spaces (see Johnson [**753**]) we have:

$$b * (m_1 * t + m_2) = (bm_1 * t + bm_2)$$
$$= (m_1 * t + m_2) * b = (m_1 * t) * b + m_2 b$$
$$= m_1 * (t * b) + m_2 b.$$

Hence, if $m_1 \neq 0$, and $b * t = t * b$, then K^+ is in the center of the quasifield, a contradiction unless the middle nucleus defines a K^+-regulus net. Hence, it follows that the only fixed components are $x = 0$, $y = xm$ for $m \in K^+$. Now derive the standard middle regulus net, and note that the group is generated from central collineations and such groups fix all Baer subplanes of the derivable net that are incident with the zero vector.

THEOREM 91.6. (Johnson and Mellinger [**782**]). The plane produces a $(q^2 + 1, q^2(q-1))$-type mixed partition using H_3.

In addition, the derived plane produces a $(q^2+1, q^2(q-1))$-type mixed partition using H_3.

Now consider the group H_2. Since the coordinate structure is 2-dimensional over its middle nucleus, let $\{1, t\}$ be a K^+-basis. Then,

$$b * (tm_1 + m_2) = (tm_1 + m_2) * b^q.$$

Since the kernel is also K^+, the coordinate structure is a left quasifield and K^+ is also the right nucleus, we have:

$$\begin{aligned} b * (tm_1 + m_2) &= (b * t)m_1 + bm_2 = (tm_1 + m_2) * b^q \\ &= (t * b^q)m_1 + m_2 b^q. \end{aligned}$$

Hence,

$$b * t = t * b^q \text{ and } m_2 = 0 \text{ or } m_1 = m_2 = 0.$$

Again, note that the group fixes all Baer subplanes of the original standard derivable net, but this time fixes exactly $q^2 - 1$ additional $y = x * m$'s outside of the derivable net.

THEOREM 91.7. (Johnson and Mellinger [**782**]). The group H_3 produces, by derivation, a mixed partition of type $(2q^2, (q^2 - 1)(q - 1))$ provided the quasifield has the required property $b * t = t * b^q$ for a basis $\{1, t\}$ over K^+, for all $b \in K^+$.

The original plane produces another mixed partition of type $(q^2 + 1, q^2(q - 1))$ which is distinct from the mixed partition of the same type produced using H_2. The two partitions are isomorphic in $PG(3, q^2)$ if and only if there is a collineation σ of the plane π such that $H_2^\sigma = H_3$ and mapping the components fixed by H_2 onto those fixed by H_3, fixing $x = 0$ and $y = 0$.

THEOREM 91.8. (Johnson and Mellinger [**782**]). Let π be a translation plane of order q^4 with kernel $K^+ \simeq GF(q^2)$. Choose a coordinate quasifield Q and assume with respect to that quasifield, the right, middle, and left (kernel) nuclei are all equal. Assume further that the translation plane is derivable with respect to a middle-nucleus net which is not a K^+-regulus net.

(1) If there is a basis $\{1, t\}$ over K^+ such that with respect to quasifield multiplication, $b * t = t * b^q$ for all $b \in K^+$ then the derived plane produces mixed partitions of $PG(3, q^2)$ of types $(q^2 + 1, q^2(q - 1)), (2q^2, (q^2 - 1)(q - 1)), (q^4 - q^2 + 2, q - 1)$.

(2) The original translation plane produces distinct mixed partitions of $PG(3, q^2)$ of types: $(q^2 + 1, q^2(q - 1)), (q^2 + 1, q^2(q - 1)), (q^4 + 1, 0)$.

(3) Furthermore, suppressing the $*$-notation, the multiplication is as follows: Let $t(t + \gamma) = tf(\gamma) + g(\gamma)$ for functions $f, g \colon K \longmapsto K$; then

$$(t\alpha + \beta)(t\delta + \gamma) = t(\alpha\delta^q f(\delta^{-q}\gamma) + \beta^q\delta) + (\alpha^q\delta g(\delta^{-q}\gamma) + \beta\gamma)$$

for all $\alpha, \beta, \delta \neq 0, \gamma \in K$.

The only known examples of quasifields with the above properties originate from the Hughes–Kleinfeld semifields. For example, the semifield plane of order 16 and kernel $GF(4)$, a plane admitting triple retraction of types $(5, 4), (8, 3)$ and $(14, 1)$ which are the types listed above when $q = 2$. In general, is every such plane a derived Hughes–Kleinfeld semifield plane?

91.1.1. Normalizing, Paired Partitions in $PG(3, q^2)$. In a mixed partition, we have at least two fixed components, say $x = 0$ and $y = 0$. Each of these components may be considered as a 2-dimensional $GF(q^2)$-subspace with respect to a given group.

If two mixed partitions produce isomorphic translation planes, we have two possible situations: (1) the partitions have different numbers, or (2) identifying the translation planes, there are two distinct sets of partition subspaces. In either case, identifying the translation plane, we see that we have a translation plane with spread in $PG(7, q)$ such that there are two fields K_1 and K_2 both isomorphic to $GF(q^2)$ and such that the groups of orders $q^2 - 1$ intersect exactly in the kernel homology group of order $q - 1$. Hence, we obtain a collineation group in the associated translation plane π of order $(q + 1)^2(q - 1)$ which contains the $GF(q)$-kernel homology group. Moreover, we are assuming that each group normalizes the other by assumption and the collineation group inherited from the partitions is a subgroup of $\Gamma L(4, q^2)$. We have noticed that the groups centralize each other provided they normalize each other.

PROPOSITION 91.9. *In a translation plane π of order q^t, t even, with kernel containing $K \simeq GF(q)$, two distinct normalizing paired partitions produce collineation groups $G_1 K^*$ and $G_2 K^*$ of π fix at least two common fixed components.*

THEOREM 91.10. (Johnson and Mellinger [**782**]). *Let π be a translation plane of order q^4 whose kernel contains K isomorphic to $GF(q)$. Assume that the plane is doubly retractive so that there are two common fixed components in the above situation, say $x = 0, y = 0$, for two distinct groups, $G_i K^*$, for $i = 1, 2$. Decompose the vector space relative to the first group $G_1 K^*$, which then acts like a scalar $GF(q^2)$-group acting on $x = 0$. On $x = 0$, the group $G_2 K^*$ must permute the $(q^4 - 1)/(q^2 - 1)$ 1-dimensional $GF(q^2)$-subspaces and hence must fix two such subspaces.*

Then, no third distinct group producing a mixed partition can be within the group generated by the first two.

THEOREM 91.11. (Johnson and Mellinger [**782**]). *If π is a multiply retractive translation plane of order q^4 for any set of distinct groups $G_i K^*$, $i = 1, 2, \ldots, t$, then there are at least two fixed components.*

COROLLARY 91.12. *For any number of distinct groups producing mixed partitions in a translation plane of order q^4 and kernel containing $GF(q)$, there are at least two common fixed components in the group generated by all groups and no third group can be in the group generated by two distinct groups. Furthermore, any two distinct groups intersect exactly in the kernel homology group of order $q - 1$.*

91.2. Triple and Quadruple Retraction.

Concerning multiple retraction, triple is essentially the end of the possibilities.

THEOREM 91.13. (Johnson and Mellinger [**782**]). *(1) If three distinct mixed partitions of $PG(3, q^2)$'s produce the same translation plane, then there are exactly two $PG(1, q^2)$'s which belong to all three mixed partitions. That is, there are exactly two fixed components in the translation plane by the three groups.*

(2) For each common $PG(1, q^2)$, there is always a homology group of the associated translation plane of order divisible by $(q + 1)$ with axis that common line and coaxis the remaining common line.

(3) For the two common components L and M, there are four Baer groups of order $q + 1$, fixing L and M. Furthermore, there are two nets of degree $q^2 + 1$ sharing L and M and each net contains exactly two of the four Baer subplanes fixed pointwise by Baer groups.

In addition, in each Baer net, there are $q - 1$ $GF(q)$-reguli that form a linear subset relative to a Baer axis of a Baer group. Each such regulus is either a $q + 1$ orbit for a given fixed-point-free group or is fixed linewise by that group.

(4) For q even, the group generated by the fixed-point-free groups may be generated by the two homology groups of order $q + 1$, any one of the Baer groups and the kernel homology group of order $q-1$, or generated by any three Baer groups and the kernel homology group.

(5) Quadruple spread retraction cannot occur.

We have given examples of semifield planes of order q^4 that produce three mixed partitions. The following questions are open:

PROBLEM 91.14. If π is a translation plane of order q^4 with kernel containing $GF(q)$ that admits triple retraction, is π a Hughes–Kleinfeld semifield plane or a derived Hughes–Kleinfeld semifield plane? If π is a semifield plane, can the plane be classified?

PROBLEM 91.15. Let π be a translation plane of order q^4 with kernel containing $GF(q)$ that produces mixed partitions of the following types:

$$(q^2 + 1, q^2(q - 1)), (2q^2, (q^2 - 1)(q - 1)), (q^4 - q^2 + 2, q - 1).$$

Is π derivable? And, if π is derivable, is it a derived semifield plane? Furthermore, if π is a derived semifield plane, is the semifield plane a Hughes–Kleinfeld plane?

Similarly, we may ask:

PROBLEM 91.16. If π is a derivable translation plane of order q^4 and kernel containing $GF(q)$ which produces mixed partitions of types

$$(q^2 + 1, q^2(q - 1)), (2q^2, (q^2 - 1)(q - 1)), (q^4 - q^2 + 2, q - 1)$$

and the derived plane produces mixed partitions of types

$$(q^2 + 1, q^2(q - 1)), (q^2 + 1, q^2(q - 1)), (q^4 + 1, 0),$$

classify π.

PROBLEM 91.17. Let π be any translation plane of order q^t, where t is odd. If $(q + 1)/(t, q + 1) < 2$, show that double retraction cannot occur.

91.3. Classification of Triply Retractive Semifield Planes.

It is possible to employ fundamental work on multiple retraction together with recent work of Jha and Johnson [634] on the 'fusion' of nuclei in semifield spreads and the work of Johnson and Mellinger [783] to completely classify the semifield planes admitting triple retraction.

DEFINITION 91.18. If a translation plane of order q^4 and kernel containing K isomorphic to $GF(q)$ retracts to a mixed partition of $PG(3, q^2)$ with k lines and n $PG(3, q)$'s, we shall say that the partition is of 'type (k, n)'.

In particular, Johnson and Mellinger [783] prove the following theorem.

THEOREM 91.19. (Johnson and Mellinger [783]). If a semifield plane of order q^4 and kernel containing $GF(q)$ produces three distinct partitions under retraction, then the kernel of the semifield is isomorphic to $GF(q^2)$ and the following statements hold:

(1) We may represent the spread in the following form:

$$x = 0, y = x \begin{bmatrix} u^q & t^q f \\ t & u \end{bmatrix}, \forall u, t \in GF(q^2), f \text{ a constant in } GF(q^2) - GF(q).$$

(2) The plane is a Hughes–Kleinfeld plane and the partition types are

$$(q^2 + 1, q^2(q - 1)), (q^2 + 1, q^2(q - 1)), (q^4 + 1, 0).$$

(3) Furthermore, the semifield plane is derivable and if π^* is the translation plane obtained by the derivation of a right- and middle-nucleus derivable net, then π^* produces the following partition types:

$$(q^2 + 1, q^2(q - 1)), (2q^2, (q^2 - 1)(q - 1)), (q^4 - q^2 + 2, q - 1).$$

(4) In addition, the semifield plane is derivable to, say, π_R^*, by a right-middle-nucleus net that is not a left-middle-nucleus net and to, say, π_L^*, by a left-middle-nucleus net that is not a right-middle-nucleus net.

The translation planes π_R^* and π_L^* both produce the partition types

$$(2q^2, (q^2 - 1)(q - 1)), (q^4 - q^2 + 2, q - 1)$$

but are non-isomorphic planes. In addition, π^* is not isomorphic to either π_R^* or π_L^*.

Note that π_R^* is a generalized Hall plane of type 2 and π_L^* is a generalized Hall plane of type 1 (a generalized Hall plane).

(5) Neither π_R^* nor π_L^* produce three partitions.

The proof of the previous theorem hinges on the results on the fusion of nuclei in semifields. The results that we have employed are special cases of the more general results of Jha and Johnson [634].

PROPOSITION 91.20. (Jha and Johnson [634]). Let π be a finite semifield plane coordinatized by a semifield plane of order q^4 with right nucleus K_R and middle nucleus K_M both isomorphic to $GF(q^2)$.

Then the line at infinity $\ell_\infty - \{(\infty)\}$, may be considered a Desarguesian affine plane coordinatized by either the right nucleus K_R or the middle nucleus K_M. In either case, we call such an affine plane, 'a Desarguesian plane at infinity'. In general, the 1-dimensional subspaces are the orbits under the particular homology group union the parallel class (0). If considered over K_R, then K_M^* fixes at least two of the 1-dimensional K_R-subspaces; two of the K_R^*-homology group orbits.

THEOREM 91.21. (Jha and Johnson [634]).
(1) Under the above assumptions, there is a coordinatizing semifield $S = (S, +, *)$ so that the right and middle nuclei are equal (we may 'fuse' the nuclei).

(2) If the right- and middle-nucleus groups share three common orbits on the line at infinity, then they share all orbits and hence act identically as scalar groups. In this latter case, the group generated by the homology groups contains a group of order $q^2 - 1$ which fixes the line at infinity pointwise. That is, the kernel is also $K_R = K_M$ and commutes with the right and middle nuclei; the right and middle nuclei define $GF(q^2)$-reguli. Furthermore, the semifield plane is Desarguesian.

(3) If the right- and middle-nucleus groups do not produce kernel-reguli (if the kernel would happen to be isomorphic to $GF(q^2)$), the two groups share exactly two common orbits on the line at infinity.

(4) If the left, middle and right nuclei are all isomorphic to $GF(q^2)$, then there is a coordinate semifield so that the three nuclei may be fused; $K_R = K_M = K_L$.

The following is taken from Biliotti, Jha, and Johnson [123]:

THEOREM 91.22. (Hughes–Kleinfeld Semifields). Suppose $a = x^{1+\theta} + xb$ has no solution for x in $F \simeq GF(h)$, where θ is an automophism of F and a, b in F. Then

$$(x + \lambda y) \circ (z + \lambda t) = (xz + aty^\theta) + \lambda(yz + (x^\theta + y^\theta b)t), \forall x, y, z, t \in F$$

defines a semifield of order h^2, and F is both its right nucleus and its middle nucleus. Conversely, if D is a finite semifield that is two-dimensional over a field F such that the middle and right nuclei of D coincide, then D is a Hughes–Kleinfeld semifield.

Now in our situation, triple retraction, we know that F is isomorphic to $GF(q^2)$, and K isomorphic to $GF(q)$ must actually commute within the semifield. This says that $\theta = q$. Moreover, we also know that the kernel (left nucleus) must be isomorphic to F as well. In this setting, we may fuse the three nuclei by Theorem 91.21, part (4). Hence,

$$\begin{aligned} k \circ ((x + \lambda y) \circ (z + \lambda t)) &= ((kx + \lambda k^\theta y) \circ (z + \lambda t)) \\ &= ((kx)z + at(ky)^\theta) + \lambda(k^\theta y)z + ((kx)^\theta + (k^\theta y)^\theta b)t), \\ &= k(xz + aty^\theta) + \lambda k^\theta(yz + (x^\theta + y^\theta b)t) \qquad \forall x, y, z, t \in F. \end{aligned}$$

This implies that

$$k^{\theta^2} b = k^\theta b \ \forall k \in F.$$

Hence, we must have $b = 0$, or $\theta = 1$, which does not occur as π is not Desarguesian.

91.3.1. Pseudo-Reguli.

In the classification of triply retractive semifield spreads of order q^4 with kernel containing $GF(q)$, we have found that there is a derivable net, which, when derived, produces three different mixed partitions of projective spaces isomorphic to $PG(3, q^2)$, depending on which of the three groups (fields) are used in the retraction process.

In Freeman [383], there is a study of how regular spreads in $PG(3, q)$ can produce partial spreads in $PG(3, q^2)$ that are termed 'pseudo-reguli'.

In our retraction process, this is arrived at in the following manner: Assume that we have a derivable vector space of order q^4 and kernel F isomorphic to $GF(q^2)$ containing K isomorphic to $GF(q)$ and assume that the Baer subplanes of the net incident with the zero vector are not all F-subspaces but are K-subspaces. This, by the way, is easy to accomplish using the construction process of 'lifting' (see Johnson [744]), whereby any quasifield coordinatizing a spread in $PG(3, q)$ may be lifted to a quasifield for a spread in $PG(3, q^2)$ having a derivable net of the indicated type.

Then, the kernel group of order $q^2 - 1$ fixes exactly two of the Baer subplanes and has $q - 1$ orbits of length $q + 1$. In this setting, we obtain a 'pseudo-regulus' in $PG(3, q^2)$. Actually, this is what might be called the 'derived' pseudo-regulus.

In our classification result of semifield spreads admitting triple-retraction, we have seen that there is a class of spreads admitting a derivable net D that produces similar possibilities.

In particular, we had three different groups, G_i, $i = 1, 2, 3$, where we allow G_1 to denote the kernel homology group. Retracting via G_1 produces the pseudo-regulus situation as mentioned above. However, the other two groups provide two other alternatives. One of the groups, say G_2, fixes all components of the derivable net and all Baer subplanes incident with the zero vector. Retraction with respect to G_2 produces a 'regulus' in the associated $PG(3, q^2)$. The other group has the property that the Baer subplanes are each fixed by G_3 but the group fixes exactly two components of the derivable net and has $q - 1$ orbits of length $q + 1$. This is exactly the way that G_1 would act on the derived net. Anyway, in this situation, the derived net produces merely a set of $q^2 + 1$ lines of the associated $PG(3, q^2)$; an original pseudo-regulus.

So, the derived partial spread of a derivable partial spread of the appropriate type can be interpreted in three ways; as a derived-pseudo-regulus, a regulus and pseudo-regulus, depending on the type of group (field) used in the retraction process.

CHAPTER 92

Transitive Baer Subgeometry Partitions.

A Baer subgeometry partition of $PG(2, q^2)$ is a partition of the points by $PG(2, q)$'s. Recently, Mathon and Hamilton [**975**] found, by computer search, several Baer subgeometry partitions (BSG's) of $PG(2, 16)$ and $PG(2, 25)$. The BSG's of $PG(2, 16)$ admit an automorphism group fixing one $PG(2, 4)$ and acting transitively on the remaining $PG(2, 4)$'s. Since BSG's seem to be quite rare, in general, this particular set of Baer subgeometry partitions is quite interesting. In fact, an open question in Baker et al. [**63**] is whether this is an example of a potentially infinite class. Furthermore, it is also asked whether such partitions can exist more generally in $PG(2m, q^2)$ for $m > 1$.

Actually, it is possible to establish some parameters of the above questions and make strong progress toward answering these questions, in general. Our method of attack is to consider the implications of such study in translation planes associated with Baer subgeometry partitions.

In particular, considering the Hamilton–Mathon BSG's, we have, by the process of 'geometric lifting' of Hirschfeld and Thas [**529**, p. 204] (also see Johnson [**755**]), established that there is a corresponding translation plane with spread in $PG(5, 4)$. Moreover, by Johnson [**755**], there is a collineation group of the associated translation plane which fixes a $GF(4)$-regulus net, that normalizes a field multiplicative group $GF(16)^*$, has one orbit of length $4^3 - 4$ of components and permutes transitively a set of $4^2 - 4$ $GF(4)$-reguli.

All of the above statements hold more generally for a translation plane associated with any Baer subgeometry partition. Using the analysis of the associated translation planes, Jha and Johnson [**629**] are able to prove the following result:

THEOREM 92.1. (Jha and Johnson [**629**]). Let π be a translation plane of order q^3 and kernel containing $GF(q)$ that arises from a BSG of $PG(2, q^2)$ admitting an automorphism group fixing one $PG(2, q)$ and transitive on the rest.

Then all planes are cubic extensions of a flag-transitive plane. Furthermore, we have the following situations:

(1) $q = 2$ and the plane is Desarguesian.

(2) $q = 4$ and the plane is Desarguesian.

(3) $q = 4$ and the plane is one of two Hamilton–Mathon planes, whose matrix spread sets are interrelated by transposition.

(4) q is odd and π is a cubic extension of a subplane π_o of order q.

In a series of articles, Hiramine, Jha, and Johnson have studied translation planes of order q^n that admit a subplane π_o of order q and a collineation group G that leaves π_o invariant and acts flag-transitively on it and furthermore acts transitively on the component tangents to π_o (see [**514**]–[**519**]). In several of the articles mentioned, 'cubic' extensions are studied. Hence, we have a translation plane of

order q^3 admitting a subplane π_o of order q and a group acting transitively on the
infinite points of π_o and transitively on the remaining infinite points. When the
group G is solvable and the spread is in $PG(5,q)$, there is a complete representation
of the spreads.

It is, in fact, possible to show that for BSG's of $PG(2,q^2)$, of the type under
consideration, the collineation group is, in fact, solvable. Hence, there is basically
a classification of the BSG's of $PG(2,q^2)$ that admit an automorphism group fixing
one $PG(2,q)$ and acting transitively on the remaining $PG(2,q)$'s.

92.1. Transitive Deficiency-One Partial Baer Subgeometry Partitions.

There is an alternative way of considering the problem; the study of 'partial'
Baer subgeometry partitions of $PG(2m,q^2)$. In particular, we are interested in
the so-called 'deficiency-one' partial Baer subgeometry partitions. Moreover, we
discuss 'transitive' deficiency-one partial Baer subgeometry partitions; deficiency-
one partial BSG's admitting a group transitive on the subplanes of the partition.

Concerning deficiency one, in general, Jha and Johnson [**629**] also prove:

THEOREM 92.2. (Jha and Johnson [**629**]).
Given a partial BSG $\mathcal{B}$ of $PG(2m,q^2)$ of deficiency one. If $m > 1$, then $\mathcal{B}$ may
be uniquely extended to a BSG.

Furthermore, for transitive deficiency-one partial BSG's admitting solvable
groups, we have:

THEOREM 92.3. (Jha and Johnson [**629**]). Let $\mathcal{P}$ be a deficiency-one BSG of
$PG(2m,q^2)$ for $m > 1$, which admits a transitive group T. Let $\mathcal{P}^+$ denote the
unique extension to a BSG.

(1) If T leaves invariant a point of the adjoined $PG(2m,q)$ and is solvable, then
$q = 2$ or 4.

(2) Either $(q, 2m + 1) = (2, 5)$ or the BSG, when it exists, cannot be classical
and the associated translation plane cannot be Desarguesian.

92.1.1. 4^3-Spreads. We note that the Desarguesian spreads of orders q^3, for q
$= 2, 4$, are, in fact, cubic extensions. The only other possibility is spreads associated
with the Hamilton–Mathon partitions. However, in this case, we may identify the
fixed regulus net of the associated translation planes with a regulus in an associated
Desarguesian spread. The automorphism group of the Hamilton–Mathon partitions
is the Galois group of $GF(2^{12})$. Our forthcoming analysis applies to show that the
group of order 12 combined with the 'inherited' group of order 5 acting on the spread
does, in fact, fix a subplane of the invariant regulus net. **Hence, all translation
planes of order 4^3 which permit retraction are cubic extensions of a
subplane of order 4.**

So, we have a group G of order $4(4^2-1)(4-1) = 2^2 \cdot 3^2 \cdot 5$ containing the $GF(4)$-
scalar group that fixes a Desarguesian subplane π_o of an invariant $GF(4)$-regulus
net which we may coordinatize by

$$\mathcal{E} : x = 0, y = x\alpha I_3 \ \forall \alpha \in GF(4).$$

We see that the 2-groups of G must act faithfully on π_o.

The group G is a subgroup of $\Gamma L(3,q^2)$ and G contains a cyclic subgroup C_{15}
of order 15 which acts sharply transitively on the non-zero vectors of π_o. We note

that C_{15} is a normal subgroup of G and fixes all of the Desarguesian subplanes of order q corresponding to $GF(q^2)$-1-dimensional subspaces.

We have seen above that any 2-group in $GL(3, 16)$ must fix exactly 4^2 points, a contradiction.

Hence, there are no linear 2-elements.

It follows that we may assume that some 2-group is isomorphic to the cyclic Galois group of order 4 generated by $z \longmapsto z^2$. Also, any such 2-group is isomorphic to a subgroup of $\Gamma L(2, 4)$ acting on π_o. We note that the unique involution σ in this group must fix a $GF(q)$-1-dimensional subspace pointwise of π_o and thus induce an elation on the translation plane π. That is, if σ is a Baer collineation then $\operatorname{Fix} \sigma$ contains fixed points exterior to the net N and hence, there must be a 2-group of order at least 8 in $\Gamma L(3, 4^2)$ giving rise to a linear involution, which we have seen cannot occur.

The group generated by the elation groups forms a group of order $2(4 + 1)$ faithfully induced on π_o since each elation group fixes each of the $1 + 4 + 4^2$ Desarguesian subplanes of $\mathcal{E}$. Thus, the group is dihedral D_5 of order $2(4 + 1)$ with the cyclic stem being $\langle C_{15}^3 \rangle$.

Suppose that there is a 3-element that induces an affine homology on π_o. Then there are at least two distinct 3-groups modulo the center $GF(4)^*$. Such groups are in $GL(2, 4)$ so the quotient in $PGL(2, 4)$ is either A_4, S_4, A_5 or dihedral of order 6 or 10. But, the Sylow 2-subgroup in $PGL(2, 4)$ is of order 2 so that we can only obtain a group of order 6 or 10.

Hence, there is a 3-group of order 3 that fixes π_o pointwise. It still follows that the group induced in $PGL(2, 4)$ is dihedral of order 10. This group of order 3 that fixes π_o pointwise commutes with any 2-group of order 4, so we obtain a cyclic group C_{12} of order 12.

Hence, we have a group isomorphic to $C_{12}GF(16)^*$ which acts on our spread.

Any 2-group in $\Gamma L(3, 4^2)$ generated as a Galois group fixes exactly 4-points. Since the involution in this group is an elation, we choose an axis as $x = 0$. We realize $GF(4) = \{0, 1, \alpha, \alpha + 1\}$ so that $\alpha^2 = \alpha + 1$. Assume that the associated field splits the $GF(4)$-irreducible polynomial $x^2 + \alpha x + 1$. Choose a generator of $GF(16)^*$ to map $x = 0 \longmapsto y = 0 \longmapsto y = x \longmapsto y = x\alpha \longmapsto y = x(\alpha + 1)$.

This generator leaves invariant every Desarguesian subplane of the net $x = 0, y = x\beta$; $\beta \in GF(4)$. Write the vector space over $GF(6)$ with vectors

$$(x_1, x_2, x_3, y_1, y_2, y_3) \qquad \text{for all } x_i,\ y_i \in GF(4) \text{ for } i = 1, 2, 3.$$

Note that

$$\pi_1 = \{(x_1, 0, 0, y_1, 0, 0); x_1, y_1 \in GF(q)\},$$
$$\pi_2 = \{(0, x_2, 0, 0, y_2, 0); x_2, y_2 \in GF(4)\}, \quad \text{and}$$
$$\pi_3 = \{(0, 0, x_3, 0, 0, y_3); x_3, y_3 \in GF(4)\}$$

are Desarguesian subplanes of the $GF(4)$-regulus net $\mathcal{E}$.

Under these conditions, it follows fairly directly that a generator which accomplishes the mapping on components has the following form:

$$\begin{bmatrix} aI_3 & aI_3 \\ \alpha aI_3 & 0_3 \end{bmatrix}$$

for some a in $GF(4)$. Since we obtain the scalar group βI_6 for all $\beta \in K$ acting as a collineation group, we may take $a = 1$.

We note that the general form for the field $GF(16)$ then becomes:

$$\begin{bmatrix} uI_3 + tI_3 & tI_3 \\ \alpha tI_3 & uI_3 \end{bmatrix} \forall u, t \in GF(4).$$

We may represent a 2-group as follows:

$$\left\langle E \colon (x_1, x_2, x_3, y_1, y_2, y_3) \longmapsto (x_1^2, x_2^2, x_3^2, y_1^2, y_2^2, y_3^2) \begin{bmatrix} I_3 & \alpha I_3 \\ 0_3 & I_3 \end{bmatrix} \right\rangle.$$

The group of order 3 which fixes a Desarguesian subplane pointwise (by the action of the kernel homology group of order 3, we may guarantee this) must also fix at least two other subplanes. We choose coordinates so that the fixed point subplane is π_1 and the other two fixed subplanes are π_2 and π_3. This forces this element to be diagonal as it fixes all components of the net $\mathcal{E}$.

Also an element h of order three on L as $PG(2,4)$ fixes a point P and permutes the five lines incident with P. Hence, h fixes at least two lines incident with P. Since there are no homologies in $PG(2,4)$, h must fix a triangle pointwise.

Hence, we have a group

$$h = \begin{bmatrix} 1 & 0 & 0 & 0 & 0 & 0 \\ 0 & a & 0 & 0 & 0 & 0 \\ 0 & 0 & b & 0 & 0 & 0 \\ 0 & 0 & 0 & 1 & 0 & 0 \\ 0 & 0 & 0 & 0 & a & 0 \\ 0 & 0 & 0 & 0 & 0 & b \end{bmatrix} ; a, b \in GF(4)^*, a \text{ and } b \text{ both not } 0 \text{ or } 1.$$

Furthermore, since h is linear, it follows that we may take $a = \alpha$ and $b = \alpha^2$.

The reader may note that h commutes with E and both E and h normalize $GF(16)^*$. Hence, we obtain the group of order $2^2 \cdot 3^3 \cdot 5$ which is a semidirect product of a cyclic group by a normal cyclic group of order 15:

$$G = \langle h \rangle \, EGF(16)^*.$$

We may choose a component $y = xM$ where

$$M = \begin{bmatrix} m_{11} & m_{12} & m_{13} \\ m_{21} & m_{22} & m_{23} \\ 1 & 0 & 0 \end{bmatrix}.$$

Then the spread is

$$x = 0, y = x\beta I, (y = xM)g \forall \; \beta \in GF(4) \; \forall g \in G/GF(4)^*.$$

Hence, by variation of M, we obtain all possible spreads. So, there are two non-Desarguesian planes; the Hamilton–Mathon planes and the spread sets above of the two planes are merely transposes of each other.

92.1.2. Desarguesian Transitive and Deficiency-One Spreads.
We consider a Desarguesian affine plane Σ of order q^{2k+1} to be represented by $GF(q^{2(2k+1)})$, where the points are the elements of the field and the components are the 1-dimensional $GF(q^{2k+1})$-subspaces. There is a subgroup of order $q^{2(2k+1)} - 1$ acting on Σ, fixing one point (say the zero vector) and acting transitively on the remaining affine points. Naturally, the group induced on the line at infinity is cyclic of order $q^{2k+1} + 1$ and acts transitively there. The stabilizer of a component has order $q^{2k+1} - 1$ and acts transitively on the non-zero vectors of each component. If we

consider the subgroup of order $q^2 - 1$ that does, in fact, act as a collineation group of Σ, the analysis of Johnson [755] shows that there is a subgroup of order $q - 1$ that fixes each component; the orbit length on the line at infinity is therefore $q + 1$, and the point orbits under this group form natural Desarguesian subplanes of order q. Hence, the orbits of length $q + 1$ form subplanes covered nets which, in fact, are regulus nets. Note that the 1-dimensional $GF(q^2)$-subspaces are the subplanes of the net incident with the zero vector. Forming the retraction as in Johnson [755], we obtain a natural Baer subgeometry partition of $PG(2k, q^2)$. Hence, we have:

THEOREM 92.4. *Any Desarguesian affine plane of order q^{2k+1} produces, by retraction, a Baer subgeometry partition of $PG(2k, q^2)$.*

Now we ask which of these Baer subgeometry partitions can admit an automorphism group that fixes one subgeometry and acts transitively on the remaining subgeometries.

We note, by the analysis of Johnson [755], that the group involved as acting on the associated translation plane (Desarguesian plane, in this case) is a subgroup of $\Gamma L(2k + 1, q^2)$. Moreover, the group is also acting on a Desarguesian affine plane so must be in $\Gamma L(2, q^{2k+1})$.

Hence, we have the following result:

THEOREM 92.5. *(Jha and Johnson [629]). Let $\mathcal{B}$ be a classical BSG of $PG(2k, q^2)$ which admits an automorphism group fixing one $PG(2k, q)$ and acting transitively on the remaining $PG(2k, q)$'s of the partition.*

(1) Then $(q, 2k + 1) \in \{(2, 3), (4, 3), (3, 3), (2, 5)\}$.

(2) Each of the BSG's of part (1) produce Desarguesian spreads of orders q^{2k+1} which admit collineation groups that fix a subplane of order q, fix one regulus net and act transitively on the remaining $(q^{2k+1} + 1)/(q + 1) - 1$ regulus nets. Hence, each such plane is a solvable n-dimensional extension of a flag-transitive plane.

(3) Note that each of the BSG's of part (1) admits an automorphism group which acts doubly transitively on the set of partition subplanes. Furthermore, these are exactly the classical BSG's that admit a doubly transitive group.

92.1.3. The Doubly Transitive BSG's. In Johnson [755], it was shown that every two-transitive BSG of $PG(2k, q^2)$ produces a Desarguesian plane, except possibly when $(q, 2k + 1) = (3, 3), (4, 3)$ or $(2, 5)$ (noting that the planes of order 8 are Desarguesian). By Dempwolff [304], the planes of order 3^3 admit such doubly transitive groups if and only if the planes are Desarguesian (see our analysis above for the 'if' part). We have seen that the planes of order $(4, 3)$ that admit doubly transitive groups are Desarguesian (by Baker et al. [63] the groups of the Hamilton–Mathon planes are not transitive on the regulus nets). Hence, this leaves the possibility that there is a doubly transitive plane on 2-reguli of order 2^5. Hence, we obtain:

THEOREM 92.6. *(Jha and Johnson [629]). Let $\mathcal{B}$ be a doubly transitive BSG of $PG(2k, q^2)$. If $(q, 2k + 1) \neq (2, 5)$ then $\mathcal{B}$ is classical and the associated translation plane is Desarguesian of order $8, 27$, or 64.*

92.2. Transitive Deficiency-One Cases in Larger Dimensions.

Now consider the possibility that there could exist BSG's of $PG(2m, q^2)$ for $m > 1$ admitting a group G that fixes one $PG(2m, q)$ and acts transitively on the remaining $PG(2m, q)$'s.

Hence, we obtain a translation plane of order q^{2m+1} with kernel containing $GF(q)$ whose spread is covered by $GF(q)$-reguli and a collineation group G in $\Gamma L(2(2m+1), q)$ which fixes one $GF(q)$-regulus and acts transitively on a set of $(q^{2m+1} + 1)/(q+1)$ other $GF(q)$-reguli.

Note that we may take the weaker hypothesis that there is a partial BSG of deficiency one with a transitive automorphism group.

To see this, we note that we obtain a translation net of order q^{2m+1} and deficiency $q + 1$.

Consider the polynomial $p(x) = x^4/2 + x^3 + x^2 + 3x/2$ and compute $p(q) = q^4/2 + q^3 + q^2 + 3q/2 < q^{2m+1}$ for $m > 1$. For polynomial inequalities, results of uniqueness of extension apply showing that there is a unique extension to a translation plane and the collineation group extends as well. For example, the reader is directed to Jungnickel [815]. We assert that the translation plane has its spread in $PG(4m + 1, q)$. We need to check that the $GF(q)$-kernel group of the original net corresponding to the deficiency-one BSG of $PG(2m, q^2)$ acts as a kernel homology group of the translation plane. Hence, there is an adjoined net of degree $q + 1$ that admits as a collineation group the group $GF(q^2)^*$ that fixes each $GF(q)$-regulus net. Each component of the adjoined net Γ is a $(2m+1)$-dimensional $GF(q)$-subspace. Suppose some element of $GF(q^2)^* - GF(q)$ leaves a component in Γ invariant. We note that $GF(q^2)^*$ is fixed-point-free so that $q^{2m+1} - 1$ is divisible by some element of order dividing $q^2 - 1$. However, $(q^{2m+1} - 1, q^2 - 1) = (q^{(2m+1,2)} - 1) = (q - 1)$. Hence, the only possible element that can leave a component L of Γ invariant has order dividing $q - 1$, which implies that the orbit length is $q + 1$ and that the translation plane has kernel containing $GF(q)$.

Now we may use the ideas of the fundamental theorem of retraction in Johnson [755] to see that the adjoined net is also a $GF(q)$-regulus net so we have an extension of the deficiency-one BSG.

We record this fact separately.

THEOREM 92.7. (Jha and Johnson [629]). Any partial BSG of $PG(2m, q^2)$ for $m > 1$ of deficiency one may be uniquely extended to a BSG.

92.2.1. Transitive Extensions. If we have a transitive deficiency-one BSG of $PG(2m, q^2)$ for $m > 1$, then there is an associated translation plane of order q^{2m+1} with kernel containing $GF(q)$ that admits a collineation group isomorphic to $GF(q^2)^*$ whose component orbits are $GF(q)$-reguli and further exists a collineation group G containing and normalizing $GF(q^2)^*$ in $\Gamma L(2m+1, q^2)$ which fixes a $GF(q)$-regulus $\mathcal{R}$ and acts transitively on the remaining components.

If there is an invariant subplane π_o in $\mathcal{R}$, then we have that π is a transitive extension of a flag-transitive plane π_o. We may apply the results of Hiramine, Jha and Johnson [517] provided the associated group is solvable. We have seen that when $m = 1$, i.e., BSG's of $PG(2, q^2)$, the group must be solvable and the associated translation plane must be a cubic extension of a subplane of order q. Thus, this might point to the general situation. In any case, it seems reasonable to make the assumption that, in general, the associated translation plane is an $(2m + 1)$-extension of a flag-transitive subplane of order q and that the group involved is solvable.

Hence, if we assume that the group is solvable, we must have either π is Desarguesian or $q = 2$ or 4 and all involutions are elations. However, we have seen, in the Desarguesian case, that other than cubic orders, the only Desarguesian possibility is when $(q, 2m + 1) = (2, 5)$. Furthermore, in this situation, the plane is a 5-dimensional extension of a subplane of order 2. Hence, applying the main result on solvable extensions (see Section 90.3), and noting that $q = 3$ requires $2m + 1$ to be even or 3, we have:

THEOREM 92.8. (Jha and Johnson [629]). Let $\mathcal{P}$ be a deficiency-one BSG of $PG(2m, q^2)$ for $m > 1$, which admits a transitive group T. Let $\mathcal{P}^+$ denote the unique extension to a BSG.

(1) If T leaves invariant a point of the adjoined $PG(2m, q)$ and is solvable, then $q = 2$ or 4.

(2) Either $(q, 2m + 1) = (2, 5)$ or the BSG, when it exists, cannot be classical and the associated translation plane cannot be Desarguesian.

PROOF. T induces G acting on the translation plane and $G/GF(q^2)^* \simeq T$ by a result of Johnson [755]. Since T leaves invariant a point, it follows that G leaves invariant a $GF(q^2)$-1-dimensional subspace. Hence, G is solvable and our previous analysis applies.   □

92.3. Dual BSG's.

Let π be a translation plane of order q^{2k+1} with kernel containing $GF(q)$ which permits spread retraction. Hence, there is an associated collineation group of order $q^2 - 1$, $GF(q^2)^*$, containing the $(q - 1)$-kernel homology group such that the component orbits are $GF(q)$-reguli. We note that the full collineation group normalizes $GF(q^2)^*$. Now let σ be any duality of the associated vector space of dimension $2(2k + 1)$ over $GF(q)$. Then, the spread components are of dimension $2k + 1$ over $GF(q)$ and, as such, we obtain another spread from the given one S, by taking $S\sigma$. The collineation group of the new spread is isomorphic to the original group (although, not necessarily permutation isomorphic).

We note that the group $GF(q^2)^*$ is fixed-point-free but fixes all one-dimensional $GF(q^2)$-subspaces. The number of hyperplane orbits under a collineation group is the same as the number of 1-dimensional subspace orbits by Dembowski [290]; hence, the group leaves invariant each $GF(q^2)$-hyperplane. Moreover, a fixed-point-free collineation in $GF(q^2)^*$, acting on the spread (a $2(2k + 1)$-dimensional vector space over $GF(q)$), has orbits of length $q + 1$ on 1-dimensional $GF(q)$-subspaces. Hence, such a fixed-point-free collineation has orbits of length $q + 1$ on $GF(q)$-hyperplanes. Taking a duality on the associated $2(2k+1)$ $GF(q)$-vector space then maintains the orbit structure on $GF(q)$-hyperplanes and $GF(q)$-1-spaces. Hence, when we dualize, we obtain a $GF(q^2)^*$-group acting on the dual space. This means that the spread obtained from a duality gives rise to a spread with the same retraction properties of the original. Hence, we obtain an associated BSG.

We note that the transpose of the translation plane associated with a BSG produces the translation plane whose retraction gives rise to the new BSG. Hence, we obtain:

THEOREM 92.9. (Jha and Johnson [629]). Given any Baer subgeometry partition $\mathcal{B}$ of $PG(2k, q^2)$. Then, there is an associated Baer subgeometry partition $\mathcal{B}^D$, called the 'dual' of $\mathcal{B}$, obtained by the transpose spread (or dual spread) of the spread in $PG(4k + 1, q)$ associated with $\mathcal{B}$.

Now consider spreads in $PG(4m - 1, q)$ that permit spread retraction; the vector space may be written over a field K isomorphic to $GF(q^2)$ which extends the indicated field $GF(q)$ as a $4m$-dimensional K-vector space, where the group $GF(q^2)^*$ is assumed to act as a collineation group with orbit lengths 1 or $q + 1$. In the transposed spread, there is a collineation group that fixes all duals of spread components for which the orbit length in the original group is 1. Hence, we obtain the same configuration of orbits of length 1 and of length $q + 1$ in the dual setting. Hence, we obtain:

THEOREM 92.10. (Jha and Johnson [**629**]). Let π be a spread in $PG(4m-1, q)$ that permits spread retraction, producing a mixed partition $\mathcal{B}$ of $PG(2m - 1, q^2)$ of δ $PG(m - 1, q^2)$'s and d $PG(2m - 1)$'s.

Then the transposed spread of π is a spread in $PG(4m - 1, q)$ that permits spread retraction, producing a corresponding mixed partition, $\mathcal{B}^D$, called the 'dual' of $\mathcal{B}$ which also has δ $PG(m - 1, q^2)$'s and d $PG(2m - 1)$'s.

CHAPTER 93

Geometric and Algebraic Lifting.

As mentioned in Chapter 35, there is a very general construction process called 'lifting' and which we further term 'algebraic lifting' (to properly distinguish from geometric lifting) which constructs from any spread in $PG(3, q)$ a derivable spread in $PG(3, q^2)$. This spread has a replaceable net which we can replace to construct a spread in $PG(7, q)$ satisfying the retraction process as presented here. We have noted that the reverse process of construction of a spread in $PG(3, q)$ from a spread of a given form in $PG(3, q^2)$ is also called 'retraction'. Hence, we might use the terms 'algebraic retraction' and 'geometric retraction' here as well to distinguish these two constructions. We can relate 'algebraic lifting' now to geometric lifting. We recall the basic construction process.

THEOREM 93.1. Let π be a translation plane with spread S in $PG(3, q)$. Let F denote the associated field of order q and let K be a quadratic extension field with basis $\{1, \theta\}$ such that $\theta^2 = \theta\alpha + \beta$ for $\alpha, \beta \in F$. Choose any quasifield and write the spread as follows:

$$x = 0, y = x \begin{bmatrix} g(t, u) & h(t, u) - \alpha g(t, u) = f(t, u) \\ t & u \end{bmatrix} \forall t, u \in F$$

where g, f and unique functions on $F \times F$ and h is defined as noted in the matrix, using the term α.

Define $F(\theta t + u) = -g(t, u)\theta + h(t, u)$. Then

$$x = 0, y = x \begin{bmatrix} \theta t + u & F(\theta s + v) \\ \theta s + v & (\theta t + u)^q \end{bmatrix} \forall t, u, s, v \in F$$

is a spread S^L in $PG(3, q^2)$ called the spread 'algebraically lifted' from S. We note that there is a derivable net

$$x = 0, y = x \begin{bmatrix} w & 0 \\ 0 & w^q \end{bmatrix} \forall w \in K \simeq GF(q^2)$$

with the property that the derived net (replaceable net) contains exactly two Baer subplanes which are $GF(q^2)$-subspaces and the remaining $q^2 - 1$ Baer subplanes form $(q - 1)$ orbits of length $q + 1$ under the kernel homology group.

Hence, we obtain a mixed partition of $(q - 1)$ $PG(3, q)$'s and $q^4 - q$ lines of $PG(3, q^2)$.

So, from any quasifield, we obtain a spread in $PG(3, q)$ which lifts and derives to a spread permitting retraction which produces a mixed partition of $(q - 1)$ $PG(3, q)$'s and $q^4 - q$ lines of $PG(3, q^2)$.

REMARK 93.2. From the mixed partition of $(q - 1)$ $PG(3, q)$'s and $q^4 - q$ lines of $PG(3, q^2)$, we construct a translation plane back which has $q^4 - q$ what might be called 'ordinary' components and $(q - 1)$ orbit's of length $q + 1$ of components each forming a $GF(q)$-regulus. Now since we 'get back', what must happen is that two of these ordinary components together with the $q-1$ reguli must form a derivable net—that is, we get the derived side—which when derived gets back to the original spread in $PG(3, q^2)$. On the other hand, if we have a mixed partition of $q-1$ $PG(3, q)$'s and $q^4 - q$ lines of $PG(3, q^2)$, it is not necessarily the case that the geometrically lifted spread is derivable.

In the text by Hirschfeld and Thas [**529**], there is given a construction of finite spreads, and hence, finite translation planes, from either Baer subgeometry partitions or mixed subgeometry partitions of a finite projective space. The partitions are the points of finite projective geometries Σ over $GF(q^2)$. When Σ is isomorphic to $PG(2m, q^2)$, the partition components are Baer subgeometries isomorphic to $PG(2m, q)$. When Σ is isomorphic to $PG(2n-1, q^2)$, $n > 1$, it is possible to have a so-called 'mixed' partition of β $PG(n-1, q^2)$'s and α $PG(2n-1, q)$'s. The configuration is such that $\alpha(q+1) + \beta = q^{2n} + 1$.

The interest in such partitions lies in the fact that they may be used to construct spreads and hence translation planes. Baer subgeometry partitions produce translation planes of order q^{2m+1} with kernel containing $GF(q)$, whereas mixed partitions produce translation planes of order q^{2n} and kernel containing $GF(q)$. The process is called 'geometric lifting' in Johnson [**755**]. Hirschfeld and Thas constructed some mixed partitions of $PG(3, q^2)$, which turn out to construct André spreads of order q^4. The main unanswered question is what spreads may be constructed using geometric lifting. For example, there does not seem to be a direct connection between spreads in $PG(3, q)$ and translation planes of order q^2 and subgeometry partitions of $PG(3, q^2)$ since partitions produce translation planes of order q^4. Or rather to directly construct a translation plane of order q^2 with spread in $PG(3, q)$ from a subgeometry partition, we would need that $n = 1$, which is prohibited in the geometric lifting construction, but see below for an attempt at such constructions when n is indeed chosen to be 1.

In Johnson [**755**], the question was considered as how to recognize spreads that have been geometrically lifted from Baer subgeometry or mixed subgeometry partitions of a finite projective space. It was found that the intrinsic character is that the translation planes have order q^t with subkernel K isomorphic to $GF(q)$ and admit a fixed-point-free collineation group (on the non-zero vectors) which contains the scalar group K^*, written as GK^*, such that GK^* union the zero mapping is a field isomorphic to $GF(q^2)$ (see Johnson [**755**] and Johnson–Mellinger [**782**]). With such a recognition theorem on such collineation groups, it is then possible to 'retract' such a translation plane or spread to construct a variety of Baer subgeometry or mixed subgeometry partitions of an associated projective space written over GK as a quadratic field extension of K. To construct spreads in $PG(3, q)$ directly requires subgeometry partitions of $PG(1, q^2)$, which of course are simply lines $PG(1, q)$'s. In this case, there will be a collineation group in the associated translation plane of order $q^2 - 1$, which fix all components; that is, the translation plane of order q^2 has $GF(q^2)$ as a kernel homology group, so only the Desarguesian plane may be so constructed.

Hence, it would seem that the only spreads that can be obtained from a mixed subgeometry partition of $PG(3, q^2)$ are those spreads corresponding to translation planes of order q^4 that admit the required 'field group' of order $q^2 - 1$. However, there is also an algebraic construction procedure for spreads which is called 'algebraic lifting' (or more simply 'lifting' in Johnson [**744**]) by which a spread in $PG(3, q)$ may be lifted to a spread in $PG(3, q^2)$. More precisely, this construction is a construction on the associated quasifields for the spread and different quasifields may produce different algebraically lifted spreads. The reverse procedure of constructing spreads in $PG(3, q)$ from certain spreads in $PG(3, q^2)$ is called 'algebraic contraction'. This material is explicated in Biliotti, Jha, and Johnson and the reader is referred to this text for additional details and information (see [**123**]).

Now apart from the name, there should be no connection between geometric lifting from a subgeometry partition of a projective space to a spread and algebraic lifting from a subspace partition of a vector space to a spread. On the other hand, when a spread in $PG(3, q)$ is algebraically lifted to a spread in $PG(3, q^2)$, it turns out that there is a suitable field group of order $q^2 - 1$, from which such an algebraically lifted spread of order q^4 can be produced from a mixed subgeometry partition. This spread in $PG(3, q^2)$ begins its existence as a spread in $PG(7, q)$, but actually has kernel isomorphic to $GF(q^2)$. This means that our original spread in $PG(3, q)$ may be constructed by a 2-step procedure from a mixed subgeometry partition of $PG(3, q^2)$. Previously, in Johnson and Mellinger [**782**], this idea of producing a spread from a series of construction methods is considered, where it is pointed out that one may algebraically lift and spread in $PG(3, q)$ to a spread in $PG(3, q^2)$, which is actually derivable. The derived translation plane produces a spread in $PG(7, q)$, and the associated plane admits a field group (the original kernel group of the spread in $PG(3, q^2)$), which may be used to produce a mixed subgeometry partition of $PG(3, q^2)$. Then the construction procedure to obtain the original spread requires a 3-step procedure: geometric lifting—derivation— algebraic contraction.

However, Jha and Johnson [**580**] note the following fundamental connection:

THEOREM 93.3. (Jha and Johnson [**580**]). Let S be any spread in $PG(3, q)$. Then there is a mixed subgeometry partition of $PG(3, q^2)$, which geometrically lifts to a spread in $PG(3, q^2)$ that algebraically contracts to S.

COROLLARY 93.4. The set of mixed subgeometry partitions of a 3-dimensional projective space, $PG(3, k^2)$, constructs all spreads of $PG(3, k)$.

PROOF. Let π be any translation plane with spread in $PG(3, q)$. Consider any matrix spread set $\mathcal{M}$, and form the algebraic lift to a spread set in $PG(3, q^2)$. This spread admits a Baer group B of order $q+1$. Let $K^*_{q^2-1}$ denote the kernel homology group of order $q^2 - 1$ of the lifted plane. We note that any lifted spread has the general form:

$$x = 0, y = x \begin{bmatrix} u & F(t) \\ t & u^q \end{bmatrix} ; u, t \in GF(q^2).$$

The Baer group has the following form

$$\left\langle \begin{bmatrix} e & 0 & 0 & 0 \\ 0 & 1 & 0 & 0 \\ 0 & 0 & 1 & 0 \\ 0 & 0 & 0 & e \end{bmatrix} ; e^{q+1} = 1, \ e \in GF(q^2)^* \right\rangle.$$

Write B as follows:

$$\left\langle \tau_a = \begin{bmatrix} a^{q-1} & 0 & 0 & 0 \\ 0 & 1 & 0 & 0 \\ 0 & 0 & 1 & 0 \\ 0 & 0 & 0 & a^{q-1} \end{bmatrix} ; a \in GF(q^2)^* \right\rangle .$$

Note that the kernel homology group $K^*_{q^2-1}$ may be written in the form:

$$\left\langle k_a = \begin{bmatrix} a & 0 & 0 & 0 \\ 0 & a & 0 & 0 \\ 0 & 0 & a & 0 \\ 0 & 0 & 0 & a \end{bmatrix} ; a \in GF(q^2)^* \right\rangle .$$

Now form the group with elements $\tau_a k_a$:

$$\left\langle \tau_a k_a = \begin{bmatrix} a^q & 0 & 0 & 0 \\ 0 & a & 0 & 0 \\ 0 & 0 & a & 0 \\ 0 & 0 & 0 & a^q \end{bmatrix} ; a \in GF(q^2)^* \right\rangle .$$

As a matrix ring

$$K = \left\langle \tau_a k_a = \begin{bmatrix} a^q & 0 & 0 & 0 \\ 0 & a & 0 & 0 \\ 0 & 0 & a & 0 \\ 0 & 0 & 0 & a^q \end{bmatrix} ; a \in GF(q^2) \right\rangle ,$$

is clearly a field isomorphic to $GF(q^2)$, whose multiplicative group acts fixed-point-free. Note that when $a \in GF(q)^*$, we obtain a kernel subgroup of order $q-1$. Hence, the associated lifted spread produces a mixed subgeometry partition. Applying the reverse construction process to algebraic lifting (algebraic contraction), we have the proof to the theorem. □

REMARK 93.5. Note the component orbits under K are the orbits under B, hence there are exactly q^2+1 components fixed by K^* and $(q^4-q^2)/(q+1) = q^2(q-1)$, $PG(3, q)$'s. Hence, there is a mixed subgeometry partition in $PG(3, q^2)$ of q^2+1 $PG(1, q^2)$'s and $q^2(q-1)$ $PG(3, q)$'s. Therefore, there is a subgeometry partition in $PG(3, q^2)$ with $q^2 + 1$, $PG(1, q)$'s and $q^2(q-1)$ $PG(3, q)$'s which geometrically lifts to the spread algebraically lifted from the spread in question in $PG(3, q)$.

It is also possible to characterize our algebraically lifted spreads using mixed subgeometry partitions.

THEOREM 93.6. (Jha and Johnson [580]). Assume that $PG(3, q^2)$ admits q^2 mixed subgeometry partitions $\mathcal{M}_i$, $i = 1, 2, \ldots, q^2$, all with $q^2 + 1$ $PG(1, q^2)$'s and with $q^2(q-1)$ $PG(3, q)$'s that pairwise share precisely one common $PG(1, q^2)$ and no common $PG(3, q)$'s. Furthermore, assume that each of these subgeometry partitions admits a collineation group H_i of order $q+1$, which fixes each $PG(3, q)$ and has orbits of length $q+1$ on the $(q^4 - 1)/(q - 1)$ points of each $PG(3, q)$. H_i also fixes all $PG(1, q^2)$'s, fixes two points of each and has orbits of length $q+1$ on the remaining points of each $PG(1, q^2)$.

(1) If the set of mixed subgeometry partitions $\mathcal{M}_i$, for $i = 1, 2, \ldots, q^2$, all geometrically lift to the same translation plane π of order q^4, then π is a translation

plane that has been algebraically lifted from a translation plane with spread in $PG(3, q)$.

(2) Conversely, an algebraically lifted plane from a plane with spread in $PG(3, q)$ produces q^2 mixed subgeometry partitions of the type listed in part (1).

CHAPTER 94

Quasi-Subgeometry Partitions.

This chapter discusses generalizations of subgeometry partitions of projective spaces. In Chapter 90, it was established that there is an equivalence between what is called 'spread retraction' and geometric lifting. We recall that in 1976, A.A. Bruen and J.A. Thas [**198**] showed that it was possible to find partitions of the points of $PG(2s - 1, q^2)$ by projective subgeometries isomorphic to $PG(s - 1, q^2)$'s and $PG(2s - 1, q)$'s. These might be called 'mixed subgeometry partitions'. Bruen and Thas showed that there is an associated translation plane of order q^{2s} and kernel containing $GF(q)$. There is another construction using Segre varieties given in Hirschfeld and Thas [**529**] which generalizes and includes Baer subgeometry partitions of $PG(2s, q^2)$ by $PG(2s, q)$'s. In this latter case, there is an associated translation plane of order q^{2s+1} and kernel containing $GF(q)$.

We pointed out that in Johnson [**755**], there is an interpretation of the above construction from the viewpoint of the translation plane. That is, starting with a translation plane of a certain type, a 'retraction' method is possible which reverses the construction and produces either a mixed or a Baer subgeometry partition of a projective space. The main consideration is that a group isomorphic to $GF(q^2)^*$ acts as a collineation group of the translation plane.

It is possible to generalize these constructions several ways. First of all, finiteness is not required. Second, to apply the group action more generally, we define what we call a 'quasi-subgeometry' of a projective space to realize that certain spreads produce 'quasi-subgeometry partitions' of projective spaces. Since we are considering only finite projective geometry, we could have a projective space isomorphic to $PG(2s - 1, q^d)$ partitioned by quasi-subgeometries isomorphic to $PG(ds/e - 1, q^e)$ for various divisors e of d. Third, it is realized that subgeometry or quasi-subgeometry partitions of a projective space need not actually produce a spread but do produce a 'generalized spread'. Furthermore, a generalized spread admitting a fixed-point-free field acting as a collineation group produces a quasi-subgeometry partition. Hence, it is possible to generalize all of this to the consideration of generalized spreads and ask how these might produce partitions of projective spaces, again considering all of this only in the finite case.

As mentioned, the fundamental device is the consideration of the nature of collineation groups acting on the generalized spread $\mathcal{S}$ that arise from extension fields F of K. Such groups define within the vector space what we call 'F_L-fans'. These fans when considered projectively are the quasi-subgeometries in question. These also have been discussed in Chapter 16 on generalized André spreads, in the finite case.

Fans are only considered from the vector space point of view and we are interested in the associated quasi-subgeometry partitions. Hence, we form a generalization and integration of the André and Bruck–Bose approaches to the study of 'spreads'.

We then have a hybrid of the vector space and projective space variations which allows a complete generalization of the theory of subgeometry partitions of projective spaces and their associated translation planes to quasi-subgeometry partitions and associated generalized spreads.

Also included is a discussion a theory of subgeometry partitions of $PG(ds/2 - 1, q^2)$ by subgeometries isomorphic to $PG(s/2 - 1, q^2)$ and $PG(s - 1, q)$ if s is even, or by $PG(s - 1, q)$'s if s is odd. This is further generalized to a consideration of arbitrary quasi-subgeometry partitions of $PG(ds/e - 1, q^e)$ by appropriate quasi-subgeometries.

In terms of examples in the finite case, it is possible to construct a variety of examples of finite quasi-subgeometry partitions from generalized André planes, using these ideas.

94.1. Subgeometries.

If Σ is a finite projective space and Π is a subset of points of Σ such that if a line ℓ of Σ intersects Π in at least two points A and B then we define a 'line' of Π to be the set of points $\ell \cap \Pi$. If the points of Π and the 'lines' induced from lines of Σ form a projective space, we say that Π is a 'subgeometry' of Σ. If Σ is a projective space over a finite field F, it is not completely clear when Π is a projective space over a field T that T may be taken as a subfield of F. However, all of the subgeometries that have been considered thus far are of this type. Furthermore, the subgeometries studied here will be assumed to be of this type. Hence, to be clear, we formulate our definition of subgeometry as follows.

DEFINITION 94.1. Let Σ be a projective space isomorphic to $PG(V^-, F)$, where F is a finite field. If Π is a subgeometry Σ that arises from a subskewfield T of F, we shall say that Π is an 'induced' subgeometry.

REMARK 94.2. (1) Let V be an F-vector space such that the points of Σ are 1-dimensional F-subspaces and lines of Σ are 2-dimensional F-subspaces. If Π is an induced subgeometry isomorphic to $PG(Z^-, T)$ then the points of Π are assumed to be exactly those 1-dimensional F-subspaces $\langle z \rangle_F$ with a generator a non-zero vector z of Z considered as a subspace of V over T and T is a subskewfield of T.

(2) In the following, any reference to 'subgeometries' will intrinsically mean 'induced subgeometries'.

Considering subgeometries in the above manner, there is a natural generalization to what we shall call 'quasi-subgeometries'.

DEFINITION 94.3. A 'quasi-subgeometry' of a projective space is a subset of points that can be made into a projective space such that the lines of the projective set are subsets of lines of the projective space.

If Σ is a projective space isomorphic to $PG(V^-, F)$, where F is a finite field and Π is a subset of Σ isomorphic to $PG(Z^-, T)$, where T is a subfield of F, if the points of Π are induced in the same manner as in the above definition of subgeometry and 'lines' of Π as a $PG(Z^-, T)$ are subsets of lines of Σ, we shall say that Π is an 'induced quasi-subgeometry'.

In the following, any reference to a 'quasi-subgeometry' shall intrinsically mean an induced quasi-subgeometry.

REMARK 94.4. In either a subgeometry or a quasi-subgeometry, we also assume that the 'lines' of the structure are induced from the set of 1-dimensional F-subspaces that lie in a 2-dimensional F-subspace and are generated by vectors of Z, (see the use of Z in the previous paragraphs).

DEFINITION 94.5. A partition of a projective space by quasi-subgeometries is called a 'quasi-subgeometry partition'. If all of the quasi-subgeometries are subgeometries, the partition is called a 'subgeometry partition'.

What we are trying to accomplish is to construct a variety of partitions of projective geometries by either subgeometries or by quasi-subgeometries.

We have seen that if the subgeometries or quasi-subgeometries are all what might be called 'half-dimensional' then the ideas of 'Bruck–Bose is André at infinity' show that there is a corresponding translation plane. However, if the quasi-subgeometries involved in a partition could be essentially chosen from any set of projective geometries, it is not at all clear that there are connections between vector space spreads and quasi-subgeometry partitions or even if there are spreads producing translation planes. In order to see what sorts of vector space spreads or generalizations of these might conceivably correspond to quasi-subgeometry partitions, we define and study 'fans' first in vector space spreads and then more generally in 'generalized spreads'. We note that all of the quasi-subgeometries that we obtain from 'fans' are induced quasi-subgeometries.

94.2. Fans.

In this section, we consider that K is a finite field. The reader is directed to Chapter 16 on fundamentals of André and generalized André planes, for example the definition of 'fan' is given there.

Let V be a K-vector space. Let D be a finite field extension of K that acts on a generalized partial spread $\mathcal{P}$ of V. For any component L of $\mathcal{P}$, then $D_L^* \cup \{0\} = D_L$ is a subfield of D since L is a K-subspace. Furthermore, L is a D_L-vector space if D^* is fixed-point-free (no non-identity element fixes a non-zero vector).

DEFINITION 94.6. If D is a 2-dimensional field extension of K, we say that the associated fan is a 'quadratic fan' provided the stabilizer of a component is not D, that the fan is not simply a single component.

LEMMA 94.7. Every quadratic fan is a K-regulus in some associated projective space over K.

94.2.1. Folding The Fan. Our definition of a 'fan' is an orbit of components under a group D^*. These components share the zero vector and are otherwise disjoint as sets. When we consider a 'folding' of a fan, we basically fold all of the components into a single one in the orbit and define a projective space arising from the fold as follows:

94.2.1.1. *The Incidence Geometry.* Let $O(L) = \eta$ be an D_L-fan, for L a component and let the stabilizer of L in D^* define a field D_L. Then, by taking the lattice of D-subspaces, we obtain a projective space $PG(L^-, D_L)$. Since D acts on the vector space V, V considered over D produces $PG(V^-, D)$ by the lattice definition. The goal is to see that $PG(L^-, D_L)$ has a subset $G(\eta)$ of points and a set of 'subsets' of lines of $PG(V^-, D)$ that forms a projective space $PG(L^-, D_L)$.

It turns out that this provides a quasi-subgeometry in $PG(V^-, D)$. To begin with, we assume that $D_L = K$ and produce a subgeometry. After this, we realize that the arguments are valid more generally and produce quasi-subgeometries.

So, for the moment, let $D_L - K$.

Hence, we define an incidence geometry $G(\eta)$ as follows: The 'points' are those D^*-orbits union the zero vector of vectors that lie in $O(L)$. Given two non-zero vectors u and v in different D^*-orbits that lie components on $O(L)$ note that the 2-dimensional D-subspace $\langle u, v \rangle_D$ is the same subspace when u and v are considered on the same component L^*. Within $\langle u, v \rangle_D$, a 'line' will be the set of 'points' as D^*-orbits arising from vectors in $\langle u, v \rangle_K$, where u and v are vectors on a component L^* of $O(L)$ that are in different D^*-orbits. The question is $\langle u, v \rangle_D \cap G(\eta) = \{ \langle \alpha_o v + \beta_o v \rangle_D ; \alpha_o, \beta_o \in K \}$?

We now utilize the model that a quadratic fan is a K-regulus.

LEMMA 94.8. Let Z be a vector space and R a regulus in $PG(Z^-, K)$ for K a field. Then, R is a partial spread and we let $N(R)$ denote the associated regulus net in the associated K-vector space Z.

Then

(a) Any pair of subplanes π_o and π_1 that share the zero vector can be embedded in a unique derivable subnet $\mathcal{D}_{\pi_o, \pi_1}$ of $N(R)$ that contains π_o and π_1.

(b) Furthermore, the only subplanes incident with the zero vector that non-trivially intersect $\pi_o \oplus \pi_1$ are in the derivable subnet $\mathcal{D}_{\pi_o, \pi_1}$.

LEMMA 94.9. If D acts on the quadratic fan, then

$$\langle u, v \rangle_D \cap G(\eta) = \{ \langle \alpha_o u + \beta_o v \rangle_D ; \alpha_o, \beta_o \in K \}.$$

Hence, we also obtain:

LEMMA 94.10. Given two distinct points A and B of $G(\eta)$, there is a unique line AB incident with A and B. This line is a subline of the line in $PG(V^-, D)$ obtained from the 2-dimensional D-space $\langle A, B \rangle_D$.

Furthermore, for any two distinct points C, D of AB, $AB = CD$.

We wish to show that $G(\eta)$ is a projective space. Since the 'point set' corresponds to the set of 1-dimensional D_L-vector subspaces on L, and the line set corresponds to the set of 2-dimensional D_L-subspaces on L, we may define the associated subgeometry based on the lattice of D_L-subspaces to realize this as a projective space isomorphic to $PG(L^-, D_L)$.

THEOREM 94.11. (Johnson [**762**]). Let η be an K-fan (a quadratic fan). Define an incidence geometry $G(\eta)$ as follows:

(i) 'Points' of $G(\eta)$ are the point-orbits of D acting on the partial spread $\mathcal{P}$ and

(ii) 'Lines' are defined by pairs of distinct 'points' A and B as follows: If $A = uD$ and $B = vD$, for u, v on same component of $O(L)$, then the line AB is the set of points arising as D-orbits of vectors in $\langle u, v \rangle_{D_L}$.

Then $G(\eta)$ is a projective space isomorphic to $PG(L^-, D_L)$ that may be considered a subgeometry of $PG(V^-, D)$.

The main result of Johnson on quadratic fans is that a vector space generalized spread that is the union of fans produces a subgeometry partition of a projective space.

THEOREM 94.12. (Johnson [**762**]). Let $\mathcal{Z}$ be a generalized spread whose components are K-subspaces with underlying vector space V over a field K. Let D be a quadratic field extension of K. Assume that D acts on $\mathcal{Z}$.

Let the orbits of D^* be denoted by $O(L)$ where L is a component of the spread of π.

For $O(L)$, let D_L denote the subfield of D leaving L invariant. If the generalized spread $\mathcal{Z}$ for π is

$$\bigcup \{ O(L); \ L \in \mathcal{Z} \},$$

let L_i be a representative for the orbit $O(L_i) = \eta_i$, for $i \in \Lambda$, an index set for the set of orbits.

Then (a) there is a subgeometry partition $\mathrm{Sub}(\mathcal{Z})$ of $PG(V^-, D)$ defined as follows:

$$\mathrm{Sub}(\mathcal{Z}) = \big\{ G(\eta_i) \simeq PG(L_i^-, D_{L_i}); \ i \in \Lambda \big\}.$$

That is, a generalized spread $\mathcal{Z}$ that is a union of D_{L_i}-fans produces a subgeometry partition of $PG(V^-, D)$ by $PG(L_i^-, D_{Li})$'s for all $i \in \Lambda$, where D_{L_i} is either K or D. (b) Each orbit of length > 1, $O(L_i)$ under D^* is a K-regulus in some projective space $PG(V^{L_i-}, K)$. Note that we are not assuming that the dimensions of L_i and L_j are equal over K.

We now take up the more general problem.

94.3. Quasi-Subgeometries.

In the general situation, a D_L-fan produces a projective geometry that sits within $PG(V^-, D)$ by defining lines to arise from the D-one dimensional subspaces originating from a vector on L; that is by

$$\{ \langle \alpha_o u + \beta_o v \rangle_D \ ; \ \alpha_o, \beta_o \in K \} \subseteq \langle u, v \rangle_D \cap G(\eta).$$

However, although this would indeed produce a projective geometry isomorphic to $PG(W_o^-, D_L)$ that sits in $PG(V^-, D)$, it may not actually be the complete set of intersection points and hence would not properly be considered a 'subgeometry'. We have defined this embedded projective geometry as a 'quasi-subgeometry', so then from any spread that is a union of D_{L_i}-fans, we would obtain a partition of $PG(V^-, D)$ by quasi-subgeometries isomorphic to $PG(W_o^-, D_{L_i})$ for $i \in \Phi$.

The analogous theorems for quasi-subgeometry partitions are as follows and note that a re-reading of the previous results reveals that the proofs are essentially the same as for the subgeometry partitions, except that a 'subline' given may not be the intersection of the point set by a given line, although it will be a subset of a line.

THEOREM 94.13. (Johnson [**762**]). Let η be a D_L-fan. Define an incidence geometry $G(\eta)$ as follows:

(i) 'Points' of $G(\eta)$ are the point-orbits of D acting on the generalized partial spread $\mathcal{P}$ and

(ii) 'Lines' are defined by pairs of distinct 'points' A and B as follows: If $A = uD$ and $B = vD$, for u, v on same component of $O(L)$, then the line AB is the set of points arising as D-orbits of vectors in $\langle u, v \rangle_{D_L}$ of the form $\langle \alpha_o u + \beta_o v \rangle_D$; $\alpha_o, \beta_o \in D_L$.

Then $G(\eta)$ is a projective space isomorphic to $PG(L^-, D_L)$ that may be considered a quasi-subgeometry of $PG(V^-, D)$.

THEOREM 94.14. (Johnson [**762**]). Let $\mathcal{Z}$ be a generalized spread $\mathcal{S}$ with underlying vector space V with components K-spaces where K is a field. Let D be a field acting on $\mathcal{Z}$.

Let the orbits of D^* be denoted by $O(L)$ where L is a component of the generalized spread.

For $O(L)$, let D_L denote the subfield of D leaving L invariant. If the generalized spread $\mathcal{Z}$ for π is

$$\bigcup\{O(L);\, L \in \mathcal{Z}\},$$

let L_i be a representative for the orbit $O(L_i) = \eta_i$, for $i \in \Lambda$, an index set for the set of orbits.

Then there is a quasi-subgeometry partition $\mathrm{Sub}(\mathcal{Z})$ of $PG(V^-, D)$ defined as follows:

$$\mathrm{Sub}(\mathcal{Z}) = \left\{ G(\eta_i) \simeq PG(L_i^-, D_{L_i});\, i \in \Lambda \right\}$$

That is, a generalized spread $\mathcal{Z}$ that is a union of D_{L_i}-fans produces a quasi-subgeometry partition of $PG(V^-, D)$ by $PG(L_i^-, D_{Li})$'s for all $i \in \Lambda$.

94.3.1. Unfolding the Fan. We think of a particular quasi-subgeometry as a 'projective' folded fan and argue that this can be unfolded into an associated vector space fan.

Let V be a K-vector space, for K a field and D be an extension field of K. Let L be a K-vector subspace such that V is a D-space and L is a D_L-space. Hence, assume that we have a (induced) quasi-subgeometry Σ_o isomorphic to $PG(L^-, D_L)$, where Σ_o is in Σ_1, and Σ_1 is the lattice of subspaces $PG(V^-, D)$. We want to realize this quasi-subgeometry as a D_L-fan in the vector space V over D. Since we assume that D_L is a subfield of D, then V is also an D_L-vector space. We also assume that D_L contains the field K.

LEMMA 94.15. Under the above assumptions and notation, there is a vector space V^+ over D defining a projective space Σ_2 as $PG(V^{+-}, D)$ so that

$$\Sigma_o \subseteq \Sigma_1 \subseteq \Sigma_2.$$

Note that Σ_o is a quasi-subgeometry of Σ_1 and Σ_1 is a projective subspace of Σ_2.

PROOF. Let Q be any 1-dimensional D-vector space, and consider the external direct sum $V \oplus Q = V^+$ as a D-vector space. Let Σ_2 denote the lattice of D-subspaces of V^+. In this way, we have embedded Σ_1 in the projective geometry Σ_2, which is $PG(V^{+-}, D)$. And, as above, Σ_o may be considered isomorphic to the lattice of subspaces of the vector space W_o over D_L:

$$PG(L^-, D_L) \simeq \Sigma_o,\ PG(V^-, D) \simeq \Sigma_1, PG(V^{+-}, D) \simeq \Sigma_2,\ \text{where}$$
$$\Sigma_o \subseteq \Sigma_1 \subseteq \Sigma_2.$$

$\square$

We now make explicit a coordinate description of the above embedding. Note that this is the reflection of our previous discussion on 'Bruck–Bose is André at infinity', found in Chapter 4 on foundations. We restate our Lemma 4.36 in terms of the field D.

LEMMA 94.16. (1) $PG(V^{+-}, D)$ is isomorphic to $PG(V, D)$.
(2) Bases may be chosen for V and V^+ so that

    (a) vectors of V^+ may be represented in the form:
        (i) $((x_i), x_\infty)$ for all $i \in \rho$, ρ an index set, where $x_i, x_\infty \in D$,
        (ii) $((x_i), 0)$ for all $i \in \rho$, ρ an index set, where $x_i \in D$ represent vectors
          in V,
    (b) regarding two non-zero 'tuples' above to be equal if and only if they are
        K-scalar multiples of each other produces the 'homogeneous coordinates'
        of the associated projective spaces $PG(V^-, D)$ and $PG(V^{+-}, D)$, and
    (c) $((x_i), 1)$ for all $i \in \rho$, ρ an index set, where $x_i \in D$ represents homogeneous
        coordinates for a subset isomorphic to $AG(V, D)$.

    (3) Furthermore, we may consider $PG(V, D)$ as the adjunction of $PG(V^-, D)$
as the hyperplane at infinity of $AG(V, D)$.

    What we try now to do is to integrate the idea that Σ_o is isomorphic to a
projective space $PG(L^-, D_L)$ while at the same time being embedded into Σ_1 the
projective space $PG(V^-, D)$ as a quasi-subgeometry. This is somewhat problematic
since we need to think simultaneously of a point of Σ_o being a 1-dimensional D_L-
vector subspace as well as a 1-dimensional D-vector subspace and we need to realize
that Σ_o is only isomorphic to $PG(L^-, D_L)$ as a quasi-subgeometry of $PG(V^-, D)$.

    LEMMA 94.17. (1) (a) Using the representation of the previous lemma, a point
of Σ_o may be represented by either a 1-dimensional D-subspace generated by
$((x_i), 0)$ or defines a 'point' of Σ_1 having homogeneous coordinate $((x_i), 1)$.

    (b) We adopt the notation $\langle (v, 0) \rangle$ for some v of L, for the first situation
and note that D-scalar multiplication may be defined as

$$\alpha(v, 0) - (\alpha v, 0).$$

    (c) Note that $(\alpha v, 1)$, for any $\alpha \in D$ and $v \in L$, is a 'point' of Σ_1 and Σ_2.
    (d) $\alpha v = \beta w$ for v, w in $L - \{0\}$ and $\alpha, \beta \in D$ if and only if $\alpha^{-1} \beta v = w$
implying that $\alpha^{-1} \beta \in D_L$.

    REMARK 94.18. We recall that the lattice $PG(L^-, D_L)$ corresponds to Σ_o .
Hence, when we consider the points of Σ_o defined using the notation $(\alpha v, x_\infty)$, for
$v \in L$, we are using the above setup. Since Σ_o is a subset of points of $PG(V^-, D)$,
there is a preimage set Σ_o^+ of 1-dimensional D-subspaces; a subset of vectors of V.
In our notation, we have a point not in $PG(V^-, D)$ of the form $(\alpha v, 1)$ for all $v \in L$
and for all $\alpha \in D$.

    We define some sets of points which will become projective quasi-subgeometries
of $PG(V^{+-}, D)$:
    Let

$$\Sigma_\alpha = \{ (\alpha v, 1), \langle v, 0 \rangle_D \, ; \, v \in L \} \text{ for fixed non-zero } \alpha \in D \}.$$

    (1) (a) Σ_α is a projective quasi-subgeometry of $PG(V^{+-}, D)$ that is isomorphic
to $PG(L^{+-}, D_L) \simeq PG(L, D_L)$.

    (b) Under the above structure of 'lines', Σ_α induces on the point set Σ_o
an isomorphic (as projective spaces) quasi-subgeometry.

    (c) $\bigcap \Sigma_\alpha = \Sigma_o = \{ \langle v, 0 \rangle_D \, ; \, v \in L \}$.
    (2) Let

$$\widehat{V} = \{ (w, 1); \, w \in V \} .$$

Then by defining $\beta(w, 1) = (\beta w, 1)$, $\widehat{V}$ is an D-vector space isomorphic to V over
D.

Also, Σ_α^- may be considered a D_L-vector subspace over D_L, that admits the scalar multiplication over $D_L \subseteq D$.

(3) Let

$$\Sigma_o^+ = \{(\alpha v, 1), \alpha \in D \text{ and } v \in L\} \subseteq \widehat{V}.$$

This set of vectors produces $\Sigma_o = PG(L^-, D_L)$ when considering Σ_o^+ as the lattice of D-subspaces of the subset Σ_o^+. Let A and B be distinct 1-dimensional D-subspaces within Σ_o^+. Hence, the 'line' of Σ_o AB containing A and B must be realized within the 2-dimensional D-subspace, $\langle A, B \rangle_D$ and must abstractly correspond to a 2-dimensional D_L-subspace of W_o. If $a, b \in L$, for $a \in A$ and $b \in B$ then the 'points' of the 'line' in the quasi-subgeometry are of the form $\langle \alpha_o u + \beta_o v \rangle_D$ for all α_o, β_o in D_L, for not both α_o and β_o zero.

Let

$$\Sigma_\alpha^- = \{ (\alpha v, 1); \text{ for fixed non-zero } \alpha \in D \text{ and for all } v \in L \}.$$

Then Σ_α^- may be made into a D_L-vector space by defining $\beta(\alpha v, 1) = (\beta \alpha v, 1)$ where $\beta \in D_L$ and $v \in L$. Under our notation, L represents (or becomes) a subset of V as Σ_1^-.

Similarly, V is a subspace of V^+ and L is a subset of V, so a subset of V^+.

94.3.2. The Associated Affine Spaces $\Sigma_\alpha^- \simeq AG(L, D_L)$. Consider the group D^*/D_L^* and let $\mathcal{C}_L = \{ \alpha_j; j \in \Omega \}$ be a coset representative set for D_L^*. Now fix $\alpha \in D - \{0\}$ and consider

$$\Sigma_\alpha^- = \{ (\alpha v, 1); v \in L \}.$$

The previous lemma shows that in our defined action of D on $\widehat{V}$, we may regard that D_L^* leaves each Σ_α^- invariant.

As an affine space, we suppress the '1' and write $(\alpha v, 1)$ as $[v]_\alpha$ to place these elements back in V (or an isomorphic copy of V).

If β_o is in D_L, then more formally, we have $\beta_o(\alpha v, 1) = (\beta_o \alpha v, 1) = (\alpha \beta_o v, 1)$. Hence, $\beta_o[v]_\alpha = [\beta_o v]_\alpha$ for all $\beta_o \in D_L$ and for all $v \in L$.

To emphasize the previous, we repeat part of the statement of the previous remark.

LEMMA 94.19. Σ_α^- is a subset of $AG(V, D)$ which may be considered isomorphic to $AG(L, D_L)$. Furthermore, the associated vector space over D_L is isomorphic to L. In addition, Σ_α is a subset of Σ_2 isomorphic to $PG(L, D_L)$.

(1) Let α_j and α_k be distinct elements in $\mathcal{C}_L$, then

$$\Sigma_{\alpha_j}^{-} \cap \Sigma_{\alpha_k}^{-} = (0, 1).$$

(2) $\left\{ \Sigma_{\alpha_j}^-; j \in \Omega \right\}$ is a set of mutually isomorphic and pairwise disjoint D_L-vector spaces.

PROOF. In this context, Σ_α^- is clearly D_L-isomorphic to L, by the mapping that maps

$$[v]_\alpha \longmapsto v.$$

Thus, we have Σ_α^- is an affine geometry isomorphic to $AG(L, D_L)$. Hence, Σ_α is a subspace of Σ_2 isomorphic to $PG(L, D_L)$.

Now assume that $(\alpha v, 1) = (\beta u, 1) \in \Sigma_\alpha^- \cap \Sigma_\beta^-$ if and only if $\alpha^{-1}\beta u = v$, and if both u and v are non-zero then this implies that $\alpha^{-1}\beta \in D_L$. Thus, two distinct elements of a coset representative set define affine spaces that are mutually disjoint as vector spaces. This completes the proof of the lemma. □

94.3.3. D Acts on the Affine Spaces Σ_α^- as D^*/D_L^*. We have previously defined an action of D^* so that:

LEMMA 94.20. (1) D^* acts transitively on $\left\{ \Sigma_{\alpha_j}^-; j \in \Omega \right\}$ and induces faithfully the group D^*/D_L^* on this set.

(2) D^* fixes $(0,1)$ and fixes Σ_o pointwise.

(3) D^* acts as a collineation group of the affine space $AG(V, D)$ that fixes the zero vector of the associated vector space V and acts as a natural scalar group.

Now realize this vector space as over K. Note that the Σ_α^-'s are naturally D_L-subspaces under the scalar action of D restricted to D_L, since each is fixed by D_L.

DEFINITION 94.21. Suppose that $V = W_o \oplus W_o$. A quasi-subgeometry of $PG(V^-, D)$ isomorphic to $PG(W_o^-, D_L)$ shall be said to have the 'congruence generating property' if and only if as vector spaces,

$$\Sigma_{\alpha_j}^- \oplus \Sigma_{\alpha_k}^- = V$$

for any distinct pair of elements $\alpha_j, \alpha_k \in \mathcal{C}_L$.

Note that there is a unique way to consider the direct sum that is dependent only on the point sets involved.

The main result from Johnson related to quasi-subgeometries is as follows:

THEOREM 94.22. (Johnson [**762**]). Let V be a vector space over a field K and assume there are field extensions D_{L_i} and D such that $K \subseteq D_{L_i} \subseteq D$, for $i \in \Lambda$, an index set, and L_i K-vector subspaces of V.

Let $\mathcal{P}$ be a quasi-subgeometry partition of $PG(V^-, D)$ by quasi-subgeometries $\mathcal{G}_i$ isomorphic to $PG(L_i^-, D_{L_i})$, so that

$$\mathcal{P} = \bigcup_{i \in \Lambda} \mathcal{G}_i.$$

(1) Then, by 'unfolding', there is a corresponding generalized spread $\mathcal{S}_\mathcal{P}$ of the vector space V such that D acts on V and $\mathcal{S}_\mathcal{P}$ consists of D_{L_i}-fans for $i \in \Lambda$.

(2) Conversely, the associated generalized spread admitting D produces, by 'folding', a quasi-subgeometry partition of $PG(V^-, D)$ consisting of $PG(L_i^-, D_{L_i})$'s on the analogous point sets of the partition.

(3) The constructed generalized spread is a spread if and only if the quasi-subgeometries generate K-subspaces that pairwise have the congruence generating property.

This is true for example if $L_i' \oplus L_j' = V$, for all distinct $i, j \in \Lambda$, for all subspaces L_i' of $O(L_i)$, the D^*-orbits.

(4) If D is a 2-dimensional field extension of K, then the quasi-subgeometries are subgeometries and there is a bijective correspondence between generalized spreads admitting D and subgeometry partitions of $PG(V^-, D)$ by $PG(L_i^-, K)$'s and $PG(L_j^-, D)$'s. The generalized spreads are unions of K-reguli and subspaces fixed by D^*.

94.4. Fans in Spreads.

In the previous sections, we have developed the connections with fans and quasi-subgeometries. Although our intention in this text is to consider finite geometries, all of these results may be viewed more generally. In the finite case, we consider the study of finite vector spaces of dimension $2ds$ over K isomorphic to $GF(q)$ that admit a fixed-point-free field group $D^* = F_d^{'*}$ of order $q^d - 1$ (i.e., $F_d = F_d^* \cup \{0\}$ is a field) that contains the field group K^* or those that could admit $D^* = F_{2d}^*$ isomorphic to $GF(q^{2d})^*$. Note that for spreads of finite dimensional vector spaces V, the dimension over a subkernel field is necessarily even dimension, say $2ds$ dimension. In this setting, the order of the associated translation plane is q^{ds}. We are interested in essentially two types of fields 'acting' as collineation groups of the translation plane. First, we consider whether it is possible that the field D could fix a component of the spread. Since a component has $q^{ds} - 1$ non-zero vectors and $D \simeq GF(q^w)$ is fixed point free, then w must divide ds in this context. So, in such a setting we take $w = d$ without loss of generality. On the other hand, assume that it would be required that D never fix a component. Since V is a $2ds$ dimensional $GF(q)$-vector space, then V would be a $2ds/w$-dimensional $GF(q^w)$-vector space. Hence, w divides $2ds$ and if it does not divide ds, then without loss of generality, we consider $w = 2d$. Hence, the fields in question could be isomorphic to $GF(q^d)$ or $GF(q^{2d})$, where $2d$ divides $2ds$ but not ds, so we assume that s is odd in the latter case to avoid reduction to the previous situation.

Hence, the q^e-fans that are under consideration have degree either $(q^d - 1)/(q^e - 1)$ where e divides d or degree $(q^{2d} - 1)/(q^e - 1)$ where e divides $2d$.

To be specific, we emphasize the definition of a specific type of fan.

DEFINITION 94.23. A 'q^e-fan' in a $2ds$-dimensional vector space over K isomorphic to $GF(q)$ is a set of $(q^w - 1)/(q^e - 1)$ mutually disjoint K-subspaces of dimension ds that are in an orbit under a field group F_w^* of order $(q^d - 1)$, where F_w contains K and such that F_w^* is fixed-point-free, and where $w = d$ or $2d$ and in the latter case s is odd.

We obtain the following corollaries.

COROLLARY 94.24. Assume a partial spread $\mathcal{Z}$ of order q^{ds} in a vector space of dimension $2ds$ over K isomorphic to $GF(q)$ admits a fixed-point-free field group F_w^* of order $(q^w - 1)$ containing K^*, for $w = d$ or $2d$ and s is odd if $w = 2d$. Then any component orbit Γ of length $(q^w - 1)/(q^e - 1)$ (a 'q^e-fan'), for e a divisor of d, produces a quasi-subgeometry isomorphic to a $PG(ds/e-1, q^e)$ in the corresponding projective space $PG(2ds/w - 1, q^w)$, considered as the lattice of F_w-subspaces of V.

COROLLARY 94.25. Let π be a translation plane of order q^{ds} and kernel containing K isomorphic to $GF(q)$ that admits a fixed-point-free field group F_w^* of order $(q^w - 1)$ containing K^*, where $w = d$ or $2d$ and s is odd if $w = 2d$.

(1) Then, for any component orbit Γ, there is a divisor e_Γ of w such the orbit length of Γ is $(q^w - 1)/(q^{e_L} - 1)$ so that Γ is a q^{e_Γ}-fan.

Hence, there is a quasi-subgeometry partition of $PG(2ds/w - 1, q^w)$ by quasi-subgeometries isomorphic to $PG(ds/e_\Gamma - 1, q^{e_\Gamma})$, for various divisors e_Γ of w.

(2) Conversely, for $w = d$ or $2d$, every quasi-subgeometry partition of $PG(2ds/w - 1, q^w)$ by quasi-subgeometries isomorphic to $PG(ds/f - 1, q^f)$, for

various divisors f of w, produces a translation plane of order q^{ds} and kernel containing K isomorphic to $GF(q)$, that admits a fixed-point-free field collineation group of order $q^w - 1$, F_w^* containing K^*. The projective quasi-subgeometries of type $PG(ds/f - 1, q^f)$ correspond to q^f-fans.

The main general result isolating particularly on the combinatorics of a finite spread case is now summarized as follows.

THEOREM 94.26. (Johnson [**762**]). Let π be a translation plane of order q^{ds} and kernel containing K isomorphic to $GF(q)$. Assume that there is a fixed-point-free collineation group GK^* such that GK is a field isomorphic to $GF(q^w)$, where $w = d$ or $2d$ and s is odd in the latter case. Let the set of divisors of w be $N = \{e_i; i = 1, \ldots, E\}$, including w and 1.

(1) Each component L has a unique maximal subfield $GF(q^f)$ within $GF(q^w)$, for $f = e_k$ for some k, such that L is a $GF(q^f)$-subspace. In this case, the orbit length of L under GK^* is $(q^w - 1)/(q^f - 1)$ and the orbit is a q^f-fan.

Let k_i denote the number of GK^*-orbits of components of length $(q^w - 1)/(q^{e_i} - 1)$; of q^{e_i}-fans, and let N^- denote the subset of N containing the divisors e_i used in the construction.

Then $\sum_{i=1}^{N^-} k_i (q^w - 1)/(q^{e_i} - 1) = q^{ds} + 1$.

(2) Consider the associated affine geometry $AG(2ds/w, q^w)$, embed in $PG(2ds/w, q^w)$ and let Δ denote the hyperplane at infinity isomorphic to $PG(2ds/w - 1, q^w)$. Consider an orbit of components under $GF(q^w)^*$ of length $(q^w - 1)/(q^{e_i} - 1)$; the various q^{e_i}-fans. Each q^{e_i}-fan will produce a quasi-subgeometry isomorphic to $PG(ds/e_i - 1, q^{e_i})$ in Δ. Hence, we obtain k_i such $PG(ds/e_i - 1, q^{e_i})$'s.

(3) If $\sum_{i=1}^{N^-} k_i (q^w - 1)/(q^{e_i} - 1) = q^{ds} + 1$, then there is a corresponding quasi-subgeometry partition of $PG(2ds/w - 1, q^w)$ by k_i $PG(ds/e_i - 1, q^{e_i})$'s for $i = 1, 2, \ldots, N$. We call this a partition of type $(k_1, \ldots, k_{N^-})$.

(4) If $PG(2ds/w - 1, q^w)$ admits a quasi-subgeometry partition by k_i $PG(ds/e_i - 1, q^{e_i})$'s for $i = 1, 2 \ldots, N^-$, for $e_i \in N^- \subseteq N$ and $N = \{e_i; i = 1, 2, \ldots, E\}$ is the set of all divisors of d, then necessarily

$$\sum_{i=1}^{N^-} k_i (q^w - 1)/(q^{e_i} - 1) = (q^{ds} + 1).$$

Furthermore, there is an associated translation plane of order q^{ds} and kernel containing $GF(q)$ admitting a collineation group GK^* such that the union with the zero mapping is a field isomorphic to $GF(q^w)$ that is fixed-point-free and there is a set of k_i mutually disjoint q^{e_i}-fans whose union is the spread for the translation plane.

We now show that there are a great variety of examples of quasi-subgeometry partitions based on net replacement. All of the associated translation planes that we shall consider are generalized André planes. It is pointed out that in the cases in question, in the following, the group acting is considered always of the type $GF(q^d)^*$.

THEOREM 94.27. (Johnson [**762**]). Let Σ be a Desarguesian affine plane of order q^{ds}. For each of the $q - 1$, André nets A_α, choose a divisor e_α of d (these divisors can possibly be equal and/or possibly equal to 1 or d). For each q-André net A_α, there is a corresponding set of k_α q^{e_α}-fans. Form the corresponding André plane $\Sigma_{(e_\alpha f_\alpha \ \forall \alpha \in GF(q))}$ obtained with spread:

$$y = x^{q^{e_\alpha f_\alpha}} m \text{ for } m^{(q^{ds}-1)/(q-1)} = \alpha, x = 0, y = 0; m \in GF(q^{ds}), (e_\alpha f_\alpha, d) = e_\alpha.$$

Then the spread $\Sigma_{(e_\alpha f_\alpha \ \forall \alpha \in GF(q))}$ is a union of $\sum_{\alpha=1}^{q-1} \left(k_{c_\alpha} = \frac{(q^{ds}-1)}{(q-1)} \frac{(q^{e_\alpha}-1)}{(q^d-1)} \right)$ q^{e_α}-fans, together with two q^d-fans $x = 0$ and $y = 0$.

There is a corresponding quasi-subgeometry partition of $PG(2s-1, q^d)$ of k_{e_α} quasi-subgeometries isomorphic to $PG(ds/e_\alpha - 1, q^{e_\alpha})$, and two $PG(s-1, q^d)$ (corresponding to $x = 0$ and $y = 0$).

94.4.1. Multiple q^e-André Replacement. Actually, a refinement of the above will produce a more general variety of partitions; however, the associated translation planes are not necessarily André planes but are certainly generalized André planes. For example, we may partition any q-André net of cardinality $(q^{ds} - 1)/(q - 1)$ into $(q^e - 1)/(q - 1)$ q^e-André nets of cardinality $(q^{ds} - 1)/(q^e - 1)$. For the q^e-André nets, the basic replacement components must be of the form $y = x^{q^{ef}} m$; however, we may choose the f's independently of each other for the $(q^e - 1)/(q - 1)$ q^e-André nets. We then may choose another divisor e_1 of d/e to produce a set of $\frac{(q^{ds}-1)}{(q^e-1)} \frac{(q^{ee_1}-1)}{(q^d-1)}$ q^{ee_1}-fans from each of the $(q^e - 1)/(q - 1)$ q^e-André nets. Furthermore, the partitioning into relative sized André nets can be continued. For example for any one of the q^e-André nets, choose a divisor e_1 of d/e and partition this q^e-net into a set of q^{ee_1}-André nets. The point is that we may choose each of these fans with possibly different component set configurations. Furthermore, it is possible to continue this partitioning and choice of component sets for the fans.

In this way, we obtain an explosion of possible subgeometry partitions as the number of possibilities is extremely large.

For example, we may require a q^e-fan for every divisor e of d. In other to obtain this, it suffices to require $q > N$, the number of divisors of d.

THEOREM 94.28. (Johnson [**762**]). If $q - 1 \geq$ the number of divisors (including d and 1) of d then there exists a generalized André translation plane obtained from a Desarguesian plane of order q^{ds} by multiple André replacement that produces a quasi-subgeometry partition of $PG(2s - 1, q^d)$ such that for every divisor e of d, there exists quasi-subgeometries of the partition isomorphic to $PG(ds/e - 1, q^e)$.

94.4.2. Recognition of André Type Partitions. The question remains, if we are given a quasi-subgeometry partition of $PG(2s-1, q^d)$ by quasi-subgeometries isomorphic to $PG(ds/e - 1, q^e)$ for a set of divisors e of d, when is the associated translation plane André or generalized André? Of course, in the translation plane, we can determine if the plane is André by consideration of its collineation group. However, this does not mean that the partition itself is André. Here we are intending this to mean that a quasi-subgeometry partition is obtained via q^e-fans of André type using André replacement in a Desarguesian affine plane. Note in all cases the translation planes associated are generalized André planes constructed by multiple André replacement.

DEFINITION 94.29. Any quasi-subgeometry partition obtained by using multiple André replacement shall be called 'André'.

So, we shall list with some open questions:

(1) When is a quasi-subgeometry (respectively, subgeometry) partition André?

(2) If a quasi-subgeometry (respectively, subgeometry) partition provides a generalized André translation plane, is the partition itself André?

(3) Do non-André quasi-subgeometry (respectively, subgeometry) partitions exist for $d > 2$? Note that when $d = 2$, we obtain simply a mixed partition and the q-fans become (are) K-reguli. Hence, non-André partitions certainly exist if $d = 2$.

We have not given examples of quasi-subgeometry partitions (respectively, subgeometry) corresponding to q^e-fan spreads, when $GF(q^{2d})$-acts. However, when $d = 1$, there are a variety of examples. In this case the q-fans are K-reguli and produce subgeometries.

(4) Do quasi-subgeometry partitions of $PG(s-1, q^{2d})$ exist by quasi-subgeometries isomorphic to $PG(ds/e - 1, q^e)$ for d and s odd?

Now we turn to our more general results involving finite generalized spreads.

94.5. Generalized Spreads and Replaceable Translation Sperner Spaces.

When there is a spread, the vector space is $2ds$ dimensional over K isomorphic to $GF(q)$ and the components have q^{ds} vectors each. We consider a more general situation where the vector space is tds-dimensional and we have a generalized spread whose components have q^{ds} vectors each. Thus, we consider 't-spreads of size (ds, q)'. Here, we also have a 'translation space' whose points are the vectors of V and whose lines are the translates of the components of the t-spread.

To see a simple example of such t-spreads, take any vector space V of dimension t over $GF'(q^{ds})$ and let S be the t-spread of V of $(q^{tds} - 1)/(q^{ds} - 1)$ 1-dimensional $GF(q^{ds})$-subspaces. Forming the 'translation Sperner space' by taking translates of these $(q^{tds} - 1)/(q^{ds} - 1)$ spread components, we obtain a Sperner space with q^{tds} total points and q^{ds} points per line.

Now assume that $ds = 2$, regard V as $2t$-dimensional over $GF(q)$ and take any proper subspace W of V of dimension 2 over $GF(q)$ which is not contained in a component of the t-spread. Hence, whenever W intersects a spread component, it must intersect in a 1-dimensional $GF(q)$-subspace. Consider the subspread induced on W of $(q^2 - 1)/(q - 1)$ 1-dimensional $GF(q)$-subspaces. This defines a subplane of the translation space that is Desarguesian of order q by taking the translates within W as lines of the subplane. We note that this subplane π_o cannot be considered a subspace of the associated affine space over $GF(q^2)$. Take the set of images of π_o under the $GF(q^2)^*$-scalar group to obtain a set of $(q + 1)$ subplanes π_i, for $i = 0, 1, \ldots, q$, that share the same set W^+ of $q + 1$ 1-dimensional $GF(q^2)$-spaces. Replace W^+ with the set $\{\pi_i; i = 0, 1, \ldots, q\}$.

Now consider the original t-spread components of 1-dimensional $GF(q^2)$-subspaces. If we consider these as 2-dimensional $GF(q)$-spaces, then the set of $(q+1)$-1-dimensional $GF(q)$-subspaces within each is a 2-spread of each original component. Hence, if we take $S - W^+$ as a partial $2t$-spread of V as a $2t$-dimensional $GF(q)$-subspace together with $\{\pi_i\}$, we obtain a 'new' $2t$-spread of size $(1, q)$ upon which $GF(q^2)$ acts. Then $GF(q^2)^*$ and has exactly one orbit of length $q + 1$, a q-fan and the remaining orbits are of length 1. Our previous analysis shows that we have a q-regulus in $PG(\pi_o \oplus \pi_1 - 1, q)$.

Thus, we see that we may fold the fan and produce a quasi-subgeometry partition of $PG(t-1, q^2)$ by quasi-subgeometries isomorphic to $PG(0, q^2)$ and $PG(1, q)$'s.

Such partitions are perhaps not all that interesting, but this gives a glimpse of what could occur. For example, there could be t-spreads of size (s,q) that admit a group isomorphic to $GF(q^2)$ without s necessarily equal to 2. Suppose that s is even and let $s/2 = s^*$, then, in this setting, there would be a partition of $PG(tds^* - 1, q^2)$ by either $PG(s^* - 1, q^2)$'s where the component is fixed by $GF(q^2)$ or $PG(2s^* - 1, q)$'s; that is, either q^2-fans or q-fans.

In the previous setting, $t = 2$ produces the subgeometry partitions of $PG(2s^* - 1, q^2)$ that correspond to spreads. Here, it still might be possible for a t-spread of size (ds,q) to involve a union of 'reguli'. Furthermore, we have seen that these $GF(q^2)$-orbits of ds-spaces are still covered by subplanes; that is, these are still subplane covered nets that become (are) reguli. We know that if A and B are two 1-dimensional $GF(q^2)$ subspaces that lie within $O(L)$, then

$$\langle A, B\rangle_{GF(q^2)} \cap O(L) = \langle \alpha_o A + \beta_o B\rangle_{GF(q^2)} \ \forall \alpha_o, \beta_o \in GF(q).$$

Hence, each such orbit induces a 'subgeometry' in $PG(tds^* - 1, q^2)$ isomorphic to $PG(2s^* - 1, q)$.

If ds is odd, however, then it still might be possible for $GF(q^2)^*$ to act on the t-spread, but there could be no fixed components. This could occur if $(q^{tds} - 1)/(q^{ds} - 1)$ is divisible by $(q+1)$.

THEOREM 94.30. (Johnson [**762**]). Let $\mathcal{S}$ be a t-spread of V of size (ds,q).

(1) (a) If s is even, let $s/2 = s^*$. Assume that $GF(q^2)$ 'acts' on $\mathcal{S}$ then there is a subgeometry partition of $PG(tds^* - 1, q^2)$ consisting of $PG(s^* - 1, q^2)$'s and $PG(2s^* - 1, q)$'s. If $(q+1)$ divides $(q^{tds} - 1)/(q^{ds} - 1)$, it is possible that there are no $PG(s^* - 1, q^s)$'s.

(b) If ds is odd, but t is even, let $t^* = t/2$ and assume that $GF(q^2)$ acts on $\mathcal{S}$ then there is a subgeometry partition of $PG(t^*ds - 1, q^2)$ by subgeometries isomorphic to $PG(s - 1, q)$'s.

(2) Conversely,

(a) Any subgeometry partition of $PG(tds^* - 1, q^2)$ by $PG(s^* - 1, q^2)$'s and $PG(2s^* - 1, q)$'s produces a t-spread of size (ds,q) that is a union of q^2-fans and q-fans and

(b) Any subgeometry partition of $PG(t^*ds-1, q^2)$ by $PG(s-1, q)$'s produce a t-spread of size (ds,q) that is a union of q-fans.

In either case (a) or (b), the t-spread corresponds to a translation Sperner space admitting $GF(q^2)^*$ as a collineation group.

(3) The t-spread is a union of components fixed by $GF(q^2)^*$ and a set of $GF(q)$-reguli each in a projective space isomorphic to $PG(2s - 1, q) = PG(4s^* - 1, q)$. **Note** that these projective spaces need not be the same for different $GF(q)$-reguli.

PROBLEM 94.31. If we have a t-spread of size (ds,q) where ds is odd, and $GF(q^2)$ acts on the t-spread, then is it possible that there is a collineation group that acts transitively (2-transitively on the d-spread components)? If $t = 2$ then a 2-spread is a spread and there would be a corresponding flag-transitive translation plane.

If we have a t-spread $\mathcal{Z}$ of size (ds,q), then we ask when $GF(q^w)$ could act on $\mathcal{Z}$. Certainly the associated vector space V of dimension tds over K isomorphic to $GF(q)$ is then a $GF(q^w)$-vector space so that w must divide tds. When $t = 2$, we isolated on fields $GF(q^d)$ or $GF(q^{2d})$, where $2d$ does not divide ds. In the t-spread case, we could analogously consider $GF(q^d)$ and $GF(t^*d)$ where t^*d does not divide ds, and t^* divides t.

THEOREM 94.32. (Johnson [**762**]). (1) Let $\mathcal{S}$ be a t-spread of V of size (ds, q), and assume that $w = d$ or t^*d where t^*d does not divide ds, and t^* divides t.

Assume that $GF(q^w)$ 'acts' on $\mathcal{S}$ then there is a quasi-subgeometry partition of $PG(tds/w - 1, q^w)$ consisting of $PG(ds/e_i - 1, q^{e_i})$'s where e_i divides w for $i \in \Lambda$.

(2) Conversely, any quasi-subgeometry partition of $PG(lds/w - 1, q^w)$ by $PG(ds/e_i - 1, q^{e_i})$'s for $i \in \Lambda$, and e_i a divisor of w, produces a t-spread of size (ds, q) that is a union of q^{e_i}-fans. The t-spread corresponds to a translation Sperner space admitting $GF(q^w)^*$ as a collineation group.

Clearly, we have barely scratched the surface of the many open questions and problems that the previous results generate. We shall be content here to list essentially one involving 'subgeometry' partitions.

PROBLEM 94.33. For any t-spread of type (s, q): (a) If s is even, determine an infinite class of subgeometry partitions of $PG(ts/2 - 1, q^2)$ by $PG(s/2 - 1, q^2)$'s and $PG(s - 1, q)$'s. (b) If t is even, determine an infinite class of subgeometry partitions of $PG(ts/2 - 1, q^2)$ by $PG(s - 1, q)$'s.

EXAMPLE 94.34. For example, for subgeometry partitions of $PG(5, q^2)$, we consider an associated vector space V of dimension $12 = ts$ over $GF(q)$, producing an t-spread where $t = 1, 2, 3, 4, 6$ or 12. We assume that such an t-spread admits $GF(q^2)$.

(i) If $t = 1$, then $s = 12$; then we have one component admitting $GF(q^2)$, so this is a trivial partition and we obtain exactly one $PG(5, q^2)$.

(ii) If $t = 2$ and $s = 6$, the projective space could be partitioned by $PG(5, q)$'s and $PG(2, q^2)$'s. In this case, a $PG(5, q)$ arises from a vector space of q^6 vectors which defines a q-fan and a $PG(2, q^2)$ produces a vector space of q^4 vectors which is fixed by $GF(q^2)$. The q-fans define reguli in $PG(11, q) = PG(2s - 1, q)$.

(iii) If $t = 3$ and $s = 4$, then the partition subgeometries are $PG(1, q^2)$ and $PG(3, q)$'s. A $PG(1, q^2)$ produces a vector space of q^4 vectors fixed by $GF(q^2)^*$ and a $PG(3, q)$ arises from a vector space of q^4-vectors which defines a q-fan. The q-fans define reguli in various $PG(7, q)$'s=$PG(2s - 1, q)$'s.

(iv) If $t = 4$ and $s = 3$, then we have a partition by $PG(2, q)$'s arising from q-fan's that define reguli in various $PG(5, q)$'s $= PG(2s - 1, q)$'s.

(v) If $t = 6$ and $s = 2$, $PG(5, q^2)$ could be partitioned by $PG(0, q^2)$'s and $PG(1, q)$'s arising from q-fans that define reguli in various $PG(3, q)$'s $= PG(2s - 1, q)$'s and a $PG(0, q^2)$ produces a vector space of q^2 vectors fixed by $GF(q^2)^*$.

(vi) If $t = 12$ and $s = 1$, $PG(5, q^2)$ could be partitions by $PG(0, q)$'s arising from q-fans that define (trivially) reguli in various $PG(1, q)$'s.

CHAPTER 95

Hyper-Regulus Partitions.

The reader is directed to Chapter 16 on generalized André planes and the constructions of Johnson using finite fans. Fundamentally, to obtain unusual subgeometry or quasi-subgeometry partitions of projective spaces, it suffices to find non-André hyper-reguli with fixed-point-free groups. Specifically, we recall the following theorems of Johnson [**760**].

THEOREM 95.1. (Johnson [**760**]). Let Σ be a Desarguesian affine plane of order q^{ds} defined by the field F isomorphic to $GF(q^{ds})$ and let F_d denote the subfield isomorphic to $GF(q^{ds})^*$. Let F^* and F_d^*, respectively, denote the associated multiplicative groups. Let e be any divisor of d.

(1) Let $\mathcal{A}_\alpha$ denote the q^e-André net $\{y = xm; m^{(q^{ds}-1)/(q^e-1)} = \alpha\}$, and let $\alpha = 1$. Next, we consider cosets of $F_d^{*s(q^e-1)}$ in $F_d^{*(q^e-1)}$. Let $\left\{ \alpha_i; i = 1, \ldots, (s, \frac{(q^d-1)}{(q^e-1)}) \right\}$ be a set of coset representatives for $F_d^{*s(q^e-1)}$.

Let $\mathcal{A}_1^{q^{e\lambda(i)}} : \{ y = x^{q^{e\lambda(i)}} m; \ m^{(q^{ds}-1)/(q^d-1)} \in \alpha_i F_d^{*s(q^e-1)} \}$, where $(d/e, \lambda(i)) = 1$, $i = 1, \ldots, (s, \frac{(q^d-1)}{(q^e-1)})$. Then, the kernel homology subgroup of Σ corresponding to F_d^* leaves $\mathcal{A}_1^{q^{e\lambda(i)}}$ invariant and has $\frac{(q^{ds}-1)}{(q^d-1)}/(s, \frac{(q^d-1)}{(q^e-1)})$ orbits of length $(q^d-1)/(q^e-1)$.

(2) $(s, \frac{(q^d-1)}{(q^e-1)}) = 1$ if and only if $\left| \mathcal{A}_1^{q^{e\lambda(1)}} \right| = \frac{(q^{ds}-1)}{(q^e-1)}$ if and only if (for arbitrary $\lambda(i)$) we have a standard q^e-André replacement.

(3) If $(\lambda(i) - \lambda(j), d/e) = d/e$ for all $i, j = 1, 2, \ldots, (s, \frac{(q^d-1)}{(q^e-1)})$ then

$$\bigcup_{i=1}^{(q^{ds}-1)/(q^d-1)} \mathcal{A}_1^{q^{e\lambda(i)}}$$

forms a generalized André replacement, admitting F_d^* as a fixed-point-free collineation group. Hence, we have a generalized André replacement that is a union of q^e-fans. If, in addition $(\lambda(i), ds/e) = 1$, for all $i = 1, \ldots, (s, \frac{(q^d-1)}{(q^e-1)})$, we obtain a hyper-regulus replacement.

(4) The generalized André replacement of (3) is not André if and only if we choose at least two of the $\lambda(i)$'s to be distinct, which is possible if there are at least two distinct integers i, j, if and only if $(s, \frac{(q^d-1)}{(q^e-1)}) \neq 1$.

COROLLARY 95.2. Choose any fixed integer f less than d/e such that $(f, d/e) = 1$. For each i, $i = 1, \ldots, (s, \frac{(q^d-1)}{(q^e-1)})$, choose any integer k_i such that $k_i d/e + f \leq ds$ and let $\lambda(i) = k_i d/e + f$. Call the corresponding generalized André replacement $\Sigma_{k_i, f}$. If $(s, \frac{q^d-1}{q^e-1}) \neq 1$ and at least two of the integers k_i are distinct then we obtain a generalized André replacement which is not André. Furthermore, the translation plane obtained by this replacement of a single André net is not an André plane.

COROLLARY 95.3. If A_α is any q^e-André net, for $\alpha \in GF(q^e) - \{0\}$, there is a generalized André replacement isomorphic to the replacement for A_1, considered as a q^e-André net, for each André net.

95.1. Mixed Multiple Hyper-Regulus Replacement.

We have previously considered partitioning André nets into parts of sizes $(q^{ds} - 1)/(q^d - 1)$ for various d's, for up to $(q - 1)$ mutually disjoint André nets. If we take different d's per André net, we have effectively partitioned the spread into a set of different sized André nets and considered replacements of each of these nets. However, our replacements restricted to the nets of the partition are never André replacements on the partitions q^d-André nets since in the André case the replacement components would necessarily be of the form $y = x^{q^d \rho(m)} m$, where our components have the form $y = x^{q^{\lambda(i)}} m$, where $(\lambda(i), d) = 1$, for various functions λ.

The previous multiple replacement method may be generalized as follows: Let $q^{ds} = q_\beta^{d_\beta s_\beta}$, where q_α is not necessarily equal to q. Although, we use a similar notation as in the previous, d_β need not be a divisor of d in this setting. Choose any set of mutually disjoint q_β-André nets of $(q^{ds} - 1)/(q_\beta - 1)$ components each for various q_β's. Choose generalized André replacements for each of these nets. Any such translation plane constructed will be a potentially new generalized André plane.

To construct spreads that are union of fans, we formally present a variation of the basic construction method.

THEOREM 95.4. (Johnson [**760**]). Let Σ be a Desarguesian affine plane of order q^{ds}. For each q-André A_α, $\alpha \in GF(q)$, choose a divisor e_α of d and partition the components of A_α into $\frac{(q^{e_\alpha} - 1)}{(q - 1)}$ pieces each of cardinality $\frac{(q^{ds} - 1)}{(q^{e_\alpha} - 1)}$.

(1) Each of the partition pieces will be q^{e_α}-André nets admitting the group F_d^*. Let $A_{\alpha,i}$ for $i = 1, \ldots, \frac{(q^{e_\alpha} - 1)}{(q - 1)}$, denote the set of q^{e_α}-André nets arising from A_α.

(2) Consider the subgroup $F_d^{*s(q^{e_\alpha} - 1)}$ of $F_d^{*(q^{e_\alpha} - 1)}$ and let

$$\left\{ \beta_{\alpha,i,j}; j = 1, \ldots, (s, \frac{q^d - 1}{q^{e_\alpha} - 1}) \right\}, \ i = 1, \ldots, \frac{(q^{e_\alpha} - 1)}{(q - 1)}, \ \alpha \in GF(q) - \{0\}$$

be a coset representative set for $F_d^{*s(q^{e_\alpha} - 1)}$ of $F_d^{*(q^{e_\alpha} - 1)}$, depending on α and i.

Let b_α be any element of F_{ds}^* such that $b_\alpha^{(q^{ds} - 1)/(q-1)} = \alpha$. Let

$$A_{\alpha,i}^{e_\alpha \lambda_{\alpha,i}(j)} = \left\{ y = x^{q^{e_\alpha \lambda_{\alpha,i}(j)}} m b_\alpha; \ m^{(q^{ds} - 1)/(q^d - 1)} \in \beta_{\alpha,i,j} F_d^{*s(q^{e_\alpha} - 1)} \right\},$$

where the functions $\lambda_{\alpha,i}$ are chosen to satisfy the conditions $(\lambda_{\alpha,i}(j), d/e_\alpha) = 1$ and $(\lambda_{\alpha,i}(j) - \lambda_{\alpha,i}(j), d/e_\alpha) = d/e_\alpha$.

Then $A_{\alpha,i}^{e_\alpha \lambda_{\alpha,i}(j)}$ is a union of $(\frac{(q^{ds} - 1)}{(q^d - 1)} / (s, \frac{(q^d - 1)}{(q^{e_\alpha} - 1)}))$, F_d^* orbits of length $\frac{(q^d - 1)}{(q^{e_\alpha} - 1)}$; a set of $(\frac{(q^{ds} - 1)}{(q^d - 1)} / (s, \frac{(q^d - 1)}{(q^{e_\alpha} - 1)}))$ q^{e_α}-fans.

Hence, $\bigcup_{i=1}^{\frac{(q^{e_\alpha} - 1)}{(q-1)}} A_{\alpha,i}^{e_\alpha \lambda_{\alpha,i}(j)}$ consists of $\frac{(q^{e_\alpha} - 1)}{(q-1)} (\frac{(q^{ds} - 1)}{(q^d - 1)} / (s, \frac{(q^d - 1)}{(q^{e_\alpha} - 1)}))$ q^{e_α}-fans.

(3) Therefore,

$$
\left\{
x = 0, y = 0, \quad \bigcup_{\alpha \in GF(q)^*} \bigcup_{i=1}^{\frac{(q^{e_\alpha}-1)}{(q-1)}} A_{\alpha,i}^{e_\alpha \lambda_{\alpha,i}(j)}
\right\} = S^{(e_\alpha; \alpha \in GF(q)^*)}
$$

is a spread defining a generalized André plane that admits F_d^* as a fixed-point-free collineation group. The spread is the union of two q^d-fans $(x = 0, y = 0)$ and $\frac{(q^{e_\alpha}-1)}{(q-1)}(\frac{(q^{ds}-1)}{(q^d-1)}/(s, \frac{(q^d-1)}{(q^{e_\alpha}-1)})) \; q^{e_\alpha}$-fans, for each of $q - 1$ possible divisors e_α of d.

For $\alpha \neq \delta$, the divisors e_α and e_δ of d are independent. For $i \neq k$, the functions $\lambda_{\alpha,i}$ and $\lambda_{\delta,k}$ are independent subject to the prescribed conditions.

PROBLEM 95.5. Completely determine the isomorphism classes of translation planes that arise from the mixed multiple generalized André replacements.

95.2. Non-André Quasi-Subgeometry Partitions.

DEFINITION 95.6. We shall say that a quasi-subgeometry partition is 'André' if it arises from a generalized André plane constructed only using André replacements.

THEOREM 95.7. (Johnson [**760**]). Let π be a translation plane of order q^{ds} and kernel containing K isomorphic to $GF(q)$. Assume that there is a fixed-point-free collineation group GK^* such that GK is a field isomorphic to $GF(q^w)$, where $w = d$ or $2d$ and s is odd in the latter case. Let the set of divisors of w be $N = \{e_i; i = 1, \ldots, E\}$, including w and 1.

(1) Each component L has a unique maximal subfield $GF(q^f)$ within $GF(q^w)$, for $f = e_k$ for some k, such that L is a $GF(q^f)$-subspace. In this case, the orbit length of L under GK^* is $(q^w - 1)/(q^f - 1)$ and the orbit is a q^f-fan. Let k_i denote the number of GK^*-orbits of components of length $(q^w - 1)/(q^{e_i} - 1)$; of q^{e_i}-fans, and let N^- denote the subset of N containing the divisors e_i used in the construction. Then $\sum_{i=1}^{|N^-|} k_i(q^w - 1)/(q^{e_i} - 1) = q^{ds} + 1$.

(2) Consider the associated affine geometry $AG(2ds/w, q^w)$, embed in $PG(2ds/w, q^w)$ and let Δ denote the hyperplane at infinity isomorphic to $PG(2ds/w-1, q^w)$. Consider an orbit of components under $GF(q^w)^*$ of length $(q^w - 1)/(q^{e_i} - 1)$; the various q^{e_i}-fans. Each q^{e_i}-fan will produce a quasi-subgeometry isomorphic to $PG(ds/e_i-1, q^{e_i})$ in Δ. Hence, we obtain k_i such $PG(ds/e_i-1, q^{e_i})$'s.

(3) If $\sum_{i=1}^{N^-} k_i(q^w - 1)/(q^{e_i} - 1) = q^{ds} + 1$, then there is a corresponding quasi-subgeometry partition of $PG(2ds/w - 1, q^w)$ by $k_i \; PG(ds/e_i - 1, q^{e_i})$'s for $i = 1, 2, \ldots, E$. We call this a partition of type $(k_1, \ldots, k_{|N^-|})$.

(4) If $PG(2ds/w-1, q^w)$ admits a quasi-subgeometry partition by $k_i \; PG(ds/e_i-1, q^{e_i})$'s for $i = 1, 2 \ldots, |N^-|$, for $e_i \in N^- \subseteq N$ and $N = \{e_i; i = 1, 2, \ldots, E\}$ is the set of all divisors of d then necessarily $\sum_{i=1}^{|N^-|} k_i(q^w - 1)/(q^{e_i} - 1) = q^{ds} + 1$. Furthermore, there is an associated translation plane of order q^{ds} and kernel containing $GF(q)$ admitting a collineation group GK^* such that the union with the zero mapping is a field isomorphic to $GF(q^w)$ that is fixed-point-free and there is a set of k_i mutually disjoint q^{e_i}-fans whose union is the spread for the translation plane.

COROLLARY 95.8. (Johnson [**760**]). In the above situation, if $d = 2$ then the quasi-subgeometries are subgeometries and the partitions are subgeometry partitions.

95.2.1. The Construction. Choose any mixed multiple generalized André replacement procedure. Certain of these replacement procedures produces a generalized André plane whose spread is the union of q^{e_i}-fans for a set of e_i's. Each of these spreads produces a quasi-subgeometry partition of the associated projective space. Clearly, there is a tremendous number of such non-André quasi-subgeometry partitions and we can determine whether these are André by the nature of the components or the replacements. In particular, we record the following results on non-André partitions, analogous to a similar theorem in [**762**].

THEOREM 95.9. If $q - 1 \geq$ the number of divisors (including d and 1) of d then there exists a generalized André translation plane obtained from a Desarguesian plane of order q^{ds} by multiple generalized André replacement that produces a quasi-subgeometry partition of $PG(2s - 1, q^d)$ such that for every divisor e of d, there exist quasi-subgeometries of the partition isomorphic to $PG(ds/e - 1, q^e)$. If, for some divisor e of d, $(s, \frac{(q^d-1)}{(q^e-1)}) \neq 1$ then there is a partition that is not André.

More specifically, we have the following theorem. We state this theorem in a more shortened version. For the complete hypothesis, the reader is re-directed back to Theorem 95.4.

THEOREM 95.10. Consider

$$
\left\{ x = 0, y = 0, \bigcup_{\alpha \in GF(q)^*} \bigcup_{i=1}^{\frac{(q^{e_\alpha}-1)}{(q-1)}} A_{\alpha,i}^{e_\alpha \lambda_{\alpha,i}(j)} \right\} = S^{(e_\alpha; \alpha \in GF(q)^*)},
$$

for $j = 1, \ldots, (s, \frac{(q^d-1)}{(q^{e_\alpha}-1)})$.

(1) Then $S^{(e_\alpha; \alpha \in GF(q)^*)}$ is a spread defining a generalized André plane that is the union of two q^d-fans ($x = 0, y = 0$) and $\frac{(q^{e_\alpha}-1)}{(q-1)}(\frac{(q^{ds}-1)}{(q^d-1)}/(s, \frac{(q^d-1)}{(q^{e_\alpha}-1)}))$ q^{e_α}-fans, for each of $q - 1$ possible divisors e_α of d.

(2) Each such spread produces a quasi-subgeometry partition of $PG(2s-1, q^d)$ by $\frac{(q^{e_\alpha}-1)}{(q-1)}(\frac{(q^{ds}-1)}{(q^d-1)}/(s, \frac{(q^d-1)}{(q^{e_\alpha}-1)}))$ quasi-subgeometries isomorphic to $PG(ds/e_\alpha - 1, q^{e_\alpha})$, for α varying over $GF(q) - \{0\}$, and two $PG(s-1, q^d)$'s. Furthermore, if for some divisor e_{α_o}, we have $(s, \frac{(q^d-1)}{(q^{e_{\alpha_o}}-1)}) \neq 1$ then the partitions obtained are not André, provided the functions $\lambda_{\alpha,i}$ are not all constant.

95.3. Non-André Subgeometry Partitions.

In the previous section, we have seen that there are a tremendous variety of non-André hyper-reguli and generalizations in Desarguesian affine planes, each of which produces a new generalized André translation plane. Furthermore, when F_d^* acts and $d = 2$, we obtain subgeometry partitions; the subgeometry partitions of $PG(2s - 1, q^2)$ by $PG(s - 1, q^2)$'s and $PG(2s - 1, q)$'s.

Hence, we see that when $d = 2$, we obtain many new families of subgeometry partitions of $PG(2s - 1, q^2)$. We restate our results on generalized André partitions when $d = 2$ so that the reader can better appreciate the subgeometry partitions that we are obtaining. The main results in this section are found in parts (3) and (4) of the following result.

THEOREM 95.11. (Johnson [**760**]). Let Σ be an affine Desarguesian plane of order q^{2s}. For each André net A_α for $\alpha \in GF(q)^*$, choose functions λ_α such that $(2, \lambda_\alpha(i)) = 1$ and $(\lambda_\alpha(i) - \lambda_\alpha(j), 2) = 2$. Let

$$\mathcal{S}_\alpha^{\lambda_a} = \bigcup \mathcal{A}_\alpha^{q^{\lambda_\alpha(i)}} : \{ y = x^{q^{\lambda_\alpha(i)}} m b_\alpha; \ m^{(q^{2s}-1)/(q-1)} \in \alpha_i F_2^{*s(q-1)} \},$$

where $\{ \alpha_i; \ i = 1, \ldots, (s, q+1) \}$ is a coset representative set for $F_2^{*s(q-1)}$ in $F_2^{*(q-1)}$.

(1) Then $\mathcal{S}_\alpha^{\lambda_a} \cup \{x = 0, y = 0\}$ is a spread for a generalized André plane admitting F_2^* as a fixed-point-free collineation group with orbits of length 1 or $(q^2 - 1)/(q - 1) = (q + 1)$.

(2) Let $\eta(2s) + 1$ equal the number of divisors of $2s$. There are at least

$$(s, q + 1)^{(s-1)\eta(ds)k} \left(1 + \sum_{k=2}^{q-1} \left(\begin{array}{c} q - 1 \\ k \end{array} \right) \right)$$

different ways of choosing André nets and generalized André replacements that possibly lead to mutually non-isomorphic translation planes that admit F_2^* as a fixed-point-free collineation group.

(3) Each of the spreads of part (2) produce subgeometry partitions of $PG(2s - 1, q^2)$ by subgeometries isomorphic to $PG(2s - 1, q)$ and $PG(2s - 1, q)$. The orbits of length $q + 1$ of F_2^* in the spread bijectively correspond to the $PG(2s-1, q)$'s and the orbits of length 1 in the spread bijectively correspond to the $PG(s - 1, q^2)$'s.

(4) Any such subgeometry partition of this type is new provided $(s, q + 1) \neq 1$ and not all of the functions λ_α are constant on the partitions of the André nets.

We have constructed a variety of generalized André planes of order q^n, for n composite, by developing a new method of replacement of André nets. When $n = ds$, we have also given variations of the constructions that produce spreads that are unions of q^e-fans, for a set of divisors e of d. Such spreads also produce quasi-subgeometry partitions of $PG(2s - 1, q^d)$ by quasi-subgeometries isomorphic to $PG(ds/e - 1, q^e)$. There are new constructions of subgeometry partitions as well when $d = 2$. We have left open a number of important questions on isomorphisms of generalized André planes and correspondingly questions on isomorphisms of the associated quasi-subgeometry partitions.

Since all of our constructions involve generalized André planes, we end by stating an open problem for arbitrary quasi-subgeometry partitions.

PROBLEM 95.12. Suppose that we have a quasi-subgeometry partition $\mathcal{P}$ of $PG(s - 1, q^d)$ by quasi-subgeometries isomorphic to $PG(ds/e_i - 1, q^{e_i})$, where $i \in \Lambda$ and e_i is a divisor of d. If d is not 2, is the corresponding translation plane $\pi_\mathcal{P}$ necessarily a generalized André plane? Find non-generalized André translation planes of order q^{ds} whose spreads are unions of q^{e_i}-fans, for $d > 2$ and $\{ e_i; i \in \Lambda \}$ a set of divisors of d.

CHAPTER 96

Small-Order Translation Planes.

In this final part of the text strictly on translation planes, we gather examples of small-order translation planes that have not been mentioned previously. Many of these translation planes and some of the known affine planes are constructed using the computer. We consider non-translation planes in Chapters 97 through 100.

96.1. Orders $2, 3, 4, 5, 7, 8, 9, 16, 25, 27, 49$.

(1) All translation planes of orders ≤ 49 are known, mostly by computer programs.

(2) All translation planes of order ≤ 8 are Desarguesian.

(3) There are two translation planes, Desarguesian and Hall, of order 9.

(4) There are eight translation planes of order 16, obtained via computer due to Dempwolff and Reifart [**308, 309**].

(5) There are 21 translation planes of order 25, by computer by Czerwinski and Oakden[**269, 270**] and by Charnes [**223**].

(6) There are seven translation planes of order 27, by computer by Dempwolff [**304, 305**]. We have seen several of these in this text. Another interesting plane is the Sherk plane of order 27.

(7) There are 1347 translation planes of order 49, by computer programs of Mathon and Royle [**976**] and Charnes and Dempwolff [**226**].

All planes of orders $16, 25, 27, 49$, their collineation groups and other details about the planes may be found on the homepage of Ulrich Dempwolff:

http : //www.mathematik.uni − kl.de/˜dempw/dempw_Plane.html

96.2. Order 64.

96.2.1. Büttner's Planes of Order 64. We have seen in Chapter 10 on semifield planes that there are a variety of planes of order q^3 admitting $SL(2, q)$. In particular, every cyclic semifield plane of order q^3 produces an associated generalized Desarguesian plane admitting $GL(2, q)$ acting canonically. Indeed, Kantor [**864**] constructs translation planes of even order q^6 admitting $SL(2, q^2)$ without using the cyclic semifield connection, but may be seen to be generalized Desarguesian planes by the action of the group $GL(2, q)$ and the characterization theorem of Jha–Johnson [**607**]. However, the translation planes of order q^3 admitting $SL(2, q)$ or even $GL(2, q)$ are not known beyond this characterization.

We list the following open problem.

PROBLEM 96.1. Determine all translation planes of order q^3 that admit $SL(2, q)$ as a collineation group.

In [**204**] Büttner studied translation planes of order 64 that admit groups generated by pairs of maximal-order cyclic 2-groups whose associated involutions do not commute and was able to classify such planes.

THEOREM 96.2. (Büttner [**204**]). Let π be a translation plane of order 64. Assume that π admits two cyclic 2-groups $\langle s \rangle$ and $\langle s' \rangle$ of order the dimension of π as a vector space over the kernel and assume that the involutions in the two groups do not commute. Then π is one of the following planes:
  (1) Desarguesian,
  (2) Lüneburg–Tits,
  (3) Kantor of order 2^6 admitting $GL(2, 2^2)$ with kernel $GF(2^3)$, or
  (4) A plane admitting $SL(2,4)$ with kernel $GF(8)$ but not admitting $GL(2,4)$.
(We shall call this the Büttner plane of order 64.)

To describe the planes, particularly the Büttner plane, we require some detailed information on the collineation group. In all of the cases, the associated spread may be considered within $PG(3,8)$. So, we are working with a vector space V_4 of dimension 4 over $GF(8)$.

Let ψ denote the semilinear mapping induced from the Frobenius automorphism of $GF(8)$, and let σ be defined by $\sigma(x) = x^4$. Let $a, b \in GF(4^3)$ and define the following group elements:

$$(a,b) = \begin{bmatrix} 1 & 0 & 0 & 0 \\ a & 1 & 0 & 0 \\ a^{\sigma+1} + b & a^\sigma & 1 & 0 \\ a^{\sigma+2} + ab + b^\sigma & b & a & 1 \end{bmatrix},$$

$$\omega = \begin{bmatrix} 0 & 0 & 0 & 1 \\ 0 & 0 & 1 & 0 \\ 0 & 1 & 0 & 0 \\ 1 & 0 & 0 & 0 \end{bmatrix}.$$

Let

$$H = \langle (a,b), \omega; a, b \in GF(8) \rangle \simeq Sz(8),$$
$$F = \langle (0,1)\omega \rangle \langle (1,0) \rangle \simeq Sz(2).$$

Then the spreads considered above admit F as a collineation group. The four involutions of F will fix a subspace of line size pointwise. Let $\mathcal{I}$ denote this set. Since the Desarguesian plane and Lüneburg–Tits planes are clearly possibilities, we shall list only the other two possibilities. The Lüneburg–Tits spread is a set of tangent lines to the Tits ovoid O in $PG(3,8)$. Let π denote the associated polarity of O. The spreads are $\mathcal{T} \cup \mathcal{L}$, where

(1) $\mathcal{L} = \{(t_1\psi^2)F\} \cup \{s_1 F\} \cup \{s_1\pi F\}$, for $t_1 = \langle (0,0,0,1), (0,1,x^3,0) \rangle (0,x)$,

  $s_1 = \langle (0,0,0,1)(0,x), (0,0,0,1)(1,x^2)\omega(0,1) \rangle$ and

(2) $\mathcal{L} = \{(t_1\psi^2)F\} \cup \{s_2 F\} \cup \{s_2\pi F\}$, for

  $s_2 = \langle (0,0,0,1)(0,x), (0,0,0,1)(1,x^2+1)\omega(0,1) \rangle.$

The plane of Kantor turns out to be type (1) and the Büttner plane is the type (2) spread. The spreads were determined with the aid of a computer.

96.3. Order 81.

In this section, we note that the work of Johnson and Prince [**807, 808**] and Jha and Johnson [**637**] shows that it is possible to determine all translation planes of order 81 that admit $SL(2,5)$ in the translation complement. We here provide a sketch of how this is accomplished.

Let π be a translation plane of order n that admits at least $n^{1/2}+1$ elations with distinct axes, in the translation complement. Let G denote the collineation group of the translation complement of π generated by the elations. All the non-solvable groups are known by Hering [**476**] and Ostrom [**1047, 1052**]. If G is non-solvable, then either n is 81 and G is $SL(2,5)$ or G is a Suzuki group $S_z(n^{1/2})$ (n square) or G is a 2-dimensional special linear group $SL(2,n^{1/2})$ or $SL(2,n)$. Furthermore, the plane π is either Lüneburg–Tits or Desarguesian or the plane has order 81 (see Fink–Johnson–Wilke [**361**] and Foulser–Johnson–Ostrom [**380**]).

If $n = 81$, and G isomorphic to $SL(2,5)$ (see Ostrom [**1052**]), then the plane admits 10 elations (corresponding to the 10 Sylow 3-subgroups). However, in this case, the plane need not be Desarguesian.

In [**1119**], Prohaska studies various configurations of circles in the Miquellian inversive plane $M(q)$. In particular, Prohaska shows that if $q = 3^r$ and 5 divides $q + 1$ then there is a set of six pairwise disjoint circles C_i in an orbit under A_5 and A_5 leaves invariant another circle C_o which is disjoint from each C_i.

Considering the circles in $M(q)$ as derivable nets in a Desarguesian plane Σ of order q^2, we have a set of mutually disjoint derivable nets $D_0, D_1, D_2, \ldots, D_6$. Furthermore, Σ admits a collineation group G isomorphic to $SL(2,5)$, which fixes D_0 and acts transitively on the remaining five derivable nets. Multiply deriving these six nets produces a translation plane that admits $SL(2,5)$ with 10 elation axes. We shall call this plane the *Prohaska* plane and denote it by π_P.

Furthermore, in the Prohaska plane, we may derive the net containing the elation axes to produce another plane admitting a collineation group isomorphic to $SL(2,5)$ but where the 3-elements now fix Baer subplanes pointwise. We call this plane the *derived Prohaska* plane and denote it by $Der(\pi_P)$.

Hence, in order to complete the determination of translation planes of order n that admit at least $n^{1/2} + 1$ elation axes, we need to consider the translation planes of order 81 that admit $SL(2,5)$ as a collineation group.

It is possible to use a combination of computer programming and combinatorial group theorem to completely determine the translation planes of order 81 that admit $SL(2,5)$ as a collineation group, generated by affine elations. It turns out that the 10 Sylow 3-subgroups of $SL(2,5)$ define a set of 10 elation axes that form a derivable net. Hence, all of the planes that we determine may be derived by the elation net producing a set of planes admitting $SL(2,5)$, where the 3-elements are Baer. There are a total of 14 planes, seven admitting elations and seven admitting Baer 3-elements. These planes are all mutually non-isomorphic. Of these 14 planes, four were previously known.

The planes admitting elations are described as follows, where the *type* is the partition $[i, j, \ldots, k]$ of 82 determined by the orbit lengths of components:

(1) One plane with kernel $GF(81)$, of type $[10, 12, 60]$; the Desarguesian plane.

(2) Five planes with kernel $GF(9)$ of type $[10, 12^6]$. One of these planes is the Prohaska plane, the rest are new. Of particular interest is that one of these planes can be constructed from a Desarguesian plane by 12-nest replacement.

(3) One plane with kernel $GF(3)$ of type $[10, 12, 60]$. In this plane, each component of the union of the orbits of length 10 and 12 is fixed by a field isomorphic to $GF(9)$, that does not fix each component in the orbit of length 60.

Then there are the seven derived planes that admit $SL(2,5)$, where the 3-elements are Baer and the Baer subplanes pointwise fixed lie on a derivable net.

The elation net R coordinatized by a field K isomorphic to $GF(9)$, defines a regulus in $PG(3, K)$. In the planes of type (2), the kernel is also K. When R is derived, we still have kernel K. The planes of the following two classes will have spreads in $PG(3, K)$:

D(1), the Hall plane, D(2), the derived Prohaska plane and four new planes. Then there is the plane D(3) derived from the plane of (3) listed above with kernel $GF(3)$. Since the relative kernel of the union of the nets of orbit lengths 10 and 12 is K, and the orbit of length 10 is a regulus in $PG(3, K)$, it follows that this plane also has kernel $GF(3)$.

The main idea is to represent the group appropriately using matrices as follows, and then the problem is to determine all spreads of an 8-dimensional vector space over $GF(3)$ which are invariant under the action of the subgroup $SL(2,5)$ of $GL(8,3)$:

$$SL(2,5) = \left\langle \begin{bmatrix} I & 0 & C & 0 \\ 0 & I & 0 & C \\ 0 & 0 & I & 0 \\ 0 & 0 & 0 & I \end{bmatrix}, \begin{bmatrix} I & 0 & 0 & 0 \\ 0 & I & 0 & 0 \\ I & 0 & I & 0 \\ 0 & I & 0 & I \end{bmatrix} \right\rangle$$

where $C = \begin{bmatrix} 0 & 1 \\ -1 & 0 \end{bmatrix}$, where I and 0 in the larger matrix denote the identity and zero 2×2 matrices, respectively.

Note that the matrices $\alpha C + \beta I$ where $\alpha, \beta \in GF(3)$ form a field $GF(9)$ with non-zero elements

$$\pm \begin{bmatrix} 1 & 0 \\ 0 & 1 \end{bmatrix}, \pm \begin{bmatrix} 0 & 1 \\ -1 & 0 \end{bmatrix}, \pm \begin{bmatrix} 1 & 1 \\ -1 & 1 \end{bmatrix}, \pm \begin{bmatrix} 1 & -1 \\ 1 & 1 \end{bmatrix}.$$

The product of the two generating matrices (which both have order 3) is

$$\begin{bmatrix} I+C & 0 & C & 0 \\ 0 & I+C & 0 & C \\ I & 0 & I & 0 \\ 0 & I & 0 & I \end{bmatrix}.$$

which has order 10.

We remark that if θ is a generator of $GF(9)$ over $GF(3)$ satisfying $\theta^2 = -1$ then the subgroup of $SL(2,9)$

$$\left\langle \begin{bmatrix} 1 & \theta \\ 0 & 1 \end{bmatrix}, \begin{bmatrix} 1 & 0 \\ 1 & 1 \end{bmatrix} \right\rangle$$

is isomorphic to $SL(2,5)$. When we represent the field $GF(9)$ by 2×2 matrices as above, the matrix C corresponds to θ. We can obtain the above representation of $SL(2,5)$ on a 4-dimensional vector space over $GF(9)$ by tensoring with the 2×2 identity matrix.

The $SL(2,5)$ action is such that the 10 elements of order 3 act as elations with the 10 axes given by the 4-spaces

$$x = 0, y = x \begin{bmatrix} \alpha C + \beta I & 0 \\ 0 & \alpha C + \beta I \end{bmatrix}; \alpha, \beta \in GF(3)$$

(using the standard way of describing components in a spread in terms of matrices). Thus, the orbit of length 10 is uniquely determined by the specified action of the $SL(2,5)$.

The computer then provides the following planes:

Let R denote the regulus of elation axes in $PG(3,9)$. Then, the opposite regulus R' is contained in the regular spread D' consisting of all the τ-invariant lines of $PG(3,9)$. Each line of R' is left invariant by the $SL(2,5)$, while the remaining lines of D' have $SL(2,5)$-orbit of length 12. There are 36 distinct such orbits each containing two lines of D'. Thus, the lines of D' outside R' are grouped into 36 pairs. It is well known that the lines of a regular spread of $PG(3,q)$, with the reguli contained in the regular spread as circles, give a model for the Miquellian inversive plane of order q [**290**]. In the present context, inversion in the circle corresponding to R' yields the 36 pairs of lines, as conjugate points of the circle.

If we fix one of these pairs, then the remaining lines of D' can be partitioned uniquely into 8 reguli, one of which is R' (the linear flock of the Miquellian plane with carrier the given pair). Replacing these reguli by their opposites gives a Desarguesian spread D which has R as a regulus. This is actually the unique Desarguesian spread admitting the given $SL(2,5)$ referred to in the previous section with orbit lengths [10, 12, 60] where the orbit of length 12 is the orbit containing the given fixed pair of lines of D'.

The linear flock of D given by the 8 opposite reguli just described consists of all the τ-invariant reguli of D. The $SL(2,5)$-orbits of these 8 reguli are as follows:

(a) A single orbit of length 1, namely $\{R\}$;

(b) A single orbit of length 6, consisting of 6 disjoint reguli whose union is the 60-orbit;

(c) Two orbits of length 12, each containing 2 reguli of the flock; each orbit is a 12-nest: the union of the reguli in each orbit is the 60-orbit and each line of the 60-orbit is contained in 2 reguli of the orbit;

(d) A single orbit of length 12, containing 2 reguli of the flock; the union of the reguli in this orbit consists of the union of the 12-orbit and 60-orbit on lines, each line of the 12-orbit being in 5 of the reguli while each line of the 60-orbit is in a unique regulus.

Thus, there is a unique τ-invariant regulus of D, which has an $SL(2,5)$-orbit of length 6 (case (b) above). Replacing the 6 disjoint reguli of this orbit by their opposites gives a Prohaska spread.

If we consider a τ-invariant regulus of D, which has an $SL(2,5)$-orbit as in case (c) above, then the computer results show that it is possible to choose (in two ways) a half-regulus in the opposite regulus that determines 5 conjugate pairs of lines of D' which, together with the starting conjugate pair of lines of D', gives a 6-clique corresponding to a plane of type $[10, 12^6]$ in the orbit of length 144 (for the action on the 720 spreads).

96.3.1. Johnson–Prince Planes of Order 81. Johnson and Prince then construct a set of seven translation planes of order 81 admitting $SL(2,5)$, generated by elations. The previously known planes are the Desarguesian and Prohaska, thus we have five new planes. Since each of these planes may be derived, there are five additional translation planes of order 81, admitting $SL(2,5)$, where the 3-elements are Baer. The derived planes are mutually non-isomorphic by the same argument as for the original planes (the group acting on the 720 spreads, described in Section 96.3, leaves the regulus of elation axes and its opposite invariant). Finally, by Foulser [**375**], Baer 3-elements and elations cannot coexist in a translation plane. Hence, we have a total of 10 new translation planes of order 81 admitting $SL(2,5)$. The five new planes where the 3-elements are elations are termed the 'Johnson–Prince planes of order 81' and the five new planes where the 3-elements are Baer are called the 'derived Johnson–Prince planes of order 81'.

REMARK 96.3. Johnson and Prince [**808**] are able to give computer-free constructions for the spreads of order 81 in $PG(3,9)$ and in so doing realize that one of the planes may be constructed using a 12-nest.

96.3.2. 12-Nest Planes.

THEOREM 96.4. (Johnson and Prince [**808**]).
(1) There is a unique translation plane of order 81 admitting $SL(2,5) \times Z_5$ that may be obtained from a Desarguesian affine plane of order 81 by 12-nest replacement.
(2) There are exactly four translation planes of order 81 and kernel $GF(9)$ that admit $SL(2,5) \times Z_5$:
  (a) The Prohaska,
  (b) 12-nest plane, where $SL(2,5)$ is generated by elations,
  (c) The derived Prohaska, and
  (d) The derived 12-nest plane, where $SL(2,5)$ is generated by Baer 3-elements.

COROLLARY 96.5. Let Σ be a Desarguesian plane of order 81 that admits $SL(2,5)$ as a collineation group. Let π_o be any subplane of a τ-invariant regulus net that sits in Γ_{60} then $SL(2,5)\pi_o$ is a partial spread of cardinality 12 that contains exactly two τ-invariant components (π_o and $\pi_o N_{SL(2,5)}(\langle\tau\rangle)$).

COROLLARY 96.6. There are exactly six mutually non-isomorphic spreads admitting $SL(2,5)$, where the 3-elements are elations, provided in the non-Desarguesian case that there are six component orbits of length 12:
(1) Desarguesian,
(2) Prohaska,
(3) Johnson–Prince spreads: (a) the 12-nest spread, and (b) three spreads arising from 24 reguli in a Desarguesian affine plane.

COROLLARY 96.7. There are exactly six mutually non-isomorphic spreads π in $PG(3,9)$ that admit $SL(2,5)$, generated by elations.

96.4. Baer Groups.

Now assume that we have $SL(2,5)$ acting on a translation plane π of order 81. We note below that by Jha and Johnson [**637**], the 3-elements are elations or Baer. Furthermore, in the Baer case, we note the following:

REMARK 96.8. If whenever there is an orbit of length 12, all orbits of components have lengths 1, 12 or 60, then the Baer axes line up into a derivable net.

It is known by computer that this is exactly the situation when there is an orbit of length 12, however, we do not have a proof of this without the use of the computer.

If we make this assumption in the dimension-2 case, we have a computer-free construction of all translation planes of order 81 with spread in $PG(3,9)$ that admit $SL(2,5)$ as a collineation group.

THEOREM 96.9. (Johnson and Prince [**807, 808**]). Let π be a translation plane of order 81 with spread in $PG(3,9)$ that admits $SL(2,5)$ as a collineation group. Acting on the vector space V_4, assume that when there is a partial spread orbit of length $12, \Gamma_{12}$, all partial spread orbits disjoint from Γ_{12} has length $1, 12$ or 60, and there is only an orbit of length 60 in the Desarguesian case.

Then π is one of the following 14 planes:

I. The 3-elements are elations and π is one of the following seven planes:

(1) Desarguesian,

(2) Prohaska,

(3) Johnson–Prince: (a) a 12-nest plane, (b) one of two planes obtained from a Desarguesian plane using 24 reguli,

II. The 3-elements are Baer and π is one of the following seven planes:

(1) Hall,

(2) Derived Prohaska,

(3) Derived Johnson–Prince 12-nest plane,

(4) The derived Johnson–Prince planes of the two planes of I(3(b)) above.

REMARK 96.10. The computer will tell us that, in fact, any translation plane of order 81 and spread in $PG(3,9)$ admitting $SL(2,5)$ does have the property that there are six orbits of components of length 12. It may be possible to prove this fact without the use of the computer, thereby completely determining the translation planes with spreads in $PG(3,9)$ in a completely analytical manner.

96.5. The Translation Planes of Order 81 Admitting $SL(2,5)$.

By Johnson and Prince [**807**] the translation planes of order 81 admitting $SL(2,5)$, where the 3-elements are elations are determined. In this situation, the set of 10 elation axes form a derivable net, so upon derivation, there are translation planes of order 81 admitting $SL(2,5)$, where the 3-elements are Baer. However, in a general situation, it would not necessarily be known that if the 3-elements were Baer, or if there were Baer that the Baer axes would be disjoint or even if they were that these axes would belong to a derivable net. In fact, we show that all 3-elements are either elation or Baer and in the Baer case, both of these latter possibilities occur.

There are exactly 14 translation planes determined by Johnson and Prince, of which exactly 10 planes are new. In this case, the subspaces fixed pointwise by 3-elements are mutually disjoint.

It is however possible to consider the more general case: Translation planes of order 81 admitting $SL(2,5)$ as a collineation group in the translation complement, but without further assumptions. Jha and Johnson are able to establish the following result.

THEOREM 96.11. (Jha and Johnson [**637**]). Let π be a translation plane of order 81 admitting $SL(2,5)$ as a collineation group of the translation complement.

(1) Then the 3-elements are either elations or Baer collineations.

(2) If the 3-elements are Baer collineations, the Baer axes are disjoint, and form a derivable subnet within the translation plane π.

(3) There are exactly 14 translation planes of order 81 admitting $SL(2,5)$ as a collineation group, as listed in Johnson and Prince [**807, 808**].

96.5.1. Foulser's Planes Admitting $SL(2,3)$. Since we are also interested in the overlap problem on a translation plane of order 3^{2r}, we assume the more general situation in the following subsection. We recall that Foulser showed that two distinct Baer p-groups for $p > 3$ cannot fix common non-zero points. However, when $p = 3$, Foulser [**375**] has determined a class of translation planes of order 81 with overlapping Baer 3-group. The construction of these planes starts with the two Baer groups with overlapping fixed-point-spaces and used the computer to construct the planes. The Baer groups generate a collineation group isomorphic to $SL(2,3)$. When two Baer groups of order 3 generate $SL(2,5)$, it is unclear from Foulser's work the exact nature of the fixed point spaces. However, it is possible to overcome this as follows.

96.5.2. Overlap Theorem.

THEOREM 96.12. (Jha–Johnson [**637**]). Let π be a translation plane of order 3^{2r} that admits a collineation group in the translation complement isomorphic to $SL(2,5)$, generated by Baer 3-elements. Then the Baer axes are mutually disjoint.

96.6. The 12-Nest Plane of Prince.

In the previous sections, planes of order 81 admitting $SL(2,5) \times Z_5$ obtained the replacement of a 12-nest. There is an analogous translation plane of order 19^2 admitting $SL(2,5) \odot Z_{20}$ due to Prince [**1110**]. The idea is to represent $SL(2,5)$ acting on an associated Desarguesian affine plane Σ and determine a 12-nest of reguli within Σ which is invariant under $SL(2,5)$. The use of $SL(2,5) \times Z_5$ applied in an manner similar to that of the 12-nest plane of order 81 of Johnson and Prince shows that the partial spread in Σ is nest-replaceable producing a 12-nest plane of order 19^2.

CHAPTER 97

Dual Translation Planes and Their Derivates.

Although this handbook considers finite translation planes, it turns out that dual translation planes obtained from translation planes of order q^2 with spread in $PG(3, q)$ are always derivable. For instance the associated dual quasifields are 2-dimensional vector spaces over the dual kernel field isomorphic to $GF(q)$. Furthermore, the derived planes are semi-translation planes and often exhibit very interesting properties, including connections to flocks of quadratic cones and the construction technique of algebraic lifting.

Dual translation planes are coordinatized by dual quasifields. Thus, if $(Q, +, \cdot)$ is a quasifield with kernel subfield K, then the corresponding dual quasifield is a right vector space under the action of K. In particular, if $(Q, +, \cdot)$ is a quasifield of order q^2 with a kernel subfield K isomorphic to $GF(q)$, then the dual quasifield with $(Q, +, *)$ is given by $a * b = b \cdot a$, a two-dimensional quasifield with respect to K acting from the right. Hence,

REMARK 97.1. If π is a spread in $PG(3, q)$ the dual of the corresponding translation plane is derivable.

In fact, we can go much further by noting that there is a bijection between quasifields $(Q, +, \cdot)$ coordinatizing a translation plane and the dual quasifields $(Q, +, *)$ coordinatizing the dual of π.

PROPOSITION 97.2. An affine translation plane of order q^2 with spread in $PG(3, q)$ produces $(q+1)(q^2+1)$ derivation sets in the set of (q^2+1) dual translation planes. Hence, there are potentially $(q^2+1)(q+1)$ mutually non-isomorphic affine planes constructed from an affine plane of order q^2 and spread in $PG(3, q)$.

PROOF. We sketch the simple proof. The dual quasifields coordinatizing the $q^2 + 1$ dual affine planes through the point at infinity are each associated with $(q^2 - 1)/(q - 1)$ derivation sets based on the orbits of the dual kerns. □

DEFINITION 97.3. Let ρ be an affine plane. ρ is said to be a 'semi-translation plane' if and only if there exists a translation group T such that each point orbit of T is an affine Baer subplane.

THEOREM 97.4. Let π be an affine dual translation plane that is derivable, where the derivation set $\mathcal{D}$ contains, as a point $(\infty)^*$, the line at infinity of the associated (affine) translation plane dual π, and let the point (∞) of the translation plane denote the line at infinity of π.
(1) Then the translation group T_∞ of dual π with center (∞) is a translation group of π which leaves the derivation set $\mathcal{D}$ invariant.
(2) The affine plane derived using $\mathcal{D}$ is a semi-translation plane.

PROOF. Notice that T_∞ fixes all lines incident with (∞) and these become the points at infinity of π. Hence, in particular, T_∞ leaves $\mathcal{D}$ invariant. Baer subplanes are replaced in the semi-translation plane by lines of the net defined by parallel classes defined by the points of the derivation set, which means, in particular, the lines incident with (∞) become Baer subplanes in the constructed affine plane. Since the translation group with center (∞) acts transitively on affine points of each of the lines incident with it, it follows that the point orbits of this translation group become Baer subplanes. $\square$

A major problem that now arises is whether the collineation group of the constructed semi-translation plane is connected to the collineation group of the dual translation plane (i.e., 'inherited'), which is, of course, connected to the collineation group of the associated translation plane. It turns out that the collineation group for the semi-translation plane is inherited or the kernel for the translation plane is $GF(2)$ and $SL(2,q)$ acts on the semi-translation plane. In particular, the relevant theorem is as follows:

THEOREM 97.5. (Johnson [**717**]). Let π_3 be an affine semi-translation plane of order $q^2 > 16$, which is derived from a dual translation plane π_2, of a translation plane π_1, by a net containing a (the) shears axis.

(a) If either q is odd or the kernel of the associated translation plane π_1 is not $GF(2)$, then the full collineation group of π_3 leaves the derivable net invariant and hence is inherited from the collineation group of the dual translation plane.

(b) If the group of π_3 is not inherited, there is a collineation group isomorphic to $SL(2,q)$ fixing an affine point P and generated by Baer involutions such that the Sylow 2-subgroups fix Baer subplanes pointwise and these $(q+1)$ Baer subplanes lie across the entire set of $q^2 + 1$ lines incident with P.

PROBLEM 97.6. Show that a semi-translation plane of order q^2 derived from a dual translation plane cannot admit $SL(2,q)$.

So, in the most common usage of the above theorem, the original spread has kernel $GF(q)$, so the collineation group is the inherited group. Since we may derive in any of $(q+1)(q^2+1)$ ways, the question is when are any two such planes isomorphic. In such a setting, two semi-translation planes are isomorphic if and only if there is a collineation of the corresponding translation plane that maps one set of q lines defining a derivation set onto the second set of q lines defining a derivation set. In particular, since the kernel homology group leaves all such derivation sets invariant and the translation group basically is not in play, we see that if the collineation group of the translation plane consists merely of the required groups (that particular translation group and the kernel homology group of order q), we will have exactly $(q+1)(q^2+1)$ mutually non-isomorphic semi-translation planes constructed from a given translation plane with spread in $PG(3,q)$.

We also note that when π is a semifield plane, the dual translation plane is also a translation plane and the derived plane then becomes another translation plane.

The proof for the above argument utilizes connecting derivable nets co-dimension two structures in an associated 3-dimensional projective space and the use of Hering's normalizer intersection theorem [**477**] and so is basically beyond the scope of this handbook, although we will discuss the form of the spreads in our chapter on transversal spreads, Chapter 47.

DEFINITION 97.7. We shall say that the translation plane is 'rigid' if and only if the full collineation group is TK^*, semi-direct product of the kernel homology group K^* by the translation group T.

THEOREM 97.8. Let π be a finite rigid translation plane with spread in $PG(3, q)$. Then there are exactly $(q^2 + 1)(q + 1)$ mutually non-isomorphic semi-translation planes constructed from π by the derivation of an associated dual translation plane.

The only known rigid translation plane with spread in $PG(3, q)$ is due to Charnes of order 17^2 [**222**] and is, in fact, translation plane arising from a slice of a Conway–Kleidman–Wilson binary ovoid. For even order, there are planes due to Kantor and Williams (see Chapter 79). These planes have kernel $GF(2)$, so admit, in fact, exactly the translation group as collineation group.

PROBLEM 97.9. Find an infinite class of rigid planes with spreads in $PG(3, q)$.

97.1. Semi-Translation Planes of Type i.

Now assume that an affine translation plane of order q^2 with spread in $PG(3, q)$ admits an elation group E of order q which leaves invariant a Baer subplane π_0. Choose a derivable net defined by an affine version of the projective extension of π_0. This constructs a semi-translation plane. Then $ET_\infty T_0$ is a group of order q^3 containing a point-line-transitivity of the projective semi-translation plane. Hence,

THEOREM 97.10. (Johnson [**731**, (7.7)]). Let π be a translation plane of order $q^2 > 16$ and kernel K isomorphic to $GF(q)$. Choose a derivable net of the dual translation plane containing a shears axis. Then the projective extension of the derived semi-translation plane admits an incident, point-line transitivity if and only if the translation plane π admits an elation group of order q which leaves invariant an affine Baer subplane of π.

Actually, it turns out that any elation group of order q either fixes 0, 1, 2 or $q + 1$ affine Baer subplanes that share the same parallel classes. These Baer subplanes necessarily share the elation axis. We call such semi-translation planes of 'type i' if and only if there is an associated elation group E fixing exactly i-affine Baer subplanes sharing the same infinite points.

We note that when a translation plane admits an elation group of order q one of whose orbits defines a regulus net, then all such orbits define regulus nets and the plane corresponds to a flock of a quadratic cone. Hence, any conical flock translation plane produces semi-translation planes of type $q + 1$. When we algebraically lift a spread in $PG(3, q)$ to a spread in $PG(3, q^2)$, we create a translation plane admitting an elation group of order q^2 in a translation plane of order q^4 that leaves invariant exactly two Baer subplanes that are incident with an affine point and share their parallel classes. Hence, algebraically lifting creates semi-translation planes of type 2. Hence, there are nice connections with type-2 and type-$(q + 1)$ semi-translation planes and known spread types or known construction techniques. There are exactly four type-1 spread constructions.

THEOREM 97.11. (Johnson [**731**, (7.8)], Semi-Translation Planes of Types $q+1$ and 2). Let π be a translation plane of order $q^2 > 16$ and kernel K isomorphic to $GF(q)$ which contains an elation group E that leaves invariant affine Baer subplanes.

(1) Then there are i type-i semi-translation planes constructed the derivation of dual translation planes by a derivable net sharing the shears axis, for $i = 1, 2, q+1$.

(2) If π is not a semifield plane, then the semi-translation planes admit exactly one incident point-line transitivity.

(3) If π corresponds to a flock of a quadratic cone, then there is an elation group E of order q, each component orbit of which defines a regulus net containing $q + 1$ Baer subplanes left invariant by E. Hence, there are $q + 1$ type-$(q + 1)$ semi-translation planes constructed.

(4) If π is a translation plane of order q^4 with spread in $PG(3, q^2)$ which has been algebraically lifted from a translation plane of order q^2 with spread in $PG(3, q)$, then there is an elation group E of order q each orbit of which defines a net containing exactly two Baer subplanes left invariant by E. Hence, there are two type-2 semi-translation planes constructed.

97.2. Semi-Translation Planes of Type 1.

We shall list the corresponding translation planes. Under appropriate coordinatization, the elation group E has the following form:

$$E = \left\langle \begin{bmatrix} I & \begin{bmatrix} u & g(u) \\ 0 & f(u) \end{bmatrix} \\ 0 & I \end{bmatrix} ; u \in GF(q) \right\rangle,$$

where I and 0 are the identity and zero 2×2 matrices.

We designate the group E by the shorthand notation $(g(u), f(u), u)$.

Description	Group	Comments
Lüneburg–Tits order $2^{2(2e-1)}$	$(u^\sigma - u, u)$ $u \in GF(2^{2e-1})$ $\sigma = 2^e$	Ostrom (see, e.g., [1046])
Kantor/Ree–Tits order $3^{2(2e-1)}$	$(u^\sigma - u, u)$ $u \in GF(3^{2e-1})$ $\sigma = 3^e$	Johnson [731]
Biliotti–Menichetti order 64	$(u^4 + u^3, u)$ $u \in GF(8)$	Johnson [731]
Jha–Johnson order 64	$(u + u^2, u)$ $u \in GF(8)$	Johnson [731]

In this chapter, we have connections with interesting semi-translation planes of order q^2 and elation groups of order q in associated translation planes of order q^2. If the associated translation plane admits an elation group of order q^2, with appropriate properties, then there is a group of order q^4 in the corresponding semi-translation plane. We consider such phenomena in Chapter 98.

CHAPTER 98

Affine Planes with Transitive Groups.

A very common theme in finite geometry is to use group theory or permutation group theory to determine or classify planes. For example, the Ostrom–Wagner theorem [**1072**] is amongst the earliest examples of such results.

THEOREM 98.1. (Ostrom–Wagner [**1072**]). Let π be a finite projective plane of order n admitting a collineation group G acting doubly transitively on the points. Then π is Desarguesian and G contains $PGL(3, n)$.

Inspired by this result, one might ask when a group transitive on the points of an affine plane π is sufficient for the plane to be a translation plane. We begin, however, by considering a rather specialized situation to illustrate what is involved. Thus, we consider a translation plane of order q^2 with spread in $PG(3, q)$ that admits a collineation group in $GL(4, q)$ of order q^2. Of course, it is possible that we have a semifield plane, which admits an elation group of order q^2. On the other hand, it is possible to have translation planes of order q^2 with non-elation groups of order q^2 as is seen in a related Chapter 48.

Now suppose a group G of order q^2 contains a normal elation group E of order q which fixes i affine Baer subplanes incident with an affine point that share all of their parallel classes. For example, if $i = q + 1$, the plane is a conical flock plane and the group of order q^2 must act transitively on the set of q reguli that share the axis $x = 0$ of the elation group of order q. Furthermore, suppose that the group G induces on $x = 0$ an elation group of order q, considering the 2-dimensional $GF(q)$-space as a Desarguesian affine plane of order q. Note that the set of q^2 affine lines will be in q orbits under the group G. The orbit of the line $x = 0$ will define a derivation set as in the chapter on derivation of dual translation planes (see Chapter 97), left invariant by the group of order q^2. Hence, the associated semi-translation plane of order q^2 will admit a collineation group of order q^4. We have seen in Chapter 97 that the semi-translation plane will admit an incident point-line transitivity. This transitivity is actually a translation group of order q^2 with fixed center. The group of order q^4 will turn out to be a collineation group that acts regularly on the affine points of the semi-translation plane and contains a translation group of order q^3. If there is a collineation group of order q^2 in $GL(4, q)$, with the properties as above, there is a semi-translation plane of order q^2 that admits a collineation group of order q^5 that contains a Baer group of order q and acts transitively on the q^4 affine points.

THEOREM 98.2. (See, e.g., Johnson [**731**]). Let π_3 be a semi-translation plane of order $q^2 > 16$, which is derived from a dual translation plane π_2 of a corresponding translation plane π_1 with spread in $PG(3, q)$. Assume that π_1 is not a semifield spread.

Then π_3 admits a collineation group that acts transitively on the affine points if and only if the associated translation plane admits a collineation group of order q^2 in $GL(4, q)$.

Thus, we see that in order to ensure that an affine plane admitting a collineation group is transitive on the points, some additional hypothesis must be made. We have mentioned the Ostrom–Wagner theorem, however, this is a projective version. A similar theorem for affine planes is:

THEOREM 98.3. (Ostrom–Wagner [**1072**]). Let π denote an affine plane admitting a collineation group acting doubly transitively on the points. Then π is a Desarguesian plane and the collineation group contains the translations.

We recall that a semi-translation plane of order q^2 admits a group of translations of order q^2, whose affine point orbits are Baer subplanes (subplanes of order q). Hence,

REMARK 98.4. If π is a translation plane with spread in $PG(3, q)$, then any affine dual translation plane derives to a semi-translation plane. If the original translation plane is not a semifield plane or the line at infinity of the dual translation plane is not the center of an affine elation group, then the semi-translation plane is not a translation plane.

REMARK 98.5. There are a wide variety of semi-translation planes. Johnson [**659**] has classified these with a method similar to the point-line transitivity classification of Lenz–Barlotti (see Barlotti [**88**]). Furthermore, the possible types of semi-translation planes obtained by the derivation of a dual translation plane are also determined by Johnson [**675**].

We will note below that the Hughes planes are semi-translation planes and these planes are also derivable. Since all known semi-translation planes are derivable, we pose the following difficult problem.

PROBLEM 98.6. Show that every semi-translation plane is derivable.

CHAPTER 99

Cartesian Group Planes—Coulter–Matthews.

The planar functions of Dembowski–Ostrom [**292**] are a potential source for cartesian group planes, planes that are (P, ℓ)-transitive for some flag (P, ℓ), without being semi-translation planes or their duals. However, for rather 'good' reasons, the theory has been dormant for many years until the discovery of the spectacular Coulter–Matthews cartesian group planes. We report here on these and related planes.

Concerning the general process, Pierce and Kallaher [**1104**] showed that either such a construction produces an affine plane with exactly one incident point-line transitive or the plane is a translation plane (semifield plane). For remarks concerning limitations of this method, see [**1104**]. Indeed, for monomial planar functions producing projective planes of order p^n that are non-translation planes, Dempwolff and Röder [**310**] completely determine the collineation group as $\Gamma L(1, p^n) \cdot GF(p^n) \times GF(p^n)$, where $0 \times GF(p^n)$ denotes the single point-line transitivity.

We begin with Dembowski's theorem.

THEOREM 99.1. (Dembowski [**288**]). Let π^+ be a projective plane of order n that admits a collineation group G of order n^2 which acts regularly on any point or line orbit. Then P contains a flag (C, a) fixed by G such that π^+ is (C, a)-transitive and the group E of (C, a)-elations is a subgroup of G. Furthermore, one of the following two situations occurs:

(1) G consists of elations all with axis a or all with center C, or

(2) The set of elations of G is E and G has exactly three point (line) orbits: the point C (line a), the points on $a - C$ (lines on C and not a), and the points of $\pi^+ - a$ (lines of π^+ not on C).

Note that if G in the above theorem is Abelian, then G induces a regular group on any point or line orbit. Now further assume that G is elementary Abelian.

DEFINITION 99.2. Let $G = H \times T$, be an elementary Abelian p-group, for p a prime, of order p^{2r}, where H and T are both subgroups of order p^r. A mapping f from H into T is a 'planar function with respect to $H \times T$' if and only if for each $u \neq 0$ of H, by defining λ_u:

$$\lambda_u(x) = f(u + x) - f(x),$$

then λ_u is bijective.

The main reason for the consideration of planar functions is in the following theorem.

THEOREM 99.3. Let f be a planar function with respect to $G = H \times T$, of order p^{2r}. Define a point-line geometry as follows: The 'points' are the elements of

$H \times T$. G acts on itself by $(x, y)g = (x + u, y + v)$, where $g = (u, v)$. The 'lines' are defined to be the sets

$$\{(c, y); c \text{ fixed in } H, \forall y \in T\} \equiv (x = c),$$

and

$$\{x, \ f(x - u) + v; \ u, v \text{ fixed}, \ u \in H, \ v \in T, \ \forall x \in H\} \equiv (y = f(x - u) + v).$$

Then the point-line geometry is a finite affine plane of order p^{2r} admitting G, and G fixes the parallel class (∞) and is transitive on the remaining parallel classes. Projectively, there is a $((\infty), \ell_\infty)$-transitivity.

PROOF. To check that G acts as a collineation group, we have $(x, y) \longmapsto (x + u, y + v)$, so $x = c$ maps to $x = c + u$. A pair $(d, f(d - u^*) + v^*)$ of points of $y = f(x - u^*) + v^*$ will map to $(d + u, f(d - u^*) + v^* + v)$, which will be incident with $y = f(x - z) + w$ if and only if

$$f(d + u - z) + w = f(d - u^*) + v^* + v.$$

Hence,

$$w = v^* + v \text{ and } z = u + u^*$$

solves the equation independently of d, implying that

$$y = f(x - u^*) + v^* \longmapsto y = f(x - (u^* + u)) + v^* + v.$$

Note that this implies immediately that there is a translation group of order p^{2r} with center (∞) obtained by taking $u = 0$.

Now in order to prove that an affine plane is obtained, we basically need only show lines of different parallel classes uniquely intersect. A parallel class is obtained by $\{x = c; c \in H\}$, or by $\{y = f(x - u) + v; v \in T\}$. Now $x = c$ will intersect $y = f(x - u) + v$ uniquely in the point $(c, f(c - u) + v)$ and $y = f(x - u) + v$ will intersect $y = f(x - u^*) + v^*$ in a point $(d, f(d - u) + v)$ if and only if

$$f(d - u) + v = f(d - u^*) + v^*$$

has a unique solution d. Let $x = d - u^*$, so that $d - u = x + (u^* - u)$ and we are considering $\lambda_{(u^* - u)}(x)$, which is bijective. Hence, there is a unique solution x such that $\lambda_{(u^* - u)}(x) = v^* - v$, namely $d - u^*$. This completes the proof. □

Let $+, \cdot$ be addition and multiplication for $GF(p^r)$, p odd, and use juxtaposition to denote multiplication of a semifield on $(GF(p^r), +, \cdot)$. Consider the following groups:

$$T \ : \ \{\tau_v : (x, y) \to (x, y + v); v \in GF(p^r)\} \text{ and}$$

$$W \ = \ \left\{\rho_u : (x, y) \to (x + u, y + xu + \frac{u^2}{2}); u \in GF(p^r)\right\}.$$

T is a translation group of order p^r with center (0) and the elements of W are products of elations with axis $x = 0$ and certain translations. Note that $\rho_u \rho_w : (x, y) \to (x + u, y + xu + \frac{u^2}{2}) \to (x + (u + w), y + xu + \frac{u^2}{2} + (x + u)w + \frac{w^2}{2})$. Since

$$xu + \frac{u^2}{2} + (x + u)w + \frac{w^2}{2} = x(u + w) + \frac{(u + w)^2}{2},$$

then $\rho_u \rho_w = \rho_{u+w}$ so that W is, indeed, a group (note we need $uw = wu$ to make this calculation, hence a commutative semifield will suffice). Form $WT \simeq W \times T$ to obtain an elementary Abelian group of order p^{2r}. Now define

$$g : g(x) = \frac{x^2}{2},$$

and notice to recapture xu, we have

$$g(x) - g(x - u) + \frac{u^2}{2} = \frac{x^2}{2} - \frac{(x-u)^2}{2} + \frac{u^2}{2} = xu.$$

Furthermore, we claim that g is a planar function since

$$g(x+u) - g(x) = \frac{(x+u)^2}{2} - \frac{x^2}{2} = xu - \frac{u^2}{2} = \lambda_u(x)$$

clearly defines a bijective function λ_u, for $u \neq 0$. Hence, we note that there is bijection between commutative semifield planes and planar functions defined as $g(x) = \frac{x^2}{2}$, where the lines of the plane become $x = c$ or $y = g(x) - g(x - u) + v$, for all $u, v \in GF(p^r)$, and the groups

$$T \quad : \quad \{\, \tau_v \colon (x,y) \to (x, y+v); v \in GF(p^r) \,\} \text{ and}$$
$$W \quad = \quad \{\, \rho_u \colon (x,y) \to (x+u, y + g(x) - g(x-u)); u \in GF(p^r) \,\}$$

define an elementary Abelian group $W \times T$ of order p^{2r}. However, this idea is more general and actually proves the following result:

THEOREM 99.4. Let g be a planar function on $GF(p^r) \times GF(p^r)$.

(1) Then there is an affine plane of order p^r with points $GF(p^r) \times GF(p^r)$ and lines $x = c, y = g(x) - g(x - u) + v; c, u, v \in GF(p^r)$ that admits the following collineation groups

$$T \quad : \quad \{\, \tau_v \colon (x,y) \to (x, y+v); v \in GF(p^r) \,\} \text{ and}$$
$$W \quad = \quad \{\, \rho_u \colon (x,y) \to (x+u, y + g(x) - g(x-u)); \ u \in GF(p^r) \,\}.$$

(2) Conversely, any commutative semifield plane admits such groups and produces a planar function g by define $g(x) = \frac{x^2}{2}$.

To reconcile this theorem with the result showing that from a planar function f on $W \times T$ produces an affine plane with lines

$$x = c, \ y = f(x - u) + v, \ c, u \in W, \ v \in T$$

we merely check that the two planes are isomorphic.

THEOREM 99.5. There is a bijection between planar functions on an elementary Abelian group of order p^{2r} and affine planes of order p^r admitting Abelian collineation groups of order p^{2r} with the elation subgroup has order p^r. The planes may be constructed either using Theorem 99.3 or isomorphically using Theorem 99.4.

99.1. The Planes of Coulter–Matthews.

Coulter and Matthews [264] provide a complete list of all known planar functions and determine two planar functions leading to new affine planes, one a commutative semifield plane discussed in Chapter 10 on semifield planes and an affine plane admitting precisely one incident point-line transitivity. The latter planes are defined as follows:

THEOREM 99.6. (Coulter and Matthews [**264**]). The function f defined on $GF(3^e)$ by $f(x) = x^{\frac{3^\alpha+1}{2}}$ is a planar function, precisely when $(\alpha, e) = 1$ and α is odd. The affine plane is not a translation plane or a dual translation plane if α is not congruent to ± 1 mod $2e$ and has exactly one incident point-line transitivity.

CHAPTER 100

Planes Admitting $PGL(3, q)$.

There are two classes of projective planes that admit $PGL(3, q)$; the Hughes planes of order q^2 and the Figueroa planes of order q^3. We consider each of these in the following two sections.

100.1. The Hughes Planes.

The following construction is valid for any right nearfield K of order q^2 which contains a field isomorphic to $GF(q)$ is its center. The reader is directed to Chapter 6 of "Subplane Covered Nets" [**753**] for more discussion on the Hughes planes.

Define a point-line geometry π^+ as follows: The 'points' of π^+ are the sets of non-zero triples $\{(x, y, z)a; \forall a \in K\}$. Considering the 3-dimensional vector space F^3 of vectors (x, y, z) over $GF(q)$, there is a cyclic group G of order $q^2 + q + 1$ acting on the points of π^+. Let $L(t)$, $t \in K^*$ denote the set of points whose 'vectors' (x, y, z) satisfy

$$x + yt + z = 0.$$

Define the 'lines' of π^+ as the sets of points of $\{L(t)g; g \in G\}$.

THEOREM 100.1. (See, e.g., Johnson [**753**, (6.4), p. 60]). The point-line geometry π^+ is a projective plane of order q^2, called the 'Hughes plane over the nearfield K'. The restriction to $GF(q)$ produces a Desarguesian projective subplane π_0^+ of order q, which is left invariant by G. Furthermore, the collineation group of π^+ contains a group isomorphic to $PGL(3, q)$ leaving invariant π_0^+.

To see how to convert to a derivable affine form, in the following theorem, let

$$L(1) \begin{bmatrix} 1 & 0 & 1 \\ 0 & 1 & 1 \\ 0 & 0 & 1 \end{bmatrix} = \{(x, y, -y, -y)a\} \begin{bmatrix} 1 & 0 & 1 \\ 0 & 1 & 1 \\ 0 & 0 & 1 \end{bmatrix} = \{(x, y, 0)a\}.$$

THEOREM 100.2. (See, e.g., Johnson [**753**, (6.6), p. 61]). Let $z = 0$ denote the line at infinity ℓ_∞ and form the affine Hughes plane. There is a representation of the affine plane with points as elements of $K \times K$ and lines as follows:

$$
\begin{aligned}
x &= c; c \in K \\
y &= x\delta + b; \delta \in GF(q),\ b \in K, \\
y &= (x - \alpha)m + \beta; \alpha, \beta \in F, m \in K - GF(q).
\end{aligned}
$$

The net with lines

$$
\begin{aligned}
x &= c; c \in K, \\
y &= x\delta + b; \delta \in GF(q),\ b \in K,
\end{aligned}
$$

is a derivable net and the derived plane is called the 'Ostrom–Rosati' plane, which has exactly one incident point-line transitivity.

100.1.1. Irregular Hughes Planes. The previous construction of the Hughes planes required that the nearfield of order q^2 has center $GF(q)$. There are three irregular nearfield of orders $11^2, 23^2, 59^2$ (non-solvable) whose centers are isomorphic to $GF(p)$, for $p = 11, 23, 59$, respectively, and hence the previous construction applies. However, the irregular nearfield planes of order 5^2 and 7^2 of order p^2 admit cyclic homology groups of order $p + 1$. Such groups defined derivable nets whose partial spreads are orbits of length $p + 1$. The analysis of Ostrom obtaining the associated affine version of the Hughes planes also works when we have a derivable net and produces Hughes planes in this setting as well. Hence, there are the so-called 'irregular Hughes planes' as order 5^2 and 7^2.

100.2. The Figueroa Planes.

From the point of view of projective plane theorem, the Figueroa planes are probably the most important class of non-Desarguesian planes. We provide the construction here.

Let π^+ be a Pappian plane coordinatized by the field K that admits a planar collineation α of order 3. Then it turns out that the points of π^+ are partitioned into three sets: The fixed points of α, the set of points P such that $P, P\alpha$ and $P\alpha^2$ are collinear and the set of points Q such that $Q, Q\alpha$ and $Q\alpha^2$ form a triangle. Similarly, we have three such sets for the lines of π^+. The Figueroa planes may be constructed by modifying the incidence relation of the set of points and lines in the triangle sets.

Let M_P denote the set of points P such that $P, P\alpha$ and $P\alpha^2$ form a triangle and similarly let M_ℓ denote the set of lines ℓ such that $\ell, \ell\alpha, \ell\alpha^2$ form the sides of a triangle. Define the involution μ interchanging M_P and M_ℓ by

$$P\mu = P\alpha P\alpha^2, \text{ and } \ell\mu = \ell\alpha \cap \ell\alpha^2.$$

Now define the new incidence I^* to modify the Pappian incidence I as follows:

$$P \ I^* \ \ell \text{ if and only if } \ell\mu \ I \ P\mu$$

for all points $P \in M_P$ and lines ℓ in M_ℓ.

The new structure becomes a projective plane, called the 'Figueroa plane over K'.

For example, if K is isomorphic to $GF(q^3)$, for any prime power q, α is the associated mapping $(x, y, z) \rightarrow (x^q, y^q, z^q)$. Figueroa initially found the plane of order 3^3 by a variant of a net replacement procedure. In a series of seminars with R. Figueroa, R. Pomareda and O. Barriga, it emerged that the construction of Figueroa is valid for all cubic extensions of finite fields. Then C. Hering and H. Schaeffer [492] and T. Grundhöfer [422] realized that there is a synthetic construction which is valid for all cubic extensions of any field. The synthetic form is that which is given here.

THEOREM 100.3. (Figueroa [355], Hering–Schaeffer [492], and Grundhöfer [422]).

The point-line geometry defined above is a projective plane. When K is isomorphic to $GF(q^3)$ the projective plane, the Figueroa plane has order q^3 and admits a collineation group isomorphic to $PGL(3, q)$ leaving invariant a Desarguesian subplane of order q.

CHAPTER 101

Planes of Order ≤ 25.

We consider non-Desarguesian planes of order ≤ 25.

101.1. Planes of Order 16.

There are a variety of interesting planes of order 16, but only one, the Mathon plane, cannot be be constructed from known translation planes, Hughes planes, etc., by derivation, dualization. In any case, we list here the known planes of order 16, recalling that there are exactly eight translation planes of this order.

The known planes of order 16, their groups and general information about the planes may be found on Gordon Royle's website:

$$http : //www.csse.uwa.edu.au/~gordon/remote/planes16/$$

101.1.1. Known Projective Planes of Order 16. We take the list of planes of order 16 as it appears in Royle's list of planes. Recall that there are exactly 8 translation planes.

PG(2,16) The Desarguesian projective plane of order 16
SEMI2 The semifield plane with kernel GF(2)
SEMI4 The semifield plane with kernel GF(4)
HALL The Hall plane
LMRH The Lorimer–Rahilly plane
JOWK The Johnson–Walker plane
DSFP The derived semifield plane
DEMP The Dempwolff plane
JOHN The Johnson plane
BBS4 Plane derived from the (dual of) SEMI4 from a derivation set of Bose–Barlotti type not on the (dual) translation line
BBH2 Plane derived from the dual of HALL from a derivation set of Bose–Barlotti type "from a line not on the regulus"
MATH The Mathon plane
BBH1 Plane derived from the dual of HALL from a derivation set of Bose–Barlotti type "from a line on the regulus"
Many details may be found in Royle's website.

101.2. Planes of Order 25.

There are 193 known projective planes of order 25 and recall that there are 21 translation planes. However, there are only two projective planes that so far do not fall into any recognizable class; the two planes of Moorhouse, which are actually derivates of each other and are obtained by a process called 'lifting quotients by involutions'. The two Moorhouse planes are lifted from a 'quotient' starting with the regular nearfield plane of order 25.

The list of planes may be obtained from the website of Eric Moorhouse and the reader is directed to his website for additional information:

http://www.uwyo.edu/moorhouse/pub/planes25/

There are 193 planes, (5 self-dual planes plus 94 dual pairs).

More details on the new planes of Moorhouse can be found in Moorhouse [**993**]

THEOREM 101.1. The known finite projective or affine planes that are not translation planes are included in the following list:

> Dual translation planes,
>
> Semi-translation planes and their duals
>
> > (Derived dual translation planes and their duals),
>
> Planar function planes of Coulter–Matthews,
>
> Hughes planes and irregular Hughes planes,
>
> Ostrom–Rosati planes and their duals
>
> > (Derived Hughes planes and their duals),
>
> Figueroa planes,
>
> Mathon planes of order 16 and the dual plane,
>
> The two Moorhouse planes of 25 (derivates of each other).

CHAPTER 102

Real Orthogonal Groups and Lattices.

We include this chapter as an appendix to equip the reader with sufficient background to read our treatment and construction of ovoids and the construction processes of Conway–Kleidman–Wilson and the generalization by Moorhouse.

DEFINITION 102.1. Let V be an n-dimensional vector space over the field R of real numbers. Let $\langle \cdot, \cdot \rangle$ denote the standard dot product of V. The group $\mathcal{O}_n(V)$ of linear transformations that preserve length and distance is said to be the 'real n-dimensional orthogonal group'. Let S be any vector subspace of V. Let $S^\perp = \{ x \in V; \langle x, s \rangle = 0 \; \forall s \in S \}$. $S^\perp$ is called the subspace 'orthogonal' to S. Note that over the R, $S \cap S^\perp = \langle 0 \rangle$ and the dimension of $S^\perp = n - \dim S$, so that $S \oplus S^\perp = V$.

Now let P be any hyperplane ($(n-1)$-dimensional vector subspace). We define a 'reflection S' to be a linear transformation such that $Sx = x$ for all $x \in P$ and $Sx = -x$ for all $x \in P^\perp$. Since $y \in V$, is $x_y + r_y$, where $x_y \in P$ and $r_y \in P^\perp$. If we choose an orthonormal basis for V, we realize that $x_y = \mathrm{Proj}_P \, y$ and $r_y = \mathrm{Proj}_{P^\perp} \, y = \langle y, r^* \rangle \, r^*$, where r^* is a unit vector in $P^\perp$. Hence, $Sy = x_y - r_y = (x_y + r_y) - 2r_y = y - 2 \langle y, r \rangle \, r$. More generally, for any r in $P^\perp$, the 'reflection along r', or the 'reflection through P' S_r is defined by $S_r x = x - 2 \frac{\langle x, r \rangle}{\langle r, r \rangle} r$ and so fixes P pointwise, where $P^\perp = \langle r \rangle$. We note that S_r is clearly an element of $\mathcal{O}_n(V)$.

DEFINITION 102.2. A subgroup G of $\mathcal{O}_n(V)$ is said to be a 'reflection group' if and only if there is a set of hyperplanes $\{ P_i; i \in \lambda \}$ such that

$$ G = \langle S_{r_i}; r_i \in P_i^\perp; i \in \lambda \rangle. $$

The 'roots' of G are the elements r_i, where $i \in \lambda$ (note that we choose one element out of each 1-dimensional subspace $P_i^\perp$) and any images of these elements under the group G (i.e., associated with conjugates of S_{r_i}). Consider the subspace $\cap_{g \in G} \{ x \in V; gx = x \} = \cap_{i \in \lambda} P_i$. If this subspace is trivial, G is said to be 'effective'. . In general, the set of all roots of G of a generating set, together with the roots of the associated conjugate reflections form what is called a 'root system' for G. .

We note (see, e.g., Grove and Benson [**415**, 4.1.2]) that if the roots of G contain a basis for V, then G is forced to be effective, and conversely, for any effective group, the roots form a basis. Furthermore, for effective groups, G is finite provided G has a finite root system (see Grove and Benson [**415**, 4.1.3]).

DEFINITION 102.3. A finite effective reflection group is said to be a 'Coxeter group'. If G is a Coxeter group, there is an element $t \in V$ such that $\langle t, r \rangle \neq 0$ for all roots r of G, and let Δ denote a root system for G. Define

$$\Delta_t^+ = \{\, r \in \Delta;\ \langle t, r\rangle > 0 \,\},$$
$$\Delta_t^- = \{\, r \in \Delta;\ \langle t, r\rangle < 0 \,\}.$$

Now choose a subset T of element of Δ_t^+, which is minimal with respect to the property that every element of Δ is a linear combination of T with non-negative coefficients. T is said to be a 't-base'. It turns out that there is a unique t-base, which also becomes a basis for V.

DEFINITION 102.4. The roots of the t-base for a root system Δ for a Coxeter group are called the 'fundamental roots' of G and the associated reflections, the 'fundamental reflections'.

Now let G be a Coxeter group with root system Δ and t-base $T = \{t_1, t_2, \ldots, t_n\}$. Then given elements r_i, r_j of T, we have $\langle r_i, r_j\rangle \,/\, \left(\langle r_i, r_i\rangle^2 \langle r_j, r_j\rangle^2\right) = -\cos(\pi/p_{ij})$, where p_{ij} is a positive integer, and furthermore, for the reflections S_{r_i} and S_{r_j}, then $\left|S_{r_i} S_{r_j}\right| = p_{ij}$ (Grove and Benson [**415**, 5.1.1]). Furthermore, a major result of Coxeter shows that $\langle S_{r_i}; S_{r_i}^2 = (S_{r_i} S_{r_j})^{p_{ij}} = 1\rangle = G$; that is, the Coxeter group may represented abstractly by a set of generators and relations.

DEFINITION 102.5. A 'lattice' (or 'lattice with respect to a basis') in an n-dimensional real vector space V is the integral span of a given basis B of V; the coefficients of the vectors are in Z.

DEFINITION 102.6. Let G be any subgroup of $\mathcal{O}_n(V)$ that leaves invariant a lattice $\mathcal{L}$ in V, then G is said to be a 'crystallographic group' (sometimes this terminology is reserved for $n = 3$). If G is the full finite subgroup of $\mathcal{O}_n(V)$ that leaves $\mathcal{L}$ invariant, then the reflection group which is the subgroup of G that is generated by reflections in G is called the 'Weyl group' $W(\mathcal{L})$ of the lattice.. The Weyl group is a Coxeter group provided it is effective.

The Coxeter groups are completely determined as are the crystallographic groups. We shall be interested in three lattices (see E_6, E_7, E_8 below) and associated Weyl groups which are, in fact, Coxeter groups, as these are related directly to constructions of translation planes.

DEFINITION 102.7. Let G be a Coxeter group and let T be a t-base for G. If $T = T_1 \cup T_2$, where T_i is non-empty and such that $T_1 \perp T_2$, we may decompose G into a direct product of Coxeter groups $G_1 \times G_2$ as follows. Let $V_i = \operatorname{Span} T_i$, so that $V = V_1 \oplus V_2$. Then $T_i \mid V_i = G_i$. Then G_i is a Coxeter groups acting on V_i, $i = 1, 2$. Furthermore, G is isomorphic to $G_1 \times G_2$. In this case, G is said to be 'reducible'; otherwise, G is 'irreducible'.

DEFINITION 102.8. There is a graph-theoretic component of this analysis, which arises as follows: Let G be a Coxeter group and let T be a t-base $\{r_i; i = 1, \ldots, n\}$. Define a graph whose vertices are the n elements r_i, $i = 1, 2, \ldots, n$ and edges are the distinct pairs (r_i, r_j). Label the edge with p_{ij} provided $(S_{r_i} S_{r_j})$ has order $p_{ij} > 2$. Then the associated labeled graph is said to be a 'Coxeter graph.' It occurs that the Coxeter graph is connected if and only if the Coxeter group is irreducible. Furthermore, the irreducible crystallographic Coxeter group have $p_{ij} \in \{2, 3, 4, 6\}$, for $i \neq j$. We follow the practice of not labeling the edge (p_i, p_j) if $p_{ij} = 3$.

It turns out that the irreducible Coxeter groups are obtained by the connected Coxeter groups.

CHAPTER 103

Aspects of Symplectic and Orthogonal Geometry.

We are interested in the methods of Kantor with regard to symplectic planes. This chapter serves as an appendix which provides the necessary background to read our treatment of orthogonal and symplectic spreads.

103.1. Bilinear Forms.

In this section, we mainly review the concepts associated with reflexive bilinear forms: isotropy, radicals, polarization, etc. We generally concentrate on those parts of the subject that have arisen in the study of symplectic and orthogonal spreads.

Let V be a vector space over a field F with basis B. Then any map $\alpha\colon B \to F$ extends uniquely to a F-linear form $\overline{\alpha}\colon V \to F$; hence, there is a natural bijective correspondence between linear forms on V and maps from the basis B to F.

The easy argument used to establish this generalizes to yield an analogous result for multilinear forms. In particular, any map $\beta\colon B \times B \to F$ extends uniquely to a bilinear form $\overline{\beta}\colon V \times V \to F$.

In the finite-dimensional case, relative to the ordered basis $B = (b_1, b_2, \ldots, b_n)$ of V, the map $\beta\colon B \times B \to F$, and hence its bilinear extension $\overline{\beta}\colon V \times V \to F$, may be identified with the matrix:

$$B := (b_{ij}), \quad \text{where} \quad b_{ij} = \beta(b_i, b_j) \, \forall i, j \in \{1, 2, \ldots, n\}.$$

It follows, by a direct computation, that the set of matrices representing a bilinear form $\overline{\beta}$, relative to all possible ordered bases B of V, coincides with a class of congruent matrices: square matrices X and Y of order n are *congruent* iff a nonsingular matrix P exists such that $P^t X P = Y$. Thus matrices X and Y represent the same bilinear form on V, relative to possibly different ordered bases, iff X and Y are congruent.

It follows that *symmetric* bilinear forms β, those satisfying the identity $\beta(x, y) = \beta(y, x)$, are represented by symmetric matrices only and conversely symmetric matrices can only represent symmetric bilinear forms. Similarly *symplectic* bilinear forms, those satisfying the identity $\beta(x, y) = -\beta(y, x)$, are precisely those bilinear forms that are represented by skew symmetric matrices: and hence must have all zeros on their main diagonal.

REMARKS 103.1. Let $\beta\colon V \times V \to F$ be a bilinear form on a vector space V over a field F.

(1) If char $F = 2$, then β is skew symmetric iff it is symmetric.
(2) If β is skew symmetric and char $F \neq 2$, then $\beta(x, x) = 0$ for all $x \in V$.
(3) If $\beta(x, x) = 0$ for all $x \in V$, then β is skew symmetric.
(4) If char $F \neq 2$, then $\beta(x, x) = 0$ for all $x \in V$ iff β is skew symmetric.

763

(5) If char $F = 2$ describe the matrices of bilinear forms β satisfying the identity $\beta(x, x) = 0 \forall x \in V$, and hence conclude that not all skew symmetric (or symmetric) matrices satisfy the condition.

The final remark is associated with the following definition.

DEFINITION 103.2. Let $\beta \colon V \times V \to F$ be a bilinear form. Then β and its matrix representations [when $\dim_F V$ is finite] are ALTERNATING if $\beta(x, x) = 0$ for all $x \in V$.

Thus, the trivial remarks above imply that in characteristic $\neq 2$, symplectic and alternating forms coincide: the corresponding matrices are just the skew symmetric matrices. But over a field F of characteristic 2, alternating bilinear forms are just the ones represented by symmetric matrices with all zeros on the main diagonal; thus all the other ('skew') symmetric matrices are *not* alternating. Relative to any bilinear form $\beta \colon V \times V \to F$, we write $a \perp b$ if $\beta(a, b) = 0$. All bilinear forms that we encounter will be of this type.

DEFINITION 103.3. A bilinear form on a vector space over a field F is REFLEXIVE if

$$a \perp b \Longleftrightarrow b \perp a \forall a, b \in V.$$

Orthogonality and related concepts are easily defined for *reflexive* bilinear forms. We summarize the related terminology:

DEFINITION 103.4. Let (V, β) be a finite-dimensional vector space equipped with a reflexive bilinear form, see Definition 103.3. Then subsets A and B of V are mutually ORTHOGONAL if:

$$a \perp b = 0 \forall a \in A, b \in B.$$

Moreover, the ANNIHILATOR or ORTHOGONAL COMPLEMENT of $A \subseteq V$,

$$A^\perp := \{\, v \in V \mid v \perp a = 0 \,\forall a \in A \,\}.$$

If $A = \{a\}$ is a singleton then $A^\perp$ is also denoted by $a^\perp$; a is ISOTROPIC if $a \in a^\perp$. If A is any subspace of V, then the subspace $A^\perp \cap A$ is called the RADICAL of A. A subspace A is called isotropic if it contains a non-zero isotropic vector and is TOTALLY isotropic if $A^\perp \supseteq A$. If $A \cap A^\perp = \mathbf{O}$ then A is NON-SINGULAR or ANISOTROPIC.

The binary form β is NON-SINGULAR if the radical of (V, β) is zero.

The importance of symmetric and alternating bilinear forms stems from the following standard result:

RESULT 103.5. [**33**, pp. 110–111]. A bilinear form over a field is reflexive iff it is symmetric or alternating.

If $(V, \cdot)$ represents a reflexive bilinear form we generally write $V = W \perp W^\perp$ if $V = W \oplus W^\perp$. Note that calling $W^\perp$ the *orthogonal complement* of W is motivated by the fact that if W is a non-singular subspace of V then $V = W \perp W^\perp$, cf. Remark 103.8(2).

The key to resolving such matters is the following lemma from elementary linear algebra concerned with orthogonal complements: we state it for the reflexive case only.

LEMMA 103.6. Let V be a finite-dimensional vector space with a reflexive bilinear form. Then

$$\dim W^\perp = \dim V - \dim W.$$

PROOF. Let $(e_1, e_2, \ldots, e_w)$ be a basis of W that extends to basis $(e_1, e_2, \ldots, e_w, e_{w+1}, \ldots, e_n)$ of V; so $x \in F^n$ is the n-tuple representing $\sum_{i=1}^n x_i w_i \in V$. Then x represents an element of $W^\perp$ iff

$$x \cdot e_i = \mathbf{o}, \quad 1 \le i \le w,$$

and the solutions of this correspond to the solution space of all $x \in F^n$ of the matrix equation

$$\left(x_1, x_2, \ldots, x_n\right) \begin{pmatrix} e_1 \\ e_2 \\ \vdots \\ e_w \end{pmatrix} = \mathbf{o}.$$

But the solution space for x of this matrix equation has rank $n-w$, as the coefficient matrix has rank w, since its rows are independent. The result follows. $\square$

We note that distinct projective points have distinct orthogonal complements:

COROLLARY 103.7. Let $(V, \cdot)$ be a finite-dimensional vector space over a field F, equipped with a reflexive bilinear form. Then

(a) For $a, b \in V^*$:
$$[a]^\perp = [b]^\perp \implies [a] = [b].$$

(b) The $\mathrm{rad}(V, \cdot) \le \frac{1}{2} \dim(V, \cdot)$.

PROOF. (a) Otherwise $[a, b]^\perp = [a]^\perp$, but the rank of the left-hand side is less than the rank of the right-hand side by the lemma. (b) Let n be the dimension of V, and assume $W = \mathrm{rad}(V, \cdot)$ has dimension $> \frac{n}{2}$. Then $\mathrm{rad}(V) \ge W \cap W^\perp > \mathbf{O}$, contradicting the non-singularity of V. $\square$

The following remarks review further elementary properties concerning orthogonality.

REMARKS 103.8. Let $(V, \cdot)$ be a vector space of rank n, over a field F, equipped with a non-singular reflexive bilinear form: thus $\mathrm{rad}(V) \ne \mathbf{O}$.

Let W be a subspace of V.

(1) V contains a non-zero isotropic vector iff at least one of the matrices congruent to $(V, \cdot)$ has a zero on its diagonal.

(2) If W is a non-singular subspace of V, i.e., $\mathrm{rad}(W) = \mathbf{O}$, then

$$V = W \perp W^\perp \forall W \le V.$$

(Apply Lemma 103.4).

(3) If $v \in V^*$ is isotropic, then $v^\perp$ is a hyperplane of V *(Apply Lemma 103.4).*

(4) If $(V, \cdot)$ is an alternating space, then $v \le v^\perp$, for all $v \in V$. Moreover, the map $v \to v^\perp$ is a bijective involution on the set of points and hyperplanes of $PG(V, F)$ that interchanges the points and hyperplanes and preserves their incidence.

103.2. Quadratic Forms and Orthogonal Decompositions.

In this section, we introduce quadratic forms and isometries between bilinear spaces and between spaces equipped with quadratic forms, called orthogonal spaces. This enables us to state the well-known Witt's lemma, a powerful tool in the investigation of bilinear and orthogonal spaces.

The homogeneous quadratic form in n variables, over any field F, gives rise to a map $Q\colon F^n \to F$, with which it may be identified. Thus a homogeneous quadratic form on F^n is, by definition, a map of type:

$$Q\colon F^n \to F$$

$$(x_1, x_2, \ldots, x_n) \mapsto \sum_{1 \le i \le j \le n} f_{ij} x_i x_j.$$

So $\beta(x,y) = Q(x+y) - Q(x) - Q(y)$ is clearly a symmetric F-bilinear form on F^n. This suggests a coordinate-free definition of a quadratic form, valid for arbitrary vector spaces, over any field.

DEFINITION 103.9. Let V be a vector space over a field F. A quadratic form is a map $Q\colon V \to F$ such that for all $x, y, v \in V$ and $\lambda \in F$:

$$Q(\lambda v) = \lambda^2 Q(v)$$

$$Q(x+y) = Q(x) + Q(y) + \beta(x,y),$$

where $\beta\colon V \times V \to F$ is a bilinear form, called the bilinear form ASSOCIATED with Q, or the bilinear form POLARIZING Q. The pair (V, Q) is an ORTHOGONAL SPACE.

The polarizing bilinear form β is thus always symmetric, and, in characteristic 2, it must be alternating—a stronger condition.

PROPOSITION 103.10. Suppose (Q, V) is a quadratic form, over a field F, polarized by β. Then the following hold.

(1) If $A\colon W \to V$ is a linear bijection, then the map

$$Q \circ A\colon W \to F$$

$$x \mapsto Q(Ax)$$

is a quadratic form on W that polarizes to the bilinear map β_A defined by:

$$\beta_A(x,y) := \beta(Ax, Ay).$$

(2) The quadratic forms of type $Q \circ A$ are said to be EQUIVALENT to Q, and this is a genuine equivalence relation on all the quadratic forms defined on the same vector space.

(3) When $\dim_F V = n$, every quadratic form on V is equivalent to a homogeneous quadratic form on F^n: any linear bijection $\alpha\colon F^n \to V$, specified by a fixed basis of V, defines a quadratic form of the required type.

Under any such identification $\alpha\colon F^n \to V$, the variety defined by corresponding homogeneous quadratic form Q_α, that is the set of projective points:

$$\{x \in F^n \mid Q_\alpha(x) = 0\},$$

is a QUADRIC on F^n.

(4) Equivalent quadratic forms on V polarize to the same congruent bilinear map.

(5) If $W \leq V$ is a subspace of V, then the restriction map $Q_W := Q^W$ is a quadratic form that polarizes to the restriction map $\beta_W := \beta^{W \times W}$. Thus, a quadratic form on V INDUCES a quadratic form on any subspace W, that polarizes to the restriction of β to $W \times W$.

(6) If char $F \neq 2$, then β is a symmetric bilinear form and

$$\forall x \in V : Q(x) = \frac{1}{2}\beta(x, x).$$

Hence over fields of characteristic $\neq 2$ quadratic forms and symmetric bilinear forms are essentially equivalent forms.

(7) If char $F = 2$ then β is an alternating bilinear form. However, as will become evident, distinct quadratic forms may polarize to the same alternating form.

(8) Suppose V is a direct sum of pairwise orthogonal spaces, relative to β; thus: $V = \oplus \sum_{i \in I} V_i$, where for distinct $i, j \in I$, $V_i \perp V_j = \{\mathbf{O}\}$, relative to β, and let Q_i denote the restriction of Q to V_i; thus β is the orthogonal sum of $(\beta_i)_{i \in I}$, where β_i denotes the polarized form of Q_i. Then:

$$Q\left(\oplus \sum_{i \in I} v_i\right) = \oplus \sum_{i \in I} Q_i(v_i).$$

We shall call the orthogonal space (V, Q) an ORTHOGONAL SUM of the family of quadratic subspaces $(V_i, Q_i)_{i \in I}$.

The converse of the last part may be stated as follows:

PROPOSITION 103.11. Suppose $(V_i, Q_i, \beta_i)_{i \in I}$ is a family of vector spaces V_i, $i \in I$, over the same field F, such that each V_i, $i \in I$, is equipped with a quadratic form Q_i, polarizing to β_i, then the external direct sum $V := \oplus \sum_{i \in I} V_i$ is assigned the DIRECT SUM quadratic form $Q = \oplus \sum_{I \in I} Q_i$ defined by

$$Q\left(\oplus \sum_{i \in I} v_i\right) := \oplus \sum_{i \in I} Q_i(v_i);$$

this Q is a genuine quadratic form on V and it polarizes to the direct sum of the family of bilinear forms $(\beta_i)_{i \in I}$. Thus (V, Q) is the direct orthogonal sum of the natural copies of V_i in V.

We introduce the notion of isometries and similarities for vector spaces equipped with bilinear forms.

DEFINITION 103.12. Let (V, f) and (W, g) be bilinear forms, on vector spaces V and W, respectively, over a common field F, and let $\phi \colon V \to W$ be a linear bijection. Then

(1) $\phi \colon V \to W$ is an ISOMETRY if:

for all x and y in V: $g(\phi(x), \phi(y)) = f(x, y)$.

(2) $\phi \colon V \to W$ is a SIMILARITY if there is a constant $\lambda_\phi \in F$ such that:

for all x and y in V: $g(\phi(x), \phi(y)) = \lambda_\phi f(x, y)$.

Since orthogonal spaces are associated with polarities, isometries and similarities between them have a natural interpretation. However, we require isometries to preserve the associated quadratic forms—is a stronger condition in characteristic 2.

DEFINITION 103.13. Let (V, Q) and (W, R) be orthogonal spaces over a field F, and let $\phi \colon V \to W$ be a linear bijection. Then

(1) $\phi \colon V \to W$ is an ISOMETRY if:

$$R(\phi(x)) = Q(x) \forall x \in V$$

(2) $\phi \colon V \to W$ is a SIMILARITY if there is a constant $\lambda_\phi \in F$ such that:

$$R(\phi(x)) = \lambda_\phi Q(x) \forall x \in V.$$

It is straightforward to check that an isometry/similarity $\phi \colon V \to W$ between orthogonal spaces, (V, Q) and (W, R), is also an isometry/similarity between (V, π_Q) and (W, π_R), where π_S denotes the polarization of quadratic form S.

We now turn to Witt's lemma, which we only state for reflexive or orthogonal spaces. We refer the reader to Aschbacher's book, [**34**, pp. 81–85] for the proof.

THEOREM 103.14. (Witt's Lemma). Let V be a reflexive bilinear form (which must therefore be symmetric or symplectic, Result 103.5) or an orthogonal space. Suppose U and W are subspaces of V and that $\alpha \colon U \to W$ be an isometry. Then α extends to an isometry of V.

Since totally singular spaces of the same dimension are isomorphic, it follows immediately:

COROLLARY 103.15. Maximal singular subspaces of V have the same dimension, called the WITT INDEX.

For any reflexive *bilinear* space (V, β) we defined its radical to be the largest subspace annihilating V, cf. Definition 103.4. We now define the radical of an *orthogonal* space (V, Q) as the subspace of *singular* points of Q that lie in the radical of the bilinear form β that polarizes Q.

DEFINITION 103.16. Let $Q \colon V \to F$ be a quadratic form, on a vector space V over a field F, with polarization β. Then $v \in V$ is SINGULAR if v is in $\mathrm{rad}(V, \beta)$ and $Q(v) = 0$. The RADICAL of (V, Q) is the subspace $\mathrm{rad}(V)$ (or $\mathrm{rad}(V, Q)$) of $\mathrm{rad}(V, \beta)$ specified by:

$$\mathrm{rad}(V) = \{ \, x \in \mathrm{rad}(\beta) \mid Q(x) = \mathbf{o} \, \}.$$

Thus, the radical of an orthogonal space (V, Q) might be a proper subspace of $\mathrm{rad}(\beta)$, where β is the associated polarization. Although, for characteristic $\neq 2$, $\mathrm{rad}(V)$ and $\mathrm{rad}(V, \beta)$ are obviously the same, in characteristic 2 it may happen that $\mathrm{rad}(V, Q) < \mathrm{rad}(V, \beta)$. However, even here, $\mathrm{rad}(V, Q)$ is at least a hyperplane of $\mathrm{rad}(V, \beta)$:

REMARK 103.17. Suppose V is a vector space over a perfect field F with characteristic 2.

A quadratic form $Q \colon V \to F$, with associated bilinear form b, is non-singular only if $V^\perp$, the radical of b, has rank ≤ 1. (In particular, any non-defective form is non-singular.)

PROOF. The restriction $Q|V^\perp$ satisfies the identities:

$$\forall x, y \in V^\perp : Q(x + y) = Q(x) + Q(y)$$
$$\forall \alpha \in F \forall x \in V^\perp : Q(\alpha x) = \alpha^2 x,$$

and so $Q|V^\perp \to F$ is an additive map and is semilinear because F is perfect (this means that $x \mapsto x^2$ is a field *automorphism* of F—rather than just an injective isomorphism). But since Q is non-singular, by definition, it cannot vanish on any non-zero subset of $V^\perp$, so Q injectively maps the F-space $V^\perp$ into the rank one F-space F, and so as an F space either $V^\perp \cong F$ or $V^\perp$ is trivial. The result follows.   □

REMARK 103.18. If the radical of quadratic form has rank 1, then the form is not always non-singular.

PROOF. For example, take the symplectic form b of type $V = H \perp [e]$ where $b|H$ is non-singular and $b|[e]$ is the zero form, where e is a non-zero vector in $V \setminus H$. Now define the quadratic form on Q so that it induces b, by specifying $Q(e) = 0$, and arbitrarily specifying Q on H, via its values on any basis of H. Thus e is a singular point on the radical of b, so Q is singular.   □

We note further the following important link between the radical of an orthogonal space and the radical of its polarization.

Let (V, Q, β) denote an orthogonal space over a field F, that polarizes to β. Then $\mathrm{rad}(V, \beta)$ is the largest subspace W of V such that the restriction $Q \colon W \to F$ is an additive homomorphism and the kernel of this homomorphism is $\mathrm{rad}(V, Q)$, the radical of the orthogonal space.

103.3. Orthogonal Basis in Characteristic $\neq 2$.

Although we shall mainly be concerned with quadratic forms of characteristic 2, to put this work in context, we discuss in this section the situation when this is not the case. Our discussion is based on Kaplansky's book [**882**].

THEOREM 103.19. Let $(V, \cdot)$ be a non-singular symmetric bilinear form over a field F of characteristic $\neq 2$. Then

(a) V contains an anisotropic vector.
(b) V contains an orthogonal basis.
(c) $(V, \cdot)$ is congruent to a diagonal matrix, with all entries non-zero.

PROOF. If all vectors are isotropic, then we have the identity

$$\mathbf{o} = (x + y) \cdot (x + y) = 2x \cdot y,$$

contradicting the non-singularity of V, and (a) follows. Choose $v \in V$ to be anisotropic, so $V = v \perp v^\perp$ and hence also $v^\perp$ non-singular by the non-singularity of V. So (b) follows by induction. Now (c) holds because the matrices associated with orthogonal bases are diagonal: the entries are non-zero since $\mathrm{rad}(V) = \mathbf{O}$.   □

The following are corollaries to the theorem.

COROLLARY 103.20. Let F be a field of characteristic $\neq 2$.

(1) Let $(V, \cdot)$ be a non-singular symmetric F-bilinear space. Every set of mutually orthogonal linearly independent anisotropic vectors B_0 extends to an orthogonal basis $B_0 \cup B_1$ of V.

(2) Every symmetric non-singular matrix over F is congruent to a non-singular diagonal matrix.

We now introduce the most important bilinear spaces: called hyperbolic planes (in vector notation) and hyperbolic lines (in projective space notion). Here, we continue to focus on the case when the characteristic $\neq 2$, but later on, Definition 103.29, we extend the definition to handle quadratic forms.

DEFINITION 103.21. Let $(V, \cdot)$ be a non-singular symmetric bilinear space over a field F of characteristic $\neq 2$. Then V is called a HYPERBOLIC PLANE (also a projective HYPERBOLIC LINE!!) if it is congruent to the diagonal matrix $\mathrm{Diag}[1, -1]$.

We now characterize the hyperbolic planes, in fields of characteristic $\neq 2$, as being the projective lines that contain non-zero isotropic vectors. The proof chosen here is based on discriminants, [**882**, Theorem 16, p. 19], and is thus not quite self-contained. However, what we really need is the alternating version of this result, for the characteristic 2 case, and for this case we can easily give the details.

LEMMA 103.22. Let $(V, \cdot)$ be a rank two non-singular symmetric bilinear space over a field F of characteristic $\neq 2$. Then

(1) If char $F \neq 2$, V is a hyperbolic plane (i.e., a projective hyperbolic line) iff it has a non-zero isotropic vector.
(2) If char $F = 2$, and the form $\cdot$ is *alternating*, then $(V, \cdot)$ is congruent to the matrix
$$\begin{pmatrix} 0 & 1 \\ 1 & 0 \end{pmatrix}.$$

PROOF. Consider first the case when char $F = 2$. The hyperbolic line, congruent to $\mathrm{diag}(1, -1)$, admits '$(1, 1)$' as an isotropic vector. Conversely, assume some $u \in V^*$ is isotropic. So by the non-singularity of $(V, \cdot)$, we may choose a $v \in V - [u]$ such that (u, v) is a basis such that $v \cdot u = 1$. Inspecting the representation of $(V, \cdot)$ on this basis shows that the discriminant of $(V, \cdot)$ is -1. But $\mathrm{Diag}[1, -1]$ has the same discriminant and the result follows, since over finite fields of odd order the discriminant determines the congruence class of symmetric *non-singular* bilinear forms.

Next consider the case when char $F = 2$, and the space is *alternating*. Now all matrices representing the form are symmetric and all vectors, a fortiori all basis vectors, are isotropic. Thus the above argument shows that on any basis the matrix of the form is
$$\begin{pmatrix} 0 & c \\ c & 0 \end{pmatrix},$$
where c is non-zero by the non-singularity hypothesis. As in the previous case we can also force $c = 1$, as desired.   □

Note that the above has an obvious analogue for the symplectic forms, but we shall not use it.

For the characteristic 2 case we shall need a slight variation of the above that applies to *quadratic forms*, and will be shown in Lemma 103.30.

An immediate inductive corollary of the lemma is:

COROLLARY 103.23. Let $(V, \cdot)$ be an n-dimensional non-singular symmetric bilinear space over a field F of characteristic $\neq 2$. Then V is an orthogonal sum of $r \leq n$ hyperbolic planes and an anisotropic space.

The uniqueness of r also holds, see [**882**, Theorem 18], e.g., by considering the null subspaces in V that have maximal rank, viz. r.

Turning to the even order version of the above, for alternating forms, we note that here there cannot be any anisotropic vectors. Thus we have the following:

COROLLARY 103.24. *Let $(V, \cdot)$ be an n-dimensional non-singular alternating bilinear space over a field F of characteristic $\neq 2$. Then V is an orthogonal sum of $r \leq n$ hyperbolic planes; in particular n is always even.*

Note that the above corollary can be formulated so as to apply to any non-singular symplectic space.

Later we shall prove an important analogue of the above for *quadratic forms*, Proposition 103.31.

103.4. Matrix Representations.

If (V, Q) is an orthogonal space over a field F of characteristic $\neq 2$, it easily follows that $Q(x) = \frac{1}{2}\beta(x, x)$ so in this case Q and β determine each other and hence Q is uniquely specified by the action of β on $B \times B$, where B is a basis of V. However, for char $F = 2$, Q is not specified completely until the restriction Q^B is explicitly specified.

The following result describes explicitly the polarizing matrix M associated with any quadratic form on F^n, F a field of any characteristic. The proof is a routine exercise.

THEOREM 103.25. *The mapping $Q \colon F^n \to F$ defined by*

$$\forall x_i \in F, i = 1, \ldots, n : Q\left(\sum_{i=1}^{n} x_i e_i\right) = \sum_{i=1}^{n} x_i{}^2 Q(e_i) + \sum_{i=1}^{n}\sum_{j>i}^{n} x_i x_j \beta(e_i, e_j),$$

is a quadratic form on F^n that polarizes to the bilinear form

$$\forall x, y \in F^n : (x, y) \to x(2S + B)y^t,$$

where $S = \operatorname{diag}(Q(e_1), Q(e_2), \ldots, Q(e_n))$, and $B = (b_{ij})$ is a symmetric matrix with an all-zero diagonal and $b_{ij} = \beta_{ij} := \beta(e_i, e_j)$ for $i > j$; more explicitly, the polarizing matrix is:

$$
2\begin{pmatrix}
Q(e_1) & 0 & 0 & \cdots & 0 \\
0 & Q(e_2) & 0 & \cdots & 0 \\
\vdots & & \ddots & & \vdots \\
\vdots & & & \ddots & \vdots \\
0 & 0 & 0 & \cdots & Q(e_n)
\end{pmatrix}
+
\begin{pmatrix}
0 & \beta_{12} & \beta_{13} & \cdots & \beta_{1n} \\
\beta_{21} & 0 & \beta_{23} & \cdots & \beta_{2n} \\
\vdots & & \ddots & & \vdots \\
\vdots & & & \ddots & \vdots \\
\beta_{n1} & \beta_{n2} & \beta_{n3} & \cdots & 0
\end{pmatrix}
$$

PROOF.

$$Q(x+y) - Q(x) - Q(y) =$$

$$\left(\sum_{i=1}^{n} (x_i + y_i)^2 Q(e_i) + \sum_{i=1}^{n} \sum_{j>i}^{n} (x_i + y_i)(x_j + y_j) \beta(e_i, e_j) \right)$$

$$- \left(\sum_{i=1}^{n} x_i^2 Q(e_i) + \sum_{i=1}^{n} \sum_{j>i}^{n} x_i x_j \beta(e_i, e_j) \right)$$

$$- \left(\sum_{i=1}^{n} y_i^2 Q(e_i) + \sum_{i=1}^{n} \sum_{j>i}^{n} y_i y_j \beta(e_i, e_j) \right)$$

$$= \left(\sum_{i=1}^{n} (2x_i y_i) Q(e_i) + \sum_{i=1}^{n} \sum_{j>i}^{n} (x_i y_j + x_j y_i) \beta(e_i, e_j) \right).$$

Thus the associated bilinear form may be represented in the matrix form xMy^t where

$$M = 2 \begin{pmatrix} Q(e_1) & 0 & 0 & \cdots & 0 \\ 0 & Q(e_2) & 0 & \cdots & 0 \\ \vdots & & \ddots & & \vdots \\ \vdots & & & \ddots & \vdots \\ 0 & 0 & 0 & \cdots & Q(e_n) \end{pmatrix}$$

$$+ \begin{pmatrix} 0 & \beta(1,2) & \beta(1,3) & \cdots & \beta(1,n) \\ \beta(2,1) & 0 & \beta(2,3) & \cdots & \beta(2,n) \\ \vdots & & \ddots & & \vdots \\ \vdots & & & \ddots & \vdots \\ \beta(n,1) & \beta(n,2) & \beta(n,3) & \cdots & 0 \end{pmatrix},$$

and since we must also have $xMy^t = yM^t x^t$, M is of the required form. □

The point of the following result is that the quadratic form considered above is generic; in particular, we have a full knowledge of the associated polarization.

THEOREM 103.26. Suppose V is a vector space with a basis $B = (e_1, e_2, \ldots, e_n)$, over any field F. Let β be any bilinear form on V and $f: B \to F$ any map.

(1) The following are pairwise equivalent:
(a) The bilinear form β is symmetric and

$$\beta(e_i, e_i) = 2f(e_i), \quad \text{for} \quad 1 \le i \le n.$$

(Thus, for char $F = 2$ the bilinear form β is any *alternating* form and $f: B \to F$ is any map, while for char $F \ne 2$, β is any *symmetric* bilinear form such that its '*diagonal* entries' are all of form $\beta(e_i, e_i) = 2f(e_i)$.)
(b) $f: B \to F$ extends to a quadratic form $Q_f: V \to F$ that polarizes to β.

(c) $f\colon B \to F$ extends to a unique quadratic form $Q_f\colon V \to F$ that polarizes to β.

(d) The quadratic form $Q_f \to F$ given by

$$\forall x_i \in F, i = 1, \ldots, n : Q_f\left(\sum_{i=1}^{n} x_i e_i\right) = \sum_{i=1}^{n} x_i{}^2 Q(e_i) + \sum_{i=1}^{n}\sum_{j>i}^{n} x_i x_j \beta(e_i, e_j),$$

is an extension of the map $f\colon B \to F$ that polarizes to β.

(2) The quadratic forms on V that polarize to a given (necessarily) symmetric β are of form Q_f above, where $f\colon B \to F$ satisfies the additional condition $2f(i) = \beta(i, i)$, for $1 \le i \le n$, but is otherwise arbitrary.

PROOF. Recall that if Q is any quadratic form, it polarizes to a symmetric bilinear form β and that for char $F = 2$, β must also be alternating (i.e., its diagonal entries associated with any basis must be zero), while for all other characteristic $\beta(e_i, e_i) = 2f(e_i)$, for all i. Thus, any of the conditions (b), (c), or (d) imply (a). Hence we may assume the hypothesis of (a), i.e., as just spelled out, to establish the other conditions.

We verify that now that if Q is a quadratic form that polarizes to the given β and such that $Q^B = f$ then $Q = Q_f$, as specified in (d), is forced: this means that (a) implies (d) and (c) and hence all the equivalences hold because (b) is a special case of (c).

We induct on n, the dimension of V, to show Q has the form claimed in (d). Any $v \in V$ may be written

$$v = \sum_{i=1}^{n} x_i e_i, \forall x_i \in F,$$

and

$$Q(v) = Q\left(x_1 e_1 + \sum_{i=2}^{n} x_i e_i\right) = x_1{}^2 Q(e_1) + Q\left(\sum_{i=2}^{n} x_i e_i\right) + x_1 \beta\left(e_1, \sum_{i=2}^{n} x_i e_i\right)$$

$$= x_1{}^2 Q(e_1) + Q\left(\sum_{i=2}^{n} x_i e_i\right) + \sum_{i=2}^{n} x_1 x_i \beta(e_1, e_i),$$

so by our tacit inductive hypothesis,

$$Q\left(\sum_{i=1}^{n} x_i e_i\right) = x_1{}^2 Q(e_1) + \sum_{i=2}^{n} x_i{}^2 Q(e_i) + \sum_{i=2}^{n}\sum_{j>i}^{n} x_i x_j \beta(e_i, e_j)$$

$$+ \sum_{j=2}^{n} x_1 x_j \beta(e_1, e_j)$$

$$- \sum_{i=1}^{n} x_i{}^2 Q(e_i) + \sum_{i=1}^{n}\sum_{j>i}^{n} x_i x_j \beta(e_i, e_j),$$

as claimed. It remains to check whether the polarizing bilinear form of this quadratic form is the given β. This follows by repeating the routine argument used in Theorem 103.25 to the matrix of Q above, which, relative to B, is clearly

$$2\,\mathrm{Diag}\,(f(e_1), f(e_2), \ldots f(e_n)) + [\beta(i, j)]_{1 \le i, j \le n},$$

where $\beta(i, i) = 2f(i)$, for all i.

Thus (a) implies Q is at most unique and corresponds to the quadratic form above. Thus (a) implies (d), and all the other equivalences easily follow, as already indicated.

The final case is a corollary of the equivalences established. □

REMARKS 103.27. (1) Let F be a field of characteristic 2, with finite basis B. There is exactly one quadratic form $Q: V \to F$ such that for a given pair of maps $(q: B \to F, b: B \times B \to F)$, where for all $i, j \in B$, $b(i,j) = b(j,i)$ and $b(i,i) = 0$, such that $Q^B = q$ and $b = \beta^{B \times B}$, where β is the alternating form polarized by Q.

(2) We note that this fails for fields of characteristic $\neq 2$.

(3) If we count the number of different quadratic forms on $GF(q)^n$, as a $GF(q)$ space, for q odd and even, they are the same, even though the polarizing matrix specifies the form only in the case when q is odd.

103.5. Hyperbolic Decomposition.

Throughout this section, V is a vector space over a field F of characteristic 2, and $Q: V \to F$ is a quadratic form on V. Thus (V, Q) is an *orthogonal* space, and we generally use multiplicative or dot-product notation to represent a bilinear form that polarizes any given quadratic form. It will often be necessary or convenient to insist that F is perfect. We shall show that orthogonal spaces over Galois fields are direct sums of hyperbolic lines (appropriately defined) and possibly one anisotropic space that cannot be orthogonally decomposed.

An orthogonal space (V, Q) is *anisotropic* if it has no non-zero singular vectors. Over such perfect fields, such spaces do not admit non-trivial orthogonal decompositions:

REMARK 103.28. An orthogonal sum of two or more non-zero anisotropic orthogonal spaces, over a perfect field F of characteristic 2, is not anisotropic.

PROOF. Suppose $V = A \perp B$ is a direct sum of quadratic spaces A and B. Choosing $a \in A^*$, $b \in B^*$ we observe that by orthogonality:

$$\forall \alpha, \beta \in F : Q(\alpha a + \beta B) = \alpha^2 Q(a) + \beta^2 Q(b),$$

and since $Q(a) \neq 0 \neq Q(b)$, the right-hand side of the identity can be made zero for $\alpha \neq \beta$ since the field F is perfect. □

Obviously, any non-singular rank one orthogonal space is an anisotropic. In the rank two case, anisotropic and isotropic orthogonal spaces exist, even over finite fields. For example, over $GF(2)$, the form $x^2 + xy + y^2$ is anisotropic but xy is isotropic. In view of the above remark, we focus on isotropic rank 2 orthogonal spaces. To classify these, we extend the definition of a hyperbolic plane (also called 'line'), Definition 103.21 to an orthogonal space.

DEFINITION 103.29. A HYPERBOLIC PLANE (or HYPERBOLIC LINE) of a quadratic form (V, Q) is a two-dimensional vector subspace of V containing a basis $[a, b]$ such that $Q(a) = Q(b) = 0$ and $a \cdot b = 1$ relative to the polarizing bilinear form.

Thus a hyperbolic plane, is just a non-degenerate rank two orthogonal space with a basis consisting of two singular vectors: the dot product can be made unity by scaling. We now show that any non-degenerate rank two *isotropic* orthogonal space is a hyperbolic line; this is an 'odd-order' version of Lemma 103.22.

LEMMA 103.30. Let (V, Q) be a non-singular orthogonal space of rank two, over any field F, such that $Q(v) = 0$ for some $v \in V^*$. Then V contains a hyperbolic line through v.

PROOF. Since $Q(v) = 0$, $v^\perp = V$ cannot hold without contradicting our assumption that Q is non-singular. Hence $\beta(v, w) \neq 0$, for some w, and by scaling v we may assume further that $\beta(v, w) = 1 = \beta(w, v)$. We wish to choose a w that satisfies the additional condition $Q(w) = 0$. So consider:

$$Q(w - \lambda v) = Q(w) - \lambda \beta(w, v) + \lambda^2 Q(v)$$
$$= Q(w) - \lambda.$$

Thus, taking $\lambda = Q(w)$, we obtain $Q(w') = 0$, where $w' = w - Q(w)v$. Moreover,

$$\beta(v, w') = \beta(v, w - Q(w)v) = \beta(v, w) - Q(w)\beta(v, v)$$
$$= \beta(v, w) = 1$$

So the subspace with basis v, w' is the required hyperbolic line. □

In Corollary 103.24 we obtained a canonical hyperbolic splitting associated with an *alternating form*. Here we do the same for an orthogonal space, i.e., the quadratic form rather than its polarization.

PROPOSITION 103.31. There exists unique non-negative integer $0 \leq r \leq n$ such that $V = H_r \perp A$ where H_r is a direct sum of hyperbolic lines and A is an anisotropic space.

PROOF. We prove existence. If V is anisotropic, the theorem is vacuous. So assume $x \in V^*$ is non-singular. Then, by definition, Q is non-singular on any plane containing x, so by the previous lemma, V contains a hyperbolic plane H—take this to be any plane containing x. Since the radical of any hyperbolic line is zero it follows that $V = H \oplus H^\perp$; as always, $H^\perp$ means the orthogonal complement of the alternating form polarizing V. Clearly, $H^\perp$ is an orthogonal space relative to the restriction $Q|H^\perp$, and $V = H \perp H^\perp$.

Proceeding inductively, the process can be repeated on $H^\perp$ until an anisotropic space arises. The desired split follows. □

Hence, any non-singular orthogonal space polarizes to an alternating form that has a matrix of type:

$$\begin{pmatrix} \mathbf{H} & \mathbf{O} \\ \mathbf{O} & \mathbf{A} \end{pmatrix} \quad \text{where } \mathbf{H} = \begin{pmatrix} 0 & 1 & 0 & 0 & 0 & 0 & 0 & 0 \\ 1 & 0 & 0 & 0 & 0 & 0 & 0 & 0 \\ 0 & 0 & 0 & 1 & & & 0 & 0 \\ 0 & 0 & 1 & 0 & \cdots & & 0 & 0 \\ \vdots & & \vdots & & \ddots & & \vdots \\ 0 & 0 & 0 & 0 & & & 0 & 1 \\ 0 & 0 & 0 & 0 & \cdots & & 1 & 0 \end{pmatrix} \quad \text{and } \mathbf{A} \text{ is anisotropic.}$$

In the case of a direct sum of hyperbolic lines, the matrix $\mathbf{H}$ above corresponds to a form of type:

$$Q\left(\sum_{i=1}^{n} x_i a_i + y_i b_i\right) = \sum_{i=1}^{n} x_i y_i$$

relative to a symplectic basis:

$$(a_1, b_1, a_2, b_2, \ldots, a_n, b_n,),$$

so reordering of this basis to

$$(a_1, a_2, \ldots, a_n, b_1, b_2, \ldots, b_n,),$$

the representation of Q on this new basis becomes the form:

$$Q\left(\sum x_i a_i + y_i b_i\right) = (x_1, x, \ldots, x_n)(y_1, y_2, \ldots, y_n)^t,$$

and we may summarize this by saying that a direct sum of hyperbolic lines may be represented as the dot product of the first n coordinates with the last n coordinates.

PROPOSITION 103.32 (Dot Product Representation of Hyperbolic Sums). Let $V = F^n \oplus F^n$; so the elements $\mathbf{z} \in V$ may be written as follows:

$$\mathbf{z} := (\mathbf{z_x}, \mathbf{z_y}), \quad \text{where} \quad \mathbf{z_x} \in F^n \oplus \mathbf{0} \quad \text{and} \quad \mathbf{z_y} \in \mathbf{0} \oplus F^n.$$

Let $a \cdot b \in F$ denote the ordinary dot product on F^n, and define the map $Q: V \to F$ by:

$$Q(\mathbf{z}) = \mathbf{z_x} \cdot \mathbf{z_y}.$$

Then $Q: V \to F$ is a quadratic form on V.

Every quadratic space that is an orthogonal sum of a finite number of non-defective hyperbolic lines may be represented over a suitable basis as a quadratic form on $F^n \oplus F^n$ such that

$$x \oplus y \in F^n \Rightarrow Q(x \oplus y) := x \cdot y.$$

We now further investigate the form Q of the proposition above. The matrix of this form is clearly

$$\begin{pmatrix} 0 & 1 \\ 1 & 0 \end{pmatrix},$$

and by an easy computation we determine that this matrix polarizes to the matrix

$$\begin{pmatrix} 0 & 1 & 0 & 0 & 0 & 0 & 0 & 0 \\ 1 & 0 & 0 & 0 & 0 & 0 & 0 & 0 \\ 0 & 0 & 0 & 1 & & & 0 & 0 \\ 0 & 0 & 1 & 0 & \cdots & & 0 & 0 \\ \vdots & & \vdots & & \ddots & & \vdots \\ 0 & 0 & 0 & 0 & & & 0 & 1 \\ 0 & 0 & 0 & 0 & & \cdots & 1 & 0 \end{pmatrix}.$$

As this matrix is non-singular, we conclude that the direct sum of alternating hyperbolic forms polarizes to a non-singular alternating form. Thus we have established:

REMARK 103.33. Every non-singular orthogonal space polarizes to a non-singular bilinear form.

103.6. Quadrics.

So far we have considered quadratic forms mainly in terms of vector spaces. In this section, we stress the geometric aspects of the subject: we regard vector spaces as being *projective spaces* and we focus on *quadrics*, rather than quadratic *forms*. Hence, almost all the results of this section are easily deduced by reformulating geometrically the results on orthogonal spaces established above.

Our treatment is closely based on [**529**, 22.1–22.5], part of the monumental three-volume treatise on projective spaces by Hirschfeld and Thas; the proofs that we omit may be found in [**529**]. To enable the reader to refer easily to this source, we have adopted the projective space notation of [**529**], and we also included a couple of items, Definition 103.39 and Lemma 103.40, that we shall never directly require but nonetheless play an important part in our source [**529**, 22.1–22.5].

Thus, when working with projective spaces, F^n generally denotes the vector space of $n+1$-tuples, over any field F, whose elements are indexed $(x_0, x_1, \ldots, x_n)$, and $\mathbf{P}(X)$ denotes the projective point associated with $X \in F^n$.

Given a homogeneous quadratic form over a finite field $F = GF(q)$:

$$f = \sum_{i=0}^{n} a_i x_i^2 + \sum_{0 \leq i < j \leq n} a_{ij} x_i x_j,$$

the associated variety $V(f)$ in $PG(n, q)$ is a QUADRIC. Thus, the quadric $V(f)$ consists of the following set of projective points of $PG(n, q)$:

$$(103.6.1) \quad V(f) = \left\{ \mathbf{P}((x_0, x_1, \ldots x_n)) \middle| \begin{array}{l} \sum_{i=0}^{n} a_i x_i^2 + \\ \sum_{0 \leq i < j \leq n} a_{ij} x_i x_j = 0, \text{ for } (x_0, x_1, \ldots x_n) \in F^n \setminus \mathbf{O} \end{array} \right\}.$$

We define some basic geometric terminology concerning quadrics.

DEFINITION 103.34. In $PG(n, F)$, F any field, let ℓ be any projective line and $\mathcal{Q}$ any quadric. The line ℓ is a SECANT to $\mathcal{Q}$ if $|\ell \cap \mathcal{Q}| = 2$ and is a TANGENT to $\mathcal{Q}$ if $|\ell \cap \mathcal{Q}| = 1$. The subspaces of maximum dimension contained in $\mathcal{Q}$ are the GENERATORS of $\mathcal{Q}$.

We state the standard properties related to the above, but only in the context of $GF(q)$.

REMARK 103.35. In $PG(n, q)$, let ℓ be a line and $\mathcal{Q}$ a quadric. Then one of the following occurs: (1) ℓ is skew to $\mathcal{Q}$; (2) ℓ is tangent to $\mathcal{Q}$; (3) $\ell \subset \mathcal{Q}$, and by Witt's lemma, Theorem 103.14, ℓ is in at least one generator of $\mathcal{Q}$.

The quadric form in, e.g., (103.6.1) is DEGENERATE if it can be reduced to a form on fewer than $n+1$ variables by a non-singular linear transformation, and in this case the variety $V(f)$ is said to be SINGULAR.

Under projectivities in $PG(n, q)$ there is one or two distinct non-singular quadrics $\mathcal{Q}_n$ according to whether n is even or odd. Although our main concern is with the n-odd case in this text, we still need to consider the possibility that both types may putatively occur as a subsystem of either type. The possible types are:

$$\mathcal{P}_n = V\left(x_0^2 + x_1 x_2 + \cdots + x_{n-1} x_n\right), \quad \text{parabolic}$$
$$\mathcal{H}_n = V\left(x_0 x_1 + x_2 x_3 + \cdots + x_{n-1} x_n\right), \quad \text{hyperbolic}$$
$$\mathcal{E}_n = V\left(\phi(x_0, x_1) + x_2 x_3 + \cdots + x_{n-1} x_n\right), \quad \text{elliptic},$$

where $\phi(x_0, x_1)$ is irreducible over $GF(q)$. A quadric projectively equivalent to a parabolic (respectively hyperbolic, elliptic) quadric is said to be parabolic (respectively hyperbolic, elliptic).

In the elliptic case, cf. [**529**, p. 7], the associated ϕ may be chosen from: $\phi(x_0, x_1) = x_0^2 + x_0 x_1 + d x_1^2$, where $d \in GF(q)$ is chosen such that ϕ is irreducible.

THEOREM 103.36. *If n is even then all non-singular quadrics are parabolic, and if n is odd, all non-singular quadrics are hyperbolic or elliptic.*

Thus, reviewing our notation, any $\mathcal{Q}_n$ is a $\mathcal{P}_n$, for n even, and for n odd any $\mathcal{Q}_n$ is an $\mathcal{H}_n$ or an $\mathcal{E}_n$. The degenerate quadrics turn out to be the CONES whose generators are all lines joining the points of a subspace (the vertex) to the points of some non-singular quadric in some complement to the vertex:

DEFINITION 103.37. Let V be the vector space with lattice $PG(n, q)$, and suppose $V = P \oplus S$ where the projective space S of rank $s + 1$ contains a non-singular quadric $\mathcal{S}_s$. Then the CONE $(P; \mathcal{S}_s)$ is the union all points that lie on the lines joining points of P, the VERTEX, to the points of $P \cup \mathcal{S}_s$:

$$(P; S) := \{ ab \mid a \in P, b \in P \cup \mathcal{S}_s \}.$$

As indicated in [**529**, p. 3], cones are just the degenerate quadrics: in particular cones are varieties. The following exercise considers important notions related to quadrics and cones.

REMARK 103.38. In $PG(n, q)$: (1) A non-degenerate quadric is a cone with an empty vertex.

(2) A degenerate quadric is a cone with a unique vertex.

We mentioned above that the generators of any quadric are the maximum-dimensional subspaces that it contains. We mention some important associated parameters:

DEFINITION 103.39. The PROJECTIVE INDEX g of a quadric is the projective dimension of its generators. If k is the dimension of the vertex of the quadric (so $k = -1$ if the quadric is non-singular), then the CHARACTER of the quadric is given by $w = 2g - k - n + 2$.

Note that the projective index is g means that the Witt index is $g + 1$, cf. Corollary 103.15. The generators of degenerate quadrics in terms of their non-singular sections may be described as follows.

LEMMA 103.40. [**529**, Lemma 22.4.3.] *The generators of a quadric $\Pi_k \cdot Q_t$, with vertex Π_k, are the subspaces of form $\Pi_k \oplus G$, where G ranges over the generators of Q_t.*

DEFINITION 103.41. Let $\mathcal{Q}_n$ be a quadric in $PG(n, q)$, associated with a vector space V of rank $n + 1$, over $F = GF(q)$. Let $f \colon V \to GF(q)$ denote a non-singular quadratic form, polarized by $g \colon V \times V \to F$, that specifies Q_n. Then the NUCLEUS of Q_n is the rad(g).

By the equivalence of quadratic and bilinear forms over $GF(q)$ of odd order, the nucleus, for non-singular quadrics, can only exist when q is even. In the non-singular-hyperbolic case, there is no nucleus even when q is even, by Remark 103.33. Recalling that for a non-singular quadric the radical of its polarizing bilinear form has rank ≤ 1, we now easily have:

RESULT 103.42. (See [**529**, Corollary 2, p. 10]). A non-singular quadric $\mathcal{Q}_n$ in $PG(n, q)$ has a nucleus, i.e., the associated quadric form polarizes to a degenerate form β, iff q and n are both even, and in this case the radical of β has rank one.

REMARK 103.43. Suppose $\mathcal{H}$ is a hyperbolic quadric in $PG(2g + 1, q)$, q even. Then for any point $y \notin \mathcal{H}$, and generator G of $\mathcal{H}$, $y^\perp$ is a hyperplane of $PG(2g+1, q)$, and the space $y^\perp \cap G$ is a hyperplane of G.

PROOF. By Result 103.42, a hyperbolic quadric must polarize to a non-singular bilinear form, so $y^\perp$ is a hyperplane, and the result holds unless $y^\perp > G$. But since G is totally singular, $y^\perp > G$ cannot hold unless $G^\perp \geq [y \cup G] > G$, and now rank $G^\perp > \text{rank } G = g + 1$, contradicting:

$$\text{rank } G^\perp = 2g + 2 - \text{rank } G = g + 1.$$

$\square$

We end with a fundamental theorem for non-singular hyperbolic quadrics, generalizing the ruling classes of a regulus. In $PG(3, q)$ the generators of a quadric (the points covered by a regulus) partition themselves into two disjoint classes such that two members are in the same class if and only if they have empty intersection. The generalization to arbitrary $PG(n, q)$ is the following:

THEOREM 103.44. Let Q be a hyperbolic quadric in $PG(n, q)$ and let $\mathcal{G}$ be its set of generators; so $n = 2g + 1$, where g is the projective index, see Definition 103.39. Define $A \sim B$, for $A, B \in \mathcal{G}$, iff the projective dimension $A \cap B$ has the same parity as g. The $\sim$ is an equivalence relation on $\mathcal{G}$ with exactly two classes: each class is called a SYSTEM OF GENERATORS, or a RULING CLASS.

PROOF. See [**529**, Lemma 22.4.10, Theorem 22.4.12, pp. 20–21]. $\square$

CHAPTER 104

Fundamental Results on Groups.

We list in this appendix various important results on groups and collineations.

104.1. The Classification of Doubly Transitive Groups.

It will be convenient to list here the classification theorem of finite doubly transitive groups.

Let v denote the degree of the permutation group.

The possibilities are as follows:

(A) G has a simple normal subgroup N, and $N \leq G \leq \operatorname{Aut} N$ where N and v are as follows:

(1) A_v, $v \geq 5$,

(2) $PSL(d, z)$, $d \geq 2$, $v = (z^d - 1)/(z - 1)$ and $(d, z) \neq (2, 2), (2, 3)$,

(3) $PSU(3, z)$, $v = z^3 + 1$, $z > 2$,

(4) $Sz(w)$, $v = w^2 + 1$, $w = 2^{2e+1} > 2$,

(5) $^2G_2(z)'$, $v = z^3 + 1$, $z = 3^{2e+1}$,

(6) $Sp(2n, 2)$, $n \geq 3$, $v = 2^{2n-1} \pm 2^{n-1}$,

(7) $PSL(2, 11)$, $v = 11$,

(8) Mathieu groups M_v, $v = 11, 12, 22, 23, 24$

(9) M_{11}, $v = 12$

(10) A_7, $v = 15$

(11) HS (Higman–Sims group), $v = 176$,

(12) .3 (Conway's smallest group), $v = 276$.

(B) G has a regular normal subgroup N, which is elementary Abelian of order $v = h^a$, where h is a prime. Identify G with a group of affine transformations $x \longmapsto x^g + c$ of $GF(h^a)$, where $g \in G_0$. Then one of the following occurs:

(1) $G \leq A\Gamma L(1, v)$,

(2) $G_0 \trianglerighteq SL(n, z)$, $z^n = h^a$,

(3) $G_0 \trianglerighteq Sp(n, z)$, $z^n = h^a$,

(4) $G_0 \trianglerighteq G_2(z)'$, $z^6 = h^a$, z even,

(5) $G_0 \trianglerighteq A_6$ or A_7, $v = 2^4$,

(6) $G_0 \trianglerighteq SL(2, 3)$ or $SL(2, 5)$, $v = h^2$, $h = 5, 11, 19, 23, 29$, or 59 or $v = 3^4$,

(7) G_0 has a normal extraspecial subgroup E of order 2^5 and G_0/E is isomorphic to a subgroup of S_5, where $v = 3^4$,

(8) $G_0 = SL(2, 13)$, $v = 3^6$.

104.2. Classification of Primitive Groups of $\Gamma L(4, q)$.

THEOREM 104.1. (Kantor and Liebler [**878**, (5.1)]).

Let H be a primitive subgroup of $\Gamma L(4, q)$ for $p^r = q$. Then one of the following holds:

781

(a) $H \geq SL(4, q)$,

(b) $H \leq \Gamma L(4, q')$ with $GF(q') \subset GF(q)$,

(c) $H \leq \Gamma L(2, q^2)$, with the latter group embedded naturally,

(d) $H \leq Z(H)H_1$, where H_1 is an extension of a special group of order 2^6 by S_5 or S_6. Here, q is odd, H_1 induces a monomial subgroup of $O^+(6, q)$, and H_1 is uniquely determined up to $\Gamma L(4, q)$-conjugacy,

(e) $H^{(\infty)}$ is $S_p(4, q)'$ or $SU(3, q^{1/2})$,

(f) $H \leq \Gamma O^{\pm}(4, q)$,

(g) $H^{(\infty)}$ is $PSL(2, q)$ or $SL(2, q)$ (many classes),

(h) $H^{(\infty)}$ is A_5. Here, H arises from the natural permutation representation of S_5 in $O(5, q)$ and $p \neq 2, 5$,

(i) $H^{(\infty)}$ is $2 \cdot A_5$, $2 \cdot A_6$ or $2 \cdot A_7$. These arise from the natural permutation representation of S_7 in $O(7, q)$,

(j) $H^{(\infty)} = A_7$ and $p = 2$,

(k) $H^{(\infty)} = S_p(4, 3)$, and $q \equiv 1 \bmod 3$. This arises from the natural representation of the Weyl group $W(E_6)$ in $O^+(6, q)$,

(l) $H^{(\infty)} = SL(2, 7)$, and $q^3 \equiv 1 \bmod 7$. Here, $H^{(\infty)}$ lies in the group $2 \cdot A_7$,

(m) $H^{(\infty)} = 4 \cdot PSL(3, 4)$, and q is a power of 9,

(n) $H^{(\infty)} = S_z(q)$, and $p = 2$.

104.3. Theorems on Collineation Groups.

104.3.1. Ostrom's Baer Trick and Ostrom Phantom.

THEOREM 104.2. (See Biliotti, Jha, and Johnson [**123**, p. 301]). Let V be a vector space of dimension $2n$ over a field $F \simeq GF(q)$, for $q = p^r$, p a prime and let S be an F-spread on V. If q is odd and A is an elementary Abelian 2-group which is a subgroup of $GL(n, F)$ such that all involutions of A are Baer then $|A|$ divides n. This result is often called 'Ostrom's Baer trick' and the associated integer n is referred to as the 'Ostrom dimension.'

LEMMA 104.3. Assume that H is an elementary Abelian collineation group of order 4 in the translation complement of a translation plane π of order q^2. If all involutions of H are Baer then q is a square.

PROOF. By Ostrom's Baer trick 4 divides the Ostrom dimension which divides $2r$, where $p^r = q$, for p a prime.   □

THEOREM 104.4. (See Johnson [**700**]). Let V be a vector space of dimension $2n$ over a field $F \simeq GF(q)$, for $q = p^r$, p a prime. Assume that a collineation σ of $GL(2n, q)$ has order dividing $q^n - 1$ but not dividing $q^t - 1$ for $t < n$. If σ fixes at least three mutually disjoint n-dimensional F-subspaces, then there is an associated Desarguesian spread Σ admitting σ as a kernel homology. Furthermore, the normalizer of $\langle \sigma \rangle$ is a collineation group of Σ.

In the situation above Σ is called an 'Ostrom phantom'.

104.3.2. Ostrom's p-Prime Theorem.

THEOREM 104.5. (Ostrom's p-Prime Theorem) (Ostrom [**1055**, (2.17)]).

Let π be a translation plane of order q^2 with spread in $PG(3, q)$. Let G be a subgroup of the linear translation complement and assume that $(p, |G|) = 1$ where $p^r = q$. Let $\overline{G} = GK/K$ where K is the kernel homology group of order $q - 1$.

Then at least one of the following holds:

(a) G is cyclic.

(b) $\overline{G}$ has a normal subgroup of index 1 or 2 which is cyclic or dihedral or isomorphic to one of $PSL(2,3)$, $PGL(2,3)$ or $PSL(2,5)$.

(c) G has a cyclic normal subgroup H such that G/H is isomorphic to a subgroup of S_4.

(d) G has a normal subgroup H of index 1 or 2. H is isomorphic to a subgroup of $GL(2,q^a)$, for some a, such that the homomorphic image of H in $PSL(2,q^a)$ is one of the groups in the list given under (b).

(e) There are five pairs of points on ℓ_∞ such that if (P,Q) is any such pair, there is an involutory homology with center P (or Q) whose axis goes through Q (or P). $\overline{G}$ has a normal subgroup $\overline{E}$ which is elementary Abelian of order 16. For each pair (P,Q) each element of $\overline{E}$ either fixes P and Q or interchanges them. $\overline{G}$ induces a transitive permutation group on these ten points.

(f) $\overline{G}$ has a subgroup isomorphic to $PSL(2,9)$ and acts in the following manner: Each Sylow 3-subgroup has exactly two fixed points on ℓ_∞. If (P,Q) is such a pair, G contains (P,OQ) and (Q,OP)-homologies of order 3. There are ten such pairs and $\overline{G}$ is transitive on these ten pairs.

(g) G has a reducible normal subgroup H not faithful on its minimal subspaces and satisfying the following conditions:

Either the minimal H-spaces have dimension two and H has index 2 in G or the minimal H-spaces have dimension 1. In the latter case, if H_0 is a subgroup fixing some minimal H-space pointwise, then H/H_o is cyclic and G/H is isomorphic to a subgroup of S_4.

104.3.3. Hering–Ostrom Theorem.

THEOREM 104.6. (Hering–Ostrom Theorem). (Hering [**476**] and Ostrom [**1047, 1052**]).

Let π denote a finite translation plane of order p^n and let E denote the collineation group in the translation complement which is generated by all affine elations.

Then one of the following holds:

(1) E is an elementary Abelian p-group,

(2) $p = 2$, and the order of E is $2t$ where t is odd,

(3) E is isomorphic to $SL(2,p^b)$,

(4) E is isomorphic to $S_z(2^c)$ and $p = 2$, or

(5) E is isomorphic to $SL(2,5)$ and $p = 3$.

Furthermore, when the group is non-solvable and not isomorphic to $S_z(2^c)$ then the net determined by the set of elation axes is a Desarguesian net.

104.3.4. Foulser–Johnson Theorem.

THEOREM 104.7. (Foulser–Johnson [**378, 379**]).

Let π denote a translation plane of order q^2 that admits a collineation group isomorphic to $SL(2,q)$.

Then π is one of the following planes:

(1) Desarguesian,

(2) Hall,

(3) Hering,

(4) Ott–Schaeffer,

(5) The Dempwolff plane of order 16.

THEOREM 104.8. (Johnson [**700**]).

Let π be a finite translation plane of order p^r which admits a collineation σ in the translation complement H_0 of order u, a prime p-primitive divisor of $p^r - 1$.

If σ fixes at least three mutually disjoint r-dimensional $GF(p)$-subspaces, then there exists a unique Desarguesian affine plane Σ consisting of the σ-invariant r-dimensional $GF(p)$-subspaces.

Furthermore, $N_{H_0}(\langle\sigma\rangle)$ is a collineation group of Σ.

104.3.5. Foulser's Theorems on Baer Groups.

THEOREM 104.9. (Foulser [**371**]).

Let π be a finite translation plane of order q^2 which contains a Baer subplane π_o incident with the zero vector. Let N_{π_o} denote the net of degree $q+1$ defined by the components of the subplane π_o. Let the kernel of π_o be isomorphic to $GF(p^a)$ where $q = p^r$ and p is a prime and assume that there are at least three Baer subplanes in N_{π_o} which are incident with the zero vector.

Then the number of subplanes of N_{π_o} incident with the zero vector is $1 + p^a$.

Furthermore, the set of Baer subplanes incident with the zero vector is isomorphic to $PG(1, p^a)$.

THEOREM 104.10. (Foulser [**375**]).

A finite translation plane of odd order p^r for p a prime cannot simultaneously admit non-trivial Baer p-collineations and non-trivial affine elations.

THEOREM 104.11. (Foulser [**375**]).

Let π be a finite translation plane of order 3^r. The following three properties cannot simultaneously hold: τ and σ are Baer 3-collineations and

(1) Fix $\tau \neq$ Fix σ,
(2) Fix $\tau \cap$ Fix $\sigma \neq 0$, and
(3) τ leaves Fix σ invariant.

104.3.6. Ostrom's Critical Deficiency Theorem.

THEOREM 104.12. (Ostrom [**1038**]).

Let N be a net of order q^2 and degree $q^2 - q$.

Then N can be extended to at most two non-isomorphic affine planes and if there are two extensions, the planes are related to each other by derivation.

104.3.7. Buekenhout, Delandtsheer, Doyen, Kleidman, Liebeck, and Saxl Theorem.

THEOREM 104.13. (Buekenhout, Delandtsheer, Doyen, Kleidman, Liebeck, and Saxl (not stated in its most general form) [**201**]).

Let π_o be a finite translation plane that admits a non-solvable flag-transitive group. Then π_o is one of the following types of planes:

(1) Hall of order 9,
(2) Hering of order 27,
(3) Lüneburg–Tits of order 2^{2r} for r odd,
(4) Desarguesian.

104.3.8. Ganley–Jha–Johnson Doubly Transitive Theorem.

THEOREM 104.14. (Ganley, Jha and Johnson [**393**]). Let π be a finite affine plane of order n not in $\{3^4, 3^6, 11^1, 19^2, 29^2, 59^2\}$ admitting a collineation group G

that has a point orbit Ω of length n and induces upon it a non-solvable doubly transitive group.

Then Ω is a line or a Baer subplane of π and the largest translation subgroup T of G acts transitively on Ω. Thus, $G = TG_0$ where G_0 is the stabilizer of any point 0 of Ω. Moreover, at least one of the following holds:

(1) π is a Desarguesian plane,

(2) π is a Hall plane,

(3) π is the Hering plane of order 25 or one of the two exceptional Walker planes of order 25,

(4) π is the Dempwolff plane of order 16.

104.3.9. Foulser–Kallaher Solvable Flag-Transitive Group Theorem.

THEOREM 104.15. (Foulser [**365**, Table II, p. 457], Foulser [**366**, p. 192 and Fig. 1 p. 201], and Foulser and Kallaher [**381**, (1.2)]). Let G be a solvable flag-transitive collineation group of a finite affine plane π of order p^r. Then, with 16 exceptions, $G \leq \Gamma L(1, p^{2r})$.

Of the 16 exceptions:

(a) 10 occur as doubly transitive groups of the Desarguesian affine planes of orders 3, 5, 7, 11, and 23.

(b) 3 occur as doubly transitive groups of the nearfield plane of order 9.

(c) 2 occur as rank 6 groups of the Desarguesian plane of order 11, and one as a rank 12 group of the Desarguesian plane of order 23.

104.3.10. Hering's Theorems on Flag-Transitive Groups.

THEOREM 104.16. (Hering [**479**, (4.4), p. 391]).

Let π_o be an affine plane of order $p^r > 8$ and $r > 1$ admitting a flag-transitive collineation group G. let u be a prime p-primitive divisor of $p^{2r} - 1$, let S denote the closure normal closure of a Sylow u-subgroup U within G_0 and let F denote the Fitting subgroup of G. Then

(a) SF/F is simple and

(b) G_o/S is isomorphic to a subgroup of $\Gamma L(1, p^{2r})$.

104.3.11. Fundamental Theorems on Quasi-Perspectivities and Reguli.

THEOREM 104.17. (Lüneburg [**961**, (37.10), p. 192]).

Let π_o be an affine plane of order p^r for p a prime. Assume that u is a p-primitive divisor of $p^r - 1$. If G is a solvable flag-transitive collineation group of π_o whose order is divisible by u then π_o is Desarguesian.

THEOREM 104.18. (Lüneburg [**961**, (49.5), p. 253]).

A collineation τ of a finite translation plane is an affine elation or a Baer collineation of order p if and only if the minimal polynomial of τ is $(x - 1)^2$.

THEOREM 104.19. (Lüneburg [**961**, (49.4), p. 253]).

Let π be a translation plane with spread in $PG(3, q)$. Let $q = p^r$ for p a prime. If τ is a collineation of π in the translation complement of order p, then the minimal polynomial is $(x - 1)^2$ or $(x - 1)^4$. If it is $(x - 1)^4$, then $p \geq 5$.

THEOREM 104.20. (Lüneburg [**961**, (48.5)]).

Let R be a regulus in $PG(3, q)$. Any Desarguesian affine plane with spread in $PG(3, q)$ that contains R admits the group which fixes the opposite regulus linewise as a collineation group.

104.3.12. Jha's Tangentially Transitive Theorem.

THEOREM 104.21. (Jha [**570**]).

A finite translation plane of order $p^{2r} \neq 16$ that admits a collineation group of order $p^r (p^r - 1)$ that fixes a Baer subplane pointwise is a generalized Hall plane and may be derived from a semifield plane whose middle nucleus contains a field isomorphic to $GF(p^r)$.

104.3.13. Liebeck's Theorem on Primitive Rank 3 Groups.

THEOREM 104.22. (Liebeck [**916**]).

Let G be a finite primitive affine permutation group of rank 3 and of degree $n = p^d$, with socle V, where $V \simeq (Z_p)^d$ for some prime p, and let G_0 be the stabilizer of the zero vector in V. Then G_0 belongs to one of the following classes (and conversely, each of the possibilities listed below does give rise to a rank 3 affine group).

(A) Infinite classes. These are:

(1) $G_0 \leq \Gamma L(1, p^d)$: all possibilities are determined in Foulser and Kallaher [**381**];

(2) G_0 is imprimitive: G_0 stabilizers a pair $\{V_1, V_2\}$ of subspaces of V, where $V = V_1 \oplus V_2$ and $\dim V_1 = \dim V_2$; moreover, $(G_o)_{V_i}$ is transitive on $V_i - \{0\}$ for $i = 1, 2$, and hence determined by Hering's theorem (and see below 104.24);

(3) Tensor product case: for some a, q with $q^a = p^d$, consider V as a vector space $V_a(q)$ of dimension a over $GF(q)$; then G_0 stabilizes a decomposition of $V_a(q)$ as a tensor product $V_1 \otimes V_2$ where $\dim_{GF(q)} V_1 = 2$; moreover, $G_0^{V_2} \rhd SL_a(q)$, or $G_o^{V_2} = A_7 < SL(4, 2)$ (and $p = q = 2$, $d = a = 8$), or $\dim_{GF(q)} V_2 \leq 3$;

(4) $G_0 \rhd SL(a, q)$ and $p^d = q^{2a}$;

(5) $G_0 \rhd SL(2, q)$ and $p^d = q^6$;

(6) $G_0 \rhd SU(a, q)$ and $p^d = q^{2a}$;

(7) $G_0 \rhd \Omega^{\pm}(2a, q)$ and $p^d = q^{2a}$, and if q is odd, G_0 contains an automorphism interchanging the two orbits of $\Omega^{\pm}(2a, q)$ on non-singular 1-spaces;

(8) $G_0 \rhd SL(5, q)$ and $p^d = q^{10}$ (from the action of $SL(5, q)$ on the skew square $\wedge^2(V_5(q))$);

(9) $G_0 \rhd \Omega_7(q).Z_{(2,q-1)}$ and $p^d = q^8$ (from the action of $B_3(q)$ on a spin module);

(10) $G_0/Z(G_0) \rhd P\Omega^+(10, q)$ and $p^d = q^{16}$ (from the action of $D_5(q)$ on a spin module);

(11) $G_0 \rhd S_z(q)$ and $p^d = q^4$ (from the embedding of $S_z(q) < Sp_4(q)$).

(B) 'Extraspecial' classes. Here, $G_0 \leq N_{GL(d,p)}(R)$ where R is an r-group, irreducible on V. Either $r = 3$ and $R \simeq 3^{1+2}$ (extraspecial of order 27) or $r = 2$ and $|R/Z(r)| = 2^m$ with $m = 1$ or 2. If $r = 2$ then either $|Z(R)| = 2$ and R is one of the two extraspecial groups R_1^m, R_2^m of order 2^{1+2m}, or $|Z(R)| = 4$, when we write $R = R_3^m$.

The possibilities are listed below in the appropriate table.

(C) 'Exceptional' classes. Here the socle L of $G/Z(G)$ is simple, and the possibilities are given in the appropriate table below.

The following remarks might be useful when reading Chapter 75 on rank 3 translation planes. In particular, consider the possible groups G that can act as a collineation group of a finite affine plane π. That is, we assume that π is a rank 3 affine plane. We know by the results of Kallaher and Liebler that π is a translation plane and that G contains the translation group of π. Hence, G is a finite primitive affine group of rank 3 and of degree p^d, with socle $V \cong (Z_p)^d$ for some prime p and G_0 is the stabilizer of the vector 0. Then G_0 is an irreducible subgroup of $GL(d,p)$. If $a \mid d$, then the group $\Gamma L(a, p^{\frac{d}{a}})$ has a natural irreducible action on V. Choose a to be minimal such that $G_0 \le \Gamma L(a, p^{\frac{d}{a}})$ in this action and put $q^a = p^d$.

LEMMA 104.23. Let $\mathcal{S}$ be a spread of V admitting G_0. Then the kernel K of $\Pi(V, \mathcal{S})$ is contained in $GF(q)$.

PROOF. Let $K = p^t$. Then $G_0 \le \Gamma L(\frac{d}{t}, p^t)$. By the minimality of a, we must have $GF(p^t) \le GF(q)$. □

Denote by $\mathfrak{D}_1$ and $\mathfrak{D}_2$ the orbits of G_0 on V, different from $\{0\}$ and by $\mathcal{S}_1$ and $\mathcal{S}_2$ the components of $\mathcal{S}$ contained in $\mathfrak{D}_1 \cup \{0\}$ and in $\mathfrak{D}_2 \cup \{0\}$, respectively (here we assume that G is not flag-transitive). Note that if $W \in \mathcal{S}$ then by Kallaher [817], $W - \{0\} = W^* \subset \mathfrak{D}_i$ for some i and $G_{0,W}$ is transitive on W^*. Since G_0 permutes the elements of $\mathcal{S}$, G_0 admits a transitive permutation representation of degree $\frac{|\mathfrak{D}_1|}{p^{\frac{d}{2}}-1}$ on $\mathcal{S}_1$ and a transitive permutation representation of degree $\frac{|\mathfrak{D}_2|}{p^{\frac{d}{2}}-1}$ on $\mathcal{S}_2$.

Now we refer to the main theorem and to Tables 12 through 14 of Liebeck.

Type of G	$n = p^d$	Subdegrees
(A1): $G_0 < \Gamma L(1, p^d)$	p^d	
(A2): G_0 imprimitive	p^{2m}	$2(p^m - 1), (p^m - 1)^2$
(A3): tensor product	q^{2m}	$(q+1)(q^m - 1), q(q^m - 1)(q^{m-1} - 1)$
(A4): $G_0 \triangleright SL_a(q)$	q^{2a}	$(q+1)(q^a - 1), q(q^a - 1)(q^{a-1} - 1)$
(A5): $G_0 \triangleright SL_2(q)$	q^6	$(q+1)(q^3 - 1), q(q^3 - 1)(q^2 - 1)$
(A6): $G_0 \triangleright SU_a(q)$	q^{2a}	$\begin{cases} \begin{cases} (q^a - 1)(q^{a-1} + 1), \\ q^{a-1}(q-1)(q^a - 1), \end{cases} & a \text{ even} \\ \begin{cases} (q^a + 1)(q^{a-1} - 1), \\ q^{a-1}(q-1)(q^a + 1), \end{cases} & a \text{ odd} \end{cases}$
(A7): $G_0 \triangleright \Omega_{2a}^\varepsilon(q)$	q^{2a}	$\begin{cases} \begin{cases} (q^a - 1)(q^{a-1} + 1), \\ q^{a-1}(q-1)(q^a - 1), \end{cases} & \varepsilon = + \\ \begin{cases} (q^a + 1)(q^{a-1} - 1), \\ q^{a-1}(q-1)(q^a + 1), \end{cases} & \varepsilon = - \end{cases}$
(A8): $G_0 \triangleright SL_5(q)$	q^{10}	$(q^5 - 1)(q^2 + 1), q^2(q^5 - 1)(q^3 - 1)$
(A9): $G_0 \triangleright B_3(q)$	q^8	$(q^4 - 1)(q^3 + 1), q^3(q^4 - 1)(q - 1)$
(A10): $G_0 \triangleright D_5(q)$	q^{16}	$(q^8 - 1)(q^3 + 1), q^3(q^8 - 1)(q^5 - 1)$
(A11): $G_0 \triangleright Sz(q)$	q^4	$(q^2 + 1)(q - 1), q(q^2 + 1)(q - 1)$

THEOREM 104.24. (Hering [479]).

Let G be a 2-transitive affine permutation group of degree p^d, with socle $V \cong (Z_p)^d$ for some prime p, and let G_0 be the stabilizer of the zero vector in V. Then G_0 belongs to one of the following classes (and conversely, each of the classes below does give a 2-transitive affine group).

(A) Infinite classes:
(1) $G_0 \le \Gamma L(1, p^d)$;
(2) $G_0 \triangleright SL(a, q)$ and $p^d = q^a$;
(3) $G_0 \triangleright Sp(2a, q)$ and $p^d = q^{2a}$;
(4) $G_0 \triangleright G_2(q)'$, $p^d = q^6$ and $p = 2$.

(B) Extraspecial classes: with the notation of part (B) of the theorem of Liebeck:

(i) $p^d = 5^2, 7^2, 11^2, 23^2, r = 2$ and $R = Q_8$, or

(ii) $r = 2$, $p^d = 3^4$ and $R = R_2^2 \simeq D_8 \circ Q_8$ and $G_0/R \leq S_5$.

(C) Exceptional classes:

(i) $G_0 \rhd SL(2,5)$ and $p^d = 9^2, 11^2, 19^2, 29^2, 59^2$ and $SL(2,5) < SL(2, p^{d/2})$,

(ii) G_0 is A_6, $p^d = 2^4$ and $A_6 \simeq S_p(4,2)'$,

(iii) G_0 is A_7, $p^d = 2^4$ and $A_7 < A_8 \simeq SL(4,2)$, or

(iv) G_0 is $SL(2,13)$, $p^d = 3^6$ and $SL(2,13) < Sp(6,4)$.

CHAPTER 105

Atlas of Planes and Processes.

We gather together in this part a listing of all know types of finite affine/projective planes and construction processes. This part of the text is intended to be a search tool and a quick reference check of the possible types of translation planes and affine planes as well as a listing of some of the more important classes of planes. We do not attempt a true classification of translation planes as we feel that it is more important to emphasize either importance of the planes in various classifications, in intrinsic construction techniques, and/or with connecting and related geometries.

Since many affine planes or translation planes are known merely by the process of construction, such as derivation or distortion, we begin with a listing of the known construction processes. All of these processes may be found in the text and are more or less listed under the 'models' index, so the reader should have no trouble tracking these down.

105.1. Construction Processes.

The following construction processes are probably a small subset that one could consider for the contruction and analysis of translation planes. We give this list to indicate the broad connections and ideas with other geometries.

> Slicing, spreading, and expanding in orthogonal geometry
>
> Projection of ovoids and Klein quadric
>
> Derivation of affine planes—multiple derivation
>
> General net replacement—hyper-regulus replacement
>
> Nest replacement
>
> Transpose, dualize
>
> Distortion of quasifields
>
> t-extension
>
> Algebraic lifting—contraction
>
> Geometric lifting—retraction
>
> Subregular lifting
>
> t-square, t-inverse—flock derivation
>
> Conical distortion (homology planes)
>
> Semifields
>
> Fusion of nuclei

Cyclic distortion of semifields

Direct products of nets

Hermitian sequences—parabolic, hyperbolic

Translation dual

105.2. Known Subregular Spreads.

By Johnson's subregular lifting theorem [656], any subregular plane of order q^2 may be lifted to a subregular plane of order q^{2s}, for any odd integer.

Name	Order	Reguli
Ostrom–Johnson–Seydel	3^6	7
Foulser–Ostrom	q^4	$q(q-1)/2$
Bruck Triple	q^2	3
Ebert Quadruple	q^2	4
Orr	9^2	81
Jha–Johnson Elation	8^2	4
Jha–Johnson 'Moved'	q^2	$\sqrt{q}$
Prohaska	q^2	$6, 7$
Ebert Quintuple	$p^2, p = 11, 19, 29, 31$	5
V. Abatangelo–Larato	q^2	$6, 7$
Bonisoli–Korchmáros–Szönyi	q^2	$5, 10, 15$
Charnes–Dempwolff	q^2	$5, 10$
Dempwolff	q^2	$5, 10$

105.3. Known Flag-Transitive Planes.

In this section, we review the set of known finite flag-transitive translation planes. We first note Wagner's theorem.

THEOREM 105.1. (Wagner [1202]). Finite flag-transitive affine planes are translation planes.

With a flag-transitive plane, the associated group G can be non-solvable or solvable. Furthermore, the non-solvable finite flag-transitive affine planes are completely determined. The following theorem is a special case of the more general result on flag-transitive designs.

THEOREM 105.2. (BDDKLS-Theorem). (Buekenhout, Delandtsheer, Doyen, Kleidman, Liebeck, and Saxl [201]). Let π be a non-solvable flag-transitive finite affine plane. Then π is one of the following types of affine planes: (1) Desarguesian, (2) Lüneburg–Tits, (3) Hering of order 27, or (4) Hall of order 9.

For completeness, we give a concrete representation for the Lüneburg–Tits planes of order 27, and Hering order 27 spreads. The Lüneburg–Tits planes are defined in Chapter 48 on translation planes of order q^2 admitting collineation groups of order q^2 and the Hering plane of order 27 is listed in Section 50.3 on indicator sets.

105.4. Flag-Transitive Planes with Spreads in $PG(3, q)$.

Here we list the classes of flag-transitive planes with spreads in $PG(3, q)$, for certain q.

> Baker–Ebert and Narayana Rao, q odd
>
> Order 49 M.L. Narayana Rao and K. Kuppuswamy Rao
>
> Order 25 Foulser

REMARK 105.3.

(1) Again, we note that all planes of orders 16, 25, 27, 49 are known by computer. These planes may be found in the homepage of Ulrich Dempwolff: http://www.mathematik.uni-kl.de/~dempw/ (or Google "homepage ulrich dempwolff"). The flag-transitive planes of order 27 are also known by the computer work of Prince [**1111**].

(2) There is no non-Desarguesian flag-transitive plane of order 16.

(3) There are exactly two flag-transitive planes of order 25, the two Foulser planes.

(4) There are exactly four flag-transitive planes of order 27, the Desarguesian, Hering and two others due to Narayana Rao, Kuppuswamy Rao, and by Narayana Rao, Kuppuswamy Rao and Satyanarayana.

Furthermore, (5) The planes of order 125 are known by the computer work of Prince [**1112**] and are given by constructions listed below.

105.5. Flag-Transitive Planes of Large Dimension.

By large dimension, we mean that associated vector space has dimension > 4. Again all of these planes appear in the text of the book, this list giving a quick indication of what planes exist.

> Order 27 Narayana Rao and Kuppuswamy Rao
>
> Order 27 Narayana Rao, Kuppuswamy Rao, Satyanarayana
>
> Order 125 Narayana Rao, Kuppuswamy Rao
>
> Desarguesian Scions, Kantor–Williams
>
> Planes of Suetake
>
> Planes of Kantor–Suetake
>
> Planes of Kantor
>
> Planes of Narayana Rao, Kuppuswamy Rao
>
> Order 81, Narayana Rao, Kuppuswamy Rao, Durga Prasad
>
> Order 81 Durga Prasad.

The planes of Narayana Rao, Kuppuswamy Rao and Durga Prasad of order 81 [**1004**], of 1988 and the general class of Narayana Rao and Kuppuswamy Rao constructed in 1989 [**1005**] predate those of Suetake [**1167**] 1991, Suetake and Kantor [**879**] 1994, and Kantor 1992, 1993 [**874, 875**], respectively. However, the general constructions of Suetake and Kantor, and Kantor include the planes of Narayana Rao, Kuppuswamy Rao and of Narayana Rao [**1005**] order, of Kuppuswamy Rao and Durga Prasad [**1004**], and Durga Prasad [**331**] of order 81, so we give, in the following, the general constructions. Similarly the planes of Durga Prasad [**331**] of order 81 are included in the constructions.

105.6. Classes of Semifields.

In Chapter 82 on symplectic spreads, we have noted the connections between commutative semifields and symplectic semifields. From a semifield flock spread, the 5th cousin will be symplectic. Most of these planes have been constructed previously. Here we provide the cross-references to those construction processes and the constructions which have not been previously mentioned.

THEOREM 105.4. *The known finite semifields and general construction processes are as follows: (the classes are not necessarily disjoint, see C and D).*

B: Knuth binary semifields

F: Flock semifields and their 5th cousins:

$\qquad F_1$: Kantor–Knuth

$\qquad F_2$: Cohen–Ganley, 5th cousin: Payne–Thas.

$\qquad F_3$: Penttila–Williams symplectic semifield order 3^5, 5th cousin, Bader, Lunardon, Pinneri flock semifield

C: Commutative semifields/symplectic semifields.

$\qquad C_1$: Kantor–Williams Desarguesian Scions (symplectic), Kantor–Williams commutative semifields

$\qquad C_2$: Ganley commutative semifields and symplectic cousins

$\qquad C_3$: Coulter–Matthews commutative semifields and symplectic cousins

D: Generalized Dickson/Knuth/Hughes–Kleinfeld semifields

S: Sandler semifields

JJ: Jha–Johnson cyclic semifields (generalizes Sandler, also of type $S(\omega, m, n)$, or p-primitive type 1, or q-primitive type 2)

$JMPT$: Johnson–Marino–Polverino–Trombetti semifields (generalizes Jha–Johnson type $S(\omega, 2, n)$-semifields)

$JMPT(4^5)$: Johnson–Marino–Polverino–Trombetti non-cyclic semifield of order 4^5

T: Generalized twisted fields

JH: Johnson–Huang 8 semifields of order 8^2

CF: Cordero–Figueroa semifield of order 3^6

General Construction Processes:

L: Algebraically lifted semifields

The algebraically lifted Desarguesian spreads are completely determined. These are also known as Cordero–Figueroa/Boerner–Lantz semifield planes for q odd or the Cardinali, Polverino, Trombetti semifield planes for q even

M: The middle-nucleus semifields by distortion-derivation

C: $GL(2, q) - q^3$ plane construction of Jha–Johnson

105.7. $GL(2, q)$-Planes of Order q^3.

Kantor order q^6 admitting $GL(2, q^2)$

Liebler order q^6 admitting $GL(2, q^2)$

Bartolone–Ostrom

Jha–Johnson, three families

105.8. Generalized André Planes (and Derived Planes).

We list here the known generalized André planes and those derived from them.

André

Multiple André (Ostrom)

Dickson nearfield

Johnson (variation on 'all except' method—new replacements of André nets)

Hiramine–Johnson order q^k admit affine homology group order $(q^k - 1)/2$

Draayer order q^k admit affine homology group order $(q^k - 1)/3$

Draayer order q^k admitting affine homology groups order $(q^k - 1)/d$,

  where d divides k

Wilke–Zemmer order 3^n, n odd (multiple André)

Narayana Rao and Zemmer

Wilke ('all except' method)

Ostrom 1st and 2nd constructions, coupled with Foulser net replacement

Foulser Q_g planes

Derived André and derived generalized André of Ostrom and Jha–Johnson

Triangle transitive generalized André of Draayer–Johnson–Pomareda

105.9. Hyper-Regulus Spreads.

In the previous section, we listed the known generalized André planes, all of which are produced by net replacement using either André replacement or Foulser-replacement or in combination with derivation. There are but a few known hyper-regulus spreads which are not obtained using either of these methods independently or in combination. These are as follows:

Culbert–Ebert cubic planes

Jha–Johnson fundamental hyper-regulus planes

Jha–Johnson multiple hyper-regulus replacement planes

105.10. Symplectic Spreads.

Symplectic semifield spreads are connected to commutative semifield spreads by the construction processes of transpose and dualization. The non-semifield symplectic spreads are quite rare and are listed below.

Suetake (includes Hering 27)

(only known symplectic, non-semifield, odd-order and odd-dimension),

The symplectic generalized twisted field spread

Lüneburg–Tits

Payne–Thas (5th cousin of Cohen–Ganley semifield flock)

Kantor–Knuth flock spreads

Penttila–Williams

(5th cousin of Bader–Lunardon–Pinneri semifield flocks order 3^{10})

Symplectic slice of Ree–Tits ovoids,

Kantor and Kantor–Williams symplectic semifield planes

105.11. Triangle Transitive Planes.

A triangle transitive plane is a finite affine translation plane admitting an autotopism group that acts transitive on the non-fixed points of each leg of the autotopism triangle. The known triangle transitive planes are:

(1) Finite generalized twisted field planes are triangle transitive precisely when the right, middle and left nuclei of the associated semifield plane coincide (see Biliotti, Jha, Johnson [121]).

(2) The Suetake planes

(3) The planes of Kantor and Williams

(4) Certain generalized André planes

(5) The nearfield planes

(6) The non-solvable triangle planes consist of the irregular nearfield planes with non-solvable groups, orders $11^2, 29^2$ or 59^2.

105.12. Chains of Reguli.

THEOREM 105.5. The following list of Bruen chains constitutes a complete list of the chains in $PG(3, q)$ for $q \leq 49$. The chains shall be denoted by a set of $(q + 3)/2$ 3-tuples, where the notation is explained previously. We shall list the chain and usually the group Aut permuting the circles. We designate the chains by the mathematician originally constructing the chain, together with the size q. For example, (Bruen(5)) denotes a chain constructed by Bruen in $PG(3, 5)$. There are exactly 19 chains.

(1)(Bruen(5)) : $q = 5$: $\{(-1, 1, 1), (0, 1, 1), (4, 1, 1), (17, 2, -1)\}$,

Aut $/ \langle e \rangle \simeq S_4$, e central involution; Bruen [195].

(2)(Bruen(7)) : $q = 7$: $\left\{ \begin{array}{c} (-1, 1, 1), (1, 1, 1), (3, 1, 1), \\ (15, 1, 1), (17, 3, -1) \end{array} \right\}$,

Aut $/ \langle e \rangle \simeq S_4$, e central involution; Bruen [195].

(3)(Korchmáros(7)) : $q = 7$:

$$\{(-1,1,1),(1,1,1),(25,1,1),(21,2,1),(45,2,1)\},$$

Aut $\simeq S_5$; Korchmáros [**903**].

(4)(V. Abatangelo(9)) : $q = 9$:

$$\left\{\begin{array}{c} (-1,0,1),(1,0,-1),(3,0,-1), \\ (18,0,1),(58,0,1),(65,0,1) \end{array}\right\},$$

Abatangelo [**7**].

(5)(Heden–Saggese(9)) : $q = 9$:

$$\left\{\begin{array}{c} (-1,0,1),(1,0,-1),(6,0,1), \\ (27,0,-1),(46,0,1),(75,0,-1) \end{array}\right\},$$

Heden and Saggese [**464**].

(6)(Capursi(11)) : $q = 11$:

$$\left\{\begin{array}{c} (-1,1,1),(0,1,-1),(7,1,1), \\ (50,1,-1),(115,3,1),(79,4,1),(103,4,1) \end{array}\right\},$$

Aut $\simeq D_{10} \times C_2$; Capursi [**213**].

(7)(Korchmáros(11)) : $q = 11$:

$$\left\{\begin{array}{c} (-1,1,1),(1,1,1),(4,2,1), \\ (34,2,1),(85,3,1),(45,5,1),(92,5,1) \end{array}\right\},$$

Aut $\simeq PSL(2,5)$; Korchmáros [**902**].

(8)(Raguso(13)) : $q = 13$:

$$\left\{\begin{array}{c} (-1,1,1),(3,1,-1),(10,1,1), \\ (165,1,-1),(158,5,-1), \\ (165,5,1),(131,6,1),(160,6,1) \end{array}\right\},$$

Aut $\simeq D_{12} \times C_2$; Raguso [**1124**].

(9)(Heden(13)) : $q = 13$:

$$\left\{\begin{array}{c} (-1,1,1),(3,1-1),(10,1,1), \\ (111,2,1),(73,3,-1), \\ (154,4,1),(0,6,1),(160,6,-1) \end{array}\right\},$$

Aut $\simeq S_4$; Heden [**458**].

(10)(Heden(17)) : $q = 17$:

$$\left\{\begin{array}{c} (-1,1,0,(0,1,1),(16,1,1), \\ (53,2,-1),(215,2,1),(26,3,1), \\ (170),(3,5,-1),(67,5,-1),(251,7,1) \end{array}\right\},$$

Aut $\simeq D_{12}$; Heden [**464**].

(11)(Baker–Ebert–Weida(17)) : $q = 17$:

$$\left\{ \begin{array}{c} (-1,1,1),(0,1,1),(27,1,-1), \\ (37,2,-1),(105,2,-1),(200,2,1), \\ (188,3,-1),(83,4,-1),(98,6,1),(225,8,1) \end{array} \right\},$$

Aut $\simeq D_{16}$; Baker, Ebert and Weida [**80**].

(12)(Heden(19)$_1$) : $q = 19$:

$$\left\{ \begin{array}{c} (-1,1,1),(1,1,1),(21,1,1), \\ (83,1,1),(125,1,1),(131,1,1), \\ (145,1,1),(4,2,-1),(119,2,-1),(92,4,1),(229,4,1) \end{array} \right\},$$

Aut $\simeq D_{12}$; Heden [**458**].

(13)(Heden(19)$_2$) : $q = 19$:

$$\left\{ \begin{array}{c} (-1,1,1),(1,1,1),(64,1,-1), \\ (226,1,-1),(247,4,1),(45,5,1), \\ (155,5,1),(293,6,-1),(355,7,-1),(170,9,1),(194,9,1) \end{array} \right\},$$

Aut $\simeq S_4$; Heden [**458**].

(14)(Heden(23)) : $q = 23$:

$$\left\{ \begin{array}{c} (-1,1,1),(0,1,-1),(4,1,-1), \\ (15,1,1),(147,1,1),(148,2,1),(192,3,-1), \\ (404,5,1),(369,6,1),(212,8,1), \\ (493,9,-1),(279,11,1),(299,11,-1) \end{array} \right\},$$

Aut $\simeq D_8$; Heden [**458**].

(15)(Heden–Saggese)(25)$_1$) : $q = 25$:

$$\left\{ \begin{array}{c} (-1,1,0,(2,0,1),(10,0,1),(470,0,1), \\ (591,26,1),(199,52,1),(342,130,1), \\ (121,156,1),(464,156,-1), \\ (572,156,1),(604,156,1),(81,182,1), \\ (205,208,-1),(610,260,1) \end{array} \right\},$$

Heden and Saggese [**464**].

(16)(Heden–Saggese)(25)$_2$) : $q = 25$:

$$\left\{ \begin{array}{c} (-1,0,1),(2,0,1),(13,0,-1), \\ (257,0,-1),(440,26,-1),(118,78,1), \\ (289,78,-1),(91,104,1),(557,104,-1), \\ (592,130,-1),(372,208,-1), \\ (513,208,-1),(278,260,-1),(1,286,1) \end{array} \right\},$$

Heden and Saggese [**458**].

$(17)(\text{Heden–Saggese})(27)) : q = 27 :$

$$\left\{ \begin{array}{c} (-1,0,1), (2,0,-1), (18,0,-1), \\ (200,0,-1), (260,0,-1), (283,28,-1), \\ (377,56,-1), (151,84,1), (420,112,-1), \\ (611,196,-1), (470,224,-1), \\ (644,224,1), (308,252,-1), \\ (58,308,-1), (350,308,-1) \end{array} \right\},$$

Heden and Saggese [458].

$(18)(\text{Heden})(31)) : q = 31 :$

$$\left\{ \begin{array}{c} (-1,1,1), (1,1,1), (113,1,1), \\ (746,2,1), (626,4,-1), (119,5,1), \\ (255,6,1), (402,6,1), (73,7,1), \\ (586,7,1), (84,8,1), (408,8,1), (72,9,1), \\ (285,10,1), (897,11,1), (707,14,1), (498,15,1) \end{array} \right\},$$

Heden [458].

$(19)(\text{Heden})(37)) : q = 37 :$

$$\left\{ \begin{array}{c} (-1,1,1), (3,1,-1), (9,1,-1), \\ (625,1,-1), (1057,1,-1), (1085,1,-1), \\ (258,2,-1), (406,2,1), (534,2,-1), \\ (183,3,1), (891,3,-1), (507,4,1), \\ (898,4,-1), (201,6,-1), (694,11,1), \\ (746,11,-1), (962,12,1), \\ (922,17,1), (424,18,1), (976,18,1) \end{array} \right\},$$

Heden [459].

105.13. Nest Planes.

The infinite classes of t-nest planes or mixed nest planes have either been constructed by Baker–Ebert (see all references to Baker and Ebert and in particular, [69], [70], and [74]) or Johnson–Pomareda [797] or more generally for arbitrary order by Johnson [750]. Also, for 12-nests by Johnson and Prince [807]. The notation refers to the group used in the construction. The reader is referred to Chapter 65 for more information.

E-nests (q-Nests)

$\langle a^{\sigma-1} \rangle$-Nests $((q+1) - \text{Nests})$

$\langle a^{\sigma+1} \rangle$-Nests $((q-1)\text{-Nests})$

Double $\langle a^{\sigma+1} \rangle$-Nests $(2(q-1)\text{-Nests})$

Mixed Nests

12-Nests

105.14. Infinite Classes of Flocks.

Since there are so many connections with flocks of quadratic cones, there are a variety of different methods of construction and hence there are competing designations for many of the infinite classes. We begin with a rough list of the infinite classes, after which, we shall add comments indicating the complexity of the situation. To add to the confusion, there are the derived flocks (the s-square and s-inverted spreads), many of which were determined after the initial discovery of the flock. Hence, we prefer to list the infinite classes in terms of their skeletons. We formulate a very rough classification of the skeletons to include 'semifield skeletons', where at least one member of the skeleton is a semifield flock, 'monomial skeletons', where at least one member of the skeleton has defining functions $(t, -f(t), g(t))$, where both f and g are monomials, 'likeable skeletons', where the conical flock spread is defined using a likeable function, 'transitive skeletons', where there is a transitive action on the members of the skeleton, 'rigid skeletons', where at least one flock member of the skeleton admits only the identify collineation group, and the 'Adelaide super skeleton', containing skeletons corresponding to the Adelaide flocks and the Subiaco flocks. Finally, the 'linear skeleton' consisting only of the linear flock or Desarguesian conical flock spread is not listed; however, the 'almost linear skeletons,' where at least one member of the skeleton contains a linear subset of $(q-1)/2$-conics consists of precisely the Fisher skeleton. The explicit constructions and comments are directly below. We provide the conical flock spread representation or the BLT-set representation, except for the Adelaide super skeleton, which is provided directly after the statement of the theorem.

THEOREM 105.6. *The infinite classes of flock skeletons are as follows:*
S: *Semifield Skeletons*:
$\quad$ S_1: The Kantor–Knuth skeleton
$\quad$ S_2: The Cohen–Ganley skeleton
M: *Monomial Skeletons*:
$\quad$ M_1: The Walker for q odd and Betten for q even Skeletons (Narayana Rao and Satyanarayana for characteristic 5)
$\quad$ M_2: The Barriga/Cohen–Ganley/Kantor/Payne skeleton
L: *Likeable Skeletons*:
$\quad$ L_1: The Kantor likeable skeleton
AL: *Almost Linear Skeletons*:
$\quad$ AL_1: The Fisher skeleton
T: *Transitive Skeletons*:
$\quad$ T_1: The Penttila skeleton
$\quad$ T_2: The Adelaide skeleton
R: *Rigid Skeletons*:
$\quad$ R_1: The Law–Penttila skeleton
A: *The Adelaide Super Skeletons*:
A_1: The Betten skeleton
A_2: The Adelaide skeleton
A_3: The Subiaco skeleton

105.15. Sporadic Flocks of Quadratic Cones.

Apart from the infinite classes, there are a number of conical flocks of small order. Actually, all flocks of quadratic cones are known from prime power orders ≤ 29, mostly by the use of the computer and the flocks of order 32 are completely known. The use of the computer is probably pushed about as far as it can, at least with current memory and speed. For example, it is estimated by Law and Penttila [**913**] that it might require eight years of computer time to determine the flocks of order 31. Still, is important to know flocks of small order as these conceivably point to infinite classes. For example, there is a flock of order 11 (i.e., in $PG(3, 11)$), whose associated translation plane q^2 admits a collineation group of order $q(q + 1)$ in the translation complement. Jha and Johnson [**632, 633**] completely determine the translation planes of order q^2 with spreads in $PG(3, q)$ that admit linear groups of order $q(q + 1)$. When q is odd, all of these planes are either conical flock planes or Ostrom derivates of conical flock planes. The classification method is one of structure, in that, in all but two sporadic cases, such planes are shown to be related to a Desarguesian plane by a single or multiple nest replacement procedure. For example, the Fisher conical flock planes are constructed from a Desarguesian affine plane by q-nest replacement. The more general concept of multiple nest replacements is developed in Jha and Johnson [**632, 633**] to analyze the planes obtained. In particular, the De Clerck–Herssens–Thas conical flock plane of order 11, originally found by computer, admits a collineation group of order $q(q + 1)$, when $q = 11$, and may be shown to be constructed from a Desarguesian plane by double-nest replacement.

In the following table, we list the conical flocks of orders q, for $q \leq 29$ and $q = 32$. This list is compiled from the work of Law and Penttila [**913**]. In that work, the term 'Kantor semifield flock' replaces our designation of 'Kantor–Knuth' flock, 'Thas–Fisher–Walker' replaces our designation of Walker for q odd or Betten for q even. The term 'Mondello' family refers to flocks determined by Penttila [**1098**] and the designation PRj refers to flocks found by Penttila and Royle [**1099**] where j refers to a group order. The group orders refer to groups acting on the associated BLT-sets. What is called a 'Kantor' monomial flock, we used the designation Barriga/Cohen–Ganley/Kantor/Payne flock(s), including both the even or odd orders. We use the term BCGKP for these flocks in both odd and even cases.

The flocks of orders $2, 3, 4$ are determined by Thas [**1177**], of orders $5, 7, 8$ by De Clerck, Gevaert and Thas [**272**], order 9 by Mylle [**995**] of order 11 and 16 by De Clerck and Herssens [**273**] (there is a class of order 11 partial flocks determined synthetically by Thas [**1187**] so when $q = 11$, the flock is called the De Clerck–Herssens–Thas flock). When $q = 13$ or 17, the flocks are determined by Penttila and Royle [**1099**]. For $q = 19, 23, 25, 27, 29$ Law and Penttila [**913**] find all flocks. Finally, when $q = 32$, the possible flocks are listed in Brown, O'Keefe, Payne, Penttila, and Royle [**170**]. In the following, we list the order q, then the possible flocks, followed by comments, if any.

$$q \; = \; 2, 3, 4, \text{linear},$$

$$q \; = \; 5, 7, 8, \text{linear or Fisher},$$

$$q \; = \; 9, \text{linear, Fisher, Kantor–Knuth},$$

$$q \; = \; 11, \text{linear, Fisher, Walker, Mondello},$$

$$q \; = \; 13, \text{linear, Fisher, two in BCGKP-skeleton},$$

$$q \; = \; 16, \text{linear, Subiaco},$$

$$q \; = \; 17, \text{linear, Fisher, Walker},$$
two BCGKP-skeleton, two in De Clerck–Herssens-skeleton,
two in Penttila–Royle-skeleton.

$$q \; = \; 19, \text{linear, Fisher, Mondello, PR20, four from PR16-skeleton},$$

q = 23, linear, Fisher, Walker, PR1152, PR24,

   two in DCH72-skeleton, two in BCGKP-skeleton, four in PR16-skeleton,

   five in PR6-skeleton.

q = 25, linear, Fisher, Kantor–Knuth, two in Kantor likeable-skeleton,

   three in PR16-skeleton, four in PR8-skeleton.

q = 27, linear, Fisher, Kantor–Knuth, two in Cohen–Ganley-skeleton,

   two in BCGKP-skeleton, seven in Law–Penttila-skeleton.

q = 29, linear, Fisher, Walker, Mondello, LP720, two in LP48-skeleton,

   five in LP8-skeleton, seven in LP6-skeleton, ten in LP3-skeleton.

q = 32, linear, Walker, BCGKP, Subiaco.

As mentioned going further than 29 by computer, except for 'nice' orders, demands extraordinary computer memory. Actually, there are two special orders, $q = 23$ and 47, that merit mention. The classification results of Jha and Johnson [633] show that any translation plane with spread in $PG(3, q)$ admitting a linear collineation group of order $q(q+1)$ is either a conical flock plane or a derived conical flock plane. Further, with the exceptions of orders 23 and 47 all of the associated translation planes may be constructed from a Desarguesian affine plane of order q^2 by a described net replacement procedure. For $q = 23$ the planes of Penttila–Royle admits a group of order $23(24)$, PR24. Moreover, there are some sporadic flocks of order 47, one of which is due to Penttila–De Clerck, which admits a group of order $47(48)$. These exceptional cases are part of the general classification results of Jha and Johnson [633].

105.15.1. Law–Penttila $27 \leq q \leq 125$. There are several other examples of BLT-sets found by computer. For example, Law and Penttila [1099], who find new examples of order q various orders. Let N denote the number of new mutually non-isomorphic BLT-sets determined. Then we have

q	N
27	1 (member of an infinite family)
29	5
31	5
37	3
41	3
43	1
47	2
49	2
53	2
59	3
125	1

The reader is directed to Law and Penttila [1099] for the description of the BLT-sets and other details.

105.15.2. Law $q = 83, 89$. In addition to the BLT-sets listed in the previous subsection, there are other examples, one each for $q = 83$ and 89 in Law [911].

105.15.3. Lavrauw–Law $q = 167$. In Lavrauw and Law [**910**], it is shown that there are two conjugacy classes of subgroups isomorphic to S_4 in $P\Gamma LO(5, q)$, for $q \equiv 5 \bmod 6$. Then a method for searching for BLT-sets admitting S_4 is given. A variety of BLT-sets are constructed, for $q = 23, 47, 71$ and a new BLT-set for $q = 167$. This method allows the possibility of continuing with the computation of new examples with the parameters given.

105.16. Semi-Translation Planes of Type 1.

We shall list the corresponding translation planes. Under appropriate coordinatization, the elation group E has the following form:

$$ E = \left\langle \begin{bmatrix} I & \begin{bmatrix} u & g(u) \\ 0 & f(u) \end{bmatrix} \\ 0 & I \end{bmatrix} ; u \in GF(q) \right\rangle, $$

where I and 0 are the identity and zero 2×2 matrices.

We designate the group E by the shorthand notation $(g(u), f(u), u)$.

Description	Group	Semi-Translation Plane
Lüneburg–Tits order $2^{2(2e-1)}$	$(u^\sigma - u, u)$ $u \in GF(2^{2e-1})$ $\sigma = 2^e$	Ostrom (see, e.g., [**1046**])
Kantor/Ree–Tits order $3^{2(2e-1)}$	$(u^\sigma - u, u)$ $u \in GF(3^{2e-1})$ $\sigma = 3^e$	Johnson [**731**]
Biliotti–Menichetti order 64	$(u^4 + u^3, u)$ $u \in GF(8)$	Johnson [**731**]
Jha–Johnson order 64	$(u + u^2, u)$ $u \in GF(8)$	Johnson [**731**]

105.17. Sporadic Planes of Order 16.

There are a variety of interesting planes of order 16, but only one, the Mathon plane, cannot be be constructed from known planes translation planes, Hughes planes, etc., by derivation, dualization. In any case, we list here the known planes of order 16, recalling that there are exactly 8 translation planes of this order.

The known planes of order 16, their groups and general information about the planes may be found on Gordon Royle's website:

http://www.csse.uwa.edu.au/~gordon/remote/planes16/

105.17.1. Projective Planes of Order 16. We take the list of planes of order 16 as it appears in Royle's list of planes. Recall that there are exactly 8 translation planes.

105.17.1.1. *PG(2,16)*. The Desarguesian projective plane of order 16
105.17.1.2. *SEMI2*. The semifield plane with kernel GF(2)
105.17.1.3. *SEMI4*. The semifield plane with kernel GF(4)
105.17.1.4. *HALL*. The Hall plane
105.17.1.5. *LMRH*. The Lorimer–Rahilly plane
105.17.1.6. *JOWK*. The Johnson–Walker plane

105.17.1.7. *DSFP.* The derived semifield plane

105.17.1.8. *DEMP.* The Dempwolff plane

105.17.1.9. *JOHN.* The Johnson plane

105.17.1.10. *BBS4.* Plane derived from the (dual of) SEMI4 from a derivation set of Bose–Barlotti type not on the (dual) translation line

105.17.1.11. *BBH2.* Plane derived from the dual of HALL from a derivation set of Bose–Barlotti type "from a line not on the regulus"

105.17.1.12. *MATH.* The Mathon plane

105.17.1.13. *BBH1.* Plane derived from the dual of HALL from a derivation set of Bose–Barlotti type "from a line on the regulus"

Many details may be found in Royle's website.

105.18. Affine Planes of Order 25.

There are 193 known projective planes of order 25 and recall that there are 21 translation planes. However, there are only two projective planes that so far do not fall into any recognizable class; the two planes of Moorhouse, which are actually derivates of each other and are obtained by a process called 'lifting quotients by involutions'. The two Moorhouse planes are lifted from a 'quotient' starting with the regular nearfield plane of order 25.

The list of planes may be obtained from the website of Eric Moorhouse and the reader is directed to his website for additional information:

http://www.uwyo.edu/moorhouse/pub/planes25/

There are 193 planes (5 self-dual planes plus 94 dual pairs).

More details on the new planes of Moorhouse can be found in Moorhouse [**993**].

105.19. Non-Translation Planes.

THEOREM 105.7. The known finite projective or affine planes that are not translation planes are included in the following list:

> Dual translation planes,
>
> Semi-translation planes and their duals
>
>   (Derived dual translation planes and their duals),
>
> Planar function planes of Coulter–Matthews,
>
> Hughes planes and irregular Hughes planes,
>
> Ostrom–Rosati planes and their duals
>
>   (Derived Hughes planes and their duals),
>
> Figueroa planes,
>
> Mathon planes of order 16 and the dual plane,
>
> The two Moorhouse planes of 25 (derivates of each other).

105.20. A Short Description—Dimension 2.

In this section, we provide a listing or summary of the known classes of finite translation planes of dimension 2. In some cases, the plane will be designated using a construction technique.

In this part, planes can be designated in one (or more) of the following ways:

(1) By matrix spread sets:

$$x = 0, y = x \begin{bmatrix} G(u,t) & F(u,t) \\ t & u \end{bmatrix} ; \forall u, t \in GF(q),$$

where $G(u,t)$, $F(u,t)$ are functions from $GF(q) \times GF(q) \longmapsto GF(q)$,

(2) By a description of ovoids on the Klein quadric (for example, the Conway–Kleidman–Wilson ovoids and the extensions by Moorhouse)

(3) By a discussion of computer programs, listing small order planes,

(4) By the use of BLT-sets (see Chapters 54 through 61), by flocks of quadratic cones or by generalized quadrangles,

(5) By a net replacement procedure ('derivation', 'multiple derivation', 't-nest replacement', 't-distortion', 't-extension', 's-square', 's-inversion', 'subregular-lifting', 'algebraic-lifting', 'transpose', 'transpose-dualization', 'dualization-transpose', 'multiple-nest replacement') or by a combination of such procedures, or

(6) By 'geometric-lifting' from subgeometry partitions.

We will indicate planes either by manner of construction or geometric significance, of which the following are important:

Distortion and derivation

Algebraic lifting

Subgeometry planes

Subregular lifting

Conic flock planes

Hyperbolic flock planes,

Nearfield planes

Partial hyperbolic flock planes

t-Nest planes

Geometric lifting

Planes of order q^2 /Groups of order q^2; likeable

Planes of order q^2 /Groups of order $q^2 - 1$

Planes of order q^2 /Group $SL(2,q)$

Planes of order q^2 /Group $Sz(q)$

$SL(2,3) \times SL(2,3)$ and $SL(2,5) \times SL(2,5)$-planes

Planes related to irregular nearfields

Ovoidal planes; Klein quadric constructions

Symplectic planes and cousins

Flag-transitive planes

Generalized André planes

Semifield planes

j-Planes

REMARK 105.8. The t-nest planes have planes with the following known parameter: $t = q - 1, q, q + 1, 2(q + 1), 12, chains$.

105.21. A Short Description—Large Dimension.

Dimension 3:

> Generalized Desarguesian Planes
>
> Long Orbit Planes
>
> Cubic Hyper-Regulus Planes
>
> Generalized André Planes
>
> Semifield Planes
>
> $j \ldots j$-Planes
>
> Cubic Extension Planes

Dimension 4:

> Regular Parallelism Planes
>
> Algebraically Lifted + Derived Planes

as well as planes of type in the dimension 3 case.
Larger Dimension:

> Hyper-Regulus Planes
>
> Generalized Hall Planes
>
> Flag Transitive Planes
>
> Long Orbit Planes
>
> $j \ldots j$-Planes
>
> Symplectic and Orthogonal Planes
>
> Semifield Planes
>
> Generalized André Planes

105.22. Planes Admitting Large Non-Solvable Groups.

105.22.1. Translation Planes of Order q^m Admitting $SL(2, q)$, $PSL(2, q)$ or $S_z(q)$.

105.22.1.1. *When $m = 1$.* When $m = 1$, the hypothesis does not require that the plane is a translation plane. Lüneburg shows that any projective plane of order q that admits $SL(2, q)$ or $PSL(2, q)$ as a collineation group must be Desarguesian.
$S_z(q)$ cannot act on a translation plane of order q.

105.22.1.2. *When $m = 2$.*
When the group is $SL(2, q)$. The translation planes of order q^2 that admit either $SL(2, q)$ or $PSL(2, q)$ as a collineation group are completely determined. The six possible classes are: (1) Desarguesian, (2) Hall, (3) Hering, (4) Ott–Schaeffer, (5) three Walker planes of order 25 of which one is Hering, (6) Dempwolff plane of order 16.

With the exception of the Dempwolff plane of order 16, all planes have kernel $GF(q)$. When the planes have spreads in $PG(3, q)$, Walker for q odd and Schaeffer for q even have determined the possible classes.

The more general situation for arbitrary kernel is considered by D.A. Foulser and N.L. Johnson in [**378, 379**].

When the group is $PSL(2, q)$. The assumptions by Foulser–Johnson do not require that the group $SL(2, q)$ act faithfully so that a rereading of these results will produce the planes of order q^2 that admit $PSL(2, q)$ as a collineation group of a translation plane. Of course, the group $PSL(2, q)$ does act on the Hughes planes of order q^2.

The only possible translation plane of odd order q^2 that admits $PSL(2, q)$ as a collineation group is the Desarguesian plane of order q^2 [**378, 379**].

105.22.1.3. *When the Group Is $S_z(q)$.* The translation planes of even order q^2 $= 2^{2(2r+1)}$ that admit $S_z(q)$ are Lüneburg–Tits planes of order q^2. This result is due independently to Ch. Hering [**480**] and R.A. Liebler [**918**].

105.22.1.4. *When $m = 3$.* When the group is $S_z(q)$: $S_z(q)$ cannot act on a translation plane of order q^3 by Büttner [**205**].

When the group is $SL(2, q)$. The translation planes of order q^3 that admit $SL(2, q)$ generated by elations are related to the "cyclic" semifield planes. We note that for the only known translation planes of order q^3 that admit $SL(2, q)$ as a collineation group, the group is generated by elations.

Kantor, Liebler, Bartolone–Ostrom, myriad planes of Jha–Johnson.

When the group is $PSL(2, q)$. This cannot occur in translation planes. Of course, the group does act on the Figueroa planes of order q^3([**355**]). Dempwolff [**297**]) has shown that when $PSL(3, q)$ acts on a projective plane of order q^3 and the elations extend (check hypothesis) then there is an invariant subplane of order q.

105.22.2. When $m = 4$. There are intimate connections between translation planes of order q^4 that admit $SL(2, q) \times Z_{1+q+q^2}$ as a collineation group. There are only three translation planes of order q^4 such a group as a collineation group. These include the two planes of order 16 that admit $SL(2, \dot{2})$. These planes may be constructed in a variety of ways. These two planes have four names attached and have been referred to as the Lorimer–Rahilly and Johnson–Walker planes of order 16. There is also a plane due to Walker of order 8^4 admitting $SL(2, 8) \times Z_{1+8+8^2}$. All of these planes are related to regular packings in $PG(3, q)$ so are of particular interest. There are new planes due to Alan Prince with regular packings in $PG(3, 5)$ and an infinite class in $PG(3, q)$ for q congruent to 2 mod 3 due to Penttila and Blair Williams and appear to contain all of the previously known examples.

There are a variety of translation planes of order q^4 that admit $SL(2, q)$ where the p-elements are either elations or Baer for $q = p^r$. In particular, there are the planes of Foulser–Ostrom–Walker.

105.22.3. Planes of order q^n admitting $PSL(3, q)$. Hughes, and Figueroa planes.

105.22.4. Sporadic Examples. Hering 27 admitting PSL(2,13), Lorimer–Rahilly, Johnson–Walker order 16 admitting PSL(2,7)$\simeq$ $GL(3, 2)$, Lüneburg $SL(2, 5) \times SL(2, 5)$ planes, Hiramine $SL(2, 5) \times SL(2, 5)$.

A_5 and A_6 planes of Ostrom, Ostrom and Mason, Mason and Shult, Nakagawa, Biliotti and Korchmáros, Charnes and Dempwolff, Dempwolff for other orders, Moorhouse. Prohaska planes and their duals of order 81 admitting $SL(2, 5)$ as well as two infinite classes of Prohaska of order q^2, all admitting A_5.

105.22.5. Solution to Design Puzzle and the Pun. At the bottom of the puzzle, you will see a Desarguesian configuration. At the top, the "gold line" indicates that this is a non-Desarguesian configuration but also somehow connected to the bottom Desarguesian configuration. That the non-Desarguesian configuration is placed on top indicates inherent importance and dominance over Desarguesian configurations.

Solution and Pun: **"This book is about non-Desarguesian projective planes, their importance in geometry and their dominance over and connection to Desarguesian projective planes."**

Bibliography

1. L.M. Abatangelo, *On the automorphism groups of β-derived planes*, J. Geom. **25** (1985), no. 1, 19–29.

2. _____, *Affine homology groups of a β-derived translation plane*, Boll. Un. Mat. Ital. B (6) **5** (1986), no. 2, 569–580.

3. _____, *Collineation groups of spreads*, Eleventh British Combinatorial Conference (London, 1987), Ars Combin. **25** (1988), no. B, 247–256.

4. _____, *On the collineation groups of a β-derived spread*, Combinatorics '88, Vol. 1 (Ravello, 1988), Res. Lecture Notes Math., Mediterranean, Rende, 1991, pp. 111–123.

5. L.M. Abatangelo and V. Abatangelo, *On Bruen's planes of order 25*, Atti Sem. Mat. Fis. Univ. Modena **31** (1982), no. 2, 348–369.

6. L.M. Abatangelo, V. Abatangelo, and G. Korchmáros, *A translation plane of order 25*, J. Geom. **22** (1984), no. 2, 108–116.

7. V. Abatangelo, *A translation plane of order 81 and its full collineation group*, Bull. Austral. Math. Soc. **29** (1984), no. 1, 19–34.

8. V. Abatangelo, M.R. Enea, G. Korchmáros, and B. Larato, *Ovals and unitals in commutative twisted field planes*, Combinatorics (Assisi, 1996), Discrete Math. **208/209** (1999), 3–8.

9. V. Abatangelo, G. Korchmáros, and B. Larato, *Transitive parabolic unitals in translation planes of odd order*, 17th British Combinatorial Conference (Canterbury, 1999), Discrete Math. **231** (2001), no. 1–3, 3–10.

10. V. Abatangelo and B. Larato, *Translation planes with an automorphism group isomorphic to* SL$(2,5)$, Combinatorics '84 (Bari, 1984), North-Holland Math. Stud., vol. 123, North-Holland, Amsterdam, 1986, pp. 1–7.

11. _____, *A group-theoretical characterization of parabolic Buekenhout-Metz unitals*, Boll. Un. Mat. Ital. A (7) **5** (1991), no. 2, 195–206.

12. _____, *A characterization of Buekenhout-Metz unitals in* PG$(2, q^2)$, *q even*, Geom. Dedicata **59** (1996), no. 2, 137–145.

13. _____, *Polarity and transitive parabolic unitals in translation planes of odd order*, J. Geom. **74** (2002), no. 1–2, 1–6.

14. _____, *Doubly β-derived translation planes*, Des. Codes Cryptogr. **28** (2003), no. 1, 65–74.

15. V. Abatangelo, B. Larato, and L.A. Rosati, *Unitals in planes derived from Hughes planes*, Graphs, designs and combinatorial geometries (Catania, 1989), J. Combin. Inform. System Sci. **15** (1990), no. 1–4, 151–155.

16. A. Aguglia and G.L. Ebert, *A combinatorial characterization of classical unitals*, Arch. Math. (Basel) **78** (2002), no. 2, 166–172.

17. K. Akiyama, *On translation planes of order q^3 that admit a collineation group of order q^3, II*, Geom. Dedicata **57** (1995), no. 2, 171–193.

18. K. Akiyama and C. Suetake, *On translation planes of order q^3 that admit a collineation group of order q^3*, Geom. Dedicata **55** (1995), no. 1, 1–57.

19. A.A. Albert, *On nonassociative division algebras*, Trans. Amer. Math. Soc. **72** (1952), 296–309.

20. _____, *Finite noncommutative division algebras*, Proc. Amer. Math. Soc. **9** (1958), 928–932.

21. _____, *On the collineation groups of certain non-desarguesian planes*, Portugal. Math. **18** (1959), 207–224.

22. _____, *Finite division algebras and finite planes*, Proc. Sympos. Appl. Math., Vol. 10, American Mathematical Society, Providence, R.I., 1960, pp. 53–70.

23. _____, *Generalized twisted fields*, Pacific J. Math. **11** (1961), 1–8.

24. _____, *Isotopy for generalized twisted fields*, An. Acad. Brasil. Ci. **33** (1961), 265–275.

25. _____, *On associative division algebras of prime degree*, Proc. Amer. Math. Soc. **16** (1965), 799–802.

26. _____, *The finite planes of Ostrom*, Bol. Soc. Mat. Mexicana (2) **11** (1966), 1–13.

27. _____, *New results on associative division algebras*, J. Algebra **5** (1967), 110–132.

28. _____, *On associative division algebras*, Bull. Amer. Math. Soc. **74** (1968), 438–454.

29. A.A. Albert and R. Sandler, *An Introduction to Finite Projective Planes*, Holt, Rinehart and Winston, New York, 1968.

30. J.L. Alperin, *Projective modules for* $SL(2, 2^n)$, J. Pure Appl. Algebra **15** (1979), no. 3, 219–234.

31. J. André, *Über nicht-Desarguessche Ebenen mit transitiver Translationsgruppe*, Math. Z. **60** (1954), 156–186.

32. _____, *Über Perspektivitäten in endlichen projektiven Ebenen*, Arch. Math. **6** (1954), 29–32.

33. E. Artin, *Geometric Algebra*, Interscience Publishers, Inc., New York-London, 1957.

34. M. Aschbacher, *Finite Group Theory*, Cambridge Studies in Advanced Mathematics, vol. 10, Cambridge University Press, Cambridge, 1986.

35. E.F. Assmus, Jr. and J.D. Key, *Translation planes and derivation sets*, J. Geom. **37** (1990), no. 1–2, 3–16.

36. L. Bader, *On generalized André planes*, Rend. Circ. Mat. Palermo (2) **35** (1986), no. 3, 448–455.

37. _____, *On the flocks of the quadratic cone*, Boll. Un. Mat. Ital. A (7) **2** (1988), no. 3, 371–375.

38. _____, *Some new examples of flocks of* $Q^+(3, q)$, Geom. Dedicata **27** (1988), no. 2, 213–218.

39. _____, *Derivation of Fisher flocks*, J. Geom. **37** (1990), no. 1–2, 17–24.

40. _____, *Flocks of cones and generalized hexagons*, Advances in Finite Geometries and Designs (Chelwood Gate, 1990), Oxford Sci. Publ., Oxford Univ. Press, New York, 1991, pp. 7–18.

41. L. Bader, A. Cossidente, and G. Lunardon, *On flat flocks*, preprint.

42. _____, *Generalizing flocks of* $Q^+(3, q)$, Adv. Geom. **1** (2001), no. 4, 323–331.

43. L. Bader and M.J. de Resmini, *On some Kantor flocks*, Combinatorics '98 (Palermo), Discrete Math. **255** (2002), no. 1–3, 3–6.

44. L. Bader, N. Durante, M. Law, G. Lunardon, and T. Penttila, *Flocks and partial flocks of hyperbolic quadrics via root systems*, J. Algebraic Combin. **16** (2002), no. 1, 21–30.

45. _____, *Symmetries of BLT-sets*, Proceedings of the Conference on Finite Geometries (Oberwolfach, 2001), Des. Codes Cryptogr. **29** (2003), no. 1–3, 41–50.

46. L. Bader, D. Ghinelli, and T. Penttila, *On monomial flocks*, European J. Combin. **22** (2001), no. 4, 447–454.

47. L. Bader and N.L. Johnson, *Subplane covered affine planes*, Geom. Dedicata **63** (1996), no. 2, 171–182.

48. L. Bader, W.M. Kantor, and G. Lunardon, *Symplectic spreads from twisted fields*, Boll. Un. Mat. Ital. A (7) **8** (1994), no. 3, 383–389.

49. L. Bader and G. Lunardon, *On the flocks of* $Q^+(3, q)$, Geom. Dedicata **29** (1989), no. 2, 177–183.

50. _____, *Generalized hexagons and BLT-sets*, Finite Geometry and Combinatorics (Deinze, 1992), London Math. Soc. Lecture Note Ser., vol. 191, Cambridge Univ. Press, Cambridge, 1993, pp. 5–15.

51. _____, *On non-hyperelliptic flocks*, European J. Combin. **15** (1994), no. 5, 411–415.

52. L. Bader, G. Lunardon, and S.E. Payne, *On q-clan geometry*, $q = 2^e$, A tribute to J. A. Thas (Gent, 1994), Bull. Belg. Math. Soc. Simon Stevin **1** (1994), no. 3, 301–328.

53. L. Bader, G. Lunardon, and I. Pinneri, *A new semifield flock*, J. Combin. Theory Ser. A **86** (1999), no. 1, 49–62.

54. L. Bader, G. Lunardon, and J.A. Thas, *Derivation of flocks of quadratic cones*, Forum Math. **2** (1990), no. 2, 163–174.

55. L. Bader, G. Marino, O. Polverino, and R. Trombetti, *Spreads of* $PG(3, q)$ *and ovoids of polar spaces*, (to appear).

56. L. Bader and C.M. O'Keefe, *Arcs and ovals in infinite K-clan geometry*, Finite Geometry and Combinatorics (Deinze, 1997), Bull. Belg. Math. Soc. Simon Stevin **5** (1998), no. 2–3, 127–139.

57. L. Bader, C.M. O'Keefe, and T. Penttila, *Some remarks on flocks*, J. Aust. Math. Soc. **76** (2004), no. 3, 329–343.

58. L. Bader and S.E. Payne, *On infinite K-clan geometry*, J. Geom. **63** (1998), no. 1–2, 1–16.

59. L. Bader and R. Trombetti, *Translation ovoids of flock generalized quadrangles*, European J. Combin. **25** (2004), no. 1, 65–72.

60. R.D. Baker, A. Bonisoli, A. Cossidente, and G.L. Ebert, *Mixed partitions of* PG(5, q), Combinatorics (Assisi, 1996), Discrete Math. **208/209** (1999), 23–29.

61. R.D. Baker, C. Culbert, G.L. Ebert, and K.E. Mellinger, *Odd order flag-transitive affine planes of dimension three over their kernel*, Special issue dedicated to Adriano Barlotti, Adv. Geom. suppl. (2003), S215–S223.

62. R.D. Baker, J.M. Dover, G.L. Ebert, and K.L. Wantz, *Hyperbolic fibrations of* $\mathcal{PG}(3, q)$, European J. Combin. **20** (1999), no. 1, 1–16.

63. _____, *Baer subgeometry partitions*, Second Pythagorean Conference (Pythagoreion, 1999), J. Geom. **67** (2000), no. 1–2, 23–34.

64. _____, *Perfect Baer subplane partitions and three-dimensional flag-transitive planes*, Special issue dedicated to Dr. Jaap Seidel on the occasion of his 80th birthday (Oisterwijk, 1999), Des. Codes Cryptogr. **21** (2000), no. 1–3, 19–39.

65. R.D. Baker and G.L. Ebert, *Enumeration of two-dimensional flag-transitive planes*, Proceedings of the Conference on Groups and Geometry, Part A (Madison, Wis., 1985), Algebras Groups Geom. **2** (1985), no. 3, 248–257.

66. _____, *Spreads and packings for a class of* $((2^n + 1)(2^{n-1} - 1) + 1, 2^{n-1}, 1)$-*designs*, J. Combin. Theory Ser. A **40** (1985), no. 1, 45–54.

67. _____, *A nonlinear flock in the Minkowski plane of order* 11, Eighteenth Southeastern International Conference on Combinatorics, Graph Theory, and Computing (Boca Raton, Fla., 1987), Congr. Numer. **58** (1987), 75–81.

68. _____, *Construction of two-dimensional flag-transitive planes*, Geom. Dedicata **27** (1988), no. 1, 9–14.

69. _____, *Nests of size* q − 1 *and another family of translation planes*, J. London Math. Soc. (2) **38** (1988), no. 2, 341–355.

70. _____, *A new class of translation planes*, Combinatorics '86 (Trento, 1986), Ann. Discrete Math., vol. 37, North-Holland, Amsterdam, 1988, pp. 7–20.

71. _____, *Intersection of unitals in the Desarguesian plane*, Proceedings of the Twentieth Southeastern Conference on Combinatorics, Graph Theory, and Computing (Boca Raton, FL, 1989), Congr. Numer. **70** (1990), 87–94.

72. _____, *On Buekenhout-Metz unitals of odd order*, J. Combin. Theory Ser. A **60** (1992), no. 1, 67–84.

73. _____, *A Bruen chain for* q = 19, Des. Codes Cryptogr. **4** (1994), no. 4, 307–312.

74. _____, *Filling the nest gaps*, Finite Fields Appl. **2** (1996), no. 1, 42–61.

75. _____, *Regulus-free spreads of* PG(3, q), Special issue dedicated to Hanfried Lenz, Des. Codes Cryptogr. **8** (1996), no. 1–2, 79–89.

76. _____, *Two-dimensional flag-transitive planes revisited*, Geom. Dedicata **63** (1996), no. 1, 1–15.

77. R.D. Baker, G.L. Ebert, K.H. Leung, and Q. Xiang, *A trace conjecture and flag-transitive affine planes*, J. Combin. Theory Ser. A **95** (2001), no. 1, 158–168.

78. R.D. Baker, G.L. Ebert, and T. Penttila, *Hyperbolic fibrations and q-clans*, Des. Codes Cryptogr. **34** (2005), no. 2–3, 295–305.

79. R.D. Baker, G.L. Ebert, and K.L. Wantz, *Regular hyperbolic fibrations*, Adv. Geom. **1** (2001), no. 2, 119–144.

80. R.D. Baker, G.L. Ebert, and R. Weida, *Another look at Bruen chains*, J. Combin. Theory Ser. A **48** (1988), no. 1, 77–90.

81. S. Ball, J. Bamberg, M. Lavrauw, and T. Penttila, *Symplectic spreads*, Des. Codes Cryptogr. **32** (2004), no. 1–3, 9–14.

82. S. Ball, A. Blokhuis, and C.M. O'Keefe, *On unitals with many Baer sublines*, Des. Codes Cryptogr. **17** (1999), no. 1–3, 237–252.

83. S. Ball and M.R. Brown, *The six semifield planes associated with a semifield flock*, Adv. Math. **189** (2004), no. 1, 68–87.

84. S. Ball and M. Lavrauw, *Commutative semifields of rank 2 over their middle nucleus*, Finite Fields with Applications to Coding Theory, Cryptography and Related Areas (Oaxaca, 2001), Springer, Berlin, 2002, pp. 1–21.

85. S. Ball and M. Zieve, *Symplectic spreads and permutation polynomials*, Finite Fields and Applications, Lecture Notes in Comput. Sci., vol. 2948, Springer, Berlin, 2004, pp. 79–88.

86. J. Banning, *Errata: "The order of a finite translation plane"*, Nieuw Arch. Wisk. (4) **8** (1990), no. 3, i.

87. _____, *The order of a finite translation plane*, Nieuw Arch. Wisk. (4) **8** (1990), no. 2, 199–207.

88. A. Barlotti, *Le possibili configurazioni del sistema delle coppie punto-retta (A, a) per cui un piano grafico risulta (A, a)-transitivo*, Boll. Un. Mat. Ital. (3) **12** (1957), 212–226.

89. _____, *Classification of finite projective planes*, Combinatorial Mathematics and its Applications (Proc. Conf., Univ. North Carolina, Chapel Hill, N.C., 1967) (R.C. Bose and T.A. Dowling, eds.), Univ. North Carolina Press, Chapel Hill, N.C., 1969, pp. 405–415.

90. _____, *Alcuni procedimenti per la costruzione di piani grafici non desarguesiani*, Confer. Sem. Mat. Univ. Bari (1971), no. 127, 17.

91. _____, *On the definition of Baer subplanes of infinite planes*, J. Geometry **3** (1973), 87–92.

92. _____, *Representation and construction of projective planes and other geometric structures from projective spaces*, Jber. Deutsch. Math.-Verein. **77** (1975), no. 1, 28–38.

93. A. Barlotti and J. Cofman, *Finite Sperner spaces constructed from projective and affine spaces*, Abh. Math. Sem. Univ. Hamburg **40** (1974), 231–241.

94. A. Barlotti and G. Lunardon, *A class of unitals in Δ-planes*, Riv. Mat. Univ. Parma (4) **5** (1979), part 2, 781–785.

95. O.E. Barriga, *On the planes of Narayana Rao and Satyanarayana*, J. Combin. Theory Ser. A **45** (1987), no. 1, 148–151.

96. _____, *A characterization of Hering's plane of order 27*, Bol. Soc. Brasil. Mat. (N.S.) **21** (1990), no. 1, 71–77.

97. O.E. Barriga and G. Mason, *Spread-invariant representations of $L_2(2^n)$ and a characterization of some Ott-Schaefer planes*, J. Algebra **137** (1991), no. 1, 233–251.

98. O.E. Barriga and R. Pomareda, *The uniqueness of Hering's translation plane of order 27*, Geom. Dedicata **19** (1985), no. 3, 237–245.

99. C. Bartolone, *The group of collineations of the plane obtained by deriving the dual of a Lüneburg plane*, Boll. Un. Mat. Ital. B (5) **18** (1981), no. 3, 921–932.

100. _____, *On some translation planes admitting a Frobenius group of collineations*, Combinatorics '81 (Rome, 1981), Ann. Discrete Math., vol. 18, North-Holland, Amsterdam, 1983, pp. 37–53.

101. C. Bartolone and F. Bartolozzi, *On a class of (R, r)-transitive planes*, Rend. Sem. Mat. Univ. Padova **61** (1979), 13–31.

102. C. Bartolone and T.G. Ostrom, *Translation planes of order q^3 which admit SL$(2, q)$*, J. Algebra **99** (1986), no. 1, 50–57.

103. A. Basile and P. Brutti, *Particular sets of strictly two-transitive permutations and their associated translation planes*, Boll. Un. Mat. Ital. A (7) **9** (1995), no. 1, 57–65.

104. _____, *Sharply 2-transitive sets of permutations and construction of some classes of translation planes*, Atti Sem. Mat. Fis. Univ. Modena **43** (1995), no. 2, 255–263.

105. _____, *Some conditions for the construction of a finite translation plane*, Boll. Unione Mat. Ital. Sez. B Artic. Ric. Mat. (8) **4** (2001), no. 2, 429–439.

106. _____, *Regular switching sets and ruled cubics of associated $\mathcal{PG}(5, q)$ translation planes*, Rend. Circ. Mat. Palermo (2) **53** (2004), no. 1, 85–92.

107. _____, *Ruled cubics in PG$(5, q)$ and fibrations of finite generalized André planes of order q^3*, Ital. J. Pure Appl. Math. (2004), no. 15, 29–34.

108. L.M. Batten and J.M. Dover, *Some sets of type (m, n) in cubic order planes*, Des. Codes Cryptogr. **16** (1999), no. 3, 211–213.

109. L.M. Batten and P.M. Johnson, *The collineation groups of Figueroa planes*, Canad. Math. Bull. **36** (1993), no. 4, 390–397.

110. M. Bernardi, *Esistenza di fibrazioni in uno spazio proiettivo infinito*, Ist. Lombardo Accad. Sci. Lett. Rend. A **107** (1973), 528–542.

111. T. Beth, D. Jungnickel, and H. Lenz, *Design Theory*, Cambridge University Press, Cambridge, 1986.

112. D. Betten, *4-dimensionale Translationsebenen mit 8-dimensionaler Kollineationsgruppe*, Geom. Dedicata **2** (1973), 327–339.

113. _____, *Sperner spaces derived from translation planes*, Proceedings of the Conference on Combinatorial and Incidence Geometry: Principles and Applications (La Mendola, 1982) (Milan), Rend. Sem. Mat. Brescia, vol. 7, Vita e Pensiero, 1984, pp. 83–88.

114. D. Betten and D.G. Glynn, *Über endliche planare Funktionen, ihre zugehörenden Schiebebenen, und ihre abgeleiteten Translationsebenen*, Results Math. **42** (2002), no. 1–2, 32–36.

115. M. Biliotti, *Il gruppo delle collineazioni dei piani di Hughes generalizzati*, Atti Sem. Mat. Fis. Univ. Modena **24** (1975), no. 2, 290–299.

116. _____, *A Dembowski generalisation of the Hughes planes*, Boll. Un. Mat. Ital. B (5) **16** (1979), no. 2, 674–693.

117. _____, *On the derived planes of Desarguesian planes*, Rend. Sem. Mat. Univ. Padova **62** (1980), 165–181.

118. _____, *On inherited groups of derivable translation planes*, Combinatorial and Geometric Structures and Their Applications (Trento, 1980), Ann. Discrete Math., vol. 14, North-Holland, Amsterdam, 1982, pp. 151–157.

119. _____, *On the collineation groups of derived semifield planes*, Math. Z. **183** (1983), no. 4, 489–493.

120. M. Biliotti, V. Jha, and N.L. Johnson, *Translation ovals in generalized twisted field planes*, preprint.

121. _____, *The collineation groups of generalized twisted field planes*, Geom. Dedicata **76** (1999), no. 1, 97–126.

122. _____, *Transitive parallelisms*, Results Math. **37** (2000), no. 3–4, 308–314.

123. _____, *Foundations of Translation Planes*, Monographs and Textbooks in Pure and Applied Mathematics, vol. 243, Marcel Dekker Inc., New York, 2001.

124. _____, *Two-transitive parabolic ovals*, J. Geom. **70** (2001), no. 1–2, 17–27.

125. _____, *Two-transitive ovals in generalized twisted field planes*, Arch. Math. (Basel) **79** (2002), no. 3, 232–240.

126. _____, *Large quartic groups on translation planes, I: Odd order: Characterization of the Hering planes*, Note Mat. **23** (2004/05), no. 1, 151–166.

127. _____, *Large quartic groups on translation planes, III: Groups with common centers*, Note Mat. **24** (2004/05), no. 2, 143–151.

128. _____, *Symplectic flock spreads in PG(3, q)*, Note Mat. **24** (2004/05), no. 1, 85–109.

129. _____, *Classification of transitive deficiency one partial parallelisms*, Bull. Belg. Math. Soc. Simon Stevin **12** (2005), no. 3, 371–391.

130. _____, *Large quartic groups on translation planes, II: Even order: characterization of the Ott-Schaeffer planes*, J. Combin. Des. **13** (2005), no. 3, 195–210.

131. _____, *Special linear group sections on translation planes*, Bull. Belg. Math. Soc. Simon Stevin **13** (2006).

132. M. Biliotti, V. Jha, N.L. Johnson, and G. Menichetti, *The collineation groups of Baer-elation planes*, Atti Sem. Mat. Fis. Univ. Modena **36** (1988), no. 1, 23–35.

133. _____, *A structure theory for two-dimensional translation planes of order q^2 that admit collineation groups of order q^2*, Geom. Dedicata **29** (1989), no. 1, 7–43.

134. M. Biliotti, V. Jha, N.L. Johnson, and A. Montinaro, *Classification of projective translation planes of order q^2 admitting a two-transitive orbit of length $q + 1$*, submitted.

135. _____, *The Hall plane of order 9—revisited*, submitted.

136. _____, *Translation planes of order q^2 admitting a two-transitive orbit of length $q + 1$ on the line at infinity*, submitted.

137. _____, *Two-transitive groups on a hyperbolic unital*, submitted.

138. M. Biliotti and N.L. Johnson, *Maximal Baer groups in translation planes and compatibility with homology groups*, Geom. Dedicata **59** (1996), no. 1, 65–101.

139. _____, *Variations on a theme of Dembowski*, Mostly Finite Geometries (Iowa City, IA, 1996), Lecture Notes in Pure and Appl. Math., vol. 190, Dekker, New York, 1997, pp. 139–168.

140. _____, *Bilinear flocks of quadratic cones*, J. Geom. **64** (1999), no. 1–2, 16–50.

141. _____, *The non-solvable rank* 3 *affine planes*, J. Combin. Theory Ser. A **93** (2001), no. 2, 201–230.

142. M. Biliotti and G. Korchmáros, *Some finite translation planes arising from A_6-invariant ovoids of the Klein quadric*, J. Geom. **37** (1990), no. 1–2, 29–47.

143. M. Biliotti and G. Lunardon, *Derivation sets and Baer subplanes in a translation plane*, Atti Accad. Naz. Lincei Rend. Cl. Sci. Fis. Mat. Natur. (8) **69** (1980), no. 3–4, 135–141.

144. M. Biliotti and G. Menichetti, *Derived semifield planes with affine elations*, J. Geom. **19** (1982), no. 1, 50–88.

145. _____, *On a generalization of Kantor's likeable planes*, Geom. Dedicata **17** (1985), no. 3, 253–277.

146. P. Biscarini, *Translation planes of order* 49, Rend. Circ. Mat. Palermo (2) **32** (1983), no. 1, 110–123.

147. I. Bloemen, *Substructures and characterizations of finite generalized polygons*, Ph.D. thesis, University of Ghent, 1995.

148. A. Blokhuis, R.S. Coulter, M. Henderson, and C.M. O'Keefe, *Permutations amongst the Dembowski-Ostrom polynomials*, Finite Fields and Applications (Augsburg, 1999), Springer, Berlin, 2001, pp. 37–42.

149. A. Blokhuis, M. Lavrauw, and S. Ball, *On the classification of semifield flocks*, Adv. Math. **180** (2003), no. 1, 104–111.

150. A. Blunck and N.L. Johnson, *Derivable pseudo-nets*, Note Mat. **21** (2002), no. 1, 107–112.

151. V. Boerner and M.J. Kallaher, *Half-Bol quasifields*, J. Geom. **18** (1982), no. 2, 185–193.

152. V. Boerner-Lantz, *A class of semifields of order* q^4, J. Geom. **27** (1986), no. 2, 112–118.

153. A. Bonisoli, *On the sharply 1-transitive subsets of* $\mathrm{PGL}(2, p^m)$, J. Geom. **31** (1988), no. 1–2, 32–41.

154. _____, *The regular subgroups of the sharply 3-transitive finite permutation groups*, Combinatorics '86 (Trento, 1986), Ann. Discrete Math., vol. 37, North-Holland, Amsterdam, 1988, pp. 75–86.

155. _____, *On resolvable finite Minkowski planes*, J. Geom. **36** (1989), no. 1–2, 1–7.

156. A. Bonisoli and A. Cossidente, *Mixed partitions of projective geometries*, Des. Codes Cryptogr. **20** (2000), no. 2, 143–154.

157. A. Bonisoli, D. Defina, and D. Saeli, *Finite translation planes arising from* $\mathrm{ASL}(1, 9)$-*invariant spreads*, Rend. Mat. Appl. (7) **10** (1990), no. 2, 327–347.

158. A. Bonisoli and C. Fiori, *Sharply 1-transitive subsets of certain permutation groups*, Geom. Dedicata **26** (1988), no. 3, 309–314.

159. A. Bonisoli and G. Korchmáros, *A characterization of the sharply 3-transitive finite permutation groups*, European J. Combin. **11** (1990), no. 3, 213–228.

160. _____, *Flocks of hyperbolic quadrics and linear groups containing homologies*, Geom. Dedicata **42** (1992), no. 3, 295–309.

161. _____, *Suzuki groups, one-factorizations and Lüneburg planes*, Discrete Math. **161** (1996), no. 1–3, 13–24.

162. A. Bonisoli, G. Korchmáros, and T. Szőnyi, *Some multiply derived translation planes with* $\mathrm{SL}(2, 5)$ *as an inherited collineation group in the translation complement*, Des. Codes Cryptogr. **10** (1997), no. 2, 109–114.

163. A. Bonisoli and P. Quattrocchi, *Incidence structures and permutation sets*, Pure Math. Appl. **5** (1994), no. 2, 127–140.

164. R.C. Bose and A. Barlotti, *Linear representation of a class of projective planes in a four dimensional projective space*, Ann. Mat. Pura Appl. (4) **88** (1971), 9–31.

165. R.C. Bose and K.J.C. Smith, *Ternary rings of a class of linearly representable semi-translation planes*, Atti del Convegno di Geometria Combinatoria e sue Applicazioni (Univ. Perugia, Perugia, 1970), Ist. Mat., Univ. Perugia, Perugia, 1971, pp. 69–101.

166. R. Brauer and C. Nesbitt, *On the modular characters of groups*, Ann. of Math. (2) **42** (1941), 556–590.

167. A.E. Brouwer and H.A. Wilbrink, *Blocking sets in translation planes*, J. Geom. **19** (1982), no. 2, 200.

168. M.R. Brown, *Ovoids of* $\mathrm{PG}(3, q)$, q *even, with a conic section*, J. London Math. Soc. (2) **62** (2000), no. 2, 569–582.

169. M.R. Brown, G.L. Ebert, and D. Luyckx, *On the geometry of regular hyperbolic fibrations*, European J. Combin. (2006), DOI 10.1016/j.ejc.2006.07.006 (online publication).

170. M.R. Brown, C.M. O'Keefe, S.E. Payne, T. Penttila, and G.F. Royle, *Spreads of $T_2(\mathcal{O})$*, α-*flocks and ovals*, Des. Codes Cryptogr. **31** (2004), no. 3, 251–282.

171. R.H. Bruck, *Some results in the theory of linear non-associative algebras*, Trans. Amer. Math. Soc. **56** (1944), 141–199.

172. ———, *Some results in the theory of quasigroups*, Trans. Amer. Math. Soc. **55** (1944), 19–52.

173. ———, *Contributions to the theory of loops*, Trans. Amer. Math. Soc. **60** (1946), 245–354.

174. ———, *Finite nets, I: Numerical invariants*, Canadian J. Math. **3** (1951), 94–107.

175. ———, *A Survey of Binary Systems*, Ergebnisse der Mathematik und ihrer Grenzgebiete. Neue Folge, Heft 20. Reihe: Gruppentheorie, Springer Verlag, Berlin, 1958.

176. ———, *Finite nets, II: Uniqueness and imbedding*, Pacific J. Math. **13** (1963), 421–457.

177. ———, *Construction problems of finite projective planes*, Combinatorial Mathematics and its Applications (Proc. Conf., Univ. North Carolina, Chapel Hill, N.C., 1967) (R.C. Bose and T.A. Dowling, eds.), Univ. North Carolina Press, Chapel Hill, N.C., 1969, pp. 426–514.

178. ———, *Circle geometry in higher dimensions*, A Survey of Combinatorial Theory (Proc. Internat. Sympos., Colorado State Univ., Fort Collins, Colo., 1971), North-Holland, Amsterdam, 1973, pp. 69–77.

179. ———, *Circle geometry in higher dimensions, II*, Geom. Dedicata **2** (1973), 133–188.

180. ———, *Construction problems in finite projective spaces*, Finite Geometric Structures and Their Applications (Centro Internaz. Mat. Estivo (C.I.M.E.), II Ciclo, Bressanone, 1972), Edizioni Cremonese, Rome, 1973, pp. 105–188.

181. ———, *The automorphism group of a circle geometry*, J. Combin. Theory Ser. A **32** (1982), no. 2, 256–263.

182. R.H. Bruck and R.C. Bose, *The construction of translation planes from projective spaces*, J. Algebra **1** (1964), 85–102.

183. ———, *Linear representations of projective planes in projective spaces*, J. Algebra **4** (1966), 117–172.

184. R.H. Bruck and E. Kleinfeld, *The structure of alternative division rings*, Proc. Amer. Math. Soc. **2** (1951), 878–890.

185. ———, *The structure of alternative division rings*, Proc. Nat. Acad. Sci. U. S. A. **37** (1951), 88–90.

186. R.H. Bruck and H.J. Ryser, *The nonexistence of certain finite projective planes*, Canadian J. Math. **1** (1949), 88–93.

187. A. Bruen, *Blocking sets in finite projective planes*, SIAM J. Appl. Math. **21** (1971), 380–392.

188. ———, *Partial spreads and replaceable nets*, Canad. J. Math. **23** (1971), 381–391.

189. ———, *Spreads and a conjecture of Bruck and Bose*, J. Algebra **23** (1972), 519–537.

190. ———, *Unimbeddable nets of small deficiency*, Pacific J. Math. **43** (1972), 51–54.

191. ———, *Subregular spreads and indicator sets*, Canad. J. Math. **27** (1975), no. 5, 1141–1148.

192. A. Bruen and J.C. Fisher, *Spreads which are not dual spreads*, Canad. Math. Bull. **12** (1969), 801–803.

193. A. Bruen and R. Silverman, *Switching sets in* PG$(3, q)$, Proc. Amer. Math. Soc. **43** (1974), 176–180.

194. A.A. Bruen, *Some new replaceable translation nets*, Canad. J. Math. **29** (1977), no. 2, 225–237.

195. ———, *Inversive geometry and some new translation planes, I*, Geom. Dedicata **7** (1978), no. 1, 81–98.

196. A.A. Bruen and J.W.P. Hirschfeld, *Applications of line geometry over finite fields, I: The twisted cubic*, Geom. Dedicata **6** (1977), no. 4, 495–509.

197. A.A. Bruen and J.A. Thas, *Flocks, chains and configurations in finite geometries*, Atti Accad. Naz. Lincei Rend. Cl. Sci. Fis. Mat. Natur. (8) **59** (1975), no. 6, 744–748.

198. ———, *Partial spreads, packings and Hermitian manifolds in* PG$(3, q)$, Math. Z. **151** (1976), no. 3, 207–214.

199. F. Buekenhout, *Existence of unitals in finite translation planes of order q^2 with a kernel of order q*, Geom. Dedicata **5** (1976), no. 2, 189–194.

200. F. Buekenhout (ed.), *Handbook of Incidence Geometry: Buildings and Foundations*, North-Holland, Amsterdam, 1995.

201. F. Buekenhout, A. Delandtsheer, J. Doyen, P.B. Kleidman, M.W. Liebeck, and J. Saxl, *Linear spaces with flag-transitive automorphism groups*, Geom. Dedicata **36** (1990), no. 1, 89–94.

202. R.P. Burn, *Finite Bol loops*, Math. Proc. Cambridge Philos. Soc. **84** (1978), no. 3, 377–385.

203. W. Büttner, *On translation planes containing* $Sz(q)$ *in their translational complement*, Geom. Dedicata **11** (1981), no. 3, 315–327.

204. _____, *Einige Translationsebenen der Ordnung* 64, Arch. Math. (Basel) **41** (1983), no. 6, 572–576.

205. _____, *On 4-dimensional translation planes admitting a Suzuki group as group of automorphisms*, J. Combin. Theory Ser. A **37** (1984), no. 1, 76–79.

206. _____, *Automorphismengruppen von Translationsebenen, die gewisse verallgemeinerte Elationen besitzen*, Rend. Sem. Mat. Univ. Padova **74** (1985), 1–5.

207. A. Caggegi, *On a class of translation planes*, Ricerche Mat. **31** (1982), no. 1, 139–153.

208. _____, *The collineations of a class of translation planes*, Proceedings of the Conference on Combinatorial and Incidence Geometry: Principles and Applications (La Mendola, 1982) (Milan), Rend. Sem. Mat. Brescia, vol. 7, Vita e Pensiero, 1984, pp. 181–203.

209. A.R. Calderbank, P.J. Cameron, W.M. Kantor, and J.J. Seidel, Z_4*-Kerdock codes, orthogonal spreads, and extremal Euclidean line-sets*, Proc. London Math. Soc. (3) **75** (1997), no. 2, 436–480.

210. P.J. Cameron, *Projective and Polar Spaces*, QMW Maths Notes, vol. 13, Queen Mary and Westfield College School of Mathematical Sciences, London, 1992.

211. P.J. Cameron and D. Ghinelli, *Tubes of even order and flat* $\pi.C_2$ *geometries*, Geom. Dedicata **55** (1995), no. 3, 265–278.

212. P.J. Cameron and N. Knarr, *Tubes in* $PG(3, q)$, European J. Combin. **27** (2006), no. 1, 114–124.

213. M. Capursi, *A translation plane of order* 11^2, J. Combin. Theory Ser. A **35** (1983), no. 3, 289–300.

214. I. Cardinali, N. Durante, T. Penttila, and R. Trombetti, *Bruen chains over fields of small order*, Discrete Math. **282** (2004), no. 1–3, 245–247.

215. I. Cardinali, O. Polverino, and R. Trombetti, *Semifield planes of order* q^4 *with kernel* F_{q^2} *and center* F_q, European J. Combin. **27** (2006), no. 6, 940–961.

216. L.R.A. Casse and D.G. Glynn, *The solution to Beniamino Segre's problem* $I_{r,q}$, $r = 3$, $q = 2^h$, Geom. Dedicata **13** (1982), no. 2, 157–163.

217. L.R.A. Casse and G. Lunardon, *On Buekenhout-Metz unitals*, Finite Geometry and Combinatorics (Deinze, 1997), Bull. Belg. Math. Soc. Simon Stevin **5** (1998), no. 2–3, 237–240.

218. L.R.A. Casse, C.M. O'Keefe, and T. Penttila, *Characterizations of Buekenhout-Metz unitals*, Geom. Dedicata **59** (1996), no. 1, 29–42.

219. M.L. Cates and R.B. Killgrove, *One-directional translation planes of order* 13, Proceedings of the Twelfth Southeastern Conference on Combinatorics, Graph Theory and Computing, Vol. I (Baton Rouge, La., 1981), Congr. Numer. **32** (1981), 173–180.

220. C. Cerroni, *Divisible designs from semifield planes*, Combinatorics '98 (Palermo), Discrete Math. **255** (2002), no. 1–3, 47–54.

221. C. Cerroni and A.G. Spera, *On divisible designs and twisted field planes*, J. Combin. Des. **7** (1999), no. 6, 453–464.

222. C. Charnes, *A nonsymmetric translation plane of order* 17^2, J. Geom. **37** (1990), no. 1–2, 77–83.

223. _____, *Quadratic matrices and the translation planes of order* 5^2, Coding Theory, Design Theory, Group Theory: Proc. Marshall Hall Conf. (Burlington, VT, 1990) (D. Jungnickel, S.A. Vanstone, K.T. Arasu, M. Aschbacher, and R. Foote, eds.), Wiley, New York, 1993, pp. 155–161.

224. _____, *A pair of mutually polar translation planes*, Ars Combin. **37** (1994), 121–128.

225. C. Charnes and U. Dempwolff, *Spreads, ovoids and* S_5, Geom. Dedicata **56** (1995), no. 2, 129–143.

226. _____, *The translation planes of order* 49 *and their automorphism groups*, Math. Comp. **67** (1998), no. 223, 1207–1224.

227. _____, *The eight-dimensional ovoids over* $GF(5)$, Math. Comp. **70** (2001), no. 234, 853–861.

228. W.E. Cherowitzo, *Hyperovals in the translation planes of order* 16, J. Combin. Math. Combin. Comput. **9** (1991), 39–55.

229. _____, *Monomial flocks of monomial cones in even characteristic*, Finite Geometry and Combinatorics (Deinze, 1997), Bull. Belg. Math. Soc. Simon Stevin **5** (1998), no. 2–3, 241–253.

230. W.E. Cherowitzo, C.M. O'Keefe, and T. Penttila, *A unified construction of finite geometries associated with q-clans in characteristic* 2, Adv. Geom. **3** (2003), no. 1, 1–21.

231. W.E. Cherowitzo and S.E. Payne, *The cyclic q-clans with* $q = 2^e$, Special issue dedicated to Adriano Barlotti, Adv. Geom. suppl. (2003), S158–S185.

232. W.E. Cherowitzo, T. Penttila, I. Pinneri, and G. F. Royle, *Flocks and ovals*, Geom. Dedicata **60** (1996), no. 1, 17–37.

233. I.V. Chuvaeva and D.V. Pasechnik, *Distance-transitive graphs of type* $q \cdot K_{q,q}$ *and projective planes*, European J. Combin. **11** (1990), no. 4, 341–346.

234. S. Çiftçi and R. Kaya, *On the Fano subplanes in the translation plane of order* 9, Doğa Mat. **14** (1990), no. 1, 1–7.

235. K.L. Clark, J.D. Key, and M.J. de Resmini, *Dual codes of translation planes*, European J. Combin. **23** (2002), no. 5, 529–538.

236. J. Cofman, *Double transitivity in finite affine and projective planes*, Atti Accad. Naz. Lincei Rend. Cl. Sci. Fis. Mat. Natur. (8) **43** (1967), 317–320.

237. _____, *Double transitivity in finite affine planes, I*, Math. Z. **101** (1967), 335–342.

238. S.D. Cohen and M.J. Ganley, *Commutative semifields, two-dimensional over their middle nuclei*, J. Algebra **75** (1982), no. 2, 373–385.

239. _____, *Some classes of translation planes*, Quart. J. Math. Oxford Ser. (2) **35** (1984), no. 138, 101–113.

240. S.D. Cohen, M.J. Ganley, and V. Jha, *On transitive automorphism groups of finite quasifields*, Arch. Math. (Basel) **35** (1980), no. 5, 406–415.

241. _____, *Transitive automorphism groups of finite quasifields*, Finite Geometries and Designs (Proc. Conf., Chelwood Gate, 1980), London Math. Soc. Lecture Note Ser., vol. 49, Cambridge Univ. Press, Cambridge, 1981, pp. 88–97.

242. J.H. Conway, P.B. Kleidman, and R.A. Wilson, *New families of ovoids in* O_8^+, Geom. Dedicata **26** (1988), no. 2, 157–170.

243. J.H. Conway and N.J.A. Sloane, *Sphere Packings, Lattices and Groups*, third ed., with additional contributions by E. Bannai, R.E. Borcherds, J. Leech, S.P. Norton, A.M. Odlyzko, R.A. Parker, L. Queen and B.B. Venkov, Grundlehren Math. Wiss., vol. 290, Springer-Verlag, New York, 1999.

244. B.N. Cooperstein, *Minimal degree for a permutation representation of a classical group*, Israel J. Math. **30** (1978), no. 3, 213–235.

245. _____, *A sporadic ovoid in* $\Omega^+(8,5)$ *and some non-Desarguesian translation planes of order* 25, J. Combin. Theory Ser. A **54** (1990), no. 1, 135–140.

246. M. Cordero, *Semifield planes of order* p^4 *that admit a p-primitive Baer collineation*, Osaka J. Math. **28** (1991), no. 2, 305–321.

247. _____, *A note on the Boerner-Lantz semifield planes*, J. Geom. **43** (1992), no. 1–2, 53–56.

248. _____, *The nuclei and other properties of p-primitive semifield planes*, Int. J. Math. Math. Sci. **15** (1992), no. 2, 367–370.

249. _____, *p-primitive semifield planes*, Combinatorics '90 (Gaeta, 1990), Ann. Discrete Math., vol. 52, North-Holland, Amsterdam, 1992, pp. 107–117.

250. _____, *Matrix spread sets of p-primitive semifield planes*, Int. J. Math. Math. Sci. **20** (1997), no. 2, 293–297.

251. M. Cordero and R.F. Figueroa, *On some new classes of semifield planes*, Osaka J. Math. **30** (1993), no. 2, 171–178.

252. _____, *Towards a characterization of generalized twisted field planes*, J. Geom. **52** (1995), no. 1–2, 54–63.

253. _____, *Transitive autotopism groups and the generalized twisted field planes*, Mostly Finite Geometries (Iowa City, IA, 1996), Lecture Notes in Pure and Appl. Math., vol. 190, Dekker, New York, 1997, pp. 191–196.

254. _____, *On the semifield planes of order* 5^4 *and dimension 2 over the kernel*, Note Mat. **22** (2003/04), no. 1, 75–81.

255. M. Cordero and G.P. Wene, *A survey of finite semifields*, Combinatorics (Assisi, 1996), Discrete Math. **208/209** (1999), 125–137.

256. M. Cordero-Vourtsanis, *The autotopism group of p-primitive semifield planes*, Ars Combin. **32** (1991), 57–64.

257. A. Cossidente, *Mixed partitions of* PG$(3, q)$, J. Geom. **68** (2000), no. 1–2, 48–57.

258. A. Cossidente, C. Culbert, G.L. Ebert, and G. Marino, *On m-ovoids of* $\mathcal{W}_3(q)$, Combinatorica (to appear).

259. A. Cossidente, G.L. Ebert, and G. Korchmáros, *A group-theoretic characterization of classical unitals*, Arch. Math. (Basel) **74** (2000), no. 1, 1–5.

260. _____, *Unitals in finite Desarguesian planes*, J. Algebraic Combin. **14** (2001), no. 2, 119–125.

261. A. Cossidente, G.L. Ebert, and G. Marino, *Commuting Hermitian varieties and the flag geometry of* PG$(2, q^2)$, Discrete Math. (to appear).

262. A. Cossidente, G.L. Ebert, G. Marino, and A. Siciliano, *Shult sets and translation ovoids of the Hermitian surface*, Adv. Geom. **6** (2006), no. 4, 523–542.

263. A. Cossidente, G. Lunardon, G. Marino, and O. Polverino, *Hermitian indicator sets*, (to appear).

264. R.S. Coulter and R.W. Matthews, *Planar functions and planes of Lenz-Barlotti class II*, Des. Codes Cryptogr. **10** (1997), no. 2, 167–184.

265. M. Crismale, $(q^2 + q + 1)$*-sets of type* $(0, 1, 2, q + 1)$ *in translation planes of order* q^2, Combinatorics '81 (Rome, 1981), Ann. Discrete Math., vol. 18, North-Holland, Amsterdam, 1983, pp. 225–228.

266. _____, *Translation planes associated with Denniston spreads*, Riv. Mat. Univ. Parma (4) **11** (1985), 299–307.

267. C. Culbert and G.L. Ebert, *Circle geometry and three-dimensional subregular translation planes*, Innov. Incidence Geom. **1** (2005), 3–18.

268. T. Czerwinski, *Finite translation planes with collineation groups doubly transitive on the points at infinity*, J. Algebra **22** (1972), 428–441.

269. _____, *The collineation groups of the translation planes of order* 25, Geom. Dedicata **39** (1991), no. 2, 125–137.

270. T. Czerwinski and D. Oakden, *The translation planes of order twenty-five*, J. Combin. Theory Ser. A **59** (1992), no. 2, 193–217.

271. E.H. Davis, *Translation planes of order* 25 *with nontrivial* $X - OY$ *perspectivities*, Proceedings of the Tenth Southeastern Conference on Combinatorics, Graph Theory and Computing (Florida Atlantic Univ., Boca Raton, Fla., 1979) (Winnipeg, Man.), Congress. Numer., XXIII–XXIV, Utilitas Math., 1979, pp. 341–348.

272. F. De Clerck, H. Gevaert, and J.A. Thas, *Flocks of a quadratic cone in* PG$(3, q)$, $q \leq 8$, Geom. Dedicata **26** (1988), no. 2, 215–230.

273. F. De Clerck and C. Herssens, *Flocks of the quadratic cone in* $PG(3, q)$, *for q small*, The CAGe Reports, no. 8, Dec. 15, 1992, pp. 1–74.

274. F. De Clerck and N.L. Johnson, *Subplane covered nets and semipartial geometries*, A collection of contributions in honour of Jack van Lint, Discrete Math. **106/107** (1992), 127–134.

275. F. De Clerck and H. Van Maldeghem, *On flocks of infinite quadratic cones*, A tribute to J.A. Thas (Gent, 1994), Bull. Belg. Math. Soc. Simon Stevin **1** (1994), no. 3, 399–415.

276. M.J. de Resmini, *Some combinatorial properties of a semitranslation plane*, Eighteenth Southeastern International Conference on Combinatorics, Graph Theory, and Computing (Boca Raton, Fla., 1987), Congr. Numer. **59** (1987), 5–12.

277. _____, *On the semifield plane of order* 16 *with kern* GF(2), Eleventh British Combinatorial Conference (London, 1987), Ars Combin. **25** (1988), no. B, 75–92.

278. _____, *On the Dempwolff plane*, Finite Geometries and Combinatorial Designs (Lincoln, NE, 1987), Contemp. Math., vol. 111, Amer. Math. Soc., Providence, RI, 1990, pp. 47–64.

279. _____, *On the derived semifield plane of order* 16, Twelfth British Combinatorial Conference (Norwich, 1989), Ars Combin. **29** (1990), no. B, 97–109.

280. _____, *On the Johnson-Walker plane*, Simon Stevin **64** (1990), no. 2, 113–139.

281. _____, *Some remarks on the Hall plane of order* 16, Proceedings of the Twentieth Southeastern Conference on Combinatorics, Graph Theory, and Computing (Boca Raton, FL, 1989), Congr. Numer. **70** (1990), 17–27.

282. _____, *On an exceptional semi-translation plane*, Advances in Finite Geometries and Designs (Chelwood Gate, 1990), Oxford Sci. Publ., Oxford Univ. Press, New York, 1991, pp. 141–162.

283. _____, *On the semifield plane of order 16 with kern GF(4)*, Combinatorics '88, Vol. 2 (Ravello, 1988), Res. Lecture Notes Math., Mediterranean, Rende, 1991, pp. 369–390.

284. _____, *On the classical semitranslation plane of order 16*, Ars Combin. **46** (1997), 191–209.

285. _____, *On the Mathon plane*, J. Geom. **60** (1997), no. 1–2, 47–64.

286. M.J. de Resmini and A.O. Leone, *Subplanes of the derived Hughes plane of order 25*, Simon Stevin **67** (1993), no. 3–4, 289–322.

287. M.J. de Resmini and L. Puccio, *Some combinatorial properties of the dual Lorimer plane*, Proceedings of the First Catania International Combinatorial Conference on Graphs, Steiner Systems, and their Applications, Vol. 1 (Catania, 1986), Ars Combin. **24** (1987), no. A, 131–148.

288. P. Dembowski, *Gruppentheoretische Kennzeichnungen der endlichen desarguesschen Ebenen*, Abh. Math. Sem. Univ. Hamburg **29** (1965), 92–106.

289. _____, *Zur Geometrie der Suzukigruppen*, Math. Z. **94** (1966), 106–109.

290. _____, *Finite Geometries*, Ergeb. Math. Grenzgeb., Band 44, Springer-Verlag, Berlin, 1968.

291. _____, *Generalized Hughes planes*, Canad. J. Math. **23** (1971), 481–494.

292. P. Dembowski and T.G. Ostrom, *Planes of order n with collineation groups of order n^2*, Math. Z. **103** (1968), 239–258.

293. U. Dempwolff, *Grosse Baer-Untergruppen auf Translationsebenen gerader Ordnung*, J. Geom. **19** (1982), no. 2, 101–114.

294. _____, *A note on the Figueroa planes*, Arch. Math. (Basel) **43** (1984), no. 3, 285–288.

295. _____, *A remark on the Figueroa planes*, Mitt. Math. Sem. Giessen (1984), no. 164, 77–81.

296. _____, *On the automorphism group of planes of Figueroa type*, Rend. Sem. Mat. Univ. Padova **74** (1985), 59–62.

297. _____, *PSL(3, q) on projective planes of order q^3*, Geom. Dedicata **18** (1985), no. 1, 101–112.

298. _____, *Involutory homologies on affine planes*, Abh. Math. Sem. Univ. Hamburg **57** (1987), 37–55.

299. _____, *Linear groups with large cyclic subgroups and translation planes*, Rend. Sem. Mat. Univ. Padova **77** (1987), 69–113.

300. _____, *A note on 2-groups and homologies on affine planes*, Boll. Un. Mat. Ital. A (7) **1** (1987), no. 3, 383–386.

301. _____, *A characterization of the generalized twisted field planes*, Arch. Math. (Basel) **50** (1988), no. 5, 477–480.

302. _____, *Involutory homologies on affine planes: the conclusion*, Abh. Math. Sem. Univ. Hamburg **59** (1989), 43–59.

303. _____, *The projective planes of order 16 admitting SL(3, 2)*, Rad. Mat. **7** (1991), no. 1, 123–134.

304. _____, *Translation planes of order 27*, Des. Codes Cryptogr. **4** (1994), no. 2, 105–121.

305. _____, *Correction to: "Translation planes of order 27" [Des. Codes Cryptogr. **4** (1994), no. 2, 105–121]*, Des. Codes Cryptogr. **5** (1995), no. 1, 81.

306. _____, *Ovoids and translation planes admitting A_5 at characteristic p > 5*, Boll. Un. Mat. Ital. A (7) **11** (1997), no. 3, 731–738.

307. U. Dempwolff and A. Guthmann, *Applications of number theory to ovoids and translation planes*, Geom. Dedicata **78** (1999), no. 2, 201–213.

308. U. Dempwolff and A. Reifart, *Translation planes of order 16 admitting a Baer 4-group*, J. Combin. Theory Ser. A **32** (1982), no. 1, 119–124.

309. _____, *The classification of the translation planes of order 16, I*, Geom. Dedicata **15** (1983), no. 2, 137–153.

310. U. Dempwolff and M. Röder, *On finite projective planes defined by planar functions*, Innov. Incidence Geom. (to appear).

311. R.H.F. Denniston, *Subplanes of the Hughes plane of order 9*, Proc. Cambridge Philos. Soc. **64** (1968), 589–598.

312. ———, *Some packings of projective spaces*, Atti Accad. Naz. Lincei Rend. Cl. Sci. Fis. Mat. Natur. (8) **52** (1972), 36–40.

313. ———, *Cyclic packings of the projective space of order 8*, Atti Accad. Naz. Lincei Rend. Cl. Sci. Fis. Mat. Natur. (8) **54** (1973), 373–377.

314. ———, *Packings of* PG(3, q), Finite Geometric Structures and Their Applications (Centro Internaz. Mat. Estivo (C.I.M.E.), II Ciclo, Bressanone, 1972), Edizioni Cremonese, Rome, 1973, pp. 193–199.

315. ———, *Spreads which are not subregular*, Glasnik Mat. Ser. III **8(28)** (1973), 3–5.

316. ———, *Some spreads which contain reguli without being subregular*, Colloquio Internazionale sulle Teorie Combinatorie (Roma, 1973), Tomo II, Atti dei Convegni Lincei, no. 17, Accad. Naz. Lincei, Rome, 1976, pp. 367–371.

317. L.E. Dickson, *Linear Groups: With an Exposition of the Galois Field Theory*, Dover Publications Inc., New York, 1958, with an introduction by W. Magnus.

318. J. Dieudonné, *Sur les Groupes Classiques*, Actualités Sci. Ind., no. 1040 = Publ. Inst. Math. Univ. Strasbourg (N.S.) no. 1 (1945), Hermann et Cie., Paris, 1948.

319. ———, *La Géométrie des Groupes Classiques*, Troisième édition, Ergeb. Math. Grenzgeb., Band 5, Springer-Verlag, Berlin, 1971.

320. J.D. Dixon and B. Mortimer, *Permutation Groups*, Graduate Texts in Mathematics, vol. 163, Springer-Verlag, New York, 1996.

321. J.M. Dover, *Some design-theoretic properties of Buekenhout unitals*, J. Combin. Des. **4** (1996), no. 6, 449–456.

322. ———, *A family of non-Buekenhout unitals in the Hall planes*, Mostly Finite Geometries (Iowa City, IA, 1996), Lecture Notes in Pure and Appl. Math., vol. 190, Dekker, New York, 1997, pp. 197–205.

323. ———, *Subregular spreads of* $\mathcal{PG}(2n + 1, q)$, Finite Fields Appl. **4** (1998), no. 4, 362–380.

324. ———, *Spreads and resolutions of Ree unitals*, Ars Combin. **54** (2000), 301–309.

325. ———, *Subregular spreads of* PG(5, 2e), Finite Fields Appl. **7** (2001), no. 3, 421–427.

326. D.E. Draayer, *Right nuclear decomposition of generalized André systems*, Adv. Geom. **2** (2002), no. 1, 81–98.

327. ———, *Translation planes admitting a pair of index three homology groups*, J. Geom. **73** (2002), no. 1–2, 112–133.

328. D.E. Draayer, N.L. Johnson, and R. Pomareda, *Triangle transitive translation planes*, Note Mat. **26** (2006), no. 1, 29–53.

329. J. Duplák, *Quasigroups and translation planes*, J. Geom. **43** (1992), no. 1–2, 95–107.

330. B.K. Durakov, *Finite translation planes that admit a collineation group that acts doubly transitively on a translation line*, Discrete Mathematics (Russian), Krasnoyarsk. Gos. Tekh. Univ., Krasnogorsk, 1996, pp. 32–39.

331. K.V. Durga Prasad, *Two flag transitive planes of order* 9^2, Bull. Pure Appl. Sci. Sect. E Math. Stat. **19** (2000), no. 1, 99–104.

332. R.H. Dye, *Partitions and their stabilizers for line complexes and quadrics*, Ann. Mat. Pura Appl. (4) **114** (1977), 173–194.

333. G.L. Ebert, *Disjoint circles: a classification*, Trans. Amer. Math. Soc. **232** (1977), 83–109.

334. ———, *Amusing configurations of disjoint circles*, Ars Combin. **6** (1978), 197–207.

335. ———, *Maximal strictly partial spreads*, Canad. J. Math. **30** (1978), no. 3, 483–489.

336. ———, *Translation planes of order* q^2: *asymptotic estimates*, Trans. Amer. Math. Soc. **238** (1978), 301–308.

337. ———, *A new lower bound for msp spreads*, Proceedings of the Tenth Southeastern Conference on Combinatorics, Graph Theory and Computing (Florida Atlantic Univ., Boca Raton, Fla., 1979) (Winnipeg, Man.), Congress. Numer., XXIII–XXIV, Utilitas Math., 1979, pp. 413–421.

338. ———, *Subregular 1-spreads of* PG(2n + 1, 2), Geom. Dedicata **14** (1983), no. 4, 343–353.

339. ———, *The completion problem for partial packings*, Geom. Dedicata **18** (1985), no. 3, 261–267.

340. ———, *Partitioning projective geometries into caps*, Canad. J. Math. **37** (1985), no. 6, 1163–1175.

341. ———, *Spreads obtained from ovoidal fibrations*, Finite Geometries (Winnipeg, Man., 1984), Lecture Notes in Pure and Appl. Math., vol. 103, Dekker, New York, 1985, pp. 117–125.

342. _____, *Nests, covers, and translation planes*, Eleventh British Combinatorial Conference (London, 1987), Ars Combin. **25** (1988), no. C, 213–233.

343. _____, *Spreads admitting regular elliptic covers*, European J. Combin. **10** (1989), no. 4, 319–330.

344. _____, *Some nonreplaceable nests*, Combinatorics '88, Vol. 1 (Ravello, 1988), Res. Lecture Notes Math., Mediterranean, Rende, 1991, pp. 353–372.

345. _____, *On Buekenhout-Metz unitals of even order*, European J. Combin. **13** (1992), no. 2, 109–117.

346. _____, *Some results on two-dimensional translation planes*, Conf. Semin. Mat. Univ. Bari (1992), no. 246, 12.

347. _____, *Buekenhout-Tits unitals*, J. Algebraic Combin. **6** (1997), no. 2, 133–140.

348. _____, *Replaceable nests*, Mostly Finite Geometries (Iowa City, IA, 1996), Lecture Notes in Pure and Appl. Math., vol. 190, Dekker, New York, 1997, pp. 35–49.

349. _____, *Partitioning problems and flag-transitive planes*, Combinatorics '98 (Mondello), Rend. Circ. Mat. Palermo (2) Suppl. (1998), no. 53, 27–44.

350. _____, *Buekenhout unitals*, Combinatorics (Assisi, 1996), Discrete Math. **208/209** (1999), 247–260.

351. _____, *Constructions in finite geometry using computer algebra systems*, Computational algebra and number theory (Milwaukee, WI, 1996), J. Symbolic Comput. **31** (2001), no. 1–2, 55–70.

352. G.L. Ebert and K. Wantz, *A group-theoretic characterization of Buekenhout-Metz unitals*, J. Combin. Des. **4** (1996), no. 2, 143–152.

353. M.R. Enea and G. Korchmáros, *Ovals in commutative semifield planes*, Arch. Math. (Basel) **69** (1997), no. 3, 259–264.

354. A.B. Evans, *Distance in finite geometries, I: Translation planes*, Geom. Dedicata **15** (1983), no. 1, 59–68.

355. R. Figueroa, *A family of not (V, l)-transitive projective planes of order q^3, $q \not\equiv 1(\mathrm{mod}\, 3)$ and $q > 2$*, Math. Z. **181** (1982), no. 4, 471–479.

356. _____, *Translation planes that admit a collineation group of order $q^2 - 1$*, Ph.D. thesis, University of Iowa, 1989.

357. _____, *A characterization of the generalized twisted field planes of characteristic ≥ 5*, Geom. Dedicata **50** (1994), no. 2, 205–216.

358. R.F. Figueroa and M. Cordero, *Semifield planes admitting a transitive autotopism group*, Proceedings of the Twenty-fifth Southeastern International Conference on Combinatorics, Graph Theory and Computing (Boca Raton, FL, 1994), Congr. Numer. **102** (1994), 91–95.

359. _____, *On the generalized twisted field planes of characteristic 2*, Geom. Dedicata **56** (1995), no. 2, 197–208.

360. J.B. Fink, N.L. Johnson, and F.W. Wilke, *A characterization of "likeable" translation planes*, Rend. Circ. Mat. Palermo (2) **32** (1983), no. 1, 76–99.

361. _____, *Transitive affine planes admitting elations*, J. Geom. **21** (1983), no. 1, 59–65.

362. J.B. Fink and M.J. Kallaher, *Simple groups acting on translation planes*, J. Geom. **29** (1987), no. 2, 126–139.

363. J.C. Fisher and N.L. Johnson, *Fano configurations in subregular planes*, Advances in Geometry (submitted).

364. J.C. Fisher and J.A. Thas, *Flocks in PG(3, q)*, Math. Z. **169** (1979), no. 1, 1–11.

365. D.A. Foulser, *The flag-transitive collineation groups of the finite Desarguesian affine planes*, Canad. J. Math. **16** (1964), 443–472.

366. _____, *Solvable flag transitive affine groups*, Math. Z. **86** (1964), 191–204.

367. _____, *A class of translation planes, $\Pi(Q_g)$*, Math. Z. **101** (1967), 95–102.

368. _____, *A generalization of André's systems*, Math. Z. **100** (1967), 380–395.

369. _____, *Collineation groups of generalized André planes*, Canad. J. Math. **21** (1969), 358–369.

370. _____, *Replaceable translation nets*, Proc. London Math. Soc. (3) **22** (1971), 235–264.

371. _____, *Subplanes of partial spreads in translation planes*, Bull. London Math. Soc. **4** (1972), 32–38.

372. _____, *Collineations of order p in translation planes of order p^r, $p > 2$*, Proceedings of the International Conference on Projective Planes (Washington State Univ., Pullman, Wash., 1973), Washington State Univ. Press, Pullman, Wash., 1973, pp. 83–90.

373. _____, *Derived translation planes admitting affine elations*, Math. Z. **131** (1973), 183–188.
374. _____, *A translation plane admitting Baer collineations of order p*, Arch. Math. (Basel) **24** (1973), 323–326.
375. _____, *Baer p-elements in translation planes*, J. Algebra **31** (1974), 354–366.
376. _____, *Planar collineations of order p in translation planes of order p^r*, Geom. Dedicata **5** (1976), no. 3, 393–409.
377. _____, *Some translation planes of order* 81, Finite Geometries and Designs (Proc. Conf., Chelwood Gate, 1980), London Math. Soc. Lecture Note Ser., vol. 49, Cambridge Univ. Press, Cambridge, 1981, pp. 114–118.
378. D.A. Foulser and N.L. Johnson, *The translation planes of order q^2 that admit SL(2, q) as a collineation group, II: Odd order*, J. Geom. **18** (1982), no. 2, 122–139.
379. _____, *The translation planes of order q^2 that admit SL(2, q) as a collineation group, I: Even order*, J. Algebra **86** (1984), no. 2, 385–406.
380. D.A. Foulser, N.L. Johnson, and T.G. Ostrom, *A characterization of the Desarguesian planes of order q^2 by SL(2, q)*, Int. J. Math. Math. Sci. **6** (1983), no. 3, 605–608.
381. D.A. Foulser and M.J. Kallaher, *Solvable, flag-transitive, rank 3 collineation groups*, Geom. Dedicata **7** (1978), no. 1, 111–130.
382. D.A. Foulser, G. Mason, and M. Walker, *On translation planes as irreducible SL(3, q)-modules, q odd*, Proceedings of the Conference on Combinatorial and Incidence Geometry: Principles and Applications (La Mendola, 1982) (Milan), Rend. Sem. Mat. Brescia, vol. 7, Vita e Pensiero, 1984, pp. 323–336.
383. J.W. Freeman, *Reguli and pseudoreguli in PG(3, s^2)*, Geom. Dedicata **9** (1980), no. 3, 267–280.
384. M. Funk, *On configurations in translation planes of positive characteristic*, Abh. Math. Sem. Univ. Hamburg **56** (1986), 119–125.
385. M.J. Ganley, *Polarities in translation planes*, Geom. Dedicata **1** (1972), no. 1, 103–116.
386. _____, *Baer involutions in semifield planes of even order*, Geom. Dedicata **2** (1974), 499–508.
387. _____, *On a paper of P. Dembowski and T. G. Ostrom: "Planes of order n with collineation groups of order n^2" (Math. Z. **103** (1968), 239–258)*, Arch. Math. (Basel) **27** (1976), no. 1, 93–98.
388. _____, *Central weak nucleus semifields*, European J. Combin. **2** (1981), no. 4, 339–347.
389. _____, *Tangentially transitive semitranslation planes*, Geom. Dedicata **12** (1982), no. 3, 287–296.
390. _____, *On likeable translation planes of even order*, Arch. Math. (Basel) **41** (1983), no. 5, 478–480.
391. M.J. Ganley and V. Jha, *On a conjecture of Kallaher and Liebler*, Geom. Dedicata **21** (1986), no. 3, 277–289.
392. _____, *On translation planes with a 2-transitive orbit on the line at infinity*, Arch. Math. (Basel) **47** (1986), no. 4, 379–384.
393. M.J. Ganley, V. Jha, and N.L. Johnson, *The translation planes admitting a nonsolvable doubly transitive line-sized orbit*, J. Geom. **69** (2000), no. 1–2, 88–109.
394. A. Gardiner, *Imprimitive distance-regular graphs and projective planes*, J. Combin. Theory Ser. B **16** (1974), 274–281.
395. H. Gevaert and N.L. Johnson, *Flocks of quadratic cones, generalized quadrangles and translation planes*, Geom. Dedicata **27** (1988), no. 3, 301–317.
396. _____, *On maximal partial spreads in PG(3, q) of cardinalities $q^2 - q + 1$, $q^2 - q + 2$*, Ars Combin. **26** (1988), 191–196.
397. H. Gevaert, N.L. Johnson, and J.A. Thas, *Spreads covered by reguli*, Simon Stevin **62** (1988), no. 1, 51–62.
398. A.M. Gleason, *Finite Fano planes*, Amer. J. Math. **78** (1956), 797–807.
399. D.G. Glynn, *A lower bound for maximal partial spreads in PG(3, q)*, Ars Combin. **13** (1982), 39–40.
400. _____, *Two new sequences of ovals in finite Desarguesian planes of even order*, Combinatorial Mathematics, X (Adelaide, 1982), Lecture Notes in Math., vol. 1036, Springer, Berlin, 1983, pp. 217–229.
401. _____, *On finite division algebras*, J. Combin. Theory Ser. A **44** (1987), no. 2, 253–266.

402. _____, *The classification of projective planes of order q^3 with cone-representations in* PG(6, q), Geom. Dedicata **66** (1997), no. 3, 343–355.

403. _____, *A general theory of cone-representations*, Boll. Un. Mat. Ital. A (7) **11** (1997), no. 3, 789–800.

404. _____, *On cone representations of translation planes*, J. Statist. Plann. Inference **58** (1997), no. 1, 33–41.

405. _____, *On non-spread representations of projective planes of order q^i in* PG(2i, q), Charlemagne and His Heritage: 1200 Years of Civilization and Science in Europe, Vol. 2 (Aachen, 1995), Brepols, Turnhout, 1998, pp. 273–286.

406. _____, *Plane representations of ovoids*, Finite Geometry and Combinatorics (Deinze, 1997), Bull. Belg. Math. Soc. Simon Stevin **5** (1998), no. 2–3, 275–286.

407. _____, *A survey of cone representations*, R. C. Bose Memorial Conference (Fort Collins, CO, 1995), J. Statist. Plann. Inference **72** (1998), no. 1–2, 265–277.

408. D.G. Glynn, C.M. O'Keefe, T. Penttila, and C.E. Praeger, *Ovoids and monomial ovals*, Geom. Dedicata **59** (1996), no. 3, 223–241.

409. D.G. Glynn and G.F. Steinke, *On conics that are ovals in a Hall plane*, European J. Combin. **14** (1993), no. 6, 521–528.

410. _____, *Pencils of translation ovals in translation planes*, Geom. Dedicata **51** (1994), no. 2, 113–121.

411. C.D. Godsil, R.A. Liebler, and C.E. Praeger, *Antipodal distance transitive covers of complete graphs*, European J. Combin. **19** (1998), no. 4, 455–478.

412. A. Gonçalves and C.Y. Ho, *On flag collineations of finite projective planes*, J. Geom. **28** (1987), no. 2, 117–127.

413. E.G. Goodaire and M.J. Kallaher, *Systems with two binary operations, and their planes*, Quasigroups and Loops: Theory and Applications, Sigma Ser. Pure Math., vol. 8, Heldermann, Berlin, 1990, pp. 161–195.

414. D. Gorenstein, *Finite Groups*, Harper & Row Publishers, New York, 1968.

415. L.C. Grove and C.T. Benson, *Finite Reflection Groups*, second ed., Graduate Texts in Mathematics, vol. 99, Springer-Verlag, New York, 1985.

416. T. Grundhöfer, *Eine Charakterisierung ableitbarer Translationsebenen*, Geom. Dedicata **11** (1981), no. 2, 177–185.

417. _____, *Reguli in Faserungen projektiver Räume*, Geom. Dedicata **11** (1981), no. 2, 227–237.

418. _____, *Die Bestimmung der Kollineationsgruppen der merkwürdigen Translationsebenen*, Geom. Dedicata **13** (1982), no. 1, 63–66.

419. _____, *Die Projektivitätengruppen der endlichen Translationsebenen*, J. Geom. **20** (1983), no. 1, 74–85.

420. _____, *Projektivitätengruppen von Translationsebenen*, Results Math. **6** (1983), no. 2, 163–182.

421. _____, *Finite subplanes and affine projectivities of translation planes*, Mitt. Math. Sem. Giessen (1984), no. 164, 179–184.

422. _____, *A synthetic construction of the Figueroa planes*, J. Geom. **26** (1986), no. 2, 191–201.

423. A. Gunawardena and G.E. Moorhouse, *The non-existence of ovoids in* $O_9(q)$, European J. Combin. **18** (1997), no. 2, 171–173.

424. M. Hall, Jr., *Projective planes*, Trans. Amer. Math. Soc. **54** (1943), 229–277.

425. _____, *Cyclic projective planes*, Duke Math. J. **14** (1947), 1079–1090.

426. _____, *Correction to "Projective planes."*, Trans. Amer. Math. Soc. **65** (1949), 473–474.

427. _____, *Uniqueness of the projective plane with 57 points*, Proc. Amer. Math. Soc. **4** (1953), 912–916.

428. _____, *Projective Planes and Related Topics*, California Institute of Technology, 1954.

429. _____, *Ovals in the Desarguesian plane of order 16*, Ann. Mat. Pura Appl. (4) **102** (1975), 159–176.

430. _____, *The Theory of Groups*, Chelsea Publishing Co., New York, 1976, reprinting of the 1968 edition.

431. M. Hall, Jr., J.D. Swift, and R. Killgrove, *On projective planes of order nine*, Math. Tables Aids Comput. **13** (1959), 233–246.

432. M. Hall, Jr., J.D. Swift, and R.J. Walker, *Uniqueness of the projective plane of order eight*, Math. Tables Aids Comput. **10** (1956), 186–194.

433. N. Hamilton, *Some inherited maximal arcs in derived dual translation planes*, Geom. Dedicata **55** (1995), no. 2, 165–173.

434. _____, *Some maximal arcs in Hall planes*, J. Geom. **52** (1995), no. 1–2, 101–107.

435. N. Hamilton and R. Mathon, *More maximal arcs in Desarguesian projective planes and their geometric structure*, Adv. Geom. **3** (2003), no. 3, 251–261.

436. N. Hamilton and T. Penttila, *A characterisation of Thas maximal arcs in translation planes of square order*, J. Geom. **51** (1994), no. 1–2, 60–66.

437. J. Hanson and M.J. Kallaher, *Finite Bol quasifields are nearfields*, Utilitas Math. **37** (1990), 45–64.

438. M.E. Harris and C. Hering, *On the smallest degrees of projective representations of the groups* PSL(n, q), Canad. J. Math. **23** (1971), 90–102.

439. R.W. Hartley, *Determination of the ternary collineation groups whose coefficients lie in the* GF(2^n), Ann. of Math. (2) **27** (1925), no. 2, 140–158.

440. V. Havel, *Zur Geometrie der Translationsebenen*, Czechoslovak Math. J **10 (85)** (1960), 432–439.

441. _____, *A characterization of affine planes over Veblen-Wedderburn systems with the right inverse property*, Rev. Roumaine Math. Pures Appl. **12** (1967), 1461–1466.

442. _____, *Nets and groupoids*, Comment. Math. Univ. Carolinae **8** (1967), 435–451.

443. _____, *One characterization of special translation planes*, Arch. Math. **3** (1967), 157–160.

444. _____, *One construction of quasifields*, Arch. Math. (Basel) **18** (1967), 33–34.

445. _____, *Ternary halfgroupoids and coordinatization*, Comment. Math. Univ. Carolinae **8** (1967), 569–580.

446. _____, *Group congruences and quasifields*, J. Algebra **8** (1968), 385–387.

447. _____, *Nets and groupoids, II*, Comment. Math. Univ. Carolinae **9** (1968), 87–93.

448. _____, *Ternary halfgroupoids and coordinatization*, Mat. Časopis Sloven. Akad. Vied **19** (1969), 102–109.

449. _____, *Zur Theorie der Zassenhausschen Verfeinerungen zweier Reihen von Zerlegungen, III: Kartesische Strukturen*, Acta Fac. Rerum Natur. Univ. Comenian. Math. (1969), no. Publ. 23, 7–11.

450. _____, *A general coordinatization principle for projective planes with comparison of Hall and Hughes frames and with examples of generalized oval frames*, Czechoslovak Math. J. **24(99)** (1974), 664–673.

451. _____, *Generalization of one Baer's theorem for nets*, Časopis Pěst. Mat. **101** (1976), no. 4, 375–378.

452. _____, *Replaceable nets and improper collineations*, Comment. Math. Univ. Carolin. **24** (1983), no. 3, 507–517.

453. V. Havel and I. Studnička, *Bemerkung über gemeinsame Beziehung zwischen kartesischen Gruppen und kartesischen Zahlensystemen*, Časopis Pěst. Mat. **103** (1978), no. 4, 384–390, 409.

454. O. Heden, *No maximal partial spread of size 10 in* PG$(3, 5)$, Ars Combin. **29** (1990), 297–298.

455. _____, *No partial 1-spread of class* $[0, \geq 2]_d$ *in* PG$(2d - 1, q)$ *exists*, Discrete Math. **87** (1991), no. 2, 215–216.

456. _____, *Maximal partial spreads and the modular n-queen problem*, Discrete Math. **120** (1993), no. 1–3, 75–91.

457. _____, *Maximal partial spreads and the modular n-queen problem, II*, Discrete Math. **142** (1995), no. 1–3, 97–106.

458. _____, *On Bruen chains*, Discrete Math. **146** (1995), no. 1–3, 69–96.

459. _____, *Another Bruen chain*, Discrete Math. **188** (1998), no. 1–3, 267.

460. _____, *Maximal partial spreads in* PG$(3, 5)$, Ars Combin. **57** (2000), 97–101.

461. _____, *A maximal partial spread of size 45 in* PG$(3, 7)$, Des. Codes Cryptogr. **22** (2001), no. 3, 331–334.

462. _____, *Maximal partial spreads and the modular n-queen problem. III*, Discrete Math. **243** (2002), no. 1–3, 135–150.

463. _____, *No maximal partial spread of size 115 in* PG$(3, 11)$, Ars Combin. **66** (2003), 139–155.

464. O. Heden and M. Saggese, *Bruen chains in* $PG(3, p^k)$, $k \geq 2$, Discrete Math. **214** (2000), no. 1–3, 251–253.

465. G. Heimbeck, *Translationsebenen der Ordnung 49 mit Scherungen*, Geom. Dedicata **27** (1988), no. 1, 87–100.

466. _____, *Translation planes of order 25 admitting a quaternion group of homologies with affine axis*, Results Math. **21** (1992), no. 3–4, 328–331.

467. _____, *Translationsebenen der Ordnung 49 mit einer Quaternionengruppe von Dehnungen*, J. Geom. **44** (1992), no. 1–2, 65–76.

468. G. Heimbeck and R. Wagner, *Eine neue Serie von endlichen Translationsebenen*, Geom. Dedicata **28** (1988), no. 1, 107–125.

469. C. Hering, *Eine Charakterisierung der endlichen zweidimensionalen projektiven Gruppen*, Math. Z. **82** (1963), 152–175.

470. _____, *Eine Bemerkung über Automorphismengruppen von endlichen projektiven Ebenen und Möbiusebenen*, Arch. Math. (Basel) **18** (1967), 107–110.

471. _____, *Zweifach transitive Permutationsgruppen, in denen 2 die maximale Anzahl von Fixpunkten von Involutionen ist*, Math. Z. **104** (1968), 150–174.

472. _____, *Eine nicht-desarguessche zweifach transitive affine Ebene der Ordnung 27*, Abh. Math. Sem. Univ. Hamburg **34** (1969/1970), 203–208.

473. _____, *A new class of quasifields*, Math. Z. **118** (1971), 56–57.

474. _____, *Eine Bemerkung über Streckungsgruppen*, Arch. Math. (Basel) **23** (1972), 348–350.

475. _____, *On 2-groups operating on projective planes*, Illinois J. Math. **16** (1972), 581–595.

476. _____, *On shears of translation planes*, Abh. Math. Sem. Univ. Hamburg **37** (1972), 258–268.

477. _____, *On subgroups with trivial normalizer intersection*, J. Algebra **20** (1972), 622–629.

478. _____, *On linear groups which contain an irreducible subgroup of prime order*, Proceedings of the International Conference on Projective Planes (Washington State Univ., Pullman, Wash., 1973), Washington State Univ. Press, Pullman, Wash., 1973, pp. 99–105.

479. _____, *Transitive linear groups and linear groups which contain irreducible subgroups of prime order*, Geom. Dedicata **2** (1974), 425–460.

480. _____, *On projective planes of type VI*, Colloquio Internazionale sulle Teorie Combinatorie (Rome, 1973), Tomo II, Atti dei Convegni Lincei, no. 17, Accad. Naz. Lincei, Rome, 1976, pp. 29–53.

481. _____, *On perspectivities in finite collineation groups of projective planes and quasiperspectivities in Chevalley groups*, Proceedings of the 5th School of Algebra (Rio de Janeiro, 1978) (Rio de Janeiro), Soc. Brasil. Mat., 1978, pp. 11–19.

482. _____, *On the structure of finite collineation groups of projective planes*, Abh. Math. Sem. Univ. Hamburg **49** (1979), 155–182.

483. _____, *Finite collineation groups of projective planes containing nontrivial perspectivities*, The Santa Cruz Conference on Finite Groups (Univ. California, Santa Cruz, Calif., 1979), Proc. Sympos. Pure Math., vol. 37, Amer. Math. Soc., Providence, R.I., 1980, pp. 473–477.

484. _____, *On shears in fixed-point-free affine groups*, Finite Geometries and Designs (Proc. Conf., Chelwood Gate, 1980), London Math. Soc. Lecture Note Ser., vol. 49, Cambridge Univ. Press, Cambridge, 1981, pp. 146–152.

485. _____, *On Beweglichkeit in affine planes*, Finite Geometries (Pullman, Wash., 1981), Lecture Notes in Pure and Appl. Math., vol. 82, Dekker, New York, 1983, pp. 197–209.

486. _____, *On homologies in fixed-point-free affine groups*, Combinatorics '81 (Rome, 1981), Ann. Discrete Math., vol. 18, North-Holland, Amsterdam, 1983, pp. 427–432.

487. _____, *On perspectivities of odd order in simple collineation groups of projective planes*, Mitt. Math. Sem. Giessen (1984), no. 164, 191–198.

488. _____, *Fibrations in free modules*, IX Latin American School of Mathematics: Algebra (Spanish) (Santiago de Chile, 1988), Notas Soc. Mat. Chile **10** (1991), no. 1, 151–163.

489. _____, *On the classification of finite nearfields*, Special issue in honor of Helmut Wielandt, J. Algebra **234** (2000), no. 2, 664–667.

490. C. Hering and C.Y. Ho, *On free involutions in linear groups and collineation groups of translation planes*, J. Geom. **26** (1986), no. 2, 115–147.

491. C. Hering, W.M. Kantor, and G.M. Seitz, *Finite groups with a split BN-pair of rank 1, I*, J. Algebra **20** (1972), 435–475.

492. C. Hering and H.-J. Schaeffer, *On the new projective planes of R. Figueroa*, Combinatorial Theory (Schloss Rauischholzhausen, 1982), Lecture Notes in Math., vol. 969, Springer, Berlin, 1982, pp. 187–190.

493. _____, *A remark on a theorem of T. G. Ostrom*, Mostly Finite Geometries (Iowa City, IA, 1996), Lecture Notes in Pure and Appl. Math., vol. 190, Dekker, New York, 1997, pp. 213–214.

494. C. Hering and M. Walker, *Perspectivities in irreducible collineation groups of projective planes, I*, Math. Z. **155** (1977), no. 2, 95–101.

495. _____, *Perspectivities in irreducible collineation groups of projective planes, II*, J. Statist. Plann. Inference **3** (1979), no. 2, 151–177.

496. A. Herzer and G. Lunardon, *Charakterisierung (A, B)-regulärer Faserungen durch Schliessungssätze*, Geom. Dedicata **6** (1977), no. 4, 471–484.

497. Y. Hiramine, *On doubly transitive permutation groups*, Osaka J. Math. **15** (1978), no. 3, 613–631.

498. _____, *Doubly transitive groups of even degree whose one-point stabilizer has a normal subgroup isomorphic to* PSL$(3, 2^n)$, Osaka J. Math. **16** (1979), no. 3, 817–830.

499. _____, *On some doubly transitive permutation groups in which* socle(G_α) *is nonsolvable*, Osaka J. Math. **16** (1979), no. 3, 797–816.

500. _____, *Automorphisms of p-groups of semifield type*, Osaka J. Math. **20** (1983), no. 4, 735–746.

501. _____, *On semifield planes of even order*, Osaka J. Math. **20** (1983), no. 3, 645–658.

502. _____, *A generalization of Hall quasifields*, Osaka J. Math. **22** (1985), no. 1, 61–69.

503. _____, *On (G, Γ, n, q)-translation planes*, J. Math. Soc. Japan **37** (1985), no. 1, 157–164.

504. _____, *On translation planes of order q^2 which admit an autotopism group having an orbit of length $q^2 - q$*, Osaka J. Math. **22** (1985), no. 3, 411–431.

505. _____, *On weakly transitive translation planes*, Osaka J. Math. **22** (1985), no. 1, 55–60.

506. _____, *On translation planes of order q^3 with an orbit of length $q^3 - 1$ on l_∞*, Osaka J. Math. **23** (1986), no. 3, 563–575.

507. _____, *Some translation planes admitting* SL$(2, 5) \times$ SL$(2, 5)$, J. Combin. Theory Ser. A **48** (1988), no. 2, 189–196.

508. _____, *A conjecture on affine planes of prime order*, J. Combin. Theory Ser. A **52** (1989), no. 1, 44–50.

509. _____, *Affine planes with primitive collineation groups*, J. Algebra **128** (1990), no. 2, 366–383.

510. _____, *On planar functions*, J. Algebra **133** (1990), no. 1, 103–110.

511. _____, *Planar functions and related group algebras*, J. Algebra **152** (1992), no. 1, 135–145.

512. _____, *On collineation groups with block orbits*, J. Combin. Theory Ser. A **64** (1993), no. 1, 137–143.

513. _____, *On finite affine planes with a 2-transitive orbit on l_∞*, J. Algebra **162** (1993), no. 2, 392–409.

514. Y. Hiramine, V. Jha, and N.L. Johnson, *Characterization of translation planes by orbit lengths*, Geom. Dedicata **78** (1999), no. 1, 69–80.

515. _____, *Cubic extensions of flag-transitive planes, II: Odd order*, Note Mat. **19** (1999), no. 2, 299–314.

516. _____, *Quadratic extensions of flag-transitive planes*, European J. Combin. **20** (1999), no. 8, 797–818.

517. _____, *Solvable extensions of flag-transitive planes*, Note Mat. **19** (1999), no. 2, 183–198.

518. _____, *Characterization of translation planes by orbit lengths, II: Even order*, Bull. Belg. Math. Soc. Simon Stevin **7** (2000), no. 3, 395–418.

519. _____, *Cubic extensions of flag-transitive planes, I: Even order*, Int. J. Math. Math. Sci. **25** (2001), no. 8, 533–547.

520. Y. Hiramine and N.L. Johnson, *Generalized André planes of order p^t that admit a homology group of order $(p^t - 1)/2$*, Geom. Dedicata **41** (1992), no. 2, 175–190.

521. _____, *Near nearfield planes*, Geom. Dedicata **43** (1992), no. 1, 17–33.

522. _____, *Characterizations of regulus nets*, J. Geom. **48** (1993), no. 1–2, 86–108.

523. _____, *Regular partial conical flocks*, Bull. Belg. Math. Soc. Simon Stevin **2** (1995), no. 4, 419–433.

524. _____, *Nets of order p^2 and degree $p + 1$ admitting* SL(2, p), Geom. Dedicata **69** (1998), no. 1, 15–34.

525. Y. Hiramine, M. Matsumoto, and T. Oyama, *On some extension of 1-spread sets*, Osaka J. Math. **24** (1987), no. 1, 123–137.

526. J.W.P. Hirschfeld, *Projective Geometries over Finite Fields*, Oxford Mathematical Monographs, The Clarendon Press, Oxford University Press, New York, 1979.

527. _____, *Finite Projective Spaces of Three Dimensions*, Oxford Mathematical Monographs, Oxford Science Publications, The Clarendon Press, Oxford University Press, New York, 1985.

528. _____, *Projective Geometries over Finite Fields*, second ed., Oxford Mathematical Monographs, The Clarendon Press, Oxford University Press, New York, 1998.

529. J.W.P. Hirschfeld and J.A. Thas, *General Galois Geometries*, Oxford Mathematical Monographs, Oxford Science Publications, The Clarendon Press, Oxford University Press, New York, 1991.

530. C.Y. Ho, *Linear transformation and collineation of finite translation planes*, J. Algebra **56** (1979), no. 1, 235–254.

531. _____, *Finite projective planes that admit a strongly irreducible collineation group*, Canad. J. Math. **37** (1985), no. 4, 579–611.

532. _____, *Involutory collineations of finite planes*, Math. Z. **193** (1986), no. 2, 235–240.

533. _____, *On multiplier groups of finite cyclic planes*, J. Algebra **122** (1989), no. 1, 250–259.

534. _____, *On the order of a finite projective plane and its collineation group*, Finite Geometries and Combinatorial Designs (Lincoln, NE, 1987), Contemp. Math., vol. 111, Amer. Math. Soc., Providence, RI, 1990, pp. 299–301.

535. _____, *Totally irregular collineation groups and finite Desarguesian planes*, Coding Theory and Design Theory, Part II, IMA Vol. Math. Appl., vol. 21, Springer, New York, 1990, pp. 127–131.

536. _____, *Some remarks on orders of projective planes, planar difference sets and multipliers*, Des. Codes Cryptogr. **1** (1991), no. 1, 69–75.

537. _____, *Subplanes of a tactical decomposition and Singer groups of a projective plane*, Geom. Dedicata **53** (1994), no. 3, 307–326.

538. _____, *Some basic properties of planar Singer groups*, Geom. Dedicata **55** (1995), no. 1, 59–70.

539. _____, *Arc subgroups of planar Singer groups*, Mostly Finite Geometries (Iowa City, IA, 1996), Lecture Notes in Pure and Appl. Math., vol. 190, Dekker, New York, 1997, pp. 227–233.

540. _____, *Finite projective planes with abelian transitive collineation groups*, J. Algebra **208** (1998), no. 2, 533–550.

541. _____, *Linear groups and collineation groups of translation planes*, J. Algebra **257** (2002), no. 2, 373–392.

542. _____, *Collineation groups of translation planes and linear groups*, Algebraic combinatorics (Kyoto, 2001), Sūrikaisekikenkyūsho Kōkyūroku (2003), no. 1299, 64–70.

543. _____, *Finite translation planes from the collineation groups point of view*, Advances in Algebra, World Sci. Publishing, River Edge, NJ, 2003, pp. 148–156.

544. _____, *Collineation groups with perspectivities*, J. Algebra **274** (2004), no. 1, 245–266.

545. _____, *Simple collineation groups with perspectivities of finite translation planes*, J. Algebra **279** (2004), no. 1, 315–325.

546. _____, *Eigenspaces of linear collineations*, Adv. Geom. **5** (2005), no. 1, 71–79.

547. C.Y. Ho and A. Gonçalves, *On collineation groups of a projective plane of prime order*, Geom. Dedicata **20** (1986), no. 3, 357–366.

548. _____, *On* PSU(3, q) *as collineation groups*, J. Algebra **111** (1987), no. 1, 1–13.

549. _____, *On totally irregular simple collineation groups*, Advances in Finite Geometries and Designs (Chelwood Gate, 1990), Oxford Sci. Publ., Oxford Univ. Press, New York, 1991, pp. 177–193.

550. _____, *On* PSL(2, q) *as a totally irregular collineation group*, Geom. Dedicata **49** (1994), no. 1, 1–24.

551. C.Y. Ho and G.E. Moorhouse, *A new characterization of the Desarguesian plane of order 11*, Proceedings of the Conference on Groups and Geometry, Part B (Madison, Wis., 1985), Algebras Groups Geom. **2** (1985), no. 4, 428–435.

552. C.Y. Ho and A. Pott, *Multiplier groups of planar difference sets and a theorem of Kantor*, Proc. Amer. Math. Soc. **109** (1990), no. 3, 803–808.
553. A.R. Hoffer, *On unitary collineation groups*, J. Algebra **22** (1972), 211–218.
554. H. Huang and N.L. Johnson, *8 semifield planes of order* 8^2, Discrete Math. **80** (1990), no. 1, 69–79.
555. D.R. Hughes, *Planar division neo-rings*, Trans. Amer. Math. Soc. **80** (1955), 502–527.
556. ———, *A class of non-Desarguesian projective planes*, Canad. J. Math. **9** (1957), 378–388.
557. ———, *Generalized incidence matrices over group algebras*, Illinois J. Math. **1** (1957), 545–551.
558. ———, *Regular collineation groups*, Proc. Amer. Math. Soc. **8** (1957), 165–168.
559. ———, *Collineation groups of non-Desarguesian planes, I: The Hall Veblen-Wedderburn systems*, Amer. J. Math. **81** (1959), 921–938.
560. ———, *Collineation groups of non-Desarguesian planes, II: Some seminuclear division algebras*, Amer. J. Math. **82** (1960), 113–119.
561. ———, *Sottopiani non-Desarguesiani di piani finiti*, Atti Accad. Naz. Lincei Rend. Cl. Sci. Fis. Mat. Natur. (8) **36** (1964), 315–318.
562. D.R. Hughes and M.J. Kallaher, *On the Knuth semifields*, Int. J. Math. Math. Sci. **3** (1980), no. 1, 29–45.
563. D.R. Hughes and E. Kleinfeld, *Seminuclear extensions of Galois fields*, Amer. J. Math. **82** (1960), 389–392.
564. D.R. Hughes and F.C. Piper, *Projective Planes*, Graduate Texts in Mathematics, vol. 6, Springer-Verlag, New York, 1973.
565. B. Huppert, *Zweifach transitive, auflösbare Permutationsgruppen*, Math. Z. **68** (1957), 126–150.
566. ———, *Endliche Gruppen, I*, Grundlehren Math. Wiss., Band 134, Springer-Verlag, Berlin, 1967.
567. A.A. Ivanov, R.A. Liebler, T. Penttila, and C.E. Praeger, *Antipodal distance-transitive covers of complete bipartite graphs*, European J. Combin. **18** (1997), no. 1, 11–33.
568. T.V.S. Jagannathan and P. Srinivasan, *Near-field-like planes*, Finite Geometries (Winnipeg, Man., 1984), Lecture Notes in Pure and Appl. Math., vol. 103, Dekker, New York, 1985, pp. 137–147.
569. J. Jakóbowski, *Conjugate points of a translation plane of order 9 and its extensibility to a Minkowski plane*, Zeszyty Nauk. Geom. **17** (1988), 85–96.
570. V. Jha, *On tangentially transitive translation planes and related systems*, Geom. Dedicata **4** (1975), no. 2/3/4, 457–483.
571. ———, *On abelian autotopism groups of finite translation planes*, Arch. Math. (Basel) **37** (1981), no. 6, 569–571.
572. ———, *On* Δ-*transitive translation planes*, Arch. Math. (Basel) **37** (1981), no. 4, 377–384.
573. ———, *On the derivability of field transitive quasifields*, J. London Math. Soc. (2) **23** (1981), no. 1, 41–44.
574. ———, *On subgroups and factor groups of* GL(n, q) *acting on spreads with the same characteristic*, Discrete Math. **41** (1982), no. 1, 43–51.
575. ———, *On spreads admitting large autotopism groups*, Finite Geometries (Pullman, Wash., 1981), Lecture Notes in Pure and Appl. Math., vol. 82, Dekker, New York, 1983, pp. 237–242.
576. ———, *On groups of Baer collineations acting on Cartesian and translation planes*, J. Algebra **88** (1984), no. 2, 361–379.
577. ———, *On translation planes which admit solvable autotopism groups having a large slope orbit*, Canad. J. Math. **36** (1984), no. 5, 769–782.
578. ———, *On finite semitransitive translation planes*, Geom. Dedicata **18** (1985), no. 1, 1–9.
579. ———, *On binary constant weight codes associated with spreads*, Twelfth British Combinatorial Conference (Norwich, 1989), Ars Combin. **29** (1990), no. A, 91–96.
580. V. Jha and N.L. Johnson, *Algebraic and geometric lifting*, Note Mat. (to appear).
581. ———, *Derivable André and generalized André planes*, J. Interdiscip. Math. (to appear).
582. ———, *A new class of translation planes constructed by hyper-regulus replacement*, J. Geom. (to appear).
583. ———, *The planes of Suetake*, J. Geom. (to appear).

584. _____, *Translation planes constructed by multiple hyper-regulus replacement*, J. Geom. (to appear).

585. _____, *Some unusual translation planes of order* 64, Arch. Math. (Basel) **43** (1984), no. 6, 566–571.

586. _____, *Translation planes of order* q^2 *that admit a collineation group of order* q^2, *II: Transitivity*, Atti Sem. Mat. Fis. Univ. Modena **33** (1984), no. 1, 161–165.

587. _____, *Coexistence of elations and large Baer groups in translation planes*, J. London Math. Soc. (2) **32** (1985), no. 2, 297–304.

588. _____, *Compatibility of Baer and elation groups in translation planes*, Finite Geometries (Winnipeg, Man., 1984), Lecture Notes in Pure and Appl. Math., vol. 103, Dekker, New York, 1985, pp. 163–170.

589. _____, *A note on finite semifield planes that admit affine homologies*, J. Geom. **24** (1985), no. 2, 194–197.

590. _____, *On spreads in* PG$(3, 2^s)$ *that admit projective groups of order* 2^s, Proc. Edinburgh Math. Soc. (2) **28** (1985), no. 3, 355–360.

591. _____, *On spreads of characteristic p admitting nonsolvable groups, whose Sylow p-subgroups are planar*, Osaka J. Math. **22** (1985), no. 2, 365–377.

592. _____, *Translation planes of order n which admit a collineation group of order n*, Finite Geometries (Winnipeg, Man., 1984), Lecture Notes in Pure and Appl. Math., vol. 103, Dekker, New York, 1985, pp. 149–162.

593. _____, *Derivable nets defined by central collineations*, J. Combin. Inform. System Sci. **11** (1986), no. 2–4, 83–91.

594. _____, *Notes on the derived Walker planes*, J. Combin. Theory Ser. A **42** (1986), no. 2, 320–323.

595. _____, *On collineation groups of translation planes of order* q^4, Int. J. Math. Math. Sci. **9** (1986), no. 3, 617–620.

596. _____, *On regular r-packings*, Note Mat. **6** (1986), no. 1, 121–137.

597. _____, *Regular parallelisms from translation planes*, Discrete Math. **59** (1986), no. 1–2, 91–97.

598. _____, *Solution to Dempwolff's nonsolvable B-group problem*, European J. Combin. **7** (1986), no. 3, 227–235.

599. _____, *Baer-elation planes*, Rend. Sem. Mat. Univ. Padova **78** (1987), 27–45.

600. _____, *Baer involutions in translation planes admitting large elation groups*, Results Math. **11** (1987), no. 1–2, 63–71.

601. _____, *A note on rational covers of spreads*, Ars Combin. **24** (1987), 175–177.

602. _____, *The odd order analogue of Dempwolff's B-group problem*, J. Geom. **28** (1987), no. 1, 1–6.

603. _____, *On the nuclei of semifields and Cofman's many-subplane problem*, Abh. Math. Sem. Univ. Hamburg **57** (1987), 127–137.

604. _____, *The centre of a finite semifield plane is a geometric invariant*, Arch. Math. (Basel) **50** (1988), no. 1, 93–96.

605. _____, *The linearity question for abelian groups on translation planes*, J. Math. Soc. Japan **40** (1988), no. 1, 77–84.

606. _____, *An analog of the Albert-Knuth theorem on the orders of finite semifields, and a complete solution to Cofman's subplane problem*, Algebras Groups Geom. **6** (1989), no. 1, 1–35.

607. _____, *A characterization of some spreads of order* q^3 *that admit* GL$(2, q)$ *as a collineation group*, Hokkaido Math. J. **18** (1989), no. 1, 137–147.

608. _____, *Nests of reguli and flocks of quadratic cones*, Simon Stevin **63** (1989), no. 3–4, 311–338.

609. _____, *A geometric characterization of generalized Desarguesian planes*, Atti Sem. Mat. Fis. Univ. Modena **38** (1990), no. 1, 71–80.

610. _____, *On the ubiquity of Denniston-type translation ovals in generalized André planes*, Combinatorics '90 (Gaeta, 1990), Ann. Discrete Math., vol. 52, North-Holland, Amsterdam, 1992, pp. 279–296.

611. _____, *Translation planes of large dimension admitting nonsolvable groups*, J. Geom. **45** (1992), no. 1–2, 87–104.

612. _____, *A characterisation of spreads ovally-derived from Desarguesian spreads*, Combinatorica **14** (1994), no. 1, 51–61.

613. _____, *Flocks of oval cones and extensions of theorems of Thas*, Note Mat. **14** (1994), no. 2, 189–197.

614. _____, *Automorphism groups of flocks of oval cones*, Geom. Dedicata **61** (1996), no. 1, 71–85.

615. _____, *Infinite flocks of a quadratic cone*, J. Geom. **57** (1996), no. 1–2, 123–150.

616. _____, *Quasifibrations*, Bull. Belg. Math. Soc. Simon Stevin **3** (1996), no. 3, 313–324.

617. _____, *Rigidity in conical flocks*, J. Combin. Theory Ser. A **73** (1996), no. 1, 60–76.

618. _____, *The doubly transitive flocks of quadratic cones*, European J. Combin. **18** (1997), no. 7, 763–778.

619. _____, *Structure theory for point-Baer and line-Baer collineations in affine planes*, Mostly Finite Geometries (Iowa City, IA, 1996), Lecture Notes in Pure and Appl. Math., vol. 190, Dekker, New York, 1997, pp. 235–273.

620. _____, *Finite doubly transitive affine planes*, J. Combin. Theory Ser. A **83** (1998), no. 1, 165–168.

621. _____, *A note on the stability of autotopism triangles in translation planes*, Note Mat. **18** (1998), no. 2, 299–302.

622. _____, *The Bella-Muro lectures on translation planes, 1997*, Quaderni del departimento di matematica dell'università di Lecce, Q. 1 (1999), 1–237.

623. _____, *Conical, ruled and deficiency one translation planes*, Bull. Belg. Math. Soc. Simon Stevin **6** (1999), no. 2, 187–218.

624. _____, *Cyclic Ostrom spreads*, Des. Codes Cryptogr. **16** (1999), no. 1, 41–51.

625. _____, *Infinite Baer nets*, J. Geom. **68** (2000), no. 1–2, 114–141.

626. _____, *Lifting quasifibrations, II: Non-normalizing Baer involutions*, Note Mat. **20** (2000/01), no. 2, 51–68.

627. _____, *Almost Desarguesian maximal partial spreads*, Des. Codes Cryptogr. **22** (2001), no. 3, 283–304.

628. _____, *Infinite flocks of quadratic cones, II: Generalized Fisher flocks*, J. Korean Math. Soc. **39** (2002), no. 4, 653–664.

629. _____, *Transitive deficiency-one Baer subgeometry partitions*, J. Combin. Theory Ser. A **98** (2002), no. 1, 127–149.

630. _____, *Transversal-free translation nets*, Des. Codes Cryptogr. **27** (2002), no. 3, 195–205.

631. _____, *Infinite maximal partial spreads*, Note Mat. **21** (2002/03), no. 2, 49–55.

632. _____, *The classification of spreads in* PG(3, q) *admitting linear groups of order* $q(q+1)$, *II: Even order*, Special issue dedicated to Adriano Barlotti, Adv. Geom. suppl. (2003), S271–S313.

633. _____, *The classification of spreads in* PG(3, q) *admitting linear groups of order* $q(q+1)$, *I: Odd order*, J. Geom. **81** (2004), no. 1–2, 46–80.

634. _____, *Nuclear fusion in finite semifield planes*, Adv. Geom. **4** (2004), no. 4, 413–432.

635. _____, *André flat flocks*, Note Mat. **24** (2004/05), no. 2, 135–141.

636. _____, *Multiple nests*, Note Mat. **23** (2004/05), no. 1, 123–149.

637. _____, *The translation planes of order 81 admitting* SL(2, 5), Note Mat. **24** (2004/05), no. 2, 59–73.

638. _____, *The Foulser-Ostrom planes*, JP J. Geom. Topol. **5** (2005), no. 3, 275–284.

639. _____, *Ostrom-derivates*, Innov. Incidence Geom. **1** (2005), 35–65.

640. _____, *Semifield flat flocks*, J. Geom. **83** (2005), no. 1–2, 71–87.

641. _____, *Transitive conical flocks of even order*, Discrete Math. **301** (2005), no. 1, 83–88.

642. _____, *Collineation groups of translation planes constructed by multiple hyper-regulus replacement*, Note Mat. **26** (2006), no. 1, 149–167.

643. _____, *Cubic order translation planes constructed by multiple hyper-regulus replacement*, Note Mat. **26** (2006), no. 1, 79–103.

644. _____, *Subregular planes admitting elations*, Des. Codes Cryptogr. **41** (2006), no. 2, 125–145.

645. _____, *Translation planes of order* q^2 *admitting collineation groups of order* q^3u *preserving a parabolic unital*, Note Mat. **26** (2006), no. 2, 105–118.

646. V. Jha, N.L. Johnson, and G.P. Wene, *On translation planes of even non-square order*, J. Combin. Inform. System Sci. **19** (1994), no. 3–4, 195–199.

647. V. Jha, N.L. Johnson, and F.W. Wilke, *On translation planes of order q^2 that admit a group of order $q^2(q-1)$; Bartolone's theorem*, Rend. Circ. Mat. Palermo (2) **33** (1984), no. 3, 407–424.

648. V. Jha and M.J. Kallaher, *On spreads admitting projective linear groups*, Canad. J. Math. **33** (1981), no. 6, 1487–1497.

649. _____, *On the Lorimer-Rahilly and Johnson-Walker translation planes*, Pacific J. Math. **103** (1982), no. 2, 409–427.

650. V. Jha and T.G. Ostrom, *On u-groups acting on vector spaces and spreads with characteristic p*, Arch. Math. (Basel) **37** (1981), no. 1, 78–84.

651. V. Jha and G.P. Wene, *The structure of the central units of a commutative semifield plane*, Finite Geometry and Combinatorics (Deinze, 1992), London Math. Soc. Lecture Note Ser., vol. 191, Cambridge Univ. Press, Cambridge, 1993, pp. 207–216.

652. _____, *An oval partition of the central units of certain semifield planes*, Combinatorics (Acireale, 1992), Discrete Math. **155** (1996), no. 1–3, 127–134.

653. N.L. Johnson, *Fano configurations in translation planes of large dimension*, Note Mat. (to appear).

654. _____, *The Hermitian ovoids of Cossidente, Ebert, Marino, Siciliano*, Note Mat. (to appear).

655. _____, *Homology groups of translation planes and flocks of quadratic cones, II: j-planes*, preprint.

656. _____, *Lifting subregular spreads*, J. Geom. (to appear).

657. _____, *Transitive parabolic unitals in semifield planes*, J. Geom. (to appear).

658. _____, *A classification of strict semi-translation planes*, Proc. Projective Geom. Conf. Univ. of Illinois (1967), 1967, pp. 52–56.

659. _____, *A classification of semi-translation planes*, Canad. J. Math. **21** (1969), 1372–1387.

660. _____, *Nonstrict semi-translation planes*, Arch. Math. (Basel) **20** (1969), 301–310.

661. _____, *Derivable chains of planes*, Boll. Un. Mat. Ital. (4) **3** (1970), 167–184.

662. _____, *Derivable semi-translation planes*, Pacific J. Math. **34** (1970), 687–707.

663. _____, *A note on semi-translation planes of class $1-5a$*, Arch. Math. (Basel) **21** (1970), 528–532.

664. _____, *On nonstrict semi-translation planes of Lenz-Barlotti class $I-1$*, Arch. Math. (Basel) **21** (1970), 402–410.

665. _____, *A note on free planar extensions*, Math. Z. **119** (1971), 281–282.

666. _____, *A note on the construction of quasifields*, Proc. Amer. Math. Soc. **29** (1971), 138–142.

667. _____, *Transition planes constructed from semifield planes*, Pacific J. Math. **36** (1971), 701–711.

668. _____, *Affine planes of order 25*, Arch. Math. (Basel) **23** (1972), 104–109.

669. _____, *A characterization of generalized Hall planes*, Bull. Austral. Math. Soc. **6** (1972), 61–67.

670. _____, *Derivation in infinite planes*, Pacific J. Math. **43** (1972), 387–402.

671. _____, *Homomorphisms of free planes*, Math. Z. **125** (1972), 255–263.

672. _____, *A class of translation planes admitting homologies with affine axes*, Arch. Math. (Basel) **24** (1973), 105–112.

673. _____, *Collineation groups of derived semifield planes*, Arch. Math. (Basel) **24** (1973), 429–433.

674. _____, *A note on generalized Hall planes*, Bull. Austral. Math. Soc. **8** (1973), 151–153.

675. _____, *Semi-translation planes derived from dual translation planes*, Proceedings of the International Conference on Projective Planes (Washington State Univ., Pullman, Wash., 1973), Washington State Univ. Press, Pullman, Wash., 1973, pp. 115–120.

676. _____, *Semi-translation planes of class $1-3a$*, Boll. Un. Mat. Ital. (4) **7** (1973), 359–376.

677. _____, *Collineation groups of derived semifield planes, II*, Arch. Math. (Basel) **25** (1974), 400–404.

678. _____, *Collineation groups of derived semifield planes, III*, Arch. Math. (Basel) **26** (1975), 101–106.

679. _____, *Distortion and generalized Hall Planes*, Geom. Dedicata **4** (1975), no. 2/3/4, 437–456.

680. _____, *Regularity derived planes*, Math. Z. **145** (1975), no. 3, 201–210.

681. _____, *The translation planes of Bruck type* {1}, Arch. Math. (Basel) **26** (1975), no. 5, 554–560.

682. _____, *Correction to: "Collineation groups of derived semifield planes, iii" (Arch. Math. (Basel) **26** (1975), 101–106)*, Arch. Math. (Basel) **27** (1976), no. 3, 335–336.

683. _____, *The geometry of* GL(2, q) *in translation planes of even order* q^2, Int. J. Math. Math. Sci. **1** (1978), no. 4, 447–458.

684. _____, *A note on the derived semifield planes of order 16*, Aequationes Math. **18** (1978), no. 1–2, 103–111.

685. _____, *On central collineations of derived semifield planes*, J. Geom. **11** (1978), no. 2, 139–149.

686. _____, *Corrigendum: "The geometry of* GL(2, q) *in translation planes of even order* q^2" *[Internat. J. Math. Math. Sci. **1** (1978), no. 4, 447–458]*, Int. J. Math. Math. Sci. **4** (1981), no. 2, 413–414.

687. _____, *On the solvability of the collineation groups of derivable translation planes*, J. Algebra **71** (1981), no. 2, 569–575.

688. _____, *Some translation planes constructed by multiple derivation*, Bull. Austral. Math. Soc. **23** (1981), no. 2, 313–315.

689. _____, *The translation planes of Dempwolff*, Canad. J. Math. **33** (1981), no. 5, 1060–1073.

690. _____, *The translation planes of Ott-Schaeffer*, Arch. Math. (Basel) **36** (1981), no. 2, 183–192.

691. _____, *On elations in semitransitive planes*, Int. J. Math. Math. Sci. **5** (1982), no. 1, 159–164.

692. _____, *On the compatibility of planar and nonplanar involutions in translation planes*, Abh. Math. Sem. Univ. Hamburg **52** (1982), 16–28.

693. _____, *The translation planes of order 16 that admit* SL(2, 4), Combinatorial and Geometric Structures and Their Applications (Trento, 1980), Ann. Discrete Math., vol. 14, North-Holland, Amsterdam, 1982, pp. 225–236.

694. _____, *Elations in translation planes of order 16*, J. Geom. **20** (1983), no. 2, 101–110.

695. _____, *The geometry of Ostrom: the contributions to geometry of T. G. Ostrom*, Finite Geometries (Pullman, Wash., 1981), Lecture Notes in Pure and Appl. Math., vol. 82, Dekker, New York, 1983, pp. 1–22.

696. _____, *Representations of* SL(2, 4) *on translation planes of even order*, J. Geom. **21** (1983), no. 2, 184–200.

697. _____, *Translation planes of characteristic p that admit* SL(2, p^r), Combinatorics '81 (Rome, 1981), Ann. Discrete Math., vol. 18, North-Holland, Amsterdam, 1983, pp. 493–509.

698. _____, *On Desarguesian extensions of elation nets*, J. Geom. **23** (1984), no. 1, 72–77.

699. _____, *The translation planes of order 16 that admit nonsolvable collineation groups*, Math. Z. **185** (1984), no. 3, 355–372.

700. _____, *Translation planes of order* q^2 *that admit* q+1 *elations*, Geom. Dedicata **15** (1984), no. 4, 329–337.

701. _____, *Foulser's covering theorem*, Note Mat. **5** (1985), no. 1, 139–145.

702. _____, *A note on net replacement in transposed spreads*, Canad. Math. Bull. **28** (1985), no. 4, 469–471.

703. _____, *Finite translation planes of dimension two*, V Latin American colloquium on algebra (Spanish) (Santiago, 1985), Notas Soc. Mat. Chile **5** (1986), no. 1, 71–107.

704. _____, *Lectures on finite translation planes of dimension two*, IX Escola de Algebra, Brazilia (1986), Sociedade Brasileira de Matemática **17** (1986), 25–52.

705. _____, *Lezioni sui Piani di Traslazione*, Quaderni del Departimento di Matematica dell'Iniversità di Lecce, Q. 3, 1986 (Italian).

706. _____, *The maximal special linear groups which act on translation planes*, Boll. Un. Mat. Ital. A (6) **5** (1986), no. 3, 349–352.

707. _____, *Translation planes and related geometric structures (ten lectures on finite translation planes)*, Notas de Curso, Departamento de Matemática, No. 27, Universidade Federal de Pernambuco, Recife, 1986, pp. 1–110.

708. _____, *On derivable Baer-elation planes*, Note Mat. **7** (1987), no. 1, 19–27.

709. _____, *Projective planes of prime order p that admit collineation groups of order* p^2, J. Geom. **30** (1987), no. 1, 49–68.

710. _____, *Semifield flocks of quadratic cones*, Simon Stevin **61** (1987), no. 3–4, 313–326.

711. _____, *Derivable nets and 3-dimensional projective spaces*, Abh. Math. Sem. Univ. Hamburg **58** (1988), 245–253.

712. _____, *A note on translation planes with large dimension*, J. Geom. **33** (1988), no. 1–2, 65–72.

713. _____, *Sequences of derivable translation planes*, Osaka J. Math. **25** (1988), no. 3, 519–530.

714. _____, *Tensor product and generalized Ott-Schaeffer planes*, Osaka J. Math. **25** (1988), no. 2, 441–459.

715. _____, *A characterization of the Hall planes by planar and nonplanar involutions*, Int. J. Math. Math. Sci. **12** (1989), no. 4, 781–785.

716. _____, *Derivation is a polarity*, J. Geom. **35** (1989), no. 1–2, 97–102.

717. _____, *The derivation of dual translation planes*, J. Geom. **36** (1989), no. 1–2, 63–90.

718. _____, *Flocks of hyperbolic quadrics and translation planes admitting affine homologies*, J. Geom. **34** (1989), no. 1–2, 50–73.

719. _____, *Maximal partial spreads and central groups*, Note Mat. **9** (1989), no. 2, 249–261.

720. _____, *Nest replaceable translation planes*, J. Geom. **36** (1989), no. 1–2, 49–62.

721. _____, *Semifield planes of characteristic p admitting p-primitive Baer collineations*, Osaka J. Math. **26** (1989), no. 2, 281–285.

722. _____, *Translation planes admitting Baer groups and partial flocks of quadric sets*, Simon Stevin **63** (1989), no. 2, 167–188.

723. _____, *Derivable nets and 3-dimensional projective spaces, II: The structure*, Arch. Math. (Basel) **55** (1990), no. 1, 94–104.

724. _____, *Derivation by coordinates*, Note Mat. **10** (1990), no. 1, 89–96.

725. _____, *Derived Fisher translation planes*, Simon Stevin **64** (1990), no. 1, 21–50.

726. _____, *Flocks and partial flocks of quadric sets*, Finite Geometries and Combinatorial Designs (Lincoln, NE, 1987), Contemp. Math., vol. 111, Amer. Math. Soc., Providence, RI, 1990, pp. 111–116.

727. _____, *Conical flocks whose corresponding translation planes are derivable*, Simon Stevin **65** (1991), no. 3–4, 199–215.

728. _____, *Derivation*, Combinatorics '88, Vol. 2 (Ravello, 1988), Res. Lecture Notes Math., Mediterranean, Rende, 1991, pp. 97–113.

729. _____, *A group-theoretic characterization of finite derivable nets*, J. Geom. **40** (1991), no. 1–2, 95–104.

730. _____, *Homology groups, nearfields and reguli*, J. Geom. **42** (1991), no. 1–2, 109–125.

731. _____, *Ovoids and translation planes revisited*, Geom. Dedicata **38** (1991), no. 1, 13–57.

732. _____, *Problems in translation planes related to flocks of quadric sets*, Notas Soc. Mat. Chile **10** (1991), no. 1, 187–199, IX Latin American School of Mathematics: Algebra (Spanish) (Santiago de Chile, 1988).

733. _____, *Reguli in translation planes defined by flocks of hyperbolic quadrics*, Ars Combin. **31** (1991), 205–210.

734. _____, *Sharply transitive sets*, Conference Proceedings 1991, Santiago, Chile, 1991.

735. _____, *Translation planes of large dimension*, Notas Soc. Mat. Chile **10** (1991), no. 1, 165–185, IX Latin American School of Mathematics: Algebra (Spanish) (Santiago de Chile, 1988).

736. _____, *Derivation of partial flocks of quadratic cones*, Rend. Mat. Appl. (7) **12** (1992), no. 4, 817–848.

737. _____, *Translation planes and related combinatorial structures*, Combinatorics '90 (Gaeta, 1990), Ann. Discrete Math., vol. 52, North-Holland, Amsterdam, 1992, pp. 297–315.

738. _____, *Translation planes covered by subplane covered nets*, Simon Stevin **66** (1992), no. 3 4, 221–239.

739. _____, *Partially sharp subsets of* $P\Gamma L(n, q)$, Finite Geometry and Combinatorics (Deinze, 1992), London Math. Soc. Lecture Note Ser., vol. 191, Cambridge Univ. Press, Cambridge, 1993, pp. 217–232.

740. _____, *Half nearfield planes*, Osaka J. Math. **31** (1994), no. 1, 61–78.

741. _____, *The classification of subplane covered nets*, Bull. Belg. Math. Soc. Simon Stevin **2** (1995), no. 5, 487–508.

742. _____, *Flocks-flocks-flocks*, Conference Proceedings, Punta de Tralca, Chile, 1995.

743. _____, *Extending partial flocks containing linear subflocks*, J. Geom. **55** (1996), no. 1–2, 99–106.

744. _____, *Lifting quasifibrations*, Note Mat. **16** (1996), no. 1, 25–41.

745. _____, *Baer groups*, Encyclopaedia of Mathematics, Supplement Volume I (M. Hazewinkel, ed.), Kluwer Academic Publishers, 1997, p. 73.

746. _____, *Flocks of infinite hyperbolic quadrics*, J. Algebraic Combin. **6** (1997), no. 1, 27–51.

747. _____, *Flocks of quadric sets*, Encyclopaedia of Mathematics, Supplement Volume I (M. Hazewinkel, ed.), Kluwer Academic Publishers, 1997, p. 254.

748. N.L. Johnson (ed.), *Mostly Finite Geometries*, In celebration of T. G. Ostrom's 80th birthday, Papers from the conference held at the University of Iowa, Iowa City, IA, March 1996, Lecture Notes in Pure and Applied Mathematics, vol. 190, Marcel Dekker Inc., New York, 1997.

749. _____, *Derivable nets may be embedded in nonderivable planes*, Groups and Geometries (Siena, 1996), Trends Math., Birkhäuser, Basel, 1998, pp. 123–144.

750. _____, *Infinite nests of reguli*, Geom. Dedicata **70** (1998), no. 3, 221–267.

751. _____, *New and old results on flocks of circle planes*, Combinatorics (Assisi, 1996), Discrete Math. **208/209** (1999), 349–373.

752. _____, *Desarguesian partial parallelisms*, J. Geom. **69** (2000), no. 1–2, 131–148.

753. _____, *Subplane Covered Nets*, Monographs and Textbooks in Pure and Applied Mathematics, vol. 222, Marcel Dekker Inc., New York, 2000.

754. _____, *Some new classes of finite parallelisms*, Note Mat. **20** (2000/01), no. 2, 77–88.

755. _____, *Retracting spreads*, Bull. Belg. Math. Soc. Simon Stevin **8** (2001), no. 3, 505–524.

756. _____, *Two-transitive parallelisms*, Des. Codes Cryptogr. **22** (2001), no. 2, 179–189.

757. _____, *Dual parallelisms*, Note Mat. **21** (2002), no. 1, 137–150.

758. _____, *Transversal spreads*, Bull. Belg. Math. Soc. Simon Stevin **9** (2002), no. 1, 109–142.

759. _____, *Beutelspacher's parallelism construction*, Note Mat. **21** (2002/03), no. 2, 57–69.

760. _____, *Hyper-reguli and non-André quasi-subgeometry partitions of projective spaces*, J. Geom. **78** (2003), no. 1–2, 59–82.

761. _____, *Parallelisms of projective spaces*, Combinatorics, 2002 (Maratea), J. Geom. **76** (2003), no. 1–2, 110–182.

762. _____, *Quasi-subgeometry partitions of projective spaces*, Bull. Belg. Math. Soc. Simon Stevin **10** (2003), no. 2, 231–261.

763. _____, *Dual deficiency one transitive partial parallelisms*, Note Mat. **23** (2004/05), no. 1, 15–38.

764. _____, *The non-solvable triangle transitive planes*, Note Mat. **23** (2004/05), no. 1, 39–46.

765. _____, *Spreads in* PG$(3, q)$ *admitting several homology groups of order* $q + 1$, Note Mat. **24** (2004/05), no. 2, 9–39.

766. _____, *Piani di Traslazione e Loro Conessioni con Flock, Quadrangoli Generalizzati, e Reti*, Quaderno 3/2005, Universitá degli Studi di Lecce, 2005, 1–137.

767. _____, *Bol planes of orders* 3^4 *and* 3^6, Note Mat. **26** (2006), no. 2, 137–146.

768. _____, *Derivable subregular planes*, Innov. Incidence Geom. **3** (2006), 89–108.

769. _____, *Homology groups of translation planes and flocks of quadratic cones, I: The structure*, Bull. Belg. Math. Soc. Simon Stevin **12** (2006), no. 5, 827–844.

770. N.L. Johnson and V. Jha, *Regulus codes*, Discrete Math. **132** (1994), no. 1–3, 97–106.

771. N.L. Johnson and M.J. Kallaher, *Transitive collineation groups on affine planes*, Math. Z. **135** (1973/74), 149–164.

772. _____, *On translation planes with affine central collineations, II*, Canad. J. Math. **28** (1976), no. 1, 116–129.

773. N.L. Johnson, M.J. Kallaher, and C.T. Long (eds.), *Finite Geometries*, Lecture Notes in Pure and Applied Mathematics, vol. 82, New York, Marcel Dekker Inc., 1983.

774. N.L. Johnson and K.S. Lin, *Embedding dual nets in affine and projective spaces*, Rend. Mat. Appl. (7) **14** (1994), no. 3, 483–502.

775. N.L. Johnson and X. Liu, *Flocks of quadratic and semi-elliptic cones*, Mostly Finite Geometries (Iowa City, IA, 1996), Lecture Notes in Pure and Appl. Math., vol. 190, Dekker, New York, 1997, pp. 275–304.

776. _____, *The generalized Kantor-Knuth flocks*, Mostly Finite Geometries (Iowa City, IA, 1996), Lecture Notes in Pure and Appl. Math., vol. 190, Dekker, New York, 1997, pp. 305–314.

777. N.L. Johnson and G. Lunardon, *On the Bose-Barlotti Δ-planes*, Geom. Dedicata **49** (1994), no. 2, 173–182.

778. _____, *Maximal partial spreads and flocks*, Des. Codes Cryptogr. **10** (1997), no. 2, 193–202.

779. N.L. Johnson, G. Lunardon, and F.W. Wilke, *Semifield skeletons of conical flocks*, J. Geom. **40** (1991), no. 1–2, 105–112.

780. N.L. Johnson, G. Marino, O. Polverino, and R. Trombetti, *On a generalization of cyclic semifield spreads*, preprint.

781. _____, *Semifield spreads of $PG(3, q^3)$ with center F_q*, submitted.

782. N.L. Johnson and K.E. Mellinger, *Multiple spread retraction*, Adv. Geom. **3** (2003), no. 3, 263–286.

783. _____, *Classification of triply-retractive semifields of order q^4*, J. Geom. **80** (2004), no. 1–2, 121–135.

784. N.L. Johnson and T.G. Ostrom, *Translation planes with several homology or elation groups of order 3*, Geom. Dedicata **2** (1973), 65–81.

785. _____, *Tangentially transitive planes of order 16*, J. Geom. **10** (1977), no. 1–2, 146–163.

786. _____, *Translation planes of characteristic two in which all involutions are Baer*, J. Algebra **54** (1978), no. 2, 291–315.

787. _____, *The geometry of $SL(2, q)$ in translation planes of even order*, Geom. Dedicata **8** (1979), no. 1, 39–60.

788. _____, *The translation planes of order 16 that admit $PSL(2, 7)$*, J. Combin. Theory Ser. A **26** (1979), no. 2, 127–134.

789. _____, *Translation planes of dimension two and characteristic two*, Int. J. Math. Math. Sci. **6** (1983), no. 1, 41–58.

790. _____, *Inherited groups and kernels of derived translation planes*, European J. Combin. **11** (1990), no. 2, 145–149.

791. _____, *Direct products of affine partial linear spaces*, J. Combin. Theory Ser. A **75** (1996), no. 1, 99–140.

792. N.L. Johnson and S.E. Payne, *Flocks of Laguerre planes and associated geometries*, Mostly Finite Geometries (Iowa City, IA, 1996), Lecture Notes in Pure and Appl. Math., vol. 190, Dekker, New York, 1997, pp. 51–122.

793. N.L. Johnson and F.C. Piper, *On planes of Lenz-Barlotti class $II - 1$*, Bull. London Math. Soc. **6** (1974), 152–154.

794. N.L. Johnson and R. Pomareda, *André planes and nests of reguli*, Geom. Dedicata **31** (1989), no. 3, 245–260.

795. _____, *A maximal partial flock of deficiency one of the hyperbolic quadric in $PG(3, 9)$*, Simon Stevin **64** (1990), no. 2, 169–177.

796. _____, *Translation planes admitting many homologies*, J. Geom. **49** (1994), no. 1–2, 117–149.

797. _____, *Mixed nests*, J. Geom. **56** (1996), no. 1–2, 59–86.

798. _____, *Translation planes with many homologies, II: Odd order*, J. Geom. **72** (2001), no. 1–2, 77–107.

799. _____, *m-parallelisms*, Int. J. Math. Math. Sci. **32** (2002), no. 3, 167–176.

800. _____, *Real parallelisms*, Note Mat. **21** (2002), no. 1, 127–135.

801. _____, *Transitive partial parallelisms of deficiency one*, European J. Combin. **23** (2002), no. 8, 969–986.

802. _____, *Parallelism-inducing groups*, Aequationes Math. **65** (2003), no. 1–2, 133–157.

803. _____, *Partial parallelisms with sharply two-transitive skew spreads*, Ars Combin. **70** (2004), 275–287.

804. _____, *Collineation groups of translation planes admitting hyperbolic Buekenhout or parabolic Buekenhout-Metz unitals*, J. Combin. Theory Ser. A (2006), DOI 10.1016/j.jcta.2006.08.006 (online publication).

805. _____, *Minimal parallelism-inducing groups*, Aequationes Math. (2006).

806. N.L. Johnson, R. Pomareda, and F.W. Wilke, *j-planes*, J. Combin. Theory Ser. A **56** (1991), no. 2, 271–284.

807. N.L. Johnson and A.R. Prince, *Orbit constructions for translation planes of order 81 admitting $SL(2, 5)$*, Note Mat. **24** (2004/05), no. 2, 41–57.

808. _____, *The translation planes of order* 81 *that admit* SL(2, 5), *generated by affine elations*, J. Geom. **84** (2006), no. 1–2, 73–82.

809. N.L. Johnson and A. Rahilly, *On elations of derived semifield planes*, Proc. London Math. Soc. (3) **35** (1977), no. 1, 76–88.

810. N.L. Johnson and R.E. Seydel, *A class of translation planes of order* 3^6, Math. Z. **135** (1973/74), 271–278.

811. N.L. Johnson and L. Storme, *Spreads corresponding to flocks of quadrics*, Mostly Finite Geometries (Iowa City, IA, 1996), Lecture Notes in Pure and Appl. Math., vol. 190, Dekker, New York, 1997, pp. 315–323.

812. N.L. Johnson and O. Vega, *Symplectic spreads and symplectically paired spreads*, Note Mat. **26** (2006), no. 2, 119–134.

813. N.L. Johnson, O. Vega, and F.W. Wilke, *j, k-planes of order* 4^3, Innov. Incidence Geom. **2** (2006), 1–34.

814. N.L. Johnson and F.W. Wilke, *Translation planes of order* q^2 *that admit a collineation group of order* q^2, Geom. Dedicata **15** (1984), no. 3, 293–312.

815. D. Jungnickel, *Maximal partial spreads and transversal-free translation nets*, J. Combin. Theory Ser. A **62** (1993), no. 1, 66–92.

816. M.J. Kallaher, *A note on Bol projective planes*, Arch. Math. (Basel) **20** (1969), 329–332.

817. _____, *On finite affine planes of rank* 3, J. Algebra **13** (1969), 544–553.

818. _____, *Projective planes over Bol quasi-fields*, Math. Z. **109** (1969), 53–65.

819. _____, *Right Bol quasi-fields*, Canad. J. Math. **21** (1969), 1409–1420.

820. _____, *A class of rank three affine planes*, Math. Z. **119** (1971), 75–82.

821. _____, *On rank* 3 *projective planes*, Pacific J. Math. **39** (1971), 207–214.

822. _____, *A note on finite Bol quasi-fields*, Arch. Math. (Basel) **23** (1972), 164–166.

823. _____, *A note on planes of characteristic three*, Bull. Austral. Math. Soc. **7** (1972), 105–111.

824. _____, *Rank* 3 *affine planes of square order*, Geom. Dedicata **1** (1973), no. 4, 415–425.

825. _____, *A survey of weak rank* 3 *affine planes*, Proceedings of the International Conference on Projective Planes (Washington State Univ., Pullman, Wash., 1973), Washington State Univ. Press, Pullman, Wash., 1973, pp. 121–143.

826. _____, *Bol quasi-fields of dimension two over their kernels*, Arch. Math. (Basel) **25** (1974), 419–423.

827. _____, *A note on Z-planes*, J. Algebra **28** (1974), 311–318.

828. _____, *A conjecture on semi-field planes*, Arch. Math. (Basel) **26** (1975), no. 4, 436–440.

829. _____, *On translation planes with affine central collineations*, Geom. Dedicata **4** (1975), no. 1, 71–89.

830. _____, *Translation planes in which the stabilizer of every line is doubly transitive*, Geom. Dedicata **5** (1976), no. 1, 97–108.

831. _____, *A note on affine planes with transitive collineation groups*, Math. Z. **153** (1977), no. 1, 89–97.

832. _____, *Semi-field-like affine planes*, Glasgow Math. J. **18** (1977), no. 2, 113–123.

833. _____, *Translation planes admitting solvable rank* 3 *collineation groups*, Geom. Dedicata **6** (1977), no. 3, 305–329.

834. _____, *A characterization of the Hall planes*, Bull. London Math. Soc. **13** (1981), no. 3, 241–243.

835. _____, *Translation planes having* PSL(2, w) *or* SL(3, w) *as a collineation group*, Finite Geometries and Designs (Proc. Conf., Chelwood Gate, 1980), London Math. Soc. Lecture Note Ser., vol. 49, Cambridge Univ. Press, Cambridge, 1981, pp. 197–204.

836. _____, *Affine Planes with Transitive Collineation Groups*, North-Holland Publishing Co., New York, 1982.

837. _____, *Quasifields with irreducible nuclei*, Int. J. Math. Math. Sci. **7** (1984), no. 2, 319–326.

838. _____, *On finite Bol quasifields*, Proceedings of the Conference on Groups and Geometry, Part A (Madison, Wis., 1985), Algebras Groups Geom. **2** (1985), no. 3, 300–312.

839. _____, *The multiplicative groups of quasifields*, Canad. J. Math. **39** (1987), no. 4, 784–793.

840. _____, *Translation planes*, Handbook of Incidence Geometry, North-Holland, Amsterdam, 1995, pp. Ch. 5, 137–192.

841. M.J. Kallaher and J.B. Fink, *Sporadic simple groups acting on finite translation planes*, Finite Geometries (Winnipeg, Man., 1984), Lecture Notes in Pure and Appl. Math., vol. 103, Dekker, New York, 1985, pp. 171–178.

842. M.J. Kallaher and G. Kelly, *Line-rank 3 affine planes*, Bull. Austral. Math. Soc. **25** (1982), no. 3, 397–403.

843. M.J. Kallaher and R.A. Liebler, *A conjecture on semifield planes, II*, Geom. Dedicata **8** (1979), no. 1, 13–30.

844. M.J. Kallaher and T.G. Ostrom, *Fixed point free linear groups, rank three planes, and Bol quasifields*, J. Algebra **18** (1971), 159–178.

845. _____, *Bol quasifields and generalized André systems*, J. Algebra **58** (1979), no. 1, 100–116.

846. _____, *Collineation groups irreducible on the components of a translation plane*, Geom. Dedicata **9** (1980), no. 2, 153–194.

847. _____, *Translation planes admitting* SL(n, w) *with* $n \geq 3$, J. Algebra **79** (1982), no. 2, 272–285.

848. _____, *Collineation groups whose order is divisible by a p-primitive divisor*, J. Geom. **24** (1985), no. 1, 77–88.

849. W.M. Kantor, *Isomorphisms of symplectic planes*, European J. Combin. (to appear).

850. _____, *On unitary polarities of finite projective planes*, Canad. J. Math. **23** (1971), 1060–1077.

851. _____, *Line-transitive collineation groups of finite projective spaces*, Israel J. Math. **14** (1973), 229–235.

852. _____, *On 2-transitive collineation groups of finite projective spaces*, Pacific J. Math. **48** (1973), 119–131.

853. _____, *On homologies of finite projective planes*, Israel J. Math. **16** (1973), 351–361.

854. _____, *Those nasty Baer involutions*, Proceedings of the International Conference on Projective Planes (Washington State Univ., Pullman, Wash., 1973), Washington State Univ. Press, Pullman, Wash., 1973, pp. 145–155.

855. _____, *2-transitive designs*, Combinatorics (Proc. NATO Advanced Study Inst., Breukelen, 1974), Part 3: Combinatorial Group Theory, Math. Centre Tracts, no. 57, Math. Centrum, Amsterdam, 1974, pp. 44–97.

856. _____, *Symplectic groups, symmetric designs, and line ovals*, J. Algebra **33** (1975), 43–58.

857. _____, *The existence of translation complements*, Geom. Dedicata **5** (1976), no. 1, 71–78.

858. _____, *On the structure of collineation groups of finite projective planes*, Proc. London Math. Soc. (3) **32** (1976), no. 3, 385–402.

859. _____, *Generalized quadrangles associated with* $G_2(q)$, J. Combin. Theory Ser. A **29** (1980), no. 2, 212–219.

860. _____, *On point-transitive affine planes*, Israel J. Math. **42** (1982), no. 3, 227–234.

861. _____, *Ovoids and translation planes*, Canad. J. Math. **34** (1982), no. 5, 1195–1207.

862. _____, *Spreads, translation planes and Kerdock sets, I*, SIAM J. Algebraic Discrete Methods **3** (1982), no. 2, 151–165.

863. _____, *Spreads, translation planes and Kerdock sets, II*, SIAM J. Algebraic Discrete Methods **3** (1982), no. 3, 308–318.

864. _____, *Translation planes of order* q^6 *admitting* SL$(2, q^2)$, J. Combin. Theory Ser. A **32** (1982), no. 2, 299–302.

865. _____, *Expanded, sliced and spread spreads*, Finite Geometries (Pullman, Wash., 1981), Lecture Notes in Pure and Appl. Math., vol. 82, Dekker, New York, 1983, pp. 251–261.

866. _____, *Non-Desarguesian planes, partial geometries, strongly regular graphs and codes arising from hyperbolic quadrics*, Combinatorics '81 (Rome, 1981), Ann. Discrete Math., vol. 18, North-Holland, Amsterdam, 1983, pp. 511–517.

867. _____, *Flag-transitive planes*, Finite Geometries (Winnipeg, Man., 1984), Lecture Notes in Pure and Appl. Math., vol. 103, Dekker, New York, 1985, pp. 179–181.

868. _____, *Generalized quadrangles and translation planes*, Proceedings of the Conference on Groups and Geometry, Part A (Madison, Wis., 1985), Algebras Groups Geom. **2** (1985), no. 3, 313–322.

869. _____, *Homogeneous designs and geometric lattices*, J. Combin. Theory Ser. A **38** (1985), no. 1, 66–74.

870. _____, *Some generalized quadrangles with parameters q^2, q*, Math. Z. **192** (1986), no. 1, 45–50.

871. _____, *Primitive permutation groups of odd degree, and an application to finite projective planes*, J. Algebra **106** (1987), no. 1, 15–45.

872. _____, *Automorphism groups of some generalized quadrangles*, Advances in Finite Geometries and Designs (Chelwood Gate, 1990), Oxford Sci. Publ., Oxford Univ. Press, New York, 1991, pp. 251–256.

873. _____, *Generalized quadrangles, flocks, and BLT sets*, J. Combin. Theory Ser. A **58** (1991), no. 1, 153–157.

874. _____, *Two families of flag-transitive affine planes*, Geom. Dedicata **41** (1992), no. 2, 191–200.

875. _____, *2-transitive and flag-transitive designs*, Coding Theory, Design Theory, Group Theory: Proc. Marshall Hall Conf. (Burlington, VT, 1990) (D. Jungnickel, S.A. Vanstone, K.T. Arasu, M. Aschbacher, and R. Foote, eds.), Wiley, New York, 1993, pp. 13–30.

876. _____, *Projective planes of order q whose collineation groups have order q^2*, J. Algebraic Combin. **3** (1994), no. 4, 405–425.

877. _____, *Commutative semifields and symplectic spreads*, J. Algebra **270** (2003), no. 1, 96–114.

878. W.M. Kantor and R.A. Liebler, *The rank 3 permutation representations of the finite classical groups*, Trans. Amer. Math. Soc. **271** (1982), no. 1, 1–71.

879. W.M. Kantor and C. Suetake, *A note on some flag-transitive affine planes*, J. Combin. Theory Ser. A **65** (1994), no. 2, 307–310.

880. W.M. Kantor and M.E. Williams, *New flag-transitive affine planes of even order*, J. Combin. Theory Ser. A **74** (1996), no. 1, 1–13.

881. _____, *Symplectic semifield planes and $\mathbb{Z}_4$-linear codes*, Trans. Amer. Math. Soc. **356** (2004), no. 3, 895–938.

882. I. Kaplansky, *Linear Algebra and Geometry: A Second Course*, revised ed., Dover Publications Inc., Mineola, NY, 2003, reprint of Chelsea Publication Company, second edition, New York, 1974.

883. O.H. Kegel and A. Schleiermacher, *Amalgams and embeddings of projective planes*, Geom. Dedicata **2** (1973), 379–395.

884. V.H. Keiser, Jr., *Finite affine planes with collineation groups primitive on the points*, Math. Z. **92** (1966), 288–294.

885. J.D. Key and K. Mackenzie, *An upper bound for the p-rank of a translation plane*, J. Combin. Theory Ser. A **56** (1991), no. 2, 297–302.

886. P. Kleidman and M. Liebeck, *The Subgroup Structure of the Finite Classical Groups*, London Math. Soc. Lecture Note Ser., vol. 129, Cambridge University Press, Cambridge, 1990.

887. E. Kleinfeld, *Techniques for enumerating Veblen-Wedderburn systems*, J. Assoc. Comput. Mach. **7** (1960), 330–337.

888. J. Klouda, *Ternary rings associated to translation plane*, Comment. Math. Univ. Carolinae **18** (1977), no. 1, 169–181.

889. D. Klucký and L. Marková, *Ternary rings with zero associated to translation planes*, Czechoslovak Math. J. **23(98)** (1973), 617–628.

890. _____, *Planar ternary rings with zero belonging to translation planes*, Grundlagen der Geometrie und algebraische Methoden (Internat. Kolloq., Pädagog. Hochsch. "Karl Liebknecht", Potsdam, 1973), Potsdamer Forschungen, Reihe B, Heft 3, Pädagog. Hochsch. "Karl Liebknecht", Potsdam, 1974, pp. 116–120.

891. _____, *Ternary rings with a left quasizero belonging to translation planes*, Sb. Prací Přírodověd. Fak. Univ. Palackého v Olomouci Mat. **19** (1980), 89–100.

892. N. Knarr, *Sharply transitive subsets of $P\Gamma L(2, F)$ and spreads covered by derivable partial spreads*, J. Geom. **40** (1991), no. 1–2, 121–124.

893. _____, *A geometric construction of generalized quadrangles from polar spaces of rank three*, Results Math. **21** (1992), no. 3–4, 332–344.

894. _____, *Translation Planes: Foundations and Construction Principles*, Lecture Notes in Mathematics, vol. 1611, Springer-Verlag, Berlin, 1995.

895. _____, *Derivable affine planes and translation planes*, Bull. Belg. Math. Soc. Simon Stevin **7** (2000), no. 1, 61–71.

896. F. Knoflíček, *On translation planes and quasifields of order* 16, Zeszyty Nauk. Geom. **16** (1987), 49–66.

897. D.E. Knuth, *A class of projective planes*, Trans. Amer. Math. Soc. **115** (1965), 541–549.

898. _____, *Finite semifields and projective planes*, J. Algebra **2** (1965), 182–217.

899. G. Korchmáros, *On translation ovals in a Galois plane of even order*, Rend. Accad. Naz. XL (5) **3** (1977/78), 55–65.

900. _____, *A class of transitive groups and finite Bol plane*, Rend. Mat. (6) **12** (1979), no. 1, 71–78.

901. _____, *The line ovals of the Lüneburg plane of order* 2^{2r} *that can be transformed into themselves by a collineation group isomorphic to the simple group* Sz(2^r) *of Suzuki*, Atti Accad. Naz. Lincei Mem. Cl. Sci. Fis. Mat. Natur. Sez. Ia (8) **15** (1979), no. 6, 293–315.

902. _____, *Example of a chain of circles on an elliptic quadric of* PG(3, q), $q = 7, 11$, J. Combin. Theory Ser. A **31** (1981), no. 1, 98–100.

903. _____, *A translation plane of order* 49 *with nonsolvable collineation group*, J. Geom. **24** (1985), no. 1, 18–30.

904. _____, *The full collineation group of the Bruen plane of order* 49, Boll. Un. Mat. Ital. B (6) **5** (1986), no. 1, 109–121.

905. K. Kuppuswamy Rao, *Finite translation planes*, Math. Student **55** (1987), no. 2–4, 96–102.

906. V. Landazuri and G.M. Seitz, *On the minimal degrees of projective representations of the finite Chevalley groups*, J. Algebra **32** (1974), 418–443.

907. B. Larato, *A characterization of the parabolic unitals of Buekenhout-Metz*, Matematiche (Catania) **38** (1983), no. 1–2, 95–98.

908. B. Larato and G. Raguso, *Collineation group of a translation plane of order* 13^2, Proceedings of the Conference on Combinatorial and Incidence Geometry: Principles and Applications (La Mendola, 1982) (Milan), Rend. Sem. Mat. Brescia, vol. 7, Vita e Pensiero, 1984, pp. 453–470.

909. _____, *Translation planes of order* 13^2, Riv. Mat. Univ. Parma (4) **10** (1984), 223–233.

910. M. Lavrauw and M. Law, *BLT-sets admitting the symmetric group* S_4, Australas. J. Combin. **33** (2005), 307–316.

911. M. Law, *Flocks, generalised quadrangles and translation planes from BLT-sets*, Ph.D. thesis, The University of Western Australia, Adelaide, 2003.

912. M. Law and T. Penttila, *Some flocks in characteristic* 3, J. Combin. Theory Ser. A **94** (2001), no. 2, 387–392.

913. _____, *Classification of flocks of the quadratic cone over fields of order at most 29*, Special issue dedicated to Adriano Barlotti, Adv. Geom. suppl. (2003), S232–S244.

914. _____, *Construction of BLT-sets over small fields*, European J. Combin. **25** (2004), no. 1, 1–22.

915. W.J. LeVeque, *Topics in Number Theory, Vol. I, II*, Dover Publications Inc., Mineola, NY, 2002, Reprint of the 1956 original [Addison-Wesley Publishing Co., Inc., Reading, Mass.], with separate errata list for this edition by the author.

916. M.W. Liebeck, *The affine permutation groups of rank three*, Proc. London Math. Soc. (3) **54** (1987), no. 3, 477–516.

917. R.A. Liebler, *Finite affine planes of rank three are translation planes*, Math. Z. **116** (1970), 89–93.

918. _____, *A characterization of the Lüneburg planes*, Math. Z. **126** (1972), 82–90.

919. _____, *On finite line transitive affine planes of rank 3*, Proceedings of the International Conference on Projective Planes (Washington State Univ., Pullman, Wash., 1973), Washington State Univ. Press, Pullman, Wash., 1973, pp. 165–179.

920. _____, *A note on the Hering-Ott planes*, J. Combin. Theory Ser. A **25** (1978), no. 2, 202–204.

921. _____, *On nonsingular tensors and related projective planes*, Geom. Dedicata **11** (1981), no. 4, 455–464.

922. _____, *Combinatorial representation theory and translation planes*, Finite Geometries (Pullman, Wash., 1981), Lecture Notes in Pure and Appl. Math., vol. 82, Dekker, New York, 1983, pp. 307–325.

923. _____, *The classification of distance-transitive graphs of type* $q \cdot K_{q,q}$, European J. Combin. **12** (1991), no. 2, 125–128.

924. _____, *Virtual derivation*, Mostly Finite Geometries (Iowa City, IA, 1996), Lecture Notes in Pure and Appl. Math., vol. 190, Dekker, New York, 1997, pp. 331–342.

925. A. Lizzio, *General ternary systems and involutory perspectivities in translation planes*, Matematiche (Catania) **33** (1978), no. 1, 138–145.

926. P. Lorimer, *A class of projective planes of cubic order*, Proc. Amer. Math. Soc. **21** (1969), 93–95.

927. _____, *The construction of semifields*, Math. Chronicle **1** (1971), 129–137.

928. _____, *Finite projective planes and sharply 2-transitive subsets of finite groups*, Proceedings of the Second International Conference on the Theory of Groups (Australian Nat. Univ., Canberra, 1973) (Berlin), Lecture Notes in Math., vol. 372, Springer, 1974, pp. 432–436.

929. _____, *A projective plane of order 16*, J. Combin. Theory Ser. A **16** (1974), 334–347.

930. _____, *An introduction to projective planes: some of the properties of a particular plane of order 16*, H.G. Forder 90th birthday volume, Math. Chronicle **9** (1980), 53–66.

931. _____, *The construction of finite projective planes*, Combinatorial Mathematics, VIII (Geelong, 1980), Lecture Notes in Math., vol. 884, Springer, Berlin, 1981, pp. 64–76.

932. _____, *Some of the finite projective planes*, Math. Intelligencer **5** (1983), no. 2, 41–50.

933. _____, *The groups A_7, A_8 and a projective plane of order 16*, Australas. J. Combin. **8** (1993), 45–51.

934. K. Lueder, *Derivability in the irregular nearfield planes*, Proceedings of the International Conference on Projective Planes (Washington State Univ., Pullman, Wash., 1973), Washington State Univ. Press, Pullman, Wash., 1973, pp. 181–189.

935. G. Luisi, *A translation plane of order 49*, Matematiche (Catania) **38** (1983), no. 1–2, 157–166.

936. G. Lunardon, *Proposizioni configurazionali in una classe di fibrazioni*, Boll. Un. Mat. Ital. A (5) **13** (1976), no. 2, 404–413.

937. _____, *A classification of the translation planes in relation to all fibrations and their associates, I*, Atti Accad. Naz. Lincei Rend. Cl. Sci. Fis. Mat. Natur. (8) **63** (1977), no. 6, 504–508.

938. _____, *A classification of translation planes in relation to all fibrations and their associates, II*, Atti Accad. Naz. Lincei Rend. Cl. Sci. Fis. Mat. Natur. (8) **64** (1978), no. 1, 59–64.

939. _____, *Derivable translation planes*, Rend. Sem. Mat. Univ. Padova **61** (1979), 271–284.

940. _____, *On regular parallelisms in* PG(3, q), Discrete Math. **51** (1984), no. 3, 229–235.

941. _____, *Planar fibrations and algebraic subvarieties of the Grassmann variety*, Geom. Dedicata **16** (1984), no. 3, 291–313.

942. _____, *Projective indicator sets and planar spreads of a finite projective space*, Boll. Un. Mat. Ital. B (6) **3** (1984), no. 3, 717–735.

943. _____, *A remark on the derivation of flocks*, Advances in Finite Geometries and Designs (Chelwood Gate, 1990), Oxford Sci. Publ., Oxford Univ. Press, New York, 1991, pp. 299–309.

944. _____, *Flocks, ovoids of* Q(4, q) *and designs*, Geom. Dedicata **66** (1997), no. 2, 163–173.

945. _____, *An approach to semifield flocks*, Rend. Circ. Mat. Palermo (2) Suppl. (1998), no. 51, 115–121.

946. _____, *Normal spreads*, Geom. Dedicata **75** (1999), no. 3, 245–261.

947. _____, *Translation ovoids*, Combinatorics, 2002 (Maratea), J. Geom. **76** (2003), no. 1–2, 200–215.

948. _____, *Blocking sets and semifields*, J. Combin. Theory Ser. A **113** (2006), no. 6, 1172–1188.

949. G. Lunardon and O. Polverino, *Blocking sets of size* $q^t + q^{t-1} + 1$, J. Combin. Theory Ser. A **90** (2000), no. 1, 148–158.

950. _____, *Blocking sets and derivable partial spreads*, J. Algebraic Combin. **14** (2001), no. 1, 49–56.

951. _____, *Translation ovoids of orthogonal polar spaces*, Forum Math. **16** (2004), no. 5, 663–669.

952. H. Lüneburg, *Charakterisierungen der endlichen desarguesschen projektiven Ebenen*, Math. Z. **85** (1964), 419–450.

953. _____, *Die Suzukigruppen und ihre Geometrien*, Springer-Verlag, Berlin, 1965.

954. _____, *Über projektive Ebenen, in denen jede Fahne von einer nicht-trivialen Elation invariant gelassen wird*, Abh. Math. Sem. Univ. Hamburg **29** (1965), 37–76.

955. _____, *Über die Anzahl der Dickson'schen Fastkörper gegebener Ordnung*, Atti del Convegno di Geometria Combinatoria e sue Applicazioni (Univ. Perugia, Perugia, 1970), Ist. Mat., Univ. Perugia, Perugia, 1971, pp. 319–322.

956. _____, *Affine Ebenen, in denen der Stabilisator jeder Geraden zweifach transitiv ist*, Arch. Math. (Basel) **24** (1973), 663–668.

957. _____, *Über einige merkwürdige Translationsebenen*, Geom. Dedicata **3** (1974), 263–288.

958. _____, *Characterizations of the generalized Hughes planes*, Canad. J. Math. **28** (1976), no. 2, 376–402.

959. _____, *On finite affine planes of rank 3*, Foundations of Geometry (Proc. Conf., Univ. Toronto, Toronto, Ont., 1974), Univ. Toronto Press, Toronto, Ont., 1976, pp. 147–174.

960. _____, *Über eine Klasse von endlichen affinen Ebenen des Ranges 3*, Colloquio Internazionale sulle Teorie Combinatorie (Rome, 1973), Tomo II, Accad. Naz. Lincei, Rome, 1976, pp. 439–446. Atti dei Convegni Lincei, No. 17.

961. _____, *Translation Planes*, Springer-Verlag, Berlin, 1980.

962. H. Lüneburg and T.G. Ostrom, *Rang-3-Ebenen mit einer Bahn der Länge 2 auf der uneigentlichen Geraden*, Geom. Dedicata **4** (1975), no. 2/3/4, 249–252.

963. D. Luyckx, *A geometric construction of the hyperbolic fibrations associated with a flock, q even*, Des. Codes Cryptogr. **39** (2006), no. 2, 281–288.

964. D.M. Maduram, *Matrix representation of translation planes*, Geom. Dedicata **4** (1975), no. 2/3/4, 485–492.

965. _____, *Transposed translation planes*, Proc. Amer. Math. Soc. **53** (1975), no. 2, 265–270.

966. A.E. Malyh, *Latin squares of order 8 from the descriptions of the Desarguesian and the translation plane of order 9*, Kombinatornyĭ Anal. Vyp. 3 (1974), 96–101.

967. A. Maschietti, *Symplectic translation planes*, Scuola Estiva di Geometrie Combinatorie, Giuseppe Tallini, Potenza, Italy, 1–6 September 2003, pp. 46–103.

968. _____, *Symplectic translation planes and line ovals*, Adv. Geom. **3** (2003), no. 2, 123–143.

969. G. Mason, *Irreducible translation planes and representations of Chevalley groups in characteristic 2*, Finite Geometries (Pullman, Wash., 1981), Lecture Notes in Pure and Appl. Math., vol. 82, Dekker, New York, 1983, pp. 333–346.

970. _____, *Orthogonal geometries over GF(2) and actions of extraspecial 2-groups on translation planes*, European J. Combin. **4** (1983), no. 4, 347–357.

971. _____, *Some translation planes of order 7^2 which admit $SL_2(9)$*, Geom. Dedicata **17** (1985), no. 3, 297–305.

972. _____, *Representation theory and translation planes*, The Arcata Conference on Representations of Finite Groups (Arcata, Calif., 1986), Proc. Sympos. Pure Math., vol. 47, Amer. Math. Soc., Providence, RI, 1987, pp. 211–222.

973. G. Mason and T.G. Ostrom, *Some translation planes of order p^2 and of extra-special type*, Geom. Dedicata **17** (1985), no. 3, 307–322.

974. G. Mason and E.E. Shult, *The Klein correspondence and the ubiquity of certain translation planes*, Geom. Dedicata **21** (1986), no. 1, 29–50.

975. R. Mathon and N. Hamilton, *Baer partitions of small order projective planes*, J. Combin. Math. Combin. Comput. **29** (1999), 87–94.

976. R. Mathon and G.F. Royle, *The translation planes of order 49*, Des. Codes Cryptogr. **5** (1995), no. 1, 57–72.

977. V.D. Mazurov, *Minimal permutation representations of finite simple classical groups: Special linear, symplectic and unitary groups*, Algebra i Logika **32** (1993), no. 3, 267–287, 343.

978. K.F. Mellinger, *Mixed partitions and spreads of projective spaces*, Ph.D. thesis, Univ. Delaware, 2001.

979. G. Menichetti, *On a Kaplansky conjecture concerning three-dimensional division algebras over a finite field*, J. Algebra **47** (1977), no. 2, 400–410.

980. _____, *n-dimensional algebras over a field with a cyclic extension of degree n*, Geom. Dedicata **63** (1996), no. 1, 69–94.

981. R. Metz, *On a class of unitals*, Geom. Dedicata **8** (1979), no. 1, 125–126.

982. H.H. Mitchell, *Determination of the ordinary and modular ternary linear groups*, Trans. Amer. Math. Soc. **12** (1911), no. 2, 207–242.

983. A. Montinaro, *Large doubly transitive orbits on a line*, J. Aust. Math. Soc. (to appear).

984. ———, *On a restricted Cofman's problem*, European J. Combin. (submitted).

985. ———, *Projective planes of order up to q^3 with a Desarguesian subplane of order q*, Bull. Belg. Math. Soc. Simon Stevin (to appear).

986. ———, *Projective planes with a doubly transitive projective subplane*, Bull. Belg. Math. Soc. Simon Stevin (to appear).

987. G.E. Moorhouse, *PSL$(2,q)$ as a collineation group of projective planes of small order*, Geom. Dedicata **31** (1989), no. 1, 63–88.

988. ———, *On the construction of finite projective planes from homology semibiplanes*, European J. Combin. **11** (1990), no. 6, 589–600.

989. ———, *Reconstructing projective planes from semibiplanes*, Coding Theory and Design Theory, Part II, IMA Vol. Math. Appl., vol. 21, Springer, New York, 1990, pp. 280–285.

990. ———, *Ovoids from the E_8 root lattice*, Geom. Dedicata **46** (1993), no. 3, 287–297.

991. ———, *Root lattice constructions of ovoids*, Finite Geometry and Combinatorics (Deinze, 1992), London Math. Soc. Lecture Note Ser., vol. 191, Cambridge Univ. Press, Cambridge, 1993, pp. 269–275.

992. ———, *Ovoids and translation planes from lattices*, Mostly Finite Geometries (Iowa City, IA, 1996), Lecture Notes in Pure and Appl. Math., vol. 190, Dekker, New York, 1997, pp. 123–134.

993. ———, *On projective planes of order less than 32*, Finite Geometries, Groups, and Computation (Pingree Park, Colorado, 2004) (A. Hulpke, R. Liebler, T. Penttila, and Á. Seress, eds.), de Gruyter Proceedings in Mathematics, Walter de Gruyter, Berlin, New York, 2006, pp. 149–162.

994. D.L. Morgan and T.G. Ostrom, *Coordinate systems of some semi-translation planes*, Trans. Amer. Math. Soc. **111** (1964), 19–32.

995. F. Mylle, *Flocks van Kwadrieken in PG$(3,q)$*, M.S. Thesis, Ghent (Belgium), 1991.

996. N. Nakagawa, *Some translation planes of order 11^2 which admit SL$(2,9)$*, Hokkaido Math. J. **20** (1991), no. 1, 91–107.

997. ———, *The Klein correspondence and some translation planes of order p^2 which admit SL$(2,9)$*, J. Sch. Sci. Eng. Kinki Univ. (1999), no. 35, 1–9.

998. M.L. Narayana Rao, *Conjecture of D. R. Hughes extended to generalized André planes*, Canad. J. Math. **22** (1970), 701–704.

999. ———, *A class of flag transitive planes*, Proc. Amer. Math. Soc. **39** (1973), 51–56.

1000. M.L. Narayana Rao and E.H. Davis, *Construction of translation planes from t-spread sets*, J. Combin. Theory Ser. A **14** (1973), 201–208.

1001. M.L. Narayana Rao and K. Kuppuswamy Rao, *A flag transitive affine plane of order 125*, Proc. Nat. Acad. Sci. India Sect. A **46** (1976), no. 1, 65–67.

1002. ———, *A new flag transitive affine plane of order 27*, Proc. Amer. Math. Soc. **59** (1976), no. 2, 337–345.

1003. ———, *A translation plane of order 25 and its full collineation group*, Bull. Austral. Math. Soc. **19** (1978), no. 3, 351–362.

1004. M.L. Narayana Rao, K. Kuppuswamy Rao, and K.V. Durga Prasad, *Flag transitive planes of dimension four over GF(3)*, J. Geom. **33** (1988), no. 1–2, 89–98.

1005. ———, *A class of d-dimensional flag transitive translation planes*, J. Geom. **34** (1989), no. 1–2, 139–145.

1006. ———, *A note on a class of translation planes of square order*, Indian J. Pure Appl. Math. **27** (1996), no. 8, 731–734.

1007. M.L. Narayana Rao, K. Kuppuswamy Rao, and V. Joshi, *A translation plane of order 25 with small translation complement*, J. Indian Inst. Sci. **66** (1986), no. 8, 533–541.

1008. ———, *A translation plane of order 49 with orbit structure 2, 16 and 32*, J. Combin. Inform. System Sci. **12** (1987), no. 1–2, 41–50.

1009. ———, *A translation plane of order 49 with 48 ideal points in a single orbit*, Indian J. Math. **31** (1989), no. 1, 89–97.

1010. ———, *A translation plane of order 7^2 with a very small translation complement*, Graphs Combin. **9** (1993), no. 3, 255–263.

1011. M.L. Narayana Rao, K. Kuppuswamy Rao, and K. Satyanarayana, *On Hering's flag transitive plane of order 27*, Bull. Austral. Math. Soc. **25** (1982), no. 1, 117–123.

1012. _____, A third flag transitive plane of order 27, Houston J. Math. 10 (1984), no. 1, 127–145.

1013. M.L. Narayana Rao, D.J. Rodabaugh, F.W. Wilke, and J.L. Zemmer, A new class of finite translation planes obtained from the exceptional near-fields, J. Combin. Theory Ser. A 11 (1971), 72–92.

1014. M.L. Narayana Rao and K. Satyanarayana, A new class of square order planes, J. Combin. Theory Ser. A 35 (1983), no. 1, 33–42.

1015. _____, Orbit structure of Rao and Davis plane of order 25, J. Combin. Inform. System Sci. 8 (1983), no. 1, 73–78.

1016. _____, Collineations in the translation complement fixing an ideal point of a translation plane, J. Combin. Inform. System Sci. 9 (1984), no. 4, 239–241.

1017. _____, On a C-plane of order 25, Bull. Austral. Math. Soc. 30 (1984), no. 1, 27–36.

1018. _____, On Sherk's plane of order 27, Linear Algebra Appl. 74 (1986), 1–9.

1019. M.L. Narayana Rao, K. Satyanarayana, and K.M.A. Rao, A note on flag transitive planes of order 27, Southeast Asian Bull. Math. 13 (1989), no. 2, 101–102.

1020. _____, A flag transitive plane of order 49 and its translation complement, Int. J. Math. Math. Sci. 14 (1991), no. 2, 339–344.

1021. _____, A flag transitive plane of order 49 and its translation complement, Combinatorial Mathematics and Applications (Calcutta, 1988), Sankhyā Ser. A 54 (1992), no. Special Issue, 367–373.

1022. M.L. Narayana Rao, K. Satyanarayana, and G. Vithal Rao, A class of translation planes of square order, Bull. Austral. Math. Soc. 30 (1984), no. 1, 59–66.

1023. _____, On a class of translation planes of square order, Discrete Math. 66 (1987), no. 1–2, 175–190.

1024. _____, On a class of derived translation planes of square order, Linear Algebra Appl. 116 (1989), 81–99.

1025. M.L. Narayana Rao and F.W. Wilke, A necessary condition that two finite quasi-fields coordinatize isomorphic translation planes, Proc. Amer. Math. Soc. 24 (1970), 124–125.

1026. M.L. Narayana Rao and J.L. Zemmer, A question of Foulser on λ-systems of characteristic two, Proc. Amer. Math. Soc. 21 (1969), 373–378.

1027. H. Neumann, On some finite non-desarguesian planes, Arch. Math. 6 (1954), 36–40.

1028. C.M. O'Keefe and T. Penttila, Characterisations of flock quadrangles, Geom. Dedicata 82 (2000), no. 1–3, 171–191.

1029. C.M. O'Keefe, T. Penttila, and G.F. Royle, Classification of ovoids in PG(3, 32), J. Geom. 50 (1994), no. 1–2, 143–150.

1030. C.M. O'Keefe and A. Rahilly, Spreads and group divisible designs, Des. Codes Cryptogr. 3 (1993), no. 3, 229–235.

1031. C.M. O'Keefe and J.A. Thas, Collineations of Subiaco and Cherowitzo hyperovals, Bull. Belg. Math. Soc. Simon Stevin 3 (1996), no. 2, 177–192.

1032. _____, Partial flocks of quadratic cones with a point vertex in PG(n, q), n odd, J. Algebraic Combin. 6 (1997), no. 4, 377–392.

1033. B. Orbán, On semi-anchors of a translation plane, J. Geometry 3 (1973), 191–197.

1034. W.F. Orr, The Miquelian inversive plane IP(q) and the associated projective planes, Ph.D. thesis, University of Wisconsin, Madison, WI, 1973.

1035. W.F. Orr, A characterization of subregular spreads in finite 3-space, Geom. Dedicata 5 (1976), no. 1, 43–50.

1036. T.G. Ostrom, Double transitivity in finite projective planes, Canad. J. Math. 8 (1956), 563–567.

1037. _____, Translation planes and configurations in Desarguesian planes, Arch. Math. 11 (1960), 457–464.

1038. _____, Nets with critical deficiency, Pacific J. Math. 14 (1964), 1381–1387.

1039. _____, Semi-translation planes, Trans. Amer. Math. Soc. 111 (1964), 1–18.

1040. _____, Collineation groups of semi-translation planes, Pacific J. Math. 15 (1965), 273–279.

1041. _____, Partially Desarguesian planes, Math. Z. 106 (1968), 113–122.

1042. _____, A characterization of generalized André planes, Math. Z. 110 (1969), 1–9.

1043. _____, Translation planes of order p^2, Combinatorial Mathematics and its Applications (Proc. Conf., Univ. North Carolina, Chapel Hill, N.C., 1967), Univ. North Carolina Press, Chapel Hill, N.C., 1969, pp. 416–425.

1044. _____, *A class of translation planes admitting elations which are not translations*, Arch. Math. (Basel) **21** (1970), 214–217.

1045. _____, *Desarguesian decompositions for planes of order p^2*, Duke Math. J. **37** (1970), 151–162.

1046. _____, *Finite Translation Planes*, Lecture Notes in Mathematics, vol. 158, Springer-Verlag, Berlin, 1970.

1047. _____, *Linear transformations and collineations of translation planes*, J. Algebra **14** (1970), 405–416.

1048. _____, *Translation planes admitting homologies of large order*, Math. Z. **114** (1970), 79–92.

1049. _____, *Collineation groups generated by homologies in translation planes*, Atti del Convegno di Geometria Combinatoria e sue Applicazioni (Univ. Perugia, Perugia, 1970), Ist. Mat., Univ. Perugia, Perugia, 1971, pp. 351–366.

1050. _____, *Classification of finite translation planes*, Proceedings of the International Conference on Projective Planes (Washington State Univ., Pullman, Wash., 1973), Washington State Univ. Press, Pullman, Wash., 1973, pp. 195–213.

1051. _____, *Homologies in translation planes*, Proc. London Math. Soc. (3) **26** (1973), 605–629.

1052. _____, *Elations in finite translation planes of characteristic 3*, Abh. Math. Sem. Univ. Hamburg **41** (1974), 179–184.

1053. _____, *Normal subgroups of collineation groups of finite translation planes*, Geom. Dedicata **2** (1974), 467–483.

1054. _____, *Recent advances in finite translation planes*, Foundations of Geometry (Proc. Conf., Univ. Toronto, Toronto, Ont., 1974), Univ. Toronto Press, Toronto, Ont., 1976, pp. 183–205.

1055. _____, *Collineation groups whose order is prime to the characteristic*, Math. Z. **156** (1977), no. 1, 59–71.

1056. _____, *Finite translation planes, an exposition*, Aequationes Math. **15** (1977), no. 2–3, 121–133.

1057. _____, *Translation planes of odd order and odd dimension*, Int. J. Math. Math. Sci. **2** (1979), no. 2, 187–208.

1058. _____, *Translation planes of dimension two with odd characteristic*, Canad. J. Math. **32** (1980), no. 5, 1114–1125.

1059. _____, *Translation planes of even order in which the dimension has only one odd factor*, Int. J. Math. Math. Sci. **3** (1980), no. 4, 675–694.

1060. _____, *Collineation groups of translation planes of small dimension*, Int. J. Math. Math. Sci. **4** (1981), no. 4, 711–724.

1061. _____, *Elementary abelian 2-groups in finite translation planes*, Arch. Math. (Basel) **36** (1981), no. 1, 21–22.

1062. _____, *Elementary abelian 2-groups on the line at infinity of translation planes*, J. Geom. **17** (1981), no. 2, 128–139.

1063. _____, *Lectures on finite translation planes*, Confer. Sem. Mat. Univ. Bari (1983), no. 191, 31 pp.

1064. _____, *New developments in finite translation planes*, Finite Geometries (Pullman, Wash., 1981), Lecture Notes in Pure and Appl. Math., vol. 82, Dekker, New York, 1983, pp. 351–360.

1065. _____, *Quaternion groups and translation planes related to the solvable nearfield planes*, Mitt. Math. Sem. Giessen (1984), no. 165, 119–134.

1066. _____, *Reducibility in finite translation planes*, Proceedings of the Conference on Groups and Geometry, Part B (Madison, Wis., 1985), Algebras Groups Geom. **2** (1985), no. 4, 455–477.

1067. _____, *Two-groups on finite translation planes*, Arch. Math. (Basel) **47** (1986), no. 6, 568–572.

1068. _____, *Correction to: "Two-groups on finite translation planes" [Arch. Math. (Basel) **47** (1986), no. 6, 568–572]*, Arch. Math. (Basel) **50** (1988), no. 6, 575–576.

1069. _____, *Higher-dimensional analogues of Klein's quadric*, Geom. Dedicata **41** (1992), no. 2, 207–217.

1070. _____, *Hyper-reguli*, J. Geom. **48** (1993), no. 1–2, 157–166.

1071. _____, *Remarks concerning linear systems with parallelism*, Mostly Finite Geometries (Iowa City, IA, 1996), Lecture Notes in Pure and Appl. Math., vol. 190, Dekker, New York, 1997, pp. 1–7.

1072. T.G. Ostrom and A. Wagner, *On projective and affine planes with transitive collineation groups*, Math. Z **71** (1959), 186–199.

1073. U. Ott, *Eine neue Klasse endlicher Translationsebenen*, Math. Z. **143** (1975), no. 2, 181–185.

1074. _____, *Endliche zyklische Ebenen*, Math. Z. **144** (1975), no. 3, 195–215.

1075. _____, *Fahnentransitive Ebenen gerader Ordnung*, Arch. Math. (Basel) **28** (1977), no. 6, 661–668.

1076. _____, *Fahnentransitive Ebenen ungerader Ordnung*, Geom. Dedicata **8** (1979), no. 2, 219–252.

1077. _____, *Some remarks on representation theory in finite geometry*, Geometries and Groups (Berlin, 1981), Lecture Notes in Math., vol. 893, Springer, Berlin, 1981, pp. 68–110.

1078. _____, *Sharply flag-transitive projective planes and power residue difference sets*, J. Algebra **276** (2004), no. 2, 663–673.

1079. T. Oyama, *On quasifields*, Osaka J. Math. **22** (1985), no. 1, 35–54.

1080. G. Panella, *Isomorfismo tra piani di traslazione di Marshall Hall*, Ann. Mat. Pura Appl. (4) **47** (1959), 169–180.

1081. D. Passman, *Permutation Groups*, W. A. Benjamin, Inc., New York-Amsterdam, 1968.

1082. S.E. Payne, *A new infinite family of generalized quadrangles*, Proceedings of the Sixteenth Southeastern International Conference on Combinatorics, Graph Theory and Computing (Boca Raton, Fla., 1985), Congr. Numer. **49** (1985), 115–128.

1083. _____, *Spreads, flocks, and generalized quadrangles*, J. Geom. **33** (1988), no. 1–2, 113–128.

1084. _____, *The Thas-Fisher generalized quadrangles*, Combinatorics '86 (Trento, 1986), Ann. Discrete Math., vol. 37, North-Holland, Amsterdam, 1988, pp. 357–366.

1085. _____, *Collineations of the generalized quadrangles associated with q-clans*, Combinatorics '90 (Gaeta, 1990), Ann. Discrete Math., vol. 52, North-Holland, Amsterdam, 1992, pp. 449–461.

1086. _____, *Collineations of the Subiaco generalized quadrangles*, A tribute to J.A. Thas (Gent, 1994), Bull. Belg. Math. Soc. Simon Stevin **1** (1994), no. 3, 427–438.

1087. _____, *A tensor product action on q-clan generalized quadrangles with $q = 2^e$*, Linear Algebra Appl. **226/228** (1995), 115–137.

1088. _____, *The fundamental theorem of q-clan geometry*, Des. Codes Cryptogr. **8** (1996), no. 1–2, 181–202, Special issue dedicated to Hanfried Lenz.

1089. _____, *Flock generalized quadrangles and related structures: an update*, Generalized Polygons (Proceedings of the Academy Contact Forum, Brussels, Belgium, October 20, 2000) (F. De Clerck, L. Storme, J.A. Thas, and H. Van Maldeghem, eds.), Universa Press, Belgium, 2001, pp. 61–98.

1090. _____, *The Law-Penttila q-clan geometries*, Finite Geometries, Dev. Math., vol. 3, Kluwer Acad. Publ., Dordrecht, 2001, pp. 295–303.

1091. S.E. Payne, T. Penttila, and I. Pinneri, *Isomorphisms between Subiaco q-clan geometries*, Bull. Belg. Math. Soc. Simon Stevin **2** (1995), no. 2, 197–222.

1092. S.E. Payne, T. Penttila, and G.F. Royle, *Building a cyclic q-clan*, Mostly Finite Geometries (Iowa City, IA, 1996), Lecture Notes in Pure and Appl. Math., vol. 190, Dekker, New York, 1997, pp. 365–378.

1093. S.E. Payne and L.A. Rogers, *Local group actions on generalized quadrangles*, Simon Stevin **64** (1990), no. 3–4, 249–284.

1094. S.E. Payne and J.A. Thas, *Conical flocks, partial flocks, derivation, and generalized quadrangles*, Geom. Dedicata **38** (1991), no. 2, 229–243.

1095. _____, *Generalized quadrangles, BLT-sets, and Fisher flocks*, Proceedings of the Twenty-Second Southeastern Conference on Combinatorics, Graph Theory, and Computing (Baton Rouge, LA, 1991), Congr. Numer. **84** (1991), 161–192.

1096. _____, *The stabilizer of the Adelaide oval*, Discrete Math. **294** (2005), no. 1–2, 161–173.

1097. G. Pellegrino and G. Korchmáros, *Translation planes of order 11^2*, Combinatorial and Geometric Structures and Their Applications (Trento, 1980), Ann. Discrete Math., vol. 14, North-Holland, Amsterdam, 1982, pp. 249–264.

1098. T. Penttila, *Regular cyclic BLT-sets*, Combinatorics '98 (Mondello), Rend. Circ. Mat. Palermo (2) Suppl. (1998), no. 53, 167–172.

1099. T. Penttila and G.F. Royle, *BLT-sets over small fields*, Australas. J. Combin. **17** (1998), 295–307.

1100. T. Penttila and L. Storme, *Monomial flocks and herds containing a monomial oval*, J. Combin. Theory Ser. A **83** (1998), no. 1, 21–41.

1101. T. Penttila and B. Williams, *Regular packings of* PG(3, q), European J. Combin. **19** (1998), no. 6, 713–720.

1102. ———, *Ovoids of parabolic spaces*, Geom. Dedicata **82** (2000), no. 1–3, 1–19.

1103. O. Pfaff, *The classification of doubly transitive affine designs*, Des. Codes Cryptogr. **1** (1991), no. 3, 207–217.

1104. D. Pierce and M.J. Kallaher, *A note on planar functions and their planes*, Bull. Inst. Combin. Appl. **42** (2004), 53–75.

1105. F.C. Piper, *Polarities in the Hughes plane*, Bull. London Math. Soc. **2** (1970), 209–213.

1106. ———, *Polarities of finite projective planes*, Atti del Convegno di Geometria Combinatoria e sue Applicazioni (Univ. Perugia, Perugia, 1970), Ist. Mat., Univ. Perugia, Perugia, 1971, pp. 373–376.

1107. ———, *Collineation groups, perspectivities and collineations of prime order*, Proceedings of the International Conference on Projective Planes (Washington State Univ., Pullman, Wash., 1973), Washington State Univ. Press, Pullman, Wash., 1973, pp. 263–273.

1108. R. Pomareda, *Hyper-reguli in projective space of dimension* 5, Mostly Finite Geometries (Iowa City, IA, 1996), Lecture Notes in Pure and Appl. Math., vol. 190, Dekker, New York, 1997, pp. 379–381.

1109. R.J. Pomareda Rodriguez, *Groups and geometry: collineations of translation planes*, Proceedings of the Eleventh Brazilian Mathematical Colloquium (Poços de Caldas, 1977), Vol. I (Portuguese) (Rio de Janeiro), Inst. Mat. Pura Apl., 1978, pp. 145–154.

1110. A.R. Prince, *A translation plane of order* 19^2 *admitting* $SL(2, 5)$, *obtained by* 12-*nest replacement*, submitted.

1111. ———, *The flag-transitive affine planes of order* 27, Combinatorics '90 (Gaeta, 1990), Ann. Discrete Math., vol. 52, North-Holland, Amsterdam, 1992, pp. 477–500.

1112. ———, *A complete classification of the flag-transitive affine planes of order* 125, J. Combin. Des. **5** (1997), no. 2, 147–153.

1113. ———, *Parallelisms of* PG(3, 3) *invariant under a collineation of order* 5, Mostly Finite Geometries (Iowa City, IA, 1996), Lecture Notes in Pure and Appl. Math., vol. 190, Dekker, New York, 1997, pp. 383–390.

1114. ———, *The cyclic parallelism of* PG(3, 5), European J. Combin. **19** (1998), no. 5, 613–616.

1115. ———, *Flag-transitive affine planes of order 64*, Designs and codes—a memorial tribute to Ed Assmus, Des. Codes Cryptogr. **18** (1999), no. 1–3, 217–221.

1116. ———, *Flag-transitive affine planes of order at most 125*, Second Pythagorean Conference (Pythagoreion, 1999), J. Geom. **67** (2000), no. 1–2, 208–216.

1117. ———, *Two new families of commutative semifields*, Bull. London Math. Soc. **32** (2000), no. 5, 547–550.

1118. ———, *Covering sets of spreads in* PG(3, q), Discrete Math. **238** (2001), no. 1–3, 131–136, Designs, codes and finite geometries (Shanghai, 1999).

1119. O. Prohaska, *Konfigurationen einander meidender Kreise in miquelschen Möbiusebenen ungerader Ordnung*, Arch. Math. (Basel) **28** (1977), no. 5, 550–556.

1120. L. Puccio and M.J. de Resmini, *Subplanes of the Hughes plane of order 25*, Arch. Math. (Basel) **49** (1987), no. 2, 151–165.

1121. W. Purpura, *Counting the generalized twisted fields*, Note Mat. (to appear).

1122. F. Radó, *Congruence-preserving isomorphisms of the translation group associated with a translation plane*, Canad. J. Math. **23** (1971), 214–221.

1123. ———, *Extension of collineations defined on subsets of a translation plane*, J. Geometry **1** (1971), 1–17.

1124. G. Raguso, *A translation plane of order* 13^2, Rend. Mat. (7) **5** (1985), no. 1–2, 51–56.

1125. A. Rahilly, *A class of finite projective planes*, Proceedings of the First Australian Conference on Combinatorial Mathematics (Univ. Newcastle, Newcastle, 1972), Univ. of Newcastle Res. Associates, Newcastle, 1972, pp. 31–37.

1126. _____, *Some translation planes with elations which are not translations*, Combinatorial Mathematics, III (Proc. Third Australian Conf., Univ. Queensland, St. Lucia, 1974), Lecture Notes in Math., vol. 452, Springer, Berlin, 1975, pp. 197–209.

1127. _____, *Spreads of lines and regular group divisible designs*, J. Combin. Math. Combin. Comput. **12** (1992), 141–151.

1128. _____, *Corrigendum to: "Spreads of lines and regular group divisible designs" [J. Combin. Math. Combin. Comput. **12** (1992), 141–151]*, J. Combin. Math. Combin. Comput. **13** (1993), 223.

1129. A.S. Raju, *A characterization of translation planes and dual translation planes of characteristic ≠ 2*, Osaka J. Math. **31** (1994), no. 1, 51–60.

1130. A. Reifart, *The classification of the translation planes of order 16, ii*, Geom. Dedicata **17** (1984), no. 1, 1–9.

1131. L.A. Rosati, *Piani proiettivi desarguesiani non ciclici*, Boll. Un. Mat. Ital. (3) **12** (1957), 230–240.

1132. _____, *Sui piani desarguesiani affini "non-ciclici"*, Atti Accad. Naz. Lincei. Rend. Cl. Sci. Fis. Mat. Nat. (8) **22** (1957), 443–449.

1133. _____, *I gruppi di collineazioni dei piani di Hughes*, Boll. Un. Mat. Ital. (3) **13** (1958), 505–513.

1134. _____, *Su una generalizzazione dei piani di Hughes*, Atti Accad. Naz. Lincei Rend. Cl. Sci. Fis. Mat. Nat. (8) **29** (1960), 303–308.

1135. _____, *Unicità e autodualità dei piani di Hughes*, Rend. Sem. Mat. Univ. Padova **30** (1960), 316–327.

1136. _____, *Su alcune varietà dello spazio proiettivo sopra un corpo non commutative*, Ann. Mat. Pura Appl. (4) **59** (1962), 213–227.

1137. _____, *Su certe varietà dello spazio proiettivo r-dimensionale sopra un corpo non commutativo*, Atti Accad. Naz. Lincei Rend. Cl. Sci. Fis. Mat. Nat. (8) **32** (1962), 907–912.

1138. _____, *Su una definizione assiomatica di determinante sopra un corpo non commutativo*, Riv. Mat. Univ. Parma (2) **3** (1962), 249–257.

1139. _____, *Su una nuova classe di piani grafici*, Atti Accad. Naz. Lincei Rend. Cl. Sci. Fis. Mat. Natur. (8) **35** (1963), 282–284.

1140. _____, *Su una nuova classe di piani grafici*, Ricerche Mat. **13** (1964), 39–55.

1141. _____, *Sui piani di Hughes generalizzati e i loro derivati*, Matematiche (Catania) **22** (1967), 289–302.

1142. _____, *Ovals of finite Desarguesian planes of even order*, Combinatorics '81 (Rome, 1981), Ann. Discrete Math., vol. 18, North-Holland, Amsterdam, 1983, pp. 713–720.

1143. _____, *Unitals in Hughes planes*, Geom. Dedicata **27** (1988), no. 3, 295–299.

1144. G.F. Royle, *An orderly algorithm and some applications in finite geometry*, Discrete Math. **185** (1998), no. 1–3, 105–115.

1145. R. Sandler, *Autotopism groups of some finite non-associative algebras*, Amer. J. Math. **84** (1962), 239–264.

1146. _____, *The collineation groups of some finite projective planes*, Portugal. Math. **21** (1962), 189–199.

1147. _____, *A note on some new finite division ring planes*, Trans. Amer. Math. Soc. **104** (1962), 528–531.

1148. K. Satyanarayana and K. Kuppuswamy Rao, *A note on a paper of M. L. Narayana Rao and E. H. Davis: "Construction of translation planes from t-spread sets" [J. Combin. Theory Ser. A **14** (1973), 201–208]*, J. Combin. Inform. System Sci. **5** (1980), no. 2, 104–106.

1149. _____, *Full collineation group of the second of Foulser's flag transitive planes of order 25*, Houston J. Math. **7** (1981), no. 4, 537–543.

1150. H.-J. Schaeffer, *Translationsebenen, auf denen die Gruppe* $SL(2, p^n)$ *Operiert*, Diplomarbeit, Univ. Tübingen, 1975.

1151. R.-H. Schulz, *Über Translationsebenen mit Kollineationsgruppen, die die Punkte der ausgezeichneten Geraden zweifach transitiv permutieren*, Math. Z. **122** (1971), 246–266.

1152. R.-H. Schulz and A.G. Spera, *Construction of divisible designs from translation planes*, European J. Combin. **19** (1998), no. 4, 479–486.

1153. _____, *Divisible designs admitting a Suzuki group as an automorphism group*, Boll. Unione Mat. Ital. Sez. B Artic. Ric. Mat. (8) **1** (1998), no. 3, 705–714.

1154. B. Segre, *Lectures on Modern Geometry*, With an appendix by Lucio Lombardo-Radice. Consiglio Nazionale delle Rierche Monografie Matematiche, vol. 7, Edizioni Cremonese, Rome, 1961.

1155. F.A. Sherk, *Indicator sets in an affine space of any dimension*, Canad. J. Math. **31** (1979), no. 1, 211–224.

1156. _____, *A translation plane of order 27 with a small translation complement*, Geom. Dedicata **9** (1980), no. 3, 307–316.

1157. _____, *Translation planes of order 16*, Finite Geometries (Pullman, Wash., 1981), Lecture Notes in Pure and Appl. Math., vol. 82, Dekker, New York, 1983, pp. 401–412.

1158. _____, *The geometry of* GF(q^3), Canad. J. Math. **38** (1986), no. 3, 672–696.

1159. _____, *Cubic surfaces in* AG($3, q$) *and projective planes of order* q^3, Geom. Dedicata **34** (1990), no. 1, 1–11.

1160. F.A. Sherk and G. Pabst, *Indicator sets, reguli, and a new class of spreads*, Canad. J. Math. **29** (1977), no. 1, 132–154.

1161. E.E. Shult, *A sporadic ovoid in* $\Omega^+(8,7)$, Proceedings of the Conference on Groups and Geometry, Part B (Madison, Wis., 1985), Algebras Groups Geom. **2** (1985), no. 4, 495–513.

1162. _____, *Nonexistence of ovoids in* $\Omega^+(10,3)$, J. Combin. Theory Ser. A **51** (1989), no. 2, 250–257.

1163. _____, *Problems by the wayside*, Discrete Math. **294** (2005), no. 1–2, 175–201.

1164. A.G. Spera, *Divisible designs associated with translation planes admitting a 2-transitive collineation group on the points at infinity*, Aequationes Math. **59** (2000), no. 1–2, 191–200.

1165. L. Storme and J.A. Thas, *k-arcs and partial flocks*, Linear Algebra Appl. **226/228** (1995), 33–45.

1166. C. Suetake, *A new class of translation planes of order* q^3, Osaka J. Math. **22** (1985), no. 4, 773–786.

1167. _____, *Flag transitive planes of order* q^n *with a long cycle* l_∞ *as a collineation*, Graphs Combin. **7** (1991), no. 2, 183–195.

1168. _____, *A family of translation planes of order* q^{2m+1} *with two orbits of length 2 and* $q^{2m+1} - 1$ *on* l_∞, Geom. Dedicata **42** (1992), no. 2, 163–185.

1169. _____, *A family of translation planes of order* q^3, J. Geom. **45** (1992), no. 1–2, 174–176.

1170. _____, *On flag-transitive affine planes of order* q^3, Geom. Dedicata **51** (1994), no. 2, 123–131.

1171. J.A. Thas, *Ovoidal translation planes*, Arch. Math. (Basel) **23** (1972), 110–112.

1172. _____, *Flocks of finite egglike inversive planes*, Finite Geometric Structures and Their Applications (C.I.M.E., II Ciclo, Bressanone, 1972), Edizioni Cremonese, Rome, 1973, pp. 189–191.

1173. _____, *Flocks of non-singular ruled quadrics in* PG($3, q$), Atti Accad. Naz. Lincei Rend. Cl. Sci. Fis. Mat. Natur. (8) **59** (1975), no. 1–2, 83–85.

1174. _____, *Construction of maximal arcs and dual ovals in translation planes*, European J. Combin. **1** (1980), no. 2, 189–192.

1175. _____, *Ovoids and spreads of finite classical polar spaces*, Geom. Dedicata **10** (1981), no. 1–4, 135–143.

1176. _____, *Semipartial geometries and spreads of classical polar spaces*, J. Combin. Theory Ser. A **35** (1983), no. 1, 58–66.

1177. _____, *Generalized quadrangles and flocks of cones*, European J. Combin. **8** (1987), no. 4, 441–452.

1178. _____, *Flocks, maximal exterior sets, and inversive planes*, Finite Geometries and Combinatorial Designs (Lincoln, NE, 1987), Contemp. Math., vol. 111, Amer. Math. Soc., Providence, RI, 1990, pp. 187–218.

1179. _____, *Maximal exterior sets of hyperbolic quadrics: the complete classification*, J. Combin. Theory Ser. A **56** (1991), no. 2, 303–308.

1180. _____, *Recent results on flocks, maximal exterior sets and inversive planes*, Combinatorics '88, Vol. 1 (Ravello, 1988), Res. Lecture Notes Math., Mediterranean, Rende, 1991, pp. 95–108.

1181. _____, *Old and new results on spreads and ovoids of finite classical polar spaces*, Combinatorics '90 (Gaeta, 1990), Ann. Discrete Math., vol. 52, North-Holland, Amsterdam, 1992, pp. 529–544.

1182. _____, *A characterization of the Fisher-Thas-Walker flocks*, Simon Stevin **67** (1993), no. 3–4, 219–226.

1183. _____, *Recent developments in the theory of finite generalized quadrangles*, Med. Konink. Acad. Wetensch. België **56** (1994), no. 1, 99–113.

1184. _____, *k-arcs, hyperovals, partial flocks and flocks*, Des. Codes Cryptogr. **9** (1996), no. 1, 95–104, Second Upper Michigan Combinatorics Workshop on Designs, Codes and Geometries (Houghton, MI, 1994).

1185. _____, *Generalized quadrangles of order (s, s^2). II*, J. Combin. Theory Ser. A **79** (1997), no. 2, 223–254.

1186. _____, *Symplectic spreads in* PG$(3, q)$, *inversive planes and projective planes*, Discrete Math. **174** (1997), no. 1–3, 329–336, Combinatorics (Rome and Montesilvano, 1994).

1187. _____, *Flocks and partial flocks of quadrics: a survey*, Second Shanghai Conference on Designs, Codes and Finite Geometries (1996), J. Statist. Plann. Inference **94** (2001), no. 2, 335–348.

1188. _____, *Ovoids, spreads and m-systems of finite classical polar spaces*, Surveys in Combinatorics, 2001 (Sussex), London Math. Soc. Lecture Note Ser., vol. 288, Cambridge Univ. Press, Cambridge, 2001, pp. 241–267.

1189. J.A. Thas, C. Herssens, and F. De Clerck, *Flocks and partial flocks of the quadratic cone in* PG$(3, q)$, Finite Geometry and Combinatorics (Deinze, 1992), London Math. Soc. Lecture Note Ser., vol. 191, Cambridge Univ. Press, Cambridge, 1993, pp. 379–393.

1190. J.A. Thas and S.E. Payne, *Spreads and ovoids in finite generalized quadrangles*, Geom. Dedicata **52** (1994), no. 3, 227–253.

1191. A.V. Vasil'ev and V.D. Mazurov, *Minimal permutation representations of finite simple orthogonal groups*, Algebra i Logika **33** (1994), no. 6, 603–627, 716.

1192. O. Veblen and J.W. Young, *Projective Geometry*, vol. 1, Blaisdell, New York–Toronto–London, 1938.

1193. _____, *Projective Geometry*, vol. 2 (by Oswald Veblen), Blaisdell, New York–Toronto–London, 1938.

1194. O. Vega, *A generalization of j-planes*, Ph.D. thesis, University of Iowa, 2006.

1195. R. Vincenti, *The collineations group of a translation plane and the collineations group of an affine geometry*, Atti Sem. Mat. Fis. Univ. Modena **28** (1979), no. 2, 342–351.

1196. _____, *Errata: "The collineations group of a translation plane and the collineations group of an affine geometry" [Atti Sem. Mat. Fis. Univ. Modena **28** (1979), no. 2, 342–351 (1980)]*, Atti Sem. Mat. Fis. Univ. Modena **29** (1980), no. 2, 385.

1197. _____, *Varieties representing Baer subplanes of translation planes of the class V*, Atti Sem. Mat. Fis. Univ. Modena **29** (1980), no. 1, 48–59.

1198. _____, *A survey on varieties of* PG$(4, q)$ *and Baer subplanes of translation planes*, Combinatorics '81 (Rome, 1981), North-Holland Math. Stud., vol. 78, North-Holland, Amsterdam, 1983, pp. 775–779.

1199. _____, *Cryptoreguli and derivable translation planes*, Proceedings of the Conference on Combinatorial and Incidence Geometry: Principles and Applications (La Mendola, 1982) (Milan), Rend. Sem. Mat. Brescia, vol. 7, Vita e Pensiero, 1984, pp. 625–633.

1200. _____, *On permutation polynomials and derivable translation planes*, Finite Fields and Applications (Augsburg, 1999), Springer, Berlin, 2001, pp. 428–436.

1201. A. Wagner, *On collineation groups of projective spaces, I*, Math. Z. **76** (1961), 411–426.

1202. _____, *On finite affine line transitive planes*, Math. Z. **87** (1965), 1–11.

1203. M. Walker, *On translation planes and their collineation groups*, Ph.D. thesis, Univ. London, 1973.

1204. _____, *A class of translation planes*, Geom. Dedicata **5** (1976), no. 2, 135–146.

1205. _____, *The collineation groups of derived translation planes*, Geom. Dedicata **5** (1976), no. 1, 87–95.

1206. _____, *The collineation groups of derived translation planes, II*, Math. Z. **148** (1976), no. 1, 1–6.

1207. _____, *A note on tangentially transitive affine planes*, Bull. London Math. Soc. **8** (1976), no. 3, 273–277.

1208. _____, *A characterization of some translation planes*, Abh. Math. Sem. Univ. Hamburg **49** (1979), 216–233.

1209. R.A. Weida, *An extension of Bruen chains*, Geom. Dedicata **30** (1989), no. 1, 11–21.

1210. S.H. Weintraub, *Spreads of nonsingular pairs in symplectic vector spaces*, J. Geom. (to appear).

1211. G.P. Wene, *Finite semifields three-dimensional over the left nuclei*, Nonassociative Algebra and Its Applications (São Paulo, 1998), Lecture Notes in Pure and Appl. Math., vol. 211, Dekker, New York, 2000, pp. 447–456.

1212. P. Wild, *Higher-dimensional ovals and dual translation planes*, Ars Combin. **17** (1984), 105–112.

1213. F.W. Wilke, *A class of translation planes and a conjecture of D. R. Hughes*, Trans. Amer. Math. Soc. **145** (1969), 223–232.

1214. _____, *Some new translation nets*, Geom. Dedicata **2** (1973), 205–211.

1215. _____, *An algorithmic search for replaceable translation nets*, Geom. Dedicata **8** (1979), no. 2, 213–217.

1216. F.W. Wilke and J.L. Zemmer, *Some new finite translation planes*, Trans. Amer. Math. Soc. **131** (1968), 378–397.

1217. M. Woltermann, *A note on doubly transitive solvable permutation groups*, Comm. Algebra **7** (1979), no. 17, 1877–1883.

1218. H. Zassenhaus, *Über endliche Fastkörper*, Abh. Math. Sem. Univ. Hamburg **11** (1935), 187–220.

1219. K. Zsigmondy, *Zur Theorie der Potenzreste*, Monatshefte Math. Phys. **3** (1892), no. 1, 265–284.

1220. Ju. N. Zvereva, *Arcs in the projective translation plane of order 9*, Kombinatornyĭ Anal. Vyp. 2 (1972), 99–102.

1221. _____, *Complete arcs in the projective translation plane of order 9*, Kombinatornyĭ Anal. Vyp. 3 (1974), 102–106.

Theorems

Models

General Index